GLENCOE

Pre-Algebra

An Integrated Transition to Algebra & Geometry

GLENCOE

McGraw-Hill

New York, New York Columbus, Ohio Woodland Hills, California Peoria, Illnois

Visit the Glencoe Mathematics Internet Site for
Pre-Algebra at

www.glencoe.com/sec/math/prealg/mathnet

You'll find:

Group Activities

Games

interNET CONNECTION links to websites relevant to Chapter Projects, Investigations, and exercises

and much more!

Glencoe/McGraw-Hill
A Division of The McGraw-Hill Companies

Send all inquiries to:
Glencoe/McGraw Hill
8787 Orion Place
Columbus, OH 43240

ISBN: 0-02-833240-7

10 027/043 06 05 04

WHY IS MATHEMATICS IMPORTANT?

Why do I need to study mathematics? When am I ever going to have to use math in the real world?

Many people, not just mathematics students, wonder why mathematics is important. ***Pre-Algebra*** is designed to answer those questions through ***integration***, ***applications***, and ***connections***.

Geometry

Did you know that algebra and geometry are closely related? Topics from all branches of mathematics, like geometry, probability, and statistics, are integrated throughout the text.

You'll learn how to use proportions to relate the measures of similar triangles. (Lesson 11-6, pages 578–583)

Movies

How did they make those dinosaurs in *Jurassic Park?* They used rational numbers! In ***Pre-Algebra***, you'll study many ways that math is used in the real world.

The size of a model is related to the size of a real object by using a rational number. (Lesson 5-1, page 224)

Biology

What does mathematics have to do with biology? Mathematical topics are connected to the other subjects that you study.

Scientists use scientific notation to describe the size of very large distances, like the distance from Earth to the Moon, and very small distances, like the diameter of a red blood cell. (Lesson 4-9, page 211)

Authors

This program builds on the mathematical background students obtain in earlier grades and prepares them to succeed in algebra. Technology is included wherever appropriate, and links between mathematics and life situations abound. Math is fun!

William Leschensky is a former teacher of mathematics at Glenbard South High School and the College of DuPage in Glen Ellyn, Illinois. He has taught mathematics at every level of the high school curriculum, served as mathematics department chairperson, and is a former textbook editor and editorial director. Mr. Leschensky received his B.A. in mathematics from Cornell College and his M.A. in mathematics from the University of Northern Iowa. He has participated in special institutes in mathematics, physics, and computer science and is a frequent speaker at local, state, and regional conferences. Mr. Leschensky is a question writer for MATHCOUNTS, a nationwide series of competitions designed to stimulate student interest and achievement in mathematics.

Wm D Leschensky

Carol Malloy is an assistant professor of mathematics education at the University of North Carolina at Chapel Hill. Dr. Malloy previously taught middle and high school mathematics. She received her B.S. in mathematics and education from West Chester State College, her M.S.T. in mathematics from Illinois Institute of Technology, and her Ph.D. in curriculum and instruction from the University of North Carolina at Chapel Hill. Dr. Malloy is an active member of numerous local, state, and national professional organizations and boards, including National Council of Teachers of Mathematics (NCTM) and Association for Supervision and Curriculum Development (ASCD). Dr. Malloy also serves as an editor of the Benjamin Banneker Association newsletter. She frequently speaks at conferences and schools on the topics of equity and mathematics for all students.

Carol E. Malloy

Mathematics teachers are excited about implementing the NCTM Standards in their mathematics classrooms. Glencoe's Pre-Algebra affords students with a strong foundation to high school mathematics through comprehensive real-world explorations, technology, and hands-on activities.

With the study of pre-algebra, you are taking your first steps toward a mathematics education that will be valuable to you throughout your lifetime. I know you will find it exciting and challenging.

Jack Price received his B.A. from Eastern Michigan University and his Doctorate in Mathematics Education from Wayne State University. Dr. Price has been active in mathematics education for over 40 years, 38 of those years at grades K through 12. In his current position as co-director of the Center for Education and Equity in Mathematics, Science, and Technology at California State Polytechnic University, he teaches mathematics and methods courses for preservice teachers and consults with school districts on curriculum change. Dr. Price is a past president of the National Council of Teachers of Mathematics, is a frequent speaker at professional conferences, conducts many in-service workshops, and is author of numerous mathematics and science instructional materials. Dr. Price is co-author on Glencoe's *Mathematics: Applications and Connections* and *Interactive Mathematics: Activities and Investigations*.

Jim Rath has 30 years of classroom experience in teaching mathematics at every level of the high school curriculum. He is a former mathematics teacher and department chairperson at Darien High School in Darien, Connecticut. Mr. Rath earned his B.A. in philosophy from the Catholic University of America and his M.Ed. and M.A. in mathematics from Boston College. He held an appointment as a Visiting Fellow in the Mathematics Department at Yale University in New Haven, Connecticut for the academic year 1994-1995. Mr. Rath is a co-author on Glencoe's *Algebra 1* and *Algebra 2*.

James N. Rath

What you learn today is a building block for what comes tomorrow. That is the reason why a student needs to do his or her best each day . . . what you learn in pre-algebra is building a strong foundation for you.

Mathematics today is a thrilling mix of present reality and investigation. Students will travel on paths of exploration to reach into the future. This textbook will make the journey a unique and exciting experience.

Yuria Alban received her B.S. in Mathematics from Armstrong State College, her M.A. in Mathematics from the University of Georgia, and her Educational Specialist degree in Computer Education from Nova University. Ms. Alban has more than twenty years of teaching experience in middle school, high school, and college mathematics. She is currently a District Mathematics Supervisor for the Dade County Public Schools in Florida. Ms. Alban is a past President of the Dade County Council of Teachers of Mathematics and has received numerous teaching awards at the local and state levels. Recently, Ms. Alban participated in U.S. Justice Department programs to assess students' needs and advise on mathematics curriculum in Guantanamo Bay, Cuba and Venezuela.

Consultants and Reviewers

Consultants

Gilbert J. Cuevas
Professor of Mathematics Education
University of Miami
Coral Gables, Florida

Dr. Luis Ortiz-Franco
Consultant, Diversity
Associate Professor of Mathematics
Chapman University
Orange, California

David Foster
Glencoe Author and Mathematics Consultant
Morgan Hill, California

Daniel Marks
Consultant, Real-World Applications
Associate Professor of Mathematics
Auburn University at Montgomery
Montgomery, Alabama

Melissa McClure
Consultant, Tech Prep
Mathematics Consultant
Teaching for Tomorrow
Fort Worth, Texas

Cleo M. Meek, Ed. D.
Mathematics Consultant
NC Dept. of Public Instruction, Retired
Raleigh, North Carolina

Jim Zwick
Mathematics Department Chairman
Guion Creek Middle School—MSD of Pike Township
Indianapolis, Indiana

Reviewers

Robert M. Allen
Mathematics Teacher
Vista Verde Middle School
Phoenix, Arizona

Bruce I. Althouse, Sr.
Mathematics Department Chairperson
Northern Lebanon Junior/Senior High School
Fredericksburg, Pennsylvania

Glenn Aston-Reese
Mathematics Department Chairperson
Trinity Middle School
Washington, Pennsylvania

Mary R. Beard
Mathematics Teacher
Southeast Guilford Middle School
Greensboro, North Carolina

Ronald L. Bow
Mathematics Teacher
Henry Clay High School
Lexington, Kentucky

Alan Brown
Registrar, Adult Education
Springfield Adult Learning Center
Springfield, Massachusetts

Sharon A. Cichocki
Secondary Math Coordinator
Hamburg High School
Hamburg, New York

Debra M. Cline
Mathematics Teacher
Atkins Middle School
Winston-Salem, North Carolina

Linda Coutts
Mathematics Coordinator
Columbia Public Schools
Columbia, Missouri

Jean E. Ellis
Mathematics Teacher
Ruddiman Middle School
Detroit, Michigan

Gary A. Graves
Coordinator of Mathematics K-12
St. Joseph Public Schools
St. Joseph, Missouri

Earlene J. Hall
Administrator
Detroit Public Schools
Detroit, Michigan

Eileen D. Harris, Ph.D.
Mathematics/Science K-12 Specialist
Charlotte County School District
Port Charlotte, Florida

Richard Hedlund
Mathematics Teacher
South Broward High School
Hollywood, Florida

James P. Herrington
Mathematics/Science Department Chairperson
O'Fallon Township High School
O'Fallon, Illinois

Deborah D. Johnson
Supervisor of Mathematics
Camden City Public Schools
Camden, New Jersey

Nancy L. Jones
Mathematics Teacher
Aycock Middle School
Greensboro, North Carolina

Stephen R. Lightner
Mathematics Department Chairperson
D.S. Keith Junior High School
Altoona, Pennsylvania

Gaye S. McKinnon
Mathematics Teacher
Columbiana Middle School
Columbiana, Alabama

Bernadette M. Meglino
Mathematics Teacher
Lake Weir High School
Ocala, Florida

Nance Marie Minnick
Mathematics Teacher
Denver Public Schools
Denver, Colorado

Rhonda C. Niemi
Mathematics Teacher
Meyzeek Middle School
Louisville, Kentucky

Georgia Thielen O'Connor
Mathematics Instructor
Stevens High School
Rapid City, South Dakota

Pamela H. Ostrosky
Mathematics Teacher
Dake Junior High School
Rochester, New York

Marcia Perry
Mathematics Teacher
Palm Beach Gardens High School
Palm Beach Gardens, Florida

Odelia Schrunk
Mathematics Department Chairperson
Clinton High School
Clinton, Iowa

Sharon Ann Schueler
Mathematics Department Chairperson
Roosevelt High School
Sioux Falls, South Dakota

Sara P. Shaw
Mathematics Teacher
Adams Middle School
Saraland, Alabama

Mary Eleanor Wood Smith
Mathematics Teacher
North Augusta High School
North Augusta, South Carolina

Diane Stilwell
Mathematics Department Chairperson
South Junior High School
Morgantown, West Virginia

Janice M. Stricko
Mathematics Teacher
Gateway Middle School
Monroeville, Pennsylvania

Jeanette Tomasullo
Assistant Principal Supervision —Mathematics
Eastern District High School
Brooklyn, New York

Ivan T. Van Dyke
Mathematics Teacher
Norfolk Junior High School
Norfolk, Nebraska

Kathy Vielhaber
Mathematics Teacher
Parkway East Middle School
Creve Coeur, Missouri

Marlene T. Vitko
Mathematics Department Chairperson
Godwin Middle School
Dale City, Virginia

Carol Welsch
Learning Coordinator
Toki Middle School
Madison, Wisconsin

Table of Contents

Introduction to the Graphing Calculator xx

Chapter 1 Tools for Algebra and Geometry 4

1-1	Problem-Solving Strategy: Make a Plan	6
1-2	Order of Operations	11
1-3	Variables and Expressions	16
1-3B	**Math Lab: Graphing Calculator Activity** Evaluating Expressions	21
1-4	Properties	22
1-5	The Distributive Property	26
1-5B	**Math Lab: Hands-On Activity** Distributive Property	31
1-6	Variables and Equations	32
1-7	**Integration: Geometry** Ordered Pairs	36
1-8	Solving Equations Using Inverse Operations	41
1-9	Inequalities	46
1-10A	**Math Lab: Hands-On Activity** Gathering Data	50
1-10	**Integration: Statistics** Gathering and Recording Data	51

Chapter 1 Highlights 57

Chapter 1 Study Guide and Assessment 58

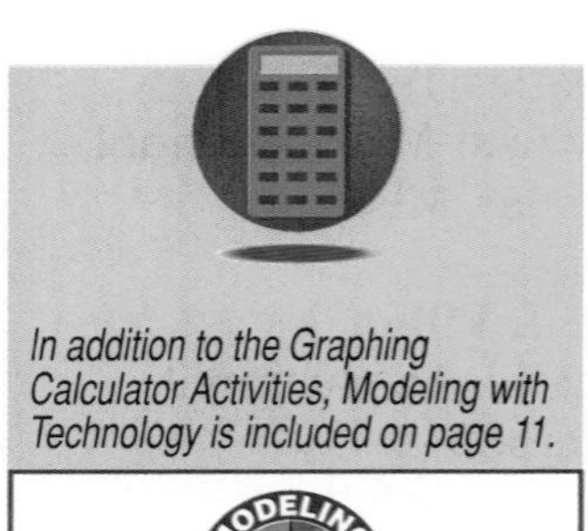

In addition to the Graphing Calculator Activities, Modeling with Technology is included on page 11.

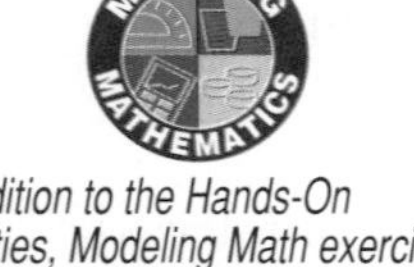

In addition to the Hands-On Activities, Modeling Math exercises are included on pages 14, 18, 24, 34, 47, and 54.

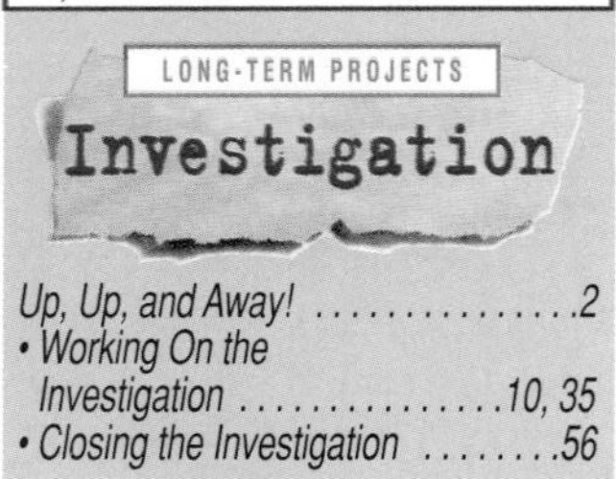

Up, Up, and Away! 2
- *Working On the Investigation* 10, 35
- *Closing the Investigation* 56

Chapter 2 Exploring Integers 64

2-1 Integers and Absolute Value 66
2-1B **Math Lab: Hands-On Activity** 71
2-2 **Integration: Geometry** The Coordinate System 72
2-2B **Math Lab: Graphing Calculator Activity** . . 77
2-3 Comparing and Ordering 78
2-4A **Math Lab: Hands-On Activity** 82
2-4 Adding Integers 83
2-5A **Math Lab: Hands-On Activity** 88
2-5 Subtracting Integers 89
2-6 Problem-Solving Strategy: Look for a Pattern 94
2-7A **Math Lab: Hands-On Activity** 98
2-7 Multiplying Integers 99
2-8 Dividing Integers 104
Chapter 2 Highlights 109
Chapter 2 Study Guide and Assessment 110
Ongoing Assessment, Chapters 1–2 114

Chapter 3 Solving One-Step Equations and Inequalities 116

3-1 Problem-Solving Strategy: Eliminate Possibilities 118
3-2A **Math Lab: Hands-On Activity** 123
3-2 Solving Equations by Adding or Subtracting 124
3-3 Solving Equations by Multiplying or Dividing 129
3-4 Using Formulas 134
3-5A **Math Lab: Hands-On Activity** 138
3-5 **Integration: Geometry** Area and Perimeter 139
3-5B **Math Lab: Graphing Calculator Activity** . . 145
3-6 Solving Inequalities by Adding or Subtracting 146
3-7 Solving Inequalities by Multiplying or Dividing 151
3-8 Applying Equations and Inequalities 156
Chapter 3 Highlights 161
Chapter 3 Study Guide and Assessment 162

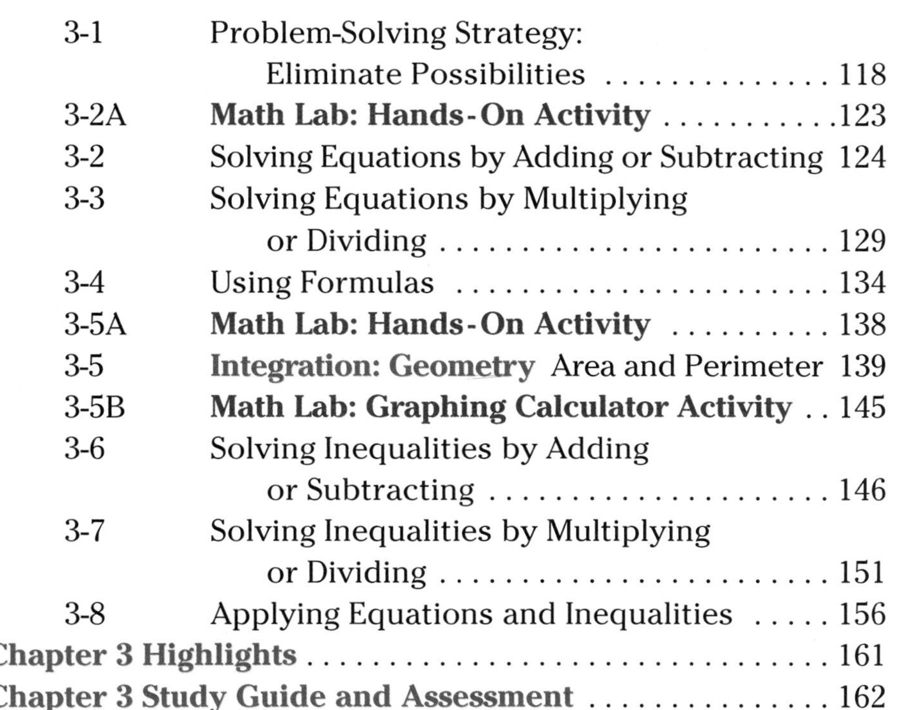

In addition to the Graphing Calculator Activities, Modeling with Technology is included on pages 83, 94, and 124.

In addition to the Hands-On Activities, Modeling Math exercises are included on pages 68, 91, 102, 120, and 135.

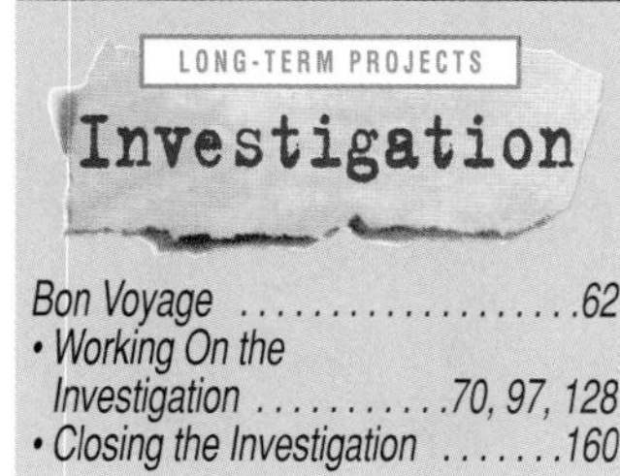

Bon Voyage 62
- *Working On the Investigation* 70, 97, 128
- *Closing the Investigation* 160

Content Integration

Why are topics like geometry and statistics in a pre-algebra book? Believe it or not, you can usually study an algebra topic from a geometric point of view and vice versa. Some examples are described below.

INTEGRATION

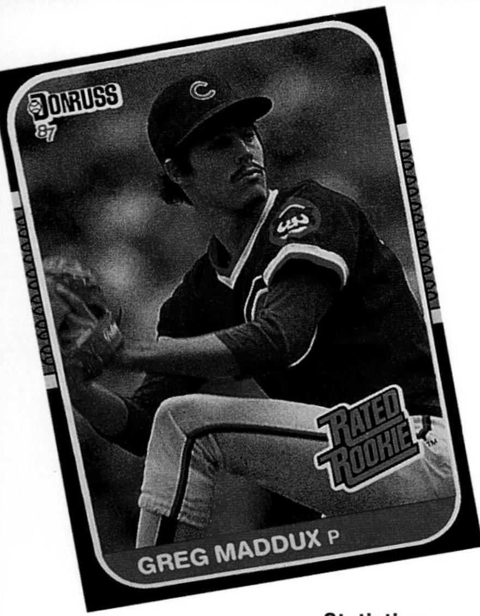

Probability You'll learn more about your chances of making that winning move in Backgammon. (Lesson 9-3, pages 440-443)

Geometry You'll use ordered pairs to develop algebraic expressions. (Lesson 1-7, pages 36-40)

Statistics
You'll learn how to use mean, median, and mode. (Lesson 6-6, pages 301-306)

Discrete Mathematics
You'll learn to represent an arithmetic sequence algebraically. (Lesson 5-9, pages 258-262)

Problem Solving
You'll learn how to solve problems by looking for a pattern. (Lesson 2-6, pages 94-97)

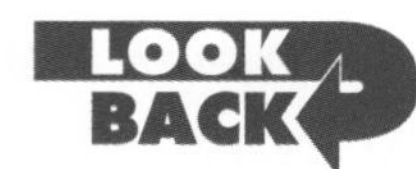

You can review dividing fractions in Lesson 6-4.

Look Back features refer you to skills and concepts that have been taught earlier in the book. Lesson 9-7, page 459

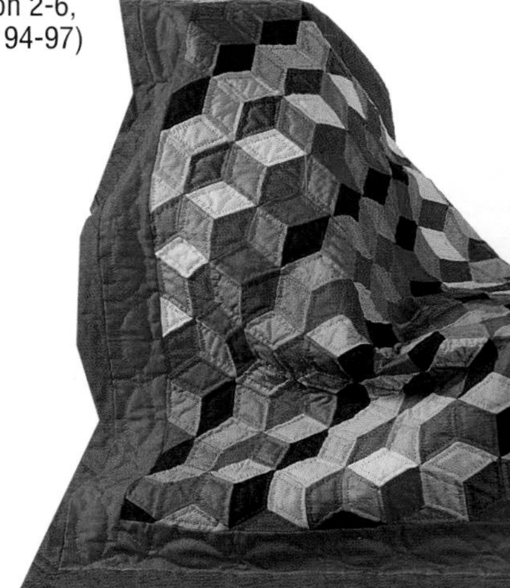

Chapter 4 Exploring Factors and Fractions 168

4-1 Factors and Monomials 170
4-2 Powers and Exponents 175
4-2B **Math Lab: Graphing Calculator Activity**
Evaluating Expressions with Exponents 180
4-3 Problem-Solving Strategy: Draw a Diagram 181
4-4 Prime Factorization 184
4-4B **Math Lab: Graphing Calculator Activity**
Factor Patterns 189
4-5 Greatest Common Factor (GCF) 190
4-6A **Math Lab: Hands-On Activity**
Equivalent Fractions 195
4-6 Simplifying Fractions 196
4-7 Using the Least Common Multiple (LCM) 200
4-8 Multiplying and Dividing Monomials 205
4-9 Negative Exponents 210
Chapter 4 Highlights 215
Chapter 4 Study Guide and Assessment 216
Ongoing Assessment, Chapters 1–4 220

Chapter 5 Rationals: Patterns in Addition and Subtraction 222

5-1 Rational Numbers 224
5-2 Estimating Sums and Differences 229
5-3 Adding and Subtracting Decimals 234
5-4 Adding and Subtracting Like Fractions ... 239
5-5 Adding and Subtracting Unlike Fractions .. 244
5-6 Solving Equations 248
5-7 Solving Inequalities 251
5-8 Problem-Solving Strategy:
Using Logical Reasoning 255
5-9 **Integration: Discrete Mathematics**
Arithmetic Sequences 258
5-9B **Math Lab: Hands-On Activity**
Fibonacci Sequence 263
Chapter 5 Highlights 265
Chapter 5 Study Guide and Assessment 266

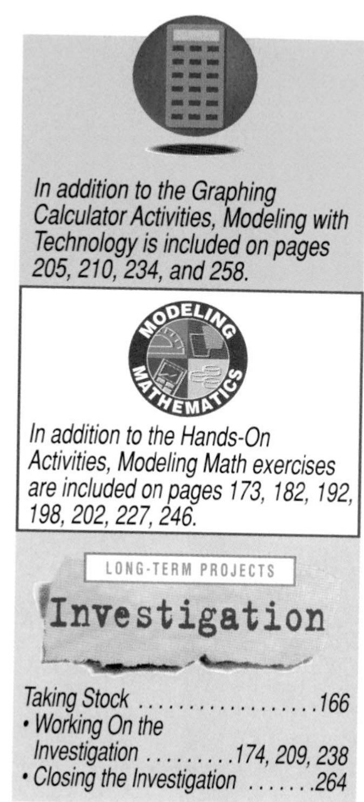

Real-Life Applications

Every lesson in this book is designed to help show you where and when math is used in the real world. Since you'll explore many interesting topics, we believe you'll discover that math is relevant and exciting. Here are some examples.

APPLICATIONS

Recycling You'll discuss recycling as you learn about solving inequalities. (Lesson 3-7, page 151)

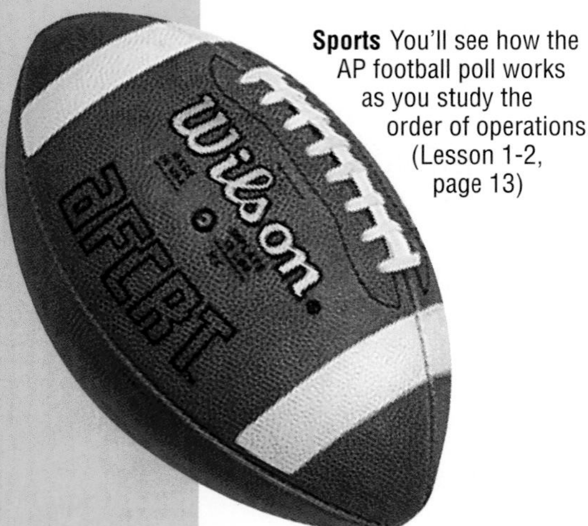

Sports You'll see how the AP football poll works as you study the order of operations. (Lesson 1-2, page 13)

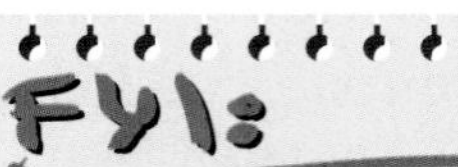

Mae Jemison was the first African-American woman to go into space. Her first mission was a cooperative effort between U.S. and Japan which focused on experiments involving space travel and biology.

FYI features contain interesting facts that enhance the applications. (Lesson 2-3, page 81)

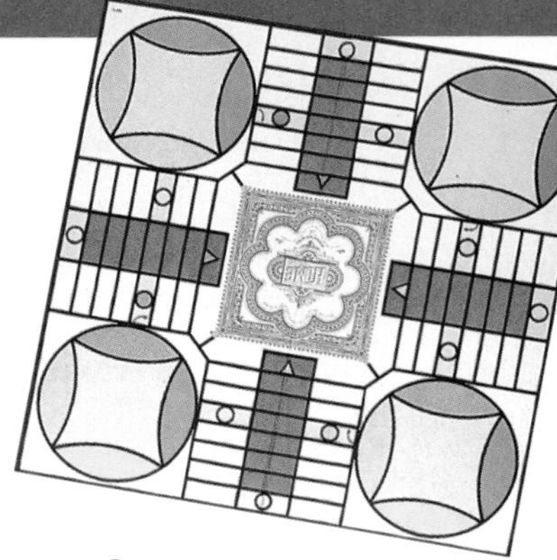

Games Rolling dice in the game of parchees which originated in India, is related t exponents. (Lesson 4-2, page 175)

When is mathematics funny? An actual reprinted *Fox Trot* comic illustrates how mathematics is a part of our society. (Lesson 1-3, page 20)

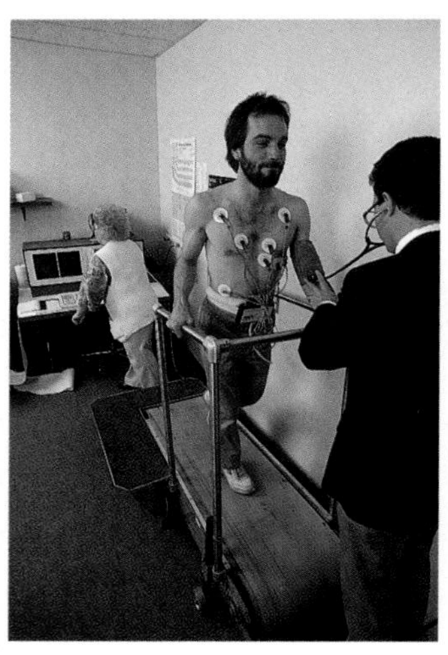

Health and Medicine You'll study a health-related application that involves heart rates and algebraic expressions. (Lesson 1-3, pages 16-17)

COOPERATIVE LEARNING PROJECT

THE SHAPE OF THINGS TO COME

Living Batteries

You'll do research about the possible future use of bacteria to generate electricity and power things like a watch. (Lesson 6-9, page 320)

Chapter 6 Rationals: Patterns in Multiplication and Division 272

6-1 Writing Fractions as Decimals 274
6-2 Estimating Products and Quotients 280
6-3 Multiplying Fractions 284
6-4 Dividing Fractions 289
6-5A **Math Lab: Hands-On Activity**
Multiplying and Dividing Decimals . . . 294
6-5 Multiplying and Dividing Decimals 295
6-6A **Math Lab: Hands-On Activity**
Mean, Median & Mode 300
6-6 **Integration: Statistics**
Measures of Central Tendency 301
6-6B **Math Lab: Graphing Calculator Activity**
Finding Mean and Median 307
6-7 Solving Equations and Inequalities 308
6-8 **Integration: Discrete Mathematics**
Geometric Sequences 312
6-9 Scientific Notation 317
Chapter 6 Highlights 321
Chapter 6 Study Guide and Assessment 322
Ongoing Assessment, Chapters 1–6 326

Chapter 7 Solving Equations and Inequalities 328

7-1 Problem-Solving Strategy: Work Backward 330
7-2A **Math Lab: Hands-On Activity**
Two-Step Equations with Cups and Counters 333
7-2 Solving Two-Step Equations 334
7-3 Writing Two-Step Equations 338
7-4 **Integration: Geometry**
Circles and Circumference 341
7-5A **Math Lab: Hands-On Activity**
Equations with Variables on Each Side 345
7-5 Solving Equations with Variables on Each Side 346
7-6 Solving Multi-Step Inequalities 351
7-7 Writing Inequalities 355
7-8 **Integration: Measurement**
Using the Metric System 358
Chapter 7 Highlights 363
Chapter 7 Study Guide and Assessment 364

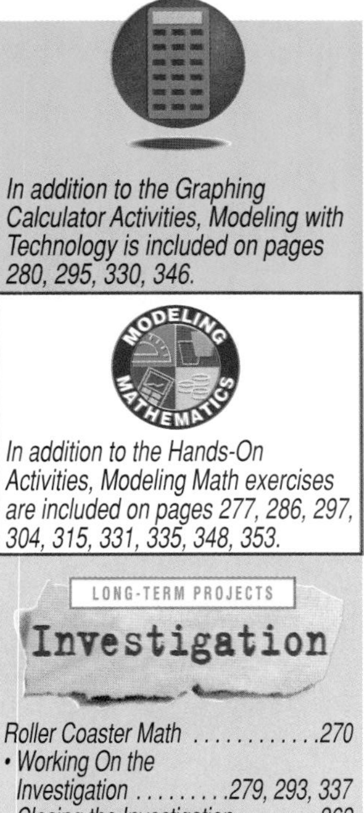

In addition to the Graphing Calculator Activities, Modeling with Technology is included on pages 280, 295, 330, 346.

In addition to the Hands-On Activities, Modeling Math exercises are included on pages 277, 286, 297, 304, 315, 331, 335, 348, 353.

Roller Coaster Math 270
• Working On the Investigation 279, 293, 337
• Closing the Investigation 362

Chapter 8 Functions and Graphing 370

8-1 Relations and Functions 372
8-2A **Math Lab: Hands-On Activity**
Scatter Plots 378
8-2 **Integration: Statistics** Scatter Plots 379
8-3 Graphing Linear Equations 385
8-3B **Math Lab: Hands-On Activity**
Graphing Parabolas 391
8-4 Equations as Functions 392
8-5 Problem-Solving Strategy: Draw a Graph . . 396
8-6 Slope 400
8-6B **Math Lab: Hands-On Activity**
Circles and Slope 405
8-7 Intercepts 406
8-7B **Math Lab: Graphing Calculator Activity**
Families of Graphs 411
8-8 Systems of Equations 412
8-9A **Math Lab: Graphing Calculator Activity**
Graphing Inequalities 417
8-9 Graphing Inequalities 418
Chapter 8 Highlights 423
Chapter 8 Study Guide and Assessment 424
Ongoing Assessment, Chapters 1–8 428

In addition to the Graphing Calculator Activities, Modeling with Technology is included on pages 379, 412, 440, and 467.

MODELING MATHEMATICS

In addition to the Hands-On Activities, Modeling Math exercises are included on pages 387, 404, 414, 434, 442, 451, 460.

LONG-TERM PROJECTS

Investigation

Stadium Stampede 368
• Working On the Investigation 384, 404, 453, 471
• Closing the Investigation 476

Chapter 9 Ratio, Proportion, and Percent 430

9-1 Ratios and Rates 432
9-2 Problem-Solving Strategy: Make a Table . . 437
9-3A **Math Lab: Hands-On Activity**
Fair and Unfair Games 439
9-3 **Integration: Probability**
Simple Probability 440
9-4 Using Proportions 444
9-4B **Math Lab: Hands-On Activity**
Capture-Recapture 448
9-5 Using the Percent Proportion 449
9-6 **Integration: Statistics**
Using Statistics to Predict 454
9-7 Fractions, Decimals, and Percents 458
9-8 Percent and Estimation 462
9-9 Using Percent Equations 467
9-10 Percent of Change 472
Chapter 9 Highlights 477
Chapter 9 Study Guide and Assessment 478

Connections

Did you realize that math can be used to model many real-world applications? Do you know that mathematics is used in biology? in history? in geography? Yes, it may be hard to believe, but mathematics is frequently used in subjects besides math.

Connections to Algebra

You'll use algebra as a tool to solve problems from math topics such as geometry and arithmetic. (Lesson 7-4, page 342)

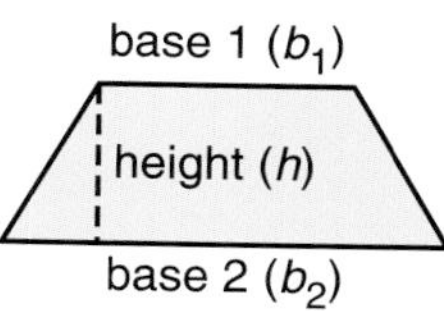

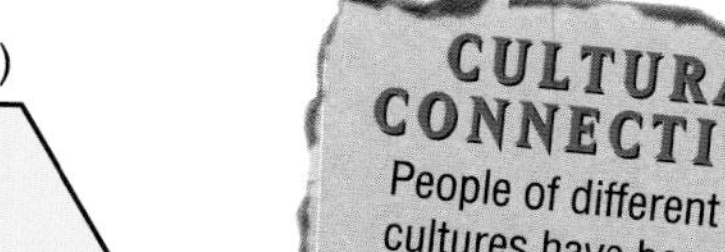

CULTURAL CONNECTIONS

People of different cultures have been playing games for centuries. One of the oldest games, Senet, made its first known appearance in Egypt in about 2686 B.C. Senet was played by people of all social classes and had both religious and secular forms.

Connections to Geometry

You'll learn how ratios can be used to describe the dimensions of the pyramids of Egypt. (Lesson 9-1, page 434)

Cultural Connections introduce you to a variety of world cultures. (Lesson 1-6, page 32)

CONNECTIONS Interdisciplinary Connections

Geography You'll study how to write and solve an equation involving the highest and lowest points in Africa. (Lesson 2-5, page 90)

Ecology You'll learn to model data about household waste as ordered pairs. (Lesson 8-1, page 372)

Music You'll solve equations that involve the number of vibrations on a guitar for a particular musical note. (Lesson 6-7, page 308)

HELP WANTED

Are you interested in the patterns and changes of the weather? For more information on this field, contact:

American Meteorological Society
45 Beacon St.
Boston, MA 02108

Help Wanted features include information on interesting careers. (Lesson 5-6, page 249)

Health You'll relate pollen count to comparing integers on a number line. (Lesson 2-3, page 78)

MATH JOURNAL

Math Journal exercises give you the opportunity to assess yourself and write about your understanding of key math concepts. (Lesson 5-2, page 231)

Chapter 10 More Statistics and Probability 484

10-1 Stem-and-Leaf Plots 486
10-2 Measures of Variation 490
10-3 Displaying Data 495
10-3B **Math Lab: Graphing Calculator Activity** Making Statistical Graphs 502
10-4 Misleading Statistics 504
10-5 Counting 509
10-6A **Math Lab: Hands-On Activity** Permutations and Combinations 514
10-6 Permutations and Combinations 515
10-7 Odds 520
10-8 Problem-Solving: Use a Simulation 524
10-8B **Math Lab: Hands-On Activity** Making Predictions 529
10-9 Probability of Independent and Dependent Events 530
10-10 Probability of Compound Events 535
Chapter 10 Highlights 539
Chapter 10 Study Guide and Assessment 540
Ongoing Assessment, Chapters 1–10 544

Chapter 11 Applying Algebra to Geometry 546

11-1 The Language of Geometry 548
11-1B **Math Lab: Hands-On Activity** Constructions 554
11-2 **Integration: Statistics** Making Circle Graphs 556
11-3 Angle Relationships and Parallel Lines 561
11-3B **Math Lab: Graphing Calculator Activity** Slopes of Parallel Lines 567
11-4 Triangles 568
11-5 Congruent Triangles 573
11-6 Similar Triangles and Indirect Measurement 578
11-7 Quadrilaterals 584
11-8 Polygons 589
11-8B **Math Lab: Hands-On Activity** Tessellations 594
11-9 Transformations 595
Chapter 11 Highlights 601
Chapter 11 Study Guide and Assessment 602

In addition to the Graphing Calculator Activities, Modeling with Technology is included on pages 490, 530, 573, and 589.

MODELING MATHEMATICS

In addition to the Hands-On Activities, Modeling Math exercises are included on pages 488, 498, 511, 552, 575, 581, 586, and 597.

LONG-TERM PROJECTS

Investigation

It's only a game 482
- *Working On the Investigation* 508, 528, 588
- *Closing the Investigation* 600

Modeling Mathematics

Using Manipulatives and Technology

One of the best ways to learn is to learn by doing. Here are two examples.

Hands-On Activities

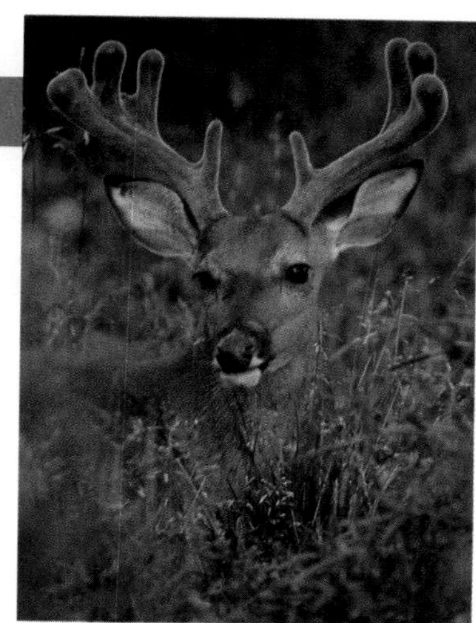

In Lesson 9-4B, you'll learn how to use the capture-recapture method of counting deer.

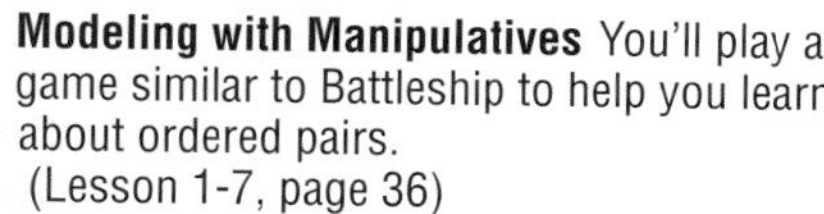

Modeling with Manipulatives You'll play a game similar to Battleship to help you learn about ordered pairs. (Lesson 1-7, page 36)

Do you know how to use high-tech machines like graphing calculators? We know that such knowledge is vital to your future success. That is why graphing calculator technology is integrated throughout this book.

There are several ways in which graphing calculators are integrated.

- **Introduction to Graphing Calculators** On page xx and page 1, you'll get acquainted with the basic features and functions of a graphing calculator.
- **Graphing Calculator Activities** In Lesson 3-5B, you'll use a program to find the area of quadrilaterals given the coordinates of the vertices.
- **Graphing Calculator Exercises** Many exercises are designed to be solved using a graphing calculator. For example, Exercise 23 in Lesson 8-2 involves making a scatter plot of keyboarding speeds.

Option	Minimum Investment	Monthly Interest Rate (Percent)	Term (Months)	Interest for Term
1	$2500	0.375	6	$56.25
2	$2500	0.45	12	$135.00
3	$5000	0.54	36	$972.00
4	$7500	0.625	60	$2812.50

- **Modeling with Technology** On page 467, Lesson 9-9 begins with an application involving a spreadsheet.

Techno Tips are designed to help you make more efficient use of technology through practical hints and suggestions.

Some calculators use a [+/–] key, called the "change sign" key instead of the [(–)] key.

(Lesson 2-4, page 83)

Chapter 12 Measuring Area and Volume 608

12-1A **Math Lab: Hands-On Activity**
Areas and Geoboards 610
12-1 Area: Parallelograms, Triangles, and Trapezoids 612
12-1B **Math Lab: Hands-On Activity**
Fractals 618
12-2 Area: Circles 619
12-3 **Integration: Probability**
Geometric Probability 623
12-3B **Math Lab: Graphing Calculator Activity**
Geometric Probability 628
12-4 Problem-Solving Strategy: Make a Model or a Drawing 629
12-5 Surface Area: Prisms and Cylinders 632
12-6 Surface Area: Pyramids and Cones 638
12-7A **Math Lab: Hands-On Activity**
Volume 643
12-7 Volume: Prisms and Cylinders 644
12-8 Volume: Pyramids and Cones 649
12-8B **Math Lab: Hands-On Activity**
Similar Solid Figures 654
Chapter 12 Highlights 655
Chapter 12 Study Guide and Assessment 656
Ongoing Assessment, Chapters 1–12 660

In addition to the Graphing Calculator Activities, Modeling with Technology is included on pages 644 and 688.

In addition to the Hands-On Activities, Modeling Math exercises are included on pages 625, 635, 651, 667, and 679.

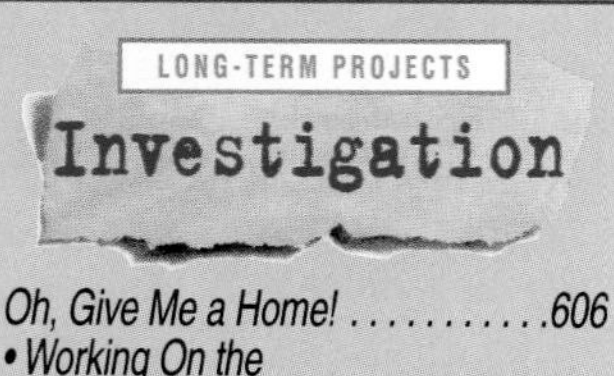

Oh, Give Me a Home! 606
- *Working On the Investigation* 637, 653, 671
- *Closing the Investigation* 698

Chapter 13 Applying Algebra to Right Triangles 662

13-1 Squares and Square Roots 664
13-2 Problem-Solving Strategy: Use Venn Diagrams 669
13-3 The Real Number System 672
13-4 The Pythagorean Theorem 676
13-4B **Math Lab: Hands-On Activity**
Graphing Irrational Numbers 682
13-5 Special Right Triangles 683
13-6A **Math Lab: Hands-On Activity**
Ratios in Right Triangles 687
13-6 The Sine, Cosine, and Tangent Ratios 688

13-6B **Math Lab: Graphing Calculator Activity**
Slope and Tangent 693
13-7 Using Trigonometric Ratios 694
Chapter 13 Highlights 699
Chapter 13 Study Guide and Assessment 700

Chapter 14 Polynomials 704

14-1 Polynomials 706
14-1B **Math Lab: Hands-On Activity**
Representing Polynomials with Algebra Tiles 710
14-2 Adding Polynomials 711
14-3 Subtracting Polynomials 715
14-4 Powers of Monomials 719
14-5A **Math Lab: Hands-On Activity**
Multiplying Polynomials 724
14-5 Multiplying a Polynomial by a Monomial 725
14-6 Multiplying Binomials 728
14-6B **Math Lab: Hands-On Activity**
Factoring 732
Chapter 14 Highlights 733
Chapter 14 Study Guide and Assessment 734
Ongoing Assessment, Chapters 1–14 738

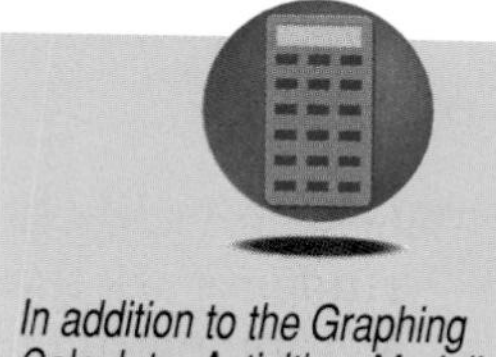

In addition to the Graphing Calculator Activities, Modeling with Technology is included on page 719.

In addition to the Hands-On Activities, Modeling Math exercises are included on pages 713, 717, 726, and 730.

Student Handbook

Extra Practice 740
Chapter Tests 774
Glossary 788
Spanish Glossary 796
Selected Answers 804
Photo Credits 830
Applications Index 834
Index 836

Introduction to the Graphing Calculator

What is it?
What does it do?
How is it going to help me learn math?

These are just a few of the questions many students ask themselves when they first see a graphing calculator. Some students may think, "Oh, no! Do we *have* to use one?", while others may think, "All right! We get to use these neat calculators!" There are as many thoughts and feelings about graphing calculators as there are students, but one thing is for sure: a graphing calculator *can* help you learn mathematics.

So what is a graphing calculator? Very simply, it is a calculator that draws graphs. This means that it will do all of the things that a "regular" calculator will do, *plus* it will draw graphs of simple or very complex equations. In pre-algebra, this capability is nice to have as you learn to graph equations and solve equations by graphing.

But a graphing calculator can do more than just calculate and draw graphs. For example, you can program it and work with data to make statistical graphs and computations. If you need to generate random numbers, you can do that on the graphing calculator. If you need to find the absolute value of numbers, you can do that too. It's really a very powerful tool—so powerful that it is often called a pocket computer. A graphing calculator can save you time and make doing mathematics easier.

As you may have noticed, graphing calculators have some keys that other calculators do not. The Texas Instruments TI-82 will be used in this text for graphing and programming. The keys located on the bottom half of the calculator are probably familiar to you as they are the keys found on basic scientific calculators. The keys located just below the screen are the graphing keys. You will also notice the up, down, left, and right arrow keys. These allow you to move the cursor around on the screen, to "trace" graphs that have been plotted, and to choose items from the menus. The other keys located on the top half of the calculator access the special features such as statistical computations and programming features.

There are some keystrokes that can save you time when using the graphing calculator. A few of them are listed below.

- Any light blue commands written above the calculator keys are accessed with the [2nd] key, which is also blue. Similarly, any gray characters above the keys are accessed with the [ALPHA] key, which is also gray.
- [2nd] [ENTRY] copies the previous calculation so you can edit and use it again.
- Pressing [ON] while the calculator is graphing stops the calculator from completing the graph.
- [2nd] [QUIT] will return you to the home (or text) screen.
- [2nd] [A-LOCK] locks the [ALPHA] key, which is like pressing "shift lock" or "caps locks" on a typewriter or computer. The result is that all letters will be typed and you do not have to repeatedly press the [ALPHA] key. (This is handy for programming.) Stop typing letters by pressing [ALPHA] again.
- [2nd] [OFF] turns the calculator off.

Some commonly-used mathematical functions are shown in the table below. As with any scientific calculator, the graphing calculator observes the order of operations.

Mathematical Operation	Example	Keys	Display
evaluate expressions	Find 2 + 5.	2 [+] 5 [ENTER]	2+5 7
exponents	Find 3^5.	3 [^] 5 [ENTER]	3^5 243
multiplication	Evaluate 3(9.1 + 0.8).	3 [×] [(] 9.1 [+] .8 [)] [ENTER]	3*(9.1+.8) 29.7
roots	Find $\sqrt{14}$.	[2nd] [√] 14 [ENTER]	√14 3.741657387
opposites	Enter −3.	[(−)] 3	−3

Graphing on the TI-82

Before graphing, we must instruct the calculator how to set up the axes in the coordinate plane. To do this, we define a **viewing window**. The viewing window for a graph is the portion of the coordinate grid that is displayed on the **graphics screen** of the calculator. The viewing window is written as [left, right] by [bottom, top] or [Xmin, Xmax] by [Ymin, Ymax]. A viewing window of $[-10, 10]$ by $[-10, 10]$ is called the **standard viewing window** and is a good viewing window to start with to graph an equation. The standard viewing window can be easily obtained by pressing ZOOM 6. Try this. Move the arrow keys around and observe what happens. You are seeing a portion of the coordinate plane that includes the region from -10 to 10 on the x-axis and from -10 to 10 on the y-axis. Move the cursor, and you can see the coordinates of the point for the position of the cursor.

Any viewing window can be set manually by pressing the WINDOW key. The window screen will appear and display the current settings for your viewing window. First press ENTER. Then, using the arrow and ENTER keys, move the cursor to edit the window settings. Xscl and Yscl refer to the x-scale and y-scale. This is the number of units between tick marks on the x- and y-axes. Xscl = 1 means that there will be a tick mark for every unit of one along the x-axis. The standard viewing window would appear as follows.

Xmin = -10
Xmax = 10
Xscl = 1
Ymin = -10
Ymax = 10
Yscl = 1

Programming on the TI-82

The TI-82 has programming features that allow us to write and execute a series of commands to perform tasks that may be too complex or cumbersome to perform otherwise. Each program is given a name. Commands begin with a colon (:), followed by an expression or an instruction. Most of the features of the calculator are accessible from program mode.

When you press PRGM, you see three menus: EXEC, EDIT, and NEW. EXEC allows you to execute a stored program, EDIT allows you to edit or change a program, and NEW allows you to create a program. The following tips will help you as you enter and run programs on the TI-82.

- To begin entering a new program, press PRGM ▶ ▶ ENTER.
- After a program is entered, press 2nd QUIT to exit the program mode and return to the home screen.
- To execute a program, press PRGM. Then use the down arrow key to locate the program name and press ENTER, or press the number or letter next to the program name.
- If you wish to edit a program, press PRGM ▶ and choose the program from the menu.
- To immediately re-execute a program after it is run, press ENTER when Done appears on the screen.
- To stop a program during execution, press ON.

While a graphing calculator cannot do everything, it can make some things easier. To prepare for whatever lies ahead, you should try to learn as much as you can. The future will definitely involve technology, and using a graphing calculator is a good start toward becoming familiar with technology. Who knows? Maybe one day you will be designing the next satellite, building the next skyscraper, or helping students learn mathematics with the aid of a graphing calculator!

LONG-TERM PROJECT

Investigation

Up, Up and Away!

Have you ever dreamed of breaking free of the ties of gravity and soaring among the clouds? Maybe you should try an ultralight airplane! An ultralight is a small one-person airplane that is very light, about the size and weight of a hang glider. Unlike a hang glider, an ultralight is propelled by a small engine and can be enclosed or open to the air.

The Federal Aviation Administration, or FAA, governs all types of aircraft in the United States. They set forth the following restrictions on ultralights.

Notice! All ultralight airplanes must:

- *be single seat aircrafts*
- *weigh 254 pounds or less*
- *have a top speed of 55 knots (63 miles per hour)*
- *stall at 24 knots (28 miles per hour)*
- *carry no more than 5 gallons of fuel*
- *never fly over towns or settlements*
- *never fly at night or around airports without special permission*
- *always yield the right of way to all other aircraft*

Special 2-seat exemptions, for training only, may weigh up to 496 pounds and carry 10 gallons of fuel.

Veteran fliers say that building and flying an ultralight is inexpensive when compared to other types of airplanes. A number of kits are available for you to put together yourself for between \$3000 and \$6000. Plan on spending 6 months to two years on construction. Flight training, while not required, is highly recommended. Instructors charge about \$600 to \$1200. Once the initial expenses of plane and training are completed, flying is very inexpensive. The engine of an ultralight usually burns only 2 to 3 gallons of fuel an hour. Many ultralights can be stored at home. If you need to store your plane in a hangar, budget about \$30 to \$90 a month.

Starting Your Investigation

Each year, the Experimental Aircraft Association, or EAA, gathers in Oshkosh, Wisconsin, for their annual convention. EAA members are concerned with all types of home-built, antique, and ultralight aircraft. Suppose you and the other members of your group are on a committee chosen by the local EAA chapter to make plans to fly the group to the convention. There are many things you must consider as you make your plans. Look over the following list of considerations.

Think About . . .

- *Will you be able to fly there in one day?*
- *How much fuel will you need? Can you make it there on one tank of fuel?*
- *How should you plan your route?*
- *Are there places to store the planes while you are at the convention?*
- *Must each person fly his or her own plane, or can two-seat ultralights be used?*

Work with your group to add to the list of questions to be considered. Then reread the information about ultralights on the previous page to find significant information. You may wish to look in some other sources.

You will continue working on this Investigation throughout Chapter 1. Be sure to keep your materials in your Investigation Folder.

For current information on ultralights, visit: **www.glencoe.com/sec/math/prealg/mathnet**

Investigation
Up, Up, and Away!

Working on the Investigation
Lesson 1-1, p. 10

Working on the Investigation
Lesson 1-6, p. 35

Closing the Investigation
End of Chapter 1, p. 56

CHAPTER 1

Tools for Algebra and Geometry

TOP STORIES in Chapter 1

In this chapter, you will:

- solve problems using the four-step problem-solving plan,
- choose the appropriate method of computation,
- use variables and algebraic properties,
- graph points in a coordinate system, and
- gather and record data.

MATH AND FASHION IN THE NEWS

The search is over . . . Jeans that fit!

***Source:** The Columbus Dispatch, December 27, 1994*

Too long, too short, too big, too small . . . While jeans may be the most popular fashion of all time, sometimes finding a pair that fits is like searching for buried treasure. But Levi Strauss, the original maker of jeans, is coming to the rescue. You can now go to a store and get a pair of jeans made to measure. A salesperson takes four measurements on you and enters them into a computer. After some fancy figuring, the computer chooses a trial pair of jeans for you to try on. Any necessary adjustments are made, and your order is placed through the computer directly to the factory. In three weeks, your dream jeans arrive. Being in fashion has never been so easy!

Putting It into Perspective

1853
Levi Strauss begins producing pants from canvas-like cloth for gold miners.

1926
First jeans with a zipper instead of buttons

1850 | 1880 | 1910

1848
Discovery of gold in California hills draws Americans westward.

1873
U.S. patent is granted for use of rivets to reinforce stress points in jeans.

EMPLOYMENT OPPORTUNITIES

Shape the Fads and Fashions of Your Generation

DO YOU LIKE dealing with people? Do you have a flair for fashion? We are looking for sharp, energetic people to work as **retail salespeople** and **clothing buyers**. Salespeople assist customers in choosing clothing and offer advice on fit, styles, colors, and fabrics. Buyers work with manufacturers to choose merchandise for our retail outlets. A high-school education is required for salespeople, college degree preferred for buyer's position. Additional training or experience is a plus.

For more information, contact:

National Retail Federation
701 Pennsylvania Avenue NW, Suite 701
Washington, D.C. 20004

interNET CONNECTION For up-to-date information on selling fashions, visit:
www.glencoe.com/sec/math/prealg/mathnet

Statistical Snapshot

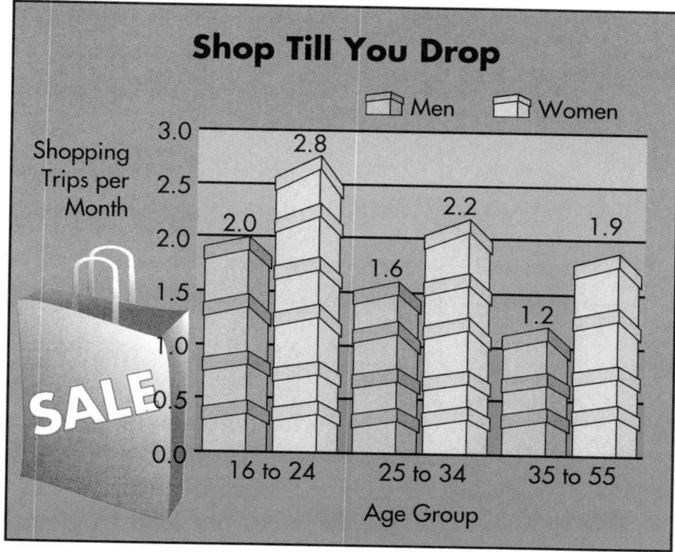

Source: *Cotton Incorporated,* 1994 survey

1939 Lady Levis, the first name-brand jeans made specifically for women introduced

1940

1965 Invention of the mini-skirt

1970

1974 Beverly Johnson, the first black model on the cover of a major fashion magazine

1986 Acid-washed jeans are introduced by an Italian jeansmaker.

1994 Made-to-measure jeans introduced

2000

1-1 Problem-Solving Strategy: Make a Plan

Setting Goals: *In this lesson, you'll use the four-step plan to solve real-life, math-related problems and choose an appropriate method of computation.*

Modeling a Real-World Application: Health and Exercise

Venus Williams studies French and Spanish so that she can talk to the tennis players who she meets from other countries. After her professional tennis career is over, Venus hopes to become a paleontologist.

At 15, Venus Williams is the newest superstar in the world of tennis. Venus learned to play tennis from her father in her home town of Compton, California. Now, as she begins her professional tennis career, she gives back to the community by giving tennis clinics at public schools near her home in Delray Beach, Florida.

Venus spends hours exercising each week to remain in top condition. When exercising, it is important to maintain a reasonable heart rate.

Based on the table below, what heart rate should Venus maintain when she exercises?

Guide to Reasonable Heart Rate (85% capacity)											
Age	20	25	30	35	40	45	50	55	60	65	70
Heart Rate (beats per minute)	174	170	166	162	157	153	149	145	140	136	132

This problem will be solved in Example 1.

Learning the Concept

You can use four steps to solve real-life, math-related problems.

1. **Explore**
   - Read the problem carefully.
   - Ask yourself questions like "What facts do I know" and "What do I need to find out?"
2. **Plan**
   - See how the facts relate to each other.
   - Make a plan for solving the problem.
   - Estimate the answer.
3. **Solve**
   - Use your plan to solve the problem.
   - If your plan does not work, revise it or make a new plan.
4. **Examine**
   - Reread the problem.
   - Ask, "Is my answer close to my estimate?"
   - Ask, "Does my answer make sense for the problem?
   - If not, solve the problem another way.

Example 1 **Health and Exercise**

Refer to the application at the beginning of the lesson. Find a reasonable heart rate for exercise for 15-year-old Venus.

Use the problem-solving plan to find the heart rate.

Explore *What facts do we know and what are we trying to find out?*

The table shows the suggested heart rates for several other ages. We are trying to find the rate for a 15 year old.

Plan *Make a plan for attacking the problem and estimate the answer.*

Since we know the heart rates for several ages, it makes sense to look for a pattern to extend to find the rate for a 15 year old. The heart rate decreases as the age increases. So the heart rate for a 15 year old should be higher than that for a 20 year old.

Solve *Use the plan to solve the problem.*

Find the differences between consecutive ages and heart rates.

+5 +5 +5 +5 +5 +5 +5 +5 +5 +5

Age (years)	20	25	30	35	40	45	50	55	60	65	70
Heart Rate (beats per minute)	174	170	166	162	157	153	149	145	140	136	132

−4 −4 −4 −5 −4 −4 −4 −5 −4 −4

Each consecutive age increases 5 years. The pattern for the heart rates is −4, −4, −4, −5. Extend this pattern as follows.

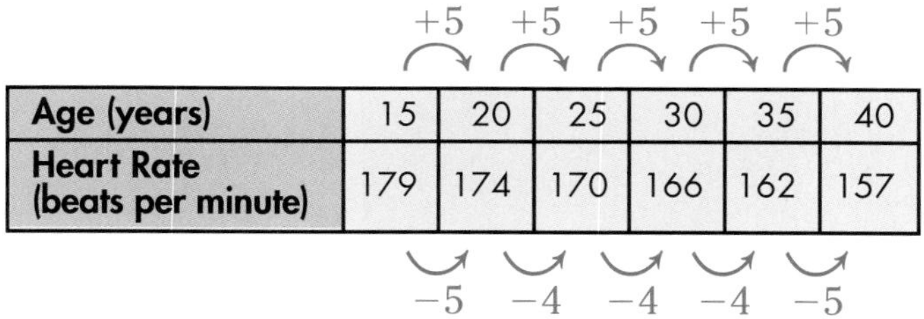

+5 +5 +5 +5 +5

Age (years)	15	20	25	30	35	40
Heart Rate (beats per minute)	179	174	170	166	162	157

−5 −4 −4 −4 −5

A reasonable heart rate for exercise for a 15 year old is 174 + 5 or 179.

Examine *Compare the answer to the problem and the estimate.*

The heart rate of 179 that we found for a 15 year old makes sense with the estimate that we made.

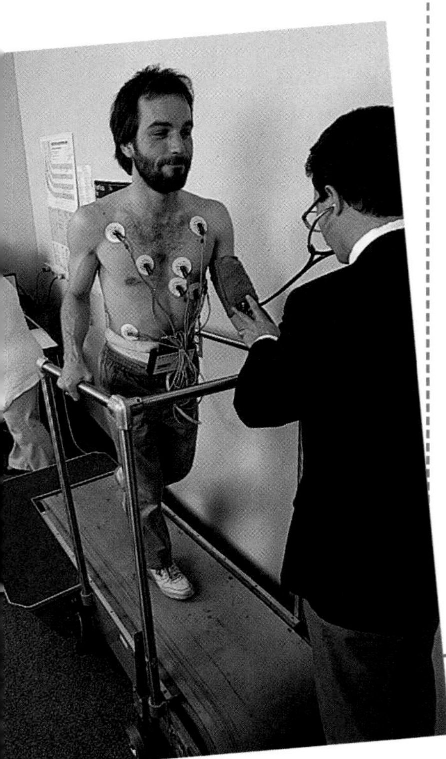

One of the important steps in solving problems is choosing the method of computation. The diagram below can help you decide which method is most appropriate.

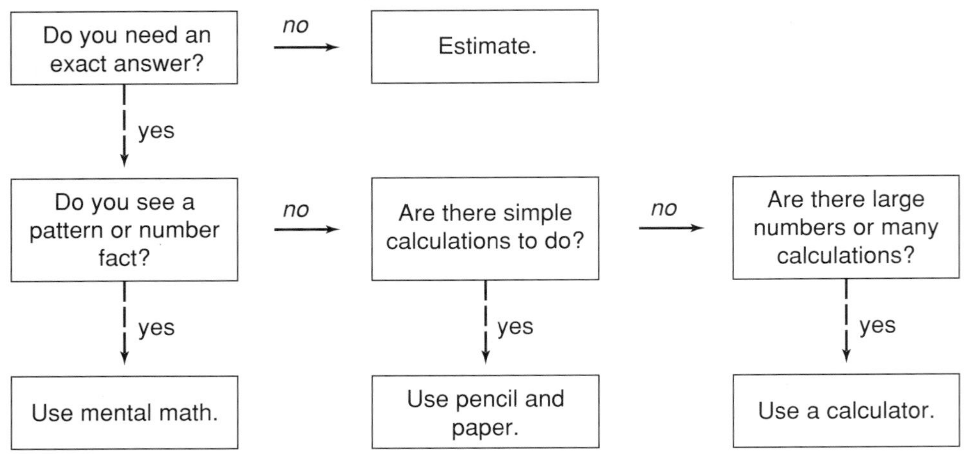

Example 2

Medicine

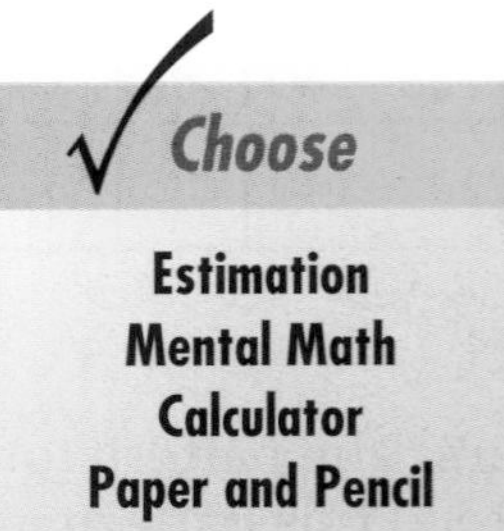

Modern medicine has made amazing advances in human transplants. The numbers of different types of transplants that were performed in the United States in 1992 are shown at the right. Find the total number of transplants performed in 1992.

Transplant	Number
heart	2172
liver	3059
kidney	10,210
heart-lung	48
lung	535
pancreas	557

Source: U.S. Department of Health and Human Services

Explore We are given the number of each type of transplant and wish to find the total.

Plan Add the numbers of transplants to find the total. Use the diagram above to find the appropriate method of computation.

Since we need an exact answer, we will not solve using estimation. There is no pattern in the numbers and there are many calculations to be performed, so using a calculator may be the best choice.

Even though we will solve using a calculator, it is a good idea to estimate first so that we know whether the answer we find is reasonable. Round each number to the nearest hundred and then add mentally.

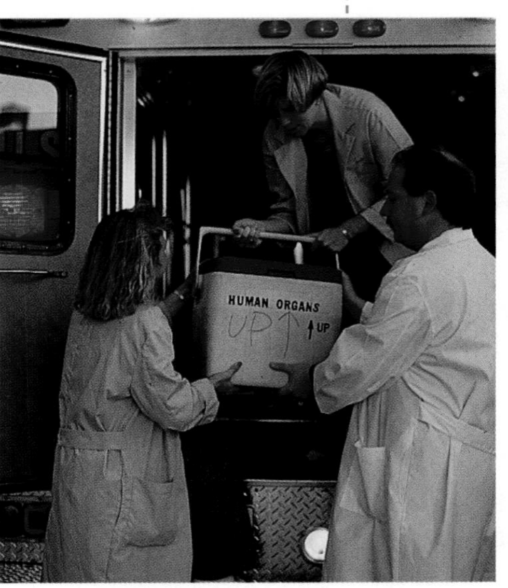

Estimate:

$$\begin{array}{rcr} 2172 & \rightarrow & 2200 \\ 3059 & \rightarrow & 3100 \\ 10{,}210 & \rightarrow & 10{,}200 \\ 48 & \rightarrow & 0 \\ 535 & \rightarrow & 500 \\ 557 & \rightarrow & \underline{600} \\ & & 16{,}600 \end{array}$$

Do you think the estimate is higher or lower than the actual total?

Solve	*Use the plan to solve the problem.* Use a calculator to find the total. 2172 [+] 3059 [+] 10210 [+] 48 [+] 535 [+] 557 [=] 16581
Examine	The total we found with the calculator is very close to our estimate, so the answer is reasonable.

Checking Your Understanding

Communicating Mathematics

Read and study the lesson to answer these questions.

1. **Describe** each step of the problem-solving plan in your own words.
2. **Explain** why it is important to examine your solution after you solve a problem.
3. **Demonstrate** how to use the diagram on page 8 to determine the best method of computation.

4. Begin your Math Journal by writing two or three sentences that describe what you like and dislike about mathematics.

Guided Practice

5. **Sales** John Diaz sells large computer software packages to accounting corporations for Object Systems, Inc. Each month, his supervisor writes a sales plan for the number of software packages that Mr. Diaz is supposed to sell. Mr. Diaz receives a salary, plus a bonus for any software packages he sells in excess of the plan. The table at the right shows the bonus amount for different levels of sales. Determine Mr. Diaz's bonus if he sells 16 software packages more than the plan.

Sales above plan	Bonus
2	$100
4	$125
6	$150
8	$175

   a. Write the *Explore* step. What do you know and what do you need to find?
   b. Write the *Plan* step. What strategy will you use? What do you estimate the answer to be?
   c. *Solve* the problem using your plan. What is your answer?
   d. *Examine* your solution. Is it reasonable? Does it answer the question?

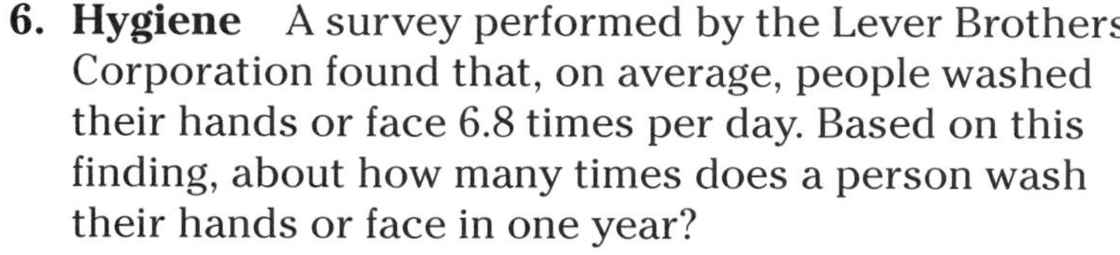

6. **Hygiene** A survey performed by the Lever Brothers Corporation found that, on average, people washed their hands or face 6.8 times per day. Based on this finding, about how many times does a person wash their hands or face in one year?
   a. Which method of computation do you think is most appropriate? Justify your choice.
   b. Solve the problem using the four-step plan. Be sure to examine your solution.
   c. Do the results of the survey imply that each person actually washes his or her hands 6.8 times each day? Explain.

Exercises: Practicing and Applying the Concept

Independent Practice

Solve each problem using the four-step plan.

7. **Money Management** The Westerville North High School Band plans to travel to Miami, Florida, to perform in the Orange Bowl Parade. Ana Vera needs to save $300 for the trip. She has $265 in her bank account and earns $20 shoveling snow and $25 babysitting for the neighbors. Does Ana have enough for the trip?
8. Find the total cost of building supplies that cost $2.34, $9.28, $92.34, $501.38 and $0.34.

Choose

Estimation
Mental Math
Calculator
Paper and Pencil

9. **Postal Service** The chart at the right shows the cost of mailing a first-class letter of different weights. How much would it cost to send a letter that weighs 8 ounces?

Weight (ounces)	Cost (1995)
1	$0.32
2	$0.57
3	$0.82
4	$1.07
5	$1.32

10. **Space Exploration** The space shuttle can carry a payload of about 65,000 pounds. A compact car weighs about 2450 pounds. About how many compact cars could be carried on the space shuttle?
11. **Geometry** Study the pattern of circles.

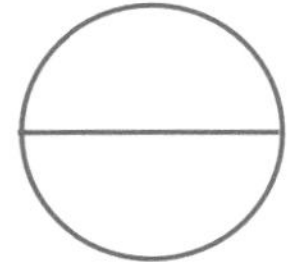
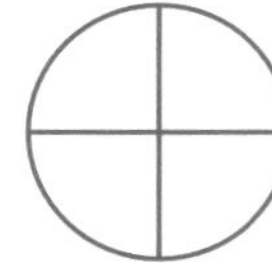

   a. Draw the next two figures in the sequence.
   b. Write a sequence of numbers for the number of pieces in each circle.
   c. Write a sequence of numbers for the number of cuts in each circle.
   d. What is the relationship between the number of cuts and the number of pieces in each circle?
12. When Gabriella divided 45,109.5 by 1479 the calculator showed 305. Is this a reasonable answer? Explain.

WORKING ON THE Investigation

Up, Up, and Away!

Refer to the Investigation on pages 2–3.

Look over your research and the list of considerations that you placed in your Investigation Folder.

- Work with the members of your group to choose one of the questions to consider first.
- Explore the question. Decide what you know and what you need to know to find the answer. You may need to do some more research.
- Make a plan for solving. If possible, make an estimate.
- Execute your plan to solve. If your plan doesn't work, make another plan and solve again.
- Examine your answer. Does it come close to the estimate? Can you justify it?

Add the results of your work to your Investigation Folder.

1-2 Order of Operations

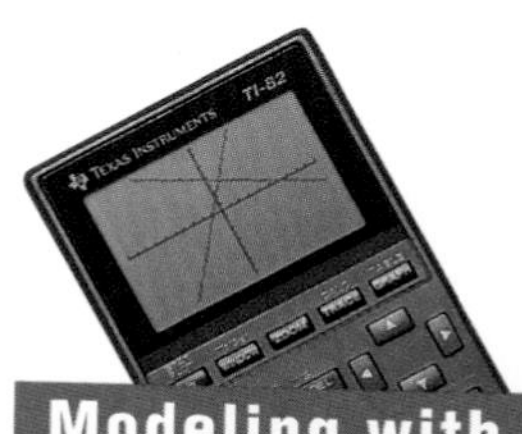

Setting Goals: *In this lesson, you'll use the order of operations to evaluate expressions.*

Modeling with Technology

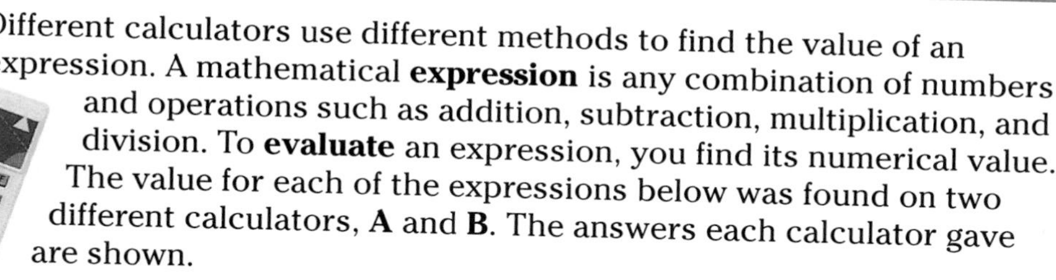

Different calculators use different methods to find the value of an expression. A mathematical **expression** is any combination of numbers and operations such as addition, subtraction, multiplication, and division. To **evaluate** an expression, you find its numerical value. The value for each of the expressions below was found on two different calculators, **A** and **B**. The answers each calculator gave are shown.

Expression	$7 + 2 \times 3$	$8 + 4 \div 2$	$12 \div 6 + 16 \div 4$	$19 - 7 + 12 \times 2 \div 8$
Calculator A	13	10	6	15
Calculator B	27	6	4.5	6

Your Turn Use a calculator to evaluate each expression above, entering the numbers and symbols in the order shown.

a. Did the calculator you used agree with the answers given by calculator A or calculator B?

b. Compare the answers you found to the ones that the rest of the class found. Did everyone agree with either calculator A or B?

c. Discuss reasons that the calculators may have given different answers. How did each calculator find the answer?

d. Which value do you think is correct for each expression? Defend your answer.

Learning the Concept

There are two methods that someone might use to find the value of an expression that contains more than one operation. Consider the expression $8 + 5 \times 2$.

Method 1

$8 + 5 \times 2 = 8 + 10$ *Multiply, then add.*
$8 + 5 \times 2 = 18$

Method 2

$8 + 5 \times 2 = 13 \times 2$ *Add, then multiply.*
$8 + 5 \times 2 = 26$

In order to avoid confusion, mathematicians established the **order of operations** to tell us how to find the value of an expression.

First: Do all multiplications and divisions from left to right.
Second: Do all additions and subtractions from left to right.

THINK ABOUT IT

Calculators that follow the order of operations are called scientific calculators. Is your calculator a scientific or non-scientific calculator?

The order of operations guarantees that each numerical expression has a *unique* value. Using the order of operations, the correct value of the expression $8 + 5 \times 2$ is 18.

Example  Find the value of each expression.

a. $7 \times 9 + 3 = 63 + 3$ *Multiply 7 and 9.*
$= 66$ *Add 63 and 3.*

b. $18 + 6 - 8 \div 2 = 18 + 6 - 4$ *Divide 8 by 2.*
$= 24 - 4$ *Add 18 and 6.*
$= 20$ *Subtract 4 from 24.*

c. $144 \div 16 + 9 \div 3 + 12 = 9 + 9 \div 3 + 12$ *Divide 144 by 16.*
$= 9 + 3 + 12$ *Divide 9 by 3.*
$= 12 + 12$ *Add 9 and 3.*
$= 24$ *Add 12 and 12.*

Many calculators allow you to evaluate expressions with parentheses. Consult your User's Guide to see if the calculator you use has this capability.

The order of operations can be changed by using grouping symbols like **parentheses**, ( ), and **brackets**, [ ]. The value of the expression is found by performing the operation in the grouping symbols first.

Example  Find the value of each expression.

a. $45 - 2 \times (15 - 3)$
$45 - 2 \times (15 - 3) = 45 - 2 \times (12)$ *Do the operations in parentheses first.*
$= 45 - 24$ *Multiply 2 by 12.*
$= 21$

b. $(22 + 3) \div (9 - 4)$
$(22 + 3) \div (9 - 4) = 25 \div 5$ *Do the operations in parentheses first.*
$= 5$ *Divide 25 by 5.*

The order for performing operations is as follows.

Order of Operations

1. **Do all operations within grouping symbols first; start with the innermost grouping symbol.**
2. **Next, do all multiplications and divisions from left to right.**
3. **Then, do all additions and subtractions from left to right.**

Example 3

Sports

In each week of the football season, the Associated Press poll ranks football teams using votes from a group of sports reporters. A team gets 25 points for each first-place vote, 24 points for each second-place vote, 23 points for each third-place vote, and so on, to one point for each twenty-fifth place vote. The team with the highest total is ranked number one. Suppose that in one week the Miami Hurricanes received 10 first-place votes, 4 second-place votes, 2 fourth-place votes, and 1 tenth-place vote.

a. Write an expression for the number of points that the Hurricanes earned.

$$\begin{pmatrix}\textit{points for}\\ \textit{10} \times \textit{first-place}\\ \textit{vote}\end{pmatrix} + \begin{pmatrix}\textit{points for}\\ \textit{4} \times \textit{second-place}\\ \textit{vote}\end{pmatrix} + \begin{pmatrix}\textit{points for}\\ \textit{2} \times \textit{fourth-place}\\ \textit{vote}\end{pmatrix} + \begin{pmatrix}\textit{points for}\\ \textit{1} \times \textit{tenth-place}\\ \textit{vote}\end{pmatrix}$$

$$(10 \times 25) + (4 \times 24) + (2 \times 22) + (1 \times 16)$$

b. Find the total number of points for the Hurricanes.

$$\begin{aligned} &(10 \times 25) + (4 \times 24) + (2 \times 22) + (1 \times 16) \\ &= 250 + 96 + 44 + 16 && \textit{Do operations in parentheses.} \\ &= 406 && \textit{Add.} \end{aligned}$$

The Hurricanes earned 406 points.

There are many ways used to indicate multiplication and division in algebra.

A raised dot or parentheses can be used to indicate multiplication.

$$5 \cdot 6 \xrightarrow{\textit{means}} 5 \times 6$$

$$7(3),\ (7)3,\ \text{or}\ (7)(3) \xrightarrow{\textit{means}} 7 \times 3$$

A fraction bar can be used to indicate division.

$$\frac{37 + 38}{30 - 5} \xrightarrow{\textit{means}} (37 + 38) \div (30 - 5)$$

Example 4 **Find the value of each expression.**

a.
$$\begin{aligned} 5(7 + 4) - 8 \cdot 4 &= 5(11) - 8 \cdot 4 \\ &= 55 - 32 && \textit{5(11) means } 5 \times 11 \textit{ and} \\ &= 23 && 8 \cdot 4 \textit{ means } 8 \times 4 \end{aligned}$$

b.
$$\begin{aligned} 5[(5 + 14) - 2(7)] &= 5[19 - 2(7)] && \textit{Do operations in innermost} \\ &= 5[19 - 14] && \textit{grouping symbols first.} \\ &= 5[5] \\ &= 25 \end{aligned}$$

c.
$$\begin{aligned} \frac{72 + 12}{35 + 7} &= (72 + 12) \div (35 + 7) && \textit{Rewrite division using} \\ &= 84 \div 42 && \textit{parentheses.} \\ &= 2 \end{aligned}$$

Checking Your Understanding

Communicating Mathematics

Read and study the lesson to answer these questions.

1. **Identify** the first operation you perform when evaluating the expression $(13 + 11) \div 4$.
2. Are the expressions $9(4 + 3)$ and $9 \cdot 4 + 3$ equivalent? Explain.
3. **Make a diagram** to show the order of steps you should follow when evaluating an expression.
4. **Write** an expression involving multiplication and subtraction in which the first step when you evaluate would be to subtract.
5. Play this game with a partner.
   - ▶ Write each of the digits 1–9 on a card and place cards face down.
   - ▶ Draw six cards from the pile. Out of the six numbers chosen, choose one to act as the answer.
   - ▶ Then work with your partner to write an expression with the other five numbers so that the result is the chosen answer.
   - ▶ You may use any of the operations—addition, subtraction, multiplication, and division—and add grouping symbols as needed. You may also rearrange the numbers.

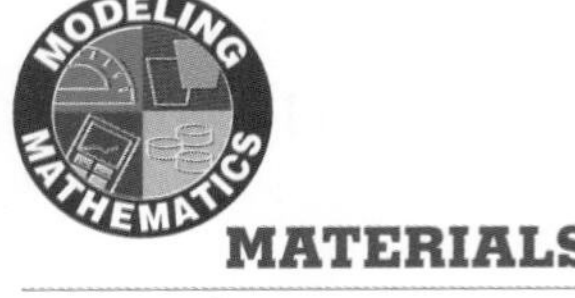

MATERIALS

index cards

pencil

Guided Practice

Name the operation that should be performed first. Then find the value of each expression.

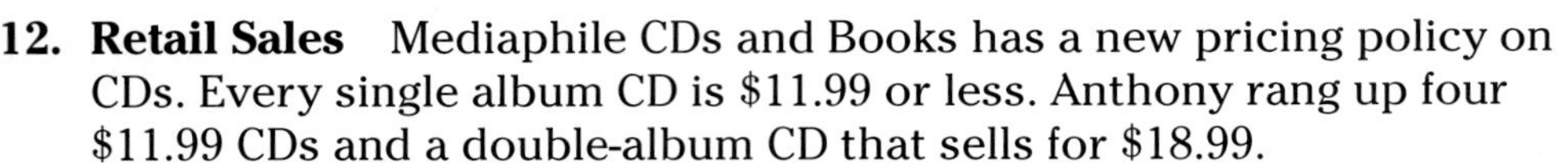

6. $9 + 2 \cdot 5$ **7.** $65 - 32 \div 8$ **8.** $33 \div 3 + 6$

9. $17 - 6(2)$ **10.** $8(2 + 4)$ **11.** $\frac{20 - 8}{9 - 5}$

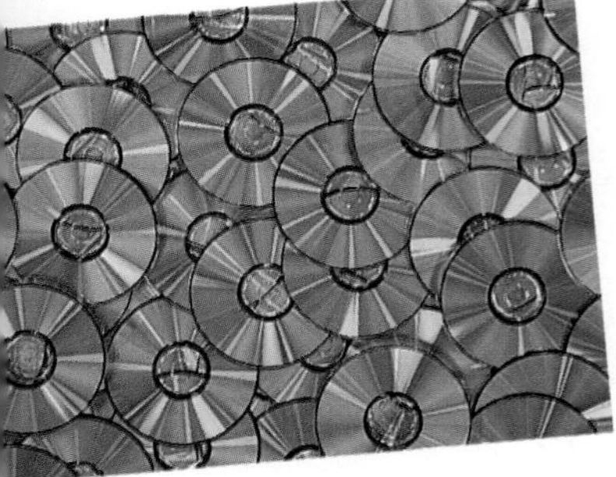

12. **Retail Sales** Mediaphile CDs and Books has a new pricing policy on CDs. Every single album CD is $11.99 or less. Anthony rang up four $11.99 CDs and a double-album CD that sells for $18.99.
    - **a.** Write an expression for the total cost of the merchandise before sales tax.
    - **b.** What was the total before sales tax was added?

Exercises: Practicing and Applying the Concept

Independent Practice

Find the value of each expression.

13. $15 \div 3 + 12 \div 4$ **14.** $3 \cdot (4 + 5) - 7$ **15.** $15 \div 5 \times 3$

16. $40 \cdot (6 - 2)$ **17.** $12(7 - 2 \times 3)$ **18.** $36 - 6 \cdot 5$

19. $\frac{18 + 66}{35 - 14}$ **20.** $\frac{16 + 8}{15 - 7}$ **21.** $\frac{2(14 - 6)}{4}$

22. $96 \div (12 \cdot 4) \div 2$ **23.** $72 \div (9 \times 4) \div 2$ **24.** $(30 \times 2) - (6 \cdot 9)$

25. $3[6(12 - 3)] - 17$ **26.** $7[5 + (13 - 4) \div 3]$

27. $4[3(21 - 17) + 3]$ **28.** $5[(12 + 5) - 3(19 - 14)]$

29. $8[(26 + 10) - 4(3 + 2)]$ **30.** $10[8(15 - 7) - 4 \cdot 3]$

Calculators

Copy each sentence below. Experiment with a calculator to find where to insert parentheses to make each sentence true. You may need to use the parentheses keys.

31. $71 - 17 + 4 = 50$ **32.** $8 - 5 \times 4 + 2 = 18$ **33.** $18 \div 3 + 6 + 12 = 14$

Critical Thinking

34. Only the 1, [+], [−], [×], [÷], [(], [)] and [=] keys on a scientific calculator are working. How can a result of 75 be reached by pushing these keys less than 20 times?

Applications and Problem Solving

35. **Business** Phyllis Sokol leads seminars on computer technology. On an upcoming trip, she will conduct a three-day seminar in New York City followed by a two-day seminar in Washington, D.C., and then a four-day seminar in Chicago. *Nation's Business* magazine estimates that a business traveler will spend $330 on food and lodging per day in New York City, $247 per day in Washington, D.C., and $229 per day in Chicago.
    a. Write an expression for the amount of money that Ms. Sokol can expect to spend on food and lodging on this trip.
    b. Find the amount of money that Ms. Sokol should expect to spend on food and lodging.
    c. List some other items that Ms. Sokol will have to pay for on her trip. Estimate the cost of each item and the total cost of the trip.

36. **Nutrition** The table below shows the number of Calories in one serving of different snack foods movie-goers love.

Snack	Calories per serving
Junior Mints (3 ounces)	360
Raisinets (2.3 ounces)	270
Buttered popcorn, medium bucket	1221
Coca-Cola (16 ounces)	205
Diet 7•Up (16 ounces)	0

    a. David and Marta ordered 2 boxes of Junior Mints, a bucket of popcorn, and four Cokes to share with friends. Find the total number of Calories in all the food.
    b. Yolanda bought two Diet 7•Ups and a box of Raisinets to share with Curtiss. How many Calories did each person consume?

37. **History** You can plant a bit of history in your yard. A non-profit conservation group called American Forests collects seeds from trees at historic homes, grows them into saplings, and sells them to the public. Each sapling costs $35 and $7 is added to each order for shipping and handling. How much would an order for seven saplings cost?

Mixed Review

38. **Sports** According to *Women's Sports and Fitness* magazine, a good swimmer can swim the length of a standard-sized pool in 24 or fewer strokes. If 1470 laps in a pool is equivalent to swimming the distance across the English Channel, about how many strokes would a good swimmer use to swim across the English Channel? (Lesson 1-1)
    a. Which method of computation do you think is most appropriate to solve this problem?
    b. Solve the problem using the four-step plan.

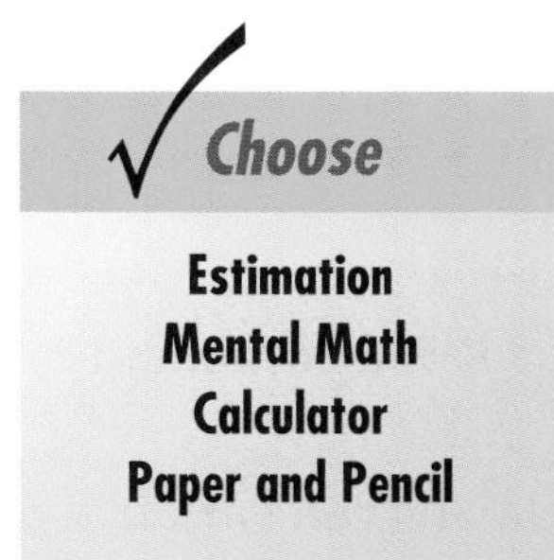

39. **Number Theory** Write a sentence or two about the pattern at the right. Find the products 44×46, 54×56, and 84×86 by extending the pattern. (Lesson 1-1)

$4 \times 6 = 24$
$14 \times 16 = 224$
$24 \times 26 = 624$
$34 \times 36 = 1224$

1-3 Variables and Expressions

Setting Goals: *In this lesson, you'll evaluate expressions containing variables and translate verbal phrases into algebraic expressions.*

Modeling a Real-World Application: Medicine

How much do you know about your hardworking heart?

- An adult heart is about the size of a clenched fist.
- In one year, a human heart beats over 30 million times. The heart of a 70 year old has beaten over 2.5 billion times!
- In an average lifetime, the heart pumps 1 million barrels of blood.

One way that people maintain good health is by monitoring their blood pressure. Blood pressure is the pressure that your blood exerts on your arteries. The *systolic pressure* is the pressure when the heart contracts, and the *diastolic pressure* is the pressure when the heart is between contractions.

As people age, their blood pressure rises. You can approximate a person's normal systolic pressure by dividing his or her age by 2 and then adding 110.

Learning the Concept

Algebra, like any language, is a language of symbols. You know the symbols for division and addition, so you can write the blood-pressure relationship as *age* ÷ 2 + 110.

In arithmetic, you could write ● ÷ 2 + 110, where ● is the person's age in years. The ● serves as a placeholder.

In algebra, we use placeholders called **variables**. Variables are usually letters. The letter x is used very often as a variable, but it is also common to use the first letter of the value you are representing. For example, we can use a to represent age in the blood-pressure relationship. The relationship can then be written as $a \div 2 + 110$.

age ÷ 2 + 110	*Words and symbols*
● ÷ 2 + 110	*Arithmetic*
$a \div 2 + 110$	*Algebra*

$a \div 2 + 110$ is called an **algebraic expression** because it is a combination of variables, numbers, and at least one operation.

Expressions like $a \div 2 + 110$ can be evaluated by replacing the variables with numbers and then finding the numerical value of the expression.

If Malcolm is 18 years old, he could estimate his normal blood pressure by evaluating the expression $18 \div 2 + 110$.

$$
\begin{aligned}
a \div 2 + 110 &= 18 \div 2 + 110 && \textit{Replace a with 18.}\\
&= 9 + 110 && \textit{Divide first, then add.}\\
&= 119
\end{aligned}
$$

Malcolm's normal systolic blood pressure is approximately 119.

Replacing a with 18 demonstrates an important property of numbers.

Substitution Property of Equality	For all numbers a and b, if $a = b$, then a may be replaced with b.

Example  **Evaluate $n + m - 34$ if $n = 18$ and $m = 49$.**

$$
\begin{aligned}
n + m - 34 &= 18 + 49 - 34 && \textit{Replace n with 18 and m with 49.}\\
&= (18 + 49) - 34 && \textit{Follow the order of operations.}\\
&= 67 - 34 && 18 + 49 = 67\\
&= 33 && 67 - 34 = 33
\end{aligned}
$$

THINK ABOUT IT

How could you rewrite $a \div 2 + 110$ using this notation?

As with numerical expression, there is special notation for multiplication and division with variables.

$3d \xrightarrow{\text{means}} 3 \times d$

$xy \xrightarrow{\text{means}} x \times y$

$7st \xrightarrow{\text{means}} 7 \times s \times t$

$\frac{q}{4} \xrightarrow{\text{means}} q \div 4$

Example  **Evaluate each expression if $x = 3$, $y = 8$, and $z = 5$.**

a.
$$
\begin{aligned}
7x - 4z &= 7(3) - 4(5) && \textit{Replace x with 3 and z with 5.}\\
&= 21 - 20 && \textit{Use the order of operations.}\\
&= 1
\end{aligned}
$$

b.
$$
\begin{aligned}
\frac{yz}{4} &= (8)(5) \div 4 && \frac{yz}{4} = yz \div 4\\
&= 40 \div 4 && \textit{Use the order of operations.}\\
&= 10
\end{aligned}
$$

c.
$$
\begin{aligned}
3x + (z + 2y) - 12 &= 3(3) + (5 + 2 \cdot 8) - 12\\
&= 3(3) + (5 + 16) - 12\\
&= 9 + 21 - 12\\
&= 18
\end{aligned}
$$

You can use variables to show patterns in expressions. For example, a babysitter may earn \$4 per hour. The chart at the right shows several possibilities for number of hours and earnings.

Hours of babysitting	Money earned
2	$4 \cdot 2$ or 8
5	$4 \cdot 5$ or 20
8	$4 \cdot 8$ or 32
h	$4 \cdot h$ or $4h$

The variable h and the algebraic expression $4h$ summarize the relationship between number of hours of babysitting and earnings.

When reading a verbal sentence and writing an algebraic expression to represent it, there are many words and phrases that suggest the operation to use. The following chart shows some common phrases and operations.

Addition	Subtraction	Multiplication	Division
plus sum more than increased by total in all	minus difference less than subtract decreased by	times product multiplied each of	divided quotient

Example 3 **Translate each phrase into an algebraic expression.**

a. ten more points than Marisa scored
Let p represent the points Marisa scored.
The words *more than* suggest addition.
The algebraic expression is $p + 10$ or $10 + p$.

b. three times last year's total
Let t represent last year's total.
The words *three times* suggest multiplication.
The algebraic expression is $3t$ or $3 \cdot t$.

Example 4

Business

Warnet Cable charges $19.95 a month for basic cable television service. Each premium channel selected costs an additional $4.95 per month. Write an expression for the cost of a month of cable service.

Let n represent the number of premium channels selected.

Since you are paying $4.95 *for each* premium channel, the total for premium channels will be $\$4.95 \times n$ or $\$4.95n$.

The word *additional* suggests addition. You will pay for basic service and the premium channels.

The algebraic expression is $19.95 + 4.95n$.

Checking Your Understanding

Communicating Mathematics

Read and study the lesson to answer these questions.

1. **Define** a variable in your own words.
2. **Write** two different expressions that are the same as $7a$.
3. If m is the number of games the Jets won, what does $m - 5$ represent?
4. Cups and counters can be used to model algebraic expressions. The cup represents the variable, and the counters are one unit each. The diagram models $2x + 3$.
   a. Use cups and counters to model the expression $4x + 2$.
   b. Model the phrase *four plus three times a number*. Then write an expression to represent the phrase.

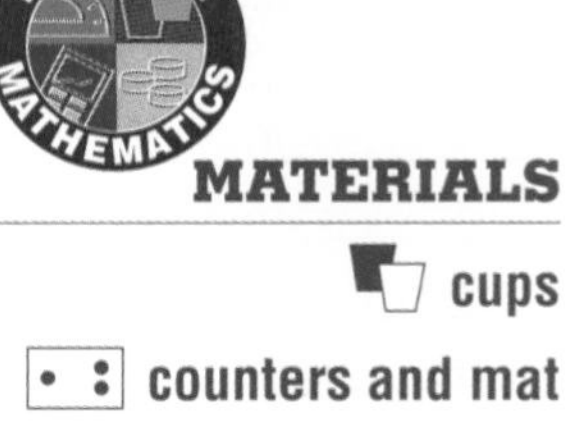

MATERIALS

cups

counters and mat

Guided Practice

Evaluate each expression if $m = 4$, $n = 11$, $p = 2$, and $q = 5$.

5. $8 + n$
6. $10 - q$
7. $3m$
8. $24 - 4q$
9. $\frac{5m}{2}$
10. $n + 110 - 2p$

Translate each phrase into an algebraic expression.

11. three more than v
12. seven less than h
13. the product of a number and 6
14. twice w
15. **Space Exploration** The force of gravity on Earth is six times greater than on the moon. As a result, objects weigh six times as much on Earth as they do on the moon.
    a. Write an expression for the weight of an object on Earth if the weight on the moon is x.
    b. A scientific instrument weighs 34 pounds on the moon. How much does the instrument weigh on Earth?

Exercises: Practicing and Applying the Concept

Independent Practice

Evaluate each expression if $a = 6$, $b = 3$, and $c = 7$.

16. $a + 22$
17. $2b - 4$
18. $ca - ab$
19. $4a - (b + c)$
20. $5c + 5b$
21. $\frac{8b}{a}$
22. $2c + 3a + 6b$
23. $\frac{6a}{b}$
24. $9a - (4b + 2c)$
25. $\frac{3(4a - 3c)}{c - 4}$
26. $12b - \frac{2c - 4}{a + 4}$
27. $\frac{15ab}{3c + 6}$

Translate each phrase into an algebraic expression.

28. a number divided by 8
29. the quotient of ninety-two and c
30. the sum of g and 8
31. p decreased by 5
32. eleven increased by f
33. the difference of u and 10
34. twice a number
35. the quotient of eighty-eight and b
36. Gil's salary plus a $500 bonus
37. three times as many fouls as Sumi
38. three times a number decreased by 18
39. two more than twice the previous month's sales

Write a verbal phrase for each algebraic expression.

40. $x + 4$
41. $16 - b$
42. $7n$
43. $v \div 5$
44. $2(a + 1)$
45. $2a + 1$

Critical Thinking

46. Write an expression for the value of a three-digit number whose hundreds, tens, and units digits are x, y, and z, respectively.

Applications and Problem Solving

47. **Meteorology** You can estimate the temperature in degrees Fahrenheit by counting the number of times a cricket chirps in one minute. Count the number of chirps, divide by 4, and then add 37.
    a. Write an expression for the temperature relationship.
    b. Find the approximate temperature when a cricket chirps 124 times in a minute.

48. **Statistics** The *mean* is a number that is often used to represent a set of numbers. To find the mean of three numbers, find the sum of the numbers and then divide by 3.

   a. Write an expression for the mean of a, b, and c.

   b. Find the mean of 51, 66, and 63.

Mixed Review

49. Find the value of the expression $6 \times 8 - (9 + 12)$. (Lesson 1-2)

50. **Retail Sales** Jamie bought four adult tickets at \$6.25 each and two children's tickets at \$3.75 each. (Lesson 1-2)

   a. Write an expression for the total cost of the tickets.

   b. Find the cost.

   c. Write your own problem involving movie tickets and solve.

51. **Patterns** Use the four-step problem-solving plan to find the next number in the sequence. (Lesson 1-1)
   1, 2, 4, 7, 11, 16, 22, __?__

52. **Art** A reproduction of Vincent van Gogh's *Sunflowers* was constructed in a wheat field near Duns, Scotland. The "painting" covered a 46,000-square foot area and contained 250,000 plants and flowers. About how many flowers and plants were planted per square foot? (Lesson 1-1)

   a. Which method of computation do you think is most appropriate for solving this problem? Explain.

   b. Solve the problem.

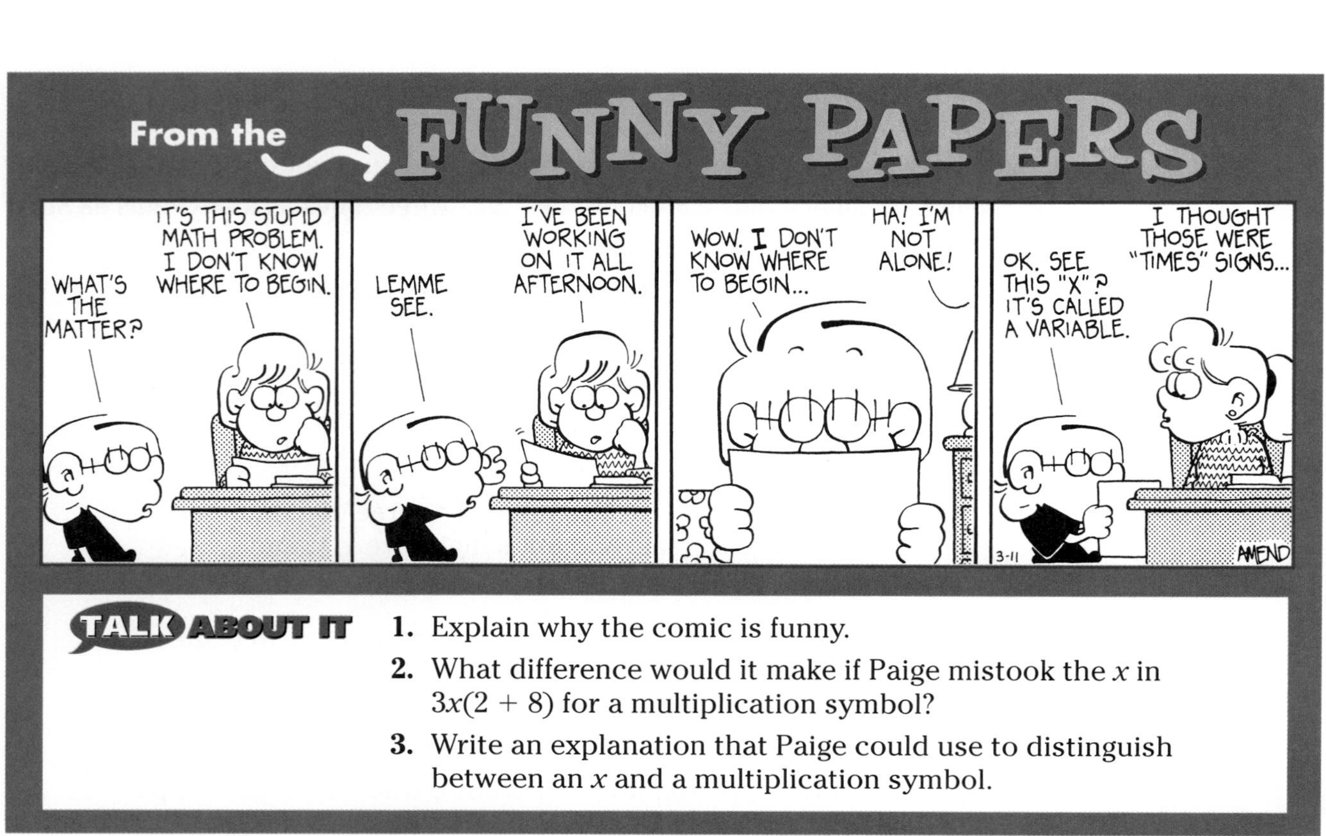

TALK ABOUT IT

1. Explain why the comic is funny.
2. What difference would it make if Paige mistook the x in $3x(2 + 8)$ for a multiplication symbol?
3. Write an explanation that Paige could use to distinguish between an x and a multiplication symbol.

1-3B

GRAPHING CALCULATOR ACTIVITY

Evaluating Expressions

An Extension of Lesson **1-3**

Graphing calculators observe the order of operations. Since you can specify the value of a variable, evaluating expressions is as simple as entering the expression with the calculator keys. The calculators also have parentheses.

Activity

Keystrokes are shown for the TI-82 graphing calculator. If you are using a different graphing calculator, consult your User's Guide.

Evaluate $3 + (2x - y)$ if $x = 7$ and $y = 5$ on a graphing calculator.

ENTER: 7 [STO▸] [X,T,θ] [2nd] [:] *Enters the value of x.*

5 [STO▸] [ALPHA] [Y] [2nd] [:] *Enters the value of y.*

3 [+] [(] 2 [X,T,θ] [−] [ALPHA] [Y] [)] [ENTER] *Enters and evaluates the expression.*

DISPLAY: 12

If you get an error message or discover that you entered the expression incorrectly, you can use the REPLAY feature to correct your error and reevaluate without reentering the expression. Follow the steps below to use the REPLAY feature.

- Press [2nd] [ENTRY] to display your expression.
- Use the arrow keys to move to the location of the correction. Then type over, use [INS], or use [DEL] to make the correction.
- Press [ENTER] to reevaluate.

Your Turn

Evaluate each expression with a graphing calculator if $a = 9$, $b = 0$, $c = 7$, and $d = 3$.

1. $5d - 1$ **2.** $b(a - c)$ **3.** $3(b + c) \div d$

4. $4a \div d$ **5.** $9a - (4d + 2c)$ **6.** $a(b + c) - 7$

7. How can you use the REPLAY feature to evaluate several expressions for variables of the same value?

8. What do you observe about the display when you press the [÷] and the [×] keys?

1-4 Properties

Setting Goals: *In this lesson, you'll identify properties of addition and multiplication and use these properties to solve problems.*

Modeling with Manipulatives

MATERIALS

counters, beans, or other small objects

Small objects like counters or beans can be used to model operations with whole numbers. The diagram at the right models the expression $5 + 3$. Counting the number of counters in the group after the additions shows that $5 + 3 = 8$.

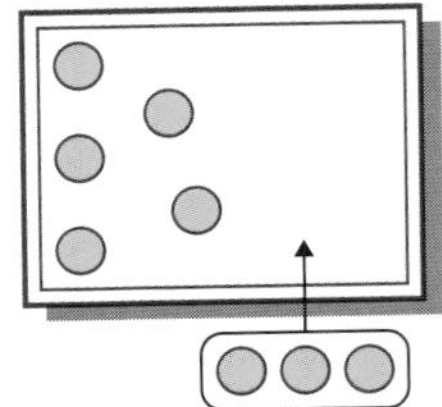

Your Turn Use counters to model $3 + 5$.

a. What is the value of $3 + 5$ that you found using the counters?

b. What do you observe about the values of $5 + 3$ and $3 + 5$?

c. Work with a partner to write and solve other problems like $a + b$ and $b + a$. Summarize your results.

d. Investigate $3 \cdot 5$ and $5 \cdot 3$ with counters by making 3 groups of 5 counters and 5 groups of 3 counters. Then explore ab and ba. What do you observe?

Learning the Concept

The charts below and on the next page summarize some of the properties of addition and multiplication.

Commutative Properties of Addition and Multiplication	
The order in which numbers are added does not change the sum. $5 + 3 = 3 + 5$ For any numbers a and b, $a + b = b + a$	The order in which numbers are multiplied does not change the product. $2 \cdot 4 = 4 \cdot 2$ For any numbers a and b, $a \cdot b = b \cdot a$
Associative Properties of Addition and Multiplication	
The way in which addends are grouped does not change the sum. $(2 + 4) + 6 = 2 + (4 + 6)$ For any numbers a, b, and c, $(a + b) + c = a + (b + c)$	The way in which factors are grouped does not change the product. $(6 \cdot 3) \cdot 7 = 6 \cdot (3 \cdot 7)$ For any numbers a, b, and c, $(a \cdot b) \cdot c = a \cdot (b \cdot c)$

Identity Properties of Addition and Multiplication	
The sum of an addend and zero is the addend. $6 + 0 = 6$ For any number a, $a + 0 = a$.	The product of a factor and one is the factor. $6 \cdot 1 = 6$ For any number a, $a \cdot 1 = a$.

Multiplicative Property of Zero
The product of a factor and zero is zero. $5 \cdot 0 = 0$ For any number a, $a \cdot 0 = 0$.

These properties can be helpful when you find sums and products mentally.

Example  **Find the value of each expression mentally.**

a. $17 + (13 + 15) = (17 + 13) + 15$ *Regroup addends.*
$= 30 + 15$
$= 45$

b. $2 \cdot 16 \cdot 5 = (2 \cdot 5) \cdot 16$ *Rearrange and group factors.*
$= 10 \cdot 16$ or 160

Example 2

On a recent tour, Elton John and Billy Joel visited Knoxville, Raleigh, Richmond, and Washington, D.C. The mileages between the cities are shown in the table below. If the equipment truck traveled from each city to the next in consecutive days, how many miles did it travel?

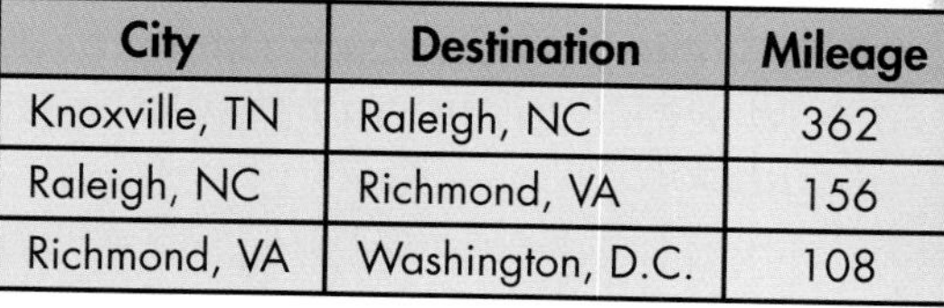

City	Destination	Mileage
Knoxville, TN	Raleigh, NC	362
Raleigh, NC	Richmond, VA	156
Richmond, VA	Washington, D.C.	108

They would travel the total of the mileages between the cities.

Rearrange and group addends so addition is easier.

$362 + 156 + 108 = (362 + 108) + 156$
$= 470 + 156$
$= 626$

Connection to Algebra

Variables represent numbers, so the properties can be used to rewrite and simplify algebraic expressions.

Example 3 **Rewrite each expression using a commutative property.**

a. $\boldsymbol{n + 6}$

$n + 6 = 6 + n$ *Change the order.*

b. $\boldsymbol{4 \cdot s}$

$4 \cdot s = s \cdot 4$

Example  **Rewrite each expression using an associative property. Then simplify.**

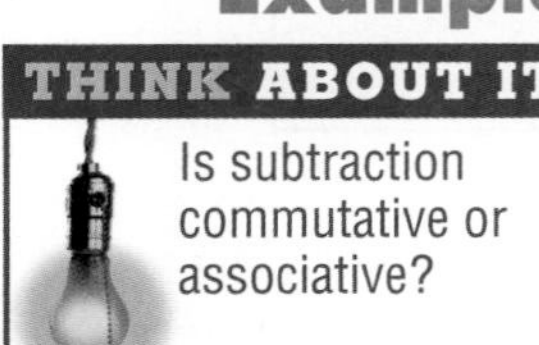
THINK ABOUT IT
Is subtraction commutative or associative?

a. $\boldsymbol{(m + 5) + 4}$

$$(m + 5) + 4 = m + (5 + 4)$$
$$= m + 9$$

b. $\boldsymbol{3(7t)}$

$$3(7t) = (3 \cdot 7)t \quad \textit{Regroup.}$$
$$= 21t$$

Example  **Evaluate $\boldsymbol{xyz}$ if $\boldsymbol{x = 2}$, $\boldsymbol{y = 8}$, and $\boldsymbol{z = 15}$ mentally.**

$xyz = 2 \cdot 8 \cdot 15$ *Replace x with 2, y with 8, and z with 15.*

$= 2 \cdot 15 \cdot 8$ *Use the Commutative property to rearrange factors.*

$= (2 \cdot 15) \cdot 8$ *Group 2 and 15 to make multiplication easier.*

$= 30 \cdot 8$

$= 240$

Checking Your Understanding

Communicating Mathematics

Read and study the lesson to answer these questions.

1. **Compare and contrast** the associative and commutative properties.
2. **Write** mathematical sentences to illustrate each of the seven properties described in the tables on pages 22 and 23.
3. **Demonstrate** how the properties can be used to perform calculations mentally.
4. Use counters to model the expressions $(2 + 5) + 6$ and $2 + (5 + 6)$.
   - **a.** What do you observe about the values of each expression?
   - **b.** Model several more expressions of the form $(a + b) + c$ and $a + (b + c)$. Explain your results in terms of one of the properties of addition.

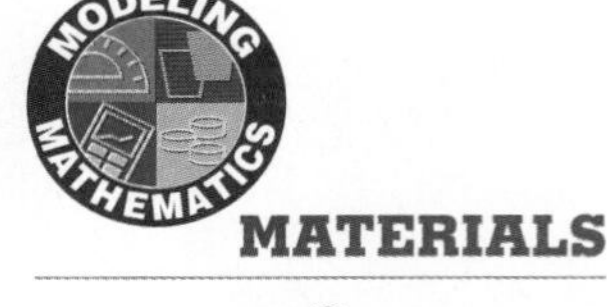

MATERIALS

counters

Guided Practice

Name the property shown by each statement.

5. $(3 + 8) + 7 = 3 + (8 + 7)$
6. $8 + 9 = 9 + 8$
7. $(5 + 7) + 2 = (7 + 5) + 2$
8. $0 + 9 = 9$

Find each sum or product mentally.

9. $12 + 9 + 8$
10. $5 \cdot 3 \cdot 8$
11. $4 + 13 + 26 + 5$

Rewrite each expression using an associative property. Then simplify.

12. $(y + 9) + 8$
13. $7(6z)$
14. $(k \cdot 8)3$
15. **Food Service** For lunch, Juana ordered an iced tea, a bowl of clam chowder, and a small Caesar salad. If the prices of the items are \$1.25, \$3.35, and \$3.75 respectively, how much should Juana's bill be before tax and tip?

Exercises: Practicing and Applying the Concept

Independent Practice

Name the property shown by each statement.

16. $22 \cdot 1 = 22$

17. $19 \cdot 12 = 12 \cdot 19$

18. $6 \cdot 33 = 33 \cdot 6$

19. $3 \cdot 7 \cdot 0 = 0$

20. $8cd = 8dc$

21. $x + (2y + z) = (2y + z) + x$

22. $0 + 9 = 9 + 0$

23. $5s + 9 = 9 + 5s$

24. $(7a)(b) = 7(ab)$

25. $8t + 6 = 6 + 8t$

26. $1m = m$

27. $7 + (2 + u) = (7 + 2) + u$

Find each sum or product mentally.

28. $16 + 7 + 14$

29. $2 \cdot 9 \cdot 20$

30. $82 + 58 + 23 + 37$

31. $18 \cdot 6 \cdot 0$

32. $7 + 99 + 123$

33. $2 \cdot 13 \cdot 5$

34. $7 + 2 + 13 + 18$

35. $6 \cdot 11 \cdot 10$

36. $129 \cdot 8 \cdot 0$

Rewrite each expression using a commutative property.

37. $5 + 9$

38. $8a + 6$

39. $9 + 18w$

Rewrite each expression using an associative property. Then simplify.

40. $(y + 7) + 6$

41. $(b \cdot 6) \cdot 5$

42. $2(8f)$

43. $13 + (11 + m)$

44. $3(2z)$

45. $(p \cdot 7) \cdot 4$

46. $(u + 8) + 16$

47. $(15 + 4w) + w$

48. $0(3x)$

Critical Thinking

49. Calculator Use a calculator to find $28 \div 7$ and $7 \div 28$.

a. Experiment with other expressions of the form $a \div b$ and $b \div a$.

b. Write a statement about division and the commutative property.

Applications and Problem Solving

50. Sports In the game of volleyball, the net is 3 feet 3 inches tall. The bottom of the net is to be set 4 feet 8 inches from the floor.

a. Write an expression for the distance from the floor to the top of the net.

b. Find the distance from the floor to the top of the net. Explain how you used the associative and commutative properties.

51. Science Mr. Rex instructed students in his class to pour acid in their flasks into water in their beakers. He warned that pouring water into acid could cause spattering and burns. These actions are not commutative. Give another example of actions that are not commutative.

Mixed Review

52. Write an expression for *the product of a number and 18.* (Lesson 1-3)

53. Evaluate the expression $8xy + 6$ if $x = 0$ and $y = 2$. (Lesson 1-3)

54. Find the value of the expression $7 \cdot (9 - 3) + 4$. (Lesson 1-2)

55. Medicine You can find the approximate number of pints of blood in your body by dividing your weight by 16. (Lesson 1-2)

a. Write an expression for the blood relationship.

b. JT weighs 144 pounds. Find the approximate number of pints of blood in JT's body.

56. Patterns Find the next number in the sequence 15, 14, 12, 9, 5, __?__. (Lesson 1-1)

1-5 The Distributive Property

Setting Goals: *In this lesson, you'll simplify algebraic expressions using the distributive property.*

Modeling a Real-World Application: Travel

HELP WANTED

Energetic people who enjoy working with people will find rewards in the travel industry. If you are interested in finding out more about this exciting field, contact:

The Travel Industry Association of America
1133 21st Street NW
Suite 800
Washington, D.C. 20036

Sunny beaches, exciting ski lodges, and exotic experiences in far-off lands await you! How do you get there? Call a travel agent! Elena Chávez is a travel agent for Carlson Travel. She helps people plan vacations and business trips by making arrangements for hotels, airline tickets, rental cars, and any other services her clients need.

Ms. Chávez received the brochure at the right for a travel package to Barbados. Her clients, Mr. and Mrs. Yoder, are considering Barbados for a summer vacation. If the round-trip airfare to Barbados from Columbus is $679, how much should Ms. Chávez tell the Yoders that the airfare and package trip would cost the couple? *This problem will be solved in Example 2.*

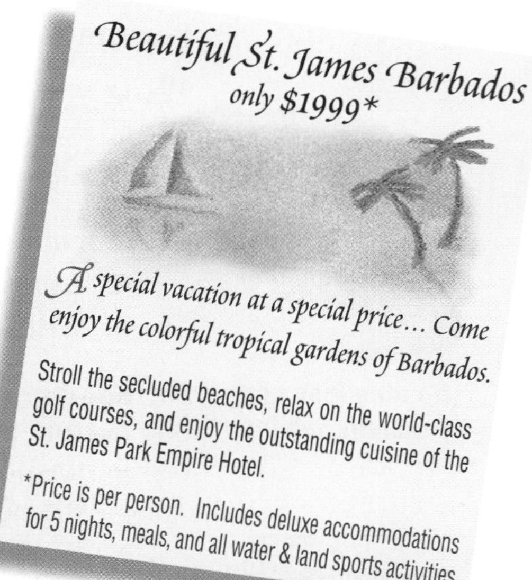

Learning the Concept

In Lesson 1-4, you learned some of the properties of addition and multiplication with whole numbers. The **distributive property** ties addition and multiplication together.

Distributive Property

In words: The sum of two addends multiplied by a number is the sum of the product of each addend and the number.

In symbols: For any numbers a, b, and c,
$a(b + c) = ab + ac$ and $(b + c)a = ba + ca$.

The expression $a(b + c)$ is read "a times the quantity b plus c" or "a times the sum of b and c."

The distributive property allows you to write expressions in different forms.

Example  Find 14 · 12 mentally using the distributive property.

$14 \cdot 12 = 14(10 + 2)$ *Replace 12 with 10 + 2.*
$= 14 \cdot 10 + 14 \cdot 2$ *Think of 140 + 28.*
$= 140 + 28$ or 168

Example 2 Refer to the application at the beginning of the lesson.

Travel

a. Write two different expressions for the price of the trip for two people.
b. Find the cost of the trip.

a. There are two methods of writing the total cost: find the total cost for one person and multiply by 2, or find the cost of each part of the trip for two people and then add.

Method 1			Method 2		
number of people	*times*	*total trip cost*	*airfare for two*	*plus*	*package for two*
2	×	(\$679 + \$1999)	2(\$679)	+	2(\$1999)

Notice that if we apply the distributive property to the expression on the left, the result is the expression on the right.

b. Evaluate either expression to find the cost of the trip.
2(\$679 + 1999) = 2(\$2678)
= \$5356
Check the result by evaluating the expression 2(\$679) + 2(\$1999).

Connection to Algebra

Refer again to the application at the beginning of the lesson. The Yoders are thinking of inviting some friends to join them on their vacation to Barbados. Ms. Chávez can use the variable p for the number of people.

If there are p people traveling, the airfare will be $679p$ dollars.
For the same number of people, the package cost will be $1999p$ dollars.

The total cost for p people could be represented as $679p + 1999p$ dollars.

The expression $679p + 1999p$ has two **terms** with the same variable. These terms are called **like terms**. Some other pairs of like terms are $6d$ and $17d$, $3st$ and $5st$, and x and $4x$. The distributive property can be helpful in simplifying expressions that have like terms. Example 3a shows how to simplify $679p + 1999p$.

THINK ABOUT IT

Are $5x$ and $5y$ like terms?

Example 3 Simplify each expression.

a. $679p + 1999p$

$679p + 1999p = (679 + 1999)p$ *p is a factor of both 679p and 1999p. Use the distributive property.*
$= 2678p$

b. $x + 12x$

$x + 12x = 1x + 12x$ *$x = 1 \cdot x$ by multiplicative identity*
$x + 12x = (1 + 12)x$ *Distributive property*
$= 13x$

An expression is in **simplest form** when it has no like terms and no parentheses.

Example  **Write each expression in simplest form.**

a. $\mathbf{16k + 9 + 5k}$

$$\begin{aligned} 16k + 9 + 5k &= 16k + 5k + 9 && \textit{Commutative property of addition} \\ &= (16 + 5)k + 9 && \textit{Distributive property} \\ &= 21k + 9 \end{aligned}$$

b. $\mathbf{m + 5(n + 7m)}$

$$\begin{aligned} m + 5(n + 7m) &= m + 5n + 5 \cdot 7m && \textit{Distributive property} \\ &= m + 5n + 35m && \textit{Substitution property} \\ &= m + 35m + 5n && \textit{Commutative property of addition} \\ &= 1m + 35m + 5n && \textit{Multiplicative identity} \\ &= (1 + 35)m + 5n && \textit{Distributive property} \\ &= 36m + 5n \end{aligned}$$

Checking Your Understanding

Communicating Mathematics

Read and study the lesson to answer these questions.

1. **Rewrite** the expression $7(5 + 3)$ in three different ways using the distributive property.
2. **Give an example** of two like terms. Explain why they are like terms.
3. **Make a table** of expressions that are and are not in simplest form. Explain how you know whether an expression is in simplest form.

MATH JOURNAL

4. Make up a method for remembering the distributive property. Write your method and a description of how it works in your journal.

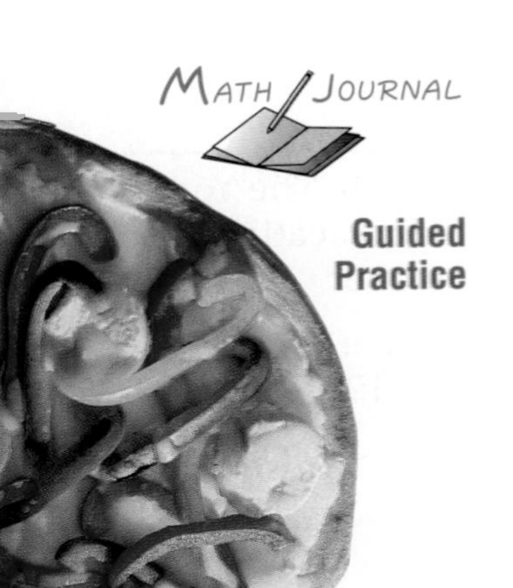

Guided Practice

Restate each expression using the distributive property. Do not simplify.

5. $5(7 + 8)$
6. $6(2 + 4)$
7. $4(x + 3)$

Simplify each expression.

8. $6 \cdot 5 + 9 \cdot 6$
9. $12a + a + 7$
10. $x + 9x$
11. $6c + c + 7$
12. $22c + 4(2 + 4c)$
13. $4ab + 6ab + 10ab$
14. **Money** Carlos works part time at The Arizona Pizzeria to save for college. He earns \$5.25 an hour plus tips.
    - **a.** Carlos worked 4 hours on Friday and 8 hours on Saturday. Write two different expressions for his wages that weekend.
    - **b.** Find the amount Carlos earned if his tips totaled \$35.50.

Exercises: Practicing and Applying the Concept

Independent Practice

Restate each expression using the distributive property. Do not simplify.

15. $3(11 + 12)$
16. $(5 + 7)x$
17. $6t + 11t$
18. $2a + 4b$
19. $2r + 12s$
20. $6x + 6y + 1$
21. $(9 + 8)v$
22. $12r + 12s$
23. $(4x + 9y)2$

Simplify each expression.

24. $6d + d + 15$ **25.** $17a + 21a + 45$ **26.** $m + m$

27. $c + 24c + 16$ **28.** $q + 4 + 11 + 4q$ **29.** $18y + 5(7 + 3y)$

30. $14(b + 3) + 8b$ **31.** $30(b + 2) + 2b$ **32.** $3(8 + a) + 7(6 + 4a)$

33. $8(9 + 3f) + f$ **34.** $x + 5x + 8(x + 2)$ **35.** $3(x + y) + 4(2x + 3y)$

Critical Thinking

36. The distributive property given on page 26 is often called the *distributive property of multiplication over addition*. Is there a distributive property of multiplication over subtraction? Evaluate several different expressions of the form $a(b - c)$ and $ab - ac$ and explain your findings.

Applications and Problem Solving

37. Engineering The foundation of a building is critical to ensuring that the building is safe. Before the foundation can be poured, an engineer calculates the volume of material to be removed. The expression used to find the volume of earth to remove for an opening of the shape shown is $\frac{d(a + b)}{2} \times L$. Find the volume of earth to remove if $a = 52$ m, $b = 48$ m, $d = 6$ m, and $L = 40$ m.

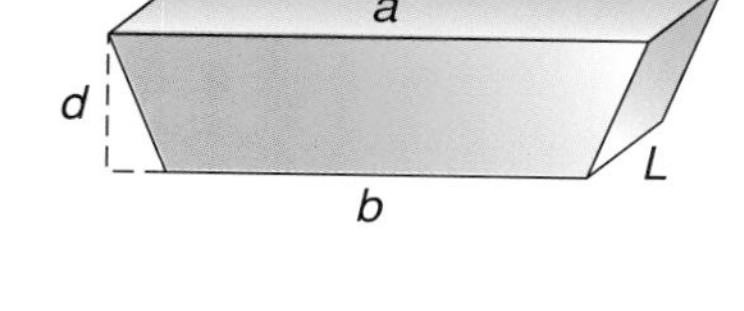

38. Business Operations Most products sold in the United States carry a universal product code (UPC) like the one at the right. The last digit of the code, the check digit, is used to ensure that the numbers of the code are entered correctly. The computer uses the steps below to verify the check digit.

1. Add the digits in the odd-numbered positions. Multiply the total by 3.
2. Add the digits in the even-numbered positions. Then add to the previous total.
3. Subtract the result from the next highest multiple of 10. For example, a result of 57 would be subtracted from 60. The answer is the check digit.

a. Write an expression for the first two steps of checking UPC 0 36800 30685 1. Be sure that you don't include the check digit in the calculation.

b. Evaluate the expression you wrote for part a. Then subtract from the next highest multiple of 10. Does the check digit verify the code?

Mixed Review

39. Rewrite $(8 + 19) + 17$ using an associative property. (Lesson 1-4)

40. Name the property shown by the statement $12 \times 1 = 12$. (Lesson 1-4)

41. Translate the phrase *the quotient of a number and 9* into an algebraic expression. (Lesson 1-3)

42. Evaluate $75 - 10n$ if $n = 7$. (Lesson 1-3)

43. **Fundraising** The local chapter of the Cancer Society sells Valentine candy to raise funds for research. Cindy sold 15 white chocolate hearts at $4.25 each, 36 milk chocolate hearts at $3.75 each, and 22 milk chocolate assortments at $7.45 each. How much money did Cindy raise? (Lesson 1-2)

44. **Geometry** Draw the next figure in the pattern. (Lesson 1-1)

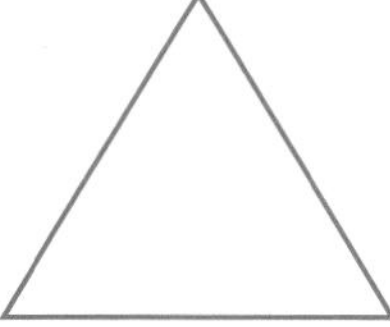  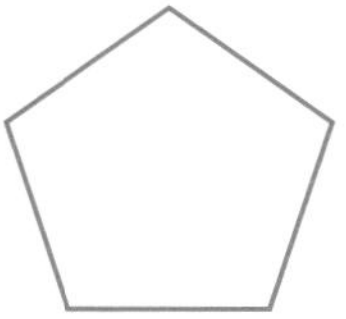

45. Bandi used a calculator to find $528 \times 7 = 3696$. Without actually calculating, determine if this answer is reasonable. (Lesson 1-1)

Self Test

1. **Number Theory** The pattern at the right is known as Pascal's Triangle. Find the pattern and complete the 6th and 7th rows. (Lesson 1-1)

1
1 1
1 2 1
1 3 3 1
1 4 6 4 1

2. **Construction** Lillehammer, Norway hosted the 1994 Winter Olympic Games. Gjovick Olympic Cavern Hall, the site of the hockey competition, was carved out of Hovdeteoppen Mountain. 29,000 truckloads of rock were removed. If a truck holds about 5 cubic yards of rock, about how much rock was removed to build the hall? (Lesson 1-1)
   a. Which method of computation do you think is most appropriate for this problem? Justify your choice.
   b. Solve the problem using the four-step plan. Be sure to examine your solution.

Find the value of each expression. (Lesson 1-2)

3. $9(2 + 5) - 4$

4. $\frac{19 - 11}{2(4)}$

Evaluate each expression if $a = 3$, $b = 8$, and $c = 12$. (Lesson 1-3)

5. $\frac{ab}{c}$

6. $8c - (b + 3a)$

Name the property shown by each statement. (Lesson 1-4)

7. $0 + c = c + 0$

8. $19 + 7 + 1 = 19 + 1 + 7$

9. **Music** Enrique's piano teacher charges $12.50 for a lesson. If Enrique has 5 lessons in one month, how much does he owe his piano teacher? Solve mentally. (Lesson 1-5)

10. Simplify $2n + 4(n + 8n)$. (Lesson 1-5)

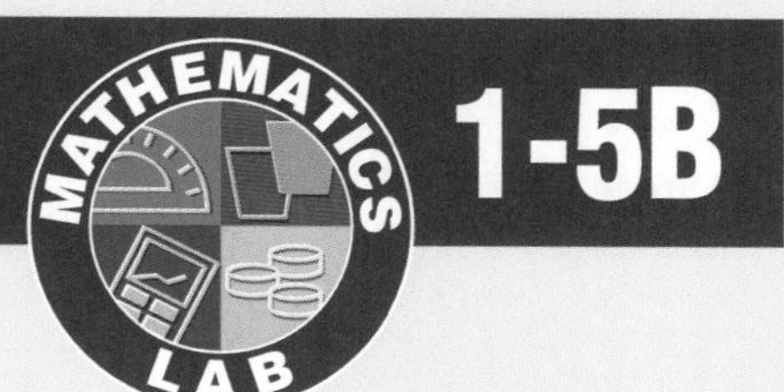

HANDS-ON ACTIVITY

1-5B Distributive Property

An Extension of Lesson **1-5**

MATERIALS

algebra tiles

Many algebraic principles can be modeled using geometry. You can look at the distributive property by modeling with algebra tiles of different sizes.

- Let 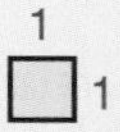

be a 1×1 square. That is, the length and width are each 1 unit. If the area of a square or a rectangle is found by multiplying the length and the width, what is the area of the square?

- Let 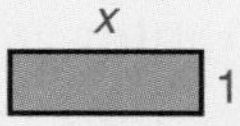

be a $1 \times x$ rectangle. The width is 1 unit, and the length is x units. What is the area of the rectangle?

Then 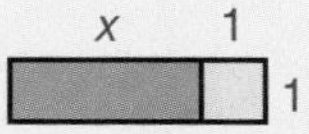

is a rectangle that has a width of 1 unit and a length of $x + 1$ units. What is the area of the rectangle?

Your Turn

Using your tiles, make rectangles with areas of $x + 2$, $2x$, $2x + 1$, and $2x + 2$ square units.

Explore

You can use geometric models to check the distributive property. Is it true that $2(x + 1) = 2x + 1$?

$2(x + 1)$ means 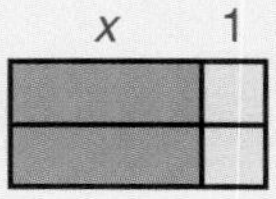

Does it matter how the tiles are arranged?

$2x + 1$ means 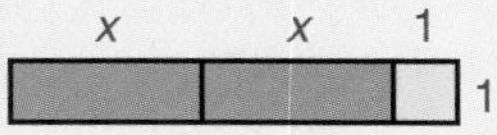

Therefore, $2(x + 1)$ is not the same as $2x + 1$. From the drawing above, $2(x + 1) = 2x + 2$.

TALK ABOUT IT

Tell whether each of the following statements is *true* or *false*. Justify your answers with algebra tiles or a drawing.

1. $2x + 3 = 6x$ **2.** $2x = x + x$ **3.** $3x + 3 = 3(x + 1)$

4. $3x + 2x = 6x$ **5.** $3x + 3 = 3(x + 3)$ **6.** $3x + 2x = x(3 + 2)$

7. You Decide Steve says that $3(x + 2) = 3x + 2$. Linh disagrees. She thinks that $3(x + 2) = 3x + 6$. Write a paragraph with drawings to explain who you think is correct.

1-6 Variables and Equations

Setting Goals: *In this lesson, you'll identify and solve open sentences.*

Modeling a Real-World Application: Entertainment

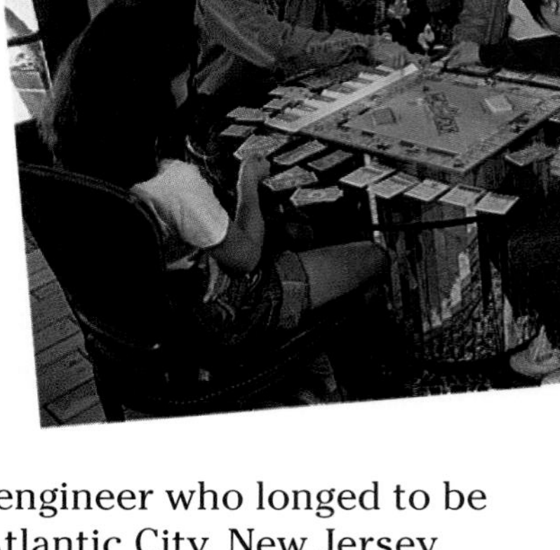

How do you like to spend a rainy Saturday afternoon? With a good book, an old movie, or how about a nice long board game? People have been playing board games for centuries and even with the advent of computer games, they still enjoy the real thing.

> **CULTURAL CONNECTIONS**
> People of different cultures have been playing games for centuries. One of the oldest games, Senet, made its first known appearance in Egypt in about 2686 B.C. Senet was played by people of all social classes and had both religious and secular forms.

Parcheesi was one of the first games to receive a copyright in the United States. It was copyrighted in 1874. Parcheesi originated in India more than 1200 years ago.

The most famous board game, Monopoly, was invented by Charles Darrow during the Great Depression. Darrow was an unemployed engineer who longed to be rich and spend his days on the boardwalks of Atlantic City, New Jersey. Darrow's game may be based on an earlier game called The Landlord's Game invented by Elizabeth J. Magie in 1904. The copyright for Monopoly was issued in 1933.

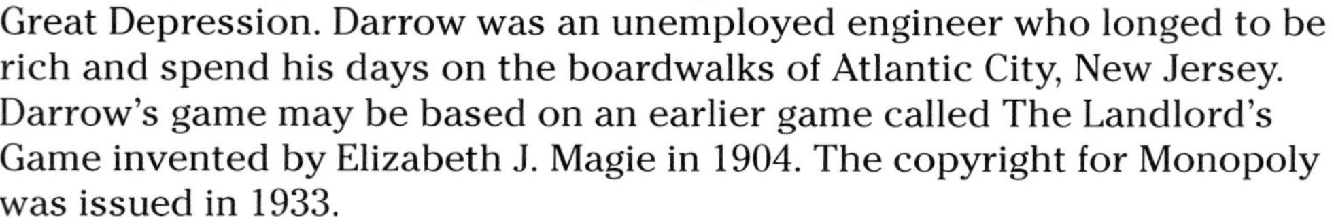

The number of years between the copyrights of Parcheesi and Monopoly can be represented by $1933 - 1874 = y$.

Learning the Concept

A mathematical sentence like $1933 - 1874 = y$ is called an **equation**. Some examples of equations are $6 + 8 = 14$, $3(6) - 9 = 9$, and $6x - 2 = 11$. Equations that contain variables are **open sentences**. When the variables in an open sentence are replaced with numbers, the sentence may be true or false.

$1933 - 1874 = y$

$1933 - 1874 \stackrel{?}{=} 67$ *Replace y with 67.*
This sentence is false because $1933 - 1874 \neq 67$.

$1933 - 1874 \stackrel{?}{=} 59$ *Replace y with 59.*
This sentence is true.

A value for the variable that makes an equation true is called a **solution** of the equation. The process of finding a solution is called **solving the equation**.

The solution of $1933 - 1874 = y$ is 59.

Example  Which of the numbers 16, 24, or 32 is the solution of $66 + x = 98$?

Replace x with each of the possible solutions to solve the equation.

$66 + 16 = 98$ *Replace x with 16.*
$82 = 98$

This sentence is false.

$66 + 24 = 98$ *Replace x with 24.*
$90 = 98$

This sentence is false.

$66 + 32 = 98$ *Replace x with 32.*
$98 = 98$

This sentence is true. The solution is 32.

THINK ABOUT IT

How could you find the solution to the equation in Example 1 without substituting each of the possible solutions?

You will learn many ways to solve equations in this course. Some equations can be solved mentally by using basic facts or arithmetic skills.

Example 2

Sports

Basketball great Shaquille O'Neal is an imposing 7' 1" tall and 300 pounds. O'Neal has appeared in Reebok commercials since 1992. When O'Neal began working with Reebok, his shoe size was 19. By 1995, O'Neal's size had risen to a whopping size 22. How many shoe sizes did Shaquille O'Neal increase from 1992 to 1995?

Since the numbers are simple, this equation can be solved mentally.

$19 + s = 22$
$19 + 3 = 22$ *Replace s with 3 since this will result in a true sentence. The solution is 3.*
$s = 3$

Shaquille O'Neal increased 3 shoe sizes from 1992 to 1995.

Equations involving products and quotients can also be found mentally.

Example 3 Solve each equation mentally.

a. $6x = 36$
$6 \cdot 6 = 36$ *Think "What number times 6 is 36?"*
$x = 6$

b. $\frac{45}{n} = 9$
$\frac{45}{5} = 9$ *Think "What number times 9 is 45?"*
$n = 5$

Checking Your Understanding

Communicating Mathematics

Read and study the lesson to answer these questions.

1. **In your own words,** define an equation.
2. **Write** an equation that is always true and another equation that is always false.

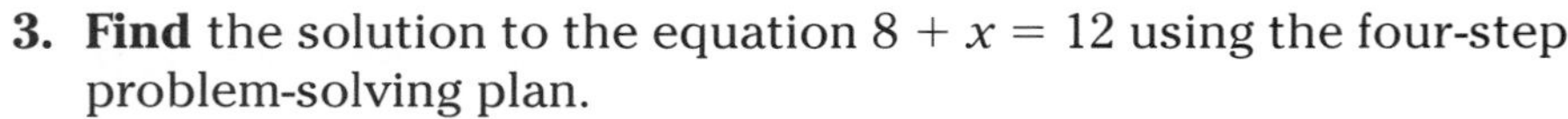

3. **Find** the solution to the equation $8 + x = 12$ using the four-step problem-solving plan.

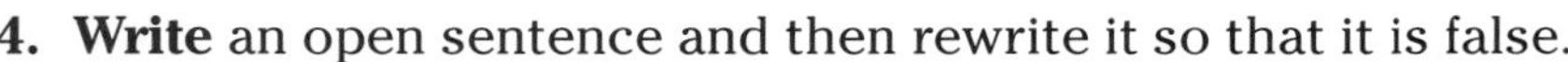

4. **Write** an open sentence and then rewrite it so that it is false.

5. You can use a TI-82 graphing calculator to solve equations if the right side of the equation is zero. You need to enter the left side of the equation, the variable you wish to solve for, and a guess at the solution. Use the keystrokes below to solve $x - 5 = 0$. We guessed that the solution was 5.

   [MATH] 0 [X,T,θ] [−] 5 [,] [X,T,θ] [,] 5 [)] [ENTER]

   The solution of the equation, 5, is displayed.

   Use a TI-82 to solve each equation below.

   a. $x - 9 = 0$ **b.** $2x = 0$ **c.** $3x - 6 = 0$

Guided Practice

Identify the solution to each equation from the list given.

6. $7 - x = 4$; 3, 4, 6
7. $y + 19 = 32$; 9, 13, 19
8. $2x + 1 = 7$; 3, 4, 5

Solve each equation mentally.

9. $t + 5 = 10$
10. $19 - q = 11$
11. $78 - r = 28$
12. $\frac{12}{x} = 3$

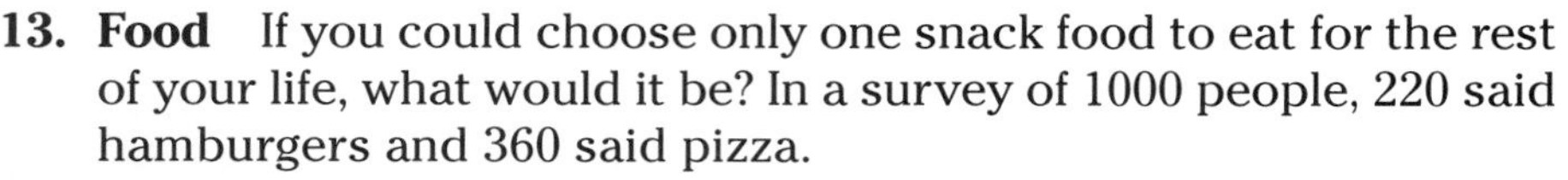

13. **Food** If you could choose only one snack food to eat for the rest of your life, what would it be? In a survey of 1000 people, 220 said hamburgers and 360 said pizza.

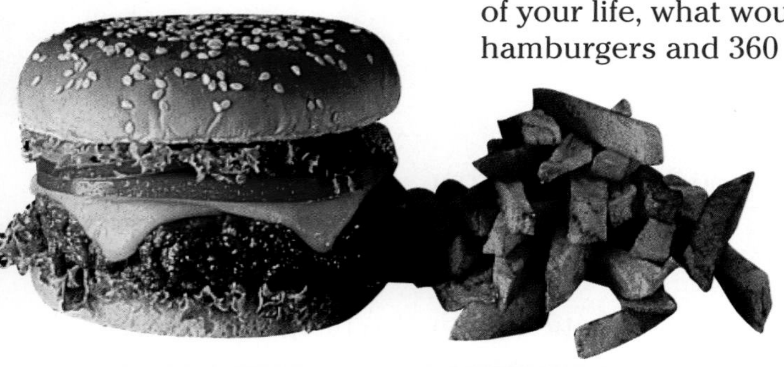

   a. Write an equation to find how many more people said pizza than hamburgers.

   b. Solve your equation.

Exercises: Practicing and Applying the Concept

Independent Practice

Identify the solution to each equation from the list given.

14. $12 - c = 8$; 2, 4, 7
15. $t + 33 = 72$; 29, 34, 39
16. $8 = \frac{16}{a}$; 1, 2, 4
17. $8 = \frac{g}{4}$; 32, 24, 16
18. $3x + 1 = 10$; 2, 3, 4
19. $7 = 5b + 2$; 0, 1, 2
20. $110 = 145 - t$; 35, 40, 45
21. $8 = \frac{48}{m}$; 3, 4, 6

Solve each equation mentally.

22. $m + 8 = 10$
23. $129 - q = 9$
24. $17 = k - 3$
25. $8x = 64$
26. $\frac{50}{h} = 5$
27. $63 = 7j$
28. $56 - s = 0$
29. $99 = 9x$
30. $\frac{72}{c} = 8$
31. $6x = 42$
32. $18 = 2w$
33. $\frac{21}{y} = 3$

Critical Thinking

34. Write two different open sentences that fit each description.
 a. The solution is 2.
 b. There is no solution that is a whole number. *Hint: The whole numbers are 0, 1, 2, 3, . . .*

Applications and Problem Solving

For the latest Dow Jones Industrial Average, visit: **www.glencoe.com/sec/math/prealg/mathnet**

35. **Finance** Stock market performance helps economists determine the health of the economy. One of the indicators of stock market performance is the Dow Jones Industrial Average. On February 21, 1995, the Dow Jones Industrial Average climbed above 4000 for the first time ever. That day, it closed at 4003.33. On February 20, 1995, the closing mark was 3973.05.
 a. Write an open sentence that describes the climb in the average.
 b. Is the solution to the open sentence 29.92, 30.28, or 33.32?

36. **Geometry** The sum of the measures of the angles in any triangle is 180°.
 a. Write an equation for the sum of the angle measures in the triangle at the right.
 b. Solve your equation for x.

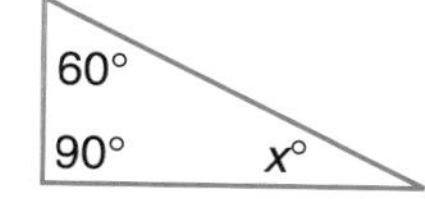

Mixed Review

37. Simplify the expression $5x + 2x + 1$. (Lesson 1-5)

38. Rewrite the expression $8(3 \cdot 6)$ using an associative property. (Lesson 1-4)

39. **Entertainment** The game Chutes and Ladders was copyrighted in the United States in 1943. Clue was copyrighted n years later. Write an expression for the year that Clue was copyrighted. (Lesson 1-3)

40. Write a verbal phrase for $y - 9$. (Lesson 1-3)

41. Evaluate $8 + 9 - 6 \cdot 2$. (Lesson 1-2)

42. **Patterns** What is the total number of rectangles, of all sizes, in the figure below? (Lesson 1-1)

WORKING ON THE

Investigation

Up, Up, and Away!

Refer to the Investigation on pages 2–3.

Look over your research and the list of considerations that you placed in your Investigation Folder at the beginning of the Investigation.

You probably listed time and fuel consumption as two of the things you need to consider as you plan your trip. You can use the formulas *speed* × *time* = *distance* and *miles traveled* ÷ *miles per gallon* = *gallons of fuel needed* to determine your time and fuel needed.

- Use a map to determine the approximate distance to Oshkosh.
- Use the information you found to write an equation involving the speed of an ultralight and the distance you will travel. Use a calculator to solve for the time needed.
- Then write an equation with the information about fuel. How many gallons of fuel will you need to make the trip?

Add the results of your work to your Investigation Folder.

1-7 Integration: Geometry
Ordered Pairs

Setting Goals: *In this lesson you'll use ordered pairs to locate points, to organize data, and to explore how certain quantities in expressions are related to each other.*

Modeling with Manipulatives

MATERIALS

grid paper

You may have played a board game called Battleship. In this game, players locate their battleships on grids like the one shown here. One player tries to locate and "sink" the other player's battleships by calling out a letter-number combination like F2. If a ship is located at this point, the player says *hit*. If not, the player says *miss*. The object of the game is to sink all of your opponent's ships before he or she sinks yours.

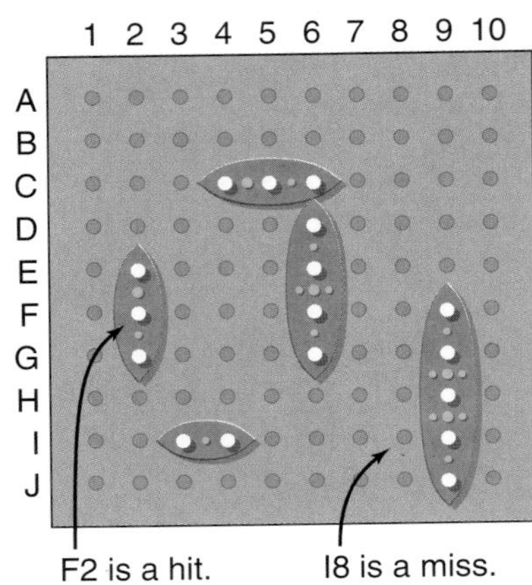

Your Turn

TALK ABOUT IT

Play a game of Battleship on grid paper.

a. Name an advantage of using a letter-number instead of a number-number combination to locate points on a grid.

b. Is the point named by C3 the same as 3C? If the locations were both named by numbers, would (3, 4) be different than (4, 3)?

c. Name a disadvantage of using a letter-number combination to locate points on a grid.

Learning the Concept

In mathematics, we can locate a point by using a **coordinate system**. The coordinate system is formed by the intersection of two number lines that meet at their zero points. This point is called the **origin**. The horizontal number line is called the ***x*-axis**, and the vertical number line is called the ***y*-axis**.

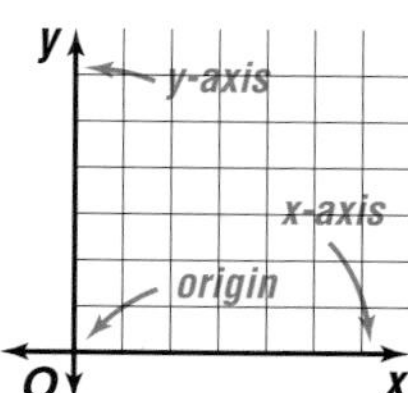

You can graph any point on a coordinate system by using an **ordered pair** of numbers. The first number in the pair is called the ***x*-coordinate**. The second number is called the ***y*-coordinate**. The coordinates are your directions to find the point.

Example  Graph each ordered pair.

a. (5, 3)

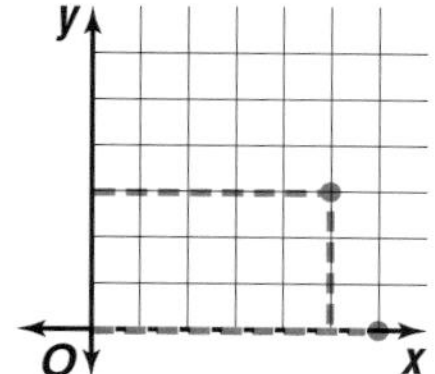

Begin at the origin. The *x*-coordinate is 5. This tells you to go 5 units right of the origin.

The *y*-coordinate is 3. This tells you to go up 3 units.

Draw a dot. You have now graphed the point whose coordinates are (5, 3).

THINK ABOUT IT

Are the numbers in an ordered pair commutative?

b. (6, 0)

Begin at the origin again. The *x*-coordinate is 6, so go 6 units right of the origin. The *y*-coordinate is 0, so you will not go up to place the dot.

Sometimes points are named by using letters. The symbol *B*(3, 4) means point *B* has an *x*-coordinate of 3 and a *y*-coordinate of 4.

The coordinate plane allows you to locate points.

Example  Name the ordered pair for each point.

a. *C*

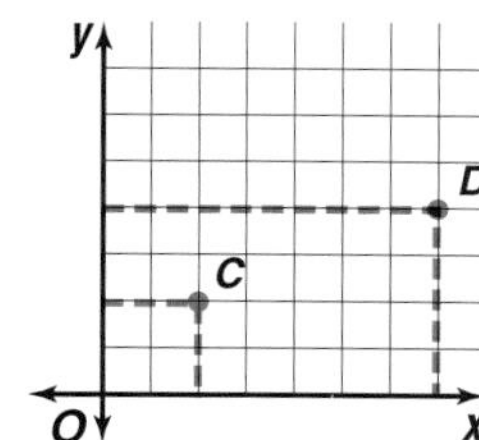

Go right on the *x*-axis to find the *x*-coordinate of point *C*, which is 2. Go up along the *y*-axis to find the *y*-coordinate, which is 2.

The ordered pair for point *C* is (2, 2).

b. *D*

The x-coordinate of *D* is 7, and the *y*-coordinate is 4.

The ordered pair for point *D* is (7, 4).

Ordered pairs can also be used to show how the value substituted for a variable is related to the value of an expression containing the variable.

Example 3

Retailing

FYI

Pocahontas was a Powhatan Indian. President Woodrow Wilson was one of her descendants.

At Movie Madness Video Store, newly-released videos rent for \$2 a day. The store paid \$18 for a copy of *Pocahontas*. If x represents the number of times the video is rented in the first month, the expression $2x - 18$ gives the amount of money the store will make on this video.

a. Make a list of ordered pairs in which the x-coordinate represents the number of rentals in the first month and the y-coordinate represents the money made for 10, 15, 20, 25, or 30 rentals.

b. Then graph the ordered pairs.

a. Evaluate the expression $2x - 18$ for 10, 15, 20, 25, and 30. A table of ordered pairs that result is shown below.

x (rentals)	$2x - 18$	y (money made)	(x, y)
10	2(10) − 18	2	(10, 2)
15	2(15) − 18	12	(15, 12)
20	2(20) − 18	22	(20, 22)
25	2(25) − 18	32	(25, 32)
30	2(30) − 18	42	(30, 42)

b.

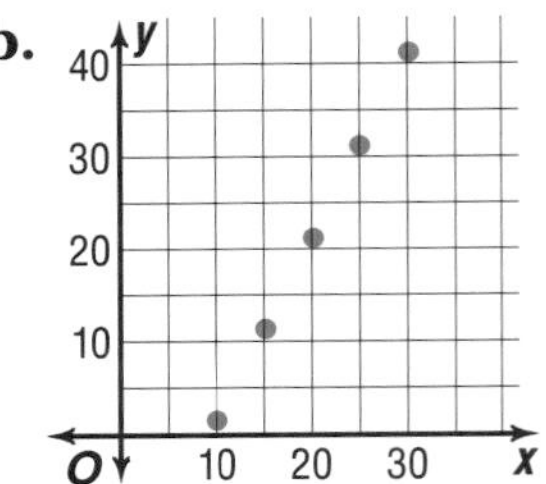

Checking Your Understanding

Communicating Mathematics

Read and study the lesson to answer these questions.

1. **Draw** a coordinate system and label the origin, x-axis, and y-axis.
2. **Tell** three different uses of ordered pairs.
3. **Describe** a coordinate system that you have seen used in everyday life.
4. **You Decide** Takala says that the points (4, 7) and (7, 4) are the same. Jamal disagrees. Who is correct and why?

Guided Practice

Use the map at the right to answer each question.

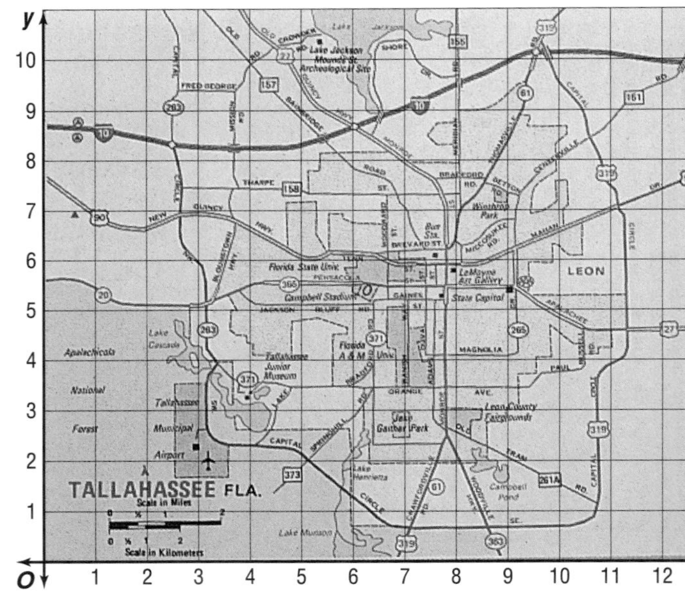

5. Which ordered pair indicates the location of the Tallahassee Municipal Airport?
6. Is (5, 6) the correct location for Campbell Stadium?
7. What location does (8, 5) indicate?
8. Which ordered pair indicates the location of the intersection of routes 319 and 10?
9. Is (3, 7) the correct ordered pair for the Lake Jackson Mounds State Archeological site?

Exercises: Practicing and Applying the Concept

Independent Practice

Use the grid at the right to name the point for each ordered pair.

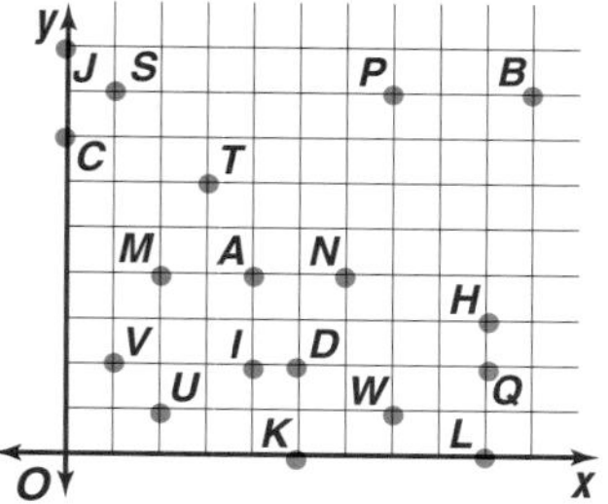

10. (5, 2)	**11.** (2, 4)
12. (1, 8)	**13.** (9, 0)
14. (4, 4)	**15.** (3, 6)
16. (9, 3)	**17.** (0, 9)

Use the grid to name the ordered pair for each point.

18. *V*	**19.** *N*	**20.** *P*
21. *K*	**22.** *W*	**23.** *Q*
24. *C*	**25.** *B*	**26.** *U*

27. **a.** Write an ordered pair for the coordinates of a point on the x-axis.

b. Write an ordered pair for the coordinates of a point on the y-axis.

c. What point lies on both the x- and y-axes?

Critical Thinking

28. **Geometry** Graph (2, 1), (2, 4), and (5, 1) on a coordinate system.

a. Connect the points with line segments. What figure is formed?

b. Multiply each number in the set of ordered pairs by 2. Graph and connect the new ordered pairs. What figure is formed?

c. Compare the two figures you drew. Write a sentence that tells how the figures are the same and how they are different.

29. **Astronomy** The table at the right contains data about the planets of our solar system. The x-value is the planet's mean distance from the sun in millions of miles. The y-value is how many years it takes each planet to complete its orbit around the Sun.

Planet	x	y
Mercury	36.0	0.241
Venus	67.0	0.615
Earth	93.0	1.000
Mars	141.5	1.880
Jupiter	483.0	11.900
Saturn	886.0	29.500
Uranus	1782.0	84.000
Neptune	2793.0	165.000
Pluto	3670.0	248.000

a. Graph each ordered pair.

b. Suppose a planet was discovered 1.5 billion miles from the sun. Use the graph to estimate how long it would take to orbit the sun.

30. **Geology** The underground temperature of rocks varies with their depth below the surface. The temperature in degrees Celsius is estimated by the expression $35x + 20$, where x is the depth in kilometers.

a. Make a list of ordered pairs in which the x-coordinate represents the depth and the y-coordinate represents the temperature for depths of 0, 2, 4, 6, and 8 kilometers.

b. Graph the ordered pairs.

31. **Traffic Control** The graph at the right shows some information about the number of cars at the traffic light at State Street and Schrock Road at two different times of a certain weekday.

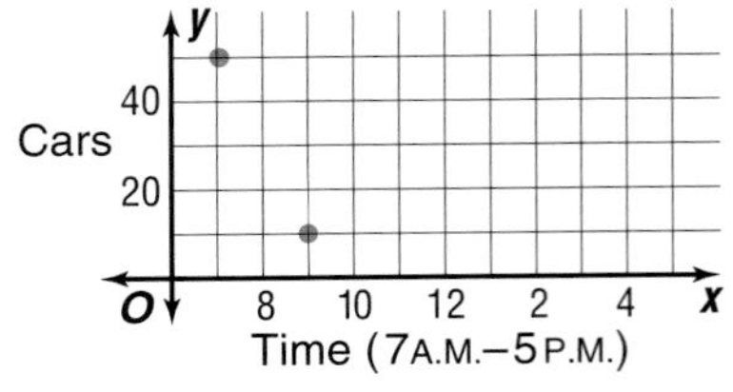

a. What information is given by the two points?

b. What do you think happened between the two times?

c. If most of the local businesses close at 5:00 P.M., copy the graph and place a point for the number of cars you think would be at the intersection at 5:15 P.M. Explain how you chose your point.

Mixed Review

32. Solve the equation $7a = 28$ mentally. (Lesson 1-6)

33. Simplify the expression $3m + 7m + 1$. (Lesson 1-5)

34. **Medicine** There are three-times more women enrolled in U.S. medical schools today than in the mid-70s. Write an expression for the number of women in medical schools today if x represents the number in the 1970s. (Lesson 1-3)

35. Evaluate the expression $4r$ if $r = 3$. (Lesson 1-3)

36. Evaluate $4c + 2a$ if $a = 2$ and $c = 7$. (Lesson 1-3)

37. Find the value of the expression $6 \cdot 3 \div 9 - 1$. (Lesson 1-2)

COOPERATIVE LEARNING PROJECT

THE SHAPE OF THINGS TO COME

The Global Positioning System

Soon you will be able to hit the open road without stopping to consult a road map. The Global Positioning System, or GPS, is becoming more available to the consumer market, with some guidance systems mounted in cars and other hand-held models as small as a candy bar. The GPS uses a system of 24 satellites to determine your exact location. The monitor receives signals from the satellites and then finds the distances to three or four different satellites to tell your location anywhere on Earth. The display tells you your latitude, longitude, altitude, and direction and speed of travel.

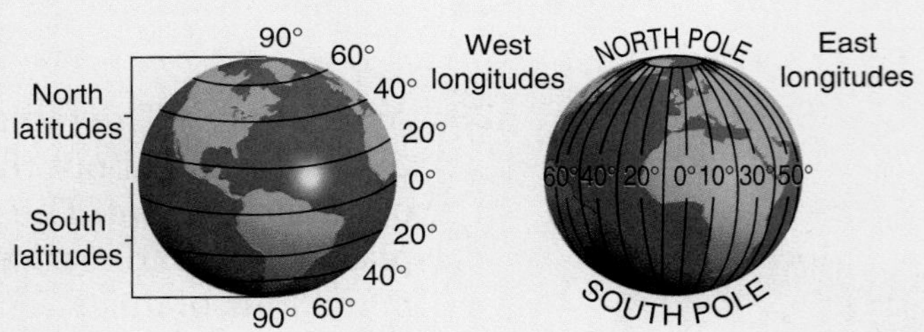

Latitude and longitude lines form a coordinate system for locating any point on Earth. Early sailors developed the latitude system from observations of the North Star. Latitude lines tell the location north or south of the equator. Longitude is used to determine east-west distance. Longitude lines tell the location east or west of the prime meridian in Greenwich, England. Both latitude and longitude are measured in degrees.

See For Yourself

Research the Global Positioning System and the latitude and longitude system.

- How does the GPS use the satellite distances to determine position?
- What are some of the ways consumers are using the GPS? What are the plans for the GPS?
- Why was the GPS first developed?
- What is the approximate latitude and longitude of your hometown?

1-8 Solving Equations Using Inverse Operations

Setting Goals: *In this lesson, you'll use inverse operations to solve equations and solve problems using equations.*

Modeling a Real-World Application: Transportation

The New York City subway system is among the oldest and longest in the world. It has 461 stations, 230 miles of track, and most trains run 24 hours a day, 365 days a year.

When the New York City subway service began in 1904, passengers bought tickets from an attendant for five cents. Another attendant tore the ticket as you passed through the turnstile. Soon the turnstiles were made to operate on coins. When the subway fare was changed to a dime in 1948, the turnstiles could still operate on a coin. However, when the fare went to fifteen cents in July of 1953, there was a problem. That's when the New York City subway token was introduced.

The "bulls-eye" tokens used in New York subways today have brass surrounding a steel center. Since the turnstiles read tokens with a magnet, using bulls-eye tokens helps reduce the use of slugs, which are worthless pieces of metal used instead of tokens. Before the bulls-eyes, New York took in about 165,000 slugs each month. Now slugs total about 20,000 a month.

In 1995, New York subways lost about $25,000 a month to people using slugs. What was the subway fare in 1995? *This problem will be solved in Example 3.*

Learning the Concept

You have solved some equations using mental math or patterns, but sometimes the solution is not easy to see. Another way to solve equations is to use **inverse operations**. Inverse operations "undo" each other. For example, to "undo" addition you would subtract. These can be shown by related sentences.

The sentences below show how addition and subtraction are related.

$$9 + 12 = 21 \quad \rightarrow \quad 9 = 21 - 12, \quad 12 = 21 - 9$$

$$9 + n = 18 \quad \rightarrow \quad 9 = 18 - n, \quad n = 18 - 9$$

Multiplication and division are also inverse operations.

$$3 \cdot 5 = 15 \quad \longleftrightarrow \quad 3 = 15 \div 5 \text{ or } 3 = \frac{15}{5}$$

$$3 \cdot 5 = 15 \quad \longleftrightarrow \quad 5 = 15 \div 3 \text{ or } 5 = \frac{15}{3}$$

$$12t = 48 \quad \longleftrightarrow \quad 12 = 48 \div t \text{ or } 12 = \frac{48}{t}$$

$$12t = 48 \quad \longleftrightarrow \quad t = 48 \div 12 \text{ or } t = \frac{48}{12}$$

You can use inverse operations to solve equations.

Example 1 Solve each equation by using the inverse operations.

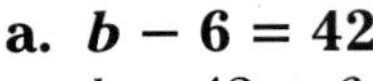

a. $\mathbf{b - 6 = 42}$

$b = 42 + 6$ *Write a related addition sentence.*

$b = 48$

b. $\mathbf{c + 16 = 54}$

$c = 54 - 16$ *Write a related subtraction sentence.*

$c = 38$ *You can check your answer by replacing c with 38 in the original equation.*

c. $\mathbf{\frac{v}{9} = 16}$

$v = 16 \cdot 9$ *Write a related multiplication sentence.*

$v = 144$

d. $\mathbf{2.5s = 37.5}$

$s = 37.5 \div 2.5$ *Write a related division sentence.*

$s = 15$ *Use a calculator.*

THINK ABOUT IT

Study each of the rewritten sentences. Why do you think each one is written with the variable alone on one side?

Using an equation to solve a problem is an important problem-solving strategy. The first step in writing an equation is to choose a variable and a quantity for the variable to represent. This is called **defining a variable**.

Example 2 The sum of 215 and a certain number is equal to 266. Find the number.

Let n represent the number. Now that you have a variable, translate the words into an equation using the variable.

215	*plus*	*number*	*equals*	*266*
215	+	n	=	266

$215 + n = 266$

$n = 266 - 215$ *Write the related subtraction sentence.*

$n = 51$ The number is 51.

Check: Since 215 + 51 = 266, the solution is correct.

Example 3

APPLICATION Transportation

Refer to the application at the beginning of the lesson. Use the four-step problem-solving plan to solve.

Explore Read the problem for the general idea. You know that about 20,000 slugs are used each month for a loss of $25,000. You need to find the subway fare. Each slug is equal to one fare. Let f represent the fare.

Plan Translate the words into an equation using the variable. The total amount of money lost to people using slugs is the number of slugs multiplied by the fare.

fare	*times*	*number of slugs*	*equals*	*total lost*
f	$\times$	20,000	=	$25,000

Solve Solve the equation for the fare.

$f \times 20{,}000 = \$25{,}000$
$f = \$25{,}000 \div 20{,}000$ *Write a related division sentence.*

25000 [÷] 20000 [=] 1.25 *Use a calculator to solve.*

The fare is $1.25.

Examine The answer is close to the estimate. Since 20,000 × $1.25 equals $25,000, the answer checks.

THINK ABOUT IT

Always estimate before you solve the equation. THINK: Since 25,000 is a little more than 20,000 the quotient will be a little over 1.

Checking Your Understanding

Communicating Mathematics

Read and study the lesson to answer these questions.

1. **Write** an addition sentence and then write a related subtraction sentence.
2. What is the inverse operation of gaining five yards in football?
3. **Explain** what it means to *define a variable*.
4. **Transportation** Refer to the application at the beginning of the lesson.
   a. Write an equation and find the amount of money lost each month if the New York Subway System was still taking in 165,000 slugs per month and the fare is $1.25.
   b. Check your answer using the words of the problem. Write a convincing argument for checking the answer of a problem against the words of the problem instead of the equation that you wrote.

Guided Practice

For each sentence, write a related sentence using the inverse operation.

5. $3 + 9 = 12$
6. $13 \cdot t = 39$
7. $16 - 12 = 4$
8. $44x \div 4x = 11$

Solve each equation using the inverse operation. Use a calculator when needed.

9. $8 + y = 22$ **10.** $77 = 7h$ **11.** $5.4 = \frac{w}{2.2}$

Translate each sentence into an equation. Then solve.

12. Twice a certain number is 6.

13. Twenty-seven more than a number is thirty-one.

Write a problem that could be solved using each equation.

14. $\$6x = \$135$ **15.** $x - 7 = 28$

16. **Sports** In April 1994, Jose Maria Olazabal won the $360,000 prize of the Masters Tournament in Augusta, Georgia. The following week, Hale Irwin took the prize of the Heritage Classic. Irwin won $135,000 less than Olazabal. How much was the Heritage Classic prize?

a. Define a variable for this problem.

b. Write an equation.

c. Solve for the amount of the Heritage Classic prize.

d. Check your solution against the words of the problem.

Exercises: Practicing and Applying the Concept

Independent Practice

For each sentence, write a related sentence using the inverse operation.

17. $14 - 3 = 11$ **18.** $23 = 17 + n$ **19.** $14 \cdot 2 = 28$

20. $\frac{x}{11} = 12$ **21.** $6h = 48$ **22.** $k \div 13 = 7$

Solve each equation using the inverse operation. Use a calculator when needed.

23. $6 + n = 28$ **24.** $g + 11 = 15$ **25.** $86 = c - 2$

26. $144 = 12h$ **27.** $8 = \frac{v}{15}$ **28.** $210 = 15w$

29. $5.56 - q = 4.73$ **30.** $3.7 = \frac{r}{5.4}$

31. $f + 15.98 = 19.75$ **32.** $6.8b = 34.68$

33. $1.5 + b = 2.1$ **34.** $\frac{p}{2} = 9$

Translate each sentence into an equation. Then solve.

35. Three times a certain number is 18.

36. Ten less than a number is 19.

37. When a number is decreased by seven, the result is 22.

38. The product of a number and eight is 88.

39. A number divided by four is 14.

40. A number divided by three is thirteen.

Write a problem that could be solved using each equation.

41. $n + 2 = 16$ **42.** $\$30n = \$90$

Critical Thinking

43. Consider the division sentence $a = \frac{5}{0}$ and the related multiplication sentence $5 = a \cdot 0$. Use these statements to explain why division by zero is not allowed.

Applications and Problem Solving

44. **Entertainment** In the movie *Maverick*, Bret Maverick worked to raise enough money to enter a card championship. Twenty players entered the tournament, each one paying the same entrance fee. The entrance fees totaled $500,000. How much did each player pay to play?
    a. Define a variable.
    b. Write an equation for the problem.
    c. Solve for the entrance fee.
    d. Check your solution against the words of the problem.

45. **Sports** *Monday Night Football* has been thrilling fans and topping the television ratings since 1969. As of the 1993 season, the San Francisco 49ers had appeared on Monday Night Football 35 times. The Miami Dolphins had appeared the most times, with 11 more appearances than the 49ers. How many times had the Dolphins appeared?

Mixed Review

46. **Geometry**
    a. Which of the points on the graph at the right has the coordinates (5, 8)? (Lesson 1-7)
    b. Write the coordinates of the point E. (Lesson 1-7)

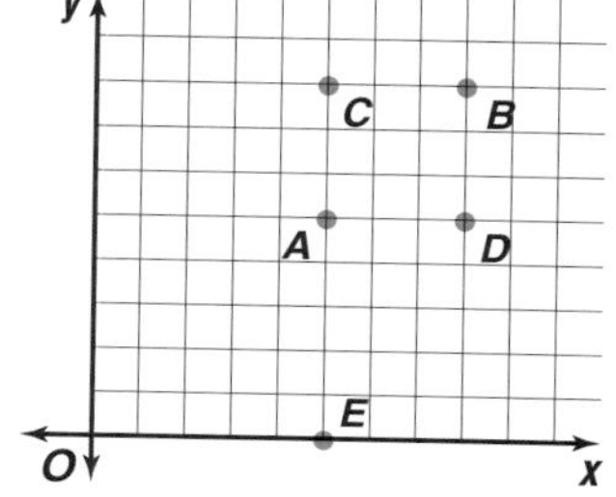

47. Which of the numbers 13, 16, or 19 is the solution of $77 + b = 96$? (Lesson 1-6)
48. Find $13 \cdot 13$ mentally using the distributive property. (Lesson 1-5)
49. Name the property shown by the statement $(9 + 4) + 5 = 9 + (4 + 5)$. (Lesson 1-4)
50. Evaluate $b + cd$ if $b = 2$, $c = 5$, and $d = 7$. (Lesson 1-3)
51. Translate the phrase *five times last year's total* into an algebraic expression. (Lesson 1-3)
52. Find the value of $(18 + 6) \div (8 - 2)$. (Lesson 1-2)
53. **Demographics** *American Demographics* magazine reported that American men drive an average of 16,500 miles each year. How many miles does an average American man drive in a week? (Lesson 1-1)
    a. Which method of computation do you think is most appropriate for this problem? Justify your choice.
    b. Solve the problem using the four-step plan. Be sure to examine your solution.

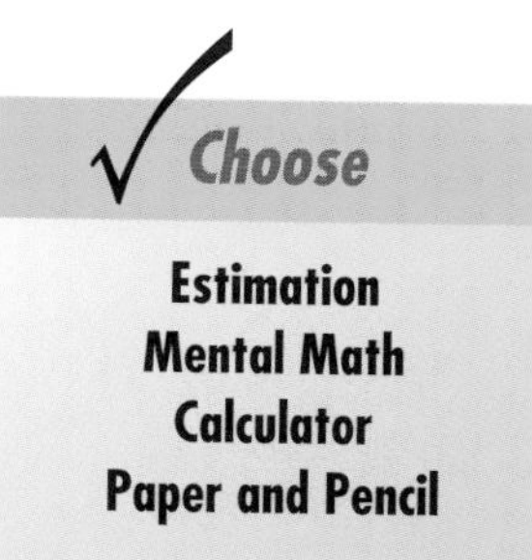

1-9 Inequalities

Setting Goals: *In this lesson, you'll write and solve inequalities.*

Modeling a Real-World Application: Health

FYI

23% of Americans say that they dream in black and white.

Insomnia, the inability to get enough sleep, affects millions of Americans each year. Researchers have come up with some suggestions on how to avoid insomnia.

- Maintain a temperature of between 68 and 75°F in your bedroom.
- Stop eating at least 2 hours before bedtime.
- Avoid reading or watching television in bed.
- Go to bed and get up at the same times each day, even when you can sleep in.

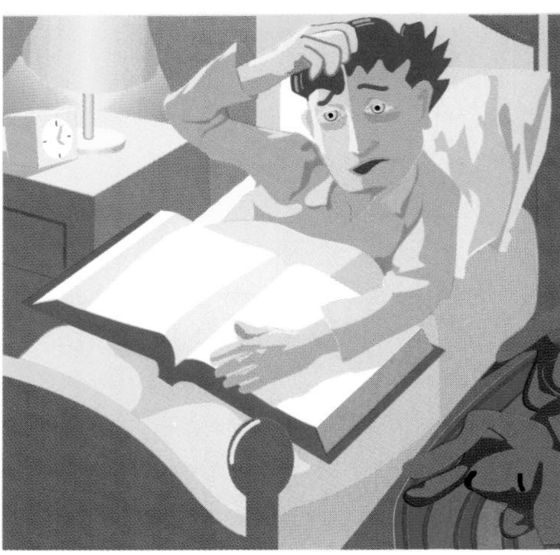

Learning the Concept

Let the variable t represent the bedroom temperature. We can write mathematical sentences to describe the temperature that physicians recommend, which is between 68°F and 75°F.

Say: t is less than 75
Write: $t < 75$

Say: t is greater than 68
Write: $t > 68$

A mathematical sentence that contains $<$ or $>$ is called an **inequality**. Inequalities, like equations, can be true, false, or open.

$18 > 13$ *This sentence is true.*
$9 < 5$ *This sentence is false.*
$t < 75$ *This sentence is open. It is neither true nor false until t is replaced with a number.*

Some inequalities contain $\leq$ or $\geq$ symbols. They are the combinations of the equals sign and the inequality symbols.

Say: n is less than or equal to 4
Write: $n \leq 4$

Say: m is greater than or equal to 8
Write: $m \geq 8$

Example  **For the given value, state whether each inequality is *true* or *false*.**

a. $r - 8 > 7, r = 12$

$12 - 8 > 7$

$4 > 7$

This sentence is false.

b. $14 \leq \frac{3s}{2} + 5, s = 10$

$14 \leq \frac{3 \times 10}{2} + 5$

$14 \leq 15 + 5$

$14 \leq 20$

This sentence is true.

Many situations in real life can be described using inequalities. The table below shows some common phrases and corresponding inequalities.

<	>	≤	≥
• less than • fewer than	• greater than • more than • exceeds	• less than or equal to • no more than • at most	• greater than or equal to • no less than • at least

Example 2

APPLICATION

Agriculture

A dozen eggs must weigh at least 30 ounces to be labeled "jumbo" by the U.S. Department of Agriculture. A dozen eggs that weigh less than 18 ounces are not supposed to be sold to consumers.

a. Write an inequality for the weight of a dozen jumbo eggs.

b. Write an inequality for the weight of eggs not sold.

a. The phrase *at least* implies a greater than or equal relationship. Let j represent the weight of a dozen jumbo eggs.
$j \geq 30$

b. Eggs that weigh less than 18 ounces per dozen are not sold. Let u represent the weight of unsold eggs.
$u < 18$

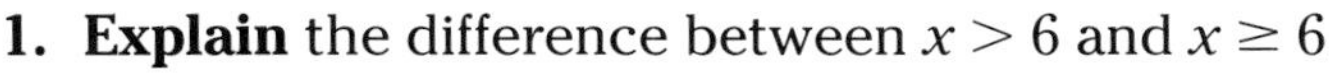

Checking Your Understanding

Communicating Mathematics

Read and study the lesson to answer these questions.

1. **Explain** the difference between $x > 6$ and $x \geq 6$.
2. **Find** a value for which $6x > 12$ is true.
3. **Compare and contrast** equations and inequalities.
4. Work with a partner.
   - ▶ Write each of the digits 0–9 on a card and place cards face down.
   - ▶ Then write each of the inequality symbols on a card and place these cards face down in a separate pile. (Mark the bottom of the symbol cards so you can tell which direction the symbol should face.)
   - ▶ Choose one card from the digits pile.
   - ▶ Then choose a card from the symbol pile.
   - ▶ Choose another card from the digits pile.
   - ▶ Determine if the inequality you created is true or false.
   - ▶ Then try to rearrange the cards to make a true inequality.
   - ▶ Shuffle the cards and draw again to create several inequalities.

MODELING MATHEMATICS

MATERIALS

index cards

Guided Practice

State whether each inequality is *true*, *false*, or *open*.

5. $9 < 3$ **6.** $8 < 8$ **7.** $12 > 4$

State whether each inequality is *true* or *false* for the given value.

8. $n + 2 > 4, n = 8$ **9.** $15 \leq 2q, q = 9$

Translate each sentence into an inequality.

10. A number is less than 6.

11. A dozen eggs must weigh 27 ounces or more to be labeled "extra large."

12. **Civics** On July 1, 1971, the 26th amendment to the U.S. Constitution was ratified. This amendment states "The right of citizens of the United States, who are 18 years of age or older, to vote shall not be denied or abridged by the United States or any state on account of age." Write an inequality showing the age of a voter.

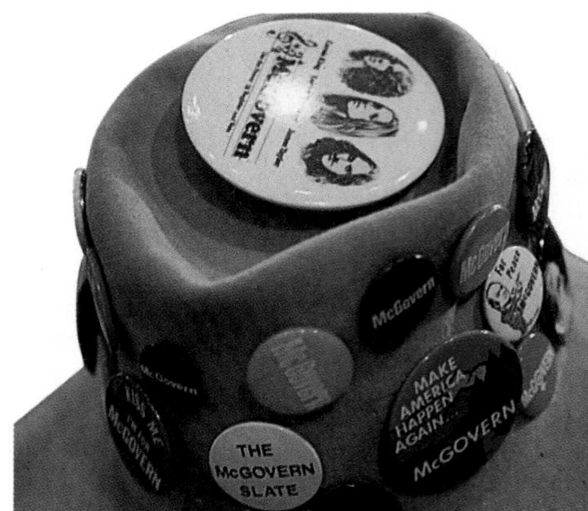

Exercises: Practicing and Applying the Concept

Independent Practice

State whether each inequality is *true*, *false*, or *open*.

13. $7 > 6$ **14.** $40 < 28$ **15.** $2n > 4$

16. $18 > 6m$ **17.** $0 \geq 0$ **18.** $j + 7 < 34$

19. $16 \leq 16$ **20.** $17 < p$ **21.** $y - 1 < 4$

State whether each inequality is *true* or *false* for the given value.

22. $22 - w > 4, w = 11$ **23.** $24 \geq 3c, c = 6$

24. $h - 9 > 6, h = 10$ **25.** $2x - 7 \leq 7, x = 7$

26. $4x + 8x < 7, x = 1$ **27.** $5z - 16 \leq 11, z = 4$

Translate each sentence into an inequality.

28. A number exceeds 24.

29. Tertius earned $100 or more.

30. At most, 75 people applied.

31. It takes no less than three years to complete law school.

32. Julita scored at least 16 points.

33. Fewer than 15 people attended.

Critical Thinking

34. Suppose $x > 6$ and $x < 18$.

a. Explain what the open sentence $6 < x < 18$ means.

b. Find a value of x that makes the sentence $6 < x < 18$ true.

Applications and Problem Solving

35. Sports On April 9, 1993, the Colorado Rockies played their first home opener in Mile High Stadium in Denver. More than 80,000 fans were in attendance. Write an inequality for the number of people who attended.

36. Finance Some companies provide 401(k) plans that help their employees save for retirement. The Internal Revenue Service sets a limit on the maximum amount that an employee may invest in a 401(k) plan in any given year. In 1995, the limitation was changed to $9240. Write an inequality for the amount that an employee may invest in a 401(k) plan.

37. Health Some of the benefits of quitting smoking are listed in the chart at the right.

a. Write an inequality for the amount of time in which a person who quits smoking can expect lung cancer death risk reduced to that of a nonsmoker.

b. Write an inequality for the amount of time it takes for a person who quits smoking to cut their risk of heart disease to half that of a smoker.

Benefits of Quitting Smoking

Within	Benefit
2 weeks to 3 months	• circulation improves • walking becomes easier • lung function increases 38%
1 year	• excess risk of heart disease is half that of a smoker
10 years	• lung cancer death risk is similar to that of a nonsmoker • precancerous cells are replaced • risk of various cancers decreases

Mixed Review

38. Charity Many airlines donate change found in seats and foreign coins that travelers no longer need to charity. TWA gives the foreign coins donated on international flights to UNICEF. The donations total $14,400 in an average year. How much does TWA collect for UNICEF in a month? (Lesson 1-8)

a. Define a variable.

b. Write an equation for the problem.

c. Solve for the monthly donations.

d. Check your solution against the words of the problem.

39. Solve $7b = 77$ mentally. (Lesson 1-6)

40. Use the distributive property to simplify $4 \cdot 8 + 4 \cdot 2$. (Lesson 1-5)

41. Rewrite the expression $8(7c)$ using an associative property. Then simplify. (Lesson 1-4)

42. Evaluate the expression $9t - 2s$ if $s = 8$ and $t = 3$. (Lesson 1-3)

43. Find the value of $\frac{84 - 12}{14 - 6}$. (Lesson 1-2)

44. Patterns Find the next number in the pattern 8, 10, 14, 22, 38, __?__ (Lesson 1-1)

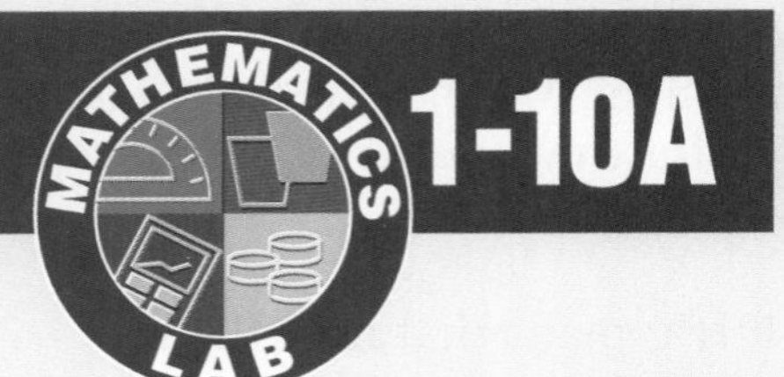

HANDS-ON ACTIVITY

1-10A Gathering Data

A Preview of Lesson **1-10**

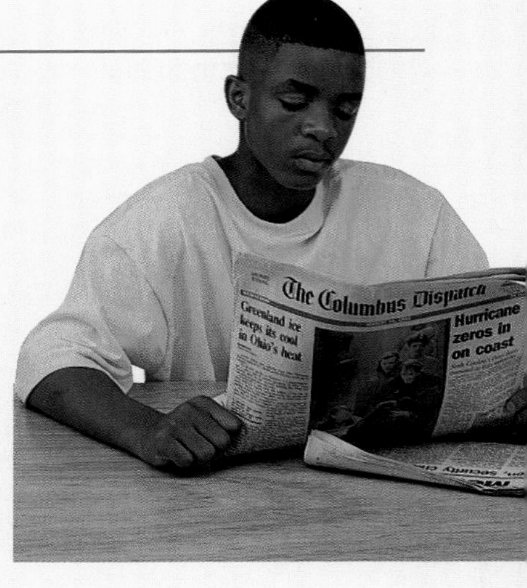

Advertisers, political parties, and media such as television stations and newspapers all want to find out about public opinion. **Demographics** is the study of the characteristics of a population. **Surveys** are often used to study a population. The results of surveys are often modeled by an equation.

Your Turn Work in small groups. Find out about the opinions of students in your school by conducting a survey. The pieces of information that you collect are called **data**.

a. Your group has been hired by a manufacturer to develop an ad campaign to convince students in your school to buy their product. Decide on a product to advertise.

b. How many people will you survey? You probably can't ask every person in your school, so you will need to ask a small group of students. A group that is used to represent a whole population is called a **sample**.

c. Write your questions.

d. Make a plan for asking students. Make sure that the group of students you will ask represents the whole school population.

e. Display your survey results in a way that is easily understood. The results of a sample survey question about television-watching habits is shown below.

How many hours of television do you watch in a week?		
Number of hours watched	**Tally**	**Frequency**
0–2	𝍸	5
3–5	𝍸 𝍸 ‖	12
6–8	𝍸 𝍸 𝍸 𝍸 ||||	24
9–11	𝍸 𝍸 𝍸 𝍸 𝍸 𝍸 |||	33
12–14	𝍸 𝍸 𝍸 𝍸 𝍸	25
15–17	𝍸		6

TALK ABOUT IT

1. Use the table above to find how many hours of television most of the students surveyed watch in a week.
2. Write a few sentences analyzing what you found out about the students in your school through your survey.
3. Explain how your advertising campaign can convince students to buy the product.
4. Prepare a report, with charts and graphs, that summarizes your findings.

Extension

5. Survey the cars in a parking lot. Determine what characteristics you will study. Then collect the data and analyze it.

1-10 Integration: Statistics
Gathering and Recording Data

Setting Goals: *In this lesson, you'll gather and record data in a frequency table or bar graph.*

Modeling a Real-World Application: Entertainment

"What do you think?" Opinion polls often ask that question. The results of an opinion poll are reported using **statistics**. Statistics involves collecting, analyzing, and presenting data.

On January 1, 1990, Epcot Center began a survey on "The Person of the Century" that will last until the year 2000. Who do you think is the most influential person of the 1900s? So far in the survey, the leaders are Albert Einstein, Martin Luther King, Jr., and Thomas Edison.

Disney's Epcot Center in Orlando, Florida, was the dream of founder Walt Disney. Disney wanted Epcot to be an open forum for people of the world to meet and talk. As part of that dream, Epcot Center conducts one of the world's largest ongoing public opinion polls. People visiting the park can gather in a 172-seat theater to participate in polls on everything from vegetables to rock videos.

The **frequency table** below shows the results of one Epcot poll on movie tastes.

What type of movie do you like most?		
	Men	Women
Comedy	39	47
Action/Adventure	40	18
Drama	6	20
Horror	2	2
Other	5	5
No response	8	8

The poll was conducted with 5174 adults at the Epcot Center. Results are shown as if groups of 100 men and 100 women were surveyed.

Learning the Concept

Obviously, it would be very difficult to ask every man and woman in the United States which type of movie they prefer. Epcot Center surveys smaller groups, called **samples**. The sample group is assumed to be representative of the entire population. However, since the group of people could not be chosen randomly but only from the group at Epcot Center, the results may not be representative.

Example 1 APPLICATION Entertainment

Refer to the frequency table at the beginning of the lesson.

a. Which type of movie did most men in the group prefer?
The greatest number of men, 40, chose action/adventure. Comedy was a close second with 39 men choosing it.

b. Overall, which type of movie is most popular?
In order to determine the type of movie most preferred, find the sum of the numbers of men and women who prefer each type.

	men	+	*women*	=	*total*
Comedy	39	+	47	=	86
Action/Adventure	40	+	18	=	58
Drama	6	+	20	=	26
Horror	2	+	2	=	4
Other	5	+	5	=	10
No response	8	+	8	=	16

Comedy is the most popular type of movie overall.

THINK ABOUT IT

The Epcot poll involved adults. How different would the results be if teens were asked? Conduct your own survey on movie preferences.

Statistical information can be displayed in a **bar graph**.

Example 2 CONNECTION Social Studies

In 1995, the Gallup Organization conducted a poll on the women that Americans admire most. The results are shown in the frequency table at the right. Make a bar graph to display the data.

Women Americans Admire	
Woman	**Votes**
Queen Elizabeth II	28
Jacqueline Kennedy Onassis	27
Mamie Eisenhower	22
Helen Keller	18
Margaret Chase Smith	18
Claire Booth Luce	17
Mother Teresa	16
Margaret Thatcher	16

Source: The Gallup Organization

Each of the categories in a bar graph has a bar to represent it. The vertical scale shows the number of votes.

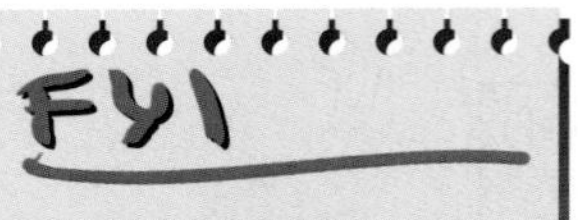

Claire Boothe Luce was an American playwright and politician noted for her sense of humor. Ms. Luce served as a U.S. Congressional Representative from Connecticut in the 1940s, as an ambassador to Italy in the 1950s, and on the Foreign Intelligence Advisory Board in the 1980s.

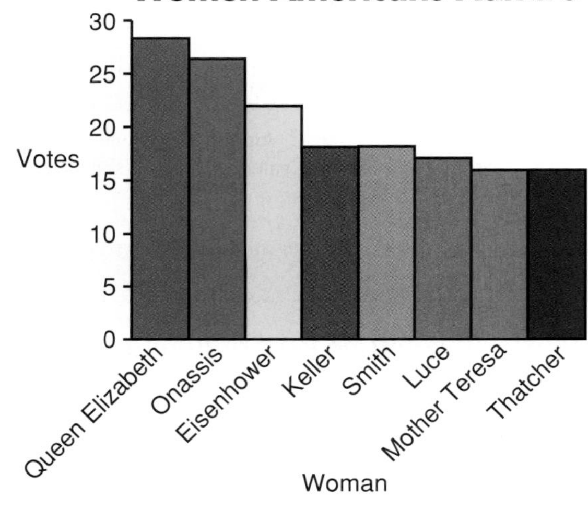

Some graphing calculators allow you to create bar graphs.

Example 3

Sports

Since the first modern Olympics in 1896, the number of events at each Summer Olympic Games has grown. The table at the right shows the number of events at recent Olympic Games. Use a TI-82 graphing calculator to make a bar graph of the data.

Olympic Year	Number of Events
1968	172
1972	196
1976	199
1980	200
1984	223
1988	237
1992	257
1996	271

Begin by accessing and clearing lists L_1 and L_2 in the calculator's memory.

Enter: [STAT] [ENTER] [▲] [CLEAR] [ENTER] [▶] [▲] [CLEAR] [ENTER]

Then enter the data. Use list L_1 for the years and L_2 for the events.

Enter: [◀] 1968 [ENTER] 1972 [ENTER] *and so on* 1996 [ENTER] [▶] 172 [ENTER] 196 [ENTER] *and so on* 271 [ENTER]

Next set the range of the data for the viewing window.

Enter: [WINDOW] [ENTER] 1968 [ENTER] 2000 [ENTER] 4 [ENTER] 0 [ENTER] 300 [ENTER] 25 [ENTER]

Now choose the type of graph and construct the graph.

Enter: [2nd] [STAT PLOT] [ENTER]

Use the arrow and [ENTER] keys to highlight "On", the bar graph, "L_1", and "L_2". Then press [GRAPH] to display the graph.

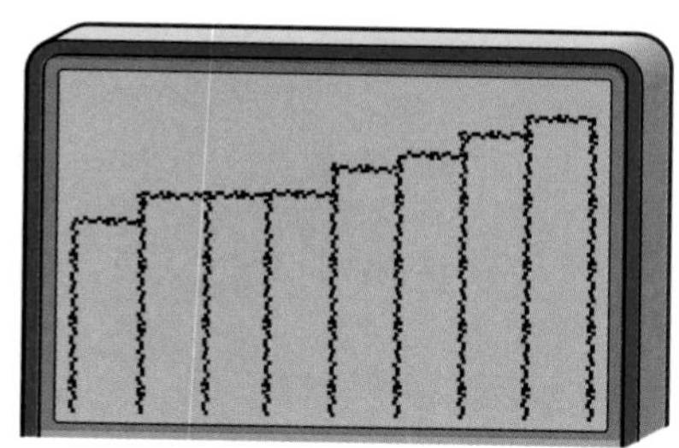

Checking Your Understanding

Communicating Mathematics

Read and study the lesson to answer these questions.

1. **Describe** the kind of information a frequency table shows in your own words.
2. **Explain** why a poll taken at Epcot Center may not be representative of the entire United States population.
3. **Statistics** Find a newspaper or magazine article that contains a graph or table. Explain how you think the results were found. Do you think the display of the information is effective? Explain.

4. **Assess Yourself** When do you use statistics in your life? Do you ever make decisions based on statistics?

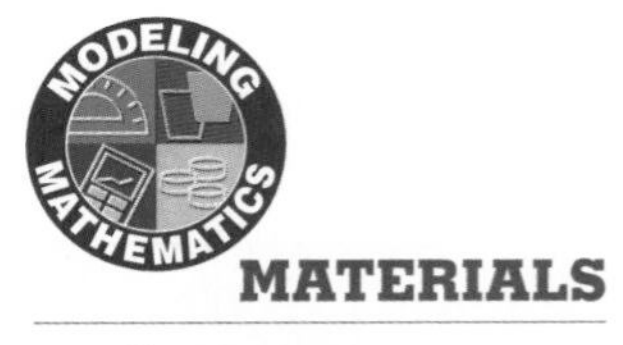

MATERIALS

M & M's or other colored candies

5. Bar graphs are excellent tools for displaying data. Examine the colors of the candies in a bag of M & Ms or a similar product. As you remove each candy from the bag, place it in a column to form a bar graph like the one at the right.

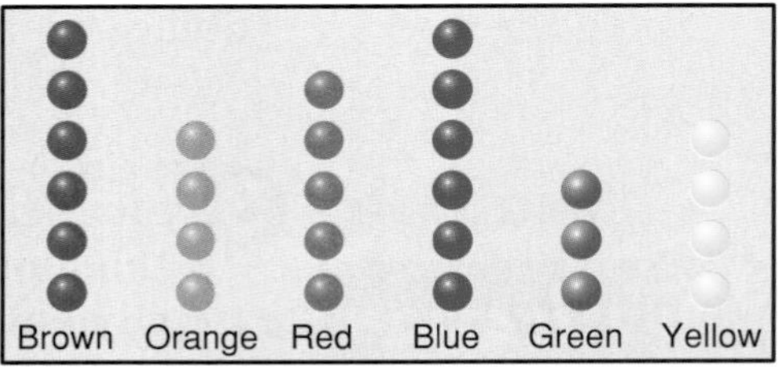

   a. Use your data to make a bar graph.

   b. How does your graph compare to the ones made by others in your class? Did each package of candy have the same color distribution and the same number of candies?

Guided Practice

The frequency table at the right contains data about the 1993 attendance at some of the major zoos in the United States.

Attendance of Selected Zoos in 1993	
Zoo	**Attendance (millions)**
Audubon (New Orleans)	0.9
Chicago	2.0
Cincinnati	1.3
Denver	1.3
Miami Metrozoo	0.8
National (Wash. D.C.)	3.0
San Diego	3.5
Toledo	0.8

6. Which zoo was visited by the most people?
7. Which zoo was visited least frequently?
8. How many people visited the Denver Zoo in 1993?
9. How many times were all of these zoos visited in 1993?
10. Make a bar graph of the zoo attendance figures.

Exercises: Practicing and Applying the Concept

Independent Practice

The table below shows the number of states with different laws about driving ages. The table shows the lowest age at which a person can receive a restricted or unrestricted driver's license.

Driver's License Laws	
Law	**States**
Restricted at 15	3
Unrestricted at 15	2
Restricted at 16	20
Unrestricted at 16	18
Restricted at 17	3
Unrestricted at 17	2
Unrestricted at 18	3

11. What is the lowest age at which someone can obtain a driver's license in any state?
12. When can a person get a driver's license in most states? Is it a restricted or unrestricted license?
13. Are all of the states and the District of Columbia represented in the table? How can you tell?
14. Make a bar graph of the data.
15. Write a sentence to describe the driver's age laws in the United States.
16. Where in the table does your state fall?

A list of the heights of the tall buildings in Seattle, Washington, is given below. Heights are given in feet.

17. Make a frequency table for the building heights. Use intervals of 100 feet.

954	740	730	722	609	605
580	574	543	514	500	498
493	487	466	456	454	409
397	389	379	371		

18. In which interval do most of the buildings fall?

19. A histogram is a special bar graph that displays the frequency of data in equal intervals. Use your frequency table to make a histogram of the data.

Critical Thinking

20. a. Find a bar graph in a newspaper or magazine.

b. Does the graph effectively communicate the data? When would a frequency distribution be more useful?

Applications and Problem Solving

21. Social Studies The heights of the U.S. presidents are given in the table at the right.

U.S. Presidents' Heights	
Height (in.)	**Frequency**
63–65	1
66–68	9
69–71	12
72–74	18
75–77	1

a. How many different presidents have there been?

b. What range of heights describes the most presidents?

c. How many presidents were exactly 6 feet tall? Explain your answer.

d. Why do you think so many presidents were or are fairly tall?

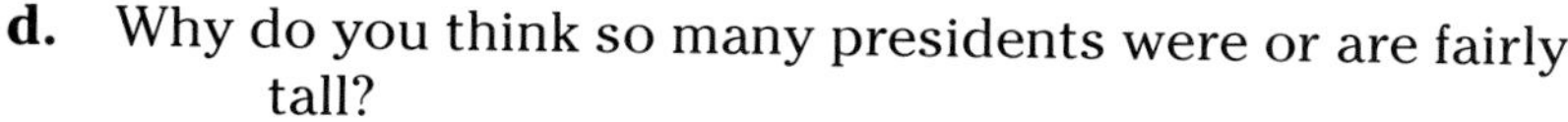

e. Research How do the heights of the presidents compare with the height of an average American man?

22. Family Activity Write your own survey question. You might ask family and friends about their favorite color, the number of people living in their home, or how many languages they speak. Make a frequency distribution and a bar graph of your findings.

Mixed Review

23. Family Management The Family Economics Research Group, which is a division of the U.S. Department of Agriculture, estimates that it costs a middle-income family $210,070 or more to raise a child to age 17. Write an inequality for the amount a middle income family will spend. (Lesson 1-9)

24. Solve the equation $t - 8 = 42$ using an inverse operation. (Lesson 1-8)

25. Geometry Graph the points $D(6, 3)$, $F(2, 2)$, $G(4, 5)$, and $H(5, 10)$ on a coordinate system. (Lesson 1-7)

26. Rewrite the expression $(t + 17) + 6$ using an associative property. Then simplify. (Lesson 1-4)

27. Evaluate each expression if $g = 9$ and $h = 4$. (Lesson 1-3)

a. $6g - 12h$

b. $\frac{gh}{3}$

CLOSING THE

Investigation

Up, Up, and Away!

Refer to the Investigation on pages 2–3.

Your chapter of the EAA is about to meet to finalize arrangements for the trip to Oshkosh. Your committee will be called upon to present its plan for the trip at the meeting next week. Complete your plan and make a presentation to the club.

- Suppose that your school represents the members of the club. Take a survey about the type of place they would like to stay on a trip. Would they like to camp? How about a hotel? Sometimes people rent rooms in their homes to travelers. Would that be an option? Use your results to choose a place to stay on your trip.
- Use the results of your study of the time and fuel consumption to choose the time of departure and to estimate the cost of the trip.
- Map out the route of the trip so that no Federal regulations are violated and the trip is as enjoyable as possible.
- Compile all of your results into a brochure to give to members interested in making the trip. Include your research and reasons for your choices.

PORTFOLIO ASSESSMENT

You may want to keep your work on this Investigation in your portfolio.

CHAPTER 1 Highlights

Vocabulary

After completing this chapter, you should be able to define each term, property, or phrase and give an example or two of each.

Algebra
algebraic expression (p. 16)
associative property of addition (p. 22)
associative property of multiplication (p. 22)
brackets (p. 12)
commutative property of addition (p. 22)
commutative property of multiplication (p. 22)
defining a variable (p. 42)
distributive property (p. 26)
equation (p. 32)
evaluate (p. 11)
evaluating an expression (p. 16)
expression (p. 11)
identity property of addition (p. 23)
identity property of multiplication (p. 23)
inequality (p. 46)
inverse operations (p. 41)
like terms (p. 27)
multiplicative property of zero (p. 23)
open sentence (p. 32)
order of operations (p. 11)
parentheses (p. 12)
simplest form (p. 28)
solution (p. 32)
solving an equation (p. 32)
substitution property of equality (p. 17)
terms (p. 27)
variable (p. 16)

Geometry
coordinate system (p. 36)
ordered pair (p. 37)
origin (p. 36)
x-axis (p. 36)
x-coordinate (p. 37)
y-axis (p. 36)
y-coordinate (p. 37)

Statistics
bar graph (p. 52)
data (p. 50)
demographics (p. 50)
frequency table (p. 51)
sample (pp. 50, 51)
statistics (p. 51)
surveys (p. 50)

Problem-Solving
method of computation (p. 8)
problem-solving plan (p. 6)

Understanding and Using Vocabulary

Choose the term that best completes each statement.

1. $8x + 6$ is an example of an __?__. (algebraic expression, inequality)
2. "For any numbers a and b, $a + b = b + a$" is a statement of the __?__ property of addition. (commutative, associative)
3. An expression is __?__ when it has no like terms and no parentheses. (an algebraic expression, in simplest form)
4. In (5, 6), 5 is the __?__. (x-coordinate, y-coordinate)
5. The inverse operation for subtraction is __?__. (addition, division)
6. A group used to represent a larger population is a __?__. (survey, sample)
7. $7y$ and $12y$ are __?__. (like terms, ordered pairs)

CHAPTER 1 Study Guide and Assessment

Skills and Concepts

Objectives and Examples

Upon completing this chapter, you should be able to:

- **use the order of operations to evaluate expressions** (Lesson 1-2)

$$\begin{aligned} 3[(7-3)+8\cdot 2] &= 3[(4)+8\cdot 2] \\ &= 3[4+16] \\ &= 3[20] \\ &= 60 \end{aligned}$$

Review Exercises

Use these exercises to review and prepare for the chapter test.

Find the value of each expression.

8. $8(16+14)$ **9.** $3(9-3\cdot 2)$

10. $14 \div 2 + 4 \cdot 3$ **11.** $12 \div 3 \cdot 4$

12. $\dfrac{14+4}{11-2}$ **13.** $\dfrac{3(2+6)}{4}$

14. $9(12-5)-11(21-8\cdot 2)$

15. $12[3(17-6)-7\cdot 4]$

- **evaluate expressions containing variables** (Lesson 1-3)

Evaluate $8x - 2y$ if $x = 3$ and $y = 7$.

$$\begin{aligned} 8x-2y &= 8(3)-2(7) \\ &= 24-14 \\ &= 10 \end{aligned}$$

Evaluate each expression if $a = 10$, $b = 8$, $c = 6$, and $d = 12$.

16. $14 - a + b$ **17.** $2b - d$

18. $d - c$ **19.** $c + (d - b)$

20. $5a + 4b$ **21.** $\dfrac{a}{b+2}$

22. $ab - cd$ **23.** $6(d - b)$

- **translate verbal phrases into algebraic expressions** (Lesson 1-3)

Forty-five fewer fans attended this football game than attended last week's game.

Let a represent last week's attendance. Then this week's attendance is $a - 45$.

Translate each phrase into an algebraic expression.

24. seven more than some number

25. the product of a number and ten

26. twice as many as last week

27. Joi's score plus 10 points extra credit

28. four more than twice last week's earnings

29. the difference of r and 12

- **use properties of addition and multiplication** (Lesson 1-4)

Name the property shown by the statement $a + 12 = 12 + a$.

commutative property of addition

Name the property shown by each statement.

30. $22 + 0 = 22$

31. $9 + 3t = 3t + 9$

32. $8 + (2 + y) = (8 + 2) + y$

33. $4(6 \cdot 3) = (4 \cdot 6) \cdot 3$

34. $0(b) = 0$

35. $1(m + 5) = m + 5$

Objectives and Examples	Review Exercises

▶ **simplify expressions using the distributive property** (Lesson 1-5)

Simplify the expression $6(8 + n) - 6$.

$$\begin{aligned} 6(8 + n) - 6 &= 6(8) + 6(n) - 6 \\ &= 48 + 6n - 6 \\ &= 48 - 6 + 6n \\ &= 42 + 6n \end{aligned}$$

Simplify each expression.

36. $7 \cdot 8 + 7 \cdot 3$ **37.** $n + 7n$

38. $7b + 12b - 10$ **39.** $16x + 2x + 8$

40. $8(5ab + 4) + 3(2ab + 2)$

41. $2(x + y) + 8(x + 3y)$

42. $10(u + 3) + 4(5 + u)$

▶ **solve open sentences** (Lesson 1-6)

Which of the numbers 8, 12, or 16 is the solution of $84 - n = 72$?

$84 - 8 = 72$
$76 = 72$ *false*

$84 - 12 = 72$
$72 = 72$ *true*

$84 - 16 = 72$
$68 = 72$ *false*

The solution is 12.

Identify the solution to each equation from the list given.

43. $15 - h = 9$; 4, 6, 7 **44.** $7 = \frac{m}{4}$; 21, 28, 35

45. $5b = 45$; 7, 9, 11 **46.** $18 - n = 18$; 18, 1, 0

47. $12 = 9x - 15$; 3, 6, 9

48. $16 = 144 \div k$; 9, 12, 16

49. $\frac{35}{r + 2} = 5$; 3, 5, 7

▶ **use ordered pairs** (Lesson 1-7)

Graph the point (7, 3).

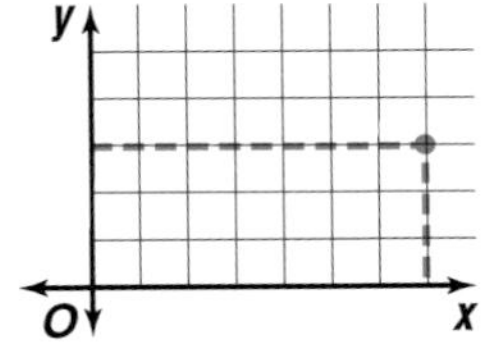

Use the grid to name the point for each ordered pair.

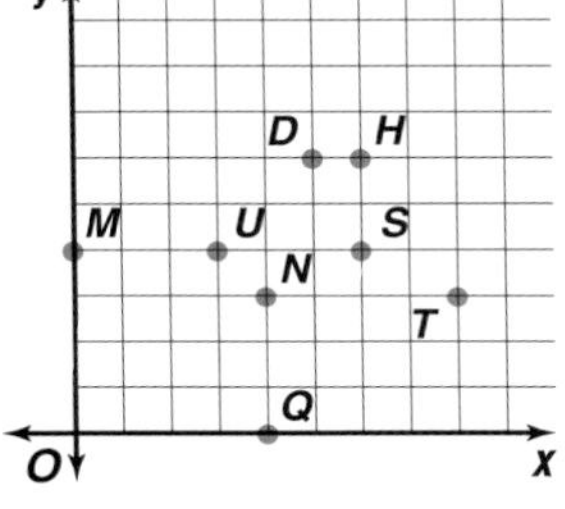

50. (8, 3)

51. (4, 3)

52. (3, 4)

Use the grid to name the ordered pair for each point.

53. *D*

54. *H*

55. *M*

▶ **solve equations using inverse operations** (Lesson 1-8)

$\frac{h}{7} = 13$

$h = 7 \cdot 13$ *Write a related multiplication sentence.*

$h = 91$

Solve each equation using the inverse operation. Use a calculator when needed.

56. $72 = 8n$ **57.** $19 - g = 7$

58. $6n = 84$ **59.** $\frac{y}{4.5} = 6.3$

60. $19.9 - x = 17.6$ **61.** $9 = \frac{w}{12}$

62. The product of a number and seven is 84. Find the number.

Objectives and Examples	Review Exercises

- **write and solve inequalities** (Lesson 1-9)

Translate "*A number is greater than or equal to fifteen.*" into an inequality. Is the inequality *true* or *false* for the value 15?

Let n represent the number.
$n \geq 15$

$15 \geq 15$ is true.

State whether each inequality is *true* or *false* for the given value.

63. $37 - z \leq 9, z = 22$

64. $5x + 8 > 0; x = 0$

65. $7c + 9c \geq 17c, c = 8$

Translate each sentence into an inequality.

66. A number exceeds 25.

67. Fewer than 60 points were scored.

68. It takes no less than seven hours to drive to Chicago from here.

- **gather and record data in a frequency table or bar graph** (Lesson 1-10)

Age	Freq.
0–10	2
11–20	8
21–30	11
31–40	6

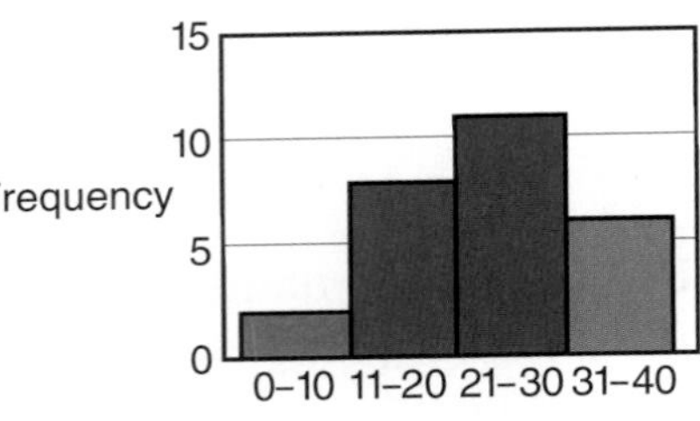

The list below shows the numbers of U.S. congressional representatives from each state for the years 1990–2000.

7	1	6	4	52	6	6	1	23	11
2	2	20	10	5	4	6	7	2	8
10	16	8	5	9	1	3	2	2	13
3	31	12	1	19	6	5	21	2	6
1	9	30	3	1	11	9	3	9	1

69. Make a frequency table of the numbers of representatives. Use intervals of 10.

70. Make a bar graph of the data.

Applications and Problem Solving

71. Space Exploration The longest flight ever made by a space shuttle ended on March 18, 1995. The astronauts aboard *Endeavour* spent 16 days, 15 hours in flight, traveling 6.9 million miles and circling Earth 262 times. About how many miles did the shuttle travel on each trip around Earth? (Lesson 1-1)

a. Which method of computation do you think is most appropriate for this problem? Justify your choice.

b. Solve the problem using the four-step plan. Be sure to examine your solution.

72. Health and Fitness One health magazine gives the equations below for finding your ideal weight, where w represents weight in pounds and h is height in inches.

males: $w = 100 + 6(h - 60)$
females: $w = 100 + 5(h - 60)$

a. Find the ideal weight of Cassie, who is 5 feet 5 inches tall. (Lesson 1-2)

b. What is Rituso's ideal weight if he is 6 feet 3 inches tall? (Lesson 1-2)

c. Rewrite each formula using the distributive property. (Lesson 1-5)

A practice test for Chapter 1 is available on page 774.

Alternative Assessment

Performance Task

Demonstrate your knowledge by giving a clear, concise solution to each problem. Be sure to include all relevant drawings and justify your answers. You may show your solutions in more than one way or investigate beyond the requirements of the problem.

Vitality magazine reports on issues of health and wellness. The table of Calories and grams of fat in different snack foods shown at the right accompanied an article on nutrition.

Snack	Calories	Fat grams
KitKat Bar (4 oz)	588	33
Strawberry Twizzlers (5 oz)	500	5
Butterfinger (4 oz)	492	21
Reeses' Peanut Butter Cups (3.2 oz)	380	22
M & M's peanut (2.6 oz)	363	20
Goobers (2.2 oz)	320	21
Skittles, fruit flavor (2.6 oz)	286	2
Plain air-popped popcorn (med. bucket)	180	trace
Coca-Cola (16 oz)	205	0
Diet Pepsi (16 oz)	<1	0

1. About how many Calories would you consume if you ate one candy item and drank a soft drink?
2. Write an expression for the number of grams of fat in two packages of peanut M & Ms and a Coca-Cola. Then find the number of Calories.
3. There are 192 Calories in a 16-ounce 7•Up.
   a. Write two expressions for the number of calories in a 7•Up and a package of candy.
   b. Find the number of calories in a 7•Up and a KitKat.
4. Explain what the entry in the Calories column for Diet Pepsi means. Could there be 6 Calories in a 16-ounce Diet Pepsi?
5. You have been asked to make a presentation on nutrition to your health class. Represent the data in the table in a different way to show the nutritious value of snack foods.

Thinking Critically

► Graph several points of the form (a, b) and (b, a). What do you observe? When are the graphs of (a, b) and (b, a) the same point? Write an equation that has no solution. Explain why it has no solution.

Portfolio

► A portfolio is representative samples of your work, collected over a period of time. Begin your portfolio by selecting an item that shows something new that you learned in this chapter.

Self Evaluation

► A good problem solver makes a plan before attacking a problem. The first step toward success is planning the route to get there.

Assess yourself. How do you start the process of solving a problem? Do you make a plan before you begin? Choose a problem from the chapter or a task in your life and make a plan for completing it. Then execute your plan and describe how the plan helped you as you worked.

LONG-TERM PROJECT

Investigation

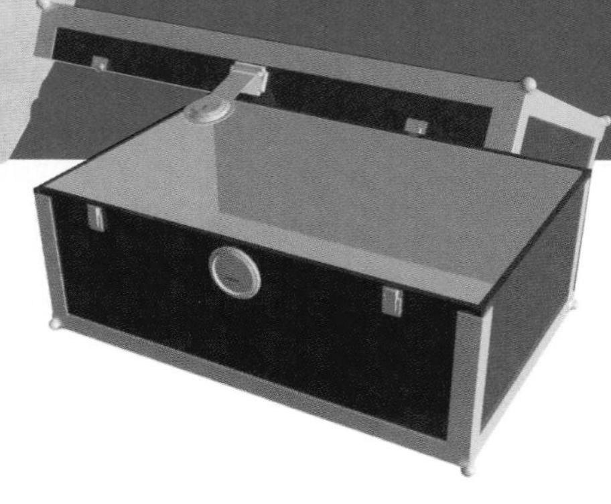

You and another member of your class have been selected to participate in a student foreign exchange program. You won't be leaving the country until June, so there is plenty of time to prepare. The first thing you should do is obtain a passport.

MATERIALS

- travel guides
- reference books or magazines

To apply for a passport, you must:

- *complete an application in person before an official designated to accept passport applications. (A parent or legal guardian must complete the application for children less than 13 years old.)*
- *provide proof of U.S. citizenship (usually a previously-issued passport or a birth certificate).*
- *submit two identical photographs that are 2 inches by 2 inches in size. (The photographs should not be more than 6 months old and must be clear, front view, full face, with a white background.)*
- *pay $30. (The fee is $55 for passports issued to persons age 18 and older.)*

Work with your partner to prepare an estimate of your pre-flight expenses. Be sure to include costs associated with obtaining your passport. Be prepared to explain your list of pre-trip expenses to the class.

Starting the Investigation

Once passports have been obtained, another concern is what items to pack. There are many things to consider as you make your plans. Look over the following list of questions to consider.

Think About . . .

- *From what airport will you depart?*
- *How much luggage can you take?*
- *Are there restrictions on items you can take into the foreign country?*
- *Are there any restrictions on items you can bring or ship back?*
- *Will you have enough luggage space to pack purchases, gifts, and souvenirs?*
- *What are the costs and quality of the communication services (phone, mail) between your final destination and your home?*
- *What's the time difference between your final destination and your home?*

Brainstorm with your partner to choose a country of interest. Use this country as your destination as an exchange student. Be prepared to share the reasoning behind your choice with the class. Then work together to add to the list of questions to be considered. You may want to write to the U.S. Customs Service for more information.

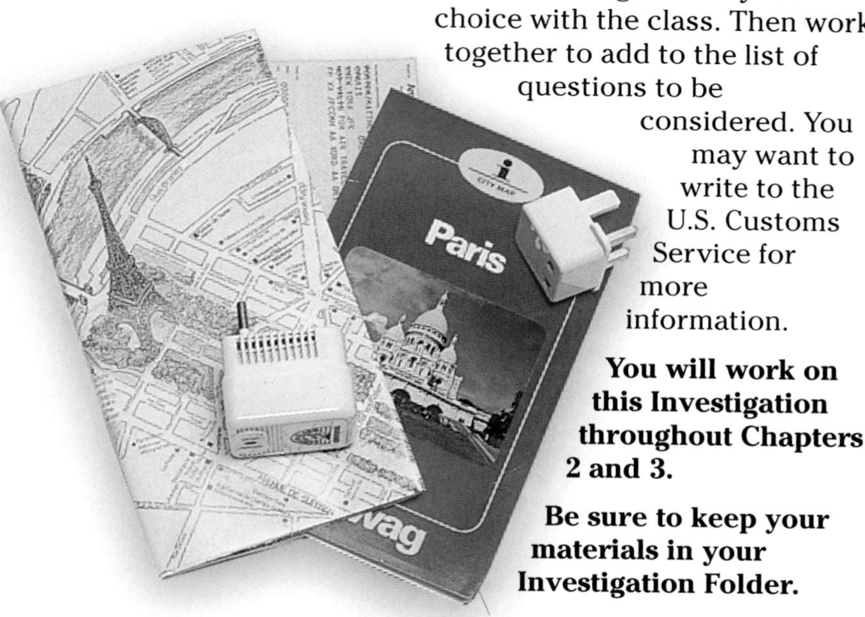

You will work on this Investigation throughout Chapters 2 and 3.

Be sure to keep your materials in your Investigation Folder.

For current information on being an exchange student, visit: **www.glencoe.com/sec/math/prealg/mathnet**

Investigation:
Bon Voyage!

Working on the Investigation
Lesson 2-1, p. 70

Working on the Investigation
Lesson 2-6, p. 97

Working on the Investigation
Lesson 3-2, p. 128

Closing the Investigation
End of Chapter 3, p. 160

CHAPTER

2 Exploring Integers

TOP STORIES in Chapter 2

In this chapter, you will:

- graph on number lines and coordinate planes,
- compare and order integers,
- add, subtract, multiply, and divide integers,
- solve problems looking for a pattern, and
- solve problems with integers.

MATH AND RECYCLING IN THE NEWS

Where does all my money go?!

Source: New York Times, May 22, 1994

It goes to the landfill. At least for now. For decades, worn out paper money has been shredded and sent to a landfill. But since landfill space is becoming scarce, the Federal Reserve is looking for ways to recycle its tired money. About 715 million bills weighing 7000 tons and having a face value of nearly $10 billion are shredded each year. The shredded bills are then pressed into bricks that weigh 2.2 pounds each. A number of companies have expressed interest in using the recycled money, including companies making roofing shingles, particle board, fuel pellets, stationery, packing material, and artwork. So where *does* all the money go? All over!

Putting It into Perspective

500 B.C. — A.D. 1400 — 1600

500 B.C.
Athens, Greece organizes the first municipal dump in the western world, 1 mile outside the city walls.

A.D. 1388
English Parliament bans waste disposal in waterways and ditches.

1690
The first paper mill in the U.S. opens near Philadelphia.

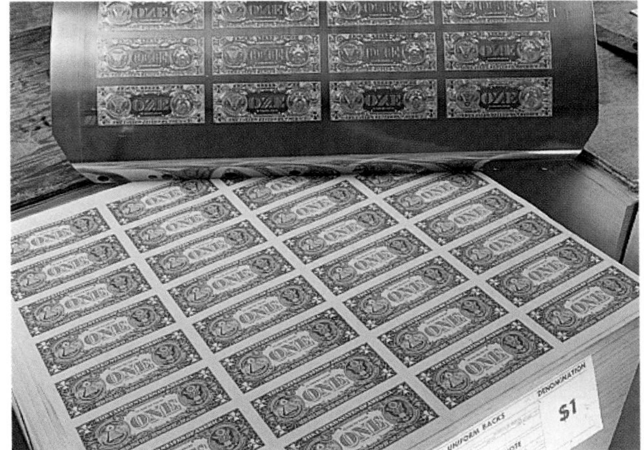

Statistical Snapshot

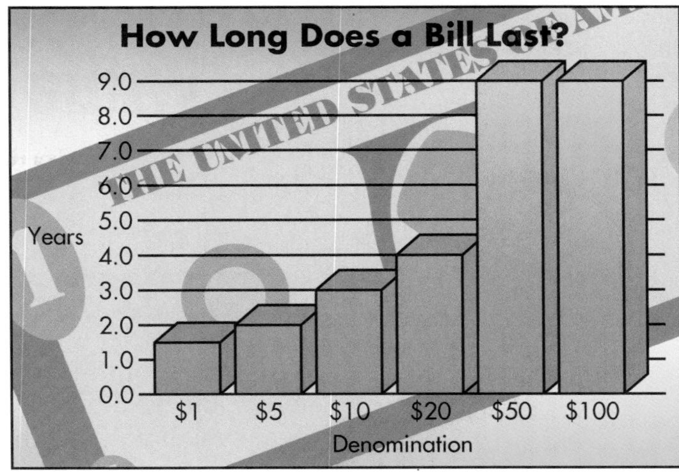

Source: Federal Reserve Bank of Los Angeles

interNET CONNECTION For up-to-date information on printing money, visit: **www.glencoe.com/sec/math/prealg/mathnet**

On the Lighter Side

1900

1904
The first two major aluminum recycling plants open in Cleveland and Chicago.

1942–45
Americans collect paper, rubber, and tin to recycle for the war effort.

1950

1970
The U.S. Environmental Protection Agency is formed.

1994
U.S. Federal Reserve begins recycling money.

2000

2-1 Integers and Absolute Value

Setting Goals: *In this lesson, you'll graph integers on a number line and find absolute value.*

Modeling a Real-World Application: Geography

The shores of the Dead Sea, between Israel and Jordan, are 1312 feet *below* sea level. If you use the number 1312 to represent 1312 feet *above* sea level, what number could be used to represent 1312 feet below sea level?

CULTURAL CONNECTIONS
For centuries, people have traveled miles to bathe in the Dead Sea. They believed that the minerals in the water were healthful. Several resorts are still in operation on the shores today.

Learning the Concept

Study the pattern of the following subtraction sentences.

$$5 - 1 = 4$$
$$5 - 2 = 3$$
$$5 - 3 = 2$$
$$5 - 4 = 1$$
$$5 - 5 = 0$$
$$5 - 6 = \underline{\ ?\ }$$

From the pattern, it appears that $5 - 6$ has an answer less than 0. Study the diagram below.

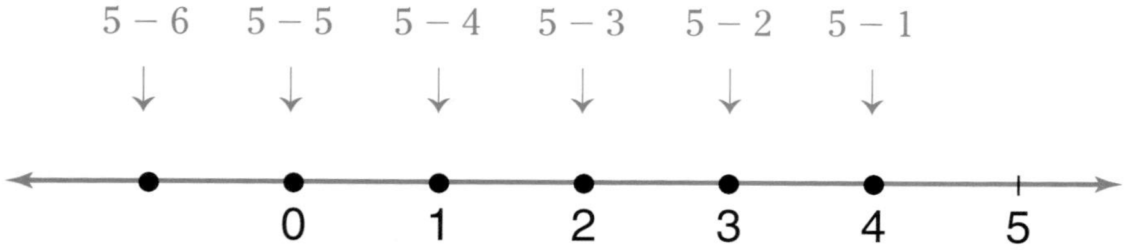

The answer for $5 - 6$ should be *one less than zero*. The number *one less than zero* is written as -1. So $5 - 6 = -1$. This is an example of a **negative** number. A negative number is a number less than zero.

You can use a negative number to write 1312 feet below sea level, the elevation of the shores of the Dead Sea. With sea level as the starting point or 0, you can express the elevation as $0 - 1312$, or -1312. So, the elevation of the Dead Sea is -1312 feet.

Negative numbers, like -1312, are members of the set of **integers**. Integers can also be represented as points on a number line.

Numbers to the left of zero are less than zero.

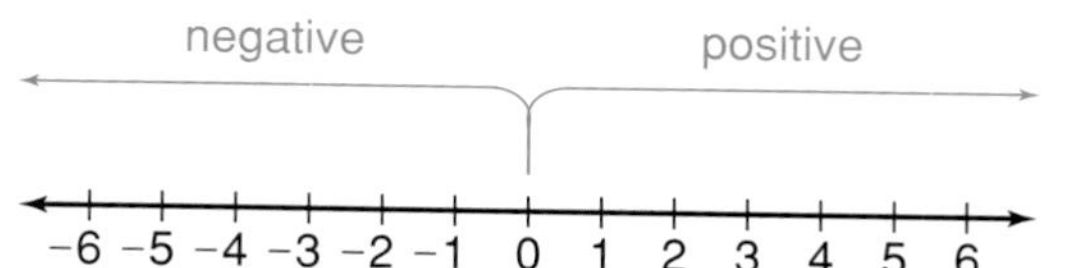

Numbers to the right of zero are greater than zero.

The numbers −1, −2, −3, . . . are called negative integers. The number negative 3 is written −3.

Zero is neither negative nor positive.

The numbers 1, 2, 3, . . . are called positive integers. The number positive 4 is written +4 or 4.

Which integer is neither positive nor negative?

This set can be written {. . ., −3, −2, −1, 0, 1, 2, 3, . . .} where . . . means continues indefinitely.

To graph a particular set of integers, locate the integer points on a number line. The number that corresponds to a point on the number line is called the **coordinate** of the point.

Example 

a. Name the coordinates of *D*, *E*, and *B*.

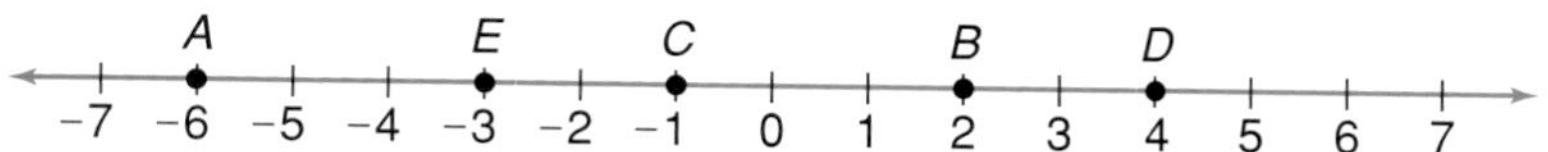

The coordinate of *D* is 4, *E* is −3, and B is 2.

b. Graph points *F*, *U*, and *N* on a number line if *F* has coordinate 1, *U* has coordinate −3, and *N* has coordinate 4.

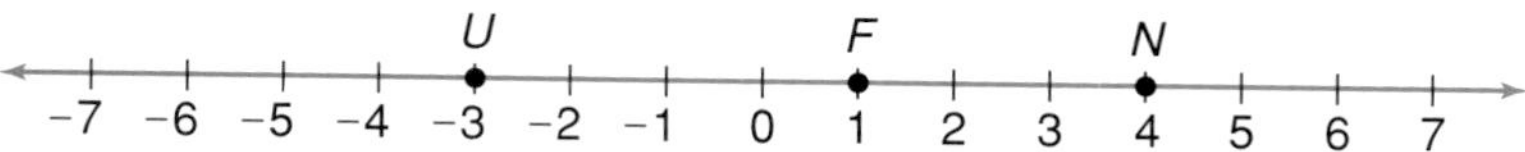

To graph a number, find its location on the number line and draw a dot. Write the letter above the dot.

Looking at the number line shown below, you can see that 4 and −4 are on opposite sides of the starting point, zero. However, they are the same distance from zero. In mathematics, we say that they have the same **absolute value**, 4.

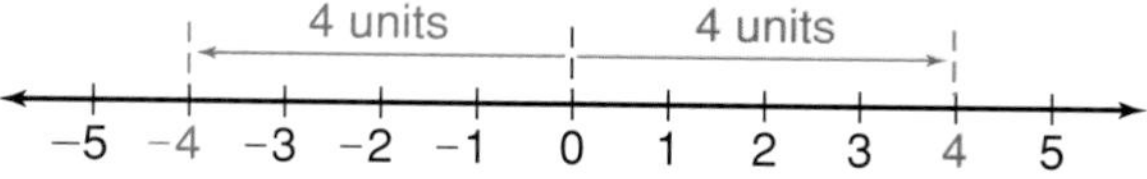

The symbol for the absolute value of a number is two vertical bars on either side of the number. $|-4| = 4$ is read *The absolute value of −4 is 4.*

Absolute Value		
	In words:	**The absolute value of a number is the distance the number is from the zero point on the number line.**
	In symbols:	$|4| = 4$ **and** $|-4| = 4$

Example 2 **Simplify.**

a. $|9| + |-9|$

$|9| + |-9| = 9 + 9$ *The absolute value of −9 is 9.*

$= 18$ *The absolute value of 9 is 9.*

b. $|13| - |-2|$

$|13| - |-2| = 13 - 2$ *The absolute value of 13 is 13.*

$= 11$ *The absolute value of −2 is 2.*

Connection to Algebra

Since variables represent numbers, you can use absolute value notation with algebraic expressions involving variables.

Example  **Evaluate the expression $|x| - 7$ if $x = -13$.**

$|x| - 7 = |-13| - 7$ *Replace x with −13.*

$= 13 - 7$ *The absolute value of −13 is 13.*

$= 6$

Checking Your Understanding

Communicating Mathematics

Read and study the lesson to answer these questions.

1. **Explain** how you would graph −7.
2. **Describe** some situations in the real world where negative numbers are used.
3. **Identify** the integers graphed on the number line at the right.

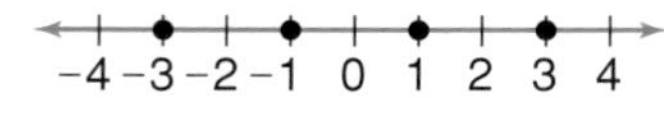

4. **Graph** two points on a number line so that the coordinates of both points have an absolute value of 5.
5. You can use counters to model integers. Each yellow counter represents a positive unit, and each red counter represents a negative unit.

MODELING MATHEMATICS

MATERIALS

counters

integer mat

Place three yellow counters on the mat to represent +3.

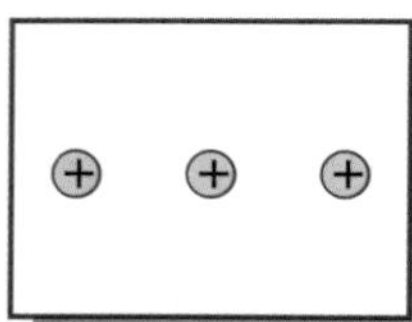

Place eight red counters on the mat to represent −8.

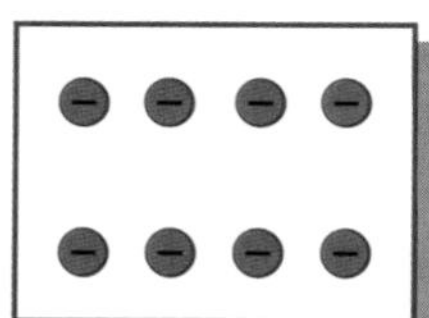

a. Model −4 using counters.   **b.** Model +5 using counters.

Guided Practice

6. Name the coordinate of each point graphed on the number line.

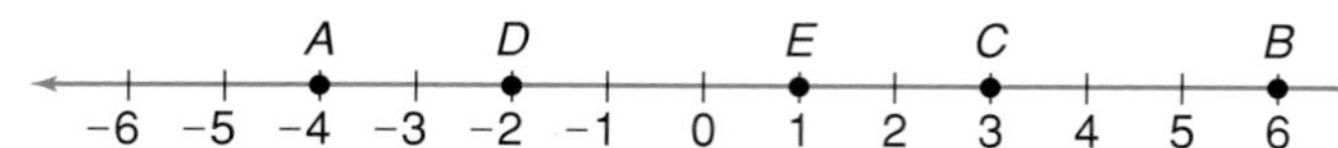

Write an integer for each situation. Then graph on a number line.

7. 5°F below zero
8. a 12-pound loss
9. 13 yards gained
10. a bank deposit of $1200

Simplify.

11. $|-5|$
12. $|8|$
13. $|13| - |-3|$
14. $|15| + |-8|$

Evaluate each expression if $d = -5$ and $r = 3$.

15. $7 + |d|$
16. $|d| - r$
17. **Scuba Diving** Two divers are descending to a coral reef off the coast of Madagascar. One diver is 15 meters below the water's surface. The other diver is 9 meters lower.
 a. Represent the depths of the divers as integers.
 b. Graph each point on a number line.

Exercises: Practicing and Applying the Concept

Independent Practice

Graph each set of numbers on a number line.

18. $\{0, -1, 3\}$
19. $\{-2, -5, 4\}$
20. $\{-4, 1, 3, 8\}$
21. $\{-3, -5, -7, -9\}$
22. Name the coordinates of each point graphed on the number line.

B D C A
-6 -4 -2 0 2 4 6 8

Write an integer for each situation.

23. a salary increase of $400
24. a bank withdrawal of $50
25. 3 seconds before liftoff
26. a gain of 9 yards
27. 5° above zero
28. a loss of 6 pounds

Simplify.

29. $|-11|$
30. $-|24|$
31. $|0|$
32. $|7|$
33. $|-5| + |3|$
34. $|-8| + |-10|$
35. $|14| - |-5|$
36. $|-15| + |-12|$
37. $|0 + 12|$
38. $|16 - 2|$
39. $-|-36|$
40. $-||-4| + |-21||$

Evaluate each expression if $a = 0$, $b = 2$, and $c = -4$.

41. $|c| - 2$
42. $12 - |b|$
43. $b + a + |c|$
44. $ac + |-30|$
45. $|c| - b$
46. $|b| \cdot |c| + |a|$

Critical Thinking

Find the next three numbers in each pattern.

47. 32, 24, 16, ?, ?, ?
48. −23, −18, −13, ?, ?, ?
49. **Logical Reasoning** Name a negative number that is not an integer.

Applications and Problem Solving

50. **Game Shows** On *Jeopardy!*, contestants earn points for each correct response and lose points for each incorrect response. Questions are valued at 100, 200, 300, 400, 500, 600, 800, or 1000 points. Describe a situation where a contestant would have a score of −300.

51. Sports Golf scores are reported in reference to *par*. A score of +2 means two strokes over par. The leaders in the 1995 U.S. Open in Southampton, New York, had the scores −2, 3, 4, 1, 4, −1, −1, −5, −4, and 1 in the final round. Graph the scores on a number line.

Mixed Review

The table shows test scores for a physical science class. Use the table to answer each question. (Lesson 1-10)

Score	Test Score Tally	Frequency
19–20	𝍸	5
17–18	𝍸 IIII	9
15–16	𝍸	5
13–14	III	3
11–12	I	1
9–10	II	2

52. How many students are in the class?

53. What scores occurred most often?

54. What scores occurred least often?

55. Make a bar graph of the data.

56. Solve $16 = z - 25$. (Lesson 1-8)

57. Geometry The area of a rectangle is given by $A = \ell w$, where ℓ is the length and w is the width. Find the area of a rectangular garden if its length is 15 feet and its width is 12 feet. (Lesson 1-6)

58. Simplify $5a + 3(7 + 2a)$. (Lesson 1-5)

59. Write an algebraic expression for the phrase *three times a number decreased by 8*. (Lesson 1-3)

60. Consumer Awareness The Columbus Association for the Performing Arts runs a summer movie series in the restored Ohio Theater. Tickets in the 1995 season were \$3.00 for adults and \$2.50 for senior citizens. (Lesson 1-2)

a. Write an expression that would help you find the total cost of fifteen adult tickets and eight senior citizen tickets.

b. Then find the total cost.

Refer to the Investigation on pages 62–63.

Materials: map or globe

Lines of longitude indicate degrees east or west of the prime meridian. The prime meridian is at 0°. The prime meridian is a semicircle passing through Greenwich, England.

A city's longitude determines its time zone. The zone labeled zero is centered at the prime meridian.

- How does the way the zones are numbered differ from the integer number line?

$37\frac{1}{2}$W $22\frac{1}{2}$W $7\frac{1}{2}$W $7\frac{1}{2}$E $22\frac{1}{2}$E $37\frac{1}{2}$E

Zone 3 | Zone 2 | Zone 1 | Zone 0 | Zone −1 | Zone −2 | Zone −3

30W 15W 0 15E 30E

- Compare your time zone to that of the country you chose at the beginning of this Investigation. How would you use this information when traveling abroad?
- Work with the members of your group to develop reasons why time zones exist.

Add the results of your work to your Investigation Folder.

HANDS-ON ACTIVITY

2-1B Statistical Line Plots

An Extension of Lesson **2-1**

MATERIALS

ruler

In the previous lesson, you saw how number lines helped to visualize integers. In statistics, data is also organized and presented on a number line. A picture of information on a number line is called a **line plot**.

Population figures from the census are used to determine the number of House of Representative members from each state of the United States.

Work with a partner.

- This chart shows the states that had a change in the number of House members after the 1990 census. Determine the change for each state. Express a decrease as a negative integer and an increase as a positive integer.

State	1980	1990	Change
Arizona	5	6	+1
California	45	52	+7
Florida	19	23	+4
Georgia	10	11	+1
Illinois	22	20	−2
Iowa	6	5	
Kansas	5	4	
Kentucky	7	6	
Louisiana	8	7	
Massachusetts	11	10	
Michigan	18	16	
Montana	2	1	
New Jersey	14	13	
New York	34	31	
North Carolina	11	12	
Ohio	21	19	
Pennsylvania	23	21	
Texas	27	30	
Virginia	10	11	
Washington	8	9	
West Virginia	4	3	

FYI

Patsy Mink, a Democrat from Hawaii, became the first Japanese American woman to serve in the U.S. Congress on January 4, 1965.

New York

alifornia

- Draw a number line. Since the largest and smallest changes were 7 and −3 for California and New York, respectively, use a scale of −4 to +8 and intervals of 1.

−4 −3 −2 −1 0 1 2 3 4 5 6 7 8

- To make the line plot, put an "x" above the number that represents the change in House members for each state. The graph below shows the first five values from the table.

TALK ABOUT IT

1. What number occurs most frequently?
2. A number that is far apart from the rest of the data is called an **outlier**. Identify any numbers that appear to be outliers.
3. Data that are grouped closely together are called a **cluster**. Identify any clusters.

Extension

4. The *net change* is the change in the total number of House members. What was the net change in the number of House members?
5. What general geographic patterns do you notice from the information in the table? In which areas of the country are states gaining or losing population?

2-2 Integration: Geometry
The Coordinate System

Setting Goals: *In this lesson, you'll graph points in all quadrants of the coordinate plane.*

Modeling a Real-World Application: Cartography

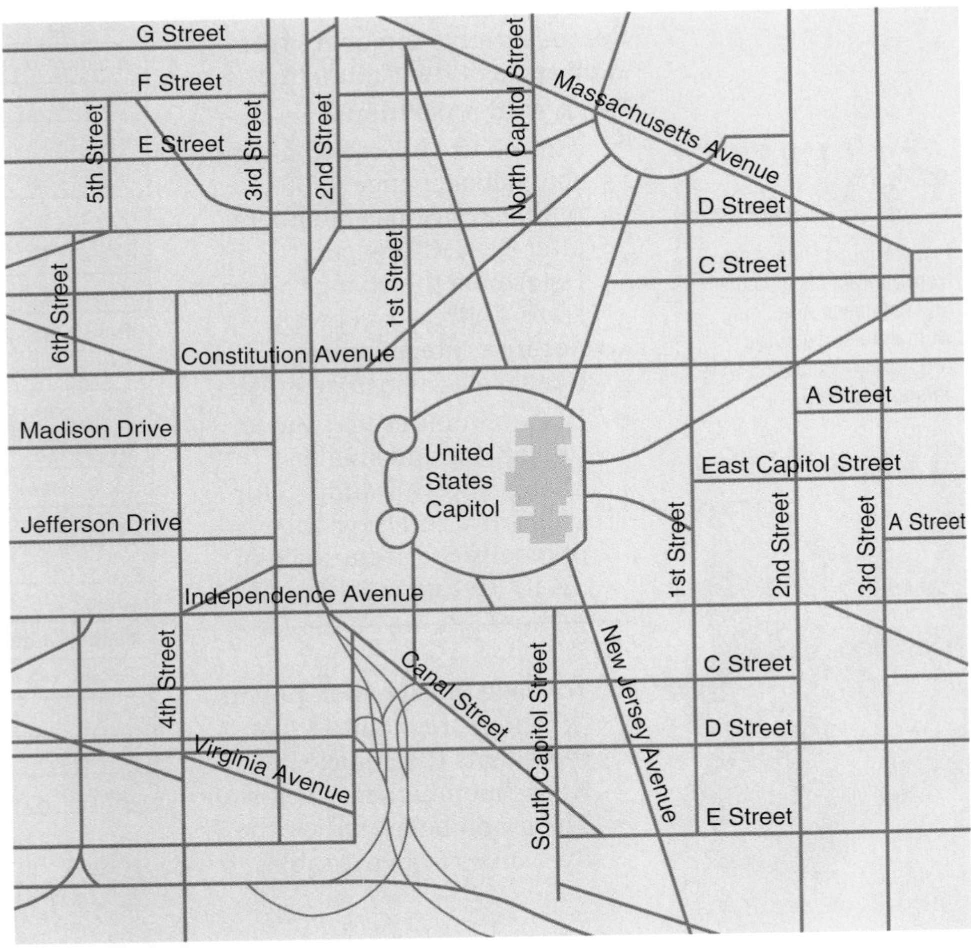

Our nation's capital has a unique system for naming its streets. The Capitol is at the center. Streets that lie east and west of the Capitol are named numerically: 1st, 2nd, 3rd, 4th, and so on. Streets that lie north and south of the Capitol are named alphabetically: A, B, C, D, . . . through W, skipping the letter J. This means that there are *two* sets each of the numerical and alphabetical street names.

Each location on the map could be indicated by a pair of numbers and letters, such as (3, D). This would represent the intersection of 3rd and D streets. However, there are *four* places where 3rd and D streets meet. The problem is resolved by separating Washington into four different sections. The sections are northeast (NE), southeast (SE), southwest (SW), and northwest (NW).

CULTURAL CONNECTIONS

Benjamin Banneker, the grandson of a slave, was part of a team charged with designing Washington, D.C. When Pierre L'Enfant resigned and took all the plans with him, Banneker reproduced the entire set of plans from memory.

Learning the Concept

You can review the coordinate system in Lesson 1-7.

A mathematical model of the city map of Washington, D.C., could be represented by a **coordinate system**. Remember that a coordinate system is formed by the intersection of two number lines, called **axes**. The axes separate the coordinate plane into four regions called **quadrants**, like the four sections of Washington, D.C.

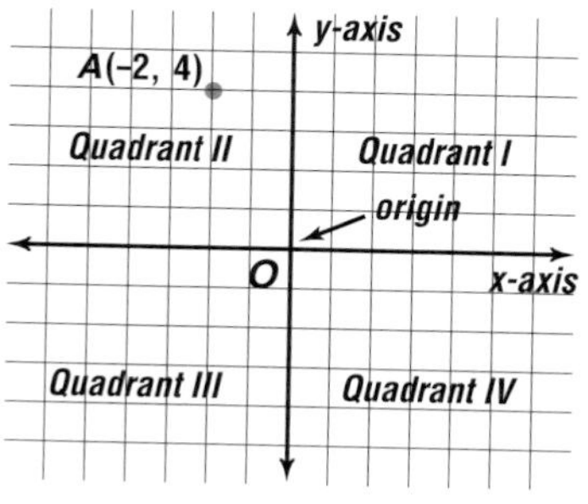

Axes is the plural of axis.

Notice that the numbers to the left of zero on the x-axis are negative as are the numbers below zero on the y-axis.

The *x-coordinate* of the ordered pair $(-2, 4)$ is -2, and the *y-coordinate* is 4.

ordered pair
$(-2, 4)$
x-coordinate ↑ ↑ *y-coordinate*

The dot at $(-2, 4)$ is the **graph** of point A.

The origin and the two axes do not lie in any quadrant.

Example 1

APPLICATION
Cartography

FYI
In 1992, Ben Nighthorse Campbell became the first Native American to serve as a U.S. Senator.

a. Write the ordered pair for NASA.
The x-coordinate is -6, and the y-coordinate is -3. Thus, the ordered pair is $(-6, -3)$.

b. What is located at (2, 3)?
Count 2 units to the right and 3 units up. The point for the Senate Offices lies at (2, 3).

c. What is located at the origin?
The origin is at (0, 0). The Capitol is at the origin.

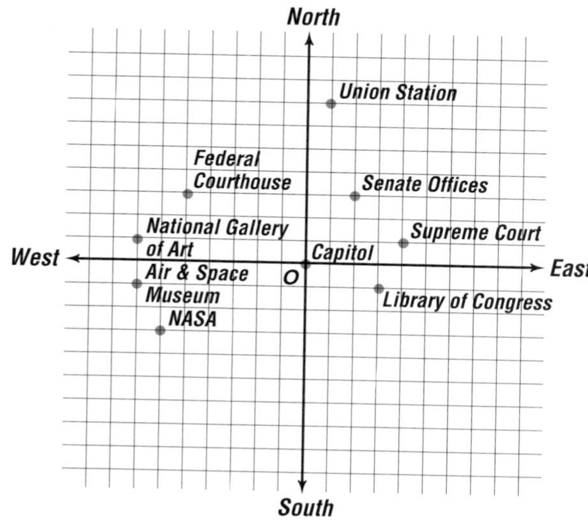

Remember that to graph a point means to place a dot at the point named by the ordered pair. This is sometimes called **plotting the point**.

 **Example** **Graph each point on a coordinate plane. Then name the quadrant in which each point lies.**

a. *D*(3, −4)

Start at the origin. Move 3 units to the right. Then move 4 units down and draw a dot. Label the dot *D*(3, −4). Point *D* is located in quadrant IV.

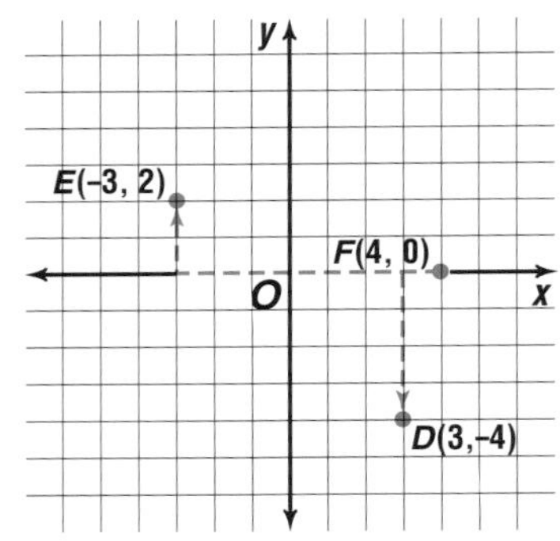

b. *E*(−3, 2)

Start at the origin. Move 3 units to the left. Then move 2 units up and draw a dot. Label this dot *E*(−3, 2). Point *E* is located in quadrant II.

c. *F*(4, 0)

Start at the origin. Move 4 units to the right.The graph is on the *x*-axis. Label this dot *F*(4, 0). Since point *F* lies on the *x*-axis, it is not in any quadrant.

Checking Your Understanding

Communicating Mathematics

Read and study the lesson to answer these questions.

1. **Draw** a coordinate system and label the origin, the *x*- and *y*-axes, and the four quadrants.
2. **Explain** why the point (4, 8) is different from the point (8, 4).
3. **Name** two ordered pairs whose graphs are *not* located in one of the four quadrants.
4. **You Decide** Danny said that if you interchange the coordinates of any point in Quadrant I, the new point still would be in Quadrant I. Cherita disagrees. She says the new point would be in Quadrant 3. Who is right? Explain.

Guided Practice

Use the coordinate grid at the right to name the point for each ordered pair.

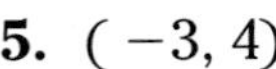

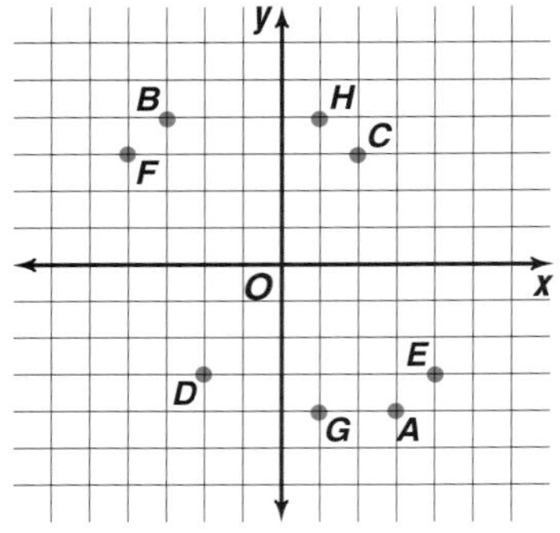

5. (−3, 4)
6. (1, 4)
7. (−2, −3)
8. (−4, 3)

On graph paper, draw coordinate axes. Then graph and label each point. Name the quadrant in which each point is located.

9. *I*(4, −7)
10. *J*(−4, 8)
11. *K*(0, 7)
12. *L*(−6, −3)
13. **Logical Reasoning** If the graph of *A*(*x*, *y*) satisfies the given condition, name the quadrant in which point *A* is located.

    a. $x > 0, y > 0$ **b.** $x < 0, y < 0$ **c.** $x < 0, y > 0$

Exercises: Practicing and Applying the Concept

Independent Practice

Name the ordered pair for each point graphed on the coordinate plane at the right.

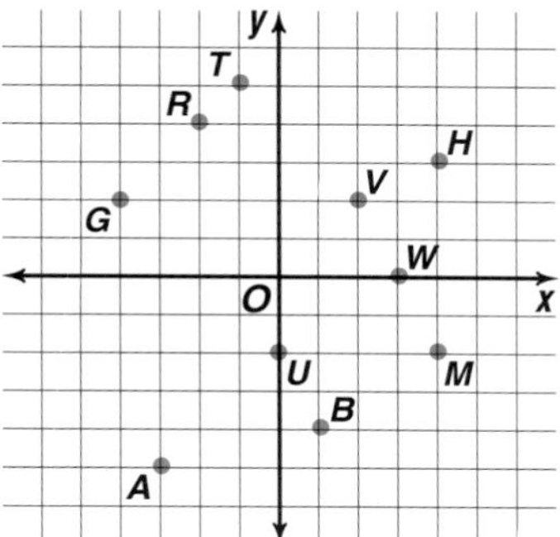

14. R
15. M
16. A
17. B
18. G
19. H
20. U
21. V
22. W
23. T

On graph paper, draw coordinate axes. Then graph and label each point. Name the quadrant in which each point is located.

24. $A(3, 5)$
25. $K(-6, 1)$
26. $M(2, -4)$
27. $B(-7, -4)$
28. $R(-1, 3)$
29. $S(3, -5)$
30. $Q(-3, -5)$
31. $D(4, 2)$
32. $E(7, -8)$
33. $F(-4, -7)$
34. $G(1, 0)$
35. $H(-7, 0)$

Graph each point. Then connect the points in alphabetical order and identify the figure.

36. $A(0, 6)$, $B(4, -6)$, $C(-6, 2)$, $D(6, 2)$, $E(-4, -6)$, $F(0, 6)$
37. $A(5, 8)$, $B(1, 13)$, $C(5, 18)$, $D(9, 13)$, $E(5, 8)$, $F(5, 6)$, $G(3, 7)$, $H(3, 5)$, $I(7, 7)$, $J(7, 5)$, $K(5, 6)$, $L(5, 3)$, $M(3, 4)$, $N(3, 2)$, $P(7, 4)$, $Q(7, 2)$, $R(5, 3)$, $S(5, 1)$

If x and y are integers, graph all ordered pairs that satisfy the given conditions.

38. $|x| < 4$ and $|y| < 3$
39. $|x| \leq 2$ and $|y| \leq 1$

Critical Thinking

40. Describe the possible location in terms of quadrants or axes, for the graph of (x, y) if x and y satisfy the condition $xy = 0$.

Applications and Problem Solving

41. **Geometry** A **vertex** of a polygon is a point where two sides of the polygon intersect.

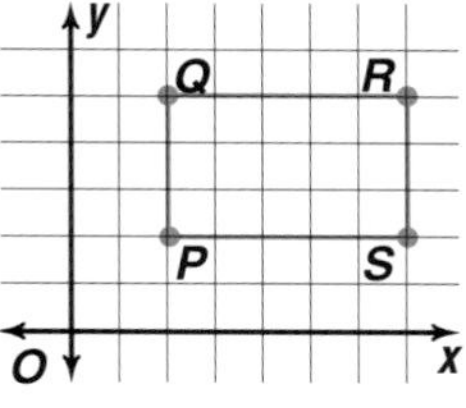

a. Identify the coordinates of the vertices in the rectangle at the right.

b. Subtract 4 from each y-coordinate. Graph the new ordered pairs. What figure is formed if the points are connected?

c. Multiply each x- and y-coordinate of the vertices of $PQRS$ by -1. Graph the new ordered pairs. Describe the figure that is formed if the points are connected.

42. **Geography** Latitude and longitude lines are used to locate places on a map or globe. Latitude is measured in degrees north or south of the equator while longitude is measured in degrees east or west of the prime meridian. When describing a location, the latitude is usually given first. For example, the location of 30°N, 90°W is New Orleans, Louisiana. Find the approximate latitude and longitude of the following locations.
   a. Sydney, Australia
   b. Porto Alegre, Brazil
   c. Johannesburg, Republic of South Africa

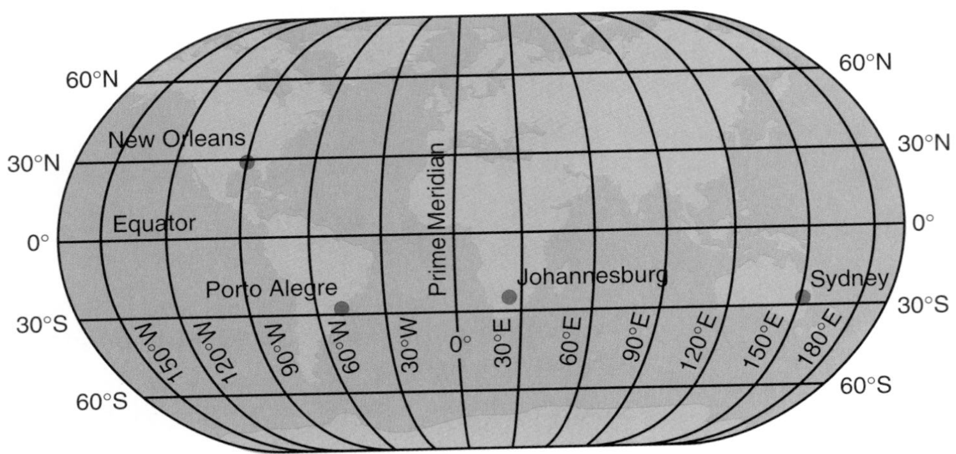

Mixed Review

43. Find the value of $|15| - |-3|$. (Lesson 2-1)

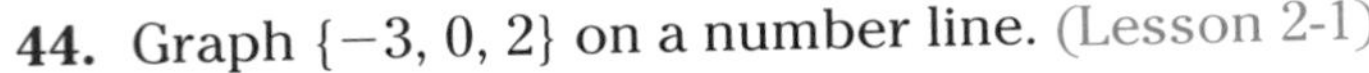

44. Graph $\{-3, 0, 2\}$ on a number line. (Lesson 2-1)

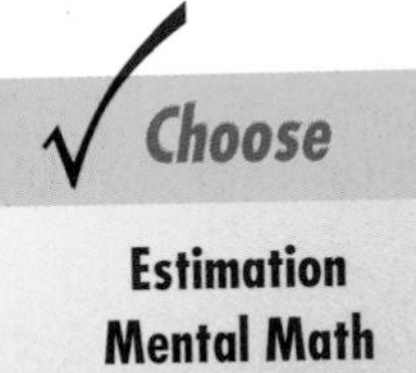

Choose
Estimation
Mental Math
Calculator
Paper and Pencil

45. **Recreation** During the last few years, in-line skating has become a popular pastime. Today there are over 500,000 people who participate in this sport. This is more than twice as many people as there were several years ago. Which inequality, $s > 250{,}000$ or $s < 250{,}000$, describes how many skaters there were several years ago? (Lesson 1-9)

46. What property is shown by $(6 \cdot 15) \cdot 3 = 6 \cdot (15 \cdot 3)$? (Lesson 1-4)

47. Evaluate each expression if $a = 4$, $b = 2$, and $c = 3$. (Lesson 1-3)
   a. $9a - (4b + 2c)$
   b. $\frac{6(a + b)}{3c}$

48. Find the value of $7[(12 + 5) - 3(19 - 14)]$. (Lesson 1-2)

49. **Sports** The Middletown Junior High School basketball team made 21 baskets to score 36 points. If two of the baskets were 3-point shots, how many baskets were worth 2 points and how many were worth 1 point? (Lesson 1-1)

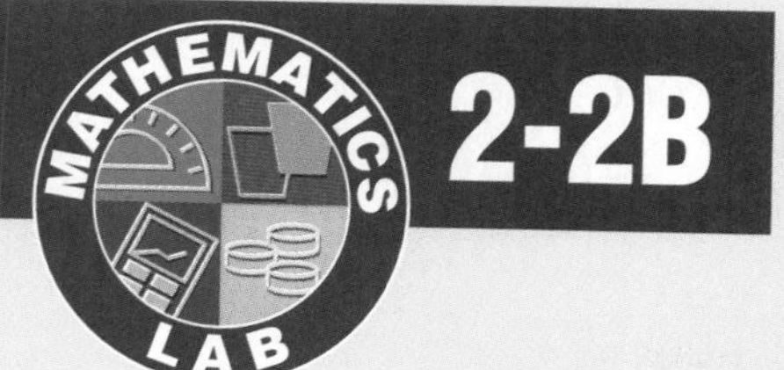

GRAPHING CALCULATOR ACTIVITY

2-2B Plotting Points

An Extension of Lesson **2-2**

You can plot points on a graphing calculator just as you do on grid paper.

Activity

Plot the ordered pairs (6, 9), (27, 13), (15, –13), (–33, 24), and (–12, –8) on a graphing calculator.

First, set the viewing window. The viewing window is the range of x- and y-values that are visible on the graphics screen. The viewing window [−47, 47] by [−31, 31], with scale factors of 10 on both axes works well for these points. This means that the scale will be from −47 to 47 on the x-axis and from −31 to 31 on the y-axis, with tick marks every 10 units.

ENTER: [WINDOW] [ENTER] [(–)] 47 [ENTER] 47 [ENTER] 10 [ENTER] [(–)] 31 [ENTER] 31 [ENTER] 10 [GRAPH]

Now draw the points.

ENTER: [2nd] [DRAW] [▶] [ENTER]

Use the arrow keys to move the cursor to the first point at (6, 9) and press [ENTER]. As you move the cursor to the next point, at (27, 13), you will see that the first point was plotted. After you position the cursor at the next point, press [ENTER] again. Continue until all points are plotted. To see the last point, press [CLEAR].

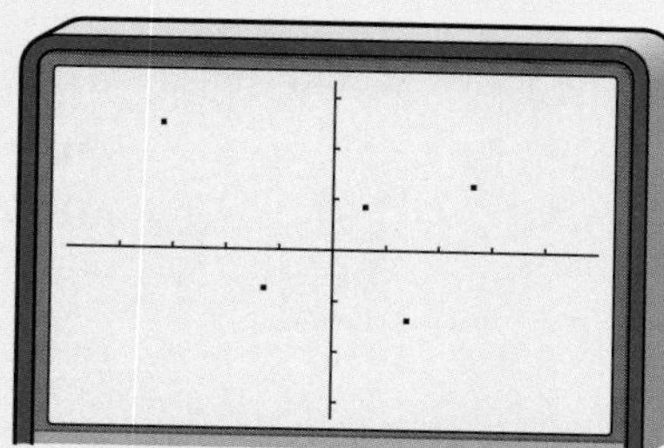

Your Turn

Clear the graphics screen by pressing [2nd] [DRAW] [ENTER]. Then graph each set of ordered pairs on a graphing calculator.

1. (8, –3), (9, 7), (–3, 4), (–7, 4), (–6, –5)
2. (–17, 22), (8, 19), (–12, 1), (–21, –7), (16, 23)
3. (–32, 4), (26, –16), (–3, –8), (–7, –11)
4. (18, –30), (39, 17), (–33, 24), (–27, 14), (32, –21)

5. The *standard viewing window* can be set automatically by pressing [ZOOM] 6.
   a. What are the minimum and maximum values of the standard viewing window?
   b. What are the minimum and maximum values of the viewing window after pressing [ZOOM] 6 [ZOOM] 8 [ENTER] ?
6. You should choose the viewing window to best display each graph.
   a. How would you set the viewing window to display a graph of Quadrant II? Quadrant IV?
   b. If the last ordered pair in Exercise 4 was (–49, 39), how could you make sure it was visible in the viewing window?

2-3 Comparing and Ordering

Setting Goals: *In this lesson, you'll compare and order integers.*

Modeling a Real-World Application: Health

Nature's way of using pollen to produce seeds in plants is remarkable. However, pollen is the enemy of millions of people who suffer from allergies. Pollen from trees or weeds can cause sneezing and eye irritation.

During allergy season, local television stations or newspapers may give daily pollen counts to warn people. As the pollen count increases, so does the discomfort level as seen moving left to right on the scale below.

Pollen Count	0–100	101–600	601–1200	over 1200
Discomfort Level	low	uncomfortable	high	very high

Learning the Concept

Just like the pollen scale, the numbers on a number line increase as you move to the right. On a number line, it is easy to determine which of two numbers is greater. –3 and 4 are graphed on the number line below.

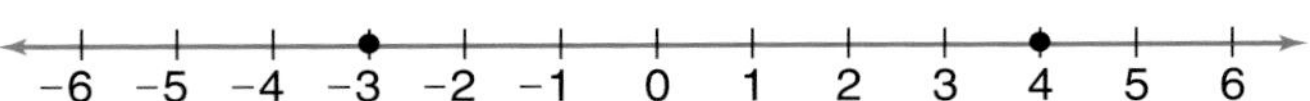

Say: 4 is greater than −3 — *The number to the right is always greater.*
Write: $4 > -3$ — *The number to the left is always less.*

You might also conclude that −3 is to the left of 4.

Say: −3 is less than 4 — *Remember, the symbol points to the lesser number.*
Write: $-3 < 4$

Remember that any mathematical sentence containing < or > is called an inequality. In inequalities, numbers are compared.

Example  **Use the integers graphed on the number line below for each question.**

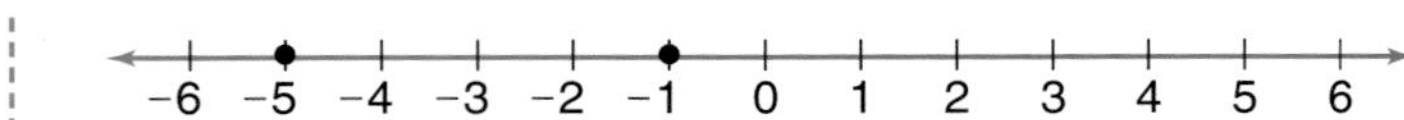

a. Write two inequalities involving −5 and −1.
Since –5 is to the left of –1, write $-5 < -1$.
Since –1 is to the right of –5, write $-1 > -5$.

b. State which number is greater.
−1 is greater since it lies to the right of –5.

c. State which number has the greater absolute value.
The absolute value of –5 is greater than the absolute value of –1 because it lies farther from the origin.

Integers are used to compare numbers in many everyday applications.

Meteorology

The chart at the right shows the record low temperatures for selected states through 1992. Order the temperatures from least to greatest.

Record Low Temperatures for Selected States (°F)	
State	**Temperature**
Alabama	−27
California	−61
Florida	−2
Georgia	−17
Hawaii	12
Indiana	−35
Kentucky	−34
Louisiana	−16
Oklahoma	−27
West Virginia	−37

Graph each integer on a number line.

−80 −60 −40 −20 0 20

Write the integers as they appear on the number line from left to right.

−61, −37, −35, −34, −27, −27, −17, −16, −2, 12 are in order from least to greatest.

Checking Your Understanding

Communicating Mathematics

Read and study the lesson to answer these questions.

1. **Write** two inequalities that show how the highest and lowest temperatures in Example 2 are related.
2. **List** the integers graphed on the number line from least to greatest.

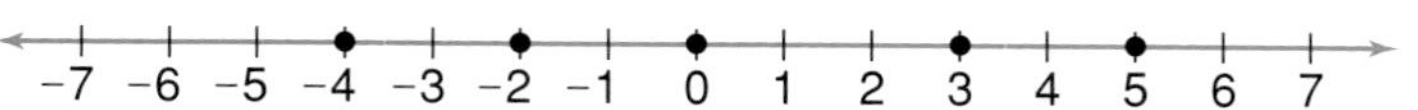

3. Graph two integers on a number line. Then write two inequalities that show how the integers are related.

4. Write a sentence that explains how to determine when one integer is less than another integer.

Guided Practice

Write an inequality using the numbers in each sentence. Use the symbols < or >.

5. 2° is warmer than −5°.
6. 5 is greater than −3.

Replace each ● with <, >, or =.

7. −19 ● −9
8. 0 ● −8
9. 4 ● |−4|
10. |−3| ● |8|
11. Order the integers {43, 0, −9, −4, 3, −88, 234} from least to greatest.

Write two inequalities for each graph.

12. −7 −6 −5 −4 −3 −2 −1 0
13. −3 −2 −1 0 1 2 3 4

14. **Sports** Patty Sheehan won the $82,500 first prize in the Rochester International LPGA tournament in June, 1995. The final round scores of the top ten finishers were −2, +1, −5, +3, −2, −1, +2, +4, 0, and +1.
    a. Order the scores from lowest to highest.
    b. If the lowest score wins, what was the winning score for this round?

Exercises: Practicing and Applying the Concept

Independent Practice

Replace each ● with $<$, $>$, or $=$.

15. 9 ● -11
16. -3 ● -5
17. $|-3|$ ● 3
18. -10 ● -3
19. -7 ● -13
20. $-|15|$ ● -15
21. -4 ● 0
22. 0 ● -2
23. 0 ● $|-8|$
24. -33 ● 5
25. -439 ● -23
26. -10 ● -3

Write an inequality using the numbers in each sentence. Use the symbols $<$ or $>$.

27. 3 m is taller than 2 m.
28. $20 is more than $15.
29. 60 kilometers per hour is slower than 75 kilometers per hour.
30. Yesterday's high temperature was 41°F. The low temperature was –3°F.
31. Today's pollen count is 565. Yesterday's count was 344.
32. Water boils at 212°F, and it freezes at 32°F.
33. One cup of green bean casserole has 265 Calories. The same amount of fresh green beans has 66 Calories.

Write two inequalities for each graph.

34. –4 –3 –2 –1 0 1 2

35. 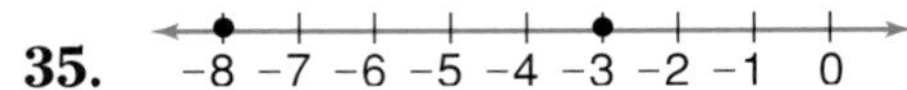

36. –100 –80 –60 –40 –20 0 20 40

37. 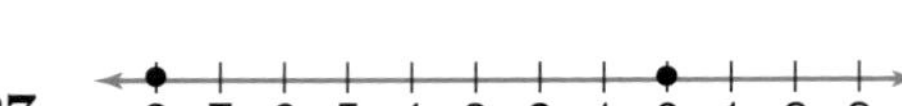

Order the integers in each set from least to greatest.

38. $\{7, 0, -5\}$
39. $\{-11, 8, -3\}$
40. $\{0, -5, -8, -1\}$
41. $\{-6, 56, 29, 1, -65\}$
42. $\{33, 9, -99, 7, 0, -4\}$
43. $\{87, -65, -53, 48, 199\}$

Critical Thinking

44. Graph all integer solutions for $|x| < 4$ on a number line.

EARTH WATCH

With the human population exploding, competition for Earth's dwindling resources is accelerating. This means that many plants and animals are in an almost impossible struggle for survival. An *endangered species* is one whose numbers have been reduced so that there is a real danger of it becoming extinct. While extinction is a natural result of changes in the environment, humans have had a hand in destroying habitat and overusing resources to make some species endangered. The table at the right shows the number of wild buffalo living in the United States in different years.

Year	Population
1851	50,000,000
1865	15,000,000
1872	7,000,000
1900	<1000
1972	30,000
1993	120,000

Source: *Audubon*

Applications and Problem Solving

45. **Chemistry** The chart at the right shows the freezing/melting points of water and selected elements.

    a. Graph this data on a number line.

    b. Which of these elements has the highest freezing/melting point?

    c. Which has the lowest freezing/melting point?

Element	Freezing/melting point (°C)
Water	0
Helium	−272
Lead	328
Mercury	−39
Oxygen	−219
Sodium	98
Tin	232

46. **Weather** The table below shows the average monthly temperatures (°F) in Fairbanks, Alaska, for the 30-year period 1961–1990. Graph these temperatures on a number line.

Jan.	Feb.	Mar.	Apr.	May	June	July	Aug.	Sept.	Oct.	Nov.	Dec.
−10	−4	11	31	49	60	63	57	46	25	3	−7

    a. In 1993, the lowest temperature in Fairbanks was −58°F on February 1. Write an inequality to compare this with the February average.

    b. The highest temperature in 1993 was 93°F on July 15. Write an inequality to compare this with the July average.

Mixed Review

On graph paper, draw and label coordinate axes. Then graph and label each point. (Lesson 2-2)

47. $L(2, -8)$

48. $M(-5, -5)$

49. **Space Exploration** About six seconds before liftoff, the three main shuttle engines start. About 120 seconds after liftoff, the solid rocket boosters burn out. Use integers to describe these events. (Lesson 2-1)

50. State whether $15 < 3m + 7$ is *true* or *false* when $m = 4$. (Lesson 1-9)

51. Restate 6×26 using the distributive property. (Lesson 1-5)

52. Evaluate $4a + b \cdot b$, if $a = 4$ and $b = 2$. (Lesson 1-3)

53. **Consumer Awareness** The Jacksonville Middle School chorale bought 30 tickets to a musical play. The total cost of the tickets is \$98. If student tickets cost \$3 each and adult tickets cost \$5 each, how many adults are going with the chorale? (Lesson 1-1)

Mae Jemison was the first African-American woman to go into space. Her first mission was a cooperative effort between U.S. and Japan that focused on experiments involving space travel and biology.

See for Yourself

There was a time when the American buffalo was in danger of extinction. Thanks to conservation, they have made a comeback.

1. The change in population from 1851 to 1865 was 15,000,000 − 50,000,000 or −35,000,000. Use a calculator to find the change in the population from each year to the next.

2. Which years showed the greatest loss in the population of buffalo? Which years showed the most growth in the population of buffalo?

3. **Research** Investigate the efforts to save the northern spotted owl. Does the program appear to be successful?

HANDS-ON ACTIVITY

2-4A Adding Integers

A Preview of Lesson **2-4**

You can use counters to model operations with integers. In this book, yellow counters will represent positive integers, and red counters will represent negative integers.

MATERIALS

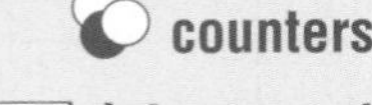

counters

integer mat

There are two important properties to keep in mind when modeling integers.

- When one positive counter is paired with one negative counter, the result is called a **zero pair**.
- You can add or remove zero pairs from a mat because removing or adding zero does not change the value of the counters on the mat.

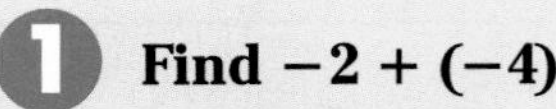

Activity 1 Find $-2 + (-4)$.

Remember that $2 + 4$ means *combine a set of two items with a set of four items.* The expression $-2 + (-4)$ means something similar. It tells you to combine a set of two negative items with a set of four negative items.

- Place 2 negative counters and 4 negative counters on the mat.

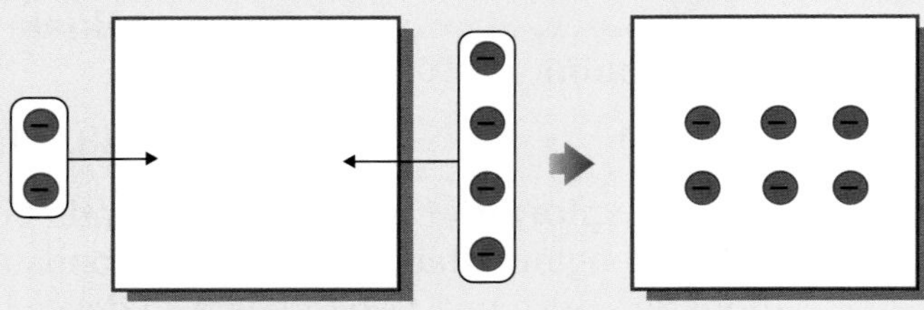

- Since there are 6 negative counters on the mat, the sum is -6. Therefore, $-2 + (-4) = -6$.

Activity 2 Find $-4 + 2$.

Remember that removing zero pairs does not change the value on the mat.

Place 4 negative counters and 2 positive counters on the mat. It is possible to remove 2 zero pairs.

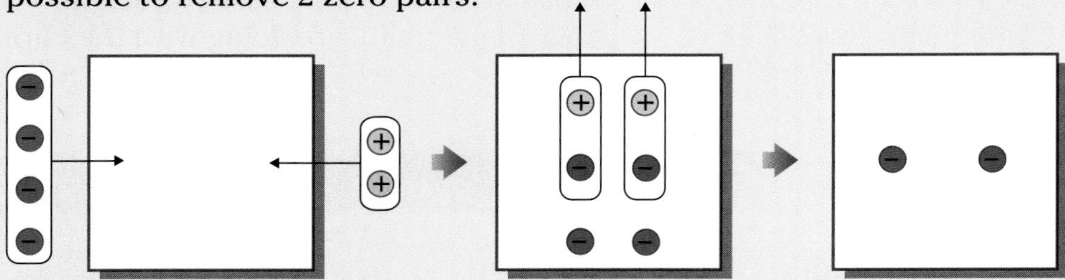

- Since there are 2 negative counters remaining, the sum is -2. Therefore, $-4 + 2 = -2$.

Your Turn **Find each sum using counters.**

1. $3 + 2$	**2.** $3 + (-2)$	**3.** $-3 + 2$	**4.** $-3 + (-2)$
5. $1 + (-4)$	**6.** $-3 + 7$	**7.** $-4 + (-4)$	**8.** $-4 + 4$

Write About It

9. Suppose you add a positive integer and a negative integer. Explain how you use a zero pair to find the sum.

2-4 Adding Integers

Setting Goals: *In this lesson, you'll add integers.*

Modeling with Technology

TECHNO TIP

Some calculators use a +/− key, called the "change sign" key instead of the (−) key.

You already know that the sum of two positive integers is a positive integer. What is the sign of the sum of two negative integers? The keystrokes for finding the sum of −5 and −8 on a scientific calculator are as follows. *Notice that the negative key is different than the subtraction key and that the negative is entered after the number.*

Enter: 5 (−) + 8 (−) = **Display:** -13

Your Turn Use a calculator to study the patterns of the sums in each column.

A	B	C	D
$4 + 8 =$	$-3 + (-4) =$	$17 + (-17) =$	$-18 + 2 =$
$12 + 3 =$	$-4 + (-6) =$	$13 + (-2) =$	$-5 + 12 =$
$29 + 17 =$	$-13 + (-2) =$	$12 + (-15) =$	$-8 + 21 =$
$25 + 93 =$	$-7 + (-3) =$	$11 + (-18) =$	$-7 + 7 =$

a. Suggest a rule for adding integers with the same signs, as in Columns A and B.

b. Suggest a rule for adding integers with different signs, as in Columns C and D.

c. Compare your rules to the rules of two of your classmates.

Learning the Concept

Another way to add integers is by using a number line.

Example  **Find $-5 + (-2)$.**

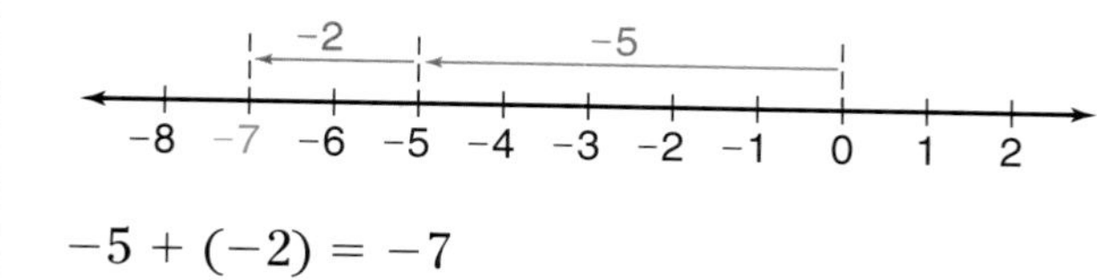

Start at zero. Move 5 units to the left. From there, move 2 more units to the left.

$-5 + (-2) = -7$

The results of this example and those in Columns A and B above suggest the following rule.

Adding Integers with the Same Signs

To add integers with the same sign, add their absolute values. Give the result the same sign as the integers.

Study Example 2 to see if you can discover the rule for adding integers that have different signs.

Example 2

Find each sum.

Recall that a zero pair is one positive and one negative counter.

a. $8 + (-4)$

Use counters. Put 8 positive counters on the mat. Add 4 negative counters. Remove all zero pairs.

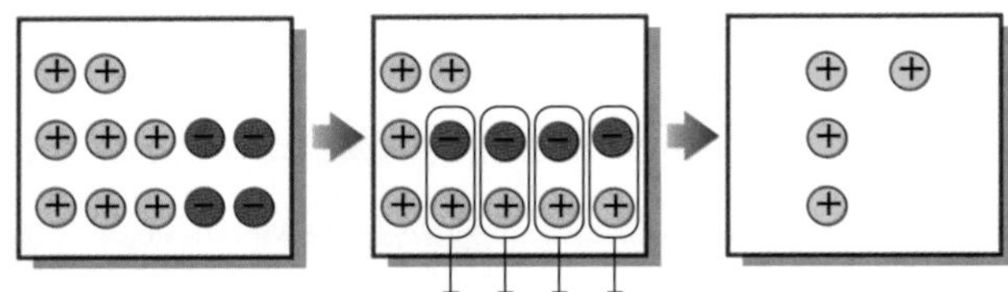

$8 + (-4) = 4$

b. $-7 + 4$

Use a number line. Start at zero. Move 7 units to the left. From there, move 4 units to the right.

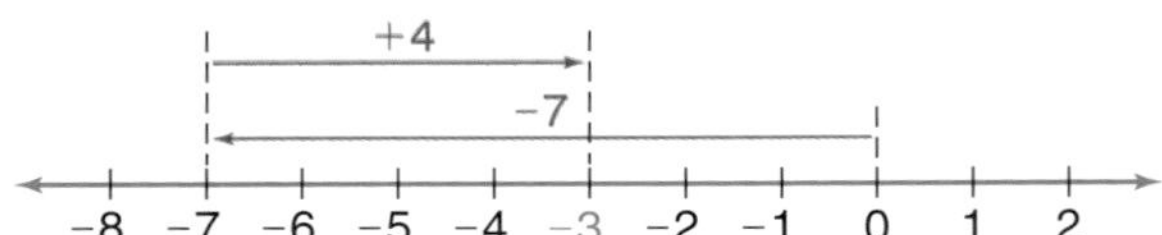

$-7 + 4 = -3$

The results in these examples and those in Columns C and D on page 83 suggest the following rule.

Adding Integers with Different Signs	To add integers with different signs, subtract their absolute values. Give the result the same sign as the integer with the greater absolute value.

Example 3

Solve each equation.

a. $x = -9 + 6$

$x = -(|-9| - |6|)$ *Subtract absolute values. The result is negative because the integer with the greater absolute value, −9, is negative.*

$x = -(9 - 6)$

$x = -3$

b. $8 + (-3) + 4 = a$

$8 + 4 + (-3) = a$ *Commutative property*

$12 + (-3) = a$ *Add 8 and 4.*

$+(12 - 3) = a$ *The result is positive because $|12| > |3|$.*

$9 = a$ *Check your result with a calculator.*

Example 4

Space Exploration

During a spacewalk on February 9, 1995, astronauts Bernard Harris Jr. and Michael Foale worked in the shadow of the shuttle *Discovery*. The temperature outside the cargo bay was −125°F. If the temperature inside the cargo bay was 111 degrees warmer, what was the temperature inside?

Explore The temperature outside the cargo bay was −125°F. Inside, it was 111 degrees warmer. You need to find the temperature inside. Let t represent the temperature inside.

Plan Translate the words into an equation using the variable. The temperature inside is the sum of −125 and 111.

FYI

Bernard Harris Jr. became the first African-American to walk in space on this *Discovery* flight.

Solve   *inside temperature* *is* *outside temperature* *plus* *111°*

$$t = -125 + 111$$

Solve the equation for t.

$t = -125 + 111$

125  [(−)] [+] 111 [=] -14   *Use a calculator to solve.*

The temperature inside the cargo bay was -14°F.

Examine The amount of increase was less than the absolute value of -125, so the temperature inside the cargo bay should still be negative. The answer is reasonable.

Connection to Algebra

You can use the rules for adding integers and the distributive property to combine like terms.

Example 5 **Simplify each expression.**

a. $-15n + 22n$

$-15n + 22n = (-15 + 22)n$ *Distributive property*
$= (7)n$ *Find the sum of -15 and 22.*
$= 7n$

b. $9x + (-18x) + 8x$

$9x + (-18x) + 8x = [9 + (-18) + 8]x$ *Distributive property*
$= -1x$ *Add 9, -18, and 8.*
$= -x$ *$-1x$ can be written as $-x$.*

Checking Your Understanding

Communicating Mathematics

Read and study the lesson to answer these questions.

1. **Explain** how to determine whether to add or subtract the absolute values to find the sum of two integers.
2. **Write** the addition sentence shown by each model.

a.

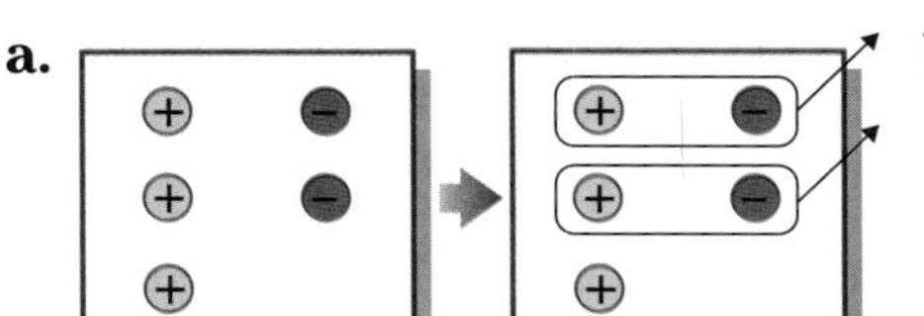

b.

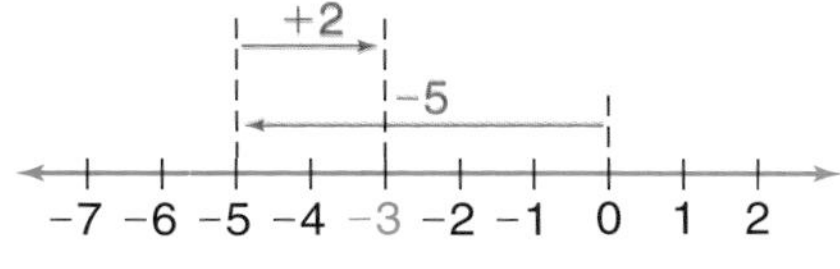

3. **Draw a model** that shows how to find the sum of 5 and -2.
4. **Explain** how to simplify $-3a + 12a + (-14a)$.

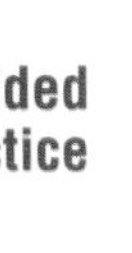

MATH JOURNAL

5. Write, in your own words, how to add pairs of integers with the same sign and with different signs. Include an example for each situation.

Guided Practice

Write an addition sentence for each model.

6.

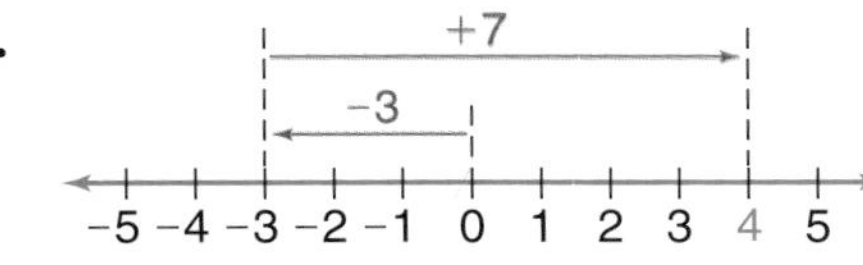

7.

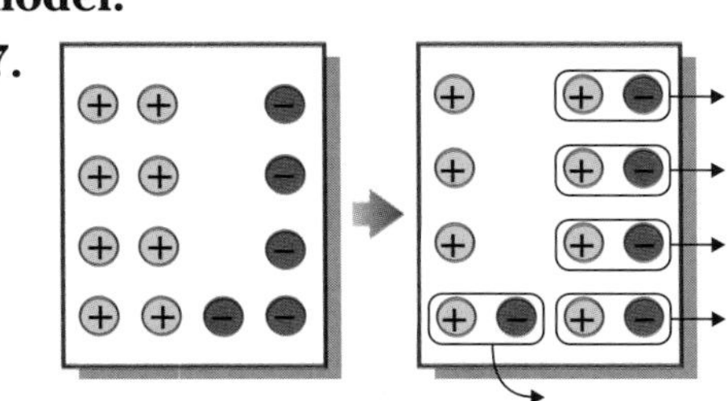

8.

-5 -4 -3 -2 -1 0 1 2 3 4 5

9.

State whether each sum is positive or negative. Then find the sum.

10. $-6 + (-9)$

11. $13 + (-4)$

12. $-8 + 12$

13. $-3 + 7 + (-8)$

Solve each equation.

14. $r = -13 + (-5) + 7$

15. $(-83) + (-21) + (-7) = z$

Simplify each expression.

16. $-3z + (-17z)$

17. $-8a + 14a + (-12a)$

18. **Golf** In golf, a score of 0 is called *even par*. A score of 3 under par is written as -3. A score of 2 over par is written $+2$ or 2. In 1995, Jose Maria Olazabal shot 6 under par, 2 over par, even par, and even par for the four rounds of the Masters Tournament. What was his final score?

Exercises: Practicing and Applying the Concept

Independent Practice

Solve each equation.

19. $18 + (-5) = x$

20. $b = -8 + (-3)$

21. $16 + (-9) = m$

22. $v = -12 + (-4)$

23. $k = 9 + (-13)$

24. $-15 + 6 = q$

25. $g = 19 + (-7)$

26. $-11 + (-15) = t$

27. $-23 + (-43) = h$

28. $-3 + 18 = n$

29. $-5 + 31 = r$

30. $y = 6 + (-16)$

31. $m = 8 + (-17)$

32. $z = -12 + (-5)$

33. $-12 + 5 = s$

Write an addition sentence for each situation. Then find the sum.

34. Maria has $500 in the bank. She owes $700 on her car.

35. In Monday night's football game, the Jacksonville Jaguars lost 7 yards on one play. They gained 12 yards on the next.

36. The research submarine *Alvin* is at 1500 meters below the sea level. It descends another 1250 meters to the ocean floor.

37. **Make Up a Problem** Write a problem that can be solved using the addition sentence $-15 + 25 = c$.

Solve each equation.

38. $m = 3 + (-11) + (-5)$

39. $k = 14 + 9 + (-2)$

40. $-18 + (-23) + 10 = c$

41. $-16 + (-6) + (-5) = a$

42. $-12 + (17) + (-7) = x$

43. $y = 47 + 32 + (-16)$

Simplify each expression.

44. $6x + (-15)x$
45. $-11y + 14y$
46. $-14k + (-7)k + 15k$
47. $8m + (-23)m$
48. $14b + (-21b) + 37b$
49. $16d + (-9d) + (-27d)$

Calculator

Use a calculator to find each sum.

50. $139 + (-316)$
51. $-249 + 7915$
52. $-3916 + (-8128)$

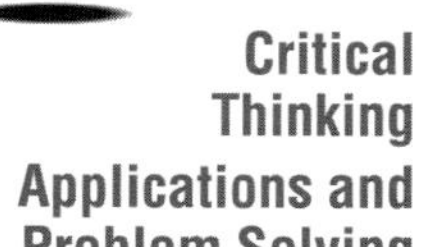

Critical Thinking

53. *True* or *false*: $-n$ names a negative number.

Applications and Problem Solving

54. **Astronomy** At noon, the average temperature on the moon is 112°C. During the night, the average temperature drops 252°C. What is the average temperature on the moon's surface during the night?

55. **Pets** The chart at the right shows the numbers of dogs of different breeds registered in the American Kennel Club in 1992 and 1993.

Breed	1992 Registration	1993 Registration
Akita	11,574	11,383
Beagle	61,051	60,661
Chow Chow	33,824	42,670
Dachshund	48,573	50,046
Labrador Retriever	124,899	120,879
Pug	15,722	16,008

a. Describe the change in the number of dogs of each breed registered from 1992 to 1993.

b. What was the total change in the number of dogs of these breeds registered from 1992 to 1993?

Mixed Review

56. Order the numbers {0, −4, −8, −3} from least to greatest. (Lesson 2-3)

57. *True* or *false*: The graph of a point whose coordinates are both negative integers would be located in Quadrant IV. (Lesson 2-2)

58. **Geography** Death Valley, California, has the lowest altitude in the United States. Its elevation is 282 feet below sea level. What integer could be used to model this information? (Lesson 2-1)

59. **Recreation** A skateboarder travels 48 meters in 16 seconds. Solve the equation $48 = r \cdot 16$ to find the speed in meters per second. (Lesson 1-8)

60. *True* or *false*: $4 + 13 = 17$ and $17 - 4 = 13$ are related sentences. (Lesson 1-8)

61. Write an equation and solve: Antonia paid $15 for a shirt on sale. It was reduced by $9. What was the regular price? (Lesson 1-8)

62. Of 19, 29, or 39, which is the solution of $3x = 87$? (Lesson 1-6)

63. Rewrite $(n + 8) + 9$ using an associative property. Then simplify. (Lesson 1-4)

64. Write a verbal phrase for the expression $2x + 3$. (Lesson 1-3)

65. Determine whether $6 \div 2 + 5 \times 4 = 32$ is *true* or *false*. (Lesson 1-2)

HANDS-ON ACTIVITY

2-5A Subtracting Integers

A Preview of Lesson **2-5**

MATERIALS

counters

integer mat

You can use counters to model the subtraction of integers.

Activity 1 **Find −6 − (−3).**

▶ Start with 6 negative counters. Remove 3 negative counters.

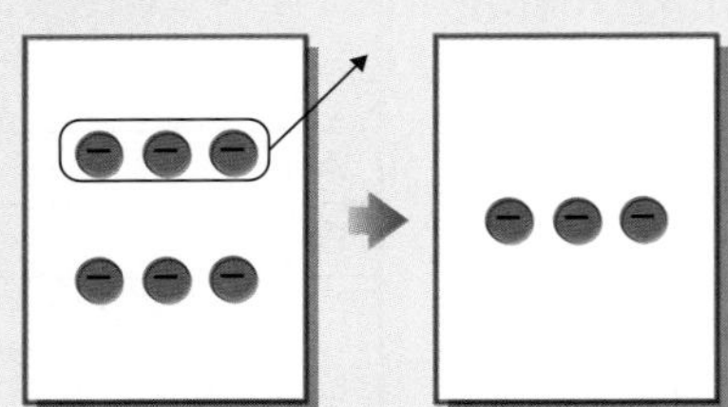

▶ Since there are 3 negative counters left, the difference is −3. Therefore, −6 − (−3) = −3.

Activity 2 **Model 4 − (−2).**

▶ Start with 4 positive counters. There are no negative counters, so you can't remove 2 negative counters. Add 2 zero pairs to the mat. Remember adding zero pairs does not change the value of the set. Now remove 2 negative counters.

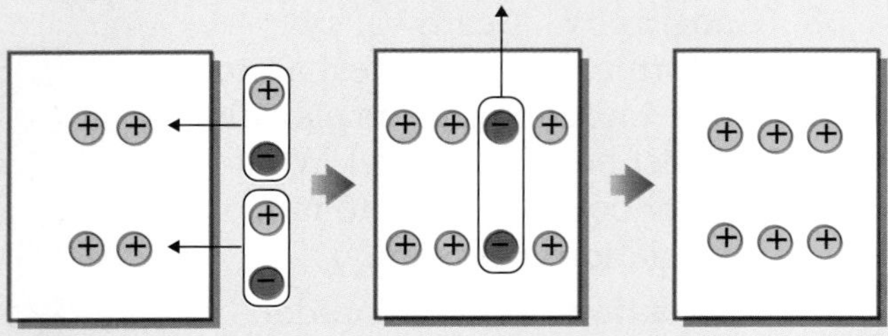

▶ Since there are 6 positive counters remaining, the difference is +6 or 6. Therefore, 4 − (−2) = 6.

Your Turn **Find each difference using counters.**

1. 2 − (−3)	**2.** 2 − 3	**3.** −2 − (−3)	**4.** −2 − 3
5. −6 − 9	**6.** 13 − 4	**7.** −8 − (−12)	**8.** −4 − (−4)

9. Compare Exercises 5–7 above with Exercises 10–12 on page 86. How are they alike? How are they different?

10. If you subtract a negative integer from a lesser negative integer, is the difference positive or negative? Justify your answer with a drawing.

2-5 Subtracting Integers

Setting Goals: *In this lesson, you'll subtract integers.*

Modeling with Manipulatives

The diagram at the right models the expression $3 - (-3)$. Counting the number of counters on the mat after the counters are removed shows that $3 - (-3) = 6$.

Use counters to find $-3 - (-2)$, $4 - 6$, and $-5 - 1$.

a. Compare the solution of $-3 - (-2)$ to the solution of $-3 + 2$.

b. Compare the solution of $4 - 6$ to the solution of $4 + (-6)$.

c. Compare the solution of $-5 - 1$ to the solution of $-5 + (-1)$.

Learning the Concept

Adding and subtracting are inverse operations that "undo" each other. Similarly, when you add opposites, like 4 and -4, the sum is 0.

You can review inverse operations in Lesson 1-8.

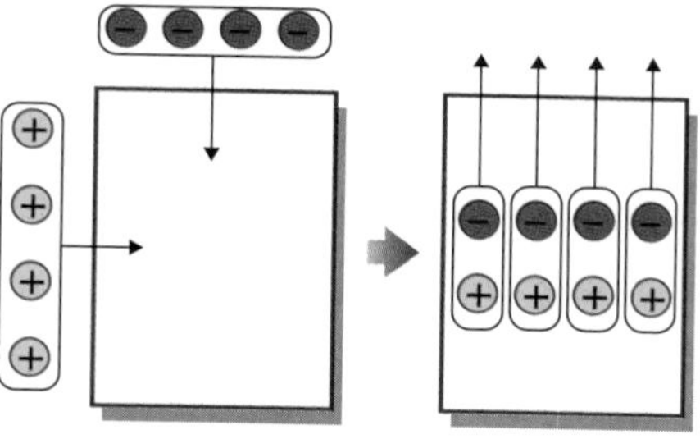

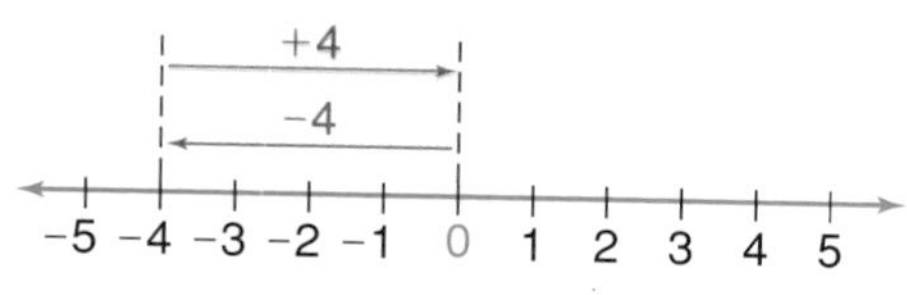

An integer and its opposite are called **additive inverses** of each other.

Additive Inverse Property

In words: The sum of an integer and its additive inverse is zero.

Arithmetic	**Algebra**
$5 + (-5) = 0$	$x + (-x) = 0$

At the beginning of the lesson, you were asked to compare the result of subtracting an integer with the result of adding its additive inverse.

Subtraction	**Addition of Additive Inverse**
$-3 - (-2) = -1$	$-3 + 2 = -1$
$4 - 6 = -2$	$4 + (-6) = -2$
$-5 - 1 = -6$	$-5 + (-1) = -6$

These examples suggest a method for subtracting integers.

Subtracting Integers

In words: To subtract an integer, add its additive inverse.

Arithmetic	**Algebra**
$3 - 5 = 3 + (-5)$	$a - b = a + (-b)$

Example  **Solve each equation.**

a. $m = 7 - 12$
$m = 7 + (-12)$ *To subtract 12, add −12.*
$m = -5$

b. $6 - (-11) = r$
$6 + 11 = r$ *To subtract −11, add 11.*
$17 = r$

c. $y = -3 - (-8)$
$y = -3 + 8$ *To subtract −8, add 8.*
$y = 5$

d. $-9 - 16 = x$
$-9 + (-16) = x$ *To subtract 16, add −16.*
$-25 = x$

You can use equations with integers to represent real-life situations.

Example 2

CONNECTION
Geography

The highest point in Africa is Kilimanjaro in Tanzania. It has an altitude of 5895 meters above sea level. The lowest point on the continent is Lake Assal in Djibouti. It is 155 meters below sea level. Find the difference between these altitudes.

Explore With sea level as the starting point, 5895 meters above sea level can be expressed as +5895 or 5895. 155 meters below sea level can be expressed as −155. Let d represent the difference.

Plan Translate the words into an equation using the variable.

difference	*is*	*altitude of Kilimanjaro*	*minus*	*altitude of Lake Assal*
d	$=$	5895	$-$	-155

Solve Solve the equation for d.

$d = 5895 - (-155)$ *To subtract −155 add its inverse, 155.*
$d = 5895 + 155$
$d = 6050$

The difference in altitudes is 6050 meters.

Examine Make a diagram to check the solution.

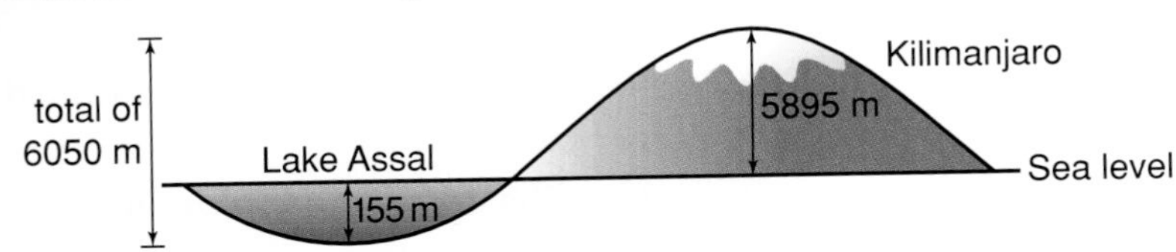

Connection to Algebra

You can use the rule for subtracting integers to evaluate expressions and combine like terms.

Example 3  **Evaluate $b - (-5)$ if $b = -13$.**

$b - (-5) = -13 - (-5)$ *Replace b with −13.*
$= -13 + 5$ *Subtract −5 by adding 5.*
$= -8$

Example 4  **Simplify $5a - 13a$.**

$5a - 13a = 5a + (-13)a$ *Add the additive inverse of 13a, −13a.*
$= [5 + (-13)]a$ *Use the distributive property.*
$= -8a$ *Add 5 and −13.*

Checking Your Understanding

Communicating Mathematics

Read and study the lesson to answer these questions.

1. **Explain** how subtraction of integers is related to addition of integers.
2. **Write** the subtraction sentence shown by each model.

a. 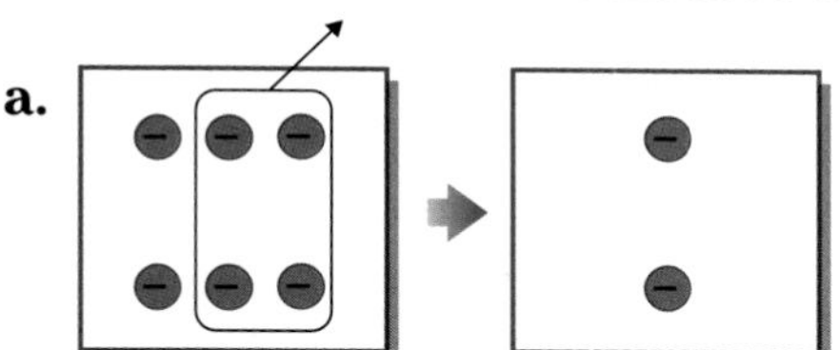

b. 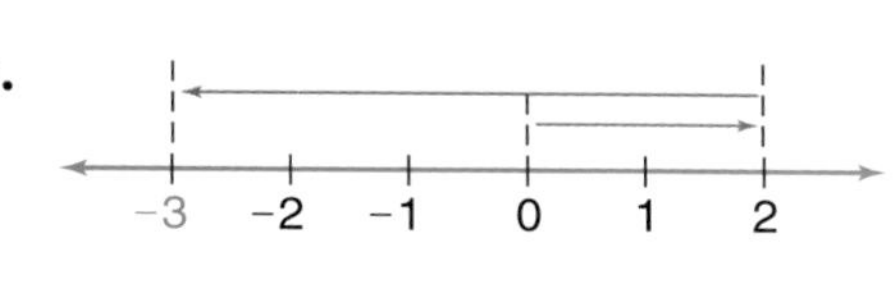

3. **Draw a model** that shows how to find $5 - (-2)$.
4. **Explain** how to simplify $12y - 14y$.
5. Using masking tape, create a large number line on the floor. Then "walk out" the solution to each problem using the following rules.
   - Stand at zero and face the positive integers.
   - To model a positive integer, move forward.
   - To model a negative integer, move backward.
   - To model subtraction, do an about-face before moving.

   a. $4 - 6$ **b.** $2 - (-3)$ **c.** $-3 - 1$ **d.** $-5 - (-4)$

MATERIALS

masking tape

Guided Practice

State the additive inverse of each of the following.

6. $+7$ **7.** -8 **8.** $-a$ **9.** $9x$

Rewrite each equation using the additive inverse. Then solve.

10. $6 - 15 = m$ **11.** $-7 - 11 = x$ **12.** $-16 - 7 = z$
13. $8 - (-3) = b$ **14.** $-24 - (-8) = r$ **15.** $-14 - (-19) = y$

Simplify.

16. $6a - 13a$ **17.** $-15x - 12x$ **18.** $-17y - (-3)y$

interNET CONNECTION

For the other windchill factors, visit: **www.glencoe.com/sec/math/prealg/mathnet**

Evaluate each expression.

19. $x - 3$, if $x = -9$ **20.** $y - (-8)$, if $y = 17$ **21.** $a - (-7)$, if $a = -15$

22. Meteorology *Windchill* factor is an estimate of the cooling effect the wind has on a person in cold weather. If the outside temperature is 10°F and the wind makes it feel like −25°F, what is the difference between the actual temperature and how cold it feels?

Exercises: Practicing and Applying the Concept

Independent Practice

Solve each equation.

23. $y = -9 - 5$ **24.** $a = -13 - 4$ **25.** $r = -8 - (-5)$
26. $x = -12 - (-7)$ **27.** $m = 12 - (-20)$ **28.** $k = 23 - (-24)$
29. $-5 - (-14) = p$ **30.** $-12 - (-16) = z$ **31.** $31 - (-6) = d$
32. $-17 - 13 = a$ **33.** $-21 - 8 = b$ **34.** $38 - (-12) = q$
35. $x = -28 - 35$ **36.** $-7 - 25 = f$ **37.** $-39 - 7 = s$

Evaluate each expression.

38. $11 - n$, if $n = 5$
39. $-15 - c$, if $c = 5$
40. $h - (-17)$, if $h = -23$
41. $r - 11$, if $r = -16$
42. $8 - m$, if $m = -12$
43. $h - (-13)$, if $h = -18$
44. $5 - b$, if $b = 8$
45. $7 - (-f)$, if $f = 19$

Simplify.

46. $5b - 17b$
47. $13x - 23x$
48. $-3a - (-5a)$
49. $18p - 3p$
50. $-28d - 17d$
51. $12cd - (-12cd)$
52. $-15xy - 18xy$
53. $7a - 14a - a$
54. $-q - (-6q) - (-7q)$

Critical Thinking

55. Explain why every integer has an additive inverse. Then identify the integer that is its own additive inverse.

Applications and Problem Solving

56. **Aviation** Steve Fossett was the first person to fly a balloon solo across the Pacific Ocean. He took off from Seoul, South Korea, on February 18, 1995, and landed in Saskatchewan, Canada, three days later. When the heaters in his gondola failed, he had to endure temperatures that ranged from 10°F to −4°F. What was the change from the highest to lowest temperatures he experienced?

57. **Accounting** Dawn Sopher is a business information analyst. She uses spreadsheet software to keep track of accounts payable. The balance of an account can be negative or positive. Purchases are subtracted from the balance, and payments are added to the balance. Find the final balance on the spreadsheet below.

	A	B	C	D
1	Date	Purchases	Payments	Balance
2	11/1			−$12,300
3	11/8	$2500		
4	11/22		$18,345	
5	11/30	$5678		

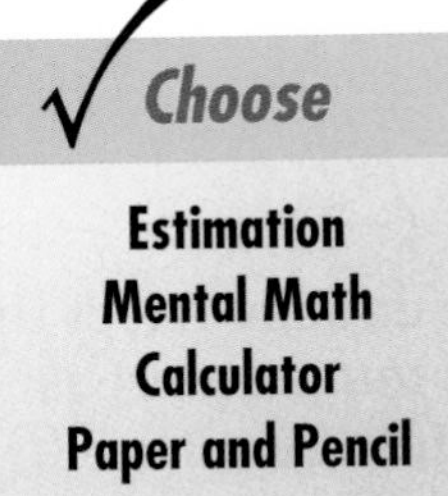

Estimation
Mental Math
Calculator
Paper and Pencil

58. **Geography** A region's time zone is determined by its longitude. The zone labeled zero is centered on the prime meridian. In the chart below, time zones to the east of the prime meridian are named by negative numbers, and time zones west of the prime meridian are named by positive integers. To find the time in another city, subtract the time zone of the other city from the time zone at your location.

Suppose you are in Miami and wish to know the time in Bombay. Find $5 - (-5)$, which is 10. Since 10 is positive, the time in Bombay is 10 hours *ahead* of Miami. So at 1:00 P.M. in Miami, it is 11:00 P.M in Bombay.

City	Time Zone
Bombay, India	−5
Honolulu, USA	11
Los Angeles, USA	8
Miami, USA	5
Paris, France	0
Rome, Italy	−1

a. If the time in Los Angeles is 2:00 P.M., find the time in Rome.

b. If the time in Miami is 11:00 A.M., find the time in Honolulu.

c. If the time in Honolulu is 3:00 A.M., find the time in Paris.

CHAPTER 2 Highlights

Vocabulary

After completing this chapter, you should be able to define each term, property, or phrase and give an example or two of each.

Algebra

absolute value (p. 67)
adding integers with different signs (p. 84)
adding integers with the same sign (p. 83)
additive inverses (p. 89)
dividing integers with different signs (p. 104)
dividing integers with the same signs (p. 105)
integers (p. 66)
multiplying integers with different signs (p. 99)
multiplying integers with the same sign (p. 100)
negative (p. 66)
subtracting integers (p. 89)
zero pair (p. 82)

Geometry

axes (p. 73)
coordinate (p. 67)
coordinate system (p. 73)
graph (p. 73)
plotting a point (p. 73)
quadrants (p. 73)
vertex (p. 75)

Problem Solving

look for a pattern (p. 94)

Statistics

cluster (p. 71)
line plot (p. 71)
outlier (p. 71)

Understanding and Using Vocabulary

Choose the letter of the term that best matches each statement or phrase.

1. an integer less than zero
2. the four regions separated by the axes on a coordinate plane
3. the number that corresponds to a point on the number line
4. an integer and its opposite
5. positive and negative whole numbers
6. a dot placed at the point named by the ordered pair
7. the distance a number is from the zero point on the number line

a. absolute value
b. additive inverses
c. coordinate
d. graph
e. integers
f. negative integer
g. quadrants

CHAPTER 2 Study Guide and Assessment

Skills and Concepts

Objectives and Examples

Upon completing this chapter, you should be able to:

- **graph integers on a number line** (Lesson 2-1)

Graph points A, B, and C on a number line if A has coordinate -2, B has coordinate 3, and C has coordinate -1.

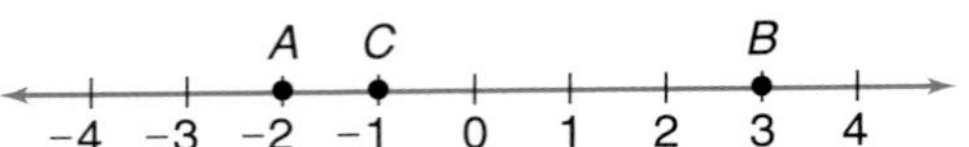

Review Exercises

Use these exercises to review and prepare for the chapter test.

Graph each set of numbers on a number line.

8. $\{-3, 6, -4, -6, 11, -1\}$

9. $\{22, 74, -30, 82, -10, -12\}$

10. $\{112, 98, -21, -41, 67\}$

11. $\{-1, -16, -4, -6, 4\}$

- **find absolute value** (Lesson 2-1)

$|-3| = 3$

$|4| - |-9| = 4 - 9 = -5$

Simplify.

12. $|-31|$

13. $|24|$

14. $-|18|$

15. $-|40|$

16. $-|6 - 2|$

17. $|-8| + |-21|$

18. $|17 - 29|$

19. $||7| - |-19||$

- **graph points in all quadrants of the coordinate plane** (Lesson 2-2)

Quadrant II | Quadrant I

(-2, 1)

O x y

Quadrant III | Quadrant IV

On graph paper, draw coordinate axes. Then graph and label each point. Name the quadrant in which each point is located.

20. $A(4, 3)$

21. $R(-1, -1)$

22. $W(-5, 3)$

23. $G(-6, 5)$

24. $M(-4, -1)$

25. $L(3, -2)$

- **compare integers** (Lesson 2-3)

Compare -4 and -6.

Since -6 is to the left of -4 on the number line, $-4 > -6$.

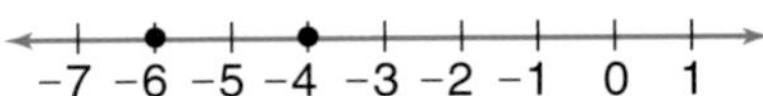

Replace each ● with <, >, or = .

26. 9 ● −9

27. −3 ● −3

28. 6 ● 13

29. −12 ● −21

30. −2 ● 0

31. −43 ● 43

Write an inequality using the numbers in each sentence. Use the symbols < or >.

32. The population of Detroit, Michigan, is 1,027,974. The population of Ann Arbor, Michigan, is 109,592.

33. The winning team in Super Bowl XXVIII won by 17 points. In Super Bowl XXVII, the winning team won by 35 points.

Objectives and Examples	Review Exercises
order integers (Lesson 2-3) Order the integers $\{-3, 5, -1, 8, -6\}$ from least to greatest –8 –6 –4 –2 –0 2 4 6 8 $-6, -3, -1, 5, 8$ are in order from least to greatest.	**Order the integers in each set from least to greatest.** **34.** $\{8, 0, -3, 6, 5\}$ **35.** $\{-16, 3, -4, 2\}$ **36.** $\{-75, 21, 4, -34, -88\}$ **37.** $\{210, -300, 124, 9, -33\}$
add integers (Lesson 2-4) $d = 4 + (-5)$ $d = -1$ $r = -4 + (-5)$ $r = -9$ $m = -4 + 5 + (-7)$ $m = 1 + (-7)$ $m = -6$	**Solve each equation.** **38.** $a = -4 + (-5)$ **39.** $8 + (-11) = s$ **40.** $-12 + 4 = t$ **41.** $k = -15 + 9$ **42.** $g = 18 + (-25)$ **43.** $h = -13 + (-2)$ **44.** $-6 + (-18) + 40 = q$ **45.** $a = 7 + (-3) + (-14)$
subtract integers (Lesson 2-5) $r = 5 - 12$ $r = 5 + (-12)$ $r = -7$ $-8p - (-11p) = -8p + 11p$ $= 3p$	**Solve each equation.** **46.** $g = -8 - 7$ **47.** $-33 - (-3) = s$ **48.** $-17 - 8 = w$ **49.** $t = 15 - (-6)$ **Simplify.** **50.** $2a - 5a$ **51.** $-12b - (-5b)$ **52.** $14m - (-12m)$ **53.** $-15r - (-21r)$
multiply integers (Lesson 2-7) $m = -7 \cdot (-4)$ $m = +(7 \cdot 4)$ $m = 28$ $-4a(3b) = (-4 \cdot 3)(a \cdot b)$ $= -12ab$	**Solve each equation.** **54.** $z = -9 \cdot 4$ **55.** $(-4)(2) = k$ **56.** $(-11)(-15) = y$ **57.** $p = -5 \cdot (-13)$ **Find each product.** **58.** $(-2)(8z)(3)$ **59.** $6 \cdot (-6c) \cdot (-d)$ **60.** $-15x(5y)$ **61.** $(-25y) \cdot (-6) \cdot (-2z)$

Objectives and Examples	Review Exercises

divide integers (Lesson 2-8)

$u = -105 \div 15$
$u = -(105 \div 15)$
$u = -7$

Evaluate $t \div 18$, if $t = -108$.

$t \div 18 = -108 \div 18$
$= -6$

Solve each equation.

62. $k = 88 \div -4$
63. $-14 \div (-2) = h$
64. $99 \div -11 = n$
65. $q = -45 \div (-3)$

Evaluate each expression.

66. $d \div 8$, if $d = 120$
67. $64 \div m$, if $m = -4$
68. $-700 \div k$, if $k = -10$
69. $z \div -12$, if $z = -228$

Applications and Problem Solving

70. **Physics** Jalisa did an experiment to see how the mass of an object attached to a spring affected the distance a spring stretched. The data are shown at the right. Graph the ordered pairs (mass, distance) in the coordinate plane. (Lesson 2-2)

Stretching of a Spring	
Mass	**Distance**
100 g	3 cm
200 g	6 cm
300 g	9 cm
400 g	12 cm
500g	15 cm

71. **Weather** A combination of wind and cold makes you feel colder than the actual temperature. According to the National Weather Service, a temperature of 20 degrees Fahrenheit with a wind of 30 miles per hour makes you feel like it is 38 degrees colder than it is. How cold does it feel when it is 20 degrees with a wind of 30 mph? (Lesson 2-4)

72. **Patterns** Find the next two integers in the pattern 1, −2, 3, −4, __?__, __?__. (Lesson 2-6)

73. **Patterns** Use the pattern at the right to find 1089 · 7. (Lesson 2-6)

1089 · 1 = 1089
1089 · 2 = 2178
1089 · 3 = 3267
1089 · 4 = 4356

74. **Physics** When an object that is traveling increases its speed it is *accelerating*, when it slows, it is *decelerating*. Acceleration and deceleration are described by positive and negative numbers respectively. Suppose a car is decelerating at a rate of 1.2 meters per second per second. This is represented by -1.2 m/s^2.
    a. How much will a car's speed change if it decelerates at a rate of -1.2 m/s^2 for 11 seconds? (Lesson 2-7)
    b. If a motorcycle slows down its rate of speed by 3.6 meters per second in 12 seconds, what is its rate of deceleration? (Lesson 2-8)

A practice test for Chapter 2 is available on page 775.

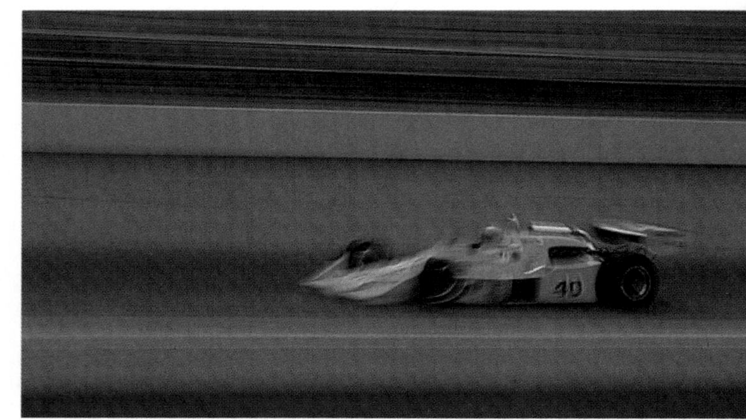

Alternative Assessment

Cooperative Learning Project

Temperature changes occur hourly, daily, weekly, and so on. For this project, you will need to be aware of this change and record it as follows.

1. Pick three cities in the world and record their temperature at the same time each day for a period of five days.
2. Make a table showing the time of day, name of the city, and the temperature for each day.
3. Describe the change in temperature for each city from day 1 to day 2, day 2 to day 3, day 3 to day 4, and day 4 to day 5.
4. On a coordinate graph, plot the daily temperature change for each city. The x-coordinate should be the first day of the series. The y-coordinate should be the change in temperature.
5. What was the total change in temperature for each city from day 1 to day 5?
6. Find the average change in temperature for each city. (Step 1: Add the change in temperature from day to day. Step 2: Divide that number by the number of temperature changes.)

Thinking Critically

- The point (a, b) is located in quadrant IV. What are all the possible values for a and b?
- Describe all the points located in each quadrant using + (for positive numbers) and − (for numbers) for the values of the x-coordinate and the y-coordinate. (*Hint*: You should have four ordered pairs as your answer.)

 **Portfolio**

- Select an item from this chapter that you feel shows your best work and place it in your portfolio.

Self Assessment

- Use the talents that you possess, for the trees would be quiet if no birds sang but the best.

Assess yourself. Are you putting your best foot forward? Does your work reflect the best that you can be? Is your work correct and presentable? Make a decision to complete your school work or a task in your life to the best of your ability. Never settle for less than your best.

CHAPTERS 1–2 Ongoing Assessment

Section One: Multiple Choice

There are ten multiple-choice questions in this section. After working each problem, write the letter of the correct answer on your paper.

1. Which is equivalent to $3 \times 8 - 6 \div 2$?

A. 3
B. 9
C. 13
D. 21

2. You have two more brothers than sisters. If you have *b* brothers, which sentence could be used to find *s*, the number of sisters you have?

A. $b = s - 2$
B. $b - 2 = s$
C. $s = b + 2$
D. $s = 2b$

3. The perimeter of a rectangle is $2(\ell + w)$, where ℓ is the length and w is the width. What is the perimeter of a rectangle with length 10 cm and width 8 cm?

A. 36 cm
B. 28 cm
C. 20 cm
D. 18 cm

4. If a diver descends at a rate of 5 meters per minute, at what depth will she be after 12 minutes?

A. +60 m
B. −17 m
C. −50 m
D. −60 m

5. If $36 - b = 20$, what is the value of *b*?

A. 56
B. 26
C. 16
D. 6

6. $7 \times (10 + 5) =$

A. $(7 + 10) + 5$
B. $(7 + 10) \times 5$
C. $(7 \times 10) + (7 \times 5)$
D. $(7 + 10) \times (7 + 5)$

Use the following graph for exercises 7 and 8.

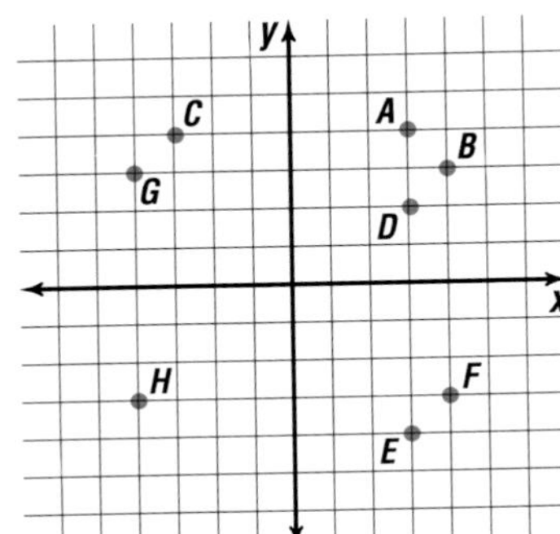

7. Which letter represents the ordered pair (4, 3)?

A. *A*
B. *B*
C. *C*
D. *D*

8. Which letter represents the ordered pair (−4, 3)?

A. *E*
B. *F*
C. *G*
D. *H*

9. The frequency table below contains students' test scores on a 25-point test.

Score	18	19	20	21	22	23	24	25
Number of Students	2	1	5	4	5	6	4	1

How many students had a score greater than 20?

A. 4
B. 5
C. 20
D. 25

10. If $a = 5$ and $b = 20$, what is the value of $2a + b$?

A. 50
B. 45
C. 40
D. 30

Section Two: Free Response

This section contains nine questions for which you will provide short answers. Write your answer on your paper.

11. Rewrite $(x + 2) + 12$ using an associative property. Then simplify.

12. Many of the roller coasters at Cedar Point in Sandusky, Ohio, require the rider to be taller than 48 inches. Write an inequality for the height requirement.

13. A bolt of fabric contains about 18 yards of fabric. You want to make T-shirts for your friends. You need 1.875 yards for each T-shirt. About how many T-shirts can you make from one bolt of fabric?

14. During a cold spell in January, 1994, the low temperatures were $-22°F$, $-20°F$, and $-18°F$. The average is the sum of these temperatures divided by 3. What was the average low temperature for those three days?

15. Translate *sixteen less than a number is twelve* into an equation and solve.

16. During one of the Chicago Bears' offensive drives, they gained 12 yards on one play. On the next play, the quarterback was sacked for a loss of 6 yards. Write and solve an addition sentence representing the net yards for these two plays.

17. Write a verbal phrase for $6u + 7$.

18. The Branson Middle School football team is 2 yards from their opponent's goal line. On the next two plays, their offense loses 17 yards and then gains 11 yards. How many yards are they from the goal line now?

19. The temperature at 6:00 A.M. was $-5°F$. What was the temperature at 8:00 A.M. if it had risen 7 degrees?

Test-Taking Tip

You can prepare for taking standardized tests by working through practice tests like this one. The more you work with questions in a format similar to the actual test, the better you become in the art of test taking.

Do not wait until the night before taking a test to review. Allow yourself plenty of time to review the basic skills and formulas that are tested. Find your weaknesses so that you can ask for help.

Section Three: Open-Ended

This section contains four open-ended problems. Demonstrate your knowledge by giving a clear, concise solution to each problem. Your score on these problems will depend on how well you do the following.

- Explain your reasoning.
- Show your understanding of the mathematics in an organized manner.
- Use charts, graphs, and diagrams in your explanation.
- Show the solution in more than one way or relate it to other situations.
- Investigate beyond the requirements of the problem.

20. Show that $7ac + 14ab + 28a = 7a(c + 2b + 4)$. Use values for a, b, c, and d to justify your answer.

21. You are at a party with 9 other guests. If everyone is to shake hands with every other guest, how many handshakes will occur?

22. If y is an integer, what is the least value that $y \cdot y$ could have? Explain your answer.

23. Rivera High School will graduate 268 seniors on June 1. The ceremony will be held in the school gymnasium. The gymnasium holds 1400 people in addition to the graduates. How many tickets should be offered to each graduate for family and friends?

CHAPTER 3

Solving One-Step Equations and Inequalities

TOP STORIES in Chapter 3

In this chapter, you will:

- solve problems by eliminating possibilities,
- solve equations and inequalities using addition, subtraction, multiplication and division, and
- use formulas to solve real-world and geometry problems.

MATH AND BRAILLE IN THE NEWS

Building Better Braille

Source: Physics Today, March, 1995

Do you ever think of math as a language? Imagine how difficult it might be for someone who cannot see letters, numbers and symbols. Physicist John Gardner and mathematician Norberto Salinas, both of whom are blind, have teamed up to develop a better method for writing mathematics for the blind using a new eight-dot code called Dots Plus. The Dots Plus system allows for four times as many symbols as the traditional six-dot system. Other new developments like speech synthesizers that "read" are opening new doors to those with visual impairments.

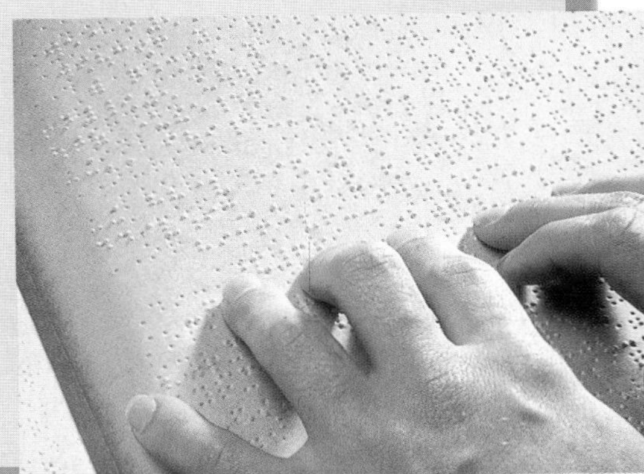

Putting It into Perspective

1800 — 1840 — 1880

1829
Louis Braille develops the Braille system, and Laura Dewey Bridgman, the first blind and deaf child to be successfully educated in the U.S., is born.

1832
The first school for the blind in the U.S. opens in Boston, MA.

1858
The American Printing House for the Blind, a major publisher, opens in Louisville, KY.

EMPLOYMENT OPPORTUNITIES

We See the Possibilities!

WE KNOW THAT new technologies are making it easier than ever for the visually impaired to choose any field of work. We are looking for talented engineers, entertainers, factory workers, teachers, social workers, computer programmers, lawyers, and anything else your imagination can conceive!

For more information, contact:
National Blindness Information Center
1800 Johnson Street
Baltimore, MD 21230

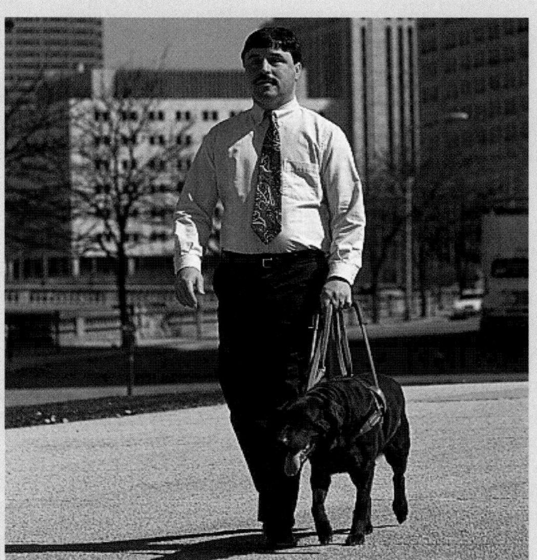

Statistical Snapshot

The Braille System

The braille cell is three dots high and two dots wide. This means that 63 different characters can be formed.

A B C D E F G H I J

The braille alphabet starts by using 10 combinations of the top 4 dots. The same 10 characters, when preceded by a special number sign, are used to express the numbers 1 to 0.

K L M N O P Q R S T

Adding the lower left-hand dot makes the next 10 letters. Adding the lower right-hand dot makes the last 5 letters of the alphabet (except *w*) and five word symbols, *below*.

U V X Y Z and for of the with

Omitting the lower left-hand dot forms nine *digraphs*, or speech sounds, and the letter *w*. This construction continues until all possible combinations have been used.

ch gh sh th wh ed er ou ow W

For up-to-date information on DotsPlus, visit:
www.glencoe.com/sec/math/prealg/mathnet

1920 1960 2000

1930
Famed blind blues singer Ray Charles is born in Albany, GA.

1944
The first eye bank is formed in New York City.

1968
Helen Keller, a famous blind and deaf writer and speaker, dies. Keller was the subject of the movie *The Miracle Worker*.

1995
Dots Plus system of writing mathematics is developed.

3-1 Problem-Solving Strategy: Eliminate Possibilities

Setting Goals: *In this lesson, you'll solve problems by eliminating possibilities.*

Modeling A Real-World Application: Education

Was there ever a question on a multiple-choice test for which you weren't quite sure of the answer? How did you go about choosing an answer? Eliminating some of the possible answers may help you choose the correct answer without having to guess.

Learning the Concept

This problem-solving strategy, **eliminating possibilities**, is helpful in solving many types of problems.

Example 1

Education

The question below appeared on a multiple choice test. Use the strategy of eliminating possibilities to choose the correct answer.

For what values of y will $2y - 4$ be equal to $2y + 6$?

a. all negative numbers
b. 0
c. all positive numbers
d. no value

Explore You know that y can represent any number. You want to find the values of y, if any, for which $2y - 4 = 2y + 6$.

Plan Look at the choices of values of y to see which ones are impossible and eliminate those.

Solve Choose a test value of y for each of the possible answers. Then test to see if $2y - 4 = 2y + 6$ for that value. If $2y - 4 \neq 2y + 6$ for that value, then we can eliminate that answer choice.

Choice	Test value for y	$2y - 4$	$2y + 6$	Is $2y - 4 = 2y + 6$?
a	-3	$2(-3) - 4 = -6 - 4$ or -10	$2(-3) + 6 = -6 + 6$ or 0	no
b	0	$2(0) - 4 = 0 - 4$ or -4	$2(0) + 6 = 0 + 6$ or 6	no
c	2	$2(2) - 4 = 4 - 4$ or 0	$2(2) + 6 = 4 + 6$ or 10	no

We can eliminate a, b, and c as possible answers. Therefore, d must be the correct choice.

Examine Study the equation $2y - 4 = 2y + 6$. Notice that $2y$ appears on both sides. Since $2y$ will have the same value for any given value of y, the equation will only be true when $-4 = 6$. But $-4 \neq 6$, so there is no value of y that makes the two expressions equal.

The process of elimination is also useful in solving some logic problems. Making a chart will help you organize the facts. The process in Example 2 is called **matrix logic**.

Example 2

Shalana, Ken, Rolon, and Erica each take their pets to Dr. McDermott for veterinary care. They each have a different pet: a dog, a cat, a gerbil, or a parrot. Use the following information to match each person with his or her pet.

- **Rolon lives next door to the person with the gerbil.**
- **Erica and Ken frequently care for the dog when its owner is out of town.**
- **Shalana cannot have a dog or cat because of allergies.**
- **Erica is teaching her pet how to talk.**

Make a chart to organize the information. Mark an X whenever you eliminate a possibility and draw a circle when you find a match.

	Shalana	Ken	Rolon	Erica
Dog				
Cat				
Gerbil				
Parrot				

Since Rolon cannot live next door to himself, the first clue tells you that he does not own the gerbil. The second clue tells us that neither Ken nor Erica owns the dog. And the third clue tells us that Shalana does not own the cat or the dog.

	Shalana	Ken	Rolon	Erica
Dog	X	X		X
Cat	X			
Gerbil			X	
Parrot				

So Rolon must own the dog. Draw a circle to indicate the match. Then mark X's for the other parts of Rolon's column.

(continued on the next page)

	Shalana	Ken	Rolon	Erica
Dog	X	X	O	X
Cat	X		X	
Gerbil			X	
Parrot			X	

Since the only pet that could possibly talk is a parrot, the fourth clue tells us that Erica must own the parrot. Draw a circle to indicate the match. Then mark Xs for each of the other choices of pets for Erica and owners for the parrot.

Complete the table to determine the answers.

	Shalana	Ken	Rolon	Erica
Dog	X	X	O	X
Cat	X	O	X	X
Gerbil	O	X	X	X
Parrot	X	X	X	O

Shalana owns the gerbil, Ken owns the cat, Rolon owns the dog, and Erica owns the parrot.

Checking Your Understanding

Communicating Mathematics

Read and study the lesson to answer these questions.

1. **Explain** why eliminating possibilities can be helpful in solving problems.
2. **Explain** how eliminating possibilities is different from guessing and checking to find the answer.

MATERIALS

25 square tiles (5 each of 5 different colors)

3. Arrange the tiles in a 5-by-5 square so that no row, column, or diagonal has more than one of the same color.
   a. Compare your arrangement with others in your class.
   b. Does the order of the first row make a difference to your final arrangement?
   c. Is there more than one way to arrange the tiles?

Guided Practice

Solve by eliminating possibilities.

4. Matt, Stacey, and Keisha each ate either cheese pizza, a hot dog, or a hamburger for dinner. Use the clues to find each person's dinner.
   - Keisha did not have any meat for dinner.
   - Stacey did not have a hot dog.
5. Use the following clues to eliminate all but the correct number.

   4235 8765 4223 6087 6444 5163
   9635 7123 6668 5228 8224 9296

   - It is an odd number that is divisible by 3.
   - It is greater than 6000 and less than 7000.
   - The sum of the digits is 21.

6. If a number is increased by 5 and the result is multiplied by 8, the product is 168. What is the original number?

a. 42 **b.** 26 **c.** 16 **d.** 128

Exercises: Practicing and Applying the Concept

Independent Practice

Solve using any strategy.

7. Study the list of numbers below. Using the clues, find the mystery number.
   - ▶ It is not divisible by 5.
   - ▶ It is even.
   - ▶ It is divisible by 3.
   - ▶ It has 2 digits.
   - ▶ The sum of its digits is 6.

6	124	76	25	7
13	252	321	60	9
24	120	15	27	8
43	561	18	78	5

8. How many packages of hot dogs balance with the ketchup?

ketchup + mustard = 4 packages ketchup = mustard + 2 packages

9. **Statistics** Trax Music Store took a survey of 100 people. Sixty-three said they like rock, 49 said they like rap, and 24 said they like jazz. Twenty-two people like rock and rap but don't like jazz, 2 like jazz and rap but not rock, and 11 like rock and jazz but not rap. If 3 people like all three types of music, how many people do not like rock, rap, or jazz?

10. If 9 less than the product of a number and −4 is greater than 7, which of the following could be that number?

a. 5 **b.** 4 **c.** −3 **d.** −5

11. What is the least positive number that you can divide by 7 and get a remainder of 4, divide by 8 and get a remainder of 5, and divide by 9 and get a remainder of 6?

12. **Business** A concession stand at the football game sold 500 candy bars for a total of $255.00. A small candy bar sells for 40¢ and a large candy bar sells for 65¢. How many of the small candy bars were sold?

a. 320 **b.** 280 **c.** 220 **d.** 77

13. **Consumerism** According to *Money* magazine, most Americans choose to buy shoes that cost $70 and last two years instead of shoes that cost $200 and last seven years. Is this good money management? Explain.

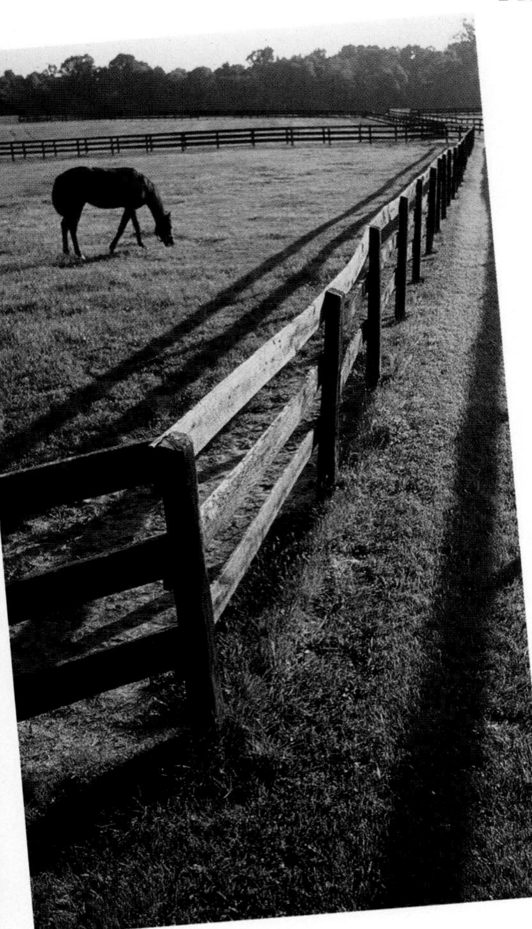

14. **Horse Riding** Five members of the Riding Club at Kingston High School went for a two-day riding trip to Manchester Lodge. They rode Sommie, Samantha, Friar, Claude, and Blaze on the first day. The stable staff wanted the club members to ride the same horses on the second day, but they could not remember who was on which horse. Use the clues below to match the rider with the horse.
    - ▶ Blaze is the fastest horse and always the first horse on the trail.
    - ▶ Friar is the most patient and gentle horse.
    - ▶ Sommie is old so she always has the lightest rider, and on the trail she liked to be near the rear.
    - ▶ Samantha and Claude are spirited and always start off together with the most experienced riders.
    - ▶ Thomas was nervous because this was his first riding trip.
    - ▶ Amber and Cynthia were the first two riders back to the stable.
    - ▶ Amber taught Cynthia how to ride, but her horse had trouble keeping up with Cynthia's horse.
    - ▶ Vernita was afraid that she was not tall enough to handle the horses she saw in the stable, so she rode behind the experienced riders.
    - ▶ Carlos rode a male horse. He started out with Amber, but stopped on the trail to check his horse's hooves. He was one of the most experienced riders, so the rest of the group knew he would catch up.

Critical Thinking

15. Three students are blindfolded and stand in a single-file line. Kieko takes three hats from a box containing three blue hats and two green hats and places one on each of the students. She tells the students how many hats of each color were in the box and removes the blindfolds. The student in the back looks at the hats on the two students in front of him and says "I don't know what color hat I am wearing." The student in the middle hears him, looks at the hat on the student in front of her and says the same thing. The student in front says "I *do* know what color hat I am wearing." What color hat is he wearing? Explain how he knows.

Mixed Review

16. Solve $x = \frac{-120}{-15}$. (Lesson 2-8)
17. **Patterns** Find the next number in the pattern 0, 1, 3, 6, 10, 15, . . . (Lesson 2-6)
18. **Sports** In Saturday's game, the Warriors lost 3 yards on a first down play. On second down, they gained 7 yards. What was the gain or loss for the two plays? (Lesson 2-4)
19. Simplify $21k + 3(k + 1)$. (Lesson 1-5)
20. Evaluate $35 - c - b$ if $b = 13$ and $c = 7$. (Lesson 1-3)

HANDS-ON ACTIVITY

3-2A Solving Equations with Cups and Counters

A Preview of Lesson **3-2**

MATERIALS

cups and counters

equation mat

You can use cups and counters as a model for solving equations.

Activity 1

An equation mat allows you to model an equation using cups and counters. The equation mat at the right represents the equation $x + 4 = 6$.

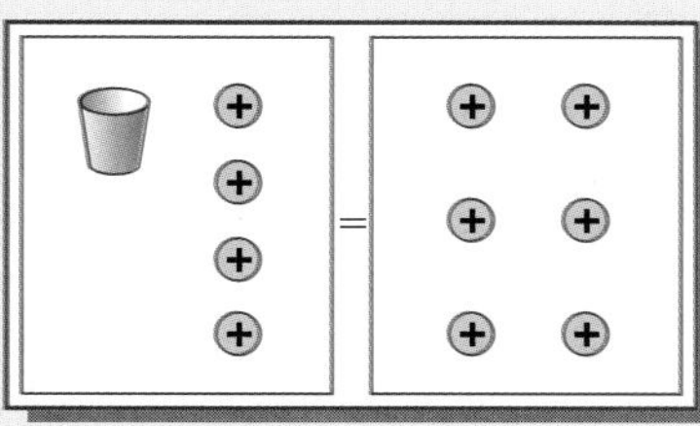

$x + 4 = 6$

To solve $x + 4 = 6$, find the value of x that makes the equation true. Think of this as finding the number of counters in the cup. To do that, get the cup by itself on one side of the mat.

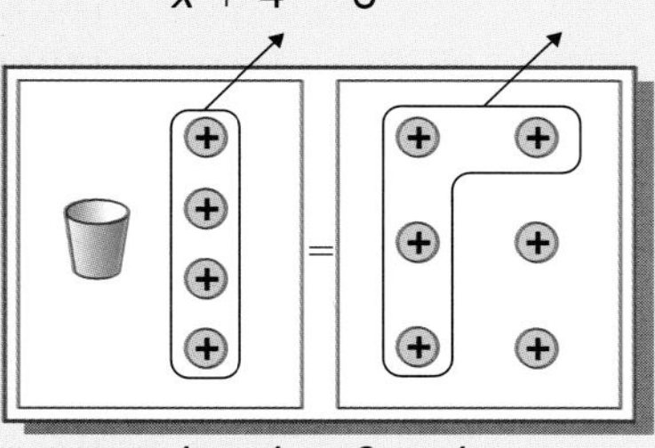

$x + 4 - 4 = 6 - 4$

If you remove four counters from each side of the mat, then you can see that the cup is equal to two counters. So, $x = 2$. Since $2 + 4 = 6$, our solution is correct.

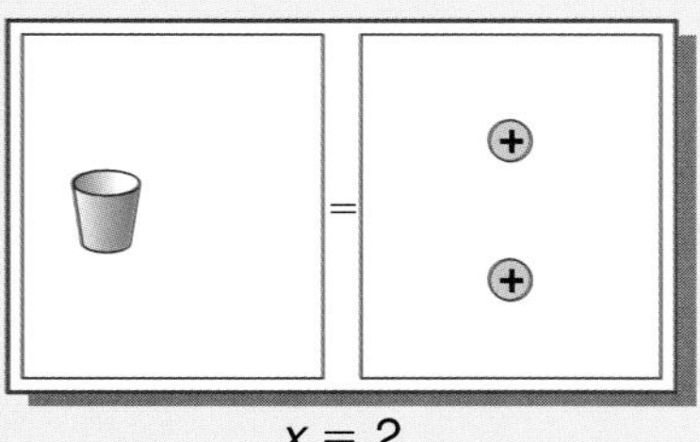

$x = 2$

Activity 2

LOOK BACK

You can review adding integers with counters in Lesson 2-4A.

Some equations involve subtraction. The mat at the right represents $7 = y - 5$. To get the cup by itself, remove 5 negative counters from each side. Since there are no negative counters on the left side, add 5 positive counters to each side to make 5 zero pairs on the right side.

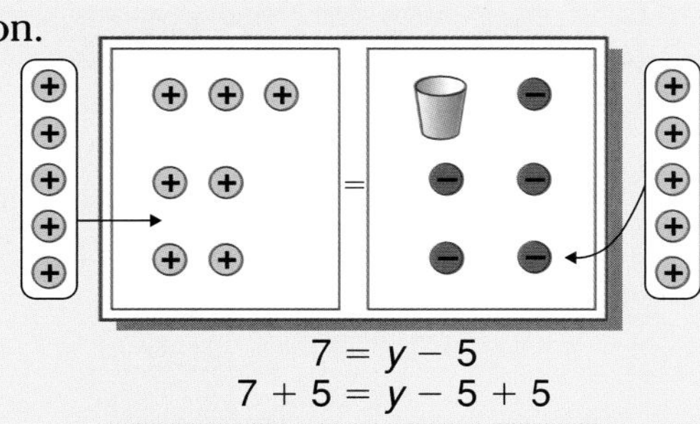

$7 = y - 5$
$7 + 5 = y - 5 + 5$

Now remove the zero pairs. The remaining cup and counters show that $y = 12$. Check by substituting 12 for y in the original equation.

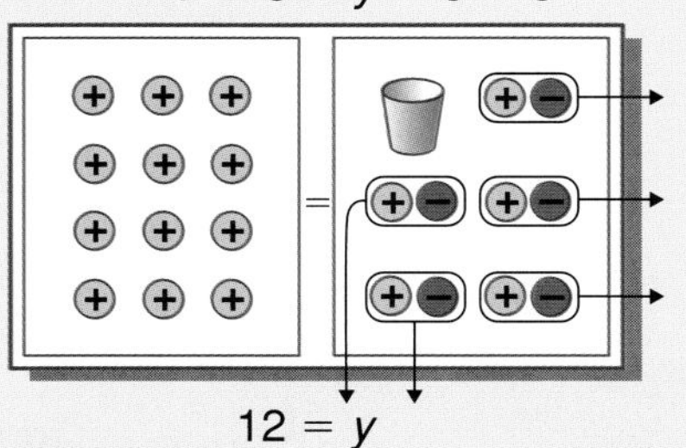

$12 = y$

Your Turn **Use cups and counters to model and solve each equation.**

1. $3 + x = 7$	**2.** $n + 6 = 9$	**3.** $y + 15 = 18$
4. $x + 6 = 10$	**5.** $t + 2 = 9$	**6.** $4 + w = 5$
7. $x - 2 = 10$	**8.** $n - 4 = 9$	**9.** $6 = q - 5$

3-2 Solving Equations by Adding or Subtracting

Setting Goals: *In this lesson, you'll solve equations by using the addition and subtraction properties of equality.*

Modeling with Technology

See page 1 for information on programming the T1-82.

Graphing calculators are computers, so you can run programs. The T1-82 program at the right allows you to enter an equation of the form $x + a = b$ or $x - a = b$ and guess the solution. Using this program and eliminating possibilities, you can solve equations.

```
PROGRAM: ONESTEP
: Disp "PRESS 1: X + A = B"
: Input "PRESS 2: X – A =
  B", N
: Input "ENTER A", A
: Input "ENTER B", B
: Lbl 1
: Input "GUESS X", X
: If N = 1 and X + A = B or
  N = 2 and X – A = B
: Then
: Disp "CORRECT"
: Stop
: Else
: Disp "TRY AGAIN"
: Goto 1
```

Your Turn Use the program to choose the solution of each equation from the possible values. Record the solutions.

$x + 5 = 9$	$x = 1, 3, 14, 4$
$x - 9 = 6$	$x = 3, 6, 15, 18$
$x + 7 = 8$	$x = 1, 7, 8, 15$
$x - 2 = 14$	$x = 2, 12, 14, 16$
$x + 10 = 15$	$x = 2, 5, 10, 20$
$x - 8 = 8$	$x = 0, 16, 1, 64$

TALK ABOUT IT

a. What do you observe about the solutions of equations like $x + a = b$?

b. What do you observe about the solutions of equations like $x - a = b$?

c. Solve $x + 3 = 12$ and $x - 5 = 15$. Use the program to verify the solutions.

Learning the Concept

Consider $x + 10 = 14$. One way to solve is to mentally find the number that makes the mathematical sentence true. Since $4 + 10 = 14$, the solution is 4.

Not all equations are easy to solve mentally. You can subtract to isolate the variable. If you subtract 10 from the left side of the equation, you must also subtract 10 from the right side to keep the equation balanced.

$$x + 10 = 14$$
$$x + 10 - 10 = 14 - 10 \quad \textit{Subtract 10 from each side.}$$
$$x + 0 = 4 \quad \textit{10 − 10 = 0 and 14 − 10 = 4.}$$
$$x = 4$$

These are **equivalent equations** because they all have the same solution, 4. We used the **subtraction property of equality** to solve this equation.

Subtraction Property of Equality

In words: If you subtract the same number from each side of an equation, the two sides remain equal.

In symbols: For any numbers a, b, and c, if $a = b$, then $a - c = b - c$.

Example 1 Solve $r + 56 = -5$.

$$r + 56 = -5$$
$$r + 56 - 56 = -5 - 56 \quad \textit{Subtract 56 from each side.}$$
$$r = -61 \quad \textit{Check your solution by replacing r with } -61.$$

Some equations can be solved by using the **addition property of equality**.

Addition Property of Equality

In words: If you add the same number to each side of an equation, the two sides remain equal.

In symbols: For any numbers a, b, and c, if $a = b$, then $a + c = b + c$.

Example 2 Solve $y - 9 = -17$.

$$y - 9 = -17$$
$$y - 9 + 9 = -17 + 9 \quad \textit{Add 9 to each side.}$$
$$y = -8 \quad \textit{Check your solution.}$$

You can often use an equation to represent a situation in real life.

Example 3 APPLICATION Entertainment

Forrest Gump **is the only one of the top ten money-making films to win an Oscar for Best Picture. It earned $317 million at the box office. This is $83 million less than *E.T. the Extra-Terrestrial*, which was the all-time top money maker. How much did *E.T.* earn?**

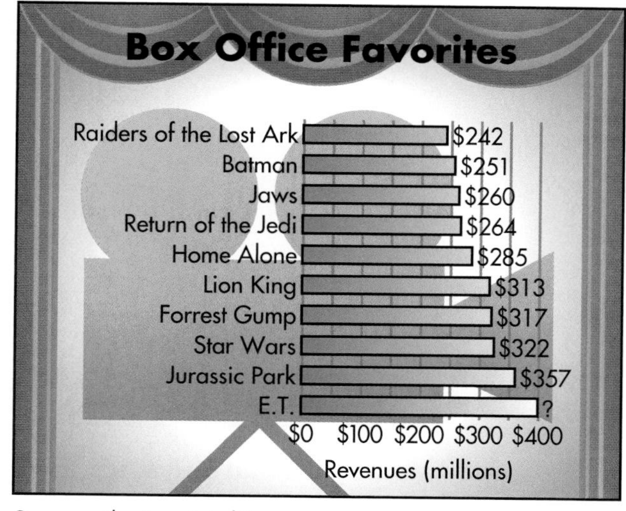

Source: The Associated Press

Write an equation.

Forrest Gump revenue	*equals*	*E.T. revenue*	*minus*	*$83 million*
317	=	r	−	83

$$317 + 83 = r - 83 + 83 \quad \textit{Add 83 to each side.}$$
$$400 = r \quad \textit{Check your solution.}$$

E.T. the Extra-Terrestrial earned $400 million at the box office.

LOOK BACK You can review additive inverses in Lesson 2-5.

Recall that to subtract an integer, you can add its inverse.

subtraction — additive inverse — addition

$8 - (-2) = 10$ $\qquad$ $8 + 2 = 10$

same result

You can also use the additive inverse to solve equations involving addition.

Example  **Solve $c + 8 = 12$.**

Method 1: Subtracting

$$c + 8 = 12$$
$$c + 8 - 8 = 12 - 8$$
$$c = 4$$

Method 2: Adding Inverse

$$c + 8 = 12$$
$$c + 8 + (-8) = 12 + (-8)$$
$$c + 0 = 4$$
$$c = 4$$

Check: $c + 8 = 12$

$4 + 8 \stackrel{?}{=} 12$ *Replace c with 4.*

$12 = 12$ ✓ The solution is 4.

Solutions to equations can be represented on a *number line*. Each solution is a **coordinate** of a point. The coordinate tells its distance and direction from the 0-point on the line. The dot marking the point is called the **graph** of the number.

The solution in the example above, 4, is represented by placing a dot above 4 on a number line.

−1 0 1 2 3 4 5 6

Example  **Solve $y - 24 = 16$. Graph the solution on a number line.**

$$y - 24 = 16$$
$$y - 24 + 24 = 16 + 24$$ *Add 24 to each side.*
$$y = 40$$ The solution is 40.

Graph the solution, 40, on a number line.

35 36 37 38 39 40 41 42 43

Checking Your Understanding

Communicating Mathematics

Read and study the lesson to answer these questions.

1. **Explain** when you use the addition property of equality to solve an equation.
2. **Tell** what property you would use to solve $x + 34 = -6$.
3. **Write** two equations that are equivalent. Then write two equations that are not equivalent.
4. **Write** an equation in the form $x - a = b$ where the solution is -12.
5. **You Decide** Ellen and Tertius solved $x - (-25) = 56$ in two different ways. Who is correct? Explain.

Ellen

$$x - (-25) = 56$$
$$x + 25 = 56$$
$$x + 25 - 25 = 56 - 25$$
$$x = 31$$

Tertius

$$x - (-25) = 56$$
$$x - (-25) + (-25) = 56 + (-25)$$
$$x = 56 - 25$$
$$x = 31$$

MODELING MATHEMATICS

MATERIALS

cups and counters

6. Use cups and counters to model the solution to $9 = w + (-5)$.

Guided Practice

Solve each equation and check your solution. Then graph the solution on a number line.

7. $w + 8 = -17$ **8.** $25 = -4 + y$ **9.** $m - (-3) = 40$

10. $y + (-7) = 9$ **11.** $k - 36 = -37$ **12.** $h - 8 = -22$

13. Entertainment *Star Wars* and its sequel, *Return of the Jedi*, were both blockbusters. Refer to the graph in Example 3 for the revenue earned by each of these movies. How much more did *Star Wars* earn?

Exercises: Practicing and Applying the Concept

Independent Practice

Solve each equation and check your solution. Then graph the solution on a number line.

14. $y + 7 = 21$ **15.** $k - 6 = 32$

16. $19 = g - 5$ **17.** $-24 = h - 22$

18. $x + 49 = 13$ **19.** $f + 34 = 2$

20. $d - (-3) = 2$ **21.** $a - (-26) = 5$

22. $12 + p = -14$ **23.** $x + 18 = 14$

24. $59 = s + 95$ **25.** $x - 27 = 63$

26. $-7 = z - (-12)$ **27.** $34 + r = 84$

28. $-234 = x + 183$ **29.** $y - 94 = 562$

30. $-591 = m - (-112)$ **31.** $846 + t = -538$

Choose the equation whose solution is graphed.

32. 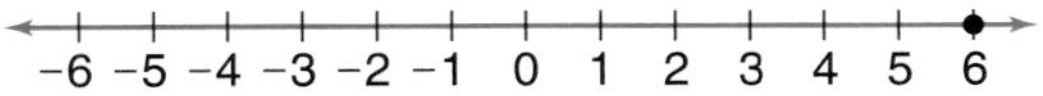

a. $x + 5 = -11$ **b.** $x - 5 = 11$ **c.** $x + 5 = 11$ **d.** $x - 5 = -11$

33. 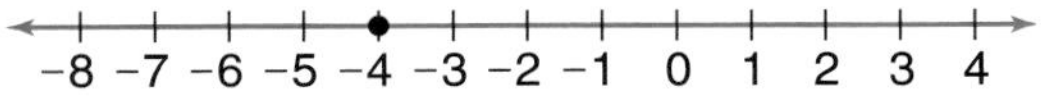

a. $-8 = t - 4$ **b.** $-8 = t + 4$ **c.** $8 = t - 4$ **d.** $8 = t + 4$

Solve each equation. Check each solution.

34. $(f + 5) + (-2) = 6$ **35.** $[y + (-3)] + 2 = 4$

36. $16 = [n - (-2)] + (-3)$ **37.** $-10 = [b + (-4)] + 2$

Critical Thinking

38. Write an equation whose solution is represented by the number line.

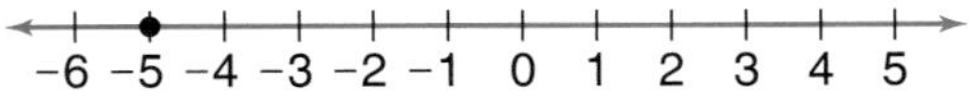

Applications and Problem Solving

39. Advertising Morris has been the official mascot for 9-Lives® Cat Food for 26 years. Actually, there have been two cats named Morris. Morris I died unexpectedly at the age of 19. If he was the official mascot for 14 years, how long has Morris II been the official mascot?

40. Meteorology Weather fronts usually move across the United States from west to east and can bring with them astounding changes in temperatures.

a. On January 23–24, 1916, the temperature in Browning, Montana, dropped 100° in 24 hours. If the temperature started at 44°F, how cold did it get?

b. In 1892, at Fort Assinaborn, Montana, the temperature rose 43° in 15 minutes. At 2:00 A.M., the temperature was −5°F. If that is when the temperature began to rise, what was the temperature at 2:15 A.M.?

41. Illustration In 1995, Charles Schulz, the creator of Peanuts comic strip, was 70 years old. He created Peanuts in 1950. How old was he when he started drawing Peanuts?

Mixed Review

42. Travel On a trip to Florida, a bus traveled 600 miles in 10 hours. At that rate, how far could the bus travel in 4 hours? (Lesson 3-1)
a. 2400 miles **b.** 400 miles **c.** 240 miles **d.** 150 miles

43. Solve $y = (-9)(18)$. (Lesson 2-7)

44. Solve $7 - 25 = h$. (Lesson 2-5)

45. Order the integers in the set $\{3, -1, 15, -3\}$ from least to greatest. (Lesson 2-3)

46. Geometry On graph paper, draw coordinate axes. Then graph and label $N(4, 6)$, $P(-3, 7)$, and $Q(4, -6)$. Name the quadrant in which each point is located. (Lesson 2-2)

47. Exercise A 150-pound person burns 384 Calories an hour jogging at a rate of 6 miles per hour. That is 120 Calories more than the same person burns playing volleyball for one hour. How many Calories does a 150-pound person burn playing volleyball for one hour? (Lesson 1-8)

a. Define a variable for this problem.

b. Write an equation.

c. Solve for the number of Calories burned in an hour of volleyball.

48. Find the value of $7 + 8 \div 2 \cdot 6 - 1$. (Lesson 1-2)

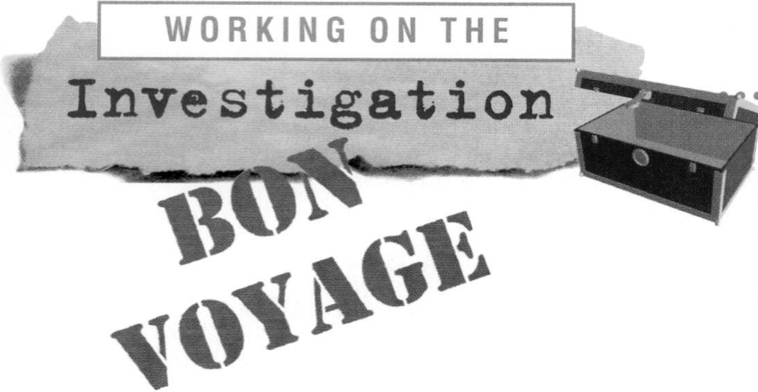

Refer to the Investigation on pages 62–63.

Look over the list of considerations that you placed in your Investigation Folder at the beginning of Chapter 2.

You probably listed weather as one thing to consider before packing clothes. You don't want to pack shorts and sandals if temperatures are going to be in the 20s—or do you?

Unless you selected Great Britain, Australia, New Zealand, or Ireland, as your destination, your hosts will refer to temperatures measured in *degrees Celsius.*

Like the Fahrenheit scale, the Celsius scale has two fixed points: the boiling point of water and the freezing point of water (or the melting point of ice).

On the Celsius scale, the boiling point of water was set at 100 and the freezing point of water was set at 0 so that the scale would have 100 equal divisions (called degrees Celsius) between them.

If you've grown up using the Fahrenheit scale, you can probably translate temperatures into "feel". You know, for example, that 85°F feels very warm.

- Work with your partner to choose at least four reference Fahrenheit temperatures that you "feel."
- Find a thermometer that has both scales on it. Use it to estimate the equivalent Celsius temperature for each Fahrenheit temperature you selected.
- If you haven't already done so, obtain environmental data for the country you and your partner selected at the beginning of the investigation. Then provide general guidelines for your wardrobe.

Add your results to your Investigation Folder.

3-3 Solving Equations by Multiplying or Dividing

Setting Goals: *In this lesson, you'll solve equations by using the multiplication and division properties of equality.*

Modeling a Real-World Application: Toys

Popular toys are often created by individuals who have a simple idea. Richard James came up with the idea for the Slinky® toy in 1943 when he watched a spring fall from his desk to the floor. He introduced the first Slinky in 1945 at the annual Toy fair. The Slinky is a wiry coil, that with a nudge, appears to walk down the stairs. After more than fifty years, the Slinky is still a "walking" success.

Millions of Slinkys have been sold over the past 50 years. They are still made using the same machines designed years ago. One Slinky is produced every 8 seconds, 24 hours a day. How many Slinkys are made each day?

Learning the Concept

To find out how many Slinkys are made in a day, translate what we know into an equation.

First find the number of seconds in a day. There are 60 seconds in a minute, 60 minutes in an hour, and 24 hours in a day.

$$\left(\frac{60\text{ s}}{\cancel{\text{min}}}\right)\left(\frac{60\ \cancel{\text{min}}}{\cancel{\text{hr}}}\right)\left(\frac{24\ \cancel{\text{hr}}}{\text{day}}\right) = 86{,}400\ \frac{\text{s}}{\text{day}}$$

Let d represent the number of Slinkys made in a day. If each Slinky takes 8 seconds to make, we can write an equation to represent the daily production of Slinkys.

$$\underbrace{8}_{\text{seconds per Slinky}} \quad \underbrace{\times}_{\text{times}} \quad \underbrace{d}_{\text{Slinkys made in a day}} \quad \underbrace{=}_{\text{equals}} \quad \underbrace{86{,}400}_{\text{seconds in a day}}$$

To solve this equation, undo the multiplication by dividing each side by the same number.

$8d = 86{,}400$

$\frac{8d}{8} = \frac{86{,}400}{8}$ *Divide each side by 8 to undo the multiplication $8 \cdot d$.*

$d = 10{,}800$

Check:

$8d = 86{,}400$

$8(10{,}800) \stackrel{?}{=} 86{,}400$ *Replace d with 10,800.*

$86{,}400 = 86{,}400$ ✓ The solution is 10,800.

Therefore, 10,800 Slinkys are made in a day.

Division Property of Equality

In words: If you divide each side of an equation by the same nonzero number, the two sides remain equal.

In symbols: For any numbers a, b, and c, where $c \neq 0$, if $a = b$, then $\frac{a}{c} = \frac{b}{c}$.

Example 1 **Solve $-68 = -4m$. Check the solution and graph it on a number line.**

$-68 = -4m$

$\frac{-68}{-4} = \frac{-4m}{-4}$ *Divide each side by −4.*

$17 = m$ *−68 ÷ (−4) = 17*

Check: $-68 = -4m$

$-68 \stackrel{?}{=} -4(17)$ *Replace m with 17.*

$-68 = -68$ ✓

The solution is 17.

13 14 15 16 17 18 19 20

Just as division can be used to solve equations in some situations, multiplication can also be used to solve equations. This makes use of the multiplication property of equality.

Multiplication Property of Equality

In words: If you multiply each side of an equation by the same number, the two sides remain equal.

In symbols: For any numbers a, b, and c, if $a = b$, then $a \cdot c = b \cdot c$.

Example 2 **Solve $\frac{d}{3} = -3$. Check the solution and graph it on a number line.**

$\frac{d}{3} = -3$

$\frac{d}{3}(3) = -3(3)$ *Multiply each side by 3 to undo the division in $\frac{d}{3}$.*

$d = -9$

Check: $\frac{d}{3} = -3$

$\frac{-9}{3} \stackrel{?}{=} -3$ *Replace d with −9.*

$-3 = -3$ ✓

The solution is −9.

−12 −11 −10 −9 −8 −7 −6 −5

Example 3 Sports

In the 1993 season, the fans at the Baltimore Orioles' Camden Yards bought 232,618 bags of peanuts at ball games. There were 80 games in the park that season.

a. On average, how many bags of peanuts were sold at each game?

b. How many bags of peanuts should the concession manager order for the 1997 season if there will be 86 games in the park that year?

The tradition of the seventh-inning stretch started in 1910, when President William Taft stood to stretch his legs during a game in Washington, D.C. The fans stood up out of respect because they thought he was leaving.

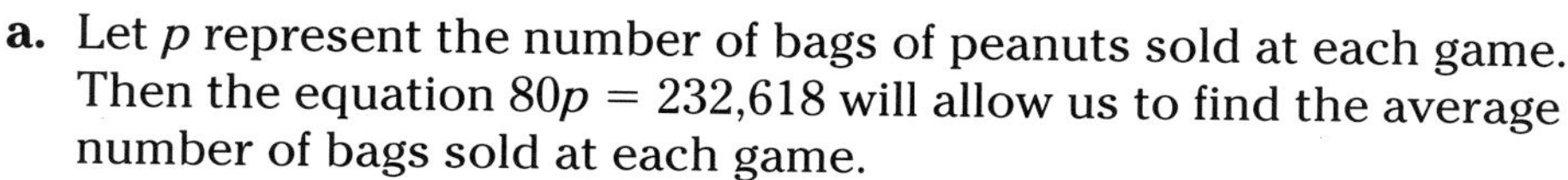

a. Let p represent the number of bags of peanuts sold at each game. Then the equation $80p = 232{,}618$ will allow us to find the average number of bags sold at each game.

$$80p = 232{,}618$$

$$\frac{80p}{80} = \frac{232{,}618}{80} \quad \textit{Divide each side by 80.}$$

$$p = \frac{232{,}618}{80} \quad \textit{Estimate: } 240{,}000 \div 80 = 3000$$

Use a calculator to find p.

232618 [÷] 80 [=] 2907.725

On average, about 2908 bags of peanuts were sold at each game.

b. You can estimate to find out how many bags of peanuts the manager should order. Round to 3000 bags per game and 90 games in the season. So, the manager should order 3000×90 or 270,000 bags of peanuts for a season with 86 games.

Checking Your Understanding

Communicating Mathematics

Read and study the lesson to answer these questions.

1. **In your own words, explain** how you would solve $\frac{x}{4} = -3$.
2. **Compare and contrast** solving equations using the multiplication property of equality and solving equations using the division property of equality.
3. **Write** an equation in the form $ax = c$ where the solution is -5.
4. In an equation of the form $\frac{x}{a} = b$, explain why a cannot be 0.

Guided Practice

Solve each equation and check your solution. Then graph the solution on a number line.

5. $5x = 45$
6. $8y = -64$
7. $-\frac{p}{3} = 4$
8. $\frac{a}{-5} = 8$
9. $-3g = -51$
10. $-2 = \frac{p}{6}$

CONNECTION

For the latest exchange rates, visit: **www.glencoe.com/sec/math/prealg/mathnet**

11. Traveling Paloma's social studies class was planning an imaginary trip to Kenya. Paloma had to find out about the rate of exchange for money in Kenya. She learned that money in Kenya is based on shillings and that the current exchange rate was one U.S. dollar for 40 shillings. If one night at a hotel cost 960 shillings, how much would that be in U.S. dollars?

Exercises: Practicing and Applying the Concept

Independent Practice

Solve each equation and check your solution. Then graph the solution on a number line.

12. $8x = 72$	**13.** $-5y = 95$	**14.** $\frac{k}{-2} = 7$
15. $-\frac{h}{7} = 20$	**16.** $86 = 2v$	**17.** $-21 = -\frac{g}{8}$
18. $\frac{a}{45} = -3$	**19.** $\frac{b}{-9} = -4$	**20.** $672 = -21t$
21. $-56 = -7p$	**22.** $-116 = -4u$	**23.** $-\frac{y}{11} = 132$
24. $\frac{f}{-34} = -14$	**25.** $\frac{y}{8} = 117$	**26.** $17r = -357$
27. $-18p = 306$	**28.** $-144 = 8x$	**29.** $-384 = -3m$
30. $\frac{b}{46} = 216$	**31.** $-71 = \frac{x}{31}$	**32.** $-171 = \frac{x}{-12}$
33. $584 = -\frac{s}{23}$	**34.** $\frac{p}{47} = 123$	**35.** $-67w = -5561$

Graphing Calculator

36. You can check your solutions to equations using a T1-82 graphing calculator. Enter one side of the equation as an expression using your solution and see if the value is the same as the other side of the equation.

For example, if $\frac{x}{5} = -2$, then $x = -10$.

Enter: [(–)] 10 [STO▸] [X,T,θ] [ENTER] *Enters the value of x.*

[X,T,θ] [÷] 5 [ENTER] -2 *Enters equation and evaluates it.*

The solution checks.

Solve each equation. Then check your solution using a graphing calculator.

a. $3x = 126$ **b.** $\frac{x}{6} = -24$ **c.** $-4x = 88$

Critical Thinking

37. Write a word problem that can be solved using the equation $4x = 13$.

Applications and Problem Solving

38. Manufacturing One Slinky is made from 80 feet of wire. If a standard spool of wire contains 4000 feet, how many Slinkys can be made from one spool?

39. **Biology** The mola is an interesting fish because it appears to have no body. Weighing more than two tons, it looks like a large head with two perpendicular fins. Female molas can have 300 million eggs. This is 6 to 15 times the number in most species of fish.

   a. Write an equation to represent 6 times the number of eggs is 300 million eggs.

   b. Write an equation to represent 15 times the number of eggs is 300 million eggs.

   c. Write a sentence that states the range of eggs produced by most species of fish.

40. **Finance** Refer to Exercise 11. On the trip to Kenya, the students in Paloma's class wanted to take a safari to the Masai Mara. They estimated that the cost of the safari would be $300 a person.

   a. Research to find out how many shillings are exchanged for a U.S. dollar today.

   b. Write an equation to find out how much the safari would cost in shillings.

   c. Solve the equation.

41. **Geometry** The perimeter of any square is 4 times the length of one of its sides. If the perimeter of a square is 72 centimeters, what is the length of each side of the square?

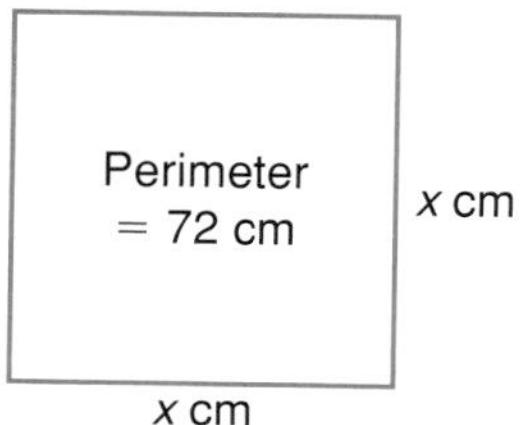

Mixed Review

42. Solve $121 = k - 34$ and check your solution. Then graph the solution on a number line. (Lesson 3-2)

43. If a number is multiplied by 4 and the result is subtracted from 52, the difference is -12. What is the original number? (Lesson 3-1)

   a. -3 b. 0 c. 12 d. 16

44. Find the product of -7 and -15. (Lesson 2-7)

45. Simplify $|-5| + |12|$. (Lesson 2-1)

46. State whether the inequality $7 > 2$ is *true*, *false*, or *open*. (Lesson 1-9)

47. Solve $7n = 70$ mentally. (Lesson 1-6)

48. **Careers** Curt Harris received a raise. He now makes $150 less than twice his old salary. Write an expression for Mr. Harris's current salary if his old salary was s. (Lesson 1-3)

49. **Shopping** Is $7 enough money to buy a loaf of bread for $0.98, a pound of hamburger for $2.29, and a pound of roast turkey for $3.29? (Lesson 1-1)

   a. Which method of computation do you think is most appropriate? Explain.

   b. Solve the problem using the four-step plan.

Estimation
Mental Math
Calculator
Paper and Pencil

3-4 Using Formulas

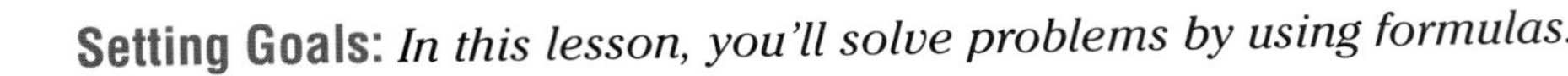

Setting Goals: *In this lesson, you'll solve problems by using formulas.*

Modeling a Real-World Application: Travel

You slide behind the wheel of a red convertible. The engine roars as you squeal away down the road. With the top down, the sound of the whipping wind drowns out the world until suddenly the engine stops.

"Just how far back was that gas station? Wish I would have figured the gas mileage on this thing before I left home."

You can use a formula to compute gas mileage.

Learning the Concept

A **formula** shows the relationship among certain quantities. The formula below can be used to find the miles per gallon achieved by a car.

number of miles driven	*divided by*	*number of gallons of gas*	*equals*	*miles per gallon*
m	$\div$	g	$=$	mpg

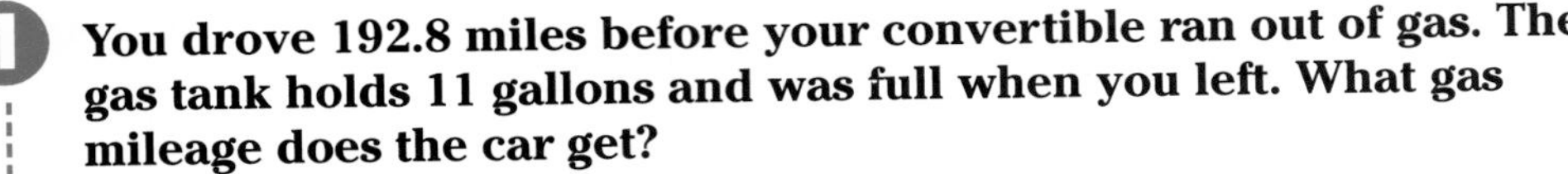

Example 1 **You drove 192.8 miles before your convertible ran out of gas. The gas tank holds 11 gallons and was full when you left. What gas mileage does the car get?**

$m \div g = \text{mpg}$

$192.8 \div 11 = \text{mpg}$ *Replace m with 192.8 and g with 11.*

192.8 [÷] 11 [=] 17.527272 *Estimate: 200 ÷ 10 = 20*

The car gets about 17.5 miles per gallon.

Connection to Geometry

There are formulas associated with almost every field of study. Example 2 uses a formula from geometry.

Example 2

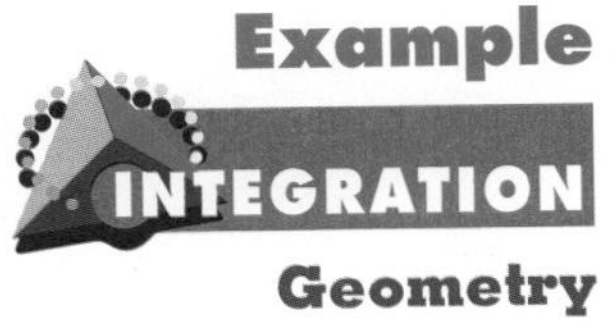

Geometry

A pentagon is a polygon with five sides. In a regular pentagon, all of the sides are the same length, and all of the angles have the same measure. The area, A, of a regular pentagon can be found using the formula $A = \frac{5}{2}sa$, where s is the length of a side of the pentagon and a is the length of the apothem. Find the area of the pentagon shown at the right.

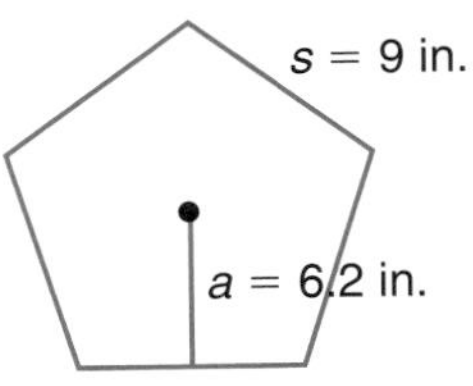

$$A = \frac{5}{2}sa$$

$$A = \frac{5}{2}(9)(6.2)$$ *Replace s with 9 and a with 6.2.*

$$A = 139.5$$

The area of the pentagon is 139.5 square inches.

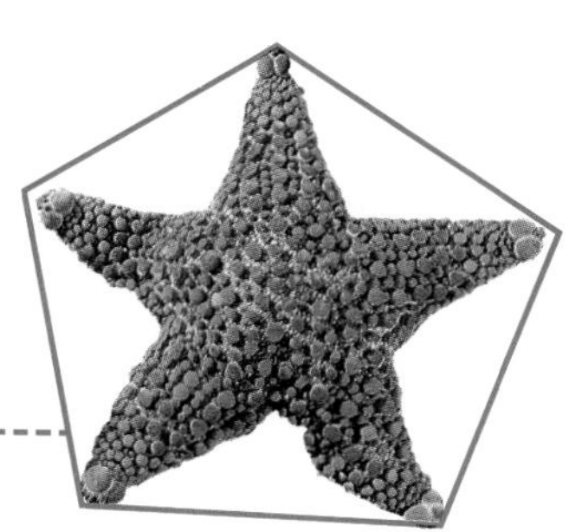

Checking Your Understanding

Communicating Mathematics

Read and study the lesson to answer these questions.

1. **Explain** how to use a formula.
2. **Write** a formula that you have used in your life.
3. **Draw** polygons with the numbers of sides in the table below. Pick a vertex and draw all the diagonals from this vertex. Count the number of triangles formed. Copy and complete the table.

MODELING MATHEMATICS

MATERIALS

ruler

Number of sides	3	4	6	8	10
Number of triangles					

Can you find a formula that will predict how many triangles will be formed in a polygon with 100 sides? 1250 sides? x sides?

Guided Practice

Solve by replacing the variables with the given values.

4. $d = rt$, if $d = 315$, $r = 45$
5. $s = l - d$, if $s = \$35$, $d = \$10$

Solve. Use the correct formula.

6. **Travel** Find the miles per gallon for a car that travels 271.7 miles on 9.6 gallons of fuel.
7. **Transportation** The formula $d = rt$ relates distance, d, rate, r, and time, t, traveled. Find the distance you travel if you drive at 55 miles per hour for 3 hours.
8. **Sports** In league bowling, each team has an equal chance of winning because team members are given a handicap based on their averages. They use a formula to find the handicap score. The formula is as follows.

   handicap score(s) = game score(g) + handicap(h) or $s = g + h$

   Monique has a game score of 145 and a handicap of 35. Dennis has a game score of 153 and a handicap of 30. Who has the higher handicap score?

Exercises: Practicing and Applying the Concept

Independent Practice

Solve by replacing the variables with the given values.

9. $d = rt$, if $r = 120$ and $t = 3$
10. $F = \frac{9}{5}C + 32$, if $C = 30$
11. $P = 5s$, if $P = 65$
12. $h = 69 + 2.2F$, if $F = 14$
13. $d = 2r$, if $r = 1.7$
14. $A = \frac{b + d}{2}$, if $b = 11$ and $d = 13$

Solve. Use the correct formula.

15. **Ballooning** What is the speed in miles per hour of a balloon that travels 56 miles in 4 hours?
16. **Transportation** A 1992 Ford Taurus can travel an average of 464 miles on one tank of gas. If the tank holds 16 gallons, how many miles per gallon does it get?
17. **Travel** The Flynn family plans to drive 600 miles from Harrisburg, Pennsylvania to Chicago, Illinois for their summer vacation. The speed limit on the highways they plan to use is 55 miles per hour. How long will the trip take if they don't exceed the speed limit?
18. **Air Travel** How long does it take an Air Force jet fighter to fly 5200 miles at 650 miles per hour?

Translate each sentence into a formula.

19. The sale price of an item, s, is equal to the list price, ℓ, minus the discount, d.
20. **Geometry** In a circle, the diameter, d, is twice the length of the radius, r.
21. **Geometry** The area, A, of a triangle is equal to the product of the length of the base, b, and the height, h, divided by two.

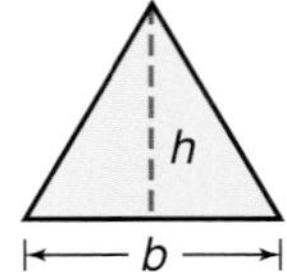

   a. What is the area of a triangle with a base of 12 inches and a height of 5 inches?
   b. What is the length of the base of a triangle with an area of 24 square centimeters and a height of 8 centimeters?

Critical Thinking

22. Measure the height and head circumference of five people. Use these measures to write a formula to predict a person's height from the size of their head.

Applications and Problem Solving

23. **Jobs** Dorinda Pinel found a job about 8 miles from home. For a few weeks, she rode the bus to and from work. But the connections were poor, and it took an hour to get home in the afternoon. She thought that she could get home earlier if she rode her bicycle to and from work. Use a calculator to answer the following questions.
   a. If she rode at an average speed of 10 miles an hour, how long would it take for her to get home?
   b. The trip to work was uphill most of the way. It took Dorinda 54 minutes, or 0.9 hours, to get to work. How fast was she riding?
24. **Physical Therapy** Physical therapists work to help people regain strength after surgery or an accident. The formula they use to determine a person's arm strength, S, is as follows.

$$S = \frac{D + P}{\frac{W}{10} + H - 60}$$

Find the arm strength of a person with the following information.

D(dips on a parallel bar)	= 4 dips
P(pull-ups)	= 8 pull-ups
W(weight in pounds)	= 120 pounds
H(height in inches)	= 60 inches

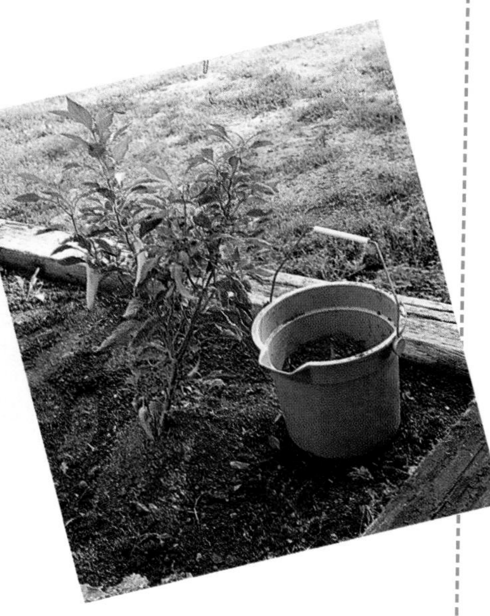

Example 3 Sumi has just come home from the plant nursery with 20 outdoor plants. The man at the plant nursery told her that each of these particular types of plants needs 24 square feet of space to grow, so that it won't be choked off by its neighbors.

a. How many square feet does Sumi need in a garden plot so all the plants have enough room?

$$\underbrace{24}_{\text{square feet per plant}} \underbrace{\cdot}_{\text{times}} \underbrace{20}_{\text{number of plants}} \underbrace{=}_{\text{equals}} \underbrace{480}_{\text{square feet}}$$

b. Sumi has decided to use old railroad ties to enclose a rectangular space along the side of her house for the plants. The space is 16 feet wide, so how long should the garden be?

$A = \ell \cdot w$	*formula for area of a rectangle*
$480 = \ell \cdot 16$	*Replace A with 480 and w with 16.*
$\frac{480}{16} = \frac{\ell \cdot 16}{16}$	*Divide each side by 16.*
$30 = \ell$	The length should be 30 feet.

Example 4

Real Estate

Real estate agents help their clients choose a home that meets their needs. One of the ways that real estate agents compare homes is by their total square footage of living space. Find the square footage of each room of the home whose floor plan is shown at the right.

Room	$\ell \times w$	Area
master bedroom	12.6×15.2	191.52
family room	15×21.3	319.5
dining room	12×9.2	110.4
living room	12.5×12.3	153.75
den/bedroom	10×12.2	122
bedroom 2	11.6×10.3	119.48

Checking Your Understanding

Communicating Mathematics

Read and study the lesson to answer these questions.

1. **Restate** the definition of perimeter in your own words.
2. **Compare and contrast** the perimeter and area of a rectangle
3. **Draw and label** a rectangle that has a perimeter of 18 inches.
4. **Explain** how to find the perimeter and area of a rectangle that is 9 cm long and 5 cm wide.

Guided Practice

Find the perimeter and area of each rectangle.

5.

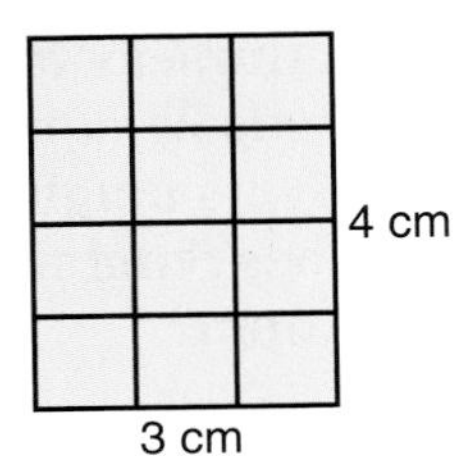

6.

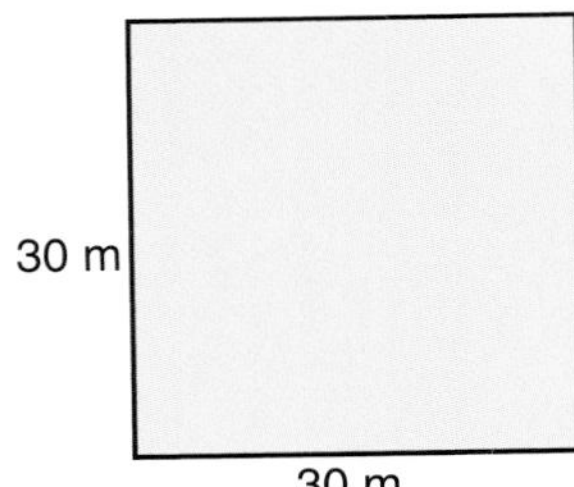

7. rectangle with length of 15 feet and a width of 6 feet

Find the missing dimension of each rectangle.

8.

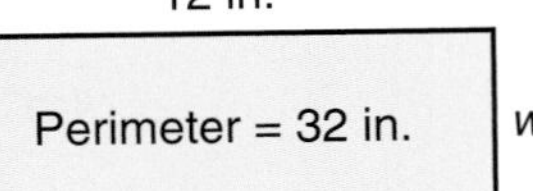

9.

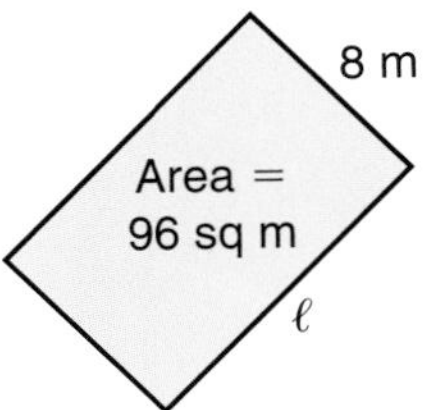

10. Community Project A neighborhood uses an empty city lot for a community vegetable garden. Each participant is allotted a space of 18 feet by 90 feet. What is the perimeter and area of each plot?

Exercises: Practicing and Applying the Concept

Independent Practice

Find the perimeter and area of each rectangle.

11.

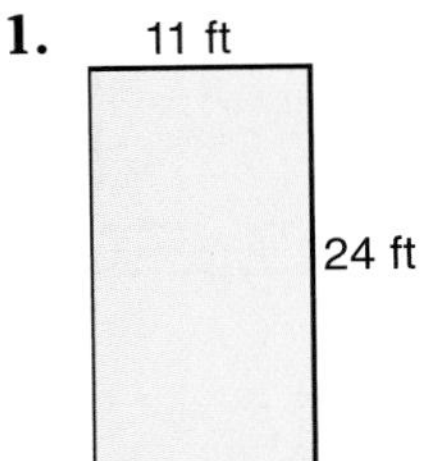

12.

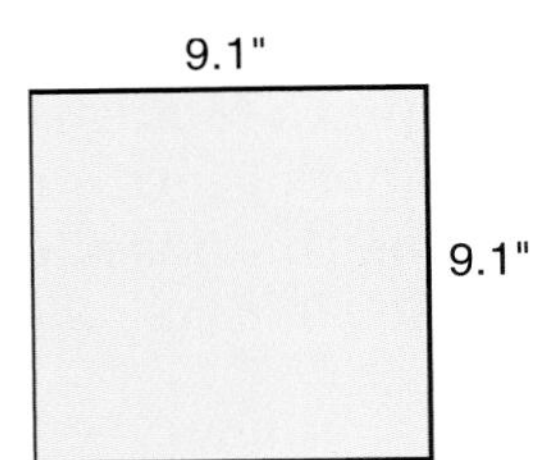

13.

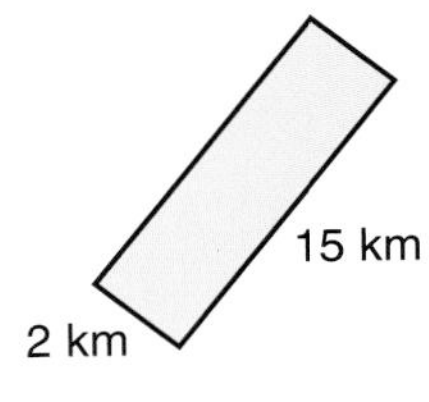

14.

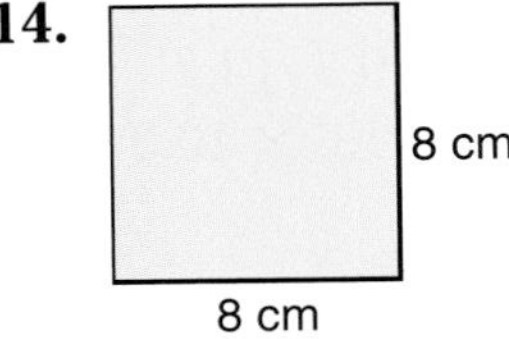

15.

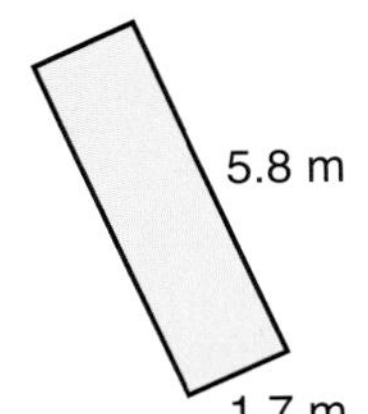

16.

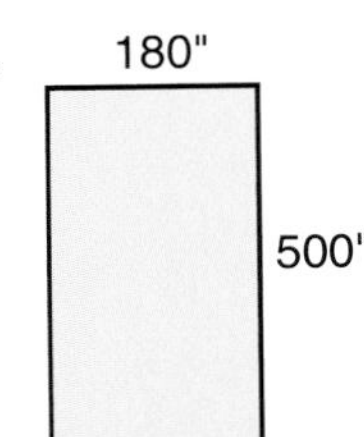

17.

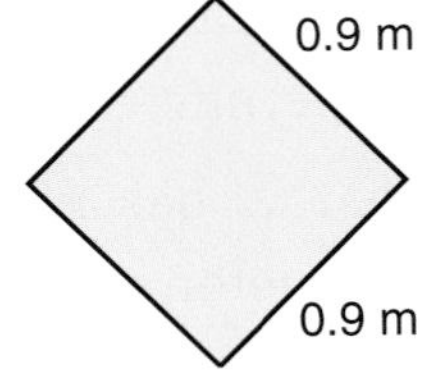

18.

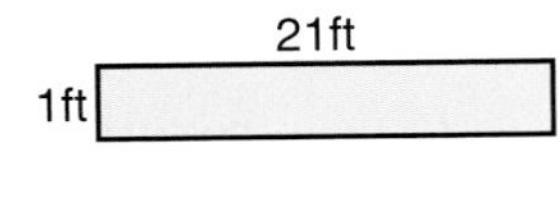

19.

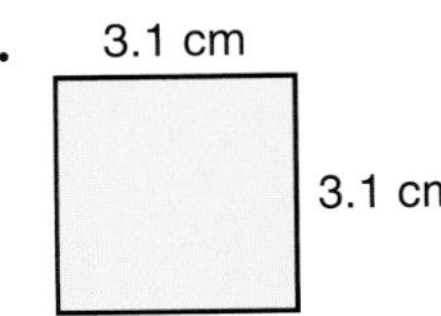

20. a rectangle that is 3.8 meters long and 1.1 meters wide

21. a square that is 4.8 millimeters on each side

22. an $8\frac{1}{2}$-inch by 11-inch rectangle

Find the missing dimension of each rectangle.

	Length	Width	Perimeter	Area
23.	7 m		24 m	35 sq m
24.		15 cm	66 cm	270 sq cm
25.	16 yd		54 yd	176 sq yd
26.	11 km		70 km	264 sq km
27.		12 ft	102 ft	468 sq ft
28.		91 yd	216 yd	1547 sq yd

Find the area of the blue part of each rectangle.

29.

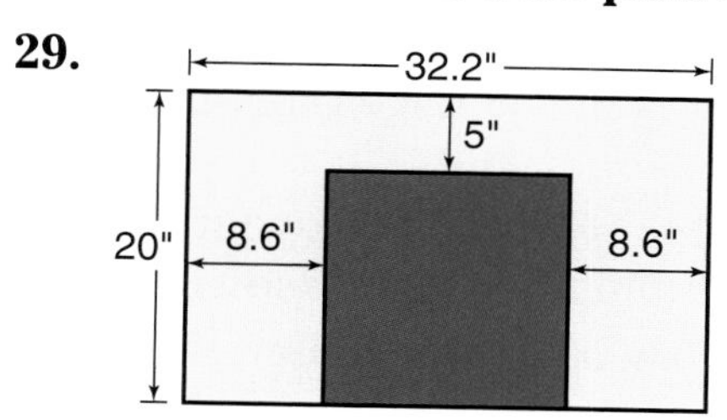

30.

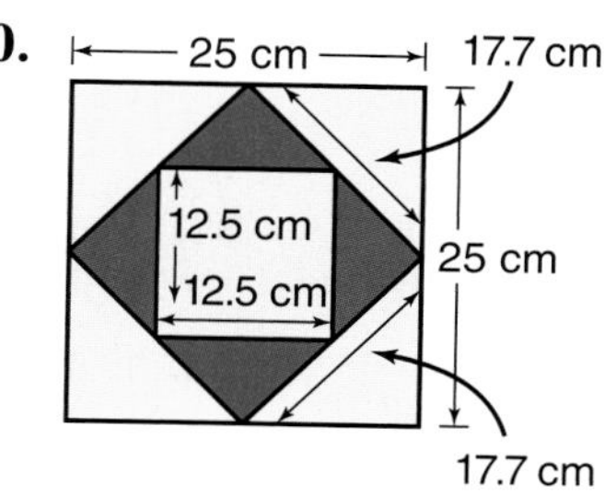

Critical Thinking

31. Copy and complete the following table that compares changes in dimension, perimeter, and area of a square.

Dimension of square	Change in each from original	Perimeter	Change in perimeter from original	Area	Change in area from original
2 by 2	—	8	—	4	—
4 by 4	multiplied by 2		multiplied by 2		multiplied by 4
6 by 6					
8 by 8					

a. Show a relationship between the length of a side of a square and its perimeter using a graph.

b. Show a relationship between the length of a side of a square and its area using a graph.

c. The change in each square from the original is called the *scale factor*. Suppose the scale factor is x. What are the new perimeter and area?

Applications and Problem Solving

32. **Landscaping** Tachiyuki Mizumoto wishes to fertilize his lawn. Fertilizer comes in bags that cover 5000 square feet. His yard is 110 by 210 feet. His house occupies a rectangle of 30 by 45 feet and the driveway is 18 by 50 feet.

a. What is the area he must fertilize?

b. How many bags should Mr. Mizumoto buy?

33. Kennels Liam raises Irish setters at his home in the country. He wants to enclose an area for a run in which his dogs can exercise.

a. If the dimensions of the run are 60 by 15 feet, how much fencing would he need?

b. What is the area enclosed?

34. Construction Jackie Rockford wishes to add a 15-by-40-foot rectangular deck to her house. However, she needs to allow for two square wells for the large trees in the deck area. The wells are 5-by-5 feet each. How much area must Jackie allow for decking materials?

35. Family Activity Suppose you are going to add a deck to the back of your home or have a garden in a community plot. Design two different plans and give the area and perimeter of each option.

Mixed Review

36. Travel Ana Perez travels 371 miles in 7 hours on a bus trip to a concert. What is the average rate of speed for the bus? (Lesson 3-4)

37. Solve $b + (-14) = 6$ and check your solution. Then, graph the solution on a number line. (Lesson 3-2)

38. Evaluate $-6ab$, if a $= -3$ and $b = -5$. (Lesson 2-7)

39. Cooking A casserole made with ground beef has 375 Calories per serving. The same recipe made with ground turkey has 210 Calories per serving. Write an inequality using the numbers and $<$ or $>$. (Lesson 2-3)

40. Solve $121 = 11x$ mentally. (Lesson 1-6)

41. Simplify $6d + 14(d + 2)$. (Lesson 1-5)

42. Evaluate $8 + 9 \cdot (7 - 4)$. (Lesson 1-2)

From the FUNNY PAPERS

1. What does “half off” usually mean at a sale?
2. What is actually half off in this sale?
3. Draw a diagram of one of the sheets. Is the sheet half of the size that it was when it was not on sale? Explain.

3-5B GRAPHING CALCULATOR ACTIVITY

Area and Coordinates

An Extension of Lesson **3-5**

You have learned about rectangles and squares, but many figures have irregular shapes. Coordinates can be used to find the area of these shapes. Aerial surveyors use this technique to find areas of plots of land. Use the TI-82 graphing calculator program below to find the area of the figure at the right.

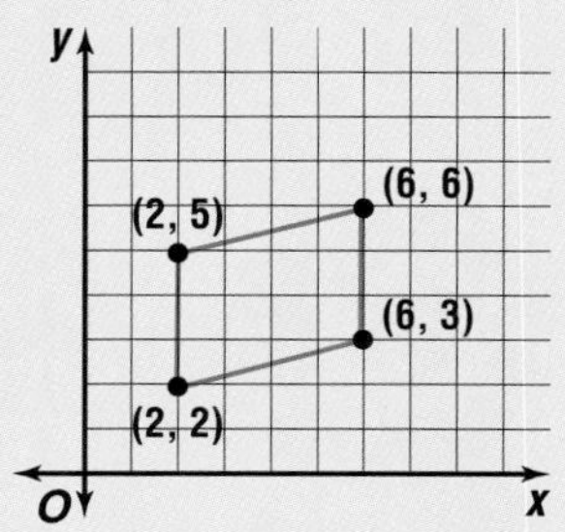

Activity

See page 1 for information on programming the TI-82.

Before you run the program, choose one vertex as the first and list the ordered pairs for the vertices in counterclockwise order. Repeat the first ordered pair at the end of the list.

(2, 2)
(6, 3)
(6, 6)
(2, 5)
(2, 2)

```
PROGRAM:AREA
: ClrList L1, L2
: 0 → D
: 0 → U
: 0 → X
: Input "VERTICES", N
: Disp "X COORDINATES"
: Input L1
: Disp "Y COORDINATES"
: Input L2
: For (X, 1, N, 1)
: L1(X)*L2(X+1)+D → D
: L2(X)*L1(X+1)+U → U
: End
: (D−U)/2 → A
: Disp "AREA ="
: Disp A
: Stop
```

The program will ask for the number of vertices in the figure. For this figure, press 4 and ENTER.

When asked for the x- and y-coordinates, enter the lists in counterclockwise order. Use braces, { }, before and after the lists.

The program will display area in square units.

Your Turn

Draw a coordinate plane and graph each figure whose vertices are listed. Then use the graphing calculator program to find the area of each figure.

1. (2, 3), (4, 1), (6, 8)
2. (−4, 2), (5, 2), (5, 5), (−4, 5)
3. (0, 4), (4, 0), (−4, 0), (0, −4)
4. (0, 0), (8, 0), (9, 5), (6, 8), (−1, 4)

Extension

5. Graph a rectangle or square on a coordinate grid and determine the coordinates of its vertices. Use the formula and the graphing calculator program to find its area. How do the methods compare?

3-6 Solving Inequalities by Adding or Subtracting

Setting Goals: *In this lesson, you'll solve inequalities by using the addition and subtraction properties of inequalities and graph the solution set.*

Modeling with Manipulatives

MATERIALS
- balance
- blocks
- paper cups

Small objects and a balance can be used to represent mathematical inequalities.

The diagram at the right shows $9 > 5$ since 9 objects are heavier than 5. The balance scale models an *inequality* because the two sides are not equal.

Let's see if the addition and subtraction properties of equality apply to inequalities.

CULTURAL CONNECTIONS
Balance scales were first used in Mesopotamia and Egypt around 2000 B.C.

First, set up your balance like the diagram. Remove 4 blocks from the left side. Remove 4 blocks from the right side. The scale returns to the position it was originally. This means that subtracting the same amount from each side does not change the inequality.

Your Turn

Work with a partner. Model the inequality $2 < 5$ and draw a picture of your model. Add 6 blocks to the side with 2. Draw a picture of what happens to the balance. Next add 6 blocks to the side with 5. Draw a picture of the result. Write down your observations.

Use a paper cup and blocks with a balance to model the inequality $x + 3 > 6$.

a. How many blocks have to be in the paper cup to model this inequality?

b. Is this the only answer?

c. Remove the 3 blocks on the side with the variable and draw a picture of what happened.

d. Next, remove 3 blocks from the other side.

e. What does this tell you about the variable? Is there more than one answer to this inequality?

Learning the Concept

To solve an inequality that involves addition, you can use subtraction just as you did to solve equations. Likewise, you can use addition to solve an inequality that involves subtraction.

Addition and Subtraction Properties of Inequalities	**In words:**	Adding or subtracting the same number from each side of an inequality does not change the truth of the inequality.
	In symbols:	For all numbers a, b, and c: 1. If $a > b$, then $a + c > b + c$ and $a - c > b - c$. 2. If $a < b$, then $a + c < b + c$ and $a - c < b - c$.

The rules are similar for $a \geq b$ and $a \leq b$.

Example 1

Solve $m + 4 < 9$. Check the solution.

$m + 4 < 9$

$m + 4 - 4 < 9 - 4$ *Subtract 4 from each side.*

$m < 5$

Check: Try -3, a number less than 5.

$-3 + 4 \overset{?}{<} 9$ *Replace m with −3.*

$1 < 9$ ✓ *This statement is true, so it checks.*

Any number less than 5 will make the statement true. Therefore, the solution is $m < 5$, all numbers less than 5.

Like solutions to equations, solutions to inequalities can be graphed. The solutions to $m < 5$ are graphed on the number line below.

The arrow represents all numbers to the left of (less than) 5. *The point, 5, is not included. Thus, the circle is open.*

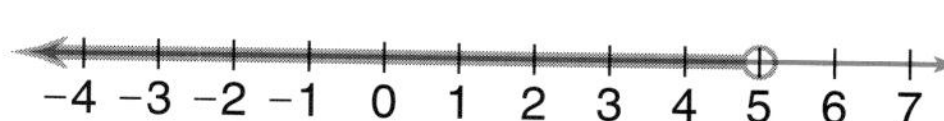

Example 2

Solve $-13 \geq x - 8$ and check the solution. Graph the solution.

$-13 \geq x - 8$

$-13 + 8 \geq x - 8 + 8$ *Add 8 to each side.*

$-5 \geq x$

Check: Try -5 and -7, numbers less than or equal to -5.

$-13 \overset{?}{\geq} -5 - 8$ $\qquad$ $-13 \overset{?}{\geq} -7 - 8$

$-13 \geq -13$ ✓ *Both are true, so they check.* $-13 \geq -15$ ✓

The solution is $-5 \geq x$. This means that the inequality is true for all numbers less than or equal to -5. The solution may also be written as $x \leq -5$. The graph of the solution set is shown below.

The point, −5, is included. Thus, the circle is solid.

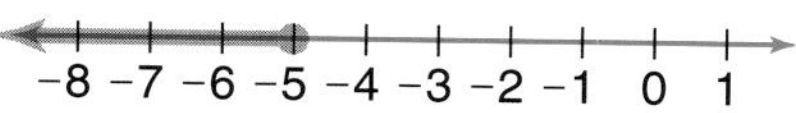

Example 3

Hobbies

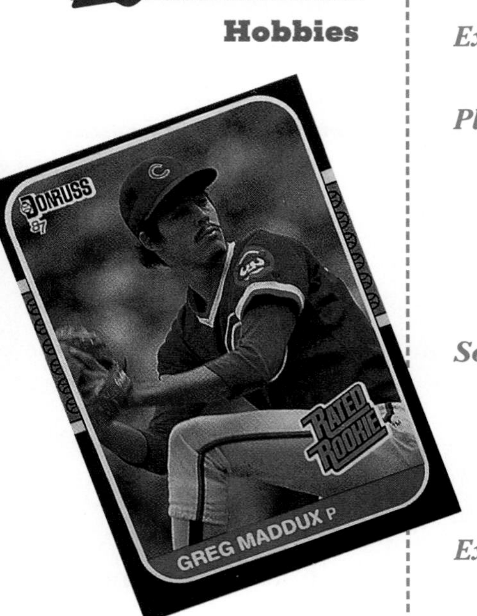

Christopher has \$30 to buy new baseball cards for his collection. He has chosen a Greg Maddux rookie card that costs \$8. What is the most he can spend on other baseball cards?

Explore We need to find out how much money he can spend on other new cards.

Plan Translate the words into an equation. Let s represent the amount he can spend on the other cards.

Cost of Maddux card	*plus*	*cost of other cards*	*must be less than or equal to*	*\$30*
8	+	s	$\leq$	30

Solve

$$8 + s \leq 30$$
$$8 - 8 + s \leq 30 - 8 \quad \textit{Subtract 8 from each side.}$$
$$s \leq 22$$

Christopher can spend no more than \$22 on other baseball cards.

Examine Suppose Christopher spent \$20, a number less than \$22, on other baseball cards. Then he would have spent \$20 + \$8 or \$28 in all. \$28 ≤ \$30, so the answer checks.

Checking Your Understanding

Communicating Mathematics

Read and study the lesson to answer these questions.

1. **Describe** how you would explain to a friend how to solve the inequality $m - 6 < 3$.
2. **Restate** in your own words the addition property of inequality.
3. **Compare and contrast** the solutions of $n + 5 = -3$ and $n + 5 > -3$.
4. **You Decide** Kelli translated the sentence *"Rachel scored no less than 8 points."* as $p < 8$. Derrick translated it as $p \geq 8$. Who is correct and why?

5. **Assess Yourself** Measure your height and a friend's height. Write an inequality that compares your heights. Suppose that each of you were to grow 3 more inches. Now write an inequality that compares your new heights.

Guided Practice

Write an inequality for each solution set graphed below.

6. −3 −2 −1 0 1 2 3 4 5
7. −6 −5 −4 −3 −2 −1 0 1

Write an inequality for each sentence. Then draw its graph.

8. Roberto spent more than \$7 for lunch.
9. Liz drove less than 75 miles today.

Solve each inequality and check your solution. Then graph the solution on a number line.

10. $z + (-5) > -3$

11. $m + 4 < -9$

12. $9 < x - 7$

13. $-6 > y - 2$

14. $k + (-1) \geq -6$

15. $18 + a \leq -13$

16. **Money** Miguel spent more than \$6 at a fast food restaurant. If he had \$25 to start, what was his remaining pocket money?

Exercises: Practicing and Applying the Concept

Independent Practice

Write an inequality for each solution set graphed below.

17. −4 −3 −2 −1 0 1 2 3 4 5 6 7

18. 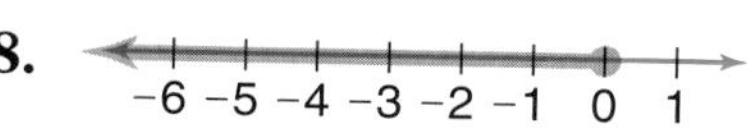

−6 −5 −4 −3 −2 −1 0 1

19. 0 1 2 3 4 5 6 7 8 9 10 11

20. −1 0 1 2 3 4 5

Write an inequality for each sentence. Then draw its graph.

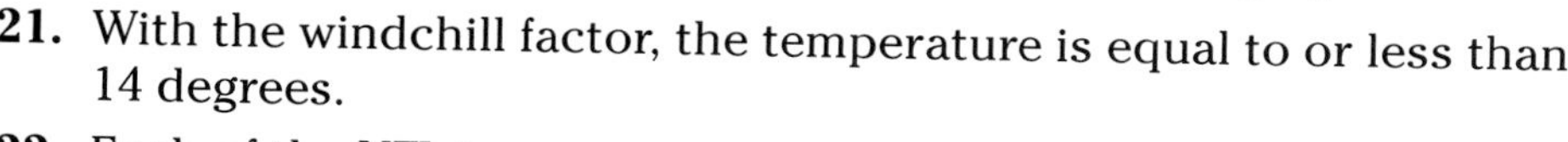

21. With the windchill factor, the temperature is equal to or less than 14 degrees.

22. Each of the NFL team rosters must have no more than 45 members at the beginning of the season.

23. Maria Vasquez is less than 5 ft. 4 in. (64 in.) tall.

24. The U.S. Constitution is more than 200 years old.

Solve each inequality and check your solution. Then graph the solution on a number line.

25. $m - 6 < 13$

26. $r \leq -12 - 8$

27. $y + 7 > 13$

28. $k + 9 \geq 21$

29. $-4 + x > 23$

30. $14 \geq a + -2$

31. $-7 + x < -3$

32. $-13 + y \geq 8$

33. $m + (-2) > -11$

34. $f + (-5) \geq 14$

35. $19 \geq z + (-9)$

36. $-41 < m - 12$

37. $73 + k < 47$

38. $33 < m - (-6)$

39. $-31 \geq x + (-5)$

40. $22 < n - (-16)$

41. $-30 \leq c + (-5)$

42. $56 > w + 72$

Critical Thinking

43. Graph the solutions for the compound inequality $y < -1$ or $y > 3$. (*Hint:* An **or** in a sentence means that either part is true.)

44. Graph the solutions for the compound inequality $x > -5$ and $x \leq 4$. (*Hint:* An **and** in a sentence means that both parts are true.)

Applications and Problem Solving

45. **Shopping** Martin is saving money to buy a new mountain bike. Bikes that he likes start at \$375. If he already has saved \$285, what is the least amount he must still save?

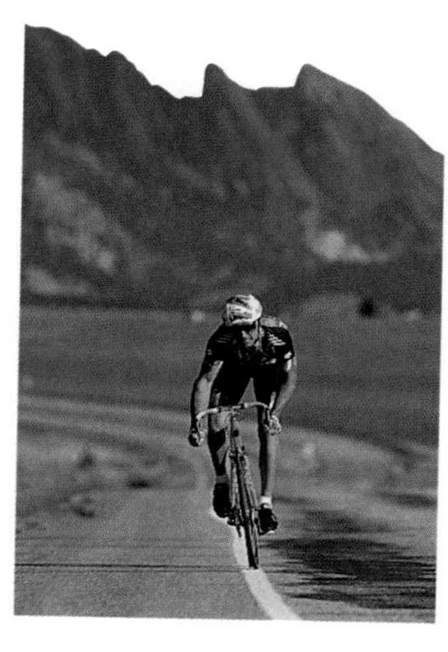

46. **Grades** On three extra credit problems on a math test, students are given up to 4 points each, provided that their scores do not exceed 100 (points). What is the minimum raw score students could have and still score 100 with the extra credit?

47. **Taxes** Residents of New York and Connecticut pay the highest taxes in the country. They must work 3 hours 9 minutes (189 minutes) of each 8-hour day to earn enough to pay their federal, state, and local taxes. The inequality $x + 52 \leq 189$ compares the time in these states with x, where x is the minimum time required in Alaska. What is the time needed in Alaska to pay off taxes?

It's Tax Time!

States where workers work the most and least to pay taxes.

Most	Hr:Min	Least	Hr:Min
Conn., N.Y.	3:09	Alaska	?
D.C., N.J.	3:01	Miss.	2:21
Hawaii	3:00	Tenn., Ala.	2:28

Source: Tax Foundation

Mixed Review

48. **Landscaping** Cherita has a rectangular garden that is 96 square feet. If the garden is 8 feet long, what is its width? (Lesson 3-5)

49. Solve $f = t - h$, if $t = 125$ and $h = 25$ by replacing the variables with the given values. (Lesson 3-4)

50. **Business** Jennifer made a \$150 down payment for soccer camp. Her unpaid balance was \$300. What is the total fee for the soccer camp? (Lesson 3-2)

51. Simplify $10a - 12a$. (Lesson 2-5)

52. Find the sum of -12 and 7. (Lesson 2-4)

53. **Communication** The table at the right shows the numbers of millions of different types of greeting cards sent in the United States each year. (Lesson 1-10)

    a. How many more cards are sent for Mother's Day than are sent for Father's Day?

    b. Make a bar graph of the data.

Occasion	Estimated number sent (millions)
Christmas	2700
Valentine's Day	1000
Easter	160
Mother's Day	150
Father's Day	101
Thanksgiving	40
Halloween	35
St. Patrick's Day	19
Jewish New Year	12
Hanukkah	11

54. Solve $\frac{36}{b} = -4$. (Lesson 1-6)

55. Name the property shown by $8 \cdot 1 = 8$. (Lesson 1-4)

3-7 Solving Inequalities by Multiplying or Dividing

Setting Goals: *In this lesson, you'll solve inequalities by using the multiplication and division properties of inequalities.*

Modeling a Real-World Application: Recycling

Environmental protection is an ever-growing concern for the world. When we pollute the environment, we are polluting *our* living space. Recycling materials is a good way to keep our habitat cleaner.

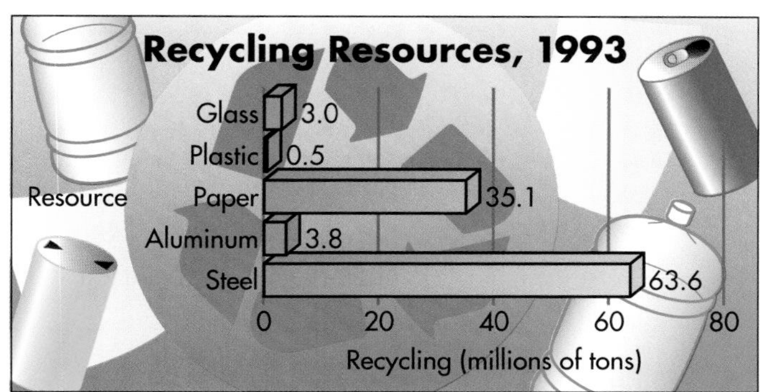

Source: Bureau of Mines

By weight, the reuse of steel and paper lead the way. The recycling of steel, by weight, is greater than that of paper.

Learning the Concept

The comparison of paper to steel recycling provides a way to look at inequalities. Rounding the weights, 35 million tons of paper being recycled is less than 64 million tons of steel being recycled. In symbols, $35 < 64$.

It is estimated that people use four times as much paper and steel as they recycle. Does the inequality between paper and steel still keep its order for the amount used?

$$35 \times 4 \stackrel{?}{<} 64 \times 4$$
$$140 < 256 \checkmark \quad \textit{The inequality is still true.}$$

Another way people can help the environment is to use as little of these products as possible. What if people cut the amount of paper and steel that they used in half? Would this affect the inequality?

$140 < 256$ *The original amount of tons of paper and steel used.*

$\frac{140}{2} \stackrel{?}{<} \frac{256}{2}$ *Each amount is divided by 2.*

$70 < 128$ ✓ *The inequality is still true.*

These examples suggest the properties on the next page.

Multiplication and Division Properties of Inequalities

In words: When you multiply or divide each side of a true inequality by a *positive* integer, the result remains true.

In symbols: For all integers a, b, and c, where $c > 0$, if $a > b$, then $a \cdot c > b \cdot c$ and $\frac{a}{c} > \frac{b}{c}$.

The rule is similar for $a < b$, $a \geq b$, and $a \leq b$.

Next, we need to consider if an inequality remains true if we multiply or divide each side by a negative number. Consider the inequality $6 < 10$.

Multiply each side by −1.		**Divide each side by −2.**	
$6 < 10$		$6 < 10$	
$6(-1) \overset{?}{<} 10(-1)$		$\frac{6}{-2} < \frac{10}{-2}$	
$-6 < -10$	**False**	$-3 < -5$	**False**

The inequalities $-6 < -10$ and $-3 < -5$ are both false. However, notice that both inequalities would be true if you reverse the order symbol. That is, change $<$ to $>$.

$-6 > -10$ and $-3 > -5$ are both true.

These examples suggest the following properties.

Multiplication and Division Properties of Inequalities

In words: When you multiply or divide each side of an inequality by a *negative* integer, you must *reverse the order symbol.*

In symbols: For all integers a, b, and c, where $c < 0$, if $a > b$, then $a \cdot c < b \cdot c$ and $\frac{a}{c} < \frac{b}{c}$.

The rule is similar for $a < b$, $a \geq b$, and $a \leq b$.

Example  **Solve each inequality and check the solution. Then graph the solution on a number line.**

a. $4x < -28$

$4x < -28$

$\frac{4x}{4} < \frac{-28}{4}$ *Divide each side by 4.*

$x < -7$

Check: Try -10, a number less than -7.

$4(-10) \overset{?}{<} -28$

$-40 < -28$ ✓ *This is true.*

The solution is $x < -7$, all numbers less than -7. The graph is shown below.

−9 −8 −7 −6 −5 −4 −3 −2 −1 0

Example 1 APPLICATION Sports

Special Olympics athlete, Andy Leonard, is one of the few American athletes who can lift up to four times his own body weight. The most weight he has lifted is 385 pounds. What is his body weight, rounded to the nearest whole number?

Explore Mr. Leonard's maximum lift is 385 pounds, which is up to four times his weight.

Plan First, estimate the solution by rounding.

$\frac{400}{4} = 100$ pounds

Translate the words into an inequality. Let w represent Mr. Leonard's body weight.

lifting weight	*is less than or equal to*	*4 times body weight*
385	$\leq$	$4w$

Solve

$385 \leq 4w$

$\frac{385}{4} \leq \frac{4w}{4}$ *Divide each side by 4.*

$96.25 \leq w$ *Use a calculator to divide 385 by 4.*

Examine Compare to the estimate. The answer appears to be reasonable. Therefore, Mr. Leonard's weight is about 96 pounds or less.

Many types of situations can be represented using an equation.

Example 

APPLICATION Entertainment

Write a realistic problem that could be solved using the equation $25x = 325$.

There is an endless number of possible situations that this equation could describe. One situation is as follows.

The pep band went on a trip to Six Flags Amusement Park. Admission was \$25 per person. If the total price of admission was \$325, how many people went to Six Flags?

Checking Your Understanding

Communicating Mathematics

Read and study the lesson to answer these questions.

1. **Explain** the advantages in using the four step problem-solving plan on word problems using equations or inequalities.
2. **List** as many word phrases as you can think of for which you would use the symbol $\leq$.
3. **You Decide** Over 2100 pounds of pancake mix were used to make the largest breakfast in 1994. Rosa solved the inequality $5b > 2100$ to find the number of 5-pound bags of mix that were used. Kiele used the inequality $5b < 2100$. Who is correct and why?

Guided Practice

Define a variable and translate each sentence into an equation or inequality. Then solve.

4. An integer increased by 12 is 28.
5. Four times some number is less than 96.
6. **Cycling** Dwain took his bike to the Leshnock Bike Shop to have the gear-shift mechanism repaired. At the shop, Dwain was told the total bill for parts and labor would be at most $63. The cost of the parts was $39. How much could Dwain expect to pay for labor?
   - **a.** What is being asked?
   - **b.** What does the phrase *at most* mean?
   - **c.** Write an equation or inequality that describes the problem.
   - **d.** Solve the problem.
7. Write a realistic problem that could be solved using the equation $169 + x = 187$.

Exercises: Practicing and Applying the Concept

Independent Practice

Define a variable and translate each sentence into an equation or inequality. Then solve.

8. The bill for labor at $13 per hour was at least $52.
9. The quotient when dividing by -3 is greater than 5.
10. The 10-km race time of 88 minutes was twice as long as the winner's time.
11. A savings account increased by $50 is now more than $400.
12. Seven times a number is less than 84.
13. The house sold for $85,000. That is at least four times the original purchase price.

Write a realistic problem that could be solved using each equation or inequality.

14. $9 + 3x = 24$
15. $b + 9 \leq 16$
16. $30m > 300$

Critical Thinking

17. Write an inequality to represent the numbers whose absolute values are less than 4. Find all the integer solutions to the inequality.

Applications and Problem Solving

Define a variable and write an equation or inequality. Then solve.

18. Brisa delivers pizza for Mama's Pizzeria. Her average tip is $1.50 for each pizza that she delivers. How many pizzas must she deliver to earn at least $20 in tips?
    - **a.** What is being asked?
    - **b.** What does the phrase *at least $20* mean?
    - **c.** Write an equation or inequality that describes the problem. Solve.
    - **d.** Can your answer be anything other than a whole number? Explain.

19. **Sports** The University of Connecticut Huskies women's basketball team was the undefeated NCAA Champion for the 1994–95 season. The team played 35 games, counting both regular and post-season play. Rebecca Lobo, one of Connecticut's star players, scored almost 600 points for the season. What was her average per game?

20. Transportation A minivan is rated for maximum carrying capacity of 900 pounds.

a. If the luggage weighs 100 pounds, what is the maximum weight allowable for passengers?

b. What is the maximum average weight allowable for 5 passengers?

21. Postage In 1971, first-class postage stamps cost 8 cents for the first ounce. In 1995, stamps for the same first-class mail cost 32 cents for the first ounce. How many times more did stamps cost in 1995 than in 1971?

22. Air Travel In 1903, the Wright brothers' first flight was at 30 mph. A Boeing 747 cruises at 600 mph. How many times faster is the flight speed of a 747?

23. Geometry The volume of a rectangular solid is found by multiplying its length, width, and depth. If the volume of the solid shown at the right is 440 cubic centimeters, what is its height?

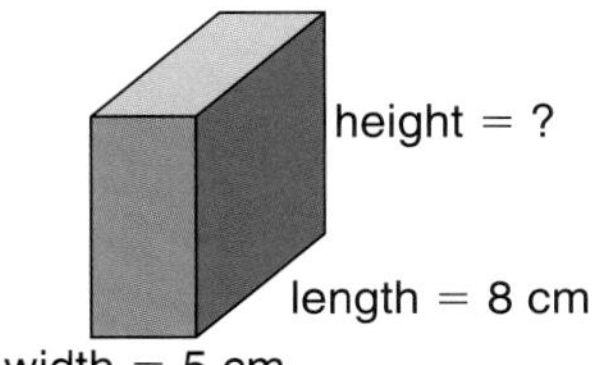

24. Agriculture Between 1987 and 1992, the number of female farmers increased 14,386 to 145,156. What was the number of female farmers in 1987?

25. Sports Danika is on her high school golf team. Her personal goal is to shoot 80. Her practice scores during the week were within 6 strokes above her desired score. What could be some of her scores?

26. Tournaments In the NBA playoffs, a team advances from the first round if they win the "best of five." Best of five means that they win more than half of the five games played between the teams. The semifinal, conference final, and final rounds are all best of seven.

a. What is the minimum number of games an NBA Champion would have to win?

b. How many games could an NBA Champion lose? Express your answer as an inequality.

NBA playoffs

Eastern Conference — Western Conference

First Round Best of Five | Conference semifinals Best of seven | Conference final Best of seven | Conference final Best of seven | Conference semifinals Best of seven | First Round Best of Five

Orlando, Boston, Charlotte, Chicago, Indiana, Atlanta, New York, Cleveland

San Antonio, Denver, Seattle, LA Lakers, Phoenix, Portland, Utah, Houston

NBA FINALS Best of seven

Mixed Review

27. Solve $-3h < -39$ and check your solution. Then graph the solution on a number line. (Lesson 3-7)

28. Science The magnification of lenses in a compound microscope, M, can be found by using the formula $E \times O = M$, where E is the power of the eyepiece lens and O is the power of the objective lens. If an eyepiece lens of a microscope has a power of 10 and its objective lens has a power of 43, what is the magnification of the microscope? (Lesson 3-4)

29. Simplify $-12y + 5y$. (Lesson 2-4)

30. Name the absolute value of -9. (Lesson 2-1)

31. Solve $\frac{t}{3} = 8$ mentally. (Lesson 1-6)

CLOSING THE

Investigation

BON VOYAGE

Refer to the Investigation on pages 62–63.

You are ready to finalize arrangements for your trip. You and your partner will be called upon to present your plan for the trip at the parents' meeting next week. Complete your research and make your presentation to the parents.

- Find the most current foreign exchange rate for the currency used in the country you chose at the beginning of the investigation. In other words, how much is *one* franc (or one krone, one peso, one rupee or one yen) worth in American dollars? These can be found in *The Wall Street Journal* and in the business section of other major newspapers.
- Explain how to find the dollar value of the foreign currency. Provide both an exact formula (that could be done on a calculator) and a simple way to estimate the value. You should be able to determine, for example, what a pair of jeans would cost in the United States if they cost 45,280 lire in Rome, Italy.
- Furnish the time zone data you found when working on the Investigation in Lesson 2-1. Work with your family to determine a convenient time to make phone calls.
- Supply environmental data for the country. Then provide the data you found when working on the Investigation in Lesson 3-2.
- Provide a short refresher course on metric units of length, weight, and liquid volume.
- Review any important business and social customs. If you can also explain the rationale behind them, it may be easier for you to remember. For example, in some Middle Eastern countries, it is considered unclean to eat or drink with your left hand!

You may wish to organize all of the information on the country you studied in a travel scrapbook to be displayed in your classroom.

PORTFOLIO ASSESSMENT

You may want to keep your work on this Investigation in your portfolio.

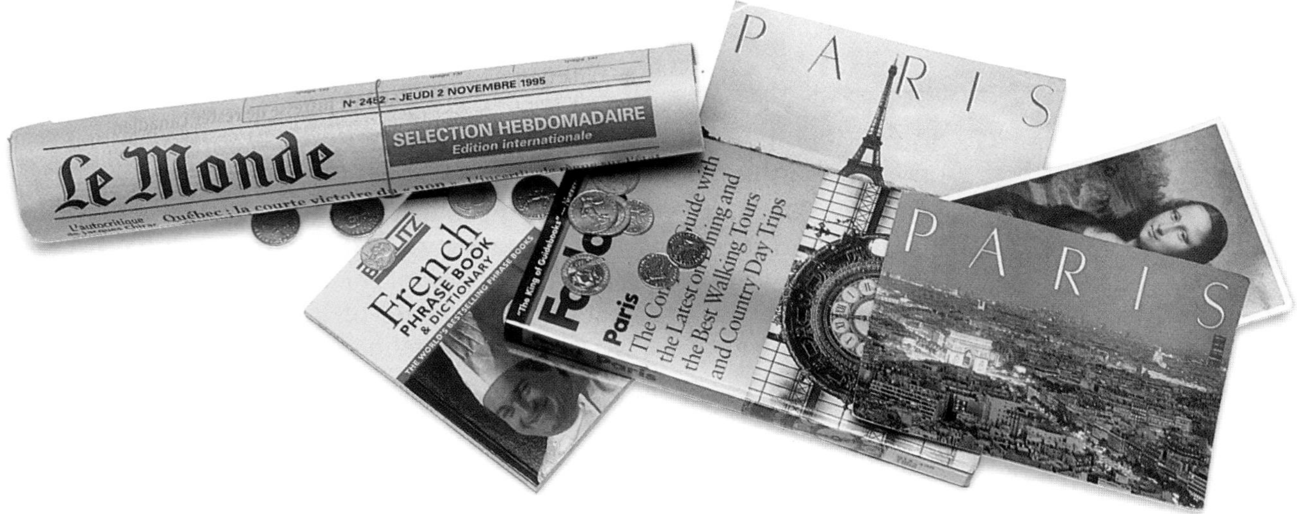

CHAPTER 3 Highlights

Vocabulary

After completing this chapter, you should be able to define each term, property, or phrase and give an example or two of each.

Algebra
addition property of equality (p. 125)
addition property of inequality (p. 147)
division properties of inequality (p. 152)
division property of equality (p. 130)
equivalent equations (p. 124)
formula (p. 134)
multiplication properties of inequality (p. 152)
multiplication property of equality (p. 130)
subtraction property of equality (p. 124)
subtraction property of inequality (p. 147)

Geometry
area (p. 138)
area of a rectangle (p. 140)
coordinate (p. 126)
graph (p. 126)
perimeter (p. 138)
perimeter of a rectangle (p. 139)

Problem Solving
eliminating possibilities (p. 118)
matrix logic (p. 119)

Understanding and Using Vocabulary

Determine whether each statement is *true* or *false*.

1. When two numbers that are additive inverses of each other are added, the sum is zero.
2. The formula *rate* · *time* = *distance* can be used to find the rate when only the distance is known.
3. The coordinates of the points on a number line are greater as you move from right to left.
4. The area of a geometric figure is the measure of the surface it encloses.
5. A solid dot on a number line indicates that the point is included in the solution set.
6. The perimeter of a geometric figure is the distance around it.
7. The division property of inequality says that if you divide both sides of an inequality by the same number, the order of the inequality symbol must be reversed.

CHAPTER 3 Study Guide and Assessment

Skills and Concepts

Objectives and Examples

Upon completing this chapter, you should be able to:

Review Exercises

Use these exercises to review and prepare for the chapter test.

- **solve equations using the addition and subtraction properties of equality.** (Lesson 3-2)

Solve $g - (-19) = 23$.

$g - (-19) = 23$ *Rewrite as addition sentence.*
$g + 19 = 23$
$g + 19 - 19 = 23 - 19$ *Subtract 19 from each side.*
$g = 4$

Solve each equation and check your solution. Then graph the solution on a number line.

8. $f + 31 = -5$
9. $r + (-11) = 40$
10. $12 = k - 7$
11. $-9 = r + (-15)$
12. $16 = h - (-4)$
13. $7 = q - 12$
14. $m - 3 = -10$
15. $z - (-51) = 36$

- **solve equations using the multiplication and division properties of equality.** (Lesson 3-3)

Solve $\frac{r}{-7} = 15$.

$\frac{r}{-7} = 15$
$(-7) \cdot \frac{r}{-7} = 15 \cdot (-7)$ *Multiply each side by (−7).*
$r = -105$

Solve each equation and check your solution. Then graph the solution on a number line.

16. $\frac{c}{6} = -29$
17. $\frac{p}{-3} = -42$
18. $840 = -28r$
19. $13b = -299$
20. $14x = 224$
21. $-\frac{y}{10} = -33$
22. $\frac{g}{-9} = 21$
23. $25d = -1775$

- **solve problems using formulas.** (Lesson 3-4)

The Herrs flew 360 miles to Tallahassee in 2 hours. What was the speed of the plane? Use $d = rt$.

$d = rt$
$360 = r(2)$ *Replace d with 360 and t with 2.*
$\frac{360}{2} = \frac{r(2)}{2}$ *Divide each side by 2.*
$180 = r$

Solve by replacing the variables with the given values.

24. $A = \ell \cdot w, A = 105, \ell = 7$
25. $I = p \cdot r \cdot t, p = 200, r = 0.06, t = 2$
26. $A = \frac{b \cdot h}{2}, A = 72, b = 6$
27. $C = 2\pi r, \pi = 3.14, r = 9$

Objectives and Examples

- find the perimeter and area of rectangles and squares. (Lesson 3-5)

Find the area and perimeter of a rectangle with length 11 cm and width 7 cm.

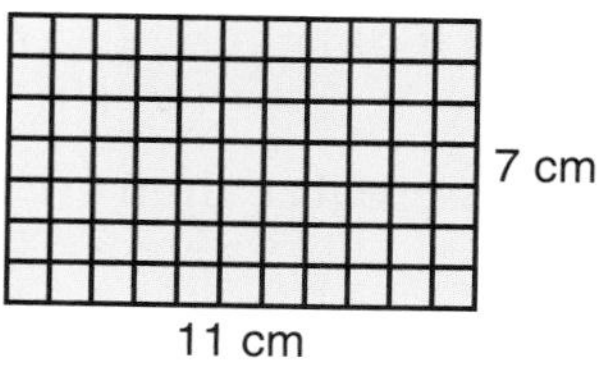

$P = 2(\ell + w)$
$= 2(11 + 7)$
$= 2(18)$ or 36 cm

$A = \ell \cdot w$
$= 11 \cdot 7$
$= 77$ sq cm

Review Exercises

Find the perimeter and area of each rectangle.

28. 4 cm, 6 cm

29. 6 in., 12 in.

30. a rectangle that is 4 meters wide and 12 meters long

31. a 15 foot-by-5 foot rectangle

32. a square with side 22 yards

Objectives and Examples

- solve inequalities by using the addition and subtraction properties and graph the solution set of the inequalities. (Lesson 3-6)

Solve $-6 < n - 6$. Then graph the solution.

$-6 < n - 6$
$-6 + 6 < n - 6 + 6$ *Add 6 to each side.*
$0 < n$

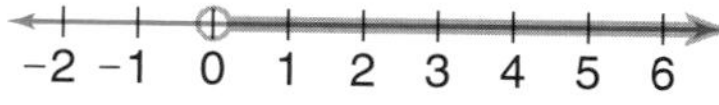

Review Exercises

Solve each inequality and check your solution. Then graph the solution on a number line.

33. $x - 2 < -14$

34. $12 + a > -9$

35. $-7 + b < -5$

36. $4 < y - 23$

37. $f + (-8) > -12$

38. $w + 9 \leq -5$

Objectives and Examples

- solve inequalities by using the multiplication and division properties. (Lesson 3-7)

Solve $-7s > 70$. Then graph the solution.

$-7s > 70$
$\frac{-7s}{-7} < \frac{70}{-7}$ *Divide each side by −7 and*
$s < -10$ *reverse the order symbol.*

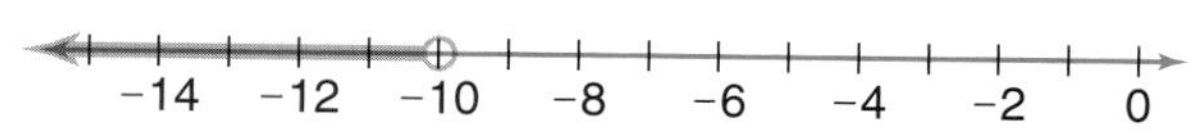

Review Exercises

Solve each inequality and check your solution. Then graph the solution on a number line.

39. $\frac{m}{-15} < 9$

40. $-72 \geq -4k$

41. $\frac{t}{12} < 11$

42. $144 < 9k$

43. $88 > -8p$

44. $\frac{s}{-22} < -7$

45. $285 \leq 19f$

46. $18 \leq \frac{r}{6}$

Objectives and Examples

▸ **solve verbal problems by translating them into equations and inequalities.** (Lesson 3-8)

After spending $17 to get into the amusement park, Marcus had $21 left. How much money did he have before he bought his ticket?

Let x be the amount of money Marcus had at the beginning.

money at start	*minus*	*ticket price*	*equals*	*money left*
x	$-$	17	$=$	21

$$x - 17 = 21$$
$$x - 17 + 17 = 21 + 17$$
$$x = 38$$

Marcus had $38 before buying his ticket.

Review Exercises

Define a variable and write an equation or inequality. Then solve.

47. Three friends went to dinner to celebrate one of them getting a job promotion. The total check was less than $36. If the cost was shared evenly, how much did each person pay for dinner?

48. Four times a number is greater than 76. What is the number?

49. A Girl Scout troop must sell more than 800 boxes of cookies to earn a trip to camp. If troop #332 has sold 576 boxes, how many boxes must they still sell to win a trip?

Applications and Problem Solving

50. Four runners represented South, Kennedy, Lincoln, and Anthony High Schools in an 880-meter race. The runner from Lincoln was faster than the runner from Kennedy. The runner from South beat the runner from Anthony, but lost to the runner from Kennedy. Which school did the runner who came in last represent? (Lesson 3-1)

a. Lincoln **b.** Kennedy
c. South **d.** Anthony

51. Statistics For many years, immigrants to the United States arrived at Ellis Island in New York. The flow of immigrants peaked in 1907 with more than 1 million. During that year, 37,807 people arrived from Germany. Find the average number of Germans who arrived daily at Ellis Island in 1907. (Lesson 3-3)

52. Physics A 110-car train goes from a velocity of 1.2 miles per hour to 6.8 miles per hour in 4 minutes. Use the formula $a = \frac{f - s}{t}$, where a is acceleration, f is final velocity, s is starting velocity, and t is time it took to make the change in velocity to find the acceleration of the train. (Lesson 3-4)

A practice test for Chapter 3 is available on page 776.

Alternative Assessment

Performance Task

Demonstrate your knowledge by giving a clear, concise solution to each problem. Be sure to include all relevant drawings and justify your answers. You may show your solutions in more than one way or investigate beyond the requirements of the problems.

	Oatmeal Creme Pies	Peanut Butter Cookies	Chocolate-Chip Cookies
Calories per serving	170		90
Grams of Fat	8 g		
Sodium	190 mg	130 mg	80 mg

Define a variable, write an equation or inequality, and then solve. Then copy and complete the table.

1. Chocolate-chip cookies have 100 Calories less than peanut butter cookies.
2. Oatmeal Creme Pies have half as much fat as peanut butter cookies.
3. Chocolate-chip cookies have 20 times more sodium than grams of fat.
4. Write a realistic problem involving the numbers given in the chart.

Nutrition Facts
Serv. Size 43 g
Serv. per package 1
Calories 220
Fat Cal. 110
* Percent Daily Values (DV) are based on a 2000 calorie diet.

Amount/serving		%DV	Amount/serving		%DV
Total fat	**12 g**	19%	**Total Carb.**	**23 g**	8%
Sat. fat	2.5 g	13%	Fiber	1 g	5%
Cholest.	5 mg	1%	Sugars	< 1 g	
Sodium	340 mg	14%	**Protein**	6 g	

Vitamin A 0% • Vitamin C 0% • Calcium 6% • Iron 10%

5. Examine the nutrition label at the left. Could you eat six servings of this product and still consume less than 5 grams of sugar? Explain.

Thinking Critically

► A rectangle is 7 inches by w inches. Suppose you increase the width by 1 inch.

a. How do the perimeter and area change?

b. If the length was 10 inches instead of 7 inches, would the change in the perimeter be different? Explain.

c. If the length was 10 inches instead of 7 inches, would the change in the area be different? Explain.

Portfolio

► Place your favorite word problem from this chapter in your portfolio and attach a note explaining why it is your favorite.

Self Evaluation

► Don't follow the paved road. Instead leave the beaten path and start a new trail. In other words, be yourself!

Assess yourself. When you try to solve a challenging problem, do you quit if the method you know doesn't work, or do you branch off and try a new way? Sometimes all you need to do is to look at a problem in a different way in order to be successful.

LONG-TERM PROJECT

Investigation

Good news: You have just inherited $10,000!

Bad news: You can't claim it until your 21st birthday.

More good news: By investing wisely, you'll have even more money in a few years!

There are several kinds of investments. You are probably most familiar with savings accounts, stocks, and bonds. Before making any kind of investment, you should learn as much as possible about investing in general.

MATERIALS

 newspaper

Some important investment terms are:

- ***Capital Gain*** *– profit made by buying something at one price and selling it at a higher price*
- ***Dividend*** *– a payment made by a corporation to its stockholders*
- ***Market Price*** *– the price per share that a stock may be purchased for or sold for at a particular time*
- ***Security*** *– a document specifying the right to property in the form of stocks or bonds*
- ***Share*** *– unit used for owning stock*
- ***Stock Exchange*** *– place where the buying and selling of stocks takes place*

Upon the advice of a relative who is a financial counselor, you decide to invest your inheritance in the stock market.

When you purchase stock, you become part owner of a corporation. The stockholders of a company share in the profits and losses of the firm. If a company's profits are high, stockholders may receive a dividend.

Guided Practice

Using divisibility rules, state whether each number is divisible by 2, 3, 5, 6, or 10.

6. 39 **7.** 118 **8.** 876 **9.** 5010

Determine whether each expression is a monomial. Explain why or why not.

10. 24 **11.** $-3m$ **12.** $4y + 2$ **13.** $5xyz$

14. Music The 78 members of the Brookhaven High School Marching Band will be performing in the Independence Day parade on Saturday. Due to the width of the street they will be marching down, the parade director has recommended that all bands march in rows of 6. Can the whole Brookhaven band be arranged in rows of 6?

Exercises: Practicing and Applying the Concept

Independent Practice

Using divisibility rules, state whether each number is divisible by 2, 3, 5, 6, or 10.

15. 111 **16.** 11,222 **17.** 5050 **18.** 4444
19. 10,505 **20.** 24,640 **21.** 117 **22.** 330
23. 49 **24.** 10,523 **25.** 434 **26.** 378

Determine whether each expression is a monomial. Explain why or why not.

27. -234 **28.** $3c - 3$ **29.** $-4b$ **30.** $6(xy)$
31. $4(3x - 1)$ **32.** v **33.** $q + r$ **34.** $166h$

35. Replace the ? in 82? so that the number is divisible by 4.

Choose

Estimation
Mental Math
Calculator
Paper and Pencil

Use mental math, paper and pencil, or a calculator to find at least one number that satisfies each condition.

36. is divisible by 2, 3, and 5

37. has four digits and is divisible by 3 and 5, but is not divisible by 10

38. has 3 digits and is not divisible by 2, 3, 5, or 10

39. has four equal digits and is divisible by 5

Critical Thinking

40. a. What is the greatest three-digit number that is not divisible by 2, 3, 5, or 10?

b. What is the least three-digit number that is not divisible by 2, 3, 5, or 10?

Applications and Problem Solving

41. Social Studies Chinese New Year, the most important holiday of the Chinese year, comes in late January or early February. The Chinese calendar follows an annual cycle, which repeats every twelve years. For example, 1996 was the Year of the Rat, so $1996 + 12$ or 2008 and $1996 - 12$ or 1984 are also the Year of the Rat.

Boar–2007 Ox–1997 Tiger–1998 Hare–1999 Dragon–2000 Snake–2001 Horse–2002 Sheep–2003 Monkey–2004 Rooster–2005 Dog–2006
1996 Rat

a. 1998 is the Year of the Tiger. How old is the youngest person born in a Year of the Tiger?

b. List five years that are Years of the Rat.

42. **Civics** Each state has a star on the U.S. flag and the stripes represent the thirteen British colonies that became the United States.
    a. How do you think the 48 stars were arranged before Alaska and Hawaii joined the Union?
    b. Suggest an arrangement for the stars if another state were to be added. Could it be rectangular?

43. **Number Theory** A perfect number is equal to the sum of its factors, except itself. There are two perfect numbers less than 30. The third perfect number is 496, since $496 = 1 + 2 + 4 + 8 + 16 + 31 + 62 + 124 + 248$.
    a. Find a perfect number between 1 and 10.
    b. Find a perfect number between 20 and 30.

Mixed Review

44. **Geography** Colorado is less than twice as large as Arkansas. Colorado is 103,730 square miles. How large is Arkansas? (Lesson 3-8)

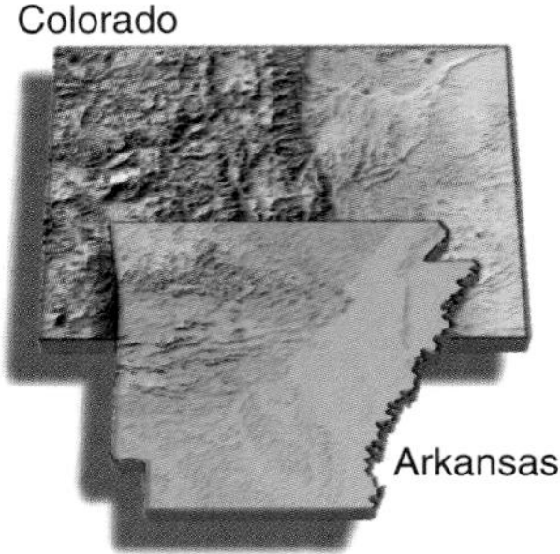

45. Solve $89 = \frac{x}{-3}$. (Lesson 3-3)
46. Solve $f = -64 \div -4$. (Lesson 2-8)
47. Solve $t = -6 + (-14) + 12$. (Lesson 2-4)
48. Solve $5b = 95$ mentally. (Lesson 1-6)
49. Evaluate $ac - 3b$ if $a = 7$, $b = 3$, and $c = 6$. (Lesson 1-3)
50. Find the value of $3 \cdot (3 + 5) \div 2$. (Lesson 1-2)

WORKING ON THE Investigation

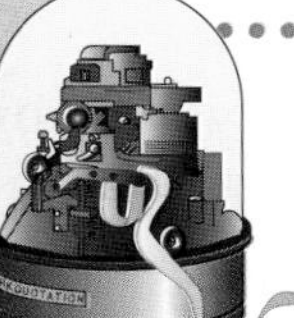

Refer to the Investigation on pages 166–167.

One of the factors to consider when investing in stock is the trends in the prices. You may see something that looks like the diagram at the right in the financial pages of a newspaper. This diagram 'charts' 30 large industrial companies whose stocks are traded on the New York Stock Exchange. This is called the *Dow Jones Industrial Average*.

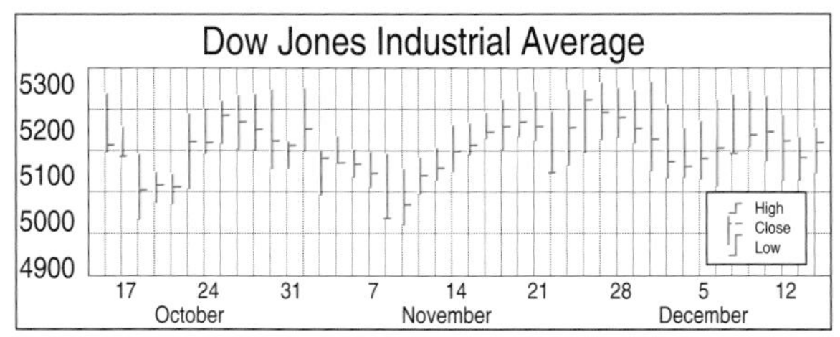

You can also chart an individual stock. When you chart a stock's activity for a day, the top of the vertical bar represents the highest price at which the stock traded, while the bottom of the bar represents the lowest price at which the stock traded. The horizontal bar indicates the closing price.

- Go to the library and use the last three months' editions of a local or national newspaper to chart the price of one of the stocks you have chosen. If the paper you are using doesn't provide the daily highs and lows, prepare a line graph of the closing prices.
- Write a brief analysis of your stocks' recent performance.

Add the results of your work to your Investigation Folder.

4-2 Powers and Exponents

Setting Goals: *In this lesson, you'll use powers in expressions and equations.*

Modeling a Real-World Application: Games

Think about India 1200 years ago. What did the people there have in common with people in the United States today? Parcheesi!

In the game of Parcheesi, or Pachisi in India, players move pieces around a board determined by tossing two dice.

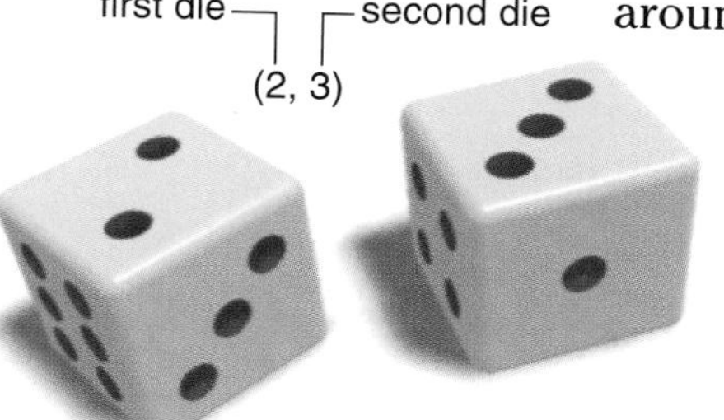

Since each die has six faces, we know that in rolling one die we can come up with six different results. How many different results are there with two dice? Make a table of the possible results.

First Die

Second Die	1	2	3	4	5	6
1	(1, 1)	(2, 1)	(3, 1)	(4, 1)	(5, 1)	(6, 1)
2	(1, 2)	(2, 2)	(3, 2)	(4, 2)	(5, 2)	(6, 2)
3	(1, 3)	(2, 3)	(3, 3)	(4, 3)	(5, 3)	(6, 3)
4	(1, 4)	(2, 4)	(3, 4)	(4, 4)	(5, 4)	(6, 4)
5	(1, 5)	(2, 5)	(3, 5)	(4, 5)	(5, 5)	(6, 5)
6	(1, 6)	(2, 6)	(3, 6)	(4, 6)	(5, 6)	(6, 6)

Six choices on the first die for each of the six choices on the second die gives us 6×6 or 36 possibilities.

Learning the Concept

The expression 6×6 can be written in a shorter way using exponents. An **exponent** tells how many times a number, called the **base**, is used as a factor. Numbers that are expressed using exponents are called **powers**.

The expression 6×6 can be written as 6^2.

base → 6^2 ← *exponent*

power

The value of any number to the first power is the number. For example, $6^1 = 6$.

The powers 7^2, 6^3, 10^4 and $(-2)^5$ are read as follows.

7^2 is seven to the second power or seven squared.
6^3 is six to the third power or six cubed.
10^4 is ten to the fourth power.
$(-2)^5$ is negative two to the fifth power.

Any number, except 0, raised to the zero power, like 5^0, is defined to be 1.

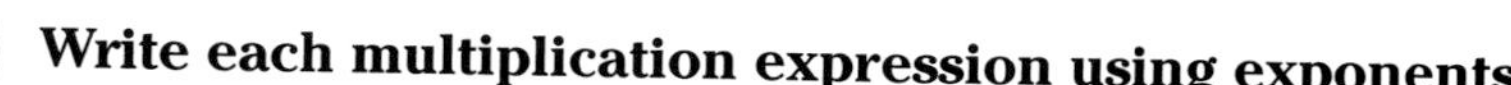

Example 1 **Write each multiplication expression using exponents.**

a. $3 \cdot 3 \cdot 3 \cdot 3$

The base is 3. It appears as a factor 4 times. The exponent is 4.

$3 \cdot 3 \cdot 3 \cdot 3 = 3^4$

b. $y \cdot y \cdot y \cdot y \cdot y$

The base is y. It appears as a factor 5 times. The exponent is 5.

$y \cdot y \cdot y \cdot y \cdot y = y^5$

c. 5

The base is 5. It appears once. The exponent is 1.

$5 = 5^1$

d. $(-2)(-2)(-2)(-2)$

The base is -2. It appears as a factor 4 times. The exponent is 4.

$(-2)(-2)(-2)(-2) = (-2)^4$

Example  **Write each power as a multiplication expression.**

a. 5^4

The base is 5. The exponent 4 means 5 is a factor 4 times.

$5^4 = 5 \cdot 5 \cdot 5 \cdot 5$

b. b^2

The base is b. The exponent 2 means b is a factor 2 times.

$b^2 = b \cdot b$

The number 12,496 is in **standard form**. You can use exponents to express a number in **expanded form**.

Example 3 **Express 12,496 in expanded form.**

To express a number in expanded form, use place value to write the value of each digit in the number.

$$12{,}496 = 10{,}000 + 2000 + 400 + 90 + 6$$
$$= (1 \times 10{,}000) + (2 \times 1000) + (4 \times 100) + (9 \times 10) + (6 \times 1)$$

Then write the multiples of 10 using exponents.

$$12{,}496 = (1 \times 10^4) + (2 \times 10^3) + (4 \times 10^2) + (9 \times 10^1) + (6 \times 10^0)$$

Recall that $10^0 = 1$.

Since powers are forms of multiplication, they need to be included in the rules for order of operations.

Order of Operations

1. **Do all operations within grouping symbols first; start with the innermost grouping symbols.**
2. ***Evaluate all powers in order from left to right.***
3. **Next do all multiplications and divisions in order from left to right.**
4. **Then do all additions and subtractions in order from left to right.**

Connection to Algebra

Follow the order of operations as you evaluate algebraic expressions.

Example  **Evaluate $3a + b^3$ if $a = 2$ and $b = 5$.**

$3a + b^3 = 3(2) + 5^3$ *Replace a with 2 and b with 5.*

$= 3(2) + 125$ *Evaluate the power.*

$= 6 + 125$ *Find the product of 3 and 2.*

$= 131$ *Find the sum.*

Many scientific formulas involve evaluating expressions containing exponents.

Example 5

APPLICATION

Space Flight

You can evaluate expressions involving exponents by using a calculator. Enter the base first, press the y^x key, and then enter the exponent.

At liftoff, the space shuttle *Discovery* has a constant acceleration, a, of 16.4 ft/s^2. The initial velocity, v, is 1341 ft/s. Use the formula $d = vt + \frac{1}{2}at^2$, where d represents distance and t represents time in seconds, to find the distance that the shuttle has traveled after 30 seconds.

$d = vt + \frac{1}{2}at^2$

$= (1341)(30) + \frac{1}{2}(16.4)(30)^2$ *Replace v with 1341, t with 30, and a with 16.4.*

Estimate: *40,000 + 8 · 900 is about 47,000.*

Evaluate this expression by using a calculator. Many calculators have a special key labeled x^2 for finding squares.

Enter: 1341 [×] 30 [+] .5 [×] 16.4 [×] 30 [x^2] [=] 47610

The shuttle has traveled 47,610 feet after 30 seconds.

Checking Your Understanding

Communicating Mathematics

Read and study the lesson to answer these questions.

1. **Explain** how to find the value of the expression *3 squared.*
2. **Write** 10^5 as a product of the same factor. What is the value of 10^5?
3. Regardless of the value of n, what can you say about 1^n? Explain.
4. **You Decide** Nichelle says that for every value of n, $n^2 > n$. Cardida disagrees. Who is correct? Explain.

MATH JOURNAL

5. **Assess Yourself** Why do you think that exponents were invented?

Guided Practice

Write each multiplication expression using exponents.

6. $10 \cdot 10 \cdot 10$
7. $m \cdot m \cdot m$
8. (3)(3)(3)

Write each power as a multiplication expression.

9. 2^3
10. $(-8)^4$
11. 10^6

Evaluate each expression if $x = 5$, $y = 9$, and $z = -3$.

12. x^4
13. $y^2 + z^3$
14. Write 598 in expanded form.
15. **Geometry** The edges of the cube shown at the right are 5 inches long.

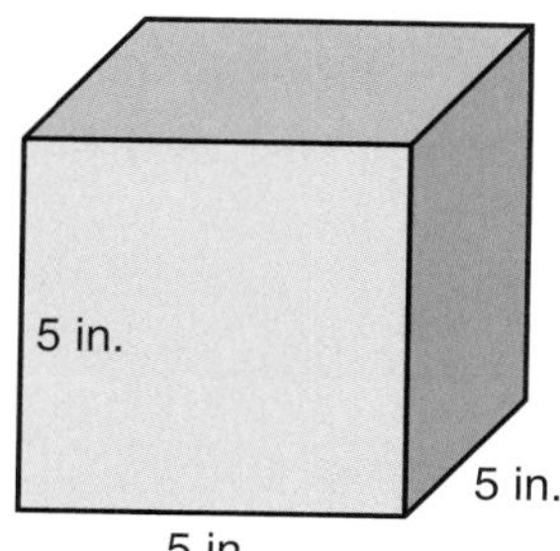

   a. The surface area of a cube is the sum of the areas of the faces. Use exponents to write an expression for the surface area of the cube.
   b. The *volume* of a cube is the amount of space that it occupies. The volume can be found by multiplying the length, width, and height of the cube. Write an expression for the volume of the cube.
   c. If you double the length of each edge of the cube, what is the effect on the surface area and volume?

Exercises: Practicing and Applying the Concept

Independent Practice

Write each multiplication expression using exponents.

16. (5)(5)(5)
17. $n \cdot n \cdot n \cdot n$
18. $1 \cdot 1 \cdot 1 \cdot 1 \cdot 1$
19. 14
20. $(p \cdot p)(p \cdot p)$
21. $\underbrace{3 \cdot 3 \cdot \cdots \cdot 3 \cdot 3}_{15 \text{ factors}}$
22. $6 \cdot 6 \cdot 6 \cdot 6 \cdot 6$
23. $b \times b \times b \times b$
24. $(-5)(-5)$

Write each power as a multiplication expression of the same factor.

25. 12^2
26. y^6
27. $(-7)^3$
28. $(-f)^2$
29. 6^{20}
30. $(x + 1)^2$
31. 1^{55}
32. q^3
33. $(-1)^4$

Evaluate each expression if $a = 3$, $b = 9$, and $c = -2$.

34. $a^4 + b$
35. $c^2 + ab$
36. $b^0 - 10$
37. $2b^2$
38. $10a^5 + c^2$
39. $a - b^2$

Write each number in expanded form.

40. 56
41. 149
42. 2053
43. Write $(2 \times 10^4) + (3 \times 10^3) + (4 \times 10^2) + (5 \times 10^0)$ in standard form.

44. **Patterns** Study the pattern at the right. Extend the pattern to find the value of each expression.

$10^0 \cdot 10^2 = 1 \cdot 100 = 100$ or 10^2
$10^1 \cdot 10^2 = 10 \cdot 100 = 1000$ or 10^3
$10^2 \cdot 10^2 = 100 \cdot 100 = 10{,}000$ or 10^4
$10^3 \cdot 10^2 = 1000 \cdot 100 = 100{,}000$ or 10^5

a. $10^5 \cdot 10^2$
b. $10^{10} \cdot 10^2$
c. $10^x \cdot 10^2$

Critical Thinking

45. **Patterns** Study the pattern of exponents at the right.

$2^4 = 16$) ÷ 2
$2^3 = 8$) ÷ 2
$2^2 = 4$) ÷ 2
$2^1 = 2$) ÷ 2
$2^0 = 1$

a. If you continue the pattern, what is the next exponent for 2?
b. What is the next value in the pattern?
c. Use a calculator to verify your results.
d. Make a conjecture about the value of 3^{-1}.

Applications and Problem Solving

46. **Electricity** Niagara Falls and Hoover Dam can produce electric energy with little pollution. As the power goes to its point of use, energy is lost to the resistance in the wires. The amount of power lost can be found by using the formula $P = I^2R$, where P is power in watts, I is current in amps, and R is resistance in ohms.
   a. The resistance of the wire leading from the source of power to a home is 2 ohms. If an electric stove causes a current of 41 amps to flow through the wire, write an equation for the power lost.
   b. Find the power lost from the wire powering the stove.

47. **Geometry** The second and third powers have special names related to geometry. Since the area of a square whose side is s units long is $s \cdot s$ or s^2, s^2 is often read as "s squared." A third power like s^3 is often read as "s cubed." Can you explain why?

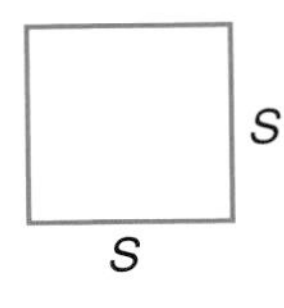

48. **Entertainment** In the movie *I.Q.*, Albert Einstein tries to make his niece Catherine Boyd fall in love with auto mechanic Ed Walters. In one scene, Ed asks Catherine how many stars she thinks are in the sky. Her answer is $10^{12} + 1$. Write this number in standard form.

49. **Literature** There is a story of a knight who slew a dragon that was destroying a small kingdom. The knight asked the king to give him reward money for a month following this pattern. The first day of the month, he would get 1 cent, the second day, 2 cents, the third day, 4 cents, and so on, continuing to double the amount for 30 days.
    - **a.** Express his reward on each of the first three days as a power of 2.
    - **b.** What is his reward on the fourth day? Express it as a power of 2.
    - **c.** How much would the knight be getting from the king on the 30th day? Express it as a power of 2.
    - **d.** Write your answer to part c in standard form.
    - **e.** Write an expression using a sum and exponents for the amount of money the knight would receive for the whole month.
    - **f.** The king agreed to the knight's request because he thought it would be a cheap reward. Was it? Explain.

Mixed Review

50. State whether 945 is divisible by 2, 3, 5, 6, or 10. (Lesson 4-1)
51. Solve and graph $-7x < 84$. (Lesson 3-7)
52. **Geometry** Find the perimeter and area of a rectangle that is 4 meters wide and 12 meters long. (Lesson 3-5)
53. **Meteorology** A maximum and minimum thermometer records both the high and low temperatures of the day. If the difference between one day's high and low was 49 degrees and the high was 67°F, what was the low temperature? (Lesson 3-2)
54. Solve $m = \frac{126}{-21}$. (Lesson 2-8)
55. Simplify $18p - 26p$. (Lesson 2-5)
56. **Education** The class of 1988 had 536 graduates, and 497 students graduated in the class of 1995. Write and solve an equation to find how many more students graduated in 1988. (Lesson 1-6)
57. Write an expression to represent the phrase *four more than the Bears scored in the last game.* (Lesson 1-3)

From the FUNNY PAPERS

1. Explain why the comic is funny.
2. Explain the differences in the formulas on the chalkboard.
3. Einstein's famous formula $E = mc^2$ relates energy, E, mass, m, and the speed of light, c. The speed of light is about 300,000 km/s. Use a calculator to find the value of c^2.

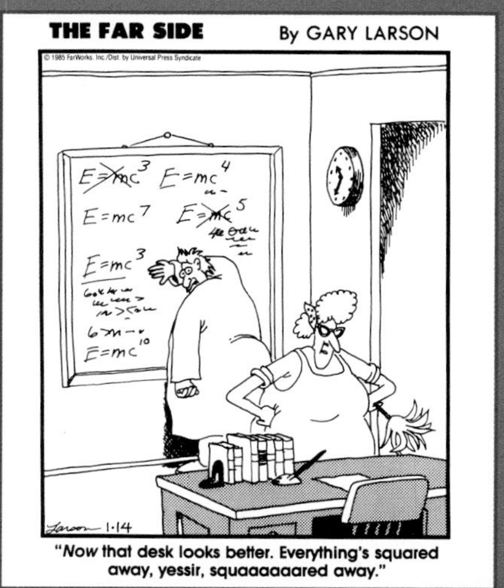

"Now that desk looks better. Everything's squared away, yessir, squaaaaaared away."

GRAPHING CALCULATOR ACTIVITY

4-2B Evaluating Expressions with Exponents

An Extension of Lesson **4-2**

You learned how to evaluate expressions with a graphing calculator in Lesson 1-3B. You can also use a graphing calculator to evaluate expressions involving exponents.

Storing the values of the variables allows you to evaluate complicated expressions accurately. Also, it will save time when you want to evaluate several expressions for the same values.

Activity 1

Evaluate $\frac{x}{(y-2)^2}$ on a graphing calculator if $x = 24$ and $y = 4$.

Keystrokes are shown for the TI-82 graphing calculator. If you are using a different graphing calculator, consult your User's Guide.

ENTER: 24 [STO▸] [X,T,θ] [2nd] [:] *Enters the value of x.*
4 [STO▸] [ALPHA] [Y] [2nd] [:] *Enters the value of y.*
[X,T,θ] [÷] [(] [ALPHA] [Y] [−] 2 [)] [x^2] [ENTER] 6 *Enters and evaluates the expression.*

Expressions involving exponents other than 2 can be evaluated using the [^] key.

Activity 2

Use a graphing calculator to evaluate $3ab^5$ if $a = -4$ and $b = 2$.

ENTER: [(−)] 4 [STO▸] [ALPHA] [A] [2nd] [:] *Enters the value of a.*
2 [STO▸] [ALPHA] [B] [2nd] [:] *Enters the value of b.*
3 [ALPHA] [A] [ALPHA] [B] [^] 5 [ENTER] -384 *Enters and evaluates the expression.*

Notice that the calculator was able to evaluate the expression without entering multiplication symbols between 3, a, and b^5.

Your Turn **Use a graphing calculator to evaluate each expression if $a = 9$, $b = -1$, $c = 7$, and $d = 3$.**

1. $7d^2 - 1$
2. $\frac{32}{(a-c)^3}$
3. $3(b + c) \div d^2$
4. $12b^4 \div d$
5. $(2c - 4d)^{10}$
6. $11 - a^2(b + c)$

7. Evaluate $12 - 4d^2$ and $(12 - 4d)^2$ if $d = 3$. Are the values the same? Explain.

4-3 Problem-Solving Strategy: Draw a Diagram

Setting Goals: *In this lesson, you'll solve problems by drawing diagrams.*

Modeling a Real-World Application: Entertainment

The game of Totolospi was invented by the Hopi Indians. You play the game with three cane dice, a counting board inscribed in stone, and a counter for each player. Each cane die can land round side up, (R), or flat side up, (F). When two players play Totolospi, each player places a counter on the nearest circle. The moves in the game are determined by tossing three cane dice.

- ▶ Toss three round sides up, (RRR), and advance 2 lines.
- ▶ Toss three flat sides up, (FFF), and advance 1 line.
- ▶ No advance with any other combination.

The first player to reach the opposite side wins.

If you are playing Totolospi, how many combinations of tosses will allow an advance?

Learning the Concept

When we are able to draw a diagram for a given situation, we often are able to understand it better. Drawing a diagram is a powerful problem-solving strategy.

Explore There are 3 dice being tossed. We need to know how many combinations of round and flat side up can be tossed.

Plan We can draw a diagram showing how each of the dice lands to determine the number of possible combinations.

Solve

First Die	Second Die	Third Die	
R	R	R (RRR)	← *advance 2*
		F (RRF)	*no advance*
	F	R (RFR)	*no advance*
		F (RFF)	*no advance*
F	R	R (FRR)	*no advance*
		F (FRF)	*no advance*
	F	R (FFR)	*no advance*
		F (FFF)	← *advance 1*

There are 8 possible outcomes. Two of them, RRR and FFF, allow you to advance. So 8 − 2 or 6 tosses allow no advance.

Examine Each die can land two different ways. So there should be $2 \cdot 2 \cdot 2$ or 8 possible tosses.

Checking Your Understanding

Communicating Mathematics

Read and study the lesson to answer these questions.

1. **Explain** why a diagram is a good strategy for problem solving.
2. Are diagrams helpful in solving *every* type of problem?
3. The diagram at the right shows a model of a five-step staircase built with concrete blocks. Use grid paper to draw a staircase with seven steps. How many blocks would be required to build a staircase with seven steps?

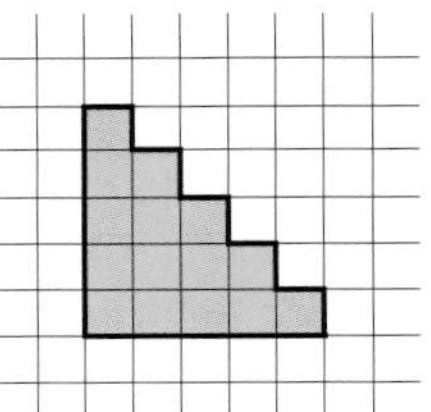

MODELING MATHEMATICS

MATERIALS

grid paper

Guided Practice

4. **Draw** a diagram to determine how many flights an airline would have to schedule if it wants to provide nonstop flights between Miami, Ft. Lauderdale, Tallahassee, Tampa, Orlando, and Daytona.

5. **Sports** The Centerville High School volleyball team is participating in the state championship. Eight teams will participate in a single-elimination tournament; that is, only winning teams will continue to play. Draw a diagram to determine the number of games that will be played.

Practicing and Applying the Concept

Independent Practice

Solve. Use any strategy.

6. The West Peoria Baseball League has 7 teams and the East Peoria League has 8 teams. Every team from the West League must play every team from the East League at least once. What is the least number of games that must be played?
7. Replace each ● with an operation symbol to make the equation true. Add grouping symbols if necessary.

   $$4 \bullet 3 \bullet 6 \bullet 3 = 45$$

8. During the first minute of a game of tug-of-war, the Red Team pulled the Blue Team forward 2 feet. During the second minute, the Blue Team pulled the Red Team forward 1 foot. During the third minute, the Red Team pulled the Blue Team forward 2 feet and during the fourth minute, the Blue Team pulled the Red Team forward 1 foot. If this pattern continues, how long will it take the Red Team to pull the Blue Team forward a total of 10 feet to win the game?
9. Meg is conditioning for the start of soccer season. She does seven minutes of stretching, followed by an 18-minute run. Then she cools down with 8 minutes of walking and 7 minutes of leg lifts. Finally, she finishes with 4 minutes of stretching. If Meg finishes her workout at 2:07 P.M., when did she begin?

10. **History** Jorge has to write a paper and complete a family tree for history class. If part of the assignment was to find the names and birthdates of the five generations that preceded him, direct relatives only, no step-parents, how many people will Jorge have to investigate?

11. Magic squares are arrays of numbers in which the sum of the numbers on each row, column, and diagonal are the same. According to legend, the most famous magic square, the *lo-shu*, was discovered on the back of a divine tortoise by China's Emperor Yu in about 2200 B.C. Copy and complete the magic square pictured at the right.

4	15		14
		12	
	3		
5		8	

12. **Geometry** A diagonal is a line segment that can be drawn from one vertex of a polygon to another vertex that does not share a side. The diagram at the right shows the two diagonals of a rectangle.

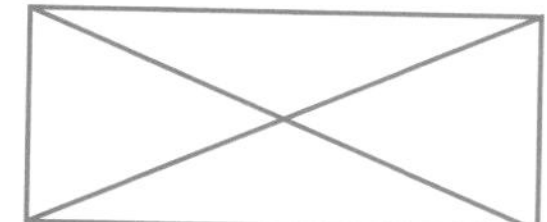

   a. A pentagon has five sides. How many diagonals does it have?
   b. How many diagonals can be drawn in a hexagon? (*Hint:* A hexagon has six sides.)
   c. Use a pattern to predict how many diagonals a heptagon, 7 sides, has. Check your answer by drawing a diagram.
   d. In your own words, write a description of the pattern.

13. Charlie traded in his old car which averaged 22 miles per gallon. The EPA sticker on the new car stated that it should average 37 miles per gallon. If Charlie drives about 12,000 miles per year and gasoline is about $1.10 per gallon, how much should he expect to save on gasoline in the first year of owning the new car?

14. At Grandma's Bakery, muffins are sold in boxes of 4, 6, or 13. You can buy 8 muffins by choosing two boxes of 4 muffins, but you can't buy 9 muffins with any combination of boxes. Find all of the numbers of muffins less than 25 that you can't buy.

Critical Thinking

15. City code requires that a party house provide 9 square feet for each person on a dance floor. If the owners of the Westman Hotel want to provide a square dance floor that is large enough for 100 people, how long should each side be?

Mixed Review

16. Write $x \cdot x \cdot x \cdot x \cdot x$ using exponents. (Lesson 4-2)
17. Solve $x - 8 \leq -17$. (Lesson 3-6)
18. **Patterns** What is the next number in the pattern 3, 3, 6, 18, . . . ? (Lesson 2-6)
19. Find the sum $5 + (-8) + 7$. (Lesson 2-4)
20. Name the property shown by $(3 \cdot 2) \cdot 8 = 3 \cdot (2 \cdot 8)$. (Lesson 1-4)

4-4 Prime Factorization

Setting Goals: *In this lesson, you'll identify prime and composite numbers and write the prime factorizations of composite numbers.*

Modeling with Manipulatives

MATERIALS

As you learned in Lesson 4-3, drawing a diagram can be very helpful as you solve problems.

Your Turn

The grid at the right shows one way of drawing a rectangle consisting of 10 squares. Use grid paper to draw as many different rectangular arrangements of 1, 2, 3, 4, 5, 6, 7, 8, 9, and 10 squares as possible.

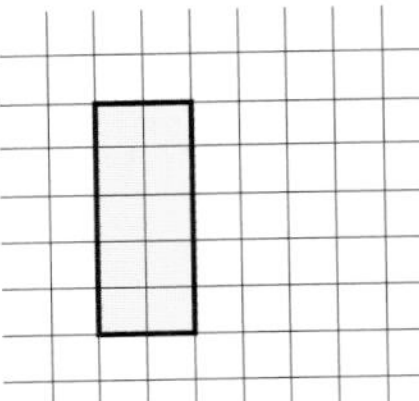

a. Which numbers of squares could only be arranged in either one row or one column?

b. Which numbers of squares could be arranged in a way other than one row or column?

c. Choose one of the numbers that could only be arranged in one row or one column. Write as many multiplication sentences with the number as the product and two whole numbers as factors as you can. What do you observe?

Learning the Concept

In the activity above, the numbers of squares that can be arranged in only one row or one column are prime numbers. A **prime number** is a whole number greater than one that has *exactly* two factors, 1 and itself.

Numbers of squares like 4, 6, 8, 9, and 10 that can be arranged in rectangles other than a row or a column are called composite numbers. A **composite number** is a whole number greater than one that has more than two factors. A composite number can always be expressed as a product of two or more primes.

- The numbers 0 and 1 are considered *neither* prime *nor* composite.
- Every number is a factor of 0, since any number multiplied by zero is zero.
- The number 1 has only one factor, itself.
- Every whole number greater than 1 is either prime or composite.

Sometimes a number can be factored in several ways. Three different ways to find all of the prime factors of 24 using diagrams called **factor trees** are shown below.

$$24 = 2 \cdot 12 = 2 \cdot 3 \cdot 4 = 2 \cdot 3 \cdot 2 \cdot 2$$

$$24 = 3 \cdot 8 = 3 \cdot 2 \cdot 4 = 3 \cdot 2 \cdot 2 \cdot 2$$

$$24 = 4 \cdot 6 = 2 \cdot 2 \cdot 2 \cdot 3$$

The factors are in a different order, but the result is the same. The order in which you factor is not important. The factoring process ends when all of the factors are prime. When a positive integer (other than 1) is expressed as a product of factors that are all prime, the expression is called the **prime factorization**.

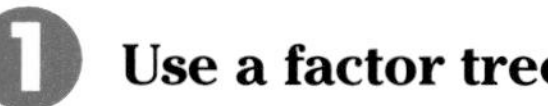

Example 1 Use a factor tree to factor 64 completely.

Begin with any two whole number factors of 64.

$64 = 8 \cdot 8$

$= 4 \cdot 2 \cdot 4 \cdot 2$ *Factor each 8.*

$= 2 \cdot 2 \cdot 2 \cdot 2 \cdot 2 \cdot 2$ *Factor each 4.*

The factorization is complete when all the factors are prime. The prime factorization of 64 is $2 \cdot 2 \cdot 2 \cdot 2 \cdot 2 \cdot 2$ or 2^6.

The prime factorization of a negative integer (other than -1) contains a factor of -1, and the rest of the factors are prime.

Example 2 Use a factor tree to factor −72 completely.

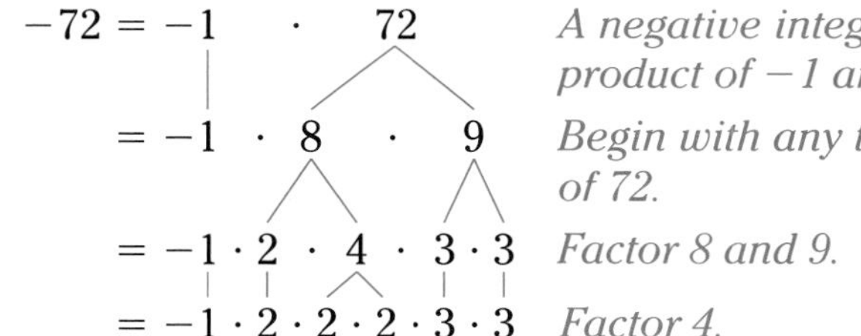

$-72 = -1 \cdot 72$ *A negative integer may be expressed as the product of −1 and a whole number.*

$= -1 \cdot 8 \cdot 9$ *Begin with any two whole number factors of 72.*

$= -1 \cdot 2 \cdot 4 \cdot 3 \cdot 3$ *Factor 8 and 9.*

$= -1 \cdot 2 \cdot 2 \cdot 2 \cdot 3 \cdot 3$ *Factor 4.*

You are finished factoring when all of the factors, other than -1, are prime. The factorization of -72 is $-1 \cdot 2 \cdot 2 \cdot 2 \cdot 3 \cdot 3$. You can write this product as $-1 \cdot 2^3 \cdot 3^2$.

You can also use a strategy called the *cake method* to find a prime factorization. The cake method uses division to find factors. For example, the prime factorization of 210 is shown below.

- Begin with the smallest prime that is a factor.
- Then divide the quotient by the smallest possible prime factor.
- Repeat until the quotient is prime.

```
    7
 5)35
3)105
2)210
```

The prime factorization of 210 is $2 \cdot 3 \cdot 5 \cdot 7$. *Use a tree diagram to check the result.*

Example 3 **Prime numbers are the key to coding and decoding information using the Rivest-Shamirs-Adleman (RSA) cryptoalgorithm. Two prime numbers p and q are chosen and the key to the code is $n = pq$. Find p and q if $n = 1073$.**

The greater the prime numbers, the more difficult it is to break the code.

Some possibilities can be eliminated by using the divisibility rules you learned in Lesson 4-1.

1073 is not divisible by 2 since the ones digit, 3, is not divisible by 2.
1073 is not divisible by 3 since $1 + 0 + 7 + 3$ or 11 is not divisible by 3.
1073 is not divisible by 5 since the ones digit, 3, is not 0 or 5.

Now use a calculator and the guess-and-check problem-solving strategy to continue checking primes.

x	1073 ÷ x	factor of 1073?
7	153.286	no
11	97.5455	no
13	82.5385	no
17	63.1176	no
19	56.4737	no
23	46.6522	no
29	37	yes

If $n = 1073$, then p and q are 29 and 37.

FYI

Dr. Arjen Lenstra and a team of volunteers spent eight months in 1993 and 1994 using calculations to factor RSA 129, a 129-digit number for encoding. The message they found was "The magic words are squeamish ossifrage."

Connection to Algebra

A monomial can also be written in factored form as a product of prime numbers, -1, and variables with no exponent greater than 1.

Example 4 **Factor each monomial completely.**

a. $\mathbf{28x^2y}$

$$28x^2y = 2 \cdot 14 \cdot x^2 \cdot y$$
$$= 2 \cdot 2 \cdot 7 \cdot x \cdot x \cdot y$$

b. $\mathbf{64ab^3}$

$$64ab^3 = 8 \cdot 8 \cdot a \cdot b^3$$
$$= 2 \cdot 4 \cdot 2 \cdot 4 \cdot a \cdot b \cdot b \cdot b$$
$$= 2 \cdot 2 \cdot 2 \cdot 2 \cdot 2 \cdot 2 \cdot a \cdot b \cdot b \cdot b$$

Checking Your Understanding

Communicating Mathematics

Read and study the lesson to answer these questions.

1. **Explain** the differences between prime and composite numbers.
2. **Find** the prime factorization of 84 by using both a factor tree and the cake method. Are the results the same? Explain.
3. **You Decide** Mansi and Wes each factored -124. Whose factorization is complete and why?

Mansi

$$\begin{aligned} -124 &= -1 \cdot 124 \\ &= -1 \cdot 4 \cdot 31 \\ &= -1 \cdot 2 \cdot 2 \cdot 31 \end{aligned}$$

Wes

$$\begin{array}{r} 31 \\ 4\overline{)124} \\ -1\overline{)-124} \end{array} \qquad -124 = -1 \cdot 4 \cdot 31$$

Guided Practice

Determine whether each number is *prime* or *composite*.

4. 17
5. 9
6. 27

Factor each number or monomial completely.

7. 38
8. 66
9. 56
10. $8x^2y^3$
11. $-42abc$
12. $25(xy)^2$

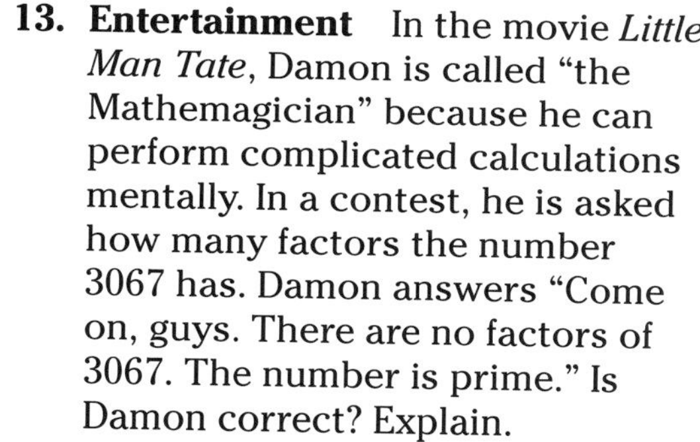

13. **Entertainment** In the movie *Little Man Tate*, Damon is called "the Mathemagician" because he can perform complicated calculations mentally. In a contest, he is asked how many factors the number 3067 has. Damon answers "Come on, guys. There are no factors of 3067. The number is prime." Is Damon correct? Explain.

Exercises: Practicing and Applying the Concept

Independent Practice

Determine whether each number is prime or composite.

14. 35
15. 19
16. 55
17. 101
18. 2

Factor each number or monomial completely.

19. 51
20. -63
21. 41
22. -95
23. 110
24. -333
25. 81
26. 59
27. 13
28. 1024
29. $42xy^2$
30. $38mnp$
31. $21xy^3$
32. $560x^4y^2$
33. $28f^2g$
34. $275st^3$
35. $210mn^3$
36. $-8a^3b^2$
37. $75m^2k$
38. $-400a^2b^3$

Critical Thinking

39. Find the least number that gives you a remainder of 1 when you divide it by 2 or by 3 or by 5 or by 7.
40. February the third is a *prime day* because the month and day (2/3) are represented by prime numbers. How many prime days are there in a leap year?

Applications and Problem Solving

41. **Cryptography** Use the formula in Example 3 to find p and q if $n = 1643$ for an RSA code.

42. **Number Theory** Mathematicians have many theories that are unproved. One theory is that there is an infinite number of *twin primes*. Twin primes are prime numbers that differ by 2, like 3 and 5. List all the twin primes that are less than 100.

43. **History** Chinese mathematician Sun-Tsŭ lived in the 1st century A.D. He studied *Chinese remainders*. This involves the remainders left when a number is divided by different primes.
   - **a.** Find the least positive integer having the remainders 2, 3, 2 when divided by 3, 5, and 7 respectively.
   - **b.** Is the least integer described in **part a** prime?

44. **Family Activity** The movie *Sneakers* involves a supposed invention that would break codes. Find the movie in a library or video store and watch it to see what the invention did.

Mixed Review

45. **Probability** If you toss a coin 4 times, how many different combinations of outcomes are possible? (Lesson 4-3)

46. Write $(-8)^7$ as the product of the same factor. (Lesson 4-2)

47. **Geometry** Find the perimeter and area of the rectangle at the right. (Lesson 3-5)

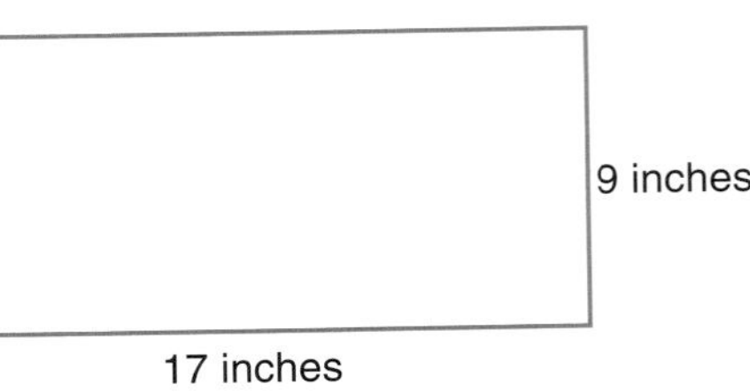

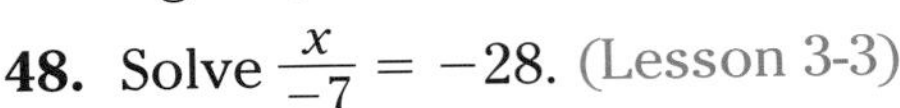

48. Solve $\frac{x}{-7} = -28$. (Lesson 3-3)

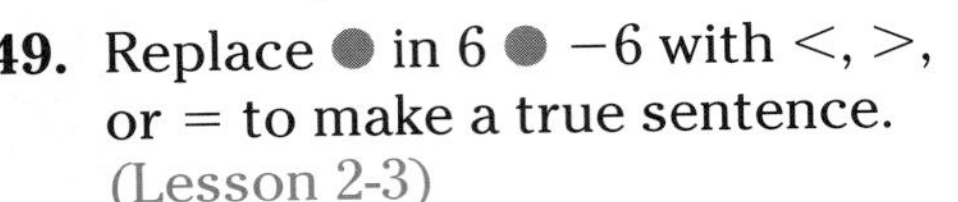

49. Replace ● in 6 ● −6 with $<$, $>$, or $=$ to make a true sentence. (Lesson 2-3)

50. Rewrite $3 \cdot (12 - 4)$ using the distributive property. (Lesson 1-5)

51. **Exercise** *Healthy Woman* magazine reports that you burn 15 additional calories each hour that you spend standing instead of sitting. Write an expression for the number of calories you burn in an hour of standing if you burn x calories per hour while sitting. (Lesson 1-3)

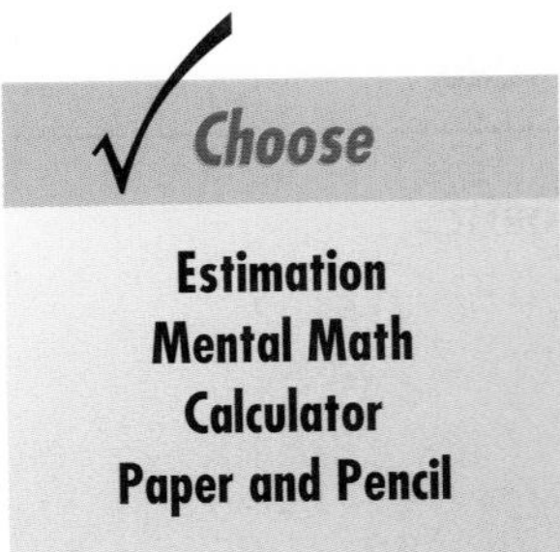

52. **Demographics** The average American earns \$1,235,720 and pays \$178,364 in taxes in his or her lifetime. The amount of money that is left in a paycheck after taxes is called take-home pay. What is the average American's lifetime take-home pay? (Lesson 1-1)
   - **a.** Which method of computation do you think is most appropriate? Justify your choice.
   - **b.** Solve the problem using the four-step plan. Be sure to examine your solution.

GRAPHING CALCULATOR ACTIVITY

4-4B Factor Patterns

An Extension of Lesson **4-4**

You already know that prime numbers have exactly two factors. Which numbers have exactly three factors? or four factors? In this Math Lab, you will investigate factor patterns.

The chart below shows the numbers 2 through 18 and their factors arranged according to the number of factors.

Exactly 2 Factors	Exactly 3 Factors	Exactly 4 Factors	Exactly 5 Factors	Exactly 6 Factors
2: 1, 2 **3:** 1, 3 **5:** 1, 5 **7:** 1, 7 **11:** 1, 11 **13:** 1, 13 **17:** 1, 17	**4:** 1, 2, 4 **9:** 1, 3, 9	**6:** 1, 2, 3, 6 **8:** 1, 2, 4, 8 **10:** 1, 2, 5, 10 **14:** 1, 2, 7, 14 **15:** 1, 3, 5, 15	**16:** 1, 2, 4, 8, 16	**12:** 1, 2, 3, 4, 6, 12 **18:** 1, 2, 3, 6, 9, 18

Your Turn Copy the chart above. Then use the TI-82 graphing calculator program below to find the factors of the numbers 19 through 50. Place each number in the correct column of the chart.

Begin entering the graphing calculator program by pressing PRGM ◀ ENTER. Then type in the program title, FACTOR. Consult your User's Guide if you need help finding the keys for the functions. When you finish entering the program, press 2nd QUIT to exit the programming mode. To run the program, press PRGM, the number next to the title FACTOR, and ENTER.

```
PROGRAM: FACTOR
: Input "ENTER NUMBER ", N
: For (D, 1, N)
: If iPart (N/D) = (N/D)
: Disp D
: End
```

1. Predict a number from 50 to 150 that can be placed in each column. Check your prediction by using the graphing calculator program.
2. Write a paragraph that describes the pattern in each column.

4-5 Greatest Common Factor (GCF)

Setting Goals: *In this lesson, you'll find the greatest common factor of two or more integers or monomials.*

Modeling a Real-World Application: History

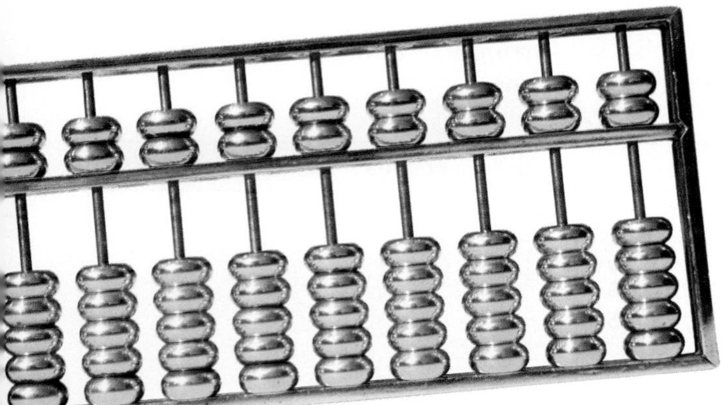

Very little of the mathematics of ancient China remains known to us today. The Chinese recorded information on bamboo, which did not last very long. Also, many of the books were destroyed during the reign of Emperor Shï Huang-ti. But the *Jiuzhang Suanshu*, or *The Nine Chapters on the Mathematical Art* survived to tell us a great deal about the mathematical accomplishments of the culture.

One of the chapters in the *Jiuzhang Suanshu* contains information on field measurement. In it, a procedure for finding the **greatest common factor (GCF)** of two numbers is described.

In Exercise 40, you will use the Chinese method to find the GCF.

Learning the Concept

The greatest of the factors of two or more numbers is called the greatest common factor. There are many methods for finding the greatest common factor. One method is to simply list the factors of each number and identify the greatest of the factors common to the numbers.

For example, consider finding the GCF of 78 and 91.

Factors of 78: 1, 2, 3, 6, 13, 26, 39, 78
Factors of 91: 1, 7, 13, 91

The common factors of 78 and 91, shown in blue, are 1 and 13. The GCF of 78 and 91 is 13.

Another method for finding the greatest common factor of two or more numbers is to find the prime factorization of the numbers and then find the product of their common factors.

Example  **Use prime factorization to find the GCF of each set of numbers.**

a. 315 and 135

First find the prime factorization of each number.

$$315 = 3 \cdot 105 = 3 \cdot 5 \cdot 21 = 3 \cdot 5 \cdot 3 \cdot 7$$

$$135 = 5 \cdot 27 = 5 \cdot 3 \cdot 9 = 5 \cdot 3 \cdot 3 \cdot 3$$

Then find the common factors.

$315 = 3 \cdot 5 \cdot 3 \cdot 7$ or $3 \cdot 3 \cdot 5 \cdot 7$ *The loops indicate each common factor.*

$135 = 5 \cdot 3 \cdot 3 \cdot 3$ or $3 \cdot 3 \cdot 5 \cdot 3$ *They are 3, 3, and 5.*

The greatest common factor of 315 and 135 is $3 \cdot 3 \cdot 5$ or 45.

b. 66, 90, and 150

Express each number as a product of prime factors.

$66 = 2 \cdot 3 \cdot 11$

$90 = 2 \cdot 3 \cdot 3 \cdot 5$ *The common factors are 2 and 3.*

$150 = 2 \cdot 3 \cdot 5 \cdot 5$

The GCF is $2 \cdot 3$ or 6.

Some real-life problems can be solved using the GCF.

Example 2 APPLICATION Hobbies

Shameka is covering the surface of an end table with equal-sized ceramic tiles. The table is 30 inches long and 24 inches wide.

a. What is the largest square tile that Shameka can use and not have to cut any tiles?

b. How many tiles will Shameka need?

a. The size of the tile is the greatest common factor of 30 and 24.

$30 = 2 \cdot 3 \cdot 5$

$24 = 2 \cdot 2 \cdot 2 \cdot 3$

The GCF of 30 and 24 is $2 \cdot 3$ or 6. So Shameka should use tiles that are 6-inch squares.

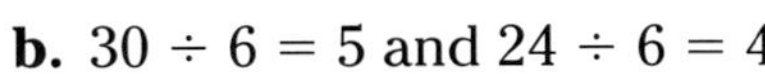

b. $30 \div 6 = 5$ and $24 \div 6 = 4$

So Shameka will need $5 \cdot 4$ or 20 tiles to cover the table.

Connection to Algebra

The product of the common prime factors of two or more monomials is their GCF.

Example 3 **Find the greatest common factor of $36x^3y$ and $56xy^2$.**

$36x^3y = 2 \cdot 2 \cdot 3 \cdot 3 \cdot x \cdot x \cdot x \cdot y$

$56xy^2 = 2 \cdot 2 \cdot 2 \cdot 7 \cdot x \cdot y \cdot y$

The GCF of $36x^3y$ and $56xy^2$ is $2 \cdot 2 \cdot x \cdot y$ or $4xy$.

Checking Your Understanding

Communicating Mathematics

Read and study the lesson to answer these questions.

1. **Explain** how to find the greatest common factor of two or more numbers.
2. **Find** the common factors of 21 and 45 by using factor trees.
3. Name two numbers whose GCF is 15.

4. In 1880, English mathematician John Venn developed the use of diagrams to show the relationships between collections of objects. The diagram at the right shows the prime factors of 16 and 28. The common factors are in both circles, so the GCF is $2 \cdot 2$ or 4. Find the greatest common factor of 36 and 48 by making a Venn diagram.

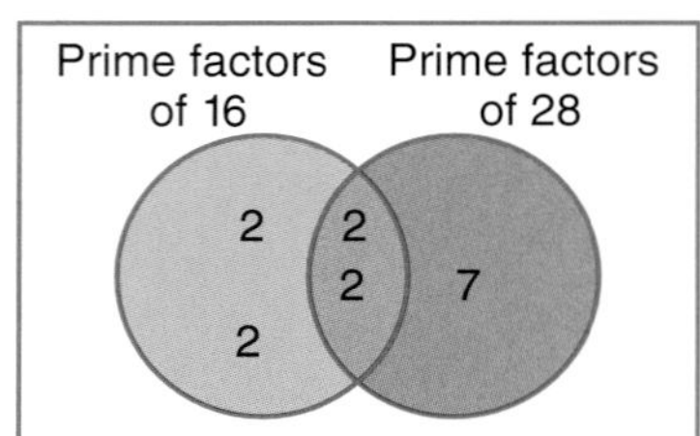

Guided Practice

Find the GCF of each set of numbers or monomials.

5. 6, 8
6. 12, 8
7. 1, 20
8. $12x$, $40x^2$
9. $-5ab$, $6b^2$
10. $15a^2b^2$, $27a^3b^3$
11. **Interior Design** Mrs. Garcia has picked out two different fabrics to use to make square pillows for her living room. One fabric comes in a width of 48 inches, and the other comes in a width of 60 inches. How long should each side of the squares for the pillows be if all the pillows are the same size and there is no fabric wasted?

Exercises: Practicing and Applying the Concept

Independent Practice

Find the GCF of each set of numbers or monomials.

12. 16, 56
13. 24, 40
14. 6, 8, 12
15. 12, 24, 36
16. 20, 21, 25
17. 108, 144
18. $14n$, $42n^2$
19. $40x^2$, $16x$
20. $24a^2$, $-60a$
21. $14b$, $-56b^2$
22. $33y^2$, $44y$
23. -18, $45mn$
24. $32mn^2$, $16n$, $12n^3$
25. $18a$, $30ab$, $42b$
26. $15v^2$, $36w^2$, $70vw$
27. $24x^3$, $36y^3$, $18xy$
28. $15r^2$, $35s^2$, $70rs$
29. $18ab$, $6a^2$, $42a^2b$

Two numbers are *relatively prime* if their only common factor is 1. Determine whether the numbers in each pair are relatively prime. Write *yes* or *no*.

30. 8 and 9 **31.** 11 and 13 **32.** 21 and 14

33. 25 and 30 **34.** 21 and 23 **35.** 9 and 12

Critical Thinking

36. Can the GCF of any set of numbers be greater than any one of the numbers? Explain.

Applications and Problem Solving

37. Patterns What is the GCF of all the numbers in the sequence 15, 30, 45, 60, 75, . . . ?

38. History Eratosthenes was a Greek mathematician who lived during the third century B.C. One of his contributions to the field was the *sieve of Eratosthenes*. Follow the steps below to complete a sieve.

MATERIALS

colored pencils or crayons

- ▶ Write the numbers 2 to 50 in a list.
- ▶ Circle 2, the first prime number. Then cross out every second number after 2.
- ▶ Using a different color, circle 3, the second prime number. Then cross out every third number after 3.
- ▶ Continue using different colors to circle prime numbers and cross out their multiples until all the numbers in the list have been circled or crossed out.

Determine the greatest common factor of 30 and 42 using the sieve. Explain how you found the GCF.

39. Carpentry Elena and her father are making shelves to store sports equipment and garden supplies in the garage. They would like to make the best use of a 48 in. by 72 in. piece of plywood. How many shelves measuring 12 in. by 16 in. could be cut from the plywood if there were no waste?

40. History Follow the steps outlined below to find the GCF of 42 and 86 by using the Chinese method mentioned at the beginning of the lesson.

- ▶ Subtract the lesser number, a, from the greater number, b.
- ▶ If the result is a factor of both numbers, it is the greatest common factor. If the result is not a factor of both numbers, subtract the result from a or subtract a from the result so that the difference is a positive number.
- ▶ Continue subtracting and checking the results until you find a number that is a factor of both numbers.

Mixed Review

41. Find the prime factorization of 3080. (Lesson 4-4)

42. Solve and graph $\frac{r}{4} \geq -14$. (Lesson 3-7)

43. Evaluate $|a| - |b| \cdot |c|$ if $a = -16$, $b = 2$, and $c = 3$. (Lesson 2-1)

44. **Sports** The table at the right shows the average cost of equipping an athlete for different sports. Make a bar graph of the costs. (Lesson 1-10)

Sport	Cost
Bowling	$100
Golf	$400
Racquetball	$60
Skiing	$500
Tennis	$75

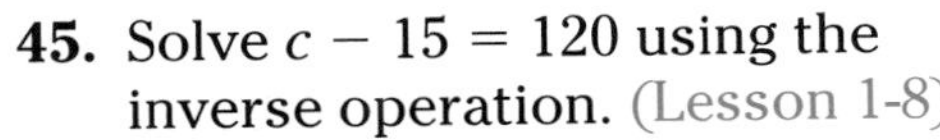

45. Solve $c - 15 = 120$ using the inverse operation. (Lesson 1-8)

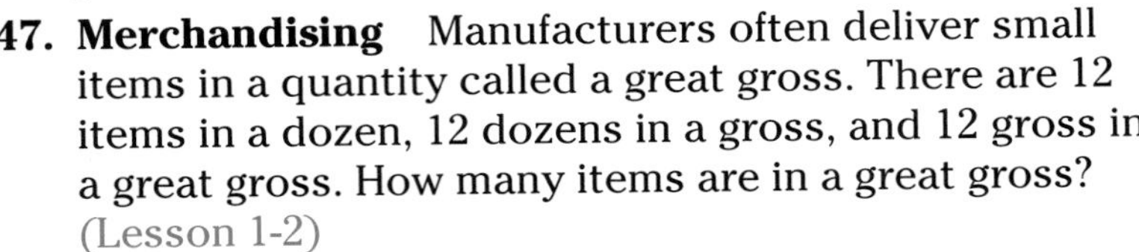

46. **Business** Book Distributors adds a $1.50 shipping and handling charge to the total price of every order. If the cost of a book is c, write an expression for the total cost. (Lesson 1-3)

47. **Merchandising** Manufacturers often deliver small items in a quantity called a great gross. There are 12 items in a dozen, 12 dozens in a gross, and 12 gross in a great gross. How many items are in a great gross? (Lesson 1-2)

48. **Automobiles** Mechanics recommend changing the motor oil in your car every 3000 miles in order to keep your car running its best. If you just changed the oil and the odometer reads 56,893 miles, at what odometer reading should you change the oil next? (Lesson 1-1)
    a. Which method of computation do you think is most appropriate? Justify your choice.
    b. Solve the problem using the four-step plan. Be sure to examine your solution.

Self Test

Using divisibility rules, state whether each number is divisible by 2, 3, 5, 6, or 10. (Lesson 4-1)

1. 117 **2.** 1002 **3.** 57

Write each product using exponents. (Lesson 4-2)

4. $9 \cdot 9 \cdot 9 \cdot 9 \cdot 9$ **5.** $2(2 \cdot 2)$ **6.** $\underbrace{k \cdot k \cdot k \cdot \ldots \cdot k \cdot k \cdot k}_{\text{28 factors}}$

7. How many cubes are in the twenty-fifth building in the sequence at the right? (Lesson 4-3)

Building 1

Building 2

Building 3

Factor each number or monomial completely. (Lesson 4-4)

8. 80 **9.** -26 **10.** $42xy^2$

Find the GCF of each set of numbers or monomials. (Lesson 4-5)

11. 42, 56 **12.** 9, 15, 24 **13.** $-18, 45xy$

4-6A

HANDS-ON ACTIVITY

Equivalent Fractions

A Preview of Lesson **4-6**

MATERIALS

ruler

It is often helpful to use a model to help you understand a concept. In this lab, you will use models to help you understand fractions.

Activity 1

Work with a partner to represent $\frac{1}{8}$ using a model.

Use a ruler to draw a rectangle and separate the rectangle into eight equal parts. Shade one part. Since one part in eight is shaded, $\frac{1}{8}$ of the rectangle is shaded.

Your Turn Draw a model to represent $\frac{1}{6}$.

1. How would you represent $\frac{2}{6}$ using a model?
2. If you shaded five parts of a rectangle with eight equal parts, what fraction is represented?

Activity 2

Compare $\frac{2}{8}$ and $\frac{3}{12}$ using models.

Draw two identical rectangles and separate one rectangle into eight equal parts. Separate the other into twelve equal parts. Shade two parts of the rectangle that has eight sections and shade three parts of the rectangle that has twelve sections.

3. What do you notice about the shaded regions in the two rectangles in Activity 2?
4. Do the fractions $\frac{2}{8}$ and $\frac{3}{12}$ name the same number? Explain.
5. Are there other fractions that name the same number? If so, draw a diagram to show an example.

Extension **Use models to determine whether each pair of fractions represents the same number.**

6. $\frac{1}{2}, \frac{5}{10}$ **7.** $\frac{1}{3}, \frac{3}{9}$ **8.** $\frac{1}{7}, \frac{2}{12}$

4-6 Simplifying Fractions

Setting Goals: *In this lesson, you'll simplify fractions using the GCF.*

Modeling a Real-World Application: Forestry

When the Rolling Stones, Eric Clapton, and the Allman Brothers need a keyboardist, they turn to Chuck Leavell. When the National Arbor Day Foundation, the world's largest tree-planting advocacy group, needs a farmer to speak on their behalf, they turn to Chuck Leavell, too.

Fifteen years ago, Mr. Leavell was a rock 'n' roller who had no experience with tree farming. His life changed when his wife's grandparents left her a 1200-acre farm in Twiggs County, Georgia. "The more I learned, the more I fell in love with it," said Leavell. Now the Leavells divide their time between their two loves, music and forestry. Their farm has now grown to 1500 acres. You can compare the original size of their farm to its current size by using a ratio.

Learning the Concept

A **ratio** is a comparison of two numbers by division. A ratio can be expressed in several ways. The expressions below all represent the same ratio.

$$2 \text{ to } 3 \qquad 2:3 \qquad \frac{2}{3} \qquad 2 \div 3$$

A ratio is most often written as a fraction in **simplest form**. A fraction is in simplest form when the GCF of the numerator and the denominator is 1.

Example 1 APPLICATION Forestry

Refer to the application at the beginning of the lesson. Write the ratio of the amount of land that Mrs. Leavell inherited to the size of the farm today in simplest form.

Mrs. Leavell inherited 1200 acres, and the farm is 1500 acres today. So the ratio is $\frac{1200}{1500}$.

First find the GCF of the numerator and denominator.

$1200 = 2 \cdot 2 \cdot 2 \cdot 2 \cdot 3 \cdot 5 \cdot 5$ *The GCF of 1200 and*

$1500 = 2 \cdot 2 \cdot 3 \cdot 5 \cdot 5 \cdot 5$ *1500 is $2 \cdot 2 \cdot 3 \cdot 5 \cdot 5$ or 300.*

Now divide the numerator and the denominator by 300 to write the fraction in simplest form.

$$\frac{1200}{1500} \overset{\div 300}{=} \frac{4}{5} \quad (\div 300)$$

Since the GCF of 4 and 5 is 1, the fraction $\frac{4}{5}$ is in simplest form.

The division in Example 1 can be represented in a different way.

$$\frac{1200}{1500} = \frac{\overset{1}{\cancel{2}} \cdot \overset{1}{\cancel{2}} \cdot 2 \cdot 2 \cdot \overset{1}{\cancel{3}} \cdot \overset{1}{\cancel{5}} \cdot \overset{1}{\cancel{5}}}{\underset{1}{\cancel{2}} \cdot \underset{1}{\cancel{2}} \cdot \underset{1}{\cancel{3}} \cdot 5 \cdot \underset{1}{\cancel{5}} \cdot \underset{1}{\cancel{5}}} = \frac{2 \cdot 2}{5} \text{ or } \frac{4}{5}$$

The slashes indicate that the numerator and denominator are both divided by $2 \cdot 2 \cdot 3 \cdot 5 \cdot 5$, the GCF.

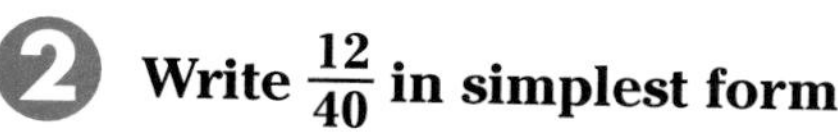

Example 2 **Write $\frac{12}{40}$ in simplest form.**

$$\frac{12}{40} = \frac{\overset{1}{\cancel{2}} \cdot \overset{1}{\cancel{2}} \cdot 3}{\underset{1}{\cancel{2}} \cdot \underset{1}{\cancel{2}} \cdot 2 \cdot 5}$$

Divide both the numerator and denominator by $2 \cdot 2$.

$$= \frac{3}{10}$$

Connection to Algebra

A fraction with variables in the numerator or denominator is called an **algebraic fraction**. Algebraic fractions can also be written in simplest form.

Example  **Simplify $\frac{16x^2y^3}{12xy^4}$. Assume that x and y are not equal to zero.**

THINK ABOUT IT

What is the GCF of $16x^2y^3$ and $12xy^4$?

$$\frac{16x^2y^3}{12xy^4} = \frac{\overset{1}{\cancel{2}} \cdot \overset{1}{\cancel{2}} \cdot 2 \cdot 2 \cdot \overset{1}{\cancel{x}} \cdot x \cdot \overset{1}{\cancel{y}} \cdot \overset{1}{\cancel{y}} \cdot \overset{1}{\cancel{y}}}{\underset{1}{\cancel{2}} \cdot \underset{1}{\cancel{2}} \cdot 3 \cdot \underset{1}{\cancel{x}} \cdot \underset{1}{\cancel{y}} \cdot \underset{1}{\cancel{y}} \cdot \underset{1}{\cancel{y}} \cdot y}$$

Divide both numerator and denominator by $2 \cdot 2 \cdot x \cdot y \cdot y \cdot y$.

$$= \frac{4x}{3y}$$

Checking Your Understanding

Communicating Mathematics

Read and study the lesson to answer these questions.

1. **Explain** what is meant by expressing a fraction in simplest form.
2. **Express** the ratio 12:14 as a fraction in simplest form.
3. **Give examples** of fractions in simplest form and fractions that are not in simplest form.
4. Why is it important to be able to simplify fractions?

5. The fractions $\frac{3}{4}$ and $\frac{6}{8}$ are graphed on the number lines at the right. Notice that the fractions have the same graph. Thus, $\frac{3}{4}$ and $\frac{6}{8}$ are equivalent. Use number lines to determine whether $\frac{3}{9}$ and $\frac{1}{3}$ are equivalent.

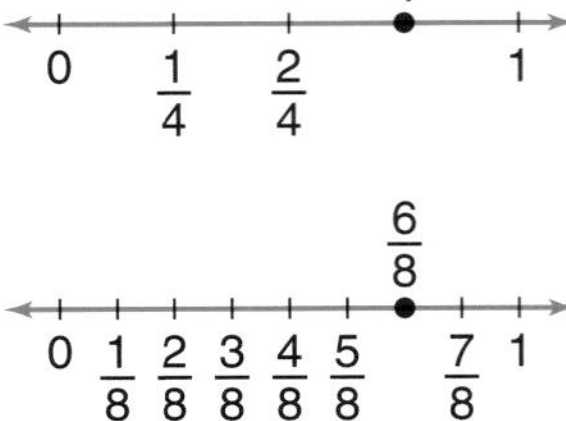

Guided Practice

Express each ratio as a fraction. Then if the fraction is not in simplest form, write it in simplest form.

6. 6 to 10 7. $1 \div 7$ 8. 20 : 100 9. eleven to fifteen

Write each fraction in simplest form. If the fraction is already in *simplest* form, write simplified.

10. $\frac{10}{37}$ 11. $\frac{15}{21}$ 12. $\frac{51}{60}$ 13. $\frac{18}{44}$

14. $\frac{x}{x^3}$ 15. $\frac{11t}{121t^2}$ 16. $\frac{8z^2}{16z}$ 17. $\frac{8ab}{15cd}$

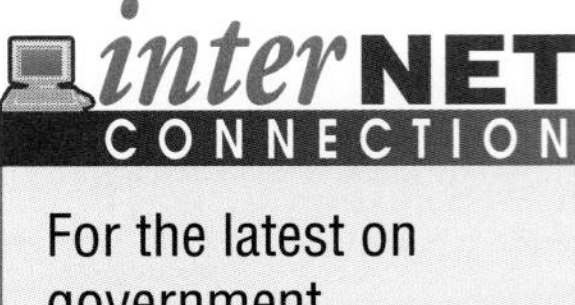

For the latest on government spending, visit: **www.glencoe.com/sec/math/prealg/mathnet**

18. **Economics** How does Uncle Sam spend a dollar? The graph at the right shows how each dollar spent by the Federal Government is used.
   a. Write the ratio of the amount spent on housing assistance to total spending as a fraction in simplest form.
   b. Write the ratio of the amount spent on the national debt to total spending as a fraction in simplest form.

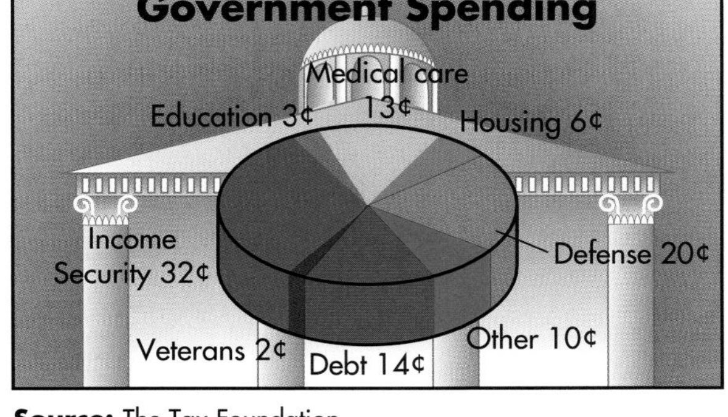

Source: The Tax Foundation

Exercises: Practicing and Applying the Concept

Independent Practice

Write each fraction in simplest form. If the fraction is already in simplest form, write *simplified*.

19. $\frac{2}{14}$ 20. $\frac{17}{20}$ 21. $\frac{34}{38}$ 22. $\frac{20}{53}$ 23. $\frac{17}{51}$

24. $\frac{9}{15}$ 25. $\frac{25}{40}$ 26. $\frac{30}{51}$ 27. $\frac{30}{37}$ 28. $\frac{16}{64}$

29. $\frac{11}{13}$ 30. $\frac{8}{27}$ 31. $\frac{124}{222}$ 32. $\frac{12m}{15m}$ 33. $\frac{12x}{15y}$

34. $\frac{40a}{42a}$ 35. $\frac{82t}{14t}$ 36. $\frac{12cd}{19ef}$ 37. $\frac{xyz^3}{x^2y}$ 38. $\frac{30x^2}{51xy}$

39. $\frac{40p^2q}{52pq^2}$ 40. $\frac{17k^2z}{51z}$ 41. $\frac{31gh}{14}$ 42. $\frac{52rst}{26rst}$ 43. $\frac{12x^2z}{23y^3}$

Critical Thinking

44. Suppose $(1 \times 10^a) + (2 \times 10^b) + (3 \times 10^c) + (4 \times 10^d) = 24{,}130$, and a, b, c and d are all positive integers. Find the value of $\frac{a + b + c + d}{16}$.

Applications and Problem Solving

45. **Transportation** According to *American Demographics* magazine, the average American worker spends 22 minutes traveling to work. What fraction of the day is this?

46. **Demographics** The United States is aging. Not only is the country getting older, but the number of older Americans increases each year. There are 2,500,000 Americans over the age of 85 and 30,000 of them are 100 or older.
    a. What is the ratio of the number of Americans age 100 or over to the number of Americans over age 85?
    b. Write the ratio in simplest terms.
    c. Explain what the ratio in part b means.

47. **Medicine** The University of California recommends using a soft gel pack to relieve muscle pain. To make a soft gel pack, fill a plastic bag with 2 ounces of rubbing alcohol and 6 ounces of water. Seal the bag and put it in a freezer. Apply to the injured area.
    a. What is the ratio of rubbing alcohol to water for the gel pack?
    b. Find a fraction in simplest form to represent the ratio.
    c. Find other amounts of rubbing alcohol and water that could be used to make a gel pack.

Mixed Review

48. Find the GCF of $42x^2y$ and $38xy^2$. (Lesson 4-5)
49. Solve $v = \ell \cdot w \cdot h$, by replacing the variables with the values $\ell = 6$, $w = 3$, and $h = 6$. (Lesson 3-4)
50. Solve $180 = 15r$. (Lesson 3-3)
51. Find the product of -4 and $-17d$. (Lesson 2-7)
52. **History** Ohio was declared a state in 1803. Hawaii became a state in 1959. Write a mathematical statement comparing the ages of these states in 1996. (Lesson 2-3)

53. Solve $\frac{144}{x} = 12$ mentally. (Lesson 1-6)
54. Evaluate $324 \div (2 \times 3 + 6)$. (Lesson 1-2)

4-7 Using the Least Common Multiple (LCM)

Setting Goals: *In this lesson, you'll find the least common multiple of two or more numbers and compare fractions.*

Modeling a Real-World Application: Astronomy

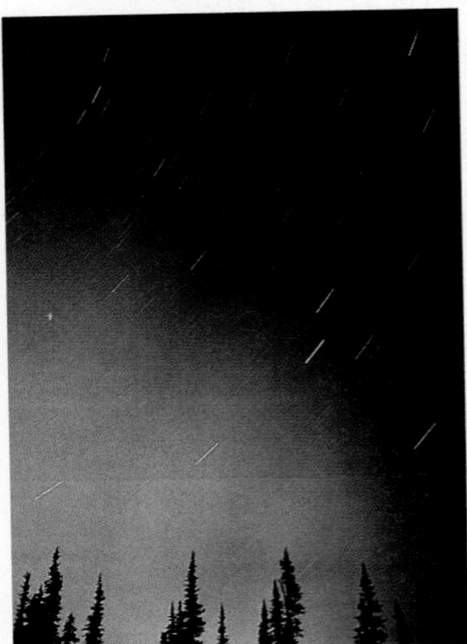

Some clear night, go to a park or out in the country and spend some time looking at the night sky. In ideal conditions, you will be able to see about 2500 stars, although there are many more than this. Some nights you will even be able to see some planets.

The chart at the right shows the number of Earth years it takes for some of the planets to revolve around the Sun. In 1982, the planets Jupiter and Saturn were in conjunction; that is, they appeared very close to each other in the sky. How often do these planets line up in these positions? *This problem will be solved in Example 1.*

Planet	Revolution Time (in Earth Years)
Earth	1
Jupiter	12
Saturn	30
Uranus	84

Learning the Concept

A **multiple** of a number is a product of that number and any whole number. *Recall that the whole numbers are 0, 1, 2, 3, . . .*

multiples of 2: **0**, 2, 4, 6, 8, **10**, 12, 14, 16, 18, **20**, . . .
multiples of 5: **0**, 5, **10**, 15, **20**, 25, 30, 35, 40, . . .

Multiples that are shared by two or more numbers are called **common multiples**. Some of the common multiples of 2 and 5, shown in blue, are 0, 10, and 20.

The least of the nonzero common multiples of two or more numbers is called the **least common multiple (LCM)** of the numbers. The least common multiple of 2 and 5 is 10.

Example 1

Astronomy

Refer to the application at the beginning of the lesson.

a. How often will Jupiter and Saturn appear close in the sky?
b. When will this happen next?
c. How often do Saturn, Jupiter, and Uranus coincide?

a. According to the table, Jupiter revolves around the Sun every 12 Earth years and Saturn every 30 years. The time required for the planets to coincide in this way again will be a multiple of 12 and 30.

multiples of 12: 0, 12, 24, 36, 48, **60**, . . .
multiples of 30: 0, 30, **60**, 90, 120, . . .

The least common multiple of 12 and 30 is 60. Jupiter and Saturn coincide every 60 years.

4-8 Multiplying and Dividing Monomials

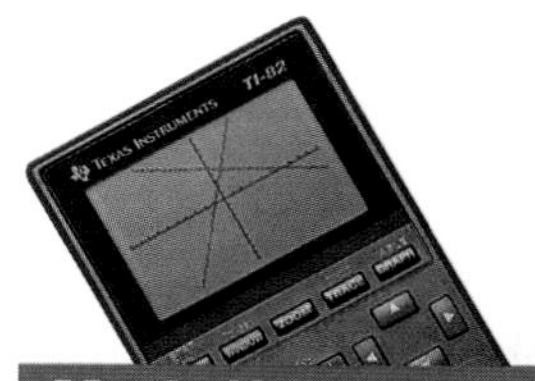

Setting Goals: *In this lesson, you'll multiply and divide monomials.*

Modeling with Technology

In dealing with large numbers and small numbers, calculators can simplify and speed our computation.

Your Turn Copy the tables below. Then use a calculator to find each product and quotient and complete the table.

Factors	Product	Product written as a power
$10^1 \cdot 10^1$		
$10^1 \cdot 10^2$		
$10^1 \cdot 10^3$		
$10^1 \cdot 10^4$		
$10^1 \cdot 10^5$		

Division	Quotient	Quotient written as a power
$10^5 \div 10^1$		
$10^4 \div 10^1$		
$10^3 \div 10^1$		
$10^2 \div 10^1$		
$10^1 \div 10^1$		

a. Compare the exponents of the factors to the exponent in the products. What do you observe?

b. Compare the exponents of the division expressions to the exponents in the quotients. What do you observe?

c. Write a rule for determining the exponent of the product when you multiply powers. Test your rule by multiplying $2^2 \cdot 2^4$ using a calculator.

d. Write a rule for determining the exponent in the quotient when you divide powers. Test your rule by dividing 8^6 by 8^4 on a calculator.

Learning the Concept

The pattern you observed in the products table above leads to the following rule for multiplying powers that have the same base.

Product of Powers

In words: You can multiply powers *that have the same base* by adding their exponents.

In symbols: For any number a and positive integers m and n, $a^m \cdot a^n = a^{m+n}$.

Example  **Find $8^4 \cdot 8^5$.**

$8^4 \cdot 8^5 = 8^{4+5}$ or 8^9

Check: $8^4 \cdot 8^5 = (8 \cdot 8 \cdot 8 \cdot 8)(8 \cdot 8 \cdot 8 \cdot 8 \cdot 8)$
$= 8 \cdot 8 \cdot 8 \cdot 8 \cdot 8 \cdot 8 \cdot 8 \cdot 8 \cdot 8$ or 8^9

Connection to Algebra

Some monomials can also be multiplied using the rule for the product of powers.

Example 2 **Find each product.**

a. $n^2 \cdot n^4$

$n^2 \cdot n^4 = n^{2+4}$ or n^6

b. $(-6x^2)(5x^3)$

$(-6x^2)(5x^3) = (-6 \cdot 5)(x^2 \cdot x^3)$ *Commutative and associative properties*
$= (-30)(x^{2+3})$ *Product of powers*
$= -30x^5$

You can also write a rule for finding quotients of powers. Study the pattern of division sentences below. Each dividend, divisor, and quotient has been replaced with a power of 2.

$2^1 = 2$
$2^2 = 4$
$2^3 = 8$
$2^4 = 16$
$2^5 = 32$
$2^6 = 64$
$2^7 = 128$
$2^8 = 256$

$16 \div 2 = 8$	$64 \div 8 = 8$	$256 \div 8 = 32$
$2^4 \div 2^1 = 2^3$	$2^6 \div 2^3 = 2^3$	$2^8 \div 2^3 = 2^5$

Look at the exponents only. Do you see a pattern?

Quotient of Powers

In words: You can divide powers *that have the same base* by subtracting their exponents.

In symbols: For any nonzero number a and whole numbers m and n, $\frac{a^m}{a^n} = a^{m-n}$.

Example  **Find each quotient.**

a. $\frac{9^7}{9^5}$

$\frac{9^7}{9^5} = 9^{7-5}$ or 9^2

Check: $\frac{9^7}{9^5} = \frac{\overset{1}{\cancel{9}} \cdot \overset{1}{\cancel{9}} \cdot \overset{1}{\cancel{9}} \cdot \overset{1}{\cancel{9}} \cdot \overset{1}{\cancel{9}} \cdot 9 \cdot 9}{\underset{1}{\cancel{9}} \cdot \underset{1}{\cancel{9}} \cdot \underset{1}{\cancel{9}} \cdot \underset{1}{\cancel{9}} \cdot \underset{1}{\cancel{9}}}$
$= 9 \cdot 9$
$= 9^2$

b. $\frac{c^8}{c^8}$

$\frac{c^8}{c^8} = c^{8-8}$
$= c^0$ or 1

Check: $\frac{c^8}{c^8} = \frac{\overset{1}{\cancel{c}} \cdot \overset{1}{\cancel{c}} \cdot \overset{1}{\cancel{c}} \cdot \overset{1}{\cancel{c}} \cdot \overset{1}{\cancel{c}} \cdot \overset{1}{\cancel{c}} \cdot \overset{1}{\cancel{c}} \cdot \overset{1}{\cancel{c}}}{\underset{1}{\cancel{c}} \cdot \underset{1}{\cancel{c}} \cdot \underset{1}{\cancel{c}} \cdot \underset{1}{\cancel{c}} \cdot \underset{1}{\cancel{c}} \cdot \underset{1}{\cancel{c}} \cdot \underset{1}{\cancel{c}} \cdot \underset{1}{\cancel{c}}}$
$= 1$

Many applications involve calculations with very large or very small numbers. The quotient of powers rule often makes these calculations less complicated.

Example 4

APPLICATION

Seismology

The intensity of an earthquake is usually measured on the Richter scale. On this scale the number designation given to an earthquake represents the intensity as a power of 10. For example, an earthquake measuring 8 on the Richter scale has an intensity of 10^7. In April, 1992, an earthquake measuring 5 on the Richter scale shook the Netherlands. Twelve days later, northern California experienced an earthquake measuring 7. How many times more intense was the California earthquake?

The Netherlands earthquake had an intensity of 10^4, and the California earthquake had an intensity of 10^6. Divide 10^6 by 10^4 to find out how many times more intense the California earthquake was.

$$\frac{10^6}{10^4} = 10^{6-4} \quad \textit{Quotient of powers}$$
$$= 10^2 \text{ or } 100$$

The California earthquake was 100 times more intense.

Communicating Mathematics

Read and study the lesson to answer these questions.

1. **Explain** how to find the product $6^6 \cdot 6^3$.
2. **Justify** the product of powers rule by multiplying 2^2 and 2^5 using the definition of exponents.
3. Make up a division problem whose solution is 4^4.

MATH JOURNAL

4. Describe, in your own words, the relationship between the rules for multiplying powers and dividing powers.

Guided Practice

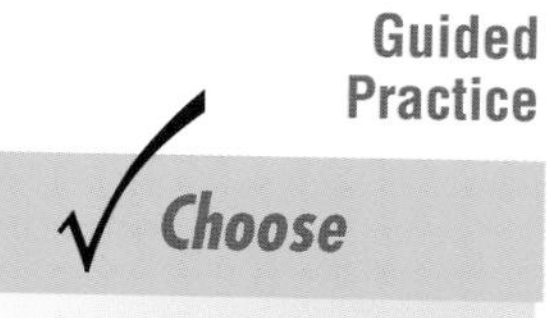

Choose

Estimation
Mental Math
Calculator
Paper and Pencil

Find each product or quotient. Express your answer in exponential form.

5. $10^4 \cdot 10^3$
6. $\frac{2^5}{2^2}$
7. $3^6 \cdot 3^4$
8. $\frac{8^4}{8^3}$
9. $8^2 \cdot 8^3 \cdot 8$
10. $m \cdot m^6$
11. $\frac{y^{11}}{y^9}$
12. $\frac{ab^4}{b^2}$
13. **Seismology** On June 6, 1994, an earthquake measuring 6.8 on the Richter scale struck southwest Colombia. Three days later an earthquake measuring 8.2 hit La Paz, Bolivia. *Approximately* how many times more intense was the earthquake in Bolivia?

Exercises: Practicing and Applying the Concept

Independent Practice

Find each product or quotient. Express your answer in exponential form.

14. $10^5 \cdot 10^5$ **15.** $w \cdot w^5$ **16.** $3^3 \cdot 3^2$ **17.** $b^5 \cdot b^2$

18. $\frac{5^5}{5^2}$ **19.** $\frac{10^{10}}{10^3}$ **20.** $\frac{z^3}{z}$ **21.** $\frac{a^{10}}{a^6}$

22. $\frac{(-2)^6}{(-2)^5}$ **23.** $(5x^3)(4x^4)$ **24.** $\frac{(-x)^4}{(-x)^3}$ **25.** $\frac{f^{20}}{f^8}$

26. $(3x^4)(-5x^2)$ **27.** $3a^2 \cdot 4a^3$ **28.** $(-10x^3)(2x^2)$

29. $a^2b \cdot ab^3$ **30.** $x^4(x^3y^2)$ **31.** $t^3t^2 \div t^4$

32. $\frac{y^{100}}{y^{100}}$ **33.** $k^3m^2 \div km$ **34.** $(-3y^3z)(7y^4)$

35. $w^5xy^2 \div wx$ **36.** $7ab(a^{16}c)$ **37.** $15n^9q^2 \div 5nq^2$

Find each missing exponent.

38. $(5^{\bullet})(5^3) = 5^{11}$ **39.** $x(x^2)(x^6) = x^{\bullet}$ **40.** $\frac{t^{\bullet}}{t} = t^{14}$

41. $3^5 \cdot 3^{\bullet} = 3^8$ **42.** $\frac{12^5}{12^{\bullet}} = 1$ **43.** $\frac{2^5}{4^2} = 2^{\bullet}$

Critical Thinking

44. If $10^n \div 10^m = 1$, what can you conclude about n and m?

Applications and Problem Solving

45. Chemistry The pH of a solution is a measure of its acidity. A low pH indicates an acid and a high pH indicates a base. Neutral water has a pH of 7; this is the dividing line between acids and bases. Each one-unit increase or decrease in the pH means that the intensity of the acid or base is changed by a factor of ten. For example, a pH of 4 is ten times more acidic than a pH of 5.

a. Acid rain is an environmental problem related to the burning of fossil fuels. Suppose the pH of a lake is 5 due to acid rain. How much more acidic is the lake than neutral water?

b. Human blood usually has a pH of about 7.4. Is it an acid or a base?

46. Biology When bacteria reproduce, they split so that one cell becomes two. The number of cells after t time periods is 2^t.

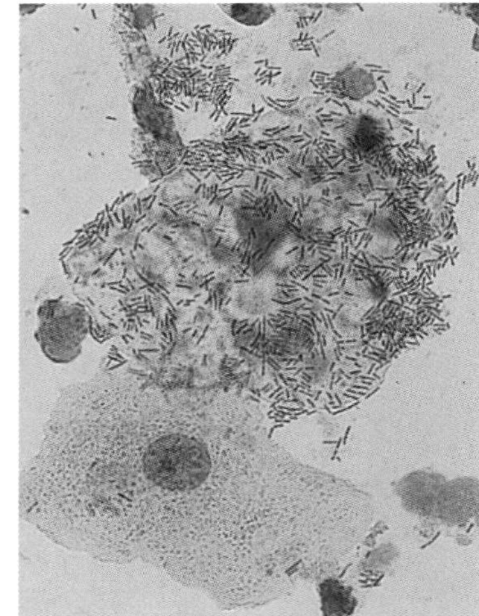

a. *E. coli* reproduce very quickly, about every 15 minutes. If there are 100 *e. coli* in a dish now, how many will there be in 30 minutes?

b. How many more *e. coli* are there in a population after 3 hours than there were after 1 hour?

Mixed Review

47. Replace ● with < or > to make $\frac{5}{8} \bullet \frac{8}{11}$ a true inequality. (Lesson 4-7)

48. Find the prime factorization of $16ab^3$. (Lesson 4-4)

49. Use divisibility rules to determine whether 298 is divisible by 2, 3, 5, 6, or 10. (Lesson 4-1)

50. **Art** A photograph measuring 9" by 11" is matted and placed in a 10" by 12" frame. Find the area of the matting. (Lesson 3-5)

51. Solve $5 + n = -2$ and check your solution. Graph the solution on a number line. (Lesson 3-2)

52. Solve $s = 40 - (-17)$. (Lesson 2-5)

53. In which quadrant is the graph of $(-3, 4)$ located? (Lesson 2-2)

54. Is $4x < 16$ *true*, *false*, or *open*? (Lesson 1-9)

55. **Music** A survey of 6500 Top-10 songs released between 1900 and 1975 showed that "I" is the most frequent first word of a song. "My" was the second most popular first word. "I" was the first word of four less than twice as many songs as the word "my." Write an expression for the number of songs that begin with "my" if x songs started with "I." (Lesson 1-3)

WORKING ON THE

Investigation

Refer to the Investigation on pages 166–167.

You can use a **spreadsheet** to find the value of your portfolio on any given day. Using a spreadsheet, it is possible to project results, make calculations, and print almost anything that can be arranged in a table.

The basic unit of a spreadsheet is called a **cell**. A cell may contain numbers, words (called labels), or a formula. Each cell is named by the column and row that describes its location. The cell C2 is the box at the intersection of column C and row 2. *Cell C2 in the spreadsheet at the right contains the number 20.*

As you know, stock prices are listed as mixed numbers. You can enter a mixed number in a spreadsheet as an expression. For example, the price of Ameritech's stock was listed as $48\frac{3}{8}$. It can be entered in cell B2 of the spreadsheet as $48 + \left(\frac{3}{8}\right)$ or 48.375.

When cells are related, a formula can be used to generate this relationship. If you type B2*C2 into cell D2, the cell will not show the formula, but the result of the multiplication of the contents of B2 and C2.

	A	B	C	D
1	Company	Price	Shares	Value
2	Ameritech	48.375	20	
3	Exxon	70.375	10	
4	GE	50	100	
5	GM	48	40	
6	IBM	109.625	10	
7	Sears	32.75	10	
8	Total Value			

- Write spreadsheet formulas for cells D3, D4, D5, D6, D7, and D8.
- Create your own spreadsheet to find the value of your portfolio. Update the value using share prices in a local or national newspaper each week. Also document any changes in your portfolio such as selling old stock holdings and buying new ones.

Add the results of your work to your Investigation Folder.

4-9 Negative Exponents

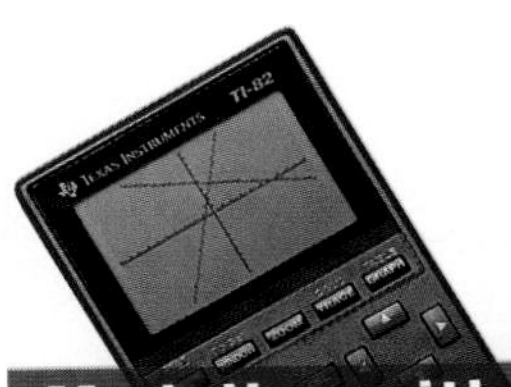

Setting Goals: *In this lesson, you'll use negative exponents in expressions and equations.*

Modeling with Technology

You can often discover important concepts by looking for a pattern.

Your Turn Copy the table at the right. Then use a calculator to complete the second column.

Expression	Value
2^6	64
2^5	32
2^4	
2^3	
2^2	
2^1	
2^0	

TALK ABOUT IT

a. Describe how successive expressions differ.

b. Describe how successive values differ.

c. If the pattern continues, what is the next expression?

d. If the pattern continues, what is the next value?

e. Use a calculator to verify that the expression you wrote for part c is equal to the value you wrote for part d.

Learning the Concept

Extending the pattern above suggests that $2^{-1} = \frac{1}{2}$. You can use the quotient of powers rule and the definition of a power to simplify the expression $\frac{x^3}{x^6}$ and write a general rule about negative powers.

Method 1: Quotient of Powers Rule

$$\frac{x^3}{x^6} = x^{3-6}$$

$$= x^{-3}$$

Method 2: Definition of Powers

$$\frac{x^3}{x^6} = \frac{\overset{1}{\cancel{x}} \cdot \overset{1}{\cancel{x}} \cdot \overset{1}{\cancel{x}}}{\underset{1}{\cancel{x}} \cdot \underset{1}{\cancel{x}} \cdot \underset{1}{\cancel{x}} \cdot x \cdot x \cdot x}$$

$$= \frac{1}{x \cdot x \cdot x}$$

$$= \frac{1}{x^3}$$

Since $\frac{x^3}{x^6}$ cannot have two different values, you can conclude that x^{-3} is equal to $\frac{1}{x^3}$. This and other examples suggest the following definition.

Negative Exponents For any nonzero number a and any integer n, $a^{-n} = \frac{1}{a^n}$.

Example  Write each expression using positive exponents.

a. 10^{-5}

$10^{-5} = \frac{1}{10^5}$ *Definition of negative exponents*

b. ab^{-3}

$ab^{-3} = a \cdot b^{-3}$

$= a \cdot \frac{1}{b^3}$ *Definition of negative exponents*

$= \frac{a}{b^3}$

Negative exponents are often used in science when dealing with very small numbers.

Example 

CONNECTION

Biology

Red blood cells are circular-shaped cells that carry oxygen through your bloodstream. A red blood cell is about 7.75×10^{-7} meters across.

a. Write the width of a red blood cell using positive exponents.

b. Write the width of a red blood cell as a decimal.

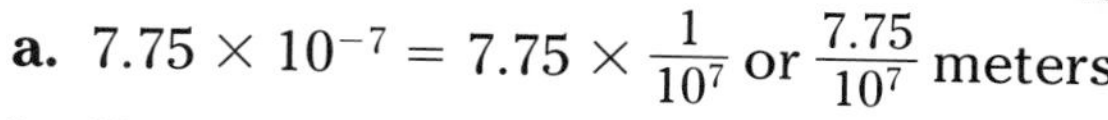

a. $7.75 \times 10^{-7} = 7.75 \times \frac{1}{10^7}$ or $\frac{7.75}{10^7}$ meters

b. Use a calculator to write the number as a decimal.

7.75 [÷] 10 [y^x] 7 [=] 0.000000775

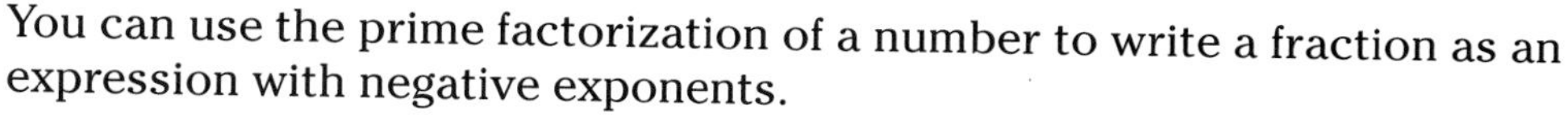

You can use the prime factorization of a number to write a fraction as an expression with negative exponents.

Example  Write each fraction as an expression using negative exponents.

a. $\frac{1}{8}$

$\frac{1}{8} = \frac{1}{2^3}$

$= 2^{-3}$ *Definition of negative exponents*

b. $\frac{x}{y^2}$

$\frac{x}{y^2} = x \cdot \frac{1}{y^2}$

$= x \cdot y^{-2}$ or xy^{-2}

Connection to Algebra

Expressions involving variables can also contain negative exponents.

Example 4 **Evaluate the expression $4x^{-3}$ if $x = 3$.**

$$4x^{-3} = 4(3)^{-3} \quad \textit{Replace x with 3.}$$
$$= 4\left(\frac{1}{3^3}\right) \quad \textit{Definition of negative exponents}$$
$$= 4\left(\frac{1}{27}\right) \quad \textit{Find } 3^3.$$
$$= \frac{4}{27} \quad \textit{Multiply.}$$

Checking Your Understanding

Communicating Mathematics

Read and study the lesson to answer these questions.

1. **Express** x^{-3} with a positive exponent.
2. **Write** a convincing argument that $2^0 = 1$ using the fact that $2^4 = 16$, $2^3 = 8$, $2^2 = 4$, and $2^1 = 2$.

3. Write in your own words the relationship between the rules for multiplying powers and dividing powers.

Guided Practice

Write each expression using positive exponents.

4. 3^{-3}
5. 15^{-5}
6. a^{-12}

Write each fraction as an expression using negative exponents.

7. $\frac{1}{10^7}$
8. $\frac{1}{4^3}$
9. $\frac{1}{9}$

Evaluate each expression.

10. 2^x if $x = -2$
11. $3n^{-2}$ if $n = 4$

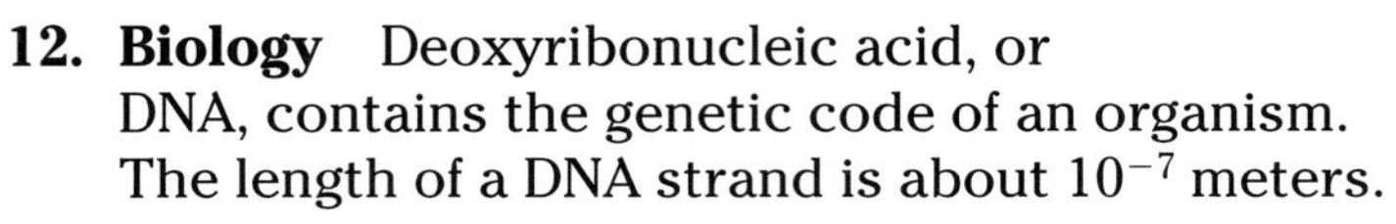

12. **Biology** Deoxyribonucleic acid, or DNA, contains the genetic code of an organism. The length of a DNA strand is about 10^{-7} meters.
    a. Write the length of a DNA strand using positive exponents.
    b. Write the length of a DNA strand as a decimal.

Practicing and Applying the Concept

Independent Practice

Write each expression using positive exponents.

13. 4^{-1}
14. 5^{-2}
15. $(-2)^{-5}$
16. x^{-1}
17. $s^{-2}t^{-1}$
18. mn^{-2}
19. $2(xy)^{-2}$
20. $\frac{1}{2^{-2}}$

Write each fraction as an expression using negative exponents.

21. $\frac{1}{5^3}$
22. $\frac{1}{a}$
23. $\frac{1}{16}$
24. $\frac{1}{x^6}$
25. $\frac{2}{3^2}$
26. $\frac{a}{b^6}$
27. $\frac{f}{g^3}$
28. $\frac{3m}{n^2}$

Evaluate each expression.

29. 3^n if $n = -3$
30. $4x^{-3}y^2$ if $x = 2$ and $y = 6$
31. $6t^{-2}$ if $t = 3$
32. $(2b)^{-3}$ if $b = -2$

Find each product or quotient. Express using positive exponents.

33. $(a^6)(a^{-3})$
34. $(b^{-10})(b^5)$
35. $\frac{c^5}{c^{-2}}$
36. $\frac{d^2}{d^3}$

Critical Thinking

37. Simplify $\frac{a^b}{a^{a-b}}$.

Applications and Problem Solving

38. **Physical Science** Electromagnetic waves are used to transmit television and radio signals. The diagram at the right shows the electromagnetic wave spectrum.
   a. Write the length of a light wave using a positive exponent.
   b. What is the range of lengths of infrared waves written in decimal form?

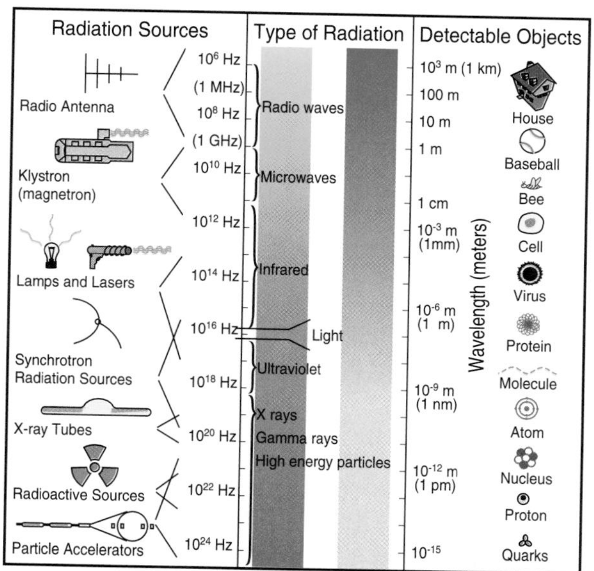

39. **Astronomy** Any two objects in space have an attraction that can be calculated by using a formula written by English scientist Henry Cavendish. The formula uses the universal gravitational constant, which is 6.67×10^{-11} Nm^2/kg^2. (N is newtons.)
   a. Write the universal gravitational constant by using positive exponents.
   b. Write the constant as a decimal.

40. **Physics** An object has energy if it is able to produce a change in itself or its surroundings. For example, people use energy to run or to digest food. Scientists use the joule as the unit in which they measure energy. A joule is 1 kg $\cdot$ m^2 $\cdot$ s^{-2}. Write this unit as a fraction using positive exponents.

Mixed Review

41. Find the product of $(10x)$ and $(-5x^3)$. (Lesson 4-8)

42. **Sports** Calida made 8 of 14 free throws in her last basketball game. How would you describe her success as a fraction in simplest form? (Lesson 4-6)

43. Write the product $7 \cdot 7 \cdot 7 \cdot 7 \cdot 7$ using exponents. (Lesson 4-2)

44. Solve $V = \frac{4}{3}\pi r^3$ for V if $\pi = 3.14$ and $r = 5$. (Lesson 3-4)

45. **Personal Finance** When Chandra and Scott got married, they put their savings into one account. Chandra had \$2230 and together they had \$4750. Write and solve an equation to find the amount of money Scott had. (Lesson 3-2)

46. Evaluate $3c - ac$ if $a = -4$ and $c = 2$. (Lesson 2-7)

47. Write and solve an equation for *eight less than a number is three.* (Lesson 1-8)

48. **Exercise** According to one medical study, you can add 21 minutes to your life for every mile that you walk. If you walked 2 miles each day for a year, how much time would you add to your life? (Lesson 1-1)
    a. Which method of computation do you think is most appropriate? Justify your choice.
    b. Solve the problem using the four-step plan. Be sure to examine your solution.

GLOBAL WARMING

You have heard about the dreadful winter suffered by General George Washington and his troops at Valley Forge. At that time, Earth was nearing the end of what has been called the "Little Ice Age." In fact, the temperatures on Earth have fluctuated from very warm to very cold many times.

Beginning in the 1980s, environmental advocates have been warning us about a global warming crisis. Carbon dioxide and other gases, called greenhouse gases, blanket Earth and keep it warm. It is said that the burning of fossil fuels adds too much carbon dioxide to the atmosphere, making Earth dangerously warm.

See for Yourself

- The carbon dioxide that is released from human activity accounts for $\frac{174}{10^7}$ of all of the greenhouse gases released each year. Write this number as a decimal.
- Some of the other greenhouse gases are chlorofluorocarbons, methane, nitrous oxide, and carbon monoxide. Research the activities that contribute to the production of these gases.
- Investigate the benefits of the greenhouse effect. Why is it necessary for life on Earth?
- Write a paragraph or two stating your opinion on the greenhouse effect. Is there any danger? If so, what should be done?

CHAPTER 4 Highlights

Vocabulary

After completing this chapter, you should be able to define each term, property, or phrase and give an example or two of each.

Algebra

algebraic fraction (p. 197)
base (p. 175)
common multiples (p. 200)
composite number (p. 184)
divisible (p. 170)
expanded form (p. 176)
exponent (p. 175)
factor trees (p. 185)
factors (p. 170)
greatest common factor (GCF) (p. 190)
least common denominator (LCD) (p. 201)
least common multiple (LCM) (p. 200)
monomial (p. 172)
multiple (p. 200)
negative exponents (p. 210)
order of operations (p. 176)
powers (p. 175)
prime factorization (p. 185)
prime number (p. 184)
product of powers (p. 205)
quotient of powers (p. 206)
ratio (p. 196)
simplest form (p. 196)
standard form (p. 176)

Problem Solving

draw a diagram (p. 181)
guess and check (p. 186)
look for a pattern (p. 171)

Understanding and Using Vocabulary

Complete each statement.

1. The greatest integer that is a factor of each of two or more integers is called their ____________.
2. Numbers that are expressed using exponents are called ______.
3. The ______ of a whole number divide that number with a remainder of zero.
4. A ________ is a whole number greater than one that has exactly two factors, 1 and itself.
5. The product of a number and any whole number is called a ______ of the original number.
6. ________ is writing numbers using place values and exponents.
7. The least common multiple of the denominators of two or more fractions is called the ____________.
8. A ______ is an integer, a variable, or a product of integers or variables.
9. A comparison of two numbers by division is called a ____.

CHAPTER **4** # Study Guide and Assessment

Skills and Concepts

Objectives and Examples	Review Exercises

After completing this chapter, you should be able to:

- **determine whether one number is a factor of another.** (Lesson 4-1)

Is 342 divisible by 2, 3, 5, 6, or 10?

It is divisible by 2 since its last digit is 2.

It is divisible by 3 because 3 + 4 + 2 or 9 is divisible by 3.

The last digit is not 5 or 0, so it is not divisible by 5.

It is divisible by 6 since it is divisible by 2 and 3.

It is not divisible by 10 since its last digit is not 0.

Use these exercises to review and prepare for the chapter test.

Using divisibility rules, state whether each number is divisible by 2, 3, 5, 6, or 10.

10. 863 **11.** 635
12. 4200 **13.** 14,577
14. 40, 180 **15.** 582
16. 5103 **17.** 1962

- **use powers and exponents in expressions and equations.** (Lesson 4-2)

Write a^5 as a product of the same factor.

$a^5 = a \cdot a \cdot a \cdot a \cdot a$

Write 3 · 3 · 3 · 3 using exponents.

$3 \cdot 3 \cdot 3 \cdot 3 = 3^4$

Write each multiplication expression using exponents.

18. $6 \cdot 6 \cdot 6 \cdot 6 \cdot 6$
19. $(c + 2)(c + 2)(c + 2)$
20. $(-4)(-4)(-4)(-4)$
21. $(1 \cdot 1) \cdot 1$

Write each power as a multiplication expression.

22. $(-16)^3$ **23.** k^9
24. $(-w)^4$ **25.** 32^{11}

- **write the prime factorization of composite numbers.** (Lesson 4-4)

Write the prime factorization of 648.

$648 = 9 \cdot 72$
$= 3 \cdot 3 \cdot 8 \cdot 9$
$= 3 \cdot 3 \cdot 2 \cdot 4 \cdot 3 \cdot 3$
$= 3 \cdot 3 \cdot 2 \cdot 2 \cdot 2 \cdot 3 \cdot 3$

Factor each number or monomial completely.

26. 120 **27.** -114
28. $630x^3$ **29.** $-825x^2y$
30. 2805 **31.** $-1827jk^3$
32. $66g^5h^2$ **33.** $550mnq^2$

Objectives and Examples

▶ **find the greatest common factor of two or more integers or monomials.** (Lesson 4-5)

Find the GCF of $42x^2$ and $110xy$.

$42x^2 = 2 \cdot 3 \cdot 7 \cdot x \cdot x$

$110xy = 2 \cdot 5 \cdot 11 \cdot x \cdot y$

The GCF is $2 \cdot x$ or $2x$.

Review Exercises

Find the GCF of each set of numbers or monomials.

34. 70, 66
35. 1092, 325
36. $100x$, $84xy$, $-76x^2$
37. $210a^3b^2$, $875a^2b^3$
38. $112j^5k$, $-144j^2$
39. $40x$, $25y$
40. $115wx$, $224wz$
41. $441ac$, $223bd$

▶ **simplify fractions using the GCF.** (Lesson 4-6)

Write $\frac{6}{45}$ in simplest form.

$\frac{6}{45} = \frac{2}{15}$ (÷ 3) *The GCF of 6 and 45 is 3.*

Simplify $\frac{15ac^2}{24ab}$.

$$\frac{15ac^2}{24ab} = \frac{\overset{1}{\cancel{3}} \cdot 5 \cdot \overset{1}{\cancel{a}} \cdot c \cdot c}{2 \cdot 2 \cdot 2 \cdot \underset{1}{\cancel{3}} \cdot \underset{1}{\cancel{a}} \cdot b}$$

$$= \frac{5c^2}{8b}$$

Write each fraction in simplest form. If the fraction is already in simplest form, write *simplified*.

42. $\frac{72}{88}$
43. $\frac{133}{140}$
44. $\frac{225}{315}$
45. $\frac{48}{66}$
46. $\frac{23a}{32b}$
47. $\frac{500x}{1000x}$
48. $-\frac{125mn}{625m}$
49. $\frac{68b^2}{105c}$

▶ **find the least common multiple of two or more numbers.** (Lesson 4-7)

Find the LCM of 18 and 15.

$18 = 2 \cdot 3 \cdot 3$

$15 = 3 \cdot 5$

The LCM is $2 \cdot 3 \cdot 3 \cdot 5 = 90$

Find the least common multiple of each set of numbers or algebraic expressions.

50. $11ab$, $6b$
51. 12, 4, 5
52. $7x^3$, $49xy$
53. $10j$, $6jk$, 3
54. $5f^4g$, $13fg^2$
55. 12, 15, 18
56. $72x^3$, $64x^5$
57. $6ab$, $18c$, $36d$

▶ **compare fractions.** (Lesson 4-7)

Which is greater, $\frac{3}{4}$ or $\frac{7}{12}$?

$\frac{3}{4} = \frac{9}{12}$ (× 3) *The LCM of 4 and 12 is 12.*

$\frac{9}{12} > \frac{7}{12}$, so $\frac{3}{4} > \frac{7}{12}$.

Replace each ● with <, >, or = to make each statement true.

58. $\frac{6}{7}$ ● $\frac{5}{14}$
59. $\frac{9}{16}$ ● $\frac{8}{15}$
60. $\frac{9}{100}$ ● $\frac{1}{9}$
61. $\frac{1}{11}$ ● $\frac{2}{23}$

Objectives and Examples

- **multiply and divide monomials.** (Lesson 4-8)

Find each product or quotient.

$$g^2 \cdot g^7 = g^{2+7} = g^9$$

$$\frac{(-3)^5}{(-3)^2} = (-3)^{5-2} = (-3)^3$$

Review Exercises

Find each product or quotient. Express your answer in exponential form.

62. $12^4 \cdot 12^7$

63. $\frac{d^9}{d^4}$

64. $n^5 \cdot n^{14}$

65. $\frac{(-4)^5}{(-4)}$

66. $(6w^3)(10w^4)$

67. $\frac{6x^{40}}{3x^{18}}$

68. $\frac{30c^2d}{15c}$

69. $22ab^3 \cdot 5a^9$

- **use negative exponents in expressions.** (Lesson 4-9)

Write $-2x^3y^{-2}$ using positive exponents.

$$-2x^3y^{-2} = -2x^3 \cdot y^{-2} = -2x^3 \cdot \frac{1}{y^2} = \frac{-2x^3}{y^2}$$

Write each expression using positive exponents.

70. $(-5)^{-3}$

71. $4c^{-2}d$

72. $\frac{1}{6^{-18}}$

73. $4x(yz)^{-2}$

74. $\left(\frac{2}{x}\right)^{-4}$

75. $\frac{3}{(f)^{-9}}$

Applications and Problem Solving

76. Business The Carpet Experts cleaning team can clean carpet in a room that is 10 feet by 10 feet in 20 minutes. Draw a diagram to find how long it would take them to do a walk-in closet that is 5 feet by 5 feet. (Lesson 4-3)

77. Sports The City Monday-Night Softball League has 7 teams. The director is making a schedule where every team plays every other team once. How many games must be scheduled? (Lesson 4-3)

78. Number Theory Numbers that have a greatest common factor of 1 are said to be **relatively prime**. Find the least two composite numbers that are relatively prime. (Lesson 4-5)

79. Literature *USA TODAY* polled Americans on the types of books that they buy. The graph below shows the results of the poll. Which type of book is more popular, travel or home improvement? (Lesson 4-7)

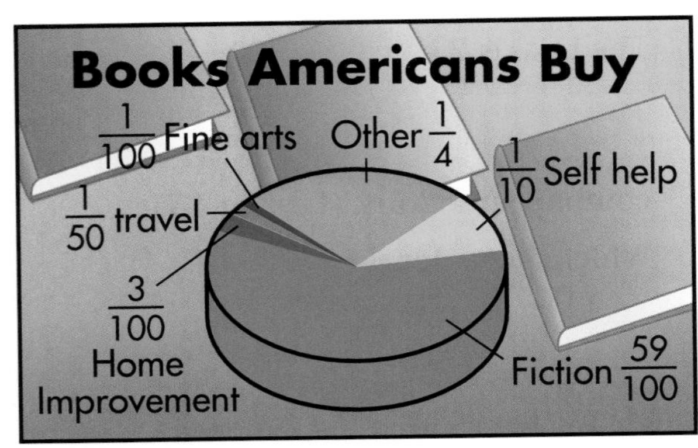

Source: *USA-Today*

A practice test for Chapter 4 is available on page 777 .

Alternative Assessment

Performance Task

Demonstrate your knowledge by giving a clear, concise solution to each problem. Be sure to include all relevant drawings and justify your answers. You may show your solutions in more than one way or investigate beyond the requirements of the problem.

A planned community housing project is providing a gardening area for its residents. The area will be laid out in individual garden plots as shown.

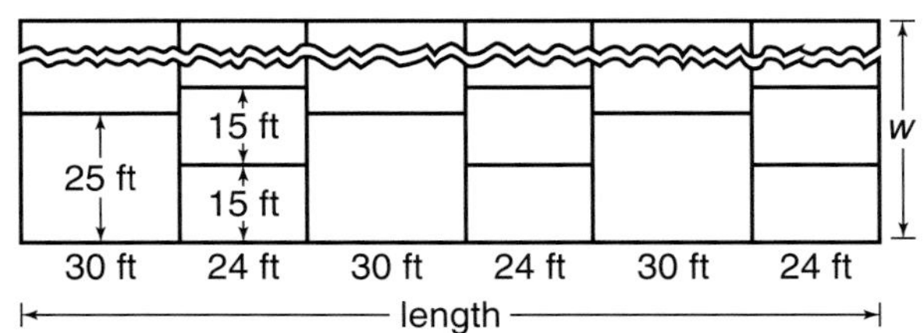

1. Explain how to find the prime factorization of a number.
2. Find the prime factorizations of 30 and 24.
3. Find the common factors of 30 and 24.
4. The garden area will be fenced. The fence posts in front are to be equally spaced along the entire length with posts at the corners of each garden plot. Would a spacing of 3 feet between posts work? How do you know?
5. What is the longest spacing between posts that can be used? What is this distance called?
6. List the multiples of the widths of each size lot.
7. If the back fence of the overall gardening area is to be straight, give at least two possible widths, w, of the area.
8. What is the shortest possible width of the gardening area? Explain.
9. If the fence posts are to be equally spaced along the width of the garden area with posts at the corners of each garden plot, what spacing would you recommend? Why?

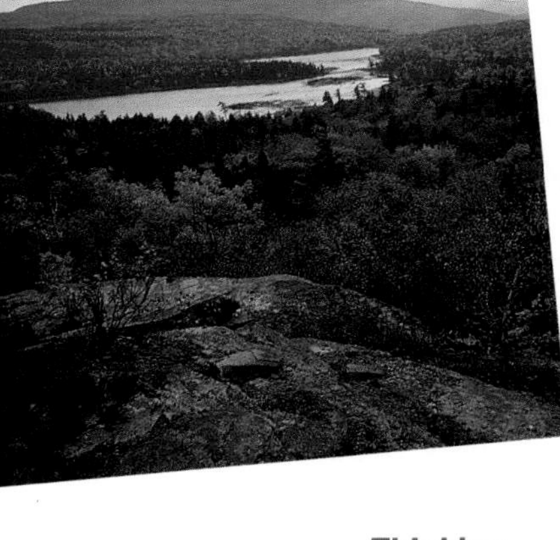

Thinking Critically

- Find the GCF and LCM of 45 and 72.
- Explain why you think the greatest common factor is called that when it is less than the numbers given.
- Explain why you think the least common multiple is called that when it is greater than the numbers given.

Portfolio

Select an item from your work in this chapter that shows your creativity and place it in your portfolio.

Self Evaluation

It isn't climbing the hill that makes you tired. It's the rock in your shoe. You are able to reach the top of the hill, if you don't let anything stop you.

Assess yourself. When you feel that rock in your shoe, what do you do? Do you stop and take it out or do you hope it moves to a place where it won't bother you? In mathematics class or in life, stop and ask a question or ask for help so you can remove the rock and continue to the top.

CHAPTERS 1–4 Ongoing Assessment

Section One: Multiple Choice

There are eight multiple-choice questions in this section. After working each problem, write the letter of the correct answer on your paper.

1. The family sizes of students in a mathematics class are recorded in the following table.

Family Size	2	3	4	5	6	7	8
Number of Students	1	4	3	6	5	2	3

How many students have fewer than 5 members in their family?

A. 6
B. 8
C. 9
D. 14

2. A group of divers needs to descend to a depth 3 times their present depth of -25 meters. At what depth do they need to be?

A. 75 meters
B. 28 meters
C. -25 meters
D. -75 meters

3. Simplify $|-12| - |-3|$.

A. -9
B. -15
C. 9
D. 15

4. An airplane descended 250 feet in 5 minutes. What was its average rate of descent?

A. 50 ft/min
B. 250 ft/min
C. 1250 ft/min
D. Not here

5. Which of the following is equivalent to 4^{-3}?

A. $3 \cdot 3 \cdot 3 \cdot 3$
B. $4 \cdot 4 \cdot 4$
C. $\frac{1}{3 \cdot 3 \cdot 3 \cdot 3}$
D. $\frac{1}{4 \cdot 4 \cdot 4}$

6. Which fraction is in simplest form?

A. $\frac{12a^2}{15ab}$
B. $\frac{15a}{50}$
C. $\frac{12d^2}{25ab}$
D. $\frac{5b^2c}{7bh^2}$

7. Which set of numbers is in order from least to greatest?

A. $\{-5, -3, 1, 14\}$
B. $\{-1, -6, 8, 11\}$
C. $\{15, -8, 1\}$
D. $\{10, 0, -5\}$

8. The distance traveled is given by the formula $d = rt$, where r is the rate of speed and t is the time spent traveling. How long would it take to ride a bike 9 miles at the rate of 6 miles per hour?

A. 15 min
B. 1.5 hours
C. 54 hours
D. 54 miles

Section Two: Free Response

This section contains eight questions for which you will provide short answers. Write your answer on your paper.

9. The temperature when you left for school this morning was 10°F. The wind speed was 20 mph, which produced a windchill of −24°F. What is the difference between the actual temperature and the windchill temperature?

10. Before she paid the electric bill, Rocio's checking account had a balance of $131. Her new account balance is $109. How much was her electric bill?

11. Juwan's yard is 10 yards wide and 20 yards long. His dad wants to build a rectangular swimming pool that has an area of 100 square feet. What are the possible dimensions of the pool?

12. Simplify $12a + 8b - 6a$.

13. A college basketball court is 94 feet long by 50 feet wide. What is the area of the court?

14. Nituna must earn 250 points to win a prize. She has already earned 176 points. Write a sentence to find p, the number of points she still needs to earn.

15. Carrie is cutting a round pizza. What is the greatest number of pieces she can get from 1 pizza with 5 straight cuts?

16. In high school football, a team can score 6 points for a touchdown, 7 points for a touchdown and an extra point, 8 points for a touchdown and a two-point conversion, 3 points for a field goal, or 2 points for a safety. Can a team have a score of 23 points? Explain.

Test-Taking Tip

You can solve many problems without much calculating if you understand the basic mathematical concepts. Always look carefully at what is asked, and think of possible shortcuts for solving the problem.

Section Three: Open-Ended

This section contains two open-ended problems. Demonstrate your knowledge by giving a clear, concise solution to each problem. Your score on these problems will depend on how well you do the following.

- Explain your reasoning.
- Show your understanding of the mathematics in an organized manner.
- Use charts, graphs, and diagrams in your explanation.
- Show the solution in more than one way or relate it to other situations.
- Investigate beyond the requirements of the problem.

17. Drina, Susan, Latisha, and Joshua are in the orchestra at school. Each student plays one of the following instruments: drums, violin, trumpet, or trombone. If each student plays a different instrument, use the following information to match each student with an instrument.
    — Drina and Joshua both play brass instruments.
    — The trumpet player and Joshua also play in the jazz band.
    — Susan lost one of her sticks last week.

18. Determine whether 435 is divisible by 2, 3, 5, 6, or 10. Justify your answers.

CHAPTER

5 Rationals: Patterns in Addition and Subtraction

TOP STORIES in Chapter 5

In this chapter, you will:

- estimate and find sums and differences of rational numbers,
- solve equations and inequalities involving rational numbers,
- solve problems using logical reasoning, and
- find terms of an arithmetic sequence.

MATH AND FITNESS IN THE NEWS

Is thin *really* in?

Source: TIME, January 16, 1995

The 1980s were not the healthy years that diet and fitness experts had hoped. According to a long-term federal study, Americans gained weight and increased their related health risks in that decade. The percentage of Americans who are seriously overweight jumped to about one-third during the 1980s. About *58 million* people in the U.S. are now classified as obese, that is they weigh at least 20% more than their ideal body weight. The percentage of overweight teenagers also rose, from 15% in the early 1970s to 21% in 1991, a 40% increase. As a nation, Americans are eating too much food, eating unhealthy food, and not exercising enough. Possible causes include heavy advertising by the food industry, eating out more often, eating more fast food, and spending more time in front of televisions and computers instead of exercising.

Putting It into Perspective

1762
The fourth Earl of Sandwich orders slices of meat and cheese served between slices of bread; the sandwich is invented.

1853
The potato chip is invented by American Indian chef George Crum in Saratoga Springs, NY.

1870
Congress creates post of Surgeon General to oversee health of U.S. Navy. Today, all public health is the concern of the Surgeon General.

On the Lighter Side

"Boy, Dad, you're sure being all you can be!"

interNET CONNECTION For up-to-date information on nutrition and weight, visit: **www.glencoe.com/sec/math/prealg/mathnet**

Statistical Snapshot

How hard do you exercise?
If you monitor how hard you exert yourself, you can estimate the number of calories you burn. The table below shows the number of calories a 130-pound person burns in an hour of different activities.

Exertion Level	Calories per hour	Type of activity
No exertion	60	sleeping, watching television reading
Almost no exertion	120	eating, doing desk work, driving washing dishes, strolling
Very light exertion	180	bowling, slowly walking walking down stairs
Light exertion	240	mopping, golfing with a cart leisurely biking, raking leaves
Moderate exertion	300	walking, riding a stationary bike playing softball, calisthenics
Vigorous exertion	360	Leisurely swimming, mowing grass shoveling light snow
Heavy exertion	420	jogging, shoveling deep snow playing tennis, casual racquetball
Extreme exertion	560 or more	sprinting, jogging uphill jumping rope

Source: *The New England Journal of Medicine*

1900 — 1950 — 2000

1904
The ice-cream cone is born at the St. Louis World's Fair when an ice cream vendor runs out of dishes and uses rolled waffles as a substitute.

1941
The United States publishes their first set of Recommended Daily Allowances (RDAs) of essential nutrients.

1994
Nutrition labels appear on all food packages as a result of U.S. federal law.

1995
Study shows alarming increase in number of overweight Americans.

5-1 Rational Numbers

Setting Goals: *In this lesson you'll identify and compare rational numbers and rename decimals as fractions.*

Modeling a Real-World Application: Movies

Were you frightened by the dinosaurs of *Jurassic Park*? Or amazed by the flight of the Enterprise in *Star Trek*? Then you have Jeff Olson to thank. Mr. Olson created miniature models for these and other blockbuster movies. Movie makers often use models to simulate situations that would be difficult or impossible to do in real life.

To create models, Mr. Olson uses a ratio called a **scale**. A scale is a comparison of the size of a model to the size of the real object.

$$\text{scale} = \frac{\textit{size of model}}{\textit{size of real object}}$$

The scale tells how much smaller or larger the model has to be than the real object. Suppose Mr. Olson was building a 15-inch model of a Bigfoot that would be 120 inches tall in real life. The scale would be $\frac{15}{120}$ or $\frac{1}{8}$. That means absolutely everything on the model is $\frac{1}{8}$ the size of the same item on the original Bigfoot. Numbers such as $\frac{1}{8}$ that can be written as fractions are called **rational numbers**.

Learning the Concept

Numbers can be organized into sets. One set is the **whole numbers**.

Whole Numbers
This set includes 0, 1, 2, 3, . . . It also includes any number that can be written as a whole number, such as $\frac{5}{5}$ or $\frac{9}{1}$.

0 $\frac{6}{6}$ 15 $\frac{9}{1}$ 27

Whole Numbers

Another set of numbers is the **integers**.

Integers
This set includes . . . , −2, −1, 0, 1, 2, . . . Notice that all whole numbers are included in the set of integers.

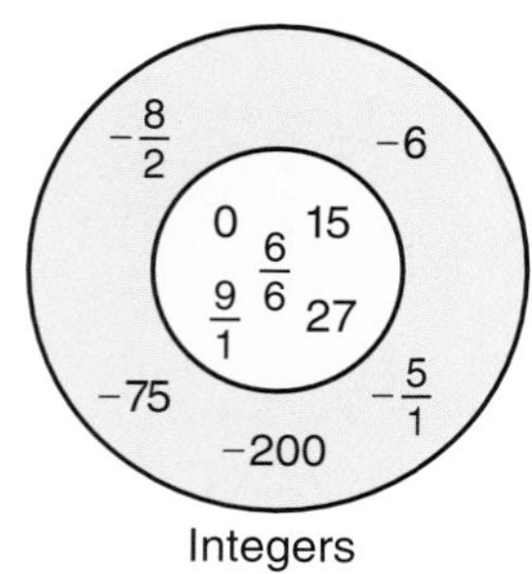

Integers

Refer to the application at the beginning of the lesson. The model of Bigfoot had a scale of $\frac{1}{8}$. Both 1 and 8 are integers, but $\frac{1}{8}$ is not. The number system is extended to include numbers like $\frac{1}{8}$. Numbers like $\frac{1}{8}$ are rational numbers.

Definition of a Rational Number	**Any number that can be expressed in the form $\frac{a}{b}$, where a and b are integers and $b \neq 0$, is called a rational number.**

Rational Numbers
This set includes common fractions, such as $\frac{1}{4}$. It also includes mixed numbers, decimals, integers, and whole numbers because all these numbers can be written in the form $\frac{a}{b}$.

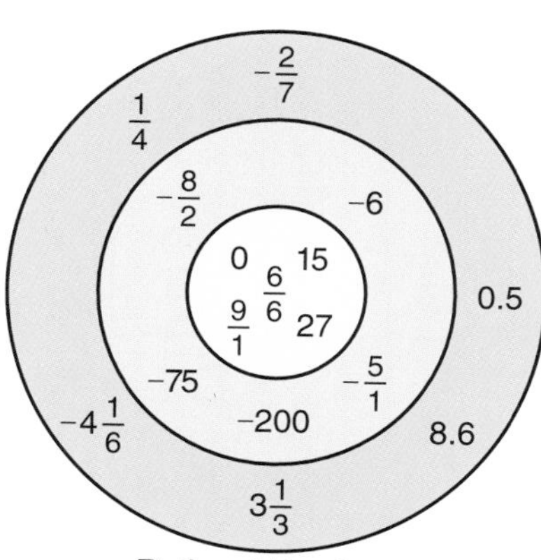

Rational Numbers

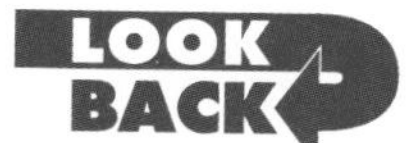

You can review simplifying fractions in Lesson 4-6.

Some decimals are rational numbers. Decimals either terminate or they go on forever. The decimal 3.74 is an example of a **terminating decimal**. Every terminating decimal can be written as a fraction with a denominator of 10, 100, 1000, and so on. For example, $0.75 = \frac{75}{100}$ or $\frac{3}{4}$, and $1.3 = 1\frac{3}{10}$ or $\frac{13}{10}$. So terminating decimals are always rational numbers.

Biology

The average adult's fingernails grow 0.125 inch each month. Write 0.125 as a fraction.

$$0.125 = \frac{125}{1000}$$

$$= \frac{1}{8}$$ *Simplify. The GCF of 125 and 1000 is 125.*

FYI

Shridhar Chillal of India has the longest fingernails in the world. On March 3, 1993, the nails on his left hand were a total of 205 inches long.

Decimals like 0.4444444444 . . . are called **repeating decimals**. Because it is inconvenient to write all of these digits, you can use the **bar notation** $0.\overline{4}$ to indicate that the 4 repeats forever. Here are some other examples.

6.23232323232323 . . .	$= 6.\overline{23}$	The digits 23 repeat.
8.4613613613613 . . .	$= 8.4\overline{613}$	The digits 613 repeat.

Repeating decimals can always be written as fractions. For example, $0.\overline{4}$ is equivalent to $\frac{4}{9}$. So, repeating decimals are always rational numbers.

Connection to Algebra

Example 2 shows how to rename repeating decimals as fractions.

Example  Express each repeating decimal as a fraction or mixed number in simplest form.

a. $0.\overline{5}$

Let $N = 0.555\ldots$ Then $10N = 5.555\ldots$ *Multiply N by 10 because one digit repeats.*

Subtract N from $10N$ to eliminate the repeating part.

$$\begin{aligned} 10N &= 5.555\ldots \quad \textit{Recall } 10N - N = 9N. \\ -\ 1N &= 0.555\ldots \\ \hline 9N &= 5 \\ N &= \frac{5}{9} \quad \textit{Divide each side by 9.} \end{aligned}$$

Therefore $0.\overline{5} = \frac{5}{9}$.

THINK ABOUT IT

What multiplier should you use to express $0.\overline{718}$ as a fraction?

b. 3.363636 . . .

Let $N = 3.363636\ldots$ Then $100N = 336.3636\ldots$. *Multiply N by 100 because two digits repeat.*

$$\begin{aligned} 100N &= 336.3636\ldots \\ -\ N &= 3.3636\ldots \\ \hline 99N &= 333 \\ N &= \frac{333}{99} \text{ or } 3\frac{4}{11} \end{aligned}$$

So $3.363636\ldots = 3\frac{4}{11}$.

Decimals that do not terminate and do not repeat such as 4.252627 . . . , cannot be written as fractions. So they are not rational numbers.

Thus, the set of rational numbers includes whole numbers, integers, fractions, terminating decimals, and repeating decimals.

Example  Name the set(s) of numbers to which each number belongs.

a. −7 -7 is an integer and a rational number.

b. $-3\frac{2}{5}$ Since $-3\frac{2}{5}$ can be written as $-\frac{17}{5}$, it is a rational number. It is not a whole number nor is it an integer.

c. 0.545556 . . . This is a non-terminating, non-repeating decimal. So it is not rational.

d. 1.469 1.469 is a terminating decimal. It is a rational number.

Like integers, rational numbers can be graphed on a number line. Examples of rational numbers graphed on a number line are shown below.

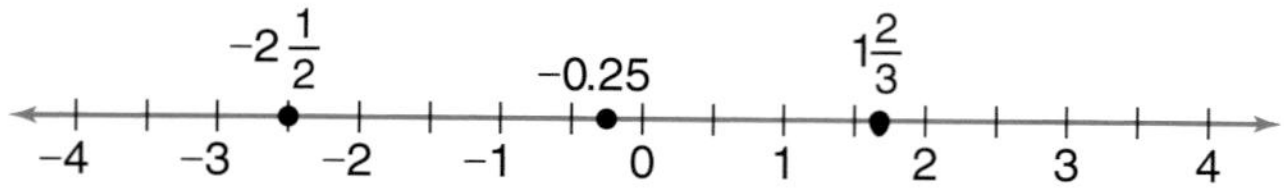

You can use a number line to compare rational numbers.

Example 4 **Use a number line to choose the greater number from each pair.**

a. $-\frac{2}{3}, -\frac{1}{2}$

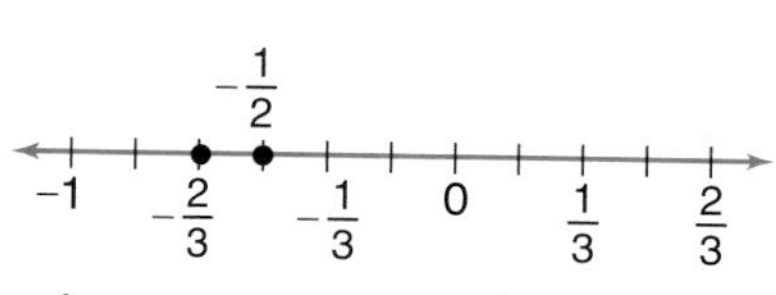

$-\frac{1}{2}$ is greater or $-\frac{1}{2} > -\frac{2}{3}$.

b. $0.3, \frac{1}{3}$

$0.3 = \frac{3}{10}$

0.3 $\frac{1}{3}$

0 $\frac{1}{10}$ $\frac{1}{5}$ $\frac{3}{10}$ $\frac{2}{5}$

$\frac{1}{3}$ is greater or $\frac{1}{3} > 0.3$.

Checking Your Understanding

Communicating Mathematics

Read and study the lesson to answer these questions.

1. **Explain** what a rational number is.
2. Are all integers also rational numbers? Explain.
3. Use $\frac{2}{3}$, $\frac{4}{5}$, and < to write a true statement.
4. **Explain** why 0.13 and $0.\overline{13}$ are rational numbers.
5. Use a ruler to draw a number line. Mark tick marks that are 1 centimeter apart from −8 to 8.
   a. Mark $3\frac{1}{5}$, $-1\frac{1}{3}$, 0.7, −5.6, $4\frac{1}{3}$, 4.3, and −2.9 on the number line.
   b. Which of the points indicates the greatest number?
   c. Which number in the list is the least?

MODELING MATHEMATICS

MATERIALS

ruler

Guided Practice

Express each decimal as a fraction or mixed number in simplest form.

6. 0.8 7. 0.05 8. −9.64 9. $0.\overline{2}$

Name the set(s) of numbers to which each number belongs.

10. −3 11. $-8\frac{3}{4}$ 12. 15 13. 0.414243 . . .

Replace each ● with <, >, or = to make each sentence true. Use a number line if necessary.

14. $\frac{3}{7}$ ● $-\frac{3}{7}$ 15. $\frac{3}{4}$ ● 0.75 16. −3.78 ● −3.88

17. **Hobbies** A model airplane has a wingspan of 27 inches. The full-sized aircraft has a wingspan of 36 feet or 432 inches.
    a. What is the scale of the model plane?
    b. Is the scale a whole number, an integer, or a rational number? Explain.

CONNECTION

For the latest on models, visit: **www.glencoe.com/sec/math/prealg/mathnet**

Exercises: Practicing and Applying the Concept

Independent Practice

Express each decimal as a fraction or mixed number in simplest form.

18. 0.4 19. −0.7 20. 0.23 21. 0.57
22. 0.48 23. 0.03 24. $0.\overline{8}$ 25. −0.333 . . .
26. −0.51 27. $0.\overline{25}$ 28. 0.375 29. $2.\overline{34}$

Name the set(s) of numbers to which each number belongs.

30. -14 **31.** $-\frac{6}{7}$ **32.** $-3\frac{7}{8}$ **33.** $\frac{35}{7}$

34. 0.12 **35.** $-3.3738\ldots$ **36.** -0.17 **37.** 3.11

38. -10 **39.** $29\frac{1}{2}$ **40.** -32.0 **41.** $0.1234\ldots$

Replace each ● with >, <, or = to make a true sentence. Use a number line if necessary.

42. $\frac{3}{4}$ ● $-\frac{3}{8}$ **43.** $-6\frac{1}{2}$ ● $-6.\overline{5}$ **44.** 9.9 ● $9.\overline{8}$

45. $\frac{3}{5}$ ● 0.6 **46.** $\frac{1}{7}$ ● $0.222\ldots$ **47.** $\frac{13}{26}$ ● $-\frac{1}{2}$

48. $0.3\overline{4}$ ● $0.\overline{34}$ **49.** $\frac{1}{11}$ ● 0.1 **50.** $-1\frac{1}{3}$ ● $-1.\overline{3}$

51. Which number is the greatest, $\frac{2}{7}$, $\frac{2}{9}$, or $\frac{2}{11}$?

52. Which number is the greatest, $-\frac{3}{5}$, $-\frac{5}{6}$, or $-\frac{6}{7}$?

Critical Thinking

53. Does $\frac{6}{2.4}$ name a rational number? Explain.

54. A machinist made a steel peg 2.37 inches in diameter for a $2\frac{3}{8}$-inch diameter hole. Will the peg fit? How do you know?

Applications and Problem Solving

55. Measurement A gauge measured the thickness of a piece of metal as 0.025 inch. What fraction of an inch is this?

56. Manufacturing A GLAD Garbage Bag has a thickness of 0.8 mils. This is 0.0008 inch. What fraction of an inch is this?

57. Engineering Ingrid Proctor-Fridia works for the United States Department of Defense checking to see that airplane parts conform to the blueprints. Blueprints are drawings of objects that show their scale.

a. The wingspan of an airplane on a blueprint is 26.9 centimeters. If the wingspan of the actual airplane is 8.07 meters, or 807 centimeters, what is the scale of the blueprints?

b. Is the scale a rational number? Explain.

Mixed Review

58. Physical Science The wavelength of visible red light is 7.50×10^{-7} meters. (Lesson 4-9)

a. Write the size of the wavelength using positive exponents.

b. Write the size of the wavelength using decimals.

59. Find the least common multiple of $4x^2$, $3x$, and 5. (Lesson 4-7)

60. Is 945 divisible by 2, 3, 5, 6, or 10? (Lesson 4-1)

61. Solve $-4b > 72$. (Lesson 3-7)

62. Solve $\frac{y}{12} = -7$. (Lesson 3-3)

63. Solve $q = \frac{323}{-17}$. (Lesson 2-8)

64. Is $17 > 40$ a true, false, or open sentence? (Lesson 1-9)

65. Food A carton of orange juice contains 64 ounces. If the nutritional information lists the serving size as 8 ounces, how many servings are in the carton? (Lesson 1-3)

66. Find the value of $24 \div (9 - 3)$. (Lesson 1-2)

5-2 Estimating Sums and Differences

Setting Goals: *In this lesson, you'll estimate sums and differences of decimals and fractions.*

Modeling a Real-World Application: Business

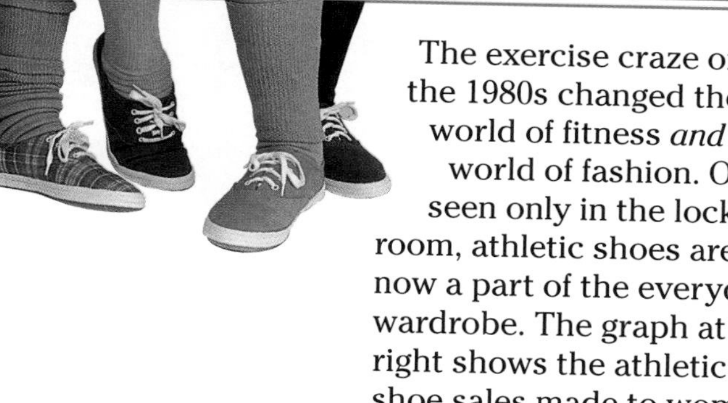

The exercise craze of the 1980s changed the world of fitness *and* the world of fashion. Once seen only in the locker room, athletic shoes are now a part of the everyday wardrobe. The graph at the right shows the athletic shoe sales made to women, men, and children in 1993. What were the total sales of athletic shoes in 1993?

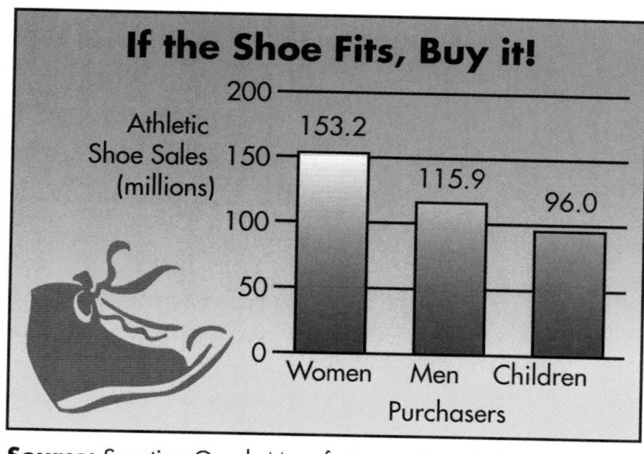

Source: Sporting Goods Manufacturers Association

Learning the Concept

Estimation is often used to provide a quick and easy answer when a precise answer is not necessary. It is also an excellent way to quickly see if your answer is reasonable or not.

To solve the problem about shoe sales, you can use *rounding* to estimate the answer. Round each number to a convenient place-value position. Often the greatest place-value position is used. Then complete the operation.

To find the total sales, round each of the sales figures to the greatest place value that they all share and then add.

$$\begin{array}{r} 153.2 \\ 115.9 \\ +\ 96.0 \\ \hline \end{array} \quad \rightarrow \quad \begin{array}{r} 150 \\ 120 \\ +100 \\ \hline 370 \end{array}$$

Round to the nearest ten.

There were approximately 370 million pairs of athletic shoes sold in 1993.

Example 1 **Estimate using rounding.**

a. 8.890 + 15.98

$8.890 + 15.98 \rightarrow 9 + 16 = 25$
8.890 + 15.98 is about 25.

b. 132.62 − 45.81

$132.62 - 45.81 \rightarrow 130 - 50 = 80$
132.62 − 45.81 is about 80.

Before you use a calculator to compute an answer, it is a good idea to estimate the answer. This helps make sure that you entered the numbers correctly.

Example 2

Consumerism

Suppose you go to Miller Cycles to buy equipment for in-line skating. You purchase a helmet for $29.89, a set of elbow pads for $19.26, knee pads for $21.67, and some energy bars for $6.03. The cashier tells you that the total is $82.88. Is this a reasonable cost?

Round each item to the nearest dollar amount.

$$\begin{array}{rcr} 29.89 & \rightarrow & 30 \\ 19.26 & \rightarrow & 19 \\ 21.67 & \rightarrow & 22 \\ +\ 6.03 & \rightarrow & +\ 6 \\ \hline & & 77 \end{array}$$

Obviously the amount charged is not correct. Probably the cashier accidentally charged you twice for the energy bars. The correct total is $76.85, which is very close to the estimate.

You can estimate sums and differences of fractions and mixed numbers by rounding also. To estimate the sum or difference of mixed numbers, round each mixed number to the nearest whole number. To estimate the sum or difference of proper fractions, round each fraction to 0, $\frac{1}{2}$, or 1.

Example 3

Estimate each sum or difference.

a. $4\frac{1}{12} + 16\frac{19}{32}$

$4\frac{1}{12} + 16\frac{19}{32} \rightarrow 4 + 17 = 21$

The sum of $4\frac{1}{12}$ and $16\frac{19}{32}$ is about 21.

b. $\frac{7}{8} + \frac{7}{16}$

$\frac{7}{8} + \frac{7}{16} \rightarrow 1 + \frac{1}{2} = 1\frac{1}{2}$

The sum of $\frac{7}{8}$ and $\frac{7}{16}$ is about $1\frac{1}{2}$.

c. $\frac{4}{5} - \frac{3}{8}$

$\frac{4}{5} - \frac{3}{8} \rightarrow 1 - \frac{1}{2} = \frac{1}{2}$

$\frac{4}{5} - \frac{3}{8}$ is about $\frac{1}{2}$.

Checking Your Understanding

Communicating Mathematics

Read and study the lesson to answer these questions.

1. **Give** two examples of numbers that round to a nearest whole number. One should round up; the other should not.
2. **State** two reasons for using estimation.
3. **You Decide** Pascual estimates that $569 - 345 = 300$. Nida estimates the difference as 220. Which is the better estimate? Explain.

4. **Assess Yourself** Write about a situation or an example in your life in which only an estimate is necessary, not an exact answer.

Guided Practice

Round to the nearest whole number.

5. 6.19
6. 12.641
7. 803.487
8. $6\frac{3}{4}$
9. $11\frac{1}{6}$
10. $33\frac{7}{24}$

Round each fraction to 0, $\frac{1}{2}$, or 1.

11. $\frac{10}{11}$
12. $\frac{1}{9}$
13. $\frac{15}{32}$

Choose the best estimate.

14. $4\frac{1}{8} - 2\frac{9}{10}$ **a.** almost 1 **b.** more than 1 **c.** more than $1\frac{1}{2}$
15. $5\frac{2}{3} + 2\frac{1}{11}$ **a.** almost 8 **b.** more than 8 **c.** less than $7\frac{1}{2}$

Estimate each sum or difference.

16. $34.32 + 19.51$
17. $159.7 - 124.8$
18. $6\frac{7}{10} - 3\frac{1}{6}$
19. $12\frac{1}{9} - 2\frac{9}{14}$
20. $1\frac{7}{12} + \frac{8}{19}$
21. $23\frac{8}{9} + 5\frac{4}{25}$

22. **Conservation** The chart at the right shows the amount of water used for average and quick showers with standard and low-flow showerheads. About how much water could be saved by changing from an average shower with a standard showerhead to a quick shower with a low-flow showerhead?

Rub-A-Dub-Dub

Gallons used by one shower daily per year:

Standard head

Average shower[1] 26,718

Quick shower[2] 13,140

Low-flow head

Average shower[1] 11,132

Quick shower[2] 5,475

(drop) = 2,000

1-National average 12.2 minutes, 2-Recommended 6.0 minutes

Source: Opinion Research Corp. for Teledyne Water Pik

Exercises: Practicing and Applying the Concept

Independent Practice

Estimate each sum or difference.

23. $12.5 + 44.8$
24. $\$20.00 - \$15.34$
25. $8.6 + 11.9$
26. $32 - 29.75$
27. $\frac{8}{9} + \frac{1}{5}$
28. $\frac{11}{15} + \frac{7}{8}$
29. $6\frac{9}{10} + \frac{2}{5}$
30. $\frac{17}{20} + 8\frac{3}{4}$
31. $\frac{4}{5} - \frac{1}{10}$
32. $\frac{7}{9} - \frac{13}{18}$
33. $18\frac{1}{8} - 12\frac{1}{2}$
34. $11\frac{89}{100} - 4\frac{1}{7}$
35. $125.8 - 22.4$
36. $16.432 + 11.910$
37. $32\frac{8}{10} - 4\frac{3}{8}$
38. $\frac{65}{131} + \frac{3}{35}$
39. $22\frac{2}{95} + 28\frac{3}{75}$
40. $145\frac{4}{5} - 121\frac{2}{15}$

Pirate Corner Store Price List	
felt-tip pen	$1.39
one inch binder	$2.89
dividers	$0.89
loose-leaf paper	$1.29
9 x 12 envelopes	$0.93
whiteout	$1.59
package of highliters	$2.79
TI-34 calculator	$13.89
composition books	
1 theme	$2.69
2 theme	$3.99
5 theme	$4.19
index tabs	$1.53
3 x 5 index cards	$0.79
ruler	$0.89
scotch tape	$1.69
can of soda or juice	$0.79
ice cream bar	$1.29
candy bars	$0.69
bag of popcorn	$1.74

The DECA club at Union High School runs a store in the school lobby as a fundraiser. Use their price list at the left to estimate each purchase price or change amount to the nearest dollar.

41. price of binder, loose-leaf paper, index cards

42. price of loose-leaf paper, dividers, white-out

43. change from \$10 for purchase of 5-theme composition book, and package of hi-liters

44. change from \$20 for purchase of TI-34 and one-inch binder

45. Is \$8.15 the correct change from \$10.00 for a ruler, ice cream bar, and a candy bar?

46. What supplies would you purchase for \$10.00, allowing yourself a candy bar? Let your goal be to get back less than \$0.50 change.

Critical Thinking

47. Make up a problem using estimation in which the answer is $12\frac{1}{2}$.

Applications and Problem Solving

48. Recycling *National Geographic* magazine estimated the amounts of steel, paper, and aluminum available for recycling in 1994. Which material accounted for the greatest amount of recycling?

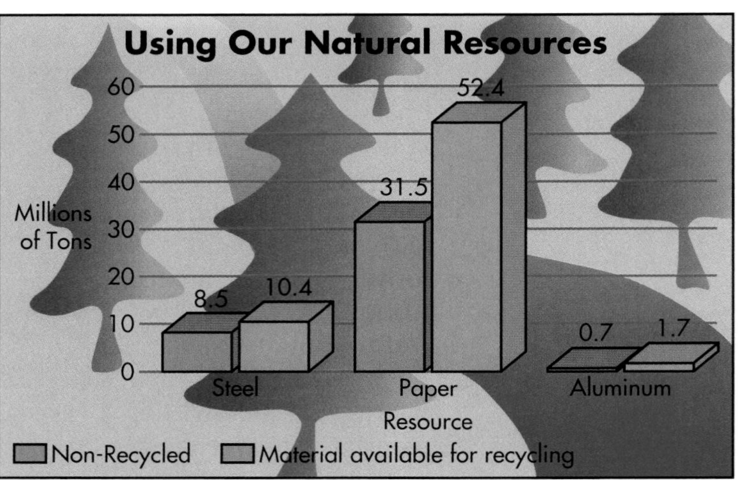

Source: *National Geographic*

EARTH WATCH

THE OZONE LAYER AND CFCs

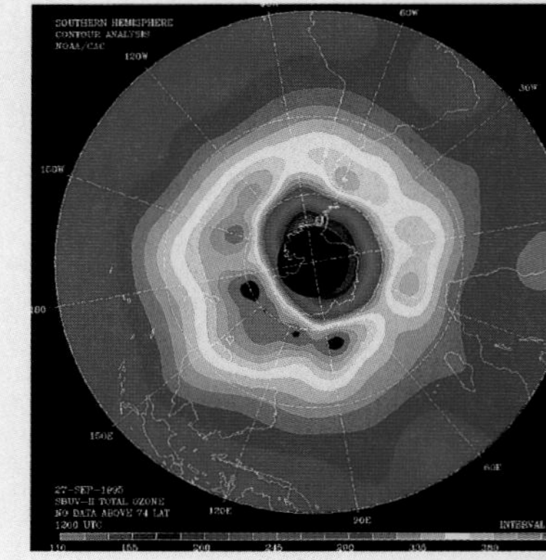

You have probably heard people speak of the "hole in the ozone layer." Did you wonder why they were concerned? Ozone is a form of oxygen that is able to block out the harmful ultraviolet radiation that radiates from the Sun. A shield of ozone, called the ozone layer, surrounds Earth about 10 to 40 kilometers above the surface. This layer protects us from the cancer-causing UVA and UVB radiation.

Chlorofluorocarbons, or CFCs, are used as coolants in air conditioners, refrigerators, and other products. Scientists have found that CFCs are able to break down ozone and destroy the protective layer above Earth.

49. **Retail Sales** Jaime bought a sweatshirt with his high school name and mascot for \$41.29 and a T-shirt with the track team logo for \$18.25. Estimate his cost to the nearest dollar.

50. **Fundraising** The Freshman class needs to earn \$500 for their annual trip to Six Flags. They made \$95.60 at a car wash, \$41.75 at a cookie and bake sale, \$150.95 at a band concert, and \$125.16 selling candy bars.
    a. Estimate their total earnings.
    b. Estimate, if necessary, the amount needed to reach their goal.

51. **Business** Pepsi-Cola reported that $\frac{3}{10}$ of the sales of their soft drinks were made through fountains and restaurants in 1994. Another $\frac{2}{5}$ of the sales were through supermarkets and other retail stores. Did the sales through these channels account for more or less than half of the total soft drink sales for Pepsi in 1994?

Mixed Review

52. Express −0.88 as a fraction in simplest form. (Lesson 5-1)
53. **Food** Tortilla chips account for $\frac{26}{100}$ of the snack chip retail sales. Write $\frac{26}{100}$ as a fraction in simplest form. (Lesson 4-6)
54. Solve $364 = w - 84$. (Lesson 3-2)
55. Divide $48 \div (-3)$. (Lesson 2-8)
56. In which quadrant is the graph of (−6, 0) located? (Lesson 2-2)
57. Is the inequality $4 < 3$ *true, false*, or *open*? (Lesson 1-9)
58. Solve $12n = 48$ mentally. (Lesson 1-6)
59. **Entertainment** When Bob and Neva went to the movies, the tickets were \$6.25 apiece. They also bought two small boxes of popcorn for \$2.25 each. (Lesson 1-5)
    a. Write two different expressions for the amount of money Bob and Neva spent.
    b. Find the amount of money they spent.
60. Find the value of $(11 + 9) \div (3 - 1)$. (Lesson 1-2)

See for Yourself

- The table at the right shows the amount of freon, the most common CFC, released into the atmosphere in different years. Estimate the amount released in the 1980s.
- A "hole" in the ozone layer is an area where the ozone is half as thick as it is in the rest of the layer. Every year, a hole appears over Antarctica. Investigate the reasons behind the appearance of the hole. Is there a similar hole over the North Pole?
- As the table of data suggests, several governments have recently passed new guidelines on the use of CFCs. What are the new guidelines? When do manufacturers have to comply? What are the expected results?

Year	Released Freon (millions of kilograms)
1980	250.8
1981	248.2
1982	239.5
1983	252.8
1984	271.1
1985	280.8
1986	295.1
1987	310.6
1988	314.5
1989	265.2
1990	216.1
1991	188.3
1992	171.1

5-3 Adding and Subtracting Decimals

Setting Goals: *In this lesson, you'll add and subtract decimals.*

Modeling with Technology

You have experience with adding and subtracting integers. Use a calculator to develop rules for adding and subtracting with decimals.

Your Turn Copy the table below. Use a calculator to find each sum or difference and complete the table.

Column A		Column B	
Sum or Difference	Answer	Sum or Difference	Answer
18 + 66		1.8 + 6.6	
56 + (−78)		5.6 + (−7.8)	
44 − 15		4.4 − 1.5	
220 − (−104)		2.20 − (−1.04)	

TALK ABOUT IT

a. Compare the sums and differences in Column A to the sums and differences in Column B.

b. Compare the answers in Column A to the answers in Column B.

c. Describe how you could find the sum of 6.8 and 9.3 if you know that 68 + 93 = 161.

Learning the Concept

Adding and subtracting rational numbers follow the same principles as addition and subtraction of integers.

Properties of Addition	Examples
Commutative Property For any rational numbers a and b, $a + b = b + a$.	$-7 + 9 = 9 + (-7)$ $-8.1 + 9.6 = 9.6 + (-8.1)$
Associative Property For any rational numbers a, b, and c, $(a + b) + c = a + (b + c)$.	$(-3.5 + 5.2) + 3.1 = -3.5 + (5.2 + 3.1)$
Identity Property For every rational number a, $a + 0 = a$ and $0 + a = a$.	$8 + 0 = 8$ $0 + -5.2 = -5.2$
Inverse Property For every rational number a, $a + (-a) = 0$.	$6 + (-6) = 0$ $15.8 + (-15.8) = 0$

You will use these properties and the rules you learned for adding and subtracting integers to add and subtract rational numbers. Estimating will help you determine the reasonableness of your answer.

Example  **Solve each equation**

a. $m = 13.2 + 11.7$ *Estimate:* $13 + 12 = 25$

$$\begin{array}{r} 13.2 \\ +11.7 \\ \hline 24.9 \end{array}$$

$m = 24.9$ *Compare with the estimate.*

b. $g = 231 - 126.7$ *Estimate:* $230 - 130 = 100$

Pencil and Paper

$$\begin{array}{r} 231.0 \\ -126.7 \\ \hline 104.3 \end{array}$$

Calculator

231 [–] 126.7 [=] 104.3

$g = 104.3$ *Compare with the estimate.*

c. $d = 8.3 + (-12.5)$ *Estimate:* $8 - 13 = -5$

$d = 8.3 - 12.5$ *The addition can be written as subtraction.*

$d = -|12.5 - 8.3|$ *The difference will be negative since* $|-12.5| > |8.3|$.

$$\begin{array}{r} 12.5 \\ -\ 8.3 \\ \hline 4.2 \end{array}$$

$d = -4.2$ *Compare with the estimate.*

You can save time by entering a sum involving a negative as a subtraction problem. For example, find $8.3 + (-12.5)$ by entering $8.3 - 12.5$.

Connection to Algebra

Simplifying expressions may involve adding and subtracting decimals.

Example  **Simplify $4.7x + 6.3x - 13.7x$.**

$$\begin{aligned} 4.7x + 6.3x - 13.7x &= (4.7 + 6.3 - 13.7)x && \textit{Distributive property} \\ &= (11.0 - 13.7)x && \textit{Associative property} \\ &= -2.7x \end{aligned}$$

You use sums and differences of rational numbers everyday when you deal with money.

Example 3

Personal Finance

Kiana uses a spreadsheet program to keep track of her money. The printout at the right shows her expenses and budgeted amounts for the month of January. Find the amount of money that Kiana is under or over budget for the month.

Item	Budget	Actual
Car payment	159.14	159.14
Car insurance	195.63	195.63
Fuel	35.00	37.32
Car repairs	150.00	135.67
Entertainment	60.00	52.44
Savings	50.00	50.00

(continued on the next page)

Find the sums of the budget and the actual amounts. Use a calculator.

$$\begin{aligned} \text{budget} &= 159.14 + 195.63 + 35.00 + 150.00 + 60.00 + 50.00 \\ &= 649.77 \\ \text{actual} &= 159.14 + 195.63 + 37.32 + 135.67 + 52.44 + 50.00 \\ &= 630.20 \end{aligned}$$

The actual total is less than the budgeted total, so Kiana is under budget by 649.77 − 630.20 or \$19.57.

Checking Your Understanding

Communicating Mathematics

Read and study the lesson to answer these questions.

1. Using the expression $12.3 + (8.3 + 2.5)$, illustrate the commutative property of addition and the associative property of addition.
2. **State** the additive inverse of 7.398.
3. **You Decide** Mariana and Leslie used different methods to solve $-3.8 - (-2.9)$. Who is correct? Explain.

Mariana	Leslie
3.8 [(−)] [−] 2.9 [(−)] [=] −.9	3.8 [(−)] [+] 2.9 [=] −.9

Guided Practice

State where the decimal point should be placed in each sum or difference.

4. $4.6 + 5.9 = 105$
5. $7.97 - 4.29 = 368$
6. $3.16 + 4.2 = 736$
7. $-8.47 + 3.56 = -491$
8. $8.17 + 9.123 = 17293$
9. $70.3 + 7.03 = 7733$

Solve each equation.

10. $a = 41.3 + 0.28$
11. $b = -9.6 + 3.2$
12. $34.2 - 43.0 = c$
13. $t = 81.9 - 38$

Simplify each expression.

14. $4.7x + 2x$
15. $5.3m - 1.4m - 8m$

16. **Transportation** Many people think that trains are a part of the past. But railroads are still a big part of the transportation industry in the United States. Use the table below to find the change in the number of passengers on each type of public transportation listed from 1993 to 1994.

Type of Transportation	1993 (billions)	1994 (billions)
Commuter rail	330.8	349.5
Heavy rail	2189.0	2279.0
Light rail	224.9	232.9
Bus/trolley	5412.1	5403.4

Source: *American Public Transit Association*

Exercises: Practicing and Applying the Concept

Independent Practice

Solve each equation.

17. $a = 4.3 + 9.8$

18. $x = 31.92 + 14.2$

19. $15.3 - 13.8 = m$

20. $85.3 - 37.07 = r$

21. $72.47 - 9.039 = b$

22. $c = 0.735 - 0.3879$

23. $-7.5 + 9.8 = x$

24. $7.4 + (-3.9) = y$

25. $z = -13.9 + (-12.5)$

26. $k = -3.91 - (-0.6)$

27. $a = -34.1 + (-17.63)$

28. $y = -18.12 - (-7.3)$

Simplify each expression.

29. $5.3m + 7m$

30. $47.9w - 31.8w$

31. $0.3y + 4.1y + 2.5y$

32. $27y - 4.7y - 13.8y$

33. $3.5x - 5 + 8x$

34. $6.93 + (3.1 + 4.07)m$

Evaluate each expression if $a = 5.3$, $b = 8.07$, $x = 21.33$, and $y = 0.7$.

35. $b + x$

36. $a - y$

37. $x - (b + y)$

38. $b + x + y$

39. $a + b - y$

40. $(x - b) + a$

Critical Thinking

41. Cassie has 45 feet of chicken wire for a fence. She wants to build a triangular pen that is 11.2 feet on one side and 9.35 feet on a second side. How much wire will she have left for the third side? Can she use all of it? Why or why not?

Applications and Problem Solving

42. Sports The physical size of the top 10 men and women tennis players has changed in 20 years. How much more do the men and women, respectively, weigh now than 20 years ago?

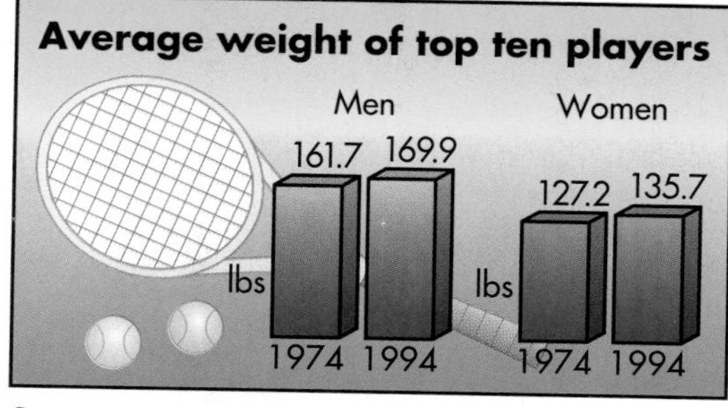

Source: *USA-Today* Research

43. Cycling Miguel Indurain of Spain won the Tour de France, the world's most prestigious bicycle race, for the fifth time in 1995. At the beginning of one of his workouts, the odometer on his bicycle reads 201.9 kilometers. If Miguel rides 176.6 kilometers that day, what will the odometer read at the end of the workout?

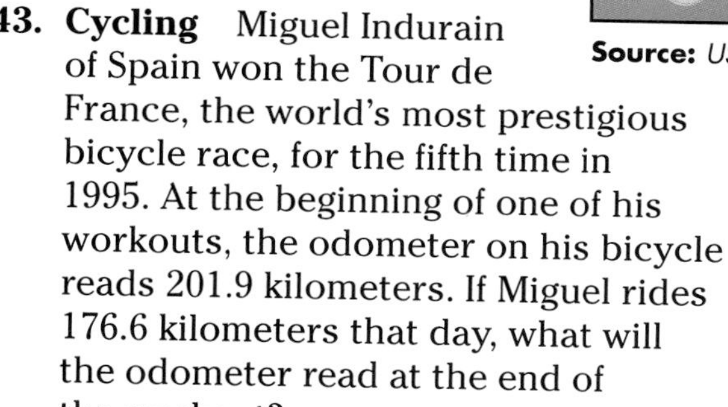

44. Recycling It is estimated that for every 13.2 million tons of glass used per year in the U.S., 2.6 million tons are recycled. How many millions of tons are simply discarded?

45. **Travel** The Department of Transportation keeps tabs on the quality of service by the airlines. In 1994, the average number of pieces of luggage lost per 10,000 passengers on major airlines was 54.4 bags. The worst month was January with 78.7 bags lost per 10,000 travelers.

a. What is the difference between the worst month and the average?

b. Why do you think January is the worst month for lost baggage?

46. **Geometry** Find the perimeter of the triangle shown at the right.

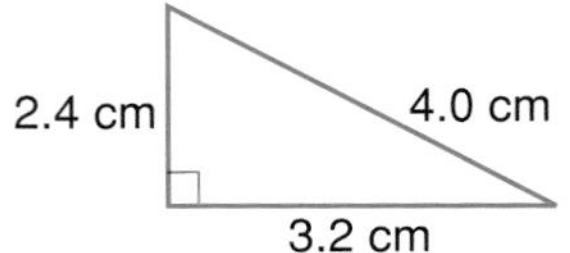

Mixed Review

47. **Personal Finance** When packing for a trip, Deandra bought toothpaste for \$1.89, suntan lotion for \$4.39, hair spray for \$1.27, and soap for \$2.04. She estimated that the \$10 she had would be enough to pay the bill. Do you agree? Explain. (Lesson 5-2)

48. Write $p \cdot p \cdot p \cdot p$ using exponents. (Lesson 4-2)

49. Solve $19 < 17 + a$. (Lesson 3-6)

50. **Fundraising** The freshman class earned \$120 selling magazines. The class account now has a balance of \$435. How much money was in the account before the magazine sale? (Lesson 3-2)

51. Solve $b = -121 \div 11$. (Lesson 2-8)

52. Replace ● with <, >, or = in $|-10|$ ● 9 to make a true sentence. (Lesson 2-3)

53. Find the value of $|8| + |-4|$. (Lesson 2-1)

54. Evaluate $2m - n$, if $m = 8$ and $n = 5$. (Lesson 1-3)

WORKING ON THE Investigation

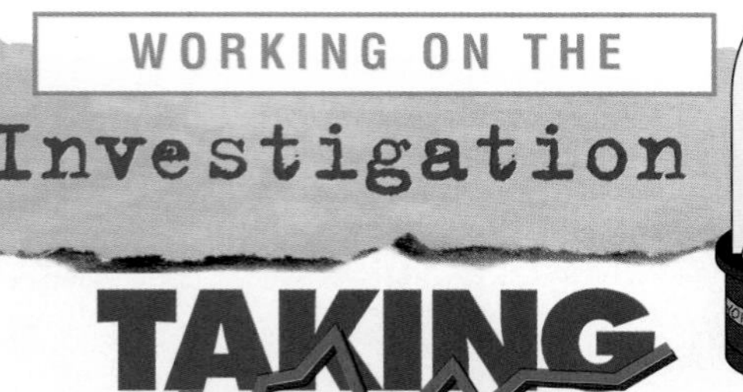

TAKING STOCK

Refer to the Investigation on pages 166–167.

To get an overall feel of how the stock market is performing, you can consult one of the market indexes. An **index** is an analysis of a selected group of individual stocks such as communications, pharmaceuticals, technology, or utilities.

There are two types of indexes. An index that is calculated by adding the price per share of all the stocks in the index and dividing by the number of stocks in the index is called a **price-weighted index**. The *Dow Jones Industrial Average* is an example of a price-weighted index.

If an index tracks the total market value, it is called a **value-weighted index**. The total market value is the product of the price per share and the total number of shares outstanding. To evaluate this type of index, the total market value is divided by the total market value on the date when the index was first started. The quotient is then multiplied by 100. The *Standard and Poor's 500* is an example of a value-weighted index.

Refer to your work with a spreadsheet on page 209 of Lesson 4-8.

- Using the stocks in your portfolio, create a spreadsheet to calculate your own price-weighted stock market index.
- Update your index each day for a five-day period and note any changes.
- Show the changes in your index by drawing a line graph of the five-day period.

Add the results of your work to your Investigation Folder.

5-4 Adding and Subtracting Like Fractions

Setting Goals: *In this lesson you'll add and subtract fractions with like denominators.*

Modeling with Manipulatives

MATERIALS

colored pencils

ruler

As you learned in Chapter 4, a fraction can be represented using a model like the one at the right. Different colors of shading can be used to show addition of fractions. This model represents $\frac{1}{6} + \frac{2}{6}$. Since 3 of the 6 sections are shaded, the sum is $\frac{3}{6}$ or $\frac{1}{2}$.

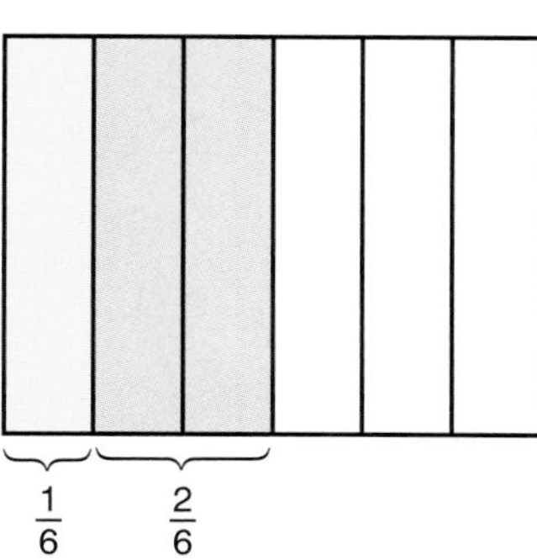

Your Turn Draw and shade a model that represents the sum of $\frac{1}{8}$ and $\frac{5}{8}$.

a. How many sections of your model are shaded? How does this compare to the sum of the numerators?

b. How many sections are there in your rectangle? How does this compare to the denominator of the addends?

c. What fraction represents the shaded sections of your rectangle?

d. Write a rule for finding the sum of fractions with the same denominator.

Learning the Concept

Adding fractions with like denominators can be described by the following rule.

Adding Like Fractions	
In words:	To add fractions with like denominators, add the numerators and write the sum over the same denominator.
In symbols:	For fractions $\frac{a}{c}$ and $\frac{b}{c}$, where $c \neq 0$, $\frac{a}{c} + \frac{b}{c} = \frac{a + b}{c}$.

When the sum of two fractions is greater than one, the sum is usually written as a mixed number in simplest form. A **mixed number** indicates the sum of a whole number and a fraction.

Example 1 **Solve each equation.**

a. $s = \frac{3}{4} + \frac{3}{4}$ *Estimate: 1 + 1 = 2*

$s = \frac{6}{4}$ *Since the denominators are the same, add the numerators.*

$s = 1\frac{2}{4}$ *Rename $\frac{6}{4}$ as a mixed number.*

$s = 1\frac{1}{2}$ *Write the mixed number in simplest form.*

b. $y = 3\frac{2}{5} + 4\frac{3}{5}$ *Estimate: 3 + 5 = 8*

$y = (3 + 4) + \left(\frac{2}{5} + \frac{3}{5}\right)$ *Use the Associative and Commutative properties to add the whole numbers and fractions separately.*

$y = 7 + \frac{5}{5}$ *Add the numerators of the fractions.*

$y = 7 + 1$ or 8 *Simplify.*

The rule for subtracting fractions with like denominators is similar to the rule for addition.

Subtracting Like Fractions

In words: To subtract fractions with like denominators, subtract the numerators and write the difference over the same denominator.

In symbols: For fractions $\frac{a}{c}$ and $\frac{b}{c}$, where $c \neq 0$, $\frac{a}{c} - \frac{b}{c} = \frac{a - b}{c}$.

Example 2 **Solve $y = \frac{7}{15} - \frac{28}{15}$. Write the solution in simplest form.**

$y = \frac{7}{15} - \frac{28}{15}$ *Since the denominators are the same, subtract the numerators.*

$y = \frac{-21}{15}$

$y = -1\frac{6}{15}$ or $-1\frac{2}{5}$ *Rename as a mixed number and simplify.*

Connection to Algebra

You can use the same rules for adding and subtracting like fractions when you evaluate algebraic expressions.

Example 3 **Evaluate $m + n$ if $m = 3\frac{1}{6}$ and $n = -1\frac{5}{6}$.**

$m + n = 3\frac{1}{6} + \left(-1\frac{5}{6}\right)$ *Replace m with $3\frac{1}{6}$ and n with $-1\frac{5}{6}$.*

$= 3\frac{1}{6} - 1\frac{5}{6}$

$= 2\frac{7}{6} - 1\frac{5}{6}$ *Rename $3\frac{1}{6}$ as $2\frac{7}{6}$.*

$= (2 - 1) + \left(\frac{7}{6} - \frac{5}{6}\right)$

$= 1\frac{2}{6}$ or $1\frac{1}{3}$

Entertainment

4 **How do you like to spend time in the great outdoors? The results of a survey by Roper Starch Worldwide on that subject are shown in the graph at the right.**

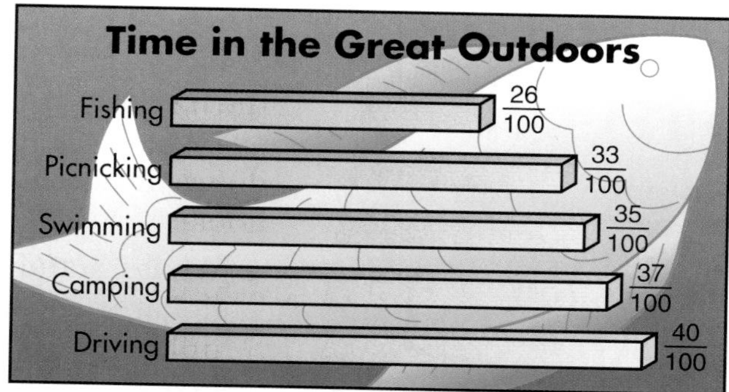

Source: Roper Starch Worldwide

a. What fraction of the people said that they enjoyed driving or swimming?

$$\frac{40}{100} + \frac{35}{100} = \frac{40 + 35}{100}$$
$$= \frac{75}{100} \text{ or } \frac{3}{4}$$

b. What fraction of the people chose camping or picnicking as favorite activities?

$$\frac{37}{100} + \frac{33}{100} = \frac{37 + 33}{100}$$
$$= \frac{70}{100} \text{ or } \frac{7}{10}$$

Checking Your Understanding

Communicating Mathematics

Read and study the lesson to answer these questions.

1. What number properties allow you to write $4\frac{2}{7} + 8\frac{3}{7}$ as $(4 + 8) + \left(\frac{2}{7} + \frac{3}{7}\right)$?
2. **Draw** and shade a model or use objects like paper plates to represent the sum of $\frac{3}{5}$ and $\frac{1}{5}$.
3. **Express** $\frac{33}{7}$ as a mixed number.

4. List several examples of like fractions and unlike fractions. Then in your own words state a simple rule for adding and subtracting like fractions.

Guided Practice

Solve each equation. Write the solution in simplest form.

5. $\frac{4}{7} + \frac{5}{7} = m$
6. $1\frac{2}{19} + \frac{17}{19} = a$
7. $\frac{19}{24} - \frac{11}{24} = r$
8. $-\frac{19}{30} + \frac{7}{30} = z$

Evaluate each expression if $x = \frac{3}{8}$, $y = \frac{7}{8}$, and $z = \frac{5}{8}$. Write in simplest form.

9. $x + z$
10. $z - y$

11. Home Maintenance The graph at the right shows the fraction of each dollar an owner spends on different expenses for a house.

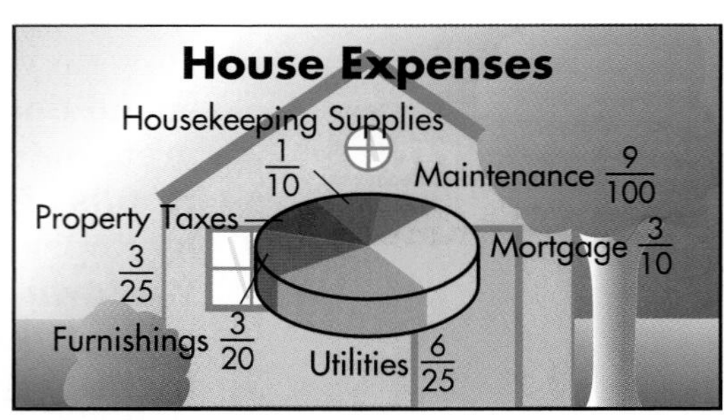

Source: Bureau of Labor Statistics

a. What fraction of each dollar is spent on utilities and property taxes?

b. How much more is spent on the mortgage than is spent on paying for housekeeping supplies?

Exercises: Practicing and Applying the Concept

Independent Practice

Solve each equation. Write the solution in simplest form.

12. $\frac{5}{7} + \frac{1}{7} = b$ **13.** $m = \frac{11}{15} - \frac{8}{15}$ **14.** $a = \frac{19}{31} - \frac{8}{31}$

15. $\frac{19}{27} - \frac{7}{27} = r$ **16.** $1\frac{13}{16} - 1\frac{5}{16} = x$ **17.** $\frac{13}{18} + \frac{11}{18} = a$

18. $m = 1\frac{9}{16} + \frac{15}{16}$ **19.** $1\frac{16}{21} - \frac{9}{21} = y$ **20.** $p = -\frac{19}{27} + \left(-\frac{7}{27}\right)$

21. $1\frac{5}{12} + \frac{11}{12} = z$ **22.** $\frac{9}{20} - \frac{17}{20} = c$ **23.** $1\frac{19}{41} + \left(-\frac{19}{41}\right) = w$

Evaluate each expression if $a = \frac{5}{12}$, $b = \frac{7}{12}$, and $c = \frac{1}{12}$. Write in simplest form.

24. $a + b$ **25.** $b - c$ **26.** $a + c$

27. $b + c$ **28.** $c - a$ **29.** $a - b$

Simplify each expression.

30. $4\frac{1}{3}x + \frac{2}{3}x - 3\frac{1}{3}x$ **31.** $-4\frac{1}{4}r + \left(-2\frac{3}{4}\right)r + 5r$

32. $5\frac{1}{7}m + \left(-3\frac{3}{7}\right)m + 1\frac{2}{7}m$ **33.** $6\frac{3}{5}a - 3\frac{4}{5}a - 2\frac{2}{5}a$

34. $5\frac{5}{6}y + 3\frac{5}{6}y - 2\frac{1}{6}y$ **35.** $3\frac{5}{9}b - 2\frac{4}{9}b - \frac{7}{9}b$

Critical Thinking

36. Write $\frac{x + y}{8}$ as the sum of two fractions.

37. Write $\frac{2x - 1}{5}$ as the difference of two fractions.

38. Find the sum of $\frac{1}{20} + \frac{2}{20} + \frac{3}{20} + \cdots + \frac{17}{20} + \frac{18}{20} + \frac{19}{20}$. Describe any pattern or shortcut you used to find the sum.

Applications and Problem Solving

39. Carpentry A carpenter is installing a countertop in a new kitchen. The countertop will be between two walls that are $54\frac{5}{8}$ inches apart. If the piece of countertop material is five feet long, how much will have to be cut off before the countertop is installed?

40. **Home Projects** The Stadlers added an exercise room to the rear of their house. They installed $\frac{3}{8}$-inch thick paneling over a layer of dry wall $\frac{5}{8}$ inch thick. How thick are the wall coverings?

41. **Sewing** Nyoko is making a linen suit. The portion of the pattern envelope that shows the amounts of fabric needed for different sizes of skirts and jackets is shown at the right. If Nyoko is making a size 6 from 45-inch fabric, how much fabric should she buy for the jacket and skirt?

SIZE	(6	8	10)
JACKET			
45"	$2\frac{5}{8}$	$2\frac{3}{4}$	$2\frac{3}{4}$
60"	2	2	2
SKIRT			
45"	$\frac{7}{8}$	$1\frac{1}{8}$	$1\frac{1}{8}$
60"	$\frac{7}{8}$	$\frac{7}{8}$	$\frac{7}{8}$

Mixed Review

42. **Shopping** Cheryl has $38.78 in her wallet. If she buys a purse for $25.59, how much will she have left? (Lesson 5-3)

43. **Economics** The Valdosta Recreation Center needs $17 million for renovations. The table at the right shows the contributions that have been made so far. (Lesson 4-6)

Contributor	Amount
City of Valdosta	$2.5 million
First One Federal Bank	$0.5 million
Montgomery County	$0.5 million
State of Georgia	$1.3 million
Recreation Center Fundraisers	$3 million
Sale of tax credits	$6.2 million

   a. Write the ratio of the amount of money the city contributed to the total amount of money needed.

   b. Write the ratio of the amount of money Montgomery County is contributing to the total amount of money received so far.

44. Solve $\frac{d}{6} < -14$. (Lesson 3-7)

45. Solve $c - 12 \geq -25$. (Lesson 3-6)

46. **Geometry** Find the missing dimension in the rectangle at the right. (Lesson 3-5)

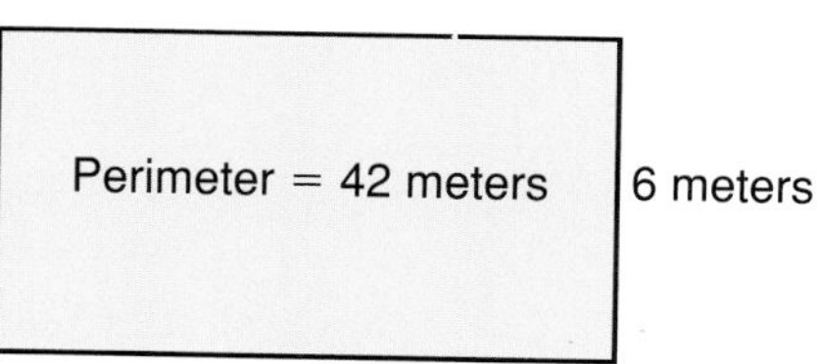

47. Simplify $-8a + (-17a) - (-3a)$. (Lesson 2-4)

48. Name the property shown by $1z = z$. (Lesson 1-4)

5-5 Adding and Subtracting Unlike Fractions

Setting Goals: *In this lesson, you'll add and subtract fractions with unlike denominators.*

Modeling with Manipulatives

MATERIALS

ruler
colored pencils

You can add or subtract fractions with unlike denominators by changing them into fractions with like denominators.

Your Turn

Make a model that represents the sum of $\frac{1}{4}$ and $\frac{1}{2}$. First shade $\frac{1}{4}$ of a rectangle like the one at the right. Then shade $\frac{1}{2}$ using a second color to represent $\frac{1}{4} + \frac{1}{2}$.

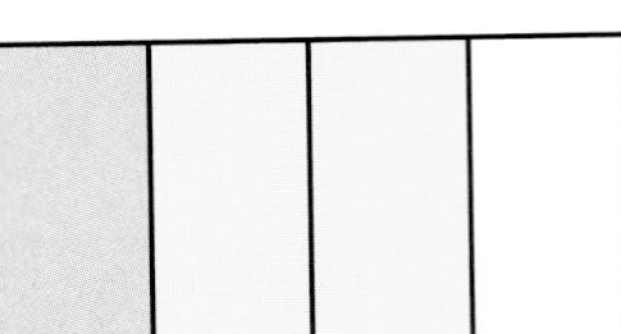

TALK ABOUT IT

a. What fraction of the model is shaded?

b. What is the least common denominator of $\frac{1}{4}$ and $\frac{1}{2}$?

c. Write an addition sentence equivalent to $\frac{1}{4} + \frac{1}{2}$ by using the least common denominator. Then find the sum.

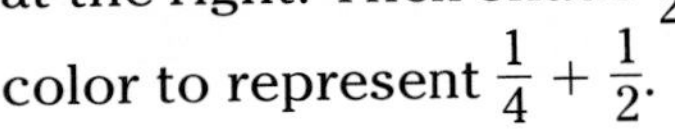

Learning the Concept

The activity leads us to the following rule for adding and subtracting unlike fractions.

Adding and Subtracting Unlike Fractions	To find the sum or difference of two fractions with unlike denominators, rename the fractions with a common denominator. Then add or subtract and simplify.

One way to rename unlike fractions before you add or subtract is to use the LCD.

Example 1 **Solve each equation. Write the solution in simplest form.**

a. $m = \frac{7}{9} + \frac{11}{12}$ *Estimate: 1 + 1 = 2*

$m = \frac{7}{9} \cdot \frac{4}{4} + \frac{11}{12} \cdot \frac{3}{3}$ *The LCD is $3 \cdot 3 \cdot 2 \cdot 2$ or 36.*

$m = \frac{28}{36} + \frac{33}{36}$ *Rename each fraction with the LCD.*

$m = \frac{61}{36}$ or $1\frac{25}{36}$

b. $\frac{1}{4} - \frac{2}{3} = a$ *Estimate:* $\frac{1}{2} - 1 = -\frac{1}{2}$

$\frac{3}{12} - \frac{8}{12} = a$ *The LCD is 12.* $\frac{1}{4} = \frac{3}{12}, \frac{2}{3} = \frac{8}{12}$

$-\frac{5}{12} = a$

To add and subtract mixed numbers with unlike denominators, rename using the LCD.

Example 2 **Solve each equation. Write the solution in simplest form.**

a. $n = 4\frac{1}{6} + 7\frac{11}{18}$ *Estimate:* $4 + 8 = 12$

$n = 4\frac{3}{18} + 7\frac{11}{18}$ *Use the LCD to rename* $\frac{1}{6}$ *as* $\frac{3}{18}$.

$n = 11\frac{14}{18}$ or $11\frac{7}{9}$

b. $7\frac{2}{3} - 9\frac{1}{12} = y$ *Estimate:* $8 - 9 = -1$

$7\frac{8}{12} - 9\frac{1}{12} = y$ *The LCD is 12. Rename the fractions.*

$7\frac{8}{12} - 8\frac{13}{12} = y$ *Rename* $9\frac{1}{12}$ *as* $8\frac{13}{12}$.

$-1\frac{5}{12} = y$

Stock market prices are expressed as mixed numbers.

Example 3

APPLICATION

Business

News that Microsoft's Windows 95 software sales were brisk after its August, 1995, debut sent stocks of technological companies up. On Monday, September 11, 1995, Oracle Systems stock rose $1\frac{3}{8}$ to close at $45\frac{1}{4}$. What was the opening price for Oracle that day?

Let p represent the opening price. Then write an equation.

$p = 45\frac{1}{4} - 1\frac{3}{8}$

$p = 45\frac{2}{8} - 1\frac{3}{8}$ *The LCD is 8. Rename* $\frac{1}{4}$ *as* $\frac{2}{8}$.

$p = 44\frac{10}{8} - 1\frac{3}{8}$ *Rename* $45\frac{2}{8}$ *as* $44\frac{10}{8}$.

$p = 43\frac{7}{8}$

The opening price was $43\frac{7}{8}$.

Checking Your Understanding

Communicating Mathematics

Read and study the lesson to answer these questions.

1. What is the first step in adding or subtracting fractions with unlike denominators?
2. **Explain** why you cannot add or subtract fractions without first finding a common denominator.

3. **Make up a problem** in which you would add $12\frac{1}{2}$ and $18\frac{3}{4}$.
4. **You Decide** Hank found the sum of $\frac{7}{9}$ and $\frac{11}{12}$ to be $\frac{18}{12}$ or $1\frac{1}{2}$. Keri found the sum to be $1\frac{25}{36}$. Who is correct? Explain.
5. **Draw** a model or use objects such as paper plates to represent the difference $\frac{3}{4} - \frac{1}{2}$.

Guided Practice

Solve each equation. Write the solution in simplest form.

6. $\frac{1}{3} + \frac{5}{6} = h$	**7.** $m = \frac{3}{4} - \frac{5}{8}$	**8.** $t = \frac{3}{5} + \frac{3}{10}$
9. $n = \frac{3}{8} - \frac{1}{2}$	**10.** $\frac{7}{12} - \frac{2}{3} = g$	**11.** $\frac{3}{7} + \frac{5}{14} = w$

12. **Travel** Chilton Research studied the factors that travelers consider when choosing an airline.
    a. What fraction of those asked said they look at cost or safety record?
    b. What is the difference between the fraction of people who said they considered time of arrival or departure and the number who chose based on the size or type of aircraft?

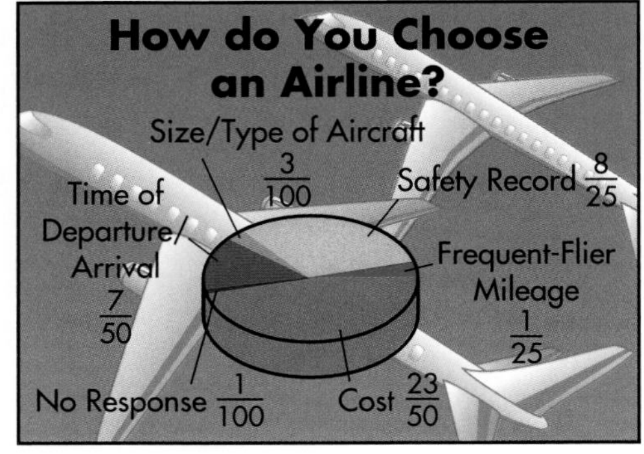

Source: Chilton Research

Exercises: Practicing and Applying the Concept

Independent Practice

Solve each equation. Write the solution in simplest form.

13. $x = \frac{1}{6} + \frac{7}{18}$	**14.** $m = \frac{4}{7} + \frac{9}{14}$	**15.** $\frac{5}{12} - \frac{1}{2} = d$
16. $y = \frac{6}{7} - \frac{5}{21}$	**17.** $r = \frac{9}{26} + \frac{3}{13}$	**18.** $b = \frac{11}{12} + \frac{3}{4}$
19. $p = 4\frac{1}{3} - 2\frac{1}{2}$	**20.** $k = \frac{7}{8} + \frac{3}{16}$	**21.** $5\frac{1}{3} - 2\frac{1}{6} = m$
22. $\frac{5}{7} - \frac{10}{21} = a$	**23.** $9\frac{3}{4} - 5\frac{1}{2} = d$	**24.** $x = 7\frac{1}{3} - 3\frac{1}{2}$
25. $y = 12\frac{3}{7} + 4\frac{1}{21}$	**26.** $19\frac{3}{8} - 4\frac{3}{4} = s$	**27.** $8\frac{9}{10} + 1\frac{1}{6} = r$
28. $b = \frac{1}{3} + \frac{2}{35}$	**29.** $\frac{7}{10} - \frac{8}{9} = c$	**30.** $18\frac{6}{7} + 2\frac{3}{5} = a$

Evaluate each expression if $x = \frac{5}{8}$, $y = -\frac{3}{4}$, and $z = 2\frac{7}{12}$. Write in simplest form.

31. $x + z$	**32.** $z + y$	**33.** $y - x$
34. $x + y + z$	**35.** $z - x + y$	**36.** $z + x - y$

Critical Thinking

37. Studies of unit fractions have been found in Greek papyrus dating back to 500 and 800 A.D. A *unit fraction* is a fraction that has a numerator of 1, such as $\frac{1}{5}$, $\frac{1}{7}$, or $\frac{1}{4}$. Greeks wrote all fractions as the sum of unit fractions. Express $\frac{2}{9}$ as the sum of two different unit fractions.

Applications and Problem Solving

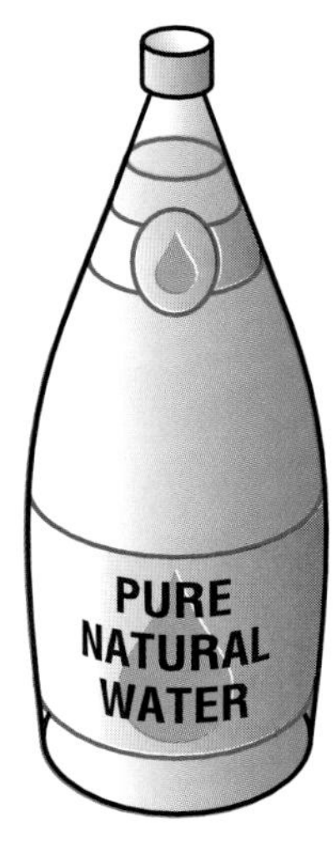

38. **Construction** The outside walls of a new home have $\frac{5}{8}$-inch drywall, $5\frac{1}{2}$ inches of insulation, $\frac{3}{4}$-inch outside wall plywood sheathing, and $\frac{7}{8}$-inch siding. How thick is the wall?

39. **Nutrition** The Beverage Marketing Corporation found that $\frac{1}{6}$ of American households bought bottled water in 1992. Only $\frac{1}{17}$ of the households bought bottled water in 1985. What fraction of the population bought bottled water in 1992 that did not in 1985?

Mixed Review

40. Simplify $5\frac{1}{3}b + 3\frac{2}{3}b - 2\frac{1}{3}b$. (Lesson 5-4)

41. Express $s^{-2}t^3$ with positive exponents. (Lesson 4-9)

42. **Boating** The U.S. Coast Guard recommends using the formula $p = \frac{\ell w}{15}$, where ℓ is the length of the boat and w is the width, to determine the number of people, p, who can safely occupy a pleasure boat. If a Daysailer is 18 feet long and 5 feet wide, how many people can occupy it? (Lesson 3-4)

43. Solve $b = -40 \div 8$. (Lesson 2-8)

44. Find $-8(3)(-2)$. (Lesson 2-7)

45. Solve $y = 9 - (-4)$. (Lesson 2-5)

46. Order the integers in the set $\{9, -3, -1, 12, -11\}$ from least to greatest. (Lesson 2-3)

47. Simplify $3x + 15x$. (Lesson 1-5)

Self Test

Express each decimal as a fraction or mixed number in simplest form. (Lesson 5-1)

1. $-0.1666\ldots$
2. 2.98
3. $1.\overline{4}$

Estimate each sum or difference. (Lesson 5-2)

4. \$50.00 − \$37.52
5. $\frac{8}{9} + \frac{1}{15}$
6. $\frac{9}{11} - \frac{5}{12}$

Solve each equation. (Lesson 5-3)

7. $r = 5.4 + 9.12$
8. $90.6 - 33.1 = k$
9. $m = 17.4 - (-13.2)$

Solve each equation. Write the solution in simplest form. (Lesson 5-4)

10. $\frac{8}{10} + \frac{3}{10} = a$
11. $b = 4\frac{6}{7} + \frac{2}{7}$
12. $8\frac{5}{35} - 2\frac{8}{35} = d$

13. **Publishing** The length of a page in a yearbook is 10 inches. The top margin is $\frac{1}{2}$ inch, and the bottom margin is $\frac{3}{4}$ inch. What is the length of the page inside the margins? (Lesson 5-5)

5-6 Solving Equations

Setting Goals: *In this lesson, you'll solve equations with rational numbers.*

Modeling a Real-World Application: Meteorology

Hurricanes can be devastating to life and property. The graph at the right compares the statistics for the five most costly hurricanes to strike the U.S. mainland. The two most costly hurricanes were Andrew and Hugo. Together, the damage they caused totaled \$32.16 billion. How much did Hurricane Andrew cost?

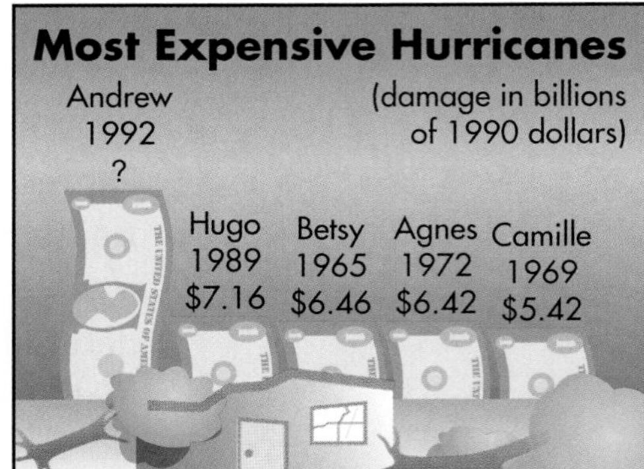

Source: National Hurricane Center

Learning the Concept

The earliest date for a hurricane to strike the U.S. was Hurricane Allison on June 5, 1995. The previous early hurricane record holder was Hurricane Alma on June 9, 1966, which like Allison, struck the Florida panhandle.

To find the cost of Hurricane Andrew, let c represent its cost and write an equation.

cost of Andrew	*plus*	*cost of Hugo*	*equals*	*total cost*
c	$+$	7.16	$=$	32.16

$$c + 7.16 = 32.16$$

$$c + 7.16 - 7.16 = 32.16 - 7.16 \quad \text{Subtract 7.16 from each side.}$$

$$c = 25$$

Hurricane Andrew cost \$25 billion.

You can solve rational number equations using the same skills you used to solve equations involving integers.

Example  **Solve each equation. Check the solution.**

a. $b + \frac{3}{5} = \frac{3}{2}$

$$b + \frac{3}{5} = \frac{3}{2}$$

$$b + \frac{3}{5} - \frac{3}{5} = \frac{3}{2} - \frac{3}{5} \quad \textit{Subtract } \frac{3}{5} \textit{ from each side.}$$

$$b = \frac{15}{10} - \frac{6}{10} \quad \textit{The LCD is 10.}$$

$$b = \frac{9}{10}$$

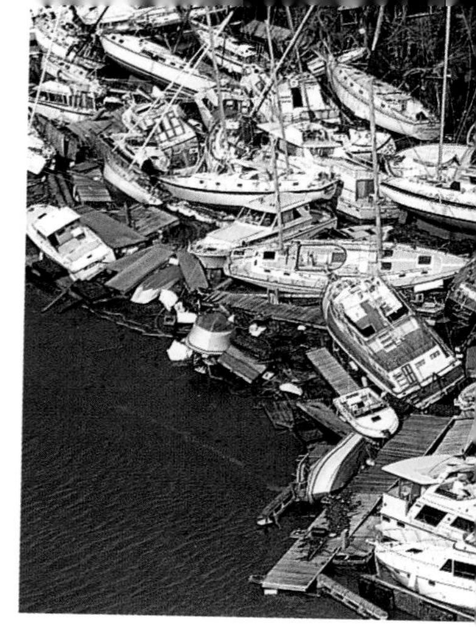

Check: $b + \frac{3}{5} = \frac{3}{2}$

$$\frac{9}{10} + \frac{3}{5} \stackrel{?}{=} \frac{3}{2}$$

$$\frac{9}{10} + \frac{6}{10} \stackrel{?}{=} \frac{3}{2}$$

$$\frac{15}{10} \stackrel{?}{=} \frac{3}{2} \checkmark$$

The solution is $\frac{9}{10}$.

b. $\boldsymbol{a - 3.2 = 2.7}$

$$a - 3.2 = 2.7$$
$$a - 3.2 + 3.2 = 2.7 + 3.2 \quad \textit{Add 3.2 to each side.}$$
$$a = 5.9$$

Check: $a - 3.2 = 2.7$

$$5.9 - 3.2 \stackrel{?}{=} 2.7$$
$$2.7 = 2.7 \checkmark$$

The solution is 5.9.

You can solve equations that represent real-world situations involving rational numbers.

Example 2

Meteorology

HELP WANTED

Are you interested in the patterns and changes of the weather? For more information on this field, contact:

American Meteorological Society
45 Beacon St.
Boston, MA 02108

Meteorologists use barometers to measure the air pressure in the atmosphere. Changes in the pressure indicate that there will be a change in the weather. As a hurricane approaches, the barometric pressure at 7:00 P.M. was 28.65 inches. It had dropped 0.37 inches from the previous reading at noon. What was the noon reading?

Let p represent the barometric pressure at noon.

pressure at noon	*dropped*	*0.37 inches*	*to*	*28.65*
p	$-$	0.37	$=$	28.65

$$p - 0.37 = 28.65$$
$$p - 0.37 + 0.37 = 28.65 + 0.37 \quad \textit{Add 0.37 to each side.}$$
$$p = 29.02$$

The reading was 29.02 inches at noon.

Checking Your Understanding

Communicating Mathematics

Read and study the lesson to answer these questions.

1. What property of equality would you use to solve the equation $x - 5.3 = 6.8$?
2. **Explain** how to solve $n - 8.9 = 6.8$.
3. **Tell** how you know that 6.9 is not a solution of $n + 8.5 = 14.7$.

Guided Practice

Solve each equation. Check your solution.

4. $\frac{7}{6} = m + \frac{5}{12}$
5. $\frac{2}{3} + r = \frac{3}{5}$
6. $7\frac{1}{2} = x - 5\frac{2}{3}$
7. $m - 4.1 = -9.38$
8. $y + 7.2 = 21.9$
9. $x - 1.5 = 1.75$
10. $a - 1\frac{1}{3} = 4\frac{1}{6}$
11. $-13.7 = b - 5$
12. $y + 3.17 = -3.17$
13. **Driver's Education** The odometer on the driver's education car read 26,375.4 miles at the start of Rachel's first in-car class. After the class, the odometer read 26,397.3 miles. How far did Rachel drive?

Exercises: Practicing and Applying the Concept

Independent Practice

Solve each equation. Check your solution.

14. $y + 3.5 = 14.9$

15. $r - 8.5 = -2.1$

16. $a - \frac{3}{5} = \frac{5}{6}$

17. $b - 1\frac{1}{2} = 14\frac{1}{4}$

18. $a + 7.1 = 4.7$

19. $b - 5.3 = 8.1$

20. $y + 1\frac{1}{3} = 3\frac{1}{18}$

21. $m + \frac{7}{12} = -\frac{5}{18}$

22. $x + \frac{5}{8} = 7\frac{1}{2}$

23. $k - \frac{3}{8} = 1\frac{3}{5}$

24. $n + 1.4 = 0.72$

25. $d - (-31.4) = 28.6$

26. $w - 0.04 = 1.2$

27. $d + (-7.03) = 0.98$

28. $5\frac{3}{10} + z = 2\frac{2}{3}$

29. $v - 4\frac{7}{9} = 8\frac{1}{6}$

30. $501.1 = y - 9.32$

31. $\frac{1}{312} + t = \frac{5}{78}$

Critical Thinking

32. Find two rational numbers, a and b, such that $a + b = ab$.

Applications and Problem Solving

33. Oil Production In 1989, the top two oil-producing states were Texas and Alaska. Together they produced 1372.2 million barrels of oil. Alaska produced 684.0 million barrels. How many barrels of oil were produced in Texas?

34. Geometry The sum of the measures of the angles in a triangle is 180°. If two of the angles measure 75.5° and 60.3°, what is the measure of the third angle?

35. Airports The graph at the right gives information on the world's busiest cargo airports for 1994. Memphis tops the list at 1.65 million metric tons. O'Hare is the world's busiest passenger airport. What is the difference in cargo handling between Memphis and O'Hare?

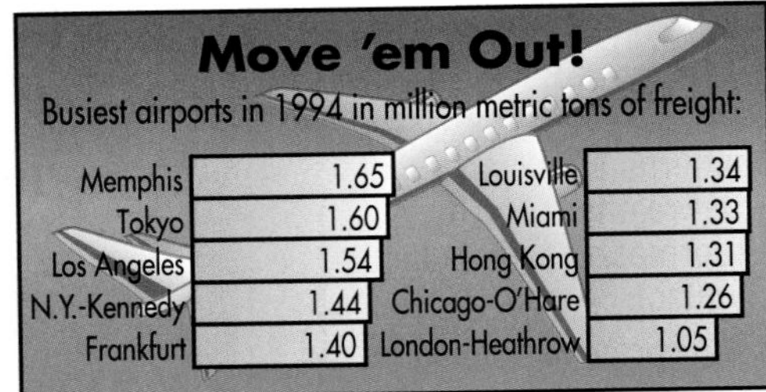

Source: Airports Council International

36. Trains As a train begins to roll, the cars are "jerked" into motion, and the movement spreads like a ripple towards the back of the train. Slack is built into the couplings so that the engine does not have to move every single car all at once. If the slack built into each coupling is 3 inches or $\frac{1}{4}$ foot, how many feet of slack is there between two freight cars and the engine?

Mixed Review

37. Solve $y = \frac{7}{8} + 4\frac{1}{24}$. Write the solution in simplest form. (Lesson 5-5)

38. Factor $420ab^3$ completely. (Lesson 4-4)

39. Solve $\frac{d}{6} \geq -14$. (Lesson 3-7)

40. Geometry Find the area of a 4" by 6" rectangle. (Lesson 3-5)

41. Solve $x + 5 = 23$. (Lesson 3-2)

42. Medicine When a baby is born, a doctor can take measurements and estimate the child's adult height. The doctor estimated that Tobi would be 5'11" tall. If Tobi is actually 5'9" tall, what is the difference between his actual height and his estimated height? (Lesson 2-5)

43. Solve $\frac{90}{x} = 6$ mentally. (Lesson 1-6)

5-7 Solving Inequalities

Setting Goals: *In this lesson, you'll solve inequalities with rational numbers.*

Modeling a Real-World Application: Sports

"So let's root, root, root, for the home team. . . ." Commentators and athletes often talk about the advantage of being the home team. Being the home team with a *new* home field may be even better.

> **CULTURAL CONNECTIONS**
> Baseball is very similar to the European game of cricket. Cricket was probably invented in England in the 1300s.

When a baseball team has a winning average of 0.600, this means they have won 0.600 or $\frac{3}{5}$ of their games. The table at the right lists the winning rates for teams in the first season in new homes. In 1995, the Colorado Rockies opened in a new park. As of May 22nd, their winning average at home was 0.600. How much would the Rockies have to raise their winning average to have the best new field record?

Home Sweet (New) Home

Field	Winning Average
Jacobs Field (1994 Indians)	0.686
SkyDome (1989 Blue Jays)	0.618
Comiskey Park (1991 White Sox)	0.568
Camden Yards (1992 Orioles)	0.531
The Ballpark (1994 Rangers)	0.492
Metrodome (1982 Twins)	0.457

Source: Major League Baseball teams

Learning the Concept

The amount that the Rockies need to raise their average should be an inequality since anything greater than the present record will give them the best new field record.

$$\underbrace{0.600}_{\text{current average}} \underbrace{+}_{\text{plus}} \underbrace{x}_{\text{increase}} \underbrace{>}_{\text{is greater than}} \underbrace{0.686}_{\text{current record}}$$

$$0.600 - 0.600 + x > 0.686 - 0.600 \quad \textit{Subtract 0.600 from each side.}$$

$$x > 0.086$$

The Rockies would have to increase their winning record by more than 0.086 to establish a new home-field advantage record.

You can solve rational number inequalities using the same skills you used to solve inequalities involving integers.

Example  **Solve each inequality. Graph each solution on a number line.**

a. $a + \frac{3}{8} > 2$

$a + \frac{3}{8} - \frac{3}{8} > 2 - \frac{3}{8}$ *Subtract $\frac{3}{8}$ from each side.*

$a > \frac{16}{8} - \frac{3}{8}$ *Rename 2 as a fraction with a denominator of 8.*

$a > \frac{13}{8}$ or $1\frac{5}{8}$

(continued on the next page)

Check: Try 2, a number greater than $1\frac{5}{8}$.

$$a + \frac{3}{8} > 2$$

$$2 + \frac{3}{8} \stackrel{?}{>} 2$$

$$2\frac{3}{8} > 2 \quad ✓ \quad \text{It checks.}$$

The solution is $a > 1\frac{5}{8}$, all numbers greater than $1\frac{5}{8}$.

b. $\mathbf{14.92 + r \geq 7.65}$

$$14.92 + r \geq 7.65$$

$$14.92 - 14.92 + r \geq 7.65 - 14.92 \quad \textit{Subtract 14.92 from each side.}$$

$$r \geq -7.27 \quad 7.65 \boxed{-} 14.92 \boxed{=} -7.27$$

Check: Try −7, a number greater than −7.27.

$$14.92 + r \geq 7.65$$

$$14.92 + (-7) \stackrel{?}{\geq} 7.65$$

$$7.92 \geq 7.65 \quad ✓ \quad \textit{It checks.}$$

The solution is $r \geq -7.27$, all numbers greater than or equal to −7.27.

THINK ABOUT IT

Are $1\frac{5}{8}$ and −7.27 solutions to the inequalities in Examples 1a and 1b, respectively? Explain.

You can translate a phrase into an inequality involving rational numbers.

Example 2

APPLICATION

Medicine

They say "If the shoe fits, wear it." But people often wear shoes that don't fit properly. Podiatrists recommend that a shoe be at least $\frac{3}{8}$ inch but no more than $\frac{1}{2}$ inch longer than your foot. If a shoe is $9\frac{1}{2}$ inches long, how long should a wearer's foot be?

Explore We know the length of the shoe. We need to find the length of the wearer's foot.

Plan This situation can be represented using two inequalities. Let f represent the length of the foot.

Solve Write the inequalities.

foot length	*plus*	*$\frac{3}{8}$ inch*	*no more than*	*$9\frac{1}{2}$ inches*
f	$+$	$\frac{3}{8}$	$\leq$	$9\frac{1}{2}$

$$f + \frac{3}{8} \leq 9\frac{1}{2}$$

$$f + \frac{3}{8} - \frac{3}{8} \leq 9\frac{1}{2} - \frac{3}{8}$$

$$f \leq 9\frac{4}{8} - \frac{3}{8}$$

$$f \leq 9\frac{1}{8}$$

foot length	*plus*	$\frac{1}{2}$ *inch*	*is no more than*	$9\frac{1}{2}$ *inches*
f	$+$	$\frac{1}{2}$	$\geq$	$9\frac{1}{2}$

$$f + \frac{1}{2} \geq 9\frac{1}{2}$$
$$f + \frac{1}{2} - \frac{1}{2} \geq 9\frac{1}{2} - \frac{1}{2}$$
$$f \geq 9$$

The foot should be between 9 and $9\frac{1}{8}$ inches long.

Examine Since $9\frac{1}{8} + \frac{3}{8} = 9\frac{1}{2}$ and $9 + \frac{1}{2} = 9\frac{1}{2}$, the answer is reasonable.

Checking Your Understanding

Communicating Mathematics

Read and study the lesson to answer each question.

1. **Compare and contrast** solving an inequality involving rational numbers to solving an inequality involving integers.
2. **Write a problem** that could be solved using the inequality $65.50 + x > 89.75$.
3. **Write** the inequality whose solution is graphed below.

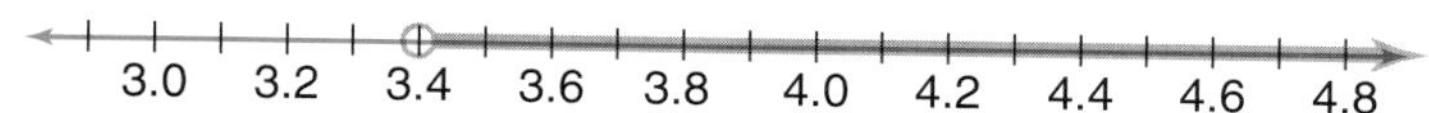

Guided Practice

Solve each inequality and check your solution. Graph the solution on a number line.

4. $m - 2\frac{3}{4} \geq \frac{7}{12}$
5. $x + \frac{5}{6} \leq 3\frac{3}{5}$
6. $\frac{5}{9} < 1\frac{1}{2} + y$
7. $x - 1.4 \leq 7.9$
8. $n - 3.7 \geq -7.2$
9. $y + 5.2 < 7.12$
10. **Computers** The number of electronic mailboxes on private networks was 2.5 million in 1990. It is predicted that more than 36 million will exist in 1996. How many mailboxes will have been created between 1990 and 1996?

Exercises: Practicing and Applying the Concept

Independent Practice

Solve each inequality and check your solution. Graph the solution on a number line.

11. $r + 7.5 \leq 13.2$
12. $x + 4.2 \geq -7.3$
13. $y + \frac{3}{4} < \frac{7}{12}$
14. $a - 2\frac{2}{3} \geq 8\frac{5}{6}$
15. $m - 3.1 < 7.4$
16. $b - 8.9 > -7.2$
17. $d - 1\frac{2}{3} < 1\frac{1}{6}$
18. $q - \frac{5}{12} \leq \frac{7}{18}$
19. $r + \frac{7}{8} \leq 2\frac{3}{4}$
20. $z - 1\frac{1}{2} \geq 4\frac{5}{9}$
21. $\frac{3}{4} \leq -1\frac{1}{2} + w$
22. $13.5 \leq 18.3 + k$
23. $h - 9.76 < 6.2$
24. $7\frac{6}{11} + f > 2\frac{14}{15}$
25. $u - 9.03 \leq 0.8$
26. $n - 289.90 < 479.21$
27. $\frac{17}{18} + t > \frac{107}{108}$
28. $0.027 + k \leq 0.00013$

Critical Thinking

29. If $\frac{1}{x} > x$, what can you say about the value of x?

Applications and Problem Solving

30. **Hobbies** Susan plans to spend no more than \$30 on model airplanes and supplies. If she buys a model for \$19.95, how much more can she spend on supplies? Express your solution as an inequality.

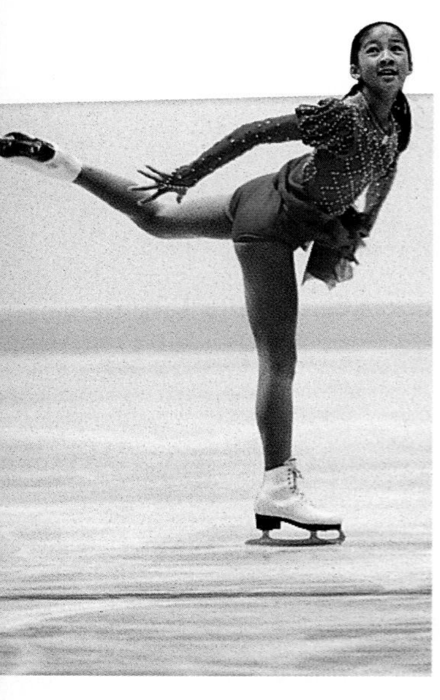

31. **Figure Skating** Michelle Kwan has scores of 9.4, 9.3, 9.8, and 8.9 in a figure skating competition. She has one more event to skate. Michelle will win the competition if her total score is greater than 47.1. What must Michelle's final event be rated in order for her to win?

32. **Family Activity** Imagine that you have won \$1000.00 in a radio contest. Choose one item that you would purchase for each person in your family including yourself. Research the prices in an advertisement or catalog. Would you have enough money? Write an inequality to show the amount of money you would have left over to put in savings.

33. **Employment** According to the Census Bureau, a high school graduate earns less than half what a college graduate earns in a lifetime.
    a. What do you think this statement means?
    b. According to the statement, the average high school graduate could earn one-tenth of what an average college graduate does. Do you think that is true? Explain.

34. **Geometry** The length of each side of a triangle is always less than the sum of the lengths of the other two sides. Find s for the triangle shown at the right.

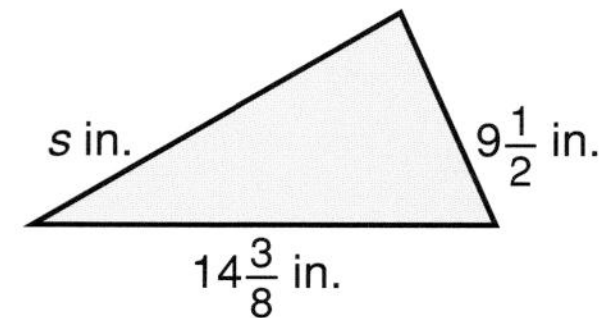

Mixed Review

35. Solve $x + 3\frac{1}{2} = 7\frac{1}{4}$. (Lesson 5-6)

36. Write $\frac{3}{a^2}$ as an expression using negative exponents. (Lesson 4-9)

37. Find the GCF of 46 and 72. (Lesson 4-5)

38. **Geometry** Find the perimeter and area of the rectangle shown at the right. (Lesson 3-5)

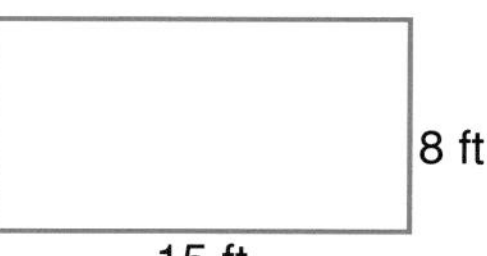

39. Simplify $2a(-3b)(5c)$. (Lesson 2-7)

40. **Sports** Plastic lenses and rubber eye pieces are used to make lightweight, shatterproof sunglasses for serious athletes. Ray-Ban's new sport-specific glasses, called Xrays, weigh less than 25 grams. Write an inequality for the weight of the glasses. (Lesson 1-9)

41. **Driving** The table at the right shows the cost of driving a car in different cities in cents per mile. The costs include insurance, depreciation of the car, license fees, taxes, fuel, oil, tires, and maintenance. (Lesson 1-10)
    a. Which of the listed cities is the most expensive for driving?
    b. What is the difference between the cost of driving in the most expensive city and in the least expensive city listed?

City	Cost (cents per mile)
Bismarck, ND	36.3
Boston, MA	49.8
Burlington, VT	36.4
Hartford, CT	48.0
Los Angeles, CA	55.8
Nashville, TN	37.1
Philadelphia, PA	49.0
Providence, RI	48.5
Raleigh, NC	37.1
Sioux Falls, SD	35.8

Source: Runzheimer International

5-8 Problem-Solving Strategy: Using Logical Reasoning

Setting Goals: *In this lesson, you'll use deductive and inductive reasoning.*

Modeling a Real-World Application: Comics

Peppermint Patty has used **inductive reasoning**. Inductive reasoning *makes* a rule after seeing several examples. Patty made a conclusion based on what happened in the past.

Example  **What is the next term in the sequence 1, 2, 4, 8, 16, . . . ?**

Notice that every number is twice the previous number. For example, 4 is twice 2 and 8 is twice 4. The next term is probably twice 16, or 32.

Suppose Patty gets back her algebra paper and finds that this time $x = 9$ and $y = 15$. Then her method of inductive reasoning did not work. Since inductive reasoning is based on past evidence only, it may sometimes fail.

The Greek letter π, called pi, represents the number 3.14159. . . . If the question on Patty's paper had been "What is the value of π?", she would know that the answer was 3.14159. . . . This is an example of **deductive reasoning**. Deductive reasoning *uses* a rule to make a conclusion.

Example 

Traffic

All octagons have eight sides. A stop sign is shaped like an octagon. How many sides does a stop sign have?

You do not need to count the sides of a stop sign. Using deductive reasoning, you can conclude that all stop signs have eight sides.

Checking Your Understanding

Communicating Mathematics

Read and study the lesson to answer these questions.

1. **Compare and contrast** deductive reasoning and inductive reasoning.
2. Which type of reasoning are you using when you look for a pattern?

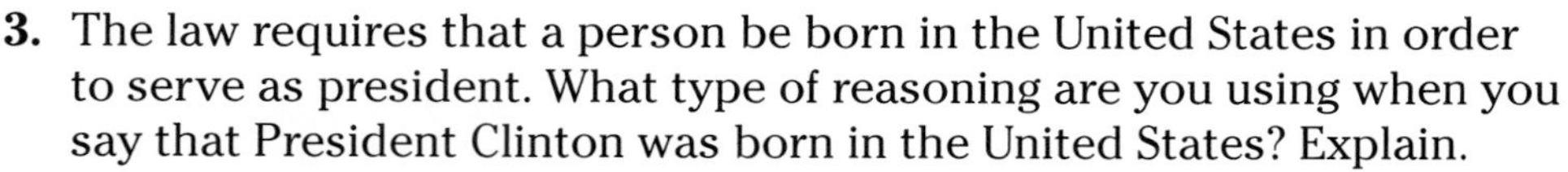

3. The law requires that a person be born in the United States in order to serve as president. What type of reasoning are you using when you say that President Clinton was born in the United States? Explain.

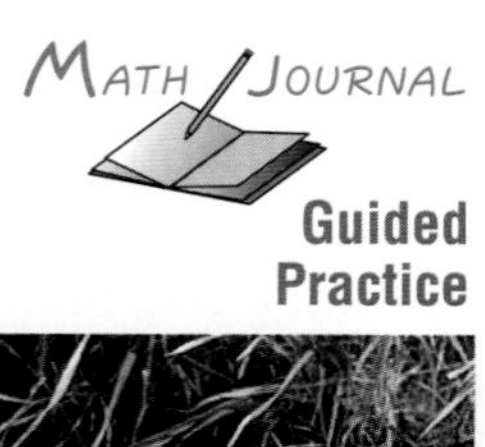

4. Make up your own examples of inductive and deductive reasoning.

Guided Practice

State whether each is an example of *inductive* or *deductive* reasoning. Explain your answer.

5. Every year for the past five years it has rained during spring break. Jamal says that it will rain during spring break this year.
6. If a student earns an "A" for each nine weeks in a class, then he or she will not have to take the final exam in that class. Stacey earned an A each nine weeks in Pre-Algebra, so she will not have to take the final.
7. The citizens of San Juan Capistrano celebrate when the swallows leave and return. For many years, the swallows have left on October 13 and returned on March 19. The citizens are predicting the swallows will return this year on March 19.

Exercises: Practicing and Applying the Concept

Independent Practice

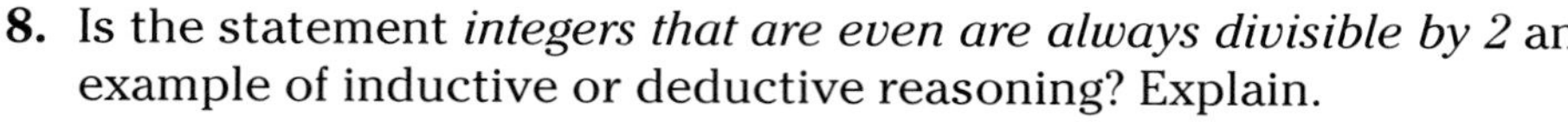

8. Is the statement *integers that are even are always divisible by 2* an example of inductive or deductive reasoning? Explain.
9. Find the next two numbers in the pattern 3, 6, 9, 12, 15, _?_, _?_.
10. The product of a number and itself is 196. Find the number.
11. **Patterns** Write the equation that you think should come next in the pattern at the right. Check your answer with a calculator.
$$1^2 = 1$$
$$11^2 = 121$$
$$111^2 = 12{,}321$$

Choose
Estimation
Mental Math
Calculator
Paper and Pencil

12. Insert one set of parentheses in $4 \times 5 - 2 + 7 = 19$ to make the equation true.
13. **School** If you do not pass your math test, then your parents will ground you for the weekend.
    a. Suppose you do not pass your math test. What will your parents do? What type of reasoning did you use?
    b. Suppose you do pass your math test. Does that mean you will definitely not be grounded? Why or why not?

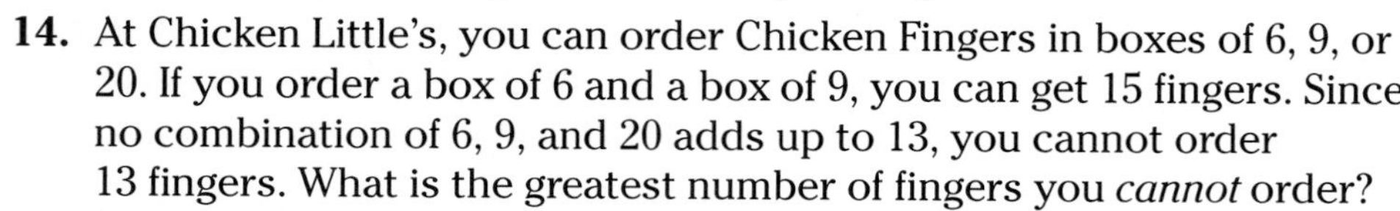

14. At Chicken Little's, you can order Chicken Fingers in boxes of 6, 9, or 20. If you order a box of 6 and a box of 9, you can get 15 fingers. Since no combination of 6, 9, and 20 adds up to 13, you cannot order 13 fingers. What is the greatest number of fingers you *cannot* order?
15. **Patterns** Study the pattern at the right.
$$11 \times 1 = 11$$
$$11 \times 2 = 22$$
$$11 \times 3 = 33$$
$$11 \times 4 = 44$$
    a. If you were to continue the pattern, how would you write 11×10?
    b. What is the actual value of 11×10?
    c. What does this tell you about inductive reasoning? Could you use inductive reasoning to prove something?

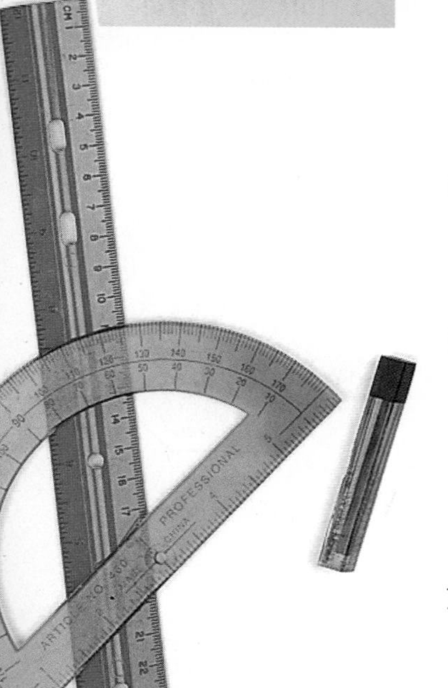

16. At Champion High School, if you take an art class you do not have to take a music class. Anastasia is taking a technical drawing course, so she does not have to take a music course. Is this an example of inductive or deductive reasoning?

17. What color piece is needed to complete the quilt at the right?

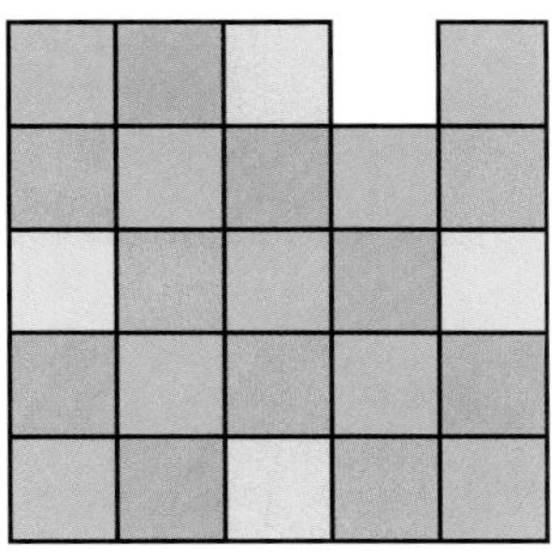

Critical Thinking

18. A set of dominoes has 28 rectangular pieces with two numbers of dots on each one. Each number, 0 through 6, is paired with every other number, including itself, on exactly one domino. The grid at the right was made by arranging dominoes and then recording the numbers. Four dominoes are shown. Copy the grid and use deductive reasoning to draw the positions of the remaining dominoes.

<table>
<tr><td>1</td><td>0</td><td>2</td><td>0</td><td>0</td><td>5</td><td>4</td><td>1</td></tr>
<tr><td>1</td><td>1</td><td>5</td><td>3</td><td>6</td><td>2</td><td>4</td><td>2</td></tr>
<tr><td>3</td><td>3</td><td>1</td><td>0</td><td>3</td><td>5</td><td>3</td><td>4</td></tr>
<tr><td>0</td><td>6</td><td>6</td><td>4</td><td>6</td><td>5</td><td>1</td><td>1</td></tr>
<tr><td>0</td><td>4</td><td>0</td><td>2</td><td>5</td><td>4</td><td>2</td><td>6</td></tr>
<tr><td>1</td><td>2</td><td>3</td><td>2</td><td>6</td><td>4</td><td>5</td><td>2</td></tr>
<tr><td>3</td><td>5</td><td>5</td><td>0</td><td>3</td><td>4</td><td>6</td><td>6</td></tr>
</table>

Mixed Review

19. Solve $z - 4.71 \leq -3.8$. (Lesson 5-7)

20. Solve $a - 9\frac{5}{6} = 2\frac{3}{24}$. (Lesson 5-6)

21. Find the GCF of 78 and 35. (Lesson 4-5)

22. **Fashion** The formula that relates men's shoe size, *s*, and foot length, *f*, in inches is $s = 3f - 21$. What is the shoe size for a man whose foot is 11 inches long? (Lesson 3-4)

23. Find $|-6| + |4|$. (Lesson 2-1)

From the FUNNY PAPERS

"Water boils down to nothing . . . snow boils down to nothing . . . ice boils down to nothing . . . everything boils down to nothing."

1. What type of reasoning did the caveman use? Explain.
2. Is the conclusion the caveman reached correct? Give an example to justify your answer.

5-9 Integration: Discrete Mathematics
Arithmetic Sequences

Setting Goals: *In this lesson, you'll find the terms of arithmetic sequences and represent a sequence algebraically.*

Modeling with Technology

Have you ever seen a flash of lightning and heard the roll of thunder and wondered how far away it was? We can generate a table to estimate the distance to a lightning bolt for which you heard the thunder. Sound travels approximately 1100 feet or $\frac{1}{5}$ of a mile per second. For every second that passes, the soundwaves of the thunder travel an additional 1100 feet.

Time (seconds)	0	1	2	3	4	5	6
Distance (feet)	0	1100	2200				
Distance (miles)	0	$\frac{1}{5}$	$\frac{2}{5}$				

Your Turn Copy the table above. Use a calculator and a pattern to complete the table.

a. How far away is the flash of lightning if the thunderclap takes 4 seconds to reach the observer?

b. What pattern do you see in the table?

c. Write a rule for finding the next numbers in the patterns.

Learning the Concept

A branch of mathematics called **discrete mathematics** deals with topics like logic and statistics. One topic of discrete mathematics is **sequences**. A sequence is a list of numbers in a certain order, such as 0, 1, 2, 3, or 2, 4, 6, 8. Each number is called a **term** of the sequence. A sequence like the distances for thunder, where the difference between any two consecutive terms is the same, is called an **arithmetic sequence**. The difference is called the **common difference**.

Arithmetic Sequence	An arithmetic sequence is a sequence in which the difference between any two consecutive terms is the same.

Example  **State whether each sequence is arithmetic. Then write the next three terms of each sequence.**

a. 1.25, 1.45, 1.65, 1.85, 2.05

Find the difference between consecutive terms in the sequence.

$$\begin{array}{ccccccccc} 1.25 & & 1.45 & & 1.65 & & 1.85 & & 2.05 \\ & +0.20 & & +0.20 & & +0.20 & & +0.20 & \end{array}$$

Since the difference between any two consecutive terms is the same, the sequence is arithmetic.

Continue the sequence to find the next three terms.

$$\begin{array}{ccccccccc} \ldots 1.85 & & 2.05 & & 2.25 & & 2.45 & & 2.65 \\ & +0.20 & & +0.20 & & +0.20 & & +0.20 & \end{array}$$

The next three terms of the sequence are 2.25, 2.45, and 2.65.

b. 1, 3, 7, 13, 21, . . .

Since there is no common difference, the sequence is not arithmetic.

$$\begin{array}{ccccccccc} 1 & & 3 & & 7 & & 13 & & 21 \\ & +2 & & +4 & & +6 & & +8 & \end{array}$$

Although the sequence is not arithmetic, it is still a sequence. Notice that the differences are in a pattern of 2, 4, 6, 8. So, the next three differences will be 10, 12, 14. The next three terms are 21 + 10 or 31, 31 + 12 or 43, and 43 + 14 or 57.

c. 13, 8, 3, −2, −7, . . .

$$\begin{array}{ccccccccccccccc} 13 & & 8 & & 3 & & -2 & & -7 & & \mathbf{-12} & & \mathbf{-17} & & \mathbf{-22} \\ & -5 & & -5 & & -5 & & -5 & & \mathbf{-5} & & \mathbf{-5} & & \mathbf{-5} & \end{array}$$

Since there is a common difference of −5, the sequence is arithmetic. The next three terms are −12, −17, and −22.

Connection to Algebra

If we know any term of an arithmetic sequence and the common difference, we can list the terms of the sequence. Consider the following sequence.

3, 7, 11, 15, 19, 23, . . .

We can write an expression that represents any term in the sequence. The first term is 3; call the first term a. The common difference between terms is 4; call the common difference d. Study this pattern.

1st term	a	3
2nd term	$a + d$	$3 + 4 = 7$
3rd term	$a + d + d$ or $a + 2d$	$3 + 2 \cdot (4) = 11$
4th term	$a + d + d + d$ or $a + 3d$	$3 + 3 \cdot (4) = 15$
⋮	⋮	
nth term	$a + (n - 1)d$	

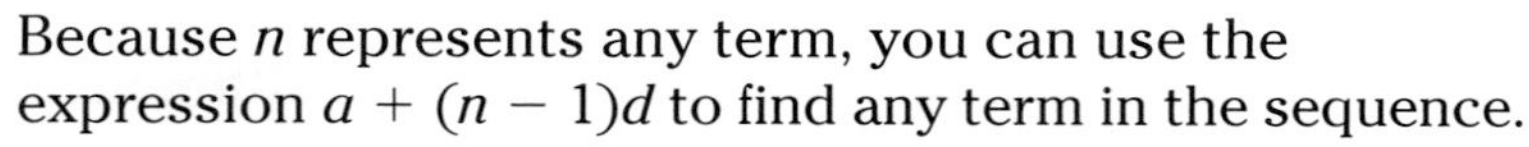
Because n represents any term, you can use the expression $a + (n - 1)d$ to find any term in the sequence.

Example 2

In 1995, the first class postage rates were raised to 32 cents for the first ounce and 23 cents for each additional ounce. A chart showing the postage for weights up to 6 ounces is shown below. What was the cost for a 10-ounce letter?

Weight (ounces)	1	2	3	4	5	6
Postage (cents)	32	55	78	101	124	147

Use the expression $a + (n - 1)d$ to find the 10th term of the postage sequence.

The first term, a, is 32. The common difference, d, is 23. Since we are looking for the tenth term, n is 10.

$a + (n - 1)d = 32 + (10 - 1)(23)$ *Substitute 32 for a, 10 for n, and 23 for d.*

$= 32 + 9(23)$ *Use the order of operations.*

$= 32 + 207$

$= 239$ The 10th term is 239.

A 10-ounce letter would cost \$2.39.

THINK ABOUT IT

Which would be easier to find the cost of a 16-ounce letter: extending the table to 16 ounces or using the formula?

Checking Your Understanding

Communicating Mathematics

Read and study the lesson to answer these questions.

1. **Give** an example of an arithmetic sequence.
2. **Tell** the relationship between consecutive terms of an arithmetic sequence.
3. **Explain** how to determine if a sequence is arithmetic.

4. Given the expression $a + (n - 1)d$, record in your Math Journal the meaning of the variables a, n, and d. Illustrate the meaning with a numerical example.

Guided Practice

State whether each sequence is arithmetic. Then write the next three terms of each sequence.

5. 2, 5, 8, 11, 14, . . .
6. 5, 9, 13, 17, 21, . . .
7. 17, 16, 14, 11, 7, . . .
8. $\frac{3}{2}, 2, \frac{5}{2}, 3, \ldots$
9. 3, 5, 8, 12, 17, . . .
10. 15, 25, 40, 60, 85, . . .
11. **On-Line Service** In 1995, America Online charged \$9.95 and \$2.95 per hour after the first five hours to use their service. Let n represent the number of hours after five.
    a. Write the sequence of prices for $n = 0$ to 5.
    b. How much would it cost if you used America Online 9 hours more than the allotted five hours in one month?

Exercises: Practicing and Applying the Concept

Independent Practice

State whether each sequence is arithmetic. Then write the next three terms of each sequence.

12. 0.75, 1.5, 2.25, 3, . . .

13. 4.5, 4.0, 3.5, 3.0, . . .

14. 0,1100, 2200, 3300, . . .

15. 1, 4, 9, 16, . . .

16. 1, 1, 1, 1, 1, . . .

17. 1, 2, 4, 8, 16, . . .

18. 7, 10, 13, 16, . . .

19. 91, 82, 73, 64, . . .

20. 83, 77, 71, 65, . . .

21. 19, 15, 11, 7, . . .

22. 0.1, 0.3, 0.9, 2.7, . . .

23. 5.47, 6.49, 7.51, 8.53, . . .

24. 10, 12, 15, 19, 24, . . .

25. 11, 14, 19, 26, 35, . . .

26. $1, \frac{1}{2}, \frac{1}{3}, \frac{1}{4}, \ldots$

27. $\frac{3}{4}, \frac{4}{5}, \frac{5}{6}, \frac{6}{7}, \ldots$

28. Write the first six terms in an arithmetic sequence with a common difference of −3. The first term is 35.

29. Given the sequence 7, 18, 29, 40, . . . , name the first term and the common difference.

30. Find the 25th even integer.

31. The first term of an arithmetic sequence is 100, and the common difference is 25. Write the first six terms of the sequence.

Critical Thinking

32. Draw the next figure in the sequence below.

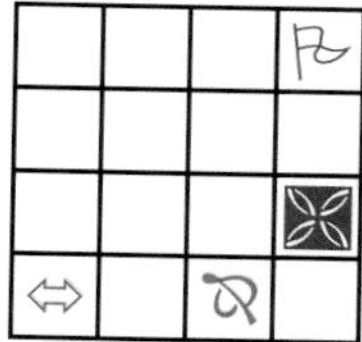
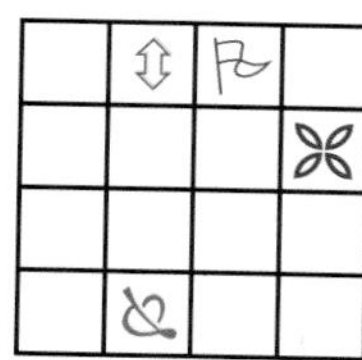
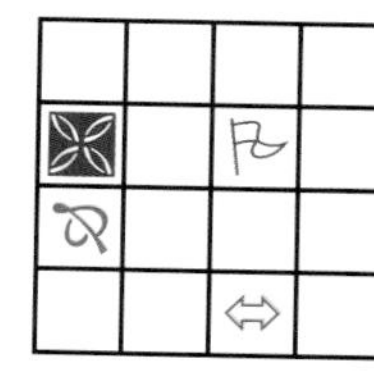
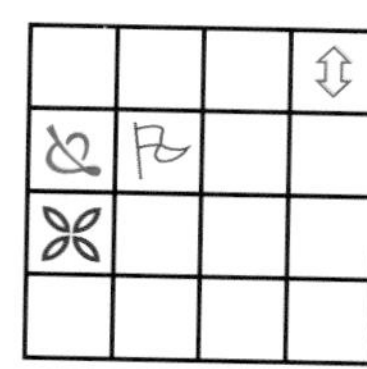
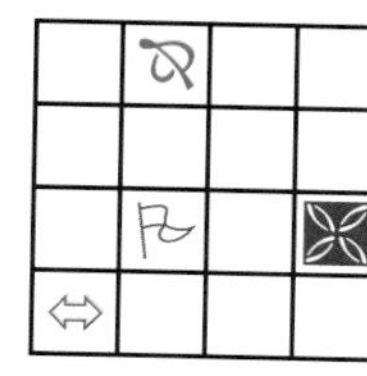

Applications and Problem Solving

33. Meteorology Refer to the application at the beginning of the lesson. Thunder can be heard from up to 10 miles away.

a. How many seconds would it take for the thunder to travel 10 miles?

b. Write a formula for finding the distance, d, in miles from lightning for which the thunder is heard t seconds after the lightning is seen.

34. Geometry Russian mathematician Sonya Kovalevsky (1850–1891) studied number sequences. One of the sequences she studied is represented by the shaded squares below.

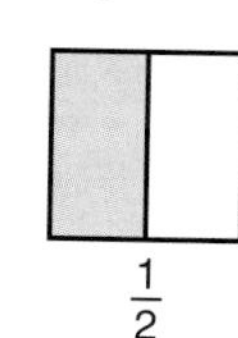
$\frac{1}{2}$ +
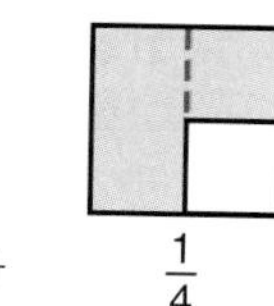
$\frac{1}{4}$ +
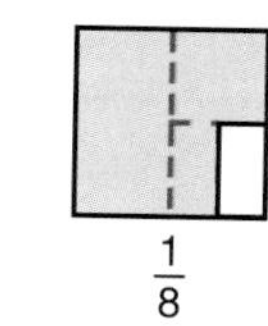
$\frac{1}{8}$ +
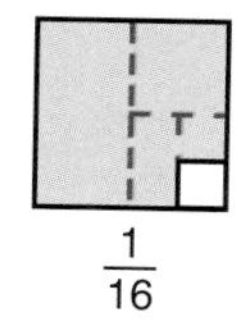
$\frac{1}{16}$ +
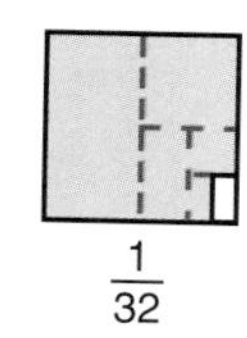
$\frac{1}{32}$

a. Write the first ten numbers of the sequence represented by the model.

b. Is the sequence arithmetic? Explain.

c. What do you think the sum of the first ten terms is close to? Do not actually add.

35. **Physics** While flying his hot-air balloon at 4000 feet, Rey dropped a sandbag in order to climb higher. A free-falling object increases its velocity by 32 feet per second every second. Use a sequence to determine how fast the sandbag will be falling at the end of 10 seconds.

36. **Meteorology** Television weather reporters frequently give the daily barometric pressure, such as 29.56 inches. This means that atmospheric pressure alone will hold up a column of mercury to a height of 29.56 inches. Centuries ago, scientists experimented with other materials such as water and discovered that atmospheric pressure would hold up a column of water more than thirty feet high.

Mercury (inches)	25	26	27	28	29	30	31	32
Water (feet)	27.2	28.33	29.46	30.59				

a. Copy and complete the table of heights above.

b. What is a in the sequence of water heights shown? What is the common difference?

c. How many feet of water are held by the pressure that holds up thirty inches of mercury?

Mixed Review

37. **Business** The long-distance telephone company that Lathan uses charges \$0.10 per minute for a long distance call. Lathan talked for one hour, so he knows his bill will be \$6.00. Is this an example of inductive or deductive reasoning? (Lesson 5-8)

38. Write $\frac{x}{125}$ as an expression with negative exponents. (Lesson 4-9)

39. Find the missing exponent in $(x^{\bullet})(x^9) = x^{15}$. (Lesson 4-8)

40. In which quadrant does the graph of $(-17, -4)$ lie? (Lesson 2-2)

41. **Statistics** Have you ever heard of "sleeping like a baby?" How about "sleeping like a koala!" A koala sleeps more than any other animal. The table at the right shows the average number of hours that different animals sleep each day. Make a bar graph of the data. (Lesson 1-10)

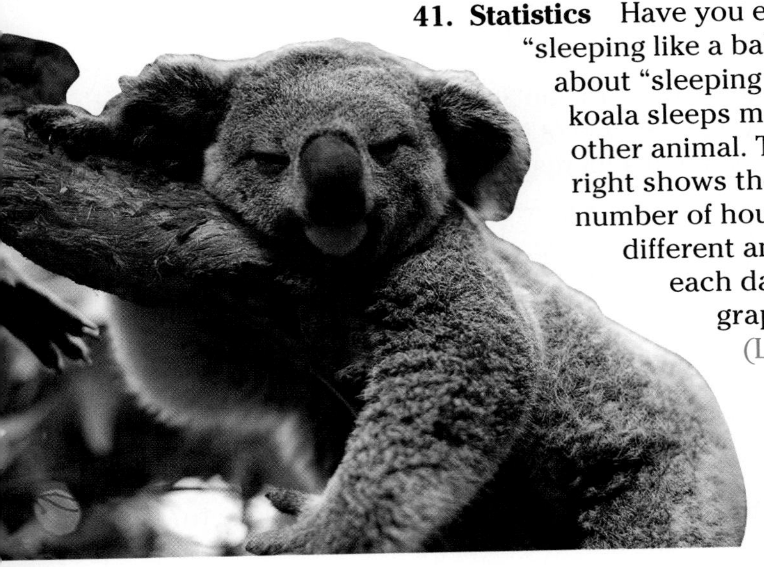

Animal	Average hours of sleep
Koala	22
Sloth	20
Armadillo	19
Opossum	19
Lemur	16
Hamster	14
Squirrel	14
Cat	13
Pig	13
Spiny anteater	12

42. Write a related sentence for $35 = 16 + m$ using an inverse operation. (Lesson 1-8)

5-9B

HANDS-ON ACTIVITY

Fibonacci Sequence

An Extension of Lesson **5-9**

As you have learned, there are sequences that are not arithmetic. One special sequence is called the **Fibonacci sequence**.

The Fibonacci sequence is not arithmetic, but its terms are found by adding. The sequence begins with 1, 1, and each term after the second is the sum of the previous two terms of the sequence.

		1 + 1	1 + 2	2 + 3	3 + 5
1,	1,	2,	3,	5,	8, . . .

Your Turn List the first ten terms of the Fibonacci sequence.

There are many patterns in the Fibonacci sequence. Here are some of the patterns.

- Every third term is divisible by 2.
- Every fourth term is divisible by 3.
- Every fifth term is divisible by 5.
- Every sixth term is divisible by 8.

TALK ABOUT IT

1. What number do you think may be a factor of every seventh term? Use a calculator to check your conjecture.
2. There is only one perfect square and one perfect cube, except for 1, in the Fibonacci sequence. Name them.
3. If you square the Fibonacci numbers and then add adjacent squares, a new sequence is formed. Describe the sequence.

Extension

4. Investigate how the Fibonacci numbers are related to a pineapple, a pine cone, and a sunflower.

CLOSING THE

Investigation

TAKING STOCK

Refer to the Investigation on pages 166–167.

Throughout this investigation, the terms *capital gain*, *closing price*, and *share* were often used.

What do these terms mean?

What do these words mean in terms of the work you did in this investigation?

Plan and give an oral presentation to describe what you know about these and any other investment terms. Your presentation should include the following.

- Examples from your portfolio to help you describe and define the investment terms you used most often in this investigation.
- A table showing how you initially invested your $10,000 and the source of the prices.
- A stock chart or line graph showing a price-weighted index of your portfolio's performance over the course of this investigation.
- An explanation of the analysis of your portfolio's performance.
- A comparison of your portfolio's performance to how the *Standard and Poor's (S&P) 500* performed over the same time period.
- A statement of what your shares would be worth if you sold all of them today.
- An explanation of what you would do differently if you had to do this investigation over again.

Extension

Schedule an interview with a financial analyst or stock broker. Make a list of questions to ask him or her.

- Be sure to ask about the requirements and duties of their job.
- Include questions about investment strategies for teenagers and how investment strategies differ depending on the age of the investor.

PORTFOLIO ASSESSMENT

You may want to keep your work on this Investigation in your portfolio.

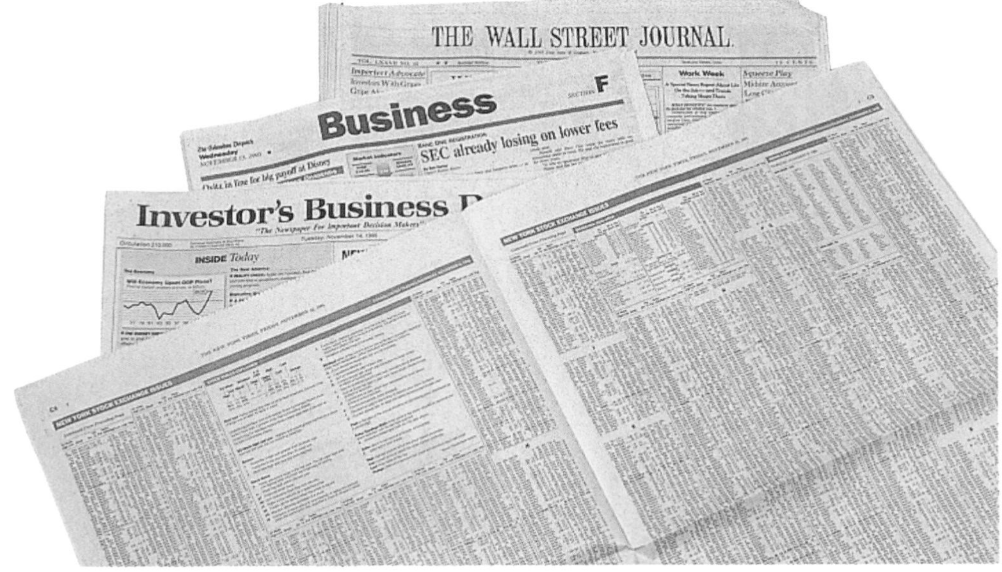

CHAPTER 5 Highlights

Vocabulary

After completing this chapter, you should be able to define each term, property, or phrase and give an example or two of each.

Algebra
adding like fractions (p. 239)
adding unlike fractions (p. 244)
associative property of addition (p. 234)
bar notation (p. 225)
commutative property of addition (p. 234)
identity property of addition (p. 234)
integers (p. 224)
inverse property of addition (p. 234)
mixed number (p. 239)
rational numbers (p. 224)
repeating decimals (p. 225)
subtracting like fractions (p. 240)
subtracting unlike fractions (p. 244)
terminating decimal (p. 225)
whole numbers (p. 224)

Geometry
scale (p. 224)

Problem Solving
deductive reasoning (p. 255)
inductive reasoning (p. 255)

Discrete Mathematics
arithmetic sequence (p. 258)
common difference (p. 258)
discrete mathematics (p. 258)
Fibonacci sequence (p. 263)
sequence (p. 258)
term (p. 258)

Understanding and Using Vocabulary

Choose the letter of the term that best matches each statement.

1. This is any number that can be expressed in the form $\frac{a}{b}$, where a and b are integers, and $b \neq 0$.
2. This type of reasoning creates a rule based on consistent examples.
3. A comparison of the size of a model to the size of the real object is called a __?__.
4. This property justifies the statement $3 + 5 = 5 + 3$.
5. This is any number which is positive, negative, or zero and whose absolute value is a whole number.
6. This type of reasoning uses a given rule to make a conclusion or a decision.
7. This property justifies the statement $4 + 0 = 4$.

A. Identity property of addition
B. rational number
C. integers
D. Commutative property of addition
E. whole numbers
F. deductive reasoning
G. Inverse property of addition
H. inductive reasoning
I. scale
J. Associative property of addition

CHAPTER 5 Study Guide and Assessment

Skills and Concepts

Objectives and Examples

Upon completing this chapter, you should be able to:

- **identify and compare rational numbers** (Lesson 5-1)

0 is a rational number, an integer, and a whole number.

0.1234567 . . . is a non-terminating, non-repeating decimal. So it is not rational.

Review Exercises

Use these exercises to review and prepare for the chapter test.

Name the set(s) of numbers to which each number belongs.

8. $\frac{9}{10}$ **9.** -4

10. 18 **11.** 0.122333 . . .

- **rename decimals as equivalent fractions.** (Lesson 5-1)

Let $N = 0.888\ldots.$ Then $10N = 8.888\ldots$

$$\begin{array}{rl} 10N &= 8.8888\ldots \\ -\quad N &= 0.8888\ldots \\ \hline 9N &= 8 \\ N &= \frac{8}{9} \\ 0.\overline{8} &= \frac{8}{9} \end{array}$$

Express each decimal as a fraction or mixed number in simplest form.

12. 3.25 **13.** -9.45

14. 0.40 **15.** -0.13

16. $0.\overline{72}$ **17.** $4.\overline{24}$

18. $0.0\overline{6}$ **19.** 8.89

- **estimate sums and differences of decimals and fractions.** (Lesson 5-2)

$48.89 + 91.21 \rightarrow 50 + 90 = 140$

$6\frac{9}{10} - 2\frac{3}{8} \rightarrow 7 - 2\frac{1}{2} = 4\frac{1}{2}$

Estimate each sum or difference.

20. $53.6 + 41.2$ **21.** $4.99 + 3.29$

22. $325.44 - 249.25$ **23.** $50.00 - 39.89$

24. $\frac{11}{12} + \frac{6}{10}$ **25.** $\frac{24}{25} + \frac{1}{11}$

26. $18\frac{1}{10} - 3\frac{1}{9}$ **27.** $24\frac{4}{7} - 22\frac{1}{6}$

- **add and subtract decimals.** (Lesson 5-3)

$f = 4.56 - 2.358$

$$\begin{array}{r} 4.560 \\ -2.358 \\ \hline 2.202 \end{array}$$

$f = 2.202$

Solve each equation.

28. $t = 8.5 + 42.25$ **29.** $9.43 - (-1.8) = p$

30. $j = -7.43 + 5.34$ **31.** $m = 17.19 - 24.87$

32. $y = 13.983 + 4.52$ **33.** $-8.52 - 9.43 = d$

Objectives and Examples

▶ **add and subtract fractions with like denominators.** (Lesson 5-4)

$n = 3\frac{3}{5} + 4\frac{4}{5}$

$n = (3 + 4) + \left(\frac{3}{5} + \frac{4}{5}\right)$

$n = 7 + \frac{7}{5}$

$n = 7 + 1\frac{2}{5}$ or $8\frac{2}{5}$

Review Exercises

Solve each equation. Write the solution in simplest form.

34. $w = \frac{4}{7} + \frac{6}{7}$
35. $t = 5\frac{2}{9} + 8\frac{5}{9}$
36. $7\frac{4}{7} + \left(-2\frac{5}{7}\right) = s$
37. $\frac{7}{8} - \frac{5}{8} = b$
38. $k = 5\frac{1}{5} - 4\frac{4}{5}$
39. $\frac{4}{15} - \frac{13}{15} = y$

▶ **add and subtract fractions with unlike denominators.** (Lesson 5-5)

$r = \frac{2}{5} + \frac{7}{8}$

$r = \frac{2 \cdot 8}{5 \cdot 8} + \frac{7 \cdot 5}{8 \cdot 5}$ *The LCD is 40.*

$r = \frac{16}{40} + \frac{35}{40}$

$r = \frac{51}{40}$ or $1\frac{11}{40}$

Solve each equation. Write the solution in simplest form.

40. $\frac{3}{7} + \frac{11}{14} = q$
41. $s = \frac{7}{15} - \frac{16}{30}$
42. $6\frac{4}{5} + 1\frac{3}{4} = h$
43. $4\frac{1}{6} - \left(-2\frac{3}{4}\right) = m$
44. $1\frac{2}{5} + \frac{1}{3} = a$
45. $9\frac{5}{12} - 4\frac{7}{18} = h$

▶ **solve equations with rational numbers.** (Lesson 5-6)

$$y + 5.7 = 3.1$$
$$y + 5.7 - 5.7 = 3.1 - 5.7$$
$$y = -2.6$$

Solve each equation. Check your solution.

46. $v - 4.72 = 7.52$
47. $s + (-13.5) = -22.3$
48. $x + \frac{3}{4} = -1\frac{2}{5}$
49. $b - \frac{5}{8} = 1\frac{3}{16}$
50. $z - (-5.8) = 1.36$
51. $k - \frac{4}{9} = 1\frac{7}{18}$

▶ **solve inequalities with rational numbers.** (Lesson 5-7)

$$p - 4\frac{9}{10} > -8\frac{6}{10}$$
$$p - 4\frac{9}{10} + 4\frac{9}{10} > -8\frac{6}{10} + 4\frac{9}{10}$$
$$p > -3\frac{7}{10}$$

Solve each inequality and check your solution. Graph the solution on a number line.

52. $w + \frac{7}{12} \geq \frac{5}{18}$
53. $f - 3\frac{1}{4} \leq \frac{7}{8}$
54. $31.6 < -5.26 + g$
55. $q - (-6.7) > 12$
56. $w - 4.32 \leq 1.234$
57. $m + 7.17 > 1.019$
58. $\frac{7}{6} + c \geq 3\frac{5}{24}$
59. $a - \left(\frac{-3}{10}\right) \leq 1\frac{1}{5}$

Objectives and Examples	Review Exercises

▶ **find the terms of arithmetic sequences and represent a sequence algebraically.** (Lesson 5-9)

2 4 8 16
+2 +4 +8

This is not an arithmetic sequence because the difference between any two consecutive terms is not the same.

The next three terms are 16 + 16 or 32, 32 + 32 or 64, and 64 + 64 or 128.

State whether each sequence is arithmetic. Then write the next three terms of each sequence.

60. 75, 90, 105, . . .

61. 45, 37, 29, 21, . . .

62. 0.0625, 0.125, 0.25, . . .

63. $\frac{10}{11}, \frac{8}{9}, \frac{6}{7}$

64. 67, 58, 49, 40, . . .

Applications and Problem Solving

65. Hobbies Tedrick is building a model yacht that is 35 centimeters long. The actual yacht is 12 meters long. (Lesson 5-1)

a. What is the scale of the model?

b. Is the scale a whole number, an integer, or a rational number? Explain.

66. Politics In the 1992 presidential election, 186.7 million people were eligible to vote. Only 113.9 million people actually voted. How many people were eligible to vote in 1992 but did not? (Lesson 5-3)

67. Carpentry Currito is making a porch swing for his house. He bought a 4-foot piece of poplar for the seat. If the finished seat on the pattern is to be $41\frac{3}{8}$ inches long, how much should Currito cut off? (Lesson 5-5)

68. Numbers that can be divided by both 2 and 3 are always divisible by 6. 18 is divisible by 2 and 3, so it is divisible by 6. Is this inductive or deductive reasoning? Explain. (Lesson 5-8)

69. Physical Science The Italian scientist Galileo discovered that there was a relationship between the time of the back-and-forth swing of a pendulum and its length. The table at the right shows the time for a swing and the length of several pendulums. (Lesson 5-9)

a. How long do you think a pendulum with a swing of 5 seconds is?

b. Describe the sequence of lengths. Is it an arithmetic sequence? Explain.

Time of swing (seconds)	Pendulum length (units)
1	1
2	4
3	9
4	16

A practice test for Chapter 5 is available on page 778.

Alternative Assessment

Performance Task

Demonstrate your knowledge by giving a clear, concise solution to each problem. Be sure to include all relevant drawings and justify your answers. You may show your solutions in more than one way or investigate beyond the requirements of the problem.

1. A house is 36 feet from front to back. On the blueprints, the drawing of the house is 18 inches from front to back.

a. What is the scale factor of the blueprints to the house?

b. What type of number is the scale factor? Explain.

2. The windows on the side of the house are $2\frac{7}{8}$ feet from the ground and $7\frac{3}{4}$ feet from the roof. The house is 15 feet from the ground to the roof.

a. Approximately how tall are the windows?

b. Exactly how tall are the windows on the side of the house?

3. The window on the front of the house is 6.72 feet wide, and there are 4.14 feet on either side of the window. What is the width of the front section of the house?

4. The house covers 1023.5 square feet. The house and the yard together cover 1 acre, or 43,560 square feet. How large is the yard?

5. The law in Genoa Township states that an individual may erect a fence as long as it is no closer than 3 inches from the property line. Mr. Owens says that the fence he is erecting 5 inches from the property line is legal. Did Mr. Owens use inductive or deductive reasoning? Explain.

6. A series of bushes is to be planted along the foundation of the house. Devise a plan for the planting that uses an arithmetic sequence.

Thinking Critically

- When you estimate an area in order to purchase wallpaper, it is wise to overestimate the area you need to cover. Give an example of a situation when it is better to estimate low.
- Do you come closer to the actual answer when you round to the nearest tenth or to the nearest whole number? Explain.

Portfolio

Review the items in your portfolio. Make a list or table of contents of the items, noting why each item was chosen and place it in your portfolio.

Self Evaluation

People who are willing to get their hands dirty can always find a sink to clean up. This is a way of saying that if you are willing to go out of your way, you can find a place to be helpful.

Assess yourself. Do you help others when they are having difficulties? Do you notice that people you have helped in the past are more likely to help you when you need some assistance?

LONG-TERM PROJECT

Investigation

MATERIALS

- ruler
- protractor
- compass
- graph paper

"Please remain seated and keep your hands and feet inside the ride at all times. Enjoy your ride."

You are a member of a group of three engineers that has been hired to design and build a new roller coaster ride. You will oversee every detail of the design of the ride from the length of its track to how riders will wait in line and board the cars. You'll even name the ride. The amusement park where the roller coaster will be built wants you to design it so that it establishes at least one new record.

Some of the existing world records are:

Longest

- *1.42 miles*—The Ultimate *at Lightwater Valley Theme Park in Ripon, Great Britain*

Tallest

- *222 feet above its footings*—Pepsi Max-The Big One *at Blackpool Pleasure Beach, Great Britain*

Fastest

- *85 mph*—Desperado *at Buffalo Bill's, Jean, NV*

Greatest Drop

- *225 feet*—Steel Phantom *at Kennywood Amusement Park, West Mifflin, PA*

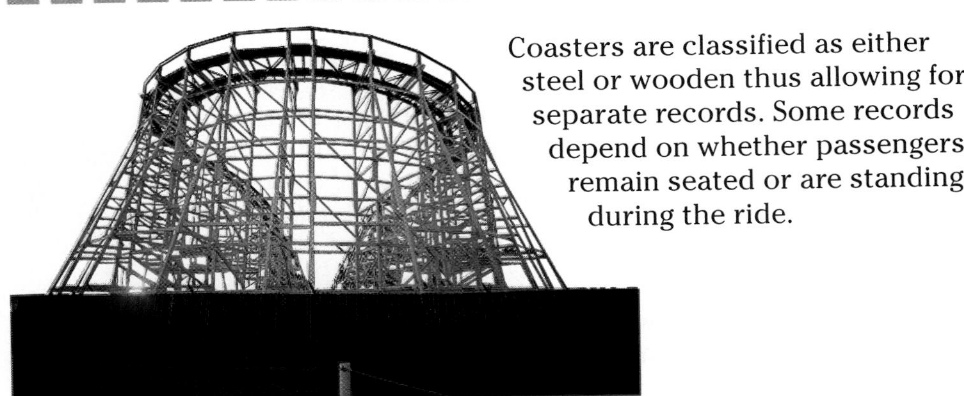

Coasters are classified as either steel or wooden thus allowing for separate records. Some records depend on whether passengers remain seated or are standing during the ride.

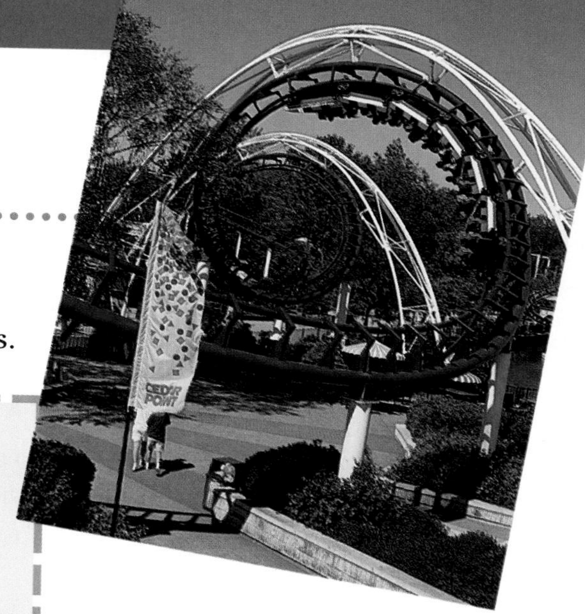

Starting the Investigation

To start the design, you will want to draw a detailed model of the path your roller coaster will take. Before you begin drawing, familiarize yourself with some roller coaster terms.

A few important roller coaster definitions are:

- ***Banked Turn***—*Turn in which the tracks are tilted to allow trains to turn at high speed.*
- ***Car***—*The part of a coaster train that carries between two and eight passengers.*
- ***Chain Lift***—*The rolling chain that carries the train to the crest of a hill.*
- ***Circuit***—*A completed journey on a coaster track.*
- ***First Drop***—*Usually the highest and fastest drop on a coaster. It most often follows immediately after the chain lift.*
- ***Gully Coaster***—*One that uses the natural terrain to give the added feeling of speed by keeping the track close to the ground.*
- ***Lift Hill***—*The hill that the train is carried up by the chain lift.*
- ***Loading Platform***—*The part of the station where passengers board the coaster trains.*
- ***Speed Dip***—*A small hill taken at high speeds usually lifting riders off their seats.*
- ***Train***—*A series of cars hooked together to travel the circuit of the coaster track.*

Since the design of a roller coaster can be very complex, you will need to include a scale drawing from several different angles including an overhead view. Be sure to include a sketch of the loading platform. Also include the number of cars in each train and how many passengers each car will carry.

Brainstorm with your group to decide what record you want to break with your design. You may wish to look in some other sources for other records and for guidance on how to design your coaster.

You will work on this investigation throughout Chapters 6 and 7. Be sure to keep your materials in your Investigation Folder.

interNET CONNECTION

For current information on roller coasters, visit:
www.glencoe.com/sec/math/prealg/mathnet

Investigation
Roller Coaster Math

Working on the Investigation
Lesson 6-1, p. 279

Working on the Investigation
Lesson 6-4, p. 293

Working on the Investigation
Lesson 7-2, p. 337

Closing the Investigation
End of Chapter 7, p. 362

CHAPTER

6 Rationals: Patterns in Multiplication and Division

TOP STORIES in Chapter 6

In this chapter, you will:

- write fractions as decimals,
- estimate and find products and quotients of rational numbers,
- find the mean, median, and mode of a set of data,
- identify and extend geometric sequences, and
- write numbers in scientific notation.

MATH AND COMPUTERS IN THE NEWS

How do you vaccinate a computer?

Source: PC Computing, September, 1994

Feed a cold, starve a fever, but what do you do for a virus? Especially if the patient has circuits and electric power! Computer viruses are destructive programs that are placed on a person's computer without their knowledge. Once inside the system, a virus manipulates or destroys the programs and data. Viruses have been a problem to computer owners for some time now, but with the dawn of the Information Superhighway, safeguards are now more important than ever.

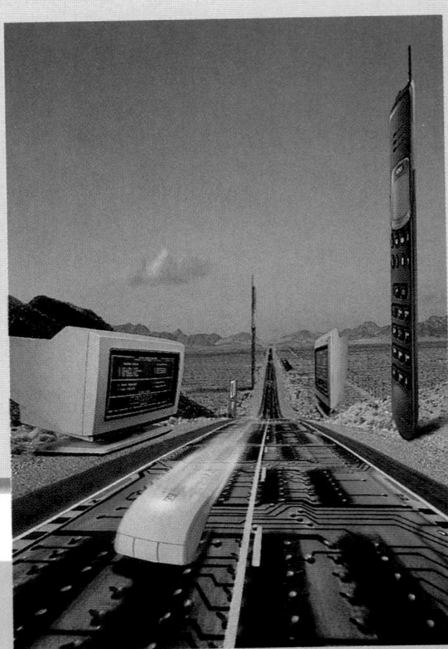

Putting It into Perspective

1850

1880

1910

1833
Charles Babbage designs the "analytical engine" with the programming help of Ada Byron Lovelace.

1888
The first successful punch-card tabulating machine is invented by an American.

On the Lighter Side

"Excuse me, I'm lost. Can you direct me to the information superhighway?"

Statistical Snapshot

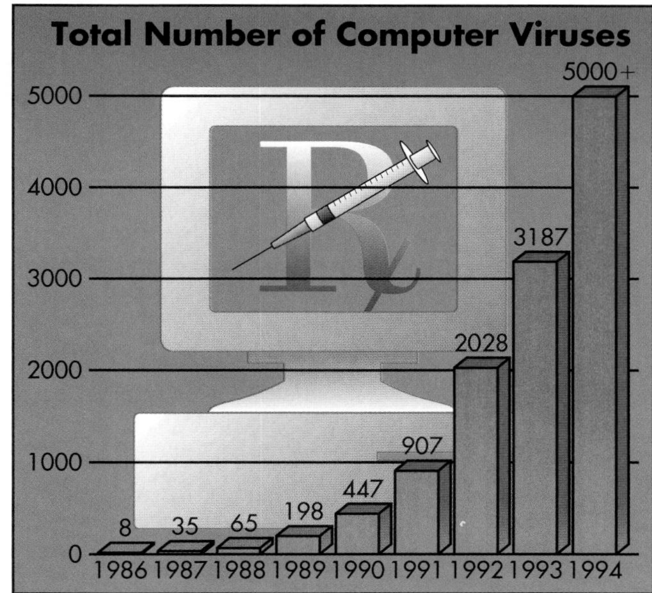

Source: Norman Data Defense Systems

interNET CONNECTION For up-to-date information on computer viruses, visit:
www.glencoe.com/sec/math/prealg/mathnet

1946
The first general purpose digital computer, the ENIAC, is constructed at the University of Pennsylvania.

1977
The Apple II is introduced as the first personal computer marketed on a large scale.

1995
The Information Superhighway, a worldwide computer network, grows.

1940 1970 2000

1968
Blockbuster movie *2001: A Space Odyssey* features the character HAL, a talking, thinking computer.

6-1 Writing Fractions as Decimals

Setting Goals: *In this lesson, you'll write fractions as terminating or repeating decimals and compare rational numbers.*

Modeling a Real-World Application: Food

"You scream, I scream, we all scream for ice cream!" What is your favorite flavor? The International Ice Cream Association polled Americans about their favorite flavors. Some of the top finishers and the approximate fractions of the votes they received are shown in the graph below. If $\frac{3}{76}$ of those surveyed chose chocolate chip as their favorite, did more people choose cookies-and-cream or chocolate chip? *This problem will be solved in Example 3.*

> **FYI**
>
> Finland produces more ice cream per citizen each year, 42 pints, than any other country in the world.

Learning the Concept

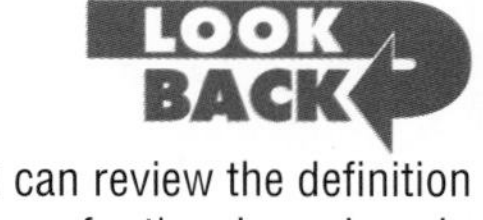

> **LOOK BACK**
>
> You can review the definition of rational numbers in Lesson 5-1.

In the case above, it is convenient to write rational numbers as decimals instead of fractions because decimals are easier to compare. Any fraction can be written as a decimal.

Consider the fraction $\frac{3}{5}$. The fraction $\frac{3}{5}$ indicates $3 \div 5$.

Method 1
Divide using paper and pencil.

$\frac{3}{5} \rightarrow$

```
   0.6
 5)3.0
  -3 0
     0
```

Method 2
Divide using a calculator.

3 ÷ 5 = .6

The fraction $\frac{3}{5}$ can be written as the decimal 0.6. This means that $\frac{3}{5}$ and 0.6 are **equivalent** rational numbers. Remember that a decimal like 0.6 is called a *terminating decimal* because the division ends, or terminates, when the remainder is zero.

Example  **Write $2\frac{3}{4}$ as a decimal.**

$2\frac{3}{4} = 2 + \frac{3}{4}$

Method 1
Use paper and pencil.

$\frac{3}{4} \rightarrow$

```
    0.75
  4)3.00
   -2 8
      20
     -20
       0
```

Consider only $\frac{3}{4}$.

Method 2
Use a calculator.

2 + 3 ÷ 4 = 2.75

Be sure your calculator follows the order of operations.

$2 + 0.75 = 2.75$

The mixed number $2\frac{3}{4}$ equals the decimal 2.75.

Not all fractions can be written as terminating decimals. Consider the fraction $\frac{2}{3}$. How can $\frac{2}{3}$ be expressed as a decimal?

Use paper and pencil.

$\frac{2}{3} \rightarrow$

```
   0.666
 3)2.000
  -1 8
     20
    -18
     20
```

Use a calculator.

2 ÷ 3 = .6666666

When you divide using paper and pencil, the remainder after each step is 2. If you continue dividing, the pattern repeats. The calculator display fills with 6s. Remember, a decimal like 0.6666666 . . . , or $0.\overline{6}$, is called a *repeating decimal.*

Some calculators round answers and others truncate answers. **Truncate** means to cut off at a certain place-value position, ignoring the digits that follow.

Example  Write $\frac{5}{11}$ as a decimal.

Use paper and pencil.

$$\frac{5}{11} \rightarrow 11\overline{)5.0000}$$

$$\begin{array}{r} 0.4545 \\ 11\overline{)5.0000} \\ \underline{-4\,4} \\ 60 \\ \underline{-55} \\ 50 \\ \underline{-44} \\ 60 \\ \underline{-55} \end{array}$$

$\frac{5}{11} = 0.\overline{45}$

Use a calculator.

5 [÷] 11 [=] .4545455

This calculator rounds.

5 [÷] 11 [=] .4545454

This calculator truncates.

Some calculators round and others truncate. Find the decimal value of $\frac{5}{11}$ on your calculator to determine whether it rounds or truncates.

Sometimes you need to compare numbers, as in the application at the beginning of the lesson. You can use what you have learned about writing fractions as decimals to compare fractions.

Example 3

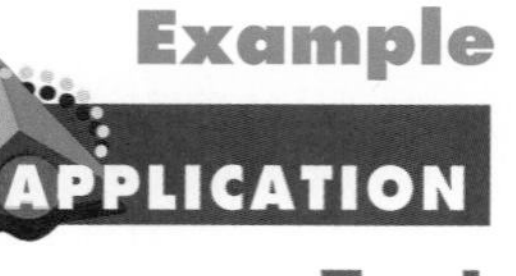

Food

Refer to the application at the beginning of the lesson. Which is more popular, chocolate chip or cookies-and-cream ice cream?

Write the fractions as decimals and then compare the decimals.

chocolate chip	cookies-and-cream
$\frac{3}{76} \rightarrow 0.0394736842$	$\frac{2}{55} \rightarrow 0.036363636$

In thousandths place, $9 > 6$.

Note that on a number line, 0.039 . . . is to the right of 0.036 . . . .

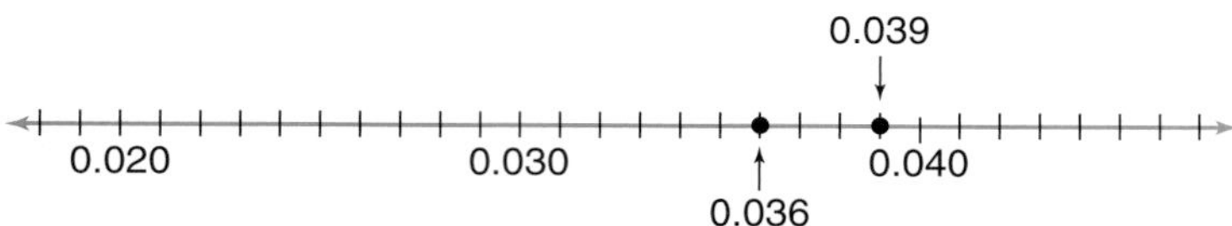

Since 0.039 . . . > 0.036 . . . , you know that $\frac{3}{76} > \frac{2}{55}$.

More people chose chocolate chip than chose cookies-and-cream.

Another way to compare two rational numbers is to write them as equivalent fractions with like denominators.

Example  Replace each ● with <, >, or = to make a true sentence.

a. $\frac{7}{8}$ ● $\frac{5}{6}$

$\frac{21}{24}$ ● $\frac{20}{24}$ *The LCM is 24.* $\frac{7}{8} = \frac{21}{24}$ *and* $\frac{5}{6} = \frac{20}{24}$

$\frac{21}{24} > \frac{20}{24}$ *Since* $21 > 20$, $\frac{21}{24} > \frac{20}{24}$.

Therefore, $\frac{7}{8} > \frac{5}{6}$.

b. $-\frac{3}{5}$ ● -0.75

-0.6 ● -0.75 *Write $-\frac{3}{5}$ as a decimal.*

$-0.6 > -0.75$ *In tenths place, $-6 > -7$.*

Therefore, $-\frac{3}{5} > -0.75$.

Verify this answer using a number line.

Checking Your Understanding

Communicating Mathematics

Read and study the lesson to answer these questions.

1. **Explain** why we use division to change a fraction to a decimal.
2. **Describe** the difference between terminating and repeating decimals. Give an example of each.
3. **Explain** the difference between 0.6 and $0.\overline{6}$. Which is greater?
4. **You Decide** Juanita says that $\frac{2}{3} > \frac{3}{5}$. Drew disagrees. Who is correct and why?
5. The grid at the right represents the number 0.45. Represent 0.5, $\frac{1}{5}$, $\frac{1}{3}$, and 0.25 by shading 10-by-10 grids.

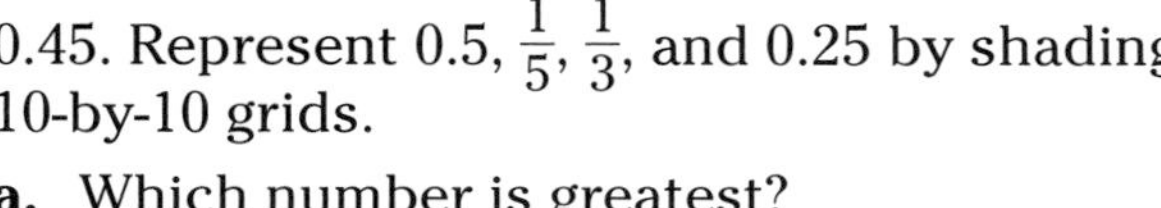

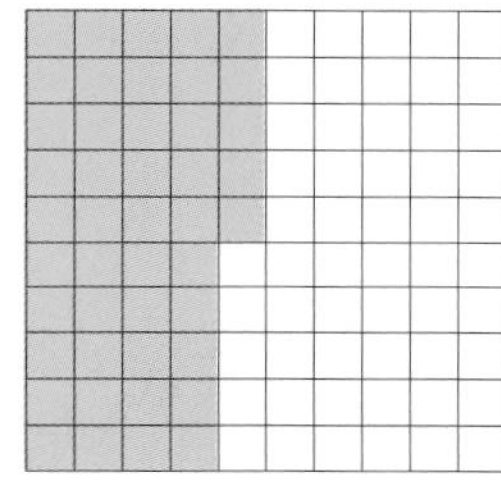

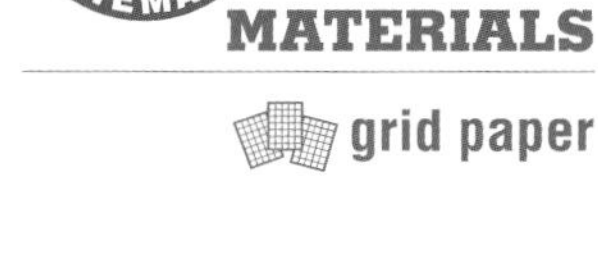

MATERIALS

grid paper

   a. Which number is greatest?
   b. Which number is least?
   c. Order the numbers from least to greatest.

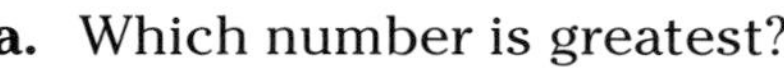

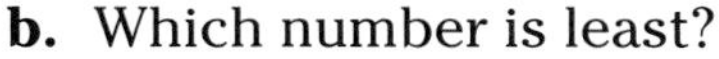

Guided Practice

Write each fraction as a decimal. Use a bar to show a repeating decimal.

6. $\frac{5}{9}$ 7. $2\frac{2}{3}$ 8. $-3\frac{1}{8}$ 9. $\frac{1}{8}$

State the greater number for each pair.

10. $\frac{1}{2}, \frac{1}{3}$ 11. $-\frac{3}{4}, -\frac{7}{8}$ 12. $\frac{3}{8}, \frac{10}{33}$ 13. $-\frac{3}{4}, -0.\overline{75}$

14. **Business** On Monday, June 12, 1995, the price of Wendy's stock closed at $16\frac{5}{8}$. The previous Friday, the stock had closed at $16\frac{3}{4}$. On which day did the stock close at a higher price?

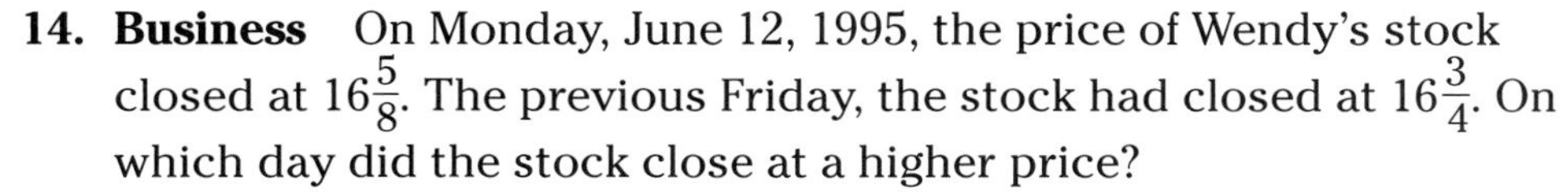

Exercises: Practicing and Applying the Concept

Independent Practice

Write each fraction as a decimal. Use a bar to show a repeating decimal.

15. $\frac{7}{10}$ 16. $-\frac{1}{3}$ 17. $\frac{1}{2}$ 18. $-1\frac{1}{4}$

19. $\frac{14}{20}$ 20. $-\frac{2}{9}$ 21. $\frac{5}{16}$ 22. $\frac{21}{25}$

23. $-3\frac{4}{11}$ 24. $\frac{23}{45}$ 25. $-4\frac{5}{16}$ 26. $\frac{31}{40}$

Replace each ● with >, <, or = to make a true sentence.

27. $5\frac{1}{5}$ ● 5.18 **28.** 7.56 ● $7\frac{12}{25}$ **29.** −3.45 ● $3\frac{2}{3}$

30. $-1\frac{1}{20}$ ● 1.01 **31.** $\frac{7}{8}$ ● $\frac{8}{9}$ **32.** $\frac{2}{5}$ ● $\frac{1}{3}$

33. $3\frac{4}{7}$ ● $3\frac{5}{8}$ **34.** $2\frac{5}{8}$ ● $2\frac{2}{3}$ **35.** $-\frac{11}{14}$ ● $-\frac{13}{16}$

36. $\frac{9}{11}$ ● $\frac{19}{23}$ **37.** $1\frac{4}{5}$ ● 1.857 **38.** $-5\frac{1}{12}$ ● −5.09

Calculators

Several different calculators' displays for 2 ÷ 3 are shown below. What would the display for 7 ÷ 9 be for each calculator?

39. .6666666 **40.** 0.6666666

41. 0.6666667 **42.** .67

Critical Thinking

43. Find the decimal equivalents for $\frac{1}{3}, \frac{1}{4}, \frac{7}{10}, \frac{4}{5}, \frac{5}{12}, \frac{4}{7}$, and $\frac{3}{25}$.

- **a.** Which are terminating decimals? Which are repeating decimals?
- **b.** Find the prime factorization of each denominator.
- **c.** What prime factors are present in the denominator of a fraction whose equivalent decimal is terminating? Write about what you observe.

44. Find a terminating and a repeating decimal between $\frac{1}{6}$ and $\frac{8}{9}$. Explain how you found them.

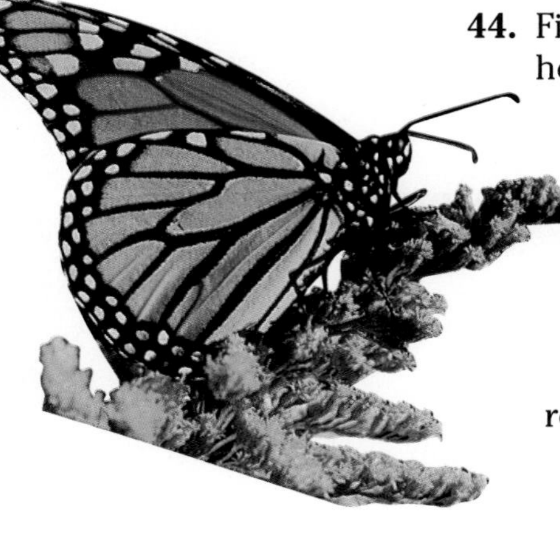

45. Zoology Monarch butterflies migrate up to 2000 miles from Canada and the northern United States to Florida, California, and Mexico. The fastest monarch butterfly can fly $\frac{1}{3}$ mile in one minute. Write $\frac{1}{3}$ as a decimal rounded to the nearest hundredth.

Applications and Problem Solving

46. Physics For best performance, 0.91 to 0.93 of a rocket's total mass should be propellant. Aerospace engineers call the following ratio the *mass fraction (MF)* and usually express it as a decimal.

$$MF = \frac{\textit{mass of propellant}}{\textit{total mass}}$$

If the total mass of a rocket is 6500 pounds and the mass of the propellant is 6000 pounds, would this ratio fall into the range for best performance? Explain.

47. **Business** The cost in dollars for one share of stock is represented by a mixed number. Find the cost of one share of the following stock for the stock market values shown below. (Round to the nearest cent.)

Nike 82 3/8	Pepsi 46 7/8	The Gap 36 3/4

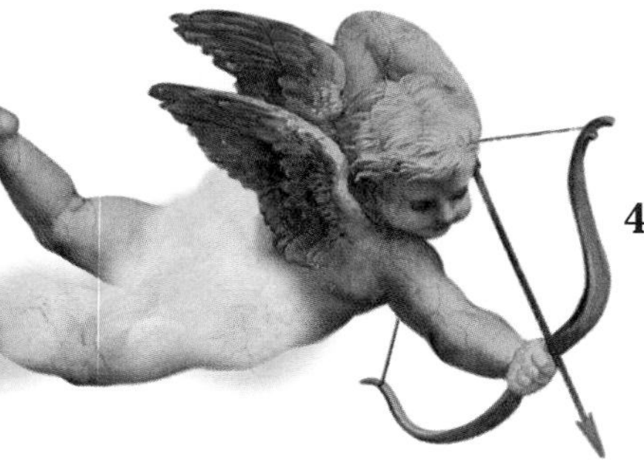

48. **Demographics** Do you believe in love at first sight? In a survey by Roper Starch Worldwide, $\frac{2}{5}$ of those asked said "It happens," $\frac{7}{25}$ said "It could happen," and $\frac{8}{25}$ said "No way." What did most people respond?

Mixed Review

49. Write the next three terms of the sequence 88, 82, 76, 70, . . . (Lesson 5-9)
50. Solve $t - 3\frac{1}{4} = \frac{7}{2}$. (Lesson 5-6)
51. Find the product of $b^5 \cdot b^6$. (Lesson 4-8)
52. Find a three-digit number that is divisible by 2, 3 and 6. (Lesson 4-1)
53. **Sewing** The Home Economics Class made a quilt measuring 4 feet by 6 feet to donate for a raffle. How much binding should they buy to trim the outer edges? (Lesson 3-5)
54. Order the integers {0, −7, 15, −6, 20} from least to greatest. (Lesson 2-3)
55. **Sports** The Cleveland Indians beat the California Angels by a score of 10 to 3. Write an inequality that would tell you how many more runs the Angels would have needed to win the game. (Lesson 1-9)
56. Solve $13 + y = 25$ by using the inverse operation. (Lesson 1-8)
57. Solve $3t = 81$ mentally. (Lesson 1-6)

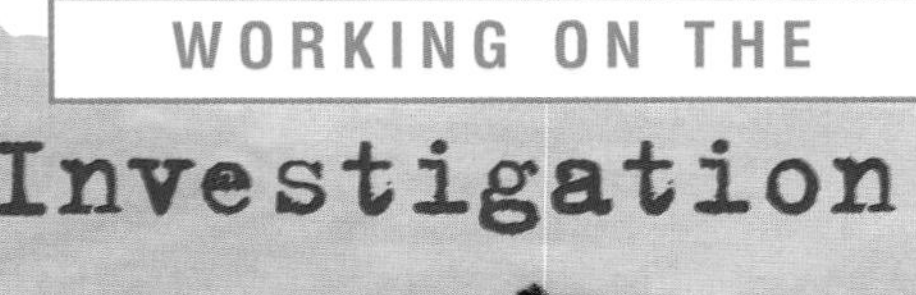

Refer to the Investigation on pages 270–271.

Begin working on the investigation by doing a group activity to determine the steepness of your lift hill.

- Build a scale model of your lift hill using a yardstick to represent the length of the track on the downhill side.
- While one member of your group holds the yardstick at the angle matching your design, another member measures how high the end of the yardstick is from the floor. Record the ratio of the height to the length of the yardstick. For example, if the height is 30 inches, you would write $\frac{30}{36}$ or $\frac{5}{6}$. Write the ratio as a decimal.
- Next, use the same method to find the ratio of an angle that is not as steep as yours and another angle that is steeper than yours.
- Write a short summary explaining the relationship between steepness and the decimal values of the ratios.

Add the results of your work to your Investigation Folder.

6-2 Estimating Products and Quotients

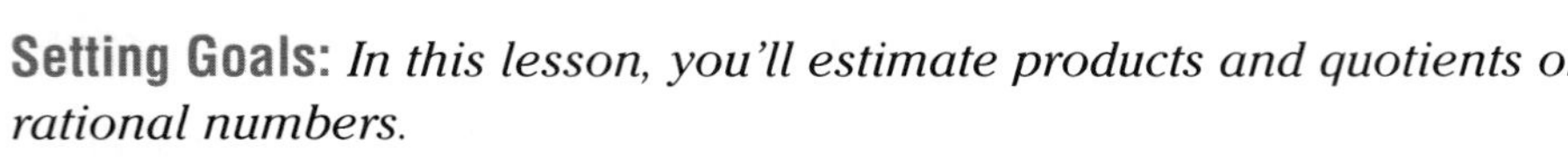

Setting Goals: *In this lesson, you'll estimate products and quotients of rational numbers.*

Modeling with Technology

In this activity, you and a partner will try to score as many points as possible by estimating decimal products. Products receive points as shown in the table below.

Your Turn Pick two numbers from the chart below and use your calculator to find the product. You may use each number only once.

Numbers:							
4.6	0.02	12.1	30.5	11.5	0.6	31.8	0.05
19.8	0.115	10.6	0.43	0.08	2.9	0.96	18.8
24.2	9.7	1.7	5.64				

Add the points for your products to find your score. You can do this activity more than once to see if you can increase your team score.

Product is between:	Points Scored
0 and 0.1	1
0.1 and 1	2
1 and 10	3
10 and 100	2
100 and 1000	1

TALK ABOUT IT

a. What strategy did you use to choose the numbers for each product?
b. How could you increase your score?

Learning the Concept

In the activity above, it is useful to know how to estimate the product of the decimals before actually multiplying them. It is also important to estimate products and quotients when using a calculator. We can make mistakes when entering numbers, and estimating will tell us if our result is reasonable.

Using rounding and **compatible numbers**, it is possible to estimate the products and quotients of rational numbers. Compatible numbers are rounded so it is easy to compute with them mentally.

$47.5 \div 5.23 \rightarrow 48 \div 6$ or 8 *Even though 5.23 rounds to 5, 6 is a compatible number because 48 is divisible by 6.*

$8 \times 98.24 \rightarrow 8 \times 100$ or 800.

Example  Use rounding or compatible numbers to estimate 67.45 ÷ 7.6.

$67.45 \div 7.6 \rightarrow 70 \div 7$ *Round 67.45 to 70 and round 7.6 to 7 since 70 and 7 are compatible numbers.*

Think: $70 \div 7 = 10$

$67.45 \div 7.6$ is about 10.

The strategy of using compatible numbers works for fractions as well as decimals.

Example  Use rounding or compatible numbers to estimate $\frac{5}{14} \times 30$.

$\frac{5}{14} \times 30 \rightarrow \frac{1}{3} \times 30$ *Replace* $\frac{5}{14}$ *with* $\frac{1}{3}$ *because* $\frac{5}{14}$ *is close to* $\frac{5}{15}$ *and* $\frac{5}{15} = \frac{1}{3}$.

Think: $\frac{1}{3} \times 30 = 10$

$\frac{5}{14} \times 30$ is about 10.

Being able to estimate products and quotients of rational numbers is very useful in everyday life.

Example 3

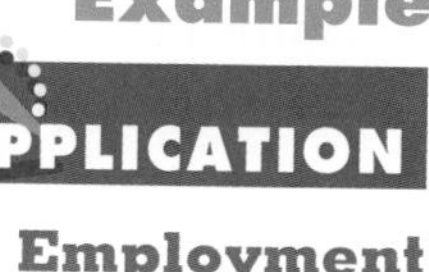

Employment

Kelly found a part-time job working at the Discus Record Store for \$5.85 an hour. The manager told Kelly that she will probably work about $7\frac{1}{2}$ hours each week. She will be paid every two weeks. Estimate Kelly's pay before taxes are deducted.

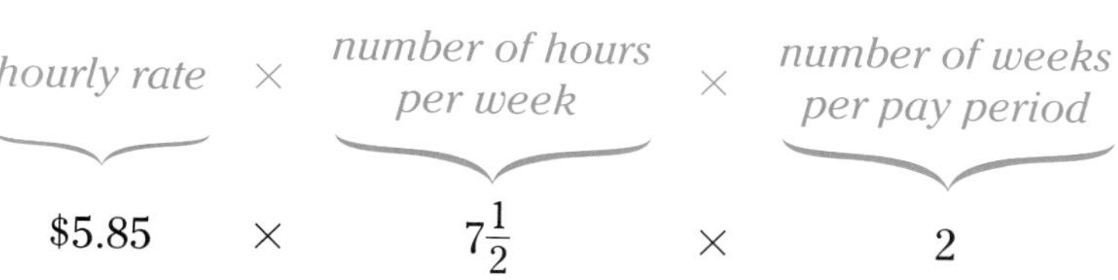

$\$5.85 \times 7\frac{1}{2} \times 2$

Think: \$5.85 rounds to \$6 and $7\frac{1}{2}$ *hours rounds to 8 hours.*

$\$5.85 \times 7\frac{1}{2} \rightarrow \$6 \times 8$ or \$48 a week

For two weeks, multiply by 2.
$\$48 \times 2 \rightarrow \$50 \times 2 = \$100$

Kelly will earn about \$100 every two weeks.

THINK ABOUT IT

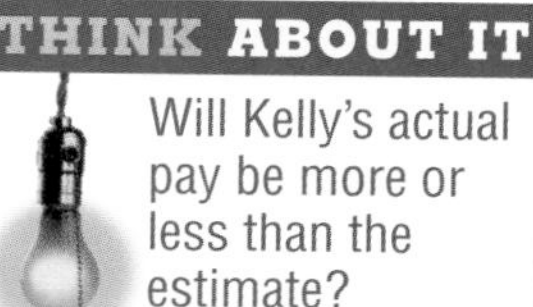

Will Kelly's actual pay be more or less than the estimate?

Checking Your Understanding

Communicating Mathematics

Read and study the lesson to answer these questions.

1. **Explain** why estimation is more important in mathematics since calculator use has become more widespread.
2. **Give an example** of two compatible numbers. Explain why they are compatible.
3. Choose the correct phrase: The product of a whole number and a fraction less than one is always (less than, greater than, equal to) the whole number.

4. **Assess Yourself** Describe a time in your daily life when estimation is useful.

Guided Practice

Choose the best estimate for each product or quotient.

5. $\left(1\frac{1}{8}\right)\left(3\frac{1}{4}\right)$ **a.** almost 3 **b.** less than 1 **c.** more than 3 **d.** none of these
6. $49 \times \frac{2}{5}$ **a.** about 20 **b.** more than 25 **c.** less than 10 **d.** none of these
7. $\$24.99 \div 4$ **a.** more than \$8 **b.** almost \$6 **c.** about \$5 **d.** none of these

Rewrite the problem using rounding and compatible numbers. Then estimate each product or quotient.

8. $40 \times \frac{2}{9}$
9. $17.8 \div 3.2$
10. $\frac{5}{12} \times 44$
11. $8.4 \div 0.95$
12. $75.4 \div 9.8$
13. $20\frac{1}{9} \times \frac{1}{4}$
14. **Geometry** The formula for the perimeter of a rectangle is $P = 2(\ell + w)$, where ℓ is the length and w is the width. Estimate the perimeter of the rectangle at the right. Is it greater than or less than 20 meters?

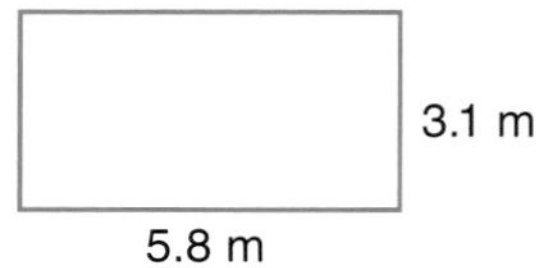

Exercises: Practicing and Applying the Concept

Independent Practice

Estimate each product or quotient.

15. 15.93×9.8
16. $2.94 \cdot 1.8$
17. $8.1 \div 2.2$
18. $47.6 \div 7.8$
19. $15.2 \div 2.7$
20. $(84.2)(3.9)$
21. $\frac{1}{4} \cdot 9$
22. $\left(\frac{1}{3}\right)(14)$
23. $\frac{3}{8} \times 17$
24. $\frac{9}{19} \times 120$
25. $\frac{11}{20} \times 41$
26. $75 \div 6\frac{7}{8}$
27. $\frac{31}{40} \times 200$
28. $146 \div 13\frac{1}{15}$
29. $43.8 \div 9.2$
30. Estimate $\frac{5}{6}$ times 10.
31. Estimate the quotient of 27.26 and 2.6.

Calculators

32. Use a calculator to find the following quotients.

a. $80 \div 0.5$	**b.** $10 \div 0.1$	**c.** $4 \div 0.25$
$6 \div 0.5$	$2 \div 0.1$	$20 \div 0.25$
$1.5 \div 0.5$	$6.5 \div 0.1$	$1.1 \div 0.25$

d. What patterns do you observe?

Use the patterns to estimate each quotient.

e. $14 \div 0.48$	**f.** $5 \div 0.12$	**g.** $12 \div 0.23$
h. $2\frac{1}{5} \div 0.25$	**i.** $21.2 \div 0.085$	**j.** $3\frac{1}{4} \div 0.51$

Critical Thinking

33. Explain how you would know if an estimated answer is greater than or less than the actual answer.

Applications and Problem Solving

34. Decorating Jennifer Green ran out of wallpaper before she finished wallpapering her room. The remaining area is 12 feet by $10\frac{1}{2}$ feet and has one window, which is $3\frac{1}{2}$ feet by $4\frac{3}{4}$ feet.

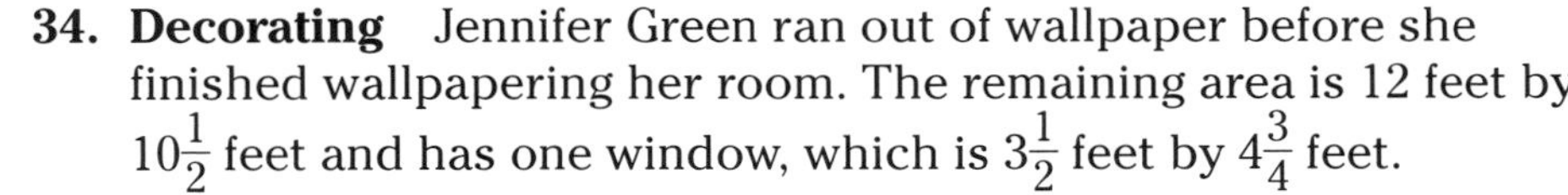

a. Estimate the amount (square feet) of wall paper Jennifer needs to cover the wall.

b. If the wallpaper comes on rolls that cover 51 square feet, how many rolls should she purchase?

35. Population In 1994, Heather Whitestone became the first hearing-impaired Miss America. In the United States, $\frac{3}{50}$ of the people are hearing impaired or deaf. If 1990 census showed that the population of the United States was 248,709,873 at that time, estimate how many people were hearing impaired.

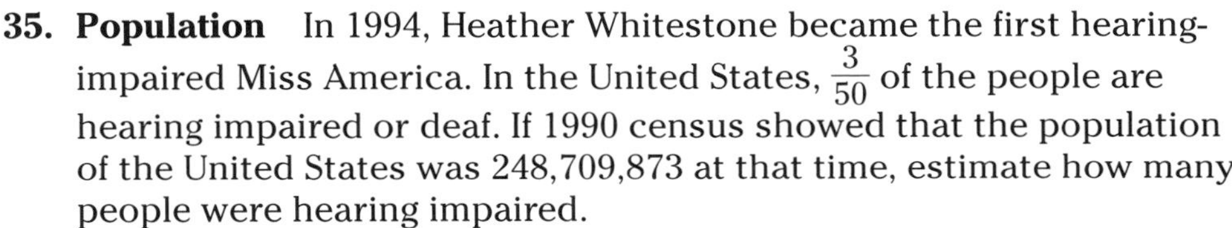

36. Stock Market Stock prices are stated as mixed numbers. On July 18, 1995, a share of IBM stock was listed as $107\frac{1}{4}$. The cost of one share of stock is \$107.25. (The fraction indicates a part of a dollar.)

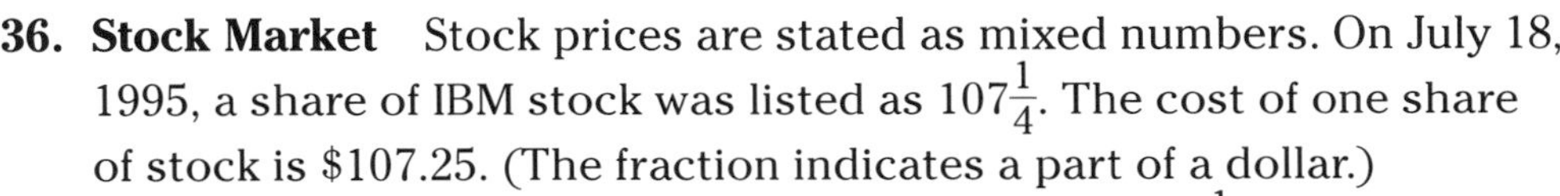

a. Estimate the cost of 37 shares of stock listed at $107\frac{1}{4}$ a share.

b. Suppose you had \$1000 to invest in IBM stock. About how many shares could you buy?

37. Biology A dolphin can swim at a speed of 37 miles per hour. A human can swim about one-eighth that speed. About how fast can a human swim?

Mixed Review

38. Write $\frac{3}{8}$ as a decimal. (Lesson 6-1)

39. Solve $\frac{6}{7} + \frac{5}{14} = s$. (Lesson 5-5)

40. Food A foot-long submarine sandwich was cut into 6 equal pieces. There are 2 pieces left. Write the ratio of pieces eaten to total pieces as a fraction in simplest form. (Lesson 4-6)

41. Solve $-14 + c = 13$. (Lesson 3-2)

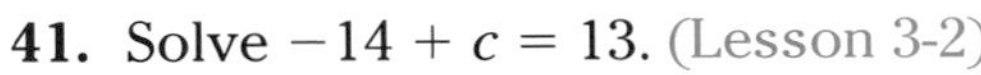

42. Find $-13 + 6$. (Lesson 2-4)

43. Simplify $|-6| - |-4|$. (Lesson 2-1)

44. Solve $\frac{r}{9} = 15$ by using the inverse operation. (Lesson 1-8)

6-3 Multiplying Fractions

Setting Goals: *In this lesson, you'll multiply fractions.*

Modeling with Manipulatives

MATERIALS

You can use grid paper to make an **area model** for multiplication of fractions. The diagram at the right models the product $\frac{2}{3} \cdot \frac{5}{6}$. Notice that the rectangle has three rows to represent thirds. Two rows are shaded yellow to represent $\frac{2}{3}$. The rectangle has six columns to represent sixths. Five columns are shaded blue to represent $\frac{5}{6}$. Notice that the overlapping area, $\frac{10}{18}$ of the rectangle, is shaded green. The product of $\frac{2}{3}$ and $\frac{5}{6}$ is $\frac{10}{18}$, or $\frac{5}{9}$.

Your Turn Use grid paper to make an area model of $\frac{3}{5} \cdot \frac{1}{4}$.

a. How did you represent $\frac{3}{5}$ in your model?

b. How did you represent $\frac{1}{4}$?

c. What is the value of $\frac{3}{5} \cdot \frac{1}{4}$ that you found using the grid paper?

d. Write a description of how an area model shows multiplication of fractions.

Learning the Concept

An area model shows what happens when you multiply fractions. You will get the same results if you multiply fractions as shown below.

$$\frac{2}{3} \cdot \frac{5}{6} = \frac{2 \cdot 5}{3 \cdot 6}$$ *Multiply the numerators and multiply the denominators.*

$$= \frac{10}{18} \text{ or } \frac{5}{9}$$ *Simplify by dividing numerator and denominator by 2.*

Multiplying Fractions

In words: To multiply fractions, multiply the numerators and multiply the denominators.

In symbols: For fractions $\frac{a}{b}$ and $\frac{c}{d}$, where $b \neq 0$ and $d \neq 0$, $\frac{a}{b} \cdot \frac{c}{d} = \frac{ac}{bd}$.

If the fractions have common factors in the numerators and denominators, you can simplify before you multiply. Compare the two methods used in Example 1.

Example  **Solve $x = \frac{5}{12} \cdot \frac{2}{9}$.**

Method 1: Use the rule.

$$x = \frac{5}{12} \cdot \frac{2}{9}$$

$$= \frac{5 \cdot 2}{12 \cdot 9} \quad \textit{Multiply.}$$

$$= \frac{10}{108} \text{ or } \frac{5}{54}$$

Method 2: Use the GCF.

$$x = \frac{5}{12} \cdot \frac{2}{9}$$

$$= \frac{5 \cdot \overset{1}{\cancel{2}}}{\underset{6}{\cancel{12}} \cdot 9}$$ *The GCF of 2 and 12 is 2. Divide 2 and 12 by 2.*

$$= \frac{5 \cdot 1}{6 \cdot 9} \text{ or } \frac{5}{54}$$

You can use the multiplication skills that you developed with integers and fractions to multiply negative fractions.

Example  **Solve each equation.**

a. $3\frac{3}{8}\left(-5\frac{1}{3}\right) = x$ — *Estimate: 3(−5) = −15.*

$\frac{27}{8}\left(-\frac{16}{3}\right) = x$ — *Rename $3\frac{3}{8}$ as $\frac{27}{8}$ and $-5\frac{1}{3}$ as $-\frac{16}{3}$.*

$\frac{\overset{9}{\cancel{27}}}{\underset{1}{\cancel{8}}}\left(\frac{-\overset{-2}{\cancel{16}}}{\underset{1}{\cancel{3}}}\right) = x$ — *Simplify.*

$\frac{-18}{1} = x$

$-18 = x$ — *Compare with the estimate.*

b. $a = \left(-1\frac{4}{9}\right)\left(-2\frac{3}{5}\right)$ — *Estimate: (−1)(−3) = 3.*

$a = \left(\frac{-13}{9}\right)\left(\frac{-13}{5}\right)$ — *Rename $-1\frac{4}{9}$ as $-\frac{13}{9}$ and $-2\frac{3}{5}$ as $-\frac{13}{5}$.*

$a = \frac{169}{45}$ or $3\frac{34}{45}$ — *The product of two rational numbers with the same sign is positive.*

Example 3 **APPLICATION**

Cooking

Manuel Mendes is the chief cook at the U.S. Military Academy at West Point, New York. He and his 96 staff members work to feed the 4500 cadets three times each day. Their sloppy joe recipe calls for $7\frac{1}{2}$ pounds of ketchup and serves 100 people. How many pounds of ketchup are needed for one meal of sloppy joes?

The recipe serves 100 people, so the staff will need to purchase enough ingredients for 4500 ÷ 100 or 45 batches of sloppy joes.

$45 \cdot 7\frac{1}{2} = \frac{45}{1} \cdot \frac{15}{2}$ — *Rename 45 as $\frac{45}{1}$ and $7\frac{1}{2}$ as $\frac{15}{2}$.*

$= \frac{45 \cdot 15}{1 \cdot 2}$

$= \frac{675}{2}$ or $337\frac{1}{2}$ — *Write as a mixed number.*

$337\frac{1}{2}$ pounds of ketchup are needed.

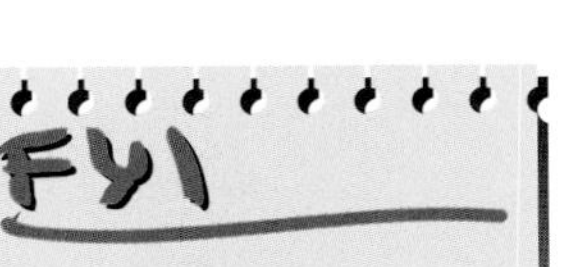

FYI

Long a staple in Latin countries, salsa has now passed ketchup in the race for most popular condiment in America.

Connection to Algebra

You can evaluate algebraic expressions involving multiplication of fractions in the same way that you evaluated expressions using integers.

Example  **Evaluate the expression ab^2 if $a = 1\frac{1}{2}$ and $b = \frac{2}{3}$.**

$ab^2 = 1\frac{1}{2} \cdot \left(\frac{2}{3}\right)^2$ *Substitute $1\frac{1}{2}$ for a and $\frac{2}{3}$ for b.*

$= 1\frac{1}{2} \cdot \frac{2}{3} \cdot \frac{2}{3}$ *Rewrite the problem without exponents.*

$= \frac{3}{2} \cdot \frac{4}{9}$ *Write $1\frac{1}{2}$ as $\frac{3}{2}$.*

$= \frac{12}{18}$ or $\frac{2}{3}$ *Multiply and write in simplest form.*

Checking Your Understanding

Communicating Mathematics

Read and study the lesson to answer these questions.

1. **Explain** how the model at the right shows the product of $\frac{3}{4}$ and $\frac{1}{2}$. What is the product?

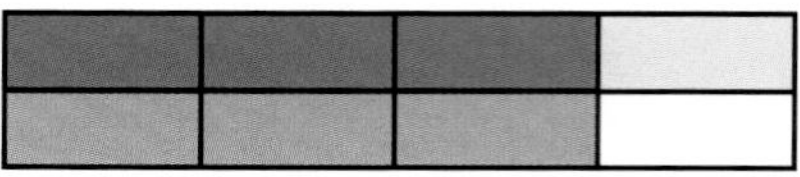

2. **You Decide** Jamal and Penny multiplied the fractions $\frac{18}{21}$ and $\frac{3}{14}$ in the following ways. Who is correct and why?

Jamal

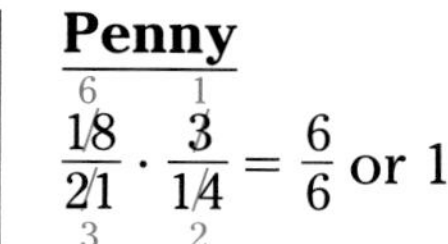

$\frac{\overset{9}{\cancel{18}}}{\underset{7}{\cancel{21}}} \cdot \frac{\overset{1}{\cancel{3}}}{\underset{7}{\cancel{14}}} = \frac{9}{49}$

Penny

$\frac{\overset{6}{\cancel{18}}}{\underset{3}{\cancel{21}}} \cdot \frac{\overset{1}{\cancel{3}}}{\underset{2}{\cancel{14}}} = \frac{6}{6}$ or 1

3. **Identify** which point shown on the number line below could be the product of the numbers graphed at *A* and *B*. Explain your reasoning.

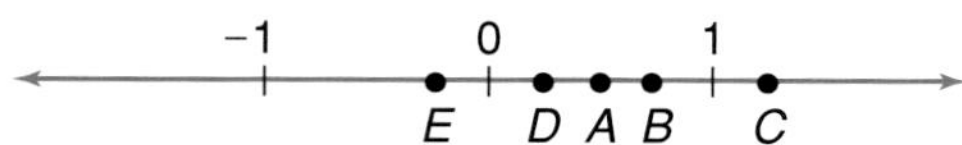

MATERIALS

construction paper
ruler

4. Fold a piece of $8\frac{1}{2}$" by 11" construction paper in half and then in half again as shown. Cut out a square from the open end of the folded paper. Unfold the paper. Then fold up the edges to form a box with no lid. Use your ruler to measure the length, width, and height of your box to the nearest sixteenth of an inch.

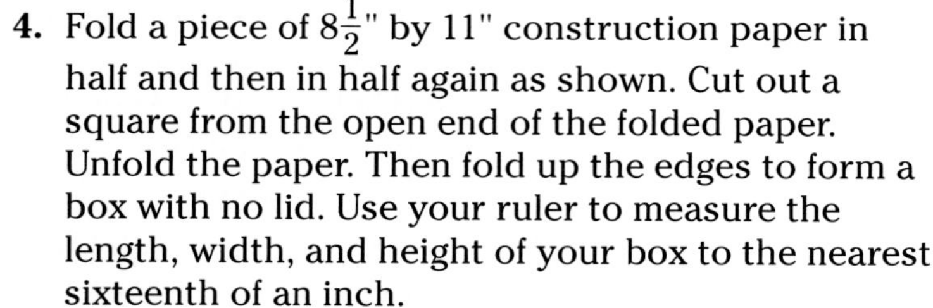

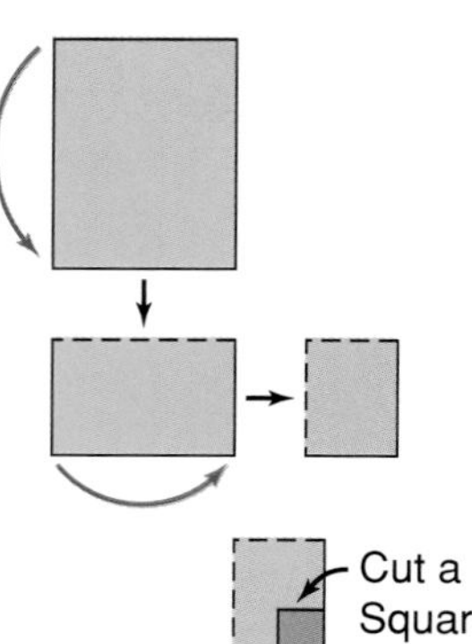

   a. Find the total area of the bottom and four sides of your box.

   b. How does the area of your box compare with your neighbor's? Why do you think they are the same or different?

Guided Practice

Choose the correct product.

5. $\frac{1}{6} \cdot \left(-\frac{3}{5}\right)$ **a.** $\frac{2}{5}$ **b.** $-\frac{2}{5}$ **c.** $\frac{1}{10}$ **d.** $-\frac{1}{10}$

6. $\left(\frac{1}{2}\right)^2$ **a.** $\frac{1}{2}$ **b.** $\frac{1}{4}$ **c.** 1 **d.** 2

Solve each equation. Write each solution in simplest form.

7. $\frac{3}{4} \cdot \frac{2}{3} = a$
8. $b = \frac{1}{2}\left(-\frac{5}{6}\right)$
9. $c = \frac{8}{12} \cdot \frac{4}{6}$
10. $x = \left(3\frac{1}{2}\right)4$
11. $m = \frac{8}{15}(-45)$
12. $\left(-\frac{4}{5}\right)^2 = p$

Evaluate each expression if $a = \frac{2}{3}$, $x = \frac{3}{5}$, and $y = \frac{5}{8}$.

13. xy
14. $2a$
15. x^2
16. ay

17. **Economics** The U.S. Congress is debating whether to replace paper one-dollar bills with coins in order to save money.

   a. It costs 8¢ to produce a one-dollar coin. The cost of producing a paper dollar is $\frac{19}{40}$ of that. How much does it cost to produce a paper dollar?
   b. An average paper dollar lasts $1\frac{1}{4}$ years. A coin lasts about $22\frac{1}{2}$ times longer. How long does a coin last?
   c. Do you think it would save money to use coins instead of paper dollars? Explain.

Exercises: Practicing and Applying the Concept

Independent Practice

Solve each equation. Write each solution in simplest form.

18. $\frac{3}{4} \cdot \left(-\frac{1}{3}\right) = b$
19. $\frac{1}{2} \cdot \frac{2}{7} = t$
20. $k = -\frac{5}{6}\left(-\frac{2}{5}\right)$
21. $d = -4\left(\frac{3}{8}\right)$
22. $(-7)\left(-2\frac{1}{3}\right) = h$
23. $c = 1\frac{4}{5} \cdot \left(-2\frac{1}{2}\right)$
24. $v = 2\frac{5}{6} \cdot 3\frac{1}{3}$
25. $s = \left(2\frac{1}{4}\right)\left(-\frac{4}{3}\right)$
26. $\left(-9\frac{3}{5}\right)\left(\frac{5}{12}\right) = y$
27. $\left(-3\frac{1}{5}\right)\left(7\frac{1}{2}\right) = w$
28. $m = (9)\left(-1\frac{5}{6}\right)$
29. $p = \left(-\frac{3}{5}\right)^2$
30. $r = \left(\frac{7}{11}\right)^2$
31. $\left(-\frac{9}{10}\right)^2 \cdot 8 = f$
32. $\left(\frac{5}{7}\right)^2 = n$
33. What is the product of $\frac{4}{9}$ and $\frac{3}{5}$?
34. What is $\frac{5}{8}$ of 36?

Evaluate each expression if $a = -\frac{1}{3}$, $b = \frac{3}{4}$, $x = 1\frac{2}{5}$, and $y = -3\frac{1}{6}$.

35. ay
36. $2a^2$
37. $b(x + a)$
38. $3a - 5x$
39. $a^2(b + 2)$
40. $-a(a - b)$

41. Use a fraction calculator and the numbers 2, 3, 4, 5, 6, and 7 to make two fractions with the greatest product. Use each digit only once. Explain your answer.

Critical Thinking

42. **Number Theory** Every whole number is a product of primes.
   a. Can you write $\frac{42}{45}$ as the product of fractions with prime numerators and denominators? You may use as many fractions as you need in the product.
   b. Is there more than one way to write the product? Explain.

Applications and Problem Solving

43. Health If you usually burn after $\frac{1}{2}$ hour exposure to sun, you can lengthen the time you are in the sun to three hours by applying SPF number 6. How long could a person who burns after $\frac{1}{4}$ hour stay in the sun using SPF 15 lotion?

44. Cooking The West Point sloppy joe recipe mentioned in Example 3 uses $3\frac{1}{4}$ ounces of cider vinegar and $1\frac{2}{3}$ ounces of garlic powder. How much of these ingredients would be needed to make sloppy joes for 4500 people?

45. Measurement People who design printed products use a system of measurement that is different from the customary and metric systems. The measurements used in this system are the *point* and the *pica*. One point equals $\frac{1}{72}$ inch, and 12 points make one pica.

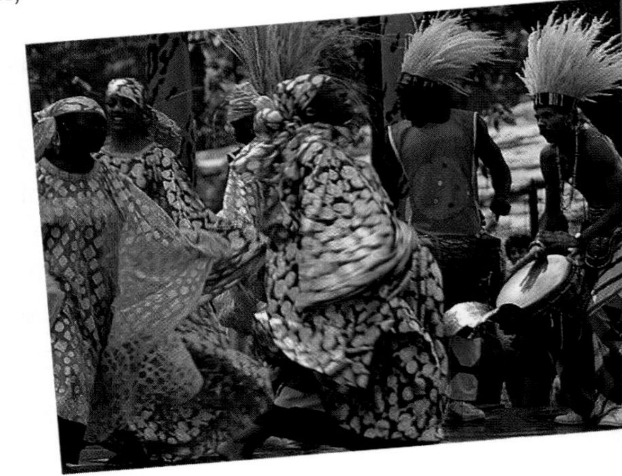

a. How many picas are in an inch?

b. The margin of a book is $\frac{3}{4}$ inches. How many picas is this?

c. If a particular kind of type has an average measure of $2\frac{3}{4}$ characters per pica, about how many characters would fit across a $5\frac{1}{2}$-inch line?

46. Statistics At Westside High School, $\frac{3}{5}$ of the students are female. The dance preferences of those students are shown in the graph at the right. What part of the students are female and interested in each kind of dance?

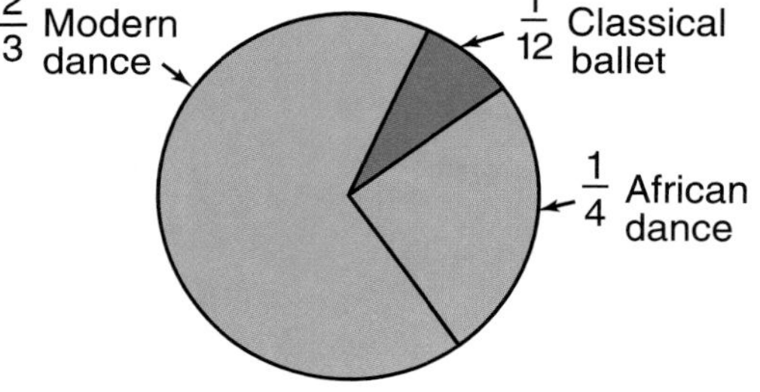

Mixed Review

47. Estimate $\frac{13}{28} \times 30$. (Lesson 6-2)

48. Consumerism At the movies, it costs $6.50 for a ticket, $1.25 for a drink, and $2.50 for popcorn. Will $10 be enough to purchase one of each? Explain. (Lesson 5-2)

49. Find the prime factorization of 72. (Lesson 4-4)

50. Solve $f + (-3) > 8$. (Lesson 3-6)

51. Find the product $-4(-x)(-5y)$. (Lesson 2-7)

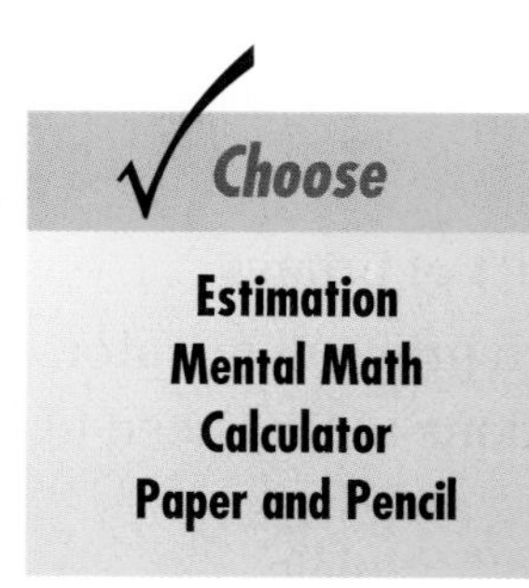

52. Geometry Supplementary angles are angles whose measures have a sum of 180°. One angle measures 57°. Solve the equation $y + 57 = 180$ to find the measure of a supplementary angle. (Lesson 1-8)

53. Find the value of $18 \div 3 + 18 \div 2$. (Lesson 1-2)

54. When Kim divided 1033.5 by 26 the calculator showed 397.5. Is this a reasonable answer? Explain. (Lesson 1-1)

6-4 Dividing Fractions

Setting Goals: *In this lesson, you'll divide fractions using multiplicative inverses.*

Modeling with Manipulatives

MATERIALS

colored pencils

One way to think of 6 ÷ 2 is to say "How many 2s are there in 6?" This way of thinking and an area model can help you divide fractions.

The drawing at the right shows a model for $\frac{3}{4} \div \frac{1}{2}$. To find the quotient, first draw a rectangle to represent one unit. Divide it into four equal parts and shade three to represent $\frac{3}{4}$. Then determine how many halves are contained in the shaded area. There is one half, and half of a second. So, $\frac{3}{4} \div \frac{1}{2} = 1\frac{1}{2}$.

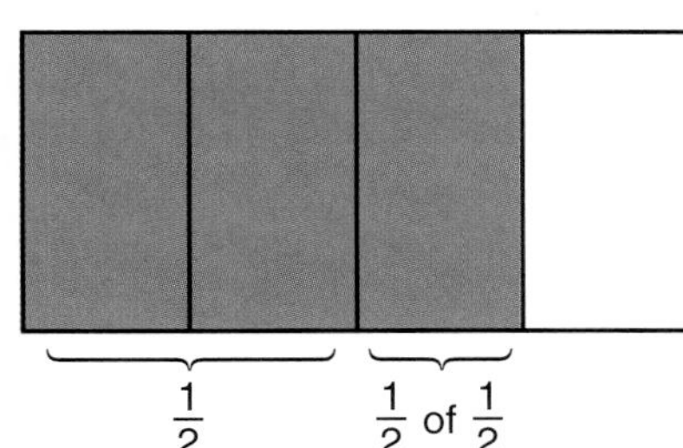

Your Turn

- Model $2\frac{2}{3} \div \frac{4}{9}$. Draw three rectangles and divide into thirds. Then shade $2\frac{2}{3}$.
- Add additional marks to show ninths in each rectangle.
- Mark as many groups of $\frac{4}{9}$ as you can.

TALK ABOUT IT

a. How many groups of $\frac{4}{9}$ did you find?

b. What is the quotient of $2\frac{2}{3} \div \frac{4}{9}$?

Learning the Concept

THINK ABOUT IT

Are $\frac{3}{4}$ and $-\frac{4}{3}$ multiplicative inverses?

You can use the same skills to divide fractions that you used to divide integers. For example, dividing 40 by 2 is the same as multiplying 40 by $\frac{1}{2}$. In each case, the solution is 20. In other words, dividing by 2 is the same as multiplying by $\frac{1}{2}$.

Notice that $2 \times \frac{1}{2} = 1$. Two numbers whose product is 1 are **multiplicative inverses**, or **reciprocals**, of each other. So, 2 and $\frac{1}{2}$ are multiplicative inverses. In the same way, $-\frac{3}{2}$ and $-\frac{2}{3}$ are multiplicative inverses because $-\frac{3}{2} \cdot \left(-\frac{2}{3}\right) = 1$.

Inverse Property of Multiplication

In words: The product of a number and its multiplicative inverse is 1.

In symbols: For every nonzero number $\frac{a}{b}$, where $a, b \neq 0$, there is exactly one number $\frac{b}{a}$ such that $\frac{a}{b} \cdot \frac{b}{a} = 1$.

Remember that dividing by 2 was the same as multiplying by $\frac{1}{2}$, its multiplicative inverse. This is true for any rational number.

Division with Fractions

In words: To divide by a fraction, multiply by its multiplicative inverse.

In symbols: For fractions $\frac{a}{b}$ and $\frac{c}{d}$ where $b, c, d \neq 0$,
$\frac{a}{b} \div \frac{c}{d} = \frac{a}{b} \cdot \frac{d}{c}$.

Example 1

Desktop Publishing

Robert G. Fernandez, a vice-president at Smith Barney financial corporation in Austin, Texas, wrote an article on money management in *Hispanic* magazine. *Hispanic* is printed on $8\frac{1}{2}$-by-11 inch pages, and the articles are printed in three columns. If the margins are $\frac{3}{8}$ inch on the left and right side of the page and there is $\frac{1}{4}$ inch of space between the columns, how wide should the columns on his computer be set?

Explore You know the total width of the page, the width of the margins, and the amount of space between each column. There are three equal columns. You need to know how wide to make each column.

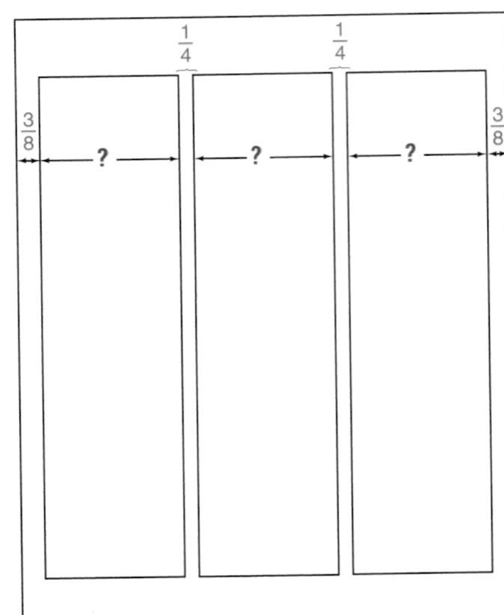

Plan Make a diagram and mark the widths you know. Subtract the known widths from the total width. Then divide the remaining width by 3.

Solve $8\frac{1}{2} - \frac{3}{8} - \frac{3}{8} - \frac{1}{4} - \frac{1}{4} = 7\frac{1}{4}$

$7\frac{1}{4} \div 3 = \frac{29}{4} \div \frac{3}{1}$ *Rename $7\frac{1}{4}$ as $\frac{29}{4}$ and 3 as $\frac{3}{1}$.*

$= \frac{29}{4} \times \frac{1}{3}$ *Multiply by the multiplicative inverse of 3, $\frac{1}{3}$.*

$= \frac{29}{12}$ or $2\frac{5}{12}$

Each column should be $2\frac{5}{12}$ inches wide.

Examine Add the widths of all the margins and the three columns. The sum should be $8\frac{1}{2}$.

$\frac{3}{8} + 2\frac{5}{12} + \frac{1}{4} + 2\frac{5}{12} + \frac{1}{4} + 2\frac{5}{12} + \frac{3}{8} \stackrel{?}{=} 8\frac{1}{2}$

$8\frac{1}{2} = 8\frac{1}{2}$ ✓ *It checks.*

You can use the division skills that you developed with integers and fractions to divide negative fractions.

Example 2 **Solve $p = -3\frac{3}{5} \div \frac{6}{7}$.**

$p = -3\frac{3}{5} \div \frac{6}{7}$ *Estimate: $-4 \div 1 = -4$*

$= -\frac{18}{5} \div \frac{6}{7}$ *Rename $-3\frac{3}{5}$ as $-\frac{18}{5}$.*

$= -\frac{18}{5} \times \frac{7}{6}$ *Multiply by the multiplicative inverse of $\frac{6}{7}$, $\frac{7}{6}$.*

$= -\frac{21}{5}$ or $-4\frac{1}{5}$ *Rename as a mixed number in simplest form. Compare to the estimate.*

Connection to Algebra

Evaluating algebraic expressions involving division of fractions uses the same procedures as evaluating algebraic expressions with integers.

Example 3 **Evaluate $a^2 \div x$ if $a = \frac{1}{3}$, and $x = 1\frac{4}{5}$.**

$a^2 \div x = \left(\frac{1}{3}\right)^2 \div 1\frac{4}{5}$ *Replace a with $\frac{1}{3}$ and x with $1\frac{4}{5}$.*

$= \left(\frac{1}{3}\right)\left(\frac{1}{3}\right) \div 1\frac{4}{5}$ *Evaluate powers first.*

$= \frac{1}{9} \div 1\frac{4}{5}$

$= \frac{1}{9} \div \frac{9}{5}$ *Rename $1\frac{4}{5}$ as $\frac{9}{5}$.*

$= \frac{1}{9} \times \frac{5}{9}$ *Multiply by the multiplicative inverse of $\frac{9}{5}$, $\frac{5}{9}$.*

$= \frac{5}{81}$

Checking Your Understanding

Communicating Mathematics

Read and study the lesson to answer these questions.

1. In your own words, what are reciprocals? Give an example.
2. How and when do you use a reciprocal in division of fractions?
3. **Write** an equation that you could use the multiplicative inverse of $\frac{3}{8}$ to solve.
4. **Give a counter example** to the statement that every rational number has a reciprocal.

5. A number is both multiplied and divided by the same rational number n, where $0 < n < 1$. Which is greater, the product or the quotient? Explain your reasoning.

Guided Practice

State whether each pair of numbers are multiplicative inverses. Write *yes* or *no*.

6. $7, \frac{1}{7}$
7. $-\frac{2}{5}, -1\frac{1}{2}$
8. $5\frac{2}{3}, \frac{17}{3}$
9. $1.4, \frac{5}{7}$

Write each division expression as a multiplication expression. Then find its value.

10. $\frac{1}{2} \div \frac{6}{7}$ **11.** $-\frac{3}{4} \div \frac{3}{4}$ **12.** $\frac{7}{9} \div \frac{2}{3}$

13. $5 \div \left(-1\frac{1}{3}\right)$ **14.** $2\frac{3}{5} \div 3\frac{6}{7}$ **15.** $-8 \div \left(-22\frac{4}{5}\right)$

Evaluate each expression.

16. $x \div y$, if $x = 1\frac{1}{2}$ and $y = \frac{1}{2}$

17. $c \div d + e$, if $c = \frac{2}{3}$, $d = 1\frac{1}{2}$, and $e = \frac{1}{6}$

18. Carpentry How many boards, each 2 feet 8 inches long, can be cut from a board 16 feet long?

Exercises: Practicing and Applying the Concept

Independent Practice

Name the multiplicative inverse for each rational number.

19. $-\frac{7}{3}$ **20.** 8 **21.** $1\frac{3}{5}$ **22.** $-3\frac{2}{7}$

23. 1.5 **24.** -0.4 **25.** $\frac{x}{y}$ **26.** $\frac{m}{n}$

Estimate the solution to each equation. Then solve. Write the solution in simplest form.

27. $r = -\frac{3}{5} \div \frac{5}{9}$ **28.** $u = -1\frac{1}{9} \div \frac{2}{3}$ **29.** $-8 \div \frac{4}{5} = t$

30. $6 \div \frac{1}{3} = m$ **31.** $2\frac{1}{4} \div \left(-1\frac{1}{2}\right) = h$ **32.** $-2\frac{3}{7} \div \left(-4\frac{4}{7}\right) = d$

33. $12 \div \frac{4}{9} = c$ **34.** $-10 \div \frac{3}{8} = f$ **35.** $z = 24 \div \frac{7}{10}$

36. $q = -2 \div \left(-\frac{1}{3}\right)$ **37.** $s = -3\frac{1}{4} \div 2\frac{1}{6}$ **38.** $7\frac{1}{2} \div 1\frac{1}{5} = n$

39. $m = -\frac{16}{7} \div \left(-\frac{12}{35}\right)$ **40.** $a = \frac{21}{30} \div \frac{7}{15}$ **41.** $12\frac{1}{4} \div \left(-\frac{14}{3}\right) = j$

Evaluate each expression.

42. $a \div b$, if $a = \frac{2}{3}$ and $b = 1\frac{1}{3}$

43. $r \div s$, if r $= -\frac{8}{9}$ and $s = \frac{7}{18}$

44. $a^2 \div b^2$, if $a = -\frac{3}{4}$ and $b = 1\frac{1}{3}$

45. $m + n \div p$, if $m = \frac{2}{3}$, $n = 1\frac{1}{3}$, and $p = \frac{1}{9}$

Critical Thinking

46. a. Divide the fraction $\frac{3}{4}$ by $\frac{1}{2}, \frac{1}{4}, \frac{1}{8}$, and $\frac{1}{12}$.

b. What happens to the quotient as the divisor changes?

c. Make a conjecture about what happens when you divide $\frac{3}{4}$ by fractions that increase in size. Then give examples that support your conjecture.

Applications and Problem Solving

47. **Business** Bertha Thangelane of Soweto, South Africa used the political changes in South Africa to start her business. She is the co-owner and managing director of RNB Creations (Pty) Ltd., which makes school uniforms. If she can make an elementary school child's uniform with $2\frac{1}{3}$ yards of fabric, how many uniforms can she make with a 50-yard bolt of fabric?

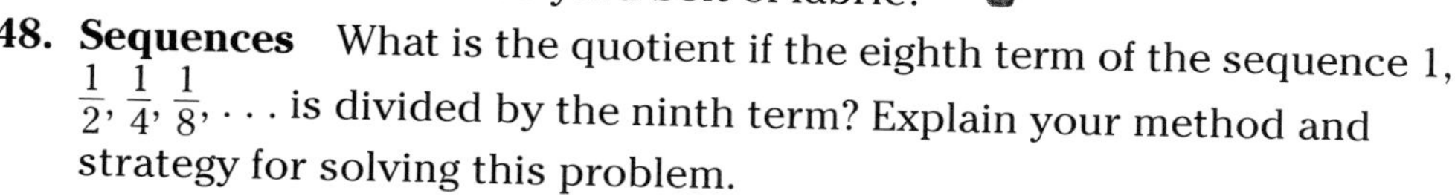

48. **Sequences** What is the quotient if the eighth term of the sequence 1, $\frac{1}{2}$, $\frac{1}{4}$, $\frac{1}{8}$, . . . is divided by the ninth term? Explain your method and strategy for solving this problem.

49. **Food** According to the U.S. Department of Agriculture, the average young American woman drinks $1\frac{1}{2}$ cans of cola each day. At this rate, how long would a 12-pack of cola last?

Mixed Review

50. Solve $\frac{3}{5} \cdot \frac{2}{7} = a$. Write the solution in simplest form. (Lesson 6-3)
51. Which number is greater, $\frac{2}{5}$ or 0.25? (Lesson 6-1)
52. **Health** Teri lost $1\frac{3}{4}$ pounds the first week of her diet. The following week she lost $3\frac{1}{4}$ pounds. How much did Teri lose? (Lesson 5-4)
53. Find the greatest common factor (GCF) of 30 and 12. (Lesson 4-5)
54. **Personal Finance** You received your paycheck on Friday. After paying your rent, which is \$395, you had \$510 left. How much was your paycheck? (Lesson 3-2)
55. **Geometry** Name the quadrant in which $(-6, 18)$ is located. (Lesson 2-2)
56. State whether $9 > 6$ is *true*, *false*, or *open*. (Lesson 1-9)
57. Simplify $h + 12h + 23$. (Lesson 1-5)

WORKING ON THE Investigation

Refer to the Investigation on pages 270–271.

Look over the scale drawings that you placed in your Investigation Folder at the beginning of Chapter 6. Make any necessary revisions to them based on what you've learned so far.

- Using the scale drawings, determine the speed of your train for each section of the circuit you designed. A typical speed down the first drop is about 60 miles an hour (88 feet per second). The length of each section should be determined by the group and will, for the most part, be dictated by the design, that is, uphill, downhill, straight, or curved.
- Determine the time it will take for your coaster train to travel each section of the circuit you designed. This can be found by dividing the length of the section of track by the speed you think the train will be traveling over that section.
- Find the total time it will take for the coaster train to complete the circuit you designed.

Add the results of your work to your Investigation Folder.

HANDS-ON ACTIVITY

Multiplying and Dividing Decimals

A Preview of Lesson **6-5**

MATERIALS

grid paper

colored pencils

In the activities below, you will explore the multiplication and division of decimals using grid paper decimal models. The whole decimal model represents 1, and each row or column represents 0.1.

Activity 1 Work with a partner.

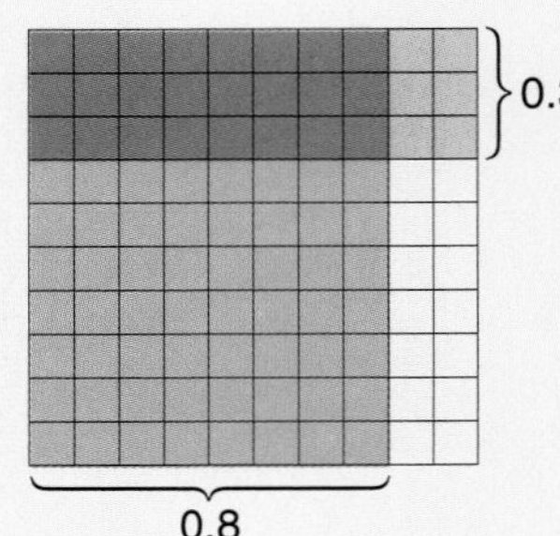

- Use a 10-by-10 decimal model to find 0.3×0.8.
- Color 3 rows of the model yellow to represent 0.3.
- Color 8 columns of the model blue to represent 0.8.

1. How many squares are green? What is the product of 0.3 and 0.8?
2. How many decimal places are there in both factors? How many are in the product? What is the relationship between the total number of decimal places in the factors and the number of decimal places in the product?
3. Use a model and a calculator to find the product 0.4×0.2. Explain your results.

Activity 2 Use decimal models to find 1.5 ÷ 0.3.

- Outline one complete model and 5 columns in the second model to represent 1.5.

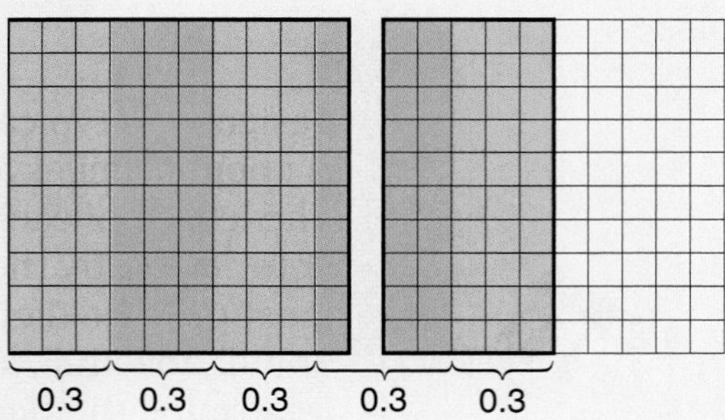

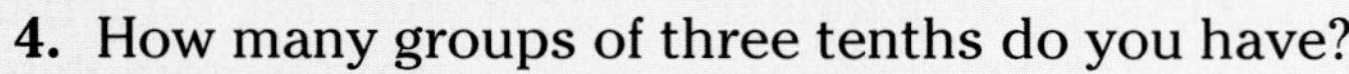

- Shade the first three columns yellow to represent 0.3.
- Shade the next three columns blue to represent another 0.3.
- Continue shading groups of three columns in alternating colors until you have filled the outline representing 1.5.

4. How many groups of three tenths do you have?
5. What is the quotient of $1.5 \div 0.3$?
6. Use decimal models to divide 2.4 by 0.4.

6-5 Multiplying and Dividing Decimals

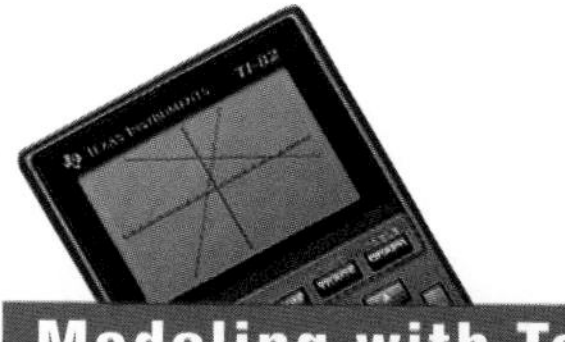

Setting Goals: *In this lesson, you'll multiply and divide decimals.*

Modeling with Technology

You can multiply decimals in the same way that you multiply whole numbers. Look for a pattern in the products below to help you write a rule for placing the decimal point in the product.

Your Turn Use a calculator to find each product.

$15.6 \times 38 =$

$15.6 \times 3.8 =$

$15.6 \times 0.38 =$

$15.6 \times 0.038 =$

a. Describe the pattern in the factors of the problems above.

b. Describe the pattern in the products.

c. What do you think 15.6×0.0038 equals? Explain your answer.

d. Write a rule for placing the decimal point in a product of two decimals.

Learning the Concept

When you multiply decimals, it is helpful to use estimation to verify the placement of the decimal point.

Example 1 **Solve each equation.**

a. $c = (3.9)(8.2)$ *Estimate: $4 \times 8 = 32$*

3.9	←	*one decimal place*
× 8.2	←	*one decimal place*
7 8		
312 0		
31.98	←	*two decimal places*

$c = 31.98$ *Compare to the estimate.*

b. $x = (-6.302)(0.81)$ *Estimate: $-6 \times 1 = -6$*

−6.302	←	*three decimal places*
× 0.81	←	*two decimal places*
6302		
504160		
−5.10462	←	*five decimal places*

$x = -5.10462$ *Compare to the estimate.*

All the properties that were true for multiplication of integers are also true for multiplication of rationals. The properties of rationals are summarized in the following chart.

Properties of Multiplication	Examples
Commutative Property For all rational numbers x and y, $x \cdot y = y \cdot x$.	$-5(-7) = -7(-5)$ $0.6 \cdot 1.5 = 1.5 \cdot 0.6$ $\frac{5}{6} \cdot \frac{8}{9} = \frac{8}{9} \cdot \frac{5}{6}$
Associative Property For all rational numbers x, y, and z, $(x \cdot y) \cdot z = x \cdot (y \cdot z)$.	$(-6 \cdot 5) \cdot 8 = -6 \cdot (5 \cdot 8)$ $(0.3 \cdot 4) \cdot 5.7 = 0.3 \cdot (4 \cdot 5.7)$ $\left(-\frac{1}{3} \cdot \frac{9}{10}\right) \cdot \frac{2}{5} = -\frac{1}{3} \cdot \left(\frac{9}{10} \cdot \frac{2}{5}\right)$
Identity Property For every rational number x, $x \cdot 1 = x$ and $1 \cdot x = x$.	$-8 \cdot 1 = -8$ $1 \cdot \frac{1}{4} = \frac{1}{4}$ $10.3 \cdot 1 = 10.3$
Inverse Property For every rational number $\frac{x}{y}$, where $x, y \neq 0$, there is a unique number $\frac{y}{x}$ such that $\frac{x}{y} \cdot \frac{y}{x} = 1$.	$6 \cdot \frac{1}{6} = 1$ $-\frac{5}{12}\left(-\frac{12}{5}\right) = 1$ $0.7 \cdot \frac{1}{0.7} = 1$

To divide by a decimal, think of the division as a fraction and write the divisor, or denominator, as a whole number. For example, consider $50 \div 0.26$.

Think of it as $\frac{50}{0.26}$.

$$50 \div 0.26 \rightarrow \frac{50}{0.26} \times \frac{100}{100} = \frac{5000}{26}$$

Since $\frac{100}{100} = 1$, the value of the fraction is unchanged. This becomes a division problem involving whole numbers.

$$\frac{50}{0.26} \rightarrow 0.26\overline{)50.00} \rightarrow 26\overline{)5000.00}$$

```
       192.30
26)5000.00
   26
    240
    234
      60
      52
       80
       78
        20
```

To the nearest tenth, the quotient is 192.3.

Example 2 **Solve $y = 42 \div (-0.8)$.** *Estimate: $42 \div -1 = -42$.*

$0.8\overline{)42.0}$ *Multiply 0.8 and 42 by 10 to get a whole number in the divisor.*

$8\overline{)420}$ with quotient 52.5 or 42 [÷] .8 [+/−] [=] -52.5

$y = -52.5$ *Compare to the estimate.*

Decimals are used every day when people work with money.

Example 3 APPLICATION Recycling

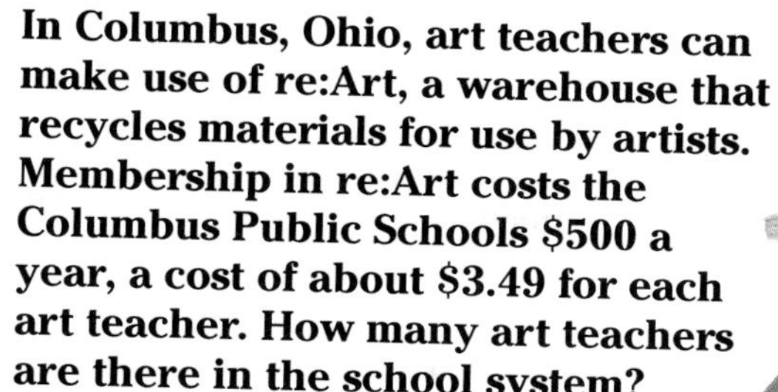

In Columbus, Ohio, art teachers can make use of re:Art, a warehouse that recycles materials for use by artists. Membership in re:Art costs the Columbus Public Schools $500 a year, a cost of about $3.49 for each art teacher. How many art teachers are there in the school system?

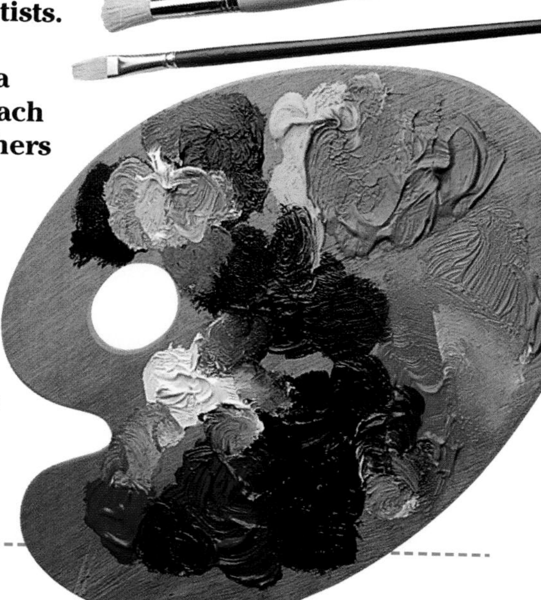

$500 ÷ $3.49 = ?
Estimate: 500 ÷ 4 = 125

500 [÷] 3.49 [=] 143.2664756

There are 143 or 144 art teachers in the system.
Check by comparing your answer to the estimate.

HELP WANTED

Many creative people find fulfillment as professional artists. If you would like to learn more, contact:

National Assn. of Schools of Art and Design
11250 Roger Bacon Dr.
Suite 21
Reston, VA 22090

Connection to Algebra

Evaluate expressions involving division of decimals in the same way you evaluate other algebraic expressions.

Example 4 **Evaluate $\frac{6m}{n}$ if $m = 0.5$ and $n = -3.2$.**

$$\frac{6m}{n} = \frac{6(0.5)}{-3.2}$$

6 [×] .5 [÷] 3.2 [(−)] [=] -0.9375

The value of the expression is −0.9375.

Checking Your Understanding

Communicating Mathematics

1. **Illustrate** or explain the rule for placement of the decimal point in (2.54)(0.067).
2. What is the first step you should take before you actually begin to multiply or divide decimals?
3. **Decide** where to place a decimal point in each factor so that the product is correct. 493 × 17 = 8.381
4. In Lesson 6-5A, you used decimal models to represent multiplication and division problems. Work with a partner to represent these multiplication and division problems.

   a. (1.2)(3) **b.** (1.2)(0.3) **c.** 4.8 ÷ 4 **d.** 4.8 ÷ 0.4

MATERIALS

grid paper

colored pencils

Guided Practice

State where the decimal point should be placed in each product or quotient.

5. (7.2)(0.2) = 144
6. (0.06)(3) = 18
7. 0.63 ÷ 0.9 = 7
8. 8.4 ÷ 0.4 = 21

Solve each equation.

9. $x = (-0.2)(-3.1)$
10. $y = (1.2)(-0.05)$
11. $s = 27 \div (-0.3)$
12. $t = 0.4 \div 2$

Evaluate each expression if $a = 15.7$, $b = 0.4$, and $c = 1.6$.

13. ac
14. $5c \div b$

Name the property shown by each statement.

15. $\left(-\frac{5}{6}\right)\left(-\frac{6}{5}\right) = 1$
16. $(-4 \cdot 5) \cdot \frac{1}{4} = [5 \cdot (-4)] \cdot \frac{1}{4}$

17. **Recycling** Sun Shares, a recycling company, is paid \$0.65 a pound for aluminum. How many pounds of aluminum would they have to collect a week to meet their weekly phone costs of \$45.50?

Exercises: Practicing and Applying the Concept

Independent Practice

Solve each equation.

18. $g = (-6.5)(0.13)$
19. $k = (0.47)(3.01)$
20. $-14.9(-0.56) = n$
21. $7.45(-0.75) = t$
22. $x = (0.001)(7.09)$
23. $b = (1.03)(-6.4)$
24. $r = 14.4 \div (0.16)$
25. $q = 0.384 \div 1.2$
26. $-85 \div (-1.7) = k$
27. $-0.51 \div 0.03 = g$
28. $s = -15.3 \div (-9)$
29. $h = 2.92 \div 0.002$
30. Find the product of 7.5 and 0.03.
31. What is the quotient of 52.64 and 9.4?

Evaluate each expression.

32. $3y^2$ if $y = 0.4$
33. ab if $a = -1.5$, $b = 10$
34. $\frac{5t}{s}$ if $t = 6.2$, $s = 2.5$
35. $n^2 \div q$ if $n = 2.2$, $q = 4$
36. $12y \div z$ if $y = 9.8$, $z = 6.4$
37. $(x + 4)(x - 2)$ if $x = 1.8$

Name the property shown by each statement.

38. $\left(2 \cdot \frac{5}{2}\right) \cdot \frac{3}{4} = \frac{3}{4}\left(2 \cdot \frac{5}{2}\right)$
39. $-\frac{8}{9} \cdot 1 = -\frac{8}{9}$
40. $0.6\left(\frac{1}{0.6}\right) = 1$
41. $(-6.5 \cdot 9.3) \cdot 2.1 = -6.5 \cdot (9.3 \cdot 2.1)$
42. $-9.3 = 1 \cdot -9.3$
43. $(0.6 \cdot 3.2) \cdot 0.3 = (3.2 \cdot 0.6) \cdot 0.3$

Calculators

Estimate. Then compute with a calculator. Round quotients to the nearest tenth if necessary.

44. $(8.01)(3.33) = w$
45. $(-0.56)(4.59) = r$
46. $-8.022(-0.03) = y$
47. $c = 90.5 \div (-8.9)$
48. $f = 93.702 \div 2.4$
49. $z = -0.36 \div (-2.1)$

Critical Thinking

50. a. Name two decimals whose sum is 2.4 and whose product is 1.44.
    b. Name two decimals whose sum is 2.7 and whose quotient is 2.

Applications and Problem Solving

51. **Consumer Economics** A 13.5 ounce box of muesli cereal costs \$2.45, and a 19.5 ounce box of frosted corn flakes costs \$3.59. Which costs less per ounce?
52. **Economics** When Lilia's softball team visited Mexico, \$1 in American money could be exchanged for 2702.7 pesos. If Lilia exchanged \$25 in American money, how many pesos did she receive?
53. **Sports** American consumers spent \$0.8 billion dollars for golf equipment in a recent year. That same year they spent four times that much on athletic shoes. How much did consumers spend on athletic shoes?

Mixed Review

54. Write $\frac{5}{11} \div \frac{10}{11}$ as a multiplication expression. Then find its value. (Lesson 6-4)
55. Solve $3\frac{4}{9} \cdot \frac{3}{4} = h$. (Lesson 6-3)
56. Solve $b - 1.6 \leq 4.3$. (Lesson 5-7)
57. Name the set(s) of numbers to which -10 belongs. (Lesson 5-1)
58. Solve $8 = -\frac{b}{11}$. (Lesson 3-3)
59. **Banking** Account #8234620 had an opening balance of \$3245. Three checks cleared on Tuesday, leaving a balance of \$2931. What was the total of the checks? (Lesson 3-2)
60. **Meteorology** The *heat index* is an estimate of the warming effect that humidity has on a person in hot weather. If the outside temperature is 86°F and the humidity makes it feel like 93°F, how much warmer does the humidity make it feel? (Lesson 2-5)
61. Name the property shown by $3 + a = a + 3$. (Lesson 1-4)
62. Find the value of $18 \div (18 - 6 \cdot 2)$. (Lesson 1-2)

Self Test

Express each fraction as a decimal. Use a bar to show a repeating decimal. (Lesson 6-1)

1. $\frac{7}{8}$
2. $-1\frac{2}{3}$
3. $-\frac{8}{11}$
4. **Business** On Tuesday, Motorola stock fell $2\frac{1}{8}$ points from the previous day's closing price. If the closing price on Monday was \$75.50, what was the closing price, in dollars, on Tuesday? (Lesson 6-1)

Estimate each product or quotient. (Lesson 6-2)

5. $\frac{2}{3} \times 25$
6. 3.85×6
7. $6.2 \div 2.1$

Solve each equation. Write each solution in simplest form. (Lessons 6-3, 6-4, and 6-5)

8. $x = \frac{1}{4}\left(-\frac{3}{8}\right)$
9. $y = -9 \div \frac{3}{8}$
10. $w = (0.5)(1.46)$

HANDS-ON ACTIVITY

6-6A Mean, Median, and Mode

A Preview of Lesson **6-6**

MATERIALS

- small boxes of raisins
- calculator

Measures of central tendency are numbers that represent a set of data. Researchers use these numbers when analyzing data.

Your Turn

Work in small groups.

- Get a small box of raisins for your group from your teacher.
- Estimate the number of raisins you think will be in the box.
- Count the number of raisins and record the number.
- Record the data for each group.
- Use a calculator to add the data. Then divide the sum by the number of groups. This number is called the **mean** of the data.
- List the data in order and mark the middle number. If there is no middle number in the set of data, circle the middle two numbers and find their mean. This number is called the **median**.
- If a number appears more often than the others in the data, this number is called the **mode**.

1. If you add the number 5 to your group's set of data, predict which measure of central tendency will be affected the most.
2. Find the mean, median, and mode of the new set of data.
3. Which measure of central tendency will be most representative of the data before adding 5? after adding 5?
4. If you add the number 400 to your group's set of data, predict which measure of central tendency will be affected the most.
5. Find the mean, median, and mode of the new set of data.
6. Which measure of central tendency will be most representative of the data after adding 400?

Extension

7. Suppose you count the number of raisins in a large box. If you add this number to your set, will it affect the mean? the median? the mode?
8. Choose your own numbers to add to your set of data. Study the effects on the mean, median, and mode.

6-6 Integration: Statistics
Measures of Central Tendency

Setting Goals: *In this lesson, you'll use the mean, median, and mode as measures of central tendency.*

Modeling a Real-World Application: Meteorology

"It's hot fun in the summertime today, folks! The high will reach a scorching 92 degrees, way above the normal of 84 for this date."

You have heard meteorologists report the normal high temperature for a given date. The normal high temperature is determined by averaging the high temperatures for a given date over several years. The high temperatures on March 1 for twelve consecutive years are recorded below. What is the normal high temperature for the date?

1985	1986	1987	1988	1989	1990	1991	1992	1993	1994	1995	1996
59°F	50°F	48°F	43°F	40°F	46°F	50°F	53°F	58°F	61°F	64°F	73°F

Learning the Concept

To analyze sets of data, researchers often try to find a number that can represent the whole set. These numbers or pieces of data are **measures of central tendency**. Three that we will study are the **mean**, the **median**, and the **mode**.

The **mean** is what people usually are talking about when they say average. It is the arithmetic average of the data. For the temperatures above, the mean is

$$\frac{59 + 50 + 48 + 43 + 40 + 46 + 50 + 53 + 58 + 61 + 64 + 73}{12} = \frac{645}{12} \text{ or } 53.75.$$

The mean is about 53.8°.
Notice that the mean may not be a member of the set of data.

Definition of Mean

The mean of a set of data is the sum of the data divided by the number of pieces of data.

The **median** is the middle number when the data are in order. Consider two cases.

Case 1: An Odd Number of Data

Write the set of data in order from least to greatest. Consider the set of data below.

8 12 22 22 25 34 35 36 55

↑ (25)

The middle number is 25.

Case 2: An Even Number of Data

If the number of data is even, there are two middle numbers. In this case, the median is the mean of the two numbers. Look back at the set of temperatures.

40 43 46 48 50 50 53 58 59 61 64 73

↑ (50) ↑ (53)

$$\frac{50 + 53}{2} = \frac{103}{2} \text{ or } 51.5$$

The median is 51.5°.

If there is an even number of data in a set, the median may not be a member of the data set.

Definition of Median

The median is the number in the middle when the data are arranged in order. When there are two middle numbers, the median is their mean.

The number or item in a data set that appears most often is called the **mode**. There can be one mode, more than one mode, or no mode in a data set.

In the temperatures above, the temperature 50° appears more times than any other (two times), so 50° is the mode.

The mode, if there is one, is always a member of the set of data. If there is no one piece of data that appears more often than the rest, the set of data has no mode.

Definition of Mode

The mode of a set of data is the number or item that appears most often.

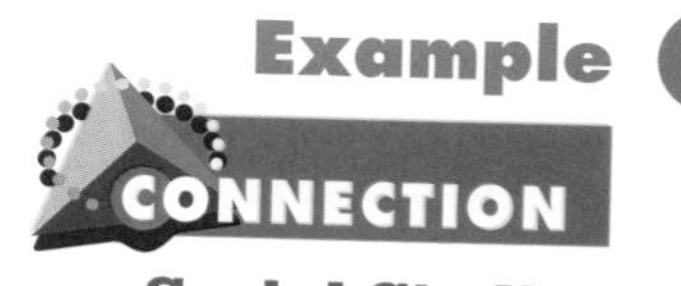

Example 1 — Connection: **Social Studies**

The table below shows the populations of several American Indian tribes as of the 1990 U.S. census. Organize the data and then find the mean, median, and mode.

Tribe	Population (thousands)
Apache	50
Cherokee	308
Chippewa	104
Choctaw	82
Creek	44
Iroquois	49
Lumbee	48
Navajo	219
Pueblo	53
Sioux	103

Median List the populations in order from greatest to least. The median is the middle number. There are 10 populations listed, so the average of the 5th and 6th populations is the median.

44 48 49 50 53 82 103 104 219 308

↑ (53) ↑ (82)

The median population is $\frac{53 + 82}{2}$ or 67.5 thousand.

Mode Since no population occurs more than the rest, there is no mode.

Mean To find the mean, add the populations and divide by 10.

$$\frac{50 + 308 + 104 + 82 + 44 + 49 + 48 + 219 + 53 + 103}{10}$$

$$= \frac{1060}{10} \text{ or } 106$$

The mean population is 106 thousand.

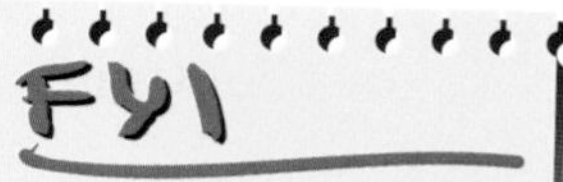

In 1985, Wilma Mankiller was elected chief of the Cherokee Nation. As chief, she placed a high priority on the economic growth of the nation.

You can use the mean, median, and mode to compare two different sets of data.

Example 2 — Application: **Entertainment**

Nielsen Media Research monitors the popularity of television programs and assigns a rating to each one. The Nielsen ratings for programs airing at 8:00 P.M. on CBS and NBC for each day of the week of July 3, 1995 are shown at the right.

CBS	NBC
7.2	6.1
5.5	5.7
7.1	4.1
8.4	9.6
5.6	5.6
5.1	5.3
9.9	12.5

a. Find the mean, median, and mode of the ratings for each network.

Begin by putting the data in order.

CBS	5.1	5.5	5.6	7.1	7.2	8.4	9.9
NBC	4.1	5.3	5.6	5.7	6.1	9.6	12.5

Median Because there are 7 pieces of data in each set, the medians are the 4th piece of data.

(continued on the next page)

CBS	The median is 7.1.
NBC	The median is 5.7.
Mode	None of the pieces of data occur more than once in either set of data. Thus, neither set has a mode.
Mean	The means are the sums of the data divided by 7.
CBS	$\frac{5.1 + 5.5 + 5.6 + 7.1 + 7.2 + 8.4 + 9.9}{7} = \frac{48.8}{7}$ or about 7.0
NBC	$\frac{4.1 + 5.3 + 5.6 + 5.7 + 6.1 + 9.6 + 12.5}{7} = \frac{48.9}{7}$ or about 7.0

b. Compare the averages to determine whether CBS or NBC programs rated better for the week.

The median rating for CBS programs was higher than for NBC programs. The means for the two networks were about the same. It appears that the popularity of the CBS programs was more consistent. NBC had more programs that were rated very high or very low.

Checking Your Understanding

Communicating Mathematics

Read and study the lesson to answer these questions.

1. Which "average" do you think best represents the data in Example 2? Explain your answer.
2. **Write** a set of data that has a mean of 7. Is the median 7 also?
3. Complete this activity with a group of three or four students. Arrange the 50 pennies in a roll in order from oldest to newest.
   a. Find the mean, median, and mode for the year the coins were minted.
   b. Based on your sample of pennies, what can you say about all pennies in circulation?
   c. Compare your results with the rest of the class and check your conclusion in **part b.**

MATERIALS

roll of 50 pennies

Guided Practice

List the data in each set from least to greatest. Then find the mean, median, and mode. When necessary, round to the nearest tenth.

4. 4, 5, 7, 3, 9, 11, 23, 37
5. 11, 45, 62, 12, 47, 8, 12, 35, 33
6. 25.98, 30.00, 45.36, 25.00, 45.36
7. 105, 116, 125, 78, 78
8. **Health** If 50 males have a resting heart rate of 59 beats a minute, 30 males have 72, 15 males have 84, and 5 males have 93, what is the mean resting heart rate for all 100 males represented?

Exercises: Practicing and Applying the Concept

Independent Practice

Solve each equation or inequality. Check your solution.

12. $7y < 3.5$
13. $-3y = 1.5$
14. $5a > -7\frac{1}{2}$
15. $\frac{1}{4}x = 4\frac{1}{8}$
16. $-\frac{1}{8}r = \frac{1}{4}$
17. $\frac{x}{3.2} < -4.5$
18. $\frac{7}{8}r \leq 1.4$
19. $-15 = 2\frac{2}{5}x$
20. $-6.7b = 14.07$
21. $\frac{p}{7.1} = -0.5$
22. $\frac{1}{3}d = -0.36$
23. $-1.3z \geq 1.69$
24. $\frac{3}{4}y > 1\frac{4}{5}$
25. $-7\frac{1}{2}x = 5\frac{1}{4}$
26. $-2h > 4.6$
27. $-0.05x = -0.95$
28. $\frac{x}{-2.5} \leq 3.2$
29. $7x < 1\frac{4}{10}$

Solve each equation or inequality and graph the solution on a number line.

30. $-\frac{1}{4}c = 3.4$
31. $\frac{4}{5}y < \frac{3}{8}$
32. $-12.6p \geq 28.98$

Critical Thinking

33. Write an inequality involving multiplication or division, using fractions, whose solution is $p < \frac{1}{3}$.

Applications and Problem Solving

34. **Geometry** The formula for the area of a triangle is $A = \frac{1}{2}bh$. A triangle has an area greater than 18 square inches. If the base of the triangle is 2.5 inches, what is the height of the triangle?

35. **Business** The Rose Department Store chain was closing one of their stores so they had to sell most of the inventory in the store. In the final markdown, they marked all items $\frac{1}{3}$ off the ticketed price. How much would a shirt that was ticketed at \$24.99 cost?

36. **Food** The National Restaurant Association polled a group of Americans about their food preferences. Less than $\frac{1}{3}$ of those polled had tried Indian or Thai food.
    a. If 332 people had tried Indian or Thai food, how large a group was polled?
    b. How large do you think the group polled was? Explain your answer.

Mixed Review

37. **Education** If you scored 93, 86, 84, 92, 85, and 93 on your first six pre-algebra tests, what is your mean test score? (Lesson 6-6)
38. Write $\frac{28}{45}$ as a decimal. (Lesson 6-1)
39. Solve $\frac{6}{7} + \frac{5}{14} = s$. (Lesson 5-5)
40. Find the LCM of $15c$ and $12cd$. (Lesson 4-7)
41. Solve $a = (-12)(3)$. (Lesson 2-7)
42. Name the property shown by $5(6 \cdot 3) = (5 \cdot 6)3$. (Lesson 1-4)

6-8 Integration: Discrete Mathematics
Geometric Sequences

Setting Goals: *In this lesson, you'll recognize and extend geometric sequences and represent them algebraically.*

Modeling a Real-World Application: Genealogy

In 1974, Alex Haley finished his book, *Roots*, which traced his family's history. His story began in 1767 with the African Kunta Kinte. Alex Haley's great-great-great-great-great-grandfather, Kunta Kinte, was brought to America to be a slave. With each preceding generation that Alex Haley researched, he found more and more ancestors.

Learning the Concept

A TV miniseries of *Roots* made in 1977 won a record-setting 9 Emmys. A follow up series about Haley's grandmother, Queen, was made in 1994.

You can use a pattern to find the number of ancestors Alex Haley had back to Kunta Kinte's generation. The number of ancestors doubles as shown below.

Generation		Ancestors	
Generation 1	*Alex Haley*	1	
			$\times 2$
Generation 2	*Parents*	2	
			$\times 2$
Generation 3	*Grandparents*	4	
			$\times 2$
Generation 4		8	
			$\times 2$
Generation 5		16	
			$\times 2$
Generation 6		32	
			$\times 2$
Generation 7		64	
			$\times 2$
Generation 8		128	

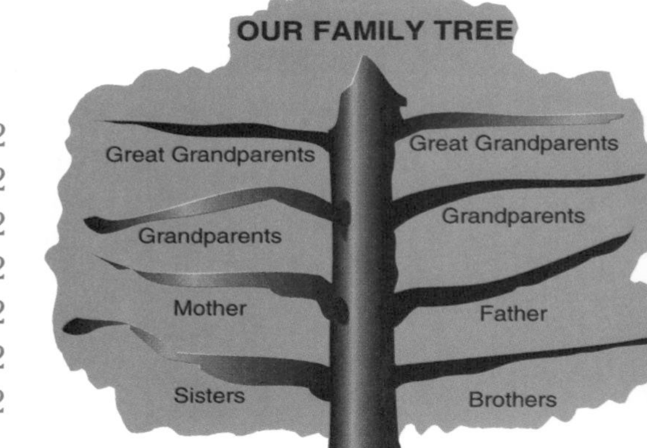

The list of numbers 1, 2, 4, 8, 16, 32, 64, 128 forms a **geometric sequence**. Each term in a geometric sequence increases or decreases by a common *factor*, called the **common ratio**. The number of ancestors doubles in each generation so the common ratio for this sequence is 2.

Geometric Sequence: **A geometric sequence is a sequence in which the ratio between any two successive terms is the same.**

Example 1 **State whether each sequence is geometric. If so, state the common ratio and list the next three terms.**

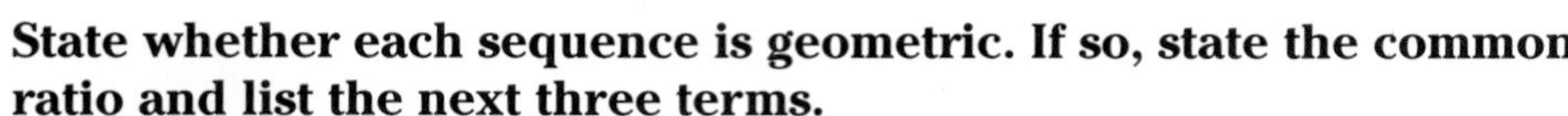

a. 3, 15, 75, 375, . . .

3, 15, 75, 375 ($\times 5$, $\times 5$, $\times 5$) *Notice* $\frac{15}{3} = 5$, $\frac{75}{15} = 5$, *and* $\frac{375}{75} = 5$.

The sequence has a common ratio so it is a geometric sequence. The common ratio is 5. The next three terms are 375×5 or 1875, 1875×5 or 9375, and 9375×5 or 46,875.

b. $-4, -2, -\frac{2}{3}, -\frac{1}{6}, \ldots$

$-4 \xrightarrow{\times \frac{1}{2}} -2 \xrightarrow{\times \frac{1}{3}} -\frac{2}{3} \xrightarrow{\times \frac{1}{4}} -\frac{1}{6}$

Since there is no common ratio, the sequence is *not* geometric.

c. $-4, 1, -\frac{1}{4}, \frac{1}{16}, \ldots$

$-4 \xrightarrow{\times\left(-\frac{1}{4}\right)} 1 \xrightarrow{\times\left(-\frac{1}{4}\right)} -\frac{1}{4} \xrightarrow{\times\left(-\frac{1}{4}\right)} \frac{1}{16}$

The sequence is geometric. The common ratio is $-\frac{1}{4}$.

$\frac{1}{16} \times \left(-\frac{1}{4}\right) = -\frac{1}{64}$ $\quad -\frac{1}{64} \times \left(-\frac{1}{4}\right) = \frac{1}{256}$ $\quad \frac{1}{256} \times \left(-\frac{1}{4}\right) = -\frac{1}{1024}$

The next three terms are $-\frac{1}{64}$, $\frac{1}{256}$, and $-\frac{1}{1024}$.

Connection to Algebra

If we know the first term of a geometric sequence and the common ratio, we can find any other term of the sequence. Consider the sequence 3, 12, 48, 192, ...

Here's how to write an expression that represents a term in the sequence. The first term is 3; call the first term a. The common ratio is 4; call the common ratio r.

Term	1st	2nd	3rd	4th	nth
Sequence	3	12	48	192	
Factors of term	3	$3 \cdot 4$	$3 \cdot 4 \cdot 4$	$3 \cdot 4 \cdot 4 \cdot 4$	$3 \cdot \underbrace{4 \cdot 4 \cdot 4 \cdot \ldots \cdot 4}_{n-1}$
Term using exponents	3	$3 \cdot 4^1$	$3 \cdot 4^2$	$3 \cdot 4^3$	$3 \cdot 4^{n-1}$
Variables	a	$a \cdot r^1$	$a \cdot r^2$	$a \cdot r^3$	$a \cdot r^{n-1}$

The exponent is one less than the number of the term.

Because n represents any term, you can use this expression to find any term in the sequence.

Example 2 **Use the expression $a \cdot r^{n-1}$ to find the eighth term in the sequence 4, 12, 36, 108, . . .**

The first term is 4, so $a = 4$. The common ratio is 3, so $r = 3$, and $n = 8$.

$a \cdot r^{n-1} = 4 \cdot 3^{8-1}$ or $4 \cdot 3^7$

4 [×] 3 [y^x] 7 [=] 8748

The eighth term is 8748.

Some real-life problems can be solved using geometric sequences.

Example 3

Sports

The NCAA Women's Basketball tournament is a single elimination tournament. After each game, the winning team progresses to the next level of competition and the loser is out of the tournament. How many teams are in the first round of play if there are 6 rounds in the tournament?

Explore With each round, half of the teams are out of the competition. So, this is a geometric sequence. If you start with the final game and work backward, each round had twice as many teams as the one after it. The first term is 2, the 2 teams who played in the final game of the tournament. The common ratio is 2 because there are twice as many games in each previous round. The number of terms is 6 because there were 6 rounds.

Plan Find out how many teams played in each round by starting with 2 and multiplying by 2 until you find the 6*th* term.

Solve 2 [×] 2 [×] 2 [×] 2 [×] 2 [×] 2 [=] 64

There were 64 teams.

Examine Another strategy is to use the formula $a \cdot r^{n-1}$

$a = 2$, $r = 2$, and $n = 6$.

$2 \cdot 2^{(6-1)} = 2 \cdot 2^5$ or 64

The solution checks.

Checking Your Understanding

Communicating Mathematics

Read and study the lesson to answer these questions.

1. **Describe** the common ratio in a geometric sequence.
2. You studied arithmetic sequences in Lesson 5-9. Compare and contrast geometric and arithmetic sequences.
3. **Give an example** of a geometric sequence with a first term of $\frac{1}{4}$.
4. **Find** the number of ancestors Alex Haley has in the generation before Kunta Kinte.

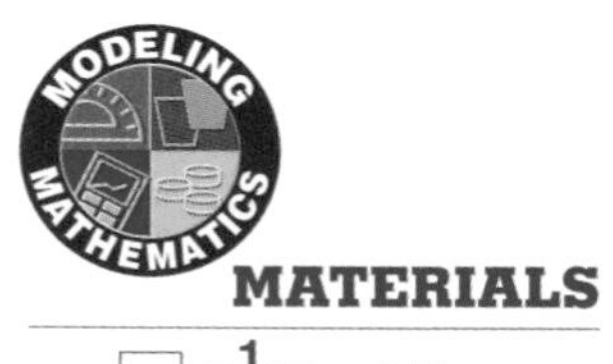

MATERIALS

☐ $8\frac{1}{2}$" by 11" paper

5. You can model a geometric sequence by folding a piece of paper and counting the number of sections.

a. Start with an $8\frac{1}{2}$"-by-11" piece of paper. There are 0 folds and 1 section. The first term of the sequence is 1.

b. Fold the paper in half. There is 1 fold and 2 sections. The second term of the sequence is 2. Continue to fold the paper in half. Keep a record of the number of folds and the number of sections.

c. How many sections result from the 5th fold?

d. Write the sequence. Does the expression $a \cdot r^{(n-1)}$ represent the terms in the sequence? Explain.

Guided Practice

State whether each sequence is a geometric sequence. If so, state the common ratio and list the next three terms.

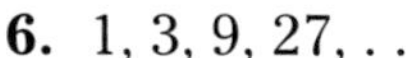

6. 1, 3, 9, 27, . . . **7.** 2, 4, 6, 8, . . . **8.** 1, −3, 5, −7, . . .

9. 81, 27, 9, 3, . . . **10.** 10, 14, 18, 22, . . . **11.** $\frac{1}{2}$, 1, 2, 4, . . .

12. Find the fourth term of a geometric sequence if $a = -5$ and $r = -2$.

13. Genealogy Find the number of ancestors in your family 11 generations ago.

Exercises: Practicing and Applying the Concept

Independent Practice

State whether each sequence is a geometric sequence. If so, state the common ratio and list the next three terms.

14. $-5, 1, -\frac{1}{5}, \frac{1}{25}, \ldots$ **15.** 24, 12, 6, 3, . . . **16.** 9, 3, −3, −9, . . .

17. $144, 12, 1, \frac{1}{12}, \ldots$ **18.** $-14, 7, -1, -\frac{1}{7}, \ldots$ **19.** $\frac{1}{2}, \frac{1}{4}, \frac{1}{8}, \frac{1}{16}, \ldots$

20. −8, 16, −32, 64, . . . **21.** 7, −14, 28, −56, . . . **22.** $25, 5, 1, \frac{1}{5}, \ldots$

23. $2\frac{1}{2}, \frac{1}{2}, -1\frac{1}{2}, \ldots$ **24.** $-\frac{1}{4}, -\frac{1}{2}, -1, -2, \ldots$ **25.** 1, 1, 2, 3, 5, . . .

26. $18, -6, 2, -\frac{2}{3}, \ldots$ **27.** $24, 6, \frac{3}{2}, \frac{3}{8}, \ldots$ **28.** 5, 7, 9, 11, 13, . . .

29. Write the first five terms in a geometric sequence with a common ratio of −2. The first term is 2.

30. Write the first four terms of a geometric sequence if $a = -6$ and $r = \frac{2}{3}$.

31. Write the first three terms of a geometric sequence if $a = 80$ and $r = \frac{5}{4}$.

32. Use the expression $ar^{(n-1)}$ to find the ninth term in the geometric sequence $-14, -\frac{7}{2}, -\frac{7}{8}, -\frac{7}{32}, \ldots$.

Critical Thinking

33. Use your calculator to find the first eight terms in the geometric sequence defined by $1 \cdot 7^n$, where n represents the number of the term. Determine the pattern that develops in the ones digits of each term. Use the pattern to find the ones digit of the 100th term.
34. Write the first five terms in a geometric sequence with a common ratio that is greater than 1. Then write the first five terms in a geometric sequence with a common ratio that is between 0 and 1.
    a. What happens to the size of the terms in a geometric sequence if the common ratio is greater than 1?
    b. What happens to the size of the terms in a geometric sequence if the common ratio is less than 1?

Applications and Problem Solving

35. **Geography** Between 1980 and 1990, the population of Florida increased from approximately 10 million to 13 million.

    a. Find the common ratio between these numbers.
    b. If the population grows in a geometric pattern over each of the next four decades, what would the population of Florida be in 2030?
36. **Investment** The face value of a high return bond increases to $1\frac{1}{10}$ of its previous value each year. If the starting value is $1250, what is the face value after 3 years?
37. **Patterns** Jacob told three friends a secret. Each of these friends told three more friends, who each told three more friends. How many people in all knew about the secret?
38. **Games** *Jeopardy* holds a tournament each year. Three people participate in each game of the competition and only the winners proceed to the next round. How many players participate if there are five rounds?

Mixed Review

39. Solve $\frac{5}{9}k = 15$. (Lesson 6-7)
40. Solve $d = 119 - 9.8$. (Lesson 5-3)
41. Find the quotient $\frac{3^3}{3^7}$. (Lesson 4-9)
42. **Geometry** The area of a square is the product of its length and width. Use exponents to write an expression for the area of a square that has sides b units long. (Lesson 4-2)
43. Solve $t = \frac{120}{24}$. (Lesson 2-8)
44. Simplify $5x + 3x + 1$. (Lesson 1-5)
45. Find the value of $6 \cdot 5 + 12 \div 3$. (Lesson 1-2)
46. **Patterns** Find the next number in the pattern 8, 9, 11, 14, 18, 23, . . . (Lesson 1-1)

6-9 Scientific Notation

Setting Goals: *In this lesson, you'll write numbers in scientific notation.*

Modeling a Real-World Application: Geology

Dr. William Rose, a geologist, specializes in the study of volcanoes and their effect on the atmosphere. He and his team studied the Augustine volcano about 280 kilometers from Anchorage, Alaska. The Augustine volcano erupted twice in 1986, once on March 27 and again on April 4.

Even though these eruptions were relatively small, on March 27th and 28th, the volcano produced 75,000,000 metric tons of fallout per day. The individual particles in the fallout were very small. The diameter of a particle was 0.000022 meter.

Learning the Concept

It's easy to make mistakes when computing with small numbers such as 0.000022 and large numbers such as 75,000,000. People who deal regularly with such numbers use **scientific notation**.

Scientific Notation	Numbers expressed in scientific notation are written as the product of a factor and a power of 10. The factor must be greater than or equal to 1 and less than 10.

To write a large positive or negative number in scientific notation, move the decimal point to the right of the left-most digit, and multiply this number by a power of ten.

To find the power of ten, count the number of places you moved the decimal point.

Example  **Write each number in scientific notation.**

a. 687,000

$687{,}000 = 6.87 \times 10^5$ *Move the decimal point 5 places to the left. Multiply by 10^5.*

b. 50,000,000

$50{,}000{,}000 = 5 \times 10^7$ *Move the decimal point 7 places to the left. Multiply by 10^7.*

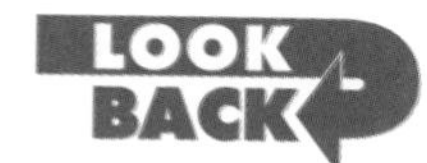

You can review negative exponents in Lesson 4-9.

To write a small positive number in scientific notation, move the decimal point to the right of the first nonzero digit, and multiply this number by a power of ten. Remember that $10^{-n} = \frac{1}{10^n}$. Count the number of places you moved the decimal point and use the negative of that number as the exponent of ten.

Example 2 **Write 0.000003576 in scientific notation.**

$0.000003576 = 3.576 \times 10^{-6}$ *Move the decimal point 6 places to the right. Multiply by 10^{-6}.*

Remember that $10^{-6} = \frac{1}{1,000,000}$.

Example 3 **Write each number in standard form.**

Use the EE key to enter numbers in scientific notation in a calculator. Enter the decimal portion, press EE, then enter the exponent.

a. 8.495×10^5

$8.495 \times 10^5 = 8.495 \times 100,000$ *Move the decimal point 5 places to the right.*

$= 849,500$ 8.49500

b. 3.08×10^{-4}

$3.08 \times 10^{-4} = 3.08 \times \left(\frac{1}{10}\right)^4$ *Remember that $10^{-4} = \left(\frac{1}{10}\right)^4$.*

$= 3.08 \times \frac{1}{10,000}$

$= 3.08 \times 0.0001$ *Move the decimal point 4 places to the left.*

$= 0.000308$ 00003.08

You can compare and order numbers in scientific notation. First compare the exponents. With positive numbers, any number with a greater exponent is greater. If the exponents are the same, compare the numbers.

Example 4 CONNECTION **Science**

The diameter of Venus is 1.208×10^4 km, the diameter of Earth is 1.276×10^4 km, and the diameter of Mars is 6.79×10^3 km. Rank these planets by diameter, from least to greatest.

$1.208 \times 10^4 < 1.276 \times 10^4$

The exponent is the same; $1.208 < 1.276$.

So Venus is smaller than Earth.

$6.79 \times 10^3 < 1.208 \times 10^4$ *$10^3 < 10^4$*

So Mars is smaller than Venus.

The rank from least to greatest is Mars, Venus, Earth.

Checking Your Understanding

Communicating Mathematics

Read and study the lesson to answer these questions.

1. In your own words, explain how to write these numbers in scientific notation.
   a. 634,930 **b.** 0.0000407
2. **Explain** the relationship between a number in standard form and the sign of the exponent when the number is written in scientific notation.

3. **Assess Yourself** What career are you interested in pursuing after high school? How might you use scientific notation in that career?

Guided Practice

State whether each number is in scientific notation. Write *yes* or *no*. If *no*, explain why not.

4. 23.45×10^5
5. 5.689×10^{-3}
6. 0.23×10^8

State where the decimal point should be placed in order to express each number in scientific notation. State the power of ten by which you should multiply. Then write the number in scientific notation.

7. 8490
8. 0.000045
9. 847.9

Write each number in standard form.

10. 5.2×10^5
11. 6.1×10^4
12. 9.34×10^{-2}
13. **Physics** The speed of light is 3.00×10^5 kilometers per second. If a light-year is the distance light travels in one Earth year, how many kilometers are in one light-year? (One Earth year = 365.25 days. Remember to convert days to seconds)

Exercises: Practicing and Applying the Concept

Independent Practice

Write each number in scientific notation.

14. 5,894,000
15. 0.0059
16. 269,000
17. 0.00015
18. 80,000,000
19. 0.0000000498
20. the possible hands in a 5-card game, 2,598,960
21. the population of Africa in 1994, 701,000,000
22. the diameter of a spider's thread, 0.001 inch

Write each number in standard form.

23. 5.6×10^{-6}
24. 1.399×10^2
25. 9.001×10^{-3}
26. 7.8×10^{11}
27. 5.985×10^{-5}
28. 4.054×10^1
29. the height of the Sears Tower, 1.454×10^3 feet
30. the diameter of a uranium atom, 3.5×10^{-8}
31. the volume of a drop of liquid, 5×10^{-5} liter
32. Write $(4 \times 10^4) + (8 \times 10^3) + (3 \times 10^2) + (9 \times 10^1) + (6 \times 10^0)$ in standard notation.
33. Write $(7 \times 10^8) + (8 \times 10^4) + (3 \times 10^0)$ in standard notation.
34. Write $(6 \times 10^0) + (4 \times 10^{-3}) + (3 \times 10^{-5})$ in standard notation.

Critical Thinking

35. In standard form, $3.14 \times 10^{-4} = 0.000314$, and $3.14 \times 10^4 = 31,400$. What is 3.14×10^0 in standard form?

Applications and Problem Solving

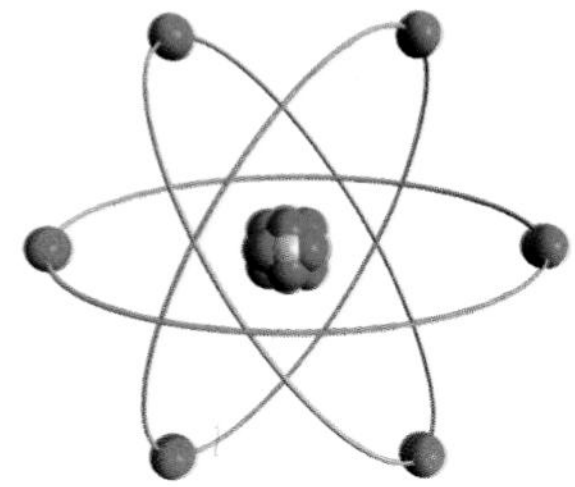

36. **Science** Everything around you is made up of atoms. Atoms are extremely small. An atom is about two millionths of an inch in diameter. Write this number in standard form and then in scientific notation.

interNET CONNECTION

For the latest volcano eruption rates, visit: **www.glencoe.com/sec/math/prealg/mathnet**

37. Geology This graph below gives the maximum eruption rate in cubic meters per second of seven volcanoes in the last century.

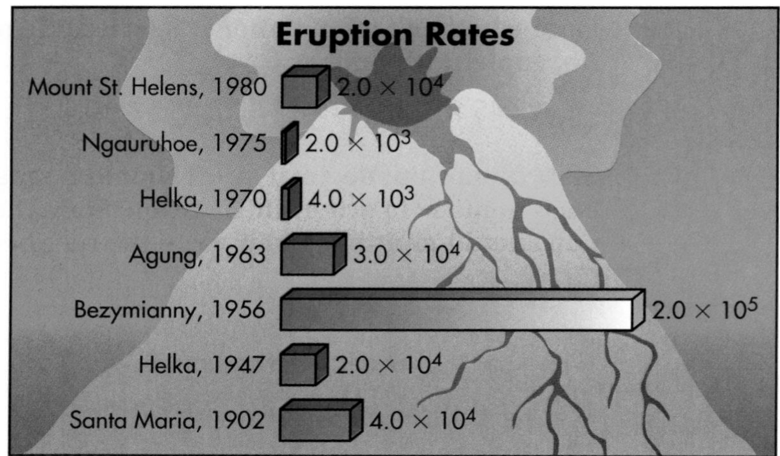

Source: University of Alaska

a. Which volcano had the greatest eruption rate?

b. Which had the least eruption rate?

c. Use standard form to rank the volcanoes in order from greatest to least eruption rate.

Mixed Review

38. State whether $9, -3, 1, -\frac{1}{3}, \ldots$ is a geometric sequence. If so, state the common ratio and list the next three terms. (Lesson 6-8)

39. Solve $x + \frac{2}{3} \geq 3$. (Lesson 5-7)

40. Solve $-3\frac{1}{5} - \left(-\frac{2}{5}\right) = a$. (Lesson 5-4)

41. Evaluate $\frac{b}{7}$, if $b = -91$. (Lesson 2-8)

42. Solve $x + 7 = 19$ mentally. (Lesson 1-6)

43. Sports Translate *three more than the Bears' score* into an algebraic expression. (Lesson 1-3)

COOPERATIVE LEARNING PROJECT

THE SHAPE OF THINGS TO COME

Living Batteries

Can you imagine taking your car to the service station for a fresh tank of sugar water? Or plugging your stereo into a lemon?! Scientists at King's College in London, England are learning to harness the energy produced by digestion to power living batteries that use sugar and bacteria.

Escherichia coli, or *E. coli*, bacteria break down carbohydrates as a part of the digestive process. This process frees up electrons, which generates electricity. As many as 100 billion, or 1×10^{11}, *E. coli* can exist in a cubic centimeter. One of the prototype living batteries is as small as a button and can run a watch for over a year.

See For Yourself

Research living batteries and the digestive system.

- How does the process of digestion work?
- A simple battery can be made using a zinc needle, a copper needle, wire, and a lemon. Place both needles into the lemon and connect them with the wire. Then place a small light bulb, like one from a flashlight, on the wire to see the result of the power flow. How does the battery work?
- How are the living batteries different than the alkaline batteries we usually use?

CHAPTER 6 Highlights

Vocabulary

After completing this chapter, you should be able to define each term, property, or phrase, and give an example or two of each.

Algebra
common ratio (p. 312)
compatible numbers (p. 280)
equivalent (p. 275)
geometric sequence (p. 312)
inverse property of multiplication (p. 289)
multiplicative inverses (p. 289)
reciprocal (p. 289)
repeating decimal (p. 275)
scientific notation (p. 317)
terminating decimal (p. 275)
truncate (p. 275)

Statistics
mean (p. 301)
measures of central tendency (p. 301)
median (p. 301)
mode (p. 301)

Geometry
area model (p. 284)

Problem Solving
work backward (p. 314)

Understanding and Using Vocabulary

Determine whether each statement is *true* or *false*. If the statement is false, rewrite it to make it true.

1. Two numbers whose product is 1 are multiplicative inverses, or reciprocals, of each other.
2. The mode is the number in the middle when the data are arranged in order.
3. Each term in a common ratio increases or decreases by a geometric sequence.
4. Mean is what people are usually talking about when they say average.
5. A decimal is called a terminating decimal because the division ends when the divisor is zero.
6. Equivalent numbers are rounded so that they fit together.
7. Numbers written in scientific notation are expressed as the product of a factor and a power of 10.
8. A calculator that truncates will display .66666666 when you divide 2 by 3. A calculator that rounds will display .66666667 when you divide 2 by 3.
9. It is possible to have a set of numerical data with no mean.
10. The inverse property of multiplication says that every number, including zero, has a multiplicative inverse. Zero is its own multiplicative inverse.

CHAPTER 6 Study Guide and Assessment

Skills and Concepts

Objectives and Examples

Upon completing this chapter, you should be able to:

- **write fractions as terminating or repeating decimals.** (Lesson 6-1)

$\frac{2}{5} \rightarrow 5\overline{)2.0}$ with quotient 0.4 $\frac{4}{9} \rightarrow 4$ [÷] 9 [=] .44444

$\frac{2}{5} = 0.4$ $\frac{4}{9} = 0.\overline{4}$

Review Exercises

Use these exercises to review and prepare for the chapter test.

Write each fraction as a decimal. Use a bar to show a repeating decimal.

11. $\frac{5}{8}$
12. $\frac{6}{9}$
13. $\frac{17}{40}$
14. $\frac{2}{25}$
15. $\frac{4}{15}$
16. $\frac{7}{12}$

- **estimate products and quotients of rational numbers.** (Lesson 6-2)

$14\frac{1}{5} \times 4\frac{8}{9} \rightarrow 14 \times 5 = 70$

$84.73 \div 4.930 \rightarrow 85 \div 5 = 17$

Estimate each product or quotient.

17. $\frac{9}{10} \cdot 21$
18. $8.36 \div 16.53$
19. $9 \div 1\frac{1}{5}$
20. 18.17×6.19

- **multiply fractions.** (Lesson 6-3)

$a = \frac{4}{3} \cdot \frac{5}{12}$

$a = \frac{\overset{1}{\cancel{4}}}{3} \cdot \frac{5}{\underset{3}{\cancel{12}}}$

$a = \frac{5}{9}$

Solve each equation. Write each solution in simplest form.

21. $-\frac{5}{9} \cdot \frac{8}{25} = n$
22. $6\frac{2}{3} \cdot \frac{1}{2} = t$
23. $w = -3\left(\frac{6}{15}\right)$
24. $-7\left(\frac{8}{21}\right) = d$
25. $g = \left(\frac{4}{5}\right)^3$
26. $q = \frac{2}{5} \cdot \frac{7}{8}$

- **divide fractions using multiplicative inverses.** (Lesson 6-4)

Evaluate $c \div d$ if $c = \frac{1}{3}$ and $d = \frac{7}{10}$.

$c \div d = \frac{1}{3} \div \frac{7}{10}$

$= \frac{1}{3} \cdot \frac{10}{7}$ *Multiply by the inverse.*

$= \frac{10}{21}$

Evaluate each expression.

27. $a \div b$, if $a = \frac{1}{4}$ and $b = \frac{2}{6}$
28. $bc \div f$, if $b = -\frac{8}{15}$, $c = 3$, and $f = 1\frac{1}{2}$
29. $\frac{x}{y}$, if $x = \frac{8}{12}$ and $y = -6$
30. $\frac{gh}{8}$, if $g = 3\frac{2}{7}$ and $h = -5\frac{1}{3}$

Objectives and Examples	Review Exercises

▸ **multiply and divide decimals.** (Lesson 6-5)

Solve $w = -5.7(3.8)$.

$$\begin{array}{r} -5.7 \\ \times 3.8 \\ \hline 456 \\ 1710 \\ \hline -21.66 \end{array}$$

−5.7 ← *one decimal place*
× 3.8 ← *one decimal place*
−21.66 ← *two decimal places*

Solve each equation.

31. $4 \times 7.07 = y$
32. $k = -1.25 \times 12$
33. $b = 62.9 \div 1000$
34. $2.65 \cdot 3.46 = c$
35. $0.7 \div 2.4 = d$
36. $-25.9 \div 2.8 = s$
37. $n = 5.3 \cdot (-9)$
38. $f = 1.3 \div 100$

▸ **find the mean, median, and mode of a set of data.** (Lesson 6-6)

Find the mean, median, and mode of the test scores 90, 88, 95, 73, and 92.

mean $\frac{90 + 88 + 95 + 73 + 92}{5} = \frac{438}{5}$ or 87.6

median 73 88 90 92 95
↑ *The median is 90.*

mode Since no scores occur more than once, there is no mode.

Find the mean, median, and mode for each set of data. When necessary, round to the nearest tenth.

39. 21, 25, 16, 18, 18, 26, 21, 17, 19, 25
40. 2.6, 3.1, 6.8, 4.9, 5.7, 3.4, 4.3
41. 0, 9, 14, 22, 26, 14
42. 5.4, 4.6, 6.2, 2.7, 8.0

▸ **solve equations and inequalities containing rational numbers.** (Lesson 6-7)

Solve $-4.7x < 14.1$.

$-4.7x < 14.1$

$\frac{-4.7x}{-4.7} > \frac{14.1}{-4.7}$ *Reverse the order symbol.*

$x > -3$

Solve each equation or inequality. Check your solution.

43. $8.67k = 78.03$
44. $6.3a \leq 52.92$
45. $-\frac{1}{4}r = 3$
46. $4.5 < -\frac{5}{8}f$
47. $-25 \geq 6.25n$
48. $3x = 55.8$
49. $8x = 42.4$
50. $2.6v > 19.76$

▸ **recognize and extend geometric sequences.** (Lesson 6-8)

State whether 768, 192, 48, . . . is a geometric sequence. If so, state the common ratio and list the next three terms.

768, 192, 48, . . .

$\times \frac{1}{4}$ $\times \frac{1}{4}$

The sequence is geometric. The common ratio is $\frac{1}{4}$. The next three terms are $48 \times \frac{1}{4}$ or 12, $12 \times \frac{1}{4}$ or 3, and $3 \times \frac{1}{4}$ or $\frac{3}{4}$.

State whether each sequence is a geometric sequence. If so, state the common ratio and list the next three terms.

51. 6.6, 5.7, 4.9, . . .
52. 3, 9, 27, . . .
53. 0.5, 2, 3.5, 5, . . .
54. $\frac{1}{4}, \frac{1}{2}$, 1, 2, . . .
55. 17, 13, 18, 14, 19, . . .

Objectives and Examples	Review Exercises

▶ **write numbers in standard form and scientific notation.** (Lesson 6-9)

Write 58,970,000,000 in scientific notation.

$58{,}970{,}000{,}000 = 5.897 \times 10^{10}$

Write 6.892×10^{-2} in standard form.

$$6.892 \times 10^{-2} = 6.892 \times \left(\frac{1}{10}\right)^2$$
$$= 6.892 \times 0.01$$
$$= 0.06892$$

Write each number in scientific notation.

56. 13,490,000 **57.** 0.00000674

58. 0.00032 **59.** 5810

Write each number in standard form.

60. 4.24×10^2

61. 5.72×10^4

62. 3.347×10^{-1}

63. 2.02×10^8

Applications and Problem Solving

64. Sports Of the medals that the United States' 1992 Summer Olympic Team earned, 0.34 were gold. $\frac{44}{112}$ of the medals earned by the Unified Team were gold. Which team had a greater portion of their medals as gold? (Lesson 6-1)

65. Economics The sales tax for many cities is $5\frac{1}{2}$%. This means that for every dollar you spend, you pay $5\frac{1}{2}$¢ in tax. If you purchase an item for $24, how much will you owe in tax? (Lesson 6-3)

66. Population Canada is divided into ten provinces. The table at the right shows the population in thousands of each province in 1991. (Lesson 6-6)

a. Find the mean population for the provinces.

b. Find the median population for the provinces.

c. Find the mode population for the provinces.

Province	Population (thousands)
Alberta	2546
British Columbia	3282
Manitoba	1092
New Brunswick	724
Newfoundland	568
Nova Scotia	900
Ontario	10,085
Prince Edward Island	130
Quebec	6896
Saskatchewan	989

67. Astronomy The first new moon discovered by *Voyager* revolves around Uranus. Called Puck, this moon makes one orbit around Uranus every 18 hours. How many times has Puck circled Uranus in 273.6 hours? (Lesson 6-7)

68. Forestry The U.S. National Park System covers a total of 80,663,217.42 acres. Write this number in scientific notation. (Lesson 6-9)

A practice test for Chapter 6 is available on page 779.

Alternative Assessment

Performance Task

Demonstrate your knowledge by giving a clear, concise solution to each problem. Be sure to include all relevant drawings and justify your answers. You may show your solutions in more than one way or investigate beyond the requirements of the problem.

Stock	Price per share
McDonalds	$39\frac{5}{8}$
Wendys	19
Sony	$55\frac{3}{8}$
Nike	$92\frac{3}{4}$

1. How much does one share of Nike stock cost in dollars?
2. How much would it cost if you bought 6 shares of Sony stock?
3. If you spent $951 buying shares of stock in McDonalds, how many shares would you have purchased?
4. Find the mean of the stock prices listed.

5. Find the median stock price.
6. If you bought 32 shares of Wendy's at $17\frac{3}{4}$ per share and sold it all at the price listed, how much money would you make?
7. Find a list of stock prices in a local newspaper. Choose ten shares of at least three stocks to "sell". Record the prices and the number of shares of each stock that you chose in a table. Then look at a paper from one week ago and record the price of each stock then.
   a. If you bought the shares last week and sold this week, how much did you make or lose?
   b. Find the gain or loss on each stock. Then find the mean, median, and mode of the numbers. If you were bragging about your financial accomplishments, which number would you tell? If you were complaining about the stock market which number would you tell? Explain.

Thinking Critically

- Why is scientific notation important?
- One meter is 1×10^2 centimeters. The approximate size of an atom is 1×10^{-10} centimeters. How many meters is that? Write the number of meters in scientific notation and standard form.

Portfolio

Select some of your work from this chapter that shows how you used a calculator or a computer and place it in your portfolio.

Self Evaluation

Assess yourself. When you first started this chapter, what did you think was going to be the most difficult lesson? Did you let the lesson win, or did you work hard and learn how to do it? How will you react when faced with a similar situation next time?

CHAPTERS 1–6 *Ongoing Assessment*

Section One: Multiple Choice

There are ten multiple-choice questions in this section. After working each problem, write the letter of the correct answer on your paper.

1. Carl bought a bike that cost \$532.16. He paid for the bike in equal payments for 12 months. Estimate: the amount of each payment was between

A. \$10 and \$20.
B. \$25 and \$35.
C. \$40 and \$50.
D. \$55 and \$65.

2. The Sanchez's added a 12-foot-by-15-foot rectangular deck to the back of their house. How much wood do they need for the railing? (The deck is attached to the house on one of the long sides.)

A. 39 feet
B. 27 feet
C. 42 feet
D. 90 feet

3. Gladys bought grocery items for the following prices: \$1.39, \$2.89, \$0.58, and \$1.19. The best estimate of the total cost is

A. \$9.
B. \$6.
C. \$5.
D. \$4.

4. If $m = 12$ and $p = 10$, what is the value of $m + mp$?

A. 120
B. 132
C. 240
D. 1440

5. If $3.2y = 80$, what is the value of y?

A. 25
B. 2.5
C. 256
D. 76.8

6. Which statement is *not* true?

A. $\frac{1}{2} > \frac{3}{8}$
B. $\frac{4}{5} < \frac{5}{6}$
C. $\frac{8}{11} < \frac{7}{13}$
D. $\frac{1}{3} < \frac{10}{13}$

7. Brent knows that to have an average of more than 85 in biology, his two exam scores must total more than 170. If his first grade was 82, which inequality could he use to find g, his second grade needed for an average greater than 85?

A. $82 + g > 170$
B. $82 > 170 + g$
C. $g + 82 < 170$
D. $\frac{82 + g}{2} > 170$

8. The mileage reading on the Garcia's car was 256.8 before they left on vacation. When the family returned, the reading was 739.4. How many miles did the Garcia family travel on their vacation?

A. 996.2 mi
B. 583.5 mi
C. 483.6 mi
D. 482.6 mi

9. Which is equivalent to $\frac{a^2}{a^4}$?

A. a^{-2}
B. a^2
C. a^6
D. a^8

10. The labor charge of repairing a car is \$42.50 per hour. If it takes 2.5 hours to repair the car, what will be the charge for labor?

A. \$812.50
B. \$106.25
C. \$13.00
D. none of these

Section Two: Free Response

This section contains eight questions for which you will provide short answers. Write your answer on your paper.

11. Solve $\frac{a}{-5} < 1$ and graph on a number line.

12. Express 0.36 as a fraction.

13. Find the GCF of $60x^2y$, $24x^3y^2$, and $84xy^4$.

14. Write the first five terms of a geometric sequence that has a common ratio of -2.

15. Carlos used $2\frac{5}{8}$ yards of material to make a shirt. He began with $3\frac{1}{4}$ yards of material. How much material did he have left?

16. The difference between the average high temperature in January and July in New York City is 37°F. The average high in July is 85°F. What is the average high in January?

17. The dimensions of Adita's yard are 25 feet long by 40 feet wide. Express the ratio of the length to the width in simplest form.

18. Ms. Clinton calculated that it would take $2\frac{5}{8}$ gallons of paint to paint one classroom at her school. She is going to paint 5 classrooms. How many gallons of paint will she need?

Test-Taking Tip

You may be able to eliminate all or most of the answer choices by estimating. Also, look to see which answer choices are not reasonable for the information given in the problem.

When questions involve variables, check your solutions by substituting values for the variables.

Section Three: Open-Ended

This section contains three open-ended problems. Demonstrate your knowledge by giving a clear, concise solution to each problem. Your score on these problems will depend on how well you do the following.

- Explain your reasoning
- Show your understanding of the mathematics in an organized manner.
- Use charts, graphs, and diagrams in your explanation.
- Show the solution in more than one way or relate it to other situations.
- Investigate beyond the requirements of the problem.

19. Give an example of either inductive or deductive reasoning. Explain why your example is either inductive or deductive reasoning.

20. The high temperatures for a week in April were 56°F, 58°F, 60°F, 63°F, 58°F, 62°F, and 70°F. What were the mean and the median temperatures? Explain the difference between the two temperatures.

21. Give an example of a decimal that neither terminates nor repeats.

CHAPTER 7

Solving Equations and Inequalities

TOP STORIES in Chapter 7

In this chapter, you will:

- solve problems by working backward,
- write and solve multi-step equations and inequalities,
- find the circumferences of circles, and
- convert measurements within the metric system.

MATH AND ENTERTAINMENT IN THE NEWS

Winning on a game show isn't all in what you know

Source: The Mathematics Teacher, May, 1994

Have you ever dreamed of hitting it big on a game show? Before you sign up, study up, on your math! Strategy is a big part of winning, and math is a big part of strategy. For example, on the show *Jeopardy!*, the winner is often determined by the last question, called *Final Jeopardy*. Before the question is asked, contestants are told the category and asked to wager any amount less than or equal to their current score. For a correct answer, the amount is added to their score. For an incorrect answer, it is subtracted. A mathematical analysis using inequalities shows that each player can carefully choose his or her wager to get the best chance of winning. So next time you play, take mathematics for $1000!

Putting It into Perspective

1928
Television recording system using aluminum gramophone records invented by John Logie Baird in London.

1955
The $64,000 Question ushers in era of big-money quiz shows on television.

1920 — 1935 — 1950

1941
The first television commercial airs on WBNT of New York.

On the Lighter Side

THE FAR SIDE By GARY LARSON

"Excuse me . . . I know the game's almost over, but just for the record, I don't think my buzzer was working properly."

Statistical Snapshot

1994's Top Ten Syndicated Programs

Rank	Show	Rating	Stations
1	Wheel of Fortune	13.6	225
2	Jeopardy!	11.5	216
3	Star Trek	9.6	244
4	Oprah Winfrey Show	8.9	234
5	Buena Vista I	8.2	162
6	Entertainment Tonight	8.1	180
7	Star Trek: Deep Space Nine	7.5	234
8	National Geographic on Assignment	7.4	188
8	Roseanne	7.4	188
10	Inside Edition	6.7	166

Source: *Brandweek*, April 3, 1995

For up-to-date information on game shows, visit:
www.glencoe.com/sec/math/prealg/mathnet

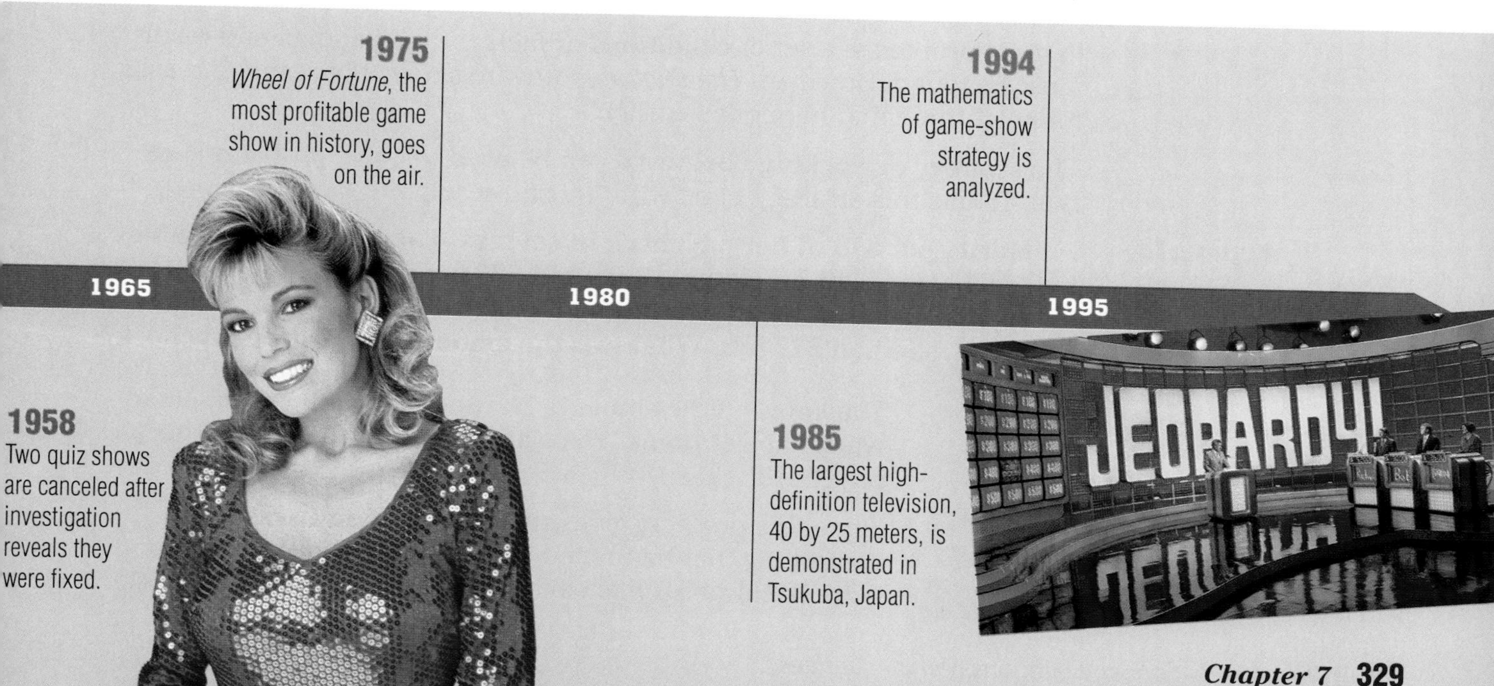

1958 Two quiz shows are canceled after investigation reveals they were fixed.

1975 *Wheel of Fortune*, the most profitable game show in history, goes on the air.

1985 The largest high-definition television, 40 by 25 meters, is demonstrated in Tsukuba, Japan.

1994 The mathematics of game-show strategy is analyzed.

7-1 Problem-Solving Strategy: Work Backward

Setting Goals: *In this lesson, you'll solve problems by working backward.*

Modeling with Technology

The computer software *The Factory* by Sunburst Communications simulates a factory assembly line. Machines rotate, punch holes, and paint a vertical stripe on geometric-shaped objects so that they match a given design.

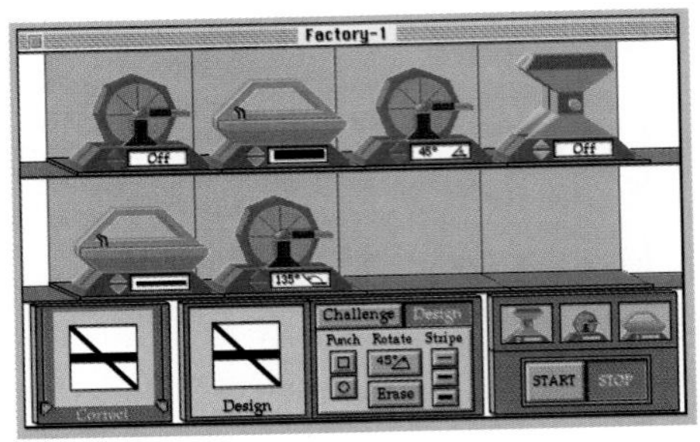

Your Turn Copy and complete the chart.

Operation	Begin with a designed square.	Rotate 45° clockwise.	Remove thin line.	Rotate 45° clockwise.	Remove thick line.
Result					

a. What process is involved in completing the chart above?

b. Look at the two squares at the right. What would you do to the first square to get the square on the far right as the end result from *The Factory*?

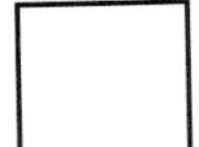 

Learning the Concept

In most problems, a set of conditions or facts is given and an end result must be found. However, *The Factory* software shows the result and asks for something that happened earlier.

The strategy of **working backward** can be used to solve problems like this. To use this strategy, start with the end result and *undo* each step.

Example **APPLICATION** **Personal Finance**

Marita put $10 of her paycheck in savings. Then she spent one-half of what was left on clothes. She paid $26 for a haircut and then spent one-half of what was left on a concert ticket. When she got home, she had $17 left. What was the amount of Marita's paycheck?

Explore You know all the amounts Marita spent and what she had when she got home. You want to find the amount of her paycheck.

Plan Since this problem gives the end result and asks for something that happened earlier, start with the result and work backward. *Undo* each step.

Solve When Marita got home, she had \$17. → 17
Undo the half that she spent on a concert ticket. → × 2 = 34
Undo the \$26 she spent on a haircut. → + 26 = 60
Undo the half she spent for clothes. → × 2 = 120
Undo the \$10 she put in savings. → + 10 = 130

The amount of Marita's paycheck was \$130.

Examine Assume that Marita started with \$130. After putting \$10 in savings, she had \$120. She spent half of \$120, so she had 120 ÷ 2 or \$60. Then she spent \$26, so she had 60 − 26 or \$34. Finally, she spent half of \$34, so she had 34 ÷ 2 or \$17. In the end, Marita would have \$17, so the answer is correct.

Checking Your Understanding

Communicating Mathematics

Read and study the lesson to answer these questions.

1. **Describe** the circumstances under which you could use the strategy of working backward.
2. **Explain** how to solve a problem by working backward.
3. **Write** a problem that can be solved by working backward. Trade problems with a classmate and solve his or her problem.

4. Create a design that could be made using *The Factory*. Record the procedure as you make it. Trade designs with a friend and reproduce your friend's design.

Guided Practice

Solve by working backward.

5. Mr. Fuentes uses half of a can of evaporated milk to make pumpkin pie. The can and the milk that is left weigh 9 ounces. If the can and the milk weighed 15 ounces before it was opened, how much does the can weigh?
6. A certain number is divided by 5, and then 1 is subtracted from the result. The final answer is 32. Find the number.

Exercises: Practicing and Applying the Concept

Independent Practice

Solve. Use any strategy.

7. Mr. and Mrs. Gentry each own an equal number of shares of IBM stock. Mr. Gentry sells one-third of his shares for \$2700. What was the total value of Mr. and Mrs. Gentry's stock before the sale?
8. A certain number is multiplied by 3 and then 5 is added to the result. The final answer is 41. Find the number.

Choose

Estimation
Mental Math
Calculator
Paper and Pencil

9. Chris has \$1.50. Half of the money he had when he left home this morning was spent on lunch. He lent Bill a dollar after lunch. How much money did Chris start with?

10. Forty-four students took the bus to the St. Louis Science Center. Each student paid \$1.25. How much did the driver collect?

11. On Monday, Fumiko told a joke to 3 of her friends. On Tuesday, each of those friends told the joke to 3 other friends. On Wednesday, each person who heard the joke on Tuesday told 3 other people. If this pattern continues, how many people will hear the joke on Saturday?

12. An ice sculpture is melting at the rate of half its weight every hour. After 8 hours, it weighs $\frac{5}{16}$ of a pound. How much did it weigh in the beginning?

13. In Canada and other countries, dates are abbreviated as "date.month.year." For example, September 19, 1991 would be abbreviated as 19.9.91 and September 5, 1995 would be 5.9.95. Dates like these, where the pattern of numbers is the same when read forward and backward, are called *palindromic dates*. What are the two palindromic dates in the 1900s that are closest together?

Critical Thinking

14. At the end of the second round of a game, Bart's and Carlos' scores were double the score each had at the end of the first round. The total number of points they gained were subtracted from Al's first round score. After the third round, Al's and Carlos' scores were double their second round scores. The total number of points they gained were subtracted from Bart's second round score. After the fourth round, Al's and Bart's third round scores doubled. The total number of points they gained were subtracted from Carlos' third round score. Now each of them has 8 points. How many points did each of them have after the first round?

Mixed Review

15. Write 6.789×10^{-7} in standard form. (Lesson 6-9)

16. **Entertainment** The table at the right shows the 1993 attendance in millions of people of the largest fairs in the United States. How many more people attended the New Mexico State Fair than attended the Los Angeles County Fair? (Lesson 5-3)

Fair	Attendance (millions)
State Fair of Texas, Dallas	3.15
State Fair of Oklahoma, Oklahoma City	1.79
New Mexico State Fair, Albuquerque	1.68
Minnesota State Fair, St. Paul	1.60
Houston (Texas) Livestock Show	1.57
Western Washington Fair, Puyallup	1.42
Los Angeles County Fair, Pomona	1.40
De Mar (California) Fair	1.11
Colorado State Fair, Pueblo	1.08
Tulsa (Oklahoma) State Fair	1.03

Source: *Top Ten of Everything*

17. Name the sets of numbers to which $\frac{42}{7}$ belongs—whole numbers, integers, or rationals. (Lesson 5-1)

18. Evaluate $ab \div (-7)$ if $a = 49$ and $b = -2$. (Lesson 2-8)

19. Find the value of $[4(3 + 8) \div (4 - 2)] + 9$. (Lesson 1-2)

HANDS-ON ACTIVITY

Two-Step Equations

A Preview of Lesson **7-2**

MATERIALS

cups

counters

equation mat

In this activity, you will extend what you know about solving one-step equations to model and solve two-step equations.

Recall from Chapter 3 that you can use cups and counters to model one-step equations. You can model two-step equations also. A two-step equation contains two operations. For example, the equation $2x + 3 = 7$ involves multiplication and addition.

Activity

Model and solve the equation $2x + 3 = 7$.

- Place 2 cups and 3 positive counters on the left side of the mat. Then place 7 positive counters on the right side of the mat.

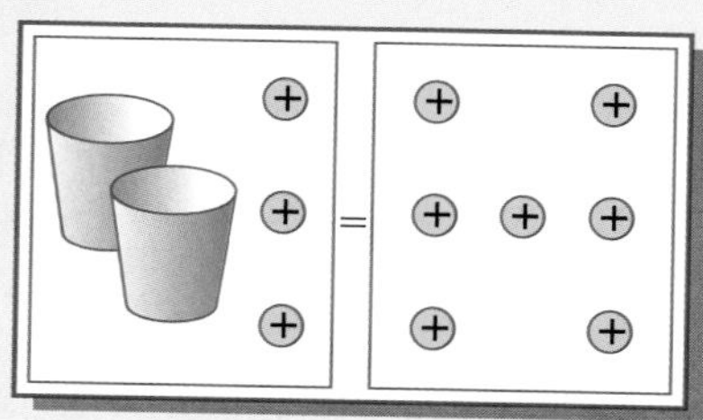

$2x + 3 = 7$

- To determine how many counters are in each cup, the goal is to isolate the cups on one side of the mat. Since the two sides of the mat represent equal quantities, you can remove an equal number of counters from each side without changing the value of the equation.

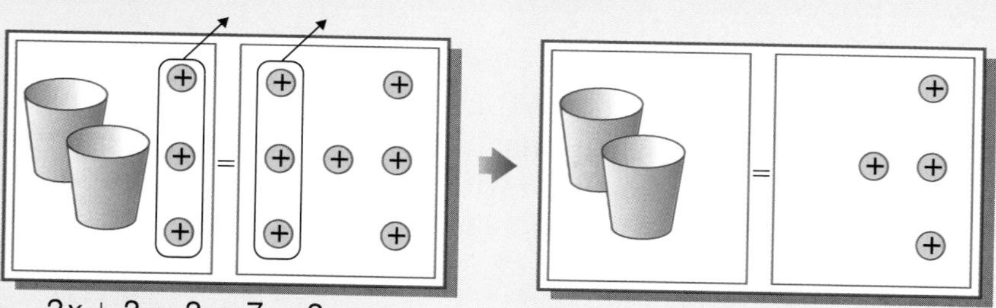

$2x + 3 - 3 = 7 - 3$ $\quad$ $2x = 4$

- Now the equation is $2x = 4$. Match an equal number of counters with each cup. Therefore, each cup must contain 2 counters, so $x = 2$.

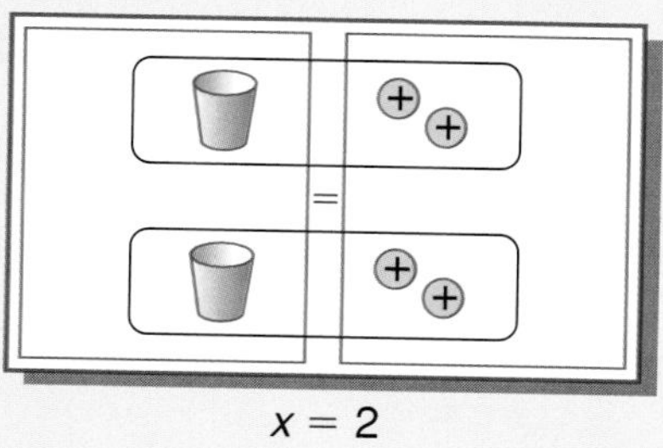

$x = 2$

Your Turn **Model each equation and solve.**

1. $2x + 2 = -4$
2. $3x + 3 = 12$
3. $2x - 2 = 6$
4. $2x - 2 = -4$
5. $4 + 2x = 8$
6. $3x - (-2) = -4$

TALK ABOUT IT

7. Explain how this activity models the work-backward strategy presented in Lesson 7-1.
8. Explain how you can use models to solve $-2x + 1 = -5$.

7-2 Solving Two-Step Equations

Setting Goals: *In this lesson, you'll solve equations that involve more than one operation.*

Modeling a Real-World Application: Consumerism

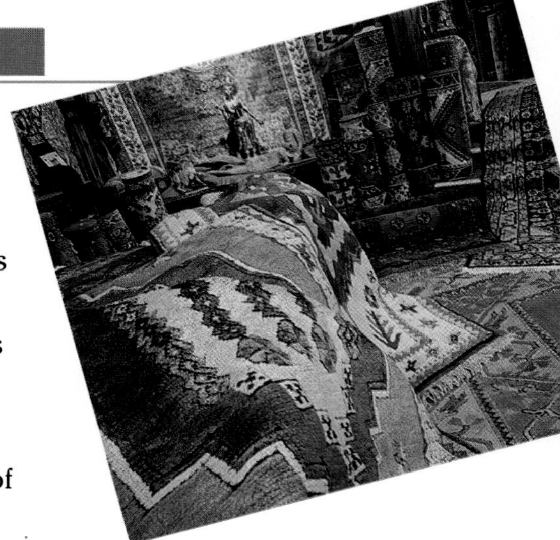

"At Carpet World, we will not be undersold! Get 18 square yards of the best stain-resistant carpet installed for just \$305.89! That *includes* the \$54 installation charge. Hurry in! At this price, the carpet will not last!"

The Jackson family is comparing prices at different carpet stores. Most carpet stores price carpet by the square yard. What price should the Jacksons use to compare the price of one square yard of carpet at Carpet World to the prices at other stores? *You will solve this problem in Exercise 36.*

Learning the Concept

To solve this problem, you will use the equation $54 + 18p = 305.89$. Some equations, like this one, contain more than one operation. To solve an equation with more than one operation, use the work-backward strategy and undo each operation.

Example 1 **Solve each equation. Check your solution.**

a. $\frac{c}{4} - 19 = 17$

$\frac{c}{4} - 19 + 19 = 17 + 19$ *Add to undo the subtraction.*

$\frac{c}{4} = 36$

$4 \cdot \frac{c}{4} = 4 \cdot 36$ *Multiply to undo the division.*

$c = 144$

Check: $\frac{c}{4} - 19 = 17$

$\frac{144}{4} - 19 \stackrel{?}{=} 17$ *Replace c with 144.*

$36 - 19 \stackrel{?}{=} 17$ *Do the division first.*

$17 = 17$ ✓ The solution is 144.

b. $-3n + 8 = -7$

$$-3n + 8 - 8 = -7 - 8 \quad \textit{Subtract 8 from each side.}$$
$$-3n = -15$$
$$\frac{-3n}{-3} = \frac{-15}{-3} \quad \textit{Divide each side by } -3.$$
$$n = 5$$

Check: $-3n + 8 = -7$
$$-3(5) + 8 \stackrel{?}{=} -7$$
$$-15 + 8 \stackrel{?}{=} -7$$
$$-7 = -7 \ \checkmark \quad \text{The solution is 5.}$$

In Chapter 3, you learned that a formula is an algebraic expression that can be used to show the relationship among certain quantities. Some formulas involve more than one operation. You can use the work-backward strategy to solve problems involving formulas.

Example 2

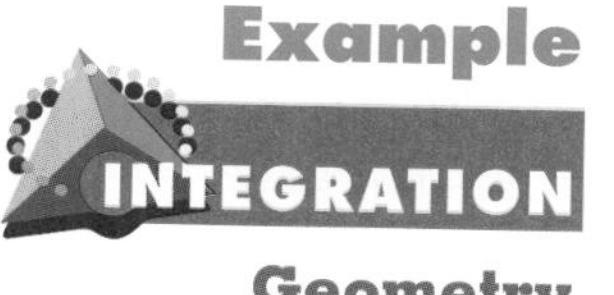

Geometry

The area A of a trapezoid can be found by multiplying the height h and one-half the sum of the lengths of the bases b_1 and b_2. The formula is $A = \frac{1}{2} \cdot h(b_1 + b_2)$. The area of the trapezoid at the right is 32 square inches. The trapezoid is 4 inches high, and the length of one of the bases is 9 inches. What is the length of the other base?

base 1 (b_1)
height (h)
base 2 (b_2)

$$A = \frac{1}{2} \cdot h(b_1 + b_2)$$
$$32 = \frac{1}{2} \cdot 4(b_1 + 9) \quad \textit{Replace A with 32, h with 4, and } b_2 \textit{ with 9.}$$
$$32 = 2(b_1 + 9)$$
$$\frac{32}{2} = \frac{2(b_1 + 9)}{2} \quad \textit{Because } 2(b_1 + 9) \textit{ means } 2 \cdot (b_1 + 9)\textit{, undo the multiplication first.}$$
$$16 = b_1 + 9$$
$$16 - 9 = b_1 + 9 - 9 \quad \textit{Subtract to undo the addition.}$$
$$7 = b_1 \quad \textit{Check the solution.}$$

The second base is 7 inches long.

Checking Your Understanding

Communicating Mathematics

Read and study the lesson to answer these questions.

1. How is the order of operations used in solving two-step equations?
2. **You Decide** Kelsey says that to solve the equation $3x + 6 = 24$, the first step should be to subtract 6 from each side. Dana says the first step should be to divide each term by 3.
   a. Which method would you use and why?
   b. Whose method would be easier to use to solve the equation $2x + 3 = 5$? Explain.
3. **Write** a problem that could be solved using the equation $3x - 5 = 15$.
4. This diagram models the equation $4 + 2x = 8$.
   a. What is the solution of $4 + 2x = 8$?
   b. Model the equation $2x - 3 = 7$. Then solve the equation.

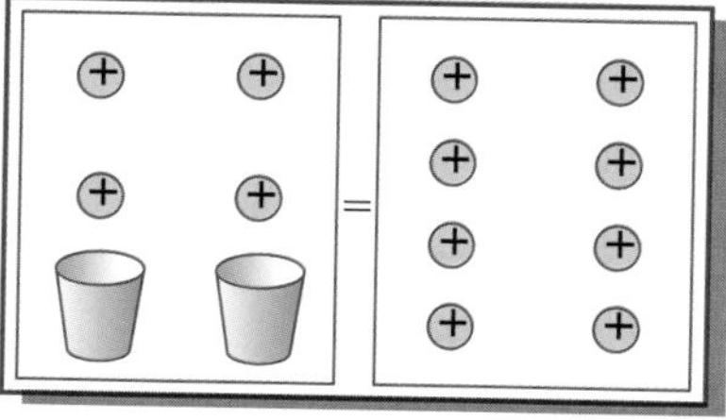

MODELING MATHEMATICS

MATERIALS

cups
counters
equation mat

Guided Practice

Solve each equation. Check your solutions.

5. $2n - 5 = 21$
6. $3 + \frac{t}{2} = 35$
7. $7.5r + 2 = -28$
8. $3(x + 5) = 9$
9. $-2.5(a + 2a) = 22.5$
10. $\frac{a + 5}{2} = 10.5$

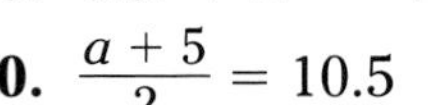

11. **Postage** Judi mailed some photographs to her friend Misae. She paid \$1.77 to send them first-class. The Post Office charges 32 cents for the first ounce and 29 cents for each additional ounce. Use the equation $32 + 29w = 177$ to find how much Judi's package weighs.

Exercises: Practicing and Applying the Concept

Independent Practice

Solve each equation. Check your solution.

12. $85 = 4d + 5$
13. $2r - 7 = 1$
14. $4 - 2b = -8$
15. $-8 - t = -25$
16. $-4y + 3 = 19$
17. $1.8 = 0.6 - y$
18. $19 = 7 + \frac{b}{7}$
19. $-12 + \frac{j}{4} = 9$
20. $-3 = -31 + \frac{c}{6}$
21. $8 = \frac{h}{-3} + 19$
22. $\frac{-4x}{3} = 24$
23. $13 + \frac{-p}{3} = -4$
24. $\frac{3}{4}n - 3 = 9$
25. $\frac{y}{3} + 6 = -45$
26. $\frac{c}{-4} - 8 = -42$
27. $\frac{n - 10}{5} = 2.5$
28. $16 = \frac{-8 + s}{-7}$
29. $\frac{-d - (-5)}{7} = 14$
30. $\frac{6 + c}{-13} = -3$
31. $\frac{n}{2} + (-3) = 5$
32. $4.7 = 1.2 - 7m$

Critical Thinking

33. **Make Up a Problem** Write a two-step equation using the numbers 2, 3, and 6, in which the solution is $-\frac{1}{6}$.

Applications and Problem Solving

34. **Medicine** Dr. Bell recommended that Marcie take eight tablets on the first day and then 4 tablets each day until the prescription was used. The prescription contained 28 tablets. Use the equation $8 + 4d = 28$ to find how many days Marcie will be taking pills after the first day.
35. **Chemistry** Chemical equations are like algebraic equations; they must be balanced on either side of the arrow. The symbol $3H_2O$ represents 3 molecules made up of 2 hydrogen and 1 oxygen atoms. Find the numbers that belong in the boxes in order for the equation $\square P + \square O_2 \rightarrow P_4O_{10}$ to be balanced.
36. **Consumerism** Refer to the application at the beginning of the lesson. How much is the carpet per square yard?
37. **Academics** Swedish scientist Alfred Nobel left a major part of his fortune to establish annual prizes to be awarded to those who have achieved the greatest common good in the fields of physics, chemistry, physiology or medicine, literature, peace, and economic science. People from the United States have won Nobel Prizes 213 times. That is 33 more than twice as many as the people from the United Kingdom, which is second in producing prize winners. How many people from the United Kingdom have won Nobel Prizes?

In 1950, Ralph Johnson Bunche became the first African-American to win the Nobel Peace Prize. Mr. Bunche worked at the U.S. State Department and the United Nations.

Mixed Review

38. **Personal Finance** Before Shantal left on her trip, she decided to plan carefully how much money she should set aside to cover the cost of the vacation. First, the plane ticket would cost \$156. Shantal decided she would use three-fourths of the remaining money in her budget for a hotel, rental car, and meals. She added \$30 for souvenirs and she wanted to have \$100 left over for emergencies. How much money should Shantal start with in her vacation budget? (Lesson 7-1)

39. Replace ● with $<$, $>$, or $=$ to make 16.43 ● $16\frac{3}{7}$ a true sentence. (Lesson 6-1)

40. Solve $c - \frac{3}{5} > 14\frac{2}{9}$. (Lesson 5-7)

41. Write $y \cdot y \cdot y$ as a product using exponents. (Lesson 4-2)

42. Simplify $|18| - |-4|$. (Lesson 2-1)

43. Is $x - (-17) > 29$ *true*, *false*, or *open* if $x = 11$? (Lesson 1-9)

Choose

Estimation
Mental Math
Calculator
Paper and Pencil

44. **Food** A survey performed by the *American Demographics* magazine found that the average American family spent \$1664 eating out in 1994. Based on this finding, about how much did the average family spend on eating out in a month? (Lesson 1-1)
    a. Which method of computation do you think is most appropriate? Justify your choice.
    b. Solve the problem using the four-step plan. Be sure to examine your solution.

WORKING ON THE Investigation

Refer to the Investigation on pages 270–271.

Look at your design of the loading platform that you placed in your Investigation Folder at the beginning of Chapter 6.

Determine how long it would take to load a coaster train by doing a whole-class experiment.

- Simulate the unloading and loading of different numbers of passengers in different numbers and sizes of cars. For example, determine how long it would take to unload and load 24 passengers in six cars or 20 passengers in ten cars. Record the time it takes from when the train stops to the time it starts moving again.
- Using the total ride time you calculated at the end of Lesson 6-4, determine how many trains you can safely operate on the track at the same time using your loading time and total ride time.

Add the results of your work to your Investigation Folder.

7-3 Writing Two-Step Equations

Setting Goals: *In this lesson, you'll solve verbal problems by writing and solving equations.*

Modeling a Real-World Application: Baseball

"It has been an interesting year for the Red Sox. Here at the 1995 All-Star Game, they stand at 10 games above 0.500. So 68 games into the season, they are atop the eastern division of the American League."

How many games had the Red Sox won before the All-Star Game?

Learning the Concept

If a baseball team is at 0.500, that means that they have won $\frac{500}{1000}$ or $\frac{1}{2}$ of their games. If a team wins more than half of their games, then their status is given as a number of games "over 0.500." At 10 over 0.500, the Red Sox have won 10 more games than they have lost. Let w represent the number of games the Red Sox have won. Then they have lost $w - 10$ games. Write an equation.

$$\underbrace{68}_{68}\ \underbrace{\text{equals}}_{=}\ \underbrace{\text{the number of wins}}_{w}\ \underbrace{\text{plus}}_{+}\ \underbrace{\text{the number of losses}}_{(w-10)}$$

Work backward to solve the equation.

$$\begin{aligned} 68 &= w + (w - 10) & \\ 68 &= 2w - 10 & \textit{Simplify.} \\ 68 + 10 &= 2w - 10 + 10 & \textit{Add 10 to each side.} \\ 78 &= 2w & \\ \frac{78}{2} &= \frac{2w}{2} & \textit{Divide each side by 2.} \\ 39 &= w & \end{aligned}$$

The Boston Red Sox had won 39 games before the All-Star Game.

Example 1 **Five more than twice some number is 19. Find the number.**

First define a variable. Let n represent the number.

$$\underbrace{\textit{five}}_{5}\ \underbrace{\textit{more than}}_{+}\ \underbrace{\textit{twice some number}}_{2n}\ \underbrace{\textit{is}}_{=}\ \underbrace{\textit{19}}_{19}$$

$$\begin{aligned} 5 - 5 + 2n &= 19 - 5 & \textit{Subtract 5 from each side.} \\ 2n &= 14 & \\ \frac{2n}{2} &= \frac{14}{2} & \textit{Divide each side by 2.} \\ n &= 7 & \text{The number is 7.} \end{aligned}$$

Example 2

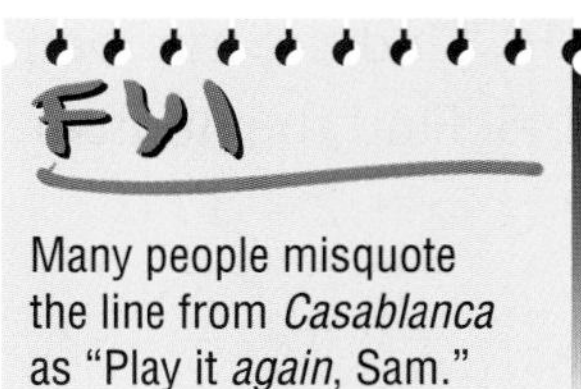

APPLICATION Entertainment

"Play it Sam." is a well known line from the classic film *Casablanca*. Rick Blaine, played by Humphrey Bogart, spoke this line to Sam, played by Dooley Wilson. Humphrey Bogart earned \$36,667 for his role in the movie. This is \$1667 more than ten times the salary Dooley Wilson earned. How much did Mr. Wilson earn for *Casablanca*?

Explore You know Mr. Bogart's salary. You are looking for Mr. Wilson's salary.

Plan Let w represent Mr. Wilson's salary. Write an equation.

Solve *36,667 is 1667 more than ten times Mr. Wilson's salary*

$$36{,}667 = 1667 + 10w$$

$$36{,}667 - 1667 = 1667 - 1667 + 10w \quad \textit{Subtract 1667 from each side.}$$

$$35{,}000 = 10w$$

$$\frac{35{,}000}{10} = \frac{10w}{10} \quad \textit{Divide each side by 10.}$$

$$3500 = w$$

Examine Since $10(3500) + 1667 = 36{,}667$, the answer is reasonable. Mr. Wilson's salary for *Casablanca* was \$3500.

FYI

Many people misquote the line from *Casablanca* as "Play it *again*, Sam."

Checking Your Understanding

Communicating Mathematics

Read and study the lesson to answer these questions.

1. **List** the steps you would use to solve a verbal problem.
2. a. **Pick** any number. Double it and then add 30. Divide the sum by 2. Subtract the original number. What do you get?
   b. Repeat the exercise with two other numbers.
   c. Write an algebraic equation to explain your answers.

Guided Practice

Define a variable and write an equation for each situation. Then solve.

3. If 17 is decreased by twice a number, the result is 5. Find the number.
4. Three times a number plus twice the number plus one is 6. What is the number?
5. **Pets** According to a survey by the Pet Food Institute, about 2682 thousand dogs in the United States can "speak." This is 354 thousand fewer than four times the number of dogs that can "sing."
   a. Let d represent the number of dogs that can "sing." Write an equation for this situation.
   b. How many dogs can "sing"?

Exercises: Practicing and Applying the Concept

Independent Practice

Match each sentence with the equation it represents.

6. Four plus three times a number is 18.
7. Eighteen equals seven plus twice some number.
8. Four times a number minus 15 is 92.
9. Ten minus three times a number is forty-five.

A. $4 + 3x = 18$
B. $4n - 15 = 92$
C. $18 = 7 + 2n$
D. $18 = 7n + 2$
E. $10 - 3x = 45$
F. $15 - 4n = 92$
G. $10x - 3 = 45$

Define a variable and write an equation for each situation. Then solve.

10. Three times a number less four is 17. What is the number?

11. The sum of a number and 6, divided by 7 is 5. Find the number.

12. Twenty more than twice a number is -30. What is the number?

13. The quotient of a number and -4, less 8 is -42. What is the number?

14. The difference between twice a number and 9 is 16. Find the number.

15. The product of 5 and a number is 10 more than 145. Find the number.

16. Write a real-life problem that could be solved using $3x + 2 = 17$.

17. **Geometry** The perimeter of the triangle at the right is 27 yards. What are the lengths of the sides of the triangle?

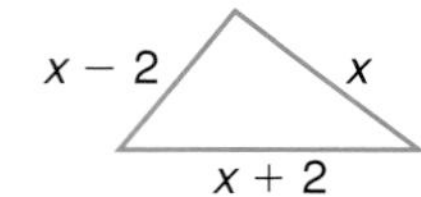

Critical Thinking

18. Numbers that are in order and differ by 1, such as 2, 3, 4, are *consecutive numbers*. If you begin with an even integer and count by two, you are counting *consecutive even integers*. Write an expression for the sum of two consecutive even integers where $2x$ is the lesser integer.

Applications and Problem Solving

19. **Fencing** Wanda used 130 feet of fence to enclose a rectangular flower garden. She also used the 50-foot wall of her house as one side of the garden. What was the width of the garden?

a. Write an equation that represents this situation.

b. Solve the equation to find the width of the garden.

20. **Running** In cross-country, a team's score is the sum of the place numbers of the first five finishers on the team. The co-captains finished first and fourth in a meet. The next three finishers on the team placed in consecutive order. The team score was 35.

a. Write an equation that represents this situation.

b. In what places did the other three members finish?

Mixed Review

21. Solve $8 + \frac{2}{5}n = 28$. (Lesson 7-2)

22. Estimate $188.76 + 14.31 - 106.5$. (Lesson 5-2)

23. Solve $y - 19 = 53$. (Lesson 3-2)

24. **Money** What is the least positive number of coins that is *impossible* to give as change for a dollar? (Lesson 3-1)

25. Name the property shown by $(2 + 5) + c = 2 + (5 + c)$. (Lesson 1-4)

1. Explain why the comic is funny.
2. Use the information in the comic to write an equation. Let y represent the number of years ago.
3. What is the value of y?

7-4 Integration: Geometry
Circles and Circumference

Setting Goals: *In this lesson, you'll find the circumference of a circle.*

Modeling with Manipulatives

MATERIALS

- 4 circular objects
- string
- metric ruler

A **circle** is the set of all points in a plane that are the same distance from a given point in the plane. The given point is called the **center**. The distance from the center to any point on the circle is called the **radius (*r*)**. The distance across the circle through its center is its **diameter (*d*)**. The **circumference (*C*)** of a circle is the the distance around the circle.

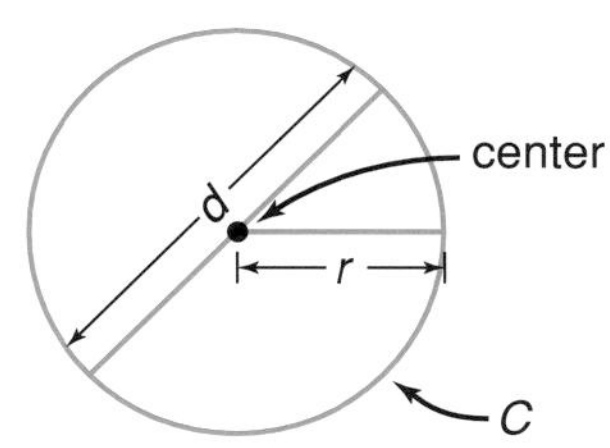

You can measure the circumference of a circle with a metric ruler.

Your Turn

- ▶ Use a metric ruler to measure the diameter of a circular object. Record your findings in a table like the one below.
- ▶ Wrap a string around the circular object once. Mark the string where it meets itself.
- ▶ Lay the string out straight and measure its length with your ruler. Record your finding. This is the circumference of the circular object.
- ▶ Use a calculator to divide the circumference by the diameter. Round to the nearest hundredth and record your finding.
- ▶ Repeat this activity with three other circular objects of various sizes.

THINK ABOUT IT

What is the relationship between the diameter and the radius of a circle?

	Object	Diameter	Circumference	$\frac{\text{Circumference}}{\text{Diameter}}$
#1				
#2				
#3				
#4				

TALK ABOUT IT

a. What is the mean of the quotients in the last column?

b. Press the π key on your calculator. How does your mean compare to this value?

Learning the Concept

The relationship you discovered in the activity above is true for all circles. The circumference of a circle is always 3.1415926... times the diameter. The Greek letter **π (pi)** stands for this number. Although π is not a rational number, the rational numbers 3.14 and $\frac{22}{7}$ are two generally accepted approximations for π.

Circumference	
In words:	**The circumference of a circle is equal to its diameter times π, or 2 times its radius times π.**
In symbols:	**$C = \pi d$ or $C = 2\pi r$**

Example 

Find the circumference of each circle described below.

a. The diameter is 8.25 cm. Use a calculator.

$C = \pi d$

$C = \pi \cdot 8.25$

[π] [×] 8.25 [=] 25.91813939

The circumference is about 25.9 cm.

b. Use $\frac{22}{7}$ for π.

$C = 2\pi r$

$C \approx \frac{2}{1} \cdot \frac{22}{7} \cdot \frac{28}{3}$ (28 and 7 divided by 7: 4 and 1)

$C \approx \frac{176}{3}$ or $58\frac{2}{3}$

The circumference is about $58\frac{2}{3}$ in.

$9\frac{1}{3}$ in.

Connection to Algebra

Sometimes you'll know the circumference of a circle and need to find its radius or diameter. In this case, you can write an equation and solve it.

Example 2

Geography

Earth's circumference is approximately 25,000 miles. If you could dig a tunnel to the center of the Earth, how long would the tunnel be?

Explore You know the circumference of Earth. You want to know the distance to the center, or the radius of Earth.

Plan Replace C with 25,000 and solve $C = 2\pi r$ for r.

Solve $25{,}000 = 2\pi r$

$\frac{25{,}000}{2\pi} = \frac{2\pi r}{2\pi}$ *Divide each side by 2π.*

$\frac{25{,}000}{2\pi} = r$ ***Estimate:*** *24,000 ÷ 6 = 4000*

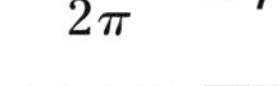

25,000 [÷] [(] 2 [×] [π] [)] [=] 3978.8736

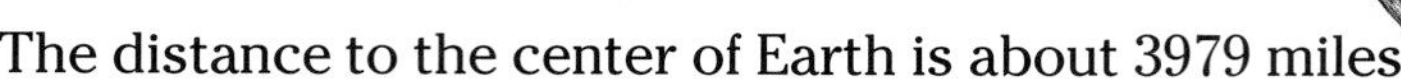

The distance to the center of Earth is about 3979 miles.

Examine The radius is about 4000 and π is about 3. Estimate the circumference by finding their product.

$2 \cdot 3 \cdot 4000 = 24{,}000$

The answer is reasonable.

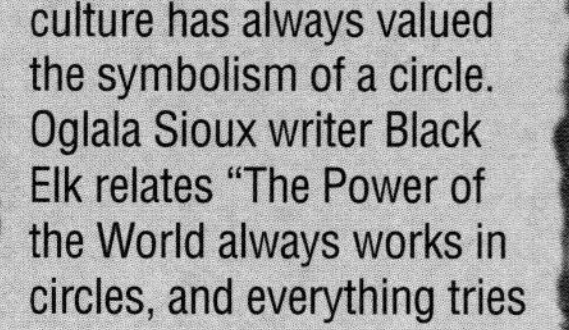

CULTURAL CONNECTIONS

The Native American culture has always valued the symbolism of a circle. Oglala Sioux writer Black Elk relates "The Power of the World always works in circles, and everything tries to be round."

Checking Your Understanding

Communicating Mathematics

Read and study the lesson to answer these questions.

1. **Explain** how to find the radius of a circle if you know its diameter.
2. **Name** two rational numbers that are approximations for π.

3. **Calculate** the circumference of a circle with a radius of 6 inches using the pi key on a calculator and each of the rational approximations.

4. A *chord* of a circle is a segment whose endpoints both lie on the circle. What is the longest chord you can draw?

5. Write a note to someone at home that explains how to estimate the circumference of a circle.

Guided Practice

Find the circumference of each circle described below.

6. 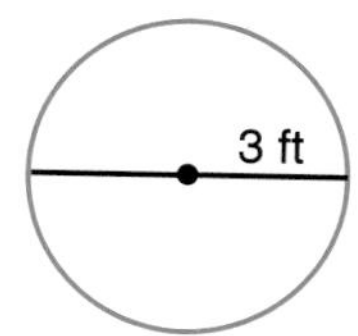

7. 

8. The diameter is $5\frac{1}{4}$ inches.

9. The radius is 1.75 meters.

10. **Recreation** A 10-speed bicycle tire has a diameter of 27 inches. Find the distance the bicycle will travel in 10 rotations of the tire.

Exercises: Practicing and Applying the Concept

Independent Practice

Find the circumference of each circle.

11. 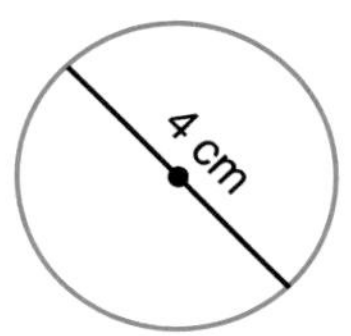

12. 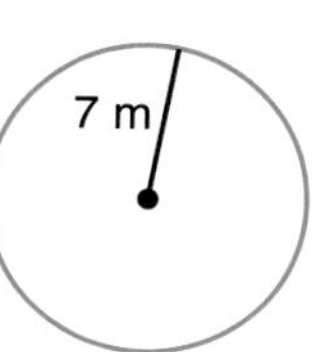

13. 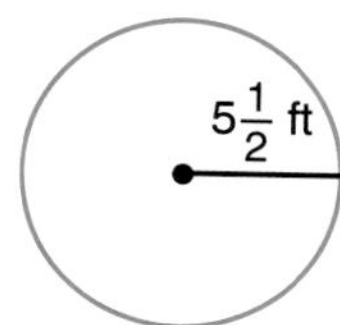

14. 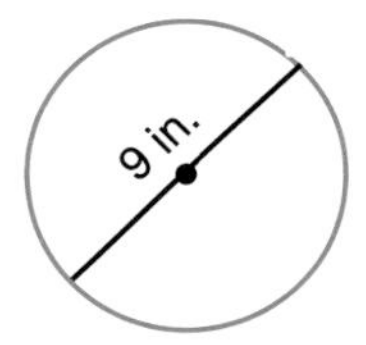

15. 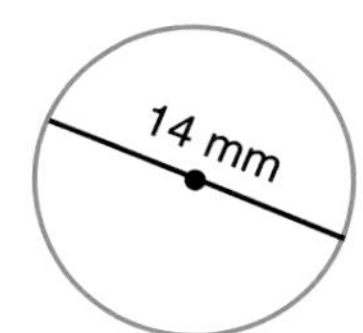

16. 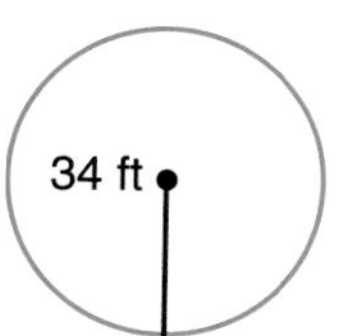

Choose
- Estimation
- Mental Math
- Calculator
- Paper and Pencil

17. The diameter is 18.8 m.

18. The radius is 0.5 meters.

19. The radius is 1.3 yd.

20. The diameter is $13\frac{1}{2}$ in.

21. The diameter is $2\frac{1}{3}$ ft.

22. The radius is $4\frac{1}{2}$ km.

Match each circle described in the column on the left with its corresponding measurement in the column on the right.

23. $C = 628$ cm
24. $r = 30$ cm
25. $2r = 28$ cm
26. $C = 47.728$ cm

A. $d = 15.2$ cm
B. $C = 87.92$ cm
C. $r = 100$ cm
D. $C = 188.4$ cm

27. The circumference of a circle is 4.082 meters. Find its radius.

28. The circumference of a circle is π units. What is the radius of this circle?

29. The radius of a circle is $\frac{7}{22}$ units. Find its circumference.

Critical Thinking

30. A *central angle* of a circle is an angle whose vertex is the center of a circle. A 90° angle intersects an arc of the circle that has a length that is $\frac{1}{4}$ the circumference. Find the length of the arc intersected by a 145° central angle of a circle that has a diameter of 8 feet.

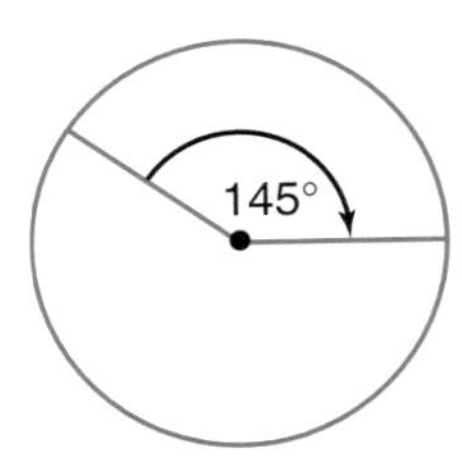

Applications and Problem Solving

31. **Entertainment** The first Ferris wheel, designed by George Ferris, was built in 1893 in Paris, France. The diameter of that Ferris wheel was 76 meters. What distance would you travel in one revolution if you rode on the first Ferris wheel?

32. **Automotive** The diameter of the wheels on many passenger cars is 14 inches. All four tires are inflated so that the weight of the car does not distort the shape of the tires. The cross section height of each 14-inch tire is 5.4 inches

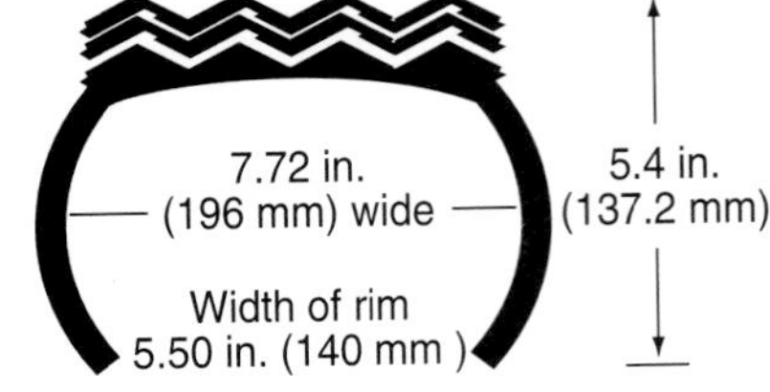

a. How far does a car with 14-inch tires travel in one complete turn of the tires?

b. How many turns of the tires would it take to travel 1 mile?

c. A well-cared-for tire will usually be able to travel about 50,000 miles before it wears out. How many turns of the tire would it take to travel 50,000 miles?

33. **Geometry** Each side of a square is 7 cm long. A circle has a diameter that is 7 cm long.

a. Which is longer, the perimeter of the square or the circumference of the circle?

b. Justify your answer to part a without doing any computation.

For the latest calculation of π, visit: **www.glencoe.com/sec/math/prealg/mathnet**

34. **Research** Mathematicians have worked for centuries to find an accurate value for π. Investigate the methods that have been used. Who was the first to calculate a value of π? How many decimal places are known now?

Mixed Review

35. **Tourism** The Louvre Museum, home to the *Mona Lisa*, is one of the most-visited museums in the world. The National Gallery of Art in Washington, D.C. is the most-visited art museum in the United States with 7.5 million visitors each year. This is 0.1 million more than twice what the second most visited museum, the Metropolitan Museum of Art, in New York, receives. How many visitors go to the Metropolitan Museum each year? (Lesson 7-3)

36. Solve $\frac{x}{-1.7} \geq 6.8$. (Lesson 6-7)

37. Factor $220pq^2$ completely. (Lesson 4-4)

38. Solve $-4 + m \geq -3$ and graph the solution. (Lesson 3-6)

39. Replace ● with <, >, or = to make $|-13|$ ● $|7|$ a true sentence. (Lesson 2-3)

40. Simplify $18z + 7(2 + 3z)$. (Lesson 1-5)

7-5A

HANDS-ON ACTIVITY

Equations with Variables on Each Side

A Preview of Lesson **7-5**

In this activity, you will use cups and counters to model and solve equations with variables on each side.

Activity

MATERIALS

cups

counters

equation mat

Model and solve the equation $2x - 4 = x - 2$.

- Place 2 cups and 4 negative counters on the left side of the mat. Then place 1 cup and 2 negative counters on the right side of the mat.

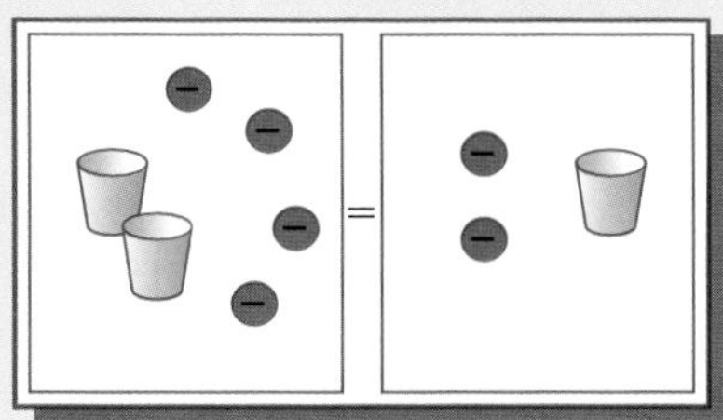

- Since the two sides of the mat represent equal quantities, you can remove an equal number of cups or counters from each side without changing the value of the equation.

- Now the equation is $x - 2 = 0$. Since there are no counters on the right, add 2 zero pairs. Then you can remove 2 negative counters from each side.

Therefore, each cup must contain 2 counters and $x = 2$.

Your Turn

Model and solve each equation.

1. $2x + 3 = x - 5$ **2.** $2x + 3 = x + 1$ **3.** $3x - 2 = x + 6$

4. $3x - 7 = x + 1$ **5.** $6 + x = 5x + 2$ **6.** $x - 1 = 3x + 7$

7. Does it matter whether you remove cups or counters first? Explain.

8. Explain how you could use models to solve $-2x + 3 = -x - 5$.

7-5 Solving Equations with Variables on Each Side

Setting Goals: *In this lesson, you'll solve equations with the variable on each side.*

Modeling with Technology

Most countries use the Celsius scale to measure temperature. In the United States, we use the Fahrenheit scale most of the time. There is one temperature which is the same on both scales. You can use the T1-82 graphing calculator program at the right to find that temperature. In this program, C represents Celsius temperature, and F represents Fahrenheit temperature.

```
PROGRAM: TEMPS
: Lbl 1
: Input "ENTER F",
  F
: (5/9)*(F−32) → C
: If F≠C
: Then
: Disp "C=", C,
  "TRY AGAIN"
: Goto 1
: End
: Disp "CORRECT,
  F=C=", C
: Stop
```

Your Turn

- ▶ Enter the program into the calculator's memory.
- ▶ Run the program.

Enter a guess at the Fahrenheit temperature that has the same value on the Celsius scale. Continue guessing and checking until you determine the temperature.

a. What temperature is the same on both scales?

b. Explain what the formula in the third line of the program does.

Learning the Concept

The formula, $F = \frac{9}{5}C + 32$, is used for finding the Fahrenheit temperature when a Celsius temperature is known. When the temperature is the same on both scales, $F = C$. Replacing F with C results in the equation $C = \frac{9}{5}C + 32$.

In $C = \frac{9}{5}C + 32$, the variable is on each side of the equal sign. Use the properties of equality to eliminate the variable from one side. Then solve the equation.

$$C = \frac{9}{5}C + 32$$

$$C - \frac{9}{5}C = \frac{9}{5}C - \frac{9}{5}C + 32 \quad \text{Subtract } \tfrac{9}{5}C \text{ from each side.}$$

$$\frac{5}{5}C - \frac{9}{5}C = 32 \quad \text{Rename } C \text{ as } \tfrac{5}{5}C. \text{ Why?}$$

$$-\frac{4}{5}C = 32$$

$$\left(-\frac{5}{4}\right)\left(-\frac{4}{5}\right)C = \left(-\frac{5}{4}\right) \cdot \frac{32}{1} \quad \text{Multiply each side by } -\tfrac{5}{4}. \text{ Why?}$$

$$C = -40$$

Check: $C = \frac{9}{5}C + 32$

$$-40 \stackrel{?}{=} \frac{9}{5}(-40) + 32 \quad \textit{Replace C with } -40.$$
$$-40 \stackrel{?}{=} -72 + 32$$
$$-40 = -40 \checkmark$$

The temperature is the same on the Celsius and Fahrenheit scales at $-40°$.

Example 1 Solve $7n + 12 = 2.5n - 2$.

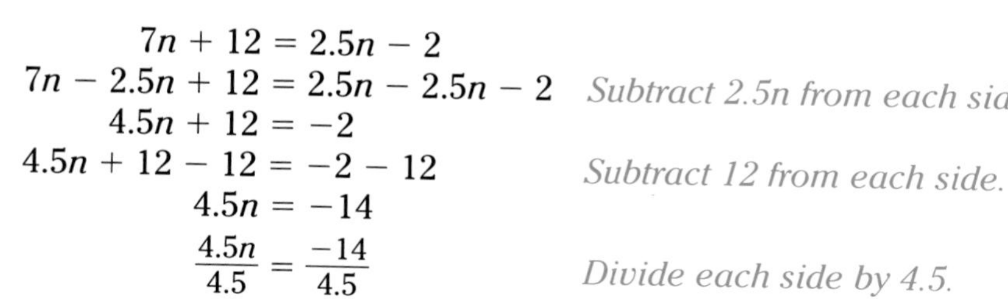

$$7n + 12 = 2.5n - 2$$
$$7n - 2.5n + 12 = 2.5n - 2.5n - 2 \quad \textit{Subtract 2.5n from each side.}$$
$$4.5n + 12 = -2$$
$$4.5n + 12 - 12 = -2 - 12 \quad \textit{Subtract 12 from each side.}$$
$$4.5n = -14$$
$$\frac{4.5n}{4.5} = \frac{-14}{4.5} \quad \textit{Divide each side by 4.5.}$$

14 [(-)] [÷] 4.5 [=] -3.11111

$$n = -3.\overline{1}$$

Sometimes a rectangle is described in terms of only one of its dimension. To find the dimensions, you will have to solve an equation that contains grouping symbols. When solving equations of this type, first use the distributive property to remove the grouping symbols.

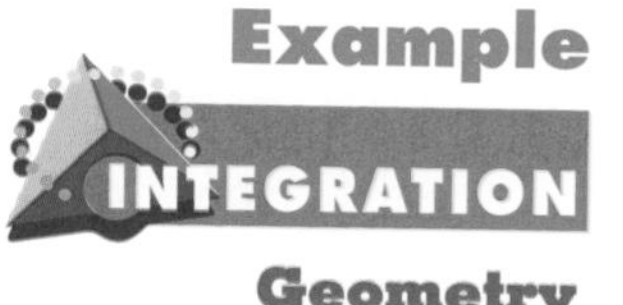

Example 2

Geometry

The perimeter of a rectangle is 74 inches. Find the dimensions if the length is 7 inches greater than twice the width.

Let w represent the width in inches. Then $2w + 7$ represents the length in inches. Recall that the formula for perimeter can be represented as $2w + 2\ell = P$.

w

2w + 7

$$2w + 2\ell = P$$
$$2w + 2(2w + 7) = 74 \quad \textit{Replace } \ell \textit{ with } 2w + 7 \textit{ and P with 74.}$$
$$2w + 4w + 14 = 74 \quad \textit{Use the distributive property.}$$
$$6w + 14 = 74 \quad \textit{Simplify.}$$
$$6w + 14 - 14 = 74 - 14 \quad \textit{Subtract 14 from each side.}$$
$$6w = 60$$
$$\frac{6w}{6} = \frac{60}{6} \quad \textit{Divide each side by 6.}$$
$$w = 10$$

The width is 10 inches. Now evaluate $2w + 7$ to find the length.

$2w + 7 = 2(10) + 7$ or 27

The length is 27 inches. *Check to see if this length and width results in the correct perimeter.*

Some equations may have *no* solution. The solution set is the **null** or **empty set**. It is shown by the symbol { } or $\varnothing$. Other equations may always be true and have every number in their solution set.

Example 3 **Solve each equation.**

a. $\mathbf{2x + 5 = 2x - 3}$

$$\begin{aligned} 2x + 5 &= 2x - 3 \\ 2x - 2x + 5 &= 2x - 2x - 3 \quad \textit{Subtract 2x from each side.} \\ 5 &= 3 \end{aligned}$$

This sentence is *never* true. The solution is ∅.

b. $\mathbf{3(x + 1) - 5 = 3x - 2}$

$$\begin{aligned} 3(x + 1) - 5 &= 3x - 2 \\ 3(x + 1) - 3 &= 3x \quad \textit{Add 2 to each side.} \\ 3x + 3 - 3 &= 3x \quad \textit{Distributive property} \\ 3x &= 3x \quad 3 - 3 = 0 \\ \frac{3x}{3} &= \frac{3x}{3} \quad \textit{Divide each side by 3.} \\ x &= x \end{aligned}$$

This sentence is *always* true. The solution set is all numbers.

Checking Your Understanding

Communicating Mathematics

MODELING MATHEMATICS

MATERIALS

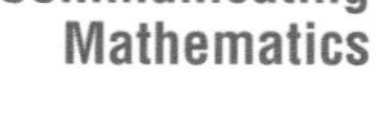

cups and counters

equation mat

Read and study the lesson to answer these questions.

1. **Describe** the steps you would take to solve $2p = 15 + 7p$.
2. **Name** the property of equality that allows you to subtract the same term from each side of an equation.
3. The diagram at the right models the equation $2x - 1 = x - 4$.

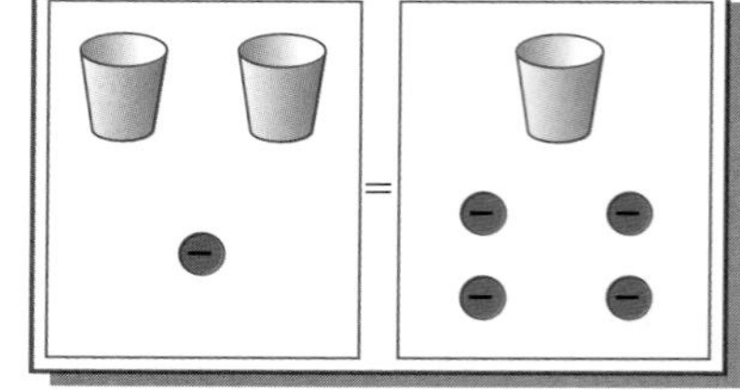

   a. Solve $2x - 1 = x - 4$.
   b. Solve $3x - 1 = x + 3$ using cups and counters.
   c. Solve $x + 5 = x + 7$ using cups and counters.

Guided Practice

Name the first two steps you should take to solve each equation. Then solve. Check your solution.

4. $12k + 15 = 35 + 2k$
5. $3x + 2 = 4x - 1$
6. $3b + 8 = -10b + b$
7. $3(a + 22) = 12a + 30$
8. $3k + 10 = 2k - 21$
9. $n + 4 = -n + 10$
10. **Geometry** The perimeter of a rectangle is 32 feet. Find the dimensions if the length is 4 feet longer than three times the width.

Exercises: Practicing and Applying the Concept

Independent Practice

Solve each equation. Check your solution.

11. $6n - 42 = 4n$
12. $8 - 3g = -2 + 2g$
13. $3(k + 2) = 12$
14. $\frac{4}{7}y - 8 = \frac{2}{7}y + 10$
15. $6 - 8x = 20x + 20$
16. $13 - t = -t + 7$
17. $3n + 7 = 7n - 13$
18. $7b - 3 = -b + 4$
19. $2 + 7(d + 1) = 9 + 7d$
20. $2a - 1 = 3.5a - 3$
21. $5m + 4 = 7(m + 1) - 2m$
22. $12x - 24 = -14x + 28$

23. $2(f - 3) + 5 = 3(f - 1)$

24. $4[z + 3(z - 1)] = 36$

25. $-3(4b - 10) = \frac{1}{2}(-24b + 60)$

26. $\frac{3}{4}a + 16 = 2 - \frac{1}{8}a$

27. $\frac{d}{0.4} = 2d + 1.24$

28. $\frac{a - 6}{12} = \frac{a - 2}{4}$

Find the dimensions of each rectangle. The perimeter is given.

29. $P = 460$ m

w

$w + 30$

30. $P = 440$ yd

w

$3w - 60$

31. $P = 110$ ft

w

$2w - 20$

Define a variable and write an equation for each situation. Then solve.

32. Twice a number is 220 less than six times the number. What is the number?

33. Fourteen less than three times a number equals the number. What is the number?

Critical Thinking

34. An apple costs the same as 2 oranges. Together, an orange and a banana cost 10¢ more than an apple. Two oranges cost 15¢ more than a banana. What is the cost for one of each fruit?

Applications and Problem Solving

35. **Census** The table at the right shows the 1990 populations and the average rates of change in population in the 1980s for Shreveport, Louisiana and Orlando, Florida. Suppose the population of each city continued to increase or decrease at these rates.

City	Shreveport, LA	Orlando, FL
1990 Population	199,000	165,000
Annual Population Change in 1980s	−700	+3700

a. Write an expression for the population of Shreveport after x years.

b. Write an expression for the population of Orlando after x years.

c. Use the expressions in parts a and b to write an equation to find the number of years until the populations are the same.

d. In how many years would the population of the two cities be the same?

36. **Cooking** Water boils at 212°F at sea level. However, as the altitude increases, the air pressure drops. The lower air pressure causes water to boil at a lower temperature. For every 550 feet above sea level, the boiling point is lowered 1°F. The boiling point rises 1°F for every 550 feet below sea level.

a. Write a formula for finding the boiling point b of water at any altitude A above sea level.

b. Denver is 5280 feet above sea level. If you were boiling an egg in Denver, how hot would the water be?

c. Find the temperature at which water would boil in Death Valley, which is 282 feet *below* sea level.

37. **Sales** Shoe World offers Veronica a temporary job during her spring break. The manager gives her a choice as to how she wants to be paid, but she must decide before she starts working.
Plan 1 \$2 an hour plus 10¢ for every dollar's worth of shoes she sells.
Plan 2 \$3 an hour plus 5¢ for every dollar's worth of shoes she sells.
    a. How much would Veronica's sales need to be in one hour to earn the same amount in either plan?
    b. If you were in Veronica's place, which method of payment would you choose? Explain your reasoning.

Mixed Review

38. **Life Science** Sequoias are the largest trees on Earth. Botanists can estimate the age of a tree based on its diameter. If the circumference of a tree is 18 inches, what is its diameter? (Lesson 7-4)

39. Solve $3x + 3 = 99$. (Lesson 7-2)
40. Solve $15 \div \frac{3}{7} = n$. (Lesson 6-4)
41. Write $\frac{36}{126}$ in simplest form. (Lesson 4-5)
42. Solve $A = \ell w$ for w if $A = 65$ and $\ell = 13$. (Lesson 3-4)
43. Solve $n - 10 = 42$ by using the inverse operation. (Lesson 1-8)
44. Solve $7m = 56$ mentally. (Lesson 1-6)

Self Test

1. **Banking** Dorinda is balancing her checkbook. Since the bank statement was mailed to her, she has written checks for \$13.85, \$19.72, and \$59.66 and made a deposit in the amount of \$497.92. If the balance she has written in her check register is \$973.88, how much should the bank statement say is in her account? (Lesson 7-1)

Solve each equation. (Lesson 7-2)

2. $12 - z = 28$
3. $9 - 4z = 57$
4. $\frac{d - 5}{7} = 14$

Define a variable and write an equation for each situation. Then solve. (Lesson 7-3)

5. Four times a number less twelve is 18. What is the number?
6. The product of a number and 6 is eight greater than 10.
7. The circumference of a circle is 6.398 yards. Find its radius. (Lesson 7-4)
8. The diameter of a circle is 16.2 inches. What is the circumference of the circle? (Lesson 7-4)
9. **Sports** A soccer field is 75 yards shorter than 3 times its width. If the perimeter is 370 yards, find the dimensions of the soccer field. (Lesson 7-5)

7-6 Solving Multi-Step Inequalities

Setting Goals: *In this lesson, you'll solve inequalities that involve more than one operation.*

Modeling with Manipulatives

MATERIALS

cups

counters

The diagram at the right models the inequality $4 + 2x > x + (-2)$. Notice that the side that is lower has the heavier weight. To solve an inequality, use the same process used for equations.

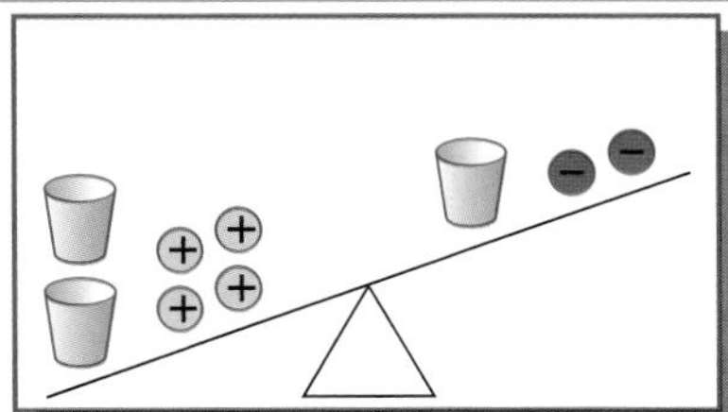

Your Turn

- Draw a diagram of the unbalanced scale on a sheet of paper and model the diagram shown above.
- Remove one cup from each side. *Perform operations on each side.*
- Place four zero pairs on the right side.
- Remove four positive counters from each side.

TALK ABOUT IT

a. State the inequality after removing one cup from each side.

b. Why did you have to place four zero pairs on the right side?

c. What is the solution to the inequality?

d. How could you check the solution?

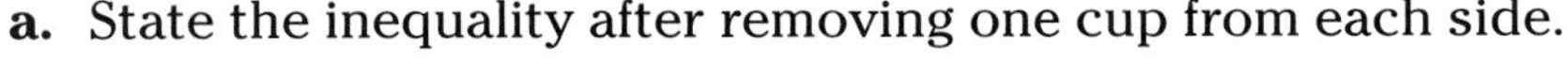

Learning the Concept

Inequalities can be solved by applying the methods for solving equations. Remember that if each side of an inequality is multiplied or divided by a negative number, the sign of the inequality should be reversed.

Example 1 **Solve each inequality and check the solution. Graph the solution on a number line.**

a. $\mathbf{7x + 29 > 15}$

$$7x + 29 > 15$$
$$7x + 29 - 29 > 15 - 29 \quad \textit{Subtract 29 from each side.}$$
$$7x > -14$$
$$\frac{7x}{7} > \frac{-14}{7} \quad \textit{Divide each side by 7.}$$
$$x > -2$$

Any number greater than -2 is a solution.

Check: Try 0, a number greater than -2.

$$7(0) + 29 \overset{?}{>} 15 \quad \textit{Replace x with 0.}$$
$$0 + 29 \overset{?}{>} 15$$
$$29 > 15 \ ✓$$

-5 -4 -3 -2 -1 0 1 2 3 4

b. $13 - 6n \le 49 - 2n$

THINK ABOUT IT

Why was the order symbol reversed?

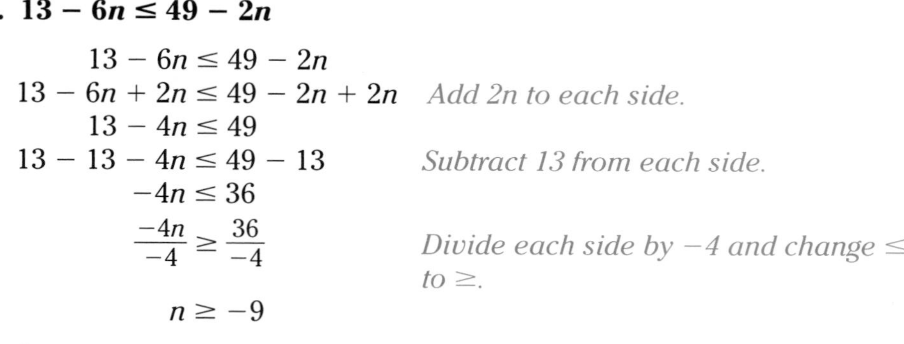

$$
\begin{aligned}
13 - 6n &\le 49 - 2n \\
13 - 6n + 2n &\le 49 - 2n + 2n && \textit{Add 2n to each side.} \\
13 - 4n &\le 49 \\
13 - 13 - 4n &\le 49 - 13 && \textit{Subtract 13 from each side.} \\
-4n &\le 36 \\
\frac{-4n}{-4} &\ge \frac{36}{-4} && \textit{Divide each side by } -4 \textit{ and change } \le \textit{ to } \ge. \\
n &\ge -9
\end{aligned}
$$

Any number greater than or equal to -9 is a solution.

Check: Try -5, a number greater than -9.

$$
\begin{aligned}
13 - 6n &\le 49 - 2n \\
13 - 6(-5) &\stackrel{?}{\le} 49 - 2(-5) && \textit{Replace n with } -5. \\
13 + 30 &\stackrel{?}{\le} 49 + 10 \\
43 &\le 59 \;\checkmark
\end{aligned}
$$

-10 -8 -6 -4 -2 0 2 4

When solving inequalities that contain grouping symbols, remember to first use the distributive property to remove the grouping symbols.

Example 2 APPLICATION Tourism

Route 1A along the Maine coast is called one of the most scenic highways in America. Every year, thousands of people go there to see the ocean and the charming New England villages. Suppose you rented a car in Portland for \$139.95 plus \$0.25 per mile for each mile over 100 miles. If you had budgeted up to \$200 for car rental in your vacation budget, how many miles could you drive and stay within the budget?

Let m represent the number of miles driven.

\$139.95 *plus* *\$0.25 per mile over 100* *is no more than* *\$200*

$139.95 \quad + \quad 0.25(m - 100) \quad \le \quad 200$

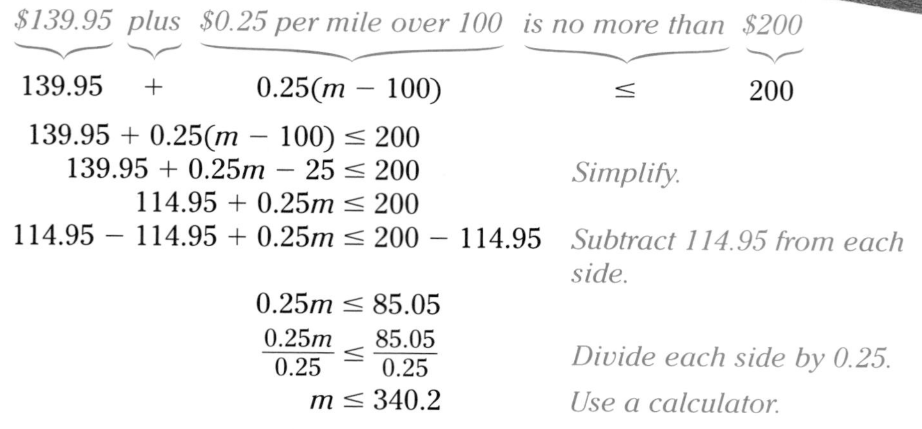

$$
\begin{aligned}
139.95 + 0.25(m - 100) &\le 200 \\
139.95 + 0.25m - 25 &\le 200 && \textit{Simplify.} \\
114.95 + 0.25m &\le 200 \\
114.95 - 114.95 + 0.25m &\le 200 - 114.95 && \textit{Subtract 114.95 from each side.} \\
0.25m &\le 85.05 \\
\frac{0.25m}{0.25} &\le \frac{85.05}{0.25} && \textit{Divide each side by 0.25.} \\
m &\le 340.2 && \textit{Use a calculator.}
\end{aligned}
$$

You could drive up to 340.2 miles and stay within your budget.

Checking Your Understanding

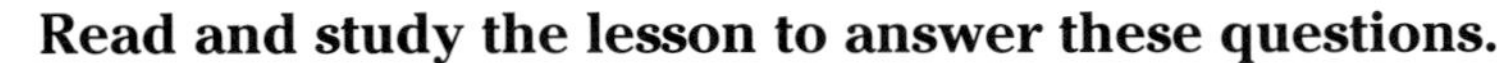

Communicating Mathematics

Read and study the lesson to answer these questions.

1. **Explain** how to check the solution for Example 2.
2. **You Decide** Tinesha says that to solve the inequality $2(2y + 3) > y + 1$, the first step should be to divide each side by 2. Kiki says the first step should be to subtract 3 from each side. Who is correct, and why?

MATERIALS

cups and counters

3. **Draw a model** that you could use to solve the inequality $2x + 3 < 3x - 7$.
4. Write an inequality for the model at the right. Then use cups and counters to solve the inequality.

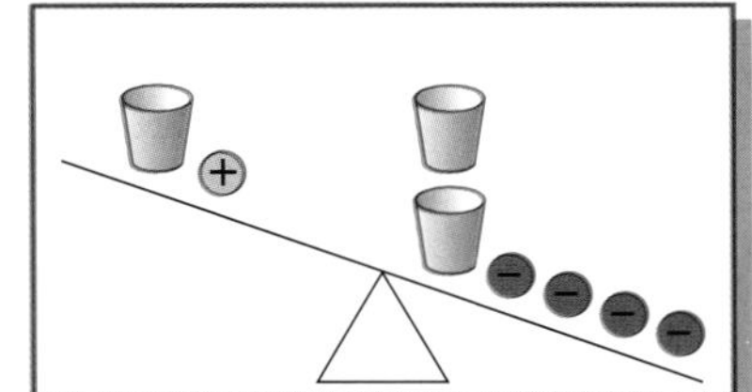

Guided Practice

Solve each inequality and check your solution. Graph the solution on a number line.

5. $3x + 2 \leq 23$
6. $18 - 2v < 16$
7. $2k + 7 > 13 - k$
8. $y + 1 \geq 5y + 5$
9. $-3(m - 2) > 12$
10. $-5 \leq \frac{x}{4} - 7$
11. **Consumer Awareness** Heather has \$90 that relatives gave her for her birthday. She needs to buy a new pair of shoes and would also like to buy two CDs. The shoes she likes cost \$65. Use the inequality $2d + 65 \leq 90$ to determine how much she can spend on each CD.

Exercises: Practicing and Applying the Concept

Independent Practice

Solve each inequality and check your solution. Graph the solution on a number line.

12. $2x + 9 > 25$
13. $-2u + 3 \leq 11$
14. $16 < 18 - 2n$
15. $2x - 5 < 2x - 9$
16. $1.2z - 2 > 7 + 0.9z$
17. $3(j - 1) \geq -12$
18. $\frac{m}{3} - 7 > 11$
19. $36 + \frac{k}{5} \geq 51$
20. $\frac{2x}{9} - 2 < -4$
21. $0.47 < \frac{t}{-9} + 0.6$
22. $-4.4 > \frac{b}{-5} - 4.8$
23. $12 - \frac{5z}{4} < 37$
24. $\frac{n - 11}{2} \leq -6$
25. $1.3x + 6.7 \geq 3.1x - 1.4$
26. $-5x + 3 < 3x + 23$
27. $-5(k + 4) \geq 3(k - 4)$
28. $8c - (c - 5) > c + 17$
29. $\frac{c + 8}{4} < \frac{5 - c}{9}$
30. $\frac{2(n + 1)}{7} \geq \frac{n + 4}{5}$

Critical Thinking

31. The sum of three times a number and 5 lies between -10 and 8. Solve the *compound inequality* $-10 < 3x + 5 < 8$ to find the solution(s). (*Hint*: Any operation must be done to every part of the inequality.)

Applications and Problem Solving

32. **Geometry** The area of trapezoid $ABCD$ is at least 320 square meters. The height of the trapezoid is 16 meters and the length of one of the bases is 15 meters. Use the inequality $16\left(\frac{15 + b}{2}\right) \geq 320$ to find the length of the other base, b.

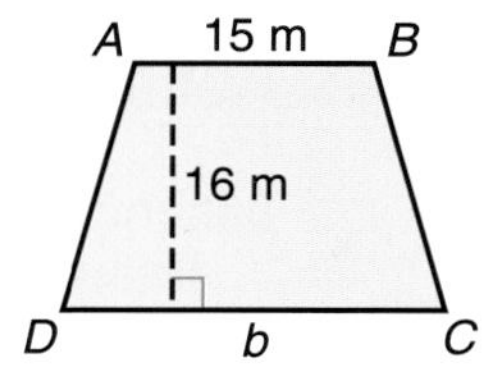

33. **Statistics** Odina's scores on the first three of four 100-point tests were 89, 92, and 82. Use the inequality $\frac{89 + 92 + 82 + s}{4} \geq 90$ to determine what scores she could receive on the fourth test and have an average of at least 90 for all the tests.

Mixed Review

34. Solve $-75 - m = 25 + 4m$. (Lesson 7-5)
35. **Patterns** Duncan received a chain letter that said he should copy the letter and send it to six friends in order to have good luck. Each of Duncan's friends sent copies to six other friends, who each sent copies to six other friends. How many people have received the chain letter after Duncan? (Lesson 6-8)
36. Simplify $7.5x - 31 + 13.78x$. (Lesson 5-3)
37. Find the missing exponent $\frac{t^{17}}{t^{\bullet}} = t^9$. (Lesson 4-8)
38. Find the value of $17 + -13 + 8$. (Lesson 2-4)
39. Translate *3 more days of summer vacation than Northridge High School* into an algebraic expression. (Lesson 1-3)

COOPERATIVE LEARNING PROJECT

THE SHAPE OF THINGS TO COME

Technology in Sports

The Olympic motto is "Faster – Higher – Stronger." Thanks to technology, athletes are accomplishing things they never dreamed were possible before.

Divers will soon be reaching new heights with diving boards that provide more spring than those of the past. Baseball players and golfers can practice their swings inside with electronic sensors that analyze their strength and accuracy. Pole vaulters will use poles made specifically for their height, weight, speed, and hold technique. And cyclists will speed by on bicycles with less drag and more stability.

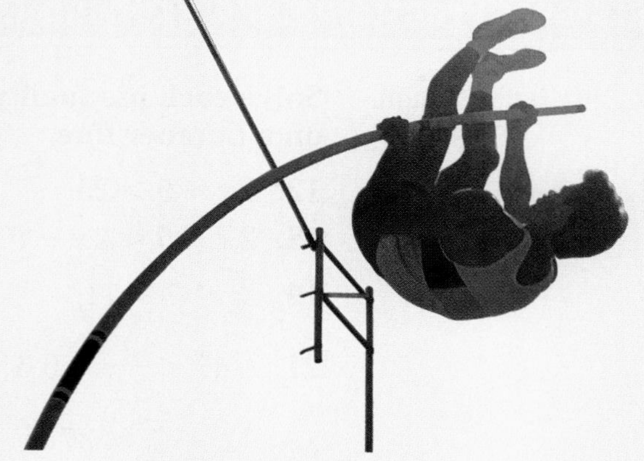

See for Yourself

- Many of the advances in sports technology have to do with making something more *aerodynamic*. Investigate what it means for something to be aerodynamic.
- In the 1980s, advances in aerodynamic technology made javelins *too* good. Research the situation. What was the final outcome?
- Choose a sport and investigate the changes that technological developments have made and will soon make in the sport.

7-7 Writing Inequalities

Setting Goals: *In this lesson, you'll solve verbal problems by writing and solving inequalities.*

Modeling a Real-World Application: Backpacking

Ahh, the great outdoors. Nearly 10 million Americans hit the trail to go backpacking each year. State and national parks are among the most popular getaways. *Women's Sports and Fitness* magazine says that three times the weight of your backpack and its contents should be less than your body weight in order for you to avoid an injury. If you weigh 120 pounds and your empty backpack weighs 15 pounds, how much should the contents of your backpack weigh? *This problem will be solved in Example 2.*

Learning the Concept

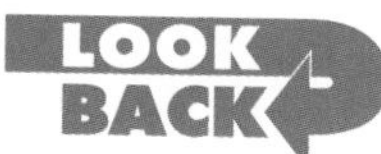

You can review writing simple inequalities in Lesson 1-9.

Verbal problems with phrases like *greater than* or *less than* describe inequalities. Other phrases that suggest inequalities are *at least*, *at most*, and *between*. Inequalities that correspond to each statement are below.

x is *at least* 3.	$x \geq 3$	*At least 3 means 3 or greater.*
y is *at most* 9.	$y \leq 9$	*At most 9 means 9 or less.*
z is *between* 2 and 4.	$2 < z < 4$	*Between 2 and 4 means greater than 2 and less than 4.*

Example 

Three times a number increased by eight is at least twenty-five. What is the number?

Let x represent the number. Write an inequality.

three times a number	*increased by*	*eight*	*is at least*	*twenty-five*
$3x$	$+$	8	$\geq$	25

$$3x + 8 \geq 25$$

$$3x + 8 - 8 \geq 25 - 8 \quad \textit{Subtract 8 from each side.}$$

$$3x \geq 17$$

$$\frac{3x}{3} \geq \frac{17}{3} \quad \textit{Divide each side by 3.}$$

$$x \geq \frac{17}{3} \text{ or } 5\frac{2}{3}$$

In many real-life problems, you will use a multi-step inequality to solve.

Example 2

Backpacking

Refer to the application at the beginning of the lesson. How much should the contents of your backpack weigh?

Explore You want to find what the backpack contents should weigh.

Plan Let c represent the weight of the backpack contents.

(continued on the next page)

Solve

$$\underbrace{3}_{\text{three times}}\left(\underbrace{15}_{\text{weight of backpack}} \underbrace{+}_{\text{plus}} \underbrace{c}_{\text{weight of contents}}\right) \underbrace{<}_{\text{is less than}} \underbrace{120}_{\text{your weight}}$$

$$3(15 + c) < 120$$

$$\frac{3(15 + c)}{3} < \frac{120}{3} \quad \textit{Divide each side by 3.}$$

$$15 + c < 40$$

$$15 - 15 + c < 40 - 15 \quad \textit{Subtract 15 from each side.}$$

$$c < 25$$

The backpack contents should weigh less than 25 pounds.

Examine Suppose the contents of your backpack weigh 20 pounds, a number less than 25. The backpack and its contents would weigh 20 + 15 or 35 pounds. 3(35) = 105 and 105 < 120, so this would be a safe weight to carry.

Checking Your Understanding

Communicating Mathematics

Read and study the lesson to answer these questions.

1. a. **Write** an inequality for *six times a number decreased by four is at most 41.*
   b. **Explain** how you would know if $4\frac{1}{2}$ is a member of the solution set to the inequality you wrote for part a.
2. **Write** an inequality for *a number is between −7 and −9.*
3. **Assess Yourself** Think of a time when you needed to use an inequality. Write an inequality about the situation.

Guided Practice

Define a variable and write an inequality. Then solve.

4. Four less than a number is at most 10. What is the number?
5. Twice a number decreased by 9 is greater than 11. Find the number.
6. You earn $2.00 for every magazine subscription you sell plus a salary of $10 a week. How many subscriptions do you need to sell each week to earn at least $40 a week?
7. You buy some candy bars at 55¢ each and one newspaper for 35¢. How many candy bars can you buy if you only have $2?
8. **Sports** Gao Min represents China in springboard diving. Suppose Gao scored 68.2, 68.9, 67.5, and 71.7 for her first four dives and has one more dive remaining. If the diver in first place has a score of 345.4, what must Gao receive on her fifth dive to overtake her?

Exercises: Practicing and Applying the Concept

Independent Practice

Define a variable and write an inequality. Then solve.

9. Four times a number increased by four is at least 16. Find the number.
10. Nine less than six times a number is at most 33. What is the number?

Write a problem that could be solved using each inequality.

11. $f + 115 \geq 260$
12. $\frac{87 + 92 + 89 + 97 + s}{5} \geq 90$

13. You can hike along the Appalachian Trail at 3 miles per hour. You will stop for one hour for lunch. You want to walk at least 18 miles. How many hours should you expect to spend on the trail?

14. Twelve times a number decreased by $\frac{1}{20}$ of the number is less than 3250. What is the number?

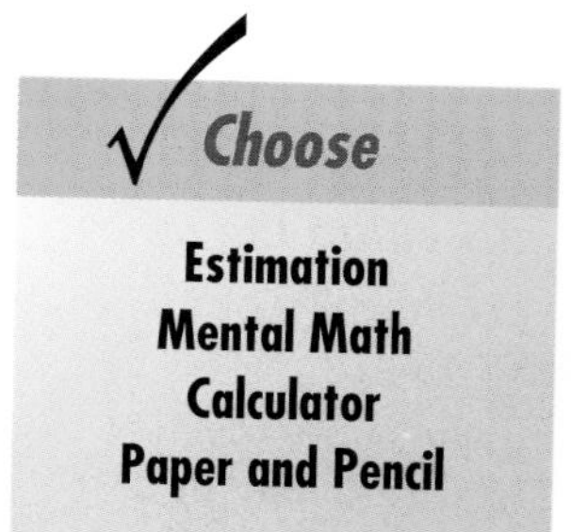

15. The sum of an integer and the next greater integer is at least 35. Find the least pair of such integers.

16. The sum of an odd integer and the next greater odd integer is more than 136. Find the least pair of such integers.

17. Three times a number t exceeds six by at least five. What are the possible values of t?

18. A number is less than the sum of twice its opposite and five. Find the number.

19. If 12.15 times an integer is increased by 9.348, the result is between 39 and 75. Find the integer.

Critical Thinking

20. A *compound inequality* is formed when two inequalities are connected by *and* or *or*. A compound inequality containing *and* is true only if *both* inequalities are true.
    a. Write an inequality equivalent to $2x - 5 < 3$ and $x + 5 > 4$ without using *and*.
    b. **Geometry** An obtuse angle is one that measures more than 90° but less than 180°. Write an inequality to represent the possible measures of angle A if you know that angle A is obtuse.

Applications and Problem Solving

21. **Geometry** The sum of the lengths of any two sides of a triangle is greater than the length of the third side. Triangle ABC has side lengths of 22 inches and 28 inches. Write a compound inequality to describe the length of the third side of triangle ABC.

22. **Agriculture** Rockwell International and John Deere Precision Farming Group are working on a system that uses satellites to help dispense the correct amount of fertilizer on crops. The basic equipment costs about $7000. Annual fees for signal reception will be about $1200.
    a. How much additional revenue would the technology have to generate for a farmer in ten years for the investment to have been a good one?
    b. One farm expert estimated that the system would produce an additional $100 per acre each year on a certain crop. How large a farm should a farmer have in order to expect to make money over a five-year period using the system?

Mixed Review

23. Solve $\frac{2}{5}x - 1 > -3$. (Lesson 7-6)

24. Solve $\frac{20}{27} - \frac{11}{27} = b$. (Lesson 5-4)

25. **Science** Neutrons and protons weigh about 1.7×10^{-24} grams each. (Lesson 4-9)
    a. Write the weight of a proton or neutron using positive exponents.
    b. Write the weight of a proton or neutron as a decimal.

26. Solve $6 \leq \frac{n}{-7}$. (Lesson 3-7)

27. Find the product of -14 and $9b$. (Lesson 2-7)

7-8 Integration: Measurement
Using the Metric System

Setting Goals: *In this lesson, you'll convert measurements within the metric system.*

Modeling a Real-World Application: Travel

You're finally here! Your dream vacation abroad. Noticing that the gas gauge is low, you pull into a gas station to buy some fuel. As you finish filling the tank, you notice that the pump stopped at 41.6!

How can the gas tank in your little rental car hold 41.6 gallons? It doesn't! The tank holds 41.6 *liters*. Most countries other than the United States use the **metric system of measurement**. They use liters to measure capacity instead of gallons.

Learning the Concept

HELP WANTED

The metric system was developed so that scientists in different countries could communicate. For more information on science-related careers, contact:

National Science Teachers Association
1840 Wilson Blvd.
Arlington, VA 22201

The metric system was developed by a group of French scientists in 1795. The **meter (m)** was defined as $\frac{1}{10,000,000}$ of the distance between the North Pole and the Equator. All units of length in the metric system are defined in terms of the meter. A prefix is added to indicate the decimal place-value position of the measurement. The chart below shows the relationships between the prefixes and the decimal place-value positions.

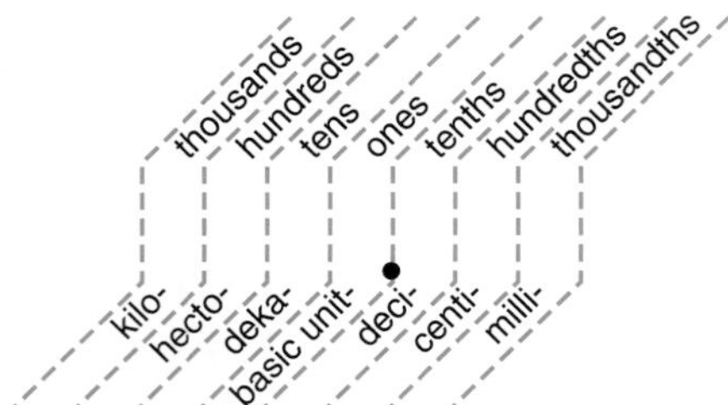

Note that each place value is 10 times the place value to its right.

Note that the value of each metric prefix is 10 times the value of the prefix to its right.

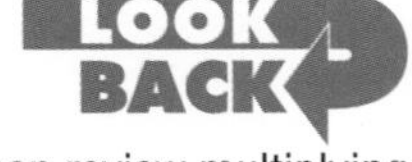

You can review multiplying and dividing by powers of ten in Lesson 6-9.

Converting units within the metric system follows the same procedure as multiplying or dividing by powers of ten.

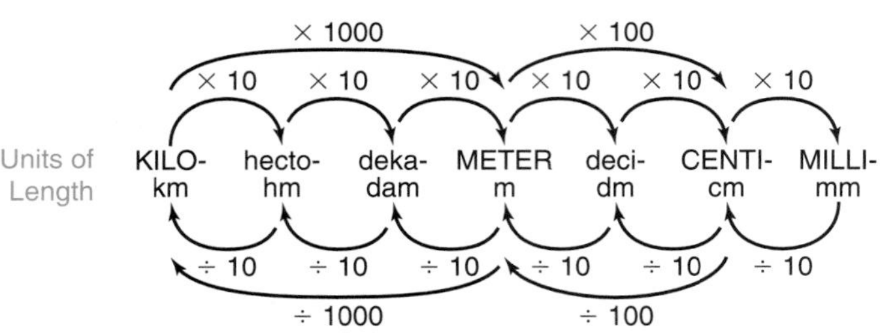

Example 

Complete each sentence.

a. 4.8 km = _?_ m

Kilometers are larger than meters. Converting to smaller units means there will be more units. Multiply by 1000.

$4.8 \times 1000 = 4800$
4.8 km = 4800 m

b. 9 mm = _?_ cm

Millimeters are smaller than centimeters. Converting to larger units means there will be fewer units. Divide by 10.

$9 \div 10 = 0.9$
9 mm = 0.9 cm

The application at the beginning of the lesson involved units of *capacity*. Capacity is the amount of liquid or dry substance a container can hold. The basic unit of capacity in the metric system is the **liter (L)**. A liter and milliliter are related in a manner similar to meter and millimeter.

1 L = 1000 mL

Example 2

Consumer Awareness

A two-liter bottle of cola sells for 99¢. A six-pack of the same cola sells for the same price. If each can contains 354 mL, which is the better value?

Explore Both packages sell for the same price. You want to determine which package gives you more cola for your money.

Plan To find the total capacity of the six-pack, multiply 354 times 6. To compare capacities, change from milliliters to liters. A milliliter is less than a liter, so divide by 1000.

$$\text{capacity of cans in liters} = \frac{(354 \text{ mL/can})(6 \text{ cans})}{1000\text{mL/L}}$$

Estimate: $350 \times 6 \div 1000 = 2.1$

Solve 354 [×] 6 [÷] 1000 [=] 2.124 The six-pack holds 2.124 L.

The six-pack is the better value since you get 2.124 liters of cola for the same price as the 2-liter bottle.

Examine The capacity we found with the calculator is very close to our estimate, so the answer is reasonable.

The *mass* of an object is the amount of matter that it contains. The basic unit of mass in the metric system is the **gram (g)**. Kilogram, gram, and milligram are related in a manner similar to kilometer, meter, and millimeter.

1 kg = 1000 g   1g = 1000 mg

 Example 3

Complete each sentence.

a. 3.2 kg = _?_ g

Kilograms are larger than grams. Larger to smaller means more units. Multiply by 1000.

$3.2 \times 1000 = 3200$
3.2 kg = 3200 g

b. 200 mg = _?_ g

Milligrams are smaller than grams. Smaller to larger means fewer units. Divide by 1000.

$200 \div 1000 = 0.2$
200 mg = 0.2 g

Checking Your Understanding

Communicating Mathematics

Read and study the lesson to answer these questions.

1. Write a formula for converting a measure in meters to an equivalent measure in centimeters. Then write a second formula for converting a measure in centimeters to an equivalent measure in meters.
2. **Show** how many times longer a kilometer is than a millimeter by using the diagram on page 358.

3. Which concept in this chapter was the most difficult for you? Why was it challenging?

Guided Practice

State which metric unit you would probably use to measure each item.

4. width of a computer screen
5. juice in a small glass
6. thickness of a coin
7. a load of lumber

Complete each sentence.

8. 0.035 m = _?_ mm
9. 40 mL = _?_ L
10. 3 kg = _?_ g
11. **Running** Last week, Ramon ran 8 laps around a 400-meter track on Monday, 4 laps on Tuesday, 8 laps on Wednesday, 4 laps on Thursday, and 8 laps on Friday. How many kilometers did he run last week?

Exercises: Practicing and Applying the Concept

Independent Practice

Write which metric unit you would probably use to measure each item.

12. distance between two cities
13. height of a 15-year-old female
14. weight of an aspirin
15. water in a washing machine

Complete each sentence.

16. 3 m = _?_ cm
17. 400 m = _?_ km
18. 30 mm = _?_ cm
19. 9400 mL = _?_ L
20. 5 L = _?_ mL
21. 600 cm = _?_ m
22. 5000 mg = _?_ g
23. 8 kg = _?_ g
24. 67 g = _?_ mg
25. How many millimeters are in a meter?
26. How many centimeters are in a kilometer?
27. How many grams are in 0.316 kilograms?
28. **Biology** The thickness of a leg of an ant is 0.035 centimeters. How many millimeters is this?
29. A bottle of ketchup holds 0.946 liters. How many milliliters does it hold?
30. How many milligrams are in 0.019 kilograms?
31. How many centimeters are in 6.032 kilometers?
32. The hair on your scalp grows about 13 millimeters per month. How long would it take for hair to grow 0.5 meter?

Critical Thinking

33. Four volumes of International Recipes are in order on a shelf. The total pages of each volume are 5 cm thick. Each cover is 5 mm thick. A bookworm started eating at page 1 of Volume I and stopped eating at the last page of Volume IV. Describe the amount the bookworm ate in terms of distance.

Applications and Problem Solving

Choose

Estimation
Mental Math
Calculator
Paper and Pencil

34. **Ecology** Every metric ton of recycled office paper saves about 19 trees. Glencoe Publishing's office in Westerville, Ohio, recycled 16,427 kilograms of paper in 1994. How many trees did this save?

35. **Geometry** The perimeter of a triangle is 1.6 meters. The length of the shortest side is 40 centimeters. Another side is 150 millimeters longer than the shortest side. Find the lengths of the other sides.

36. **Refreshments** To mix a punch, Lavonna starts with 3 liters of lemon-lime soda and adds 1893 milliliters of cranberry juice.
   a. Estimate how many liters are in the punch bowl when the punch is completed.
   b. Use your estimate to find how many 250-mL glasses of punch can be served.

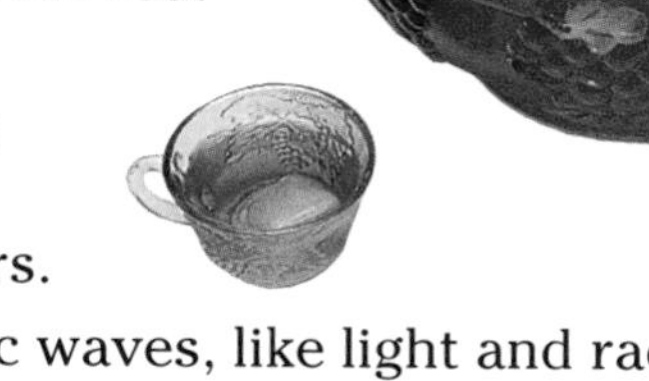

37. **Photography** Most cameras use 35-millimeter film. This means that the film is 35 millimeters wide. Write the width of the film in meters.

38. **Physical Science** Electromagnetic waves, like light and radar waves, are measured in angstroms and microns. An angstrom is 10^{-10} meters, and a micron is 10^{-6} meters.
   a. Write the length of an angstrom and a micron in millimeters.
   b. The visible portion of the electromagnetic spectrum, including white light, has waves that are approximately 3900 to 7700 angstroms long. Write the lengths of these waves in millimeters.
   c. The waves used for radio and television transmission are 1 to 10 meters long. Write these lengths in angstroms.

39. **Family Activity** Find at least five metric measures in your home. Take measurements or find some on packages of food or other products. Write each measure as you found it and in one other metric unit of measure.

Mixed Review

40. The sum of three integers, x, $x + 1$, and $x + 2$, is at least 216. Find the three smallest such integers. (Lesson 7-7)

41. Solve $7b - 39 > 2(b - 2)$. (Lesson 7-6)

42. Solve $2z - 500 = 100 - z$. (Lesson 7-5)

43. Evaluate $\frac{5x}{y}$ if $x = 31.74$ and $y = 9.2$. (Lesson 6-5)

44. **Agriculture** The chart at the right shows the numbers of millions of 42-pound units of different types of apples produced in 1992. What is the mean, median, and mode of the numbers of apples? (Lesson 6-6)

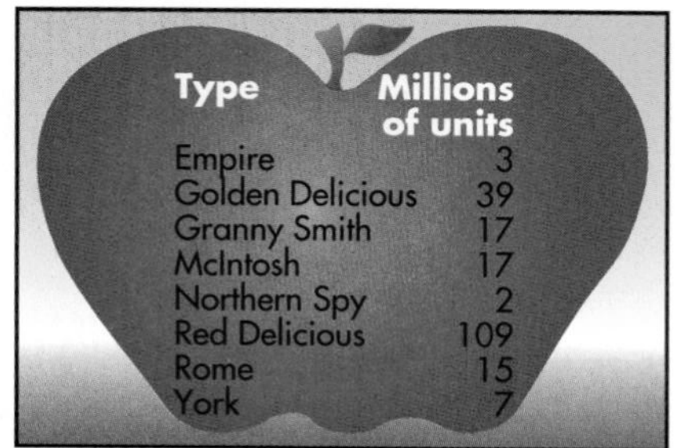

Type	Millions of units
Empire	3
Golden Delicious	39
Granny Smith	17
McIntosh	17
Northern Spy	2
Red Delicious	109
Rome	15
York	7

Source: International Apple Institute

45. Solve $\left(3\frac{1}{4}\right)\left(-6\frac{2}{5}\right) = r$. (Lesson 6-3)

46. Solve $15x = 45$. (Lesson 3-3)

CLOSING THE

Investigation

Refer to the Investigation on pages 270–271.

You are ready to begin construction of your new roller coaster ride. The amusement park's information director is preparing a press release to announce when the new ride will open.

Using information from your design and the data from your class experiments, write a report for the information director. Your report should include the following.

- The name of your roller coaster and an explanation of why you chose it.
- Any records and track data such as height, speed, and so on.
- Ride data such as how long a ride lasts and how many passengers will be carried per hour.
- Any special-interest items like providing entertainment for people waiting in line.

Plan and give an oral presentation to describe what you know about roller coasters and designing roller coasters. Your presentation could include the following.

- A brief history of roller coasters.
- Examples from your portfolio to help you describe and define the roller coaster terms you used most often in this investigation.
- A table showing how you determined the most efficient number of passengers and trains for your design.
- An explanation of what you would do differently if you had to do this Investigation over again.

Extension

Build a scale model of your roller coaster using an Erector Set®, or wooden craft sticks.

PORTFOLIO ASSESSMENT

You may want to keep your work on this Investigation in your portfolio.

CHAPTER 7 Highlights

Vocabulary

After completing this chapter, you should be able to define each term, property, or phrase and give an example or two of each.

Algebra

empty set (p. 347)

null set (p. 347)

Geometry

center (p. 341)

circle (p. 341)

circumference (p. 341)

diameter (p. 341)

pi, π (p. 341)

radius (p. 341)

Measurement

gram (p. 359)

liter (p. 359)

meter (p. 358)

metric system (p. 358)

Problem Solving

working backward (p. 330)

Understanding and Using Vocabulary

Determine whether each statement is *true* or *false*.

1. The basic unit of capacity in the metric system is the gram.
2. The distance around a circle is its circumference.
3. In the metric system, the meter is the basic unit used for measuring distance.
4. A radius of a circle is twice as long as its diameter.
5. The null set, or the empty set, contains no elements.
6. A segment from any point on a circle to the center of the circle is a radius.
7. A formula for the circumference of a circle is $C = \pi d$.

CHAPTER 7 Study Guide and Assessment

Skills and Concepts

Objectives and Examples

Upon completing this chapter, you should be able to:

- **solve equations that involve more than one operation.** (Lesson 7-2)

$$\begin{aligned} 3x + 6 &= 27 \\ 3x + 6 - 6 &= 27 - 6 \quad \textit{Subtract 6 from each side.} \\ 3x &= 21 \\ \frac{3x}{3} &= \frac{21}{3} \quad \textit{Divide each side by 3.} \\ x &= 7 \end{aligned}$$

Review Exercises

Use these exercises to review and prepare for the chapter test.

Solve each equation. Check your solution.

8. $54 = 7x - 9$ **9.** $\frac{g}{3} + 10 = 14$

10. $12 = 7 - m$ **11.** $\frac{x}{4} - 17 = -2$

12. $24 - y = -9$ **13.** $4d - 8 = -88$

14. $-2f - 37 = -11$ **15.** $2.9 = 3.1 - t$

- **solve verbal problems by writing and solving equations.** (Lesson 7-3)

One more than twice a number is 17. What is the number?

one → 1, *more than* → +, *twice a number* → $2n$, *is* → =, *17* → 17

$$\begin{aligned} 1 + 2n &= 17 \\ 1 - 1 + 2n &= 17 - 1 \\ 2n &= 16 \\ \frac{2n}{2} &= \frac{16}{2} \\ n &= 8 \end{aligned}$$

Define a variable and write an equation for each situation. Then solve.

16. The sum of six times a number and 4 is 52. What is the number?

17. Twice a number less 7 is 19. Find the number.

18. Six hundred is 15 more than 15 times a number. What is the number?

19. Vashawn earns \$50 a week less than twice his old salary. If he earns \$450 a week now, what was his old salary?

- **find the circumference of a circle.** (Lesson 7-4)

Find the circumference of a circle with a diameter 12.3 cm long.

$C = \pi d$

$C = \pi(12.3)$

$C \approx 38.6$ cm

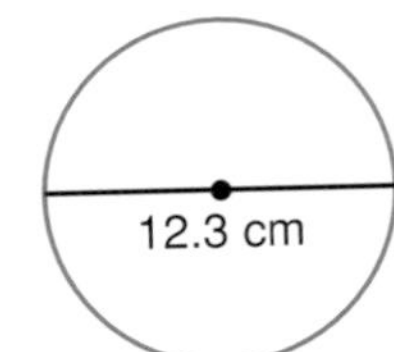

Find the circumference of each circle shown or described below.

20.

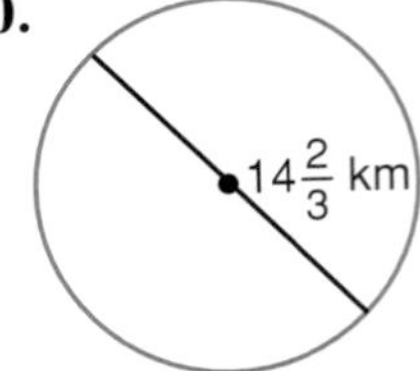

21. The radius is $3\frac{1}{2}$ inches.

22. The diameter is 18.45 centimeters.

23. The radius is 4.9 meters.

Objectives and Examples

► **solve equations with the variable on each side.** (Lesson 7-5)

$$3t - 1 = t + 8$$
$$3t - t - 1 = t - t + 8 \quad \textit{Subtract t from each side.}$$
$$2t - 1 = 8$$
$$2t - 1 + 1 = 8 + 1 \quad \textit{Add 1 to each side.}$$
$$2t = 9$$
$$\frac{2t}{2} = \frac{9}{2} \quad \textit{Divide each side by 2.}$$
$$t = \frac{9}{2} \text{ or } 4\frac{1}{2}$$

Review Exercises

Solve each equation. Check your solution.

24. $6n - 8 = 2n$
25. $4 + 2w = 8w + 16$
26. $4m + 11 = 6 - m$
27. $5(3 - k) = 20$
28. $5(q + 2) = 2q + 1$
29. $\frac{2}{7}h - 10 = h - 20$
30. $\frac{x - 5}{6} = \frac{x - 11}{2}$

► **solve inequalities that involve more than one operation.** (Lesson 7-6)

$$5b - 10 < 7b + 4$$
$$5b - 5b - 10 < 7b - 5b + 4 \quad \textit{Subtract 5b from each side.}$$
$$-10 < 2b + 4$$
$$-10 - 4 < 2b + 4 - 4 \quad \textit{Subtract 4 from each side.}$$
$$-14 < 2b$$
$$\frac{-14}{2} < \frac{2b}{2} \quad \textit{Divide each side by 2.}$$
$$-7 < b$$

Solve each inequality and check your solution. Graph the solution on a number line.

31. $5n + 4 > 34$
32. $-7m - 12 < 9$
33. $\frac{r}{7} - 11 \leq 3$
34. $16f > 13f + 45$
35. $0.52c + 14.7 > 2.48c$
36. $\frac{a - 5}{6} \leq \frac{a + 9}{8}$

► **solve verbal problems by writing and solving inequalities.** (Lesson 7-7)

Six times a number less 12 is at least twenty-two. What is the number?

six times a number	*less*	*12*	*is at least*	*twenty-two*
$6x$	$-$	12	$\geq$	22

$$6x - 12 \geq 22$$
$$6x - 12 + 12 \geq 22 + 12 \quad \textit{Add 12 to each side.}$$
$$6x \geq 34$$
$$\frac{6x}{6} \geq \frac{34}{6} \quad \textit{Divide each side by 6.}$$
$$x \geq \frac{34}{6} \text{ or } 5\frac{2}{3}$$

Define a variable and write an inequality for each situation. Then solve.

37. Eight times a number increased by two is at least 18. What is the number?
38. Twelve less than three times a number is at most 27. Find the number.
39. The sum of an integer and the next greater integer is more than 47. What is the integer?
40. Four times a number s exceeds 6 by at least 8. What are the possible values of s?

Objectives and Examples	Review Exercises
▶ **convert measurements within the metric system.** (Lesson 7-8) 4.5 km = ? m Kilometers are larger than meters, so there will be more units. There are 1000 meters in a kilometer. $4.5 \times 1000 = 4500$ 4.5 km = 4500 m	**Complete each sentence.** **41.** 6 mm = ? m **42.** 8.3 g = ? kg **43.** 43 L = ? mL **44.** 560 mg = ? g **45.** 3 km = ? cm **46.** 40 mL = ? L **47.** 88 m = ? cm

Applications and Problem Solving

48. Patterns "One potato, two potato, . . ." says Alejándro as the children decide who is "it." At the beginning, six children were standing in a circle and each round of the rhyme eliminated the eighth child. Alejándro counts clockwise and each round begins with the person next to the one who was eliminated. Let A represent Alejándro and B through F the rest of the children clockwise around the circle. If E is declared "it," where did Alejándro begin the first rhyme? (Lesson 7-1)

49. Health Doctors recommend that you not exceed a safe heart rate when you exercise. The formula $r = \frac{4(220 - a)}{5}$, gives the desirable heart rate, r, in beats per minute for a person a years old. Find the maximum safe heart rate for a 44-year old. (Lesson 7-2)

50. Music Before valves were invented in the late 19th century, musicians used to adjust the scale of their brass instruments by adding a crook. A crook is a metal piece of tubing inserted between the mouthpiece and the main tubing of the instrument. If a circular crook has a diameter of 3.4 inches, how long was the piece of metal used to make it? (Lesson 7-4)

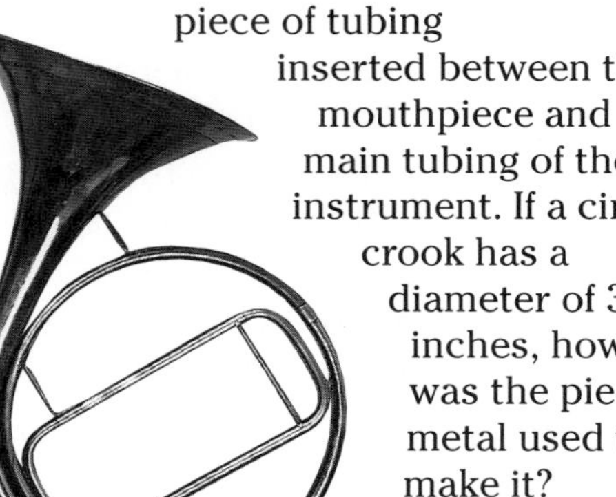

51. Aviation The pilot of a small plane wants to avoid going above the cloud ceiling that is currently at 12,000 feet. If the plane is ascending at a rate of 630 feet per minute, how long can the pilot ascend and still stay at least 1500 feet below the ceiling? (Lesson 7-7)

A practice test for Chapter 7 is available on page 780.

Alternative Assessment

Performance Task

Demonstrate your knowledge by giving a clear, concise, solution to each problem. Be sure to include all relevant drawings and justify your answers. You may show your solutions in more than one way or investigate beyond the requirements of the problem.

1. The table at the right shows the top five biggest money-losing films.

Film	Loss (millions)
The Adventures of Baron Münchhausen	48.1
Ishtar	47.3
Hudson Hawk	47.0
Inchon	44.1
The Cotton Club	38.1

Source: *Top Ten of Everything*

   a. *The Adventures of Baron Münchhausen* lost \$10.3 less than twice what the British film *Raise the Titanic* lost. How much did *Raise the Titanic* lose?

   b. The sum of what *Billy Bathgate* and *The Cotton Club* lost is \$5.1 million more than twice what *Billy Bathgate* lost. How much did *Billy Bathgate* lose?

   c. *Rambo III* lost just over \$6.3 million more than half what *Ishtar* lost. How much did *Rambo III* lose? Write your answer as an inequality.

   d. Use the numbers in the table to write a problem that could be solved using a two-step equation or inequality.

2. Terrica is cutting fabric to make a tablecloth for a round table in her living room. The top of the table is 56.25 centimeters across. The table is 1.2 meters tall.

   a. What is the circumference of the top of the table?

   b. What should Terrica make the diameter of the tablecloth be if it goes to the floor on all sides?

   c. If Terrica is going to put lace around the edge of the tablecloth, how much should she buy?

Thinking Critically

Two formulas for the circumference of a circle are $C = 2\pi r$ and $C = \pi d$. If you could only remember one of them, could you always find the circumference of a circle? Explain.

Portfolio

Select one of the assignments from this chapter that you found especially challenging and place it in your portfolio.

Self Evaluation

The only place where success comes before work is in the dictionary. In other words, if you want to succeed, you must first work!

Assess yourself. Are you willing to put forth the effort to succeed? Describe a situation that was difficult for you when you worked hard and succeeded or didn't work hard and failed. What did you learn from the situation? What would you do the same or differently if you could do it again?

LONG-TERM PROJECT

Investigation

You are the stadium manager for the West Prairie State University, home of the Mighty Mustangs. You oversee the operations of Mustang Stadium and help plan team-spirit activities for the home football games. The university has a large stadium with a seating capacity of 80,000. The stadium has two levels, an upper level and a lower level.

- *In each level, there are 26 sections. Each section has about the same number of seats.*
- *The sections are labeled A through Z, with section A situated at the center of the north end zone of the field.*
- *The rest of the sections are labeled in alphabetical order going around the stadium in a clockwise direction.*
- *In each section of the lower level, there are 90 rows numbered from 1 to 90. They are in sequential order starting with row 1 next to the field.*
- *Each section of the upper level has only 75 rows numbered from 1 to 75. Each seat in the row is also numbered.*

The following ticket stub shows a seat for the big game.

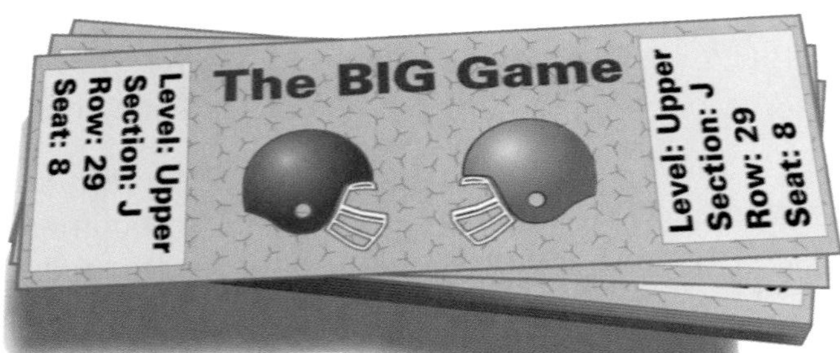

Various groups come to your office for research, advice, and permission for different projects. The spirit squad has been working on keeping the team spirit high this year for the football games. Also, the athletic director has been working on ticket pricing for stadium seating.

Starting Your Investigation

To start the research for either the spirit squad or athletic director, you will want to draw a model of the stadium. Since every row in every section of the stadium seats about the same number of people (plus or minus one person), how many seats are in a row? Explain your answer. What is the seating capacity of an upper section in the stadium? What is the seating capacity of a lower section?

On your model, label all the sections and levels in the stadium. Also, draw an enlargement of section A (both upper and lower levels), and label the number of each row in the section. Also, choose one row and label all the seat numbers.

Describe where the best seats are in the stadium. If you could sit wherever you want during the big game, what seat would you pick?

Think About . . .

The spirit squad wants to do the wave during a time-out. The wave is a type of stadium cheer where the people in each section of the stadium stand, raise their hands, and cheer in a sequence around the stadium. The spirit squad asks you a few questions:

- *How long does it take the crowd to do one rotation around the stadium?*
- *How many times can the wave go around the stadium during a two-minute time-out?*
- *If the wave is done for two minutes during each of six time-outs in the game and for three minutes during each quarter and half-time, how many times would the average fan have to stand up and sit down during a game?*

Reread the above questions throughout this Investigation. As you come up with other questions, write these new questions in the list also.

inter NET CONNECTION

For current information on football stadiums, visit: **www.glencoe.com/sec/math/prealg/mathnet**

Investigation
Stadium Stampede

Working on the Investigation
Lesson 8-2, p. 384

Working on the Investigation
Lesson 8-6, p. 404

Working on the Investigation
Lesson 9-5, p. 453

Working on the Investigation
Lesson 9-9, p. 471

Closing the Investigation
End of Chapter 9, p. 476

You will continue working on this Investigation throughout Chapters 8 and 9.

Be sure to keep your materials in your Investigation Folder.

CHAPTER

8 Functions and Graphing

TOP STORIES in Chapter 8

In this chapter, you will:

- determine whether relations are functions,
- graph data in scatter plots,
- graph linear equations using slope and intercepts,
- solve systems of equations, and
- graph inequalities.

MATH AND TRANSPORTATION IN THE NEWS

Finding your way from here to there

Source: Popular Electronics, October, 1994

The days of the map and compass may be coming to an end. Technology is making it easier to find your way in a car. Some electronics companies and car manufacturers are making on-board navigation systems to assist drivers. The car navigation systems rely on computers, databases of map information, and satellites in the Navstar Global Positioning System (GPS). Geometry and precise timing allow the computer to tell you a route to your destination and give visual displays, and computer-generated voice instructions guide you as you go. Although this system is new to consumers in the United States, it has been a popular accessory on cars sold in Japan for a number of years.

1200	1800	1850

1150
Chinese and Mediterranean navigators use magnetic compasses to guide ships.

1769
Steam-powered car is built by French engineer Nicolas-Joseph Cugnot.

On the Lighter Side

thomas

"George and I haven't traveled by car since the map-folding incident of '84."

Statistical Snapshot

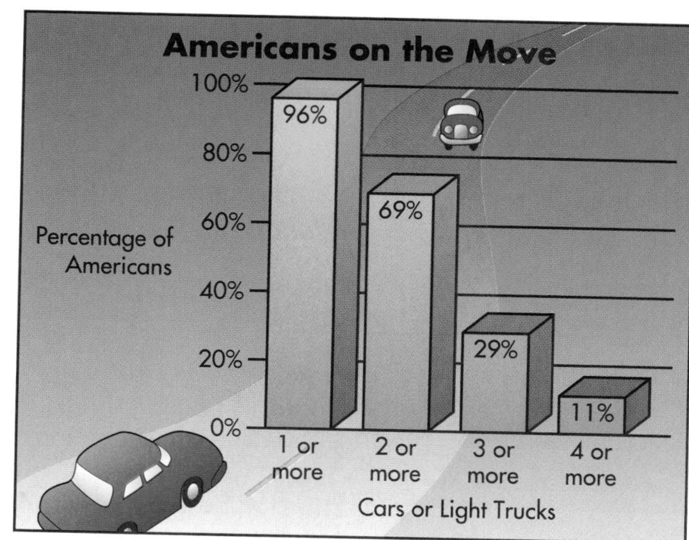

Source: *Ameripoll,* Maritz Marketing Research

interNET CONNECTION For up-to-date information on the Global Positioning System, visit:
www.glencoe.com/sec/math/prealg/mathnet

1978
The U.S. Government launches the first satellite of the Navstar system.

1994
Oldsmobile is first in North America to offer an onboard navigation system.

1900 — 1950 — 2000

885
ottlieb Daimler and
arl Benz develop
rerunners of modern
asoline engines.

1971
Leon Sullivan becomes the first African-American to serve on the board of directors of General Motors.

8-1 Relations and Functions

Setting Goals: *In this lesson, you'll use tables and graphs to represent relations and functions.*

Modeling a Real-World Application: Ecology

Have you ever thought you couldn't make a difference in landfills by recycling? Think again! The table at the right shows how the United States compares to other nations in terms of the amount of annual household waste. These five countries produce the most waste in the world.

Country	Annual Domestic Waste (thousands of tons)	Equivalent per Person (pounds)
France	15,500	634
Great Britain	15,816	620
Japan	40,225	634
United States	200,000	1925
West Germany	20,780	741

Source: *Encarta*, 1994

This data could also be displayed using a set of ordered pairs. Each first coordinate would be the annual domestic waste, and each second coordinate would be the equivalent per person.

{(15,500, 634), (15,816, 620), (40,225, 634), (200,000, 1925), (20,780, 741)}

Learning the Concept

A **relation** is a set of ordered pairs, like the set shown above. The set of first coordinates is called the **domain** of the relation. The set of second coordinates is called the **range** of the relation. So, the domain of the relation above is {15,500, 15,816, 40,225, 200,000, 20,780}, and the range is {634, 620, 1925, 741}. *Notice that 634 is listed only once.*

THINK ABOUT IT

What is the domain and range of this relation?

A relation can also be modeled by a table or a graph. For example, the relation $\{(-3, 1), (0, -2), (4, -2)\}$ can be represented in each of the ways shown at the right.

Table

x	y
−3	1
0	−2
4	−2

Graph

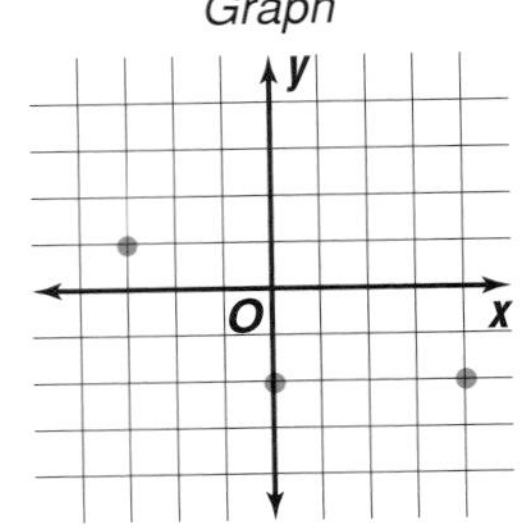

Example 1 **Express the relation shown in the graph below as a set of ordered pairs. Then determine the domain and range.**

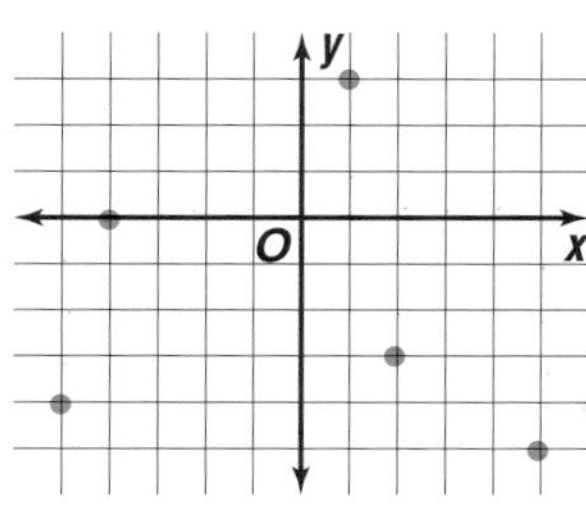

The set of ordered pairs for the relation is $\{(1, 3), (5, -5), (-4, 0), (2, -3), (-5, -4)\}$.

The domain is $\{1, 5, -4, 2, -5\}$.

The range is $\{3, -5, 0, -3, -4\}$.

The relation in Example 1 is a special type of relation called a **function**.

Definition of a Function

A function is a relation in which each element of the domain is paired with exactly one element in the range.

Example 2 **Graph each relation. Then determine whether each relation is a function.**

a.

x	6	3	6	−2
y	0	8	−1	3

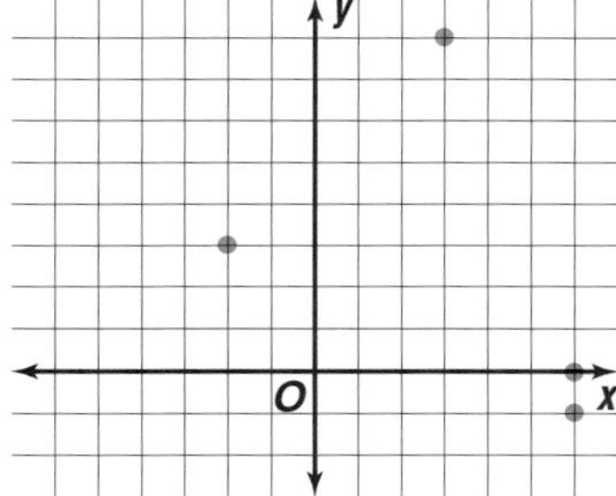

This relation is not a function since the element 6 in the domain is paired with two elements, 0 and −1, in the range.

b.

x	−3	−2	0	2
y	1	−2	1	3

Since each element of the domain is paired with exactly one element in the range, this relation is a function.

You can use the **vertical line test** to determine if a relation is a function. The relation $\{(-2, 5), (0, 4), (4, 2), (6, 1), (8, 0)\}$ is graphed at the right. Place a straightedge at the left of the graph to represent a vertical line. Slowly move the straightedge to the right across the graph.

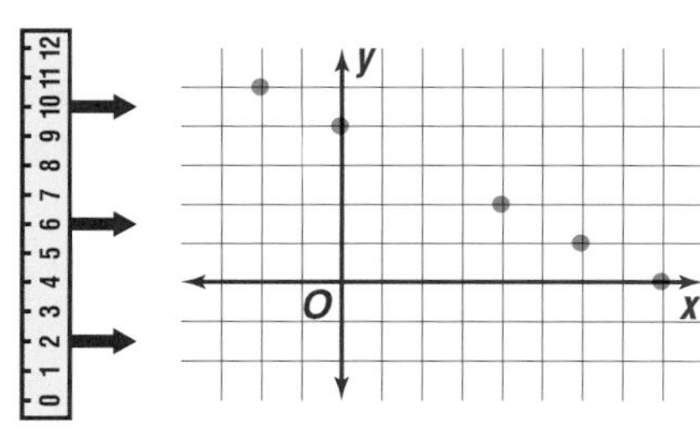

For each value of *x*, this vertical line passes through no more than one point on the graph. This is true for *every* function.

Example 3 — Probability

The graph on the left shows the possible outcomes when two six-sided dice are tossed. The graph on the right shows those outcomes when the faces on the dice show a sum of 7. Is either relation a function? Explain.

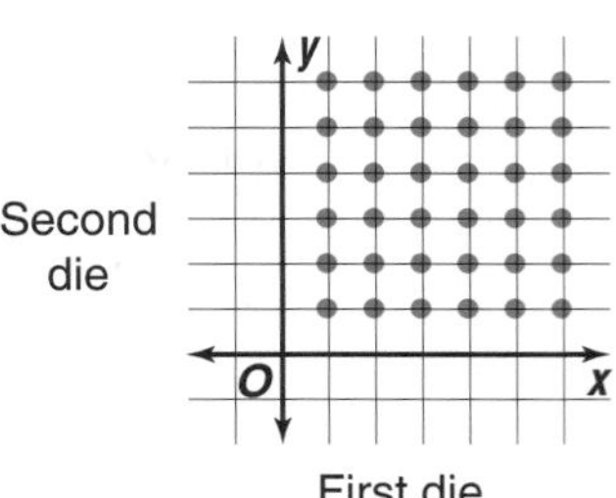

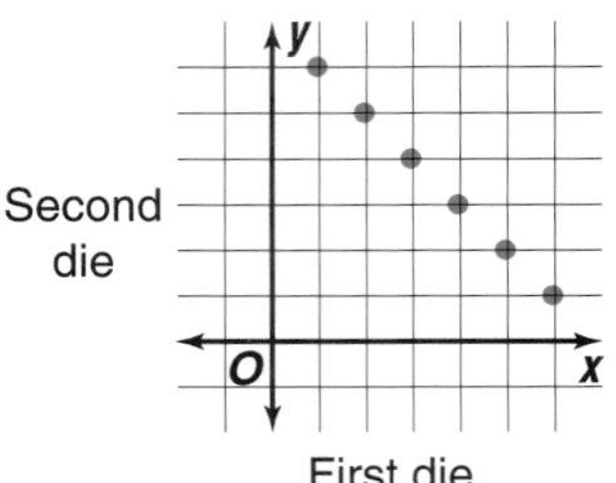

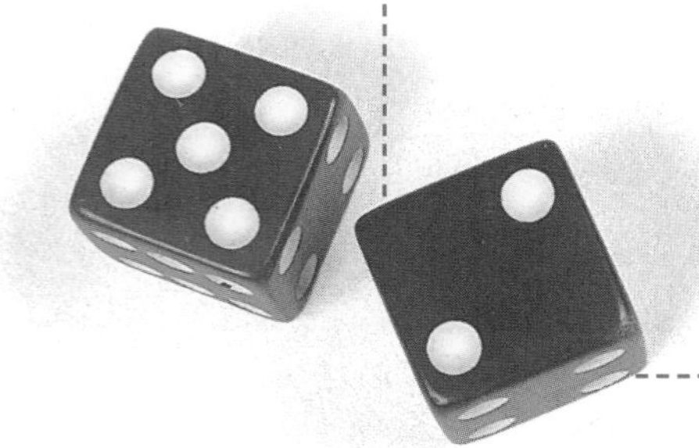

The relation graphed on the left is *not* a function since it does not pass the vertical line test.

The relation graphed on the right is a function since any vertical line passes through no more than one point of the graph of the relation.

Checking Your Understanding

Communicating Mathematics

Read and study the lesson to answer these questions.

1. **Describe** three different ways to show a relation. Give an example of a relation shown in each way.
2. **Explain** the difference between the domain and the range of a relation.
3. **You Decide** Marita says that $\{(9, 3), (0, 3), (-5, 0), (-1, 11)\}$ is a function. Terrica disagrees. Who is correct and why?
4. **Graph** the relation described by the annual waste figures in the application at the beginning of the lesson.

Guided Practice

State the domain and range of each relation.

5. $\{(-1.3, 1), (4, -3.9), (-2.4, 3.6)\}$
6. $\left\{\left(-\frac{1}{2}, -\frac{1}{4}\right), \left(1\frac{1}{2}, -\frac{2}{3}\right), \left(3, -\frac{2}{5}\right), \left(5\frac{1}{4}, 6\frac{2}{7}\right)\right\}$

Express the relation shown in each table or graph as a set of ordered pairs. Then state the domain and range of the relation.

7.

x	y
−1	3
0	6
4	−1
7	2

8.

x	y
2	4
0	2
−2	0
−4	2

9. 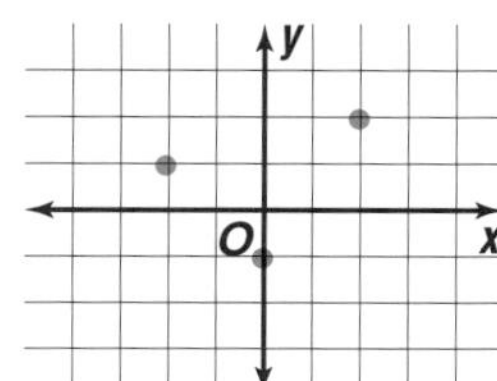

10. 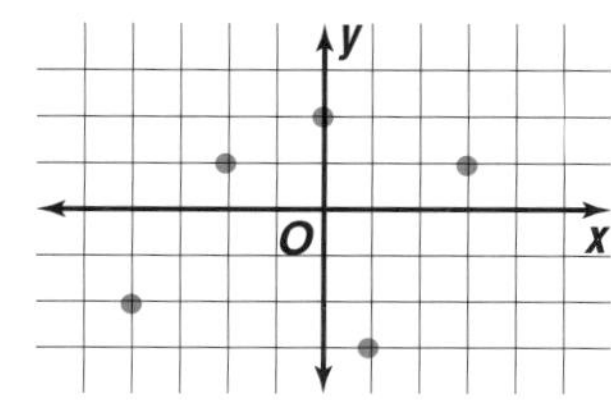

Determine whether each relation is a function.

11. $\{(5, -2), (3, 2), (4, -1), (-2, 2)\}$

12. 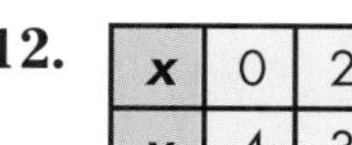

x	0	2	0
y	4	3	8

13. 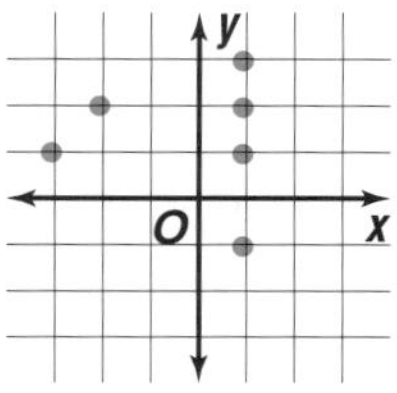

14. 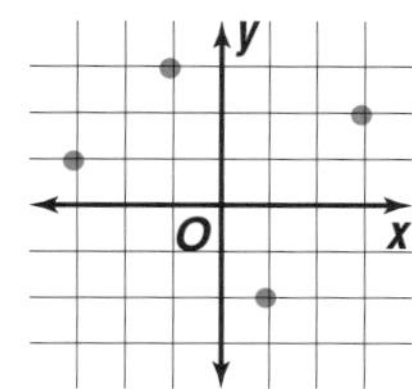

15. **Food** Lisle buys a dozen bagels for a morning meeting. She knows that the staff prefers blueberry bagels over plain bagels. If she buys at least twice as many blueberry bagels as plain bagels, write a relation to show the different possibilities. (*Hint:* Let the relation consist of ordered pairs of the form (blueberry bagels, plain bagels).)

Exercises: Practicing and Applying the Concept

Independent Practice

Write the domain and range of each relation.

16. $\{(5, -4), (-2, 3), (5, -1), (2, 3)\}$

17. $\{(-1, 6), (4, 2), (2, 36), (1, 6)\}$

18. $\{(3.1, 2), (7, -4.4), (-3.9, -8.8)\}$

19. $\{(1.4, 3), (-2, 9.6), (4, 4), (6, -2.7)\}$

20. $\left\{\left(\frac{2}{5}, \frac{3}{4}\right), \left(98\frac{3}{5}, 37\frac{1}{2}\right), \left(-4, -\frac{7}{12}\right)\right\}$

21. $\left\{\left(-\frac{1}{2}, \frac{1}{3}\right), \left(4\frac{2}{3}, -17\right), \left(-12\frac{3}{8}, 66\right)\right\}$

Express the relation shown in each table or graph as a set of ordered pairs. Then state the domain and range of the relation.

22.

x	y
0	5
2	3
1	−4
−3	3
−1	−2

23.

x	y
−4	−2
−2	1
0	2
1	−3
3	1

24.

x	y
−1	5
−2	5
−2	4
−2	1
−6	1

25.

x	y
5	4
2	8
−7	9
2	12
5	14

26. 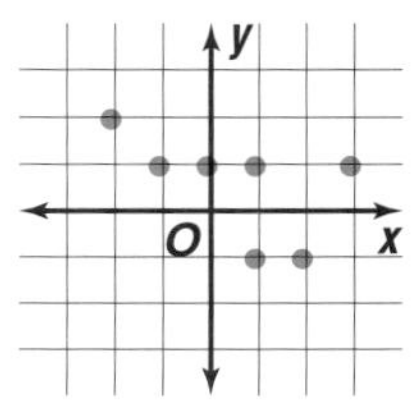

27. 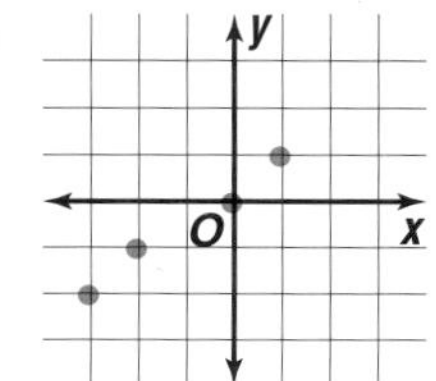

28. 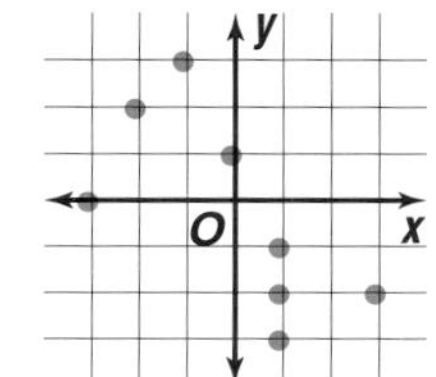

Determine whether each relation is a function.

29.

x	1	2	3
y	3	4	5

30.

x	3	5	7
y	−4	−6	−8

31.

x	1	2	1
y	3	5	−7

32.

x	−2	0	2
y	3	1	3

33. {(4, 0), (6, 0), (8, 0)}

34. {(0, 4), (0, 6), (0, 8)}

35. {(5, 4), (5, −6), (5, 4), (0, 4)}

36. {(−2, 3), (4, 7), (24, −6), (5, 4)}

37.

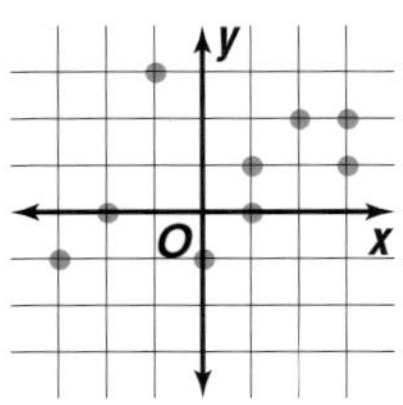

38.

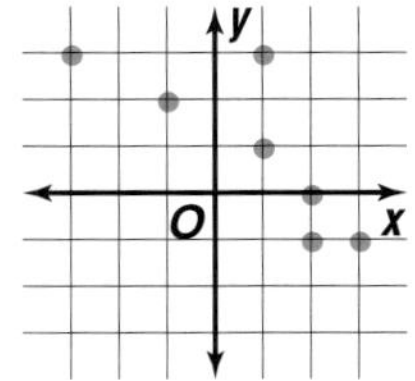

39.

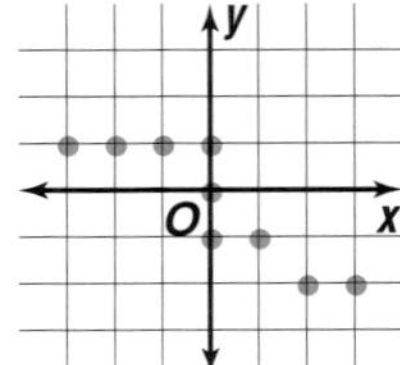

Critical Thinking

The *inverse* of any relation is obtained by switching the coordinates in each ordered pair of the relation. Determine whether the inverse of each relation is a function.

40. {(4, 0), (5, 1), (6, 2), (6, 3)}

41. {(5, 5), (−3, −1), (−2, 1), (−4, 0)}

Applications and Problem Solving

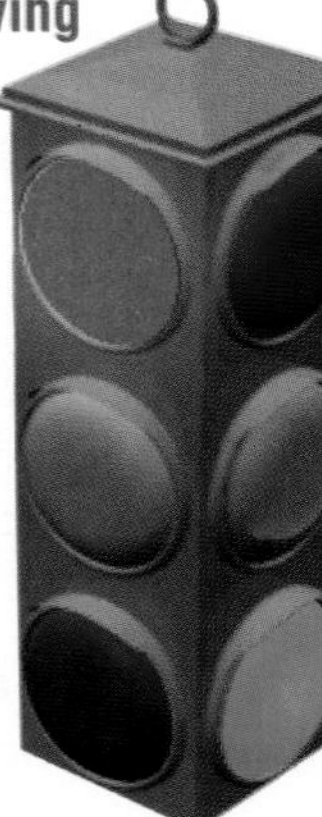

42. Traffic Control The Traffic Engineering Division for Tallahassee studies the flow of traffic at certain intersections to determine if traffic signals are needed. The table at the right shows the actual count of cars and trucks that passed through a particular intersection on one weekday.

Time	Cars	Trucks
3:00P.M.–3:30P.M.	1330	650
3:30P.M.–4:00P.M.	2000	600
4:00P.M.–4:30P.M.	2100	630
4:30P.M.–5:00P.M.	2300	650
5:00P.M.–5:30P.M.	2550	750
5:30P.M.–6:00P.M.	2220	700

a. Use the information in the second and third columns of the table to illustrate a relation that is a function.

b. Use the information to illustrate a relation that is not a function. Explain.

EARTH WATCH

SAVING THE WHALES

Fishing and whaling have long been important industries in many countries. In recent years, environmentalists have become concerned with the effect that fishing and whaling have had on the populations of these animals. One of the problems associated with protecting whales and other marine animals has been deciding who owns the ocean in which they live. Exclusive Economic Zones, or EEZs, are being established so that countries have rights and responsibilities for parts of the oceans near their shores. Nations can impose quotas or bans on fishing in their EEZ.

Year	Fish catch (metric tons)
1970	66,969,420
1972	63,855,330
1974	68,185,310
1976	71,509,720
1978	73,422,560
1980	75,587,120
1982	80,050,870
1984	87,685,160
1986	96,666,970
1988	103,149,600
1990	101,755,000

Source: World Resources Institute

Choose

Estimation
Mental Math
Calculator
Paper and Pencil

43. Shipping Rates When you order something from a mail-order company, you usually pay a shipping fee. The chart at the right relates shipping costs to the total price of merchandise ordered from *Lands' End*.

Total Price of Merchandise	Shipping Costs
\$0.00–\$30.00	\$4.25
\$30.01–\$70.00	\$5.75
\$70.01 and over	\$6.95

a. What is the shipping cost for an order of merchandise totaling \$75?

b. For what price of merchandise is the shipping cost \$5.75?

c. Does the chart represent a function? Explain.

Mixed Review

44. Consumer Awareness Which has a greater capacity, a 2-liter bottle of cola or a six-pack of cans each containing 354 mL of the same cola? (Lesson 7-8)

45. Solve $-2.2 < \frac{b}{-10} - 2.4$. (Lesson 7-6)

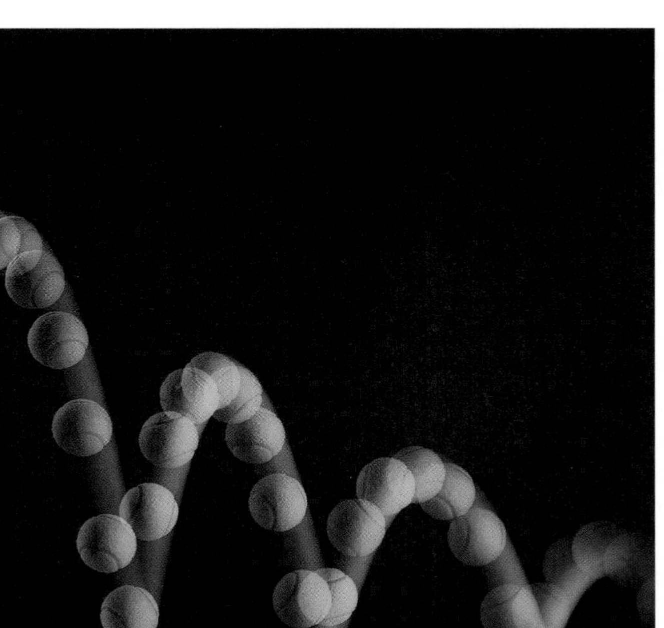

46. Physics A ball rebounds $\frac{2}{3}$ of its height after every fall. If it is dropped from a height of 48 feet, how high will it bounce at the end of the third bounce? (Lesson 6-8)

47. Solve $z - \frac{2}{5} \geq -2$. (Lesson 5-7)

48. Simplify $(-7y^2)(3y^3)$. (Lesson 4-8)

49. Geometry Find the perimeter of the rectangle shown below. (Lesson 3-5)

6 miles
2 miles

50. Three subtracted from some number is equal to -7. What is the number? (Lesson 2-5)

51. Write a verbal phrase for $2(x + 3)$. (Lesson 1-3)

See for Yourself

The table at the bottom of page 376 shows the number of tons of fish caught in the world for selected years.

1. Does the table represent a function?
2. What trend do you see in the size of the catch over these 20 years?
3. Research the United States' EEZ. Are there bans or quotas on fishing? How has the state of the marine life changed since the EEZ was established?

HANDS-ON ACTIVITY

8-2A Scatter Plots

A Preview of Lesson **8-2**

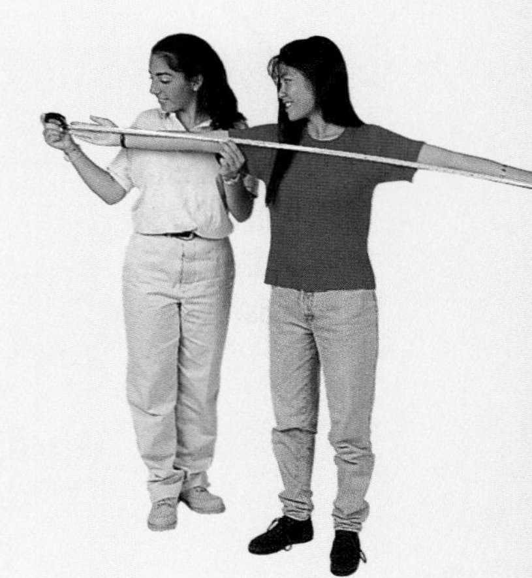

MATERIALS

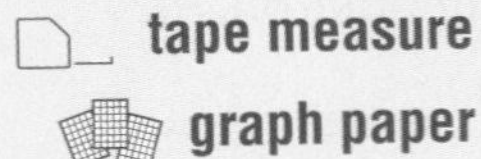

tape measure

graph paper

Cap size is related to the circumference of your head, and your shoe size is related to the length of your foot. Often it is not easy to establish whether there is a relationship between pairs of numbers by simply looking at them. Graphing ordered pairs on a coordinate system is one way to make it easier to "see" if there is a relationship.

Your Turn Work with a partner.

Name	Shoe Length (in.)	Armspan (in.)

- Measure, to the nearest inch, the length of your partner's shoe and your partner's armspan. Record your data in a table like the one shown at the right.
- Extend the table and combine your data with that of your classmates.
- On a piece of graph paper, draw a large coordinate axes. Label the axes by writing *Shoe Length (in.)* along the horizontal axis and *Armspan (in.)* along the vertical axis.
- Number the vertical lines, called tick marks, on each axis to make an appropriate scale for the data collected.
- Plot a point for each ordered pair (*shoe length, armspan*) using the data your class collected.

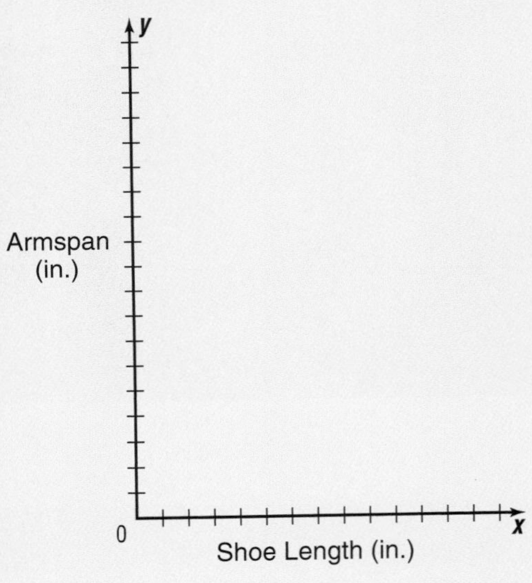

The graph you made is called a **scatter plot** of the data you collected.

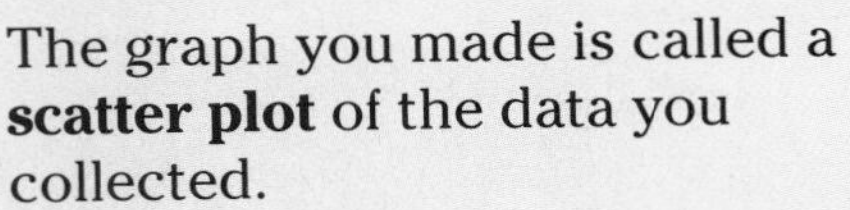

1. Describe the scatter plot.
2. Is there a relationship between shoe length and armspan? If so, write a sentence or two that describes the relationship.
3. Why do you suppose this kind of graph is called a scatter plot?

Extension

4. Using similar methods, determine if there is a relationship between another pair of lengths; for example, the circumferences of your head and wrist, or the circumference of your neck and your armspan.

8-2 Integration: Statistics
Scatter Plots

Setting Goals: *In this lesson, you'll construct and interpret scatter plots.*

Modeling with Technology

The sales and advertising costs for ten of the largest U.S. industrial corporations are shown in the chart below. Use a TI-82 graphing calculator to create a scatter plot of the data. A **scatter plot** is a graph that shows the general relationship between two sets of data.

Company	Sales (millions of dollars)	Advertising Costs (millions of dollars)
General Motors	132,775	1333.6
Ford	100,786	794.5
IBM	65,096	185.5
General Electric	62,202	250.7
Philip Morris	50,157	2024.1
Chrysler	36,897	756.6
Procter & Gamble	29,890	2165.6
PepsiCo	22,084	928.6
Eastman Kodak	20,577	686.6
Dow Chemical	19,177	186.6

Source: *World Almanac,* 1995

Your Turn

Before you create a scatter plot, clear the statistical memories.

Enter: [STAT] 4 [L1] [,] [2nd] [L2] [ENTER]

Next, enter the sales in L1 and the advertising costs in L2.

Enter: [STAT] [ENTER] *Accesses the statistical lists.*
132775 [ENTER] 100786 [ENTER] . . . 19177 [ENTER]
[▶] 1333.6 [ENTER] 794.5 [ENTER] . . . 186.6 [ENTER]

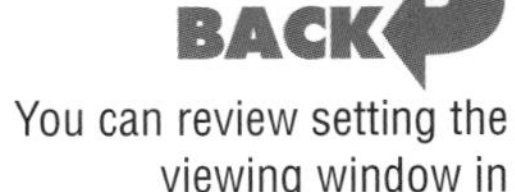

You can review setting the viewing window in Lesson 2-2B.

After the data is entered, set the range for the graph. Set a viewing window of [0, 150,000] by [0, 2500] with a scale factor of 10,000 on the x-axis and 250 on the y-axis.

Create the scatter plot by pressing [2nd] [STAT PLOT] [ENTER] and then using the arrow and [ENTER] keys to highlight "On", the scatter plot, L1 as the Xlist, L2 as the Ylist, and • as the mark. Press [GRAPH] to see the scatter plot.

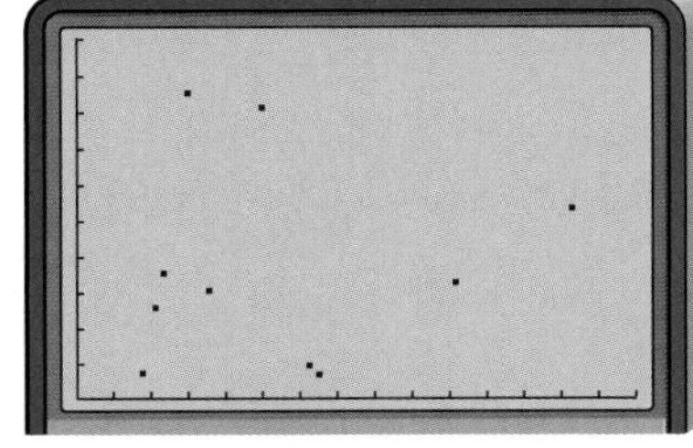

a. Does it appear that there is any relationship between sales and advertising?

b. Are there any clusters of points that are somewhat separated from the other points? If there are, describe the characteristics of each cluster.

Learning the Concept

Study the scatter plots below.

If the points appear to suggest a line that slants upward to the right, there is a **positive relationship**.

If the points appear to suggest a line that slants downward to the right, there is a **negative relationship**.

If the points seem to be random, then there is **no relationship**.

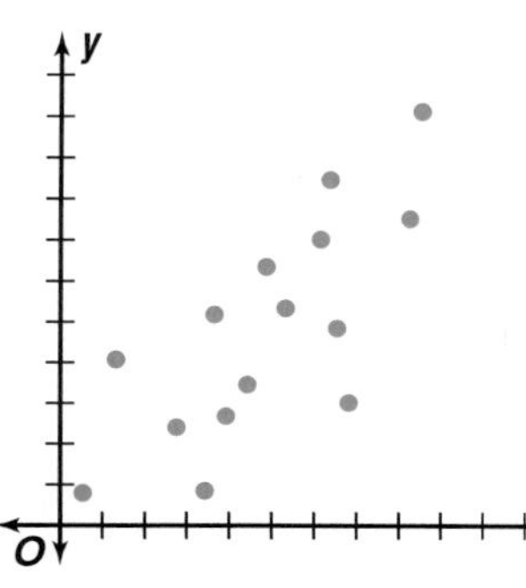

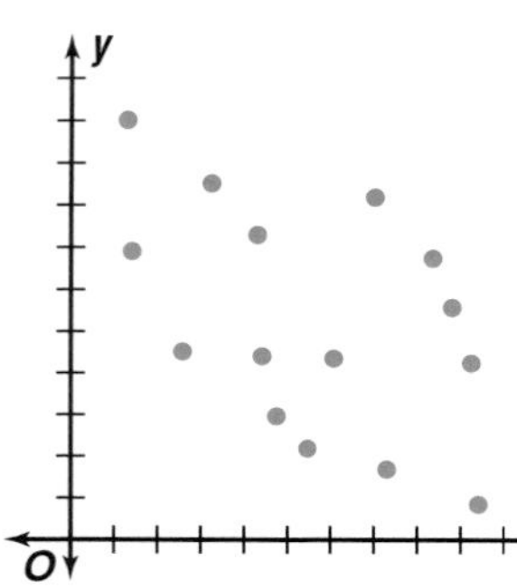

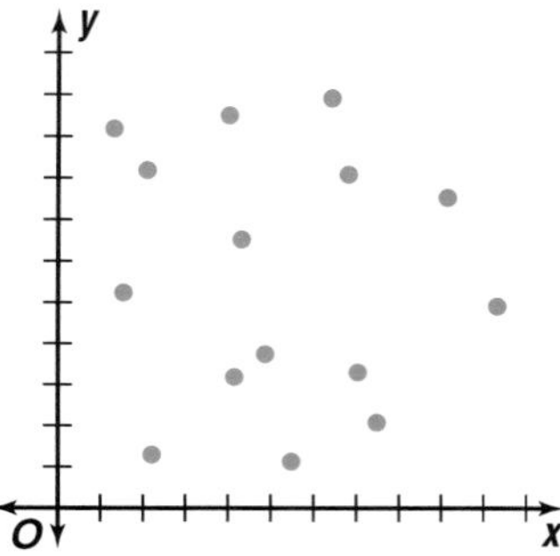

Look at the scatter plot of sales and advertising costs on the graphing calculator. The points in the scatter plot are spread out. There appears to be no relationship between the amount spent on advertising and the amount of sales.

Besides establishing a relationship, scatter plots can be used to study trends or make predictions.

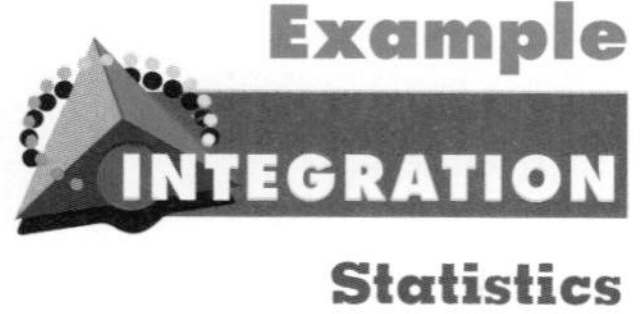

Statistics

Make a scatter plot of the following data. Then determine if there is a *positive*, *negative*, or *no* relationship. What trends do you see?

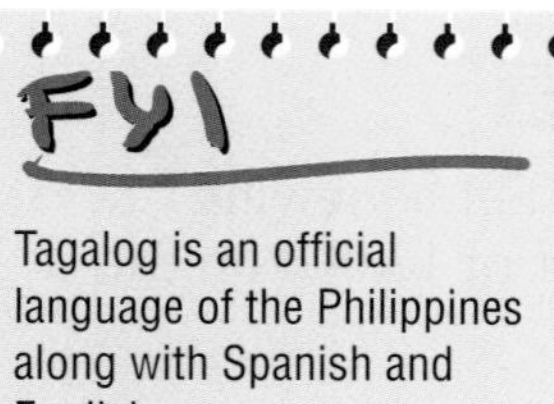

Tagalog is an official language of the Philippines along with Spanish and English.

Languages Other than English and Spanish Spoken at Home by Americans

Language used at home	Total speakers over 5 years old	
	1980	1990
French	1,572,000	1,703,000
German	1,607,000	1,547,000
Italian	1,633,000	1,309,000
Chinese	632,000	1,249,000
Tagalog	452,000	843,000
Polish	826,000	723,000
Korean	276,000	626,000
Vietnamese	203,000	507,000
Portuguese	361,000	430,000

Source: Bureau of the Census, U.S. Dept. of Commerce

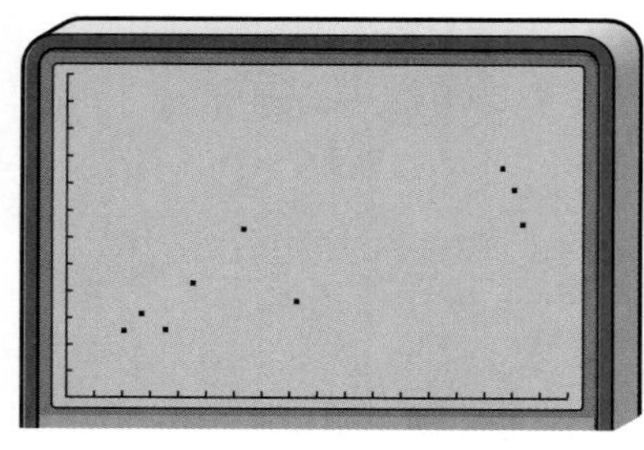

Since the points in the scatter plot are in a pattern that seems to slant upward to the right, the scatter plot shows a *positive* relationship. In general, the greater the number of speakers of a language in 1980, the greater the number of speakers of the same language in 1990.

Checking Your Understanding

Communicating Mathematics

Read and study the lesson to answer these questions.

1. **Write** a sentence explaining how a scatter plot can be used.
2. **Describe** how to draw a scatter plot for two sets of data.
3. **Draw** a scatter plot with 15 points that shows a negative relationship.
4. **Write** a short paragraph or two describing what you can determine about the data in the scatter plot shown at the right.

Relationship of Value of Cars and Age of Cars

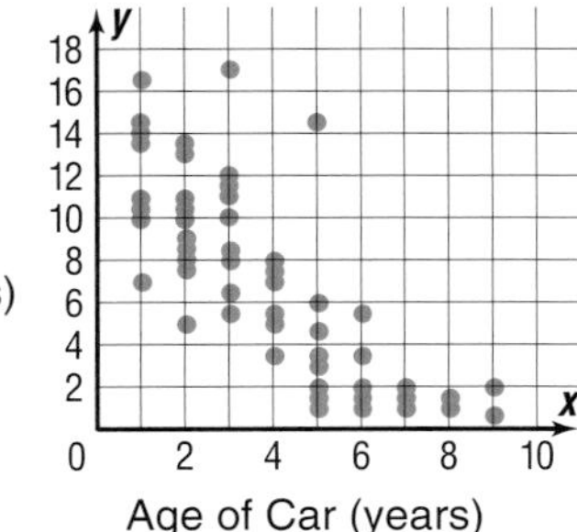

Guided Practice

What type of relationship, *positive*, *negative*, or *none*, is shown by each scatter plot?

5.

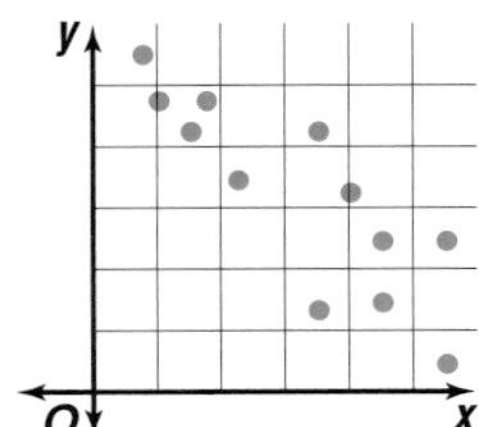

6.

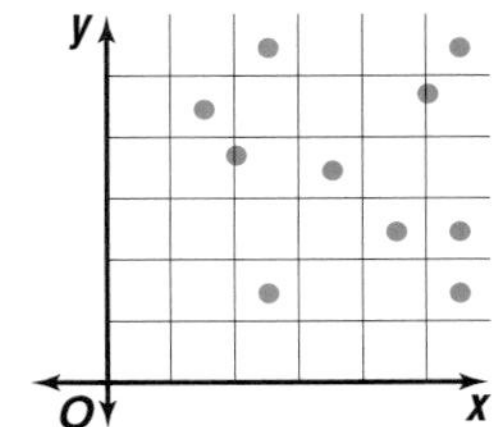

7.

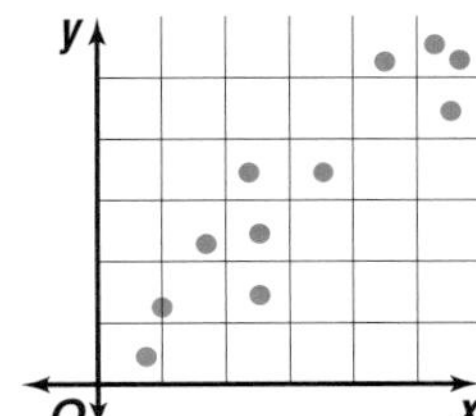

Determine whether a scatter plot of the data for the following might show a *positive*, *negative*, or *no* relationship. Explain your answer.

8. weight, month of birth
9. playing time, points scored
10. **Track and Field** The scatter plot at the right compares the winning time to the year of the race for the women's Olympic 100-meter dash.

Women's Olympic 100-Meter Dash Times

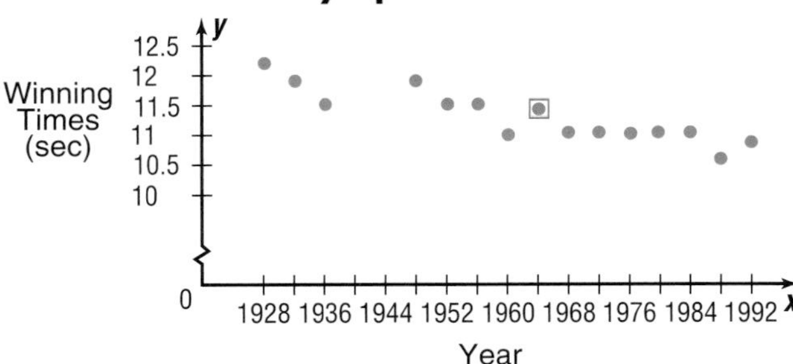

a. What year and speed does the point with the box around it represent?
b. Find the point(s) that represent(s) the slowest winning time ever run in the Olympic 100-meter dash.
c. If you had trained in the 1950s for the 100-meter race, what time would have been competitive?
d. What conclusion can you make from the scatter plot?
e. Use the scatter plot to predict the winning time in the 1996 Olympics in Atlanta, Georgia. Check your prediction.

Exercises: Practicing and Applying the Concept

Independent Practice

What type of relationship, *positive*, *negative*, or *none*, is shown by each scatter plot?

11. 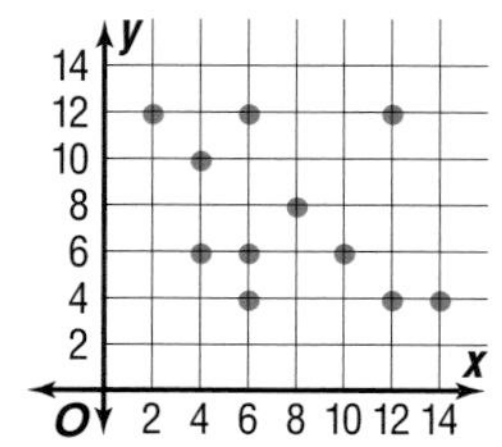

12. 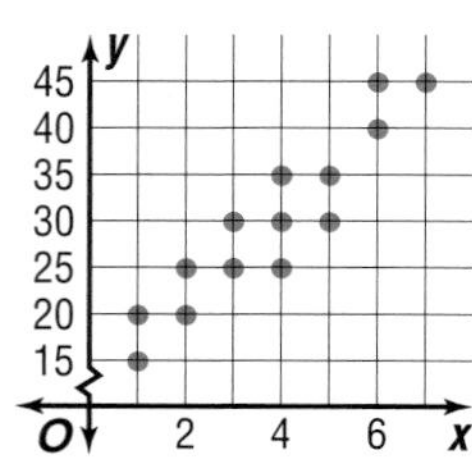

13. 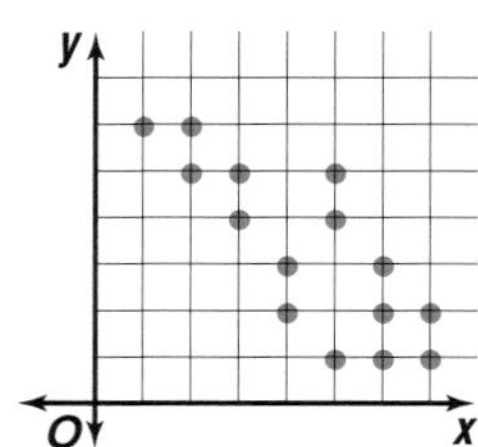

14. A scatter plot of study time and test scores for a class is shown below.

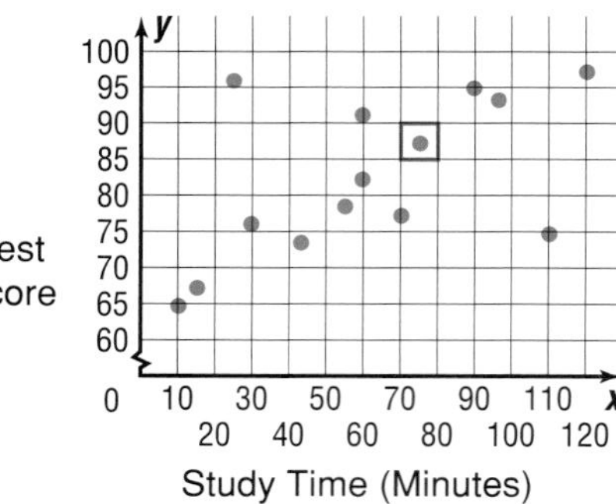

a. What study time and test score does the point with the box around it represent?

b. Do the data show a relationship between study time and test scores? If so, is it positive or negative?

c. Where on the plot are the points for people who studied quite a bit and had a fairly high test score?

d. How many students are in the class?

Determine whether a scatter plot of the data for the following might show a *positive*, *negative*, or *no* relationship. Explain your answer.

15. height, weight

16. armspan of student, test scores

17. hair color, height

18. outside temperature, heating bill

19. weight of car, miles per gallon

20. amount of sales, years of experience

21. month of birth, test scores

22. head circumference, height

Graphing Calculator

23. Keyboarding The table below shows the keyboarding speeds in words per minute (wpm) of 12 students. Use a graphing calculator to create a scatter plot of the data.

Experience (weeks)	4	7	8	1	6	3	5	2	9	6	7	10
Speed (wpm)	38	46	48	20	40	30	38	22	52	44	42	55

a. Do the data show a relationship between speed and experience? If so, is it positive or negative?

b. Use this scatter plot to estimate the keyboarding speed of a student with 12 weeks experience.

c. Estimate the experience level of a student whose speed is 33 wpm.

d. What conclusion can you make from the scatter plot?

Critical Thinking

24. **Sales** A scatter plot of monthly insulated boot sales and monthly skiing accidents in West Virginia shows a positive relationship.
   a. Why might this be true?
   b. Does a positive relationship necessarily mean that one factor causes the other? Explain.

Applications and Problem Solving

interNET CONNECTION

For the latest college costs, visit: **www.glencoe.com/sec/math/prealg/mathnet**

25. **Education** The table below shows the average tuition for public colleges in the United States for the years 1985 to 1994.

Year	1985	1986	1987	1988	1989
Cost per Year ($)	1386	1536	1651	1726	1846

Year	1990	1991	1992	1993	1994
Cost per Year ($)	2035	2159	2410	2610	2784

   a. Make a scatter plot of the data.
   b. Do the data show a relationship between year and cost per year? If so, what type of relationship is it?

26. **Geography** Use an almanac to find the area and the average depth of the ten largest oceans and seas.
   a. Construct a scatter plot to show the relationship between the average depth and area.
   b. What can you determine about the relationship between area and depth of oceans from the scatter plot?

27. **Sports** The number of points scored and rebounds made by members of the 1994 National Basketball Association champion Houston Rockets is given in the table below.
   a. Make a scatter plot of the data to show the relationship between the rebounds and points.
   b. Does the scatter plot show any relationship?
   c. Could you predict the number of points a player would have if you were given the number of rebounds for that player? Explain.
   d. Do you think that the position of a player affects the number of rebounds that player will make? Explain.
   e. Do you think the amount of time played affects the number of points scored by a player? Why or why not?

Player	Rebounds	Points
Olajuwon	955	2184
Thorpe	870	1149
Maxwell	229	1023
Smith	138	906
Jent	15	31
Horry	440	803
Elie	181	626
Cassell	134	440
Brooks	102	381
Herrera	285	353
Robinson	10	25
Bullard	84	226
Petruska	31	53
Cureton	12	4
Riley	59	88

28. **Geometry** Make a table with ten possible lengths and widths of rectangles with perimeters of 24 centimeters. Then make a scatter plot of the data. Does there appear to be a positive, negative, or no relationship between length and width?

Mixed Review

29. Determine the domain and range of the relation {(8, 1), (4, 2), (6, −4), (5, −3), (6, 0)}. (Lesson 8-1)

30. Solve $12x - 24 = -14x + 28$. (Lesson 7-5)

31. **Statistics** Find the mean, median, and mode for the data set 121, 130, 128, 126, 130, 131. (Lesson 6-6)

32. **Patterns** Write the next three terms of the sequence 5, 6.5, 8, 9.5. (Lesson 5-9)

33. **Biology** Cicadas are sometimes called 17-year locusts because they emerge from the ground every 17 years. The number of a certain type of caterpillar peaks every 5 years. If the peak cycles of the caterpillars and the cicadas coincided in 1990, what will be the next year in which they will coincide? (Lesson 4-7)

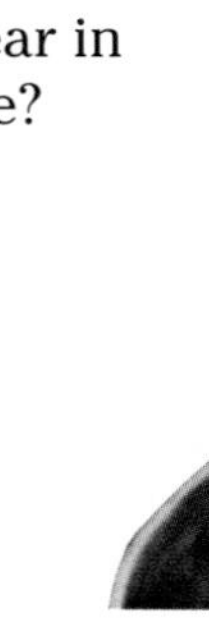

WORKING ON THE Investigation

Refer to the Investigation on pages 368–369.

Begin working on the Investigation by doing a class experiment. Use a stopwatch to time one student doing the wave. The timer says "go" and starts the stopwatch. The student stands up and then sits down. At this point, the timer stops the stopwatch. Record one person and the time in a table.

Next, time two students doing the wave (each student represents a different section in the stadium). The timer says "go" and starts the stopwatch. The first person stands and then sits. As the first person sits, the second person stands and then sits. The timer stops the stopwatch. Record two people and the time in the table.

Continue the experiment with three people (representing three sections), then eight people, then eleven people, then fourteen people, and so on until you have determined a wave time using every student in the class.

Your table of the class should contain all the data including the number of people (sections) and their corresponding wave times. Use a scatter plot to illustrate the data. The horizontal axis will be the number of people in the wave, and the vertical axis will be the wave time.

Does this data show a relationship between the number of sections and the wave time? Does this scatter plot show a *positive*, *negative*, or *no relationship*?

Write a paper answering the questions raised by the spirit squad on page 369.

Add the results of your work to your Investigation Folder.

8-3 Graphing Linear Relations

Setting Goals: *In this lesson, you'll find solutions for relations with two variables and graph the solutions.*

Modeling a Real-World Application: Science

Did you know that a cricket is nature's thermometer? You can use the equation $t = c + 40$ to find the temperature t if you know the number of chirps c a cricket makes in 15 seconds.

Chirps in 15 seconds	$c + 40$	Temperature (°F)
21	21 + 40	61
23	23 + 40	63
31	31 + 40	71
50	50 + 40	90
47	47 + 40	87

The chart shows the relationship between various numbers of chirps in 15 seconds and the temperature. The domain is {21, 23, 31, 50, 47}, and the relation is {(21, 61), (23, 63), (31, 71), (50, 90), (47, 87)}. The relation can also be shown in a graph or scatter plot like the one shown at the right.

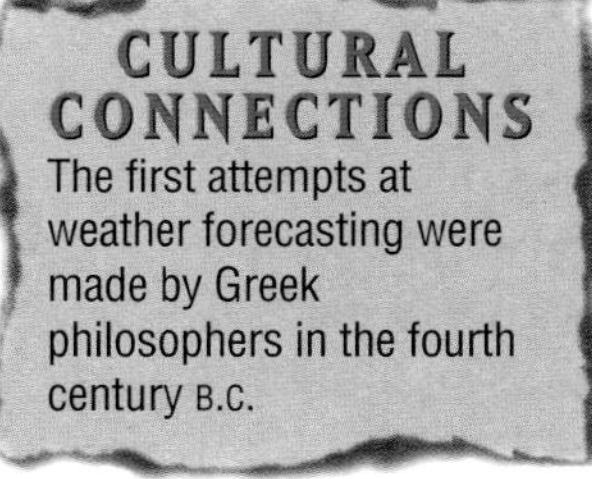

CULTURAL CONNECTIONS
The first attempts at weather forecasting were made by Greek philosophers in the fourth century B.C.

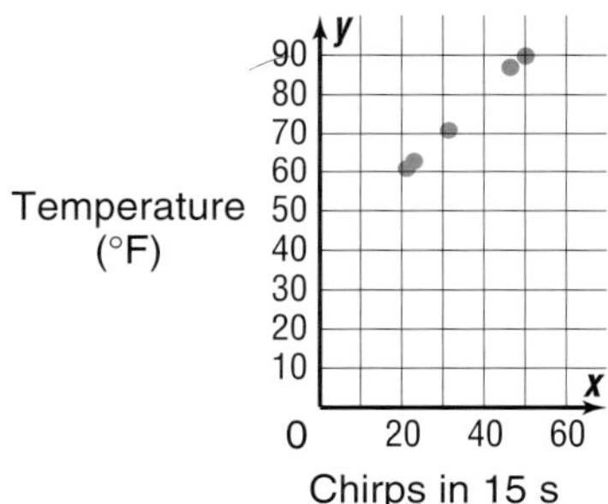

Learning the Concept

Recall that solving an equation means to replace the variable so that a true sentence results. The solutions of an equation with two variables are ordered pairs.

To find a solution of such an equation, choose any value for x, substitute that value into the equation, and find the corresponding value for y.

It is often convenient to organize the solutions in a table.

Example 

a. Find four solutions for the equation $y = -2x + 1$. Write the solutions as ordered pairs.
b. Graph the equation $y = -2x + 1$.

a. Choose four convenient values for x. Substitute each value for x in the expression $-2x + 1$.

Do the computation to find y.

Four solutions are $(-2, 5)$, $(0, 1)$, $(1, -1)$, and $(3, -5)$.

x	$-2x + 1$	y
−2	−2(−2) + 1	5
0	−2(0) + 1	1
1	−2(1) + 1	−1
3	−2(3) + 1	−5

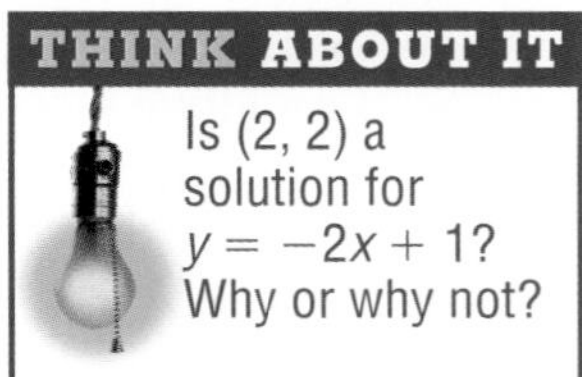

b. First, graph the ordered pairs you found in part a.

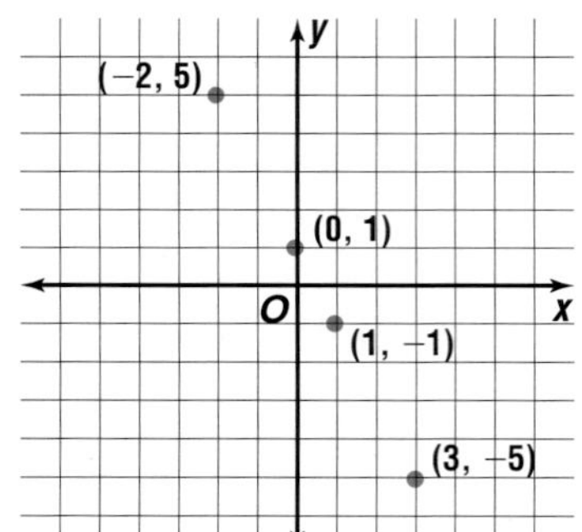

Then draw the line that contains these points.

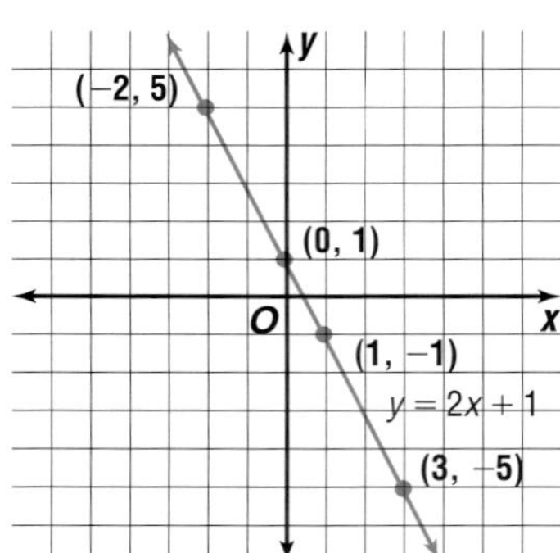

The ordered pair for every point on the line in part b of Example 1 is a solution of $y = -2x + 1$. It appears from the graph that $(-1.5, 4)$ and $(2, -3)$ are also solutions. You can check this by substitution.

Check $(-1.5, 4)$.
$y = -2x + 1$
$4 \stackrel{?}{=} -2(-1.5) + 1$
$4 = 4$ ✓

Check $(2, -3)$.
$y = -2x + 1$
$-3 \stackrel{?}{=} -2(2) + 1$
$-3 = -3$ ✓

An equation like $y = -2x + 1$ is called a **linear equation** because its graph is a straight line. An equation with two variables has an infinite number of solutions.

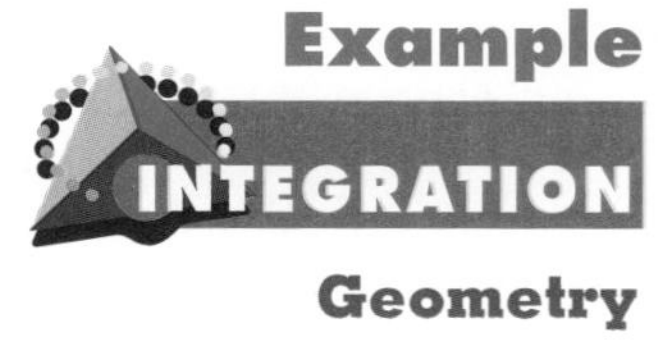

In any right triangle, the sum of the measures of the two acute angles is 90°. If x and y represent the measures of the two acute angles, then the equation $x + y = 90$ models this condition.

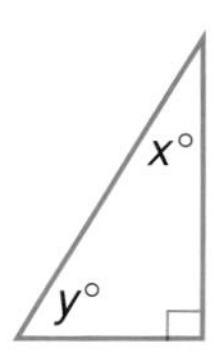

a. Graph the equation.

b. Use the graph to name another solution of the equation.

c. Is (−10, 100) a solution of the equation?

a. First solve the equation for y in terms of x.

$x + y = 90$
$y = 90 - x$ *Subtract x from each side.*

Now make a table of ordered pairs that satisfy the equation. Since any line can be defined by only two points, only two ordered pairs are needed. Finding a third ordered pair is a good idea to check the accuracy of the first two.

x	90 − x	y	(x, y)
45	90 − 45	45	(45, 45)
10	90 − 10	80	(10, 80)
70	90 − 70	20	(70, 20)

Then graph the ordered pairs and connect them with a line.

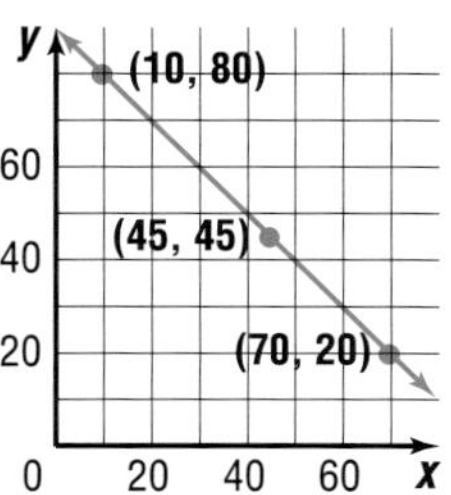

b. Since (50, 40) is on the graph, it is another solution of the equation.

c. Since $-10 + 100 = 90$, the ordered pair is a solution of the equation. However, since angles in triangles cannot have negative measure, the domain for this particular situation should not include negative numbers. *Should the domain include 0?*

Checking Your Understanding

Communicating Mathematics

Read and study the lesson to answer these questions.

1. **Determine** which ordered pair, (10, 20), (8, 4), or (300, 600), is a solution of $y = 0.5x$.
2. **Explain** why a linear equation has an infinite number of solutions.
3. **Indicate** which equation is graphed at the right.
   a. $x - y = 3$
   b. $x + y = 3$
   c. $y = x + 3$
   d. $y = -x - 3$

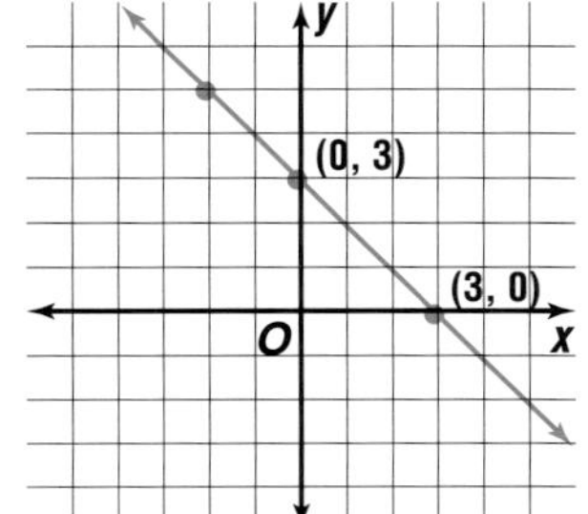

4. **Define** a linear equation.
5. **Describe** how you would graph the equation $x + y = 3$.

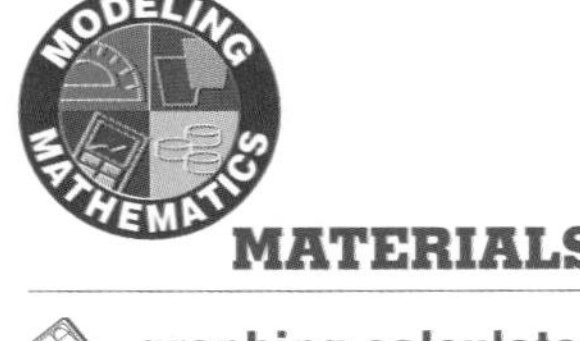

MATERIALS

graphing calculator

6. You can graph linear equations on a graphing calculator. First, clear all the STAT PLOTS. Next, press [Y=] to access the Y= list. Use the [CLEAR] key to remove any equations that are already in the list. Then enter the equation $y = 3x - 2.1$ in as function Y_1.

   Enter: 3 [X,T,θ] [−] 2.1

   Press [ZOOM] 6 to select the standard viewing window and complete the graph.

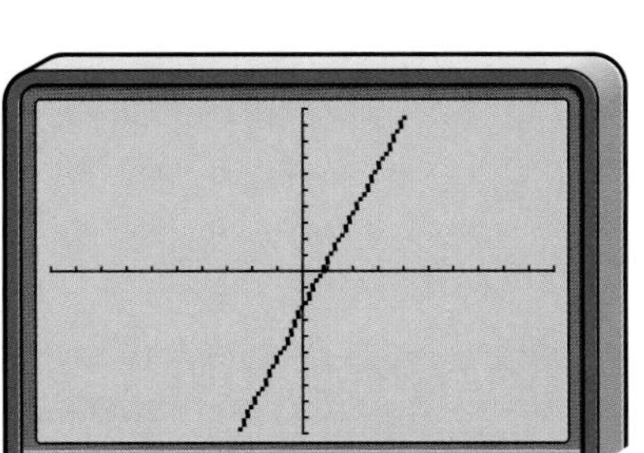

   Graph each equation in the standard viewing window.
   a. $y = 13 - 5.5x$
   b. $y = 2x + 3.7$

Guided Practice

7. Which ordered pair(s) is a solution of $x - 3y = -7$?
   a. (2, 4) **b.** (2, −1) **c.** (2, 3) **d.** (−1, 2)

Copy and complete the table for each equation. Then use the results to write four solutions for each equation. Write the solutions as ordered pairs.

8. $y = x + 4$

x	$x + 4$	y
-1	$-1 + 4$	
0		
1		
3		

9. $y = 1.5x - 2$

x	$1.5x - 2$	y
-2	$1.5(-2) - 2$	
0		
2		
4		

Find four solutions for each equation. Write the solutions as ordered pairs.

10. $y = x - 7$ **11.** $y = 5x + 2.8$ **12.** $y = -2x$

Determine whether each relation is linear.

13.

x	-2	-1	0	1	2
y	4	5	-1	2	0

14.

x	-1	-1	-1	-1	-1
y	-7	-3	0	4	12

Graph each equation.

15. $y = x - 7$ **16.** $y = 2.5x + 2$ **17.** $y = -2$ **18.** $y = \frac{1}{3}x$

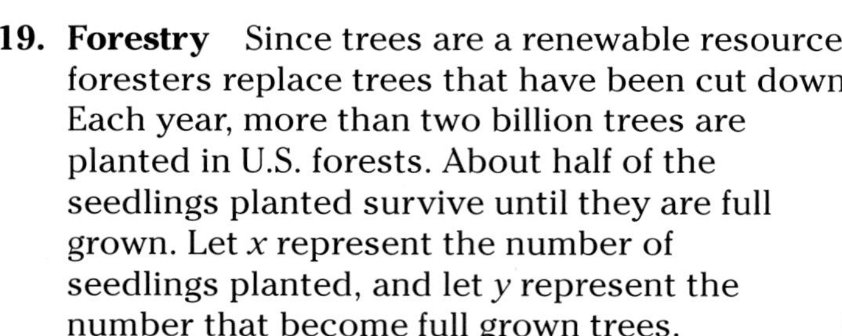

19. Forestry Since trees are a renewable resource, foresters replace trees that have been cut down. Each year, more than two billion trees are planted in U.S. forests. About half of the seedlings planted survive until they are full grown. Let x represent the number of seedlings planted, and let y represent the number that become full grown trees.

FYI

The American Forest Council estimates that nearly 70 percent of the forest land that existed in the U.S. when European settlers first landed in the early 1600s is still forested.

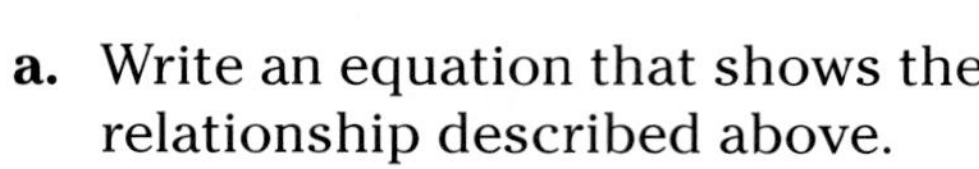

a. Write an equation that shows the relationship described above.

b. Make a table and find four solutions of the equation. Write the solutions as ordered pairs.

c. Why do negative values for x not make sense?

d. Graph the equation.

Exercises: Practicing and Applying the Concept

Independent Practice

Which ordered pair(s) is a solution of the equation?

20. $4x + 2y = 8$ **a.** $(2, 0)$ **b.** $(0, 2)$ **c.** $(0.5, -3)$ **d.** $(1, -2)$

21. $2a - 5b = 1$ **a.** $(-2, -1)$ **b.** $(2, 1)$ **c.** $(7, 3)$ **d.** $(-7, -3)$

22. $3x = 8 - y$ **a.** $(3, 1)$ **b.** $(2, 2)$ **c.** $(4, -4)$ **d.** $(8, 0)$

Find four solutions for each equation. Write the solutions as ordered pairs.

23. $y = 2.5x$ **24.** $y = 4x$ **25.** $y = 3x + 7$

26. $y = x$ **27.** $y = -x - 4$ **28.** $y = 10x - 1$

29. $x + y = 1$ **30.** $2x + y = 10$ **31.** $y - x = 1$

32. $y = \frac{1}{3}x$ **33.** $y = -\frac{1}{2}x + 3$ **34.** $y = \frac{2}{3}x - 1$

35. $x + y = 0$ **36.** $y = 5$ **37.** $x = -2$

Determine whether each relation is linear.

38. $3x + y = 20$ **39.** $y = x^2$ **40.** $y = 5$

Graph each equation.

41. $y = x + 1.5$ **42.** $y = -\frac{1}{2}x$ **43.** $y = \frac{2}{3}x + 1$

44. $y = x$ **45.** $2x + y = 5$ **46.** $x - y = 4$

47. $x + y = 2$ **48.** $y = 5$ **49.** $x = -2$

Translate each sentence into an equation. Then find four solutions for each equation. Write the solutions as ordered pairs. Graph each equation.

50. Some number is four more than a second number.

51. Some number is one-fourth a second number.

52. Four times a number minus a second number is 8.

53. A solution of $y = cx + 1$ is $(2, 5)$. Find the value of c.

54. A solution of $2x + ay = 4$ is $(-1, 2)$. Find the value of a.

55. Find a linear equation that has the following solutions.
$\{(0, 0), (1, -3), (2, -6), (-1, 3)\}$

Critical Thinking

56. Find an ordered pair that is a solution for both $x + y = 15$ and $x - y = -1$.

Applications and Problem Solving

57. Geometry The formula for finding the perimeter of a square with sides s units long is $P = 4s$.

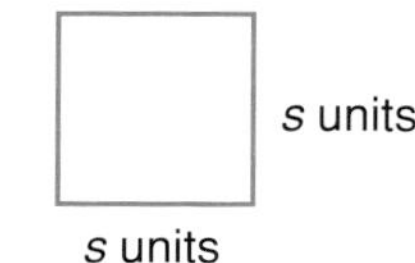

a. Find five ordered pairs of values that satisfy this condition.

b. Draw the graph that contains these points.

c. Why do negative values of s make no sense?

58. Physics As a thunderstorm approaches, you see lightning as it occurs, but you hear the accompanying sound of thunder a short time afterward. The distance y, in miles, that sound travels in x seconds, is given by the equation $y = 0.21x$.

a. Create a table to find five ordered pairs of values that relate the time it takes to hear thunder and the distance from the lightning.

b. Graph the equation.

c. How far away is lightning when the thunder is heard 2.5 seconds after the light is seen?

59. Weather The freezing point for water is 0°C and 32°F, and the boiling point is 100°C and 212°F. The relationship between Fahrenheit and Celsius temperatures is linear.

a. Write two ordered pairs that relate these equivalent Celsius and Fahrenheit temperatures.

b. Draw the graph that contains these two points.

c. How could you use this graph to find other equivalents of other Celsius and Fahrenheit temperatures?

d. Use the graph to estimate two equivalent Celsius and Fahrenheit temperatures.

Mixed Review

60. Health What relationship (*positive*, *negative*, or *none*) do you think a scatter plot of Calorie intake and weight gained might show? (Lesson 8-2)

61. Solve $2a - 5 > 17$. (Lesson 7-6)

62. Express 0.004976 in scientific notation. (Lesson 6-9)

63. Estimate 2.49×1.9. (Lesson 6-2)

64. Carpentry When remodeling a living room, a carpenter needed a $14\frac{3}{4}$-inch piece of molding. How much remained after she cut it from a 36-inch piece of molding? (Lesson 5-5)

65. Find $\frac{7^4}{7^2}$. Express the quotient using an exponent. (Lesson 4-8)

66. Geometry The perimeter of any square is 4 times the length of one of its sides. If the perimeter of a square is 56 inches, what is the length of each side of the square? (Lesson 3-5)

67. Graph and label $A(-2.5, 4)$ and $B\left(-1\frac{1}{2}, -6\right)$. (Lesson 2-2)

68. Solve $4b = 36$ mentally. (Lesson 1-6)

69. Simplify $5n + 9n$. (Lesson 1-5)

70. Find the value of $4[12(22 - 19) - 3 \cdot 6]$. (Lesson 1-2)

From the FUNNY PAPERS

In the United States, we measure temperature in degrees Fahrenheit, but most other countries use degrees Celsius.

1. Is it really cooler at "Celsius's Place"? Explain.
2. Use the graph you made in Exercise 59 to find the temperature in degrees Fahrenheit if it is 36°C.

"LET'S GO OVER TO CELSIUS'S PLACE. I HEAR IT'S ONLY 36° OVER THERE."

34. Anthropology When a human skeleton is found, an anthropologist uses the length of certain bones to determine the height of the living person. A *femur* is the bone from the knee to the hip. The height h, in centimeters, of a female with a femur of length x is estimated by $h = 61.412 + 2.317x$.

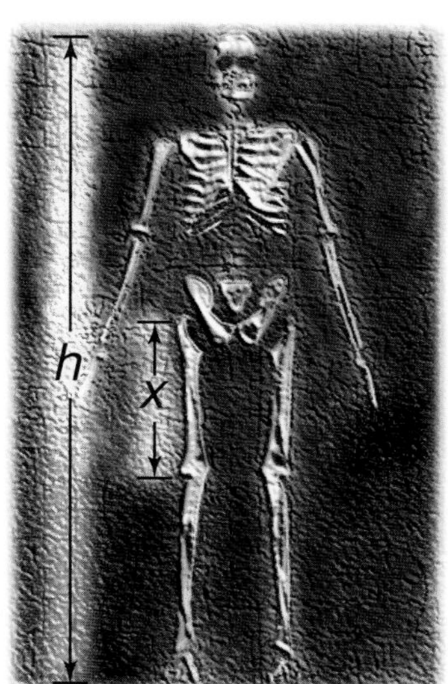

a. Graph the equation $h = 61.412 + 2.317x$.

b. Is the relation described by $h = 61.412 + 2.317x$ a function? Explain.

c. A woman's femur measuring 49 cm is found in some ruins. What was the height of the person?

Mixed Review

35. Mining The deepest mine in the world is located in Carletonville, South Africa. The temperature of the walls of the mine varies with their depth. The temperature in degrees Fahrenheit, y, is estimated by $y = 18x + 66.5$, where x is the depth in kilometers. (Lesson 8-3)

a. Create a table to find five ordered pairs of values that relate the depth of a mine and the temperature of its walls.

b. Graph the equation.

c. What would the temperature of the walls be if the mine is 3.6 kilometers deep?

36. Geometry Find the circumference of a circle with a radius of 14 yards. (Lesson 7-4)

37. Find the value of $\left(-\frac{3}{4}\right)^2$. (Lesson 6-3)

38. Write $\frac{15rs^2}{50rs}$ in simplest form. (Lesson 4-6)

39. Solve $-3a \geq 18$ and graph the solution. (Lesson 3-7)

40. Geography The table at the right shows the number of miles of seacoast in selected states. Make a bar graph of the data. (Lesson 1-10)

State	Miles of Seacoast
Alaska	5580
Delaware	28
Florida	1350
Georgia	100
Oregon	296
Maine	228
Mississippi	44
Texas	367

8-5 Problem-Solving Strategy Draw a Graph

Setting Goals: *In this lesson, you'll solve problems by using graphs.*

Modeling a Real-World Application: Air Travel

Imagine you are sitting in your summer clothes and sandals eating lunch and it's 74 degrees below zero! It happens every day on high-flying airplanes that are far from the heat of Earth's surface.

Altitude (thousands of feet)	Temp. (°F)
5	41
10	23
15	4
20	−15
25	−33

The temperature at different altitudes above sea level when the temperature at sea level is 60°F is shown in the table at the left.

The table does not include an altitude for −74°. At what altitude would a jet be flying when the thermometer has this reading? *This problem will be solved in Example 1.*

Learning the Concept

A graph represents the same information as its equation. But, because a graph is visual, it allows you to see patterns that may not be obvious from the equation. A graph is a powerful tool in problem solving.

Example 1

Air Travel

Refer to the application at the beginning of the lesson. Find the altitude where the temperature is −74°F. Assume that the temperature decreases at a constant rate.

Explore You need to estimate an altitude that is not in the table. You know the temperature at five different altitudes and that the temperature decreases at a constant rate.

Plan One way to solve this problem is to graph the given information. Then read the graph to find the altitude that corresponds to a temperature of −74°F.

Solve Let the horizontal axis of the graph represent altitude. Let the vertical axis represent the temperature. Graph each ordered pair (*altitude, temperature*) using the pairs of values taken from the table. Then draw the line that contains these points.

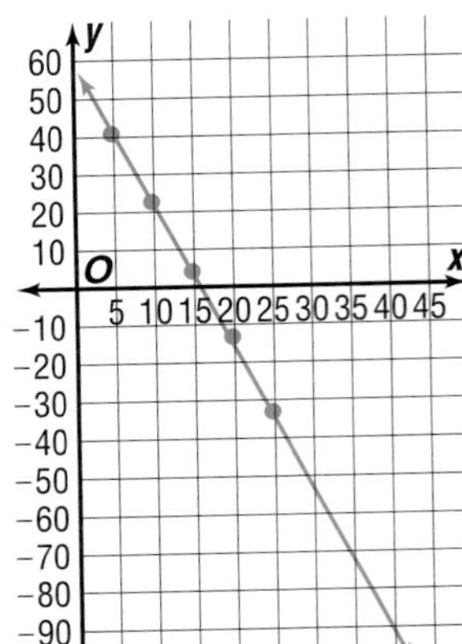

An estimate of $-74°$ would be about halfway between -70 and -80 on the vertical axis. Estimate the altitude by reading the corresponding value on the horizontal axis. The ordered pair $(36, -74)$ appears to be on the graph, so $-74°$ corresponds to about 36,000 feet.

Examine Is the solution reasonable? The lowest temperature in the table, $-33°$, corresponds to an altitude of 25,000 feet. The temperature drops about 19 degrees for each 5,000 feet increase in altitude. $-74 - (-33°) = -41°$ and -41 is about $2(-19)$. Therefore, the altitude will be about $2 \times 5{,}000$ or 10,000 feet higher. Since $25{,}000 + 10{,}000 = 35{,}000$, a solution of 36,000 is reasonable.

You can use a TI-82 graphing calculator to plot points and construct a line without erasing the points. Then you can use the trace function to answer the question.

Example 2

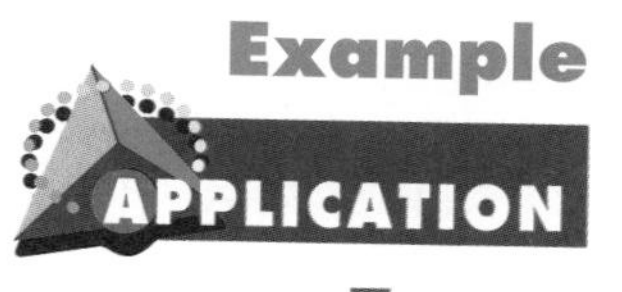

Energy

The ordered pairs (50.7, 16.47) and (99.92, 25.42) represent data from two rows of a table comparing cubic meters of natural gas used and the cost. What would the bill for 140 cubic meters of natural gas be?

Use the scatter plot procedure on page 379 to plot the two ordered pairs using □ as the mark.

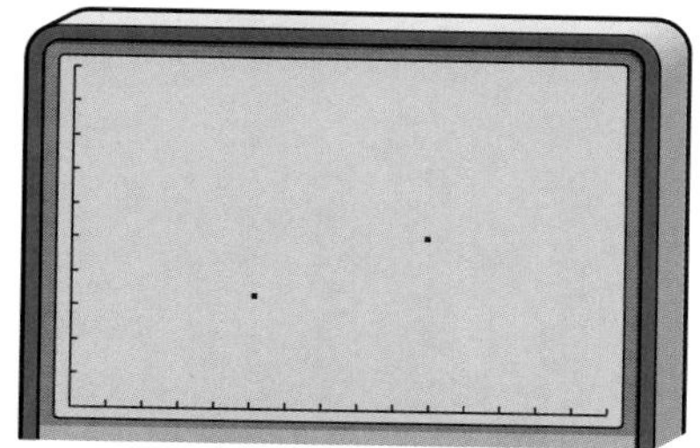

The calculator can draw the graph of the line that passes through the points.

Enter: STAT ▶ 5 ENTER Y=
VARS 5 ▶ ▶ 7 GRAPH

Press TRACE and the up arrow key, to trace the line. Then press the right arrow key until the cursor gets to the point on the line where the x-coordinate is about 140. The corresponding value of y is the cost for 140 cubic meters of gas.

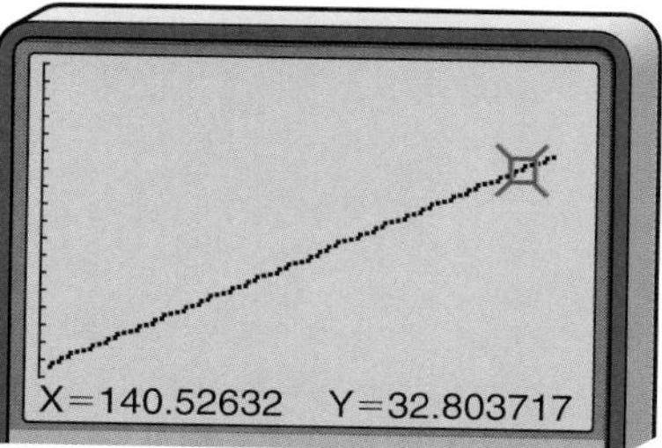

The cost will be about $32.80.

Checking Your Understanding

Communicating Mathematics

Read and study the lesson to answer these questions.

1. **Explain** how a graph can provide the same information as a table and at the same time provide more information.
2. **Describe** how you would make a graph using a table of data.

Guided Practice

3. **Estimate** the corresponding value for each item using the graph you created in Example 2.
   a. cost: $21.50
   b. gas used: 30 cubic meters

4. **Forestry** An area has been reforested. After five years, one of the trees is 2 meters tall. After six more years, the same tree is 3.2 meters tall. Name two ordered pairs that can be used to graph the information.

5. **Oceanography** The average atmospheric pressure at sea level is 14.7 pounds per square inch. As divers go deeper into the ocean, the pressure increases as shown in the chart. Use a graph to estimate the pressure at 10,000 feet below sea level. Assume that the rate is constant.

Depth (feet)	Pressure (lb/in^2)
(Sea level) 0	14.7
500	237
1500	683
4500	2019

Exercises: Practicing and Applying the Concept

Independent Practice

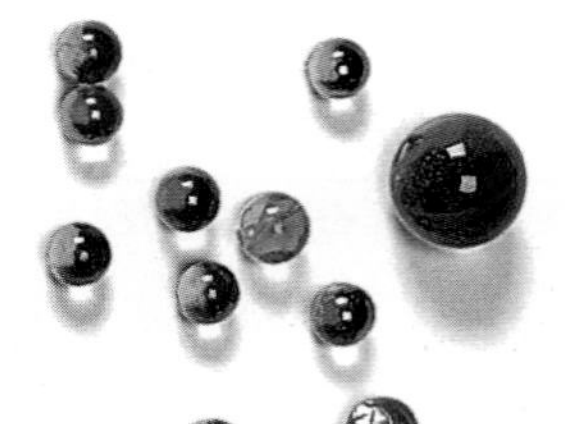

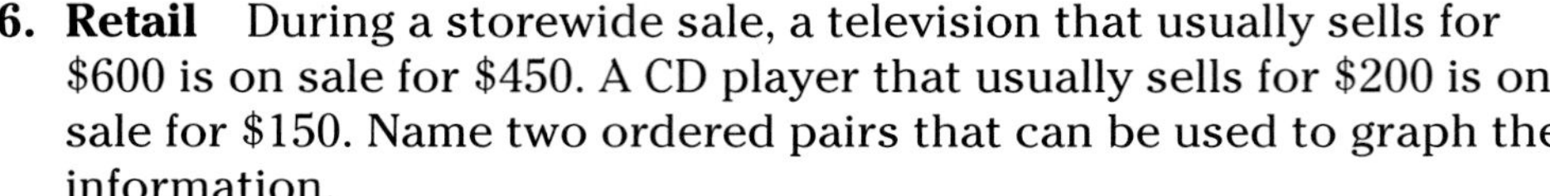

6. **Retail** During a storewide sale, a television that usually sells for $600 is on sale for $450. A CD player that usually sells for $200 is on sale for $150. Name two ordered pairs that can be used to graph the information.

Use a graph to solve each problem. Assume that the rate is constant in each problem.

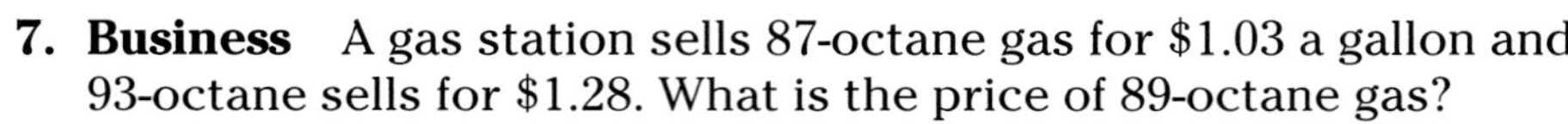

7. **Business** A gas station sells 87-octane gas for $1.03 a gallon and 93-octane sells for $1.28. What is the price of 89-octane gas?

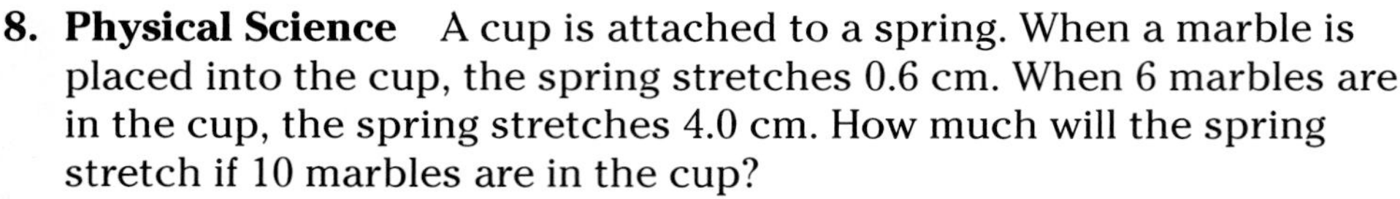

8. **Physical Science** A cup is attached to a spring. When a marble is placed into the cup, the spring stretches 0.6 cm. When 6 marbles are in the cup, the spring stretches 4.0 cm. How much will the spring stretch if 10 marbles are in the cup?

9. **Energy** Natural gas companies use *degree days* to determine the amount of gas to charge for on a calculated bill. In September, there were 300 degree days, and a homeowner used 85 cubic meters of natural gas. In January, there were 700 degree days, and the homeowner used 265 cubic meters of gas. How much gas would be used if there are 450 degree days?

Choose

Estimation
Mental Math
Calculator
Paper and Pencil

Solve using the graph. Explain your method.

10. Tamoko knows that her car can travel about 30 miles on one gallon of gasoline. The graph shows this relationship.
    a. Tamoko is starting on a 250-mile trip. If she has 6 gallons of gasoline in her car, will she have to buy more gasoline sometime during the trip?
    b. If so, approximately when?

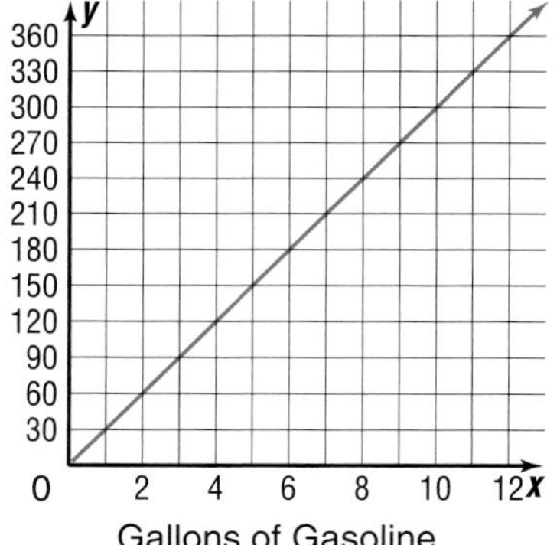

11. **Health** Steve used a chart to see if he was close to the average weight for his height. The chart gave a weight of 130 pounds for a male 60 inches tall. The average weight for a male 66 inches tall was 143 pounds. Steve is 6 feet tall. What should Steve's weight be?

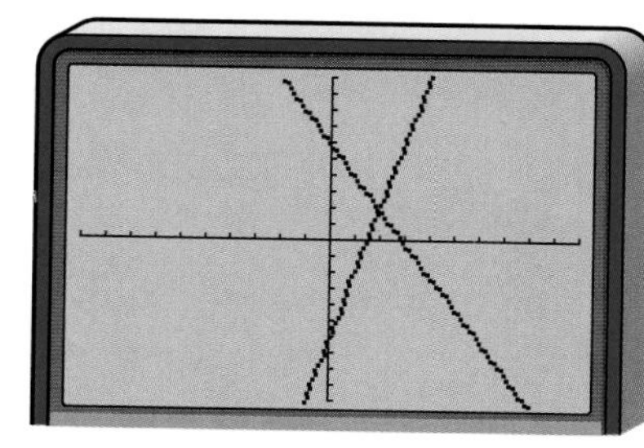

Critical Thinking

12. The graphs of $f(x) = 4x - 6$ and $g(x) = -2x + 6$ are shown at the right. Find an ordered pair that is a solution for both equations. Justify your answer.

Mixed Review

13. If $g(x) = 2x - 1$, find $g\left(\frac{5}{2}\right)$. (Lesson 8-4)

14. **Geometry** The area of a trapezoid can be found using $A = h \cdot \frac{1}{2}(b_1 + b_2)$. A trapezoid has an area 64 square inches, a height of 8 inches, and one base 7 inches long. Find the length of the other base. (Lesson 7-2)

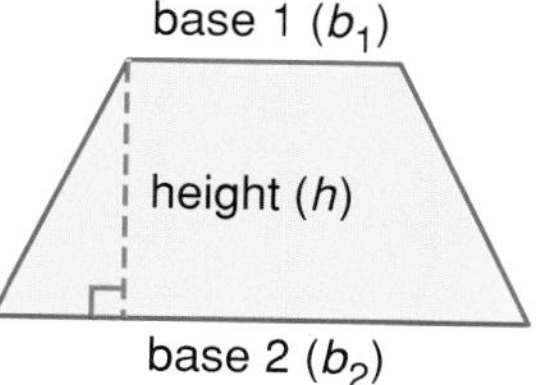

15. Solve $y = 4\frac{3}{4} - 5\frac{1}{6}$. (Lesson 5-5)

16. Evaluate $(a^2 - b)^2$ if $a = 2$ and $b = 4$. (Lesson 4-2)

17. Is the quotient of -42 and 7 positive or negative? (Lesson 2-8)

Self Test

Write the domain and range of the relation. Then determine whether each relation is a function. (Lesson 8-1)

1.

x	4	1	3	6
y	2	3	3	4

2. $\{(4, 2), (-3, 2), (8, 2), (8, 9), (7, 5)\}$

Use the scatter plot to answer each question. (Lesson 8-2)

3. What kind of relationship, if any, is shown by the scatter plot?

4. What does the point with the triangle around it represent?

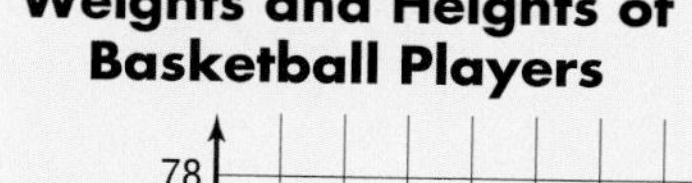

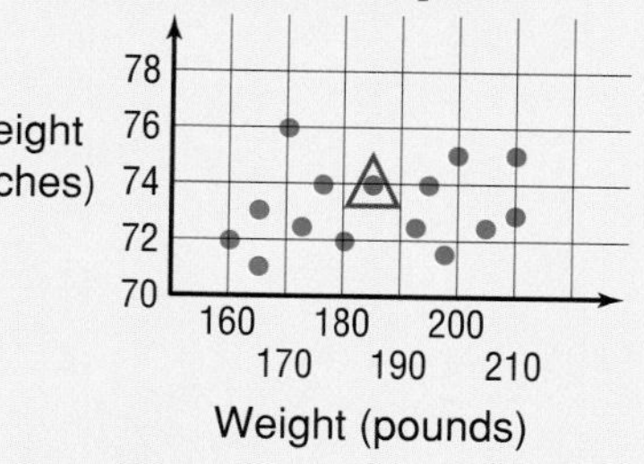

Find four solutions for each equation. Write the solutions as ordered pairs. Then draw the graph. (Lesson 8-3)

5. $y = 4x - 2$

6. $y = \frac{1}{2}x + 5$

Given $f(x) = 9x - 4$, find each value. (Lesson 8-4)

7. $f(6)$

8. $f(0)$

9. $f(-3)$

10. **Physical Science** At 20°C, sound travels 172 meters in 0.5 seconds and 206 meters in 0.6 seconds. Name two ordered pairs that could be used to graph this information. (Lesson 8-5)

8-6 Slope

Setting Goals: *In this lesson, you'll find the slope of a line.*

Modeling a Real-World Application: Construction

Rick Hansen is in the Guinness Book of World Records for the longest wheelchair journey. He traveled just under 25,000 miles in 26 months, through 4 continents and 34 countries.

In many buildings, the wheelchair ramps are built next to the walls with stairways in the middle. If you were to compare the handrail on the ramp to the handrail on the stairway, you would notice that the handrail on the ramp is not as steep as the one for the stairs.

The steepness of the handrails depends on the vertical change and the horizontal change. It can be expressed as a ratio.

$$\text{steepness} = \frac{\text{vertical change}}{\text{horizontal change}}$$

Learning the Concept

In mathematics, the **slope** m of a line describes its steepness. The vertical change is called the **change in *y***, and the horizontal change is called the **change in *x***.

$$\text{slope or } m = \frac{\text{change in } y}{\text{change in } x}$$

Example  **Find the slope of the line graphed below.**

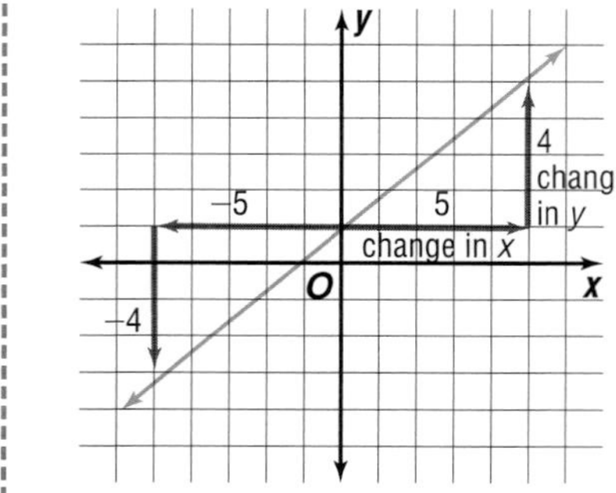

In Quadrant I, the change in y is 4, and the corresponding change in x is 5. Therefore, the slope of the line is $\frac{4}{5}$. Is the slope of the line the same in Quadrant III?

$$\frac{\text{change in } y}{\text{change in } x} \text{ or } \frac{-4}{-5} \text{ or } \frac{4}{5}$$

The slopes are the same.

The slope of a line can be determined by using the coordinates of any two points on the line. The change in y can be found by subtracting the y-coordinates. Likewise, the change in x can be found by subtracting the corresponding x-coordinates.

$$\text{slope} = \frac{\text{change in } y}{\text{change in } x} \text{ or } \frac{\text{difference in } y \text{ coordinates}}{\text{difference in } x \text{ coordinates}}$$

Example  **Find the slope of the line that contains $A(-2, 5)$ and $B(4, -5)$. Then graph the line.**

$$\text{slope} = \frac{\text{difference in } y\text{-coordinates}}{\text{difference in } x\text{-coordinates}}$$

$$\text{slope of line } AB = \frac{5 - (-5)}{-2 - 4}$$

$$= \frac{10}{-6} \text{ or } -\frac{5}{3}$$

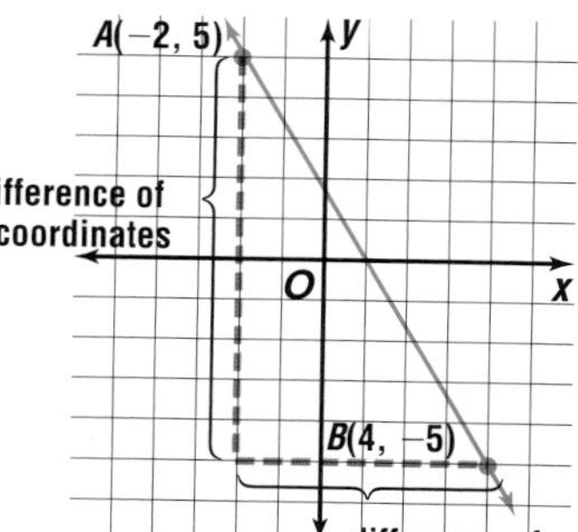

The slope of the line is $-\frac{5}{3}$.

Graph the two points and draw the line. Use the slope to check your graph by selecting any point on the line and then go down 5 units and right 3 units. This point should also be on the line.

Sometimes the vertical change is referred to as the *rise*, and the horizontal change is referred to as the *run*. You can remember slope as *rise over run*.

$$\text{slope} = \frac{\textit{rise}}{\textit{run}}$$

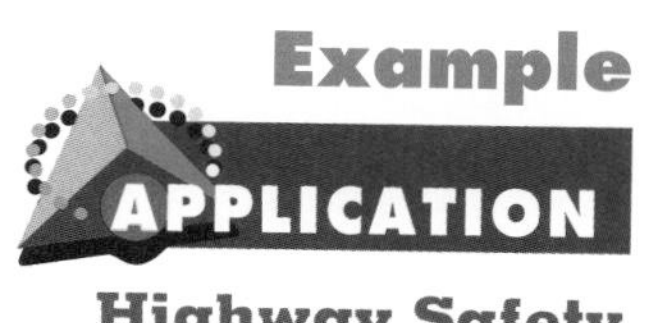

Highway Safety

Example 3 **Signs indicating the slope of an upcoming hill are often posted in hilly areas. Find the slope of a road that increases 633.6 feet in 2 miles.**

$$\text{slope} = \frac{\textit{rise}}{\textit{run}} = \frac{\textit{vertical change}}{\textit{horizontal change}}$$

Since 1 mile equals 5280 feet, 2 miles equals 10,560 feet.

$$\text{slope} = \frac{633.60}{10,560}$$

$$= \frac{3}{50}$$

The slope of the road is $\frac{3}{50}$. This means that the road rises 3 feet vertically for every 50 feet traveled horizontally.

THINK ABOUT IT

Why is the slope of a horizontal line 0?

Why does a vertical line have no slope?

Connection to Geometry

The figure at the right shows the graphs of two lines that will never intersect. These lines are **parallel**. Is there a special relationship between the slopes of parallel lines?

$$\text{slope of line } CD = \frac{3 - (-3)}{1 - (-1)} = \frac{6}{2} \text{ or } 3$$

$$\text{slope of line } EF = \frac{3 - 0}{3 - 2} = \frac{3}{1} \text{ or } 3$$

The slopes are equal.

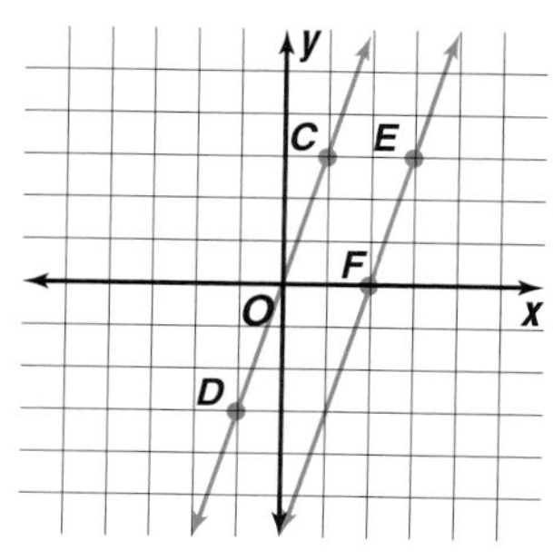

Checking Your Understanding

Communicating Mathematics

Read and study the lesson to answer these questions.

1. **Explain** what it means when a line has a slope of 3.

2. Match all the positions of the drawing of a ski slope below with its corresponding slope as shown in the table at the right.

Ski Surface	Slope
A. vertical drop	negative
B. downhill	none
C. flat	positive
D. uphill	0

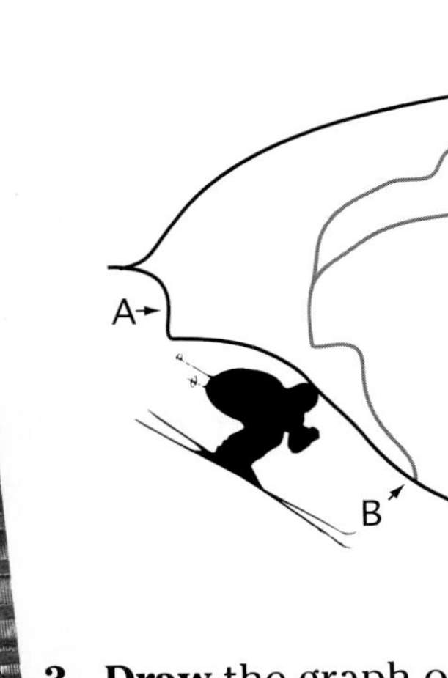

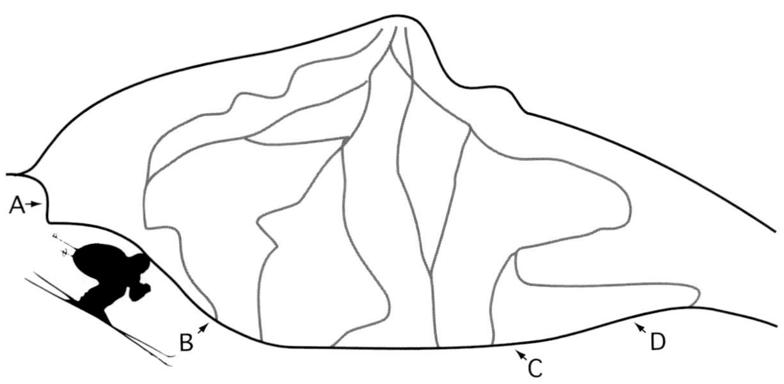

3. **Draw** the graph of a line that has a slope of $\frac{2}{3}$.

4. **Demonstrate** how to find the slope of a line when you know the coordinates of two points on the line.

5. Write a note to someone at home that describes how to tell the difference between the graph of a line with a positive slope and the graph of a line with a negative slope. Include drawings in your note.

Guided Practice

Find the slope of each line.

6. 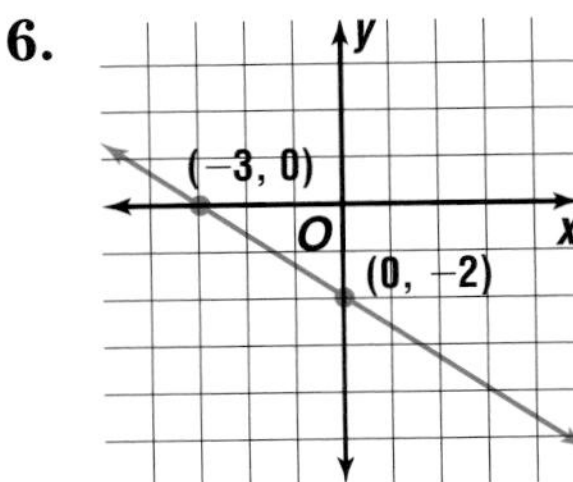

7. 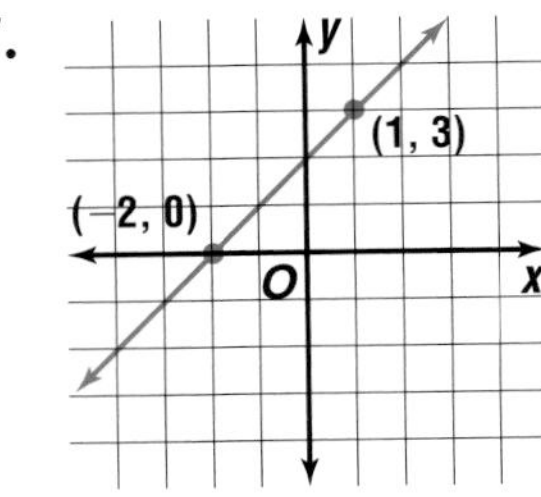

8. 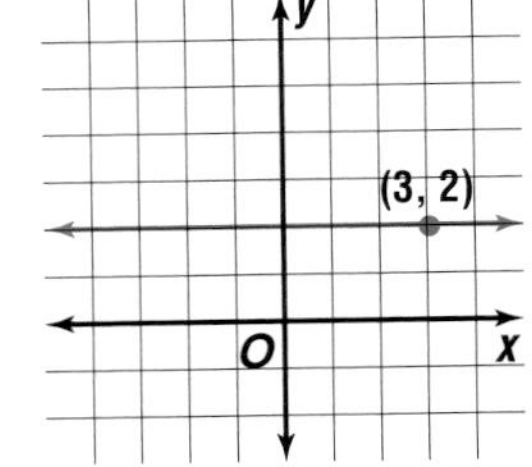

Find the slope of the line that contains each pair of points.

9. $R(9, -2), S(3, -5)$
10. $T(14, 3), U(-11, 3)$
11. $V(-1, -2), X(2, -5)$
12. $B(-6, -4), C(-8, -3)$

13. **Painting** A ladder that reaches a height of 16 feet is placed 4 feet away from the wall. What is the slope of the ladder?

Exercises: Practicing and Applying the Concept

Independent Practice

Determine the slope of each line named below.

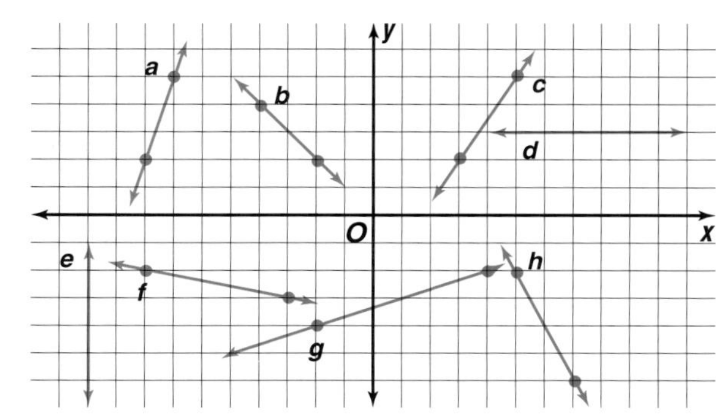

14. a
15. b
16. c
17. d
18. e
19. f
20. g
21. h
22. Look at lines b, f, and h above. These lines have negative slopes. Make a conjecture about how the absolute value of the slope of the line affects how steeply it slants.
23. **Statistics** What does a scatter plot that shows a positive relationship share with the graph of a line that has a slope of 1?

Find the slope of the line that contains each pair of points.

24. $C(3, 4)$, $D(4, 6)$
25. $E(-3, 6)$, $F(-5, 9)$
26. $G(2, 3)$, $H(-1, 3)$
27. $J(1, -3)$, $K(5, 4)$
28. $M(7, -4)$, $N(9, -1)$
29. $P(5, -2)$, $Q(4, -3)$
30. $Y(0, 0)$, $Z(0.5, 0.75)$
31. $D(5, -1)$, $E(-3, -4)$
32. $F\left(\frac{3}{4}, 1\right)$, $G\left(\frac{3}{4}, -1\right)$
33. $H\left(3\frac{1}{2}, 5\frac{1}{4}\right)$, $J\left(2\frac{1}{2}, 6\right)$

Critical Thinking

34. **Geometry** The figure at the right shows two lines that are perpendicular. Parallel lines have the same slope. Make a conjecture about the slopes of perpendicular lines.

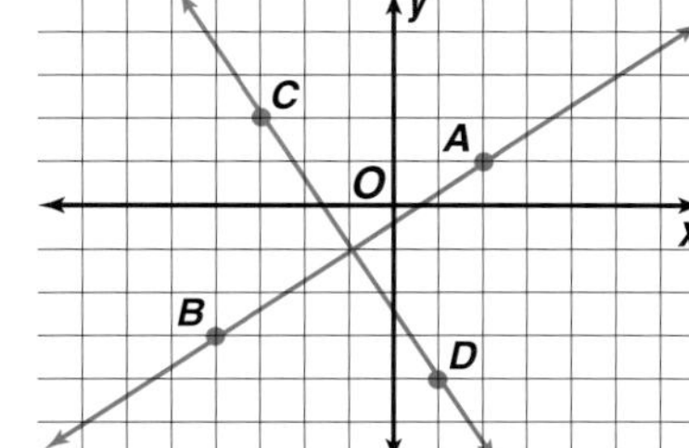

Applications and Problem Solving

35. **Carpentry** In a stairway, the slope of the handrail is the ratio of the riser to the tread. If the tread is 12 inches long and the riser is 8 inches long, what is the slope of the handrail?
36. **Statistics** The graph at the right shows how fruit juice consumption changed from 1990 through 1994.
    a. Describe the section of the graph that shows when the greatest increase occurred.
    b. Describe what no change in consumption would look like on the graph.
    c. What does a section with a negative slope mean?

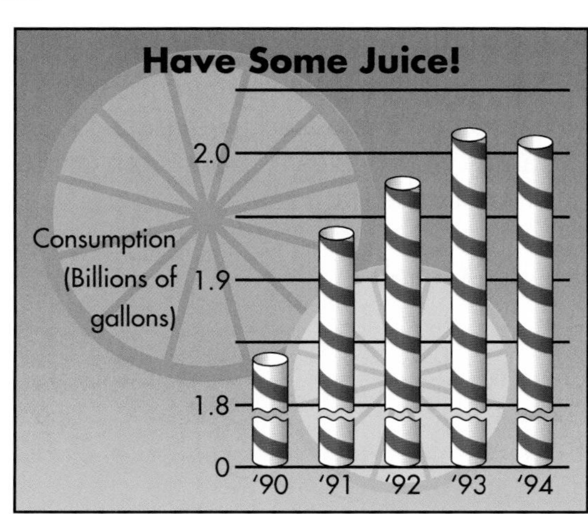

Source: Beverage Marketing Corp.

37. **Driving** The western entrance to the Eisenhower Tunnel in Colorado is at an elevation of 11,160 feet. The roadway has a downward slope of 0.00895 toward the eastern entrance, and its horizontal distance is 8941 feet. What is the elevation of the eastern end of the tunnel?

Mixed Review

38. **Travel** After 2 hours, Andrea checks the odometer in her car. She has traveled 160 kilometers. After 3 more hours, she has traveled a total distance of 400 kilometers. Use a graph to find the number of hours it will take for Andrea to travel a total distance of 560 kilometers. Assume that the rate is constant. (Lesson 8-5)
39. **Food Preparation** Suppose you are in charge of baking a ham for your family picnic. The directions tell you to bake the ham at 325° for 12 minutes for each pound. The ham weighs 7.45 pounds. To the nearest minute, how long should you bake the ham? (Lesson 6-5)
40. Round 24.692 to the nearest tenth. (Lesson 5-2)
41. Is the number 57 *prime* or *composite*? (Lesson 4-4)
42. Solve $a + (-7) = 8$. (Lesson 3-2)
43. Find the value of $6(5 + 9) \div 7$. (Lesson 1-2)
44. **Problem Solving** After half of the people at a Spanish Club meeting left, one third of those remaining began to plan the club's Cinco de Mayo (5th of May) celebration. The other 18 people were cleaning the room. How many people attended the meeting? (Lesson 1-1)

WORKING ON THE

Investigation

Refer to the Investigation on pages 368–369.

- Transcribe the data from your table in Lesson 8-2 onto a coordinate grid using the same type of axes as your scatter plot. Connect the data points.
- If you compared a graph of the data from another class, how could you tell if 11 sections move faster than 20 sections? Explain.
- Below are three graphs of wave cheer data for a stadium with 26 sections.
- Describe each graph separately. Explain what you think happened. Are all these graphs reasonable?

Add the results of your work to your Investigation Folder.

Graph A

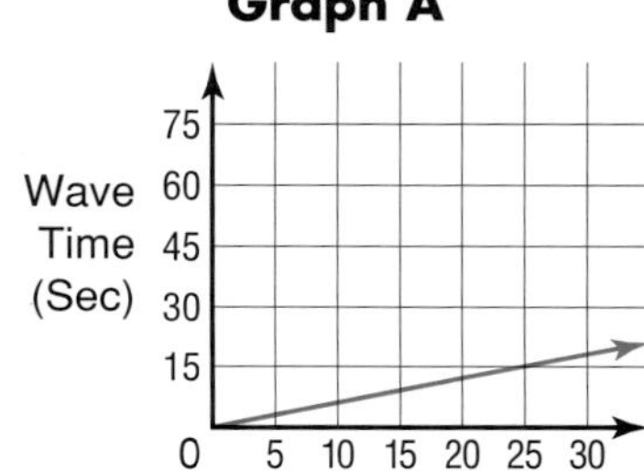

Graph B

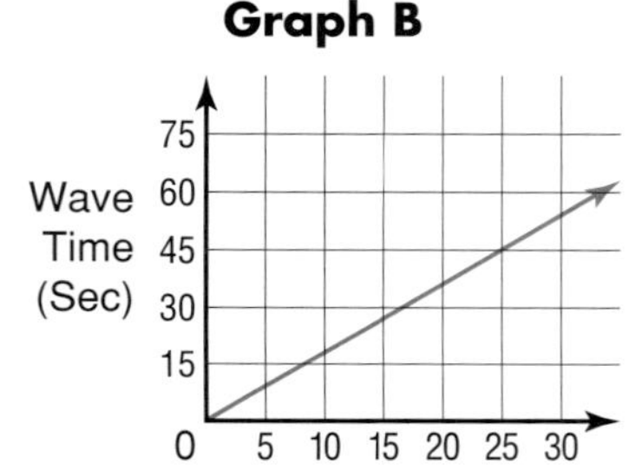

Graph C

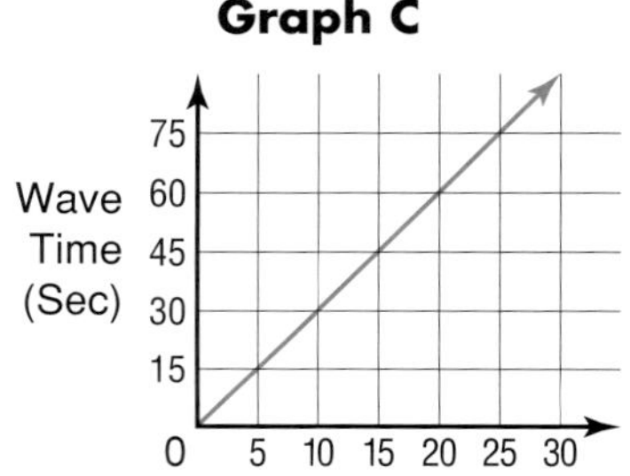

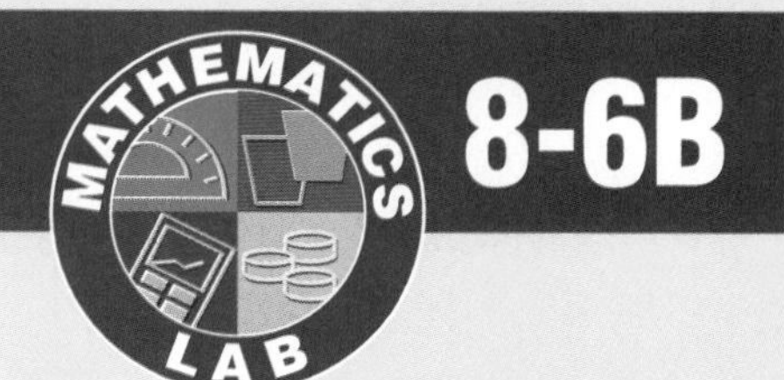

HANDS-ON ACTIVITY

8-6B Circles and Slope

An Extension of Lesson **8-6**

MATERIALS

- graph paper
- scissors
- tape
- cylindrical objects
- straightedge

Work with a partner.

- Copy the coordinate axes on a sheet of graph paper as shown at the right. Cut off the bottom of the paper so that the horizontal axis becomes the edge of the paper.

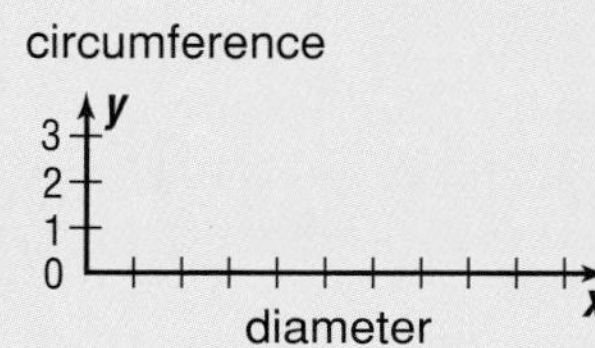

- Affix the graph paper to a wall or chalkboard.
- Hold the cylindrical base of an object against the graph paper so its left edge is at the origin and half of the cylindrical object is on either side of the horizontal axis. Mark the right side of the base where it touches the horizontal axis.

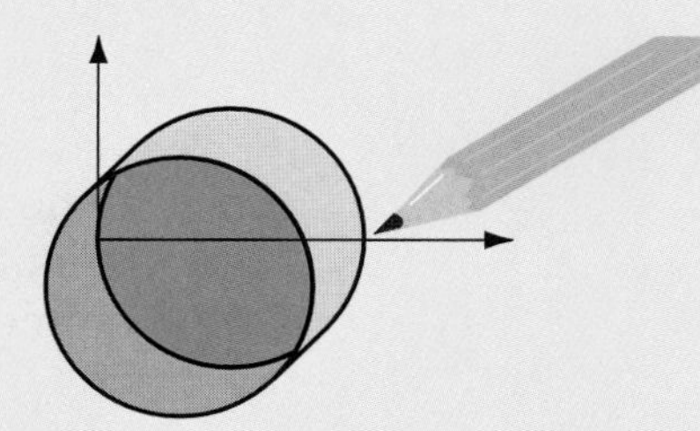

- Roll the cylindrical object up one complete rotation. The mark on the cylindrical object will be at the right side. Mark this point on the graph.

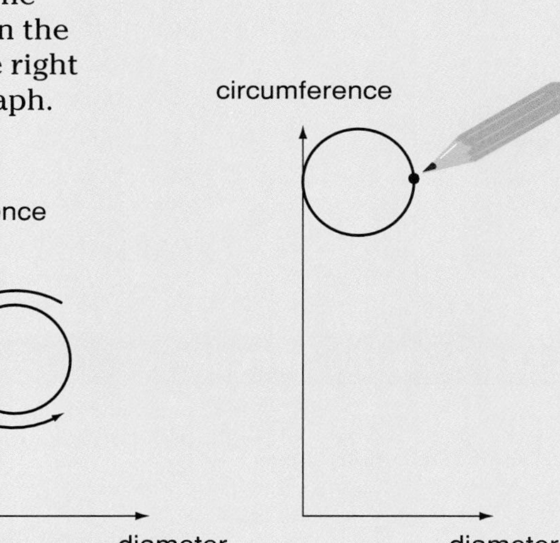

circumference
diameter

Start

circumference
diameter

Rolling

All the way around

- Repeat this procedure with a different cylindrical object.
- Write the ordered pair for each point.
- Draw a line through the two points marked on the graph paper.

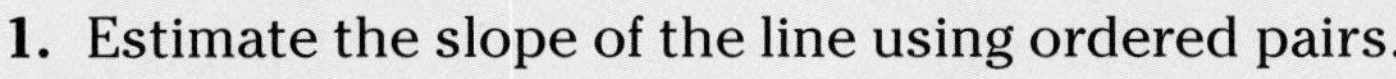

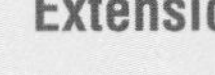

1. Estimate the slope of the line using ordered pairs.
2. Determine the slope of the line.
3. Where would you expect a point to be if you repeated the above procedure with a third cylindrical object?

Extension

4. Refer to Lesson 7-4 on circles and circumference. What should the slope of the line be? Explain your reasoning.

8-7 Intercepts

Setting Goals: *In this lesson, you'll graph a linear equation using the x- and y-intercepts or using the slope and y-intercept.*

Modeling a Real-World Application: Sports

On your mark, get set, go! Sports analysts have found that the world record for the 10,000-meter run has been decreasing steadily since 1940. The trend can be described by the equation $y = 30.18 - 0.07x$, where x is the number of years since 1940 and y is the record time. The graph of the equation is a line that crosses the x-axis at approximately (431, 0) and crosses the y-axis at (0, 30.18). The points where a graph crosses the axes are called **intercepts**.

Learning the Concept

> **THINK ABOUT IT**
>
> What kind of line only crosses the x-axis? What kind of line only crosses the y-axis?

The graph of a linear function may cross either the x-axis, the y-axis, or both axes.

The ***x*-intercept** is the x-coordinate of the point where the graph crosses the x-axis.

The ***y*-intercept** is the y-coordinate of the point where the graph crosses the y-axis.

The line graphed at the right crosses both axes.

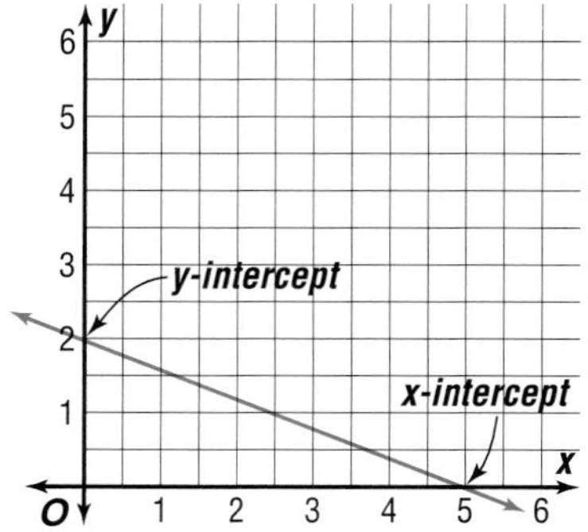

***x*-intercept** The line graphed above crosses the x-axis at (5, 0). Therefore the x-intercept is 5. Note that the corresponding y-coordinate is 0.

***y*-intercept** The line above crosses the y-axis at (0, 2). So the y-intercept is 2. Note that the corresponding x-coordinate is 0.

The y-intercept and the x-intercept can be used to graph a linear equation.

Example 1 Graph $y = \frac{3}{2}x - 6$ using the x- and y-intercepts.

Step 1 Find the x- and y-intercepts.

To find the x-intercept, let $y = 0$.

$$y = \frac{3}{2}x - 6$$
$$0 = \frac{3}{2}x - 6$$
$$0 + 6 = \frac{3}{2}x - 6 + 6$$
$$6 = \frac{3}{2}x$$
$$4 = x$$

The x-intercept is 4.
The ordered pair is (4, 0).

To find the y-intercept, let $x = 0$.

$$y = \frac{3}{2}x - 6$$
$$y = \frac{3}{2}(0) - 6$$
$$y = -6$$

The y-intercept is -6.
The ordered pair is $(0, -6)$.

Step 2 Graph the intercepts and draw the line that contains them.

Step 3 Check by choosing some other point on the line and determine whether its ordered pair is a solution of $y = \frac{3}{2}x - 6$. Try $(2, -3)$.

$$y = \frac{3}{2}x - 6$$
$$-3 \stackrel{?}{=} \frac{3}{2}(2) - 6$$
$$-3 \stackrel{?}{=} 3 - 6$$
$$-3 = -3 \checkmark$$

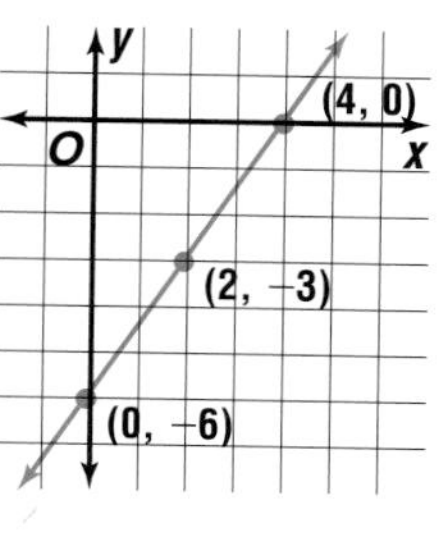

In Example 1, you may have noticed that the y-intercept for the equation was easy to find. Is there a quick way to find the slope? Use the intercepts to find the slope.

$$\text{slope} = \frac{0 - (-6)}{4 - 0} = \frac{6}{4} \text{ or } \frac{3}{2}$$

Notice that $\frac{3}{2}$ is the coefficient of x in the equation $y = \frac{3}{2}x - 6$. When a linear equation is written in the form $y = mx + b$, it is in **slope-intercept form**.

$$y = mx + b$$

↑ slope ↑ y-intercept

You can use the slope-intercept form of an equation to graph a line quickly.

Example 2 APPLICATION Child Care

The cost for a child to attend The Learning Station is \$19 a day plus a registration fee of \$30. Make a graph that can be used to find the cost of any length of attendance.

Explore You know it costs \$19 a day and there is a \$30 registration fee. Let d represent the number of days. Then $C = 19d + 30$ represents the total cost, C, of attendance.

(continued on the next page)

THINK ABOUT IT

What does the ordered pair (1, 49) represent?

Plan The y-intercept is 30 so the graph contains the point at (0, 30).

The slope is 19 or $\frac{19}{1}$.

Solve $\frac{19}{1} = \frac{\text{change in } y}{\text{change in } x}$

Starting at (0, 30), go to the right 1 unit and up 19 units. This will be the point at (1, 49). Then draw the line that contains (0, 30) and (1, 49).

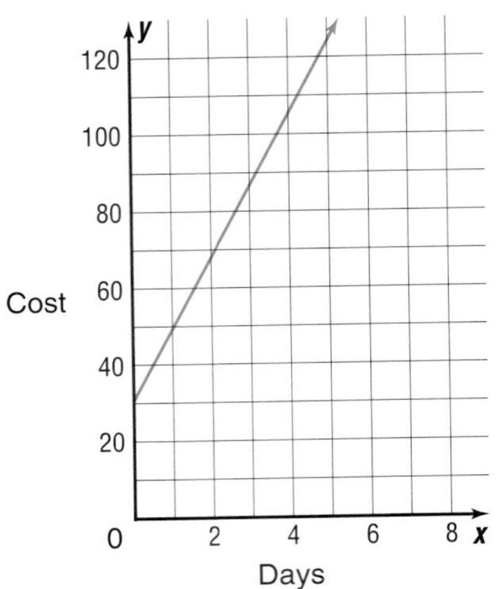

Examine Substitute the coordinates of a third point on the line into the equation $C = 19d + 30$ to determine whether it is a solution of the equation.

Checking Your Understanding

Communicating Mathematics

Read and study the lesson to answer these questions.

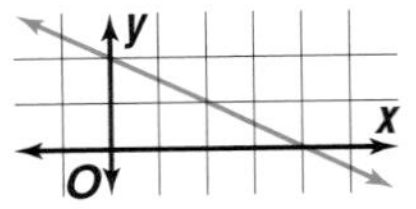

1. **Record** all the information you can determine from the graph shown at the right.
2. **Explain** how to graph a line if the x-intercept is -3 and the y-intercept is 1.
3. **Explain** how to find the x- and y-intercept for $y = -\frac{1}{2}x + 8$.
4. **Draw** a graph of a line that has a y-intercept, but no x-intercept. Describe your line.
5. **Explain** what the intercepts of the graph of the equation involving 10,000-meter run times described at the beginning of the lesson mean.
6. **Illustrate** how to graph a line if you know its y-intercept and its slope.

7. **Assess Yourself** Design a personal fitness program for yourself.
   a. Choose an activity from the chart and decide how many hours a week you will do the activity.
   b. Make a graph that shows the number of Calories burned for 0–10 hours.
   c. Use the graph to estimate how many Calories you will burn each week.

Activity	Calories burned per hour
Swimming	288
Walking	300
Jogging	654
Bicycling	174
Aerobics	546
Weight training	756

Guided Practice

State the x-intercept and the y-intercept for each line.

8.

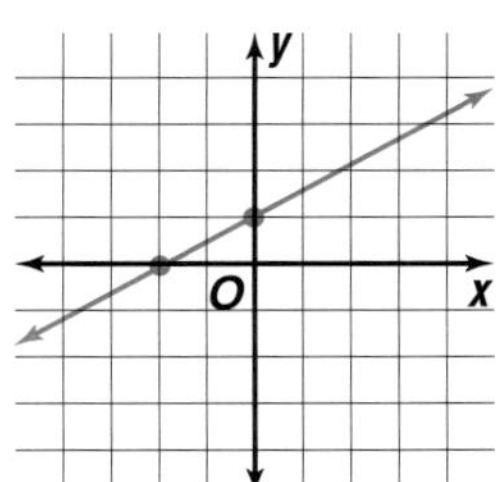

9.

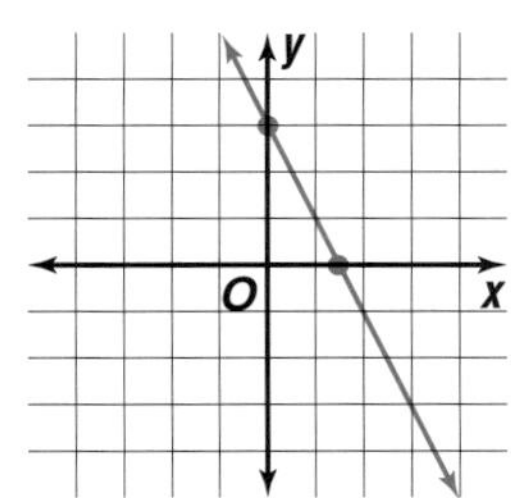

Find the x-intercept and the y-intercept for the graph of each equation. Then graph the line.

10. $y = x - 8$

11. $y = x + 6$

12. $y = -2x - 1$

13. $y = 6 - 9x$

Graph each equation using the slope and y-intercept.

14. $y = 2x + 1$

15. $y = -7x - 4$

16. Travel Tom Wishnock is taking a long trip. In the first hour, he drives only 40 miles because of heavy traffic. After that, he averages 60 miles an hour. The equation $y = 60x + 40$, where x represents the number of hours that Tom spent driving 60 miles per hour, and y represents the total distance traveled.

a. Name the x-intercept and the y-intercept of the graph of $y = 60x + 40$.

b. Graph the equation.

c. How long will it take Tom to finish his 400-mile trip?

Exercises: Practicing and Applying the Concept

Independent Practice

State the x-intercept and the y-intercept for each line.

17. a

18. b

19. c

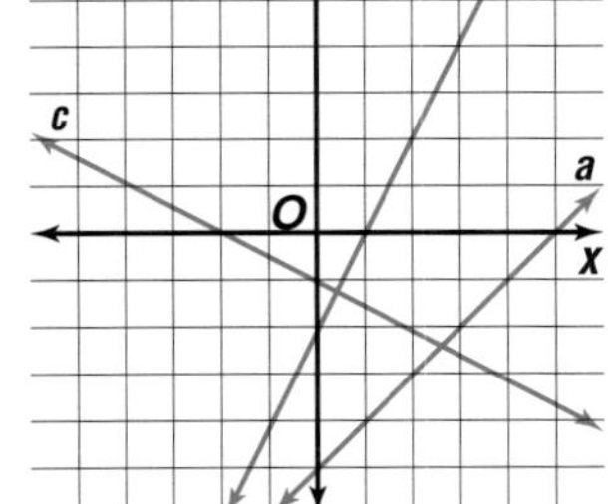

20. d

21. e

22. f

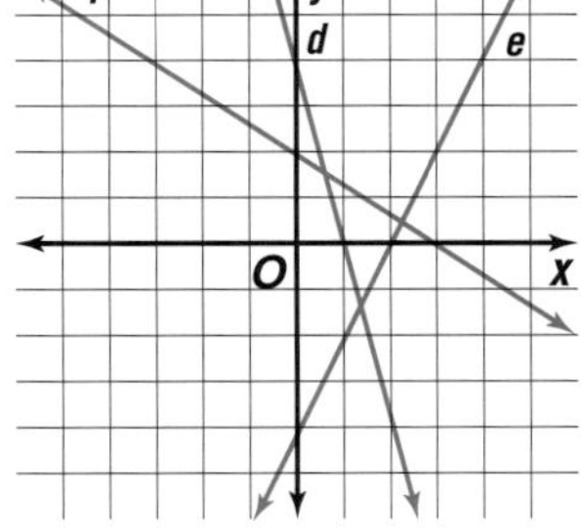

Use the x-intercept and the y-intercept to graph each equation.

23. $y = x - 2$

24. $y = x + 1$

25. $y = x - 4$

26. $y = 2x - 1$

27. $y = 3x + 4$

28. $y = -5x + 10$

29. $y = \frac{1}{2}x - 5$

30. $y = 3 - 0.5x$

31. $y = \frac{1}{3}x + 2$

32. $y = \frac{2}{3}x + 4$

33. $y = 3x - 4$

34. $y = -5x + 6$

Graph each equation using the slope and y-intercept.

35. $y = \frac{2}{3}x + 3$ **36.** $y = \frac{3}{4}x + 4$ **37.** $y = -\frac{3}{4}x + 4$

38. $-4x + y = 6$ **39.** $-2x + y = 3$ **40.** $3y - 7 = 2x$

Given the slope of a line and its y-intercept, graph the line.

41. slope = 3, y-intercept = 1 **42.** slope = $-\frac{4}{3}$, y-intercept = -1

Critical Thinking

43. Explain why you cannot graph the equation $y = 2x$ by using intercepts only. Then draw the graph.

Applications and Problem Solving

44. Business Suppose you have a lawn-mowing service. You charge a flat fee of \$1.50 per mowing for gas, plus a fee of \$4.50 per hour for labor. The equation $y = 4.5x + 1.5$ represents the total fee, y, for mowing a lawn that takes x hours to mow.

a. Graph this equation.

b. What does the y-intercept represent?

c. What is your total fee for a lawn that takes 3 hours to mow?

45. Aviation The equation $a = 24{,}000 - 1500t$, where t is the time in minutes and a is the altitude in feet, represents the steady descent of a jetliner.

a. Graph this equation.

b. Name the x-intercept of the graph.

c. What does the x-intercept represent?

Mixed Review

46. Find the slope of a line that contains the points $S(-7, -3)$ and $T(-4, -5)$. (Lesson 8-6)

47. Statistics Refer to the graph at the right. Let x represent the number of visitors to Harper's Ferry National Park. The expression $2x + 100{,}000$ is an estimate of the number of visitors at which park? (Lesson 7-5)

Visiting the Past

Most visited in 1989	Visitors
Gettysburg National Military Park, Pa.	1,352,728
Chickamauga and Chattanooga National Military Park, Ga.	851,534
Manassas National Battlefield Park, Va.	767,138
Kennesaw Mountain National Battlefield Park, Ga.	762,422
Harper's Ferry National Historical Park, W. Va.	624,168

48. Solve $y = 15 - 8.7$. (Lesson 5-3)

49. Find the GCF of 39 and 65. (Lesson 4-5)

Estimation
Mental Math
Calculator
Paper and Pencil

50. Solve $-41 > r - (-8)$. (Lesson 3-6)

51. Simplify $-3a + 12a + (-14a)$. (Lesson 2-4)

52. Sports The Mets are losing to the Astros by a score of 5 to 3. Write an inequality that would tell how many runs would give the Mets the lead. (Lesson 1-9)

53. What property allows you to say that $(9 + 10) + 7 = 9 + (10 + 7)$? (Lesson 1-4)

GRAPHING CALCULATOR ACTIVITY

8-7B Families of Graphs

An Extension of Lesson **8-7**

MATERIALS

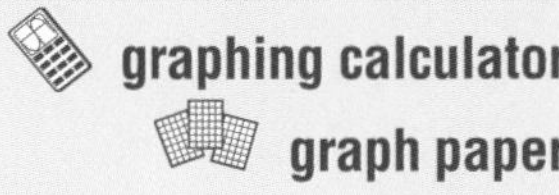

graphing calculator
graph paper

A **family of graphs** is a group of graphs that displays one or more similar characteristics. Many linear graphs are related because they have the same slope or the same y-intercept as other functions in the family. You can graph several functions on the same screen and observe if any family traits exist.

Activity 1

Work with a partner.

- The TI-82 graphing calculator can graph up to nine functions at one time. Graph $y = 1.5x$, $y = 1.5x + 2$, $y = 1.5x + 4$, and $y = 1.5x - 4$ on the same screen in the standard viewing window.
- Begin by clearing the graphics screen. [Y=] Press and clear any equations in the list using the arrow and the [CLEAR] keys.
- Enter each equation into the Y= list.
- **Enter:** [Y=] 1.5 [X,T,θ] [ENTER] 1.5 [X,T,θ] [+] 2 [ENTER] 1.5 [X,T,θ] [+] 4 [ENTER] 1.5 [X,T,θ] [−] 4

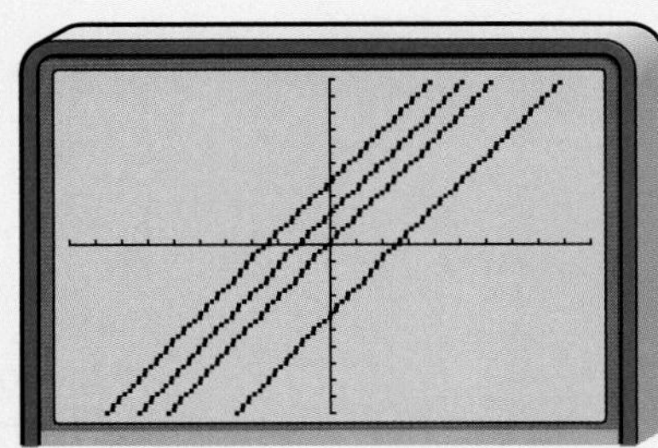

- The standard viewing window can be selected automatically from the zoom menu. Press [ZOOM] 6 and the graphs will appear automatically.

TALK ABOUT IT

1. Describe the family of graphs.

Activity 2

Work with a partner.

- Clear the graphics screen. Then graph $y = 4x$ and $y = -4x$ on the same screen in the standard viewing window. Sketch the graphs on a piece of graph paper.
- Then graph $y = -\frac{1}{4}x$ on the same screen. Sketch the graph on the graph paper.
- Finally graph $y = -6x$ and $y = -1.5x$ on the same screen. Add the graphs to the graph paper.

TALK ABOUT IT

2. Explain the difference between the graphs of $y = 4x$ and $y = -4x$.
3. Write the equation of a line whose graph is between the graph of $y = -4x$ and $y = -6x$.

Extension

4. Clear the screen and graph $y = 2x$, $y = 2x + 1$, $y = 2x + 2$, and $y = 2x + 3$. How are the graphs the same? How do they differ?

8-8 Systems of Equations

Setting Goals: *In this lesson, you'll solve systems of linear equations by graphing.*

Modeling with Technology

Between 1984 and 1994, the number of cellular phone users doubled more than 7 times.

The Use and Cost of Cellular Phones

Subscribers (millions): 1984 — .125; 1994 — 16.070 (axis 0, 4, 8, 12, 16)

Average Cost*: 1984 — $275; 1994 — $72 (axis 0, $100, $200, $300)

* in constant 1992 dollars

Source: Technology Futures, Inc.

In a questionnaire concerning phone service, potential customers were asked whether they would prefer to pay \$10 a month and 10¢ for each local call or \$5 a month and 20¢ for each local call. Is there a certain number of calls where the two payment plans result in the same total cost?

To solve this problem, let n represent the number of local calls made in a month and let c represent the monthly cost. You can write two equations to represent the situation above.

Plan 1 $c = 10 + 0.1n$ **Plan 2** $c = 5 + 0.2n$

Your Turn You can use a graphing calculator to solve the problem quickly. Enter each equation into the calculator and use the table feature to find the value of n for which the value of n is the same.

Enter: [Y=] 10 [+] .1 [X,T,θ] [ENTER] 5 [+] .2 [X,T,θ] [ENTER] [2nd] [TABLE]

The table on the calculator will show the values of n in the column marked X. Values of c for the plans 1 and 2 will be in columns Y_1 and Y_2 respectively.

a. Describe each output.

b. Locate the row where the outputs are the same. Explain what the identical outputs mean and why it happened.

Learning the Concept

The equations $c = 10 + 0.1n$ and $c = 5 + 0.2n$ together are called a **system of equations**. The **solution** to this system is the ordered pair that is a solution of both equations. Refer to the outputs in the situation above. What is the solution of the system of equations?

Another method for solving a system of equations is to graph the equations on the same coordinate plane. The coordinates of the point where the graphs intersect is the solution of the system of equations.

Example 1 **Solve the system of equations $y = 2x$ and $y = -x + 3$ by graphing.**

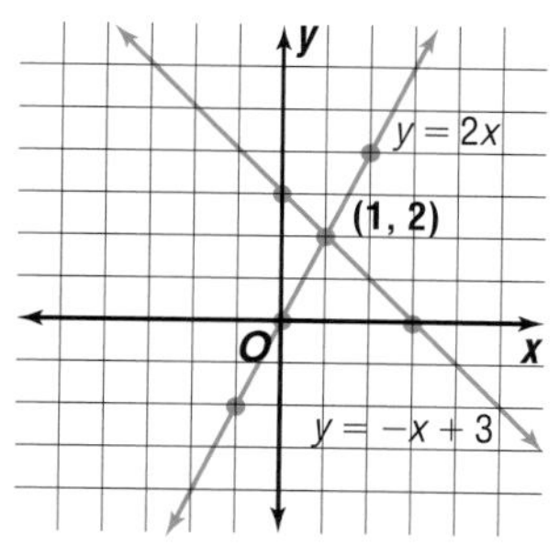

Graph $y = 2x$ and $y = -x + 3$ on the same coordinate plane.

The lines intersect at (1, 2). Therefore, the solution to the system of the equations is (1, 2).

To check, substitute the coordinates into each equation to see whether it is a solution to both equations.

Check:

$y = 2x$
$2 \stackrel{?}{=} 2(1)$
$2 = 2$ ✓

$y = -x + 3$
$2 \stackrel{?}{=} -(1) + 3$
$2 = 2$ ✓

Solving systems of equations is also used to solve problems.

Example 2

Communication

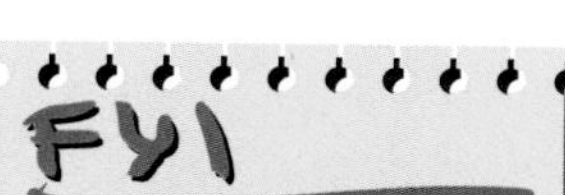

The first test of a cellular phone was in Chicago in 1978.

Refer to the application at the beginning of the lesson. Solve the system of equations $c = 10 + 0.1n$ and $c = 5 + 0.2n$ by graphing.

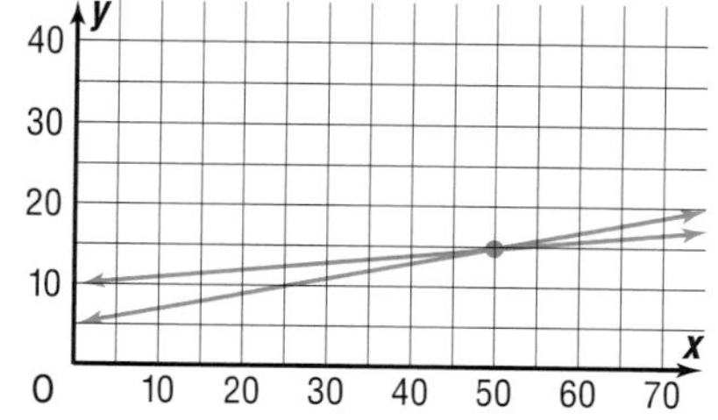

The graphs intersect at the point (50, 15).

Check:

$c = 10 + 0.1n$
$15 \stackrel{?}{=} 10 + 0.1(50)$
$15 = 15$ ✓

$c = 5 + 0.2n$
$15 \stackrel{?}{=} 5 + 0.2(50)$
$15 = 15$ ✓

The solution is (50, 15). This means that two payment plans result in the same total cost, \$15, when 50 local calls are made.

As you can see in the examples above, when the graphs of two linear equations intersect in exactly one point, the system has exactly one ordered pair as its solution. It is also possible for the two graphs to be parallel lines or to be on the same line. When the graphs are parallel, the system of equations does not have a solution. When the graphs are the same line, the system of equations has infinitely many solutions.

Example 3 **Solve each system of equations by graphing.**

a. $y = x + 3$ and $y = x + 4$

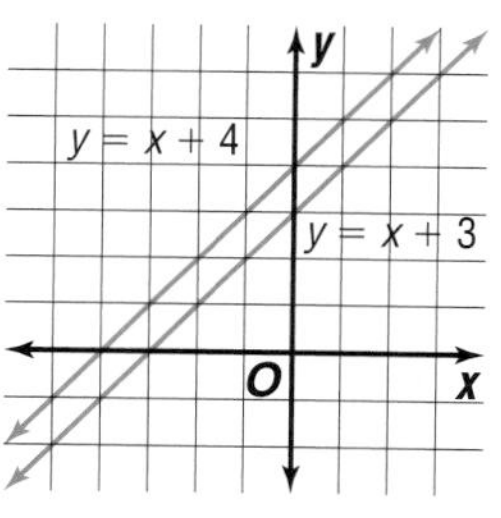

The graphs of the equations are parallel lines. Since they do not intersect, there is no solution to this system of equations.

Notice that the two lines have the same slope but different *y*-intercepts.

b. $y = -2x + 3$ **and** $3y = -6x + 9$

Each equation has the same graph. Any ordered pair on the graph will satisfy both equations. Therefore, there are infinitely many solutions to this system of equations.

Notice that the graphs have the same slopes and intercepts.

$y = -2x + 3$
$3y = -6x + 9$

Checking Your Understanding

Communicating Mathematics

Read and study the lesson to answer these questions.

1. **Explain** what is meant by a system of equations and describe its solution.
2. **Compare and contrast** the two local calling plans at the beginning of the lesson. Which plan would you prefer? Explain.
3. **Show** why there is no solution for the system $x + y = 4$ and $y = 8 - x$.
4. **Create and graph** a system of equations that has one solution, another system that has no solution, and a third that has infinitely many solutions.

MODELING MATHEMATICS

MATERIALS

geoboard

geobands

5. **Geometry** You can use a geoboard to model the first quadrant of the coordinate plane and geobands to model graphs of equations. Let the row of pegs on the bottom represent the x-axis and the column of pegs on the left represent the y-axis. The graph of the equation $y = x + 1$ is shown at the right.

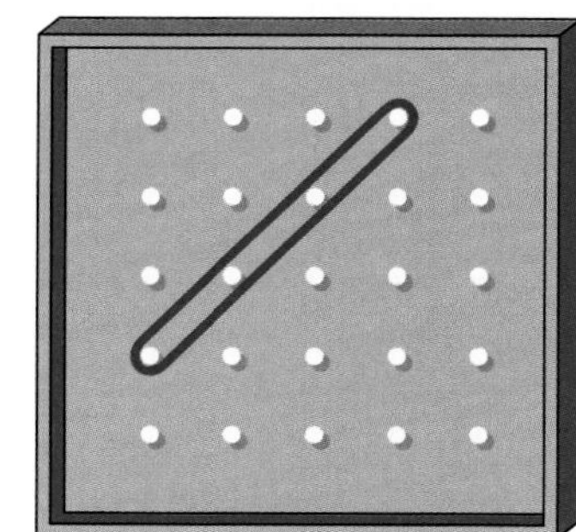

   a. Make the "line" shown at the right on a geoboard.
   b. Use a geoband to show the graph of $y = -x + 3$ on the same geoboard.
   c. What is the coordinate of the peg that is enclosed by both geobands?
   d. What is the solution to the system of equations $y = x + 1$ and $y = -x + 3$?

Guided Practice

State the solution of each system of equations.

6.

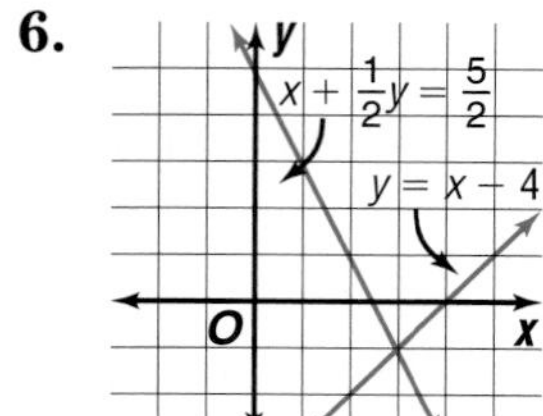

7.

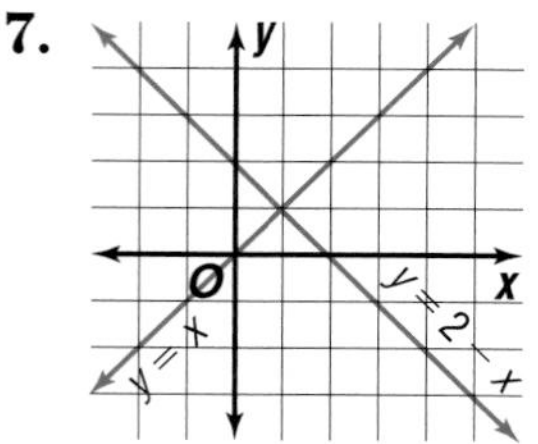

8.

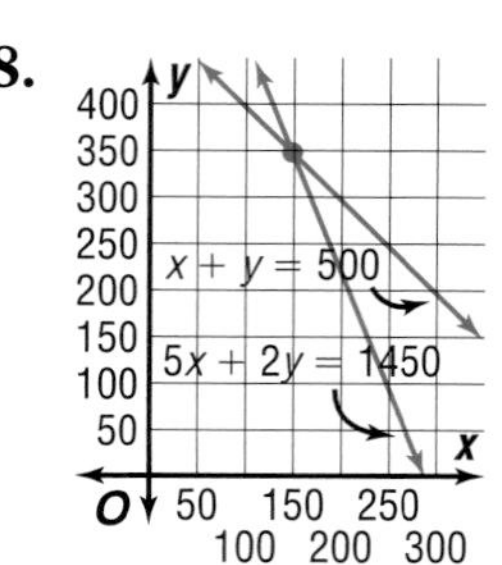

Use a graph to solve each system of equations.

9. $y = x - 8$
$y = -2x - 2$

10. $y = x + 6$
$y = 6 - 9x$

11. $y = -\frac{1}{2}x + 2$
$y = 3x - 5$

12. $y = \frac{3}{8}x - \frac{1}{2}$
$y = \frac{3}{8}x - \frac{21}{8}$

13. Art Kirima bought 12 feet of framing material to make a rectangular frame for her oil painting. She uses all the framing material with no waste and the length is twice the width.

a. Write a system of equations that represents this situation.

b. Solve this system of equations by graphing.

c. Explain what the solution means.

Exercises: Practicing and Applying the Concept

Independent Practice

The graphs of several equations are shown at the right. State the solution of each system of equations.

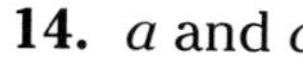

14. a and c

15. b and c

16. c and d

17. b and d

18. a and d

19. a and the x-axis

20. b and the y-axis

21. a, b, and d

22. b, c, and the y-axis

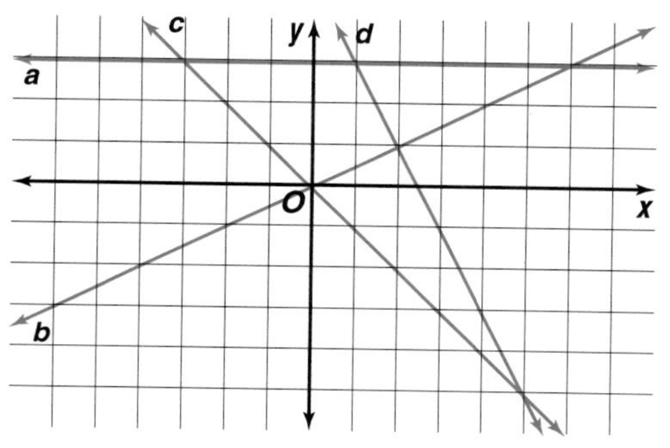

Use a graph to solve each system of equations.

23. $y = x + 2$
$y = -2x + 17$

24. $y = x$
$y = 3$

25. $y = -x + 2$
$y = 2x + 5$

26. $y = -2x - 5$
$y = -2x - 8$

27. $y = x + 6$
$y = 3x$

28. $y = -x$
$y = 4x$

29. $y = -5x - 2$
$y = 2x + 12$

30. $y = -x$
$y = 0.5x + 3$

31. $y = 3x - 8$
$y = 2x - 3$

32. $\frac{1}{2}x + \frac{1}{3}y = 6$
$y = \frac{1}{2}x + 2$

33. $\frac{2}{3}x + \frac{1}{4}y = 4$
$x = -\frac{3}{8}y + 6$

34. $3.4x + 6.3y = 4.4$
$2.1x + 3.7y = 3.1$

Critical Thinking

35. The solution for the system of equations $Ax + y = 6$ and $Bx + y = 7$ is $(1, 3)$. What are the values of A and B?

Applications and Problem Solving

36. Geometry The graphs of the equations $y = 2$, $3x + 2y = 1$, and $3x - 4y = -29$ contain the sides of a triangle.

a. Graph the three equations on the same coordinate plane.

b. Find the coordinates of the vertices of the triangle.

37. Business The Howell Company has fixed costs of \$900 per week. Each item produced by the company costs \$2 to manufacture and can be sold for \$5. If x is the number of items produced each week, the cost of producing them can be represented by the equation $y = 900 + 2x$. The weekly income from selling the items can be represented by $y = 5x$.

a. Solve this system of equations by graphing.

b. The *break-even point* is the point at which the income is the same as the cost of producing the goods or services. Name the break-even point of the system of equations and explain what it means.

38. Human Growth Jordan Gossell was born six weeks premature. He weighed 3 pounds at birth and gained an average of 2 pounds per month. Paige Plesich was born on her due date. She weighed 7 pounds at birth and gained an average of 1 pound per month.

a. Let a represent each baby's age in months, and let w represent each baby's weight in pounds. Write a system of equations that represents this situation.

b. Solve this system of equations by graphing.

c. Explain what the solution means.

39. Law Enforcement A police K-9 unit spots a prowler 60 meters away. The dog runs toward the prowler at a speed of 15 meters per second, and the prowler runs away at a speed of 5 meters per second.

a. Write a system of equations that represents this situation. (*Hint*: Let x represent time and y represent distance.)

b. Solve this system of equations by graphing.

c. Explain what the solution means.

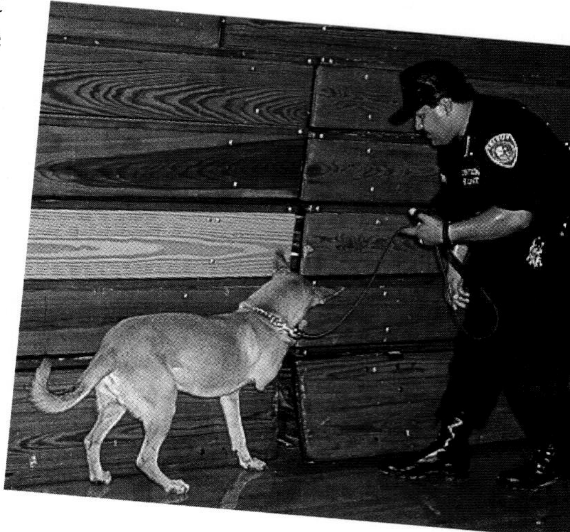

Mixed Review

40. Graph $y = -3x + 2$ by using the x-intercept and the y-intercept. (Lesson 8-7)

41. If 4.05 times an integer is increased by 3.116, the result is between 13 and 25. Use an inequality to find the number. (Lesson 7-6)

42. Publishing A page of type is to be divided into three columns. If the page is $6\frac{3}{4}$ inches wide, how many inches wide is each column? (Lesson 6-4)

43. Personal Finance Leya has \$575.29 in her checking account. She deposits her paycheck for \$125.90. The same day, she receives bills for \$397.28 and \$225.40. If she pays both bills, how much is in her checking account at the end of the day? (Lesson 5-3)

44. Solve $-15 + t \leq 12$. Check your solution. (Lesson 3-6)

45. Evaluate $|-17| - |3|$. (Lesson 2-1)

46. Simplify $4(c + 3)$. (Lesson 1-5)

GRAPHING CALCULATOR ACTIVITY

8-9A Graphing Inequalities

A Preview of Lesson **8-9**

MATERIALS

graphing calculator
graph paper

The graph of a linear function separates the coordinate plane into two regions, one above the line and one below it. For the graph of $y = 2x + 5$, shown at the right, points in the region *above* the line are represented by the inequality $y > 2x + 5$. Points in the region *below* the line are represented by the inequality $y < 2x + 5$. You can use a graphing calculator to investigate the graphs of inequalities.

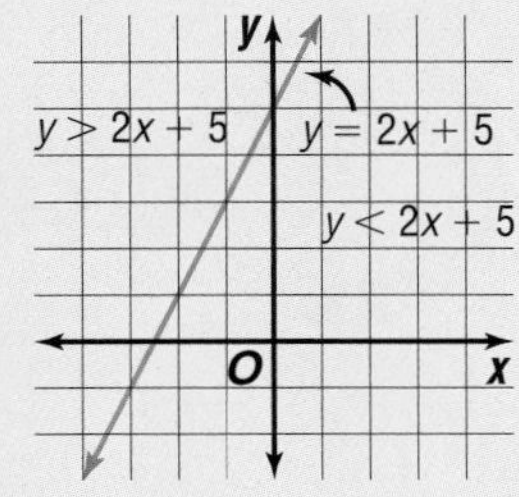

Your Turn

Work with a partner.

▶ Graph $y < 2x + 5$ in the standard viewing window.

Enter: 2nd DRAW 7 (–) 10 , 2 X,T,θ + 5) ZOOM 6 CLEAR ENTER.

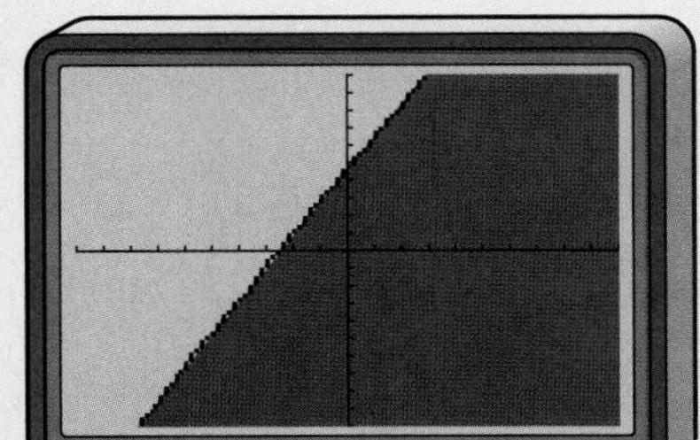

The TI-82 shades between two functions. The first function defines the lower boundary; in this case, Ymin, or −10. The second function defines the upper boundary; in this case, the function $y = 2x + 5$.

Notice that the graph is a shaded region. This indicates that all ordered pairs in that region satisfy the inequality $y < 2x + 5$.

▶ Clear the drawing that is currently displayed.

Enter: 2nd DRAW ENTER.

▶ Now graph $y > 2x + 5$ in the standard viewing window.

Enter: 2nd DRAW 7 2 X,T,θ + 5 , 10) ENTER.

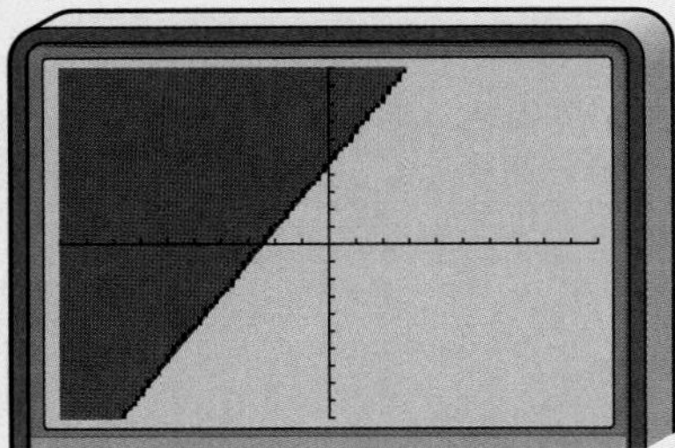

The lower boundary is $y = 2x + 5$, and the upper boundary is Ymax or 10.

TALK ABOUT IT

1. What is similar about the two graphs?
2. How does the graph of $y < 2x + 5$ differ from the graph of $y > 2x + 5$?
3. How do the graphs differ from the graph of $y = 2x + 5$?
4. Write the keystroke sequence that would graph the following inequalities. Then use a graphing calculator to graph each inequality.
   a. $y < x + 4$
   b. $y > 2x - 1$
   c. $y > -x + 3$
   d. $y < -2x + 2$

8-9 Graphing Inequalities

Setting Goals: *In this lesson, you'll graph linear inequalities.*

Modeling with Manipulatives

MATERIALS

graph paper

The graph of $y = x + 2$ is shown at the right. Copy the axes and the graph on a piece of graph paper.

Your Turn

Graph the following ordered pairs on your coordinate system.

(3, 4) (3, −2) (−2, −1)
(−1, 0) (4, 4) (0, −3)

a. Where do these points lie in the plane in relation to the graph of $y = x + 2$?

b. Which sentence, $y = x + 2$, $y > x + 2$, or $y < x + 2$, is true for all of the ordered pairs you graphed?

c. Which sentence, $y = x + 2$, $y > x + 2$, or $y < x + 2$, is true for the points located above the graph of $y = x + 2$?

d. How many points belong to the graph of $y < x + 2$?

Learning the Concept

To draw the graph of an inequality such as $y > 2x - 3$, first, draw the graph of $y = 2x - 3$. This graph is a line that separates the coordinate plane into two regions. In the graph at the right, one of the regions is shaded yellow, and the other is shaded blue. The line is the **boundary** of the two regions.

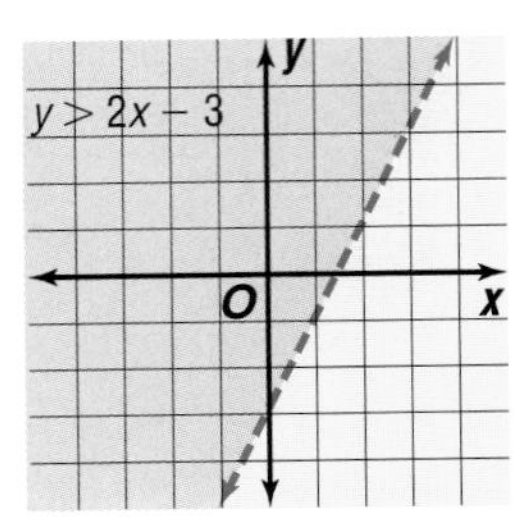

To determine which region is the solution to $y > 2x - 3$, test a point in the blue region. For example, you can test the origin, (0, 0).

$$y > 2x - 3$$
$$0 \overset{?}{>} 2(0) - 3$$
$$0 > -3 \quad \checkmark$$

Since $0 > -3$ is true, (0, 0) *is* a solution of $y > 2x - 3$.

Points on the boundary are *not* solutions of $y > 2x - 3$, so it is shown as a dashed line. Thus, the graph is all points in the blue region.

Now consider the graph of $y \leq 2x - 3$. The origin, (0, 0), is not part of the graph of $y \leq 2x - 3$ since $0 \leq -3$ is not true.

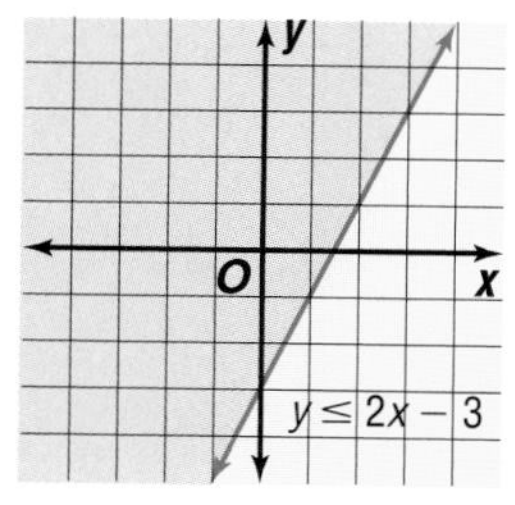

Since the inequality $y \leq 2x - 3$ means $y < 2x - 3$ or $y = 2x - 3$, points on this boundary *are* part of the solution of $y \leq 2x - 3$. Therefore, the boundary is shown as a solid line on the graph.

Thus, the graph is all the points in the yellow region and the line $y = 2x - 3$.

Example  **Graph $y < -2x + 2$.**

Graph the equation $y = -2x + 2$. Draw a dashed line since the boundary is *not* part of the graph. The origin, (0, 0), is part of the graph since $0 < -2(0) + 2$ is true. Thus, the graph is all points in the region below the boundary. Shade this region.

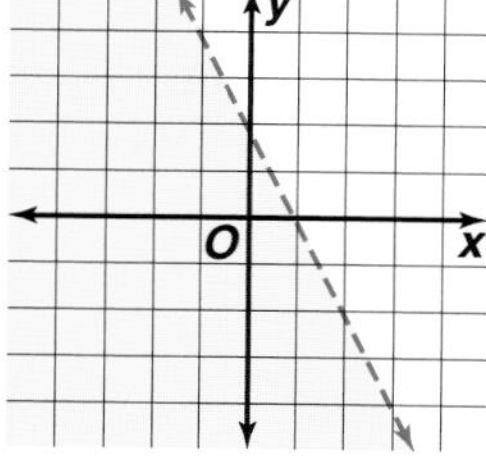

You may have to solve some inequalities for y first to find a solution to the system graphically.

Example 2

APPLICATION

Sales

Tickets for *West Side Story* are $66.50 for floor and $47.50 for balcony. In order to cover the expenses for the show, at least $66,500 per show must be made from ticket sales.

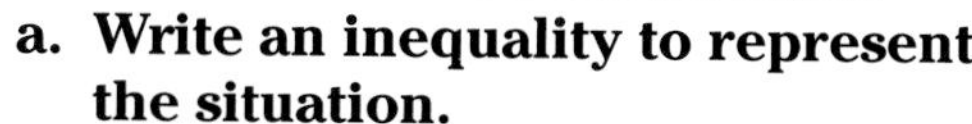

a. Write an inequality to represent the situation.

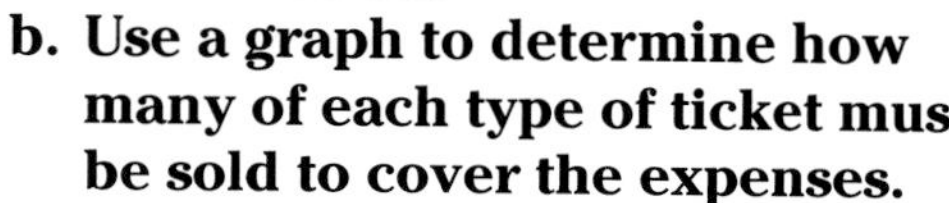

b. Use a graph to determine how many of each type of ticket must be sold to cover the expenses.

c. List three possible solutions.

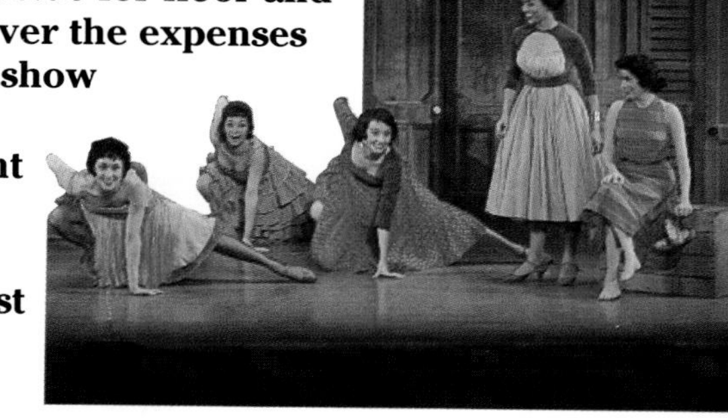

a. Let x represent the number of $47.50 tickets sold, and let y represent the number of $66.50 tickets sold.

sales of $47.50 tickets	*plus*	*sales of $66.50 tickets*	*is greater than or equal to*	*$66,500*
$47.5x$	$+$	$66.5y$	$\geq$	$66{,}500$

b. To graph the inequality, first solve for y.

$$47.5x + 66.5y \geq 66{,}500$$

$$66.5y \geq 66{,}500 - 47.5x \quad \textit{Subtract 47.5x from each side.}$$

$$y \geq 1000 - \frac{5}{7}x \quad \textit{Divide each side by 66.5}$$

Graph $y = 1000 - \frac{5}{7}x$ as a solid line since the boundary is part of the graph. The origin is not part of the graph since $0 \geq 1000 - \frac{5}{7}(0)$ is not true. Thus, the graph is all the points in the region that do not contain the origin.

c. Any point in the shaded region represents a solution. However, fractional or negative solutions do not make sense because fractional or negative numbers of tickets cannot be sold. One solution is (500, 700). This corresponds to selling five hundred \$47.50 tickets and seven hundred \$66.50 tickets for a total cost of 47.5(500) + 66.5(700) or \$70,300. (420, 700) and (400, 800) are also solutions.

Checking Your Understanding

Communicating Mathematics

Read and study the lesson to answer these questions.

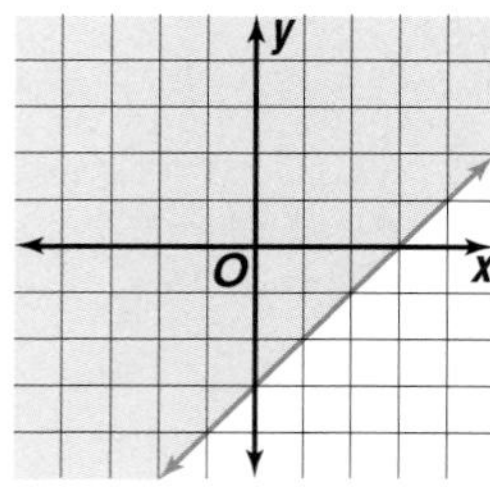

1. **Write** an inequality that describes the graph shown at the right.
2. **You Decide** Jeff says that inequalities with $<$ and $\leq$ symbols have dashed lines and those with $>$ and $\geq$ symbols have solid lines. Sanura disagrees. Who is correct and why?
3. **Write** an inequality whose graph is all points below the line $y = 9x + 4$.

4. **Describe** the location of solutions that result in a total cost of exactly \$66,500 on the graph in Example 2.
5. **Explain** how to determine which side of the boundary line to shade when graphing an inequality.

Guided Practice

State whether the boundary is included in the graph of each inequality.

6. $y \geq x - 13$ 7. $y \leq 1$ 8. $y > 3x$

Determine which ordered pair(s) is a solution to the inequality.

9. $y < 2x + 1$ **a.** $(-2, 2)$ **b.** $(4, -1)$ **c.** $(3, 1)$
10. $4y > -3x - 2$ **a.** $(-1, -1)$ **b.** $(2, -2)$ **c.** $(-2, 4)$

Determine which region is the graph of each inequality.

11. $y > x - 1$ 12. $y \leq \frac{2}{3}x + 2$ 13. $y < 5x$

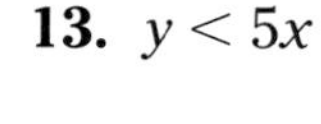

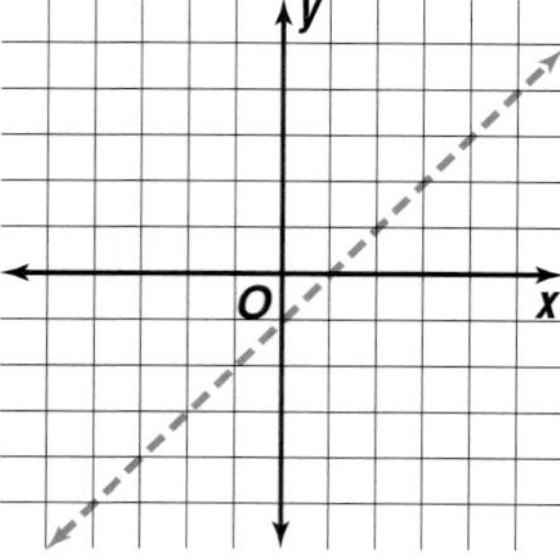

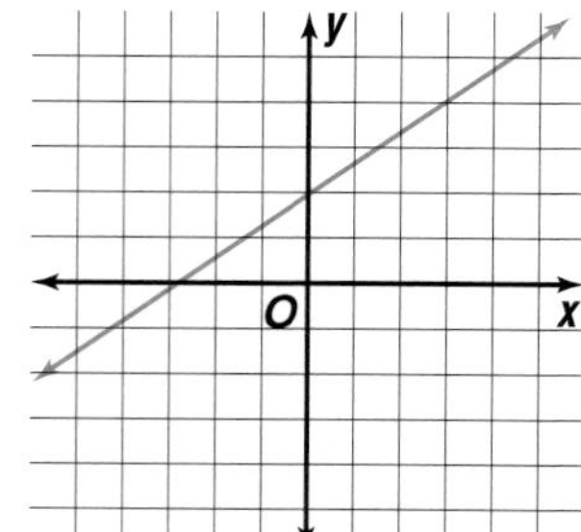

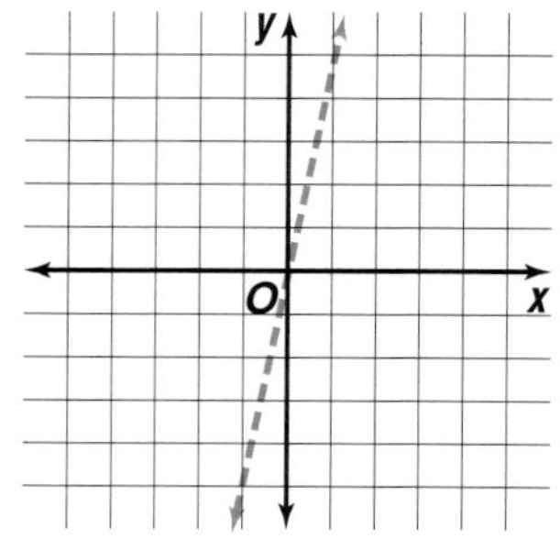

Graph each inequality.

14. $y < x + 2$ 15. $y \geq x$ 16. $y < -2x$ 17. $y \geq 3$

18. Personal Finance Silvina wants to earn at least \$23 this week. Her father has agreed to pay her \$2.25 an hour to weed the garden and \$5.00 to mow the lawn.

a. Graph the inequality $2.25h + 5n \geq 23$ where h is the number of hours she spends weeding the garden and n is the number of times she mows the lawn.

b. If Silvina mows the lawn once, what is the least number of hours she will need to spend weeding the garden in order to make at least \$23?

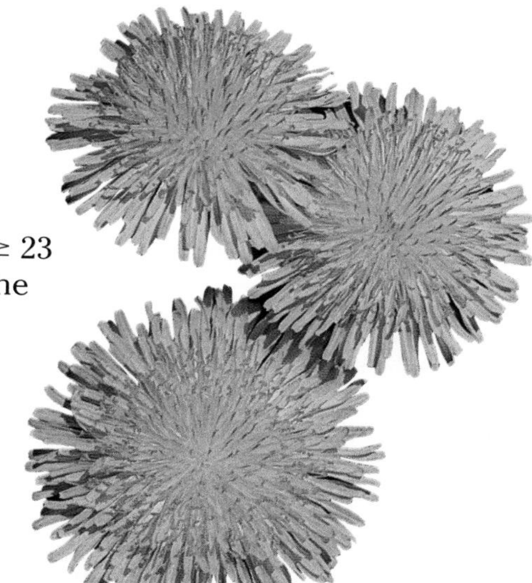

Exercises: Practicing and Applying the Concept

Independent Practice

Determine which ordered pair(s) is a solution to the inequality.

19. $y < 3 - x$	**a.** $(-2, 2)$	**b.** $(4, -1)$	**c.** $(3, 1)$
20. $3y > -1 - 2x$	**a.** $(2, 1)$	**b.** $(-5, 1)$	**c.** $(1, 1)$
21. $4y - 8 < 0$	**a.** $(0, 2)$	**b.** $(2, 5)$	**c.** $(-2, 0)$
22. $3x + 4y \geq 17$	**a.** $(1, 1)$	**b.** $(4, 2)$	**c.** $(-3, 7)$
23. $2x \geq y - 8$	**a.** $(5, 12)$	**b.** $(4, 8)$	**c.** $(-3, -1)$
24. $5x + 1 \leq 3y$	**a.** $(-2, 3)$	**b.** $(4, 7)$	**c.** $(2, 3)$

Determine which region is the graph of each inequality.

25. $y > 2x - 6$ **26.** $y \geq 4 - 3x$ **27.** 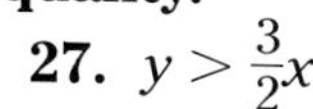 $y > \frac{3}{2}x$

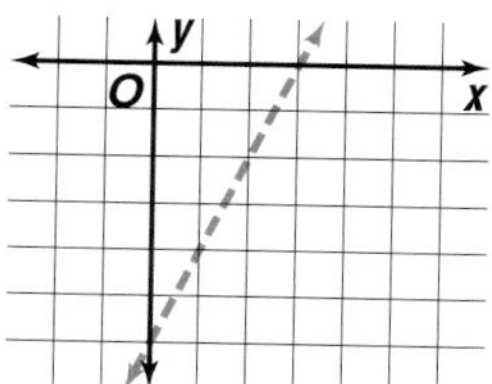

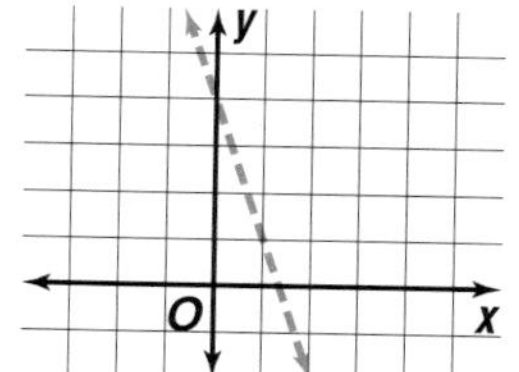

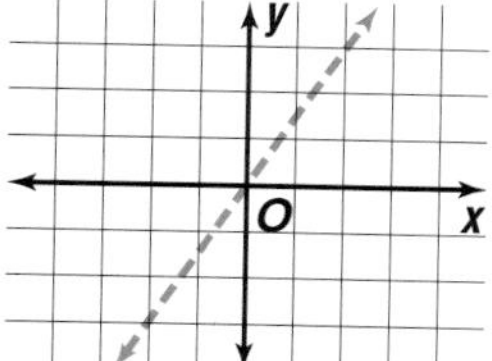

Graph each inequality.

28. $y < 2$	**29.** $y \geq -2$	**30.** $x \leq 2$	**31.** $x > -4$
32. $x + y > 2$	**33.** $x + y \geq -3$	**34.** $y \leq 0.5x + 5$	**35.** $x + y < 1$
36. $y > \frac{1}{4}x - 8$	**37.** $y > -x - 6.5$	**38.** $y \leq \frac{1}{3}x$	**39.** $y \geq 4.5x - 0.5$
40. $-y < -x$	**41.** $4y + x \leq 16$	**42.** $2x \geq 3y$	**43.** $-x > -y$

Critical Thinking

44. The solution to a system of inequalities is the set of all ordered pairs that satisfies *both* inequalities.

a. Write a system of inequalities for the graph shown at the right.

b. Check your answers by substituting into the inequalities.

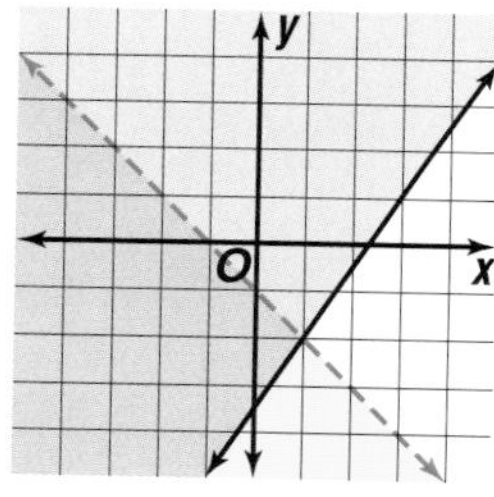

Applications and Problem Solving

45. Business For a certain business to be successful, its monthly overhead O must be at least \$3000 less than its monthly sales S.

a. Write an inequality to represent this situation.

b. Graph the inequality.

c. Do points above or below the boundary line indicate a successful business? Justify your answer.

d. Would negative numbers make sense in this problem? Explain.

e. List two solutions.

46. Pets Biologists have found that for fish to thrive, the total length L of all the fish (in inches) in a tank must be no more than 3.75 times the number of quarts Q of water in the tank.

a. Write an inequality to represent this situation.

b. Graph the inequality.

c. List two solutions where at least five 2-inch fish are in a tank. Explain what each solution means.

47. Consumer Awareness You are shopping for cassettes and CDs. Cassettes cost \$7, CDs cost \$14, and you have \$28 to spend.

a. Use a graph to determine how many cassettes and CDs you can buy.

b. List two solutions and explain what they mean.

48. Family Activity Investigate your family's eating habits and the U.S. recommended daily allowances (RDAs) of nutrients. Write inequalities for the recommended amounts of at least three nutrients. Does your diet measure up?

Mixed Review

49. Graph $y - 4x = 3$ and $y = x$. Then find the solution of the system of equations. (Lesson 8-8)

50. Recreation Angie's bowling handicap is 7 less than half her average. Her handicap is 53. What is Angie's bowling average? (Lesson 7-3)

51. Order $\frac{3}{5}$, 0.63, and 6.02×10^{-1} from least to greatest. (Lesson 6-1)

52. Weather The people of Waynesburg, PA, celebrate July 26 as Rain Day, because it almost always rains on that day. Is this an example of inductive or deductive reasoning? (Lesson 5-8)

53. Use divisibility rules to determine if 38 is divisible by 2, 3, 5, 6, or 10. (Lesson 4-1)

54. Solve $6b > -31.2$. (Lesson 3-7)

55. Solve $\frac{96}{-8} = m$. (Lesson 2-8)

56. State whether $3x - 7 > 5$ is *true*, *false*, or *open*. (Lesson 1-9)

CHAPTER 8 Highlights

Vocabulary

After completing this chapter, you should be able to define each term, property, or phrase and give an example of two of each.

Algebra

boundary (p. 418)
domain (p. 372)
function (p. 373)
functional notation (p. 393)
linear equation (p. 386)
parabola (p. 391)
range (p. 372)
relation (p. 372)
solution of system (p. 412)
system of equations (p. 412)
vertical line test (p. 373)

Statistics

scatter plot (p. 378)
negative relationship (p. 380)
no relationship (p. 380)
positive relationship (p. 380)

Geometry

change in x (p. 400)
change in y (p. 400)
family of graphs (p. 411)
intercepts (p. 406)
parallel lines (p. 401)
slope (p. 400)
slope-intercept form (p. 407)
x-intercept (p. 406)
y-intercept (p. 406)

Problem Solving

draw a graph (p. 396)

Understanding and Using Vocabulary

Choose the term from the list above that best completes each statement or phrase.

1. A ________ is a relation in which each element of the domain is paired with exactly one element in the range.
2. The set of first coordinates of a relation is called the ________.
3. The ________ is the y-coordinate of the point where a graph crosses the y-axis.
4. The set of second coordinates of a relation is called the ________.
5. A ________ is a set of ordered pairs.
6. ________ lines are always the same distance apart.
7. The ________ is the x-coordinate of the point where the graph crosses the x-axis.
8. In an equation in slope-intercept form, the b represents the ________.

CHAPTER 8 Study Guide and Assessment

Skills and Concepts

Objectives and Examples

Upon completing this chapter, you should be able to:

- **use tables and graphs to represent relations and functions.** (Lesson 8-1)

Table

x	y
1	2
6	11
5	11

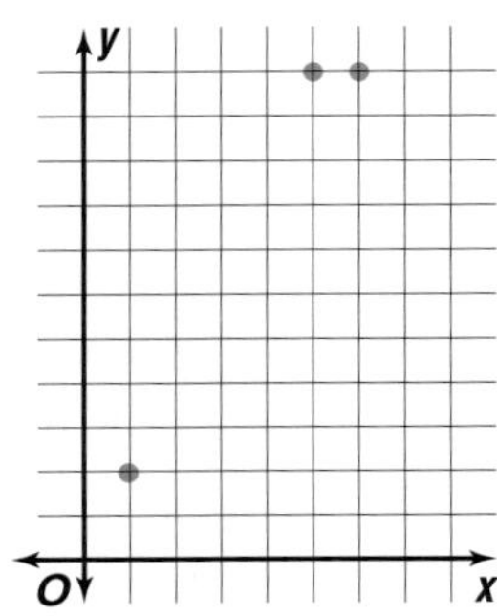

The domain is {1, 6, 5}.

The range is {2, 11}.

Since each element of the domain is paired with exactly one element of the range, this is a function.

Review Exercises

Use these exercises to review and prepare for the chapter test.

Write the domain and range of each relation. Then determine whether each relation is a function.

9. {(7, 17), (9, 19), (0, 18), (6, 40)}

10. {(4, 5), (4, 10), (4, 22), (4, 36)}

11. {(6, 2), (8, 2), (15, 2)}

12.

x	1	18	18	35	27
y	2	4	6	4	2

13. y, O, x

- **construct and interpret scatter plots** (Lesson 8-2)

Relation of years of experience and amount of money earned

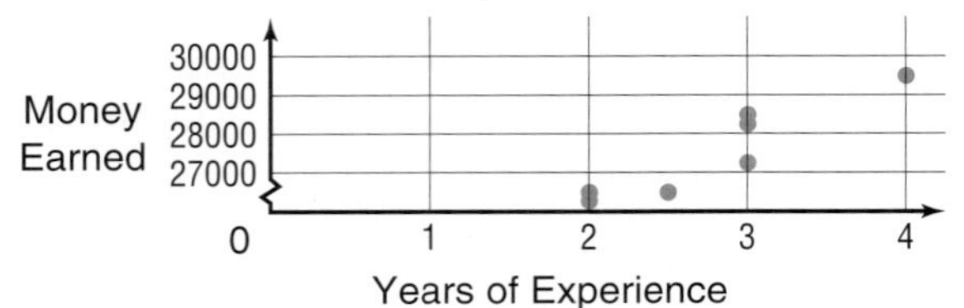

The points in the scatter plot are in a pattern that slants upward to the right. So it shows a positive relationship.

14. What type of relationship, *positive*, *negative*, or *none*, is shown by the scatter plot below?

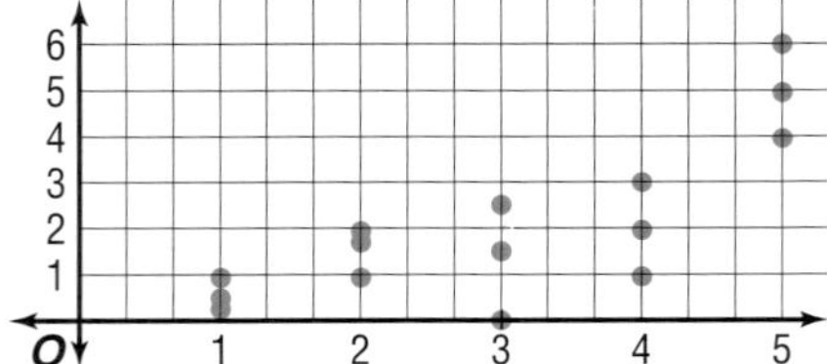

Determine whether a scatter plot of the data might show a *positive*, *negative*, or *no* relationship. Explain your answer.

15. size of TV set, amount of TV watched

16. number of small children, amount of glass displayed

17. age of a person, volume of the radio

18. amount of income, value of home

19. years of marriage, number of kids

Objectives and Examples

▶ **find solutions for relations with two variables and graph the solution** (Lesson 8-3)

Find four solutions for $y = 2.2x - 1$. Then graph the equation.

x	2.2x – 1	y
0	2.2 (0) – 1	–1
1	2.2 (1) – 1	1.2
2	2.2 (2) – 1	3.4
3	2.2 (3) – 1	5.6

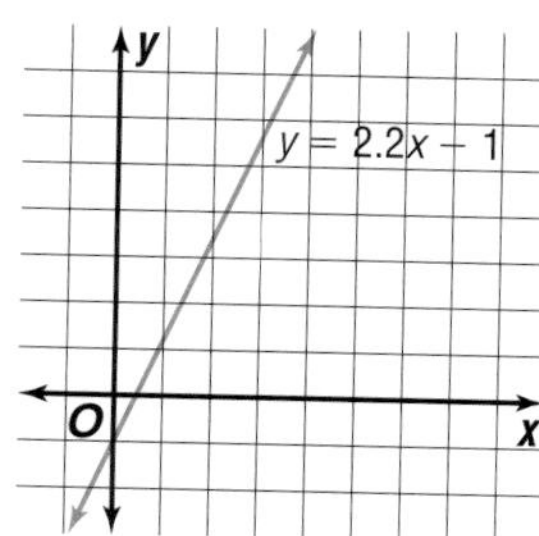

Four solutions are (0, –1), (1, 1.2), (2, 3.4), (3, 5.6).

▶ **find functional values for the given function** (Lesson 8-4)

$$g(x) = 2x^2 - 3$$

$g(3) = 2 \cdot 3^2 - 3$
$= 2(9) - 3$
$= 18 - 3$
$= 15$

$g(-1) = 2 \cdot (-1)^2 - 3$
$= 2(1) - 3$
$= 2 - 3$
$= -1$

▶ **find the slope of a line** (Lesson 8-6)

Find the slope of the line that contains $A(4, 4)$ and $B(0, 1)$.

$$\text{slope} = \frac{4 - 1}{4 - 0} = \frac{3}{4}$$

▶ **graph a linear equation using the *x*- and *y*-intercepts or using the slope and the *y*-intercept** (Lesson 8-7)

$y = 4x - 1$

slope
$m = 4$

***y*-intercept**
$y = 4(0) - 1$
$y = -1$
$(0, -1)$

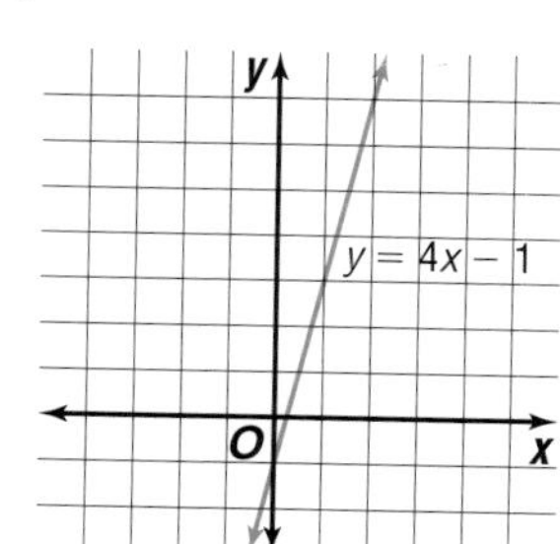

Review Exercises

Find four solutions for each equation. Write the solution as ordered pairs. Then graph each equation.

20. $y = x - 1$
21. $y = 3x + 2$
22. $x + y = 5$
23. $2x + y = 7$
24. $x - 2y = -1$
25. $x = 7$
26. $y = 1$
27. $y = \frac{1}{2}x$

Given $f(x) = x + 5$ and $g(x) = 3x^2 + 1$, find each value.

28. $f(2)$
29. $f(0)$
30. $f(-6)$
31. $g(1)$
32. $g(5)$

Find the slope of the line that contains each pair of points.

33. $A(-3, 5), B(2, 7)$
34. $J(4, 1), K(2, 6)$
35. $W(4, 0), X(7, -3)$
36. $A(6, 18), B(6, 40)$
37. $P(0, 2), Q(3, 2)$
38. $W(3, 8), X(1, 5)$

Use the slope and the *y*-intercept to graph each equation.

39. $y = 2x + 1$
40. $y = \frac{2}{3}x - 3$
41. $y = -\frac{1}{2}x + 1$
42. $y = 4$
43. $x = 7$
44. $x + 2y = 6$
45. $y = 2 - \frac{1}{2}x$
46. $5x - y = 9$

Objectives and Examples

▸ **solve systems of linear equations by graphing** (Lesson 8-8)

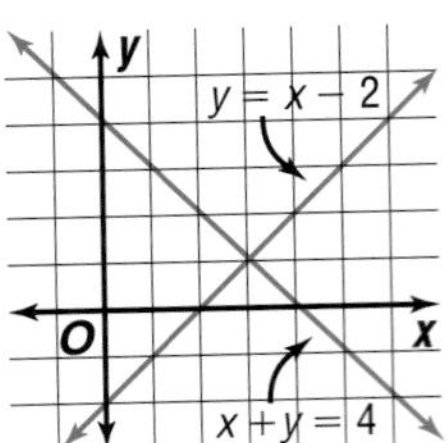

$y = x - 2$
$x + y = 4$

The lines intersect at (3, 1). So the solution to the system of equations is (3, 1).

Review Exercises

Use a graph to solve each system of equations.

47. $x + y = 8$
$x - y = -8$

48. $y = 2x - 4$
$2y = 4x - 8$

49. $y = -x + 1$
$y = -x + 2$

50. $y = \frac{1}{2}x$
$y = \frac{3}{2}x - 2$

51. $y = 3 - x$
$y = 5 - x$

52. $x + 2y = 4$
$-2x - 2y = -10$

▸ **graph linear inequalities** (Lesson 8-9)

Graph $y > \frac{1}{3}x + 2$.

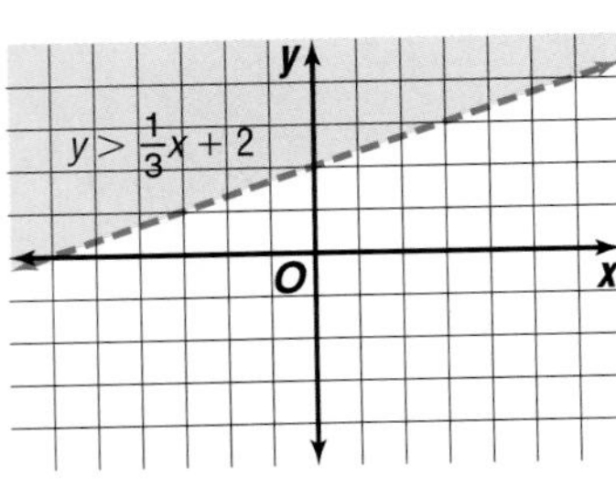

The boundary is dashed since the line is not part of the graph. The origin (0, 0) is not part of the graph since $0 < \frac{1}{3}(0) + 2$.

Graph each inequality.

53. $y < x - 2$
54. $y \geq -2x + 1$
55. $y > \frac{1}{2}x$
56. $y \leq -x + 8$
57. $y \leq 3x - 1$
58. $y > \frac{1}{4}x + 2$

Applications and Problem Solving

59. **Nutrition** This table shows nutritional values of selected foods. (Lesson 8-2)

a. Draw a scatter plot using Calories from fat as the x values and Calories from carbohydrates as the y values.

b. Does the graph show a positive, negative, or no relationship?

Food	Percentage of Calories from fat	Percentage of Calories from carbohydrate
Apple (medium, delicious)	9	89
Bagel (plain)	6	76
Banana (medium)	2	93
Bran Muffin (large)	40	53
Fig Newtons (four cookies)	18	75
PowerBar	8	76
Snickers bar (regular size)	42	51
Ultra Slim-Fast bar	30	63

60. **Living Expenses** An investment company charged \$495 a month to rent a 2-bedroom townhouse in 1993. In 1995, the rate was \$515 per month. Use a graph to determine how much the townhouse rent will be in 2000. Assume that the rate is constant. (Lesson 8-5)

61. **Sports** In the 1992–1993 season, Dino Ciccarelli played 82 games and scored 41 goals. The next season, he played 66 games and scored 28 goals. Use a graph to determine about how many goals Dino would score in a 76-game season. Assume that the rate is constant. (Lesson 8-5)

A practice test for Chapter 8 is available on page 781.

Alternative Assessment

Performance Task

Demonstrate your knowledge by giving a clear, concise solution to each problem. Be sure to include all relevant drawings and justify your answers. You may show your solutions in more than one way or investigate beyond the requirements of the problem.

The table below lists the movies with the top five box-office revenues for 1993.

Rank/Title	Distributor	U.S. Gross Totals (millions)	Foreign Gross Totals (millions)
1. *Jurassic Park*	Universal	\$338	\$530
2. *The Fugitive*	Warner Brothers	\$179	\$170
3. *Aladdin*	Buena Vista	\$118	\$185
4. *The Bodyguard*	Warner Brothers	\$45	\$248
5. *Indecent Proposal*	Paramount	\$107	\$151

Source: *Variety*, January 3–9, 1994

1. Make a graph of the information using the rank as the x-coordinate and the world gross totals as the y-coordinate. (*Hint*: The world total is the sum of U.S. and foreign revenues.)
2. Give the domain of the relation.
3. Give the range of the relation.
4. Does the chart represent a function?
5. Do you think that the film's distributor affected the amount of money it made?
6. The marketing manager of a film company has determined that a new film must make \$700 million in sales between U.S. and foreign sales in order to break even.
   a. Graph an inequality to show this situation.
   b. Would the films listed in the table have broken even if this was their goal?

Thinking Critically

- Give two examples of relations that could be graphed on a scatter plot.
- Choose one of the relations you wrote. Develop sample data and graph the relation. Then determine whether it shows a positive, negative, or no relation.

Portfolio

Select one of the assignments from this chapter that you found especially challenging. Revise your work if necessary and place it in your portfolio. Explain why you found it to be a challenge.

Self Evaluation

Never accepting a challenge, you will never feel the excitement of victory. Remember how happy you were when your team won a difficult game? Experience the feeling firsthand in math class.

Assess yourself. Did you find a lesson in the chapter more difficult than the others? How many times did you have to do it in order to solve the problems? How did you feel when you succeeded? What advice would you give to someone faced with a similar challenge?

CHAPTERS 1–8 Ongoing Assessment

Section One: Multiple Choice

There are eleven multiple-choice questions in this section. After working each problem, write the letter of the correct answer on your paper.

1. If $4x - 8 = 28$, what is the value of x?
   - **A.** 5
   - **B.** 9
   - **C.** 20
   - **D.** 36
2. An investor bought 80 shares of stock for \$2160 and later sold them for \$1920. Which integer represents the profit or loss per share of stock?
   - **A.** −3
   - **B.** −240
   - **C.** 240
   - **D.** 3
3. What is the y-intercept for the graph of $y = x + 2$?
   - **A.** −2
   - **B.** 1
   - **C.** 0
   - **D.** 2
4. A group of divers are at a level of −20 meters. If they descend 15 meters more, at what level will they be?
   - **A.** 5 m
   - **B.** −5 m
   - **C.** 35 m
   - **D.** −35 m
5. How is the product $5 \times 5 \times 5$ expressed in exponential notation?
   - **A.** 5^2
   - **B.** 3^5
   - **C.** 5^3
   - **D.** 5^5
6. A mail-order card company charges 50¢ for each greeting card plus a handling charge of \$1.50 per order. Which sentence could be used to find n, the number of cards ordered, if the total charge was \$9?
   - **A.** $0.5n + 1.5 = 9$
   - **B.** $0.50n + 1.50n = 9$
   - **C.** $9 = 50n + 1.50$
   - **D.** $9 = (0.5 + 1.50)n$
7. Which number should come next in this pattern?

   0.5, 2, 3.5, 5, . . .
   - **A.** 8
   - **B.** 7.5
   - **C.** 7
   - **D.** 6.5
8. Which is the solution of the system of equations $x + y = 4$ and $y = -3x$?
   - **A.** (6, 2)
   - **B.** (1, 3)
   - **C.** (−2, 6)
   - **D.** (3, 1)
9. One red blood cell is about 7.5×10^{-4} centimeters long. What is another way to express this measure?
   - **A.** 0.000075 cm
   - **B.** 0.00075 cm
   - **C.** 0.0075 cm
   - **D.** 0.075 cm
10. If $3x + 4(x + 1) = 5x - 8$, what is the value of x?
    - **A.** −6
    - **B.** −4.5
    - **C.** −2
    - **D.** 6
11. Express 65,000 in scientific notation.
    - **A.** 6.5×10^3
    - **B.** 6.5×10^4
    - **C.** 65×10^3
    - **D.** 0.65×10^5

Section Two: Free Response

This section contains twelve questions for which you will provide short answers. Write your answer on your paper.

12. John has \$2.50. He spent one third of the money he had this afternoon at lunch. His friend Lakesha borrowed \$2, and he spent 50¢ on a pop. How much money did John have when he left this morning?
13. Lee wants to set up a baseball tournament. Each team should have 12 different players. How many people does he need to have 8 teams?

14. Yolanda has \$50. She bought jeans for \$25. How much money can she spend for each of 4 gifts and have \$1 left for parking?

15. The price of Kodak's stock this morning was $87\frac{7}{8}$. At the end of today, the price had increased by $\frac{7}{8}$. What was the closing price of Kodak's stock?

16. When practicing during the summer on their own, members of the Lincoln High track team must run at least 18 miles a week. If Jaime wants to run the distance in no more than 6 days each week, write and solve an inequality representing how to find the number of miles he needs to run each day.

17. Mei used a chart to see if she was close to the average weight for her height. The chart gave a weight of 118 pounds for a female 60 inches tall. The average weight for a female 64 inches tall was 134 pounds. Mei is 5 feet 8 inches tall. What should Mei's weight be?

18. How many boards, each $3\frac{1}{2}$ feet long, can be cut from a board 21 feet long?

19. Graph the inequality $y \geq 2x + 2$.

20. How can the product $4 \times 4 \times 4 \times 4 \times 4 \times 4$ be written in exponential notation?

21. A contractor is going to paint a room with four walls, each 8 feet tall and 12 feet wide. Each wall will be given two coats of paint. If each can of paint will cover up to 400 square feet, how many cans will the job take?

22. The City Marathon had 3542 participants this year. 1054 runners did not finish the race. How many runners finished the City Marathon this year?

23. A computer catalog offers discount components that customers can add to their computers. Kim chose a printer for \$300, a modem for \$150, and a CD-ROM drive for \$225. If the tax, shipping, and handling was \$71.50, how much was the total bill?

Test-Taking Tip

Most standardized tests have a time limit, so you must use your time wisely. Some questions will be much easier than others. If you cannot answer a question within a few minutes, go on to the next one. If there is still time left when you get to the end of the test, go back to the questions that you skipped.

Take time out after several problems to freshen your mind.

Section Three: Open-Ended

This section contains two open-ended problems. Demonstrate your knowledge by giving a clear, concise solution to each problem. Your score on these problems will depend upon how well you do the following.

- Explain your reasoning.
- Show your understanding of the mathematics in an organized manner.
- Use charts, graphs, and diagrams in your explanation.
- Show the solution in more than one way or relate it to other situations.
- Investigate beyond the requirements of the problem.

24. Explain the difference between a relation and a function using an example of each.

25. Make a scatter plot of the areas and populations of the states and territories of Australia as shown in the table below.

State/Territory	Area (thousands of square miles)	Population (thousands)
New South Wales	310	5732
Victoria	88	4244
Queensland	667	2977
Western Australia	975	1586
South Australia	380	1401
Tasmania	26	452
Australia Capital	1	280
Northern Territory	520	175

Source: *World Almanac*, 1995

CHAPTER 9

Ratio, Proportion, and Percent

TOP STORIES in Chapter 9

In this chapter, you will:

- write expressions for ratios, decimals, fractions, and percents,
- find simple probability,
- use ratios and proportions to solve problems,
- use statistics to predict, and
- solve problems by making a table.

MATH AND GENETICS IN THE NEWS

Bone fragments contain what might be dinosaur DNA

Source: USA Today, November 18, 1994

Maybe the movie *Jurassic Park* is not as far from reality as you might think . . . Scientists have found what they believe are genes of a dinosaur.

The genetic material is the first to be taken directly from ancient bones. Researchers believed that complex DNA molecules were too fragile to survive in dinosaur bone for millions of years. But, the team of scientists who studied the bone believe that the dinosaur died in a bog and was covered by peat so that the delicate bones were preserved.

The genetic code found in the bone was 30% similar to birds, 30% similar to reptiles, and 30% similar to mammals. The discovery may help settle whether modern birds are related to ancient dinosaurs.

Some scientists are skeptical about the discovery. If others can repeat the experiment, the conclusions may become accepted.

Putting It into Perspective

1780 | 1820 | 1860

1833
First appearance of a time scale separating prehistoric time into eras of life development.

1859
First oil well is drilled by "Colonel" Edwin Drake in Titusville, Pennsylvania.

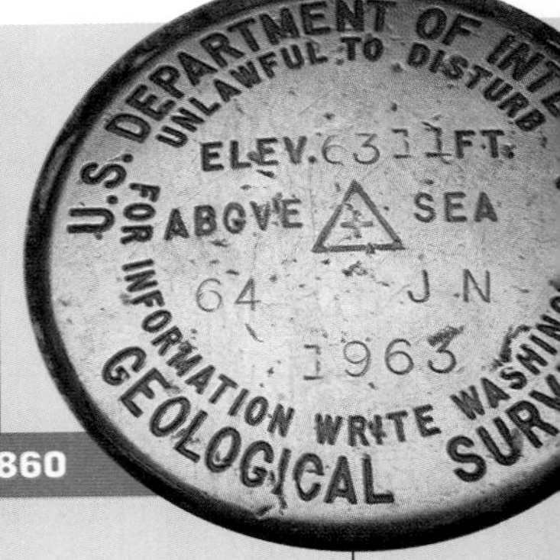

1879
The U.S. Geological Survey established.

EMPLOYMENT OPPORTUNITIES

Find Your Place in the Future of Prehistoric Life

Costa Rica • Galapagos • Australia • Alaska • Baja Mongolia • New Zealand

DO YOU LIKE traveling in the most exotic and untamed parts of the world? Career opportunities abound for those interested in the study of the remains of plants and animals preserved in Earth's rocks. Choose your specialty: as a **vertebrate paleontologist**, you will study extinct fishes, amphibians, reptiles, birds, and mammals; **invertebrate paleontologists** deal with fossil insects and shells; and the study of fossil plants is the concentration of the **paleobotanist**. College degree and related experience required.

For more information, contact:
The Paleontological Society
New Mexico Bureau of Mines and Mineral Resources
Socorro, NM 87801

Statistical Snapshot

GEOLOGIC TIME SCALE

ERAS		PERIODS	MAJOR EVENTS
	Present		
Cenozoic Era		Quaternary	Ice Ages/active faulting in west
Cenozoic Era		Tertiary	Modern topography Oldest fossil humans
	65 million years ago		
Mesozoic Era		Cretaceous	Major mountain-building in Rockies Giant dinosaurs
Mesozoic Era		Jurassic	Oldest fossil birds
Mesozoic Era		Triassic	Beginning of opening of the Atlantic Ocean
	225 million years ago		
Paleozoic Era		Permian	Last major mountain-building in Appalachians and Quachitas
Paleozoic Era		Pennsylvanian	Vast coal swamps in North America
Paleozoic Era		Mississippian	
Paleozoic Era		Devonian	Oldest forest
Paleozoic Era		Silurian	Oldest fish
Paleozoic Era		Ordovician	Major mountain-building in New England and Eastern North America
Paleozoic Era		Cambrian	Seas covered most of North America
	570 million years ago		
Precambrian	Origin of Earth ±4.5 billion years ago		

(Eras shown on the left bar: Cenozoic, Mesozoic, Paleozoic, Precambrian)

interNET CONNECTION For up-to-date information on dinosaurs, visit: **www.glencoe.com/sec/math/prealg/mathnet**

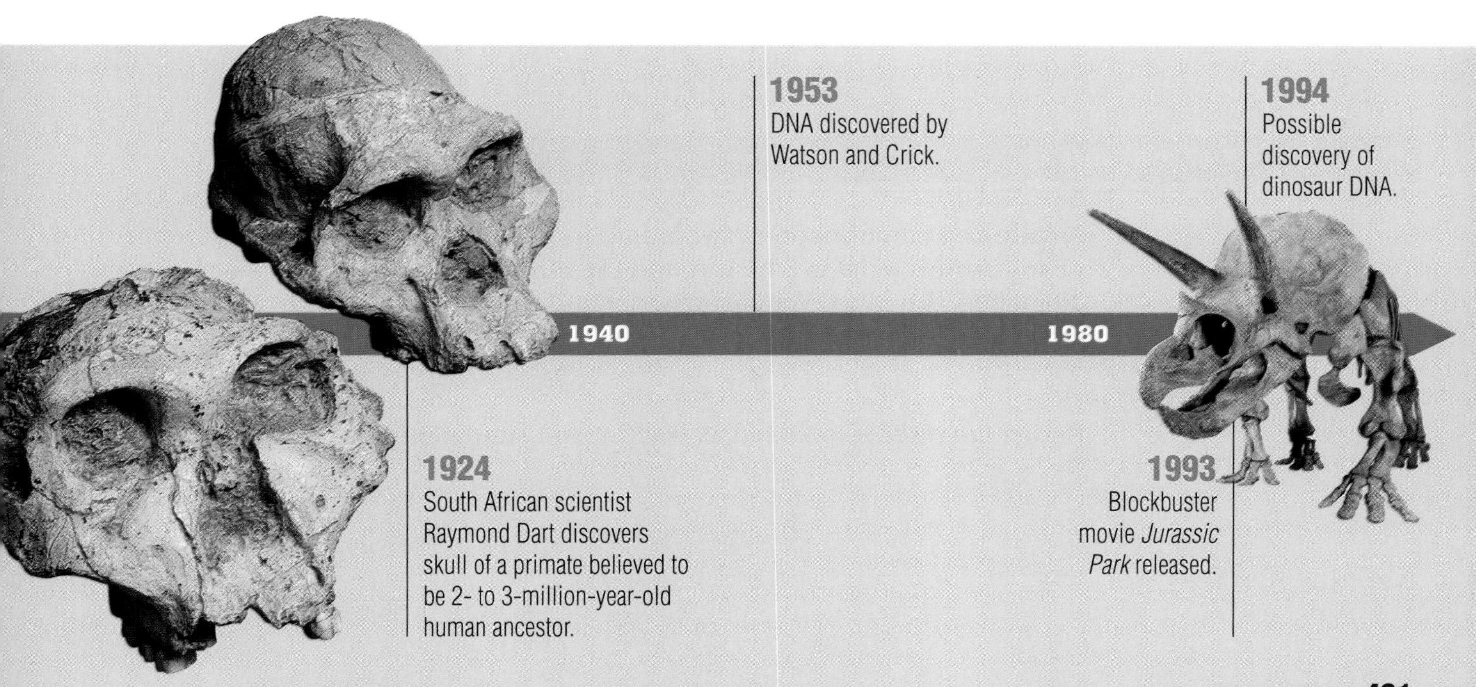

9-1 Ratios and Rates

Setting Goals: *In this lesson, you'll write ratios as fractions in simplest form and determine unit rates.*

Modeling with Manipulatives

MATERIALS

- tape measure
- grid paper
- ruler

People come in many different shapes and sizes. But with a closer look, you can find a pattern in some body measurements.

Your Turn

Work with a partner. Use a tape measure to find the distance around your wrist and around your head to the nearest $\frac{1}{2}$ inch. Copy the table below and record the data for you and your partner. Graph the data as ordered pairs on a graph similar to the one below.

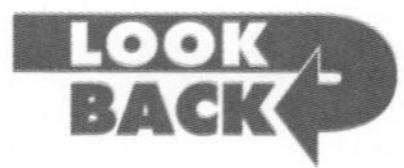

You can review scatter plots in Lesson 8-2 and slope in Lesson 8-6.

	Head Measurement	Wrist Measurement
You		
Your Partner		

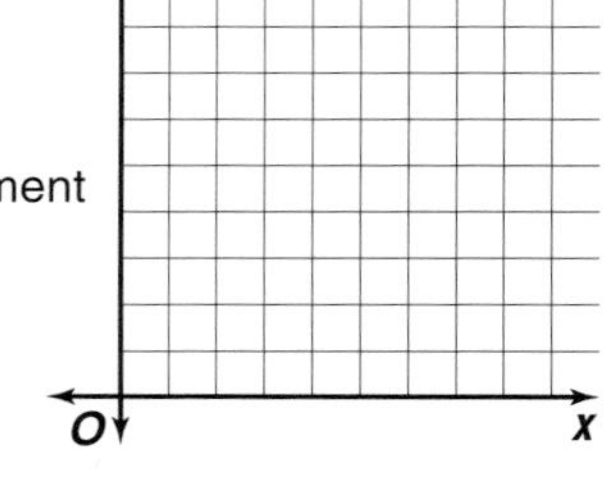

Draw the line between the two points. Then find the slope of the line.

a. Divide each wrist circumference by the corresponding head circumference and record your answers. This is the wrist to head *ratio*. How do the ratios compare to the slope of the line?

b. How do your results compare to the results of other groups?

Learning the Concept

A **ratio** is a comparison of two numbers by division. If the circumference of someone's wrist is 5 inches and the circumference of his or her head is 20 inches, the ratio comparing wrist and head measurements could be written as follows.

5 to 20   5:20 $\frac{5}{20}$

Ratios are often expressed as fractions in simplest form or as a decimal.

$$\frac{5}{20} = \frac{1}{4}$$ (÷ 5) *The GCF of 5 and 20 is 5.*

or 5 [÷] 20 [=] .25

Example 1 **Express each ratio as a fraction in simplest form and as a decimal.**

a. 6 free throws made out of 10 attempts

$$\frac{6}{10} = \frac{3}{5} \quad (\div 2)$$

6 [÷] 10 [=] *0.6*

$\frac{3}{5}$ or 0.6 of the free-throws were made.

b. \$375 in commission to \$9375 in sales

$$\frac{375}{9375} = \frac{\overset{1}{\cancel{3}} \cdot \overset{1}{\cancel{5}} \cdot \overset{1}{\cancel{5}} \cdot \overset{1}{\cancel{5}}}{\underset{1}{\cancel{3}} \cdot \underset{1}{\cancel{5}} \cdot \underset{1}{\cancel{5}} \cdot \underset{1}{\cancel{5}} \cdot 5 \cdot 5} \text{ or } \frac{1}{25}$$

375 [÷] 9375 [=] *0.04*

$\frac{1}{25}$ or 0.04 of the sales is commission.

A **rate** is a special ratio. It is a comparison of two measurements with different units of measure, like miles and gallons or cents and grams. The ratio $\frac{89 \text{ cents}}{15 \text{ ounces}}$ compares the price and size of a can of soup. To find the price per ounce, simplify the rate so that the denominator is 1 ounce. A rate with a denominator of 1 is called a **unit rate**.

Example 2 APPLICATION Consumerism

Many grocery stores list the unit price of items on the shelf tag so that customers may compare similar items. The cost of a 12-ounce box of Cheerios is \$3.29. A store brand cereal that is similar to Cheerios costs \$4.89 for an 18-ounce box. Which cereal has a lower per-ounce cost?

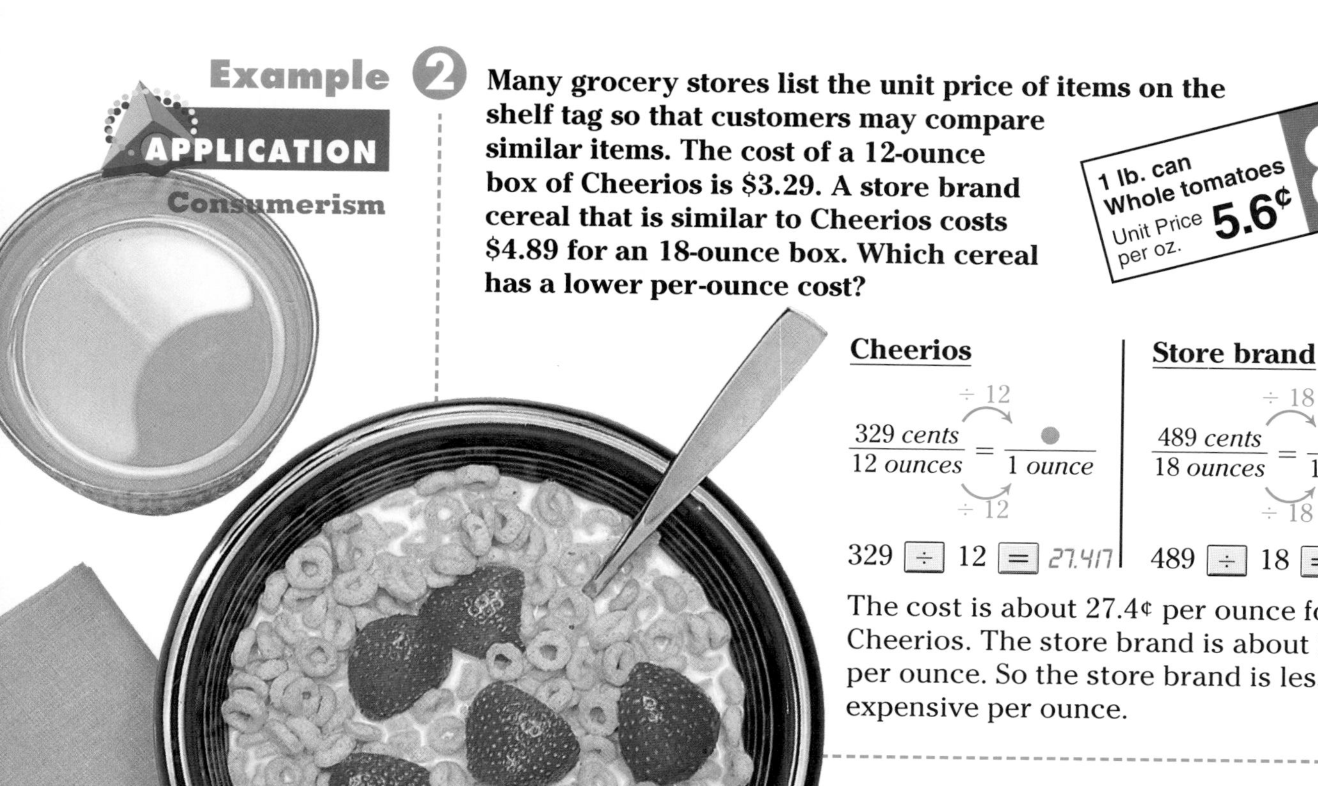

Cheerios

$$\frac{329 \text{ cents}}{12 \text{ ounces}} = \frac{\bullet}{1 \text{ ounce}} \quad (\div 12)$$

329 [÷] 12 [=] *27.417*

Store brand

$$\frac{489 \text{ cents}}{18 \text{ ounces}} = \frac{\bullet}{1 \text{ ounce}} \quad (\div 18)$$

489 [÷] 18 [=] *27.167*

The cost is about 27.4¢ per ounce for Cheerios. The store brand is about 27.2¢ per ounce. So the store brand is less expensive per ounce.

Connection to Geometry

The group of Greek mathematicians called the Pythagoreans studied the geometric figures and objects in nature that displayed the **golden ratio**. The great painter, Leonardo da Vinci, and many other artists and architects have used the golden ratio to create beautiful designs for over 4000 years. You can find the value of the golden ratio by comparing measurements of the Pyramid of Khufu.

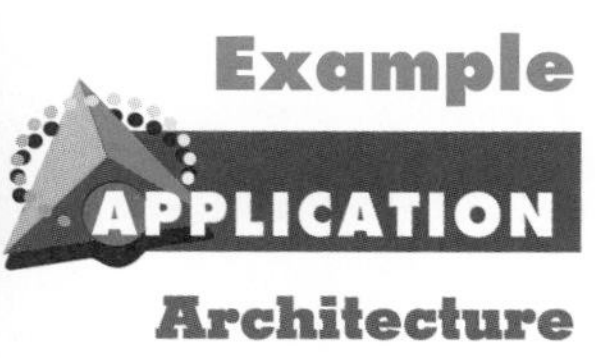

Example 3 APPLICATION **Architecture**

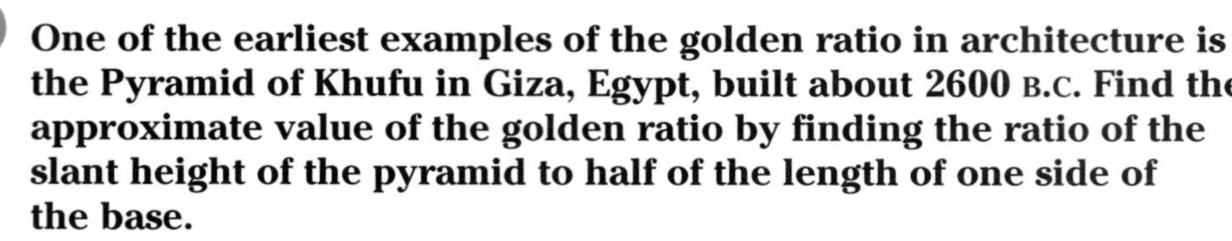

One of the earliest examples of the golden ratio in architecture is the Pyramid of Khufu in Giza, Egypt, built about 2600 B.C. Find the approximate value of the golden ratio by finding the ratio of the slant height of the pyramid to half of the length of one side of the base.

$$\frac{\textit{slant height}}{\textit{half of base length}} = \frac{611.54}{\frac{1}{2}(756.08)} = \frac{611.54}{378.04}$$

611.54 [÷] 378.04 [=] 1.617659507

The golden ratio is approximately 1.618.

slant height 611.54 ft

base 756.08 ft

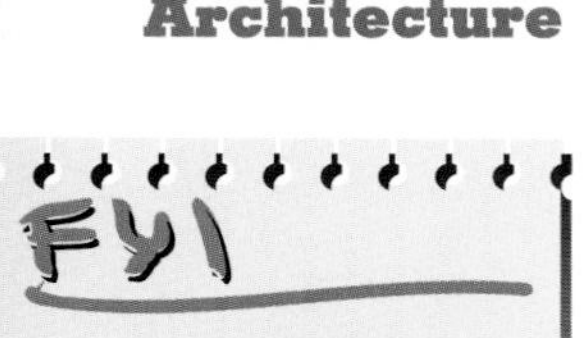

FYI

The Pyramid of Khufu was the largest structure of its day. It is the tomb of Pharaoh, Khufu, who died in 2567 B.C.

Checking Your Understanding

Communicating Mathematics

Read and study the lesson to answer these questions.

1. **Compare and contrast** ratios and rates.
2. **Write** a ratio that compares the number of students in your school to the number of teachers. Then express this ratio as a unit rate.
3. **Describe** and give an example of a unit rate.

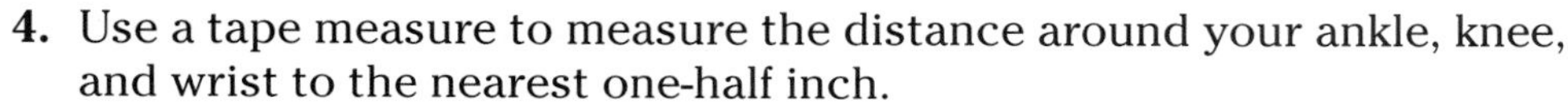

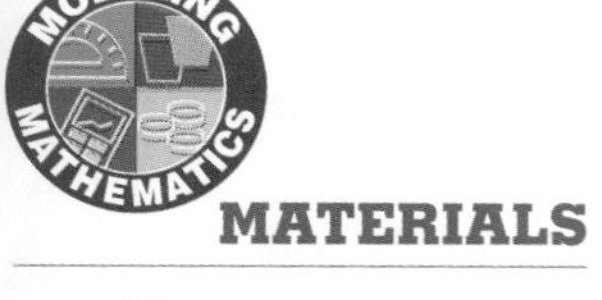

MATERIALS

tape measure

4. Use a tape measure to measure the distance around your ankle, knee, and wrist to the nearest one-half inch.
   a. Write a knee/ankle ratio, wrist/knee ratio, and ankle/wrist ratio.
   b. Compare the ratios in part a to those of a classmate.

Guided Practice

Express each ratio or rate as a fraction in simplest form.

5. 100 to 500
6. 325:25
7. 150 miles in 6 gallons
8. 98 cents for 12 ounces
9. Juan Gonzalez' 43 home runs in 1992 to 46 home runs in 1993

Express each ratio as a unit rate.

10. 12 pounds lost in 5 weeks
11. $2.29 for a 6 pack of cola
12. 99¢ for a dozen
13. 6 inches of rain in 24 hours
14. **Consumer Awareness** *Garbage* magazine reported that the average American commuter burns 190 gallons of gasoline a year driving about 4000 miles to and from work. How many miles per gallon does the average commuter get in his or her car?

Exercises: Practicing and Applying the Concept

Independent Practice

Express each ratio or rate as a fraction in simplest form.

15. 13 out of 169 **16.** 125:50 **17.** 64 to 16
18. 351:117 **19.** 156 to 84 **20.** 35 out of 149
21. 2 girls to 15 boys **22.** 9 Fords to 7 Chevrolets
23. 5 out of 7 oranges **24.** 11 pennies out of 99 pennies
25. 24 marbles to 56 marbles **26.** 15 successes in 35 attempts

√ **Choose**

Estimation
Mental Math
Calculator
Paper and Pencil

Express each ratio as a unit rate.

27. 399.3 miles on 16.5 gallons **28.** $9.60 for 12 pounds
29. 100 meters in 10.5 seconds **30.** 27 inches of snow in 9 hours
31. $210 for 30 tickets **32.** 550 mL for $3.30

Critical Thinking

33. Arrange the digits 1 to 9 into two numbers whose ratio is 1:5.

Applications and Problem Solving

34. Architecture A golden rectangle is a rectangle in which the ratio of the length to the width is the golden ratio.

- **a.** Find a golden rectangle in the photo of the Parthenon at the right.
- **b.** Many everyday objects are golden rectangles. For example, credit cards are approximately golden rectangles. Measure a rectangular object. Is it close to a golden rectangle?

35. Consumer Awareness Best buys in grocery stores are generally found by comparing unit rates such as cents/oz.

- **a.** Which is the better buy, a 16-oz bag of candy at $2.49 or a 32-oz bag at $3.69? How do you know?
- **b.** Are there instances in which the better buy might not necessarily be the one you choose to purchase?

36. Family Activity Visit a grocery store. Choose a product such as laundry detergent or pretzels. Find the unit price of a number of different brands. Which one would you choose and why?

37. Patterns Many sequences of numbers follow a pattern in the ratios of a term to the previous term.

- **a.** The sequence 3, 6, 12, 24, 48, . . . is geometric. Find the ratio between each pair of consecutive terms. What do you observe?
- **b.** Recall that the sequence 1, 1, 2, 3, 5, 8, 13, 21, . . . is called the Fibonacci sequence. The ratios between consecutive Fibonacci numbers approach a famous number. Find several ratios and make a conjecture about what number the ratios approach.

38. Travel The distance between two cities on a map is 5 centimeters. The mileage chart with the map shows that the cities are 10 miles apart. The ratio of an actual distance to the distance on a map is called the scale of miles. What is the scale of miles on the map?

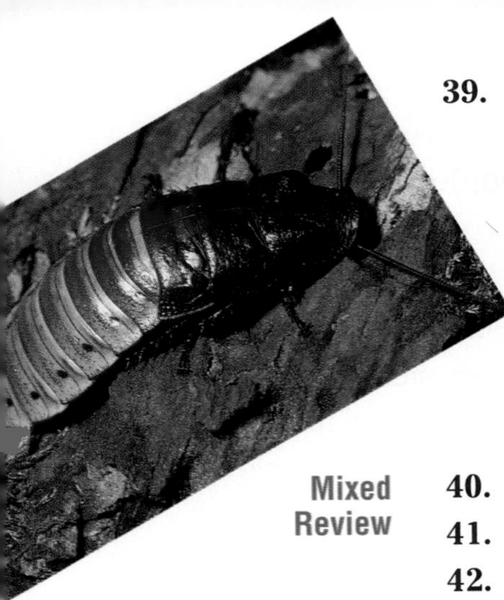

39. **Zoology** On your mark, get set, go! The fastest human can run about 26 miles per hour. That may sound fast, but many animals can leave us in the dust.

   a. A pronghorn antelope is the fastest mammal over long distances. It can run 4 miles in about 6.9 minutes. What is a pronghorn's speed in miles per hour? (*Hint:* Remember that 1 hour is 60 minutes.)

   b. The fastest insect, a tropical cockroach, can scurry up to 300 feet in a minute. What is its speed in feet per second?

Mixed Review

40. Graph $y \leq 2x - 1$. (Lesson 8-9)

41. Solve $7(a + 2) = 5a - 8$. (Lesson 7-5)

42. **Geography** The table at the right shows the area in thousands of square miles of the oceans of the world. Find the mean, median, and mode of the areas. (Lesson 6-6)

43. Factor 456 completely. (Lesson 4-4)

44. Solve $x - 9 < 19$. Graph the solution on a number line. (Lesson 3-6)

45. Find $-8 + (-5)$. (Lesson 2-4)

46. **Pets** In 1988, there were 205 million pets in the United States. By 1995, that number had grown to 235 million. Write an equation for the change in the number of pets from 1988 to 1995. Then solve. (Lesson 1-8)

Ocean	Area
Pacific Ocean	64,186
Atlantic Ocean	33,420
Indian Ocean	28,351
Arctic Ocean	5106
South China Sea	1149
Caribbean Sea	971
Mediterranean Sea	969
Bering Sea	873
Gulf of Mexico	582
Sea of Okhotsk	537
Sea of Japan	391
Hudson Bay	282
East China Sea	257
Andaman Sea	218
Black Sea	196
Red Sea	175
North Sea	165
Baltic Sea	148
Yellow Sea	114
Persian Gulf	89
Gulf of California	59

Source: *World Almanac,* 1995

EARTH WATCH

KEEPING WATCH OVER OUR OCEANS

You have probably seen news reports about oil spills in the ocean and the effects on the sea life. But oil is just one of the substances that can threaten our oceans.

When dealing with a spill, the first step is for an oceanographer to determine what the material is. One way of doing that is to find its **density.** A material's density is the ratio of its mass to its volume.

$$density = \frac{mass}{volume}$$

Each material has a unique density. So you can compare the density of a sample to the known density of a material to help identify the sample.

See for Yourself

- The density of ethyl alcohol is 0.789 grams per milliliter. An oceanographer has a sample from a chemical spill that has a mass of 136 grams and a volume of 191 milliliters. Is the sample ethyl alcohol? Explain.
- A material that has a lower density will float on one that has a higher density. A 250-milliliter sample of seawater has a mass of 260 grams. Would a spill of ethyl alcohol float on seawater? Justify your answer.
- **Research** ocean spills. Does spilled crude oil float? What chemicals are spilled most often? Describe the methods used to clean up oil and other chemical spills.

9-2 Problem-Solving Strategy: Make a Table

Setting Goals: *In this lesson, you'll solve problems by making a table.*

Modeling a Real-World Application: Business

On average, 5¢ of every dollar an American spends on food is spent in a vending machine. That's a lot of nickels!

A fruit machine accepts dollars and all the fruit costs 65 cents. If the machine gives only nickels, dimes, and quarters, what combinations of coins could it be programmed to give as change for a dollar?

Learning the Concept

You can make a table to show all of the possible combinations of coins. A table allows you to organize information in an understandable way.

Explore The machine will give back $\$1.00 - 0.65$ or 35 cents in change in a combination that can include nickels, dimes, or quarters.

Plan Organize data in a table using different combinations of nickels, dimes, quarters. Start with the most quarters.

Solve

quarters	dimes	nickels
1	1	0
1	0	2
0	3	1
0	2	3
0	1	5
0	0	7

Examine The total for each combination of the coins is 35 cents. Only nickels, dimes, and quarters are used.

Checking Your Understanding

Communicating Mathematics

Read and study the lesson to answer these questions.

1. **Explain** how a table can help you solve problems.
2. What kinds of problems are best solved by making a table or chart?

Guided Practice

Solve by making a table.

3. How many ways can you make change for a 50-cent piece using only nickels, dimes, and quarters?
4. **Number Theory** How many different ways are there to add prime numbers to make a sum of 12? Show all the ways.

5. An octahedron has faces numbered 1 to 8. If two are thrown and the faces down are added, how many ways can you roll a sum less than 8?

Exercises: Practicing and Applying the Concept

Independent Practice

Solve. Use any strategy.

6. Emily had 40 baseball cards. She traded 7 cards for 5 from Dani. She traded 3 more for 4 from Ted and 2 for 1 from Anita. Finally, she traded 11 cards for 8 from Shawna. How many cards does Emily have now?
7. A penny, a nickel, a dime, and a quarter are in a purse. How many amounts of money are possible if you grab two coins at random?
8. Lisa is challenged to find her way through a maze. Each time she walks through a correct opening, she receives \$1. In each correct aisle, she is given a reward equal to the total she already has. The correct path passes through 8 openings and 7 aisles. If Lisa takes the correct path, how much money will she have at the end?
9. Ms. Torres sold 100 shares of stock for \$5975. When the price per share went down \$5, she bought 200 more shares. When the price per share went back up \$3, she sold 100 shares. How much did Ms. Torres gain or lose in these transactions?
10. Copy the figure at the right. Place the numbers 1 through 12 at the dots so that the sum of each of the six rows is 26.
11. **Patterns** Find the next number in the pattern 0, $-1, -3, -6, -10, -15, \ldots$

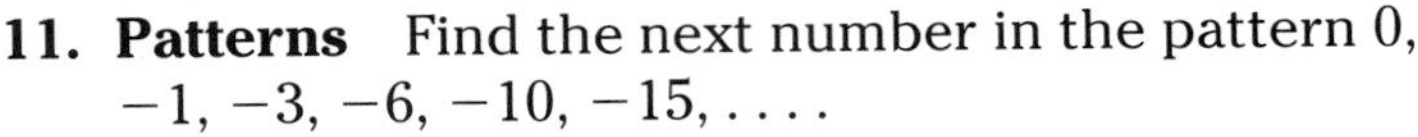

12. Tokala has 5 coins with a total value of \$1.05. He cannot make change for a dollar, 5 cents, 10 cents, 25 cents, or 50 cents. What five coins does he have?

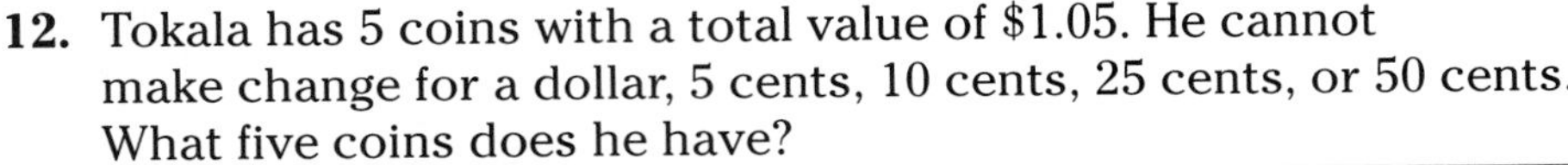

13. **Geometry** If the sides of the rectangle are whole numbers of inches, how many different combinations of measures are possible?

Area = 48 in^2, w, ℓ

Critical Thinking

14. A programmer is making a program that will create calendars. The program will contain calendar pages with just the days of the week and the dates, no months. The user will insert the name of the month before printing the calendar. If a page will exist for every arrangement of dates in a month, how many pages will need to be designed?

Mixed Review

15. **Aviation** The first non-stop, around-the-world flight was made by Capt. James Gallagher in March 1949. The 23,452-mile flight was made in 94 hours. What was the average rate in miles per hour? (Lesson 9-1)
16. Solve $3v + 7 < 19$. (Lesson 7-6)
17. Estimate $90.23 + 89.53$. (Lesson 5-2)
18. Determine whether $6c - d$ is a monomial. Explain why or why not. (Lesson 4-1)
19. Write an inequality using the numbers in the following sentence. *The Montreal Canadiens have won the Stanley Cup 24 times, and the Detroit Red Wings have won it 7 times.* (Lesson 2-3)

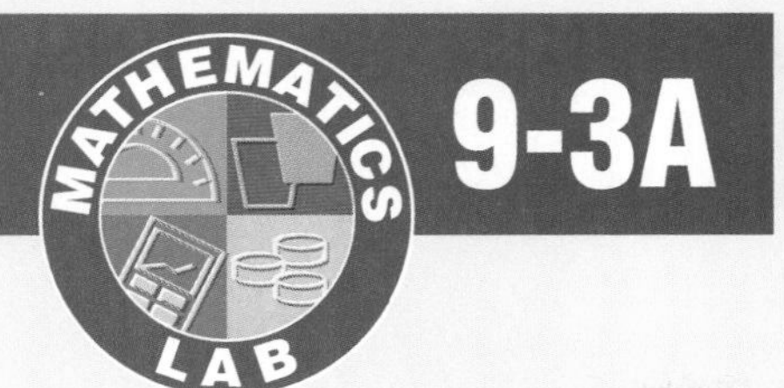

HANDS-ON ACTIVITY

9-3A Fair and Unfair Games

A Preview of Lesson **9-3**

MATERIALS

dice or number cubes

Have you ever heard someone say "Hey, no fair!" while you were playing a game? Most games are designed to be fair; that is, each player has an equal chance of winning. If the game is weighted in some way so that one player has a better chance of winning than the other players, the game is unfair.

Suppose you toss a coin 50 times. Each time the coin shows a head, you win and each time the coin shows tails, you lose.

Wins	Losses
𝍸 𝍸 𝍸 𝍸 III	𝍸 𝍸 𝍸 𝍸 𝍸 II

Since there are approximately the same number of wins as losses, the game of tossing a coin seems to be fair.

Your Turn

Work with a partner and play games A and B. Copy the table below and record your results in it.

	Win	Lose
Game A		
Game B		

Roll two dice 50 times.

Game A Find the sum of the numbers on the dice. Record an even sum as a win and record an odd sum as a loss.

Game B Find the product of the numbers on the dice. Record an even product as a win and record an odd product as a loss.

1. Are the games fair or unfair? Explain.
2. If one of the games is unfair, modify the rules in order to make it fair.

Extension

3. Enriqueta is blindfolded. She picks a marble out of a bag containing 6 blue marbles and 6 red marbles. If the marble is red, she wins. If the marble is blue, she loses. Do you think this is a fair game? Why or why not?
4. Two dice are rolled. If both dice show the same number, you win. If the dice show different numbers, you lose. Do you think this is a fair game? Why or why not?

9-3 Integration: Probability
Simple Probability

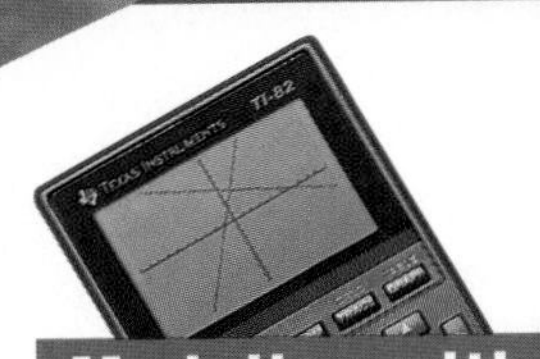

Setting Goals: *In this lesson, you'll find the probability of simple events.*

Modeling with Technology

In the fall of 1994, WSNY radio station in Columbus, Ohio, ran a contest called "Lucky Bucks". If you had a dollar bill that met the criteria announced by the DJ, you could call in to win \$100, \$1000, or even more. Suppose the DJ said, "If you have a dollar with a serial number ending in 7 or 8, be the first caller and you win!"

Your Turn

Use the graphing calculator program at the right to generate random numbers between 0 and 9 to represent the last digit of your serial number. Generate 50 different digits and record the number of times you qualify to win. Enter 0 as the least integer and 9 as the greatest.

```
PROGRAM: RANDNUM
: ClrHome
: Input "LEAST INTEGER", S
: Input "GREATEST INTEGER", L
: Input "NUMBER OF VALUES", N
: 0 → B
: Lbl 1
: B + 1 → B
: int ((L-S + 1) rand + S) → R
: Disp R
: Pause
: If A ≠ B
: Goto 1
```

TALK ABOUT IT

a. How many times did you qualify?

b. Compare results with your classmates. How many times did someone in the class qualify?

c. List the digits that qualify to win and the digits that are possible.

d. Compare $\frac{\textit{number of times qualified to win}}{\textit{number of times played}}$ and $\frac{\textit{number of digits that qualify}}{\textit{number of possible digits}}$.

e. If you had 10 dollar bills, do you think at least one would qualify? Explain.

f. Predict the outcome for 1 dollar bill and for 100 dollar bills.

Learning the Concept

Probability is the chance some event will happen. It is the ratio of the number of ways an event can occur to the number of possible outcomes.

Definition of Probability

$$\text{Probability} = \frac{\textit{number of ways a certain outcome can occur}}{\textit{number of possible outcomes}}$$

Example 1 **What is the probability that you will roll an even number when you roll a number cube?**

There are 3 possible even outcomes—2, 4, and 6.
There are 6 possible outcomes—1, 2, 3, 4, 5, and 6.

$$P(\text{even}) = \frac{\textit{number of even outcomes}}{\textit{number of outcomes}} = \frac{3}{6} \text{ or } \frac{1}{2}$$

The set of all possible outcomes is called the **sample space**. The sample space in Example 1 was {1, 2, 3, 4, 5, 6}. When you flip a coin, the sample space is heads or tails, or {heads, tails}.

Example 2 — APPLICATION: Games

In the game of Backgammon, you get to roll again if you roll doubles. What is the probability of rolling doubles?

Make a table showing all of the combinations that you could roll on a pair of dice as ordered pairs.

	1	2	3	4	5	6
1	(1, 1)	(1, 2)	(1, 3)	(1, 4)	(1, 5)	(1, 6)
2	(2, 1)	(2, 2)	(2, 3)	(2, 4)	(2, 5)	(2, 6)
3	(3, 1)	(3, 2)	(3, 3)	(3, 4)	(3, 5)	(3, 6)
4	(4, 1)	(4, 2)	(4, 3)	(4, 4)	(4, 5)	(4, 6)
5	(5, 1)	(5, 2)	(5, 3)	(5, 4)	(5, 5)	(5, 6)
6	(6, 1)	(6, 2)	(6, 3)	(6, 4)	(6, 5)	(6, 6)

In the sample space, there are 6 outcomes that show doubles and 36 possible outcomes. So the probability of rolling doubles is $\frac{6}{36}$ or $\frac{1}{6}$.

Example 3

In *Scrabble*, each player chooses seven letter tiles. The 100 tiles are in the distribution shown at the right. Find the probability of each choice.

a. an E if no tiles have been chosen

b. an M if 28 tiles have been chosen and 1 of them was an M

c. an X if 42 tiles have been chosen and 1 of them was an X

Number of tiles	Letters
1	J, K, Q, X, Z
2	B, C, F, H, M, P, V, W, Y, blank
3	G
4	D, L, S, U
6	N, R, T
8	O
9	A, I
12	E

CULTURAL CONNECTIONS

Versions of *Scrabble* in languages other than English have different ratios of letters. For example, in the Dutch version there are 18 Es, 10 Ns, and 2 Js.

a. There are 12 E tiles and 100 tiles in all.

$P(\text{choosing an E}) = \frac{12}{100}$ or $\frac{3}{25}$

b. There are 2 M tiles. So, if one has been chosen, 1 remains. There are 100 − 28 or 72 tiles from which to choose.

$P(\text{choosing an M}) = \frac{1}{72}$

c. There is only one X tile, so no Xs remain. The probability of choosing an X is 0.

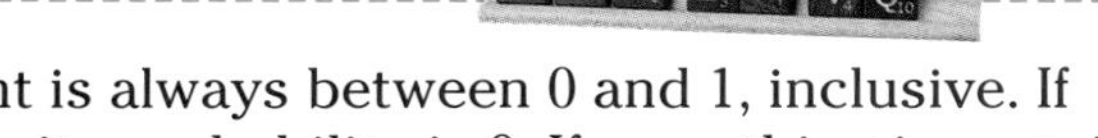

The probability of an event is always between 0 and 1, inclusive. If something cannot happen, its probability is 0. If something is certain, its probability is 1.

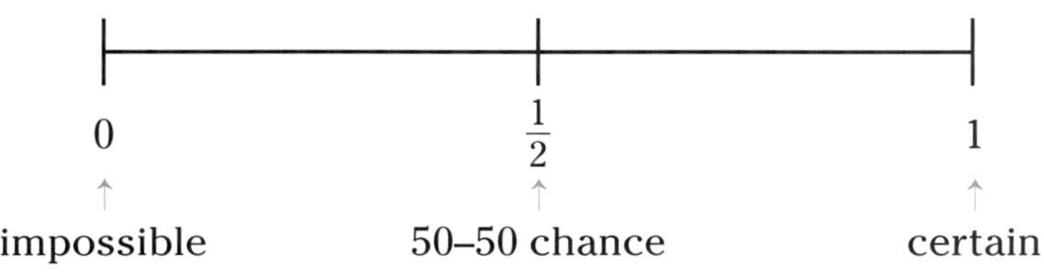

Checking Your Understanding

Communicating Mathematics

Read and study the lesson to answer these questions.

1. **Define** probability and sample space in your own words.
2. Give an example of an event with a probability of $\frac{1}{6}$.
3. **Compare and contrast** an event with a probability of 0 and an event with a probability of 1.
4. **You Decide** Four coins are tossed. Manuel says it will be fair if he wins if the majority of the coins are heads and Paige wins if the majority of the coins are tails. Paige doesn't think the game is fair. Who is correct and why?

MATERIALS

three coins

5. **a.** Use coins to help determine the sample space of the toss of three coins. Make a drawing, chart, or table to show your answer.
   b. What is the probability of two heads?
   c. What is the probability of at least one tail?

Guided Practice

State the probability of each outcome.

6. Today is the 30th of February.
7. Randomly turn to a month in a calendar and it is June.
8. A 3 is rolled on a die.

A spinner like the one at the right is used in a game. Determine the probability of spinning each outcome if the spinner is equally likely to land on each section.

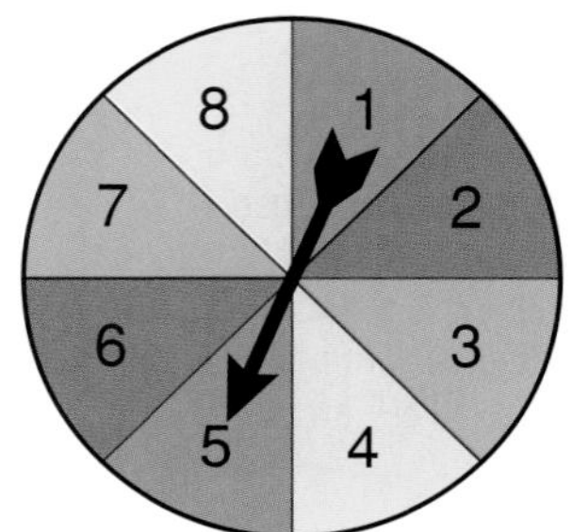

9. a four
10. a one
11. an even number
12. a two or a four
13. not a three
14. less than a six

15. **Government** In 1994, 1529 of the 7424 state legislators in the United States were women. If you chose a state legislator at random for an interview, what is the probability that you would choose a man?

Exercises: Practicing and Applying the Concept

Independent Practice

There are 5 red marbles, 7 green marbles, 4 black marbles, and 8 blue marbles in a bag. Suppose you select one marble at random. Find each probability.

16. P(red)
17. P(orange)
18. P(blue)
19. P(not green)
20. P(green or black)
21. P(not yellow)

A die is rolled. Find each probability.

22. P(odd)
23. P(even)
24. P(prime)
25. P(greater than 6)

Use the chart of possible outcomes when two dice are rolled in Example 2 to find each probability.

26. P(both odd)

27. P(both even)

28. P(both prime)

29. P(sum greater than 11)

30. P(sum of 7)

31. P(sum of 1)

Graphing Calculator

32. Refer to the application at the beginning of the lesson. What is the probability that you would have qualified to win the Lucky Bucks contest if the DJ asked for dollars that had serial numbers ending in 24? (*Hint:* Use the graphing calculator program on page 440 with numbers between 0 and 99 to simulate the situation.)

Critical Thinking

33. **Complementary events** are two events in which either one or the other must take place, but they cannot both happen at the same time. The sum of their probabilities is 1. Use this definition to solve the following problem. If a batter usually hits 3 out of every 10 pitches, what is the probability that she will *not* hit the next pitch?

Applications and Problem Solving

34. **Demographics** The graph at the right shows the population of the United States by age. Suppose that you are taking a telephone poll to determine how popular a political candidate is with different groups of voters.

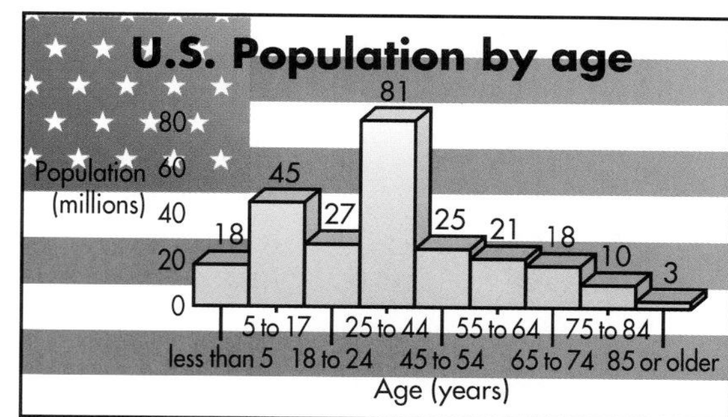

Source: U.S. Census Bureau

a. What is the probability that a randomly chosen person is old enough to vote? (*Hint:* In the United States you must be at least 18 years old to be able to vote.)

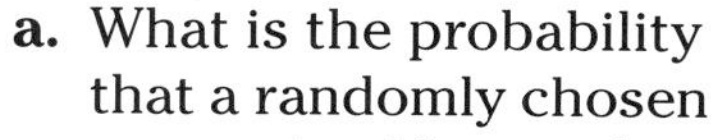

b. Find the probability that a randomly-chosen person is 65 years old or older.

35. **Space Exploration** What is the probability that humans will explore the moon on foot? What is the probability that humans will explore the Sun on foot? Explain.

36. **Family Activity** Devise an experiment using objects in your home such as tossing coins, or choosing objects out of a bag. Write each possible outcome and its probability. Then perform the experiment at least 20 times. How do the probabilities compare with the results?

Mixed Review

37. **Number Theory** How many whole numbers less than 124 are divisible by 2, 3, and 5? (Lesson 9-2)

38. Find the slope of a line that passes through $A(3, 9)$ and $B(1, -3)$. (Lesson 8-6)

39. Solve $6x + 3 = -9$. (Lesson 7-2)

40. Solve $n - 9.06 = 2.31$. (Lesson 5-6)

41. **Geometry** Find the perimeter and area of a 5.1-by-2.6-meter rectangle. (Lesson 3-5)

9-4 Using Proportions

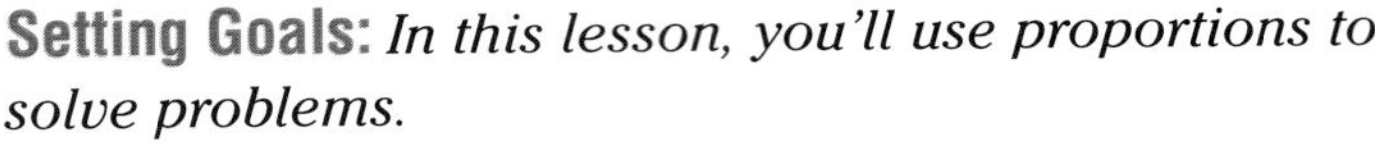

Setting Goals: *In this lesson, you'll use proportions to solve problems.*

Modeling a Real-World Application: Economics

Imagine paying almost $20 for a 2-liter bottle of Coca-Cola. That's the kind of price that the Russian people have been paying since President Yeltsin opened their markets to free trade in 1992.

The average monthly wage for a Russian is $110, while the average American takes home about $2197 each month. So if a Russian pays $0.44 for a loaf of bread, that is like an American paying $8.79!

Purchase	Russian cost (U.S. dollars)	Equivalent U.S. cost
eggs (10)	0.69	13.78
blue jeans	48.00	958.56
Coca-Cola (2 liter bottle)	0.95	18.98
gasoline (1 gallon)	1.80	35.96
McDonald's Big Mac	2.08	41.53
Heinz ketchup (17-ounce bottle)	2.31	46.13
American ice cream (2 quarts)	9.24	184.52
oranges (2 pounds)	1.57	31.35

Source: J. Patrick Lewis

The table at the right shows the cost in U.S. dollars for different items purchased in Russia and the equivalent price in terms of the average American income. What is the U.S. equivalent cost for a bottle of aspirin that costs $6.72 in Russia? *This problem will be solved in Example 3.*

Learning the Concept

You can use a proportion to solve problems that relate to ratios. A **proportion** is a statement of equality of two or more ratios. Consider a general proportion $\frac{a}{b} = \frac{c}{d}$.

$$\frac{a}{b} = \frac{c}{d}$$

$$\frac{a}{b} \cdot bd = \frac{c}{d} \cdot bd \quad \textit{Multiply each side by b and d to eliminate the fractions.}$$

$$ad = cb$$

The products ad and cb are called the **cross products** of a proportion. One way to determine if two ratios form a proportion is to check their cross products.

Property of Proportions

In words: The cross products of a proportion are equal.

In symbols: If $\frac{a}{b} = \frac{c}{d}$, then $ad = bc$. If $ad = bc$, then $\frac{a}{b} = \frac{c}{d}$.

Example  **Use cross products to determine if each pair of ratios forms a proportion.**

a. $\frac{1}{4}, \frac{4}{16}$

$$\frac{1}{4} \stackrel{?}{=} \frac{4}{16}$$
$$1 \cdot 16 \stackrel{?}{=} 4 \cdot 4$$
$$16 = 16$$
So, $\frac{1}{4} = \frac{4}{16}$.

b. $\frac{1.5}{5.0}, \frac{3}{9}$ *Use a calculator.*

1.5 [×] 9 [=] 13.5

5 [×] 3 [=] 15

$$13.5 \neq 15$$
So, $\frac{1.5}{5.0} \neq \frac{3}{9}$.

You can use cross products to solve proportions.

Example  **Solve each proportion.**

a. $\frac{c}{35} = \frac{3}{7}$

$c \cdot 7 = 35 \cdot 3$ *Cross products*

$7c = 105$ *Multiply.*

$\frac{7c}{7} = \frac{105}{7}$ *Divide each side by 7.*

$c = 15$

The solution is 15.

b. $\frac{10}{8.4} = \frac{5}{d}$

$10 \cdot d = 8.4 \cdot 5$ *Cross products*

8.4 [×] 5 [÷] 10 [=] 4.2 *Use a calculator.*

Thus, $d = 4.2$.

Proportions can be used to solve many real-life problems.

Example 3 **Refer to the application at the beginning of the lesson. Find the U.S. equivalent cost for a bottle of aspirin.**

Economics

Explore We know the average monthly income for a Russian and an American and the Russian cost for the aspirin. We need to find the U.S. equivalent cost.

Plan Write and solve a proportion with the Russian and American incomes and costs. The table shows that the U.S. equivalents are about 20 times the Russian costs. So an estimate of the U.S. equivalent is 20×7 or \$140.

Solve Let a represent the U.S. cost for aspirin.

$$\frac{\text{Russian income}}{\text{Russian cost}} = \frac{\text{American income}}{\text{American cost}}$$
$$\frac{110}{6.72} = \frac{2197}{a}$$
$$110 \cdot a = 6.72 \cdot 2197$$ *Find the cross products.*

6.72 [×] 2197 [÷] 110 [=] 134.2167273

The equivalent American cost is about \$134.22.

Examine The cost is close to the estimate, so it is reasonable.

Connection to Geometry

Proportions are used to relate the size of figures with the same shape.

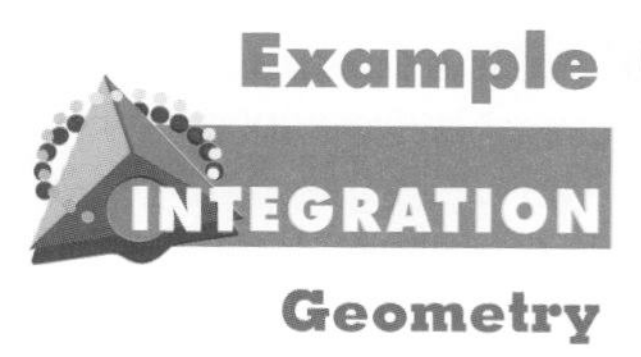

Example 4

INTEGRATION

Geometry

After setting up the proportion, you can estimate to help you decide if your solution is reasonable.

$\frac{1}{4}x \approx 2 \cdot 20$

$\frac{1}{4}x \approx 40$

$x \approx 160$

Blueprints for a house show a room that is $2\frac{5}{12}$ inches long. The key to the blueprints says that $\frac{1}{4}$ inch on the blueprints represents 18 inches in the house. How long is the actual room?

Let x represent the actual room length.

$$\begin{array}{lcl} \textit{blueprints} \rightarrow & \dfrac{\frac{1}{4}\text{ inch}}{2\frac{5}{12}\text{ inches}} = \dfrac{18\text{ inches}}{x\text{ inches}} & \leftarrow \textit{actual} \\ \textit{blueprint length} \rightarrow & & \leftarrow \textit{actual length} \end{array}$$

$$\frac{1}{4} \cdot x = 2\frac{5}{12} \cdot 18 \quad \textit{Find the cross products.}$$

$$\frac{1}{4}x = 43\frac{1}{2}$$

$$4 \cdot \frac{1}{4}x = 4 \cdot 43\frac{1}{2} \quad \textit{Multiply each side by 4.}$$

$$x = 174$$

The room is 174 inches or $14\frac{1}{2}$ feet long.

Checking Your Understanding

Communicating Mathematics

Read and study the lesson to answer these questions.

1. **Identify** the relationship between ratios and proportions.
2. **State** in your own words how to solve a proportion.
3. **Describe** three career areas that might use proportions.
4. **Demonstrate** how cross products can be used to tell whether two fractions could form a proportion.
5. **Assess Yourself** Explain how you use proportions in your life.

MATH JOURNAL

Guided Practice

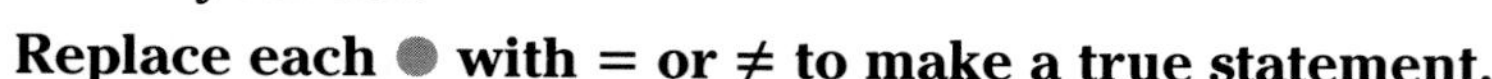

Replace each ● with = or ≠ to make a true statement.

6. $\frac{2}{5}$ ● $\frac{4}{10}$
7. $\frac{6.25}{5}$ ● $\frac{2.5}{2}$

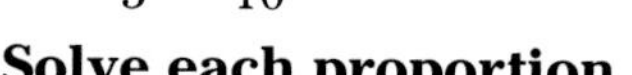

Solve each proportion.

8. $\frac{2}{y} = \frac{10}{20}$
9. $\frac{5}{7} = \frac{n}{10.5}$
10. $\frac{3}{a} = \frac{18}{24}$
11. $\frac{7}{16} = \frac{x}{4.8}$

Write a proportion that could be used to solve for each variable. Then solve.

12. 3 pounds for \$1.50
    x pounds for \$4.50
13. 2.7 liters at m dollars
    3 liters at \$7.00
14. **Cooking** A recipe calls for $4\frac{1}{2}$ cups of flour for 72 cookies. How many cups of flour would be needed for 48 cookies?

Exercises: Practicing and Applying the Concept

Independent Practice

Replace each ● with = or ≠ to make a true statement.

15. $\frac{16}{5}$ ● $\frac{4}{2}$
16. $\frac{15}{2}$ ● $\frac{18}{2.4}$
17. $\frac{1}{3}$ ● $\frac{8.6}{25.3}$
18. $\frac{2.1}{3.5}$ ● $\frac{3}{7}$
19. $\frac{2}{1}$ ● $\frac{15}{7.5}$
20. $\frac{3}{8}$ ● $\frac{2.4}{0.64}$

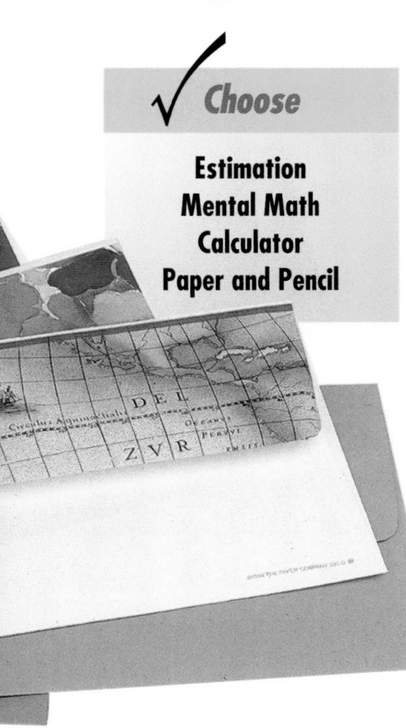

Solve each proportion.

21. $\frac{5}{m} = \frac{25}{35}$ 22. $\frac{r}{3.5} = \frac{7.2}{9}$ 23. $\frac{9.6}{t} = \frac{1.6}{7}$ 24. $\frac{6.4}{0.8} = \frac{8.1}{y}$

25. $\frac{5.1}{1.7} = \frac{7.5}{t}$ 26. $\frac{2.4}{1.6} = \frac{s}{3.4}$ 27. $\frac{18}{12} = \frac{24}{g}$ 28. $\frac{7}{45} = \frac{x}{9}$

29. $\frac{7}{16} = \frac{x}{4.8}$ 30. $\frac{3.5}{6.2} = \frac{7.35}{b}$ 31. $\frac{3.2}{4.8} = \frac{6.8}{y}$ 32. $\frac{t}{0.2} = \frac{9.4}{2}$

Write a proportion that could be used to solve for each variable. Then solve the proportion.

33. 5 liters at \$6.15
x liters at \$8.00

34. 625 bushels for 5 acres
250 bushels for y acres

35. 8 boxes in 2 crates
z boxes in 5 crates

36. 100 candies in 3 bags
300 candies in d bags

37. 25 envelopes in 5 boxes
m envelopes in 25 boxes

38. 12 pens in one package
30 pens in p packages

Solve each proportion.

39. $\frac{8}{28} = \frac{2x}{7}$ 40. $\frac{1.4}{4} = \frac{0.28}{m}$ 41. $\frac{5}{4} = \frac{x + 5}{16}$ 42. $\frac{15}{m - 3} = \frac{3}{14}$

Critical Thinking

43. Find two integers such that the ratio of the difference to the sum is 1:7 and the ratio of the sum to the product is 7:24.

Applications and Problem Solving

44. **Photography** In simple cameras, like the one shown, light from a subject passes through a lens and makes an image on film. The image and subject are always in proportion.

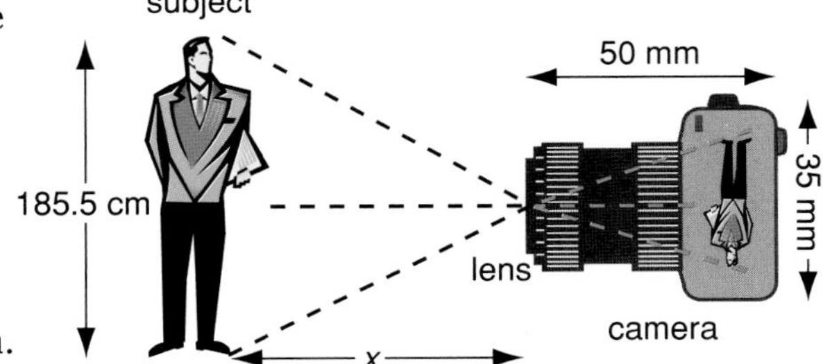

$$\frac{\textit{image size}}{\textit{subject size}} = \frac{\textit{image distance from lens}}{\textit{subject distance from lens}}$$

Find the distance represented by x in the diagram.

45. **Models** The *Preussen* was the largest sailing ship ever built. Built in Germany in 1902, it could carry 7300 metric tons of cargo. The *Preussen* was 433 feet long and 54 feet wide. How wide should a model that is 26 inches long be? Use estimation to check your answer.

Mixed Review

46. **Games** Lin needs to get a jack or higher when she cuts a deck of cards in order to win the deal. If an ace is the highest card possible, what is the probability that Lin will get the deal? (Lesson 9-3)

47. Find the slope of the line that contains $Q(5, 6)$ and $S(2, -2)$. (Lesson 8-6)

48. Solve $-8k - 21 = 75$. (Lesson 7-2)

49. Write 9,412,000 in scientific notation. (Lesson 6-9)

50. Write $0.\overline{7}$ as a fraction in simplest form. (Lesson 5-1)

51. Represent 4^{-3} using positive exponents. (Lesson 4-9)

52. **Patterns** Find the next number in the pattern 1, 2, 4, 5, 7, 8, 10, 11, . . . (Lesson 2-6)

53. Write an expression to represent *\$200 more than the budget.* (Lesson 1-3)

HANDS-ON ACTIVITY

9-4B Capture-Recapture

A Extension of Lesson **9-4**

MATERIALS

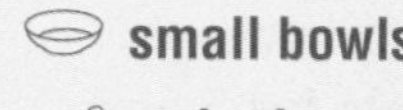

small bowls

dry beans

Did you know that there is a way to estimate how many deer are in a forest? Often naturalists want to know such a population, but it would be impossible or impractical to make an actual count.

One method of estimating a population is the **capture-recapture** technique. In this activity, you will model this technique using dry beans as "deer" and a bowl as the "forest."

Your Turn

- CAPTURE Fill a small bowl with dry beans. Grab a small handful of beans. Mark each bean with an X on both sides. Count the "tagged" beans and record this number in a chart like the one below. This number is the original number captured. Return the "tagged" beans to the bowl and mix well.

Original Number Captured: ___		
Sample	Recaptured	Tagged
A		
B		
C		
J		
Total		

- RECAPTURE Grab another small handful of beans. Count the total number of beans you grabbed. This number is the number *recaptured.* Count the number of "tagged" beans. Record these numbers. This is sample A. Return all the beans to the bowl and mix.
- Repeat RECAPTURE nine more times for samples B through J. Find the total tagged and the total recaptured.
- Use the proportion shown below to estimate the number of beans in your bowl.

$$\frac{\textit{original number captured}}{\textit{number in bowl}} = \frac{\textit{total tagged in samples}}{\textit{total recaptured}}$$

1. Why is it a good idea to base your prediction on several samples instead of just one sample?
2. What would happen to your estimate if some of your tags fell or wore off?
3. Count the number of beans in your bowl. How does your estimate compare to the actual number?
4. **Research** Investigate how researchers tag deer and estimate the total number of deer in a forest. Explain how the activity with the bowl and beans simulates their system.

38. **Personal Finance** Interest on savings accounts is often figured every quarter of a year. One quarter, Louam's savings account earned $54.84 in interest. This is 2% of her savings.

   a. Find Louam's savings.

   b. The interest in Louam's account is compounded each quarter, so the interest for a quarter is based on the original balance plus the interest earned the previous quarter. Find the interest Louam will earn next quarter if no deposits or withdrawals are made.

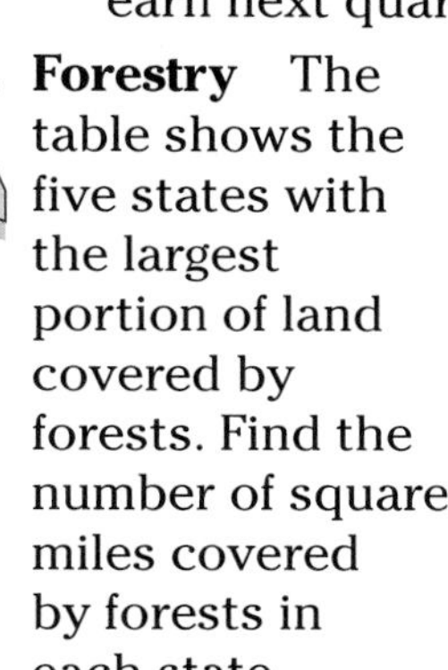

39. **Forestry** The table shows the five states with the largest portion of land covered by forests. Find the number of square miles covered by forests in each state.

State	Percent of land covered by forests	Area of state (square miles)
Maine	89.9%	35,387
New Hampshire	88.1%	9351
West Virginia	77.5%	24,231
Vermont	75.7%	9615
Alabama	66.9%	52,423

Source: The U.S. Forest Service

Mixed Review

40. **Jewelry** Jewelers use karats to describe the amount of gold in a piece of jewelry. A piece of jewelry that is 24-karat gold is all gold. How many grams of gold are in a 15-gram piece of jewelry that is labeled as 18-karat gold? (Lesson 9-4)

41. Express the ratio *6 out of 8 free throws* in simplest form. (Lesson 9-1)

42. Solve the system $y = 8x - 1$ and $3x + y = 21$. (Lesson 8-8)

43. **Personal Finance** Yolanda plans to spend no more than $15 on birthday presents for her brother. She bought a tape for $7.21 and a keychain for $3.57. What is the most money Yolanda can spend on other things for her brother? (Lesson 7-1)

44. Evaluate the expression $4b$ if $b = 3.2$. (Lesson 6-5)

45. Find the GCF of 65 and 105. (Lesson 4-5)

46. In which quadrant does the graph of $(9, -5)$ lie? (Lesson 2-2)

WORKING ON THE Investigation

Refer to the Investigation on pages 368–369.

The ticket prices of the seats are divided into three categories.

Deluxe seats	$33.50
Spectator seats	$22.75
Stadium seats	$15.20

The university has asked you to determine which seats should be designated deluxe, spectator, and stadium. To earn enough gross income, 32% of the seats need to be deluxe, and only 18% should be stadium seats.

They have asked you to prepare a report showing the location of the seats, the price of each seat, the number of seats in each category, and the gross income from a sellout crowd. Prepare this report. Be sure to include any charts or drawings that support your reasoning.

Add the results of your work to your Investigation Folder.

9-6 Integration: Statistics
Using Statistics to Predict

Setting Goals: *In this lesson, you'll use a sample to predict the actions of a larger group.*

Modeling a Real-World Application: Consumerism

How important is the brand name when you choose a product? The International Mass Retail Association surveyed consumers 8 to 17 years old on that subject. The results of the survey are shown in the graph at the right.

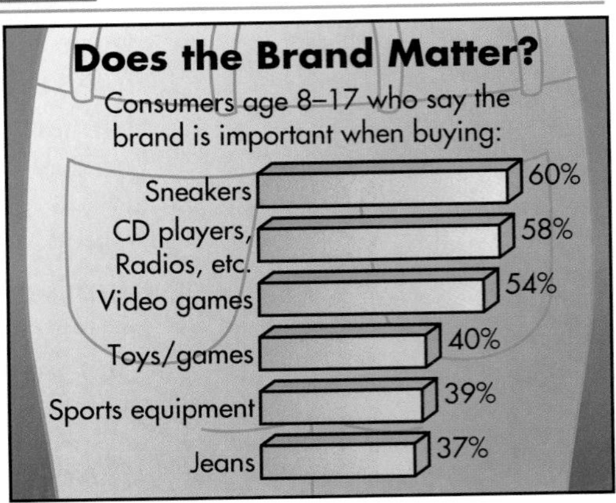

Source: International Mass Retail Association

Learning the Concept

You can use the preferences or behavior of a group to make predictions about a larger or smaller group that is similar. For example, the survey on brand buying was taken using only a portion of the consumers between 8 and 17 years old. But, the results were used to make statements about all consumers in that age group.

Example — APPLICATION — Consumerism

Refer to the application at the beginning of the lesson. How many of the 423 9th-grade students at Kilbourne High School would you expect to say that they consider brand name when buying jeans?

Use the problem-solving plan to find the expected number.

Explore The graph shows that 37% of the consumers surveyed considered brand name when they purchased jeans. We need to know how many of the freshman class students consider brand name when buying jeans.

Plan Write and solve a proportion using the percent proportion. Estimate that since 37% is a little over $\frac{1}{3}$, the solution should be a little over 423 ÷ 3, or 140.

Solve

$$\frac{\text{percent considering brand name}}{100} = \frac{\text{number of Kilbourne freshmen who consider brand name}}{\text{number of Kilbourne freshmen}}$$

$$\frac{37}{100} = \frac{x}{423}$$

$$37 \cdot 423 = 100 \cdot x \quad 37 \;[\times]\; 423 \;[\div]\; 100 \;[=]\; 156.51$$

$$x = 156.51$$

Expect 157 students to consider brand when buying jeans.

Examine The solution is close to the estimate, so it is reasonable.

Example 2

APPLICATION

Entertainment

interNET CONNECTION

For the latest Nielsen ratings, visit: **www.glencoe.com/sec/math/prealg/mathnet**

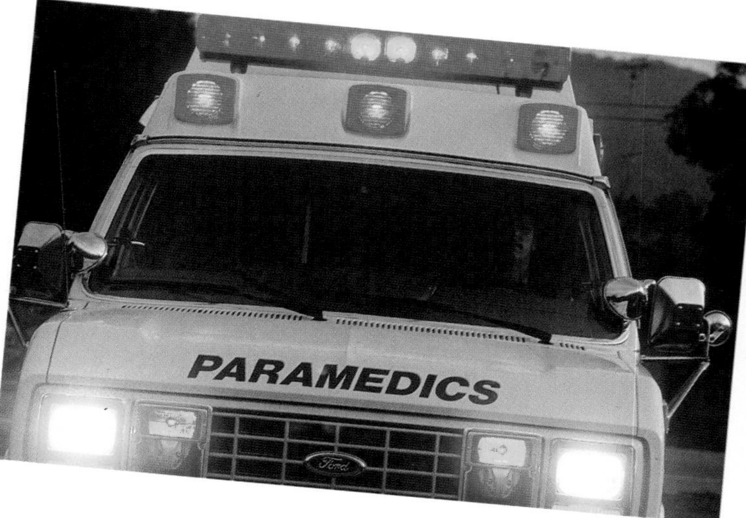

The NBC television program *ER* began the 1995 television season as the number one show in the Nielsen ratings. In the first week of the season, it attracted 41% of the Nielsen viewers. If about 91 million viewers watched television that night, about how many were watching *ER*?

Use a proportion.

$$\frac{x}{91} = \frac{41}{100}$$

$$x \cdot 100 = 91 \cdot 41 \quad \textit{Use a calculator.}$$

$$x = 37.31$$

About 37.31 million viewers watched *ER* that night.

When taking surveys and using the results to predict the actions of a larger population, be sure that your sample is random and is large enough to represent the population. If the sample is too small, it is likely that the items or people that it includes are not typical of the population. For example, suppose you were told that 4 out of 5 athletes chose a particular type of shoe for basketball. The claim would not be meaningful if only 5 athletes were surveyed.

Checking Your Understanding

Communicating Mathematics

Read and study the lesson to answer these questions.

1. **Explain** how to use the results of a survey to predict the characteristics of a population.
2. **Name** the most important features of a sample.
3. **You Decide** Refer to the application at the beginning of the lesson. Tamika estimates that 152 of the 253 students in her class consider brand name when choosing sneakers. Gloria estimates only 121 of them do. Who is correct and why?

Guided Practice

Use the survey on favorite drinks to answer the questions.

4. How large is the sample?
5. If this represents a group of 3600, what percent is the sample?
6. What fraction chose fruit drink? What percent is this?
7. Name a location that would not be a good place to conduct this survey.

Favorite Drink	
lemon lime	17
cola	10
root beer	25
fruit drink	12
ginger ale	8

Exercises: Practicing and Applying the Concept

Independent Practice

The owner of a *Dairy Dream* store surveyed 100 people at her store about their favorite ice cream flavor. The results are in the table below.

Favorite Flavors	
vanilla	20
chocolate	25
strawberry	10
peanut butter fudge	35
mint chocolate chip	10

8. Give some reasons why this is a valid sample and some reasons why it is not.

9. What percent of the people preferred peanut butter fudge?

10. If there were 800 customers on a Saturday, about how many would prefer chocolate?

The results of a 1992 survey by Independent Sector on the reasons that teens spend time as volunteers are shown in the table below.

Reason	Percent	Reason	Percent
to help others	47	for a friend	20
enjoy the work	38	religion	19
lots of free time	25	past experience	10
to learn	24	other	7
		don't know	2

11. If you asked 500 teens who volunteer why they do, how many would you expect to answer because they have free time?

12. Out of 250 teen volunteers, how many would you expect volunteer because they enjoy the work?

13. If you surveyed 25 volunteers at a homeless shelter on reasons for volunteering, would you expect the same results as this survey? Explain.

14. Research Make a frequency table showing the hair color for students in your class. Use the data to predict the number of students in your school with each hair color. Choose a time and place to observe at least 50 students and record their hair color. Compare that data to your predictions. What are your findings?

Critical Thinking

15. Thirteen percent of the letters used in English words are Es. If you were guessing the letters in a two-letter word of a puzzle on *Wheel of Fortune*, would you guess E? Explain.

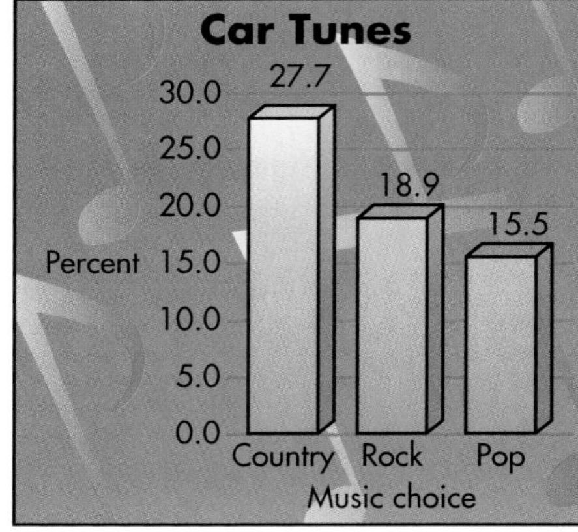

Source: 20/20 Research

Applications and Problem Solving

16. Music A *20/20 Research* survey asked what music people listen to in the car.

a. How many of the 445 cars in a traffic jam do you predict have rock playing?

b. If you owned a store that specialized in car stereos, what type of music would you have playing? Explain your choice.

17. Fashion *Cotton Incorporated* surveyed 3600 people about clothing.

a. 1872 people said they would rather be dressed too casually for an occasion than be dressed too fancy. What percent of the population do you think would feel the same?

b. What percent of the population would say they were slow to change with the fashions if 2160 of those surveyed said that?

18. **Housekeeping** A *TIME* magazine poll showed that 76% of women and 46% of men surveyed made their bed that morning.

   a. If you used the results of the *TIME* poll, about how many of the 346 females and 362 males at Centennial High school would you predict made their beds this morning?

   b. The *TIME* poll questioned adults. Do you think that a poll of teens would have the same results? Explain.

Mixed Review

19. **Food** The average American eats 10.28 pounds of chocolate in a year. If the total annual candy consumption of the average American is 17.86 pounds, what percent of the candy is chocolate? (Lesson 9-5)

20. **Patterns** State whether the sequence 2, 3.5, 5, 6.5, 8, 9.5, 11, . . . is arithmetic. Then write the next three terms. (Lesson 5-9)

21. If 15 more than the product of a number and -2 is greater than 12, which of the following could be the number? (Lesson 3-1)

   A. 6   **B.** 4   **C.** 1.5   **D.** 0

22. **Geometry** Use the coordinate grid at the right to name the point for the ordered pair $(-4, 2)$. (Lesson 2-2)

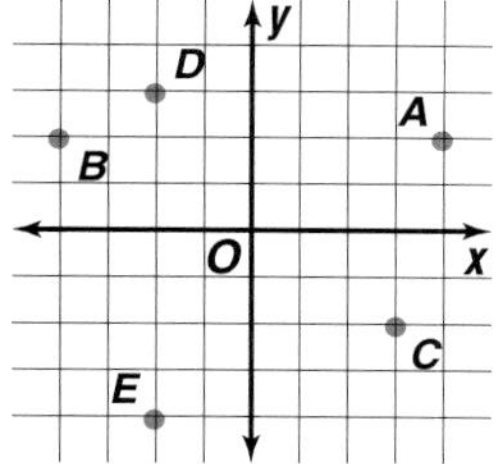

23. State whether $14 \geq 2b + 4$ is true or false if $b = 3$. (Lesson 1-9)

Self Test

Express each ratio as a unit rate. (Lesson 9-1)

1. 300 feet in 5 minutes
2. 54 inches of rain in 4 months
3. How many ways can you make change for a $50 bill using only $5, $10, and $20 bills? (Lesson 9-2)

There are 3 blue pencils, 5 green pencils, 2 black pencils, and 6 red pencils in a drawer. Suppose you grab one pencil at random. Find each probability. (Lesson 9-3)

4. $P(\text{green})$
5. $P(\text{blue or red})$

Solve each proportion. (Lesson 9-4)

6. $\frac{m}{7} = \frac{25}{35}$
7. $\frac{1.5}{2} = \frac{s}{2.4}$

Use the percent proportion to solve each problem. (Lesson 9-5)

8. Find 75% of 400.
9. 250 is 500% of what number?

Destination	Percent
Florida	35
California	30
Hawaii	20
Nevada	14
New York	11
Colorado	9
Arizona	8
Texas	6
South Carolina	6
Washington	6

Source: Travel Industry Assn.

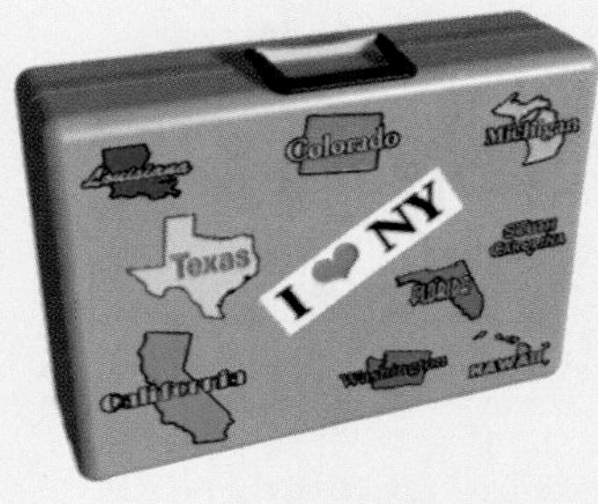

The Travel Industry Association polled 1500 adults on their choice of fall vacation destinations. (Lesson 9-6)

10. How many of those surveyed chose New York as a destination?
11. If a travel agency books vacations for 300 families this fall, how many should they expect will go to Florida?

9-7 Fractions, Decimals, and Percents

Setting Goals: *In this lesson, you'll express decimals and fractions as percents and vice versa.*

Modeling with Manipulatives

MATERIALS

grid paper

You can compare quantities using fractions, percents, or decimals. Each of these can be rewritten in the other two forms. The model at the right represents one unit. Since 36 of the 100 cells are shaded, $\frac{36}{100}$ or 0.36 of the model is shaded. Using the definition of percent, we could also say that 36% of the model is shaded.

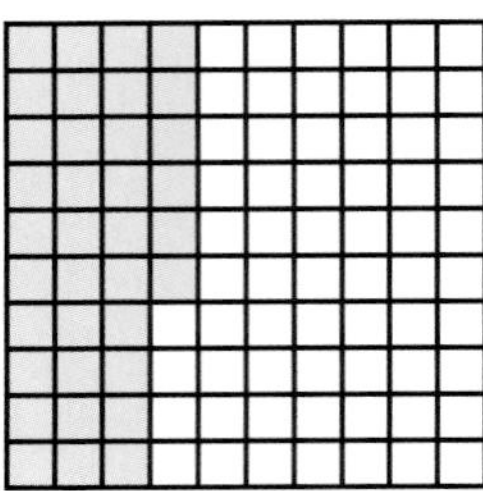

Your Turn

Make models to represent $\frac{1}{2}$, 0.42, and 8%.

a. How could you write $\frac{1}{2}$ as a decimal and a percent?

b. Write 0.42 as a percent. What do you observe about the decimal point?

c. Write a procedure for writing a decimal as a percent and for writing a percent as a fraction.

Learning the Concept

You have written fractions as decimals and percents. Fractions, decimals, and percents are all different names that represent the same number.

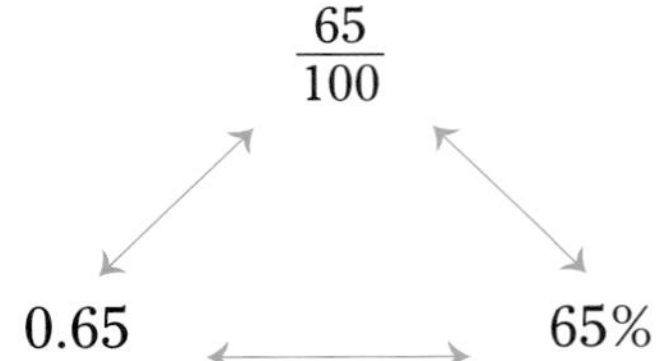

To express a decimal as a percent, write the decimal as a fraction with 1 as the denominator. Then write that fraction as an equivalent fraction with 100 as the denominator.

Example 1 **Express each decimal as a percent.**

a. 0.36 → $\frac{0.36}{1} = \frac{36}{100}$ or 36% (× 100 numerator and denominator)

b. 0.07 → $\frac{0.07}{1} = \frac{7}{100}$ or 7% (× 100 numerator and denominator)

c. 0.004 → $\frac{0.004}{1} = \frac{0.4}{100}$ or 0.4% (× 100 numerator and denominator)

d. 1.8 → $\frac{1.8}{1} = \frac{180}{100}$ or 180% (× 100 numerator and denominator)

THINK ABOUT IT

How could you convert a decimal to percent or a percent to a decimal mentally?

Compare each decimal in Example 1 with its equivalent percent. Notice that, in each case, the decimal point moved two places to the right.

To express a fraction as a percent, first write the fraction as a decimal by dividing numerator by denominator. Then write the decimal as a percent.

Example 2 **Express each fraction as a percent.**

a. $\frac{3}{5} = 0.60 \rightarrow$ 3 [÷] 5 [=] 0.6 $\rightarrow \frac{60}{100} = 60\%$

b. $\frac{7}{4} = 1.75 \rightarrow$ 7 [÷] 4 [=] 1.75 $\rightarrow \frac{175}{100} = 175\%$

c. $\frac{2}{500} = 0.004 \rightarrow$ 2 [÷] 500 [=] 0.004 $\rightarrow \frac{0.4}{100} = 0.4\%$

To express a percent as a fraction, write the percent in the form $\frac{r}{100}$ and simplify.

Example  **Write each percent as a fraction in simplest form.**

LOOK BACK

You can review dividing fractions in Lesson 6-4.

a. $15\% = \frac{15}{100}$

$= \frac{3}{20}$

b. $87\frac{1}{2}\% = \frac{87\frac{1}{2}}{100}$ *The fraction bar indicates division.*

$= 87\frac{1}{2} \div 100$

$= \frac{\overset{7}{\cancel{175}}}{2} \times \frac{1}{\underset{4}{\cancel{100}}}$

$= \frac{7}{8}$

When you want to express a percent as a decimal, write the percent in the form $\frac{r}{100}$ and then write as a decimal.

Example  **Write each percent as a decimal.**

a. $17\% = \frac{17}{100}$

$= 0.17$

b. $250\% = \frac{250}{100}$

$= 2.50$

In Example 4, notice that the decimal point was moved two places to the left to get the equivalent decimal.

Cosmetics

5 **In a 1995 *Self* magazine survey on favorite perfumes, 26% of those asked used one certain perfume all the time while one-twentieth said that they never use perfume. Which was the larger group?**

Write each number as a decimal to compare.

$26\% = \frac{26}{100}$

$= 0.26$

$\frac{1}{20} = 0.05$ 1 [÷] 20 [=] 0.05

Since 0.26 is greater than 0.05, the group that said they use the same perfume each day was larger.

CULTURAL CONNECTIONS

Perfume had its beginning in ancient incense. Greeks and Romans learned about perfumes from the Egyptians. In the 1200s, crusaders took perfumes from Palestine to England and France.

Checking Your Understanding

Communicating Mathematics

Read and study the lesson to answer these questions.

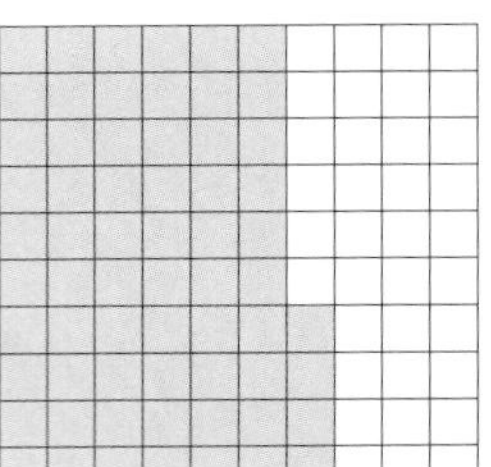

1. **Describe** how you would change 24.7% to a decimal.
2. **Write** a fraction, a decimal, and a percent to represent the model at the right.
3. **In your own words, explain** how you can tell if a fraction is greater than 100% or less than 1%.
4. Put 25 beans in a bag. Have your partner grab a handful of beans. Express the portion of the beans that were removed as a fraction, a percent, and a decimal. Repeat. Which handful was a greater portion of the beans?

MATERIALS

dry beans

paper bag

Guided Practice

Express each decimal as a percent.

5. 0.37 6. 0.475 7. 1.03

Express each fraction as a percent.

8. $\frac{27}{100}$ 9. $\frac{36}{500}$ 10. $\frac{6}{5}$

Express each percent as a fraction.

11. 25% 12. $33\frac{1}{3}\%$ 13. 0.3%

Express each percent as a decimal.

14. 35% 15. $24\frac{1}{2}\%$ 16. 0.4%

17. **Sports** In a survey, 48 of 120 people said they preferred hockey as a spectator sport.
    a. What fraction is this?
    b. What percent is this?

Exercises: Practicing and Applying the Concept

Independent Practice

Express each decimal as a percent.

18. 0.71 19. 0.03 20. 0.543 21. 2.37
22. 0.004 23. 1.32 24. 0.035 25. 0.0004

Express each fraction as a percent.

26. $\frac{3}{16}$ 27. $\frac{7}{12}$ 28. $\frac{5}{4}$ 29. $\frac{7}{3}$
30. $\frac{4}{300}$ 31. $\frac{5}{9}$ 32. $1\frac{1}{4}$ 33. $\frac{3}{40}$

Choose
Estimation
Mental Math
Calculator
Paper and Pencil

Express each percent as a fraction.

34. 36% 35. 125% 36. 58% 37. 34.5%
38. 22% 39. $37\frac{1}{2}\%$ 40. $66\frac{2}{3}\%$ 41. $16\frac{1}{3}\%$

Express each percent as a decimal.

42. 35% **43.** 75% **44.** 93% **45.** 39.8%

46. 42.7% **47.** 0.4% **48.** 127% **49.** 235.4%

Choose the greatest number in each set.

50. $\left\{\frac{3}{5}, 0.75, 80\%, 6 \textit{ to } 8\right\}$

51. $\left\{22\%, 0.022, \frac{1}{11}, 2 \textit{ out of } 13\right\}$

Write each list of numbers in order from least to greatest.

52. $\frac{1}{2}$, 31%, 0.05

53. 16%, $\frac{1}{4}$, 0.067

Critical Thinking

54. A grocery store carries three different-sized bottles of contact lens solution. Size A is 50% more expensive than size C, but it contains 20% more solution than size B. Size B contains 50% more solution than size C, but it costs 25% more than size A. Which bottle is the most economical choice?

Applications and Problem Solving

55. Retail Sales A pair of basketball shoes is listed at $33\frac{1}{3}\%$ off. What fraction off is this?

56. Health Pediatricians and obstetricians care for newborn babies and their mothers.

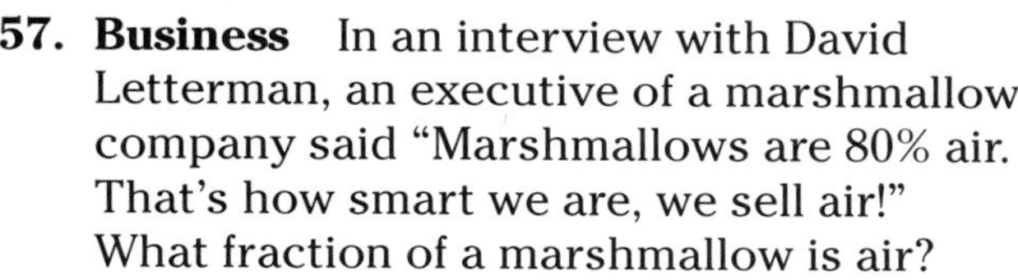

a. Pediatricians estimate that 3 babies out of every 1000 are likely to contract a cold their first month. What percent is this?

b. Obstetricians have observed that only 4% of babies are born on their due date. What fraction of babies are born on their due dates?

57. Business In an interview with David Letterman, an executive of a marshmallow company said "Marshmallows are 80% air. That's how smart we are, we sell air!" What fraction of a marshmallow is air?

Mixed Review

58. Cosmetics In the *Self* survey, 34% of those asked prefer floral-scented perfume. Of 50 customers at a department store cosmetics counter, how many would you expect to choose a floral perfume? (Lesson 9-6)

59. Find the *x*- and *y*-intercept of the graph of $y = 8x - 12$. (Lesson 8-7)

60. Solve $\frac{b}{-2} - 12 \leq 11$. (Lesson 7-6)

61. Find the product of $\frac{4}{9}$ and $\frac{5}{12}$. (Lesson 6-3)

62. Solve $x + 14.7 < 51.2$. (Lesson 5-7)

63. Find the product $n^4 \cdot n^{10}$. (Lesson 4-8)

64. Science In the metric system, the prefix *giga* means 10^9. Write 10^9 as a product of the same factor. (Lesson 4-2)

65. Choose the value of *h* if $\frac{h}{0.7} = -2.8$. (Lesson 3-1)

A. −4 **B.** −2.8 **C.** −1.96 **D.** 0.7

66. Solve $r = -9889 \div -319$. (Lesson 2-8)

67. Solve $4 + x = 12$ using an inverse operation. (Lesson 1-8)

9-8 Percent and Estimation

Setting Goals: *In this lesson, you'll use percents to estimate.*

Modeling a Real-World Application: Health

"You are what you eat." So if you want to be healthy, you need to eat healthful food. Nutritionists suggest that people get no more than 30% of their daily Calorie intake from fat. Do any of the popcorn snacks shown in the table at the right contain about 30% fat Calories or less? *This problem will be solved in Example 2.*

Snack	Calories per serving	Calories from fat per serving
Jiffy Pop	140	63
Newman's Own Light	110	27
Pop Secret Original Butter	142	90
Pop Secret by Request	108	18
RedenBudders Light	100	36
Screaming Yellow Zonkers Glazed	130	36
Smartfood White Cheddar Cheese	160	90

Source: *Vitality*

Learning the Concept

Many times when you are working with percents, an exact answer is not needed. In cases like this, you can estimate. We could estimate the percent of the model that is shaded.

Nine of the 20 squares are shaded.

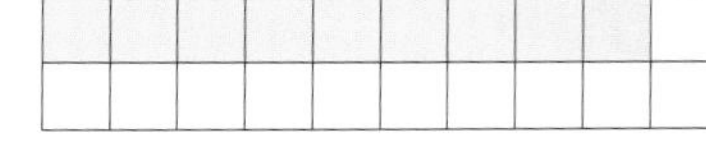

$\frac{9}{20}$ is about $\frac{10}{20}$ or $\frac{1}{2}$

Since $\frac{1}{2} = 50\%$, about 50% of the model is shaded.

There are three methods you can use to estimate a percentage. The table below shows how to estimate 30% of 657 using each method.

Fraction Method	1% Method	Meaning of Percent Method
30% is about 33% or $\frac{1}{3}$.	1% of 660 is 6.6 or about 7.	30% means 30 for every 100 or 3 for every 10. 657 has 6 hundreds and about 6 tens.
$\frac{1}{3}$ of 660 is 220.	30 times 7 is 210.	$(30 \times 6) + (3 \times 6) = 198$
Estimate: 220	Estimate: 210	Estimate: 198

Use a calculator to find the exact amount.

657 [×] 30 [%] [=] 197.1

The actual percentage is very close to all of the estimates.

Example 1

Estimate.

a. 60% of 996
1% of 996 is 9.96 or about 10. So 60% of 996 is about 60×10 or 600.

b. 8% of $58
8% is about 10% or $\frac{1}{10}$. $\frac{1}{10}$ of $58 is $5.80, so 8% of $58 is about $6.00

c. 0.5% of 795
0.5% is half of 1%.
795 is almost 800.
1% means 1 out of 100.
So 1% of 800 is 8, and $\frac{1}{2}$ of 8 is 4.
0.5% of 795 is about 4.

d. 109% of 62
109% is more than 100%, so 109% of 62 is more than 62.
109% is almost 110%.
$110\% = 100\% + 10\%$
$62(100\% + 10\%) = 62 + 6.2 = 68.2$
109% of 62 is about 68.

Example 2

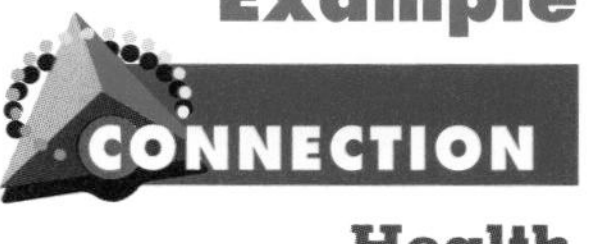

Health

Refer to the application at the beginning of the lesson. Do any of the popcorn snacks have less than 30% of their Calories from fat?

Use the meaning of percent method to estimate 30% of the Calories for each snack. 30% means 30 for every 100 Calories and 3 for every 10 Calories.

FYI

The average American eats 71 quarts of popcorn each year.

Snack	Calories	Estimate of 30% of Calories	Fat Calories
Jiffy Pop	140	$30 + 3(4) = 42$	63
Newman's Own Light	110	$30 + 3(1) = 33$	27
Pop Secret Original Butter	142	$30 + 3(4) = 42$	90
Pop Secret by Request	108	$30 + 3(1) = 33$	18
RedenBudders Light	100	$30 + 3(0) = 30$	36
Screaming Yellow Zonkers Glazed	130	$30 + 3(3) = 39$	36
Smartfood White Cheddar Cheese	160	$30 + 3(6) = 48$	90

Compare the estimate of 30% of the total calories to the number of calories from fat in each snack. Newman's Own Light, Pop Secret by Request, and Screaming Yellow Zonkers Glazed all have about 30% or less of their calories from fat.

Estimating percents is a useful skill for everyday situations.

Example 3

Consumerism

A tip of 15% is standard for good service in a restaurant. If Simone wants to leave a tip of about 15% on a dinner check of $23.85, how much should she leave?

Estimate: $23.85 is about $24.
10% of $24 is $2.40.
5% of $24 is $1.20. *5% is half of 10%.*
15% is about 2.40 + 1.20 or $3.60.

Simone should leave about $3.60 as a tip.

Checking Your Understanding

Communicating Mathematics

Read and study the lesson to answer these questions.

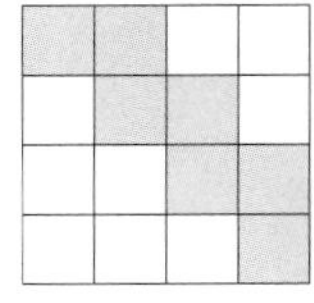

1. **Explain** how you would estimate 27% of 198.
2. **Describe** in your own words one method for estimating using percents.
3. **Estimate** the percent of the model at the right that is shaded.
4. **You Decide** Once a month Angeni and Darlene meet for lunch. The food tax where they live is $5\frac{1}{4}$%. Each of them leaves about a 15% tip. Angeni always finds her tip by taking 10% of the total and then adding that amount to half of that amount. Darlene finds her tip by multiplying the tax by 3. Who is correct and why?

5. **Describe** a situation that would best use each kind of method for estimating using percents.

Guided Practice

Choose the best estimate for the percent shaded.

6. 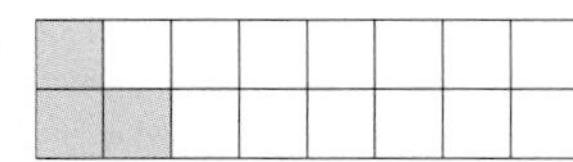

   a. 10% **b.** 25% **c.** 55%

7. 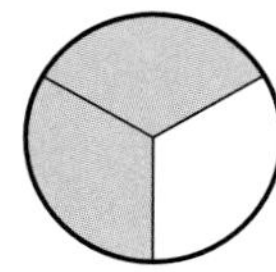

   a. 33% **b.** 50% **c.** 66%

Write the fraction, mixed number, or whole number you could use to estimate.

8. 24%
9. 79%
10. 145%

Estimate.

11. 45% of 430
12. 112% of 14.5
13. 0.6% of 325

Estimate each percent.

14. 8 out of 30
15. 9 out of 19
16. $\frac{5}{7}$

17. **Retail Sales** Davon bought a pair of shoes on sale for $29. The regular price was $50.
    a. For about what percent of the selling price did he buy them?
    b. About what percent-off was the sale?

Exercises: Practicing and Applying the Concept

Independent Practice

Choose the best estimate.

18. 39% of 300	**a.** 1.2	**b.** 12	**c.** 120
19. 47% of 605	**a.** 3	**b.** 30	**c.** 300
20. $\frac{1}{3}$% of 240	**a.** 0.8	**b.** 8	**c.** 80
21. 129% of 400	**a.** 50	**b.** 500	**c.** 5000

Write the fraction, mixed number, or whole number you could use to estimate.

22. 67%	**23.** 98%	**24.** 18%	**25.** $2\frac{1}{2}\%$
26. 148%	**27.** $8\frac{5}{9}\%$	**28.** 0.8%	**29.** 119%

Estimate.

30. 47% of 84	**31.** 28% of 390	**32.** $8\frac{1}{2}\%$ of 55
33. 98% of 98	**34.** 126% of 198	**35.** 0.9% of 514
36. 15% of $34	**37.** 116% of 18	**38.** 0.05% of 1180

Estimate each percent.

39. 12 out of 15	**40.** 8 out of 35	**41.** 39 out of 79
42. 57 out of 176	**43.** 9 out of 95	**44.** 13 out of 68
45. 1 out of 9	**46.** 3 out of 200	**47.** 7 out of 2445

Critical Thinking

48. In a controversial vote, 40 percent of the Democrats and 92.5 percent of the Republicans voted yes. If all the members of the assembly voted and 68 percent of the voters voted yes, what is the ratio of Democrats to Republicans?

Applications and Problem Solving

49. Nutrition Estimate the percent of fat Calories of each entree in the chart at the right.

Entree	Calories (3.5 oz)	Calories from fat
shrimp	90	7
lobster	90	15
chicken wing, with skin	290	171
ham	220	99
ground chuck	222	81

Source: *The Fat Counter*

50. Research The land area of Georgia is about 60% of the land area of Great Britain.

a. Find the size (in square miles) of Great Britain.

b. Estimate the size of Georgia.

51. History The table below shows the populations of the American colonies in 1780. If the total population of the colonies was 2,780,000 in 1780, estimate what percent of the total each colony had.

Colony	Population (thousands)	Colony	Population (thousands)
Connecticut	206.7	North Carolina	270.1
Delaware	45.4	Pennsylvania	327.3
Georgia	56.1	Plymouth and Massachusetts	268.6
Kentucky	45.0	Rhode Island	52.9
Maine counties	49.1	South Carolina	180.0
Maryland	245.5	Tennessee	10.0
New Hampshire	87.8	Vermont	47.6
New Jersey	139.6	Virginia	538.0
New York	210.5		

Source: Bureau of the Census

52. Nutrition A McDonald's Big Mac has 560 Calories, and 288 of these are fat Calories.

a. About what percent of the Calories are fat Calories?

b. If a person on a 2000-Calorie-per-day diet eats a Big Mac, how many Calories and fat Calories remain for the rest of the day?

Mixed Review

53. Express 64% as a fraction. (Lesson 9-7)

54. **World Facts** On German highways, called autobahns, the speed limit is 130 kilometers per hour. What is the speed limit in meters per hour? (Lesson 7-8)

55. **Patterns** State whether 8, 12, 18, 27, 40.5, . . . is a geometric sequence. If so, state the common ratio and list the next three terms. (Lesson 6-8)

56. Solve $\frac{9}{16} + \frac{11}{16} = y$. (Lesson 5-4)

57. Solve $5x = 105$ and check your solution. Then graph the solution on a number line. (Lesson 3-3)

58. Simplify $|-21|$. (Lesson 2-1)

59. **Entertainment** The table at the right shows the number of radios per thousand people in several countries. Make a bar graph of the data. (Lesson 1-10)

Country	Radios per thousand people
United States	2091
Bermuda	1710
United Kingdom	1240
Australia	1144
Finland	984
New Zealand	902
Virgin Islands (U.S.)	884
France	866
Sweden	842
Canada	828

Source: Duncan's American Radio, Inc.

60. Translate the phrase *twice the sum of a number and 7* into an algebraic expression. (Lesson 1-3)

From the FUNNY PAPERS

TALK ABOUT IT

1. Explain why the comic is funny.
2. If the student missed six out of ten questions on the quiz, how many questions did he answer correctly?
3. What is the student's real percent score on the quiz?
4. Write a sentence or two to explain to the father how to estimate the percent score on a quiz.

9-9 Using Percent Equations

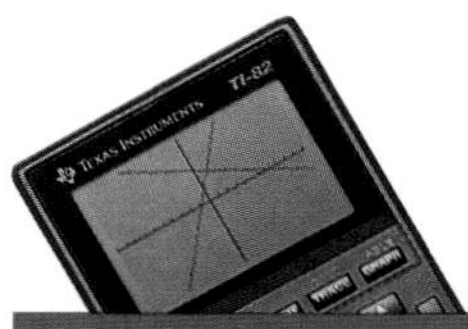

Setting Goals: *In this lesson, you'll solve percent problems using percent equations.*

Modeling with Technology

Maria Morales has saved some money that she wishes to invest. Her banker has provided the terms, interest rates, and minimum investments on the certificates of deposit, or CDs, that they offer. Maria used the information to prepare the spreadsheet below to compare the options.

Option	Minimum Investment	Monthly Interest Rate (percent)	Term (months)	Interest for Term
1	\$2500	0.375	6	
2	\$2500	0.45	12	
3	\$5000	0.54	36	
4	\$7500	0.625	60	

The **principal** is the amount of money in the account. The amount of **interest** earned in a month can be found by using the percent proportion $\frac{\text{interest}}{\text{principal}} = \frac{\text{interest rate}}{100}$. Then multiply the monthly interest by the number of months to find the total interest earned.

Your Turn

TALK ABOUT IT

Use a calculator or spreadsheet to find the interest on each investment.

a. Which option earns the most? What factors contribute to this?

b. Let I represent interest, p the principal, r the interest rate as a decimal, and t the time in months. Write a formula for finding the interest on any account.

Learning the Concept

As you discovered in the activity above, you can find the interest earned on an account using the formula $I = prt$ if you write the rate as a decimal. Writing the rate as a decimal in the percent proportion also allows you to write it as an equation and solve percent problems more quickly.

$$\frac{P}{B} = \frac{r}{100}$$

$$\frac{P}{B} \cdot B = \frac{r}{100} \cdot B \quad \textit{Multiply each side by B.}$$

$$P = \frac{r}{100} \cdot B$$

Remember $\frac{r}{100}$ is the rate. Let R represent the decimal form of $\frac{r}{100}$.

Therefore, $P = R \cdot B$ or Percentage = Rate · Base.

The form $P = R \cdot B$ is usually easier to use when the rate and base are known.

Example 

a. Find 36% of 65. ***Estimate:*** *$\frac{1}{3}$ of 60 is about 20.*

$$\underbrace{\text{What number}}_{P}\ \underbrace{\text{is}}_{=}\ \underbrace{36\%}_{0.36}\ \underbrace{\text{of}}_{\times}\ \underbrace{65?}_{65}$$ *Write in $P = R \cdot B$ form.*

0.36 [×] 65 [=] *23.4*

23.4 is 36% of 65.

b. 15 is what percent of 125?

$15 = R \cdot 125$ *Replace P with 15 and B with 125.*

$\frac{15}{125} = \frac{125R}{125}$ *Divide each side by 125.*

$0.12 = R$ *Use a calculator.*

15 is 12% of 125.

Example 

APPLICATION

Charity

The Nature Conservancy is among the largest charities involved in protecting the environment. They reported spending $198,722,500 on conservation in 1993. If this is 72.5% of the money they received, how much money did they receive?

The Nature Conservancy®

$$\underbrace{198{,}722{,}500}_{198{,}722{,}500}\ \underbrace{\text{is}}_{=}\ \underbrace{72.5\%}_{0.725}\ \underbrace{\text{of}}_{\times}\ \underbrace{\text{what number?}}_{B}$$ *$P = R \cdot B$*

$\frac{198{,}722{,}500}{0.725} = \frac{0.725B}{0.725}$ *Divide each side by 0.725.*

$274{,}100{,}000 = B$ *Use a calculator.*

The Nature Conservancy raised $274,100,000 in 1993.

One of the common uses of percents in everyday life is **discounts**.

Example 

Retail Sales

The regular price of a pair of jeans is $45.95. If there is a 25% discount and a 6% sales tax (based on the discounted price), how much will the jeans cost?

You can find the sale price in one of two ways.

Method 1

First, find the amount of the discount.

25% of 45.95 = d

0.25 [×] 45.95 [=] *11.475*

The discount is $11.49.

Then, subtract to find the discount price.

45.95 [−] 11.49 [=] *34.46*

The discount price is $34.46.

Method 2

First, subtract the percent discount from 100%.

Since 100% − 25% = 75%, the discount price will be 75% of the original price.

Then, multiply to find the discount price.

0.75 [×] 45.95 [=] *34.4625*

The discount price is $34.46.

Next, find the price including the sales tax.

Method 1

Find the amount of the tax.
6% of 34.46 = t
0.06 [×] 34.46 [=] 2.0676
The tax is $2.07.

Then, add to find the total price.
34.46 [+] 2.07 [=] 36.53

The total price is $36.53.

Method 2

First, add the percent of tax to 100%.

Since 100% + 6% = 106%, the total price will be 106% of the discount price.

Then, multiply to find the total price.
1.06 [×] 34.46 [=] 36.5276

The total price is $36.53.

The jeans will be $36.53 after the discount and the sales tax.

Checking Your Understanding

Communicating Mathematics

Read and study the lesson to answer each question.

1. **Explain** how r and R are different.
2. **Describe** two ways of finding the price of a $60 item after a 25% discount.
3. Under what conditions might you use $P = R \cdot B$ instead of $\frac{P}{B} = \frac{r}{100}$?
4. If $R = 0.32$, what is r?
5. **You Decide** Miguel and Jalisa are trying to find the monthly interest on a credit card balance of $2500 if the monthly percentage rate is 1.5%. Miguel solves using $P = (1.5)(2500)$. Jalisa uses $P = (0.015)(2500)$. Who is correct? Explain.

Guided Practice

Solve each problem by using the percent equation, $P = R \cdot B$.

6. 75 is 50% of what number?
7. 15% of what number is 30?
8. 18 is what percent of 60?
9. What is 24% of 72?

Find the discount or interest to the nearest cent.

10. $450 TV, 33% off
11. $315 suit, 15% off
12. $3500 at 12% annually for $2\frac{1}{2}$ years
13. **Travel** About 28% of the Japanese tourists who travel abroad each year visit the United States. If 12 million Japanese people went abroad in 1992, about how many visited the United States?

Exercises: Practicing and Applying the Concept

Independent Practice

Solve each problem by using the percent equation, $P = R \cdot B$.

14. 15 is what percent of 100?
15. What is 20% of 135?
16. What number is 43% of 15?
17. 28 is 40% of what number?
18. 15% of what number is 39?
19. 42% of 75 is what number?
20. 45 is what percent of 150?
21. 42 is what percent of 14?
22. 110% of what number is 880?
23. 5 is what percent of 300?
24. 2.04% of 256 is what number?
25. 18.6 is 15.5% of what number?
26. $33\frac{1}{3}$% of 420 is what number?
27. 500% of what number is 100?

Choose

Estimation
Mental Math
Calculator
Paper and Pencil

Find the discount or interest to the nearest cent.

28. \$49.95 sweat suit, 28% off

29. \$199.99 tools, 35% off

30. \$299.99 VCR, 40% off

31. \$1250 refrigerator, 15% off

32. \$500 at 0.3% a month for 15 months

33. \$250 at $8\frac{1}{2}$% annually for 2 years

34. \$1250 at 1.5% monthly for 7 months

35. \$945 at 3.5% annually for 15 months

Critical Thinking

36. Is x% of y always equal to y% of x? Give examples to support your answer.

Applications and Problem Solving

37. Medicine Doctors divide blood into eight different types. Of the approximately 249 million Americans, 37.4% have type O+. How many Americans have O+ blood?

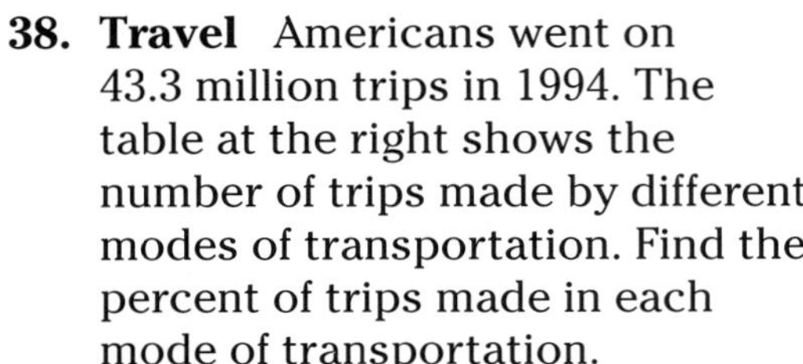

38. Travel Americans went on 43.3 million trips in 1994. The table at the right shows the number of trips made by different modes of transportation. Find the percent of trips made in each mode of transportation.

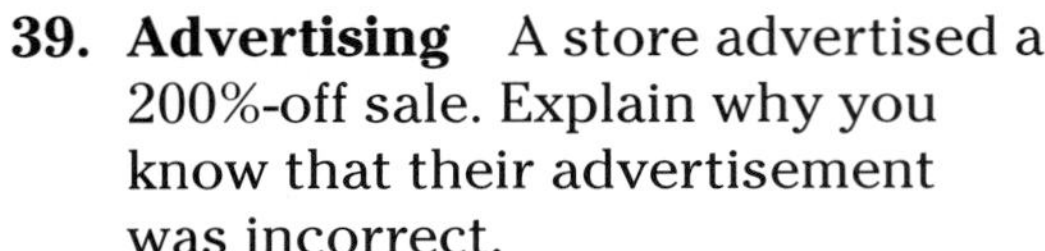

Mode of Transportation	Number of Trips (millions)
car, not rented	25.7
airplane	12.0
car, rented	3.3
motor home	1.0
bus	0.7
train	0.6

Source: *American Demographics*

39. Advertising A store advertised a 200%-off sale. Explain why you know that their advertisement was incorrect.

40. Business Sales associates are often paid in **commission**. A commission is usually given as a percent of the amount of the associate's sales. If Computer Generation pays its associates 15% commission, how much commission would be earned from the sale of a \$2300 computer?

41. Retail Jackee Johnson receives a 10% discount on all merchandise at the store where she works. She is buying a \$439 television that is on sale at 33% off.

a. Find Jackee's price for the television.

b. Does it matter which discount is taken first, the 33% sale price or the 10% employee discount?

c. Could the discount percents be added together first and then taken off or must they be taken off one at a time? Give an example to support your answer.

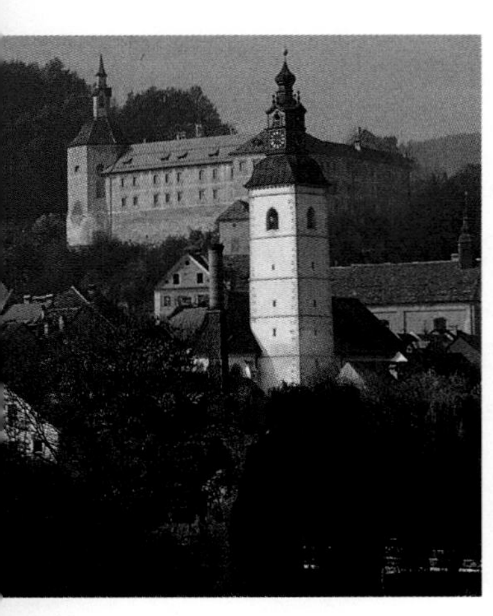

42. Business Many car dealers offer special interest rates as incentives to attract buyers. How much interest would you pay for the first month of a \$5500 car loan with a monthly interest rate of 0.24%?

43. Economics When the prices of products in a country rise over time, the country is experiencing inflation. The rate of inflation in the United States is about 5% per year. In 1993, *USA TODAY* reported "The rate of inflation in Yugoslavia is 20% a day. In other words, prices double every five days." Are the two statements made in the article the same? Explain.

Mixed Review

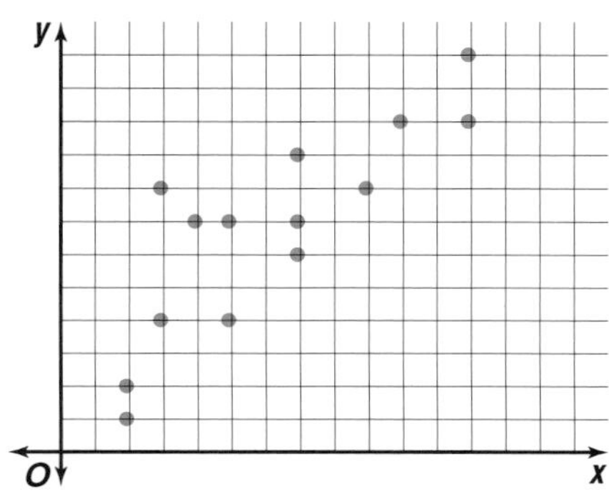

44. Estimate 68% of 210. (Lesson 9-8)

45. **Statistics** Does the scatter plot at the right show a positive relationship, a negative relationship, or no relationship? (Lesson 8-2)

46. Determine whether {(6, 1), (8, 4), (−5, 5), (3, 4)} is a function. (Lesson 8-1)

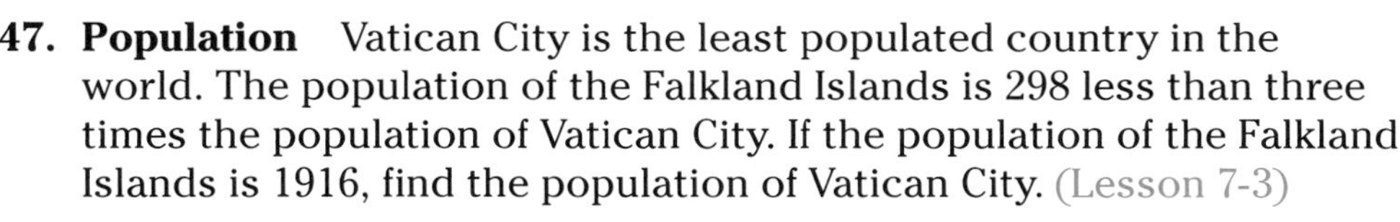

47. **Population** Vatican City is the least populated country in the world. The population of the Falkland Islands is 298 less than three times the population of Vatican City. If the population of the Falkland Islands is 1916, find the population of Vatican City. (Lesson 7-3)
    a. Write an equation that represents this situation.
    b. Solve the equation to find the population of Vatican City.

48. Write $\frac{5}{16}$ as a decimal. (Lesson 6-1)

49. Find the GCF of 150 and 345. (Lesson 4-5)

50. **Art** Pablo Picasso is the artist with the most paintings that have sold for over $1 million, with 148 paintings. This is six more than the number of paintings by Auguste Renoir that have sold for more than $1 million. (Lesson 3-2)
    a. Write an equation to represent the number of Auguste Renoir paintings that have sold for more than $1 million.
    b. Find the number of paintings by Auguste Renoir that have sold for more than $1 million.

51. Find 12 + (−5). (Lesson 2-4)

WORKING ON THE Investigation

Refer to the Investigation on pages 368–369.

The Athletic Director of the university called you into her office. She said that the university needs to increase the gross income from the football games. This could be done by either raising the prices of seats or increasing the number of deluxe and/or spectator seats. Conference regulations require that the stadium has at least 18% of its seats priced at under $16.00.

The Athletic Director asked you to prepare three different options to increase the gross income from a game by 15%. Prepare three options and state the advantages and disadvantages of each option.

Add the results of your work to your Investigation Folder.

9-10 Percent of Change

Setting Goals: *In this lesson, you'll find percent of increase or decrease.*

Modeling a Real-World Application: Electricity Rates

Rate Hike Announced

Source: New York Times, April 22, 1995

The Consolidated Edison Company of New York announced that they will be raising rates on May 1, 1995. The increase will generate $55.1 million. That represents a 1.1% increase. A typical customer in an apartment that uses 300 kilowatt-hours a month will see an increase in his or her monthly bill from $50.52 to $51.26. A Westchester customer using 450 kilowatt-hours a month will pay $71.05 instead of $70.05.

Find the percent of increase for each customer and compare your answers to the 1.1% amount the company stated. *This problem will be solved in Example 3.*

Learning the Concept

Newspapers often report changes in prices and rates as a **percent of change**. The percent of change is the ratio of the amount of change to the original amount.

Example  **Find the percent of change from $120 to $135.**

There are two methods you can use to find the percent of change.

Method 1

Step 1: Subtract to find the amount of change.
$135 - 120 = 15$

Step 2: Solve the percent equation. Compare the amount of increase to the original amount.

$$P = R \cdot B$$
$$15 = R \cdot 120$$
$$\frac{15}{120} = \frac{120R}{120}$$
$$0.125 = R$$

The percent of change is 12.5%.

Method 2

Step 1: Divide the new amount by the original amount.

135 [÷] 120 [=] 1.125

Step 2: Subtract 1 from the result and write the decimal as a percent.

$1.125 - 1 = 0.125$ or 12.5%

The percent of change is 12.5%.

When an amount increases, like in Example 1, the percent of change is a **percent of increase**. When the amount decreases, the percent of change is negative. You can state a negative percent of change as a **percent of decrease**.

Example  **Find the percent of decrease from a population of 257 thousand to 243 thousand.**

Method 1

Step 1: Subtract.
243 − 257 = −14

Step 2: Solve the percent equation. Compare the amount of change to the original amount.

$$P = R \cdot B$$
$$-14 = R \cdot 257$$
$$\frac{-14}{257} = \frac{257R}{257}$$
$$-0.054 = R$$

The percent of change is −5.4%.
The percent of decrease is 5.4%.

Method 2

Step 1: Divide the new amount by the original amount.

243 [÷] 257 [=] .945525

Step 2: Subtract 1 from the result and write the decimal as a percent.

0.946 − 1 = −0.054 or −5.4%

The percent of change is −5.4%.
The percent of decrease is 5.4%.

Example 3

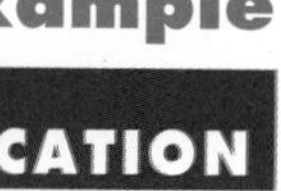

Energy

Refer to the application at the beginning of the lesson. Find the percent increase for each customer.

Apartment bill

Use Method 1.

Step 1: Subtract. \$51.26 − \$50.52 = \$0.74 ***Estimate:*** $\frac{1}{50} = 2\%$

Step 2: Solve the percent equation.

$$P = R \cdot B$$
$$0.74 = R \cdot 50.52$$
$$\frac{0.74}{50.52} = \frac{50.52R}{50.52} \quad \textit{Divide each side by 50.52.}$$
$$0.0146 = R$$

The percent of increase is about 1.5%.

Westchester bill

Use Method 2. ***Estimate:*** *$\frac{71}{70}$ will be a little over 100%.*

Step 1: Divide the new amount by the original amount.
71.05 [÷] 70.05 [=] 1.014276

Step 2: Subtract 1 from the result and write the decimal as a percent.
1.014 − 1 = 0.014 or 1.4%

The percent of increase is 1.4%.

The apartment bill rose 1.5%, and the Westchester bill rose 1.4%, not the 1.1% estimated by the power company.

THINK ABOUT IT

Could it be that the 1.1% increase in total revenue and a greater increase in customer prices could both be correct? Explain.

Checking Your Understanding

Communicating Mathematics

Read and study the lesson to answer each question.

1. **Explain** how to find a percent of change.
2. **Estimate** the percent of change if a temperature goes from 20°C to 25°C.
3. **Name** the amount used as the base when you find a percent of change.
4. The Cardinals football team scored 14 points against the Vikings. The following week, the Cardinals scored 21 points in their game against the Wolves. The coach said the team had improved their score 150%. Is that correct? Explain.

MATH JOURNAL

5. Write about the least difficult and most difficult concepts for you in this chapter.

Guided Practice

State whether each percent of change is a percent of increase or a percent of decrease. Then find the percent of increase or decrease. Round to the nearest whole percent.

6. old: $10
   new: $16
7. old: 25 thousand people
   new: 22 thousand people
8. old: 120 pounds
   new: 150 pounds
9. old: 59°F
   new: 55°F

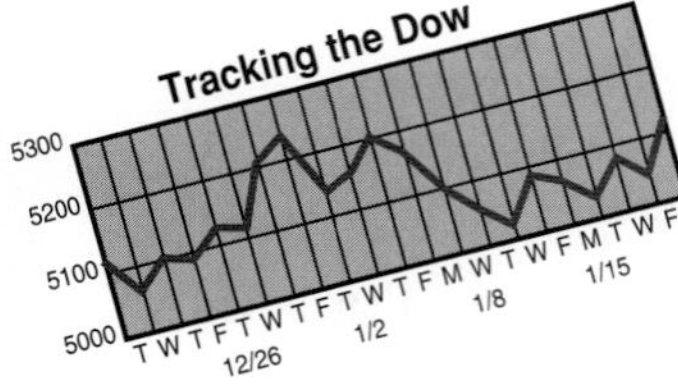

10. **Economics** October 19, 1987 is called Black Thursday because the Dow had its biggest decline, 508.32 points, on that day. If the average was 1738.42 points at closing, what was the percent of decrease of the average that day?

Exercises: Practicing and Applying the Concept

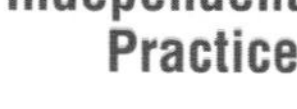

Independent Practice

State whether each percent of change is a percent of increase or a percent of decrease. Then find the percent of increase or decrease. Round to the nearest whole percent.

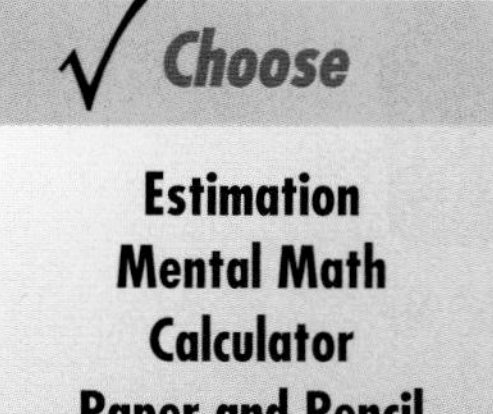

11. old: 142 pounds
    new: 114 pounds
12. old: $0.59
    new: $0.30
13. old: 425 people
    new: 480 people
14. old: $48
    new: $44
15. old: 96 minutes
    new: 108 minutes
16. old: 14.5 liters
    new: 12.5 liters
17. old: $228
    new: $251
18. old: 29.5 ounces
    new: 26.5 ounces
19. old: $39.99
    new: 42.59
20. old: 124 minutes
    new: 137 minutes
21. old: $106
    new: $250
22. old: 4329.80 points
    new: 4351.50 points

Critical Thinking

23. A tennis outfit is marked down 50% and then reduced at the cash register another 30%.
    a. Is this a total reduction of 80%? Why or why not?
    b. If the outfit is not 80% off, find the actual discount.
24. Explain why a 10% increase followed by a 10% decrease is less than the original amount if the original amount was positive.

Applications and Problem Solving

25. **Consumer Awareness** Most cars depreciate, or lose value, each year. Kiku paid \$7800 for her 1994 Geo Metro. The "blue book" that tells the value of used cars lists the value of a Metro like Kiku's as \$6225. What was the percent of decrease of the value of the car?

26. **Merchandising** A retail store buys a coat for \$85 and marks the price up 65% for sale. Later the price is reduced to \$110.
    a. If the coat sells for \$110, what percent of the selling price is profit?
    b. What was the percent of discount from the original selling price?

27. **Postage** When the U.S. Postal Service changed its rates in 1995, the prices of 1-ounce first class letters and two-pound priority mail packages increased. If the price of a priority package was increased at the same percent of increase as a first class letter, what would priority mail have cost?

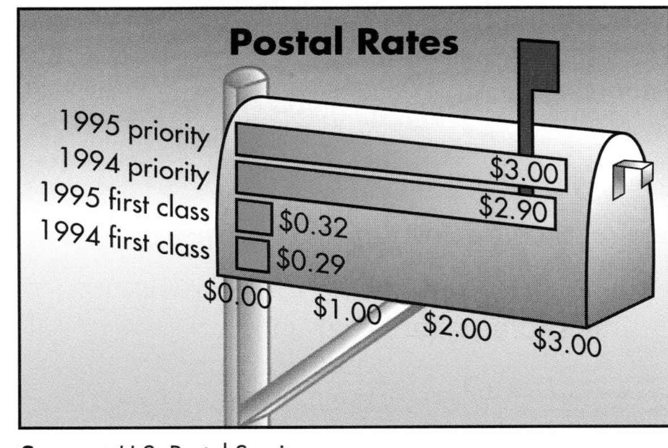

Source: U.S. Postal Service

28. **Manufacturing** Sometimes manufacturers increase their revenues by selling a product at the same price, but decreasing the quantity sold. For example, a can of fruit juice was changed so that a can which used to hold 6 ounces now contains 5.5 ounces.
    a. What was the percent of decrease?
    b. If the cans were sold in a 6-pack for \$1.59, what was the percent of increase in price per ounce?

29. **Pets** Looking for an unusual pet? How about a hedgehog! They are becoming so popular that some breeders have reported a 250% increase in sales in recent years. If a breeder sold 50 hedgehogs one year before the increase, how many should he or she expect to sell in a year now?

Mixed Review

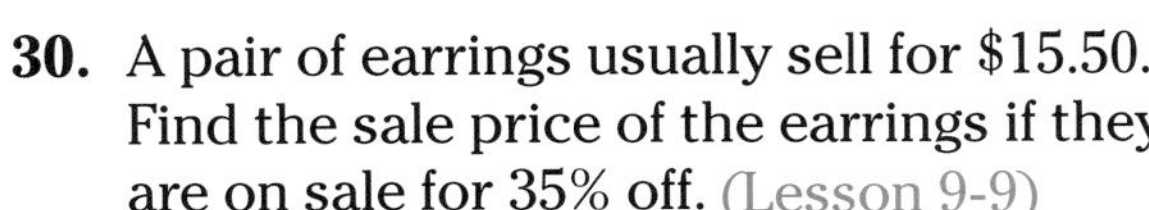

30. A pair of earrings usually sell for \$15.50. Find the sale price of the earrings if they are on sale for 35% off. (Lesson 9-9)

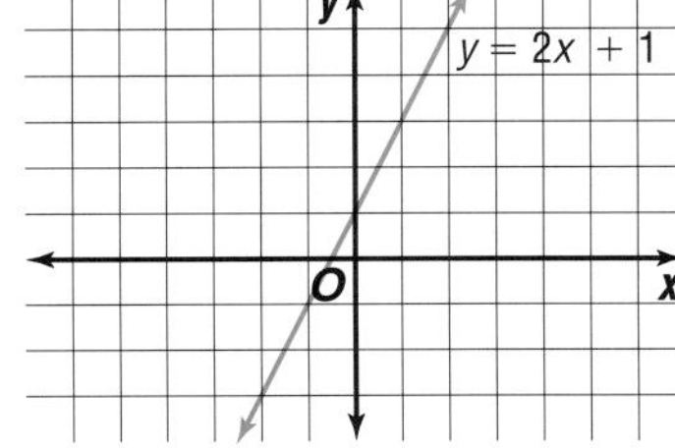

31. Is the equation whose graph is shown at the right a function? Explain. (Lesson 8-4)

32. **Energy** The United States consumes about 4,921,000,000 barrels of oil each year. The annual production of crude oil in the United States is about half of what is consumed. About how much oil is produced in the United States each year? (Lesson 6-2)

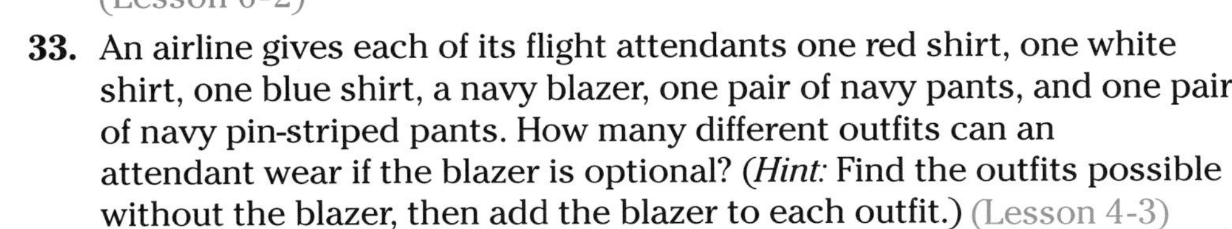

33. An airline gives each of its flight attendants one red shirt, one white shirt, one blue shirt, a navy blazer, one pair of navy pants, and one pair of navy pin-striped pants. How many different outfits can an attendant wear if the blazer is optional? (*Hint:* Find the outfits possible without the blazer, then add the blazer to each outfit.) (Lesson 4-3)

34. Solve $n = 14 - (-3)$. (Lesson 2-5)

CLOSING THE

Investigation

Refer to the Investigation on pages 368–369.

The spirit squad has an idea for the homecoming game that will increase fan spirit. They want to pass out pompoms in the university colors. There are several items that they have questions about and they have asked you to prepare a presentation for their next meeting.

- How many pompoms should they pass out? Should they go to every fan or only a certain percent of them? How should they determine that percent and which fans should get them?
- How should the pompoms be passed out? Should they be passed out at all of the gates, certain gates, or placed on random seats prior to the game?
- Can a sponsor be obtained to pay for the pompoms? Would the sponsor pay for all of the pompoms, only a certain percent of them, a specified dollar amount, or a certain number of them? If the sponsor cannot cover all the costs, how will the remaining costs be paid?
- Do you charge for the pompoms, and if so, how much do you charge for each one? What would be a good price? At what point will the cost be too much for the fans to want to buy one?
- Of what type of material should the pompoms be made? Should recycling be considered when determining the material? If so, should containers be at the gates for fans to recycle the pompoms on their way out of the stadium after the game?

Reread the list of questions. Are there other issues that need to be addressed? Prepare a written report for the committee.

PORTFOLIO ASSESSMENT

You may want to keep your work on this Investigation in your portfolio.

CHAPTER 9 Highlights

Vocabulary

After completing this chapter, you should be able to define each term, property, or phrase and give an example or two of each.

Algebra

base (p. 449)
commission (p. 470)
cross products (p. 444)
discount (p. 468)
interest (p. 467)
percent (p. 449)
percentage (p. 449)
percent of change (p. 472)
percent of decrease (p. 473)
percent of increase (p. 473)
percent proportion (p. 450)
principal (p. 467)
property of proportions (p. 444)
proportion (p. 444)
rate (p. 433, 449)
ratio (p. 432)
unit rate (p. 433)

Geometry

golden ratio (p. 434)

Probability

capture-recapture (p. 448)
probability (p. 440)
sample space (p. 441)

Problem Solving

make a table (p. 437)

Understanding and Using Vocabulary

Determine whether each statement is *true* or *false*.

1. A rate is a special ratio.
2. Percentage is a ratio that compares a number to 1.
3. In the percent proportion $\frac{16}{32} = \frac{50}{100}$, 16 is the base.
4. A sample space is the area of a closed figure.
5. When you use the percent equation $P = R \cdot B$, R represents $\frac{r}{100}$ where r is the rate.
6. A ratio is a comparison of two numbers by multiplication.
7. The property of proportions allows you to compare two proportions.

CHAPTER 9 Study Guide and Assessment

Skills and Concepts

Objectives and Examples

Upon completing this chapter, you should be able to:

▶ **write ratios as fractions in simplest form and determine unit rates.** (Lesson 9-1)

Express 15 to 21 as a fraction in simplest form.

$$\frac{15}{21} \underset{\div 3}{\overset{\div 3}{=}} \frac{5}{7}$$

Express $1.29 for a dozen as a unit rate.

$$\frac{129 \text{ cents}}{12 \text{ items}} \underset{\div 12}{\overset{\div 12}{=}} \frac{\bullet}{1 \text{ item}}$$

129 [÷] 12 [=] *10.75*

The unit rate is 10.75 cents.

Review Exercises

Use these exercises to review and prepare for the chapter test.

Express each ratio as a fraction in simplest form.

8. 10 red to 18 blue
9. 287 yards to 315 yards
10. 15 out of 90
11. 600 to 1000

Express each ratio as a unit rate.

12. 339.2 miles on 10.6 gallons
13. $1.78 for 2 pounds
14. $425 for 17 tickets
15. 142.5 miles in 2.5 hours

▶ **find the probability of simple events.** (Lesson 9-3)

Find the probability of rolling a number that is a factor of 6 on a die.

There are four ways to roll a factor of 6 –1, 2, 3, and 6, and six possible outcomes –1, 2, 3, 4, 5, and 6.

$P(\text{factor of } 6) = \frac{4}{6}$ or $\frac{2}{3}$

There are 6 cans of cola, 4 cans of fruit punch, 9 cans of iced tea, and 2 cans of lemonade in a cooler. Suppose you choose a can randomly. Find each probability.

16. P(cola)
17. P(fruit punch or iced tea)
18. P(not carbonated)
19. P(cola or lemonade)
20. P(iced tea)
21. P(neither cola nor fruit punch)

▶ **use proportions to solve problems.** (Lesson 9-4)

If 2 pounds of pecans cost $13.90, how much will 5 pounds cost?

$\frac{13.90}{2} = \frac{c}{5}$ *Write a proportion.*

$13.90 \cdot 5 = 2 \cdot c$ *Write cross products.*

$\frac{69.50}{2} = \frac{2c}{2}$ *Divide each side by 2.*

$c = 34.75$

Five pounds will cost $34.75.

Solve each proportion.

22. $\frac{8}{6} = \frac{z}{14}$ **23.** $\frac{5.1}{1.7} = \frac{7.5}{a}$

Write a proportion that could be used to solve for each variable. Then solve.

24. 20 grams for $5.60
x grams for $9.80

25. 88.4 miles on 3.4 gallons
161.2 miles on g gallons

Objectives and Examples

▶ **use the percent proportion to solve problems involving percents.** (Lesson 9-5)

Sixteen is what percent of 20?

$$\frac{P}{B} = \frac{r}{100} \rightarrow \frac{16}{20} = \frac{r}{100}$$
$$16 \cdot 100 = 20 \cdot r$$
$$\frac{1600}{20} = \frac{20r}{20}$$
$$80 = r$$

16 is 80% of 20.

Review Exercises

Use the percent proportion to solve each problem.

26. What is 40% of 5?

27. Nineteen is what percent of 25?

28. What is 120% of 50?

29. What is 0.1% of 4000?

▶ **use a sample to predict the actions of a large group.** (Lesson 9-6)

Of the students surveyed, 43% chose pizza as their favorite lunch. How many of the 1294 students in the school would choose pizza?

$$\frac{x}{1294} = \frac{43}{100}$$
$$x \cdot 100 = 1294 \cdot 43$$
$$x = 556.42 \quad \textit{Use a calculator.}$$

About 556 students would choose pizza.

The results of a survey on dessert preferences are shown in the table at the right.

Dessert	Number
cheesecake	7
chocolate cake	48
apple pie	37
ice cream	18

30. How many of the 459 cafeteria customers would choose apple pie?

31. If you asked 500 people which of these desserts they prefer, how many would choose cheesecake?

▶ **express decimals and fractions as percents and vice versa.** (Lesson 9-7)

Express 0.32 as a percent.

$$\frac{0.32}{1} = \frac{32}{100} \text{ or } 32\%$$
(× 100 applied to numerator and denominator)

Express each decimal as a percent.

32. 0.92 **33.** 0.0056

Express each fraction as a percent.

34. $\frac{63}{100}$ **35.** $\frac{113}{200}$

Express each percent as a fraction.

36. 90% **37.** 65%

Express each percent as a decimal.

38. 45% **39.** 235%

▶ **use percents to estimate.** (Lesson 9-8)

Estimate 12% of 54.

12% is about 10%.
10% of 54 is 5.4.
12% of 54 is about 5.4.

Choose the best estimate.

40. 2% of 180 **a.** 36 **b.** 3.6 **c.** 0.36

41. 198% of 5 **a.** 10 **b.** 1.0 **c.** 0.10

42. 8% of 420 **a.** 42 **b.** 5 **c.** 400

43. 73% of 80 **a.** 60 **b.** 40 **c.** 80

44. 352% of 20 **a.** 8 **b.** 40 **c.** 70

45. 0.6% of 620 **a.** 2 **b.** 6 **c.** 12

Objectives and Examples

▸ **solve percent problems using the percent equation.** (Lesson 9-9)

1.8 is 4% of what number?

$$P = R \cdot B$$
$$1.8 = 0.04 \cdot B \quad \textit{Replace P with 1.8 and R with 0.04.}$$
$$\frac{1.8}{0.04} = \frac{0.04B}{0.04}$$
$$45 = B$$

1.8 is 4% of 45.

Review Exercises

Solve each problem by using the percent equation, $P = R \cdot B$.

46. 54 is 150% of what number?
47. 16% of what number is 3.2?
48. 63 is what percent of 105?
49. 2600 is 65% of what number?
50. 16 is what percent of 2?
51. What percent of 16 is 5?

Objectives and Examples

▸ **find percent of increase or decrease.** (Lesson 9-10)

Find the percent of increase from 159 pounds to 168 pounds.

Subtract: $168 - 159 = 9$
Use the percent equation.

$$9 = R \cdot 159 \quad P = R \cdot B$$
$$\frac{9}{159} = \frac{R \cdot 159}{159}$$
$$0.0566 = R$$

The percent of increase is about 5.7%.

Review Exercises

State whether each percent of change is a percent of increase or a percent of decrease. Then find the percent of increase or decrease. Round to the nearest whole percent.

52. old: \$0.60
new: \$0.75
53. old: 16 ounces
new: 15.8 ounces
54. old: 18.4 grams
new: 17.5 grams
55. old: \$15.00
new: \$11.75
56. old: \$84
new: \$118
57. old: 210 pounds
new: 185 pounds

Applications and Problem Solving

58. **Records** Nippondenso's Micro-Car is listed as the world's smallest car by The Guinness Book of World Records. It is a model of Toyota's first passenger car, the 1936 Model AA sedan. A part of the actual car that is 1 meter long is 0.001 m long on the model. Write the scale of the model as a fraction in simplest form. (Lesson 9-1)

59. **Games** A domino has two squares on its face. Each of the two spaces is marked with 1, 2, 3, 4, 5, or 6 dots, or it is blank. A complete set of dominoes includes one domino for each possible combination of dots and blanks. Doubles, like 5 dots in each space, are included. How many dominoes are in a complete set? (Lesson 9-2)

60. **Safety** Fifty-eight percent of the adults surveyed said they wore their seat belts all the time. If there are 185 million adults in the United States, how many American adults wear their seat belts all the time? (Lesson 9-6)

A practice test for Chapter 9 is available on page 782.

Alternative Assessment

Cooperative Learning Project

Work as a group.

1. Develop a multiple-choice questionnaire on any topic. Include at least five questions with three or more choices. Survey 20% of your class for their opinions.
2. Record the data you gather in a table, graph, or scatter plot.
3. What fraction of the people surveyed chose each possible answer? Write a sentence about the probability of randomly choosing a person who chose a certain answer.
4. List the sample space of each question on the questionnaire.
5. Predict the number of people in your class who would choose a certain answer on your questionnaire. Explain how you made your prediction.
6. Write a statement about the percent of your survey group that chose a certain answer.
7. Present the results of your survey in an interesting way. You may wish to make a poster or brochure, including tables or graphs.

Thinking Critically

- ▶ When you write a decimal as a percent, the first step is to write it as a fraction with a denominator of 1. Then you multiply by 100. Can you write one step that could replace these two?
- ▶ A weather forecaster said "There is a 50% chance of rain on Saturday and a 50% chance of rain on Sunday. So there is a 100% chance of rain this weekend." Do you think this is correct? Explain.

 **Portfolio**

Select one of the assignments from this chapter that you found especially challenging and place it in your portfolio.

Self-Evaluation

"It is the province of knowledge to speak and it is the privilege of wisdom to listen." This quote by Oliver Wendell Holmes is saying that it is wise to be a good listener.

Assess yourself. Do you listen well? Studies show that people remember about 20% of what they hear. How much of what you hear do you remember? How can you remember more of what you hear?

LONG-TERM PROJECT

Investigation

It's Only A Game

What do you do in your spare time? In ancient times people must have had spare time too because archaeologists have found ancient artifacts which they think are game boards and pieces. A board game believed to be about 4500 years old was found at Ur, a city of ancient Sumer (now in Iraq).

The game of chess is believed to have originated in India 1400 years ago. Another game that has been around for some time is dominoes. Historians do not agree on how old dominoes are or who invented them. They believe they were invented in China and spread to Europe in the 1300s.

Just as the playing of games appears to be a natural instinct of the human race, the analysis of games would appear to be a natural instinct of many mathematicians. Games are most commonly classified according to the equipment they use.

The major types of games are:

- *board games*
- *card games*
- *tile games*
- *target games*
- *dice games*
- *table games*
- *paper and pencil games*
- *electronic games*

Starting the Investigation

Work with your group to design a game. Ask friends, teachers, relatives, and neighbors what their favorite game is and why it is their favorite. Describe, in detail, the qualities of each game that led them to their choice. Determine what qualities are common to the favorite games. Then determine which of these qualities you will use in the design of your game.

Think About . . .

- *What will be the age group of the players who will play your game?*
- *For how many players is the game designed?*
- *What is the object of the game?*
- *Will the game be designed for educational purposes? (Will it teach a player something?)*
- *Where will the game be played? (Is it a party game like charades? a street game, such as hopscotch? a lawn game, such as croquet? or a TV game show?)*
- *What type of game do you want to design? Will it be skill-based, pure chance, or a combination of both?*
- *How will players determine who goes first?*
- *How will players take a turn?*
- *What are the basic rules?*
- *Are there any special rules? What are they based on?*
- *When is the game over? What determines the winner?*

Investigation
It's Only a Game

Working on the Investigation
Lesson 10-4, p. 508

Working on the Investigation
Lesson 10-8, p. 528

Working on the Investigation
Lesson 11-7, p. 588

Closing the Investigation
End of Chapter 11, p. 600

You will work on this Investigation throughout Chapters 10 and 11.

Be sure to keep your materials in your Investigation Folder.

For current information on game design, visit:
www.glencoe.com/sec/math/prealg/mathnet

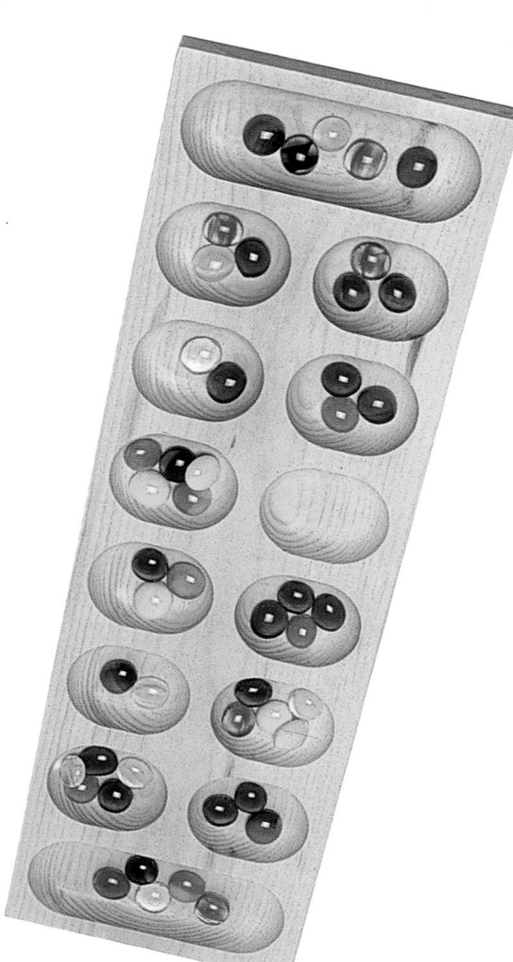

CHAPTER 10

More Statistics and Probability

TOP STORIES in Chapter 10

In this chapter, you will:

- display data in stem-and-leaf plots and box-and-whisker plots,
- find measures of variation,
- recognize misleading data,
- count outcomes,
- find permutations and combinations, and
- find probabilities and odds.

MATH AND MARKETING IN THE NEWS

Television ads take aim at young viewers

Source: American Demographics, May, 1995

Many people reach for the TV remote when the commercials come on. But advertisers hope you'll stay tuned, and then go out and buy! The products that are advertised heavily on network television are used primarily by young consumers. Companies compete fiercely for *your* dollars. They hope that if they win you as a customer now, you will be a lifetime customer.

Marketing decisions like when and how to advertise are usually based on surveys and statistics. Attitudes, buying habits, and television viewing patterns are all studied by advertisers as they make decisions. They have found that selling to today's young people is very different from selling to their parents. Advertisers hope that the Nielsen television rating system will soon be updated to give more information about the television viewing habits of consumers of different ages.

Putting It into Perspective

950
Town criers give product information and direct people to shops in European countries.

1400

1440
German inventor Johannes Gutenberg invents movable printing type that makes printing posters, handbills, and newspapers possible.

1600

1704
The first newspaper advertisement in the American colonies appears in *The Boston News-Letter.*

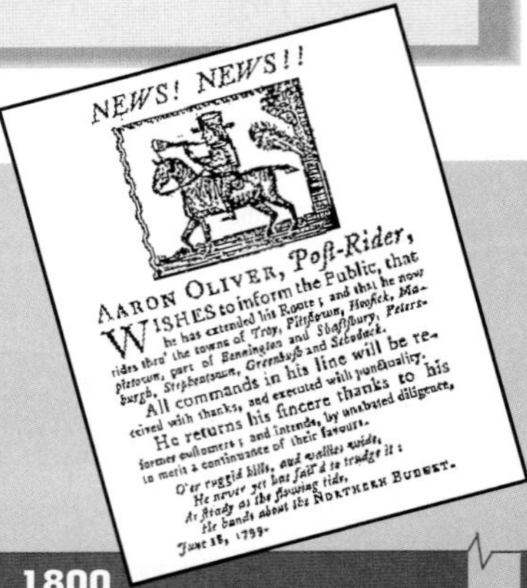

1800

1841
The first advertising agency in the United States opens in Philadelphia.

EMPLOYMENT OPPORTUNITIES

The Exciting World of Advertising

ARE YOU CREATIVE? Do you like working with people and ideas? We want you to write advertisements for national magazines, newspapers, and television. If you prefer analyzing data and developing strategies, we want you! Exciting opportunities await you in the field of marketing research and management. Positions for writers and researchers exist in advertising agencies and in marketing departments of larger corporations. College education in journalism, liberal arts, business, marketing, or commercial art needed.

For more information, contact:

American Advertising Federation
1101 Vermont Ave. NW, Suite 500
Washington, D.C. 20005

Statistical Snapshot

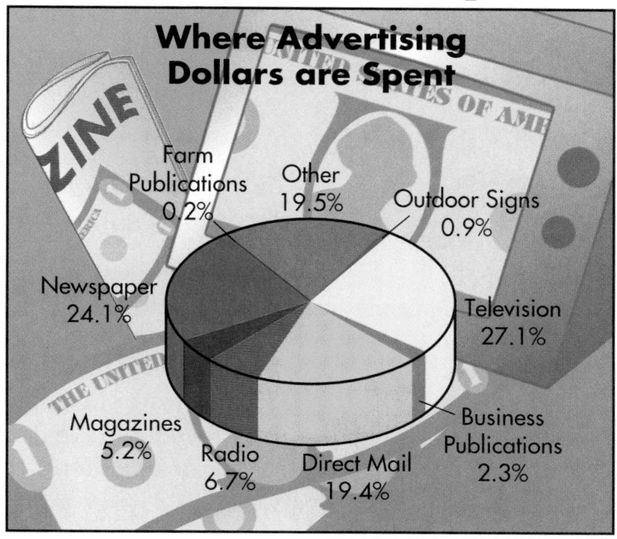

Source: McCann-Erickson, Inc.

interNET CONNECTION For up-to-date information on advertising, visit:
www.glencoe.com/sec/math/prealg/mathnet

1950 **1975** **2000**

1952
Dwight D. Eisenhower uses advertising executives to help plan his successful presidential campaign.

1969
Gail Fisher is the first African-American to have a speaking role in a nationally-televised commercial.

1971
The U.S. Federal Government bans cigarette advertising on radio and television.

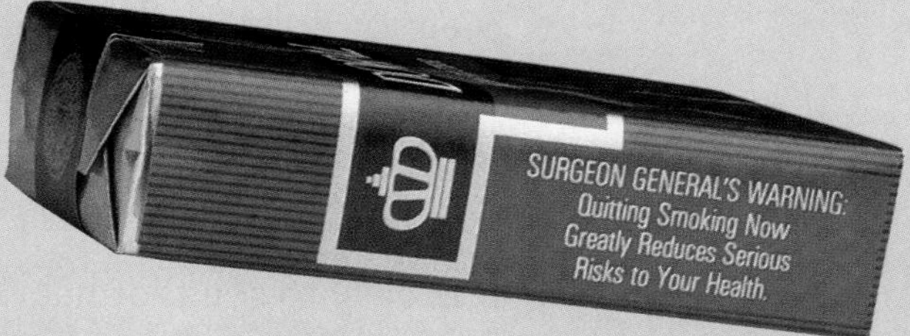

1995
Advertisers consider updating Nielsen system of television ratings.

10-1 Stem-and-Leaf Plots

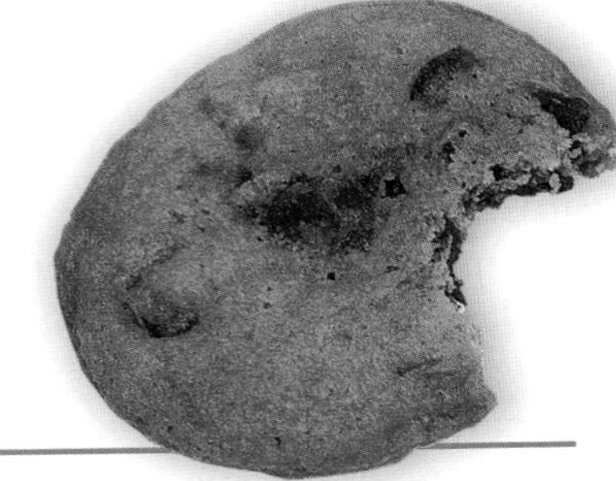

Setting Goals: *In this lesson, you'll display and interpret data in stem-and-leaf plots.*

Modeling a Real-World Application: Consumerism

In a taste test reported in *Consumer Reports* and *Zillions* magazines, 35 different chocolate-chip cookies were rated according to flavor, freshness, and texture. For the 27 cookies that were rated good, very good, or excellent, the cost per 30-gram serving is shown below.

10¢	11¢	58¢	10¢	25¢	49¢	24¢	46¢	16¢
13¢	21¢	34¢	16¢	29¢	28¢	13¢	14¢	14¢
20¢	18¢	18¢	21¢	13¢	22¢	11¢	23¢	12¢

If you were asked to organize this set of data and present it in an easy-to-read format, you could construct a **stem-and-leaf plot**.

Learning the Concept

In a stem-and-leaf plot, the greatest place value common to all the data values is usually used for the **stems**. The next greatest place value forms the **leaves**. Follow these steps to construct a stem-and-leaf plot from the data above.

Step 1 Find the least and greatest price.

The least price is 10¢, and the greatest price is 58¢.

Step 2 Find the stems.

The least price, 10¢, has a 1 in the tens place. The greatest price, 58¢, has a 5 in the tens place. Draw a vertical line and write the digits in the tens places from 1 to 5 to the left of the line.

```
1|
2|
3|
4|
5|
```

Step 3 Put the leaves on the plot.

Record each of the prices on the plot by pairing the units digit, or leaf, with the corresponding stem. For example, 10 is plotted by placing the units digit, 0, to the right of the stem 1.

```
1|0 1 0 6 3 6 3 4 4 8 8 3 1 2
2|5 4 1 9 8 0 1 2 3
3|4
4|9 6
5|8
```

Step 4 Rearrange the leaves so they are ordered from least to greatest.

Step 5 Include an explanation or key of the data.

```
1|0 0 1 1 2 3 3 3 4 4 6 6 8 8
2|0 1 1 2 3 4 5 8 9
3|4
4|6 9
5|8        1|3 means 13¢.
```

It is easy to see that nearly all of the prices are between 10¢ and 29¢ because those rows have the most leaves.

Sports

In the last ten games of the 1995 NBA playoffs, the points scored by Hakeem Olajuwon of the Houston Rockets were 27, 41, 43, 20, 42, 39, 31, 34, 31, and 35.

a. Make a stem-and-leaf plot of this data.

The stems are 2, 3, and 4.

Stem	Leaf
2	0 7
3	1 1 4 5 9
4	1 2 3

2|0 means 20 points.

b. In which interval do most of the points scored occur?

From the leaves, you can see that Olajuwon scored in the 30–39 point interval more times than in the other intervals.

A **back-to-back stem-and-leaf plot** is used to compare two sets of data. In this type of stem-and-leaf plot, the leaves for one set of data are on one side of the stem and the leaves for the other set of data are on the other side of the stem. Two keys to the data are needed.

Nutrition

A nutritional analysis was done for several fast food restaurants. The amount of fat, in grams, of their burgers and chicken, is shown below.

Burgers	10, 15, 20, 26, 19, 33, 36, 30
Chicken	9, 13, 13, 19, 18, 20, 15, 15

a. Make a back-to-back stem-and-leaf plot of the data.

The stems are 0, 1, 2, and 3.

Burgers		Chicken
	0	9
9 5 0	1	3 3 5 5 8 9
6 0	2	0
6 3 0	3	

6|3 = 36 *2|0 = 20*

b. If a lower amount of fat is better, which type of fast food is better?

Chicken is better because it has fewer amounts in the 20–39 range than the burgers.

Checking Your Understanding

Communicating Mathematics

Read and study the lesson to answer these questions.

1. **Explain** why you might use a stem-and-leaf plot.
2. **List** the numbers in the stem for the stem-and-leaf plot at the right.
3. **Explain** how a stem-and-leaf plot is similar to a bar graph. How is it different?
4. **Describe** the information you can; obtain from the stem-and-leaf plot in Example 1.

Stem	Leaf
3	0
4	1 5 7
5	3 4 4 7
6	0 1 8

3|0 = 30

Work with a partner to complete the following activity.

5. Choose a topic that interests you. Research or gather data about the topic. Topics might include: heights of U.S. presidents, ages of U.S. presidents at inauguration, points scored by basketball team members, time it takes to get to school or do homework.

a. Construct a stem-and-leaf plot of the data you collected.

b. Write two or three statements that describe the data.

Guided Practice

State the numbers you would use for each stem in a stem-and-leaf plot of each set of data. Then make the plot.

6. 65, 82, 73, 91, 95, 86, 78, 69, 80, 88

7. 5.7, 5.4, 6.8, 6.3, 7.1, 8.5, 7.5, 6.9, 6.7, 7.7

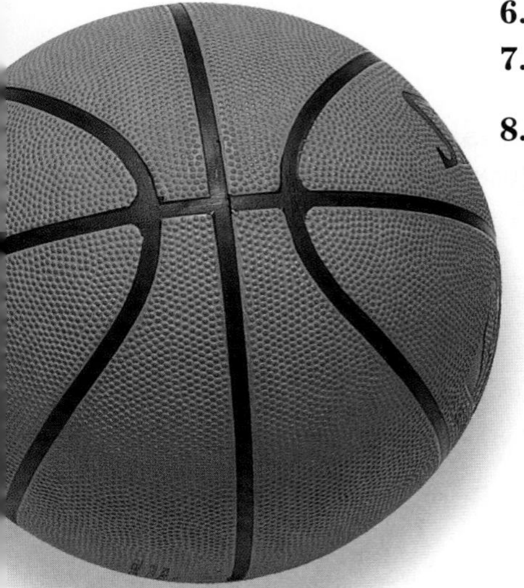

8. Basketball The stem-and-leaf plot at the right shows the scores of the Knox Junior High School's girl's basketball team for the games in the 1996 season.

Knox		Opponents
3	2	
6 3 1	3	6 7
9 8 6 5 4 4 2	4	0 3 4 5 5 6
9 8 8 6 6 2	5	1 1 3 5 6 6 7
5 0	6	1 3 5 6
0	7	2

3|2 = 23 *7|2 = 72*

a. What were the highest and lowest scores for the Knox team?

b. What were the highest and lowest scores of their opponents?

c. What is the median of each set of scores?

Exercises: Practicing and Applying the Concept

Independent Practice

Make a stem-and-leaf plot of each set of data.

9. 64, 60, 72, 61, 73, 80, 68, 70, 65, 67, 70, 80

10. 23, 14, 25, 36, 45, 34, 32, 21, 39, 28, 48, 19, 39, 45, 48, 48, 51

11. 42, 23, 9, 21, 7, 11, 14, 6, 40, 5, 9, 45, 12

12. 83, 94, 54, 92, 85, 54, 96, 89, 75, 117, 68, 99, 116

13. 11.2, 10.4, 12.6, 12.3, 11.1, 14.7, 9.2, 10.8, 12.9, 11.2, 11.7, 13.3, 13.8, 12.8

14. Architecture The *World Almanac* lists 15 tall buildings in New Orleans, Louisiana. The number of floors in each of these buildings is listed in the chart at the right.

51	47	33
53	42	28
45	33	28
39	32	25
36	31	23

a. Make a stem-and-leaf plot of the data.

b. What is the greatest number of floors?

c. What is the least number of floors?

d. How many buildings had 34 or less floors?

e. How many buildings were in the 30–39 floor range?

15. **Retail Sales** The table below lists the prices, in dollars, for pairs of running shoes.

Men's	83, 70, 70, 75, 70, 82, 70, 125, 82, 45, 70, 120, 110, 133, 100
Women's	124, 70, 72, 70, 65, 90, 60, 80, 55, 70, 67, 45, 55, 75, 85

a. Make a back-to-back stem-and-leaf plot of the prices for men's and women's shoes.

b. What information about the relationship between the prices for men's and women's shoes does it present?

Critical Thinking

16. Describe any restrictions when using a stem-and-leaf plot.

17. Is it possible to estimate the mean from a stem-and-leaf plot? Explain.

Applications and Problem Solving

18. **Meteorology** The normal January temperatures of 16 southern cities are shown in the table at the right.

a. Make a stem-and-leaf plot of the temperatures.

b. If you were touring the south in January, what range of temperatures would you expect?

Normal January Temperatures

City	Temp	City	Temp
Asheville, NC	37	Memphis, TN	67
Atlanta, GA	42	Mobile, AL	51
Birmingham, AL	42	New Orleans, LA	37
Charleston, SC	49	Norfolk, VA	52
Jackson, MS	46	Raleigh, NC	40
Jacksonville, FL	53	Richmond, VA	40
Miami, FL	38	Savannah, GA	49
Knoxville, TN	40	Tampa, FL	60

Source: *The World Almanac*, 1995

19. **Engineering** The table below lists the temperatures at launch time for the first 23 space shuttle launches and how many of those times the O-rings on the solid rocket motors experienced thermal distress or no distress.

Temp (°F)	66	70	69	68	67	72	73	57	63	78	53	75	81	76	79	58
Distress	0	2	0	0	0	0	0	1	1	0	1	1	0	0	0	1
No Distress	1	2	1	1	3	1	1	0	0	1	0	1	1	2	1	0

The last launch of the space shuttle *Challenger* was on January 28, 1986. The temperature was 31°F at the time of the launch.

a. Make a back-to-back stem-and-leaf plot of the temperatures for distress and no distress.

b. What information about the relationship between the temperatures of distress and no distress does your back-to-back stem-and-leaf plot make?

c. If the temperature on the day of the launch was 31°F, would you expect the O-rings to experience thermal distress? Explain.

Mixed Review

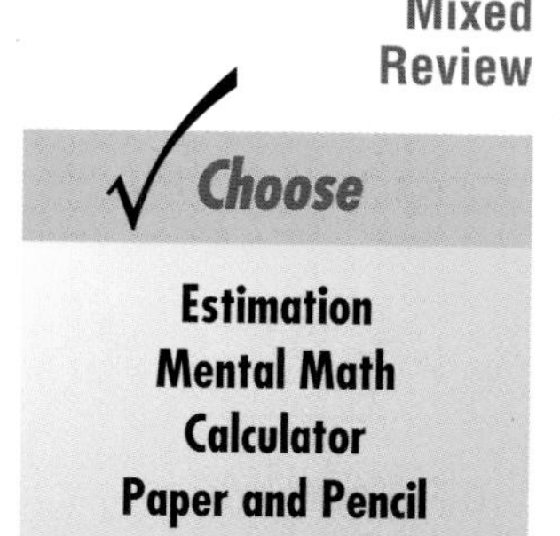

20. **Economics** The Consumer Price Index (CPI) represents the relative costs of goods and services. If the CPI is 233.2 in April and 236.4 in May, what is the percent of increase from April to May? (Lesson 9-10)

21. Express 45% as a fraction in simplest form. (Lesson 9-7)

22. Solve $x - 3.4 = 9.2$. Check your solution. (Lesson 5-6)

23. If 5 more than the product of a number and -2 is greater than 10, which of the following could be that number? (Lesson 3-1)

a. 3 **b.** 2 **c.** -1 **d.** -4

24. Find the value of $|-9| + 14$. (Lesson 2-1)

10-2 Measures of Variation

Setting Goals: *In this lesson, you'll use measures of variation to compare data.*

Modeling with Technology

In Chapter 6, you learned that the mean, median, and mode indicate the center of a set of data. Sometimes measures of center don't fully describe a set of data. In the back-to-back stem-and-leaf plot of bicycle helmet prices at the right, both models have the same mean price, the same median price, and the same mode price. But, it is easy to see that all the prices are not the same.

Adult		Youth
	2	5
7 2 1 1	3	4 7 8
3 3 3	4	2 3 3 8
0	5	

1 | 3 = $31 *2 | 5 = $25*

You can describe the two lists of helmet prices more accurately by using measures of variation to describe the spread of each set of data. You can use a TI-82 graphing calculator to find common measures of variation.

Your Turn

- ▶ First, clear the statistical memory.

  Enter: STAT 4 2nd L1 , 2nd L2 ENTER

- ▶ Enter the adult prices in the L1 list and the youth prices in the L2 list.

  Enter: STAT ENTER 31 ENTER 31 ENTER . . . 50 ENTER ▶ 25 ENTER 34 ENTER . . . 48 ENTER

- ▶ Find the median of the lower and upper halves of the adult prices.

  Enter: STAT ▶ ENTER 2nd L1 ENTER

  Press the down arrow to scroll through the list. Record the values for Q_1 and Q_3. Then find the difference between the greatest value, max X, and the least value, min X.

- ▶ Find the median of the lower and upper halves of the youth prices.

  Enter: STAT ▶ ENTER 2nd L2 ENTER

  Record the values for Q_1 and Q_3. Then find the difference between the greatest and least values.

a. How do the differences between the greatest and least values of each list compare?

b. How do the medians of the lower half (Q_1) and upper half (Q_3) of each set of data compare?

Learning the Concept

When you found the difference between the greatest and least values of the helmet prices, you calculated a measure of variation called the **range**.

Definition of Range — The range of a set of numbers is the difference between the least and the greatest number in the set.

Example 1 **Determine the range of each set of data.**

a. 12, 17, 16, 23, 14, 18, 11, 21

First, list the data in order.

11 12 14 16 17 18 21 23

The range is 23 − 11 or 12.

b.

12 13 14 15 16 17 18 19

Since the line plot is organized from the least data to the greatest, the range is the difference in the first and last data values of the plot.

The range is 19 − 12 or 7.

The TI-82 graphing calculator uses Q_1 to represent the lower quartile and Q_3 to represent the upper quartile.

In a large set of data, it is helpful to separate the data into four equal parts called **quartiles**. Recall that the median of a set of data separates the data in half. The median of the lower half of a set of data is called the **lower quartile** and is indicated by LQ. Likewise, the median of the upper half of the data is called the **upper quartile** and is indicated by UQ. Quartiles are used in another measure of variation called the **interquartile range**.

Definition of Interquartile Range

In words: The interquartile range is the range of the middle half of a set of numbers.

In symbols: Interquartile range = UQ − LQ

Example 2

Olympics

For the latest on Olympic medals, visit: **www.glencoe.com/sec/math/prealg/mathnet**

In the thirteen summer Olympic Games from 1936–1992, athletes from the United States earned the following numbers of gold medals.

Year	1936	1948	1952	1956	1960	1964	1968
Gold Medals	24	38	40	32	34	36	45

Year	1972	1976	1980	1984	1988	1992
Gold Medals	33	34	0	83	36	37

a. **Find the range and interquartile range for the number of gold medals.**

First, list the data from least to greatest. Then find the median.

0 24 32 33 34 34 36 36 37 38 40 45 83

↑ Median (36)

(continued on the next page)

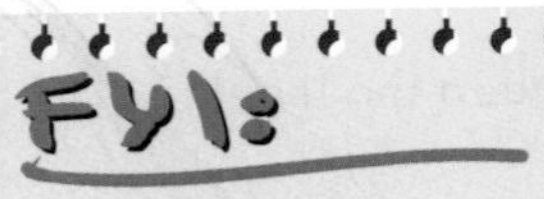

FYI:

The Summer Olympics were boycotted by the United States and 63 other countries in 1980. The USSR and 13 other Eastern Bloc nations boycotted the 1984 Summer Olympics.

The upper quartile is the median of the upper half and the lower quartile is the median of the lower half.

0 24 32 33 34 34 36 36 37 38 40 45 83

$LQ = \frac{32 + 33}{2}$ or 32.5   Median $UQ = \frac{38 + 40}{2}$ or 39

The interquartile range is 39 − 32.5 or 6.5.

b. Which measure of variation better summarizes the performance of United States athletes, the range or the interquartile range?

Since the range represents values for years in which the Summer Olympics were boycotted, the interquartile range is more representative of the variability of this set of data.

Checking Your Understanding

Communicating Mathematics

Read and study the lesson to answer these questions.

1. **Compare and contrast** the measures of variation and the measures of central tendency.
2. **Write** a list of at least twelve numbers that has a range of 50 and an interquartile range of 10.
3. What can you say about a group of test scores if it has a small interquartile range?

Guided Practice

Find the range of each set of data.

4. 63, 82, 71, 65, 92, 86, 80, 95, 78, 89
5.

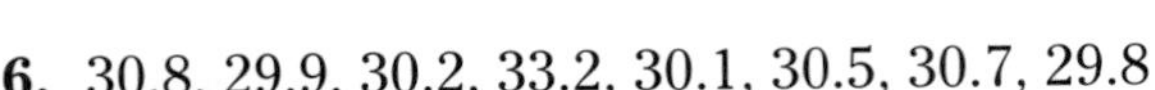

6. 30.8, 29.9, 30.2, 33.2, 30.1, 30.5, 30.7, 29.8

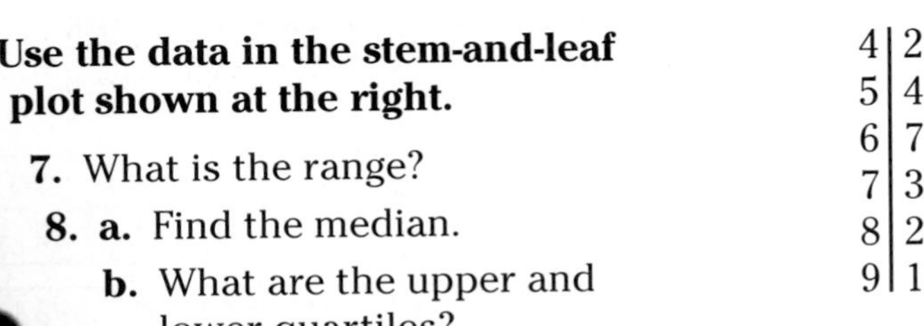

Use the data in the stem-and-leaf plot shown at the right.

Stem	Leaf
4	2 3 9
5	4 5
6	7
7	3 8
8	2
9	1 3 4

$4|2 = 42$

7. What is the range?
8. **a.** Find the median.
   b. What are the upper and lower quartiles?
   c. Find the interquartile range.
9. **Sales** Discworld's sales of Gloria Estefan's new CD over a two-week period are shown in the chart at the right.
   a. Organize the data into a line plot or a stem-and-leaf plot.
   b. What is the median?
   c. Find the upper and lower quartiles.
   d. What is the interquartile range?

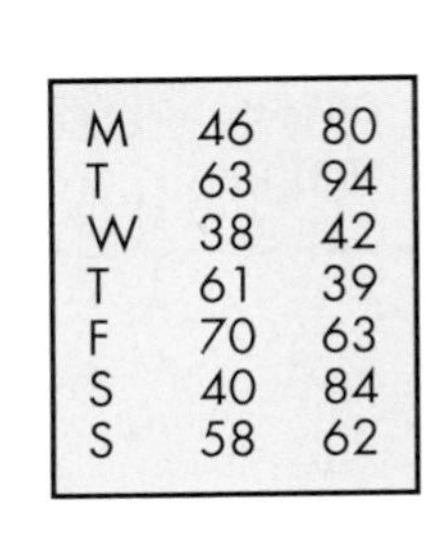

M	46	80
T	63	94
W	38	42
T	61	39
F	70	63
S	40	84
S	58	62

Exercises: Practicing and Applying the Concept

Independent Practice

Find the range, median, upper and lower quartiles, and the interquartile range for each set of data.

10. 56, 45, 37, 43, 10, 34, 29

11. 30, 90, 40, 70, 50, 100, 80, 60

12. 19°, 21°, 18°, 17°, 18°, 22°, 36°

13. 8, 11, 23, 7, 2, 4, 16, 2, 4

14. 135, 170, 125, 174, 136, 145, 180, 156, 188

15. 211, 225, 205, 207, 208, 213, 180, 200, 210, 229, 199

16.

Stem	Leaf
5	9
6	0 0 1 2 2 3 5
7	1 2

5|9 = 59¢

17.

Stem	Leaf
7	3 5
8	3 8 9
9	0 0 1 2 4 4 7 7 8
10	3 6
11	4

7|3 = 7.3 cm

18.

Stem	Leaf
25	0 3 7 9
26	1 3 4 5 5 6
27	1 5 6 6 9
28	1 2 3 5 8
29	2 5 6 9

29|2 = $292

19.

World's Busiest Airports in 1994

City	Passengers
Atlanta	54,100,000
Chicago-O'Hare	66,400,000
Dallas-Ft. Worth	52,600,000
Denver	33,100,000
Frankfurt	35,100,000
London Heathrow	51,700,000
Los Angeles	51,100,000
Miami	30,200,000
N.Y.-Kennedy	28,800,000
San Francisco	34,600,000

Source: Airports Council International

20.

Top 10 states with largest Asian-Indian population in 1990

State	Population
California	159,973
Florida	31,457
Illinois	64,200
Maryland	28,330
Michigan	23,845
New Jersey	79,440
New York	140,985
Ohio	20,848
Pennsylvania	28,396
Texas	55,795

Source: U.S. Census Bureau

21. Home Runs Hit by League Leaders, 1960–1995

National League		American League
	2	2
	2	
1	3	2 2 2 2 3
9 9 8 8 8 7 7 7 6 6 6 5	3	6 6 7 9 9 9
4 4 4 3 1 0 0 0 0 0	4	0 0 0 0 1 2 3 3 4 4 4 4
9 9 8 8 8 7 7 6 6 5 5	4	5 5 6 6 8 9 9 9 9
2 2	5	0 1
	5	
	6	1
	6	

2|5 means 52 *6|1 means 61*

Graphing Calculator

Work with a partner to describe the variation in the populations of capital cities of the U.S.

22. Each person should find the population of half of the state capitals. Enter the data in a TI-82 graphing calculator. Use the sort command to place the populations in order. Find the median, range, and interquartile range of the data. Use this information to describe the population of the capital cities.

Critical Thinking

23. a. Produce a set of at least ten test scores that has a median of 60 and an interquartile range of 20.
    b. Produce a set of at least ten test scores that has a median of 60 and an interquartile range of 50.
    c. What conclusions can you draw about the two sets of test scores in parts a and b from comparing the measures of variation?

Applications and Problem Solving

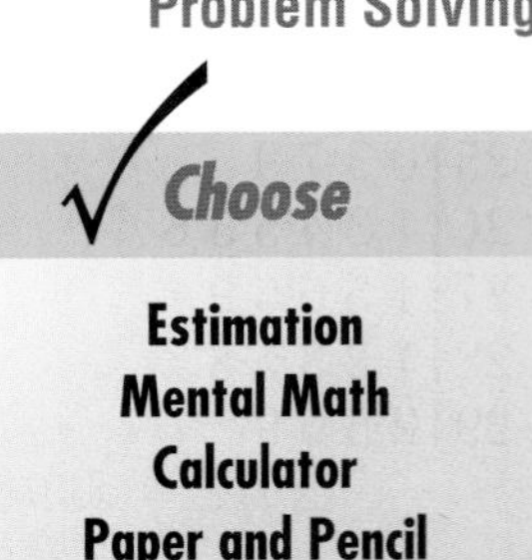

24. **Meteorology** The table at the right lists the average winter and summer precipitation for selected cities around the world.
    a. Construct a back-to-back stem-and-leaf plot for the precipitation amounts.
    b. Find the mean and median for each set of data.
    c. How do the means compare?
    d. How do the medians compare?
    e. How do the interquartile ranges compare?
    f. Which month has more consistent precipitation? Explain.

January and July Precipitation (in.)

City	Jan.	July
Atlanta, GA	4.9	4.7
Berlin, Germany	1.9	3.1
Dallas, TX	1.7	2.0
Dublin, Ireland	2.7	2.8
Indianapolis, IN	2.7	4.3
Kinshasa, Zaire	5.3	0.1
Miami, FL	2.1	6.0
New Orleans, LA	5.0	6.7
Oklahoma City, OK	5.0	6.7
Oslo, Norway	1.7	2.9
Paris, France	1.5	2.1
Rome, Italy	3.3	0.4
San Diego, CA	2.1	0.0
San Francisco, CA	4.5	0.0
Santiago, Chile	0.1	3.0
Singapore	9.9	6.7
Sydney, Australia	3.5	4.6
Washington, DC	2.8	3.9

Source: *Universal Almanac 1995*

25. **World Cultures** Many North American Indians hold conferences called powwows where they celebrate their culture and heritage through various ceremonies and dances.

    ▶ The ages of participants in a Menominee Indian powwow were: 20, 18, 12, 13, 14, 72, 65, 23, 25, 43, 67, 35, 68, 13, 56.

    ▶ The ages of observers of the powwow were: 43, 55, 70, 63, 15, 41, 9, 42, 75, 25, 16, 18, 51, 80, 75, 39, 23, 55, 50, 54, 60, 43.

    a. Find the range of each group.
    b. Find the interquartile range for each group.
    c. Write a paragraph describing the participants.
    d. Write a paragraph comparing the participants with the observers.

Mixed Review

26. **Statistics** The ages of the first twenty people into the museum on Saturday were 17, 9, 12, 25, 8, 39, 27, 14, 29, 40, 36, 8, 15, 41, 28, 29, 30, 31, 29, and 11. Make a stem-and-leaf plot for the data. (Lesson 10-1)

27. Replace the ● in $\frac{2}{3}$ ● $\frac{5}{8}$ with $<$ or $>$ to make a true sentence. (Lesson 6-1)

28. **Computers** Shantia plans to spend no more than \$50 on new software. If she buys a program for \$19.95, how much more can she spend? Express your solution as an inequality. (Lesson 5-7)

29. Find the quotient of c^2d^5 and cd. (Lesson 4-8)

30. *True* or *false*: The graph of a point with one negative coordinate and one positive coordinate is located in Quadrant III. (Lesson 2-2)

10-3 Displaying Data

Setting Goals: *In this lesson, you'll use box-and-whisker plots, pictographs, and line graphs to display data.*

Modeling a Real-World Application: Safety

Students at Horace Mann Middle School were concerned about the speed of traffic on the street in front of their school. The student council asked the city's Traffic Engineering Division to place two radar speed sensors with large digital displays near the street. One display allowed vehicles traveling west to see their speed, and the other was positioned for vehicles traveling east to see their speed.

Sample data on the speeds (in mph) of vehicles traveling west were 25, 35, 27, 22, 34, 40, 20, 19, 23, 25, and 30. Speeds of vehicles traveling east were 26, 22, 31, 36, 22, 27, 15, 50, 32, 29, and 30.

The student council voted to ask the city council to erect a stop sign at either end of the block. When they made their presentation, they wanted to make a display of the data so that the city council members would understand it quickly and easily. They chose a **box-and-whisker plot** because it summarizes data using the median, the upper and lower quartiles, and the highest and lowest, or extreme, values.

Learning the Concept

The speeds of the vehicles traveling west are displayed in the stem-and-leaf plot at the right. The median is marked by a box, and the quartiles are circled. The upper extreme is 40, and the lower extreme is 19.

Stem	Leaf
1	9
2	0 (2) 3 5 [5] 7
3	0 (4) 5
4	0

1 | 9 = 19

Here's how to construct a box-and-whisker plot to display the data.

Step 1 Draw a number line for the range of the speeds. Above the number line, mark points for the extreme, median, and quartile values.

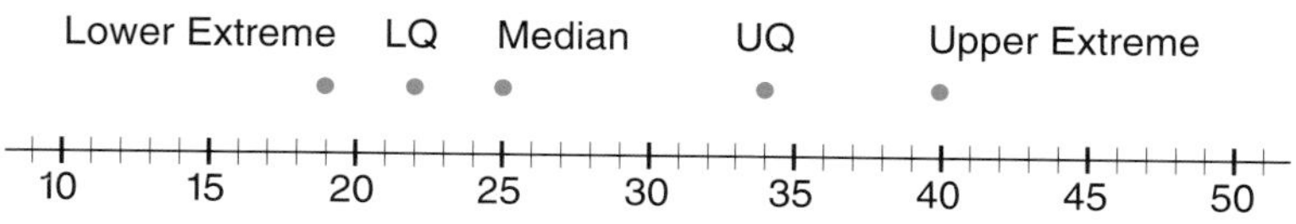

Step 2 Next, draw a *box* that contains the quartile values. Draw a vertical line through the median value. Finally, extend the *whiskers* from each quartile to the extreme data points.

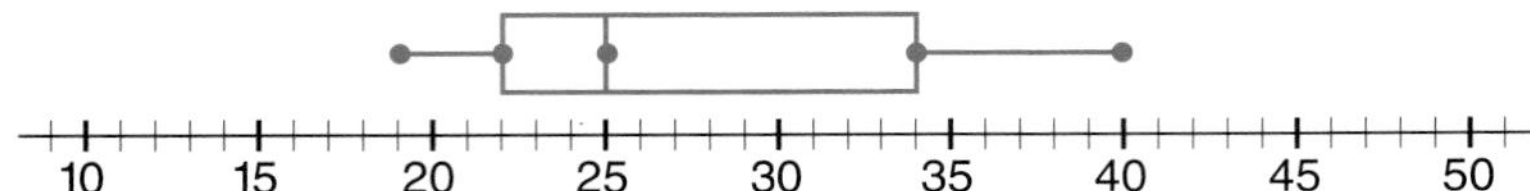

This display, the box-and-whisker plot, separates the data into fourths, each having an equal amount of data. For example, the whisker from 19 to 22 contains 25% of the data.

Example  1

Use the box-and-whisker plot below to answer the questions.

Ages of Volunteers at the Homeless Shelter

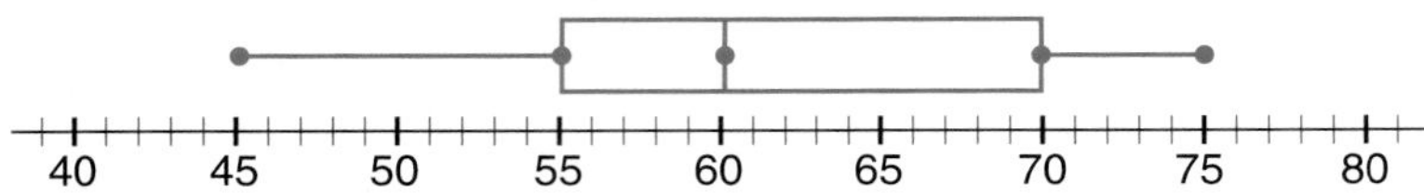

a. What is the age of the youngest volunteer? 45

b. Half of the volunteers are under what age? 60

c. What percent of the volunteers are over 70? 25%

Sometimes the data will have such great variation that one or both of the extreme values will be far beyond the other data. Data that are more than 1.5 times the interquartile range from the quartiles are called **outliers**.

Example  2

APPLICATION

Safety

Refer to the application at the beginning of the lesson.

a. Draw a box-and-whisker plot for the speed data of the cars traveling east.

15 22 22 26 27 29 30 31 32 36 50

Step 1 The median is 29, the lower quartile is 22, and the upper quartile is 32. Draw a box to show the median and quartiles.

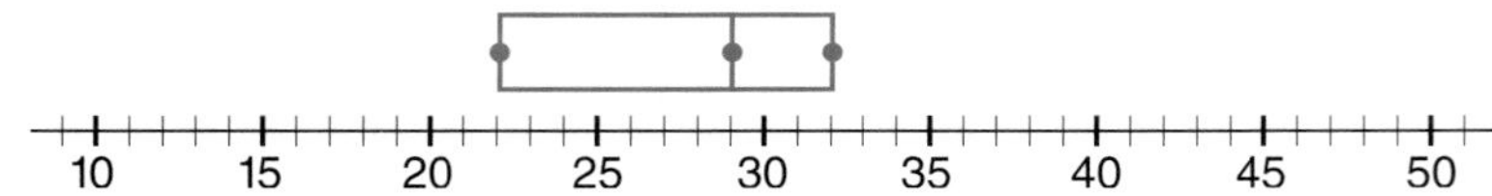

Step 2 The interquartile range is 32 − 22 or 10. So, data more than 1.5 times 10 from the quartiles are outliers.

$$1.5(10) = 15$$

Find the limits of the outliers.

Subtract 15 from the lower quartile $22 - 15 = 7$

Add 15 to the upper quartile. $32 + 15 = 47$

The limits for the outliers are 7 and 47. There is one outlier in the data, 50. Plot the outlier with an asterisk. Draw the whiskers to the extreme data that are not outliers.

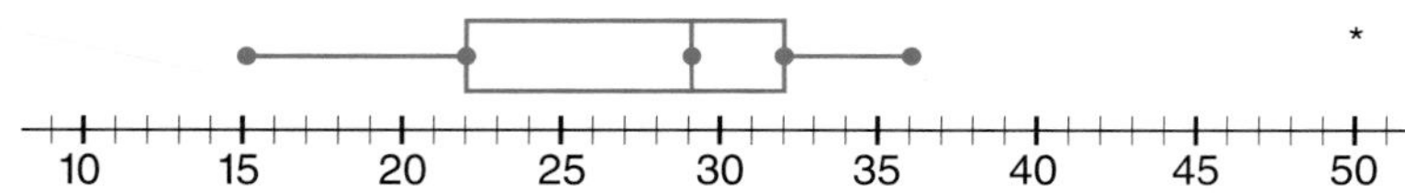

b. The speed limit in front of the school is 20 mph. Make a double box-and-whisker plot of the east and west speeds to convince the city council that a stop sign is needed.

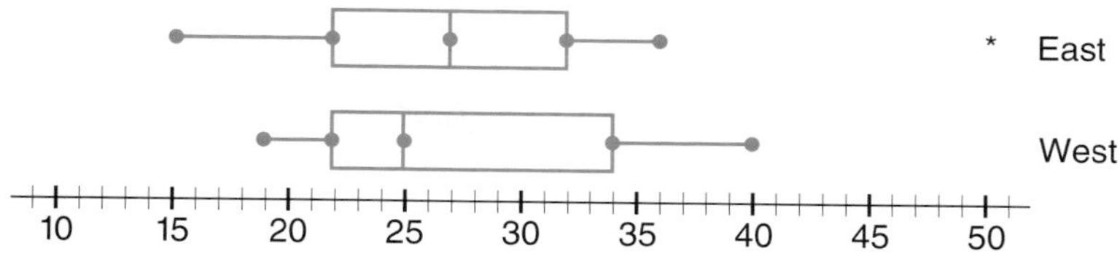

Almost all of the speeds of vehicles traveling west exceed the speed limit, and more than three-fourths of the vehicles traveling east exceed the speed limit. Stop signs at both ends of the block should slow traffic because cars would have to stop twice.

Other statistical graphs that are often used to present data include *line graphs*, *pictographs*, *circle graphs*, and *comparative graphs*.

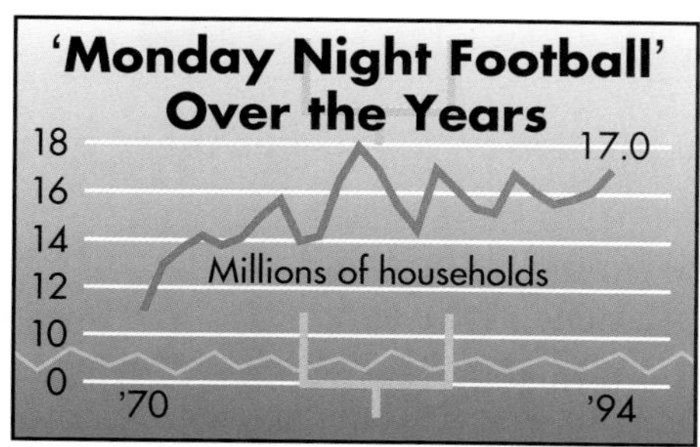

Source: ABC

Line graphs usually show how values change over a period of time. The line graph at the left shows the number of viewers of *Monday Night Football* from 1970 to 1994.

Pictographs use pictures or illustrations to show how specific quantities compare. Each symbol represents a convenient number of items to display the data. The pictograph at the right shows how much different age groups in the United States spend on sporting apparel.

Source: Sporting Goods Manufacturing Association

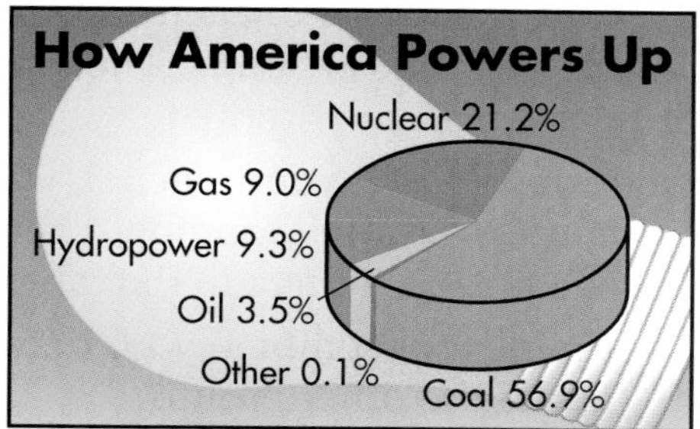

Source: Energy Information Administration

Circle graphs show how parts are related to the whole. The circle graph at the left shows how electricity is generated in the United States.

You can review bar graphs in Lesson 1-10.

Comparative graphs like a double-bar graph are used to show trends. They are usually used to compare results of similar groups. The graph at the right shows how the temperatures in the winter of 1995 compared to the 100-year average.

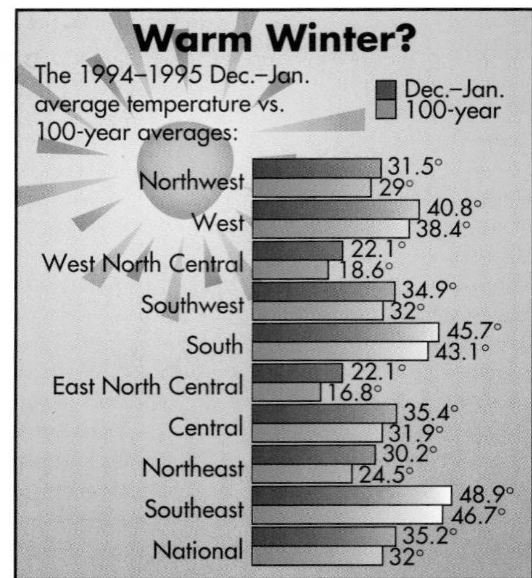

Source: National Climactic Data Center

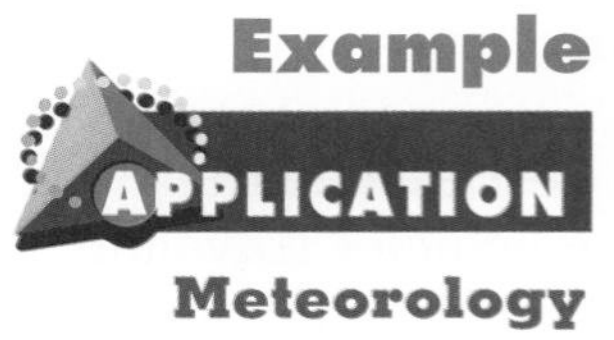

Meteorology

Example 3 ***The Herald Sun*** **wants to display the high temperatures of the past week. Should they use a box-and-whisker plot, line graph, pictograph, circle graph, or a double-bar graph?**

Since the data would show how values change over a period of time, a line graph would give the reader a clear picture of what temperatures were and the changes in temperature.

Checking Your Understanding

Communicating Mathematics

Read and study the lesson to answer these questions.

1. **Explain** how to find the median of the data shown in the box-and-whisker plot at the right.
2. **Describe** the data represented by the box.
3. **Write** the five pieces of information you can learn from the plot.
4. **Describe** how to determine if there are any outliers in a set of data.
5. **Explain** how the information you can learn from a set of data shown in a box-and-whisker plot is different from what you can learn from the same set of data shown in a stem-and-leaf plot.

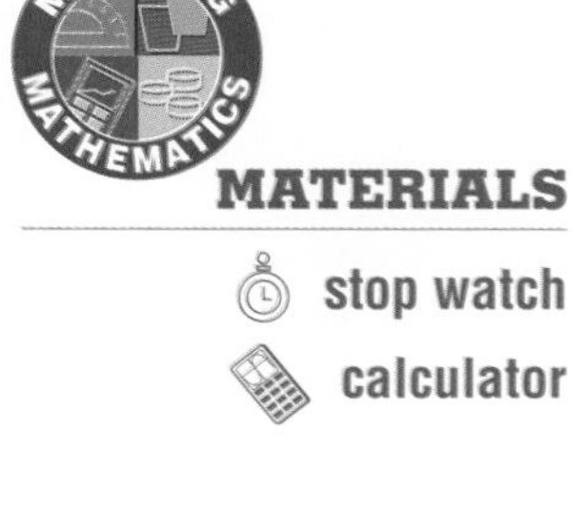

MATERIALS

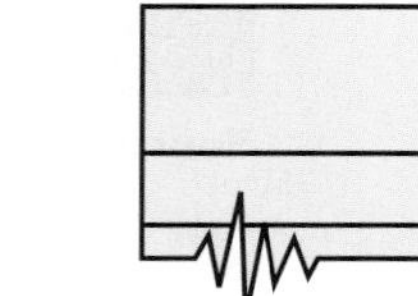

Work in groups of 4 or 5 to complete the following activity.

6. Make a table like the one shown below.

Name	Heartbeats per Minute	
	Resting Rate	Active Rate

a. Find the resting heart rate of each member of your group by taking your pulse for 15 seconds. Then multiply by 4 to find the beats per minute and record the product in your table.

b. In an area designated by your teacher, walk at a brisk rate for 5 minutes without stopping. Then take the pulses again and record them in the table.

c. Extend your table and combine your data of heart rates with that of the other groups.

d. Display the class data in a box-and-whisker plot.

Guided Practice

7. Use the box-and-whisker plot below to answer each question.

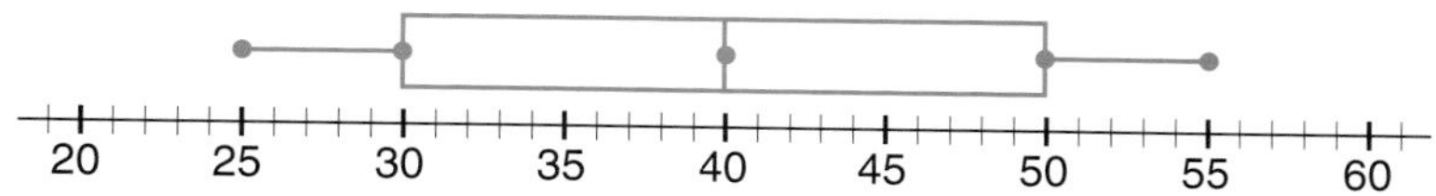

a. What is the median?
b. What is the upper quartile?
c. What is the lower quartile?
d. What is the interquartile range?
e. What are the extremes?
f. What are the outliers, if any?

8. **Education** From 1987 to 1996, Richton High School had the following number of graduates.

145, 165, 134, 173, 201, 204, 194, 185, 205, 198

a. Find the range, median, upper and lower quartiles, and the interquartile range for the data.
b. What are the limits for the outliers?
c. Use the data to make a box-and-whisker plot.

Exercises: Practicing and Applying the Concept

Independent Practice

9. The box-and-whisker plot below shows the heights in inches of the eleven girls on the Towne High girls' volleyball team.

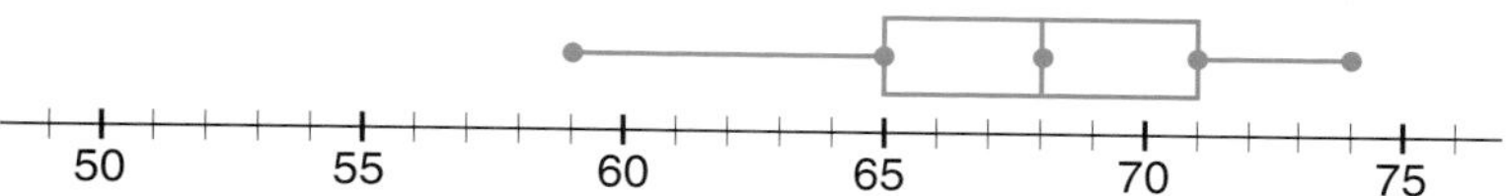

a. What is the range?
b. What is the median?
c. What is the upper quartile?
d. What is the lower quartile?
e. What is the interquartile range?
f. Are there any outliers?

10. Use the stem-and-leaf plot at the right to answer each question.

a. Make a box-and-whisker plot of the data.
b. What is the median?
c. What is the upper quartile?
d. What is the lower quartile?
e. What is the interquartile range?
f. What are the extremes?
g. What are the limits for outliers?
h. What are the outliers, if any?

Stem	Leaf
4	0 2
5	1 5 9
6	3 5 7 8
7	2 3 7 8
8	0 1

4 | 0 = 40

11. Use the box-and-whisker plot below to answer each question.

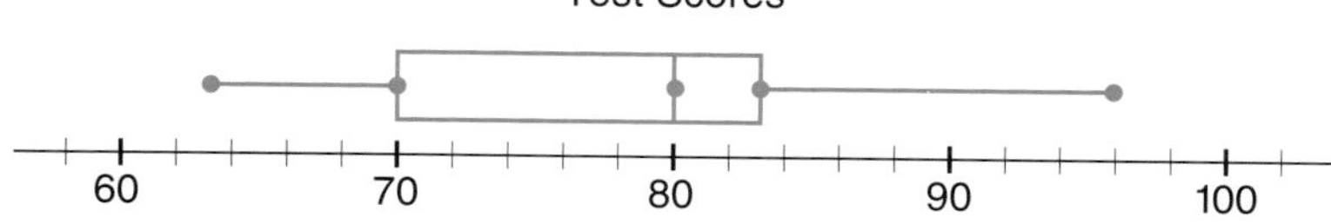

a. What percent of the test scores are below the median?
b. What percent of the test scores are represented by the box?
c. What percent of the test scores are below the lower quartile?
d. What percent of the test scores are represented by each whisker?
e. Why isn't the median in the middle of the box?
f. Why is one whisker longer than the other?

For each of the following sets of data, determine whether a bar graph, box-and-whisker plot, comparative graph, line graph, or pictograph is the best way to display the data. Explain your reasoning.

12. The number of people who have different kinds of pets.

13. Points scored by basketball players in 15 games.

14. Number of movies rented from a video store by different age groups.

15. Shoe size of students in your class.

16. Gender of teachers in your school.

17. Students SAT scores by grade level.

18. The chart at the right shows the number of sunny days for each city.

Sunny Days in 1993

City	Number of Days
Albany, NY	60
Burlington, VT	60
Dallas, TX	120
Des Moines, IA	80
Indianapolis, IN	68
Jacksonville, FL	90
Kansas City, MO	100
Lexington, KY	69
Mobile, AL	104
New Orleans, LA	99
Salt Lake City, UT	130

Source: *The World Almanac*, 1995

a. Make a bar graph to represent the number of sunny days.

b. Make a pictograph to represent the number of sunny days.

c. Describe any advantages the bar graph has over the pictograph.

d. Describe any advantages the pictograph has over the bar graph.

Critical Thinking

19. In making a box-and-whisker plot of the weights of the 23 juniors on the varsity football team, Randy found the median to be 170 pounds, the upper quartile to be 178 pounds, the lower quartile to be 168 pounds, and there were no outliers. Then he discovered four more weights: 200, 175, 169, and 167. When he included these in the data, what happened to the box-and-whisker plot? (*Hint:* You may want to sketch two plots.)

20. Is it possible to have a box-and-whisker plot with only one whisker? Explain your answer.

Applications and Problem Solving

21. Teaching Mr. Juarez has two classes of Physical Science. He made the stem-and-leaf plot of the scores for the last test shown at the right.

First Period		Second Period
8 0	2	
8 5	3	
5	4	3 5
0	5	2 4 5 6
0 0	6	0 0 0 8
5 2	7	2 5 8
9 8	8	1 5
8 6 1	9	

1|9 = 91 *8|1 = 81*

a. Make a double box-and-whisker plot for each class. (Use the same number line for both; make one plot above the other.)

b. What is similar about the data in the two plots?

c. What is different about the data in the two plots?

d. Which class had scores that were more evenly spread over the range? Explain.

22. **Consumerism** The average retail prices for one gallon of unleaded gasoline from 1984–1993 are shown in the chart below.

Year	1984	1985	1986	1987	1988
Price	\$1.21	\$1.20	\$0.92	\$0.95	\$0.95
Year	1989	1990	1991	1992	1993
Price	\$1.02	\$1.16	\$1.14	\$1.13	\$1.11

a. Make a box-and-whisker plot of the prices.
b. What is the median?
c. Are there any outliers? If so, why do you think they occurred?
d. Draw a line graph of the data.
e. Does the line graph present a better picture of the data than the box-and-whisker plot? Explain your answer.

23. **Employment** The chart below shows the percent of men and women who held two jobs from 1970 to 1995.

Percent Who Hold Down Two Jobs						
Year	1970	1975	1980	1985	1990	1995
Men	7.0	5.8	5.8	5.9	6.4	5.9
Women	2.2	2.9	3.8	4.7	5.9	5.9

a. Make a comparative bar graph to show the percentage of men and women who hold two jobs.
b. What changes have occurred in men's employment?
c. What changes have occurred in women's employment?
d. In general, what conclusion can you make from the data?

Mixed Review

24. **Statistics** Find the range, median, upper and lower quartiles, and the interquartile range for the set of data in the stem-and-leaf plot shown at the right. (Lesson 10-2)

Words Typed Per Minute

Stem	Leaf
4	0 2
5	1 5 9
6	3 5 7 8
7	2 3 7 8
8	0 1

4|0 = 40

25. Write the domain and range of the relation $\{(5, -3), (-1, 4), (4.5, 0), (-1, -3)\}$ (Lesson 8-1)

26. **Measurement** How many grams are in 0.046 kilograms? (Lesson 7-8)

27. Estimate the product of 0.25 and \$27.98. (Lesson 6-2)

28. **Nutrition** One cup of skim milk has 80 Calories, and one cup of whole milk has 150 Calories. If you drink one cup of milk per day, about how many Calories would be saved each year by switching from whole milk to skim? (Lesson 1-1)

a. Which method of computation do you think is most appropriate? Justify your choice.
b. Solve the problem using the four-step plan. Be sure to examine your solution.

GRAPHING CALCULATOR ACTIVITY

10-3B Making Statistical Graphs

An Extension of Lesson **10-3**

You can use a graphing calculator to construct statistical graphs. A line graph, box-and-whisker plots, and a special kind of bar graph, called a **histogram**, can be created using a TI-82 graphing calculator.

Activity 1

Use a TI-82 graphing calculator to construct a line graph of the data on bicycle sales.

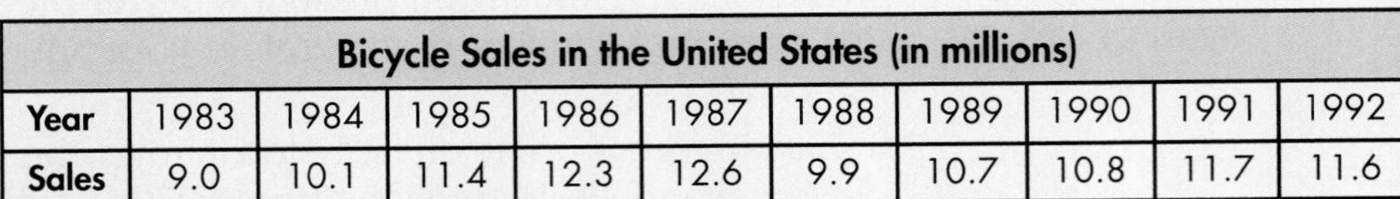

Bicycle Sales in the United States (in millions)										
Year	1983	1984	1985	1986	1987	1988	1989	1990	1991	1992
Sales	9.0	10.1	11.4	12.3	12.6	9.9	10.7	10.8	11.7	11.6

You can review setting the range in Lesson 2-2B and clearing the statistical memory in Lesson 6-6B.

Before you enter any data to graph, you must clear the statistical memory, clear the Y= list, and set the range. Use the range values [1983, 1993] by [0, 15] with a scale factor of 1 on both axes.

Now enter the data into the memory and draw the graph.

Enter: STAT ENTER — *Accesses the statistical lists.*

1983 ENTER . . . 1992 ENTER — *Enters the years in L_1.*

▶ 9 ENTER . . . 11.6 — *Enters the sales in L_2.*

Now choose the type of statistical plot you would like to create.

Enter: 2nd STAT PLOT ENTER — *Accesses the menu for the first statistical graph.*

Use the arrow and ENTER keys to highlight your selections for the graph. Select "On", the line graph icon, "L_1" as the Xlist, "L_2" as the Ylist, and "•" for the marks.

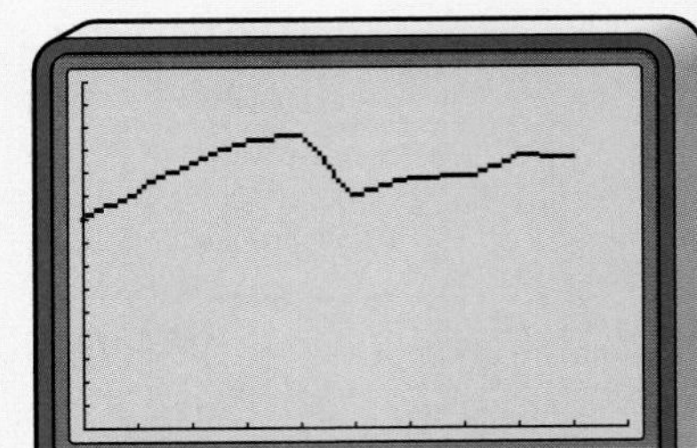

To create the graph, press GRAPH.

Your Turn

1. Use a TI-82 graphing calculator to create a line graph of the data below. (Enter 1, 2, 3, . . . , 13 in list L_1 to represent the months.)

Unemployment Rate, 1993–1994													
Month	Nov.	Dec.	Jan.	Feb.	Mar.	Apr.	May	June	July	Aug.	Sep.	Oct.	Nov.
Percent	6.5	6.4	6.7	6.5	6.5	6.4	6.0	6.0	6.1	6.1	5.9	5.8	5.6

2. Based on your graph, do you think the unemployment rate for December, 1994 increased or decreased? Explain.

A *histogram* is a special bar graph that displays the frequency of data that has been organized into equal intervals. The intervals cover all possible values of data. Therefore, there are no spaces between the bars of the histogram.

Activity 2 **Use a TI-82 graphing calculator to create a histogram of the number of graduates for each of the states.**

Public High School Graduates in 1991 (in thousands)

39.0	5.5	31.3	25.7	234.2	31.3	27.3	5.2	87.4	60.1
9.0	12.0	103.3	58.6	28.6	24.4	35.8	33.5	13.2	39.0
52.1	88.2	46.5	23.7	46.9	9.0	16.5	9.4	10.1	67.0
15.2	133.6	62.8	7.6	107.5	33.0	24.6	104.8	7.7	33.1
7.1	44.8	174.3	22.2	5.2	58.4	42.5	21.1	49.3	5.7

When you set the range for a histogram, you determine the number of bars (equal intervals) you want to graph. Use the range values [0, 240] by [0, 25] with a scale factor of 30 on the x-axis and 5 on the y-axis. These settings will create $240 \div 30$ or 8 bars.

Now enter the data into the memory and draw the graph.

Enter: STAT ENTER *Accesses the statistical lists.*

39.0 ENTER 5.5 ENTER 31.3 ENTER . . . 5.7 ENTER

Enter: 2nd STAT PLOT ENTER *Accesses the menu for the first statistical graph.*

Use the arrow and ENTER keys to highlight "On", the histogram icon, "L1" as the Xlist, and "1" as the frequency. Press GRAPH to view the histogram.

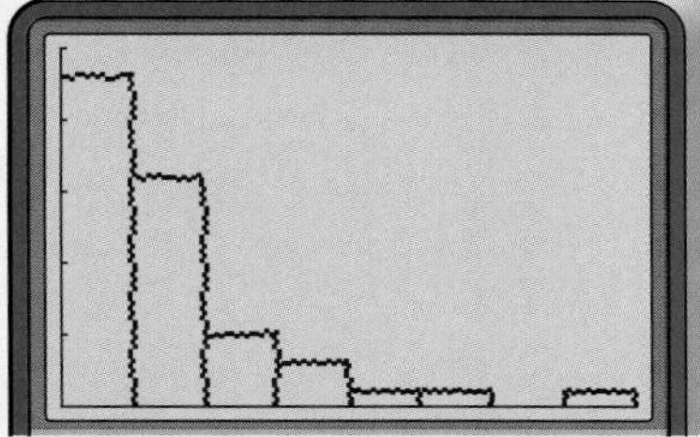

You can make a box-and-whisker plot of the same data.

Enter: 2nd STAT PLOT ENTER

Select the same plot and use the arrow and ENTER keys to highlight the box-and-whisker plot icon. Press GRAPH to view the box-and-whisker plot.

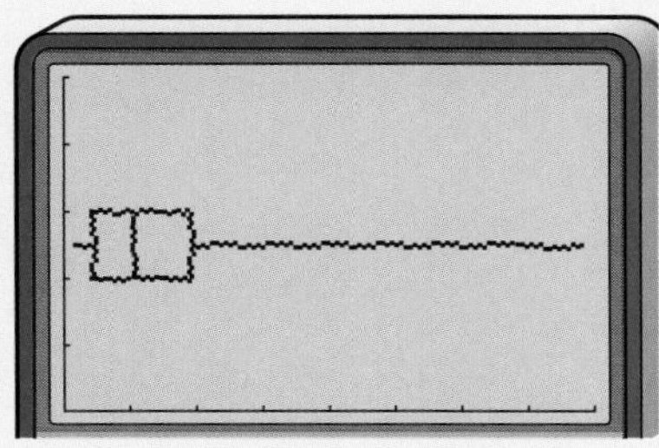

Your Turn

3. Use a TI-82 graphing calculator to create a histogram and a box-and-whisker plot for the heights of the United States presidents.

Heights of U.S. Presidents (in inches)

74	66	74.5	64	72	67	73	66	66	72	68	68	69	70
72	76	70	68.5	68.5	70	74	71	66	66	68	74	71	72
69	71	74	70	70	73	75	71.5	72	70	73	74	74.5	

4. If the range of values for a histogram is 35 and you want to create 7 bars, what scale factor should you enter? Explain.

10-4 Misleading Statistics

Setting Goals: *In this lesson, you'll recognize when statistics are misleading.*

Modeling a Real-World Application: Education

Mr. Snyder's sociology class was studying birthrates in the United States. For a project on the history of birthrates from 1960 to 1990, Maria and Jennifer made the two graphs shown below.

- Do both graphs show the same information?
- Which graph suggests a baby boom from 1975–1990? Which graph suggests a fairly steady birthrate?

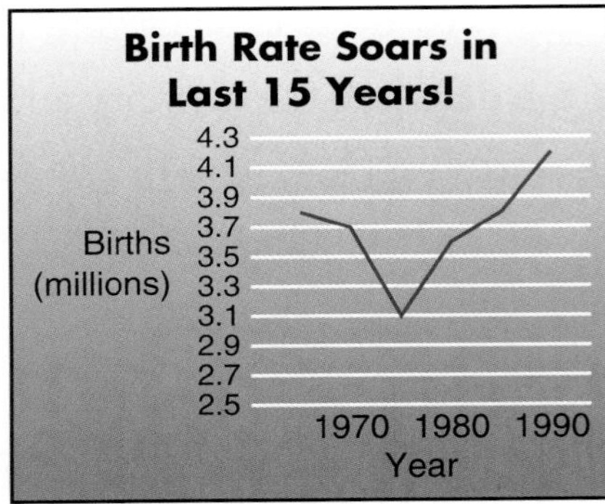

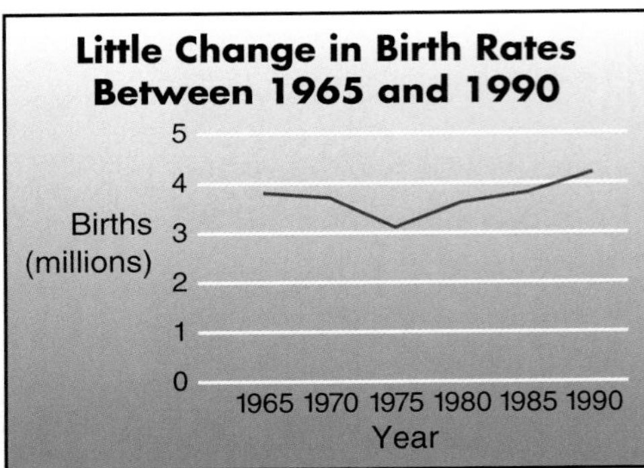

What is the same and what is different about the two graphs?

Learning the Concept

The data used in the graphs above are the same. However, different scales were used to make the data appear to support a particular point of view. The soaring graph at the left is a result of an extended vertical axis and a shortened horizontal axis. A small "break" in the vertical axis is used to show that the axis is not to scale between zero and 2.7. However, it is visually misleading when compared to the complete scale of the graph at the right.

Entertainment

Example 1

The graphs at the right display data about movie attendance in the United States.

a. Is the top graph misleading? Explain.

The graph is misleading because there is no title and there are no labels on either scale. Also, the vertical axis does not include zero.

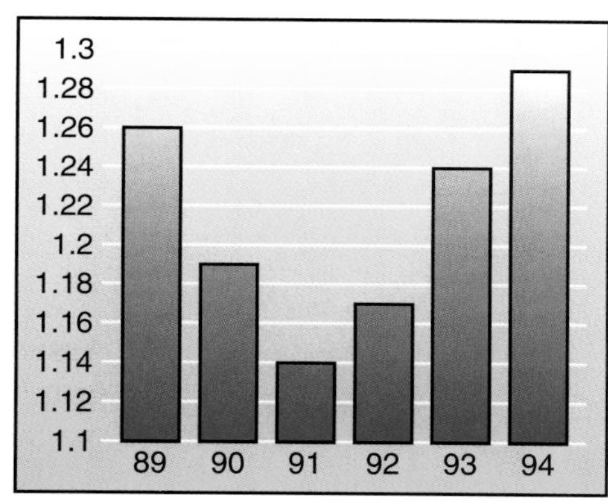

b. Is the bottom graph misleading? Explain.

The graph is not misleading. All the necessary information is present. The scales are labeled, the distance between the units on each axis is uniform, and the vertical axis includes 0.

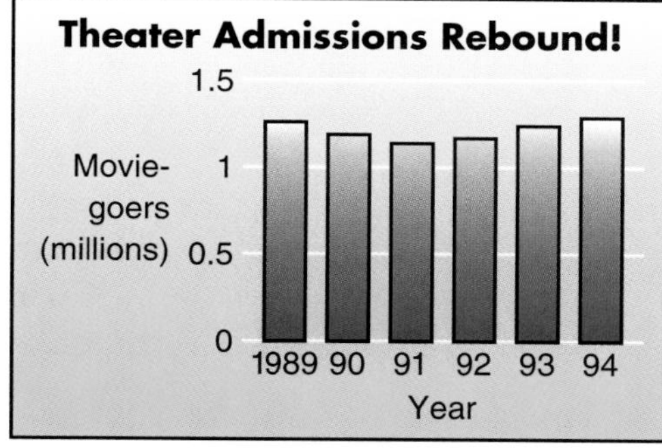

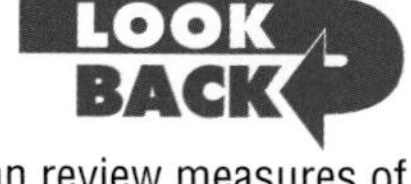

You can review measures of central tendency in Lesson 6-6.

Another way statistics can be misleading is by using the inappropriate measure of central tendency.

Statistics

Example 2

After a semester exam in science, Mr. Myers displayed the results to the class. The scores are shown in the stem-and-leaf plot at the right. Mr. Myers told the principal that the average grade on the exam was 77. Ryan told his parents that the class average was 93. Tai told her grandmother the class average was 80. How can all three people think they are correct?

Scores

Stem	Leaf
9	01333
8	012345
7	023
6	0123579

$9|0 = 90$

Each person used a different measure of central tendency to describe the average grade for the class. Mr. Myers used the mean and Tai used the median. Ryan's use of the mode is misleading because 93 is also the highest score on the semester exam.

Checking Your Understanding

Communicating Mathematics

Read and study the lesson to answer these questions.

1. **Describe** two ways graphs can be misleading.
2. **List** some things you might question when reading the results of a survey on employee salaries.

3. **You Decide** Khandi says that the graph at the right is misleading. Carlos disagrees. Who is correct and why?

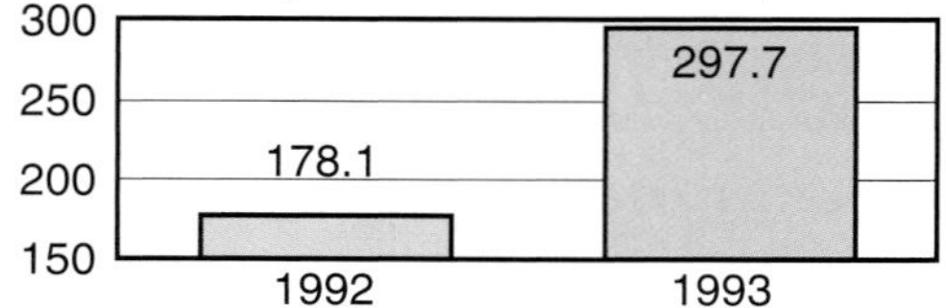

4. **Assess Yourself** Do you understand how to determine whether graphs are misleading? Find three graphs in a newspaper or magazine. Determine if the graphs are misleading and explain why or why not.

Guided Practice

5. **Refer to the graph at the right.**

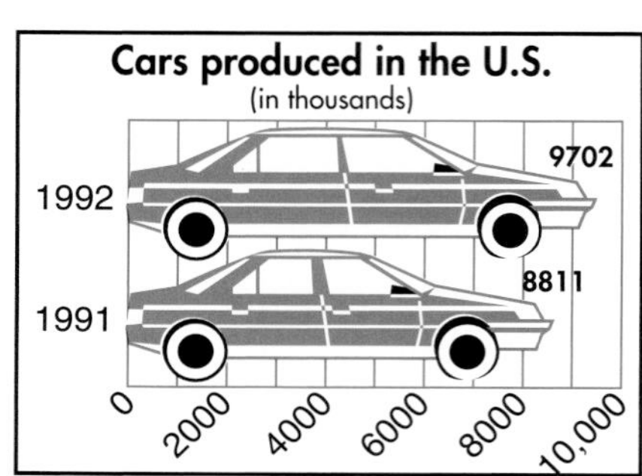

   a. What is the ratio of the number of cars produced in 1992 to the number of cars produced in 1991?
   b. How does this ratio compare with the ratio of the sizes of the cars on the graph?
   c. Is this a misleading graph? Explain.

6. **Statistics** Roberta made two graphs of her social studies grades for each grading period.

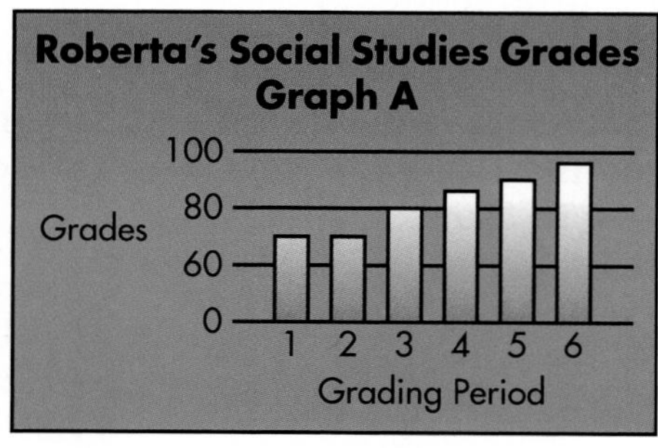

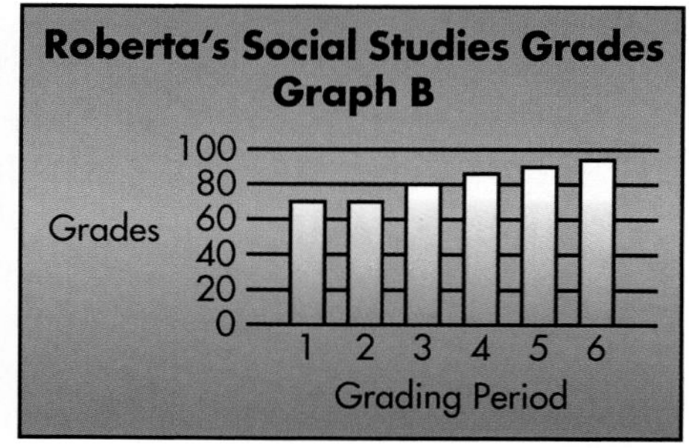

   a. Explain why these graphs made from the same data look different.
   b. If Roberta wanted to convince her parents that her social studies grades have improved dramatically, which graph would she probably show her parents? Explain.

Exercises: Practicing and Applying the Concept

Independent Practice

7. The houses in the Cambridge section of Anderson were assessed for school taxes. The values are shown in the frequency table at the right.

Value	Number
$150,000	2
$130,000	5
$115,000	4
$95,000	10
$85,000	20
$75,000	2

   a. Which measure of central tendency would you use to find the average cost of a house? Why?
   b. Which average might a real estate agent use with a business executive? Explain.

Use the chart to describe each category. Write the average you used.

	Tashelle	Kim	Dawn	Sita	Bill
Age	40	24	33	28	41
Favorite Entertainment	sports	concerts	theater	movies	sports
Height (cm)	185	163	165	157	193
Salary	$34,700	$18,000	$30,500	$25,500	$47,000

8. average age

9. average height

10. average salary

11. average favorite entertainment

12. How can you describe the "average" person in this group?

13. Find the mean, median, mode, and range of the salaries at each company in the table at the right. Which company would you rather work for? Explain your reasoning.

	Frequency	
Salary	**Company A**	**Company B**
$15,000	18	
$20,000	4	24
$25,000	4	6
$50,000	2	1
$100,000	1	

Critical Thinking

14. Find a graph in a magazine or newspaper. Redraw the graph so that the data will appear to show different results.

Applications and Problem Solving

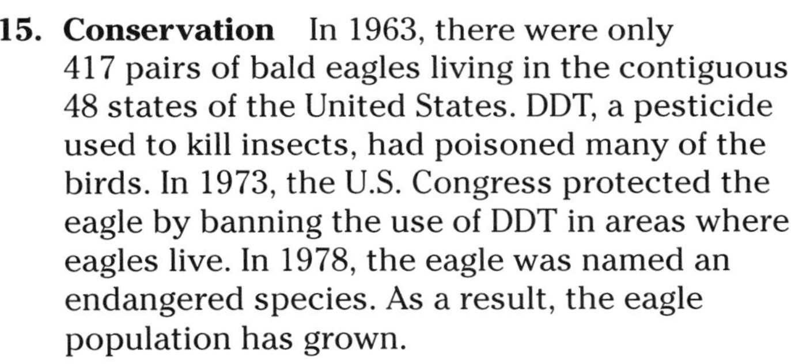

15. Conservation In 1963, there were only 417 pairs of bald eagles living in the contiguous 48 states of the United States. DDT, a pesticide used to kill insects, had poisoned many of the birds. In 1973, the U.S. Congress protected the eagle by banning the use of DDT in areas where eagles live. In 1978, the eagle was named an endangered species. As a result, the eagle population has grown.

a. Do the data in the table represent the growth of the eagle population since the policy changes were made in the United States? Explain.

b. Draw a graph that you might use to accurately represent the data.

Eagle Pairs in Contiguous 48 States of the U.S.

Year	Number of Pairs
1988	2475
1989	2680
1990	3020
1991	3391
1992	3747
1993	4016

Source: U.S. Fish and Wildlife Service

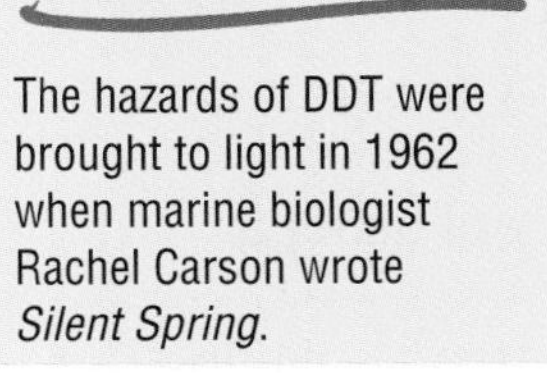

FYI

The hazards of DDT were brought to light in 1962 when marine biologist Rachel Carson wrote *Silent Spring.*

16. Employment The salaries of all the employees of Grenwich Bottling Company are shown in the frequency table at the right. They are negotiating a new contract and want a raise because they learned that the average salary for all hourly employees was only $24,000. The president informed them that the average salary for employees was $31,000.

Salary	Number of Employees
$150,000	1
$85,000	2
$60,000	2
$35,000	8
$24,000	25
$15,000	5
$12,000	3

a. Which "average" do you think the union would want to use in its negotiations? Explain.

b. If you were called in to mediate the negotiations, what recommendation would you make and why?

Mixed Review

17. The box-and-whisker plot below shows the amount students spent on entertainment for one month. (Lesson 10-3)

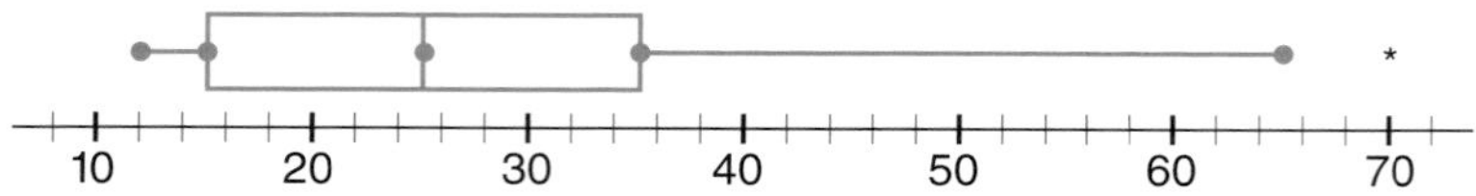

a. What is the range?
b. What is the median?
c. What is the upper quartile?
d. What is the lower quartile?
e. What is the interquartile range?
f. What are the extremes?
g. What are the limits for outliers?
h. What are the outliers, if any?

18. Which ordered pair(s) is a solution of $3x + 2y = 12$? (Lesson 8-3)

a. (4, 0) **b.** (0, 4) **c.** (5, −1.5) **d.** (−5, 1.5)

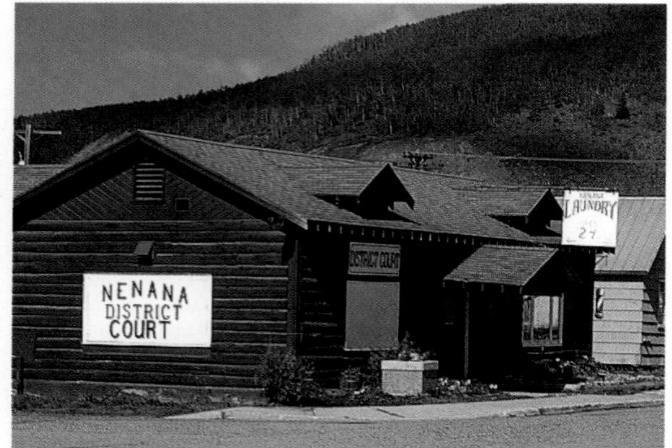

19. Geography Multipurpose buildings are common in small towns in Alaska, since the population is small. Between 1980 and 1990, the population of Alaska increased from about 400,000 to 550,000. (Lesson 6-8)

a. Find the ratio of the 1980 population to the 1990 population.

b. If the population grows in a geometric pattern over each of the next four decades, what would the population of Alaska be in 2030?

20. Measurement A millimeter is $\frac{1}{1000}$ of a meter or 10^{-3} meters. A kilometer is 1000 meters. How many millimeters are in a kilometer? (Lesson 4-8)

21. Patterns Find the next two integers in the pattern 0, −2, 2, −4, 4, . . . (Lesson 2-6)

Refer to the Investigation on pages 482–483.

Test your game by having the members of your group play the game several times. Play the game enough times to develop some different strategies. Have each member of your group use a different strategy. Then have players who are in the intended age group play the game several times.

Use the results of your test to:

- determine the best strategy for your game
- revise any rules
- display the scores or winners using a statistical graph
- determine what makes someone playing the game a "good" player
- make an advertisement for your game

Add the results of your work to your Investigation Folder.

10-5 Counting

Setting Goals: *In this lesson, you'll use tree diagrams or the Fundamental Counting Principle to count outcomes.*

Modeling a Real-World Application: Advertising

Bruegger's Fresh Bagel Bakery makes 9 types of bagels and has 10 flavors of cream cheese. They wanted to advertise the number of different bagel with cream cheese combinations that are possible. How many different possible choices are there? *This problem will be solved in Example 1.*

Learning the Concept

To solve this problem, it may be helpful to solve a simpler problem. Suppose Bruegger's Bagel Bakery offers chicken, tuna, and vegetable sandwiches, on plain or onion bagels. How many possible sandwiches are there?

You can draw a diagram to find the number of possible combinations or **outcomes**.

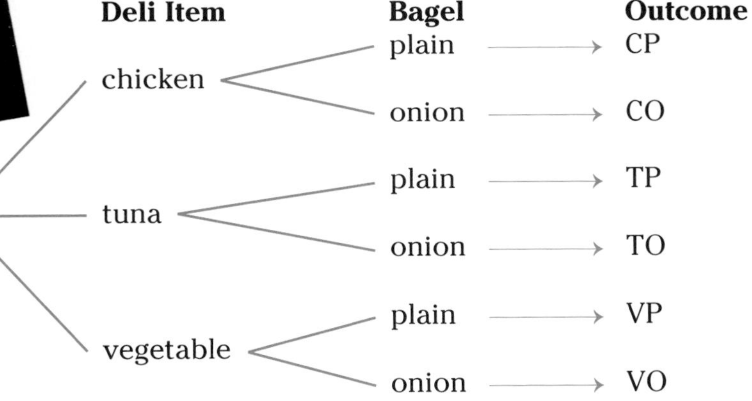

There are 6 possible choices, or outcomes.

The diagram above is called a **tree diagram**. Notice that the product of the number of deli items and the number of bagels, 3×2, also equals 6. Thus, you can find the number of possible outcomes by multiplying. This principle is called the **Fundamental Counting Principle**.

Fundamental Counting Principle	**If event M can occur in m ways and is followed by event N that can occur in n ways, then the event M followed by event N can occur in $m \cdot n$ ways.**

Example 1 APPLICATION **Advertising**

Refer to the application at the beginning of the lesson. How many different bagel with cream cheese combinations can Bruegger's Bagel Bakery advertise are possible?

Explore You know that there are 9 types of bagels and 10 flavors of cream cheese. You want to know how many different possible choices there are.

Plan Use the Fundamental Counting Principle.

Solve

$$\left(\begin{array}{c}\textit{number of types}\\ \textit{of bagels}\end{array}\right) \times \left(\begin{array}{c}\textit{number of flavors}\\ \textit{of cream cheese}\end{array}\right) = \left(\begin{array}{c}\textit{number of}\\ \textit{possible outcomes}\end{array}\right)$$

$$9 \times 10 = 90$$

There are 90 possible outcomes.

Examine Make a list or draw a tree diagram to verify the solution.

Connection to Algebra

If you already know the total number of possible outcomes, you can write and solve an equation to find the number of ways the events can occur.

Example 2 INTEGRATION **Algebra**

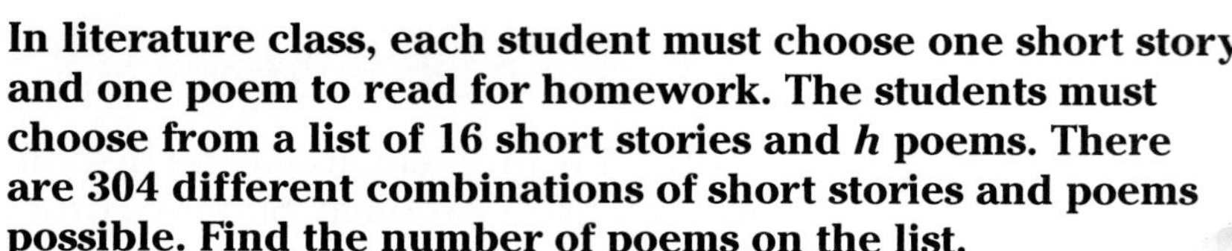

In literature class, each student must choose one short story and one poem to read for homework. The students must choose from a list of 16 short stories and *h* poems. There are 304 different combinations of short stories and poems possible. Find the number of poems on the list.

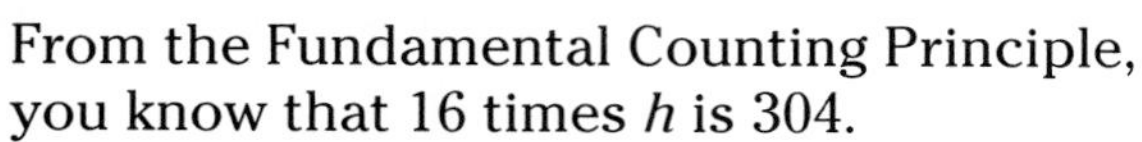

From the Fundamental Counting Principle, you know that 16 times h is 304.

$$16h = 304$$

$$\frac{16h}{16} = \frac{304}{16} \quad \textit{Divide each side by 16.}$$

304 [÷] 16 [=] 19

There are 19 poems on the list.

Example 3 INTEGRATION **Probability**

A die is rolled twice.

a. How many outcomes are possible?

$$\left(\begin{array}{c}\textit{number of outcomes}\\ \textit{for the first roll}\end{array}\right) \times \left(\begin{array}{c}\textit{number of outcomes}\\ \textit{for the second roll}\end{array}\right) = \left(\begin{array}{c}\textit{number of}\\ \textit{possible outcomes}\end{array}\right)$$

$$6 \times 6 = 36$$

There are 36 possible outcomes.

b. What is the probability of rolling a five and then a six?

Only one outcome can be a five and then a six.

$$P(\text{five and six}) = \frac{\textit{number of favorable outcomes}}{\textit{number of possible outcomes}} = \frac{1}{36}$$

The probability of rolling a five and then a six is $\frac{1}{36}$.

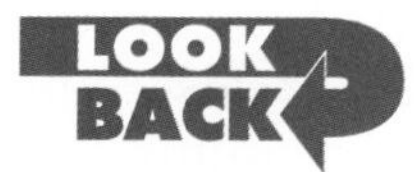

LOOK BACK

You can review probability in Lesson 9-3.

Checking Your Understanding

Communicating Mathematics

Read and study the lesson to answer these questions.

1. a. **Write** a problem that corresponds to the tree diagram at the right.
   b. **List** the outcomes illustrated in the tree diagram at the right.

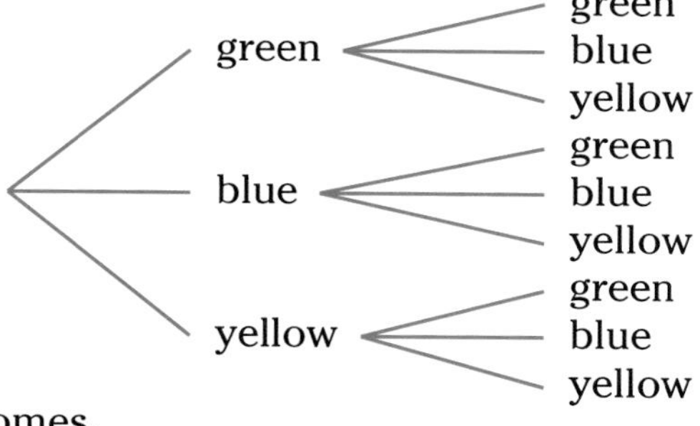

2. **Compare and contrast** using the Fundamental Counting Principle and using a tree diagram to find numbers of outcomes.

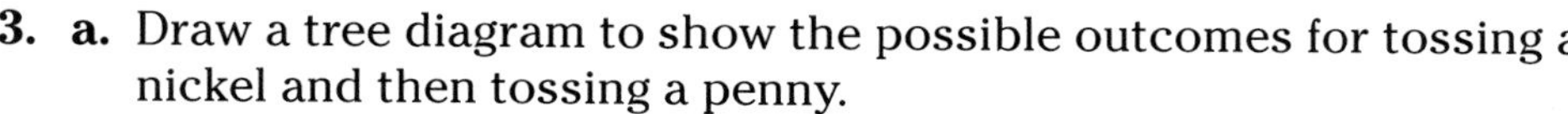

MATERIALS

 a penny and a nickel

3. a. Draw a tree diagram to show the possible outcomes for tossing a nickel and then tossing a penny.
   b. What is the probability of tossing two heads?
   c. Make a frequency table listing the outcomes. Toss a nickel and then a penny and tally the outcome in your frequency table. Repeat for 25 trials.
   d. What fraction of the time did you toss two heads?
   e. Your answer to part b is the *theoretical probability* of tossing two heads. Your answer to part d is the *experimental probability* of tossing two heads. How does the experimental probability compare to the theoretical probability?
   f. Toss the coins for 25 more trials. Compare the probabilities again. Did the experimental probability change? If so, how?

Guided Practice

4. The spinner at the right is spun twice.
   a. Draw a tree diagram that represents the situation.
   b. How many outcomes are possible?
   c. How many outcomes show red and blue?
   d. What is the probability of two oranges?

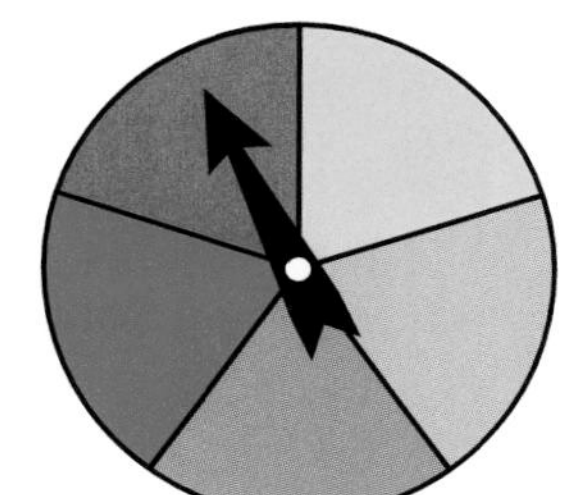

5. A coin is tossed, and a die is rolled.
   a. How many outcomes are possible?
   b. What is the probability of heads and a 4?
6. Three coins are tossed. How many outcomes are possible?
7. **Fashion** Reina has three silver necklaces, three pairs of silver earrings, and two silver bracelets. How many combinations of the three types of jewelry are possible?

Exercises: Practicing and Applying the Concept

Independent Practice

Draw a tree diagram to find the number of outcomes for each situation.

8. Each four-sided die is rolled once.

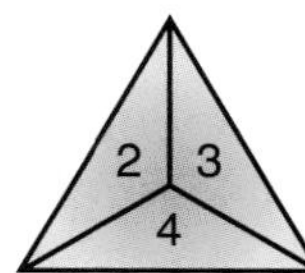

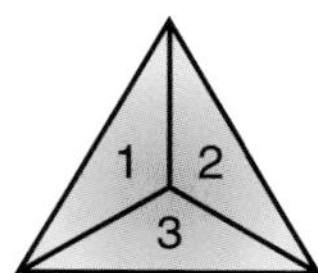

9. Each spinner is spun once.

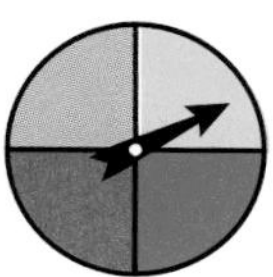

10. The continental breakfast at Myriad Hotel has toast, muffin, or bagel with coffee, milk or juice. How many combinations of one bread and one beverage are possible?

11. Jarred had five different poses to choose for his senior picture and five different frames. How many different ways could he choose a picture and a frame?

Find the number of possible outcomes for each event.

12. Four coins are tossed.

13. A die is rolled twice.

14. Two coins are tossed and a die is rolled.

15. School sweatshirts come in two colors, gray or white, and in four sizes, small, medium, large, and extra large.
    a. How many outcomes are possible?
    b. What is the probability of selecting a large gray sweatshirt at random?

16. A sedan comes with two or four doors, a four or six cylinder engine, and six exterior colors. How many different autos are possible?

17. A quiz has five true-false questions. How many outcomes for answering the five questions are possible? (*Hint:* One outcome is FFTFF.)

18. An eight-sided die is rolled four times.
    a. How many outcomes are possible?
    b. What is the probability of getting four 7s?

19. A coin is tossed thirty times.
    a. How many outcomes are possible?
    b. What is the probability of getting heads every time?

Critical Thinking

20. A test has all true-false questions. If there are x questions on the test, write an expression for the number of possible answer keys.

Applications and Problem Solving

21. **Telecommunication** Telephone numbers such as 219-555-1212 are made up of a three-digit area code (219), a three-digit exchange (555), and a four-digit extension (1212). The digits of an exchange cannot be 0 or 1, so it appears that there are $8 \cdot 8 \cdot 8$ or 512 exchanges in an area code. However, there are 26 exchanges that are not available in each area code because of special numbers like 911, 555, and 800. There are actually only 486 exchanges available in each area code.
    a. How many extensions are possible in one exchange?
    b. How many telephone numbers are available in each area code?
    c. Many local telephone companies ran out of numbers during the 1990s. They had to extend the three-digit exchange to include 0 and 1 in the second and third digits. (Zero and one still cannot be used as the first digit.) How many new numbers became available in each area code by allowing 0 and 1 in the second and third digits?

22. **Business** The Yogurt Oasis advertises that there are 1512 ways to enjoy a one-topping sundae. They offer six flavors of frozen yogurt, six different serving sizes, and several toppings. How many toppings do they offer?

Mixed Review

23. **Real Estate** A builder has house styles that are priced at $80,000, $95,000, $92,000, and $170,000. Which "average" price gives the best description of the houses that the builder offers? Why? (Lesson 10-4)

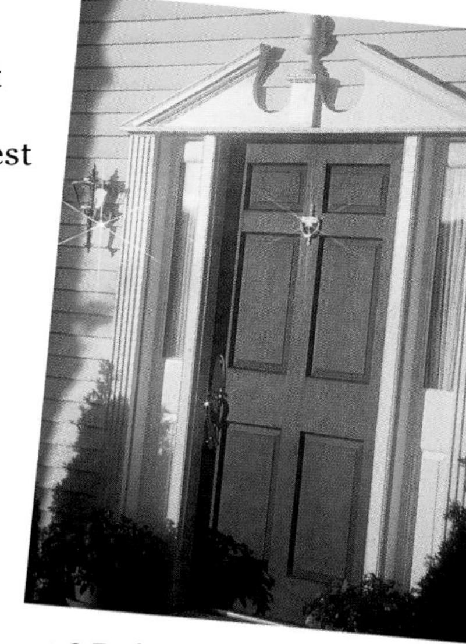

24. **Finance** Use the formula $I = prt$ to find the interest to the nearest cent on $500 at 4.25% simple interest for 6 months. (Lesson 9-9)
25. Given $f(x) = 2x + 4$, find the value of $f(-5)$. (Lesson 8-4)
26. Solve $2(k - 3) \leq 8$ and check your solution. Then graph the solution on a number line. (Lesson 7-6)
27. Sarah earned less than $20 baby-sitting at $3.50 an hour. How many hours did Sarah baby-sit? Define a variable and translate the sentence into an inequality. Then solve. (Lesson 3-8)

Self Test

The table below shows the normal monthly precipitation, in inches, for selected cities in the United States. Use this data to complete Exercises 1–5.

City	Jan.	Feb.	Mar.	Apr.	May	June	July	Aug.	Sept.	Oct.	Nov.	Dec.
Birmingham, AL	5.1	4.7	6.2	5.0	4.9	3.7	5.3	3.6	3.9	2.8	4.3	5.1
Louisville, KY	2.9	3.3	4.7	4.2	4.6	3.5	4.5	3.5	3.2	2.7	3.7	3.6
Miami, FL	2.0	2.1	2.4	2.9	6.2	9.3	5.7	7.6	7.6	5.6	2.7	1.8
Oklahoma City, OK	1.0	1.3	2.1	2.9	5.5	3.9	3.0	2.4	3.4	2.7	1.5	1.2

1. Make a stem-and-leaf plot of the precipitation for Birmingham. (Lesson 10-1)
2. Find the range and interquartile range of the precipitation for Oklahoma City. (Lesson 10-2)
3. What are the extreme measures of precipitation for Louisville? (Lesson 10-2)
4. Construct a box-and-whisker plot of the precipitation for Miami. (Lesson 10-3)
5. Determine whether a bar graph, box-and-whisker plot, comparative graph, line graph, or pictograph, is the best way to display the data. Explain your reasoning. (Lesson 10-3)

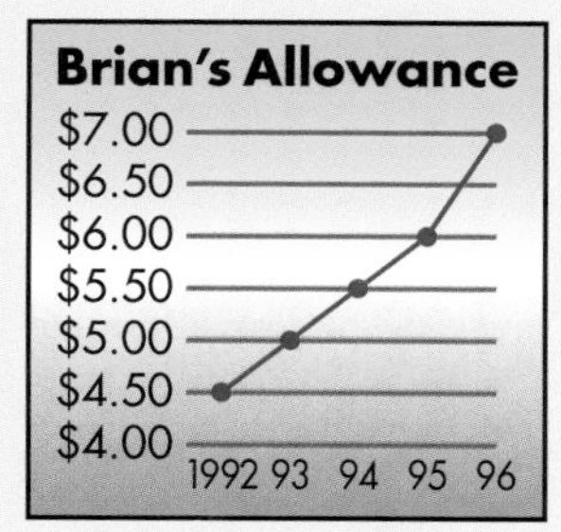

Use the graphs at the right to complete Exercises 6–7. (Lesson 10-4)

6. Explain why these graphs made from the same data look different.
7. Which graph would Brian show to his parents if he wants an increase in his allowance? Explain.
8. Wrapping paper comes in red, pink, green, or blue with white or yellow ribbon. Draw a tree diagram that illustrates how many combinations are possible. (Lesson 10-5)
9. Mark has three pairs of shorts, four shirts, and two pairs of running shoes. How many outcomes are possible? (Lesson 10-5)
10. Three coins are tossed. What is the probability that all 3 coins land heads up? (Lesson 10-5)

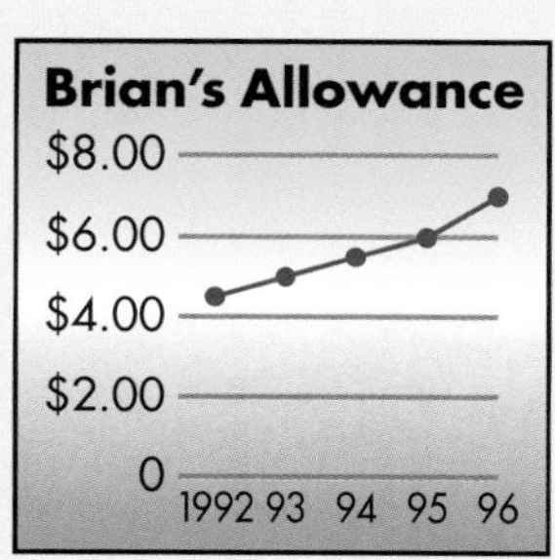

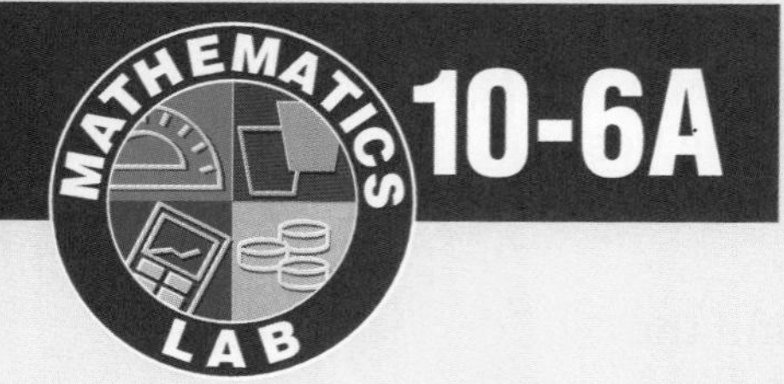

HANDS-ON ACTIVITY

10-6A Permutations and Combinations

A Preview of Lesson **10-6**

MATERIALS

6 index cards

markers

An arrangement of names, objects, or people in a particular order is called a **permutation**. For example, if you use a triangle, square, and circle, you can arrange them in the following permutations.

△□○ △○□ □○△ □△○ ○□△ ○△□

Activity 1

Work with a partner.

Write the letters M, A, T, H on four cards, one letter per card. Shuffle the cards and place them face down. Choose three of the cards and turn them face up. For example, you may have chosen the cards shown at the right. Record the "word" shown; for example, ATM.

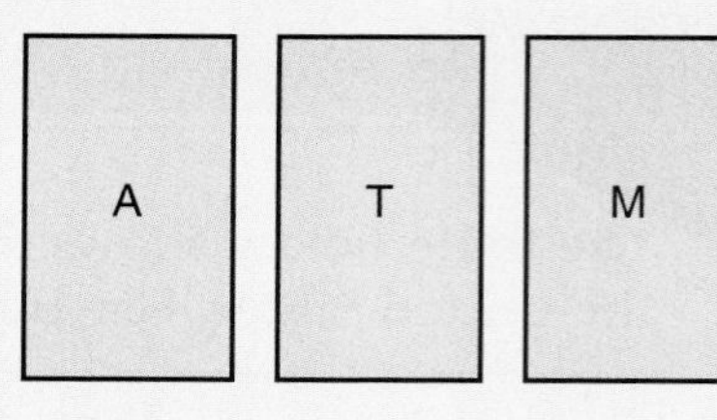

- Rearrange the three cards to make another "word." Record that word.
- Continue rearranging the cards and recording the words until you have listed all the 3-letter words you can with the cards chosen. Record the total number.
- Now use all four cards. Arrange the cards to form all the 4-letter "words" that you can. How many arrangements are there?

Activity 2

The student council at Northern Potter Junior High School is planning to have a sundae sale. They plan to offer six different toppings.

peanuts	chocolate chips	chocolate syrup
raisins	strawberries	butterscotch syrup

The price depends on the *combinations* of toppings chosen.

Your Turn **Work in groups of two or three.**

- Write the names of the six toppings on six index cards.
- To make a sundae, select any pair of cards. Make a list of all the different combinations of two toppings. Note that the order is not important; that is, *peanuts*, *raisins*, is the same as *raisins*, *peanuts*.
- Suppose that the student council decides to offer a deluxe sundae with four toppings on it. Make a list of all the different sundaes that are possible if four toppings of the six are used.

TALK ABOUT IT

1. In Activity One, suppose that you had five cards with five different letters. How many five-letter words do you think can be formed?
2. What is the difference between a combination (like a combination of toppings) and a permutation?

10-6 Permutations and Combinations

Setting Goals: *In this lesson, you'll use permutations and combinations.*

Modeling a Real-World Application: Swimming

The swim team coach at Clearwater High School is preparing the lineup for the state finals in the 400-meter freestyle relay. She must choose four swimmers from the five that qualified at the district championships. As she looks over the five names, she wonders how many arrangements are possible.

She reasons that any of the 5 swimmers can start the race. Once that swimmer is selected, there are 4 swimmers left who can swim second, 3 who can swim third, and 2 who can swim fourth.

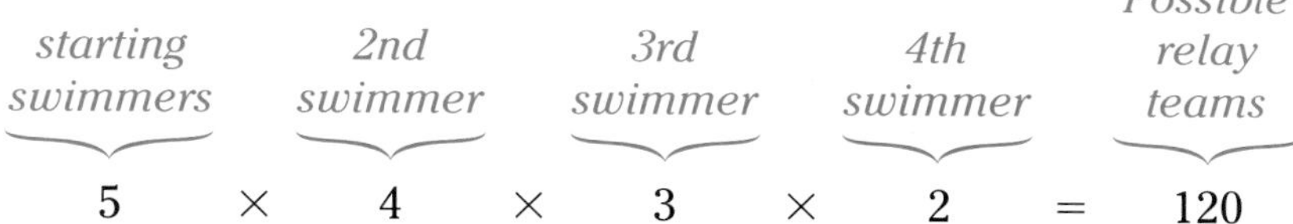

Using the Fundamental Counting Principle, as shown above, she finds there are 120 possible relay teams.

Learning the Concept

An arrangement or listing in which order is important is called a **permutation**. For the example above, the symbol $P(5, 4)$ represents the number of permutations of 5 things taken 4 at a time.

$$P(5, 4) = 5 \cdot 4 \cdot 3 \cdot 2$$

If one of the five Clearwater swimmers became ill and was unable to swim, the number of relay teams can be expressed as $P(4, 4)$.

$$P(4, 4) = 4 \cdot 3 \cdot 2 \cdot 1$$

The product $4 \cdot 3 \cdot 2 \cdot 1$ can be written using the mathematical notation $4!$. The symbol $4!$ is read *four* ***factorial***. The expression $n!$ means the product of all counting numbers beginning with n and counting backward to 1. We define $0!$ as 1.

Example  Find each value.

a. *P*(8, 3)

$P(8, 3)$ represents the number of permutations or orders of 8 things taken 3 at a time.

$$P(8, 3) = \underbrace{\textit{choices for 1st position}} \quad \underbrace{\textit{choices for 2nd position}} \quad \underbrace{\textit{choices for 3rd position}}$$

$$= 8 \cdot 7 \cdot 6$$

$$= 336$$

b. 6!

$6! = 6 \cdot 5 \cdot 4 \cdot 3 \cdot 2 \cdot 1$

$= 720$

Sometimes order is not important. For example, *pepperoni, mushrooms, onions* is the same as *onions, pepperoni, mushrooms* when you buy a pizza.

Arrangements or listings where order is not important are called **combinations**. To find the number of different three-topping pizzas when there are eight toppings to choose from, divide the number of permutations, $P(8, 3)$, by the number of different ways three things can be ordered. There are 3! or $3 \cdot 2 \cdot 1$ ways to order three toppings. The symbol $C(8, 3)$ means the number of combinations of 8 things taken 3 at a time.

$$C(8, 3) = \frac{P(8, 3)}{3!} = \frac{8 \cdot 7 \cdot 6}{3 \cdot 2 \cdot 1} \text{ or } 56$$

It is possible to make 56 different pizzas with three toppings when there are eight toppings to choose from.

Example 

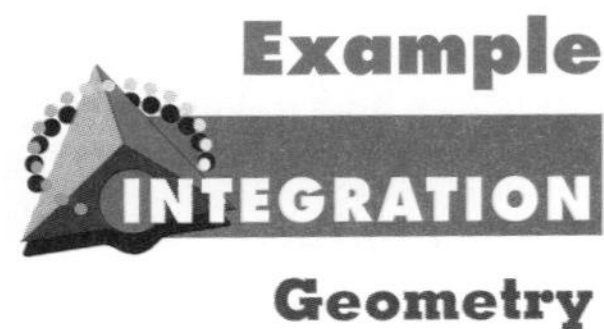

Geometry

Polygons can be drawn with any number of sides and vertices. Find the number of line segments that can be drawn between two vertices of a hexagon.

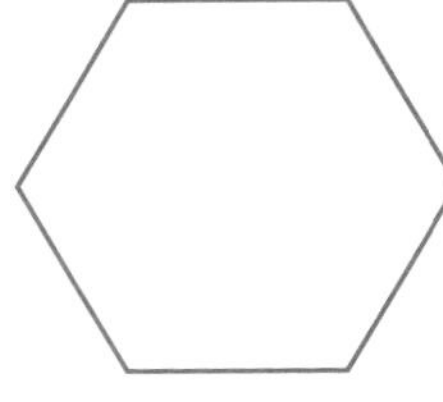

Explore A hexagon has 6 vertices.

Plan Each line segment connects two vertices, so you must find the combination of 6 vertices taken 2 at a time.

Solve

$$C(6, 2) = \frac{P(6, 2)}{2!}$$

$$= \frac{6 \cdot 5}{2 \cdot 1} \text{ or } 15$$

There are 15 line segments that can be drawn.

Examine Draw a hexagon and all the line segments and see if there are 15. Be sure to count the sides of the hexagon.

Checking Your Understanding

Communicating Mathematics

Read and study the lesson to answer these questions.

1. **Explain** what $P(4, 3)$ means.
2. **Compare** $6 \cdot 5 \cdot 4$ and $6!$.
3. **Write** an expression to represent the number of five-person basketball teams that could be formed from all the students in your math class.

4. Write a short summary explaining the difference between a permutation and a combination of 4 things taken 3 at a time. Draw models to illustrate your written explanation.

Guided Practice

Tell whether each situation represents a *permutation* or *combination*.

5. six people remaining in a game of musical chairs
6. first, second, third place, and honorable mention awards of six students who are finalists in a science fair

How many ways can the letters of each word be arranged?

7. WE
8. STUDY
9. MATH

Find each value.

10. $P(6, 3)$
11. $5!$
12. $C(5, 4)$
13. $\frac{7!3!}{5!0!}$
14. How many combinations of 6 different flowers can you choose from one dozen different flowers?
15. **Music** The Riverside High School Orchestra has been invited to play three pieces at the opening of the Summer Music Festival. The orchestra has eight pieces that they could play. How many different programs can they play?

Exercises: Practicing and Applying the Concept

Independent Practice

Tell whether each situation represents a *permutation* or *combination*.

16. four books in a row
17. six CDs from a group of twenty
18. a team of 6 players from 12
19. ten people in a line to buy tickets
20. position of students at the twelve computers in the computer center
21. ten wooden giraffes placed on display from a large collection

How many ways can the letters of each word be arranged?

22. SHE
23. BOUGHT
24. SANDWICH

Find each value.

25. $P(6, 6)$
26. 7!
27. 9!
28. $C(6, 6)$
29. 11!
30. $P(5, 1)$
31. $C(7, 3)$
32. $P(8, 4)$
33. 0!
34. $C(8, 4)$
35. $C(10, 2)$
36. $P(12, 12)$
37. $\frac{7!3!}{5!1!}$
38. $\frac{10!0!}{6!2!}$
39. $\frac{5!2!}{4!3!}$
40. $\frac{9!5!0!}{11!}$

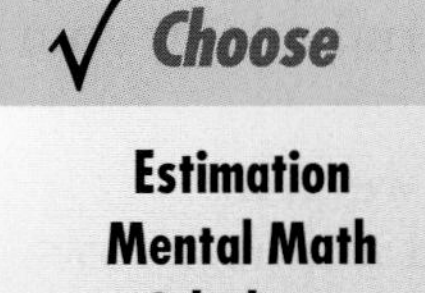

Estimation
Mental Math
Calculator
Paper and Pencil

41. How many different 6-player starting squads can be formed from a volleyball team of 15 players?
42. How many 3-digit numbers can you write using only the digits 1, 2, and 3 exactly once in each number?
43. How many ways can 4 members of a family be seated in a theater if the mother is seated on the aisle?
44. How many different 5-card hands is it possible to deal from a standard deck of 52 cards?

Critical Thinking

45. There are 720 ways for three students to win first, second, and third place in a debating match. How many students were competing?
46. Ten people are running for chair and vice-chair of the Social Studies Club. After they are selected, three directors will be elected from the remaining candidates. How many different ways can the offices be filled?

Applications and Problem Solving

47. **United Nations** The United States has brought a procedural measure to the United Nations Security Council for a vote. Nine votes are necessary for the measure to pass. There are 5 permanent members (China, France, Russian Federation, United Kingdom, and the United States) and 10 nonpermanent members on the Security Council. How many combinations of votes are possible if the measure passes with 9 votes?
48. **Geometry** Twelve points are marked on a circle. How many different line segments can be drawn between any two of the points?

EARTH WATCH

THE NEW GOLD RUSH

You have probably studied the gold rush that occurred in the 1840s and 1850s in the western United States. The news of an easy life of riches made people from all over the world move to the west in search of gold. This westward movement made the United States spread from coast to coast.

But did you know there is a new gold rush occurring in the western United States right now? A new mining process is prompting entrepreneurs to go in search of gold once again.

In the first gold rush, the process of mining was a physical process that was done on a small scale by individual miners. Today, large mining companies use a chemical process called cyanide heap-leaching. This process requires removing millions of tons of rock from the land. Then the rock is pulverized and piled into mounds. For months, the rock is sprinkled with a cyanide solution that draws the gold out of the rock fragments.

49. License Plates North Carolina issues general license plates with three letters followed by four numerals. (The first numeral cannot be 0.) Numerals can repeat, but letters cannot. How many license plates can North Carolina generate with this format?

Mixed Review

50. Draw a tree diagram to find the number of outcomes for rolling two identical number cubes. (Lesson 10-5)

51. A temperature of 32° Fahrenheit corresponds to a temperature of 0° Celsius. A temperature of 100° Celsius corresponds to a temperature of 212° Fahrenheit. About what temperature in degrees Celsius corresponds to a temperature of 0° Fahrenheit? (Lesson 8-5)

52. Geometry The perimeter of a rectangle is 38 meters. Find the dimensions if the length is 5 meters less than twice the width. (Lesson 7-5)

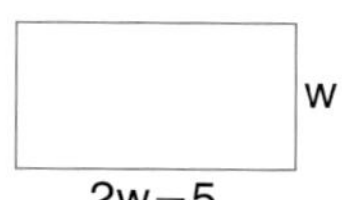

53. Solve $5.4c = -13.5$. Check your solution. (Lesson 6-7)

54. Estimate $1\frac{3}{5} - \frac{5}{12}$. (Lesson 5-2)

55. Evaluate $5x^{-2}$ if $x = 2$. (Lesson 4-9)

56. Travel Juan and his friends are traveling 120 miles to attend a concert. (Lesson 3-4)

a. If they average 50 mph, how long will the trip take?

b. If road construction on the trip limits their average speed to 40 mph, how much longer will the trip take?

57. Simplify $3x - 12x$. (Lesson 2-5)

58. Name the property shown by the statement $(3 + 6) + 8 = 3 + (6 + 8)$. (Lesson 1-4)

See for Yourself

- Many of the residents of Colorado and other western states are opposed to the new method of gold mining. They say that the cyanide heap-leaching process frees poisonous heavy metals such as lead, mercury, and arsenic. These metals may then escape into streams and groundwater. Research the effects of these metals on waterways. What are the possible dangers to humans?
- In the United States, the mining of hard-rock minerals such as gold, copper, and silver is regulated by the General Mining Act that was passed in 1892. The act exempts miners from paying royalties to the government and has no provisions for environmental protection or restoration. Investigate the proposals being made to Congress that would update the laws concerning mining.
- Write a letter to your congressional representative giving your opinion of the situation. Include reasons for what you believe and what, if anything, you think should be done.

10-7 Odds

Setting Goals: *In this lesson, you'll find the odds of a simple event.*

Modeling with Manipulatives

MATERIALS

- graph paper
- straight pin

On a piece of graph paper, copy the diagram shown at the right. Shade squares gray or black, as shown.

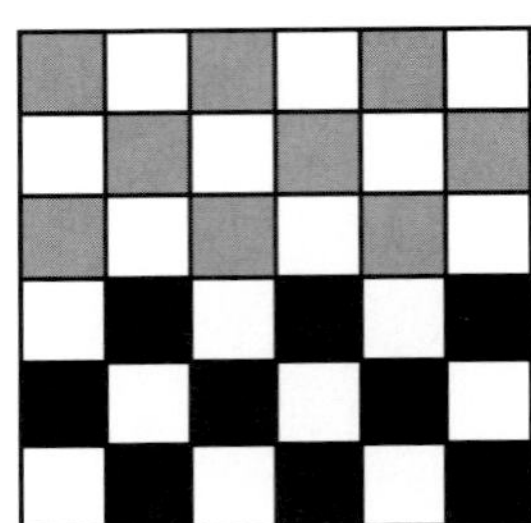

Your Turn

► Hold a straight pin about six inches above the center of the diagram. Drop the pin 50 times and record the color of the square where the point of the pin lands. If you miss the diagram, drop the pin again.

a. Would you say that the pin is more likely to land on a white square or a shaded square? Explain?

b. How many ways can the point of the pin land on a black square?

c. How many ways can the point of the pin not land on a gray square?

Learning the Concept

The probability that the point of the pin will land on a gray square is $\frac{9}{36}$ or $\frac{1}{4}$. The probability that the point of the pin will *not* land in a gray square is $\frac{27}{36}$ or $\frac{3}{4}$.

Another way to describe the chance of an event's occurring is with **odds**. The odds in favor of an event is the ratio that compares the number of ways the event can occur to the ways that the event *cannot* occur.

Definition of Odds

The odds in favor of an outcome is the ratio of the number of ways the outcome can occur to the number of ways the outcome cannot occur.

Odds in favor = number of successes: number of failures

The odds against an outcome is the ratio of the number of ways the outcome cannot occur to the number of ways the outcome can occur.

Odds against = number of failures: number of successes

Example 1 **A bag contains 5 green marbles, 2 yellow marbles, and 3 blue marbles. What are the odds of drawing a green marble from the bag?**

There are 10 − 5 or 5 marbles that are not green.

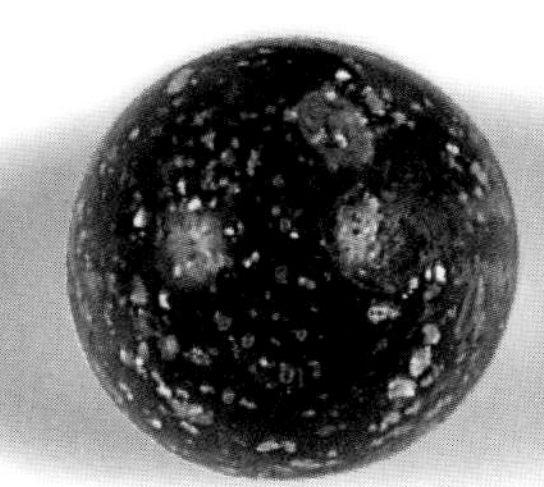

Odds of drawing a green marble

$$= \underbrace{\begin{matrix}\textit{number of ways of}\\ \textit{drawing green marble}\end{matrix}}_{5} : \underbrace{\begin{matrix}\textit{number of ways of}\\ \textit{drawing other marbles}\end{matrix}}_{5}$$

$= 1:1$ *This is read "1 to 1."*

The odds of drawing a green marble are 1:1.

Sometimes when finding odds, you must first find the total number of possible outcomes. This can involve finding permutations or combinations.

Example 2

APPLICATION

Finance

The state of Florida has a *Fantasy 5* drawing three times a week in which 5 numbers out of 39 are drawn at random. The proceeds from the lottery help to finance education in the state. What are the odds of winning the *Fantasy 5* jackpot?

Order is not important. You can find the odds using combinations.

There is only one winning combination, so the number of ways to succeed is 1. There are $C(39, 5)$ ways to draw 5 numbers from a group of 39. So the total number of possible outcomes is $C(39, 5)$ or 575,757. This means that the number of failures (not drawing all 5 numbers) must be 575,757 − 1 or 575,756.

odds of drawing 5 correct numbers
= number of successes : number of failures
= 1 : 575,756 *This is read "1 to 575,756."*

The odds of winning the *Fantasy 5* jackpot are 1: 575,756.

Checking Your Understanding

Communicating Mathematics

Read and study the lesson to answer these questions.

1. **Explain** how to find the odds of an event occurring.
2. **You Decide** Ade says that the probability of getting two in one roll of a die is 1 out of 6. Sharla says that the odds of getting two in one roll of a die are 1:5. Who is correct and why?
3. Describe a situation in real life that uses odds.

Guided Practice

Find the odds of each outcome if a die is rolled.

4. a number greater than 5
5. a multiple of 2
6. a number less than 4
7. not a 3

8. **Statistics** The United States Census Bureau reported that the number of unmarried adults nearly doubled from 1970 to 1993. The report stated that 58% of unmarried adults had never been married.
   a. What are the odds that an unmarried adult was never married?
   b. In a group of 200 unmarried adults, how many would you estimate had never been married?

Exercises: Practicing and Applying the Concept

Independent Practice

Find the odds of each outcome if the spinner at the right is spun.

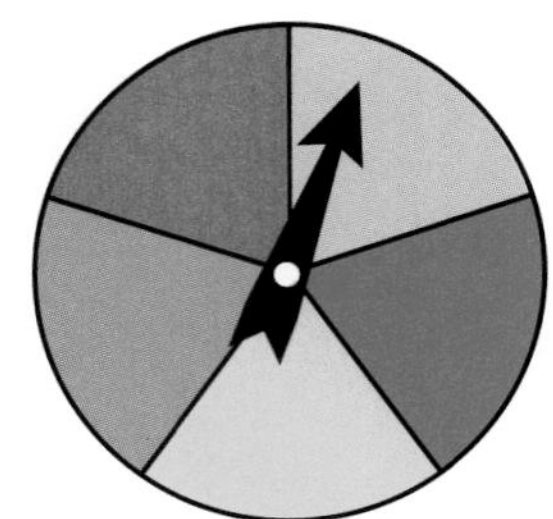

9. blue
10. purple
11. not green
12. red or orange

Find the odds of each outcome if a pair of fair dice are rolled.

13. an even sum
14. an odd sum
15. a sum that is a multiple of 3
16. a sum less than 2
17. a composite number
18. a prime number greater than 9
19. *not* a sum of 7
20. *not* a sum of 11 or 12
21. a sum of 8 with a 6 on one die
22. an even sum or a sum less than 6
23. *neither* an odd sum *nor* a sum greater than 8

Five cards are selected from a standard deck of 52 playing cards.

24. What are the odds of selecting a black king?
25. What are the odds of *not* selecting a heart *or* a seven?

Critical Thinking

26. A red die and a blue die are rolled. What are the odds that the number showing on the red die is less than the number showing on the blue die?

Applications and Problem Solving

27. **Entertainment** The Durham Bulls baseball team in Durham, North Carolina, invited the community to submit promotion ideas for the opening of their new stadium. One suggestion was that the team call out the last four-digit numbers of a telephone number and award a prize to the first person with those digits in their telephone number to reach the press box. Find the odds that the four digits called out are the last four digits of your telephone number.
28. **Television** The odds in favor of a person in North America appearing on television sometime in their lifetime is 1:3. If there are 32 students in your class, predict how many will appear on television.

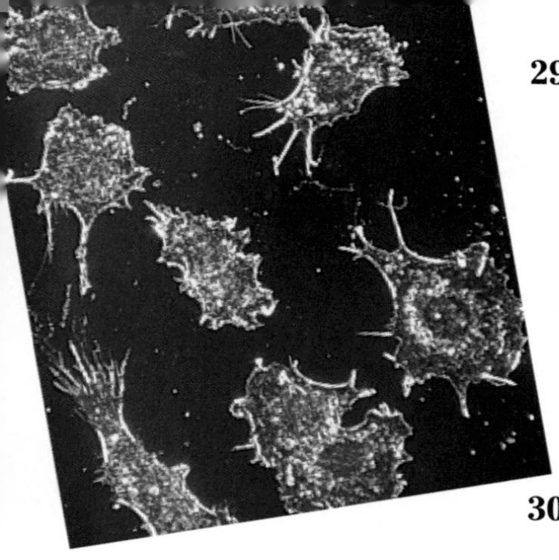

29. **Medicine** A worker at an aluminum plant in Washington state filed a workers' compensation suit against his employer because he developed cancer. He claimed that 8 of 90 workers in his shop had developed the same type of cancer. The medical odds of a person developing this type of cancer is normally 1:1000.
    a. Based on the medical odds, how many people in his shop should have developed the cancer?
    b. Based on the number of workers and the number of cases of cancer in this shop, estimate the number of people who would develop this cancer if 1000 people worked in this shop.
30. **Family Activity** During two hours of watching television or listening to the radio with your family, keep track of the commercials. Record the categories, such as food, cars, cosmetics, and so on, of each commercial break. Find the odds that a randomly-selected commercial is from each category.

Mixed Review

31. Find the value of 8!. (Lesson 10-6)
32. **Sports** The heights (in inches) of members of a volleyball team are 64, 60, 72, 61, 73, 80, 68, 70, 65, 67, 70, and 80. (Lesson 10-1)
    a. Construct a stem-and-leaf plot of the data.
    b. What percent of the heights are between 60 and 70 inches? Why do you think that is true?
33. A penny, a nickel, a dime, and 2 quarters are in a purse. Without looking, Andie picks out three coins. What different amounts of money could she have chosen? (Lesson 9-2)
34. **Geometry** Find the circumference of the circle shown at the right. (Lesson 7-4)

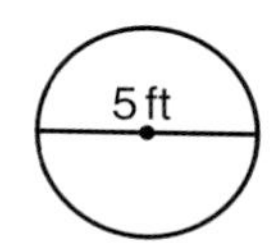

35. Replace ● with $<$, $>$, or $=$ to make $\frac{1}{6}$ ● 0.16 a true statement. Use a number line if necessary. (Lesson 6-1)
36. State whether the inequality $c + 5 < 16$ is *true*, *false*, or *open*. (Lesson 1-9)

From the FUNNY PAPERS

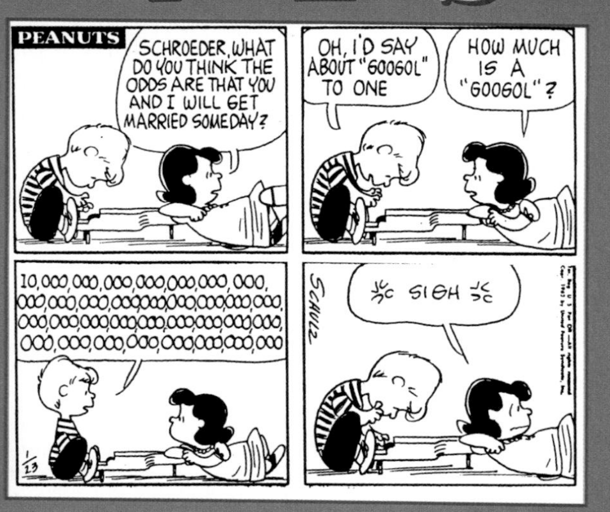

1. Express a googol in scientific notation.
2. Do you think the odds are expressed correctly here? Why or why not?
3. Research who defined the term *googol*.

10-8 Problem-Solving Strategy: Use a Simulation

Setting Goals: *In this lesson, you'll investigate problems using simulation.*

Modeling with Manipulatives

MATERIALS

8 pennies

cup

The Paiute ('pi- (y)ut) Indians of Pyramid Lake in northwestern Nevada invented the Walnut Shell Game. This game was played with 8 walnut shell halves filled with pitch and powdered charcoal. A player tosses a basket tray containing the walnut shell halves and scores one point if three or five of the walnut shells land with the flat side up. No point is scored for any other combination. The winner is the first player to reach an agreed-on sum. On average, how many times would you have to toss the basket before you get 10 points?

Your Turn Work with a partner to simulate the walnut shell game.

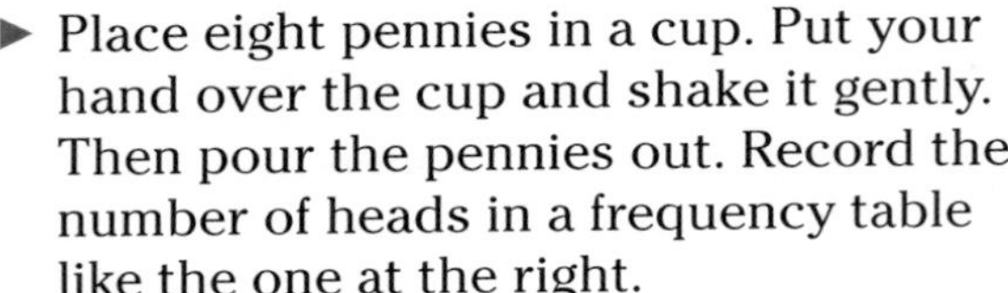

- ▶ Place eight pennies in a cup. Put your hand over the cup and shake it gently. Then pour the pennies out. Record the number of heads in a frequency table like the one at the right.
- ▶ Repeat the process until the sum of the frequencies for three heads and five heads is 10.
- ▶ Compare your results to other pairs of students.

Outcome	Tally	Frequency
0 heads		
8 heads		

CULTURAL CONNECTIONS
Sarah Winnemucca, a member of the Paiute tribe, is famous for her lectures on the U.S. government's treatment of her people.

a. Do you think this process is a reasonable approximation of the walnut shell game? Explain.

b. What is an advantage of finding an answer by acting it out?

Learning the Concept

A simulation is an application of the Acting It Out problem-solving strategy.

When you solve a problem by modeling a situation, like you did in the activity above, you are doing a **simulation**. A simulation acts out the event so that you can see outcomes.

You can conduct a simulation of the outcomes of many problems by using manipulatives such as dice, a coin, or a spinner. The manipulative, such as a die, a coin, a spinner, or a combination of them, should have the same number of outcomes as the number of possible outcomes.

Real-World Event	Simulation
having a baby boy or baby girl	flip a coin
history of making 2 free throws for every 3 attempted	spinning a 3-section spinner with 2 sections the same color
win one of twelve prizes	roll a die and flip a coin

Example 1 APPLICATION Education

A quiz has 10 true-false questions. The correct answers are T, F, F, T, F, F, T, T, T, F. You need to correctly answer 7 or more questions to pass the quiz. Is tossing a coin to decide the answers a good strategy for taking the quiz?

Explore The quiz has ten questions. Since two choices are available for each answer, tossing a coin is used to simulate guessing the answers.

Plan Toss a coin and record the answer for each question. Write T (true) if tails shows and write F (false) if heads shows. Repeat the simulation three times. Shade the cells with the correct answers.

Solve

Answers	T	F	F	T	F	F	T	T	T	F	Number Correct
Simulation 1	F	T	F	F	T	F	F	F	T	F	4
Simulation 2	F	T	F	F	F	T	T	F	T	F	5
Simulation 3	F	F	T	T	T	F	T	T	T	T	6

Since none of the simulations results in a passing grade, this is probably not a good way to take the quiz.

Examine Try some more simulations to confirm the results.

Sometimes it is necessary to create a manipulative for a simulation.

Example 2 APPLICATION Sports

The Cougars are one point behind in their basketball game. Carmen is fouled as time runs out. If she misses the foul shot, the Cougars lose. If Carmen makes the foul shot, the score is tied and she gets another shot. If she makes the first and second shots, the Cougars win. Carmen has a record of making three out of every four free throws.

a. Conduct a simulation for 25 trips to the foul line.

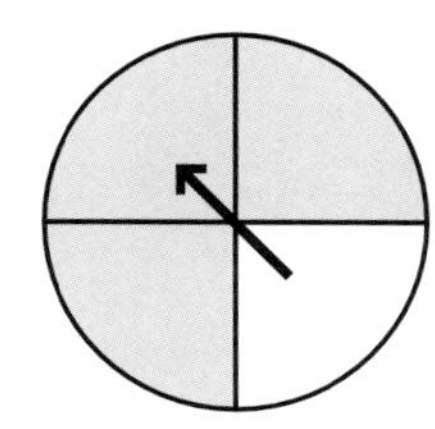

Construct a spinner with four equal sections like the one shown at the right. Spin the spinner and record the results of several trips to the foul line. A sample simulation for 25 trips to the foul line is shown in the chart below.

Misses the first shot (0 points)	Makes the first shot, misses the second (1 point)	Makes both shots (2 points)
IIII	IIII	𝍸 𝍸 𝍸 II

b. What is the probability that Carmen makes both shots and the Cougars win the game?

The experiment shows that Carmen will make 2 points $\frac{17}{25}$ or 68% of the time. Based on this simulation, the probability that Carmen scores 2 points and the Cougars win the game is 68%.

Checking Your Understanding

Communicating Mathematics

Read and study the lesson to answer these questions.

1. **Explain** what a simulation is.
2. **Tell** whether the results of a simulation will be exactly the same as the theoretical probability. Explain your reasoning.
3. **Write** an explanation of how you would modify the spinner in Example 2 to simulate a player who has a record of making one basket out of every three tries.

Guided Practice

4. Using a device other than a coin, conduct three more simulations for the problem in Example 1. Do your simulations verify the results obtained while using a coin?
5. **Football** Faud Reveiz is a placekicker for the NFL's Minnesota Vikings. In 1994, he made 80% of the field goals he attempted when the distance was between 40 and 49 yards.
   a. How would you simulate his next four field goal attempts between 40 and 49 yards?
   b. Conduct a simulation for 15 field goal attempts between 40- and 49-yards long.
   c. What is the probability that his next two field goal attempts between 40 and 49 yards will be successful?

Exercises: Practicing and Applying the Concept

Independent Practice

Solve. Use a simulation to act out the problem.

6. A fast-food restaurant includes prizes with children's meals. During the summer promotional campaign, six different prizes were available. There is an equally likely chance of getting each prize each time.
   a. Use a die to simulate this problem. Let each number represent one of the prizes. Conduct a simulation until you have one of each number.
   b. According to your simulation, how many children's meals must be purchased in order to get all six different prizes?
7. **Business** Mr. Namura runs a small gourmet restaurant. He has 12 tables. It is a popular restaurant and anyone wanting to eat at 7:00 on Saturday evening must have reservations. Mr. Namura knows that one out of six reservations usually does not show, so he takes reservations for 14 tables for 7:00.
   a. Describe a way to simulate the number of tables that will be filled at 7:00.
   b. Conduct your simulation for part a ten times.
   c. From your simulations, how many times is the restaurant overbooked? Are all 12 tables always filled?

Choose

Estimation
Mental Math
Calculator
Paper and Pencil

Solve. Use any strategy.

8. Unit 2 of a science book starts on page 126 and ends on page 241. How many pages are in the unit?
9. Antonio received a birthday gift of money from his grandmother. He loaned one dollar to his younger brother and spent half of the remaining money. The next day he earned \$6. Later, he lost $\frac{1}{6}$ of his money, but he still had \$15 left. How much money did his grandmother send him?
10. There are three traffic lights along the route that Yoshie rides from school to her home. The probability that any one of the lights is green is 0.4, and the probability that it is not green is 0.6. If the three lights operate independently, estimate the probability that these lights will all be green on Yoshie's way home from school.
11. **Hockey** There are 16 hockey teams in a single elimination tournament.
    a. How many games will be played during the tournament?
    b. How many ways can the teams win first and second place?
12. The diagram below shows a system of power lines that bring electricity from a power plant to three towns. Recently it has been very windy. For the last 10 days, each line has been knocked down about half the time due to the high winds. What is the probability that Altoona will still be able to get power on any given day even if some of the lines are down?

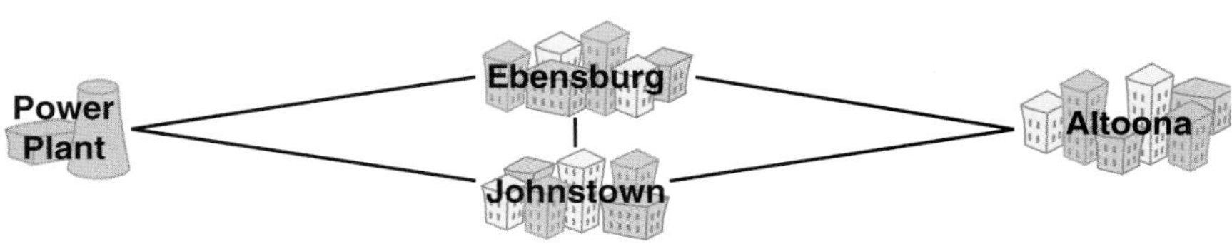

Graphing Calculator

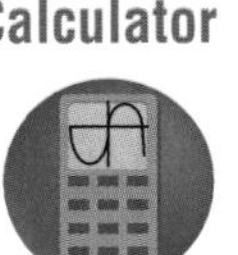

13. When you have a situation involving numbers that cannot be generated from an object like a coin or a spinner, you can use the random number generator function on a computer or a graphing calculator. A random number generator produces numbers that have no pattern.

    The graphing calculator program below will generate 20 random numbers between 1 and the number you enter for T. In order to use the program, you must first enter the program into the calculator's memory. To access the program memory, use the following keystrokes.

    Enter: PRGM ▶ ▶ ENTER

    Run the program. Press PRGM and choose the program from the list by pressing the number next to its name.

    Press the enter key twenty times. The calculator will display a number from 1 to 6 each time you press the enter key.

```
PROGRAM: RANDOM
: Input T
: For(N,1,20)
: int((T*rand) + 1) →D
: Disp D
: Pause
: End
```

    a. Run the program to simulate 20 spins of a spinner that has 7 equal sections.
    b. How would you alter the program to generate 50 random numbers?

Critical Thinking

14. In a survey of 240 students, 100 students said they play soccer, and 120 students said they play basketball. If 40 students play both sports, how many students do not play either basketball or soccer?

Mixed Review

15. Probability Find the odds of stopping on 2 if the spinner at the right is spun. (Lesson 10-7)

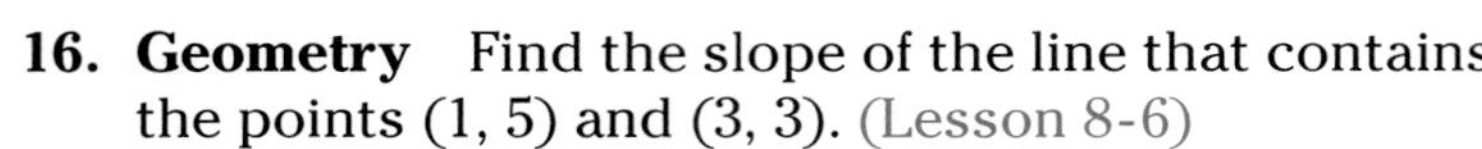

16. Geometry Find the slope of the line that contains the points (1, 5) and (3, 3). (Lesson 8-6)

17. Sales Jeff bought a used 10-speed bike for \$20 more than one-half its original price. He paid \$110 for the bike. (Lesson 7-3)

a. Write an equation that represents this situation.

b. What was the original price of the bike?

18. Write r^4 as the product of the same factor. (Lesson 4-2)

19. Meteorology The temperature outside was 20°F. The windchill factor made it feel like -15°F. Find the difference between the real temperature and the apparent temperature. (Lesson 2-5)

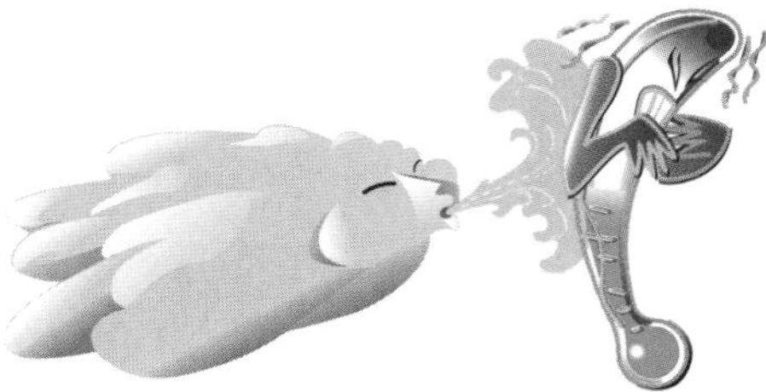

Refer to the Investigation on pages 482–483.

Look over the list of favorite games that you placed in your Investigation Folder at the beginning of this Investigation. Some people may have said that their favorite games are TV game shows, such as *The Price is Right.*

For this part of the Investigation, your group has been chosen to test a new TV game show called *Don't Lose It All!* This game is for two players. Each player takes a turn spinning two identical wheels at the same time. Each wheel is divided into six equal regions. Five of the regions say "Win \$25" and the sixth region says "Lose Your Cash". The object of the game is to accumulate the most money by the end of the show.

A player's turn ends if one or both of the wheels stop on "Lose Your Cash" or the player decides to let his or her opponent spin. On a spin, one of three things can occur.

1. Both wheels stop on "Win \$25" and the player is awarded \$50.
2. One wheel stops on "Lose Your Cash," and one stops on "Win \$25." In this case, a player loses any money made on this turn but keeps money from any previous turns.
3. Both wheels stop on "Lose Your Cash" and the player loses all the money won so far.

- Create two spinners to simulate the wheels in the game. Play the game enough times to develop some different strategies. Have each member of your group use a different strategy and play the game several more times to determine the best strategy.
- Determine a way to simulate the game your group is designing.

Add the results of your work to your Investigation Folder.

10-8B HANDS-ON ACTIVITY Making Predictions

An Extension of Lesson **10-8**

MATERIALS

- paper bag
- 10 marbles of 3 different colors

In this activity, you will conduct and interpret probability experiments. Probability calculated by making observations or experiments is called **experimental probability**. Experimental probability is an estimate based on the **relative frequency**. Relative frequency is the number of times a certain event actually happened.

Activity  Work with a partner.

- Get a bag of marbles from your teacher. You will try to determine how many marbles there are of each color.
- Draw one marble from the bag, record its color, and replace it in the bag. Repeat this process 10 times.
- Determine the relative frequency for each color marble.

1. Based on the relative frequencies, which color marble do you think is most prevalent in the bag? Predict the colors of the marbles in your bag.
2. Based on your experiment, what is the experimental probability of drawing each color marble?

$$\text{experimental probability} = \frac{\text{relative frequency of color}}{\text{total number of draws}}$$

Activity  Repeat both steps above for 20, 30, 40, and 50 draws.

TALK ABOUT IT

3. Is it possible to have a certain color marble in the bag and never draw that color? Is this situation more likely to happen if you make two draws, ten draws, or fifty draws? Explain.
4. Compute the experimental probability of drawing each color for each number of draws. Write a paragraph that describes how the experimental probability changed as you increased the number of draws.
5. Predict the colors of the marbles in your bag. Did your prediction change from the prediction you made in Exercise 1? Explain why or why not. Open the bag and check your prediction against the marbles.

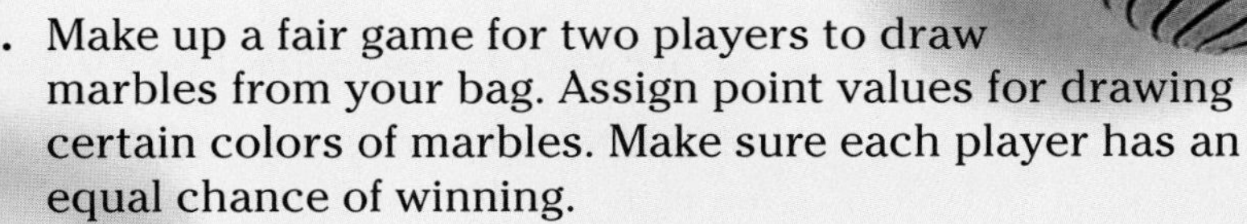

6. Make up a fair game for two players to draw marbles from your bag. Assign point values for drawing certain colors of marbles. Make sure each player has an equal chance of winning.

10-9 Probability of Independent and Dependent Events

Setting Goals: *In this lesson, you'll find the probability of independent and dependent events.*

Modeling with Technology

The television meteorologist said that there was a 50% chance of rain on Saturday and a 25% chance of rain on Sunday. Janice was upset because the ninth grade at her high school was planning their class picnic for Saturday and the rain date was Sunday. She thought that the probability for rain on Saturday *and* Sunday was 75%. Kiona said that the probability of rain on both days was just 12.5%.

You can use the TI-82 graphing calculator program at the right to simulate this situation. Recall that you must first enter the program into the calculator's memory. Access the program memory using the following keystrokes.

Enter: PRGM  ▶

The program generates random numbers for 50 weekends. The first use of the random number function simulates the probability of rain on Saturday. The second use of the random number function simulates the probability of rain on Sunday.

```
:PROGRAM:RAIN
:0→S: 0→R
:Disp "PROBABILITY",
 "OF RAIN ON"
:Input "SATURDAY", A
:Input "SUNDAY", B
:For (N,1,50)
:int ((100/A*rand)
 +1)→T
:If T=1
:S+1→S
:End
:For (M,1,50)
:int ((100/B*rand)
 +1)→G
:If G=1
:R+1→R
:End
:(S/50)*(R/50)→F
:F*100→F
:Disp "PERCENT OF
TIMES", "IT RAINS
BOTH", "DAYS IS", F
:Stop
```

Your Turn Run the program several times and record the percent of times that it rains on both days for each 50-weekend simulation.

a. Are your results closer to what Janice thought or to what Kiona said?

b. Run the program several times for a weekend when the probability of rain is 100% for Saturday and 20% for Sunday. What do you observe?

Learning the Concept

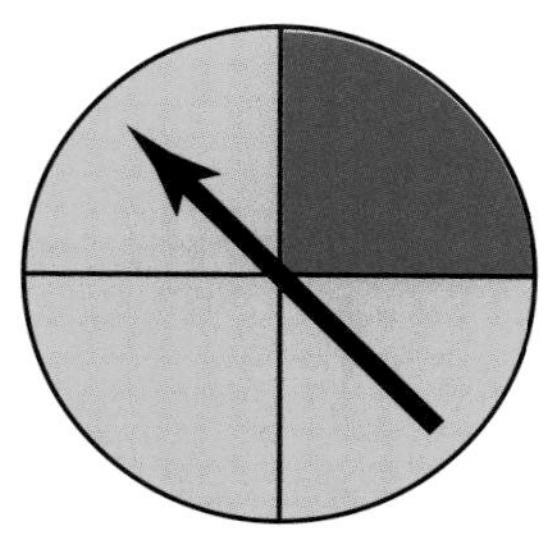

You could also simulate the situation above using manipulatives. This could be done by tossing a coin and spinning the spinner at the right 50 times each. Tossing a head has the same probability as the chance of rain on Saturday and spinning blue on the spinner has the same probability as the chance of rain on Sunday. Does the outcome of tossing the coin affect the outcome of spinning the spinner?

The weather situation in the activity above is an example of **independent events**. Events are independent when the outcome of one event does *not* influence the outcome of a second event.

Using the Fundamental Counting Principle, you know that there are $2 \cdot 4$ or 8 possible outcomes of tossing a coin and spinning the spinner. Of the 8 outcomes, there is one pair (heads, blue) that matches the meteorologist's prediction (50% chance of rain on Saturday, 25% chance of rain on Sunday). Determine the probability.

$$P(\text{heads, blue}) = \frac{\textit{number of ways heads and blue can occur}}{\textit{number of possible outcomes}}$$
$$= \frac{1}{8}$$

Note that $P(\text{heads})$ is $\frac{1}{2}$, $P(\text{blue})$ is $\frac{1}{4}$, and $\frac{1}{2} \cdot \frac{1}{4} = \frac{1}{8}$, which is $P(\text{heads, blue})$.

This suggests that the probability of two independent events is the product of the probability of each event.

Probability of Two Independent Events

In words: The probability of two independent events can be found by multiplying the probability of the first event by the probability of the second event.

In symbols: $P(A \text{ and } B) = P(A) \cdot P(B)$

Example 1 Application — Games

In the game of Monopoly, you go to jail if you roll three doubles in a row. What is the probability of rolling three doubles in a row?

The events are independent since each roll of the dice does not affect the outcome of the next roll.

There are six ways to roll doubles, (1, 1), (2, 2), and so on, and there are 36 ways to roll two dice. Thus, the probability of rolling doubles on a toss of the dice is $\frac{6}{36}$ or $\frac{1}{6}$.

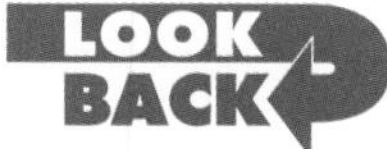

You can review the sample space of rolling two dice on page 441.

$$P\left(\begin{matrix}\textit{doubles on}\\ \textit{three rolls}\end{matrix}\right) = P\left(\begin{matrix}\textit{doubles on}\\ \textit{roll one}\end{matrix}\right) \cdot P\left(\begin{matrix}\textit{doubles on}\\ \textit{roll two}\end{matrix}\right) \cdot P\left(\begin{matrix}\textit{doubles on}\\ \textit{roll three}\end{matrix}\right)$$
$$= \frac{1}{6} \cdot \frac{1}{6} \cdot \frac{1}{6}$$
$$= \frac{1}{216}$$

The probability of rolling all doubles in three rolls of the dice is $\frac{1}{216}$.

Sometimes the outcome of one event affects the outcome of a second event. When this happens, the events are called **dependent events**. For example, suppose you choose a card from a deck, keep it, and then your friend chooses a card. Since you did not replace your card, your friend has fewer cards from which to choose. These events are dependent.

Probability of Two Dependent Events

In words: If two events, *A* and *B*, are dependent, then the probability of both events occurring is the product of the probability of *A* and the probability of *B* after *A* occurs.

In symbols: $P(A \text{ and } B) = P(A) \cdot P(B \text{ following } A)$

Example  **Bag lunches are prepared for 50 students on a field trip. There are three different desserts in the bags. Twenty bags contain a brownie, 15 contain a chocolate chip cookie, and 15 contain an oatmeal raisin cookie. The bags are arranged in random order. Connie and Tale were the first and second in line. What is the probability that Connie and Tale both select a bag containing a chocolate chip cookie?**

For Connie, $P(\text{chocolate chip}) = \dfrac{\textit{number containing chocolate chip}}{\textit{total number of lunches}}$

$= \dfrac{15}{50}$ or $\dfrac{3}{10}$

Assume that Connie chose a bag containing a chocolate chip cookie.

For Tale, $P(\text{chocolate chip}) = \dfrac{\textit{number containing chocolate chip left}}{\textit{number of lunches left}}$

$= \dfrac{14}{49}$ or $\dfrac{2}{7}$

$P(\text{chocolate chip cookies for both}) = \dfrac{3}{10} \cdot \dfrac{2}{7}$

$= \dfrac{6}{70}$ or $\dfrac{3}{35}$

The probability that both students get a chocolate chip cookie is $\dfrac{3}{35}$.

Checking Your Understanding

Communicating Mathematics

Read and study the lesson to answer these questions.

1. **Explain** the difference between independent and dependent events.
2. **Write** an example of two independent and two dependent events. List each sample space.
3. **Describe** how to find the probability of the second of two dependent events.

Guided Practice

Determine whether the events are independent or dependent. Explain.

4. rolling a die and then spinning a spinner
5. drawing a marble from a bag and then drawing a second marble without replacing the first one

The chart below lists the number and type of two soft drinks found in a tub filled with ice. Two cans are selected at random. Find the probability of each outcome.

Soda	Regular	Diet
Coca-Cola	3	4
Mountain Dew	7	6

6. a Diet Coke and a regular Mountain Dew
7. a regular Coke and a regular Mountain Dew
8. a diet Mountain Dew and a regular Mountain Dew
9. a regular Coke and a diet Mountain Dew
10. Do the probabilities in Exercises 6–9 represent dependent or independent events? Explain.
11. **Economics** In 1993, 44% of full-time, hourly wage earners in the United States were women. If 4% of these women earn the minimum wage, what is the probability that a full-time, hourly wage earner chosen at random for a survey is a woman who is paid the minimum wage?

Independent Practice

Determine whether the events are independent or dependent. Explain.

12. choosing a card from a deck of playing cards and then choosing a second card without replacing the first card
13. rolling a die twice
14. selecting a name from the Raleigh telephone book and a name from the Miami telephone book
15. tossing a coin and spinning a spinner
16. as team captain, selecting someone to be on your basketball team and then selecting a second player

A die is rolled and the spinner is spun. Find each probability.

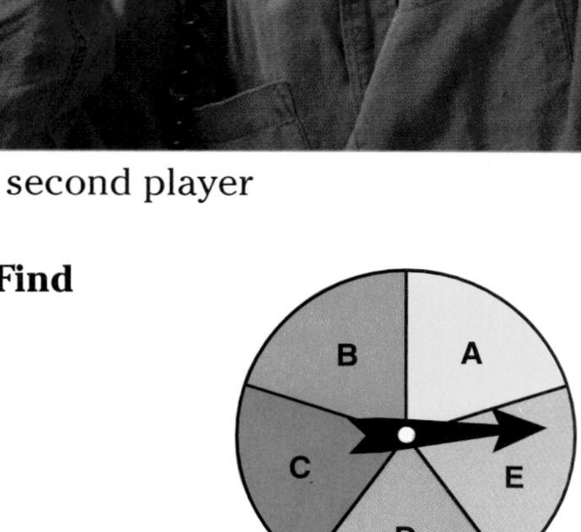

17. P(1 and D)
18. P(an odd number and C)
19. P(a composite number and a vowel)
20. P(an even number and a consonant)
21. P(a 6 and F)

In a bag there are 5 red marbles, 2 green marbles, 4 yellow marbles, and 3 blue marbles. Once a marble is drawn, it is not replaced. Find the probability of each outcome.

22. two green marbles in a row

23. a red and then a yellow marble

24. two red marbles in a row

25. a green and then a blue marble

26. a red marble, a green marble, and then a blue marble

27. three yellow marbles in a row

Critical Thinking

28. There are 9 marbles in a bag. Some of the marbles are red, some are white, and some are blue. The probability of selecting a red marble, a white marble, and then a blue marble is $\frac{1}{21}$.

a. How many of each color are in the bag?

b. Explain why there is more than one correct answer to part a.

Applications and Problem Solving

Choose

Estimation
Mental Math
Calculator
Paper and Pencil

29. TV Game Shows On *The Price is Right*, contestants can win a car if they draw the digits in the price of the car from a bag that contains the five digits in the price of the car and three strikes. Once a number or strike is drawn, it is not replaced.

a. What is the probability that the first draw is a strike?

b. A contestant has drawn three digits in the price of the car without drawing a strike. What is the probability that the next two draws will be strikes?

30. Ambulance Service Mercy Hospital has two ambulances that are each available for emergencies 80% of the time.

a. In an emergency situation, what is the probability that both ambulances will be available?

b. What is the probability that neither will be available?

c. Explain why the probabilities for parts a and b do not have a sum of 1.

31. Aerospace Design The electronic telemetry system on a spacecraft has three motion sensors, a signal conditioner, and a transmitter. The probability of failure for each motion sensor and the signal conditioner is 0.0001, and the probability of failure for the transmitter is 0.001. The systems run independently. What is the probability of *success* for the telemetry system?

Mixed Review

32. Sports Colin kicks extra points for the football team. He makes 75% of his attempts for extra points. How could you simulate the results of the next six attempts? (Lesson 10-8)

33. Consumer Awareness Keandre is shopping for new tennis shoes. He finds a pair that are priced at $49.99 on a sale rack marked 20% off. Estimate the amount of savings and the sale price. (Lesson 9-8)

34. Evaluate $c - a$ if $a = \frac{2}{9}$ and $c = \frac{5}{9}$. Write the result in simplest form. (Lesson 5-4)

35. Solve $14 = b - (-12)$ and check your solution. Then graph the solution on a number line. (Lesson 3-2)

36. State whether the product of $-3(5)(-2)$ is positive or negative. Then find the product. (Lesson 2-7)

37. Write an expression for *twice the sum of a number and 3*. (Lesson 1-3)

10-10 Probability of Compound Events

Setting Goals: *In this lesson, you'll find the probability of mutually exclusive events or inclusive events.*

Modeling with Manipulatives

MATERIALS

colored pencils

Suppose you are playing a game of *Monopoly* with three friends. It's your turn and you need to roll a 6 or 10 to avoid landing on properties owned by the other players and having to pay them rent. What is the probability that you can avoid paying rent?

Your Turn ▶ The chart at the right is a partial list of the possible sums when two dice are rolled. Copy and complete the chart. Circle the combinations with a sum of 6 with a colored pencil. Then circle those with a sum of 10 using a second color.

Second Die (columns); First Die (rows)

+	1	2	3	4	5	6
1	2	3	4	5	6	7
2	3	4	5			
3	4	5				
4	5					
5	6					
6	7					

TALK ABOUT IT

a. What is the probability the sum of the dice will be 6?

b. What is the probability the sum of the dice will be 10?

c. What is the probability the sum of the dice will be 6 or 10?

d. What do you notice about the probabilities in parts a, b, and c?

Learning the Concept

When two events cannot happen at the same time, like rolling a 6 or a 10, they are said to be **mutually exclusive**. The probability that one of two mutually exclusive events will occur is the sum of their individual probabilities.

Probability of Mutually Exclusive Events

In words: The probability of one or the other of two mutually exclusive events can be found by adding the probability of the first event to the probability of the second event.

In symbols: $P(A \text{ or } B) = P(A) + P(B)$

Example 1

THINK ABOUT IT

What is the probability of two events both occurring when they are mutually exclusive?

The spinner at the right is spun. What is the probability that the spinner will stop on yellow or on 2?

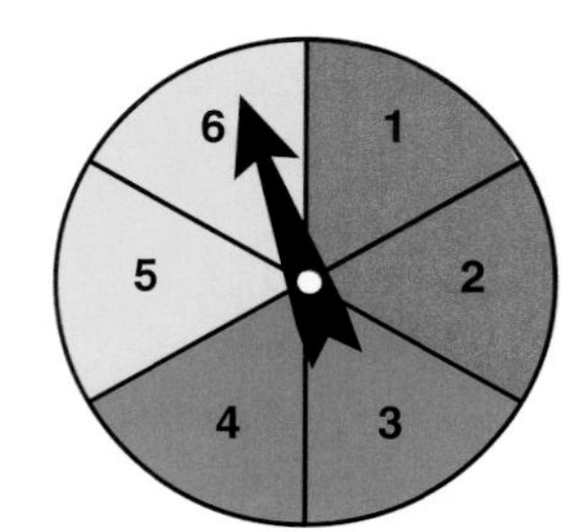

The events are mutually exclusive because the spinner cannot stop on both yellow and 2 at the same time.

$$P(\text{yellow or } 2) = P(\text{yellow}) + P(2)$$
$$= \frac{1}{3} + \frac{1}{6}$$
$$= \frac{3}{6} \text{ or } \frac{1}{2}$$

The probability that the spinner stops on yellow or 2 is $\frac{1}{2}$.

When two events are **inclusive**, they can happen at the same time. For example, from a standard deck of 52 playing cards, it is possible to draw a card that is *both* a queen and a red card. What is the probability of drawing a queen or a red card?

P(queen)	P(red card)	P(red queen)
$\frac{4}{52}$	$\frac{26}{52}$	$\frac{2}{52}$
1 queen in each suit	*13 hearts and 13 diamonds*	*queen of hearts and queen of diamonds*

The probability of drawing a red queen is counted twice, once for a queen and once for a red card. To find the correct probability, you must subtract P(red queen) from the sum of P(queen) and P(red card).

$$P(\text{queen or red card}) = \frac{4}{52} + \frac{26}{52} - \frac{2}{52} \text{ or } \frac{7}{13}$$

The probability of drawing a queen or a red card is $\frac{7}{13}$.

Probability of Inclusive Events

In words: The probability of one or the other of two inclusive events can be found by adding the probability of the first event to the probability of the second event and subtracting the probability of both events happening.

In symbols: $P(A \text{ or } B) = P(A) + P(B) - P(A \text{ and } B)$

Example 2

Business

An auto dealer finds that of the cars coming in for service, 70% need a tune up, 50% need a new air filter, and 35% need both. What is the probability that a car brought in for service needs either a tune-up or a new air filter?

Since it is possible to get a tune-up *and* a new air filter, these events are inclusive.

$P(\text{tune-up}) = \frac{70}{100}$ $\quad$ $P(\text{air filter}) = \frac{50}{100}$

$P(\text{tune-up and air filter}) = \frac{35}{100}$

$$P(\text{tune-up or air filter}) = \frac{70}{100} + \frac{50}{100} - \frac{35}{100} \text{ or } \frac{85}{100}$$

The probability that a car needs a tune-up or air filter is $\frac{85}{100}$ or 85%.

Checking Your Understanding

Communicating Mathematics

Read and study the lesson to answer these questions.

1. **Write** an example of two mutually exclusive events in your own life.
2. **Describe** the difference between mutually exclusive and inclusive events.
3. **Explain** why you must subtract the probability of both events happening when finding the probability of two events that are not mutually exclusive.

Guided Practice

Determine whether each event is *mutually exclusive* or *inclusive*.

4. A student is selected at random from a group of male and female students in the seventh, eighth, or ninth grade.
   a. P(male or ninth grader) **b.** P(male or female)
   c. P(female or eighth grader) **d.** P(seventh or eighth grader)
5. The spinner at the right is spun. Find each probability.
   a. P(C or D)
   b. P(E or vowel)
   c. P(B or consonant)
6. **Advertising** In April 1995, a poll asked 1225 adults across the United States whether they liked a certain television commercial for a long-distance service. Those polled were also asked how effective they thought the commercial was in helping to sell the service. The results of the second question are shown in the graph at the right.

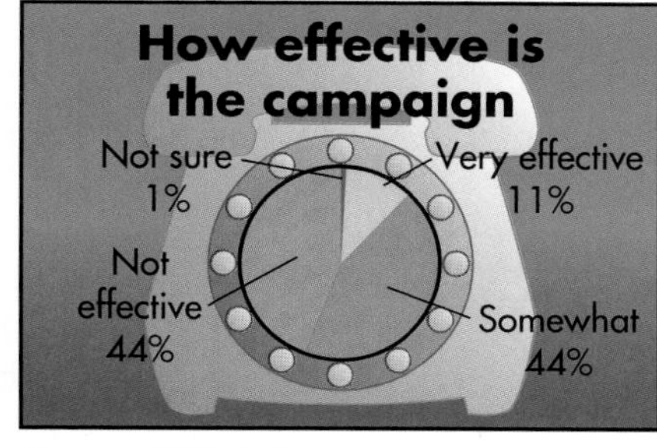

Source: *USA-Today*

   a. If a person is selected at random, what is the probability that he or she would say the commercial was not effective or somewhat effective?

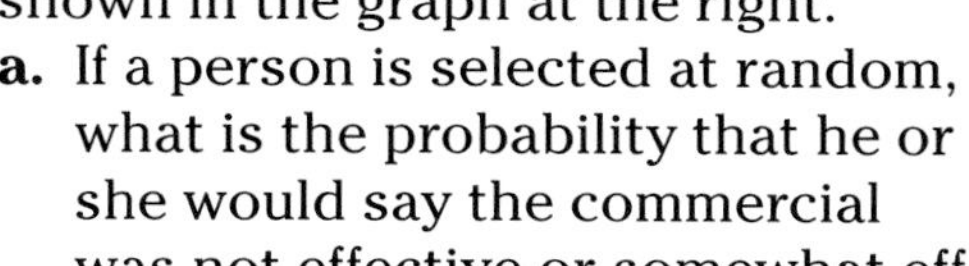

   b. If a person is selected at random, what is the probability that he or she would say the commercial was somewhat or very effective?

Practicing and Applying the Concept

Independent Practice

Determine whether each event is *mutually exclusive* or *inclusive*. Then find the probability.

7. An eight-sided die is tossed.
   a. P(even or less than 5) **b.** P(6 or odd)
   c. P(4 or prime) **d.** P(greater than 4 or an 8)
8. A card is drawn from the cards at the right.
   a. P(6 or even)
   b. P(9 or even)
   c. P(odd or even)
   d. P(5 or less than 3)
   e. P(4 or greater than 3)

Suppose $P(\text{A}) = \frac{1}{5}$, $P(\text{B}) = \frac{2}{3}$, and A and B are mutually exclusive.

9. What is P(A and B)?

10. What is P(A or B)?

Suppose $P(\text{A}) = \frac{1}{5}$, $P(\text{B}) = \frac{2}{3}$, and A and B are independent.

11. What is P(A and B)?

12. What is P(A or B)?

Two cards are drawn from a standard deck of cards. Find each probability.

13. P(both aces or both jacks)

14. P(both red or both face cards)

15. A bag contains six blue marbles and three red marbles. A marble is drawn, it is replaced and another marble is drawn. What is the probability of drawing a red marble and a blue marble in either order?

An eight-sided die is tossed 40 times. Determine how many times you would expect each outcome.

16. a 5 or a 7

17. a 6 or a 3

18. an even number or a 5

19. a prime number or an even number

20. a multiple of 3 or a multiple of 2

21. a multiple of 2 or a multiple of 4

Critical Thinking

22. Design and draw a spinner containing the numbers 1, 2, 3, and 4 so that the probability of spinning a 3 or a 4 is $\frac{2}{3}$.

Applications and Problem Solving

23. **Economics** Thirty-one percent of minimum-wage workers are between 16 and 19 years old. Twenty-two percent of minimum-wage workers are between 20 and 24 years old. If a person who makes minimum wage is selected at random, what is the probability that he or she will be between 16 and 24 years old?

24. **Government** Florida has 23 members in the United States House of Representatives. Eight members are Democrats, and 15 are Republicans. Of the 17 men in the house, 4 are Democrats. The Speaker of the House wants to choose a representative from Florida at random to serve on the agriculture committee. What is the probability that the representative will be a woman or a Democrat?

Mixed Review

25. **Probability** Suppose you choose a card from a deck of playing cards, returning the card, and then choose another card from the deck. Are these independent or dependent events? Explain. (Lesson 10-9)

26. **Consumer Awareness** Unit prices are often displayed for products at grocery stores. Suppose a 13-ounce box of breakfast cereal costs \$3.69 and a unit price of 0.95¢ per ounce is displayed. Is this the correct unit price? Explain your answer. (Lesson 9-1)

27. Graph $y = x - 3$. (Lesson 8-3)

28. Write an inequality for "the number of students in each class is less than or equal to 28." Then draw a graph of the inequality. (Lesson 3-6)

29. Find $15 \cdot 20$ mentally using the distributive property. (Lesson 1-5)

CHAPTER 10 Highlights

Vocabulary

After completing this chapter, you should be able to define each term, property, or phrase and give an example or two of each.

Discrete Mathematics
combination (p. 516)
factorial (p. 515)
Fundamental Counting Principle (p. 509)
outcomes (p. 509)
permutation (pp. 514, 515)
tree diagram (p. 509)

Probability
dependent events (p. 532)
experimental probability (p. 529)
inclusive (p. 536)
independent events (p. 531)
mutually exclusive (p. 535)
odds (p. 520)
relative frequency (p. 529)
simulation (p. 524)

Statistics
back-to-back stem-and-leaf plot (p. 487)
box-and-whisker plot (p. 495)
circle graph (p. 497)
comparative graph (p. 498)
histogram (p. 502)
interquartile range (p. 491)
leaves (p. 486)
line graph (p. 497)
lower quartile (p. 491)
outliers (p. 496)
pictograph (p. 497)
quartile (p. 491)
range (p. 490)
stem-and-leaf plot (p. 486)
stems (p. 486)
upper quartile (p. 491)

Problem Solving
use a simulation (p. 524)

Understanding and Using Vocabulary

Choose the letter of the term that best matches each statement or phrase.

1. two events that can happen simultaneously
2. the difference between the least number and the greatest number in a set of numbers
3. an arrangement or listing in which order is important
4. the median of the upper half of a set of data
5. two events that cannot happen at the same time
6. the range of the middle half of a set of data
7. an arrangement or listing where order is not important

A. mutually exclusive
B. relative frequency
C. inclusive
D. upper quartile
E. interquartile range
F. range
G. combination
H. outliers
I. permutation

CHAPTER 10 Study Guide and Assessment

Skills and Concepts

Objectives and Examples

Upon completing this chapter, you should be able to:

- **display and interpret data in stem-and-leaf plots.** (Lesson 10-1)

Display the data 46, 32, 59, 42, 51, 30, 49 in a stem-and-leaf plot.

3 | 2 0
4 | 6 2 9
5 | 9 1   *5|9 = 59*

Review Exercises

Use these exercises to review and prepare for the chapter test.

Make a stem-and-leaf plot of each set of data.

8. 3, 4, 6, 8, 17, 19, 7, 21, 40, 43

9. 10.8, 11.0, 9.8, 10.7, 11.5, 10.7, 9.5, 11.9

10. 18, 16, 20, 25, 30, 23, 30, 29, 21, 15, 18

11. 120, 111, 133, 148, 117, 112, 121, 146

- **use measures of variation to compare data.** (Lesson 10-2)

4 | 1 2 0 3
5 | 9 1 8
6 | 6 5   *6|6 = 6.6*

Range: 6.6 − 4.0 or 2.6
Median: 4.0 4.1 4.2 4.3 5.1 5.8 5.9 6.5 6.6
↑ median (5.1)
Lower Quartile: 4.15
Upper Quartile: 6.2
Interquartile Range: 6.2 − 4.15 or 2.05

Find the range, median, upper and lower quartiles, and the interquartile range.

12. 23 45 16 51 47

13. 6 10 15 21 28 36

14.
9 | 9 7 5
10 | 1 3 6
11 | 2 9 8 3   *11|3 = 113*

15. 205 236 217 214 200 298 240

16.
11 | 0 6 0
12 | 8 5 5
13 | 7 1 3   *12|8 = 12.8*

- **use box-and-whisker plots to display data.** (Lesson 10-3)

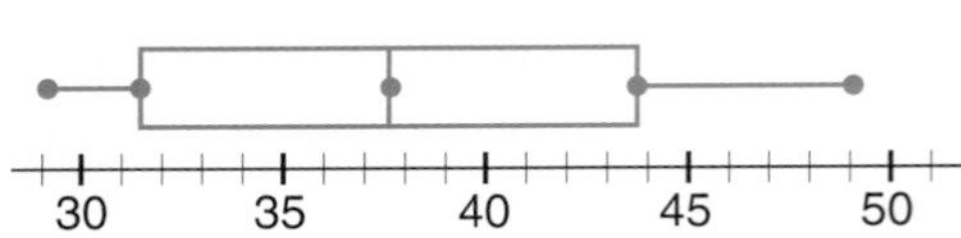

Range: 49 − 29 or 20
Median: 37.5
Upper Quartile: 43.75
Lower Quartile: 31.25
Interquartile Range: 43.75 − 31.25 or 12.50
Extremes: 29 and 49

17. Use the stem-and-leaf plot below to answer each question.

2 | 4 5 5 7 9
3 | 0 3 4 6 8
4 | 1 1 3 5
5 | 4 4   *2|4 = 24*

a. Make a box-and-whisker plot of the data.

b. What is the median?

c. What are the outliers, if any?

Objectives and Examples

▶ **recognize when statistics are misleading.** (Lesson 10-4)

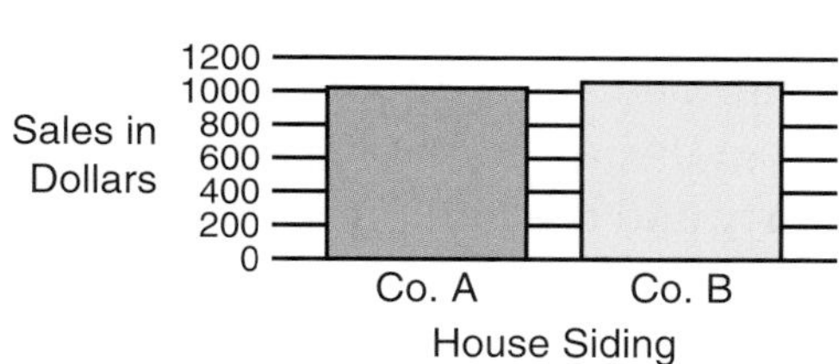

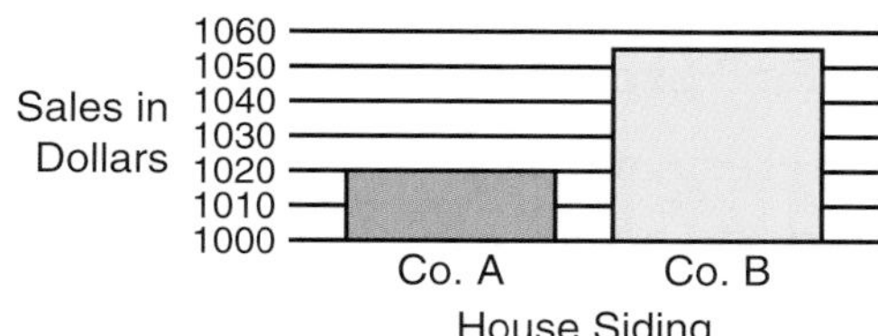

The second graph is misleading because the vertical axis does not include zero.

Review Exercises

Use the following chart to describe each category. Write the type of average you used.

	Tina	Steve	Ryan	Katie
Favorite Sport	football	softball	soccer	soccer
Number of classes	9	10	5	3
Number of family members	3	5	12	6

18. average favorite sport

19. average number of classes.

20. average number of family members

21. How can you describe the "average" person in this group?

▶ **use tree diagrams or the Fundamental Counting Principle to count outcomes.** (Lesson 10-5)

How many ways can you choose a shirt and tie?

$$\underbrace{\text{shirt choices}}_{4} \cdot \underbrace{\text{tie choices}}_{3} = 12$$

There are 12 ways to choose a shirt and tie.

Find the number of possible outcomes for each event.

22. An eight-sided die is rolled twice.

23. A multiple-choice quiz has 5 questions each with 4 choices.

24. A 6-sided die and a coin are tossed.

25. A card is drawn and a coin is tossed.

▶ **use permutations and combinations.** (Lesson 10-6)

Seven people are being arranged for a portrait. Does this represent a permutation or a combination?

This is a permutation because the order that the photographer chooses the people determines how the picture looks.

Tell whether each situation represents a *permutation* or a *combination*.

26. five people in a row

27. 32 cassette tapes from a group of 47

Find each value.

28. $P(7,4)$

29. $5!$

30. $C(8, 2)$

31. $\frac{6!3!}{4!2!}$

▶ **find the odds of a simple event.** (Lesson 10-7)

Green	Red	White	Blue

The odds of picking the green box is 1:3.

Find the odds of each outcome if a card is selected from a standard deck of cards.

32. a queen

33. not a face card

34. an ace or six

35. a red card

Objectives and Examples	Review Exercises

- **find the probability of independent and dependent events.** (Lesson 10-9)

Suppose you are playing a carnival game where each person chooses a plastic duck from a pond. Ten ducks are marked with numbers for prizes. Once a prize is given, the number is removed, and there is one less prize to be won. Is two people winning prizes an example of a dependent or independent event?

It is dependent because the first person winning affects the probability that the second person will win.

Determine whether the events are independent or dependent. Explain.

36. choosing a recipe from a box and then choosing another without replacement
37. on the first turn you buy a Monopoly property and then on the second turn you buy a property

A bag contains 5 red, 3 yellow, and 3 purple marbles. Once a marble is drawn, it is not replaced. Find each probability.

38. P(two green marbles)
39. P(a red and a yellow)
40. P(two yellow)
41. P(a purple and a red)

- **find the probability of mutually exclusive events or inclusive events.** (Lesson 10-10)

Suppose you roll two dice. Are the events *rolling a six* and *rolling a sum of less than seven* mutually exclusive or inclusive?

Since you cannot roll a six and a sum of less than seven in one roll, the events are mutually exclusive.

Determine whether each event is mutually exclusive or inclusive. Then find the probability.

42. P(rolling a six or sum greater than ten)
43. P(rolling an even number or sum is odd)

A die is tossed. Find each probability.

44. P(3 or 4)
45. P(4 or even)

Applications and Problem Solving

46. **Transportation** The numbers of seats in commonly-used passenger airplanes are listed below. Make a box-and-whisker plot of the data. (Lesson 10-3)

398	390	288	281	266	254
248	221	186	185	149	148
144	141	131	124	113	112
100	97	72			

A practice test for Chapter 10 is available on page 783.

47. **Games** Bruno's Restaurant is giving out game cards as a promotional contest. There are eight different cards available. How many different ways could you get cards on two trips to Bruno's? (Lesson 10-5)
48. **Sports** A soccer team plays 16 games during the season. The team manager predicts that the team has a 75% chance of winning each game. Describe a simulation to determine the probability of the team winning three games in a row. (Lesson 10-8)

Alternative Assessment

Cooperative Learning Activity

Work with your group to write an article for your school newspaper entitled "Are you average?"

- Choose the type of information you think students in your school would find interesting. For example, you might investigate students' heights, the numbers of languages they speak, what kind of pets they own, or how long they have lived in the area.
- Next, conduct a survey in the school cafeteria to gather data.
- Present the results of your research in an interesting way. Include some graphs and tables. Be sure to explain what the results mean in the article.
- State the probability that a randomly-chosen student at your school possesses one or more characteristics. For example, how likely is it that a person speaks two languages and owns a hedgehog?

Thinking Critically

In 1994, *First for Women* magazine reported that the average American is a white woman who is 32.7 years old. She is married with children, owns a home, and has two television sets and a VCR.

- Does this describe anyone you know?
- Explain what the magazine meant by "the average American."

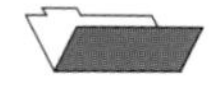 **Portfolio**

Select an item from your work in this chapter that shows how your understanding of the concepts improved and place it in your portfolio.

Self-Evaluation

"We always have time enough, if we but use it aright."
—Johann Wolfgang Von Goethe

Assess yourself. Do you use your time wisely? Or do you find yourself rushing to complete projects and not giving your best effort? Make a plan for the time that you have to complete the tasks you do each week such as home chores, school work, and community activities. Are there ways you could spend your time more wisely?

CHAPTERS 1–10 Ongoing Assessment

Section One: Multiple Choice

There are ten multiple-choice questions in this section. After working each problem, write the letter of the correct answer on your paper.

1. Theo wants to lose 2 pounds per week for 8 weeks. What will the change in his weight be?

A. -6 pounds **B.** -8 pounds
C. $+8$ pounds **D.** -16 pounds

2. The high temperatures (°F) for January 15th during the past 10 years are 15°, 18°, 25°, 38°, 6°, 25°, 10°, 30°, 18°, and 25°. What are the lower and upper quartiles?

A. 18°, 25° **B.** 15°, 25°
C. 21.5°, 30° **D.** 10°, 30°

3. Which fraction is less than $\frac{5}{12}$?

A. $\frac{1}{2}$ **B.** $\frac{3}{4}$
C. $\frac{5}{8}$ **D.** $\frac{2}{6}$

4. If $m + 2.1 = 8$, what is the value of m?

A. 5.9 **B.** 8
C. 6.9 **D.** 10.1

5. Mr. Garza has 4 black ties, 3 gray ties, 2 blue ties, and 1 brown tie in his closet. If he selects one tie without looking, what is the probability that it will be blue?

A. $\frac{1}{2}$ **B.** $\frac{1}{5}$
C. $\frac{1}{10}$ **D.** $\frac{2}{8}$

6. Angie's Place sells made-to-order sandwiches. There are 3 kinds of bread, 3 kinds of cheese, and 4 kinds of meat. How many different combinations of bread, cheese, and meat (one of each) can be ordered?

A. 10 **B.** 18
C. 12 **D.** 36

7. If $5y - 4 = 3y + 12$, what is the value of y?

A. 1 **B.** 4
C. 2 **D.** 8

8. The number of students in Sam's aerobics class increased from 15 to 24. What was the percent of increase?

A. 37.5% **B.** 62.5%
C. 60% **D.** 135%

9. A coin is tossed and a die is rolled. What is the probability of tossing tails and rolling a 5?

A. $\frac{1}{12}$ **B.** $\frac{1}{4}$
C. $\frac{1}{10}$ **D.** $\frac{1}{3}$

10. Which ordered pairs are solutions for $y = 2x - 3$?

A. $(-3, -9), (2, 1), (4, 5)$
B. $(-3, 0), (-1, -5), (3, 3)$
C. $(-3, 3), (-2, 1), (-1, -1)$
D. $(-2, -7), (-1, 1), (0, -3)$

Section Two: Free Response

This section contains eight questions for which you will provide short answers. Write your answer on your paper.

11. Barbara needs at least $7\frac{1}{8}$ pounds of fertilizer for her lawn. She has $4\frac{3}{4}$ pounds. How much more does she need?

12. Juanita drove 350 miles and used 14 gallons of gasoline. Express her gas mileage as a unit rate.

13. The circumference of the circle outlining the boundary of where a discus thrower competes is 17.27 feet. What is the diameter of the circle?

14. Find the slope of a line that contains $(3, -4)$ and $(-2, 2)$.

15. As a team, the Eagles made 24 shots in 40 times at the free throw line. What percentage did they make?

16. Matt's parents have agreed to help him buy a car. They will purchase the car and he will make no-interest payments to them over the next three years. Matt can make a down payment of $1500. He cannot afford monthly payments over $200. What is the most he can pay for the car?

17. When rolling a 6-sided die, what are the odds of rolling a number divisible by 3?

18. In how many different ways can four students be arranged in a row?

Test-Taking Tip

As part of your preparation for a standardized test, review basic definitions and formulas such as the ones below.

A number is *prime* if it has no factors except itself and 1. For example, 7 is a prime number. The factors for 8 are 1, 2, 4, and 8. Therefore, 8 is not prime since it has factors other than itself and 1.

The area of a circle with radius r units is πr^2.

Section Three: Open-Ended

This section contains two open-ended problems. Demonstrate your knowledge by giving a clear, concise solution to each problem. Your score on these problems will depend on how well you do the following.

- Explain your reasoning.
- Show your understanding of the mathematics in an organized manner.
- Use charts, graphs, and diagrams in your explanation.
- Show the solution in more than one way or relate it to other situations.
- Investigate beyond the requirements of the problem.

19. Describe how you would calculate the sale price of any item. Use a specific example to demonstrate your method.

20. Collect information from your classmates on how long it takes them to get to school. Construct an appropriate statistical graph to display the data. Explain your choice of graph.

CHAPTER 11

Applying Algebra to Geometry

TOP STORIES in Chapter 11

In this chapter, you will:

- use some basic terms of geometry,
- construct a circle graph,
- identify the relationships of intersecting, parallel, and skew lines,
- identify properties of congruent and similar figures,
- classify and draw polygons, and
- identify and draw transformations.

MATH AND SAILING IN THE NEWS

It's Magic at the America's Cup

Source: Sail, July, 1995

"Over the past two-and-a-half years, we've done everything we could to prepare for this moment." said *Black Magic 1* team leader Peter Blake. Determination and technology helped Blake and his New Zealand team capture the America's Cup, the most coveted prize in yachting. The team won five races in a row in the final round to earn the victory. The New Zealanders used the lessons they learned in their previous losses to build two boats especially for the competition. Computers and mathematics helped them study the way that the boat and the sails interact with the air and the water. The boat's co-designer says there are literally "hundreds of little refinements." Some of the major refinements are the canoe-style hull, a carbon-fiber mast, innovative sail rigging, and a thin carbon-steel rudder.

Putting It into Perspective

1850 1880 1910

1851 U.S. schooner *America* defeats 17 British yachts to win the Hundred Guinea Cup.

1857 The Hundred Guinea Cup is renamed the America's Cup and becomes an international competition.

1900 Sailing is made an Olympic sport.

EMPLOYMENT OPPORTUNITIES

Set Sail for Your Future

THERE IS MORE to sailing than just wind and water. And even the most skilled sailor depends upon the boat. The beauty of sailing begins with the designer. If you enjoy creating something from an idea, and are good at visualizing and analyzing, your skills are needed as a **designer**, **engineer**, or a **naval architect**. These jobs require a thorough knowledge of mathematics, especially geometry, and computer skills. A college degree is usually required, but a two-year technical degree with experience will be considered.

For more information, contact:
Society of Naval Architects and Marine Engineers
601 Pavonia Avenue, Suite 400
Jersey City, NJ 07306

Statistical Snapshot

1995 America's Cup Race Course

Leg	1	2	3	4	5	6	Total
Length (nautical miles)	3.275	3.275	3.000	3.000	3.000	3.000	18.550

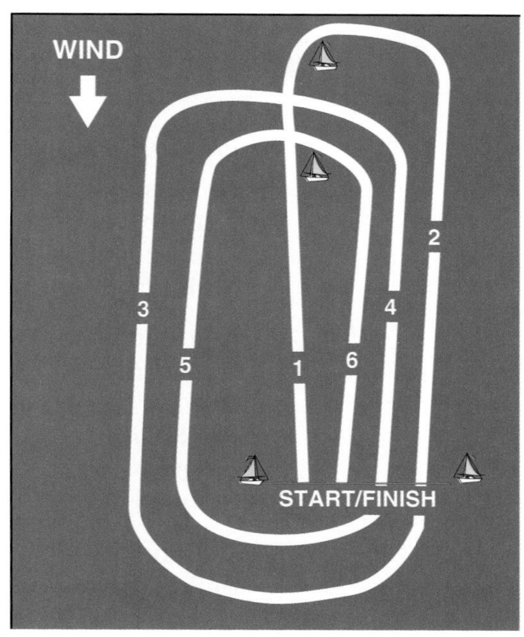

interNET CONNECTION For up-to-date information on boat making, visit:
www.glencoe.com/sec/math/prealg/mathnet

1980 — 1990 — 2000

1983
The America's Cup is won by a non-American yacht, *Australia II*, for the first time in 25 competitions.

1992
Martin Stephan and Art Prince become the first African Americans to sail in the America's Cup race.

1995
Black Magic I wins the America's Cup.

1995
The first all-female crew competes to represent the United States as America's Cup defender.

11-1 The Language of Geometry

Setting Goals: *In this lesson, you'll identify points, lines, planes, rays, segments, angles, and parallel, perpendicular, and skew lines. You'll also classify angles as acute, right, or obtuse.*

Modeling a Real-World Application: Construction

Students in Todd Scholl's construction trades classes in Fairplay, Colorado, don't just study. They put their learning into practice by building real houses. In fact, they built the town's Silverheels Middle School, a four-classroom structure.

Before construction of a building can begin, the corners of the building site must be level. A laser interferometer can determine the heights of the corners of the site with great accuracy. A laser is visible because light is being reflected from particles in the air.

We can use ideas from the construction of buildings to model geometric ideas.

Learning the Concept

A point has no dimensions.

Each particle visible in the laser beam has a specific location in space and suggests the idea of a geometric **point**. A point is a specific location in space with no size or shape. A point can be represented by a dot and named with a capital letter.

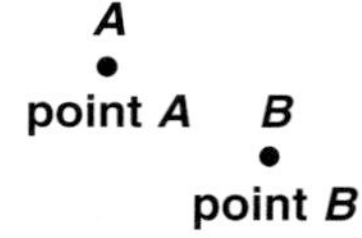

A line has one dimension, length.

A laser beam travels in a straight path. In geometry, a straight path is called a **line**. A line is a collection of points that extends indefinitely in two directions. Arrowheads are used to show that a line has no endpoints. A line can be named with a single lowercase letter or by two points on the line.

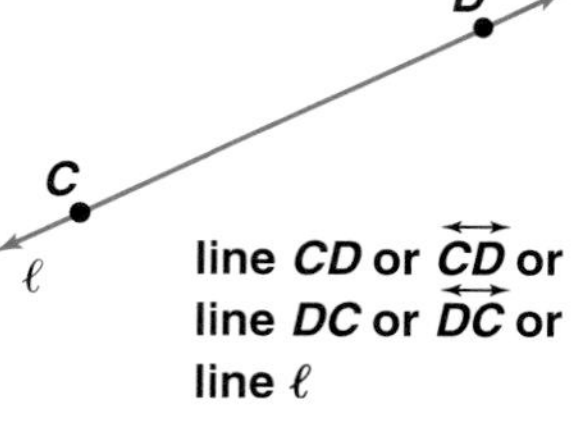

line *CD* or $\overleftrightarrow{CD}$ or line *DC* or $\overleftrightarrow{DC}$ or line ℓ

THINK ABOUT IT

Can you measure the length of a line? Can you measure a line segment?

Each wooden beam used in the construction of a building site is a model of a **line segment**. A line segment is part of a line containing two endpoints and all points between the endpoints. A line segment is named by its endpoints.

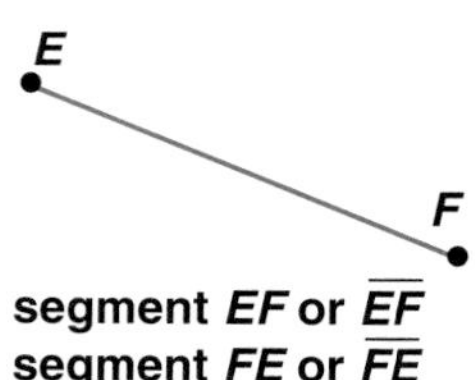

segment *EF* or $\overline{EF}$
segment *FE* or $\overline{FE}$

The path of a laser beam as it begins at the generator and extends in one direction is a model of a **ray**. A ray is a portion of a line that extends from one point infinitely in *one* direction. A ray is named using the endpoint first and then any other point on the ray.

ray *DF* or $\overrightarrow{DF}$

Example 1 **Name two points, a line, two rays, and a line segment in the figure at the right.**

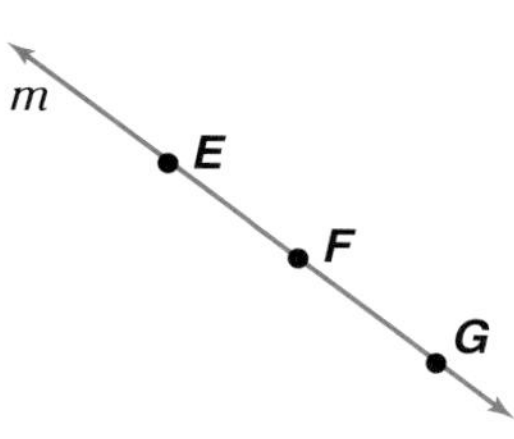

Two of the points are E and F.

We can use a single lowercase letter or any two of the labeled points to name the line. So we can name it m, $\overleftrightarrow{EF}$, $\overleftrightarrow{EG}$, or $\overleftrightarrow{FG}$.

Two of the rays are $\overrightarrow{FG}$ and $\overrightarrow{FE}$.

One of the line segments shown is $\overline{EG}$.

When two rays have the same endpoint, they form an **angle**. The common endpoint is called the **vertex**, and the rays are called **sides** of the angle. An angle is named using three points with the vertex as the middle letter. If no other angle shares the same vertex, the angle is named with just the vertex letter or with a numeral.

- angle ABC or $\angle ABC$, angle CBA or $\angle CBA$,
- angle B or $\angle B$,
- angle 1 or $\angle 1$

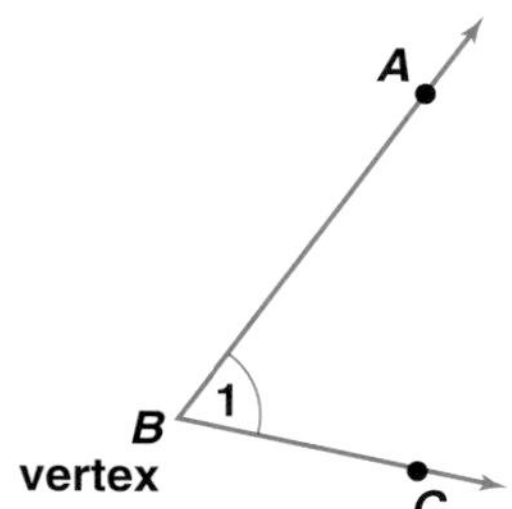

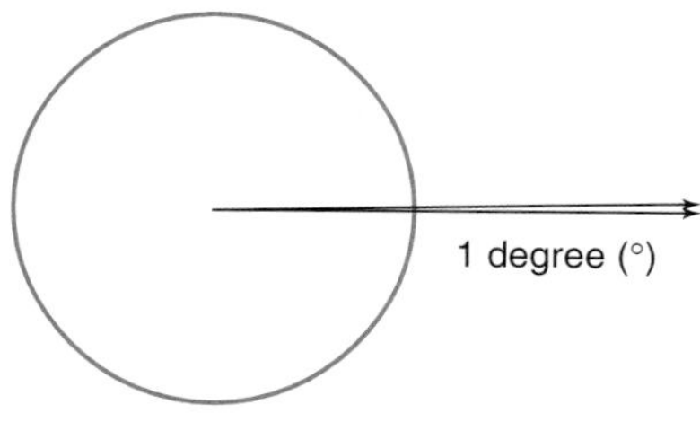

The most common unit of measure for angles is the **degree**. A circle can be separated into 360 arcs of the same length. An angle has a measurement of one degree if its vertex is at the center of the circle and its sides contain the endpoints of one of the 360 equal arcs.

You can use a **protractor** to measure angles.

Example 2 **Use a protractor to measure $\angle PQR$.**

Place the protractor so the center mark is at Q (the vertex of $\angle PQR$) and the straightedge aligns with $\overrightarrow{QR}$ (one side of the angle). Use the scale that begins with 0 at $\overrightarrow{QR}$. Read where $\overrightarrow{QP}$ (the other side of the angle) crosses this scale.

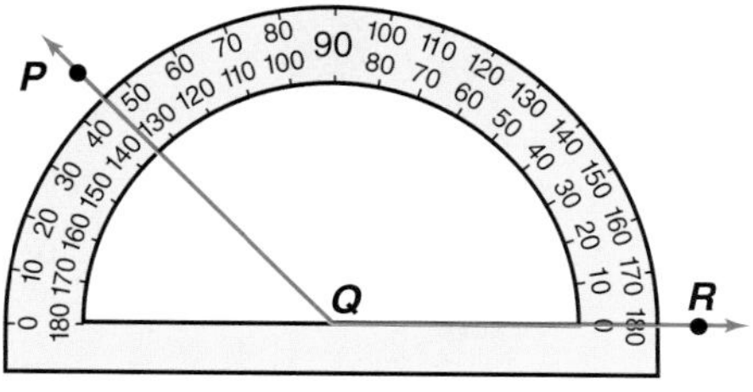

The measure of angle PQR is 135 degrees. Using symbols, $m\angle PQR = 135°$.

You can also use a protractor to draw an angle of a given measure.

Example  Draw an angle Y that measures 50°.

Use a straightedge to draw a ray.

Place the protractor with the center mark at Y and the 0-mark aligned with the ray.

Use the scale that begins with 0 at the ray, find the mark for 50, and draw a dot.

Use a straightedge to draw the other side of the angle. Angle Y measures 50°.

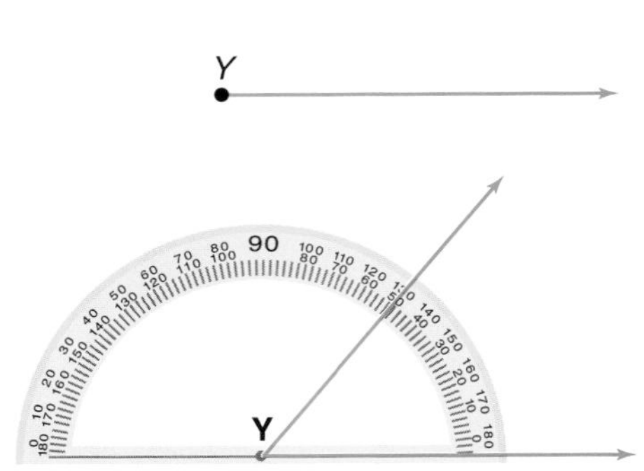

You can classify angles by their degree measure. **Acute** angles have measures greater than 0° and less than 90°. **Right** angles have measures of 90°. **Obtuse** angles have measures greater than 90° but less than 180°. **Straight** angles measure 180°.

Example 4

APPLICATION

Baseball

According to Ted Williams, one of baseball's greatest hitters, the best way to strike a baseball is at a right angle. If you swing too early or too late, the ball will probably go foul. Classify each angle as *acute*, *right*, or *obtuse*.

a.

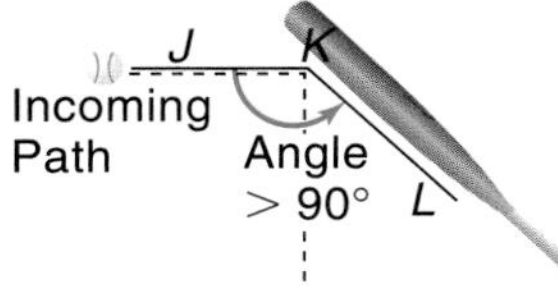

b.

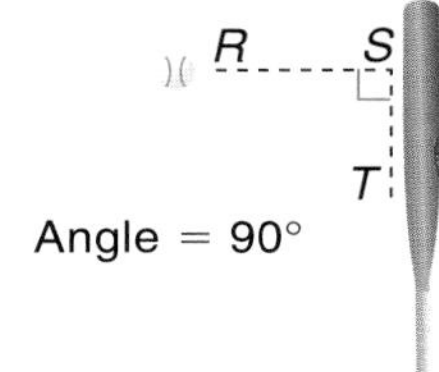

The small square indicates that $\angle RST$ is a right angle.

c.

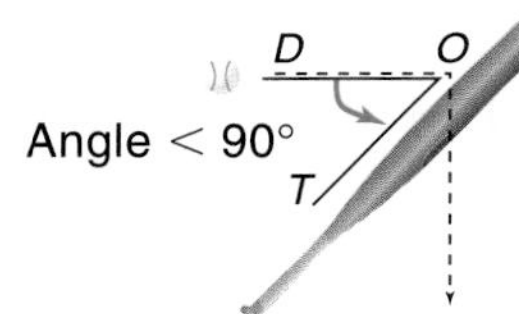

a. $m\angle JKL > 90°$. So, $\angle JKL$ is obtuse. *The swing is too early.*

b. $m\angle RST = 90°$. So, $\angle RST$ is right.

c. $m\angle DOT < 90°$. So, $\angle DOT$ is acute. *The swing is too late.*

A plane has two dimensions, length and width, but no thickness.

The level building site described in the application on page 548 is a model of part of a **plane**. A plane is a flat surface with no edges, or boundaries. A plane can be named by a single uppercase script letter or by using any three points of the plane. *The three points must not lie on the same line.*

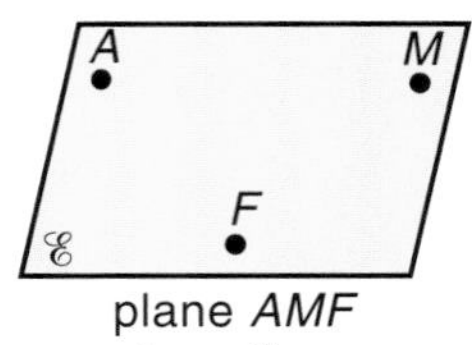

plane AMF
plane $\mathscr{E}$

Lines that lie in the *same plane* either intersect or are parallel.

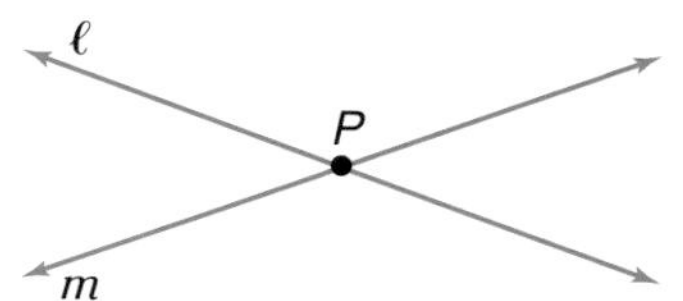

Lines ℓ and m **intersect** at point P.

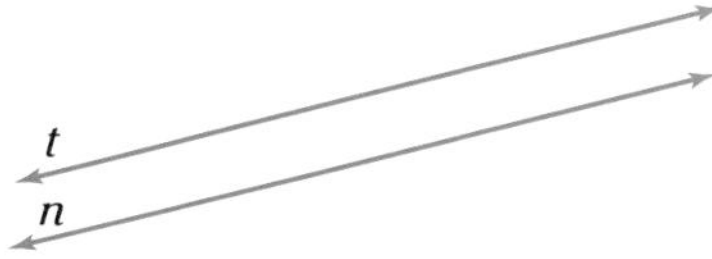

There is no point of intersection of lines t and n. Lines t and n are **parallel**. Using symbols, $t \parallel n$.

Two lines that intersect to form a right angle are **perpendicular**. Rays and line segments can also be perpendicular. The symbol $\perp$ means *is perpendicular to*.

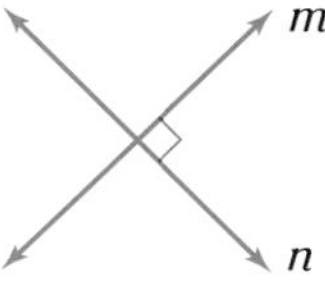

$m \perp n$

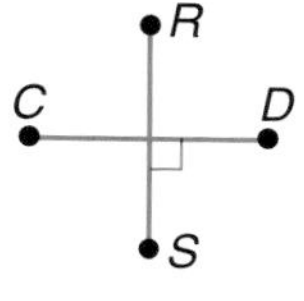

$\overline{RS} \perp \overline{CD}$

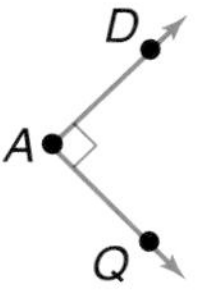

$\overrightarrow{AD} \perp \overrightarrow{AQ}$

If two lines do not intersect and are not in the same plane, they are called **skew lines**. The roof line along the front of a rectangular building and the line of the base of the side of the building are an example of skew lines.

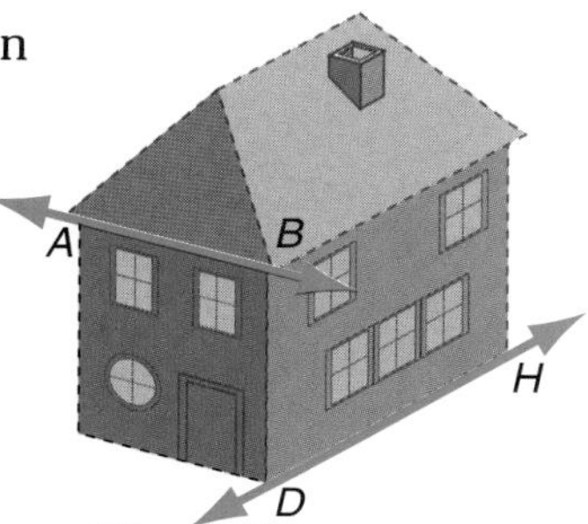

$\overleftrightarrow{AB}$ and $\overleftrightarrow{DH}$ are skew.

Example 5 **Identify each pair of line segments as *intersecting*, *parallel*, or *skew*.**

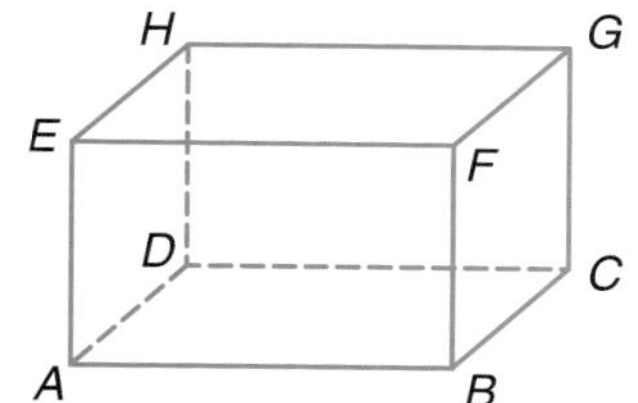

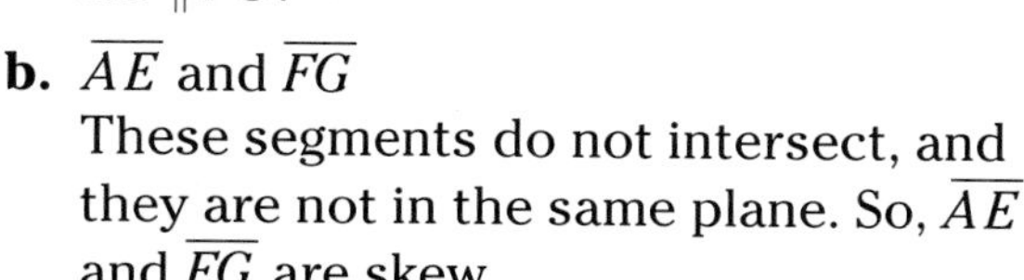

a. $\overline{EH}$ and $\overline{FG}$

These segments are in the same plane (EFG), and they do not intersect. So, $\overline{EH} \parallel \overline{FG}$.

b. $\overline{AE}$ and $\overline{FG}$

These segments do not intersect, and they are not in the same plane. So, $\overline{AE}$ and $\overline{FG}$ are skew.

c. $\overline{BC}$ and $\overline{CD}$

These segments are in the same plane (ABC), and they intersect at point C.

Checking Your Understanding

Communicating Mathematics

Read and study the lesson to answer these questions.

1. **Tell** how many endpoints each of the following has.
   a. a line **b.** a line segment **c.** a ray
2. **Compare and contrast** parallel lines and skew lines.
3. **Demonstrate** how a protractor is used to measure angles and how it is used to draw angles.

4. **Draw** and label a diagram to show a right angle whose sides are $\overrightarrow{QP}$ and $\overrightarrow{QR}$.
5. **Explain** the difference between an acute angle and an obtuse angle.

MATERIALS

toothpicks

gumdrops

6. You can use gumdrops and toothpicks to model geometric concepts. A gumdrop represents a point, and a toothpick represents a portion of a line. The figure at the right shows a model of a line segment. Use gumdrops and toothpicks to model each of the following.

a. an angle **b.** a ray **c.** a line **d.** intersecting lines

Guided Practice

Tell whether each model suggests a point, a line, a plane, a ray, or a line segment. Explain your answer.

7. a wall
8. the tip of a needle

Use the figure at the right to name an example of each term.

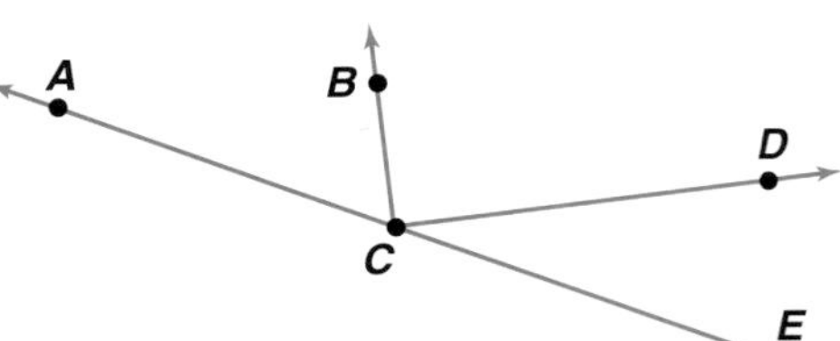

9. point
10. line segment
11. ray
12. line

In the figure above, use a protractor to find the measure of each angle. Then classify the angle as *acute*, *right*, or *obtuse*.

13. $\angle ACB$
14. $\angle BCD$

Use a protractor to draw angles having the following measurements.

15. $78°$
16. $145°$

17. **Photography** Cameras are often mounted on tripods to give stability. Why do tripods give stability?

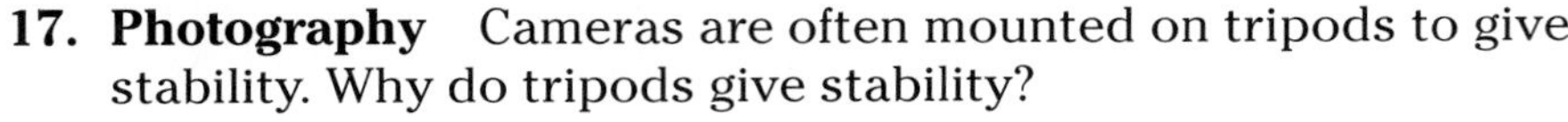

Exercises: Practicing and Applying the Concept

Independent Practice

Tell whether each model suggests a point, a line, a plane, a ray, or a line segment. Explain your answer.

18. a pencil
19. a CD-ROM
20. a telephone wire
21. an arrow's path

Draw and label a diagram to represent each of the following.

22. point T
23. $\angle Q$
24. line ℓ
25. $\overline{XY}$
26. plane RST
27. $\overrightarrow{EF}$
28. $\angle BOY$
29. $\overleftrightarrow{AB}$
30. lines ℓ and m intersect
31. $\overleftrightarrow{AB} \parallel \overleftrightarrow{CD}$
32. $\overleftrightarrow{XY} \perp \overleftrightarrow{MN}$
33. lines a and b are skew

In the figure at the right, use a protractor to find the measure of each angle. Then classify the angle as *acute*, *right*, or *obtuse*.

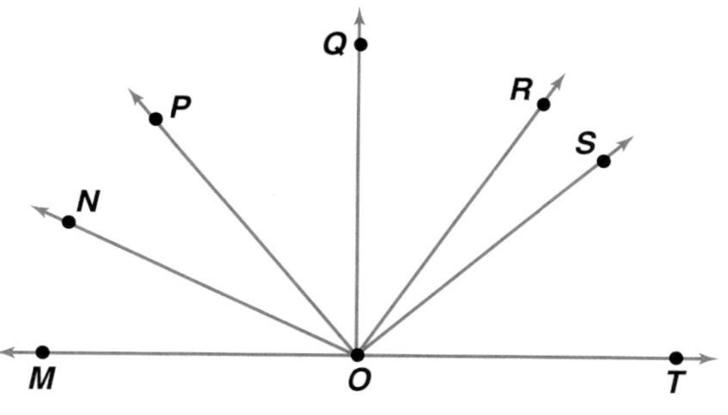

34. $m\angle MOQ$

35. $m\angle NOT$

36. $m\angle TOP$

37. $m\angle QON$

38. $m\angle POS$

39. $m\angle ROS$

Use a protractor to draw angles having the following measurements. Classify each angle as *acute*, *right*, or *obtuse*.

40. 112° **41.** 47° **42.** 95° **43.** 16° **44.** 162° **45.** 90°

Critical Thinking

46. ***True* or *false*. Explain your answers.**

a. A line is part of a line segment.

b. The intersection of any two rays is always a point.

Applications and Problem Solving

47. **Space Travel** The space shuttle approaches the runway at an angle of 18°, six times as steep as that of a commercial jet. At what angle does a commercial jet make its approach? Draw a diagram showing both approach angles.

48. **Aviation** Airplane flight paths can be described using angles and compass directions. The path of a particular airplane is described as 36° west of north. Draw a diagram that represents this path.

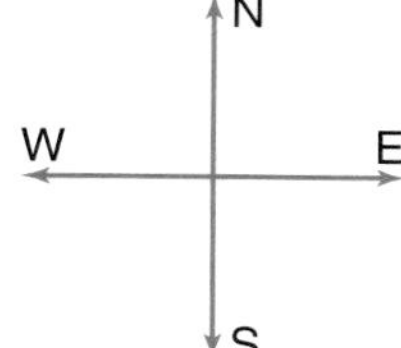

49. **Patterns** At certain times of the day, the hands of a clock form a right angle. In twelve hours, how many times do the hands form a right angle?

Mixed Review

50. **Probability** A die is tossed. Is P(even or prime) *mutually exclusive* or *inclusive*? Find the probability. (Lesson 10-10)

51. Replace the ● with = or ≠ to make $\frac{1.5}{2}$ ● $\frac{1.8}{2.4}$ true. (Lesson 9-4)

52. **Health** What relationship (*positive*, *negative*, or *none*) do you think a scatter plot of number of students with colds and season of year (spring, summer, fall, winter) might show? Explain. (Lesson 8-2)

53. **Number Theory** The sum of an even integer and the next greater even integer is more than 98. Find the least pair of such integers. (Lesson 7-7)

54. Solve $n - 6.2 \leq 5.8$. Graph the solution on a number line. (Lesson 5-7)

55. A volleyball team has 6 members. Suppose each member shakes hands with every other member. How many handshakes take place? (Lesson 4-3)

56. **Geometry** Find the perimeter and area of a rectangular floor that is 12 feet wide by 15 feet long. (Lesson 3-5)

57. Evaluate $\frac{d}{8}$ if $d = -96$. (Lesson 2-8)

58. Solve $3d = 24$ mentally. (Lesson 1-6)

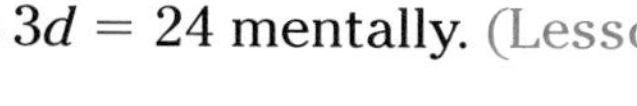

HANDS-ON ACTIVITY

11-1B Constructions

An Extension of Lesson **11-1**

MATERIALS

 compass

ruler

protractor

When two line segments have the same length, we say they are **congruent**. Similarly, two angles with the same measure are congruent. You can use a compass and straightedge to **construct** congruent segments and angles. A **straightedge** is any object you can use to draw a straight line, such as an index card or a ruler. You use a **compass** to draw a circle or part of a circle.

Activity  Construct a line segment congruent to a given line segment.

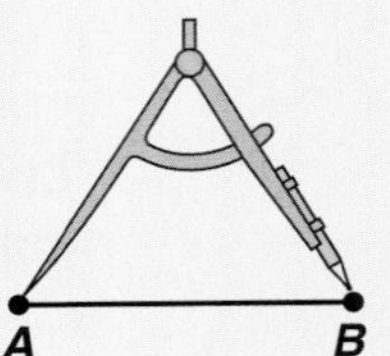

First, draw a segment and label it $\overline{AB}$. This will be the given segment.

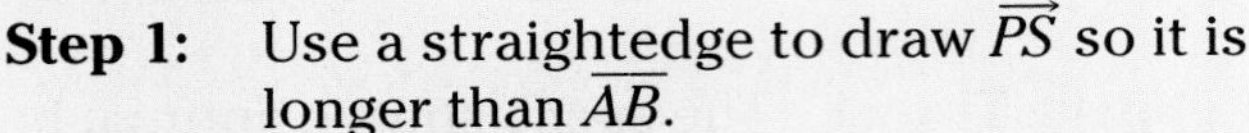

Step 1: Use a straightedge to draw $\overrightarrow{PS}$ so it is longer than $\overline{AB}$.

Step 2: Place the steel tip of the compass at A and the writing tip at B.

Step 3: Keep the same setting on the compass and place the steel tip at P. Draw an arc that intersects $\overrightarrow{PS}$ at Q. $\overline{PQ}$ is congruent to $\overline{AB}$. In symbols, $\overline{PQ} \cong \overline{AB}$.

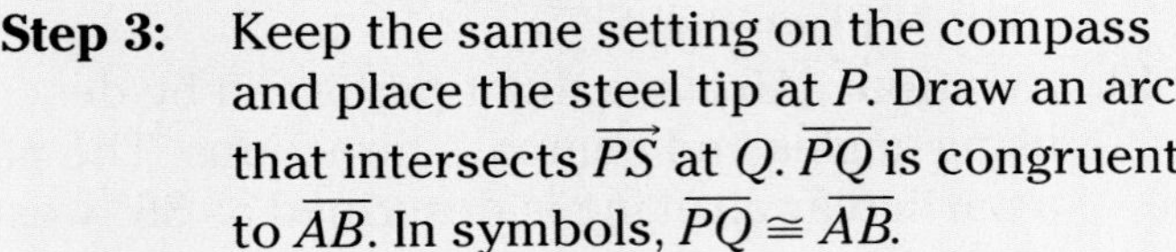

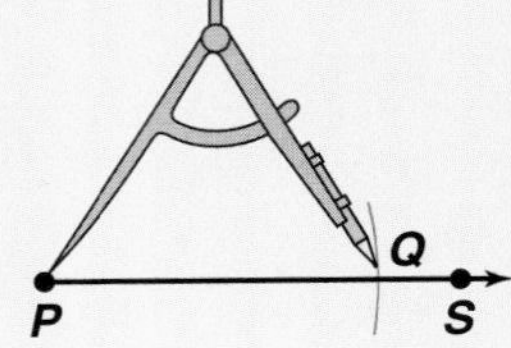

Activity  Construct an angle that is congruent to a given angle.

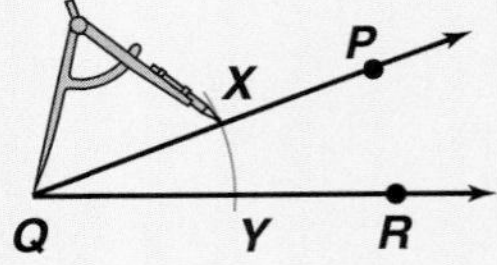

First, draw an angle and label it $\angle PQR$. This will be the given angle.

Step 1: Use a straightedge to draw $\overrightarrow{ST}$.

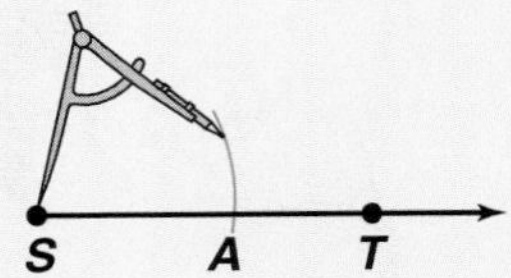

Step 2: Place the steel tip of the compass at Q. Draw an arc that intersects both sides of $\angle PQR$ to locate points X and Y.

Step 3: Keeping the same compass setting, place the steel tip at S and draw an arc that intersects $\overrightarrow{ST}$. Label the intersection point A.

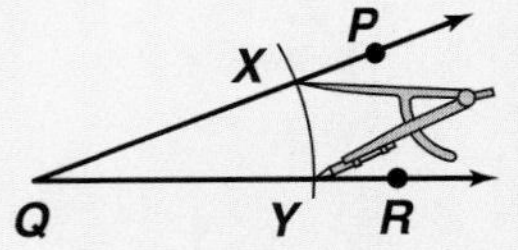

Step 4: Place the compass so one tip is at X and the other is at Y. Keep that setting and place the steel tip at A. Draw an arc that intersects the arc you drew in step 3. Label the intersection point M.

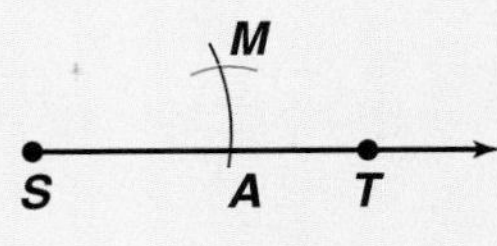

Step 5: With a straightedge, draw $\overrightarrow{SM}$. $\angle MST$ is congruent to $\angle PQR$. In symbols, $\angle MST \cong \angle PQR$.

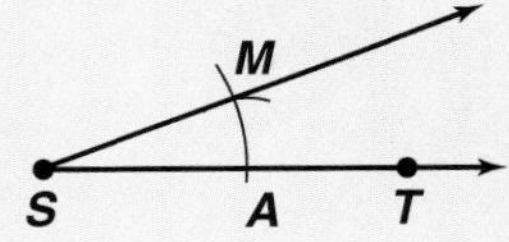

TALK ABOUT IT

1. How does the compass work in copying a line segment?
2. Explain how the compass works in copying an angle.

Activity 3 Construct a line segment that bisects a given line segment.

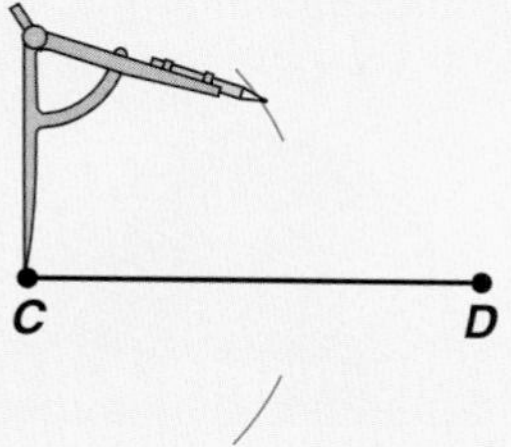

Draw a segment and label it $\overline{CD}$. This is the given line segment.

Step 1: Place the steel tip of the compass at C. Extend the compass until the writing tip is more than halfway to D and draw an arc "above" and "below" $\overline{CD}$ as shown.

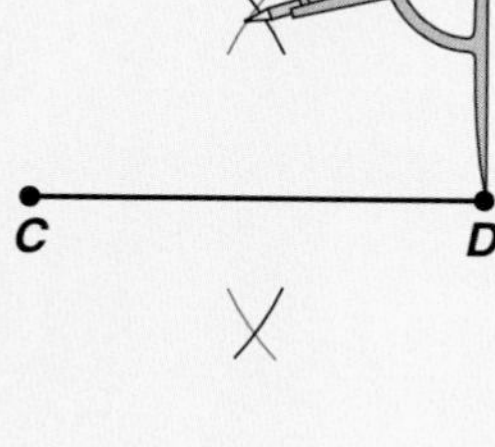

Step 2: With the same setting on the compass, place the steel tip at D and draw two arcs as shown to locate two new points, Q and R.

Step 3: With a straightedge, draw the line segment determined by these two new points. This segment will intersect $\overline{CD}$ at M.

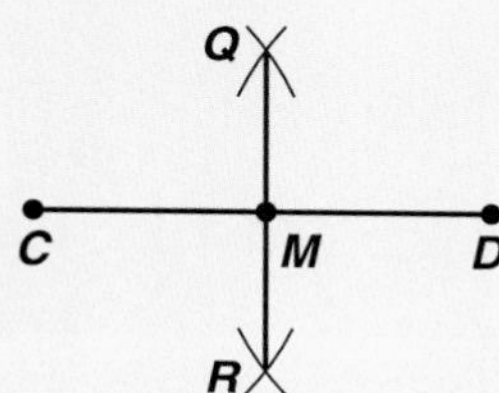

TALK ABOUT IT

3. Use a ruler to measure $\overline{CM}$ and $\overline{MD}$. This construction *bisects* a segment. What do you think *bisects* means?
4. In this construction, $\overline{QR}$ is perpendicular to $\overline{CD}$. What type of angles are formed by perpendicular segments?

Activity 4 Construct a ray that bisects a given angle.

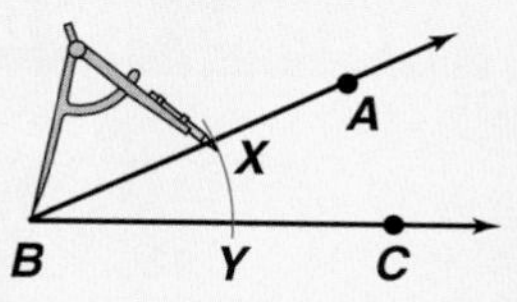

Draw an angle and label it $\angle ABC$. This is the given angle.

Step 1: Place the steel tip of the compass at B and draw an arc that intersects both sides of the angle. Label the points of intersection X and Y.

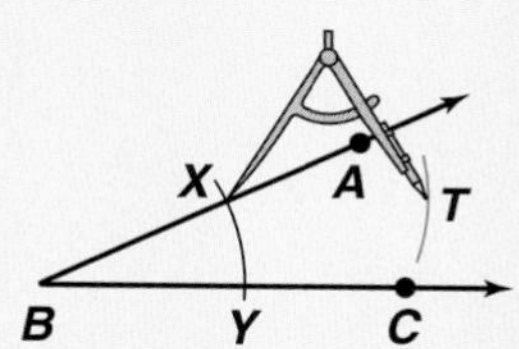

Step 2: Place the steel tip of the compass at X and open the compass to a length more than halfway to Y. Draw an arc in the interior of $\angle ABC$.

Step 3: Keeping the same compass setting, place the steel tip at Y and draw an arc that intersects the arc you drew in Step 2. Label the point of intersection T.

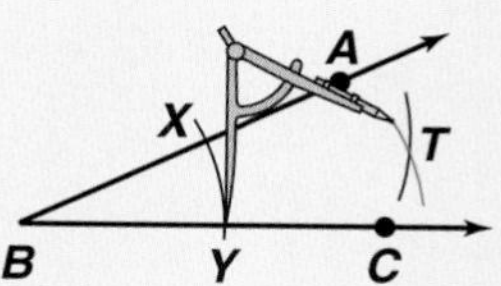

Step 4: With a straightedge, draw $\overrightarrow{BT}$. $\overrightarrow{BT}$ is the bisector of $\angle ABC$.

5. Use a protractor to measure $\angle ABT$ and $\angle TBC$. What is true about their measures?
6. Why do we say BT is the bisector of $\angle ABC$?

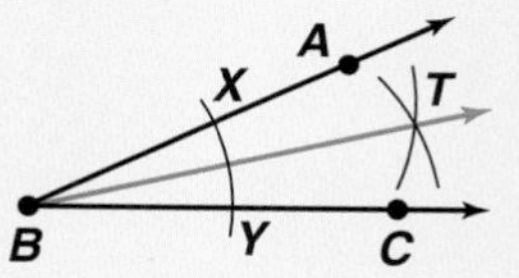

11-2 Integration: Statistics
Making Circle Graphs

Setting Goals: *In this lesson, you'll make a circle graph to illustrate data.*

Modeling a Real-World Application: Weather

Where Lightning Strikes

Location	Percent
Fields, Ballparks	28%
Under trees	17%
Bodies of Water	13%
Near Heavy Equipment	6%
Golf Courses	4%
Telephone Poles	1%
Other / Unknown	31%

Source: National Weather Service

Have you ever been caught in a thunderstorm and wondered where the safest place to be is? The chart at the left indicates where lightning strikes most often.

A circle graph can be used to display this data. What angles would you use for each section?

Learning the Concept

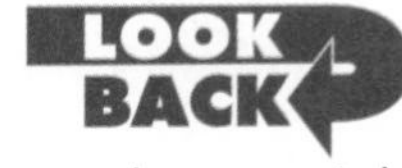

You can review percents in Lesson 9-5.

A **circle graph** is used to compare parts of a whole in a way that helps you visualize the information. As you know, there are 360° in a circle. So, you can multiply to find the number of degrees in each section of the circle graph.

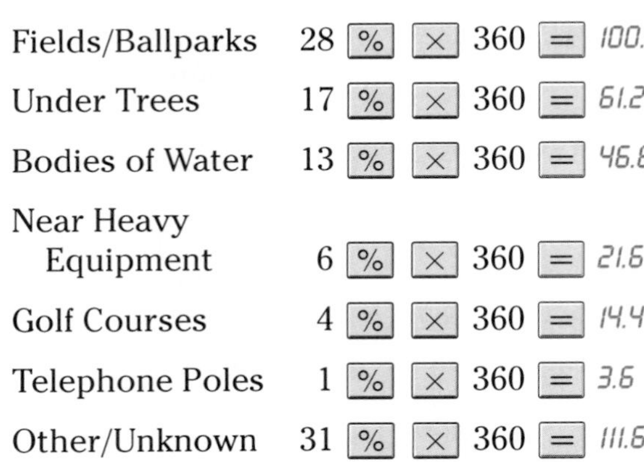

Fields/Ballparks	28	%	×	360	=	100.8
Under Trees	17	%	×	360	=	61.2
Bodies of Water	13	%	×	360	=	46.8
Near Heavy Equipment	6	%	×	360	=	21.6
Golf Courses	4	%	×	360	=	14.4
Telephone Poles	1	%	×	360	=	3.6
Other/Unknown	31	%	×	360	=	111.6

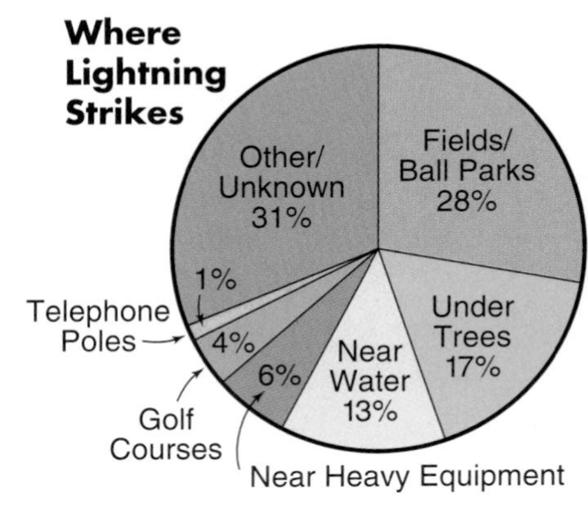

THINK ABOUT IT

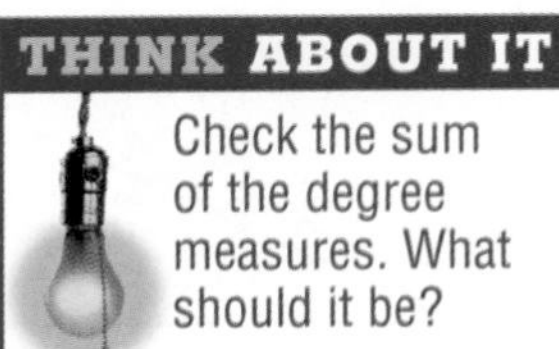

Check the sum of the degree measures. What should it be?

Example 1 shows how to make a circle graph for data that has not been expressed as a percent.

Example CONNECTION

History

You can use spreadsheets and word processing programs to construct circle graphs from data you enter.

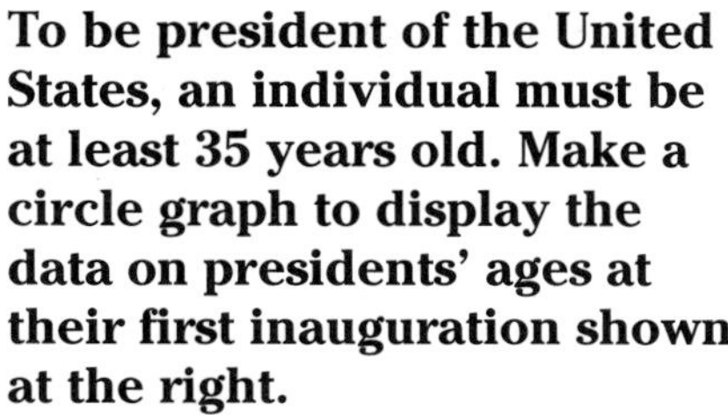

To be president of the United States, an individual must be at least 35 years old. Make a circle graph to display the data on presidents' ages at their first inauguration shown at the right.

President's Age on Inauguration Day	
Age Interval	**Number of Presidents**
35–43	2
44–52	14
53–61	19
62–70	6

Explore You know the number of presidents in each age interval.

Plan Find the total number of presidents. Find the ratio of the number in each interval to the total. Then find the number of degrees for each interval. Finally, draw the graph.

Solve

Step 1: Find the total number of presidents.

$2 + 14 + 19 + 6 = 41$

Step 2: Find the ratio that compares the number in each interval to the total.

35–43: 2 ÷ 41 = 0.048780

44–52: 14 ÷ 41 = 0.341463

53–61: 19 ÷ 41 = 0.463415

62–70: 6 ÷ 41 = 0.146341

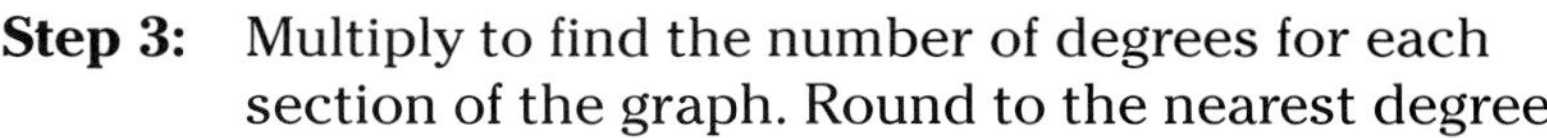

Step 3: Multiply to find the number of degrees for each section of the graph. Round to the nearest degree.

35–43: 0.048780 × 360 = 17.5608 → 18°

44–52: 0.341463 × 360 = 122.92668 → 123°

53–61: 0.463415 × 360 = 166.8294 → 167°

62–70: 0.146341 × 360 = 52.68276 → 53°

Step 4: Draw a circle graph. Use a compass to draw a circle.

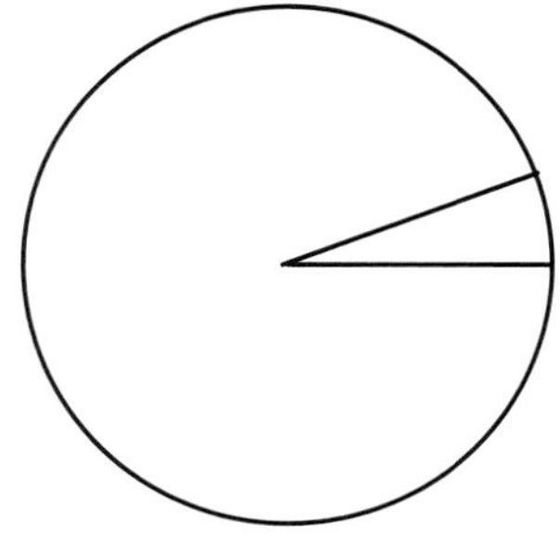

Start with the least number of degrees, in this case 18°. Use a protractor to draw an angle of 18°.

Repeat for the remaining sections. Label each section and give the graph a title.

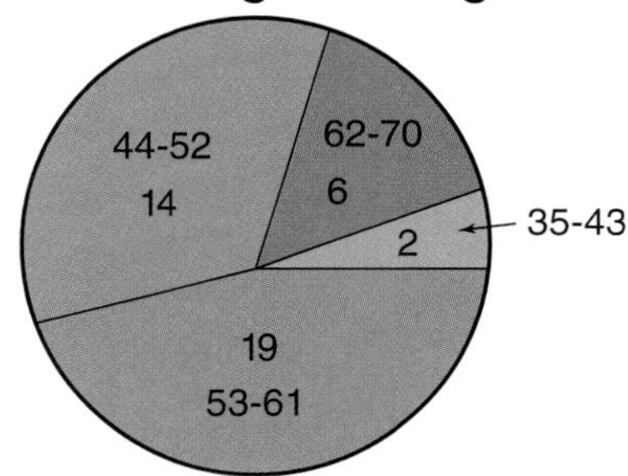

Examine Add the degree measures of each section. The total is 361°. With rounding, this is reasonable.

Checking Your Understanding

Communicating Mathematics

Read and study the lesson to answer these questions.

1. **Explain** how a circle graph differs from a bar graph and from a line graph. When is it appropriate to display data in a circle graph?
2. **Tell** the steps to make a circle graph if your data is in percent form.
3. **Explain** how to find the circle graph angles for data that is not in percent form.

4. Find a circle graph in a newspaper or magazine and make a copy to put in your journal. Write a short paragraph that explains the data displayed by the graph.

Guided Practice

Use the circle graph at the right to answer each question.

Elements of Earth's Crust

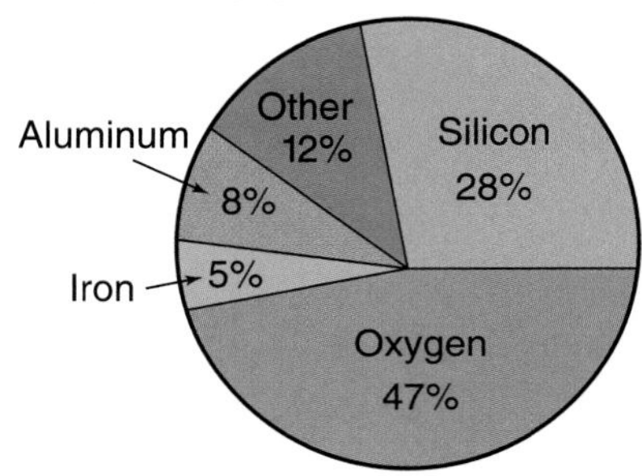

5. Which element makes up approximately $\frac{1}{4}$ of the earth's crust?
6. How much of Earth's crust is composed of oxygen and silicon?
7. The table below shows the Vera family's monthly budget.

Monthly Budget			
Category	**Amount ($)**	**Percent**	**Angle (°)**
Housing	640		
Food	400		
Transportation	450		
Insurance	140		
Savings	90		
Misc.	360		
Totals			

a. What total would you expect in the percent column?
b. What total would you expect in the degree column?
c. Copy and complete the table. Round degrees to the nearest tenth.
d. Make a circle graph to display the data. Use spreadsheet or word processing software if it is available.

8. **Probability** What is the probability of tossing a head when you flip a fair coin? What would a circle graph for all the possible results of this experiment be like?

Exercises: Practicing and Applying the Concept

Independent Practice

9. Count the months of the year that are in each of these three categories: 28 or 29 days long, 30 days long, and 31 days long. Make a circle graph to display your findings.

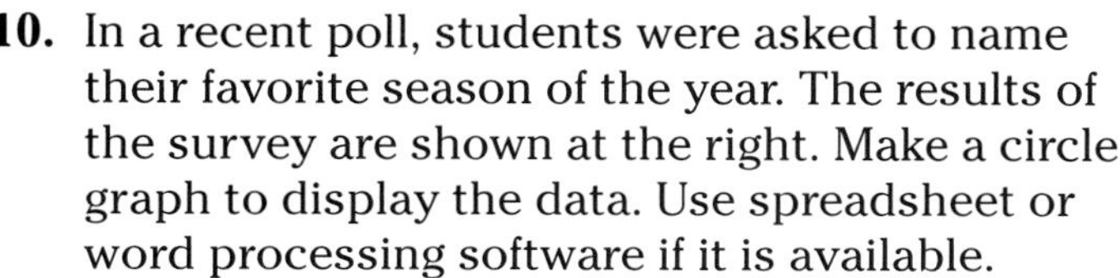

10. In a recent poll, students were asked to name their favorite season of the year. The results of the survey are shown at the right. Make a circle graph to display the data. Use spreadsheet or word processing software if it is available.

Season Preference	
Summer	27%
Autumn	22%
Winter	18%
Spring	33%

11. **Probability** Three red marbles and 5 green marbles are placed in a bag. In an experiment, you are to select one marble. If the marble you select is red, you will win a silver dollar. If the marble you select is green, you will lose.
    - **a.** What is the probability that you will win?
    - **b.** Make a circle graph that shows the probability of winning and losing.

Critical Thinking

12. Suppose someone drew the graph at the right to display information on the audience share during the NBC show *Friends* on Thursday, October 5, 1995.
    - **a.** Does the graph accurately represent the data it presents? Why or why not?
    - **b.** Why might someone have drawn the graph this way?

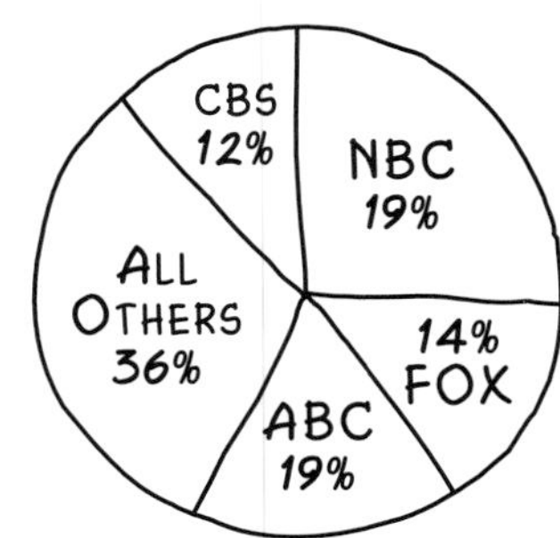

Applications and Problem Solving

Make a circle graph to display each set of data.

13. **Geography**

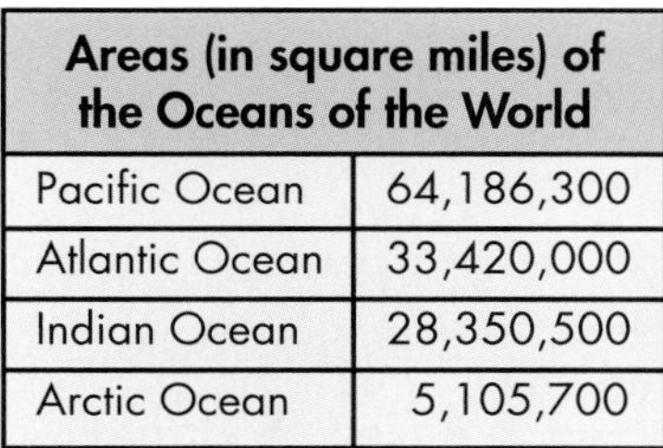

Areas (in square miles) of the Oceans of the World	
Pacific Ocean	64,186,300
Atlantic Ocean	33,420,000
Indian Ocean	28,350,500
Arctic Ocean	5,105,700

14. **Biology**

Chemical Composition of the Human Body	
Oxygen	65%
Carbon	18%
Hydrogen	10%
Nitrogen	3%
Other	4%

15. **Landscaping** A seed company prepares a grass seed mixture that is designed specifically for the Black Hills of South Dakota where the growing season is very short. The following mixture is used.

Black Hills Reclamation Mix	
30% VNS Kentucky Bluegrass	25% VNS Timothy
10% Regar Bromegrass	10% Alsike Clover
10% Lincoln Smooth Brome	10% VNS Tall Fescue
5% Medium Red Clover	

    - **a.** Make a circle graph to display the data.
    - **b.** Find the amount of each seed that would be needed to make 250 pounds of the mixture.

16. **Family Activity** Choose a set of data to collect, such as colors of cars in a parking lot, or the number of commercials in TV commercial breaks and record the data in a frequency table.
    - **a.** Make a circle graph and a bar graph to display the data.
    - **b.** Which graph better represents the data and why?

Mixed Review

17. Draw and label a diagram to represent parallel lines *a* and *b*. (Lesson 11-1)

18. **Business** The phone company assigns the first three digits of a phone number based on the geographic area of the home or business. The remaining four digits are selected from a data base. If any digit from 0 to 9 can be used in any position, how many different combinations of the last four digits are possible? (Lesson 10-6)

19. Graph $y = \frac{1}{2}x - 4$ by using the slope and y-intercept. (Lesson 8-7)

20. The distance from Kuri's house to school is about 1800 feet. About what part of a mile is this? (Lesson 6-5)

21. Write a problem that could be solved using the equation $3x = 25$. (Lesson 1-8)

COOPERATIVE LEARNING PROJECT

THE SHAPE OF THINGS TO COME

Virtual Reality

Imagine yourself caught up in a daydream so real that you can see and hear the sights and sounds of some faraway reality. Perhaps you are walking through a dense forest. Soon you will be able to simply slip on a headset and enter such a world—a world of *virtual reality*, without ever leaving your room.

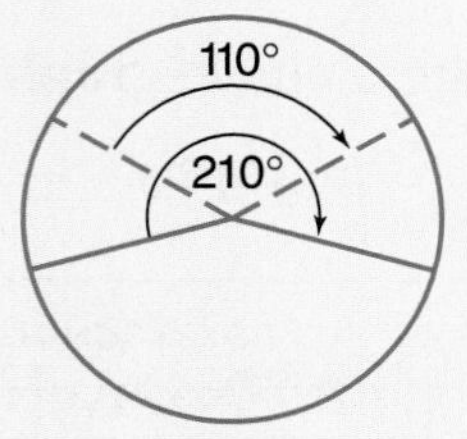

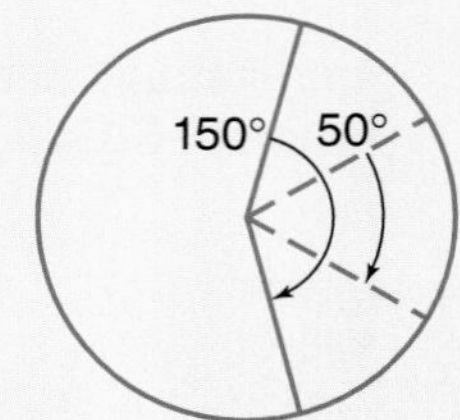

Virtual reality is a three-dimensional computer world. Viewing the images of your virtual reality headset is similar to watching a living, moving graph. Moving your head up and down, the first dimension, allows you to scan whatever images are along the vertical axis. Similarly, moving your head left and right, the second dimension, allows you to scan whatever images are along the horizontal axis. Finally, special lenses cover the screen of your headset and allow you to scan distant objects, the third dimension.

By moving your head from side to side, but not turning your body, your horizontal field of view is about 210°. When you move your head up and down, your vertical field of view is about 150°. Today's virtual reality headsets can only provide a horizontal field of view of about 110° and a vertical field of view of about 50°. But, they should catch up in a few years.

See for Yourself

Research virtual reality systems and computer graphics.

- How does the main computer of a virtual reality system know the direction in which you are looking at any given time?
- Identify and name the three dimensions of a virtual reality system.
- How is a virtual reality screen similar to a living, moving graph?
- With the help of two partners, how can you find the number of degrees of your own horizontal and vertical fields of view?
- Why is the reality of virtual reality headsets referred to as "virtual"?

11-3 Angle Relationships and Parallel Lines

Setting Goals: *In this lesson, you'll identify the relationships of vertical, adjacent, complementary, supplementary angles, and angles formed by two parallel lines and a transversal.*

Modeling with Manipulatives

MATERIALS

protractor

Notice that the writing lines on notebook paper are parallel lines. You can use these lines to investigate angles and parallel lines.

Your Turn

- ▶ Use the lines on notebook paper to draw two parallel lines.
- ▶ Then draw another line that intersects the parallel lines.
- ▶ Use a protractor to measure all the angles formed when this line crosses the parallel lines. Make a sketch that includes the measure of each angle.
- ▶ Repeat this activity using a different line.

TALK ABOUT IT

a. Not including straight angles, how many angles are formed when a line intersects two parallel lines?

b. What do you notice about the measures of the angles?

c. Describe the pairs of angles that appear to have the same measure.

d. What do you notice about measures of angles that are side by side?

Learning the Concept

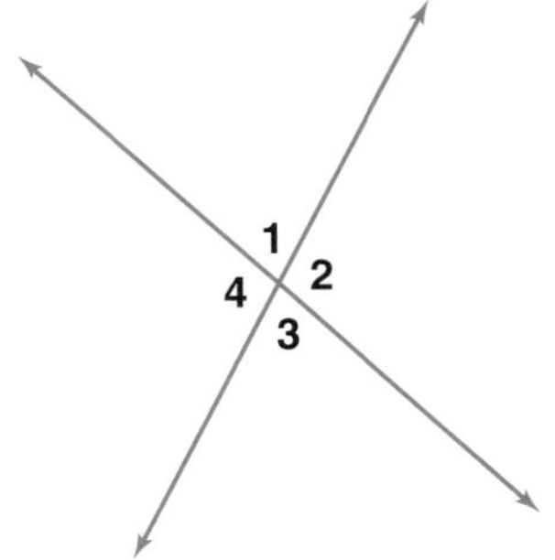

In the activity above, you discovered some of the special relationships that *pairs* of angles can have.

When two lines intersect, they form two pairs of "opposite" angles called **vertical angles**. In the figure at the left, angles 1 and 3 are vertical angles as are angles 2 and 4. If you use a protractor to measure the angles, you will discover that the angles in each pair have the same measure. Angles with the same measure are **congruent**. Vertical angles are always congruent.

When two angles in a plane have the same vertex, share a common side, and do not overlap, they are called **adjacent angles**. In the figure at the right, $\angle 1$ and $\angle 2$ are adjacent angles.

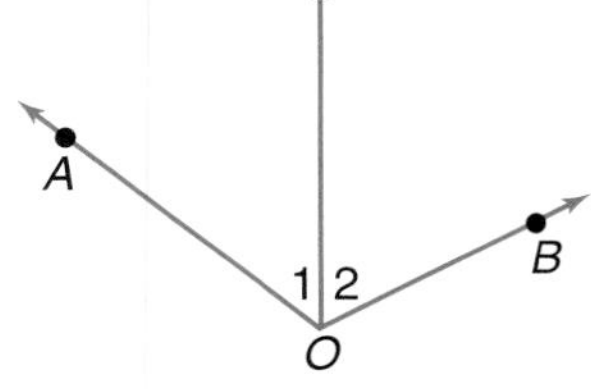

$m\angle AOB$ means the degree measure of angle AOB.

We can find the measure of $\angle AOB$ by adding the measures of $\angle 1$ and $\angle 2$. That is, $m\angle AOB = m\angle 1 + m\angle 2$.

If the sum of the measures of two angles is 90°, the angles are **complementary**.

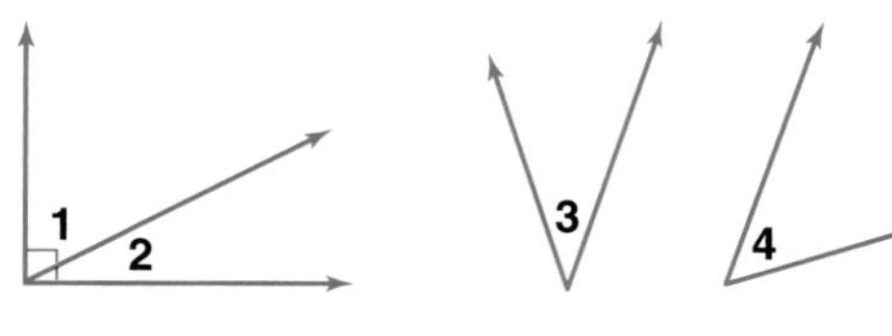

$m\angle 1 + m\angle 2 = 90°$ $m\angle 3 + m\angle 4 = 90°$

If the sum of the measures of two angles is 180°, they are **supplementary**.

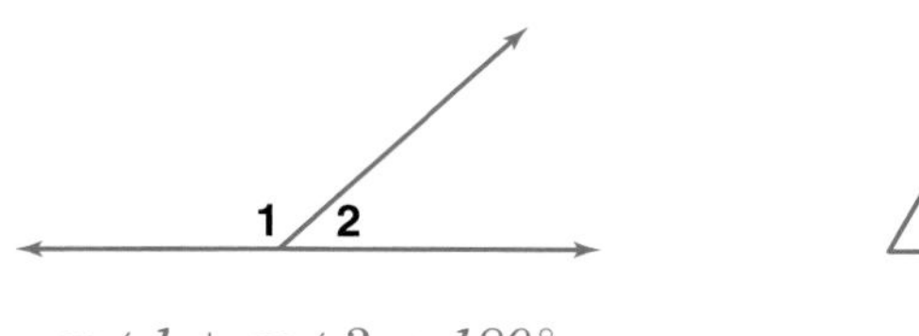

$m\angle 1 + m\angle 2 = 180°$ $m\angle 3 + m\angle 4 = 180°$

Example 1 **Use angles *A*, *B*, and *C* to solve each problem.**

a. Angles *A* and *B* are complementary. If $m\angle A = 70°$, find $m\angle B$.

$m\angle A + m\angle B = 90$ *Definition of complementary angles*
$70 + m\angle B = 90$ *Substitute.*
$m\angle B = 20$ *Subtract 70 from each side.*

The measure of $\angle B$ is 20°.

b. Angles *A* and *C* are supplementary. Find $m\angle C$.

$m\angle A + m\angle C = 180$ *Definition of supplementary angles*
$70 + m\angle C = 180$ *Replace $m\angle A$ with 70.*
$m\angle C = 110$ *Subtract 70 from each side.*

The measure of $\angle C$ is 110°.

Connection to Algebra

Geometric relationships can be expressed using the tools of algebra.

Example 2 INTEGRATION Algebra

Angles *PQR* and *RQT* are complementary. If $m\angle PQR = x + 5$ and $m\angle RQT = x - 9$, find the measure of each angle.

$m\angle PQR + m\angle RQT = 90$ *Definition of complementary angles*
$(x + 5) + (x - 9) = 90$ *Substitute.*
$2x - 4 = 90$ *Combine like terms.*
$2x = 94$ *Add 4 to each side.*
$x = 47$ *Divide each side by 2.*

Now replace *x* with 47 in the expression for each angle.

$m\angle PQR = x + 5 = 47 + 5$ or 52 $m\angle RQT = x - 9 = 47 - 9$ or 38

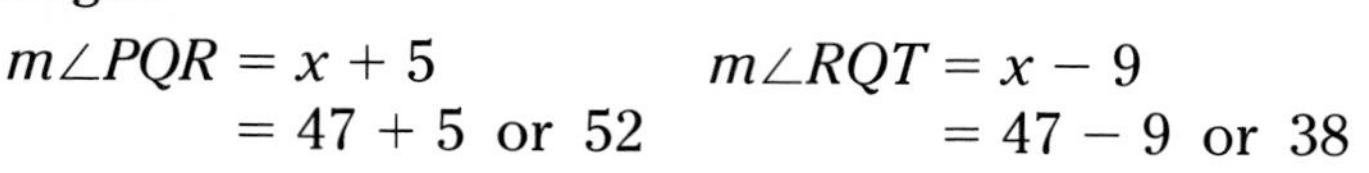

The measure of $\angle PQR$ is 52°, and the measure of $\angle RQT$ is 38°.

Check: $52° + 38° = 90°$ The angles are complementary.

THINK ABOUT IT

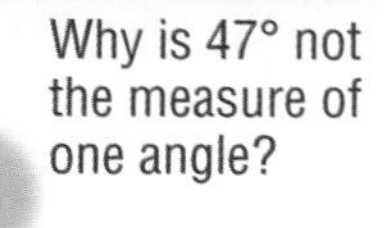

Why is 47° not the measure of one angle?

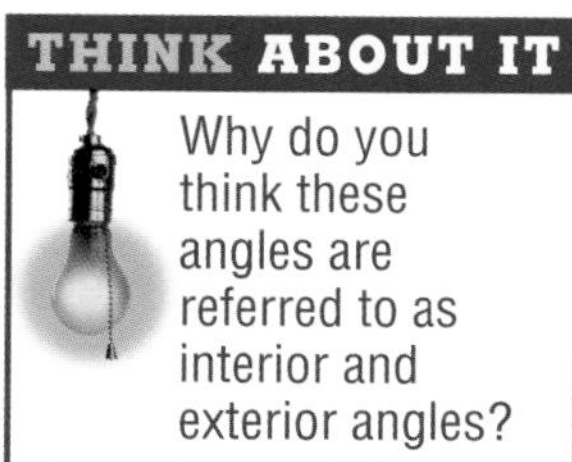

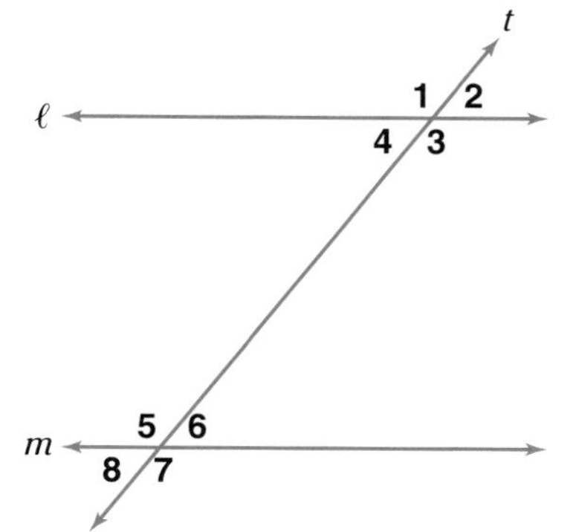

As you discovered in the modeling activity at the beginning of the lesson, when two parallel lines are intersected by a third line, called a **transversal**, eight angles are formed. Four are *interior* angles and four are *exterior* angles.

Interior angles: $\angle 3$, $\angle 4$, $\angle 5$, $\angle 6$
Exterior angles: $\angle 1$, $\angle 2$, $\angle 7$, $\angle 8$

When we study the relationship between the different angles, we can pair the angles as follows.

- **alternate interior angles:** $\angle 4$ and $\angle 6$, $\angle 3$ and $\angle 5$
  Alternate interior angles are nonadjacent interior angles found on opposite sides of the transversal.
- **alternate exterior angles:** $\angle 1$ and $\angle 7$, $\angle 2$ and $\angle 8$
  Alternate exterior angles are nonadjacent exterior angles found on opposite sides of the transversal.
- **corresponding angles:** $\angle 1$ and $\angle 5$, $\angle 2$ and $\angle 6$, $\angle 3$ and $\angle 7$, $\angle 4$ and $\angle 8$
  Corresponding angles are angles that have the same position on two different parallel lines cut by a transversal.

In the activity at the beginning of the lesson, you may have discovered the following relationships.

Parallel Lines Cut by a Transversal

Corresponding angles are congruent.
Alternate interior angles are congruent.
Alternate exterior angles are congruent.

Example 

A transversal crosses two parallel lines so that $m\angle 4 = 75°$.

a. Which other angles also have a measure of 75°?

b. Find the measures of the other angles.

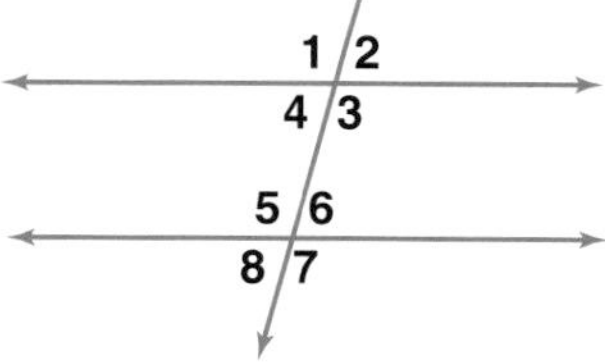

a. Since $\angle 2$ and $\angle 4$ are vertical angles, they are congruent. So $m\angle 2 = 75°$.
Since $\angle 4$ and $\angle 6$ are alternate interior angles, they are congruent. So, $m\angle 6 = 75°$.
$\angle 4$ and $\angle 8$ are corresponding angles so they are congruent. Thus, $m\angle 8 = 75°$.
Angles 2, 4, 6, and 8 have a measure of 75°.

b. Both angles 1 and 3 are supplementary to $\angle 4$ so the sums of each of their measures and the measure of angle 4 is 180°.
$180 - 75 = 105$. So, $m\angle 1 = 105°$ and $m\angle 3 = 105°$.
Since $\angle 7$ and $\angle 1$ are alternate exterior angles, they are congruent. So, $m\angle 7 = 105°$.
$\angle 5$ and $\angle 1$ are corresponding angles so they are congruent. Thus, $m\angle 5 = 105°$.

Angles 1, 3, 5, and 7 have a measure of 105°.

Example 4 — APPLICATION: Carpentry

A carpenter uses a protractor and a plumb line (a string with a weight attached) to measure the angle between a slanted ceiling and the wall. If $m\angle WXY = 65°$, find $m\angle XYZ$.

$\angle WXY$ is supplementary to $\angle WXV$.

$m\angle WXY + m\angle WXV = 180$

$65 + m\angle WXV = 180$

$m\angle WXV = 115$

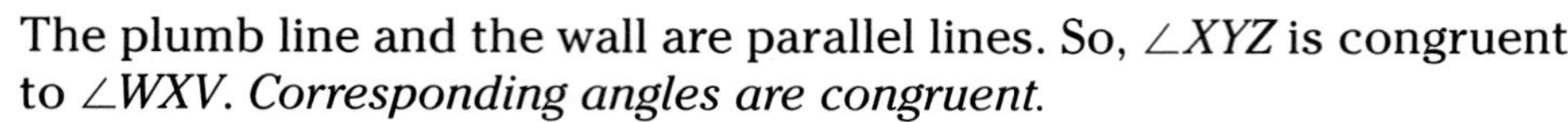

The plumb line and the wall are parallel lines. So, $\angle XYZ$ is congruent to $\angle WXV$. *Corresponding angles are congruent.*

$m\angle XYZ = m\angle WXV$

$m\angle XYZ = 115$

The measure of $\angle XYZ$ is 115°.

Checking Your Understanding

Communicating Mathematics

Read and study the lesson to answer these questions.

1. **Compare and contrast** vertical angles and adjacent angles.
2. **Draw** $\overline{MN} \parallel \overline{PQ}$ and transversal $\overline{RS}$. Number the angles and name a pair of each.
   a. corresponding angles
   b. alternate interior angles
   c. alternate exterior angles
   d. adjacent angles
   e. adjacent, supplementary angles

Draw and label a diagram to show each pair of angles and describe their characteristics.

3. $\angle WXY$ and $\angle YXZ$ are adjacent angles.
4. $\angle GHI$ and $\angle IHJ$ are adjacent, complementary angles.
5. $\angle KLM$ and $\angle MLN$ are adjacent, supplementary angles.

Guided Practice

6. Identify the figure in which angles 1 and 2 are supplementary.

a. 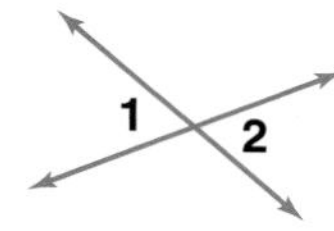

b. 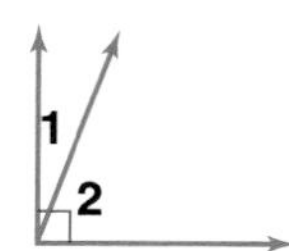

c. 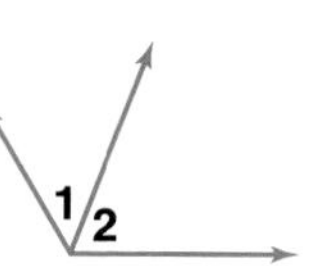

d. 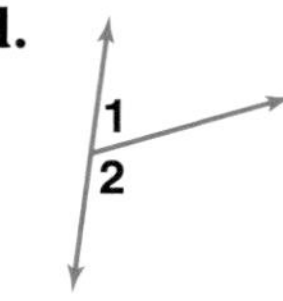

Find the value of *x* in each figure.

7. 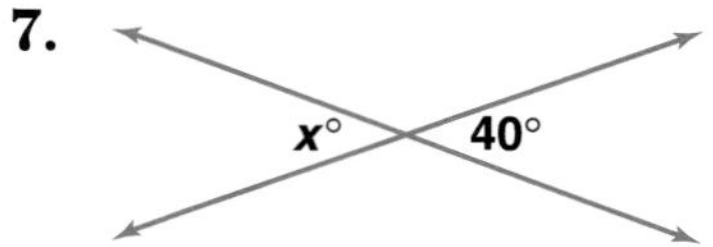

8. 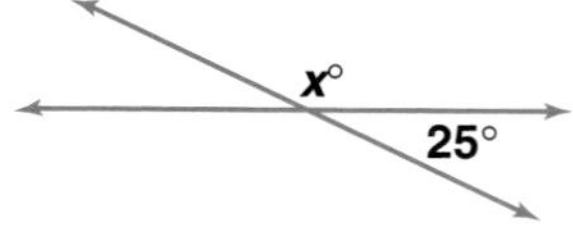

9. Angle C and $\angle D$ are complementary. Find $m\angle C$ if $m\angle D$ is 75°.
10. Angle P and $\angle Q$ are supplementary. Find $m\angle P$ if $m\angle Q$ is 90°.
11. Angles X and Y are supplementary. $m\angle X = 2x$ and $m\angle Y = 4x$. Find the measure of each angle.
12. Angles F and G are complementary. $m\angle F = x + 8$ and $m\angle G = x - 10$. Find the measure of each angle.

Refer to the diagram at the right to complete Exercises 13–15.

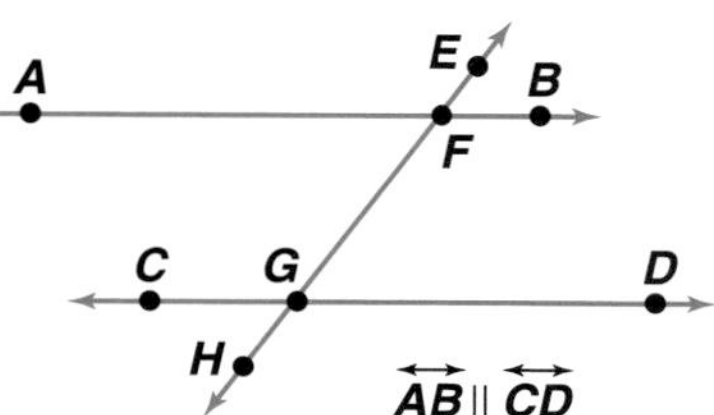

13. Name the interior angles.
14. Name the exterior angles.
15. **a.** Use a protractor to find the measure of one angle.
    b. Use the measure from **part a** to find all the other angle measures.
16. **Public Transit** The angle at the corner where two streets intersect is 125°. If a bus cannot make a turn at an angle of less than 70°, can bus service be provided on a route that includes turning that corner in both directions? Explain.

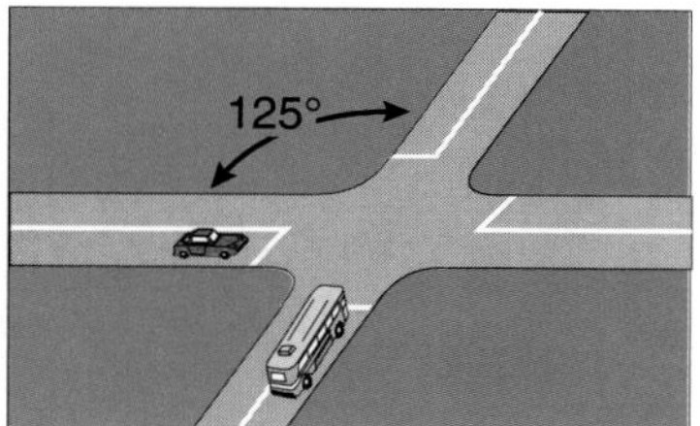

Exercises: Practicing and Applying the Concept

Independent Practice

Find the value of x in each figure.

17.

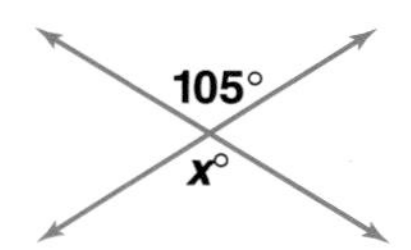

18.

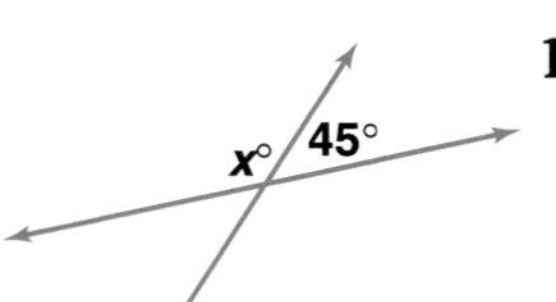

19.

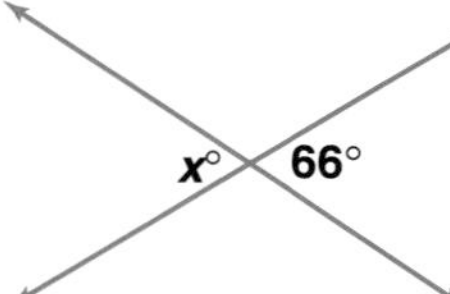

20.

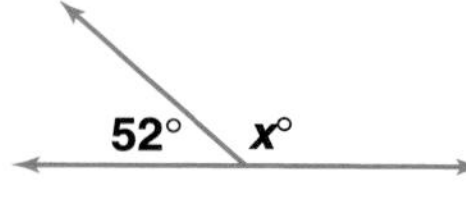

21.

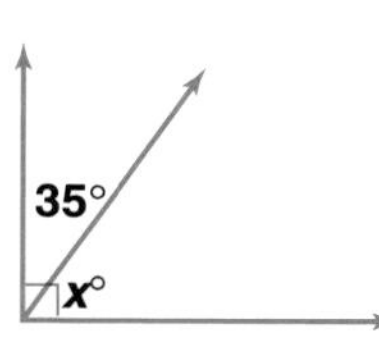

22.

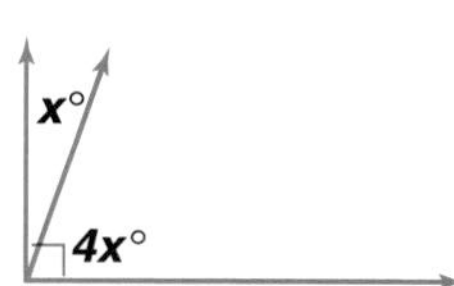

Find the measure of each angle.

23.

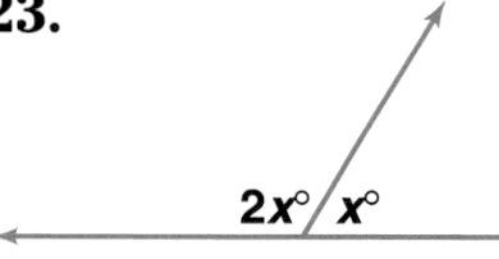

24.

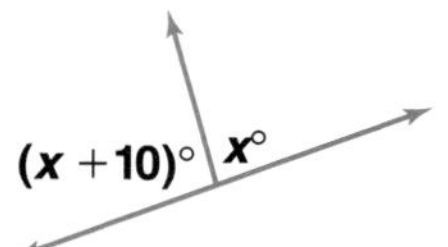

25.

In the figure at the right, $\ell \parallel m$. If the measure of $\angle 1$ is 47°, find the measure of each angle.

26. $\angle 2$
27. $\angle 3$
28. $\angle 5$
29. $\angle 4$
30. $\angle 7$
31. $\angle 8$

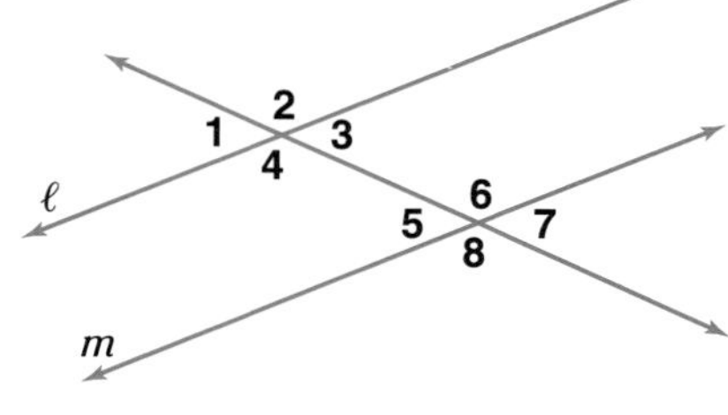

Each pair of angles is complementary. Find each angle measure.

32. $m\angle A = 2x + 15, m\angle B = x - 3$
33. $m\angle R = x - 9, m\angle S = x + 17$

Each pair of angles is supplementary. Find the measure of each angle.

34. $m\angle T = 2x + 17$ and $m\angle S = 5x - 40$
35. $m\angle F = 3x + 40$ and $m\angle G = 2x + 10$

In the figure at the right, $\ell \parallel m$. Find the value of x for each of the following. *The figure is not drawn to scale.*

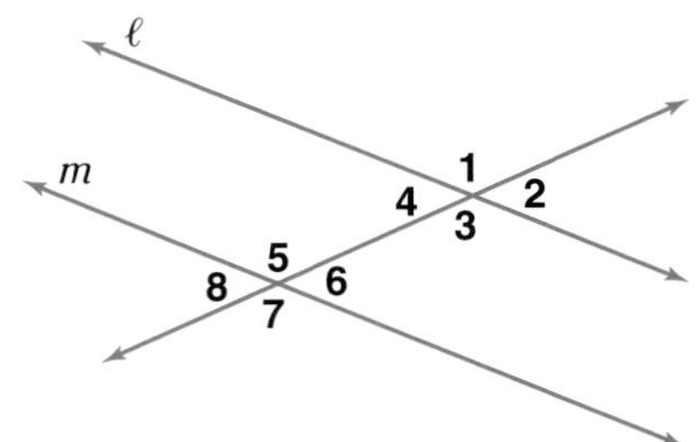

36. $m\angle 1 = 6x - 20$ and $m\angle 2 = x + 40$

37. $m\angle 4 = 3x + 17$ and $m\angle 5 = 2x + 13$

38. $m\angle 6 = 9x - 10$ and $m\angle 7 = 5x + 36$

39. The measure of the supplement of an angle is 15° less than four times the measure of the complement. Find the measure of the angle.

Critical Thinking

Draw a diagram for each situation. Use a protractor to measure the angles formed and answer each question.

40. If two parallel lines are cut by a transversal, how are the interior angles on the same side of the transversal related? Explain.

41. If a transversal is perpendicular to one of two parallel lines, is it perpendicular to the other parallel line also? Explain.

Applications and Problem Solving

42. Physics The drawing at the right shows a beam of light reflected from the surface of a flat mirror. The angle of incidence (the angle between the incident ray and the normal line) and the angle of reflection (the angle between the reflected ray and the normal line) are congruent. The normal line is perpendicular to the surface of the mirror. Suppose a beam of light strikes the mirror at a 52° angle to the mirror. What is the measure of the angle between the incident ray and the reflected ray?

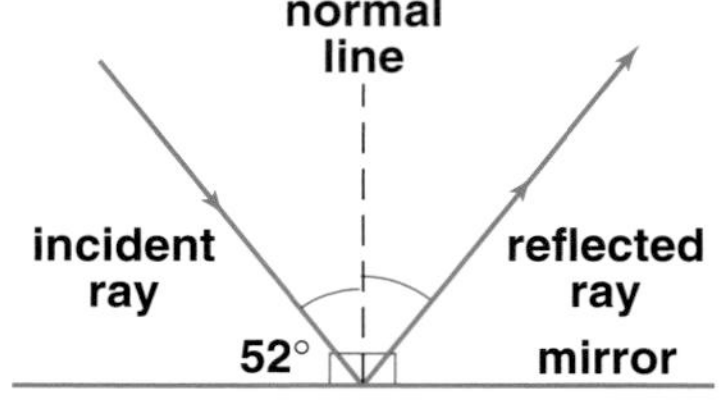

43. Carpentry A standard stair rail is designed to make an angle of 45° with the floor. The first vertical post and the lower rail also form an angle of 45°. If the upper rail is parallel to the lower rail, what angle should it make with the first vertical post?

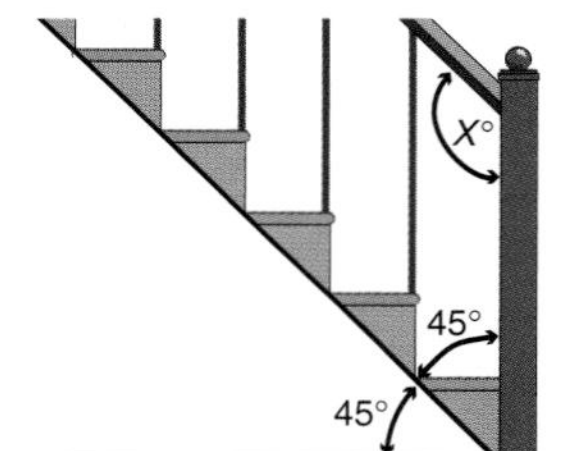

Mixed Review

44. Retail Sales 17% of Foodtown's sales come from the produce department. Find the measure of the angle to represent produce sales for a circle graph showing Foodtown's sales. (Lesson 11-2)

45. Probability Two dice are thrown. Find $P(\text{sum} > 8)$. (Lesson 9-3)

46. Biology A certain bacteria doubles its population every 12 hours. After 3 full days, there are 1600 bacteria. How many bacteria were there at the beginning of the first day? (Lesson 7-1)

47. Write 56,780 in scientific notation. (Lesson 6-9)

48. Write $\frac{12}{54}$ in simplest form. (Lesson 4-6)

49. Simplify $9b + (-16)b$. (Lesson 2-4)

50. Evaluate $|-5| - |2|$. (Lesson 2-1)

51. Solve $8 + n = 13$ mentally. (Lesson 1-6)

GRAPHING CALCULATOR ACTIVITY

Slopes of Parallel Lines

An Extension of Lesson **11-3**

MATERIALS

You can determine whether two lines are parallel by finding their slopes. In Lesson 8-6, you learned that parallel lines have the same slope. So, if two lines have the same slope, then they are parallel. The graphing calculator program at the right finds the slope of a line when you enter the coordinates of two points on the line.

```
PROGRAM:SLOPE
: Disp "ENTER"
: Disp "COORDINATES"
: Disp "POINT 1"
: Input A
: Input B
: Disp "POINT 2"
: Input C
: Input D
: If A = C
: Then
: Disp "SLOPE UNDEFINED"
: Stop
: End
: (D-B)/(C-A) → M
: Disp "SLOPE = ", M
```

Your Turn

Work with a partner.

- ▶ Enter the program in a TI-82 graphing calculator.
- ▶ Find the coordinates of two points on line m below, Point 1 and Point 2.

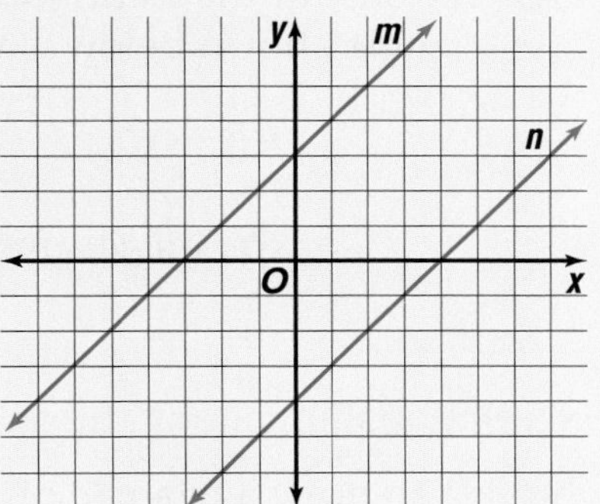

- ▶ Enter the coordinates in the program. Record the slope of line m.
- ▶ Now find the coordinates of two points on line n above.
- ▶ Enter the coordinates in the program. Record the slope of line n.

1. How does the program calculate the slope of the line?
2. How do the slopes of lines m and n compare?
3. Based on their slopes, are lines m and n parallel?

Your Turn

- ▶ Now draw two different lines that you think are parallel on graph paper.
- ▶ Find the coordinates of two points on each line.
- ▶ Enter the coordinates in the program to find the slope of each line.
- ▶ Repeat the steps above for two lines that you think are *not* parallel.

TALK ABOUT IT

4. What were the results for the lines that you drew?
5. Suppose you wanted to draw two lines that were parallel. How could you draw the lines to be sure that they were parallel?

11-4 Triangles

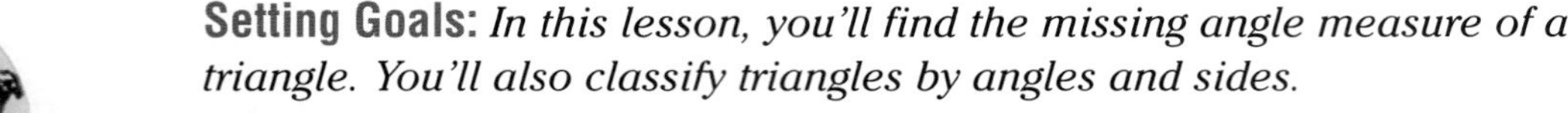

Setting Goals: *In this lesson, you'll find the missing angle measure of a triangle. You'll also classify triangles by angles and sides.*

Modeling a Real-World Application: Engineering

Civil engineers who design and build railroad bridges often choose a *truss* bridge to span canyons and rivers. A truss bridge can span more than 1000 feet and requires less building material than other types of bridges.

Learning the Concept

> **HELP WANTED**
>
> Civil engineers use geometry to make beautiful and functional designs. For more information on this field, contact:
>
> American Society of Civil Engineers
> 345 East 47th Street
> New York, NY 10017

Trusses get their strength from braces that form triangles. Because triangles are rigid, they are strong. A **triangle** is formed by three line segments that intersect only at their endpoints. A triangle can be named by its vertices. The triangle shown below is triangle *PQR*, or $\triangle PQR$.

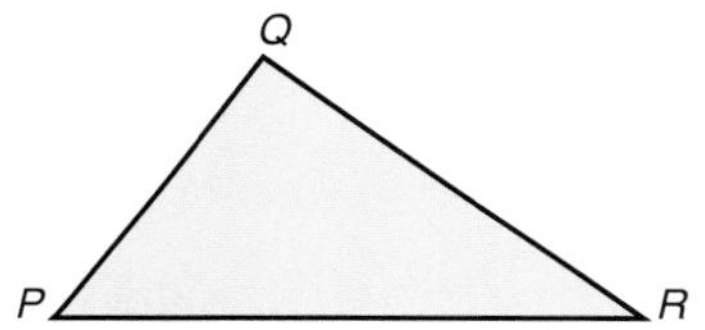

The vertices are P, Q, and R.
The sides are $\overline{PQ}$, $\overline{QR}$, and $\overline{PR}$.
The angles are $\angle P$, $\angle Q$, and $\angle R$.

There is a special relationship among the angles of a triangle.

1. Use a straightedge to draw a triangle on a piece of paper. Cut out the triangle and label the vertices *A*, *B*, and *C*.
2. Fold the triangle so that the point *C* lies on $\overline{AB}$ and the fold is parallel to $\overline{AB}$. Label $\angle C$ as $\angle 2$.
3. Fold again so point *A* meets the vertex of $\angle 2$. Label $\angle A$ as $\angle 1$.
4. Finally, fold so point *B* also meets the vertex of $\angle 2$. Label $\angle B$ as $\angle 3$.

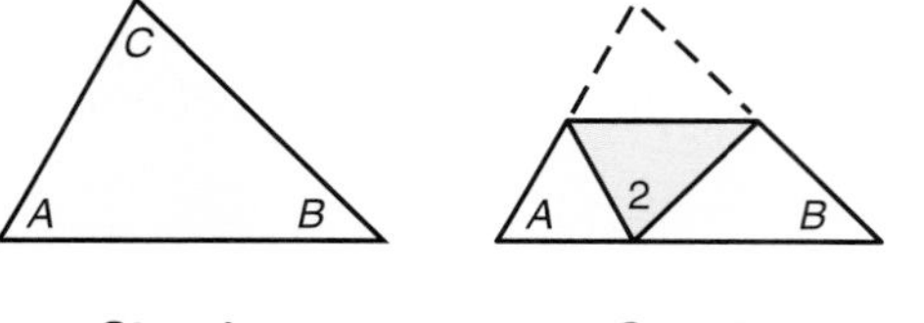

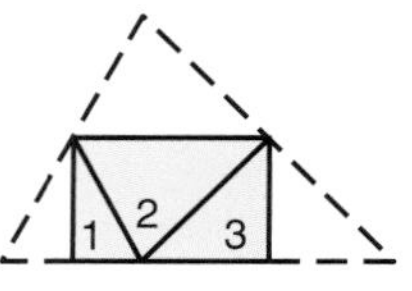

Step 1　Step 2　Step 3　Step 4

What do you notice about the sum of the measures of angles 1, 2, and 3? This activity suggests the following relationship.

> **Angles of a Triangle**　The sum of the measures of the angles of a triangle is 180°.

Example  **In $\triangle CAT$, the measure of $\angle C$ is 47°, and the measure of $\angle A$ is 59°. Find the measure of $\angle T$.**

Estimate: *50 + 60 = 110 and 180 − 110 = 70, so $m\angle T$ is about 70°.*

$m\angle C + m\angle A + m\angle T = 180$ *The sum of the measures of the angles of a triangle is 180°.*

$47 + 59 + m\angle T = 180$ *Replace $m\angle C$ with 47 and $m\angle A$ with 59.*

$106 + m\angle T = 180$ *Add 47 + 59.*

$m\angle T = 74$ *Subtract 106 from each side.*

The measure of $\angle T$ is 74°.

Connection to Algebra

The relationships of the angles in a triangle can be represented using algebra.

Example 

Algebra

The measures of the angles of a certain triangle are in the ratio 2:3:4. Find the measure of each angle.

Let $2x$ represent the measure of one angle, $3x$ the measure of a second angle, and $4x$ the measure of the third angle.

$2x + 3x + 4x = 180$ *The sum of the measures is 180°.*

$9x = 180$ *Combine like terms.*

$x = 20$ *Divide each side by 9.*

$2x = 40$, $3x = 60$, and $4x = 80$.

The measures of the angles are 40°, 60°, and 80°. *Check by adding.*

A triangle can be classified by its angles. Every triangle has two acute angles. The third angle can be used to classify the triangle.

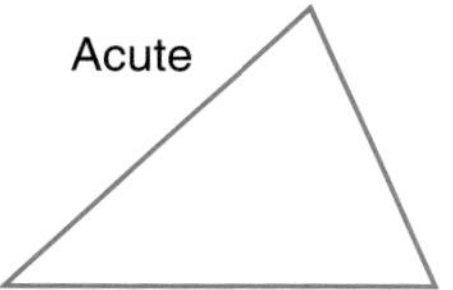

An **acute triangle** has three acute angles.

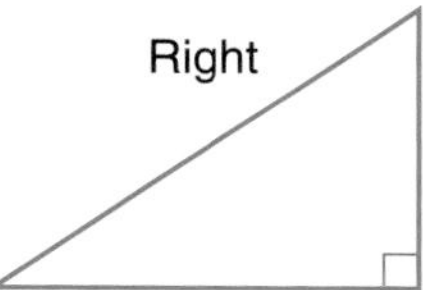

A **right triangle** has one right angle.

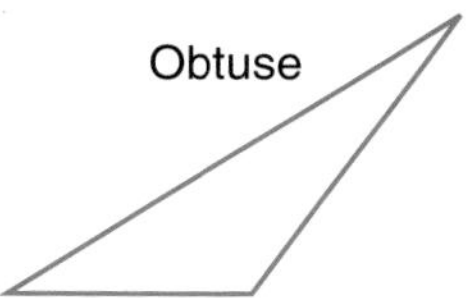

An **obtuse triangle** has one obtuse angle.

Triangles can also be classified by the number of congruent sides.

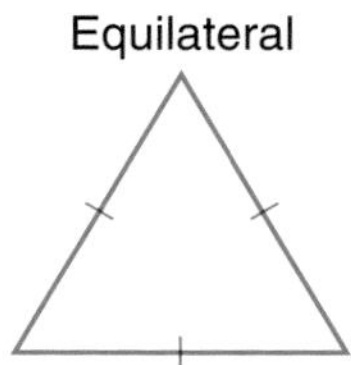

An **equilateral triangle** has three congruent sides.

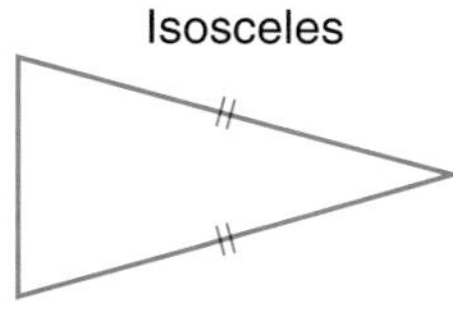

An **isosceles triangle** has at least two congruent sides.

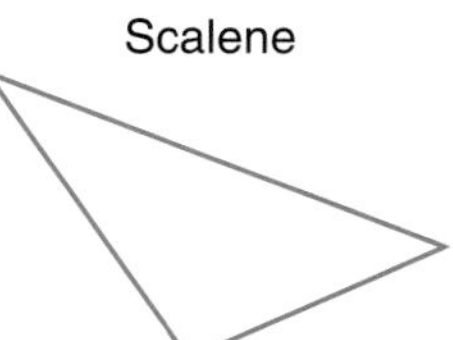

A **scalene triangle** has no congruent sides.

Checking Your Understanding

Communicating Mathematics

Read and study the lesson to answer these questions.

1. If you know the measures of two of the angles in a triangle, how can you find the measure of the third angle?

An A-frame house is shaped like an isosceles triangle. Give a real-world example of the following triangles.

2. obtuse
3. equilateral
4. right

5. Make up memory tools you can use to remember the names of triangles that are classified by sides or angles. Record the tool in your journal.

Guided Practice

Find the value of x. Then classify each triangle as *acute*, *right*, or *obtuse*.

6. 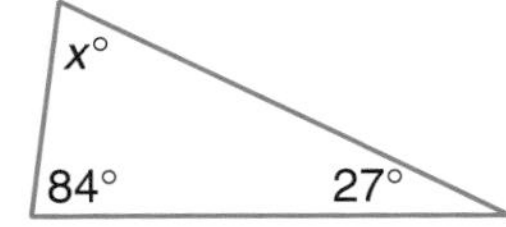

7. 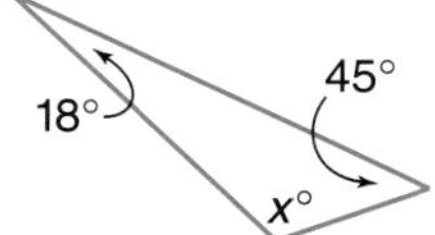

8. **Algebra** The measures of the angles of a certain triangle are in the ratio 1:3:5. Find the measures of each angle.

Use a ruler to determine the number of congruent sides in each triangle. Then classify the triangle as *scalene*, *isosceles*, or *equilateral*.

9. 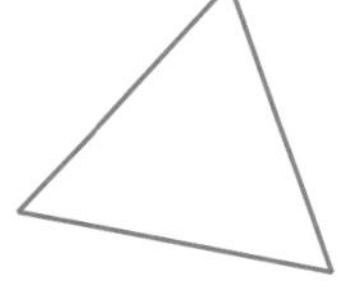

10. 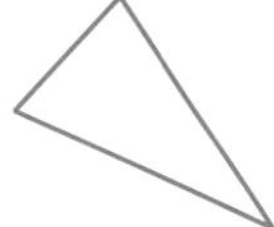

11. **Construction** Find the measure of the missing angle, $x°$, in the roof truss shown at the right.

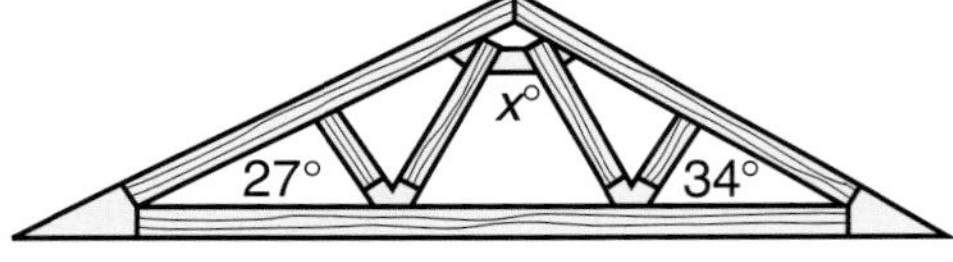

Exercises: Practicing and Applying the Concept

Independent Practice

Find the value of x. Then classify each triangle as *acute*, *right*, or *obtuse*.

12. 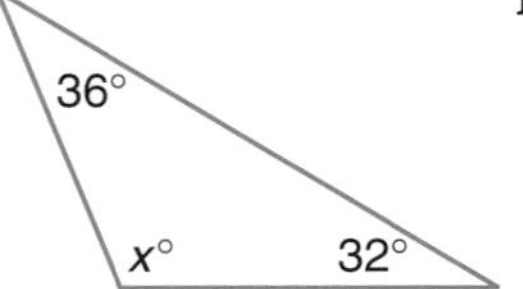

13. 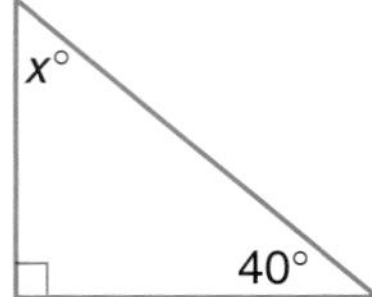

14. 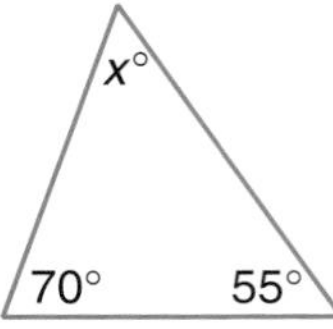

Choose
Estimation
Mental Math
Calculator
Paper and Pencil

15. The measures of the angles of a triangle are in the ratio 2:3:5. Find the measure of each angle.
16. The measures of the angles of a triangle are in the ratio 1:4:7. Find the measure of each angle.

Use a ruler to determine the number of congruent sides in each triangle. Then classify the triangle as *scalene*, *isosceles*, or *equilateral*.

17. 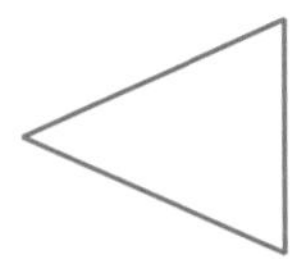

18. 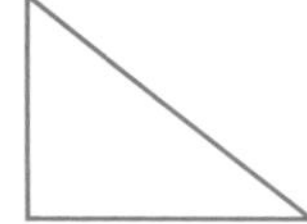

19. 

Use a protractor to determine the measures of the angles in each triangle. Then classify the triangle as *acute*, *right*, or *obtuse*.

20. 

21. 

22. 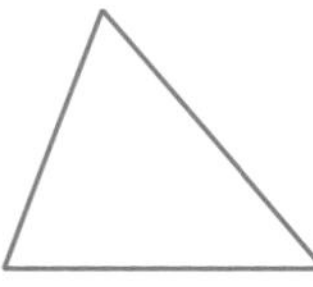

Use the figure at the right to solve each of the following.

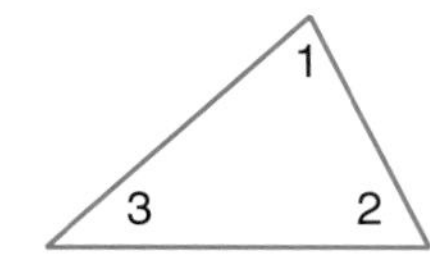

23. Find $m\angle 1$ if $m\angle 2 = 40°$ and $m\angle 3 = 55°$.

24. Find $m\angle 1$ if $m\angle 2 = 60°$ and $m\angle 3 = 60°$.

25. Find $m\angle 1$ if $m\angle 2 = 81°$ and $m\angle 3 = 74°$.

26. Find $m\angle 2$ if $m\angle 1 = 45°$ and $m\angle 3 = 75°$.

27. Find $m\angle 2$ if $m\angle 1 = 47°$ and $m\angle 3 = 48°$.

Find the measures of the angles in each triangle.

28. 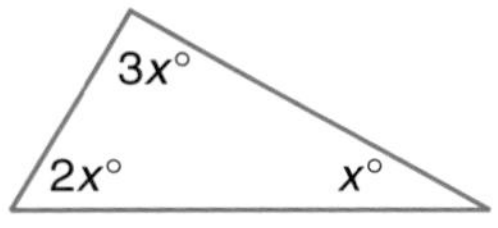

29. 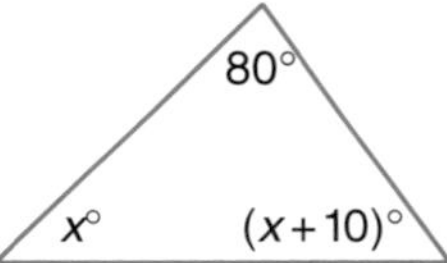

30. 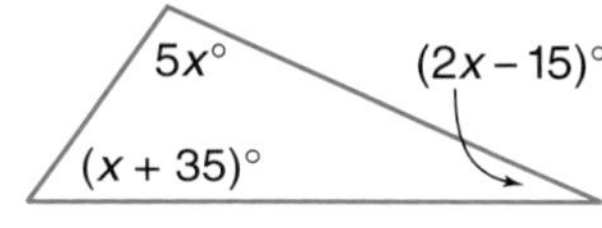

Critical Thinking

31. Explain how you could use a piece of string to convince someone that a given triangle was:

a. scalene **b.** isosceles **c.** equilateral

32. Are the acute angles of a right triangle complementary? Explain.

Applications and Problem Solving

33. Civil Defense The Civil Defense program is a civilian organization created to help people when a disaster occurs. The Civil Defense symbol is shown at the right. If the triangle pictured in the symbol is an equilateral triangle, and all of the angles are congruent, what is the measure of each angle?

34. Number Theory Numbers that can be represented by a triangular arrangement of dots are called *triangular numbers*. The first three triangular numbers are 1, 3, and 6. Find the next four triangular numbers and draw the triangular arrangement of dots for each one.

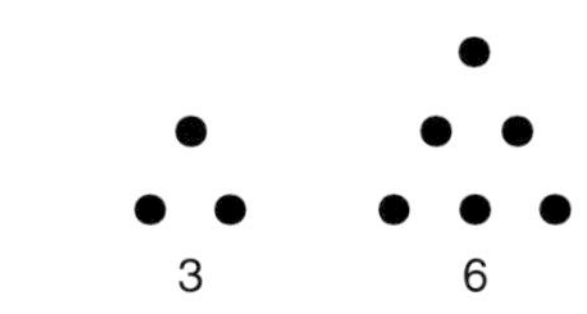

35. **Public Utilities** Above-ground electric wires and transformers, are mounted on utility poles. A support cable, called a guy wire, is sometimes attached to give the pole stability. If the guy wire makes an angle of 65° with the ground, what is the measure of the angle formed by the guy wire and the pole? Assume that the ground is level.

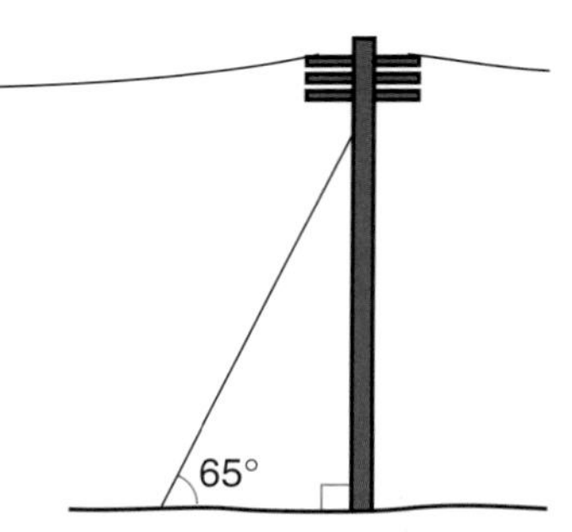

Mixed Review

36. Angles X and Y are supplementary. If $m\angle X = 35°$, find $m\angle Y$. (Lesson 11-3)

37. **Probability** Find the odds of rolling a sum less than 6 if a pair of dice are rolled. (Lesson 10-7)

38. Graph $y > x + 3$. (Lesson 8-9)

39. **Sewing** A curtain pattern says to measure the length of the window and add 4 inches for curtain length. Then it says to add 4 inches for a heading, $2\frac{1}{2}$ inches for a rod casting, and 6 inches for a hem. Mary's window is $42\frac{3}{4}$ inches long. How long should the fabric be? (Lesson 5-5)

40. Replace each ● with $<$, $>$, or $=$. (Lesson 2-3)

a. -8 ● -14 **b.** $-|20|$ ● -20 **c.** 0 ● $|-2|$

41. Simplify $4(x + 3)$. (Lesson 1-5)

Self Test

Use a protractor to draw an angle with the given measurements. Classify each angle as *acute*, *right*, or *obtuse*. (Lesson 11-1)

1. 60°
2. 115°
3. 20°
4. **Retail** At Recordtown, 42% of sales are from CDs, 28% are from tapes, 18% are from equipment, and 12% are from other items. Make a circle graph of this data. (Lesson 11-2)

In the figure at the right, ℓ is parallel to m. If the measure of $\angle 3$ is 34°, find the measure of each angle. (Lesson 11-3)

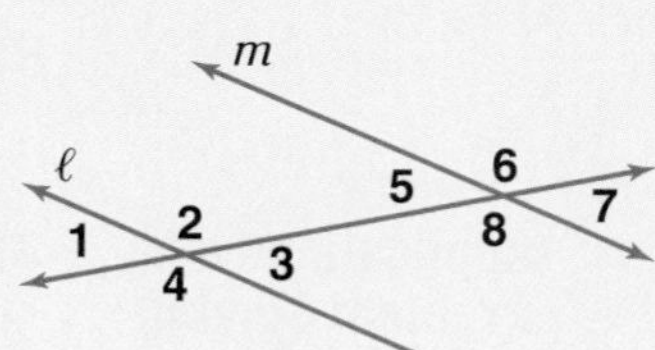

5. $\angle 1$
6. $\angle 4$
7. $\angle 5$

Find the measure of each angle using the following information. (Lesson 11-3)

8. $\angle A$ and $\angle B$ are complementary. $m\angle A = x + 25$; $m\angle B = 2x - 10$
9. $\angle F$ and $\angle G$ are supplementary. $m\angle F = 3x - 50$; $m\angle G = 2x - 20$

Find the value of x. Then classify each triangle as *acute*, *right*, or *obtuse*. (Lesson 11-4)

10. 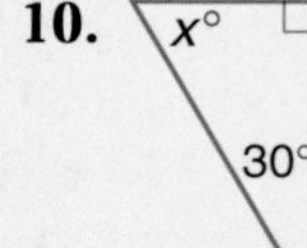

11. 42°, x°, 38°

12. 65°, x°, 52°

11-5 Congruent Triangles

Setting Goals: *In this lesson, you'll identify congruent triangles and corresponding parts of congruent triangles.*

Modeling with Technology

You can use a graphing calculator to plot points and draw line segments connecting them. By plotting three points and the connecting segments, you can graph triangles.

On the TI-82, you will use the line feature from the draw menu. When you enter two points, the calculator graphs the line segment between them.

Your Turn

Graph $\triangle XYZ$ with vertices $X(10, 5)$, $Y(30, 5)$, $Z(5, 25)$ and $\triangle X'Y'Z'$ with vertices $X'(-15, 2)$, $Y'(5, 2)$, and $Z'(-20, 22)$ using the steps below.

- Press ZOOM 6 ZOOM 8 ENTER. *Sets the screen to integer values.*
- Press 2nd DRAW 2.
- Use the arrow keys to move the cursor to (10, 5). The coordinates at the bottom of the screen will help you find the point. Press ENTER.
- Then move the cursor to (30, 5). Press ENTER twice.
- Move the cursor to (5, 25) and press ENTER twice. Then move the cursor back to (10, 5) to complete the triangle. Press ENTER.
- Repeat the third, fourth, and fifth steps above using the coordinates of $\triangle X'Y'Z'$ to draw the second triangle.

a. How do the size and shape of $\triangle XYZ$ and $\triangle X'Y'Z'$ compare?

b. If you placed $\triangle X'Y'Z'$ on top of $\triangle XYZ$, which angles would match up?

c. Which sides would match up?

Learning the Concept

Figures that have the same size and shape are **congruent**. The triangles shown below are congruent. Slash marks are used to indicate which *sides* are congruent and arcs are used to indicate which *angles* are congruent. Remember, the symbol $\cong$ means *is congruent to*.

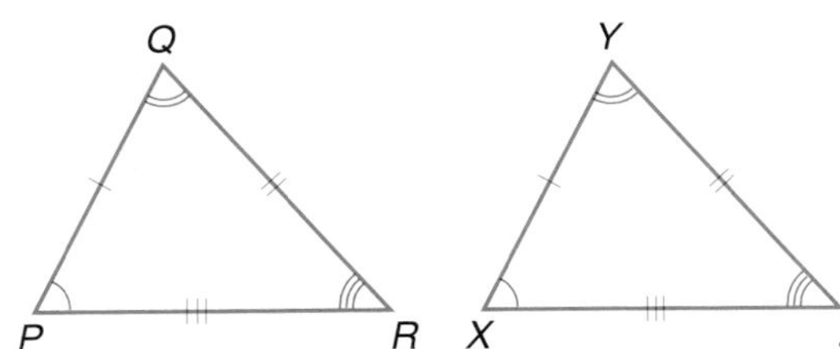

Congruent Angles	Congruent Sides
$\angle P \cong \angle X$	$\overline{PQ} \cong \overline{XY}$
$\angle Q \cong \angle Y$	$\overline{QR} \cong \overline{YZ}$
$\angle R \cong \angle Z$	$\overline{PR} \cong \overline{XZ}$

Parts of congruent triangles that "match" are called **corresponding parts**. For example, in the triangle above, $\angle P$ corresponds to $\angle X$, and $\overline{PQ}$ corresponds to $\overline{XY}$.

Corresponding Parts of Congruent Triangles

If two triangles are congruent, their corresponding sides are congruent and their corresponding angles are congruent.

When you write $\triangle PQR \cong \triangle XYZ$, the corresponding vertices are written in order. So, $\triangle PQR \cong \triangle XYZ$ means that vertex P corresponds to vertex X, vertex Q corresponds to vertex Y, and vertex R corresponds to vertex Z.

Example  **If $\triangle COD \cong \triangle ATM$, name the congruent angles and sides.**

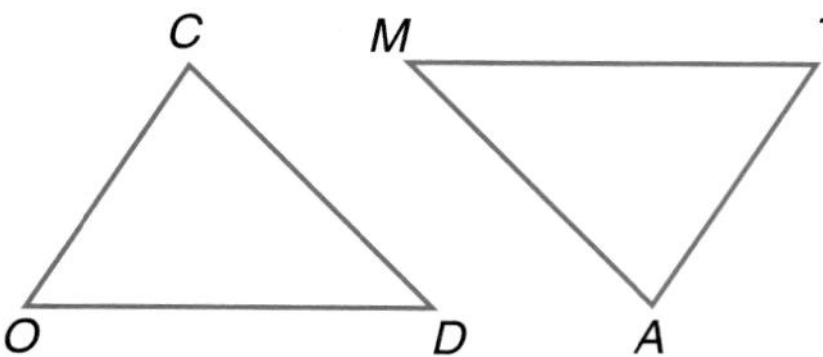

First, name the three pairs of congruent angles by looking at the order of the vertices.

$\angle C \cong \angle A$
$\angle O \cong \angle T$
$\angle D \cong \angle M$

Since C corresponds to A and O corresponds to T, $\overline{CO}$ corresponds to $\overline{AT}$. Similarly $\overline{OD}$ corresponds to $\overline{TM}$, and $\overline{CD}$ corresponds to $\overline{AM}$. The congruent sides are as follows.

$\overline{CO} \cong \overline{AT}$, $\overline{OD} \cong \overline{TM}$, $\overline{CD} \cong \overline{AM}$

You can draw conclusions about congruent triangles based on the congruence statements.

Example  **The corresponding parts of two congruent triangles are marked in the figure. Write a congruence statement for the two triangles.**

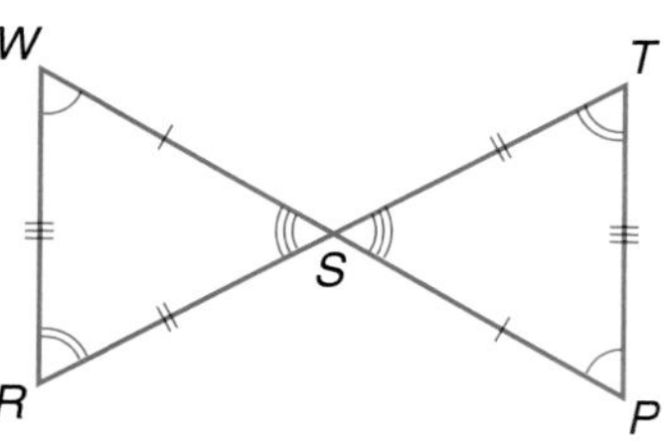

List the congruent angles and sides.

$\angle W \cong \angle P$	$\overline{PS} \cong \overline{WS}$
$\angle R \cong \angle T$	$\overline{TS} \cong \overline{RS}$
$\angle WSR \cong \angle PST$	$\overline{TP} \cong \overline{RW}$

The congruence statement can be written by matching the vertices of the congruent angles. $\triangle TPS \cong \triangle RWS$.

THINK ABOUT IT

How would you know that $\angle WSR \cong \angle PST$ even if they were not marked?

You can use corresponding parts to find measures of angles and sides in a figure that is congruent to a figure with known measures, as shown in Example 3.

Roof trusses allow a roof to withstand the stress of heavy loads. A roof truss for a particular building is made so that the angle on the left measures 35°. The triangular sides of the truss are to be congruent as marked below.

a. What should the measure of angle *TSR* be?

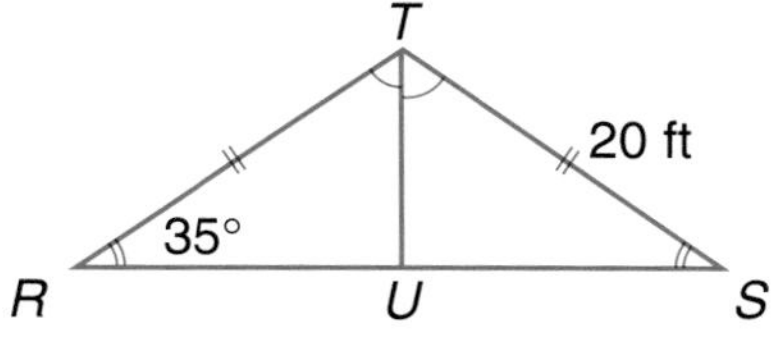

$\angle R \cong \angle S$
$\angle R$ has a measure of 35°,
so $\angle S$ also has a measure of 35°.

b. What should the measure of side *TR* be?

$\overline{TR} \cong \overline{TS}$

$\overline{TS}$ has a given measure of 20 feet. So $\overline{TR}$ also has a measure of 20 feet.

Checking Your Understanding

Communicating Mathematics

Read and study the lesson to answer these questions.

1. **Tell** what it means when one triangle is congruent to another triangle.
2. **Name** the side that corresponds to $\overline{ST}$ and name the angle that corresponds to $\angle T$ if $\triangle RST \cong \triangle WXY$.
3. **Tell** which figure appears to be congruent to △.

a. 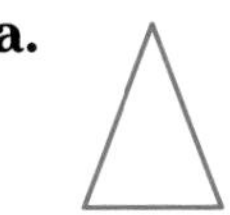 **b.**  **c.**  **d.** none of these

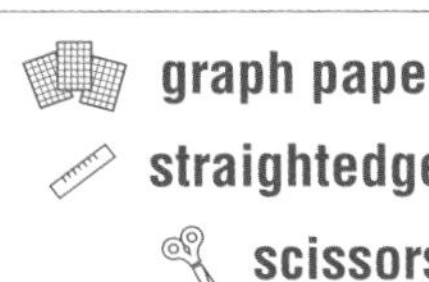

MATERIALS

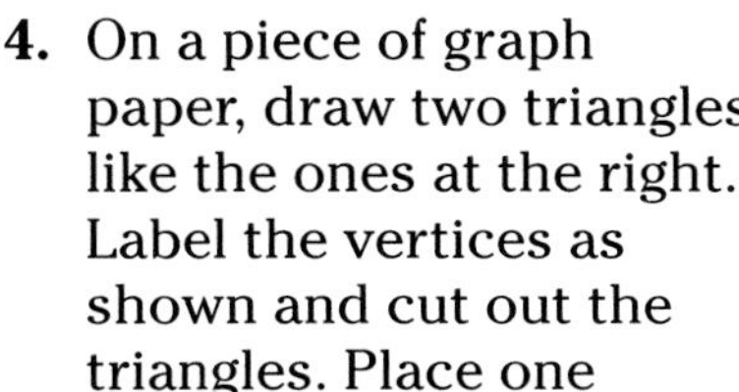

4. On a piece of graph paper, draw two triangles like the ones at the right. Label the vertices as shown and cut out the triangles. Place one triangle over the other and turn until the corresponding parts match. Next, slide the top triangle to the right until the two triangles are side-by-side. Now the corresponding parts are in the same relative positions and you can easily identify them.

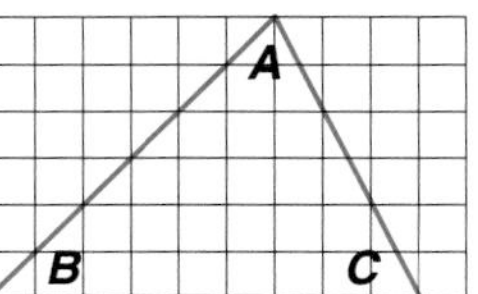

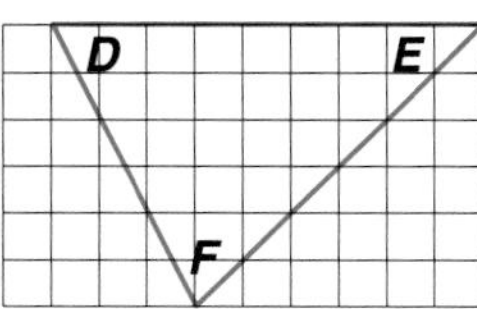

   a. Write a congruence statement for the two triangles.
   b. Name congruence statements for the congruent angles and sides.

Guided Practice

5. Complete the congruence statement for the triangles shown at the right. Then name the corresponding parts.

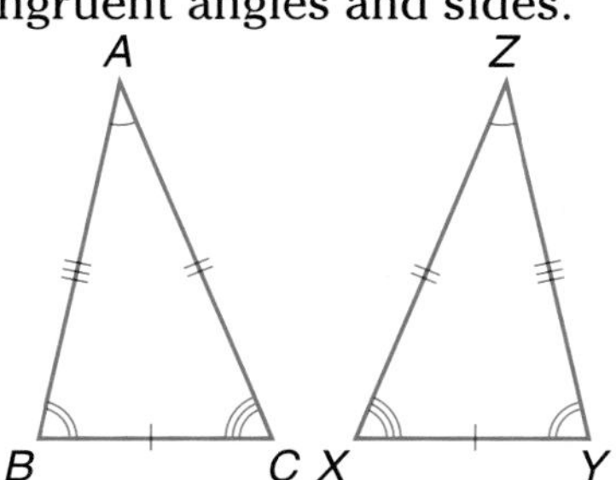

a. $\triangle ABC \cong$ ___

b. $\overline{AB} \cong$ ___ **c.** $\angle Y \cong$ ___

d. $\angle C \cong$ ___ **e.** $\overline{BC} \cong$ ___

f. $\overline{XZ} \cong$ ___ **g.** $\angle Z \cong$ ___

6. Write a congruence statement for the triangles shown below.

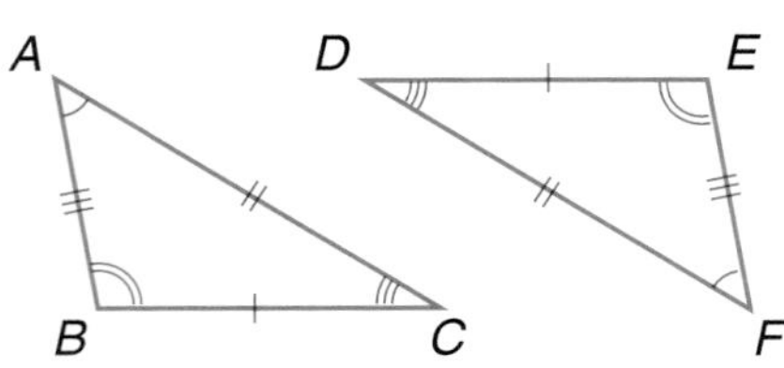

7. If $\triangle DEF \cong \triangle HEG$, what is the measure of $\overline{EH}$?

a. 6 b. 8

c. 10 d. not enough information

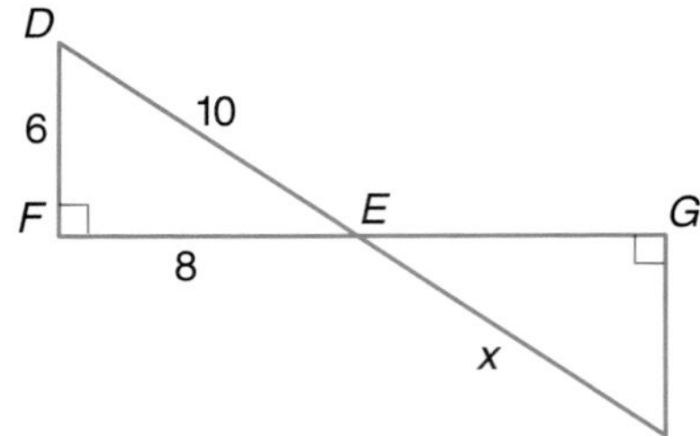

8. **Landscaping** The fence around a triangular garden is 12 meters long. A second garden is congruent to the first. How much fence is needed for the second garden?

Exercises: Practicing and Applying the Concept

Independent Practice

Complete the congruence statement for each pair of congruent triangles. Then name the corresponding parts.

9. 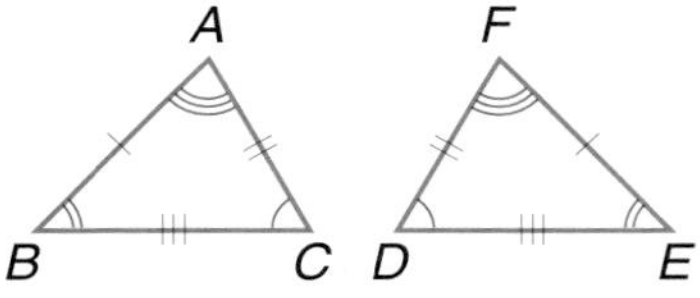

10. 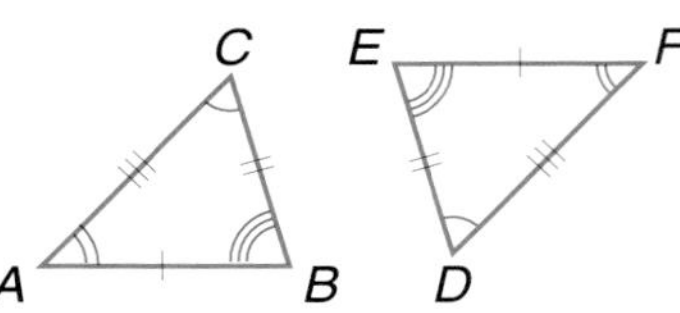

11. 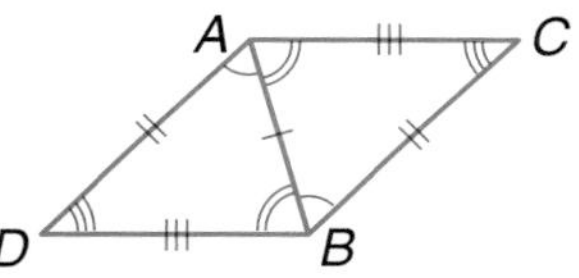

12. 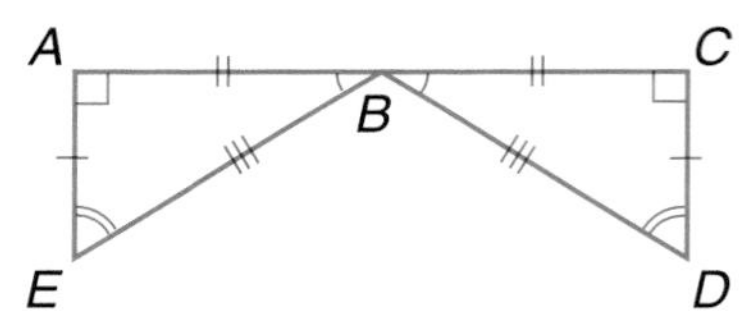

Write a congruence statement for each pair of congruent triangles.

13. 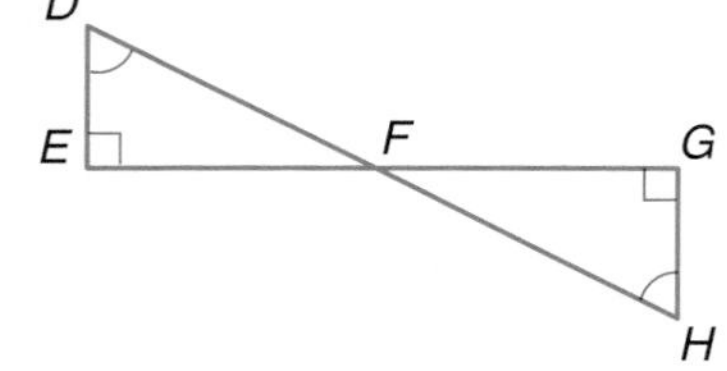

14. 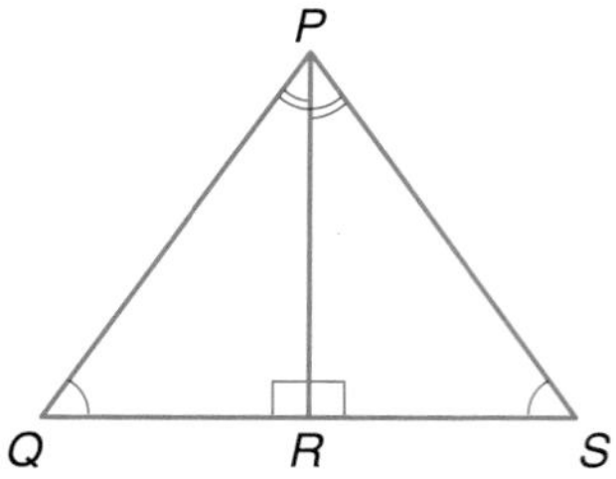

If $\triangle RST \cong \triangle MAP$, name the part congruent to each angle or segment given. (*Hint*: Make a drawing.)

15. $\overline{RS}$ 16. $\angle P$ 17. $\overline{MP}$ 18. $\overline{ST}$ 19. $\angle R$ 20. $\angle A$

Find the value of x for each pair of congruent triangles.

21. 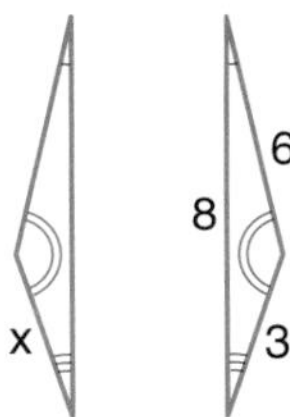

22. 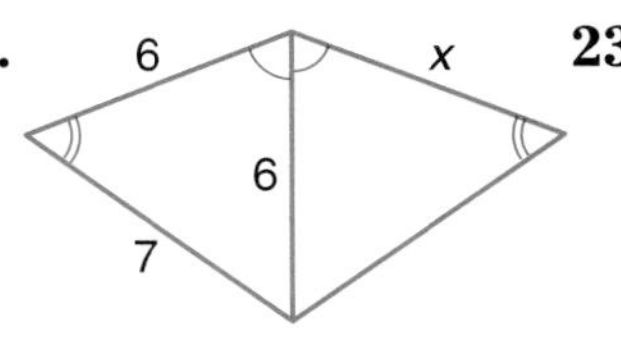

23. 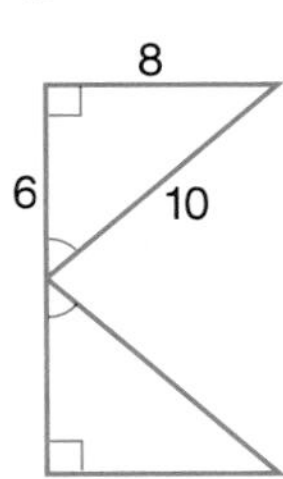

Critical Thinking

24. If two triangles have three pairs of congruent, corresponding angles, are the triangles congruent? Explain your answer by making a drawing.

25. In the figure at the right, two pairs of overlapping triangles appear to be congruent. Write a congruence statement for each pair.

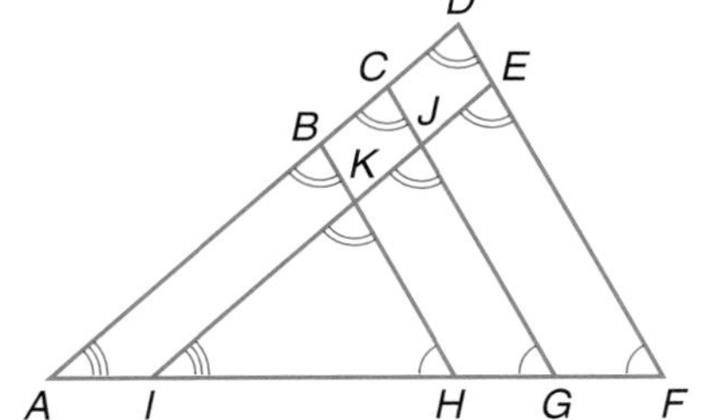

Applications and Problem Solving

26. Construction One pattern for a roof truss is shown at the right. Use the labels in the figure and name the triangle that seems to be congruent to each triangle listed below.

a.  $\triangle ABC$ **b.**  $\triangle BDC$ **c.** $\triangle ADC$

27. Geometry Two triangles are congruent and the perimeter of one triangle is 5 feet. What is the perimeter of the second triangle?

28. Recreation Make a paper airplane of your own design from an $8\frac{1}{2}$-by-11 inch piece of paper. When you have finished, open the paper and observe the pattern made by the folds. Look for pairs of congruent triangles. Use a protractor to measure angles if you wish. Label the vertices and write congruence statements for the pairs of congruent triangles.

Mixed Review

29. In $\triangle PQR$, $m\angle P = 45°$, $m\angle Q = 40°$. Find the measure of $\angle R$ and classify the triangle as *acute*, *right*, or *obtuse*. (Lesson 11-4)

30. Consumer Awareness A shirt that normally sells for $45 is on sale for 15% off. What is the amount of savings? (Lesson 9-5)

31. Solve $\frac{a}{5} - 6 = 19$. (Lesson 7-2)

32. Name the multiplicative inverse of $1\frac{3}{5}$. (Lesson 6-4)

33. Eliminate Possibilities Liz, Renee, and Pablo each either brought a sack lunch, bought a plate lunch, or bought from the snack bar for lunch. Use the clues to find each person's lunch. (Lesson 3-1)

- Liz did not buy a lunch.
- Renee did not have a plate lunch.

11-6 Similar Triangles and Indirect Measurement

Setting Goals: *In this lesson, you'll identify corresponding parts and find missing measures of similar triangles. You'll also solve problems involving indirect measurement by using similar triangles.*

Modeling with Manipulatives

MATERIALS

graph paper

ruler

protractor

Have you ever used a copy machine to make a reduction or enlargement of something? The original and the copy are the same shape, but different sizes. In mathematics, we call this a **dilation.** You can use graph paper to investigate the relationship of two such figures.

Your Turn

Work with a partner.

- ▶ Draw and label a triangle on a coordinate plane.
- ▶ Multiply the coordinates of each vertex by 2. Draw the new triangle.

a. Measure the angles in the two triangles. How do they compare?

b. Use a ruler to measure the sides of the two triangles. Write a ratio comparing the measures of corresponding sides. What do you notice?

c. Repeat the activity by multiplying the coordinates of the original triangle by $\frac{1}{2}$. Are the results the same?

Learning the Concept

The triangles you drew in the activity were **similar**. Figures that have the same shape but not necessarily the same size are similar figures.

As with congruent triangles, write the corresponding vertices of similar triangles in the same order.

In the figure below, $\triangle ABC$ is similar to $\triangle PQR$. This is written as $\triangle ABC \sim \triangle PQR$ in symbols. The symbol $\sim$ means *is similar to.*

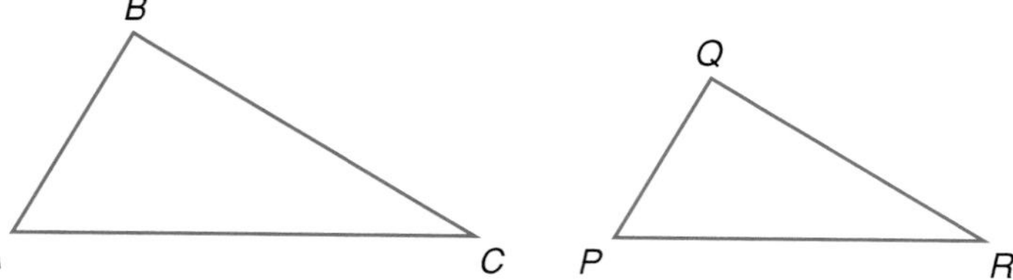

Measure each angle. Compare the corresponding angles. This example and the activity at the beginning of the lesson suggest the following.

Corresponding Angles of Similar Triangles

If two triangles are similar, then the corresponding angles are congruent.

It is also true that if the corresponding angles of two triangles are congruent, then the triangles are similar. *You will use this property in Example 2.*

The corresponding sides of congruent triangles also have a special relationship. In the figure below, $\triangle LMN \sim \triangle RST$.

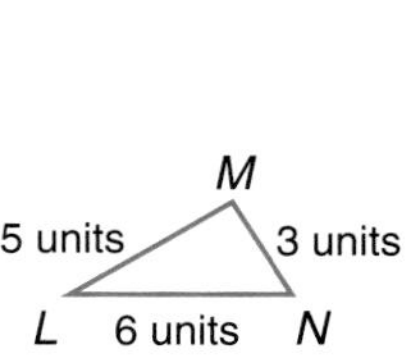

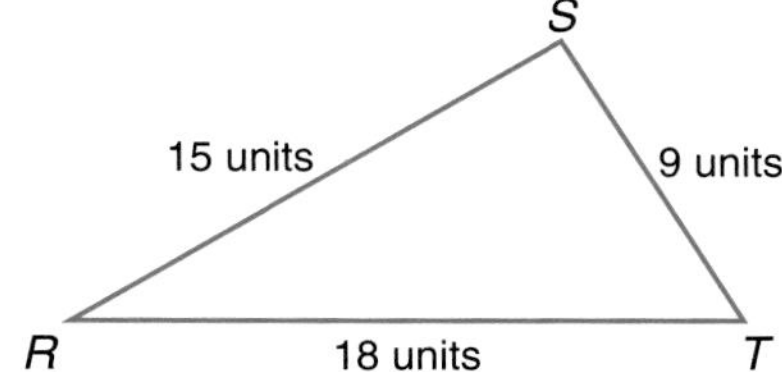

We can compare the measures of corresponding sides by using ratios.

$m\overline{LN}$ means the measure of segment LN.

$$\frac{m\overline{LN}}{m\overline{RT}} = \frac{6 \text{ units}}{18 \text{ units}} \text{ or } \frac{1}{3} \qquad \frac{m\overline{MN}}{m\overline{ST}} = \frac{3 \text{ units}}{9 \text{ units}} \text{ or } \frac{1}{3} \qquad \frac{m\overline{LM}}{m\overline{RS}} = \frac{5 \text{ units}}{15 \text{ units}} \text{ or } \frac{1}{3}$$

Notice that the ratios are equivalent. So, the corresponding sides are proportional to each other. This example and the activity at the beginning of the lesson seem to suggest the following.

Corresponding Sides of Similar Triangles

If two triangles are similar, then their corresponding sides are proportional.

It is also true that if the corresponding sides of two triangles are proportional, then the triangles are similar.

You can use proportions to find the measures of the sides of similar triangles when some measures are known.

Example  **If $\triangle GHI \sim \triangle JKL$, find the value of x.**

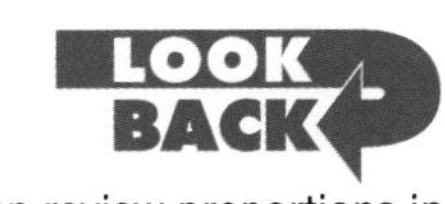

You can review proportions in Lesson 9-4.

Write a proportion using the known measures.

$\frac{m\overline{GH}}{m\overline{JK}} = \frac{m\overline{HI}}{m\overline{KL}}$ *Corresponding sides are proportional.*

$\frac{12}{4} = \frac{9}{x}$

$12x = 4 \cdot 9$ *Find the cross products.*

$12x = 36$

$x = 3$ *Divide each side by 12.*

The measure of $\overline{KL}$ is 3 cm.

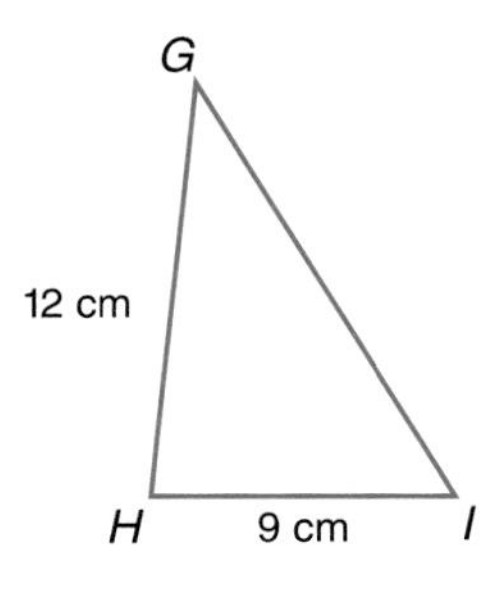

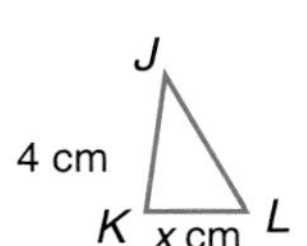

Example 2 Use the information in the figure below to find the value of *x*.

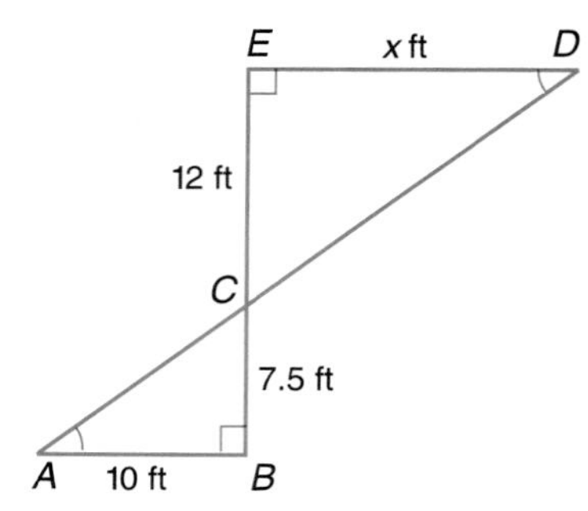

Since vertical angles are congruent, $\angle ACB \cong \angle DCE$.
Angles A and D are marked congruent.
Angles B and E are congruent because they are both right angles.

So, $\triangle ABC \sim \triangle DEC$, because their corresponding angles are congruent.

$$\frac{m\overline{AB}}{m\overline{DE}} = \frac{m\overline{BC}}{m\overline{EC}}$$ *Corresponding sides of similar triangles are proportional.*

$$\frac{10}{x} = \frac{7.5}{12}$$

$$10 \cdot 12 = 7.5x$$ *Find the cross products.*

$$120 = 7.5x$$

$$\frac{120}{7.5} = \frac{7.5x}{7.5}$$ *Divide each side by 7.5.*

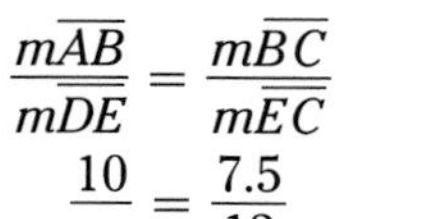

120 [÷] 7.5 [=] 16

$$16 = x$$

The measure of $\overline{ED}$ is 16 feet.

You can use what you know about similar triangles to find the measure of objects that are too large to measure directly, as shown in Example 3. This kind of measurement is called **indirect measurement**.

Example 3

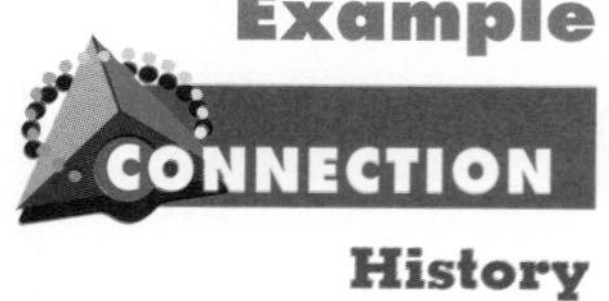

History

In ancient Egypt, mathematicians used a technique called *shadow reckoning* to determine the heights of tall objects, such as the pyramids. The height of a staff and the length of its shadow are proportional to the height of an object and the length of its shadow. The objects and their shadows form two sides of similar triangles. Use shadow reckoning to find the height of the flagpole in the figure below.

Explore The meterstick in the drawing is like the staff.

Plan In the figure, the shadow of the meterstick is 0.7 meters long, and the shadow of the flagpole is 5.6 meters long. Write a proportion comparing corresponding sides of the similar triangles.

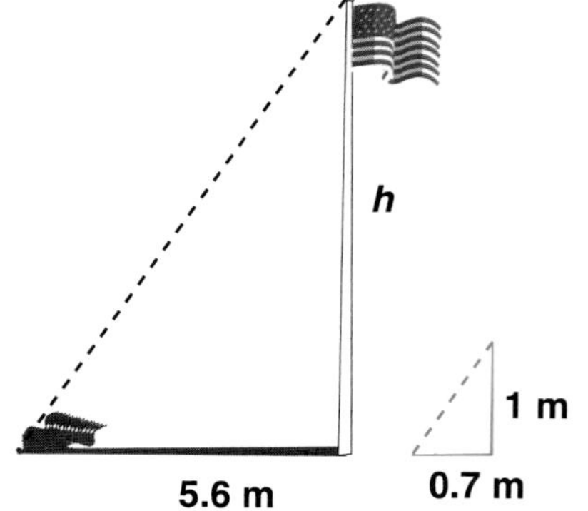

THINK ABOUT IT

How could you use shadow reckoning to measure the height of your school?

Solve $\frac{\text{length of meterstick shadow}}{\text{length of flagpole shadow}} = \frac{\text{length of meterstick}}{\text{length of flagpole}}$

$$\frac{0.7}{5.6} = \frac{1}{h}$$

$$0.7h = 5.6 \quad \textit{Find the cross products.}$$

$$\frac{0.7h}{0.7} = \frac{5.6}{0.7} \quad \textit{Divide each side by 0.7.}$$

$$h = 8$$

The flagpole is 8 meters tall.

Examine You can estimate to check your answer. The meterstick is about $1\frac{1}{2}$ times its shadow, so the flagpole is about $1\frac{1}{2}$ times its shadow.

Checking Your Understanding

Communicating Mathematics

Read and study the lesson to answer these questions.

1. **Explain** what it means for one figure to be similar to another.
2. **You Decide** Shawn thinks the triangles below are similar. Marta thinks they are not. Who is correct and why?

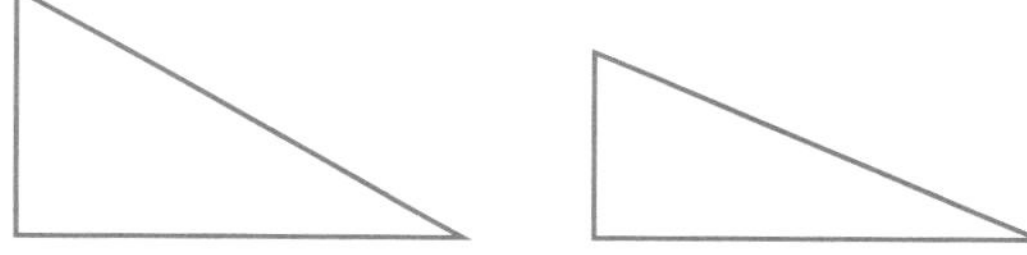

3. **Draw** two similar triangles. Label the vertices and name the corresponding sides and angles.
4. **Explain** how you can find the height of an object that is too large to measure directly.
5. **Make Up a Problem** Write a problem involving the use of similar triangles.

MATERIALS

TI-82 graphing calculator

6. Use a TI-82 graphing calculator to graph $\triangle ABC$ with vertices $A(0, 3)$, $B(15, 3)$, and $C(15, 15)$ and $\triangle AXY$ with vertices $A(0, 3)$, $X(30, 3)$, and $Y(30, 27)$. Follow the steps in the procedure on page 573, using the coordinates given in this problem. Do the triangles appear to be similar?

Guided Practice

7. Refer to the similar triangles at the right.

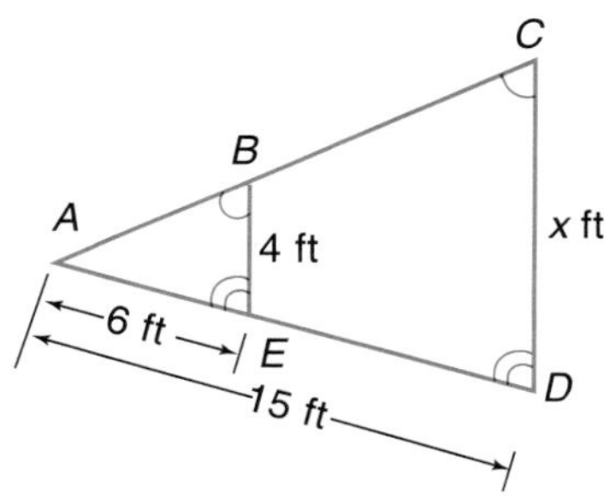

   a. List three proportions that can be written for $\triangle ABE$ and $\triangle ACD$.
   b. What angle of $\triangle CAD$ corresponds to $\angle BAE$?
   c. What side corresponds to $\overline{BE}$?
   d. Find the value of x.

Write a proportion to find each missing measure x. Then find the value of x.

8.

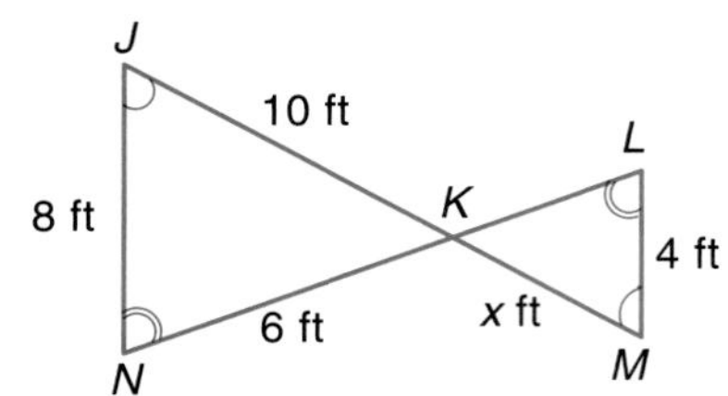

9.

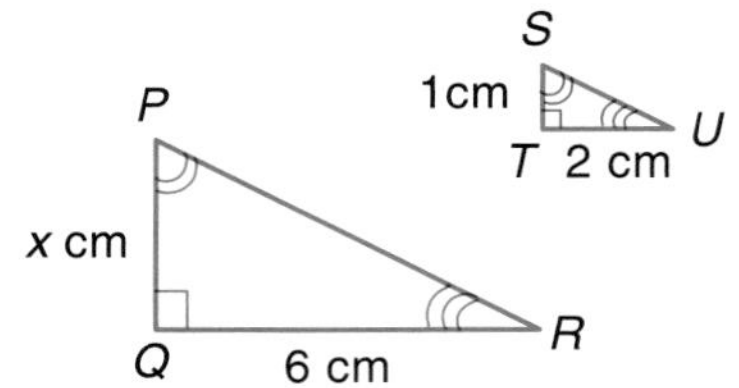

10. Given: 2 / 3 / 4 triangle. Which of the following is similar to the given triangle?

a. 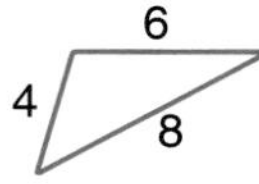

b. 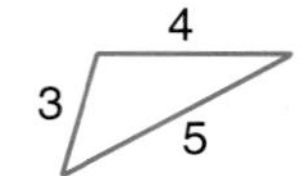

c. 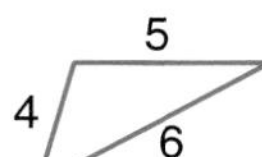

d. none of these

11. Civil Engineering The city of Marion plans to build a bridge across Brandon Lake. Use the information in the diagram at the right to find the distance across Brandon Lake.

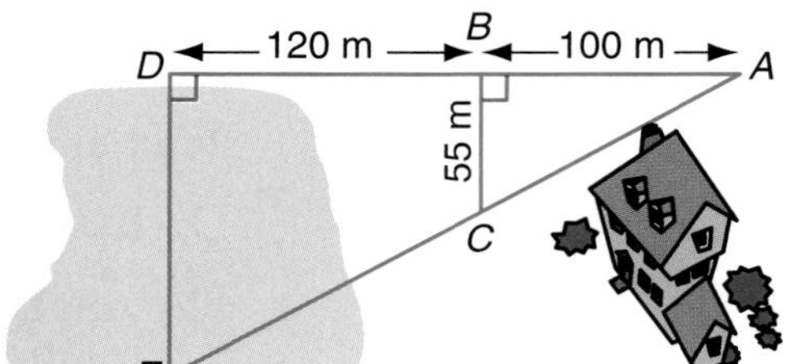

Exercises: Practicing and Applying the Concept

Independent Practice

Write a proportion to find each missing measure x. Then find the value of x.

12.

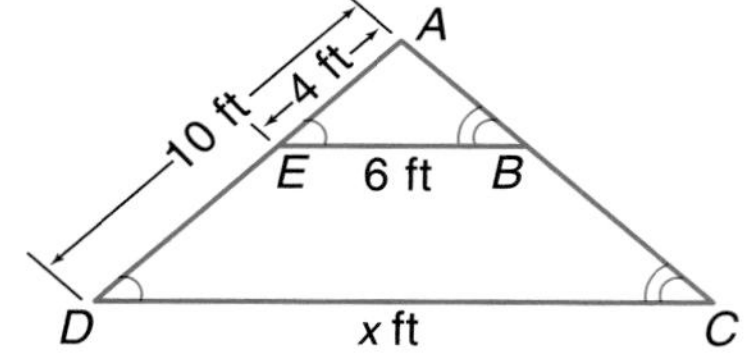

13.

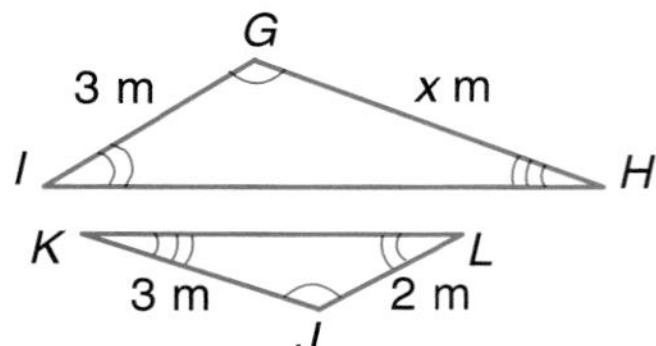

14.

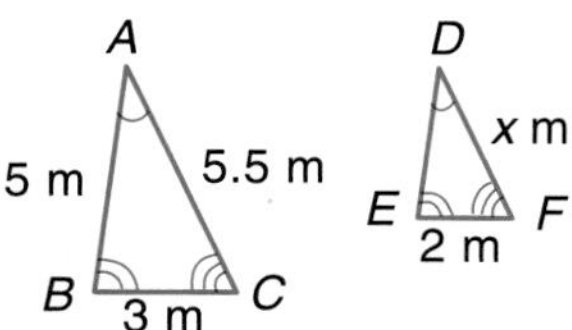

15.

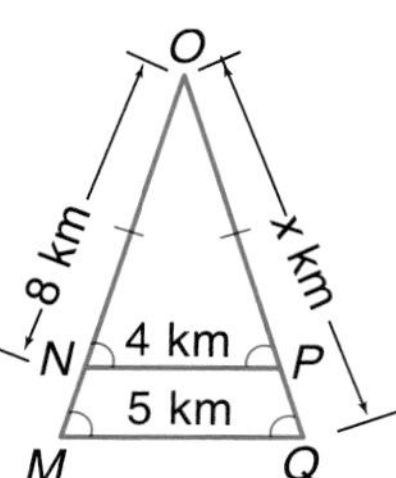

16.

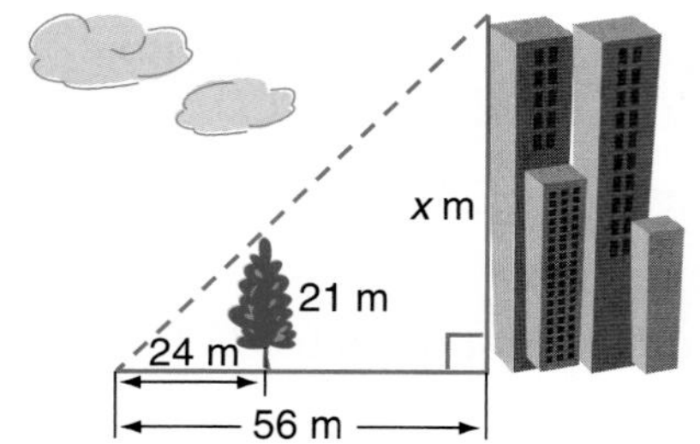

17.

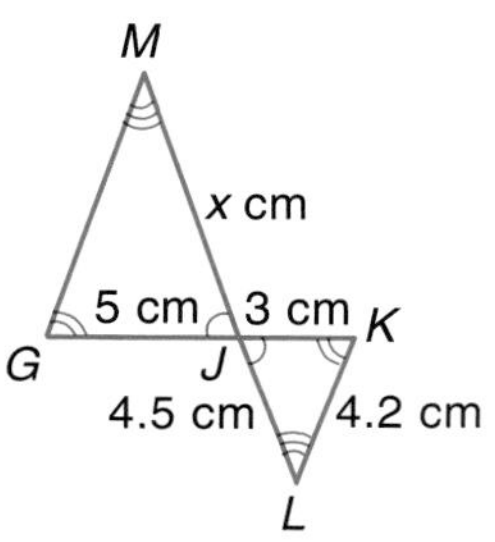

18. On a coordinate grid, graph $A(6, 6)$, $B(2, 2)$, $C(8, 2)$, $D(0, 0)$, and $E(9, 0)$. Draw $\triangle ABC$ and $\triangle ADE$.

   a. Are the triangles similar? How do you know?

   b. Do $\overline{BC}$ and $\overline{DE}$ appear to be parallel?

   c. Will two triangles with a common angle and the sides opposite that angle parallel always be similar? Explain.

Critical Thinking

19. a. Are all congruent triangles also similar? Explain.

   b. Are all similar triangles also congruent? Explain.

20. **Photography** A photo negative is 52.5 mm wide by 35 mm high. The print is made 15.24 centimeters wide. How many times greater is the area of the print than the area of the negative?

Applications and Problem Solving

21. **Construction** Find the length of the brace in the roof rafter shown at the right.

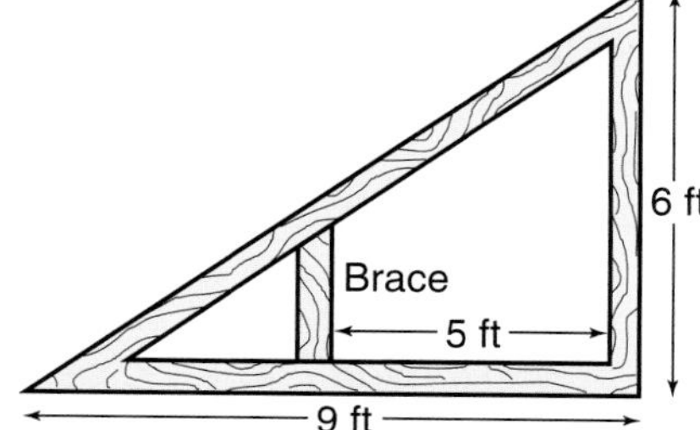

22. **History** Refer to Example 3. The largest known pyramid is Khufu's pyramid. At a certain time of day, a yardstick casts a shadow 1.5 ft long, and the pyramid casts a shadow 241 ft long. Draw a sketch that shows the two similar triangles and use shadow reckoning to find the height of the pyramid. Estimate to check your answer.

23. **Surveying** A surveyor needs to find the distance across a river and draws the sketch shown at the right. Explain how the surveyor determined the sketch and the known measurements. Then find the distance across the river.

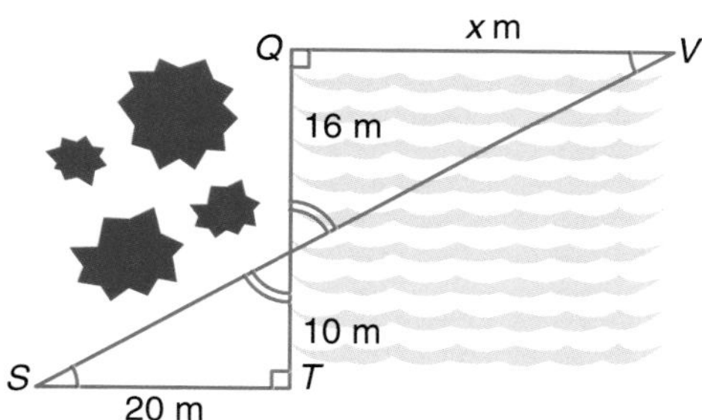

Mixed Review

24. **Geometry** Two triangles are congruent, and the area of one triangle is 15 square feet. What is the area of the second triangle? (Lesson 11-5)

25. A quiz has 5 true-false questions. How many outcomes for giving answers to the five questions are possible? (Lesson 10-5)

26. Solve the system of equations $y = 3x$ and $y = x + 4$ by graphing. (Lesson 8-8)

27. **Statistics** Find the mean, median, and mode for the set of data below. When necessary, round to the nearest tenth. (Lesson 6-6)
26, 32, 54, 34, 36, 48, 39, 40, 54, 89, 78, 45

28. Find the GCF of $14n^2$, $22p^2$, and $36n^2p$. (Lesson 4-5)

11-7 Quadrilaterals

Setting Goals: *In this lesson, you'll find the missing angle measure of a quadrilateral, classify quadrilaterals, and explore similar quadrilaterals.*

Modeling a Real-World Application: Quilting

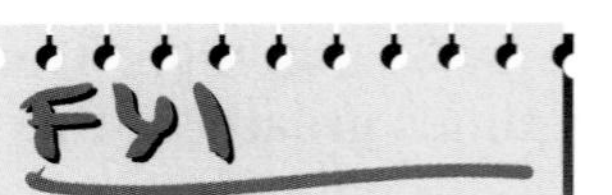

FYI

The world's largest quilt was made by 7000 citizens of North Dakota for the state's centennial celebration in 1989. It was 85 feet by 134 feet.

Quilting in America dates back to colonial days. Early quilts were often made using only square or rectangular patches. Later designs used other geometric shapes. The quilt shown at the right was made using a standard one-patch pattern and silk cloth. It is called "Tumbling Blocks" and is on display in the Howell House in Philadelphia, Pennsylvania. Notice how the four-sided patches and the design give the illusion of three dimensions.

Learning the Concept

When you name a quadrilateral, you can begin at any vertex. But it is important to name consecutive vertices in order.

Squares and rectangles are simple quadrilaterals. A **quadrilateral** is a closed figure formed by four line segments that intersect only at their endpoints. As with triangles, a quadrilateral can be named by its vertices. The quadrilateral shown below can be named quadrilateral *ABCD*.

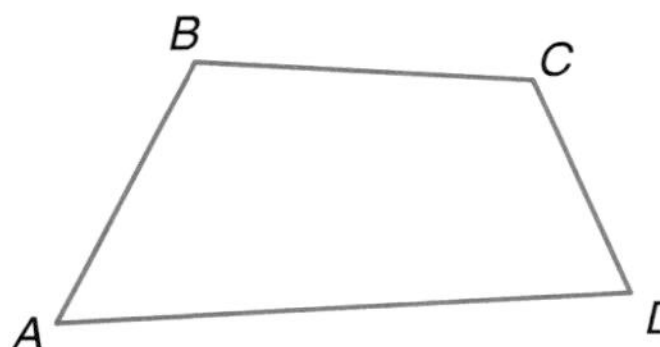

The vertices are A, B, C, and D.
The sides are $\overline{AB}$, $\overline{BC}$, $\overline{CD}$, and $\overline{AD}$.
The angles are $\angle A$, $\angle B$, $\angle C$, and $\angle D$.

If you use a protractor to measure each interior angle of quadrilateral *ABCD*, you will find that $m\angle A = 60°$, $m\angle B = 115°$, $m\angle C = 120°$, and $m\angle D = 65°$. The sum of the measures of the angles of *ABCD* is 360°.

This example suggests a relationship of the angles of a quadrilateral.

Angles of a Quadrilateral	The sum of the measures of the angles of a quadrilateral is 360°.

Example 1

Algebra

Find the value of *x*. Then find the missing angle measures.

The sum of the measures of the angles is 360°.

$72 + 96 + x + 2x = 360$

$168 + 3x = 360$ *Combine like terms.*

$3x = 192$ *Subtract 168 from each side.*

$x = 64$ *Divide each side by 3.*

So $m\angle Q = 64°$, and $m\angle R = 2(64)$ or $128°$.

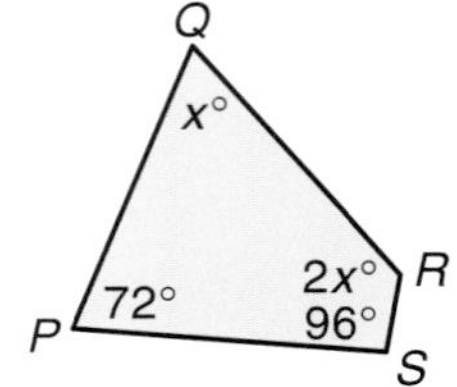

There are many kinds of quadrilaterals. One way to identify them is to look for pairs of parallel sides.

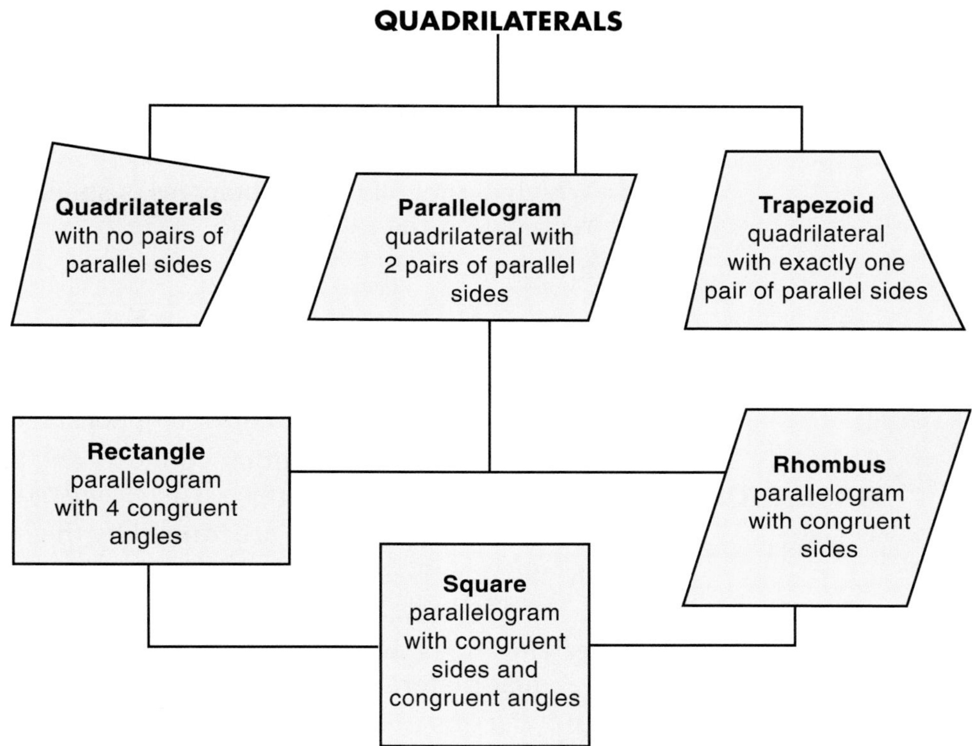

In Lesson 11-6, you learned various properties of similar triangles. These same properties also hold for similar quadrilaterals or any pair of similar figures.

Corresponding Angles and Sides of Similar Figures

If two figures are similar, then the angles of one figure are congruent to the corresponding angles of the other figure. If two figures are similar, then their corresponding sides are porportional.

A **scale drawing** is similar to the actual object, but it is generally either smaller or larger. The scale is the ratio of the lengths on the drawing to the actual lengths of the object.

Example  **On a scale drawing of a room, 1 inch represents 2 feet. In the scale drawing, the room is 11 inches long. How long is the actual room?**

The scale is the ratio 1 in. to 2 ft or $\frac{1\text{in.}}{2\text{ft}}$. *This is often written 1 in. = 2 ft.*

Write a proportion.

$$\frac{1\text{in.}}{2\text{ft}} = \frac{11\text{ in.}}{x}$$

$1x = 2(11)$ *Find the cross products.*

$x = 22$

The actual room is 22 feet long.

Checking Your Understanding

Communicating Mathematics

Read and study the lesson to answer these questions.

1. **Draw** an example of each kind of quadrilateral named in this lesson. Give a real-world example of each.
2. **You Decide** Susan says that all squares are similar. Lynn says they are not. Who is correct? Explain.
3. Which of the following rectangles is similar to the rectangle at the right? Why?

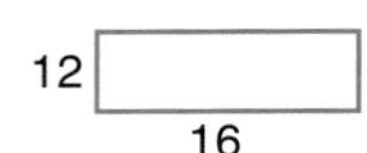

a. 

b. 

c. 

MATERIALS

 straightedge

4. Work with a partner.
   - **a.** Use a straightedge to draw a quadrilateral.
   - **b.** Draw a line segment that connects any two nonconsecutive vertices. This is a *diagonal*.
   - **c.** How many triangles are formed by the diagonal?
   - **d.** What is the sum of the measures of the angles of a triangle? of two triangles?
   - **e.** How does this activity show that the measures of the angles of a quadrilateral have a sum of 360°?

Guided Practice

5. Find the value of x.

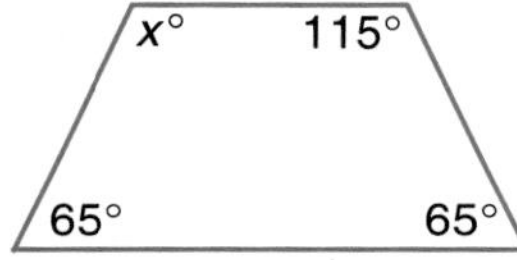

6. Find the value of x. Then find the missing angle measures.

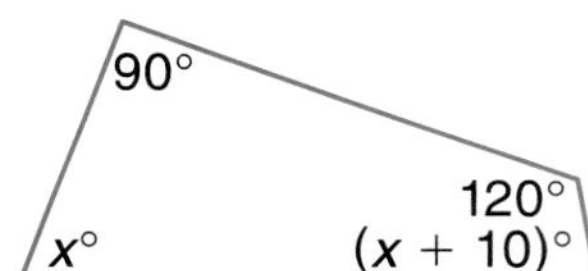

List every name that can be used to describe each quadrilateral. Indicate the name that *best* describes the quadrilateral.

7. 

8. E F H G

9. **Interior Design** The scale of the drawing at the right is 1 cm = 0.5 m. What is the length of the actual room?

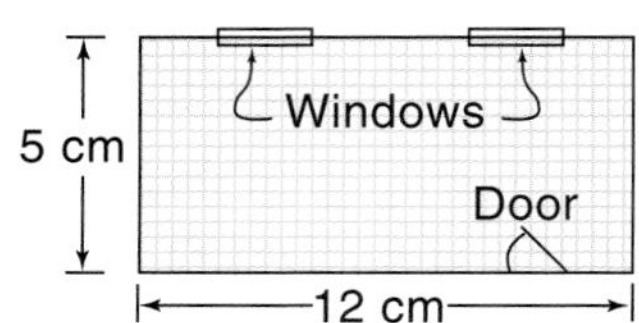

Exercises: Practicing and Applying the Concept

Independent Practice

Find the value of *x*.

10. 150° 30° 30° x°

11. 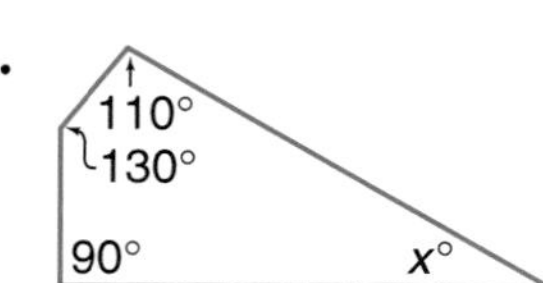

12. 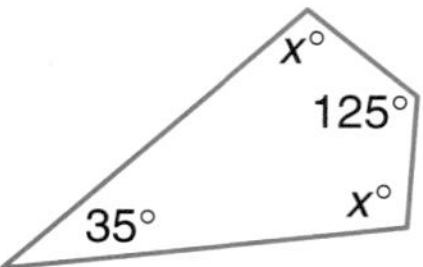

Estimation
Mental Math
Calculator
Paper and Pencil

Find the value of x. Then find the missing angle measures.

13.  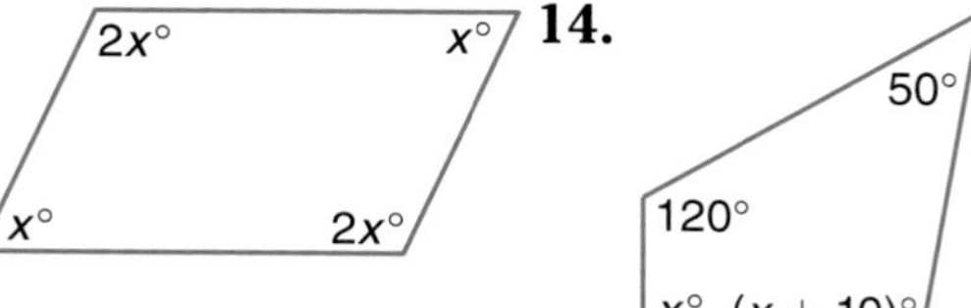

14.

15. 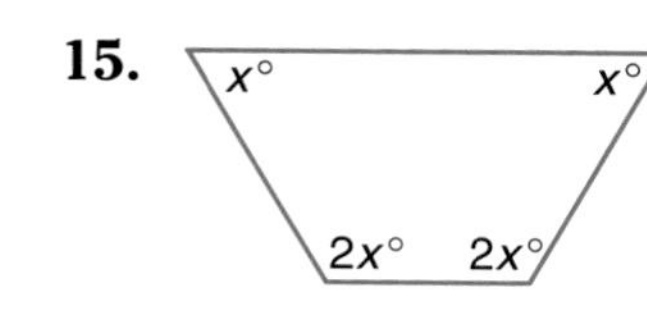

Classify each quadrilateral using the name that *best* describes it.

16. **17.** 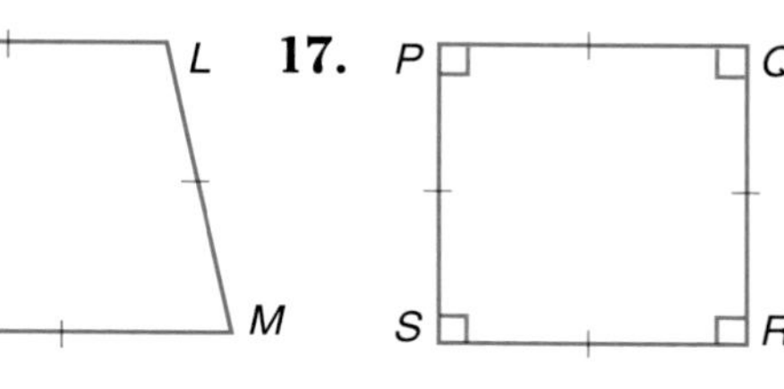

18. 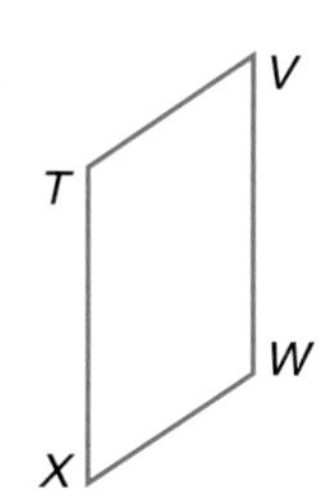

Make a drawing of each quadrilateral. Then classify each quadrilateral using the name that *best* describes it.

19. In quadrilateral $WXYZ$, $m\overline{WX} = 3$ in., $m\overline{XY} = 5$ in., $m\overline{YZ} = 3$ in., and $m\overline{WZ} = 5$ in. Angles W, X, Y, and Z are right angles.

20. In quadrilateral $CDEF$, $\overline{CD}$ and $\overline{EF}$ are parallel, and $\overline{CF}$ and $\overline{DE}$ are parallel. Angle C is not congruent to angle D.

21. In quadrilateral $JKLM$, $m\angle J = 90°$, $m\angle K = 50°$, $m\angle L = 90°$, and $m\angle M = 130°$.

Determine whether each statement is *always*, *sometimes*, or *never* true.

22. A rectangle is a parallelogram.

23. A rectangle is a square.

24. A rhombus is a square.

25. A trapezoid is a parallelogram.

26. A square is a parallelogram.

27. A square is a rhombus.

28. Use grid paper to draw each figure. Then enlarge each figure by making each segment twice the length of the original. How does the area of the enlargement compare to the area of the original figure?

a. 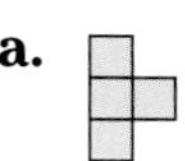

b. 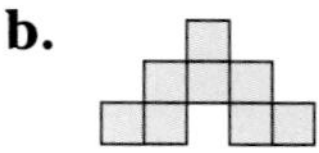

c. 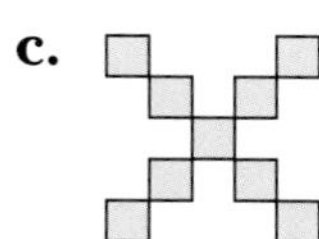

Critical Thinking

29. In an **equilateral** figure, all sides have the same measure. In an **equiangular** figure, all angles have the same measure.

a. Is it possible for a quadrilateral to be equi*lateral* without being equi*angular*? Explain.

b. Is it possible for a quadrilateral to be equiangular without being equilateral? Explain.

30. One angle of a parallelogram measures 40°. What are the measures of the other angles?

Applications and Problem Solving

31. **Sports** Name four sports with playing fields that are quadrilaterals. Find the dimensions of each field and classify the shape.

For other house blueprints, visit: **www.glencoe.com/sec/math/prealg/mathnet**

32. Construction In a house blueprint, the actual height of 8 feet is represented by 2 inches. If the actual length of the house is 60 feet, what is the length of the house in the blueprint?

33. Make a Drawing A basketball court is 84 feet by 50 feet. Make a scale drawing on $\frac{1}{4}$-inch grid paper. Use the scale $\frac{1}{4}$ inch = 6 feet.

34. Publishing Kiona's school newspaper uses 3-inch columns. If a $3\frac{1}{2}$-by-5 inch vertical photograph is reduced to fit in one column, how long will the reduced photograph be?

Mixed Review

35. Find the value of x in the $\triangle DEC$ at the right. (Lesson 11-6)

36. Find the measure of $\angle A$ in $\triangle ABC$ at the right. (Lesson 11-4)

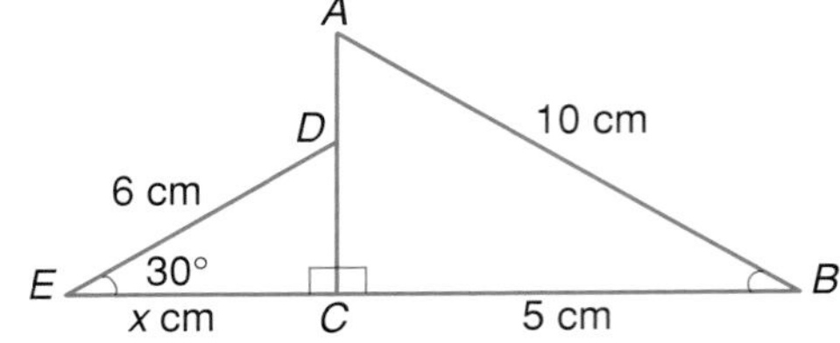

37. Express 450 miles on 20 gallons of gas as a unit rate. (Lesson 9-1)

38. Geography The distance around Earth at the equator is about 25,000 miles. Find the approximate diameter of Earth at the equator. (Lesson 7-4)

39. Viñita noticed the school cafeteria served pizza on the last five Mondays. Viñita decides that the cafeteria always serves pizza on Monday. Is this an example of *inductive* or *deductive* reasoning? (Lesson 5-8)

40. Use mental math, paper and pencil, or a calculator to find at least one number that has four digits and is not divisible by 2, 3, or 5. (Lesson 4-1)

Refer to the Investigation on pages 482–483.

Look over the list of favorite games that you placed in your Investigation Folder at the beginning of Chapter 10. Some people may have said that their favorite games are target games, such as darts or horseshoes.

For this part of the Investigation, your group has been hired to design a new target game using Velcro® balls. The new board should be square and divided into several rectangular regions. Some regions should be marked with an X, and other regions marked with a Y.

- Work with the members of your group to design the board. If a Velcro ball is thrown at random at your board, the game should meet the following requirements.
  1. Player A scores points if the ball lands in a region marked X, and player B scores points if the ball lands in a region marked Y.
  2. The game should be fair.
- Write a report explaining how your board meets the requirements. Be sure to use concepts from Chapter 10 in your explanation.

Add the results of your work to your Investigation Folder.

11-8 Polygons

Setting Goals: *In this lesson, you'll classify polygons and determine the sum of the measures of the interior and exterior angles of a polygon.*

Modeling with Technology

MATERIALS

computer
LOGO software
Your Turn

If you were to walk a path in the shape of a rectangle, you would make a 90° turn at each corner. In LOGO, the turtle "walks" a path following directions you give. You can use LOGO to make paths of various shapes.

Work with a partner. Type the following commands to make the shapes shown to the right of each set of commands. Sketch each figure and mark the angles. Clear the screen before typing the second set of commands.

FD 100 RT 60
FD 100 RT 90
FD 173 RT 120
FD 173
What angle must you turn to return the turtle to its starting position?

FD 100 RT 60
FD 80 RT 60
FD 80 RT 60
FD 100 RT 90
FD 139
What angle must you turn to return the turtle to its starting position?

TALK ABOUT IT

a. In each figure, compare the angle of turn (the exterior angle) to the adjacent angle inside the figure (the interior angle). What pattern do you see?

b. How could you use this pattern to draw figures using LOGO?

c. In each figure, find the sum of the turning (exterior) angles including the final turn to return the turtle to its starting position. What do you notice?

Learning the Concept

THINK ABOUT IT

Can you name a simple, closed figure that is NOT a polygon?

The figures you drew above were simple closed figures. A simple, closed figure can be traced in a continuous path without tracing any point other than the starting point more than once. A **polygon** is a simple, closed figure in a plane that is formed by three or more line segments, called **sides**. The segments meet only at their endpoints. These points of intersection are called **vertices** (plural of **vertex**).

These plane figures are polygons.

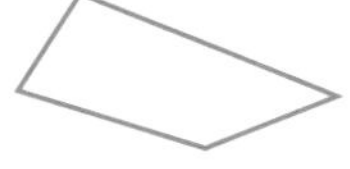

These plane figures are not polygons.

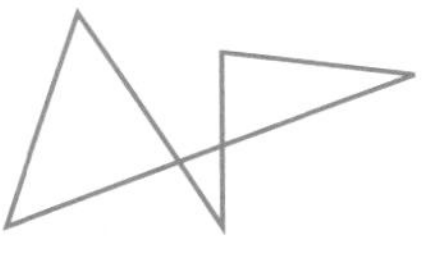

We can classify polygons by the number of sides. Some of the more common polygons are shown below.

A polygon with n sides is called an ***n-gon****.*

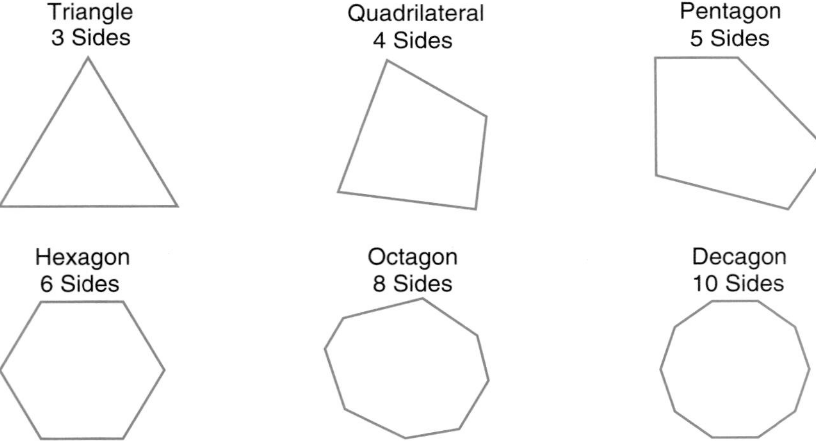

A **diagonal** is a line segment that joins two nonconsecutive vertices. You can draw diagonals in any polygon with more than three sides. In the polygons shown below, all possible diagonals from one vertex are shown. The table shows the number of diagonals drawn, and the number of triangles formed in each polygon.

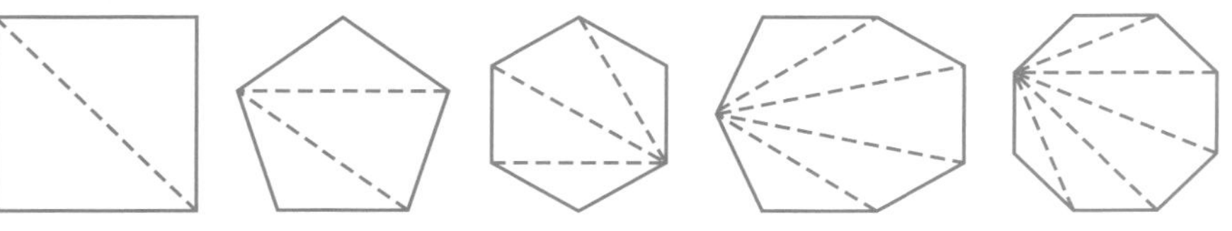

Number of:	Quadrilateral	Pentagon	Hexagon	Heptagon	Octagon
Sides	4	5	6	7	8
Diagonals	1	2	3	4	5
Triangles	2	3	4	5	6

Connection to Algebra

Compare the number of sides to the number of triangles. What pattern do you see? The number of triangles is always 2 less than the number of sides. You can use diagonals and the property of the sum of the angles of a triangle to find the sum of the measures of the angles of any polygon without using a protractor.

Sum of the Interior Angle Measures in a Polygon

If a polygon has n sides, then $n - 2$ triangles are formed, and the sum of the degree measures of the interior angles of the polygon is $(n - 2)180$.

Example 

a. Find the sum of the measures of the interior angles of a pentagon by drawing diagonals.

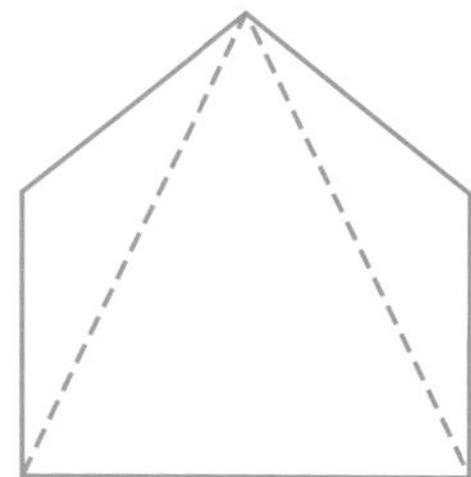

Draw all of the diagonals from one vertex.

Three triangles are formed.

The sum of the measures of the interior angles of a triangle is 180°, so the sum of the measures of the interior angles of a pentagon is 3(180) or 540°. *Check by using the formula.*

b. Find the sum of the measures of the interior angles of a hexagon by using the formula.

A hexagon has 6 sides. Therefore, $n = 6$.

$(n - 2)180 = (6 - 2)180$ *Replace n with 6.*

$= 4(180)$ or 720 *You can check this solution by drawing diagonals.*

The sum of the measures of the interior angles of a hexagon is 720°.

A **regular** polygon is a polygon that is **equilateral** (all sides are congruent) and **equiangular** (all angles are congruent). Since the angles of a regular polygon are congruent, their measures are equal.

Example 2

Architecture

The Pentagon in Washington, D.C., is named for its unusual shape, a regular pentagon. In a scale model of the Pentagon, what is the measure of each angle?

The scale model and the actual building are similar figures. In similar figures, the angles are congruent. From Example 1a, you know that the sum of the measures of the interior angles of a pentagon is 540°. Since the Pentagon is regular, you can find the degree measure of one angle by dividing 540 by 5.

540 [÷] 5 [=] 108

The measure of each angle in a regular pentagon is 108°.

The angles we considered in Examples 1 and 2 are "inside" the polygon. These are the **interior angles**. When a side of a polygon is extended, a special angle is formed, called an **exterior angle**.

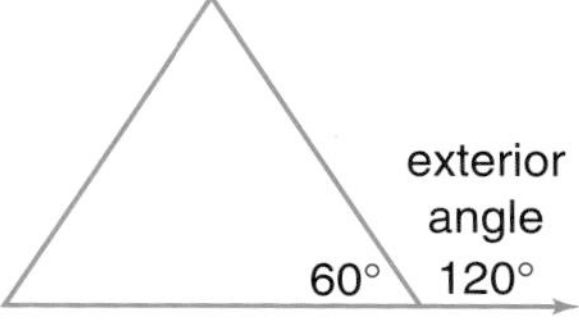

In the activity at the beginning of the lesson, you discovered that the interior and exterior angles of a polygon are supplementary. You also discovered that the sum of the exterior angles of a polygon is 360°. You can use these ideas for another way to find the measure of one angle of a regular polygon.

Example  Find the measure of each interior angle in a regular hexagon.

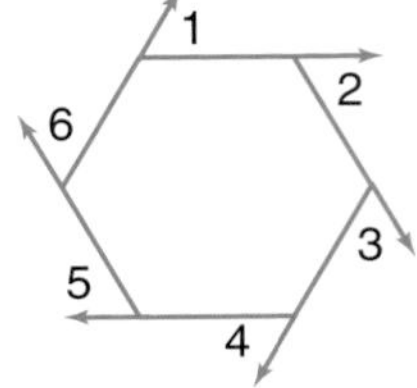

Sketch a regular hexagon.

Draw an exterior angle at each vertex.

There are 6 exterior angles. The sum of the measures of the exterior angles is 360°.

360 [÷] 6 [=] 60 *Measure of each exterior angle.*

180 − 60 = 120 *Interior and exterior angles are supplementary.*

The measure of each interior angle in a regular hexagon is 120°.

Checking Your Understanding

Communicating Mathematics

Read and study the lesson to answer these questions.

1. **Explain** why a square is a regular polygon.
2. **Explain** the relationship between the number of sides in a polygon and the number of triangles formed by the diagonals.
3. **Tell** the difference between equiangular and equilateral. Give an example of a figure that is equiangular or equilateral but not both.
4. **Draw** $\triangle ABC$ with exterior angle BCD.

MATH JOURNAL

5. **Assess Yourself** How can knowing the properties of geometric shapes help you in your daily life?

Guided Practice

Classify each polygon below and determine whether it appears to be *regular* or *not regular*.

6. 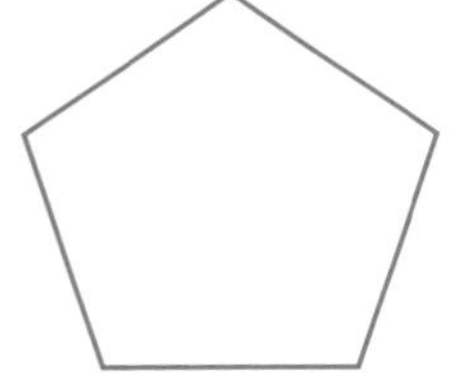

7. 

8. Find the sum of the measures of the interior angles of an octagon.
9. Find the measure of each exterior angle and each interior angle in a regular nonagon (9 sides).
10. **Scale Models** Refer to Example 2. The scale of the model of the Pentagon is to be 1 in. = 50 ft. Each side of the Pentagon is about 1 mile (5280 ft) long. What length should each side of the scale model be?

Exercises: Practicing and Applying the Concept

Independent Practice

Find the sum of the measures of the interior angles of each polygon.

11. heptagon
12. decagon
13. dodecagon
14. nonagon
15. 15-gon
16. 25-gon

Find the measure of each exterior angle and each interior angle of each regular polygon.

17. regular hexagon
18. regular decagon
19. regular dodecagon
20. regular 20-gon

Estimation
Mental Math
Calculator
Paper and Pencil

21. The measure of one angle of a regular polygon with n sides is $\frac{180(n-2)}{n}$. Find the measure of an interior angle in a regular triangle.

Find the perimeter of each regular polygon.

22. equilateral triangle with sides 27 ft long
23. regular quadrilateral with sides 18 cm long
24. regular pentagon with sides 3 m long
25. regular hexagon with sides 4.5 in. long
26. regular octagon with sides $5\frac{3}{4}$ yd long

Critical Thinking

27. Trace the dot pattern shown at the right. Without lifting your pencil from the paper, draw four line segments that connect all the points.

Applications and Problem Solving

28. **Driver's License Exam** Part of a driver's license exam includes identifying road signs by color and by shape. The shapes are polygons and some are regular polygons. Identify the shape of each road sign pictured below and tell what it means.

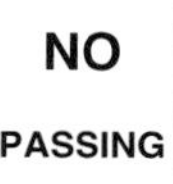

29. **Manufacturing** Some cafeteria trays are designed so that four people can place their trays around a square table without bumping corners. Each tray is shaped like the one shown at the right. The top and bottom of the tray are parallel.

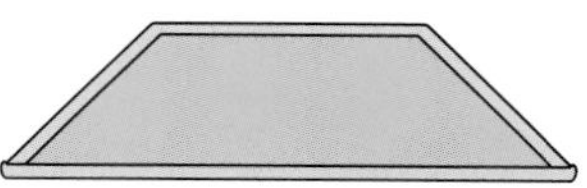

   a. What shape does the tray have?
   b. Find the measure of each angle of the tray so that the trays will fit side-to-side around the table.
   c. What polygon will be formed at the center of the table if four trays are placed around the table with their sides touching?

Mixed Review

30. **Geometry** A lot for a new house is shaped like a trapezoid, with two right angle corners. If the third corner is an 80° angle, what is the measure of the fourth angle? (Lesson 11-7)
31. **Probability** A die is rolled and the spinner is spun. What is P(an even number and a vowel)? (Lesson 10-9)

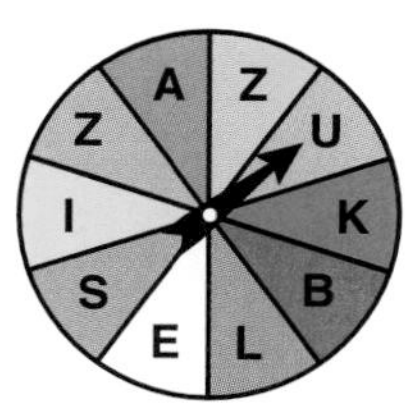

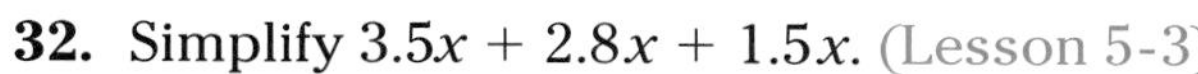

32. Simplify $3.5x + 2.8x + 1.5x$. (Lesson 5-3)
33. Write 1604 in expanded form. (Lesson 4-2)
34. Solve the equation $-6 - 8 = c$. (Lesson 2-5)

HANDS-ON ACTIVITY

11-8B Tessellations

An Extension of Lesson **11-8**

MATERIALS

- tracing paper
- cardboard
- scissors

The repetitive pattern of regular polygons shown at the right is an example of a **tessellation**. In a tessellation, the polygons fit together with no holes or gaps.

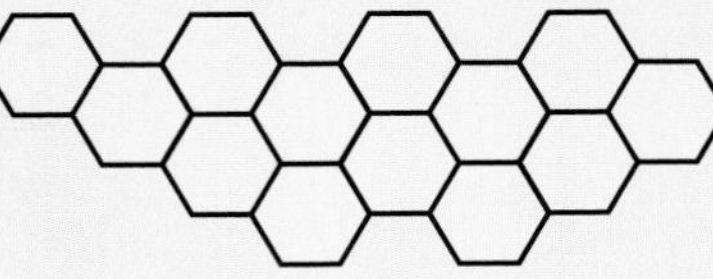

In this Math Lab, you will investigate tessellations with regular polygons.

Your Turn

Work with a partner.

- ► Trace each regular polygon shown. Cut each shape from the tracing paper and trace it onto a piece of cardboard. Cut each shape from the cardboard.
- ► Use each cardboard piece to try to draw a tessellation for each regular polygon.
- ► For each polygon, calculate the sum of the interior angle measures. Then determine the measure of one angle in each polygon. Record your results in a table.

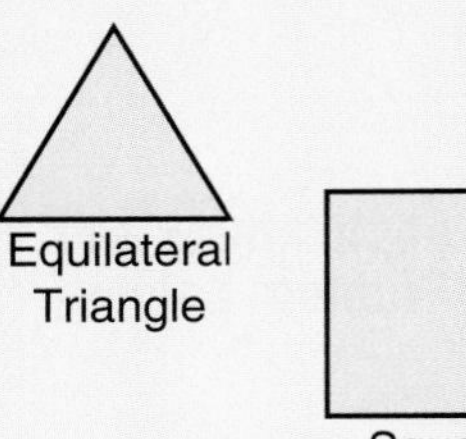

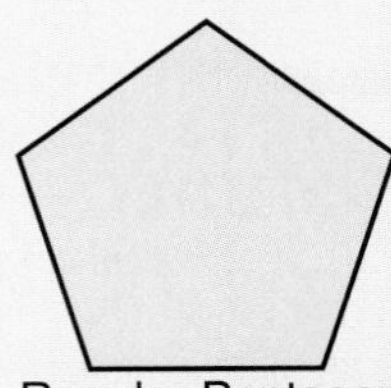

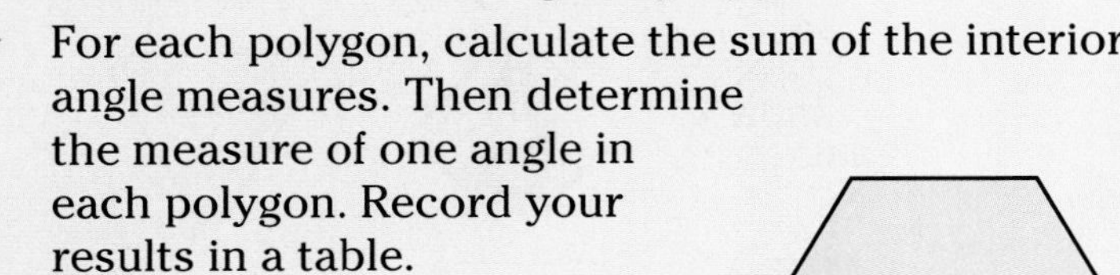

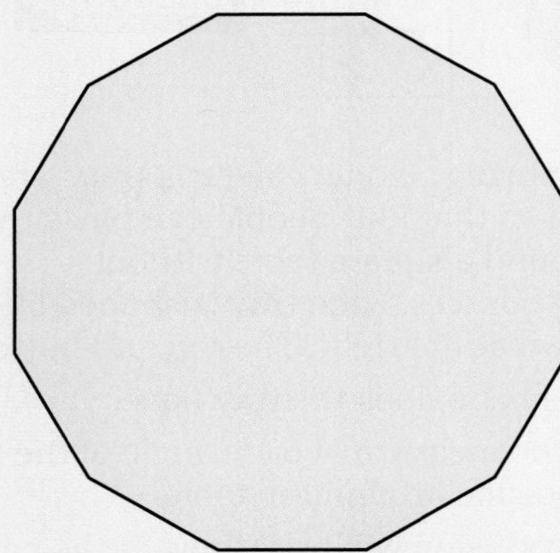

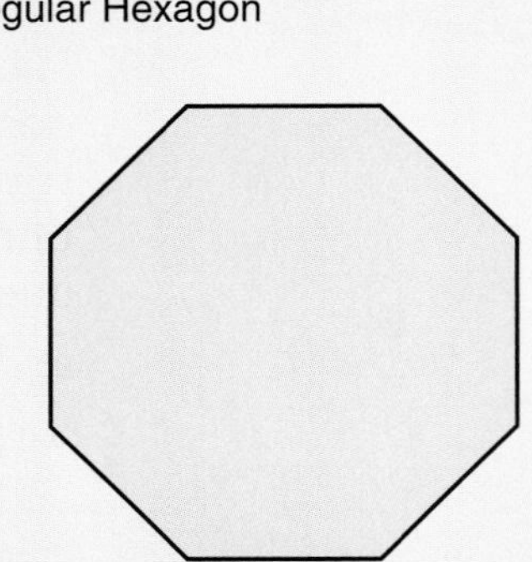

TALK ABOUT IT

1. How can you tell from the measure of one angle whether or not a single polygon will tessellate?

Your Turn

Try to make tessellations from the combinations below.

a. square and octagon
b. square, triangle, and hexagon
c. square, triangle, and dodecagon
d. another combination you choose

2. How can you tell whether a combination of polygons will tessellate?

11-9 Transformations

Setting Goals: *In this lesson, you'll identify and draw reflections, translations, and rotations. You'll also identify and draw symmetric figures.*

Modeling with Manipulatives

MATERIALS

cardboard, scissors

If you look around, you will probably see geometric shapes in repeating patterns. Floors and ceilings are often tiled in patterns. In the activity below, you will investigate one way to create a pattern.

Your Turn

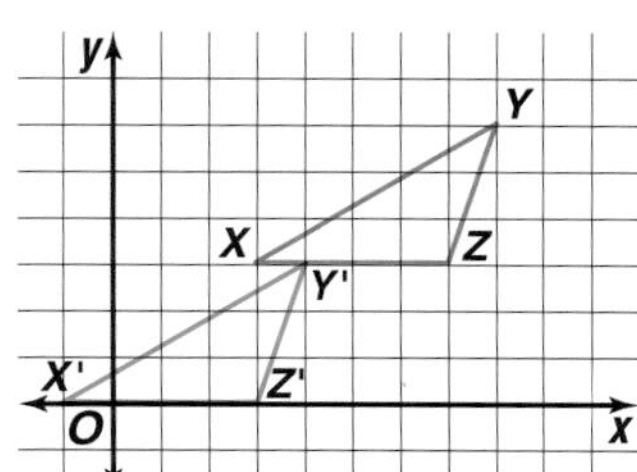

- Cut a triangle out of cardboard. Place the triangle on a coordinate grid as shown, so one side lies parallel to the *x*-axis. Trace the triangle on the paper. Label the vertices as shown. Write the coordinates of the vertices.
- Slide the cutout triangle 4 units left and 3 units down as shown. Trace the triangle. Label the vertices X', Y', and Z', so they correspond to the first triangle. Write the coordinates of the vertices of $\triangle X'Y'Z'$.

How are the coordinates of $\triangle XYZ$ and $\triangle X'Y'Z'$ related?

Learning the Concept

Transformations are movements of geometric figures. A **translation** is one type of transformation. In a translation, you "slide" a figure horizontally, vertically, or both, as you did in the activity.

Two other types of transformations are rotations and reflections.

In a **rotation**, you turn the figure around a point. Computer drawing packages allow the user to create drawings. After a drawing is created, a command called "rotate right" can be used to rotate the figure as shown at the right.

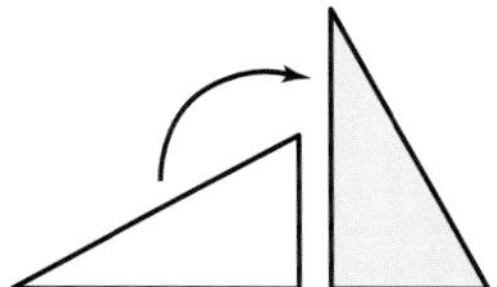

In a **reflection**, you "flip" a figure over a line. You can use a geomirror to draw the reflection of a figure. Each diagram below shows a figure and its reflection over a line. Place a geomirror on the dashed line in each figure and verify each reflection.

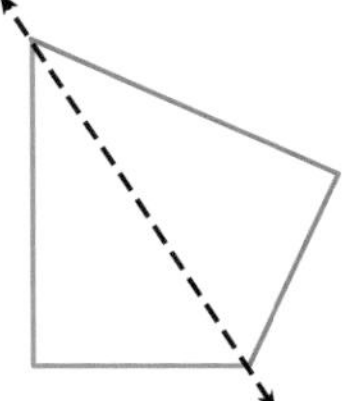

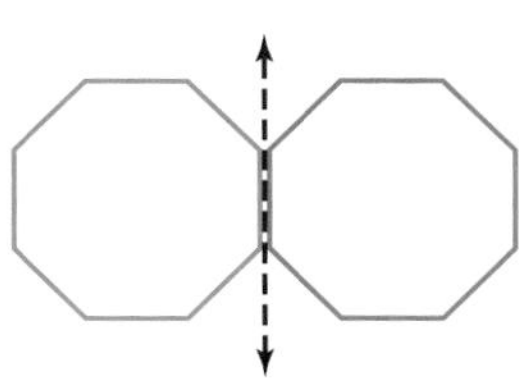

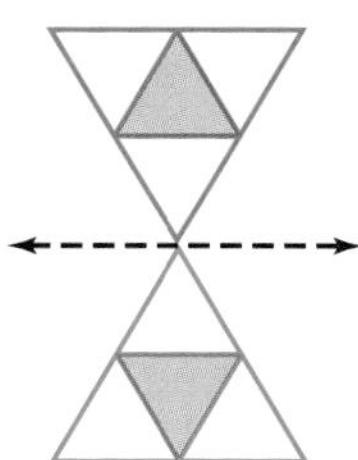

How could you describe the figure and its reflection?

Example 1 Tell whether each transformation is a translation, a rotation, or a reflection.

a.

b.

c.

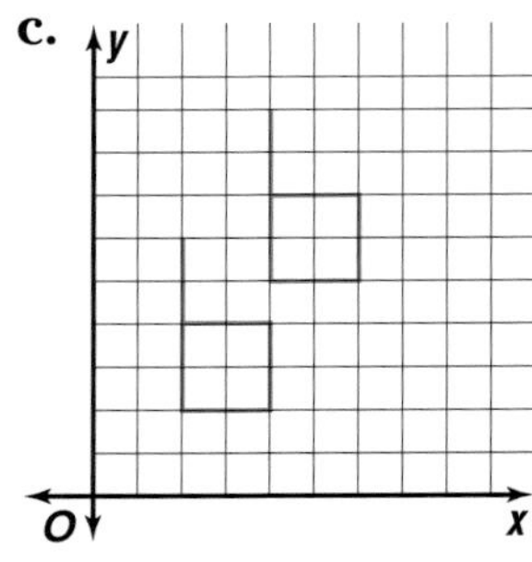

a. The figure is *rotated* about point A.

b. The figure is *reflected* about the y-axis.

c. The figure is *translated* 2 units horizontally and 3 units vertically.

THINK ABOUT IT

How many lines of symmetry does a square have? How many lines of symmetry does a circle have?

When you draw a figure and its reflection, you create a figure that is **symmetric**. The line where you placed the geomirror is called a *line of symmetry*. Some common items with line symmetry are shown below.

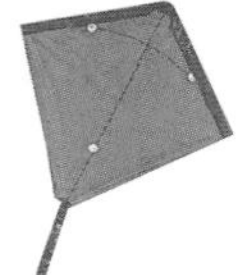

Some of the items shown above have one line of symmetry. Others have several lines of symmetry. Each line of symmetry separates the figure into *two* congruent parts.

Example 2 Draw all lines of symmetry for each figure.

a.

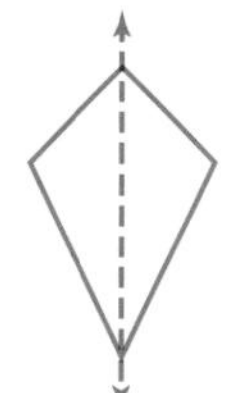

A kite has one line of symmetry.

b.

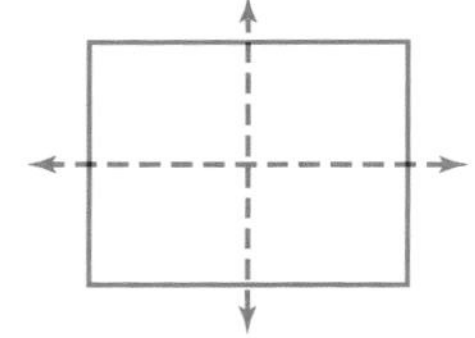

A rectangle has two lines of symmetry.

c.

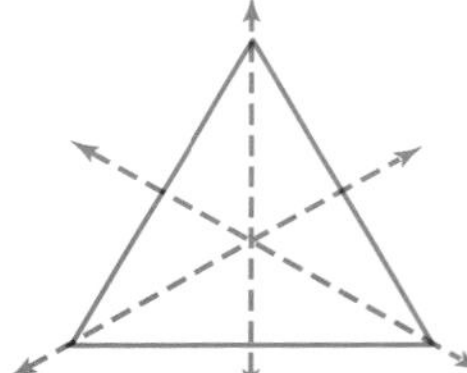

An equilateral triangle has three lines of symmetry.

Artists have used symmetry to create beautiful designs for centuries.

Example 3

World Culture

In 1895, excavators found Hopi pottery among the ruins of the pueblo known as Sikyatki. The pattern below was found on the exterior of a Sikyatki food bowl. Describe the transformations used to make the pottery design.

The original design is translated to the right and rotated 180° or reflected over a horizontal line.

Checking Your Understanding

Communicating Mathematics

Read and study the lesson to answer these questions.

1. **Compare and contrast** a translation, a reflection, and a rotation of the same figure.
2. **Show** how to complete the figure so that $\overleftrightarrow{XY}$ is a line of symmetry.

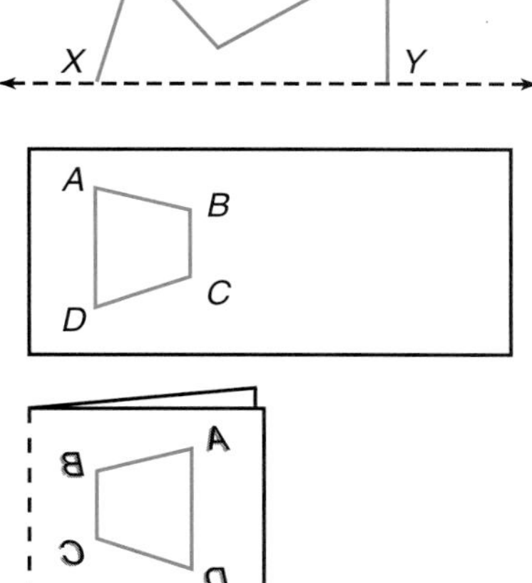

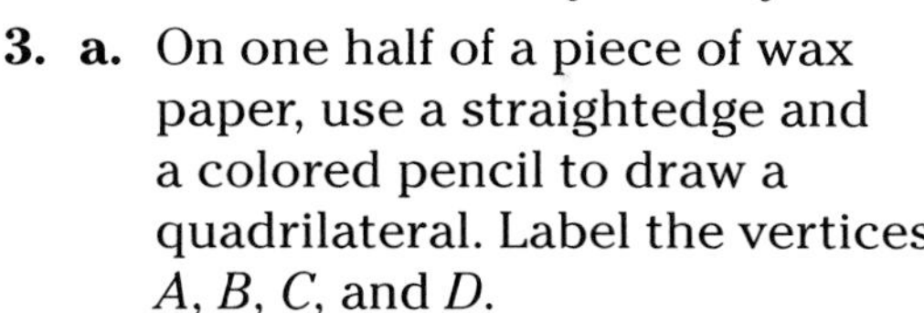

MATERIALS

wax paper

colored pencils

ruler

3. a. On one half of a piece of wax paper, use a straightedge and a colored pencil to draw a quadrilateral. Label the vertices A, B, C, and D.
   b. Fold the wax paper in half and turn it over so $ABCD$ is on the bottom side. Use a different color pencil to trace $ABCD$.
   c. Label the vertices A', B', C', and D' so that A corresponds to A', B to B', and so on.
   d. Unfold the paper. The fold line represents the line of reflection and $A'B'C'D'$ is a reflection of $ABCD$.
   e. Use a ruler to measure $\overline{AA'}$, $\overline{BB'}$, $\overline{CC'}$, and $\overline{DD'}$. At what point does the line of reflection intersect each segment?

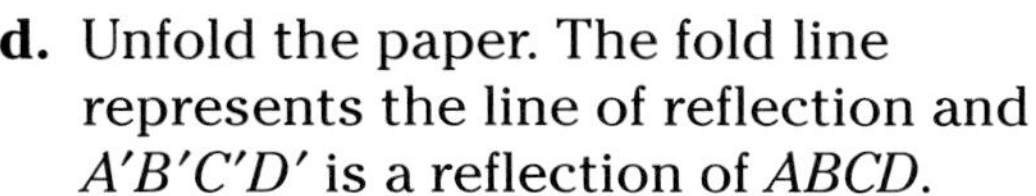

Guided Practice

Tell whether each geometric transformation is a translation, a reflection, or a rotation. Explain your answer.

4.

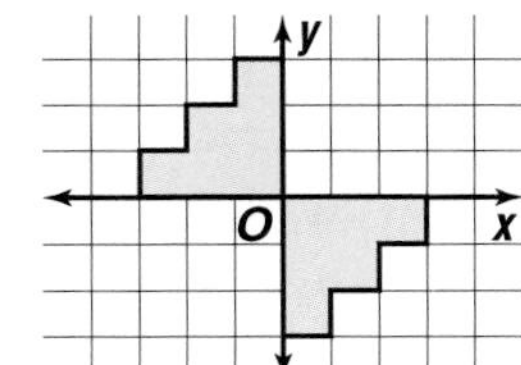

5.

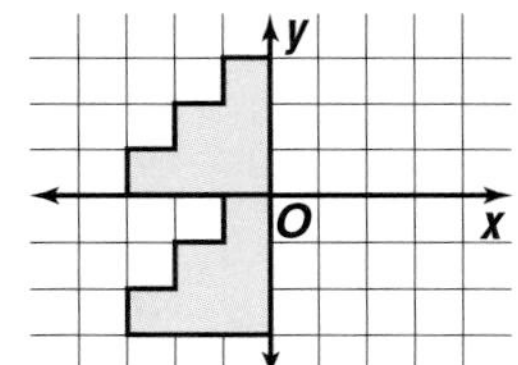

Trace each figure. Draw all lines of symmetry.

6.

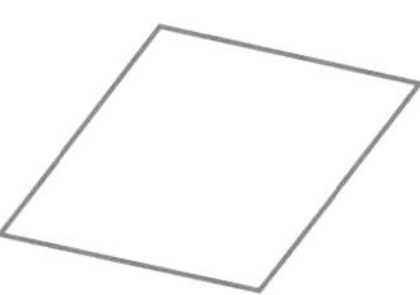

7.

8. **Art** A sample of a Navaho weaving is shown below. Describe the transformations and lines of symmetry that exist in the weaving.

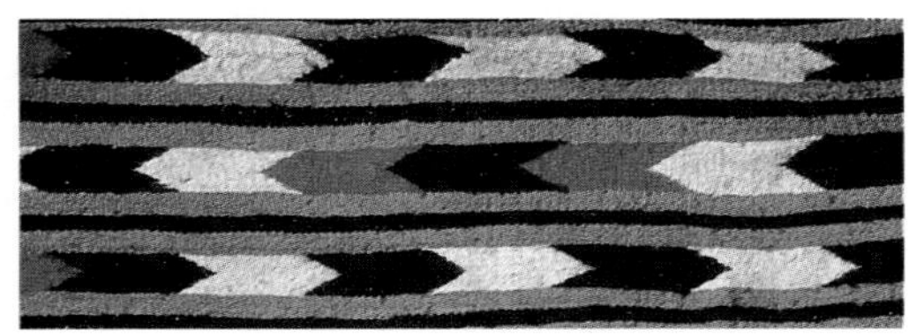

Exercises: Practicing and Applying the Concept

Independent Practice

Tell whether each geometric transformation is a *translation*, a *reflection*, or a *rotation*. Explain your answer.

9. 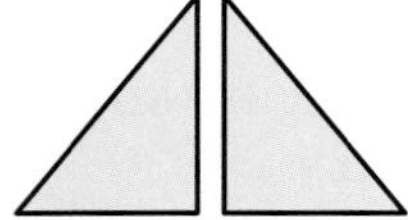

10. 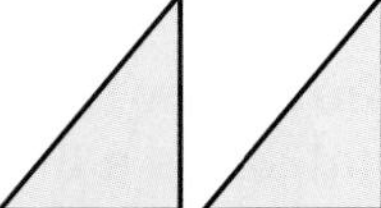

11. 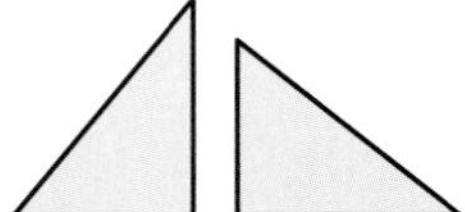

12. A translation moves $\triangle PQR$ 3 units to the right and 3 units down to form $\triangle P'Q'R'$. Write the coordinates of each vertex of $\triangle P'Q'R'$.

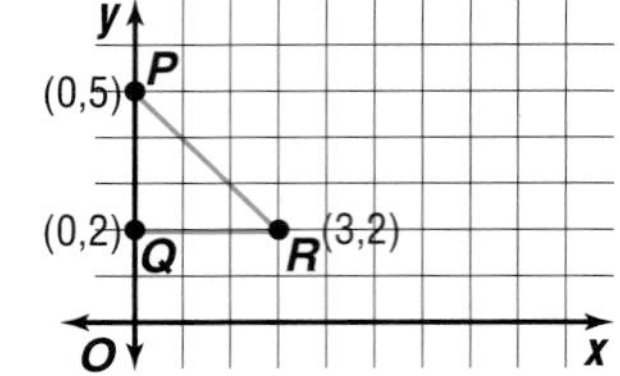

13. If the first figure below is rotated about point R, which could be the resulting figure?

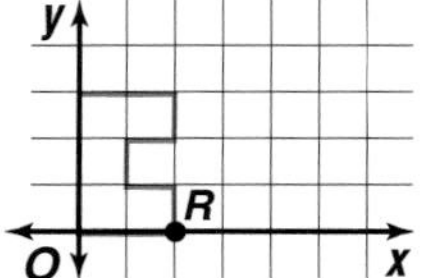

a. 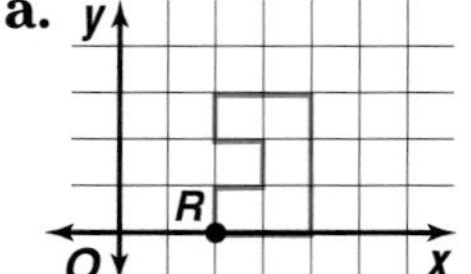

b.

c. 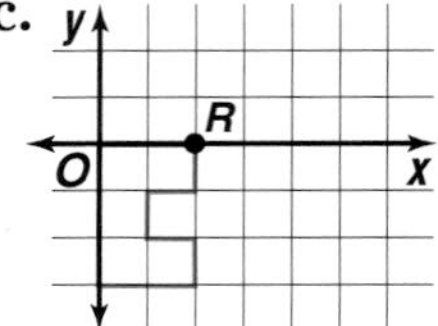

Trace each figure. Draw all lines of symmetry.

14. 

15. 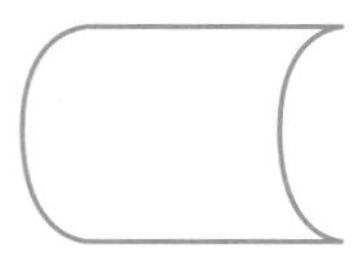

16. 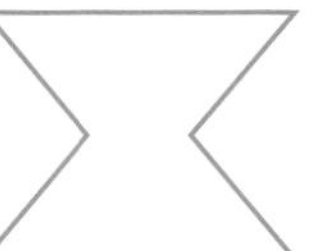

17. Graph $A(-1, 3)$, $B(4, 4)$, and $C(2, 1)$ on a coordinate plane. Draw $\triangle ABC$ and graph its reflection if the x-axis is the line of reflection.

18. Graph $P(-5, 1)$, $Q(-3, 4)$, and $R(-1, 3)$ on a coordinate plane. Draw $\triangle PQR$ and graph its reflection if the y-axis is the line of reflection.

19. Graph $R(3, -1)$, $S(6, 2)$, and $T(1, 4)$ on a coordinate plane. Draw $\triangle RST$ and translate the triangle 5 units to the left and 4 units down.

Critical Thinking

20. Write a formula to find the number of lines of symmetry for any regular polygon with n sides. (*Hint:* Sketch several regular polygons and draw all the lines of symmetry. Record the results in a table and look for a pattern.)

Application and Problem Solving

21. **Communication** International code flags can be used by sailors to send messages at sea. Describe the kinds of symmetry that are displayed by the flags and tell which flags have each kind of symmetry.

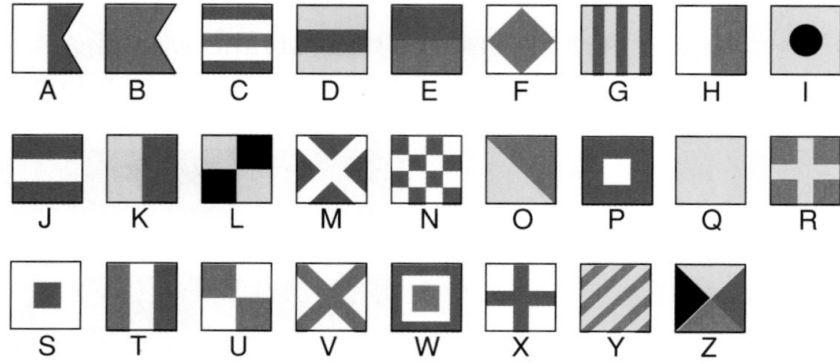

B • H E F

22. Games Miniature golf is played with a putter and a golf ball on a carpeted "green" surrounded by side rails. On some holes there is a straight path to the hole. Others are "dog legs", and it is not possible to aim directly at the hole. You can, however, get a "hole in one" by mentally reflecting the hole across the line of the wall and aiming for the reflection.

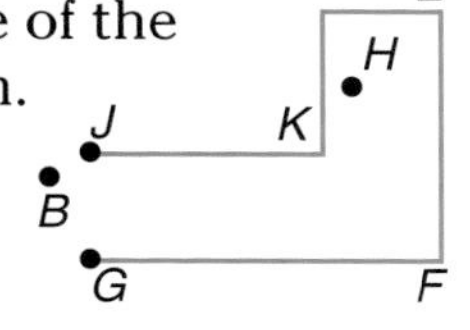

a. Trace the figure at the top right. Draw a path using a reflection that will allow the ball to reach the hole in one stroke.

b. A more difficult "dog leg" is shown at the bottom right. Trace the figure and find the path of the ball for a "hole in one."

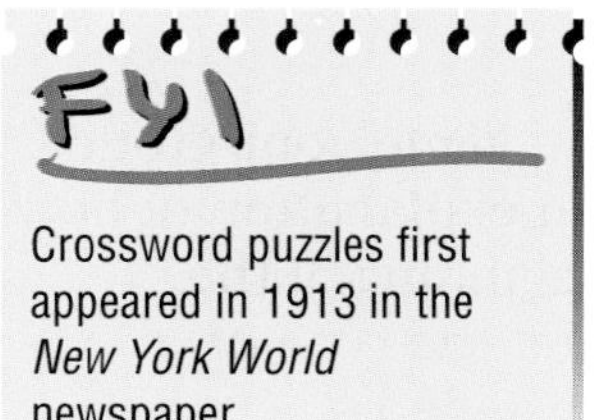

FYI

Crossword puzzles first appeared in 1913 in the *New York World* newspaper.

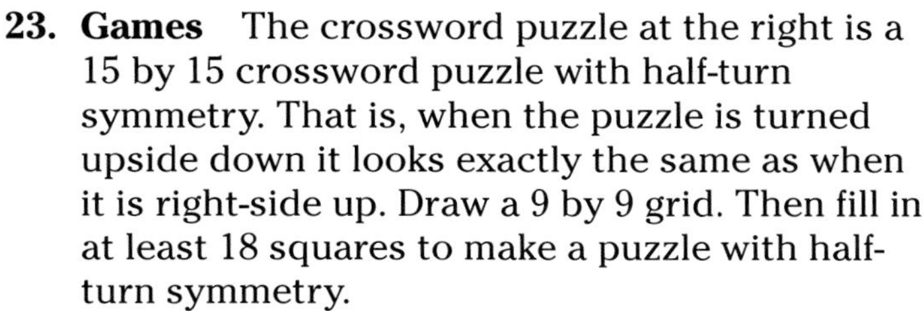

23. Games The crossword puzzle at the right is a 15 by 15 crossword puzzle with half-turn symmetry. That is, when the puzzle is turned upside down it looks exactly the same as when it is right-side up. Draw a 9 by 9 grid. Then fill in at least 18 squares to make a puzzle with half-turn symmetry.

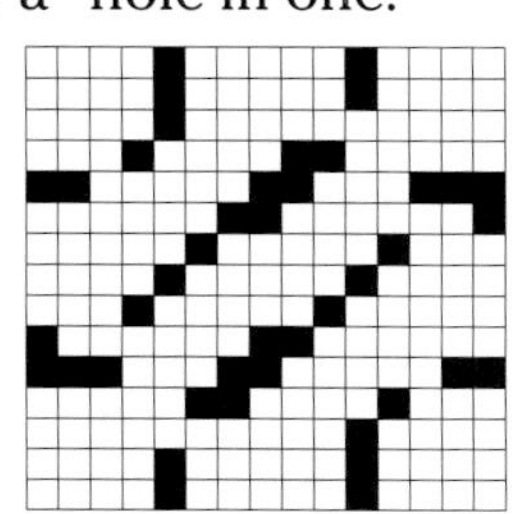

Mixed Review

24. Find the sum of the measures of the angles of a 15-gon. (Lesson 11-8)

25. Health The average birth weight for full-term female babies is 7 pounds. The average weight of a one-year old girl is 21 pounds. What is the percent of change in weight in the first year? (Lesson 9-10)

26. Solve $\frac{2}{5}x \leq -0.16$. (Lesson 6-7)

27. Solve $-84 > 4t$. Then graph the solution on a number line. (Lesson 3-7)

28. Write the coordinates for a point on the x-axis. (Lesson 1-7)

From the FUNNY PAPERS

TALK ABOUT IT

1. Explain why the comic is funny.
2. What is shown in the second frame of the comic?
3. What calculations did Jason have to make to get the ball to hit Paige?

CLOSING THE

Investigation

Refer to the Investigation on pages 482–483. Add the results of your work to your Investigation folder.

Plan and give an oral presentation to describe what you know about games and strategy. Your presentation should include the following.

- A report of the survey you conducted at the beginning of the Investigation.
- A report of the results of your work at the end of Lesson 10-4. Include any revisions that were the direct result of the test and an explanation of why those revisions were made.
- A report on how your group determined the best strategy for *Don't Lose It All!* at the end of Lesson 10-8.
- A model of the board your group created at the end of Lesson 11-7 and a brief explanation of the accompanying game.
- A final version of the game your group designed. Include an explanation of how the survey at the beginning of the Investigation affected the type of game your group designed.

Extension

Contact a computer game manufacturer and inquire about how a computer game is created and how long it takes to create it.

PORTFOLIO ASSESSMENT

You may want to keep your work on this Investigation in your portfolio.

CHAPTER 11 Highlights

Vocabulary

After completing this chapter, you should be able to define each term, property, or phrase and give an example or two of each.

Geometry

acute angle (p. 550)
acute triangle (p. 569)
adjacent angles (p. 561)
alternate interior angles (p. 563)
alternate exterior angles (p. 563)
angle (p. 549)
complementary (p. 562)
congruent (pp. 554, 561, 573)
construct (p. 554)
corresponding angles (p. 563)
corresponding parts (p. 573)
degree (p. 549)
diagonal (p. 590)
equiangular (p. 591)
equilateral (p. 591)
exterior angle (p. 591)
indirect measurement (p. 580)
interior angle (p. 591)
intersect (p. 551)
isosceles triangle (p. 569)
line (p. 548)
line segment (p. 548)
obtuse angle (p. 550)
obtuse triangle (p. 569)
parallel (p. 551)
parallelogram (p. 585)
perpendicular (p. 551)
plane (p. 550)
point (p. 548)
polygon (p. 589)
quadrilateral (p. 584)
ray (p. 548)
rectangle (p. 585)
reflection (p. 595)
regular polygon (p. 591)
rhombus (p. 585)
right angle (p. 550)
right triangle (p. 569)
rotation (p. 595)
scale drawing (p. 585)
scalene triangle (p. 569)
sides (pp. 549, 589)
similar (p. 578)
skew lines (p. 551)
straight angle (p. 550)
square (p. 585)
supplementary (p. 562)
symmetric (p. 596)
tessellation (p. 594)
transformation (p. 595)
translation (p. 595)
transversal (p. 563)
trapezoid (p. 585)
triangle (p. 568)
vertex (pp. 549, 589)
vertical angles (p. 561)

Statistics

circle graph (p. 556)

Understanding and Using Vocabulary

Choose the letter of the term that best completes each statement.

1. A(n) _?_ angle measures between 0° and 90°.
2. Two angles are _?_ if the sum of their measures is 90°.
3. When a _?_ intersects two parallel lines, the corresponding angles are congruent.
4. When a line is perpendicular to another line, the angles formed are _?_ angles.
5. A(n) _?_ angle measures between 90° and 180°.
6. A parallelogram with four congruent sides is a _?_.
7. Two angles are _?_ if the sum of their measures is 180°.

a. obtuse
b. acute
c. vertical
d. trapezoid
e. rhombus
f. right
g. transversal
h. complementary
i. supplementary

CHAPTER 11 Study Guide and Assessment

Skills and Concepts

Objectives and Examples

Upon completing this chapter, you should be able to:

- **identify points, lines, planes, rays, segments, angles, and parallel, perpendicular and skew lines.** (Lesson 11-1)

Draw and label a diagram to represent ∠*ABC*.

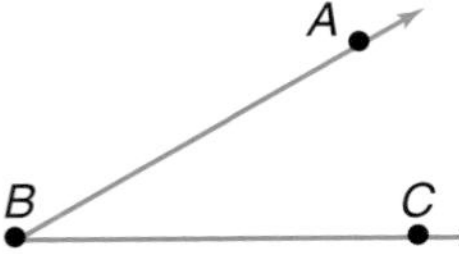

Review Exercises

Use these exercises to review and prepare for the chapter test.

Draw and label a diagram to represent each of the following.

8. point *R*
9. ray *TS*
10. plane *XYZ*
11. acute angle *JKL*
12. line *n*
13. $\overline{EF}$
14. $\overleftrightarrow{HG} \parallel \overleftrightarrow{KL}$
15. lines *w* and *y* are skew

- **make a circle graph to illustrate data.** (Lesson 11-2)

This circle graph compares the amount of land and water to the whole surface of Earth.

30  × 360  108

70 % × 360 =  252

Earth's Surface

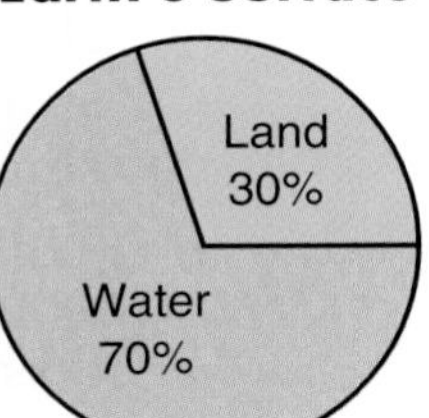

16. The table below shows the types of jams and jellies that were entered in state fairs in 1994. Make a circle graph of the data.

Type	Entries
Strawberry	266
Raspberry	266
Blackberry	176
Grape	165
Plum	131
Other	592

17. Make a circle graph of the number of hours each week that you study, talk on the telephone, watch television or listen to music, and sleep.

- **identify the relationships of vertical, adjacent, complementary, and supplementary angles, and angles formed by two parallel lines and a transversal.** (Lesson 11-3)

Find the measure of each labeled angle.

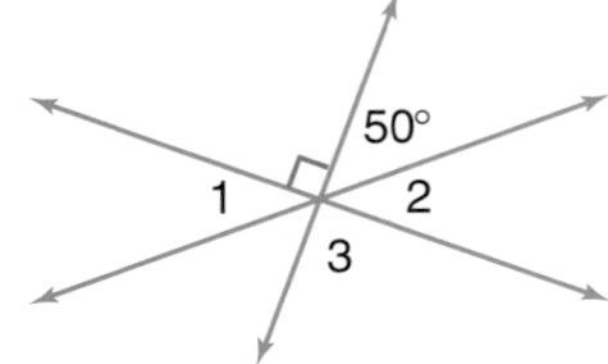

$m\angle 1 = 180 - 90 - 50$ or $40°$
$m\angle 2 = m\angle 1$ or $40°$
$m\angle 3 = 90°$

In the figure below, *a* ∥ *b*. Find the measure of each angle.

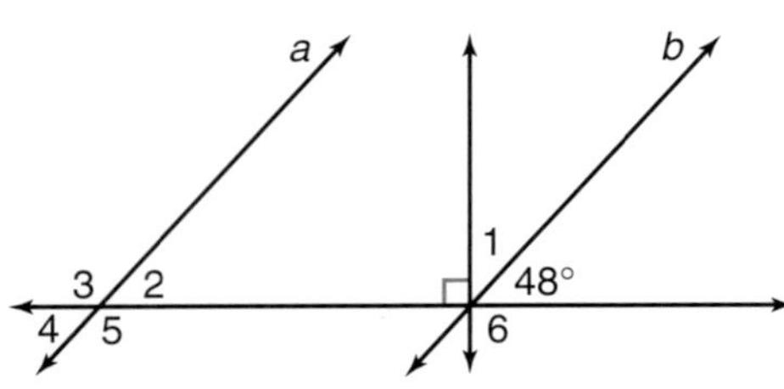

18. ∠1
19. ∠2
20. ∠3
21. ∠4
22. ∠5
23. ∠6

Objectives and Examples

▶ **find the missing angle measure of a triangle.** (Lesson 11-4)

Find $m\angle 1$ if $m\angle 2 = 50°$ and $m\angle 3 = 80°$. Classify the triangle as *acute*, *right*, or *obtuse*.

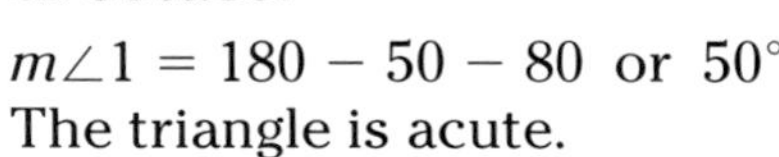
$m\angle 1 = 180 - 50 - 80$ or $50°$

The triangle is acute.

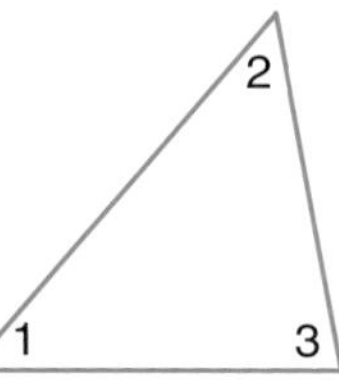

Review Exercises

Use the figure below to solve each problem.

24. Find $m\angle 1$ if $m\angle 2 = 35°$ and $m\angle 3 = 90°$. Classify the triangle as *acute*, *right*, or *obtuse*.

25. Find $m\angle 2$ if $m\angle 1 = 42°$ and $m\angle 3 = 46°$. Classify the triangle as *acute*, *right*, or *obtuse*.

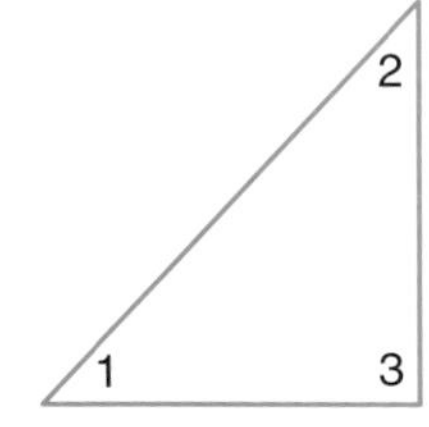

▶ **identify congruent triangles and their corresponding parts.** (Lesson 11-5)

If $\angle ABC \cong \angle FED$, name the parts congruent to angle A and segment AC.

$\angle A$ is congruent to $\angle F$. $\overline{AC}$ is congruent to $\overline{FD}$.

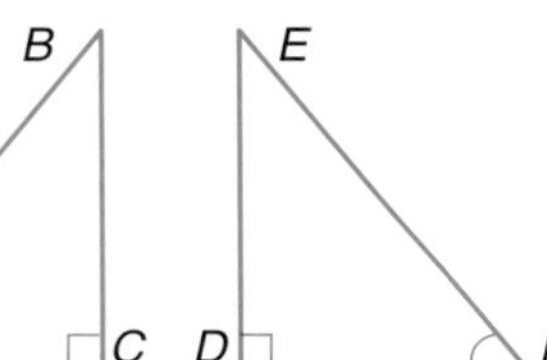

If $\triangle XYZ \cong \triangle ABC$, name the part congruent to each angle or segment.

26. $\angle X$

27. $\angle B$

28. $\overline{YZ}$

29. $\overline{AC}$

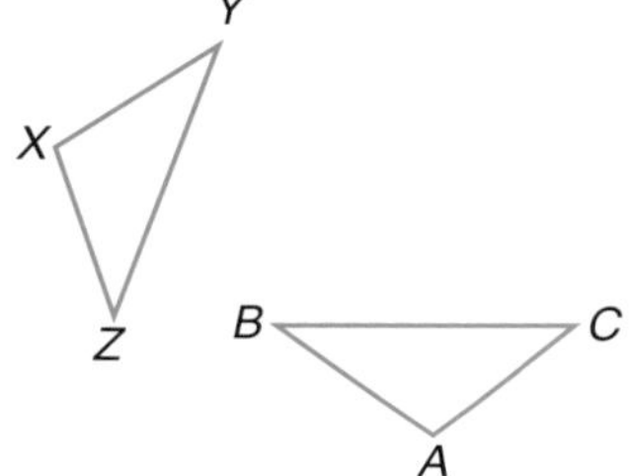

▶ **identify corresponding parts and find missing measures of similar triangles.** (Lesson 11-6)

Find the value of x in the similar triangles.

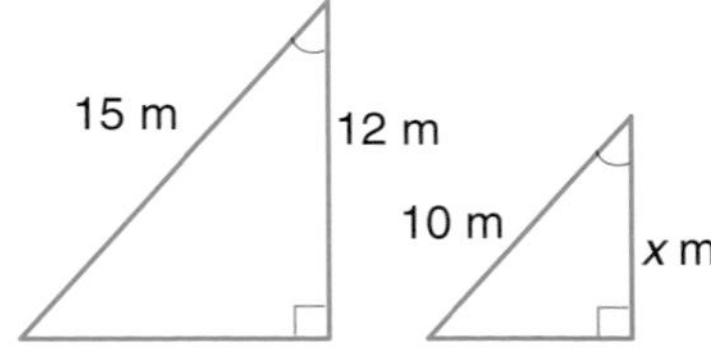

$$\frac{15}{10} = \frac{12}{x}$$
$$120 = 15x$$
$$8 = x$$

Find the value of x in each pair of similar triangles.

30.

x ft
6 ft
7 ft
21 ft

31.

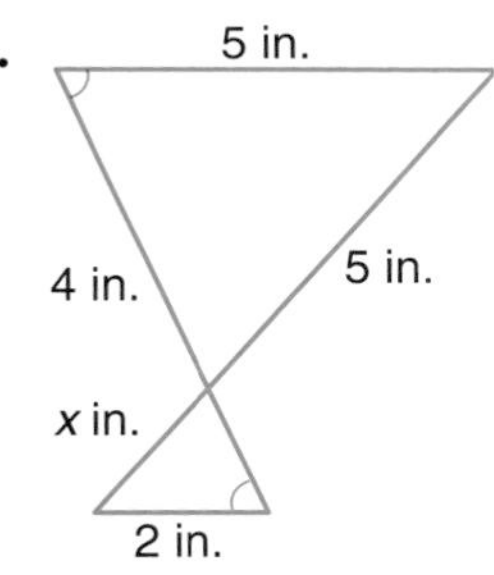

▶ **find the missing angle measure of a quadrilateral and classify quadrilaterals.** (Lesson 11-7)

Find the missing angle measures.

x° 135° 75° 2x°

$$75 + 135 + x + 2x = 360$$
$$210 + 3x = 360$$
$$3x = 150$$
$$x = 50$$

The measures are 50° and $50 \cdot 2$ or 100°.

Find the missing angle measures in each figure.

32.

33.

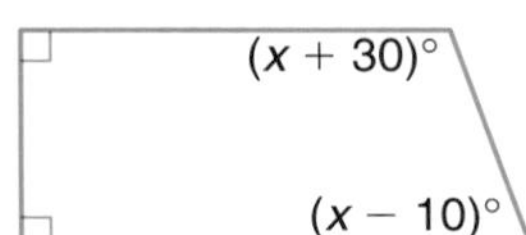

Objectives and Examples

▶ **classify polygons and determine the sum of the measures of the interior and exterior angles of a polygon.** (Lesson 11-8)

Find the sum of the measures of the angles of an octagon.

$(n - 2)180 = (8 - 2)180$
$= 6(180)$
$= 1080$

Review Exercises

34. Find the sum of the degree measures of the angles of a hexagon.

35. Find the measure of each interior angle in a regular decagon.

36. Find the measure of an exterior angle in a regular pentagon.

▶ **identify and draw reflections, translations, and rotations and identify and draw symmetric figures.** (Lesson 11-9)

Draw all lines of symmetry for the figure below.

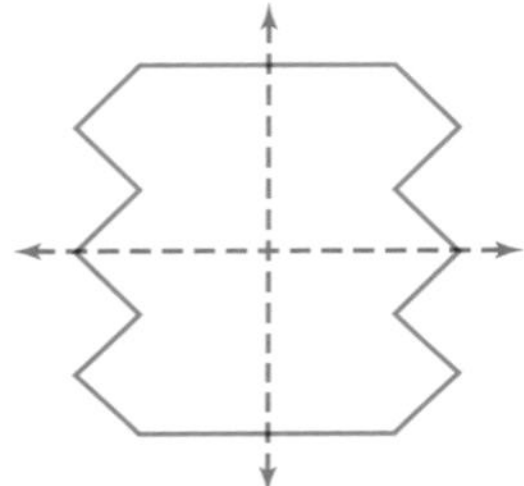

Trace each figure. Draw all lines of symmetry.

37. 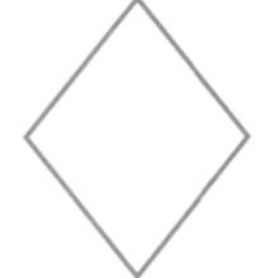

38. 

Tell whether each transformation is a *translation*, a *reflection*, or a *rotation*. Explain your answer.

39. 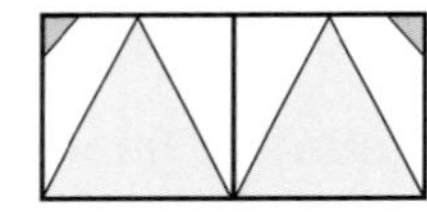

40. 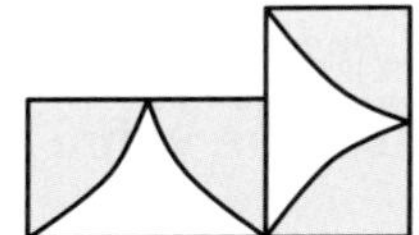

Applications and Problem Solving

41. Statistics Make a circle graph to display the data in the chart. (Lesson 11-2)

Water Use in the U.S.	
Agriculture	36%
Public Water	8%
Utilities	33%
Industry	23%

42. Home Maintenance When a 6-foot ladder is placed against the side of a house at an angle that is reasonable for climbing, it touches the side of the house at a point 5 feet above the ground. If a 15-foot ladder is placed at the same angle to the house, can it be used to reach a point 12 feet above the ground? How high will it reach? (Lesson 11-6)

A practice test for Chapter 11 is available on page 784.

Alternative Assessment

Cooperative Learning Project

Scale Drawings Measurements of angles and geometric shapes are used to make scale drawings. For this project, you will need to choose a real-life space and complete the following activities.

1. Choose an indoor or outdoor area of which to make a scale drawing. For example, you might choose a room or the exterior of a house.
2. Make a sketch of the area, marking details such as walls, doors, windows, electrical outlets, and any other details in the space.
3. Make a plan for measuring the details of the space. Consider how you will measure the angles in the space as well as the distances.
4. Make your measurements and record them on the sketch.
5. Select an appropriate scale for your drawing.
6. Prepare a final scale drawing from the information on your sketch.

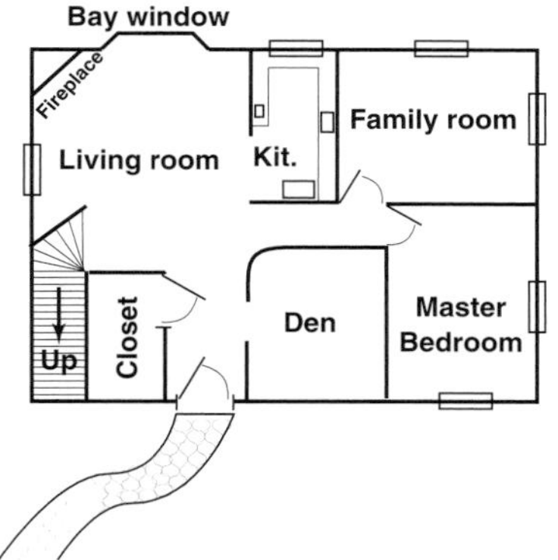

Thinking Critically

- A quadrilateral has two pairs of parallel sides. What kind of figure could it be?
- A triangle has angles measuring x, $2x$, and $3x$ degrees. What kind of triangle is it?
- The sum of the degree measures of the angles in a regular polygon is $(n - 2)180$. Explain how you can remember this formula.

Portfolio

Select your favorite activity from this chapter and write a note explaining what you learned from the activity.

Self Evaluation

Honeycombs are made in the shape of hexagons because hexagons are very strong structures. They distribute the weight and do not collapse.

Assess yourself. How do you distribute the weight of your tasks over the time you have been allowed? Do you wait for the last minute and then rush to finish, or do you spread out your work so you can give it your best attention? Write a paragraph about how you can improve the way you use your time.

LONG-TERM PROJECT

Investigation

OH GIVE ME A HOME

MATERIALS

- ruler
- protractor
- graph paper
- cardboard or foam board
- glue
- paint (optional)

Look around the room where you are sitting. What factors do you think had to be considered before any walls were built? If you did not know any mathematics, would you be able to construct the building that you are sitting in now?

Few things require the use of mathematics more than the construction of buildings. Even the most complex buildings are made up of arrangements of geometric shapes.

Suppose you are a member of a group of three architects that has been hired to design a home. You will oversee every detail of the design from selecting the location to planting the last shrub in the yard.

To create an attractive and efficient building, you must bring harmony to the following elements.

- ***Appearance*** *–The outside appearance of a structure is determined not only by its shape but also by the choice of building materials. Many architects have skillfully blended the rough textures of wood and stone with the smoothness of glass and metal.*
- ***Durability*** *–A structure should be well built so that it can withstand all kinds of weather and provide protection from temperature extremes without expensive maintenance.*
- ***Function*** *–Every structure is designed for a certain purpose. To be functional, a building should fulfill the needs of its users.*

Starting the Investigation

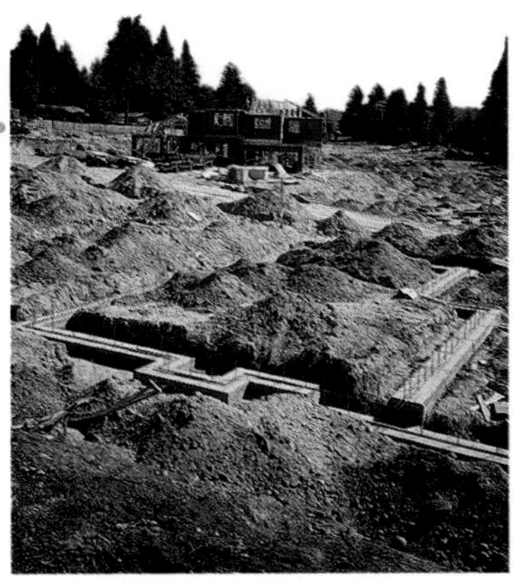

When someone decides to build a house, the first thing they must do is choose a *lot*, or piece of land. Work with your group to select a location for the dwelling you will design and build. Then check the zoning laws and building codes for that location.

As a group, identify at least four different dwellings in your community. Before you begin, you should familiarize yourself with some styles of houses by referring to an encyclopedia, or consulting with a Realtor. Describe the physical structure of the house and state its purpose. For example, Description: brick 1-story; Purpose: home for a retired couple. Include a sketch of each building.

An *elevation* is a drawing of a building's exterior. An elevation shows the location and size of features such as doors and windows. Typically, four elevation drawings are done, one for each of the four sides of the building. They also show what types of materials will be used on the outside of the building. Choose a style of dwelling that your group would like to build. Then sketch its front elevation.

For current information on architecture, visit:
www.glencoe.com/sec/math/prealg/mathnet

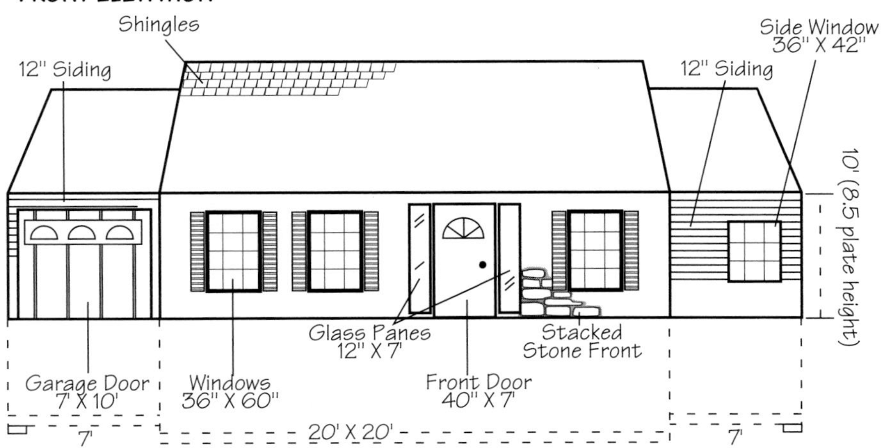

Working on the Investigation
Lesson 12-5, p. 637

Working on the Investigation
Lesson 12-8, p. 653

Working on the Investigation
Lesson 13-2, p. 671

Closing the Investigation
End of Chapter 13, p. 698

You will work on this Investigation throughout Chapters 12 and 13. Be sure to keep your materials in your Investigation Folder.

CHAPTER 12

Measuring Area and Volume

TOP STORIES in Chapter 12

In this chapter, you will:

- find the area of polygons and circles,
- investigate the relationship between area and probability,
- find the surface area and volume of prisms, cylinders, pyramids, and cones, and
- solve problems by making a model or drawing.

MATH AND ARCHAEOLOGY IN THE NEWS

An Amazing Find

Source: U.S. News & World Report, May, 29, 1995

For years, archaeologists believed that the major discoveries had already been made in Egypt's Valley of Kings. But Kent Weeks, an archaeologist from the American University in Cairo, believed there was something still hidden.

When the Egyptian government proposed adding some roadways to accommodate the tourist traffic, Weeks asked them to put off construction so he could check his hunch. After seven summers of digging, Weeks was proved right. He and his research team have discovered the largest and most elaborate tomb in Egypt. It may hold the remains of as many as fifty of the sons of Ramses II, the pharoah who ruled Egypt from 1279 to 1213 B.C. A large hall that is 50 feet square and 12 feet high opens to a T-shaped corridor with many chambers.

Many of the riches of the tomb have been stolen through the centuries, but the knowledge that it holds will be valuable to Egyptologists. Many rooms are yet to be explored, and there may be another level beneath this find.

Putting It into Perspective

1799 The Rosetta Stone is discovered during Napoleon's occupation of Egypt.

1800

1822 Jean François Champollion publishes research that enables archaeologists to translate ancient Egyptian writings.

1850

1879 Prehistoric wall paintings are found in a cave at Altamira, Spain.

EMPLOYMENT OPPORTUNITIES

Open a Window to the Past

DO YOU HAVE dreams of being a real-life Indiana Jones? Careers in archaeology are not always as dangerous as his adventures. Archaeologists study the remains of human cultures to learn what life was like and to better understand history. **Field archaeologists** travel to all parts of the world to search for artifacts. Many other archaeologists conduct research or teach in universities. A master's degree is required, a doctorate is preferred.

For more information, contact:
Society for American Archaeology
900 2nd Street NE, No. 12
Washington, D.C. 20002

Statistical Snapshot

Cross Section of the Great Pyramid of Khufu

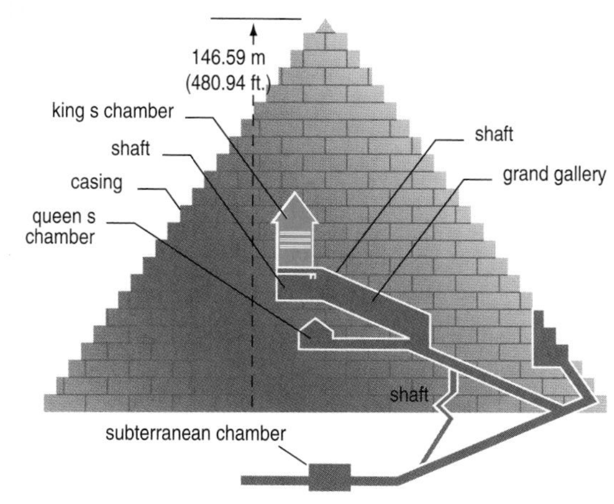

interNET CONNECTION For up-to-date information on archaeology, visit:
www.glencoe.com/sec/math/prealg/mathnet

1952 British archaeologist Dame Kathleen Kenyon proves Jericho, Jordan, to be one of the oldest known human communities.

1981 Hit movie *Raiders of the Lost Ark* features archaeologist character Indiana Jones.

1900 1950 2000

1922 Discovery of King Tutankhamen's tomb.

1995 Kent Weeks announces discovery of major tomb in Egypt's Valley of Kings.

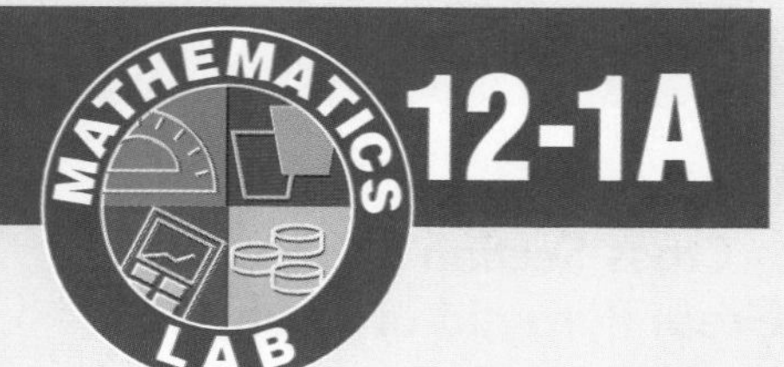

HANDS-ON ACTIVITY

12-1A Areas and Geoboards

A Preview of Lesson **12-1**

MATERIALS

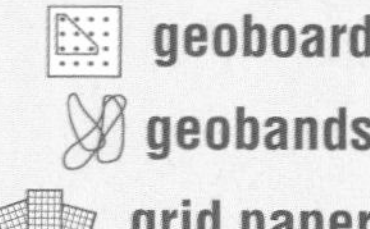

geoboard
geobands
grid paper

In Lesson 3-5, you learned a formula to find the area of a rectangle. But sometimes you need to find the area of shapes that are not rectangular. In this lab, we'll use a geoboard to find the areas of other polygons.

Activity 1

Make these figures on a geoboard. If one square represents one square unit, find the area of each figure.

a. 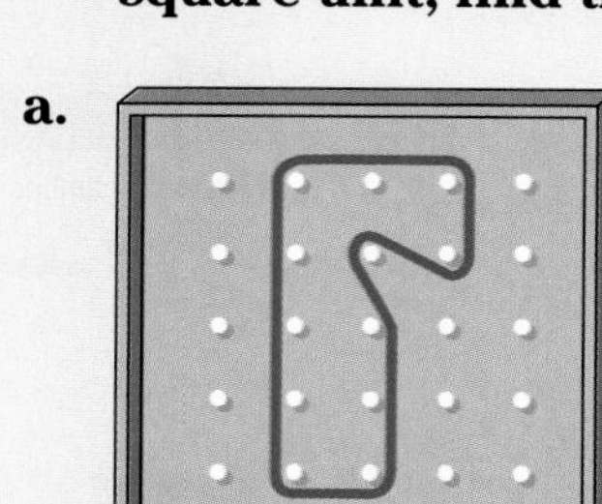

b. 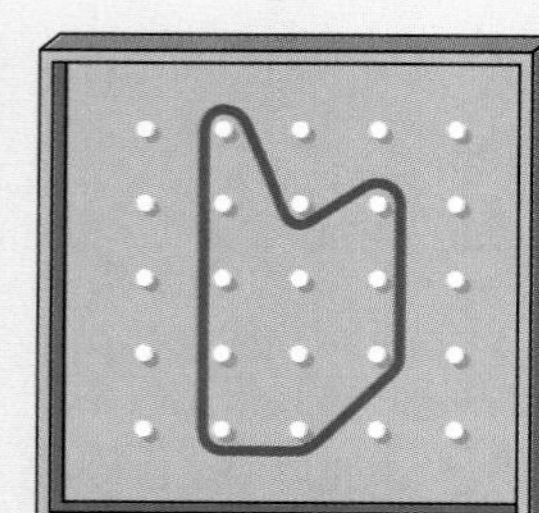

c. 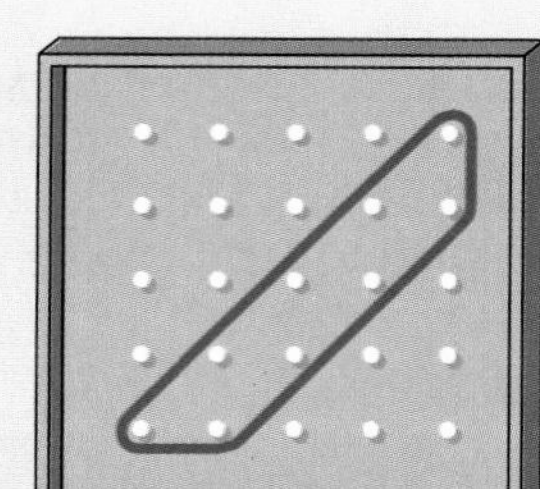

d. 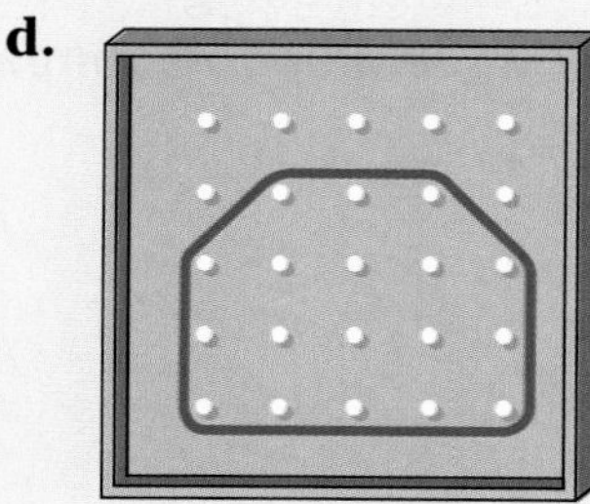

e. 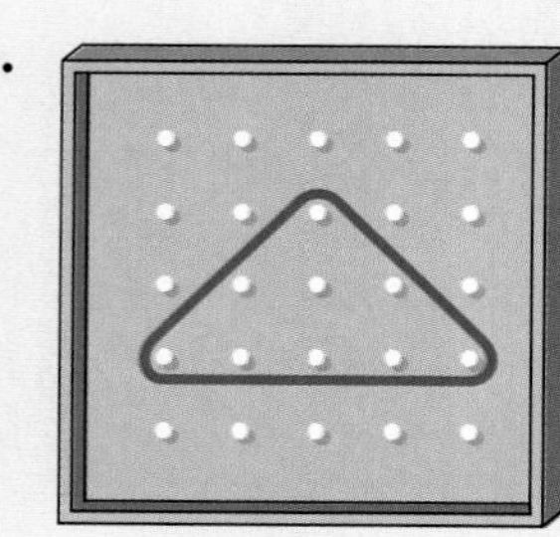

f. 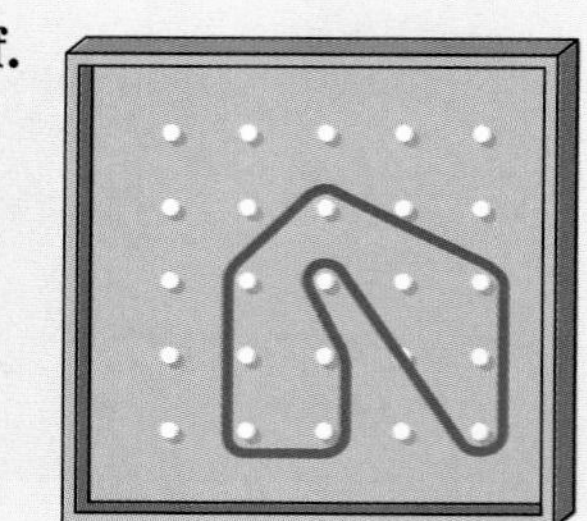

g. Make a figure on the geoboard and ask your partner to find its area.

TALK ABOUT IT

1. How did you find the areas of the polygons in a–f?
2. Was your partner's polygon made up of whole- and half-squares, or did it include other parts of squares?

Activity 2

Look at the triangle below. Suppose you make another triangle just like it to form a rectangle. You can find the area of the rectangle. Then divide by 2 to find the area of each triangle.

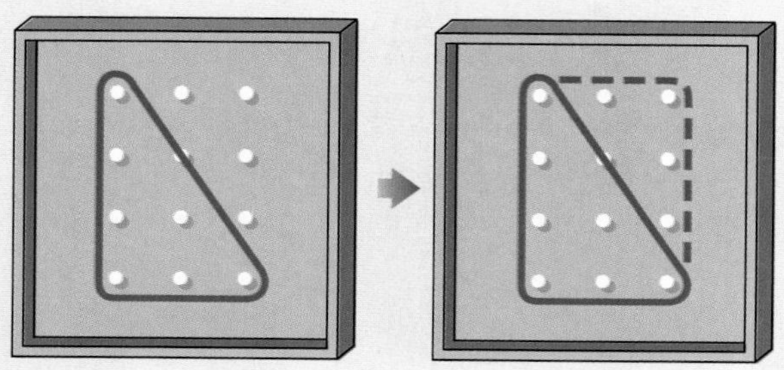

The area of the rectangle is 6 square units. So, the area of each triangle is 3 square units.

Your Turn **Find the area of each triangle.**

h. i. 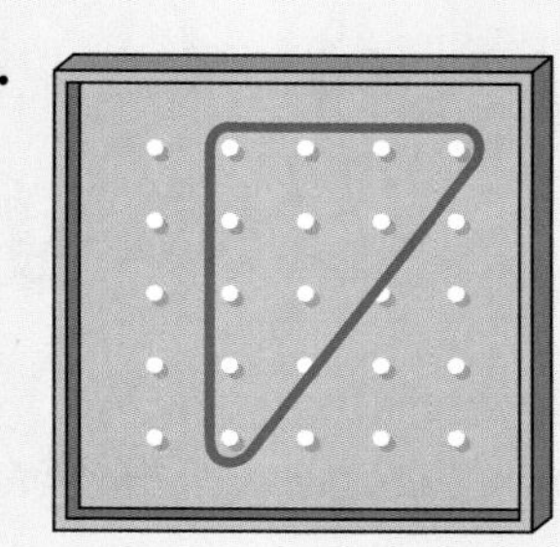 j. 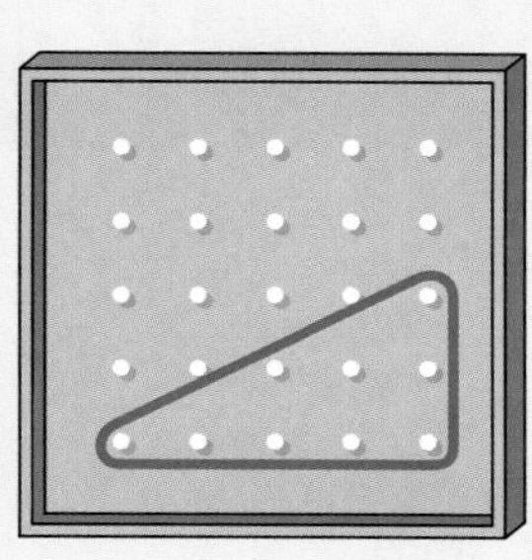

k. Make a right triangle on your geoboard and ask your partner to find the area.

3. Explain how you found the areas of the triangles.

Extension Here is another strategy for finding areas. Make the triangle at the right on a geoboard. Now build a rectangle around it as shown in step 2. You can subtract to find the area of the original triangle.

The large rectangle has an area of 16 square units. The area of triangle *a* is half of 4 or 2 square units. The area of triangle *b* is half of 12 or 6 square units. The area of the original triangle is $16 - 2 - 6$ or 8 square units.

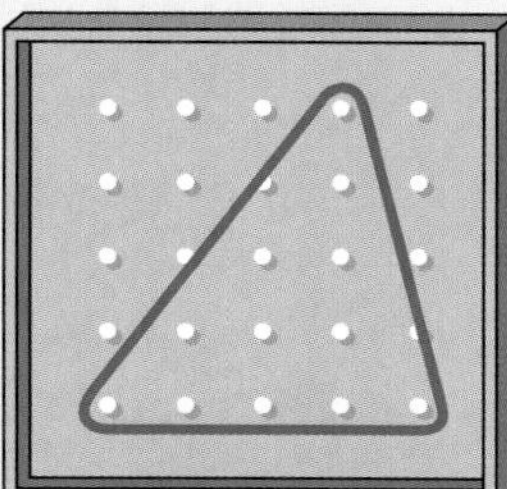

Step 1

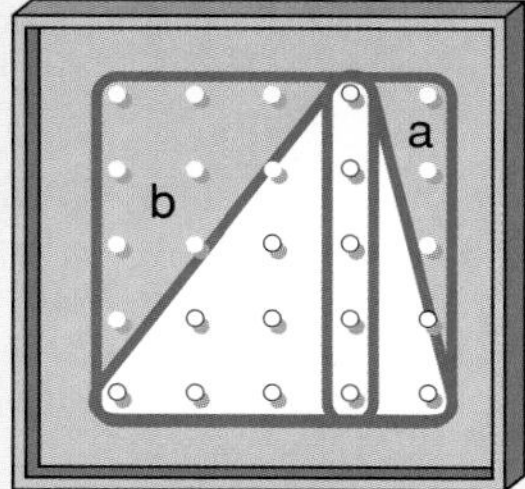

Step 2

Your Turn **Find the area of each polygon.**

l. 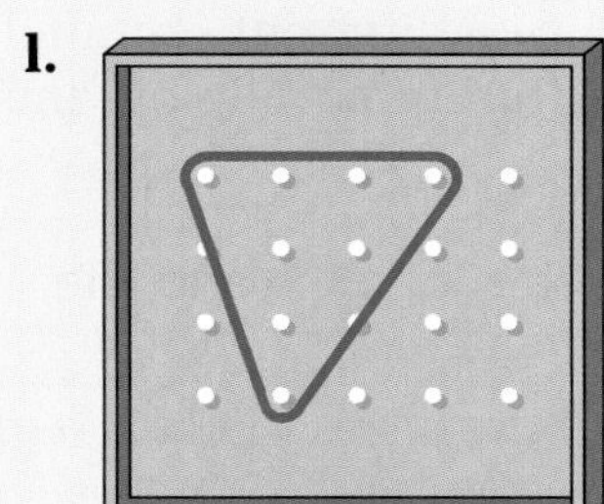 m. 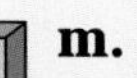 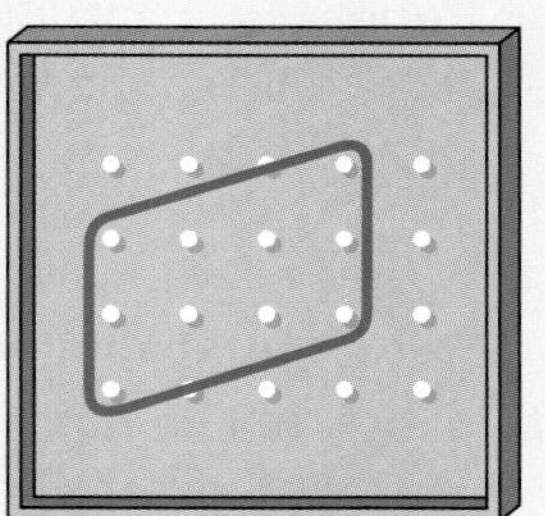 n. 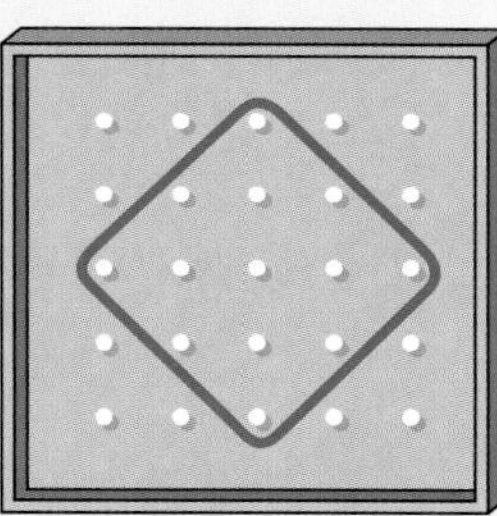

o. 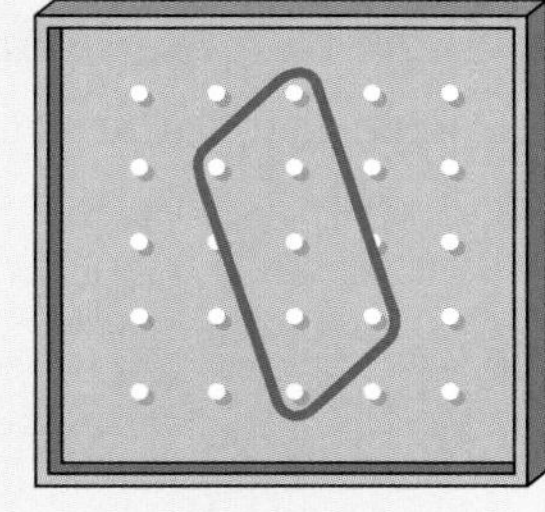 p. 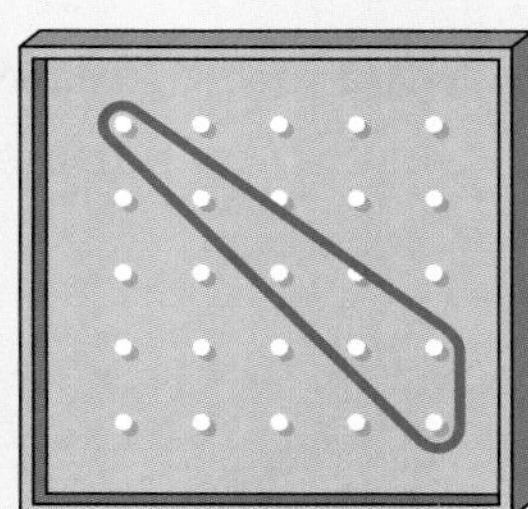 q. 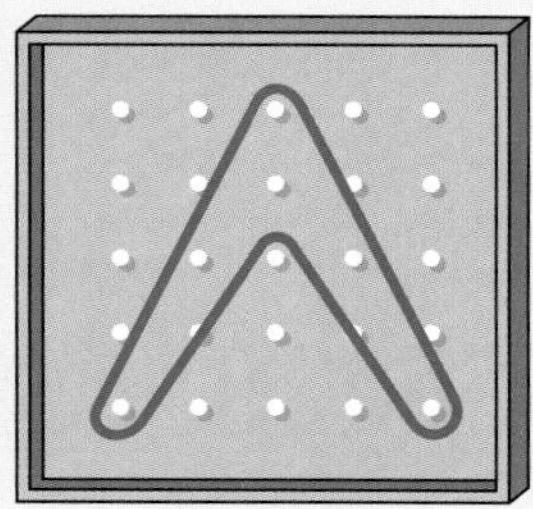

r. Make a polygon on your geoboard and ask your partner to find its area.

4. How did you find the areas of the polygons?

12-1 Area: Parallelograms, Triangles, and Trapezoids

Setting Goals: *In this lesson, you'll find areas of parallelograms, triangles, and trapezoids.*

Modeling with Manipulatives

MATERIALS

- centimeter grid paper
- scissors
- tape

Remember Transformers™? Several years ago, Transformers were one of the hottest toys around. A Transformer starts out as one toy and can be changed into a different toy by moving its parts. In the activity below, you will transform one geometric shape into another.

Your Turn Work with a partner.

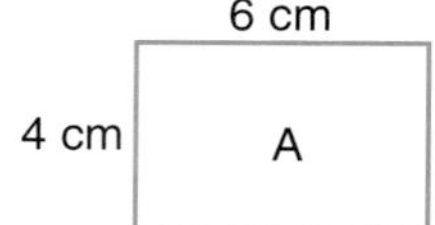

- ▶ On centimeter grid paper, draw a rectangle that is 4 cm high and 6 cm long. Cut out the rectangle. Trace it on a piece of plain paper and label the measures on the sides as shown. Label it rectangle A.

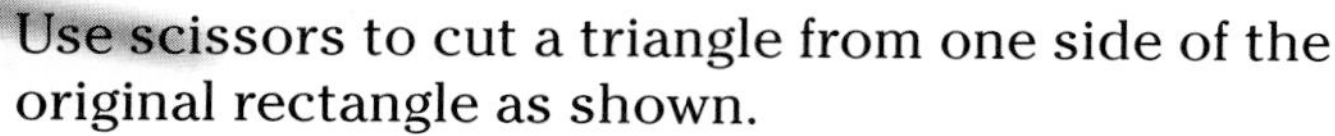

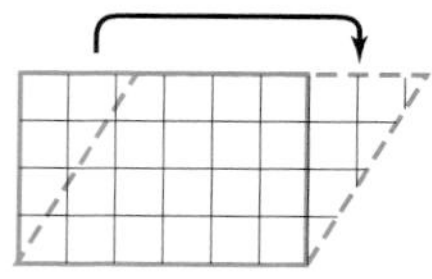

- ▶ Use scissors to cut a triangle from one side of the original rectangle as shown.
- ▶ Tape the triangle to the other side of the rectangle to form a parallelogram. Trace the parallelogram on the plain paper next to the rectangle. Label it parallelogram A.

- ▶ Repeat this activity with two different rectangles (B and C), making two different parallelograms (B and C).

a. Compare the area of each rectangle and the parallelogram made from it.

b. How do the measures of the sides of each rectangle relate to the parallelogram formed?

c. How do you find the area of a rectangle?

d. What is the area of a rectangle that is 10 cm long and 2 cm high?

e. Describe a parallelogram with the same area.

Learning the Concept

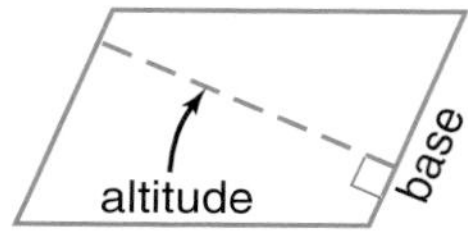

The modeling activity above suggests that you can find the area of a parallelogram by multiplying the measures of the base and the height. The base can be any side of the parallelogram. The height is the length of an **altitude**, a line segment perpendicular to the base with endpoints on the base and the side opposite the base.

23. parallelogram: base $3\frac{1}{4}$ ft; height, 2 ft
24. triangle: base 0.8 km; height, 2 km
25. trapezoid: height, 3.5 m; bases 10 m and 11 m
26. parallelogram: base, 25 yd; height, 11 yd
27. In Exercise 20, you may have found the area of the square by finding the areas of two triangles and adding them. Show that this is the same as $A = \frac{1}{2}d^2$ where d is the length of a diagonal.

Critical Thinking

28. Explain or demonstrate how the formula for the area of a trapezoid can be used to find the areas of parallelograms and triangles.

Applications and Problem Solving

29. **Consumerism** Carpeting is sold in several standard widths, but since room sizes are not standard, there frequently is wasted carpeting, for which the home owner must pay.
    a. If you are carpeting a room that is 14 feet by 22 feet, what is the area of the room?
    b. If the carpet you have chosen is sold in 12 foot and 15 foot widths, which size will result in less waste?
30. **Geography** The state of Tennessee is shaped almost like a parallelogram. The distance along the north boundary is 447 miles, and the north-south distance is 116 miles. Estimate the area of the state.

31. **Pet Care** In many communities, pets must be kept in a fenced area, or on a leash. If you had 48 feet of fencing to fence a rectangular area, describe the shape you would choose in order to give your pet the greatest amount of area. Justify your answer.
32. **Advertising** Use the formula in Exercise 27 to determine whether the claim about the viewing area in the advertisement below is accurate. Remember that television screens are measured diagonally and that the diagonals of a square are congruent.

Mixed Review

33. **Geometry** How many lines of symmetry does a regular pentagon have? (Lesson 11-9)
34. Estimate 26% of 120. (Lesson 9-8)
35. Solve $n = -9.45 \div -4.5$. (Lesson 6-5)
36. Write $(-4)(-4)(-4)$ using exponents. (Lesson 4-2)
37. Rewrite $3 + (x + 4)$ using the associative and commutative properties. Then simplify. (Lesson 1-4)

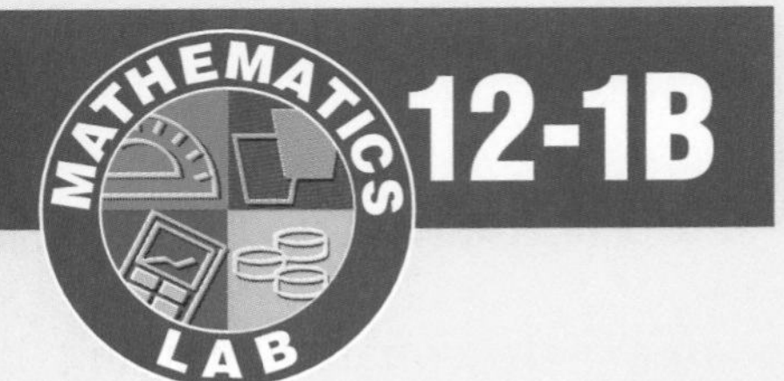

HANDS-ON ACTIVITY

12-1B Fractals

An Extension of Lesson **12-1**

MATERIALS

- ruler
- compass
- protractor

The design shown at the right is called the Koch snowflake. It is made up of three rotated copies of the Koch curve, introduced by Helge von Koch, a mathematician who worked on fractal geometry. Fractals are self-similar. That is, if you look at any part of a fractal, you will see a miniature replica of the larger design. A tree of broccoli is an example of a fractal in nature. Each branch of the tree is a miniature replica of the entire tree.

Your Turn

Work with a partner to construct a Koch curve.

a. Use a ruler to draw a line segment about 6 inches long. This is stage 0 of the curve.

b. Use the ruler to separate the line into three congruent segments.

c. Set the compass to the length of one segment. Then draw arcs from each end of the middle segment as shown at the right. Use the point where the arcs intersect to draw an equilateral triangle. Then erase the base of the triangle. This is stage 1.

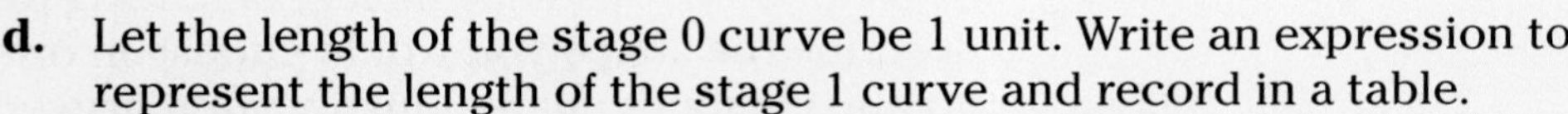

d. Let the length of the stage 0 curve be 1 unit. Write an expression to represent the length of the stage 1 curve and record in a table.

e. Repeat step c on the sides of stage 1 to produce a stage 2 figure. Again determine the length and record in the table.

f. Repeat step c once more to produce a stage 3 figure. Determine the length and record in the table.

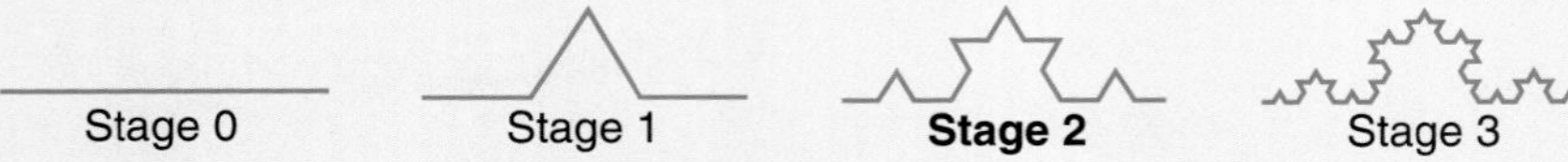

TALK ABOUT IT

1. Why do you think the Koch curve is considered to be self-similar?
2. What happens to the length in each stage?
3. Suppose the curve is continued indefinitely. Make a conjecture about the length.
4. Research fractal geometry. Then write your own definition of a fractal.

Extension

5. The Koch curve is produced by repeating step c above an infinite number of times. The Koch snowflake is made from three rotated copies of the curve. Stage 0 is an equilateral triangle. Use your constructions to produce a stage 1, stage 2, and stage 3 snowflake. Suppose the area of the stage 0 triangle is 1 square unit. Determine the area of each stage. Then make a conjecture about the area of the Koch snowflake.

12-2 Area: Circles

Setting Goals: *In this lesson, you'll find areas of circles.*

Modeling with Manipulatives

MATERIALS

- centimeter grid paper
- compass
- ruler
- protractor

Pizza, anyone? Most pizzas are round and come in different sizes, measured by diameter. Have you ever wondered whether one size would give you more pizza for your money? To find out, you need to find the area of a circle. This activity will help you understand how to find the area of a circle.

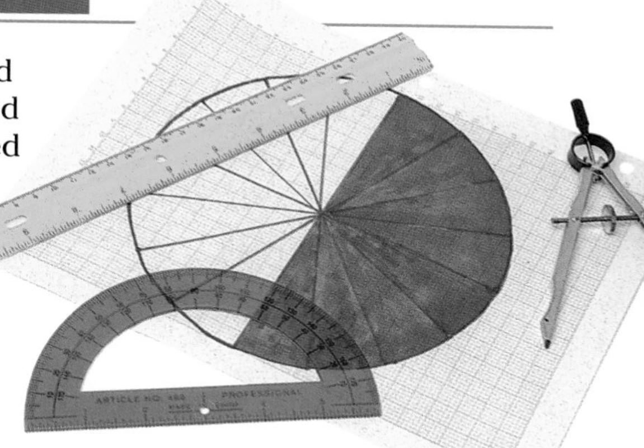

Your Turn

Work with a partner.

- ▶ On centimeter grid paper, use a compass to construct a circle with a diameter of 20 centimeters. Do this by putting the point of the compass at the corner of a grid square and adjusting the compass so it is 10 centimeters wide.
- ▶ Count the grid squares and the parts of grid squares that lie inside the circle. Use your judgment on how to count each partial grid square. Record this estimate of the area of the circle.
- ▶ With a protractor, straightedge, and pencil, separate the circle into 16 congruent parts as shown in the photograph above. Shade one half of the circle as shown.
- ▶ Cut out the parts and arrange them in a side-by-side pattern as shown. When you have finished, you should have a shape that looks like a parallelogram.

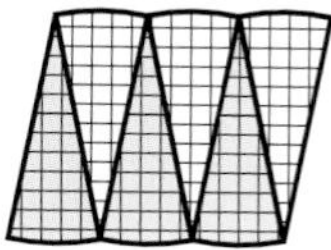

TALK ABOUT IT

a. What part of the circle makes up the base of the parallelogram?

b. What part of the circle is the height of the parallelogram?

c. Use a ruler to find the approximate measure of the base and height.

d. What is the approximate area of the parallelogram?

e. Compare to the estimate you found by counting grid squares.

Learning the Concept

In the activity, you separated the circle into parts and made a figure that looks like a parallelogram. The circle has the same area as that figure.

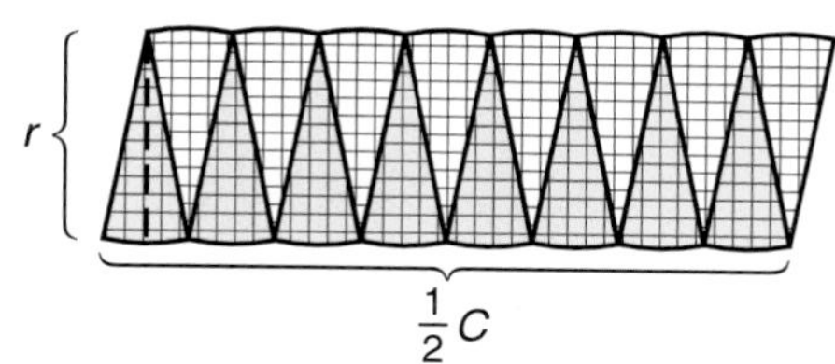

Connection to Algebra

You can use the formula for the area of a parallelogram to find the formula for the area of a circle.

You can review circumference in Lesson 7-4.

$A = b \times h$

$A = \left(\frac{1}{2} \times C\right) \times r$ *The base of the parallelogram is one-half the circumference.*

$A = \left(\frac{1}{2} \times 2\pi r\right) \times r$ *Remember, $C = 2\pi r$.*

$A = \pi \times r \times r$

$A = \pi r^2$ *The formula is $A = \pi r^2$.*

Area of a Circle

In words: If a circle has a radius of r units, then the area A is $\pi \cdot r^2$ square units.

In symbols: $A = \pi r^2$

Example 1 **Find the area of each circle. Round to the nearest tenth.**

THINK ABOUT IT

You can estimate the area of a circle by squaring the radius and multiplying by 3. So, the area in Example 1 is about $4 \times 4 \times 3$ or 48 cm².

a.

$A = \pi r^2$ *Formula for area of a circle*

$A = \pi \cdot 4^2$ *The radius is 4 in. Estimate: $3 \cdot 16 = 48$*

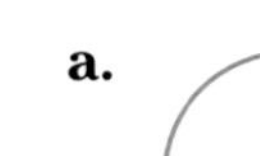

π [×] 4 [x^2] [=] 50.26548246

$A \approx 50.3$ *Compare with the estimate.*

The area is about 50.3 in².

b.

$A = \pi r^2$

$A = \pi \cdot 3^2$ *The radius is one-half the diameter, so $r = \frac{1}{2}(6)$ or 3.*

$A \approx 28.3$ ***Estimate:** $3 \times 9 = 27$*

The area is about 28.3 cm².

Example 2 **APPLICATION** **Consumerism**

A 10-inch, two-topping pizza sells for $5.99 and a 14-inch, two-topping pizza sells for $11.19. Which is a better buy, one 14-inch or two 10-inch pizzas?

Use $A = \pi r^2$ to find the area of each size of pizza.

10-inch	14-inch
$A = \pi r^2$	$A = \pi r^2$
$A = \pi \cdot 5^2$	$A = \pi \cdot 7^2$
$A \approx 79$	$A \approx 154$

The area of the 14-inch pizza is almost twice as much as the 10-inch. So two 10-inch pizzas are about the same amount of pizza for 2($5.99) or $11.98, as one 14-inch pizza for $11.19. It appears that the 14-inch pizza is the better buy.

Checking Your Understanding

Communicating Mathematics

Read and study the lesson to answer these questions.

1. **Explain** how to find the area of a circle if you know the measure of the radius.
2. **Estimate** the area of the circle at the right by counting grid squares. Then find the area using the formula. Compare the difference of the two areas to the actual area to find the percent of error.

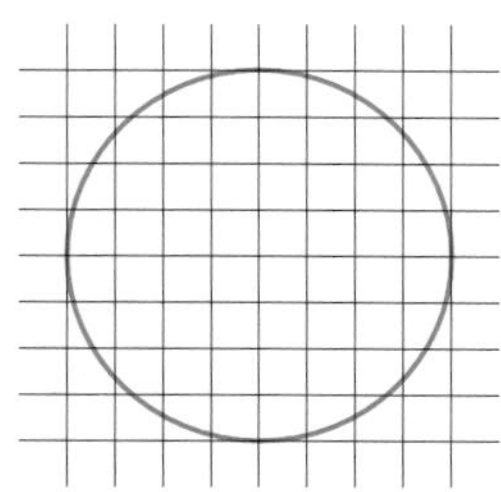

3. **You Decide** Joyce says the area of a circle with a diameter of 10 cm is about 75 cm^2. Sanjay says the area is about 300 cm^2. Use estimation to determine who is correct. Explain.

Guided Practice

Find the area of each figure. Round to the nearest tenth.

4. 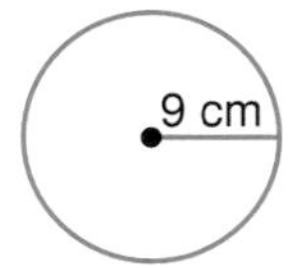

5. 

6. 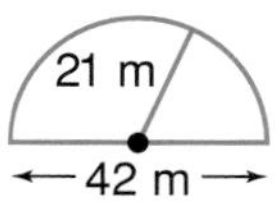

7. **City Planning** The circular region inside the streets at DuPont Circle in Washington, DC, is 250 feet across. How much area does the grass and sidewalk cover?

Exercises: Practicing and Applying the Concept

Independent Practice

Find the area of each circle. Round to the nearest tenth.

8. 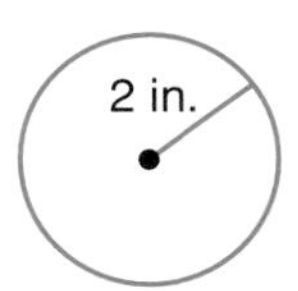

9. 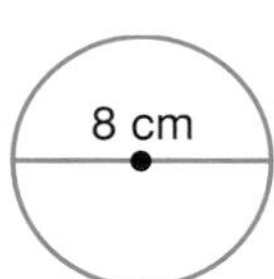

10. 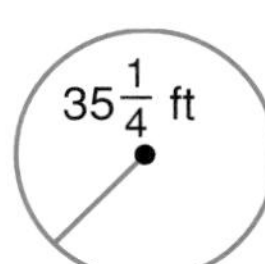

11. 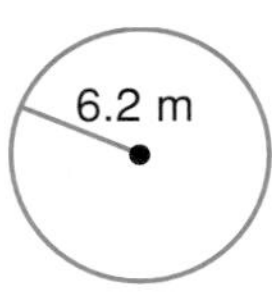

12. 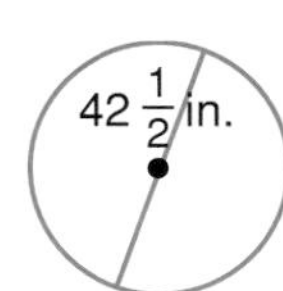

13. 

14. radius, 23 cm
15. diameter, 14.6 in.
16. radius, 24.6 m

Find the area of each figure. Round to the nearest tenth.

17. 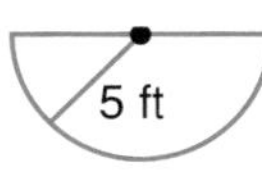

18. 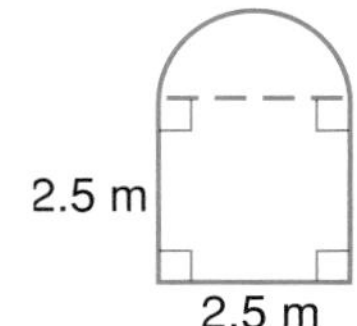

19. 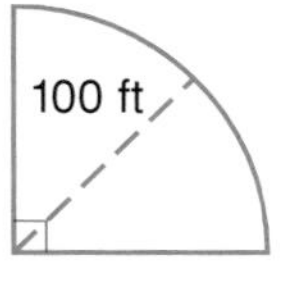

Critical Thinking

20. The area and the circumference of a circle are the same number.
    a. What is the radius of the circle?
    b. What is unique about this radius?

Applications and Problem Solving

21. **History** Stonehenge, an ancient monument in England, may have been used as a calendar. The stones are arranged in a circle 30 meters in diameter. Find the area of the circle.

22. **Landscaping** Hiroshi needs to match some pipe used in the sprinkling system where he works. He doesn't want to cut the pipe to find the diameter, but he can measure the circumference. He finds that it is about 6.25 inches. What is the diameter of the pipe?

23. **Architecture** The largest dome built in ancient times is the Roman Pantheon. The diameter of its dome is 42.7 m. The Houston Astrodome has a diameter of 216.4 m. About how many times more area does the Astrodome cover than the Pantheon?

24. **Sports** The Athletic Booster Club for a high school is raising funds for artificial grass for the football playing area in the infield of the track. The grass will cost \$50 per square yard, including installation. To the nearest dollar, how much money must they raise to pay for the grass?

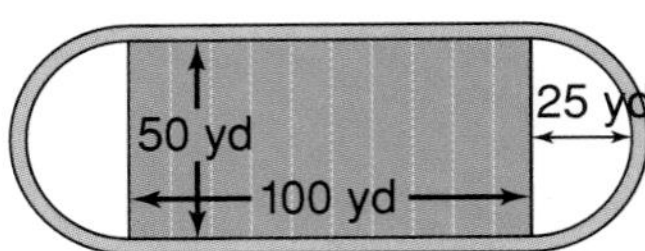

25. **Public Safety** A town has installed a tornado warning system. The sound emitted can be heard for a 2-mile radius. Find the area that will benefit from the system.

Mixed Review

26. **Real Estate** A farmer's land is in the shape of a trapezoid. The north boundary is 220 yards long, the south boundary is 176 yards long, and the east-west distance is 1320 yards. How much land does the farmer have? (Lesson 12-1)

27. Solve $\frac{15}{c} = \frac{8}{9.6}$. (Lesson 9-4)

28. Solve $3x - 5 > 3x + 4$. (Lesson 7-6)

29. Express $0.\overline{4}$ as a fraction in simplest form. (Lesson 5-1)

30. Solve $9c = -54$. Graph the solution on a number line. (Lesson 3-3)

From the FUNNY PAPERS

1. What is the formula the man is trying to write?
2. What does the formula represent?
3. Make up a way of remembering the correct formula.

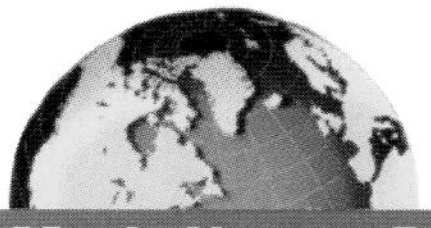

12-3 Integration: Probability
Geometric Probability

Setting Goals: *In this lesson, you'll find probabilities using area models.*

Modeling a Real-World Application: Games

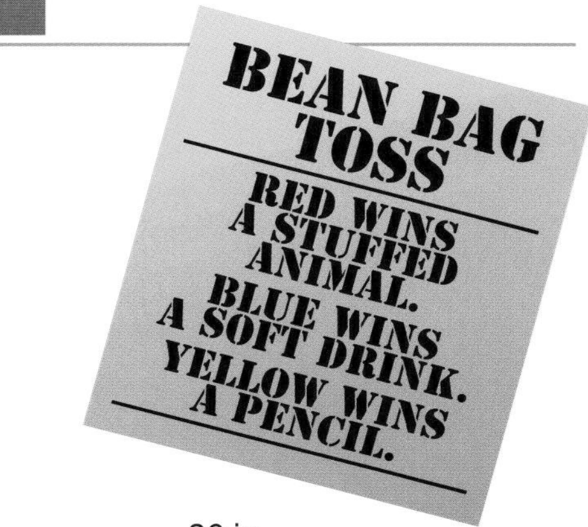

In a game at the Johnson Middle School Carnival, you toss a beanbag onto a target like the one shown at the right. You get a prize if your beanbag lands in one of the colored circles. Let's assume that all the beanbags land on the board, and that any beanbag is equally likely to land any place on the board. What is the probability that a beanbag will land in one of the circles? *This problem will be solved in Example 2.*

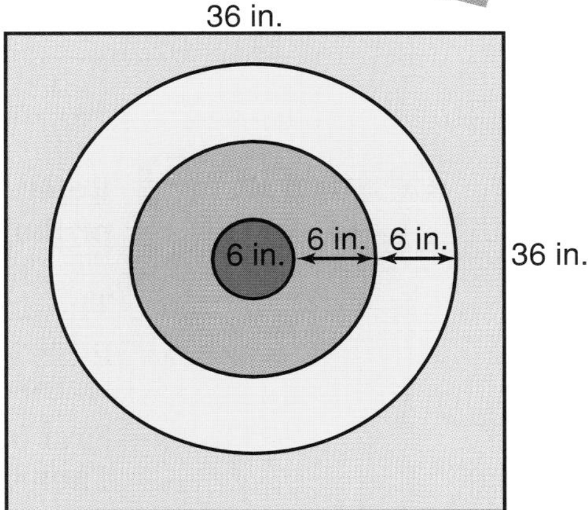

Learning the Concept

Geometric probability uses ideas about area to find the probability of an event. In Lesson 9-3, you calculated probabilities for spinners. A dartboard is similar to a spinner. On the dartboard below, there are ten equally likely outcomes, five of which are shaded. The probability of a dart landing in a shaded section is $\frac{5}{10}$ or $\frac{1}{2}$.

In this lesson, for problems involving dartboards, we will assume that the dart will land on the dartboard and that it is equally likely to land anywhere on the dartboard.

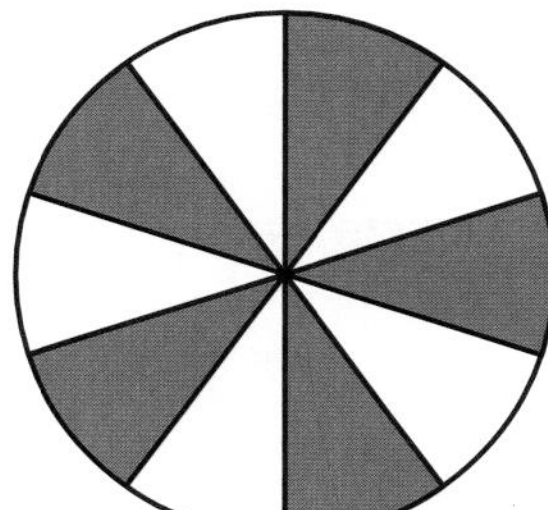

$P(\text{shaded}) = \frac{5}{10}$ or $\frac{1}{2}$

You can also relate probability to the areas of other geometric shapes.

Example  **The figure at the right represents a dartboard.**

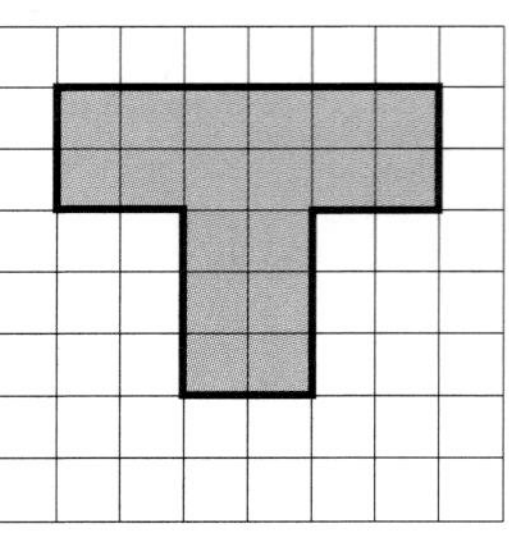

a. Find the probability of landing in the shaded region.

b. Suppose you threw 50 darts. How many would you expect to land in the shaded region?

a. You can count squares to find the area of the shaded region and the area of the whole target.

$$P(\text{shaded}) = \frac{\text{shaded area}}{\text{area of target}}$$
$$= \frac{18}{64} \text{ or } \frac{9}{32}$$

b. Let n represent the darts landing in the shaded region.

$\frac{n}{50} = \frac{9}{32}$ *Write a proportion.*

$50 \cdot 9 = 32n$ *Find the cross products.*

$450 = 32n$

$14.1 \approx n$ *About 14 darts should land in the shaded region.*

Example  **Refer to the application at the beginning of the lesson. What is the probability of winning a prize on any given toss?**

The area within the circles represents the outcomes of winning a prize and the total area of the board (including the circles) represents all the possible outcomes.

First find the area of the entire board and the area within the circles. Then write a fraction to represent the probability of winning.

The board is a 36-inch by 36-inch square.

$A = s^2$ *Formula for area of a square*

$A = 36^2$ *Replace s with 36.*

$A = 1296$

The board has an area of 1296 in^2.

The area for winning a prize is a circle with a radius of 15 in. *Why?*

$A = \pi r^2$ *Formula for area of a circle*

$A = \pi(15)^2$ *Replace r with 15. Estimate: 3 × 225 = 675*

$A \approx 707$

The area for winning a prize is about 707 in^2.

So, $P(\text{winning a prize}) = \frac{\text{winning area}}{\text{total area}}$

$\approx \frac{707}{1296}$

≈ 0.546 *Use a calculator.*

The probability of winning is about 55% or $\frac{11}{20}$.

Checking Your Understanding

Communicating Mathematics

Read and study the lesson to answer these questions.

1. Refer to the application at the beginning of the lesson. Tell how you would find the probability of winning a soft drink or a pencil.
2. **Draw** a dart board for which the probability of landing on a shaded area is $\frac{1}{3}$.
3. Work with a partner.

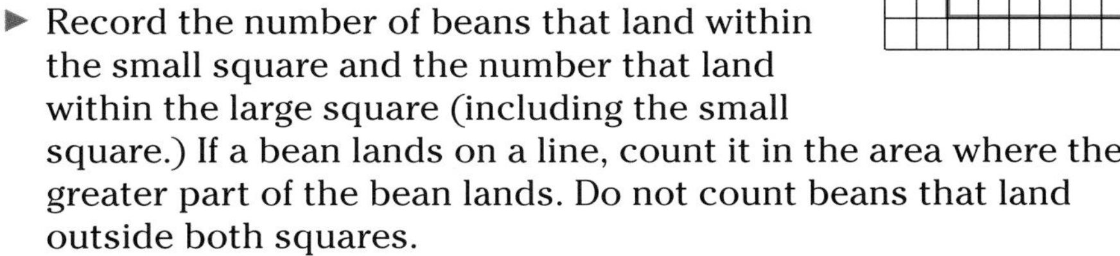

MODELING MATHEMATICS

MATERIALS

grid paper

dry beans

- ▶ Outline two squares as shown at the right.
- ▶ Hold 25 beans about 3 inches above the paper and drop them onto the paper.
- ▶ Record the number of beans that land within the small square and the number that land within the large square (including the small square.) If a bean lands on a line, count it in the area where the greater part of the bean lands. Do not count beans that land outside both squares.
- ▶ Repeat three more times.
- ▶ Find the ratio $\frac{\text{number of beans in small square}}{\text{number of beans in large square}}$.
- ▶ Find the ratio $\frac{\text{area of small square}}{\text{area of large square}}$.
- ▶ Repeat the activity or combine results with other pairs.
- ▶ Compare the two ratios. What conclusion can you make about the relationship of area and probability?

Guided Practice

Each figure represents a dartboard. Find the probability of landing in the shaded region.

4.

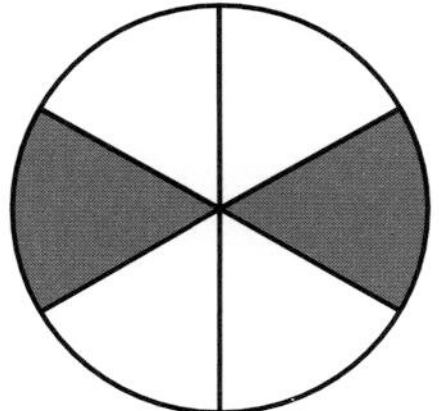

5.

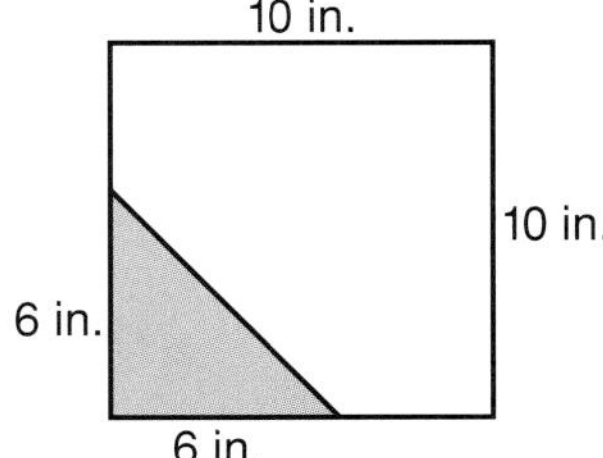

6. Suppose you throw 50 darts at the target in Exercise 4. Predict how many darts will land in the shaded region.
7. Suppose you throw 50 darts at the target in Exercise 5. Predict how many darts will land in the shaded region.
8. **Skydiving** A sky diver parachutes onto a 100-yard by 100-yard square field with a tree in each corner of the field. The diver will not get tangled in tree branches if she stays at least 6 yards away from the tree trunk. Assume the diver lands within the field and that she is equally likely to land anywhere in the field. What is the probability of a clear landing?

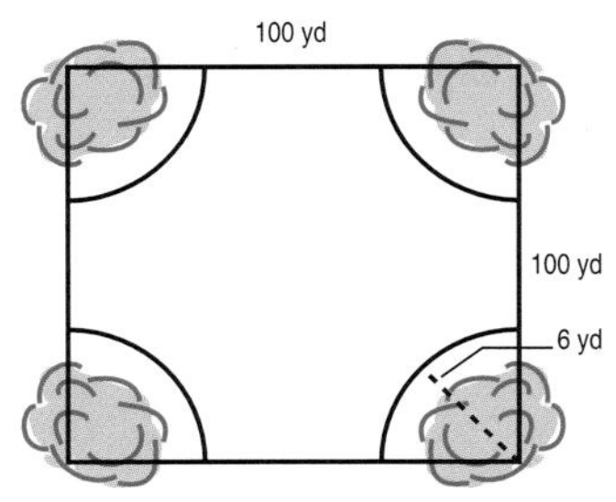

Exercises: Practicing and Applying the Concept

Independent Practice

Each figure represents a dartboard. Find the probability of landing in the shaded region.

9. 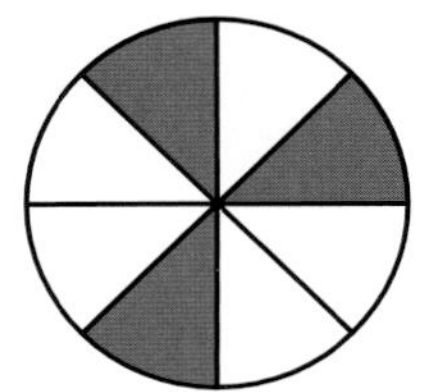

10. 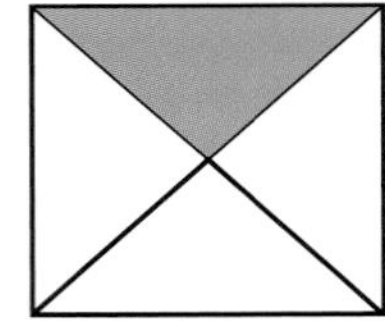

11. 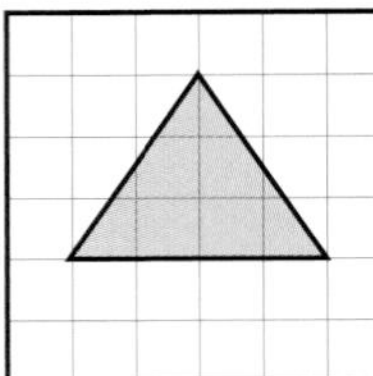

12. 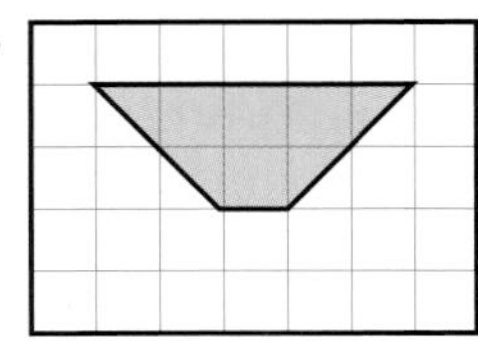

13. 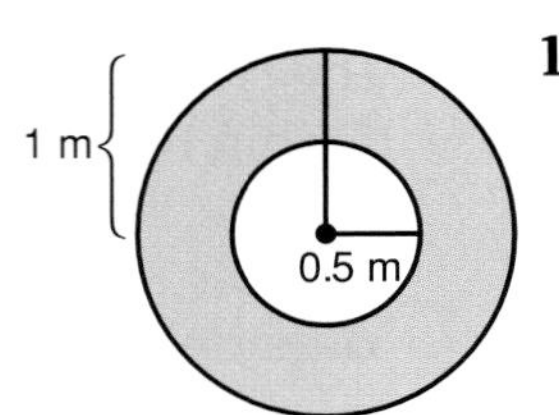

14. 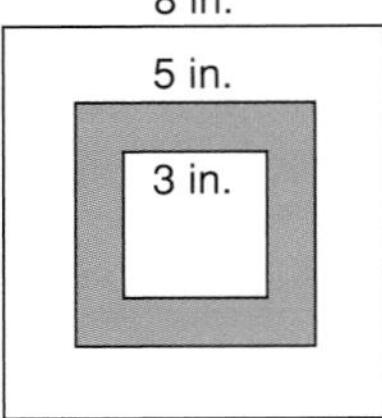

15. Suppose you throw 100 darts at the target in Exercise 9. Predict how many darts will land in the shaded region.

16. Suppose you throw 75 darts at the target in Exercise 12. Predict how many darts will land in the shaded region.

Critical Thinking

17. Games At the Johnson Middle School Carnival, you can win a CD by tossing a quarter onto a grid board so that it doesn't touch a line. The sides of the squares of the grid are 40 millimeters long and the radius of a quarter is 12 millimeters. What is the probability of winning? (*Hint:* Find the area where the center of the coin could land so that the edges don't touch a line.)

Applications and Problem Solving

18. Games Refer to the application at the beginning of the lesson. What is the probability of winning a stuffed animal?

19. Mining A geologist told Sam Antonio that there is oil under 500 acres of his 17,500 acre ranch. If Sam randomly starts drilling, what is the probability that he will strike oil?

20. Computers Computers can be used to simulate problem situations using a random number generator. Suppose a computer program randomly selects two real numbers between 0 and 2. Find the probability that the sum of the two numbers is:

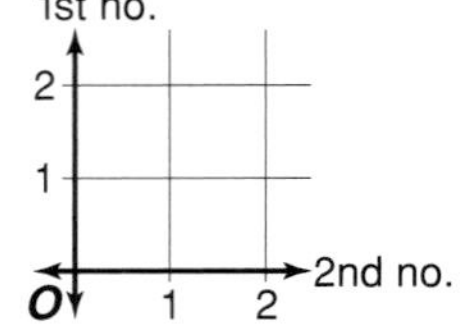

a. less than 2

b. greater than 1

c. between 1 and 2

(*Hint:* Use a coordinate grid.)

Mixed Review

21. **Geometry** Estimate the area of a circle with a radius of 12 cm. (Lesson 12-2)

22. **Geometry** The measures of the angles of a triangle are in the ratio 1:3:5. Find the measure of each angle. (Lesson 11-4)

23. **Science** An inclined plane is a slanted surface, which may be used for raising objects to higher places. Find the slope of the inclined plane pictured at the right. (Lesson 8-6)

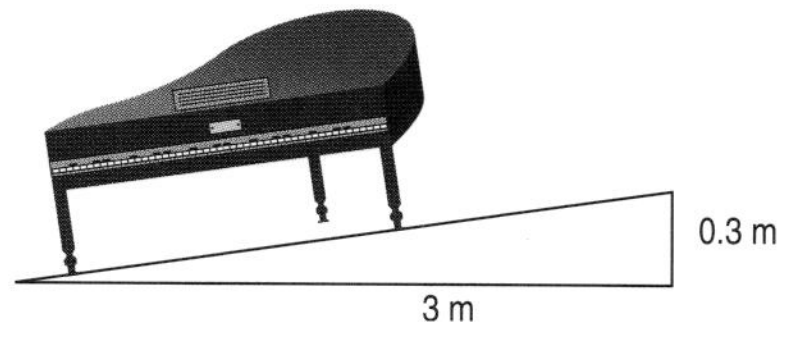

24. Evaluate xy if $x = \frac{1}{5}$ and $y = \frac{3}{5}$. (Lesson 6-3)

25. Write an inequality using the numbers in the sentences below. Use the symbols $<$ or $>$. (Lesson 2-3)
Water boils at 212°F. It freezes at 32°F.

COOPERATIVE LEARNING PROJECT

THE SHAPE OF THINGS TO COME

Digital Cameras

Wouldn't it be great to have a camera that allowed you to take a snapshot and instantly view it, before it is developed? Or a camera that, when you viewed a snapshot, allowed you to zoom in on some interesting detail? Such cameras are just now becoming available. They are called digital cameras, cost under $1000, and are about the size of today's film-loaded cameras. Digital cameras can even be plugged into a TV. When you are satisfied with your snapshots, you can transfer them to a videotape, save them on a floppy disk, or print them out on a video printer for placement in an album. By the year 2005, it is expected that the price of digital cameras will come down to about $300.

Digital cameras, like digital computers and digital televisions, produce images by treating the screen as a two-dimensional grid. Each square on the grid is called a pixel. The greater the number of pixels, the higher the "resolution" of the image, and the sharper the image becomes.

To help you understand how image resolution works, think about drawing a circle. But, you can only draw one *straight* line in any given grid square, or pixel. Compare figures 1 and 2. Figure 2 has twice as many squares along each axis.

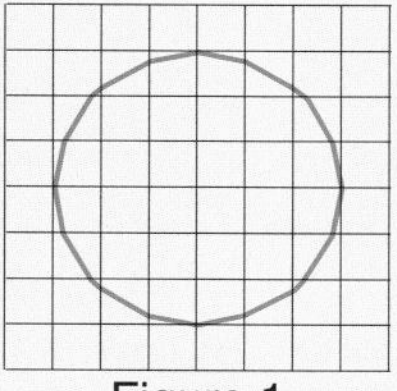
Figure 1

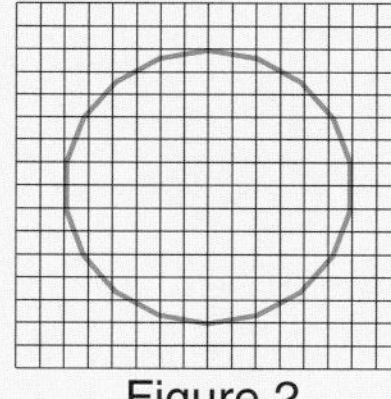
Figure 2

See for Yourself

Research digital photography and the new digital cameras.

- What are several advantages of digital cameras over today's film-loaded cameras?
- How will you be getting your digital-camera snapshots developed by the year 2005?
- Notice that in figure 2, the drawing comes closer to looking like a circle. As you increase the number of grid squares, you achieve higher resolution and come closer and closer to a sharp, circular image. Reproduce the grid squares of figure 2 on graph paper. Then increase the number of grid squares and, by drawing only one straight line per grid square, see how close you can come to drawing the image of a circle.

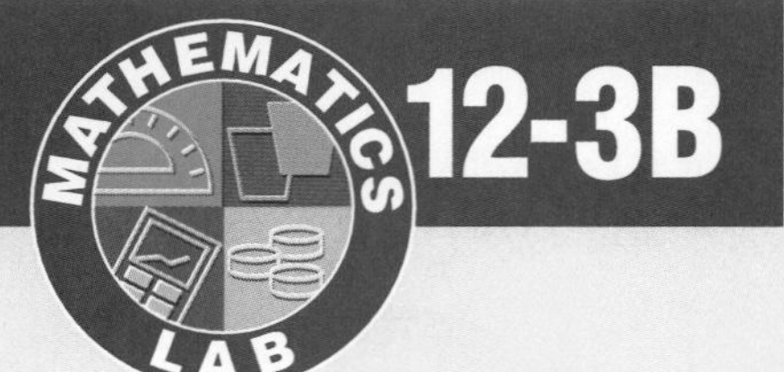

GRAPHING CALCULATOR ACTIVITY

12-3B Geometric Probability

An Extension of Lesson **12-3**

You can use a T1-82 graphing calculator to experiment with area and probability. The program below generates random ordered pairs.

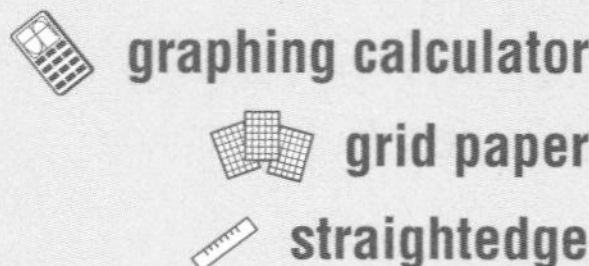

MATERIALS

- graphing calculator
- grid paper
- straightedge

```
PROGRAM:RANDPRS
: ClrHome
: Fix 1
: Input "LEAST X", A
: Input "GREATEST X", B
: Input "LEAST Y", C
: Input "GREATEST Y", D
: Input "NUMBER OF PAIRS", P
: 0 →I
: Lbl 1
: I +1 → I
: (B–A)rand+A →X
: Disp X
: (D–C)rand+C →Y
: Disp Y
: Disp " "
: Pause
: If I≠P
: Goto 1
: Stop
```

Your Turn Work with a partner.

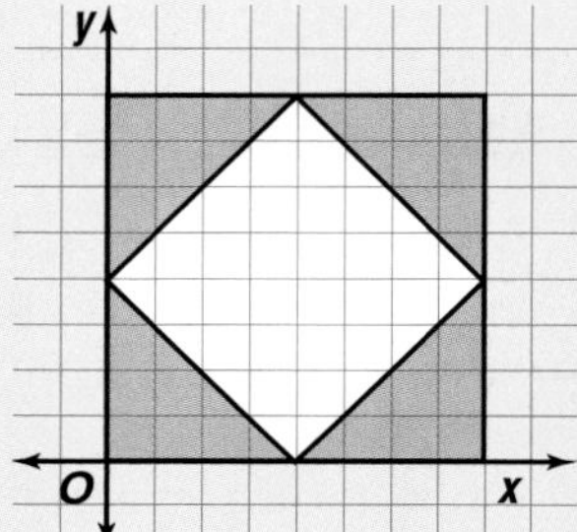

- ► Use a straightedge to draw a square like the one shown at the right. Shade the figure as shown.
- ► Enter the program above into a graphing calculator.
- ► Use the program to generate 50 ordered pairs. Enter 0 as the least value and 8 as the greatest value for both x and y.
- ► Graph each pair on the grid paper and keep a tally of how many points are in the shaded and unshaded areas.
- ► Write a ratio comparing the points in the shaded area to the total number of points.
- ► Find the areas of the small square and the large squares. Write a ratio comparing the area of the small square to the area of the large square.

1. How do the two ratios you wrote compare?
2. Repeat the activity above to generate and graph another 50 ordered pairs. Combine the results with the results from the first run and write the ratio.
3. What effect does more data have on the results?
4. Predict the results of repeating the activity using the figure at the right. Then check your prediction by running the program and graphing the results.

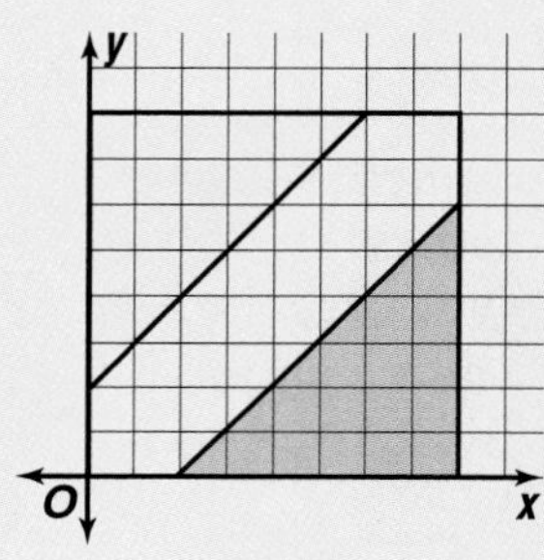

12-4 Problem-Solving Strategy: Make a Model or a Drawing

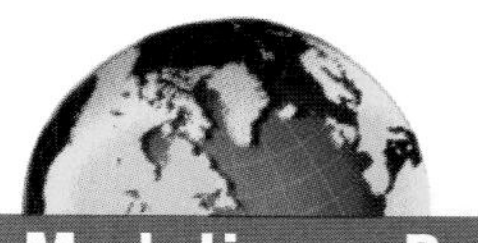

Setting Goals: *In this lesson, you'll solve problems using a model or a drawing.*

Modeling a Real-World Application: Pet Care

June has attached a 20-foot rope to the corner of a 10-foot by 10-foot shed to tie up her horse Harry. How much grazing area does Harry have?

Learning the Concept

Without a drawing, it is difficult to find the area or even visualize the shape of the horse's grazing area. Many problems, especially those that involve geometry, are easier if you make a model or a drawing.

Explore You know the dimensions of the shed and the length of the rope. You need to find the grazing area.

Plan Make a drawing to illustrate the problem. Notice that you have sections of circles. The largest section is $\frac{3}{4}$ of a circle with a radius of 20 feet. The smaller sections are each $\frac{1}{4}$ of a circle with a radius of 10 feet. You can find the areas of each section and add to find the total grazing area.

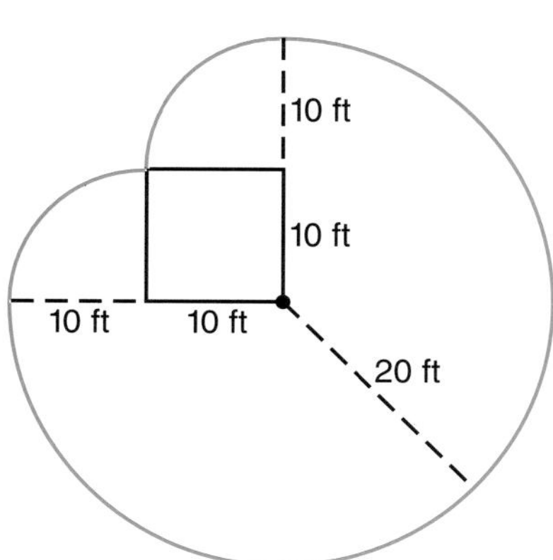

THINK ABOUT IT

How do you know that the large section is three-fourths of a circle and the small sections are one-fourth of a circle?

Solve Area of the large section: $A = \pi r^2 \times \frac{3}{4}$

Estimate: $3 \times 400 \times \frac{3}{4} = 900$

$A = \pi \times 20^2 \times \frac{3}{4}$

$A \approx 942$

Area of each smaller section: $A = \pi r^2 \times \frac{1}{4}$

Estimate: $3 \times 100 \times \frac{1}{4} = 75$

$A = \pi \times 10^2 \times \frac{1}{4}$

$A \approx 79$

Total area = 942 + 2(79) or 1100 *There are 2 smaller sections.*
The horse has a grazing area of about 1100 ft^2.

Examine An answer of 1100 ft^2 seems reasonable, compared to your estimate.

CULTURAL CONNECTIONS

Horses were introduced to North America by the Spanish in 1519. The horses they left behind were probably the ancestors of the American wild horses.

Checking Your Understanding

Communicating Mathematics

Read and study the lesson to answer these questions.

1. Refer to the application at the beginning of the lesson. Draw a picture of the problem if the rope was 25 feet long. Then explain how the problem is different.
2. How can making a model or drawing help in problem solving?

3. Think of a real-life problem where making a model or drawing can help you solve the problem. Write about the problem and its solution. Include a drawing.

Guided Practice

Solve by making a model or drawing.

4. Refer to the application at the beginning of the lesson. Suppose the rope is only 15 feet long. Find the grazing area.
5. **Framing** A painting 15 cm by 25 cm is bordered by a mat that is 3 centimeters wide. The frame around the mat is 2 centimeters wide. What is the area of the picture including the frame and mat?

Exercises: Practicing and Applying the Concept

Independent Practice

Solve. Use any strategy.

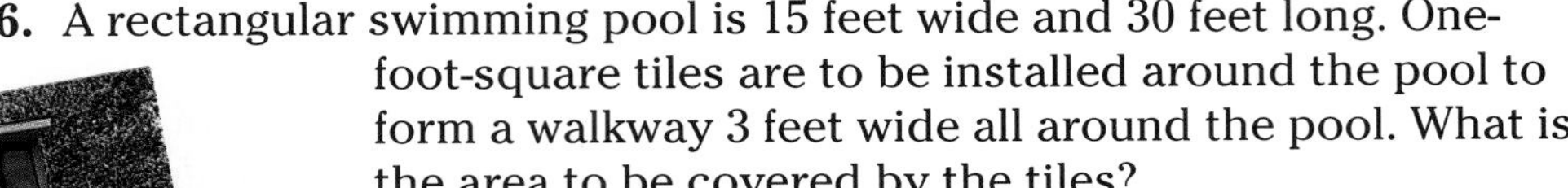

6. A rectangular swimming pool is 15 feet wide and 30 feet long. One-foot-square tiles are to be installed around the pool to form a walkway 3 feet wide all around the pool. What is the area to be covered by the tiles?

7. There are five ways to place four one-inch squares together so that any two adjoining sides match exactly.

These match: 

These do not match: 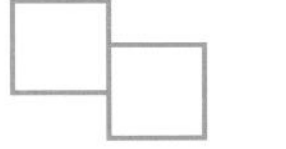 

   a. Find the five ways.
   b. Find the perimeter of each figure.
   c. Describe the shape or shapes for which the perimeter is greatest.
   d. Describe the shape or shapes for which the perimeter is least.
   e. How many ways can you arrange five squares? Do their perimeters differ?

LOOK BACK

You can review perimeter in Lesson 3-5.

8. Find the sum of the first 100 *even* positive numbers.
9. The perimeter of a rectangle is 32 meters. Its area is 48 square meters. What are the dimensions of the rectangle?
10. **Sports** A circular track has a diameter of 400 feet. Les starts out jogging around the track at a rate of 600 feet per minute. One minute later, Antonio starts jogging at the same place. He goes in the same direction and jogs at 700 feet per minute. Will Antonio pass Les on the first, second, third, or fourth quarter of the track?
11. Sandy has 36 identical cubes. Each edge of a cube is 1 inch long. How can Sandy arrange the cubes to have the smallest surface area?

12. **Geometry** Various views of a solid figure are shown below. The edge of one block represents one unit of length. A dark segment indicates a break in the surface.
    a. Use the views to sketch the solid.

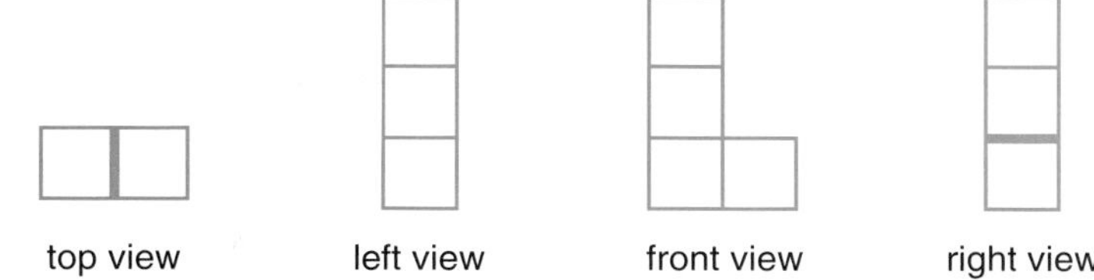

    b. Choose a solid object from your home and draw the top, left, front, and right views.
13. Ms. Warren sold 100 shares of stock for \$2475. When the price per share went down \$4, she bought 200 more shares. When the price per share went back up \$3, she sold 100 shares. How much did Ms. Warren gain or lose in her transactions?
14. Keiko's little brother fills an 8-quart pail with sand for a sand castle. He also has empty 3-quart and 5-quart pails. How could he use these to divide the sand into two equal portions?
15. Three different views of the same cube are shown below. What figure is on the side opposite the X?

Critical Thinking

16. The figures you drew in Exercise 7 are called *nets*. Some nets can be folded to form solid figures. Draw all of the nets containing five one-inch squares that could be folded into open boxes. Mark the square that would be the bottom of the box.

Mixed Review

17. **Probability** The figure at the right represents a dartboard. What is the probability of landing in the shaded region? (Lesson 12-3)
18. Find the value of $P(5, 4)$. (Lesson 10-6)
19. **Wages** You earn \$8 an hour and a \$12 bonus. What is the least number of hours you must work to earn at least \$100? (Lesson 7-7)
20. Write 2^{-3} using positive exponents. (Lesson 4-9)
21. **Statistics** The graph at the right shows the number of millions of acres of federally-owned land in several states. (Lesson 1-10)
    a. Which state has the greatest number of acres that are federally owned?
    b. What is the difference between the number of acres owned by the Federal Government in California and Arizona?

For the latest on federally-owned land, visit: **www.glencoe.com/sec/math/prealg/mathnet**

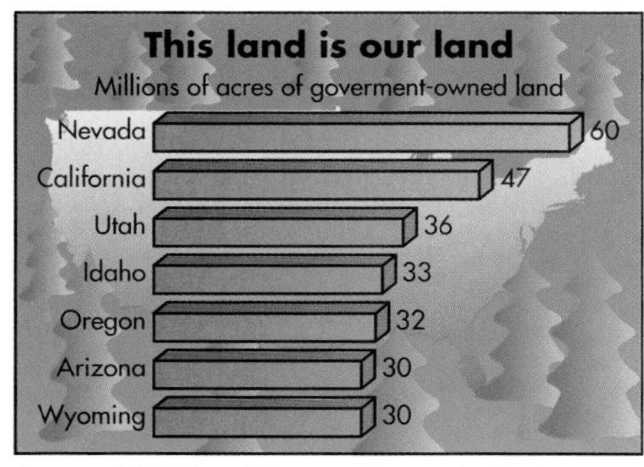

Source: USDA Natural Resources Conservation Service, National Resources Inventory

12-5 Surface Area: Prisms and Cylinders

Setting Goals: *In this lesson, you'll find surface areas of triangular and rectangular prisms and circular cylinders.*

Modeling a Real-World Application: Postal Service

Maria was buying stamps at a U. S. Post Office when she noticed shipping boxes on display. The price list is shown at the right.

Size (length × width × height)	Price
8 in. × 8 in. × 8 in.	$1.25
15 in. × 10 in. × 12 in.	$2.00
20 in. × 14 in. × 10 in.	$2.50

Maria wondered if the price of each box was related to the amount of material needed to make the box. Is the ratio of price to material about the same for each box? *You will solve this problem in Exercise 27.*

Learning the Concept

THINK ABOUT IT

Rectangles *ABCD* and *EFGH* are bases. Could another pair of regions be the bases of this prism? Does a base always need to be on the bottom?

In geometry, solids like the shipping box shown above are called **prisms**. A prism is a solid figure that has two parallel congruent sides, called **bases**. A prism is named by the shape of its bases. The shipping box and the prism shown at the right are **rectangular prisms**.

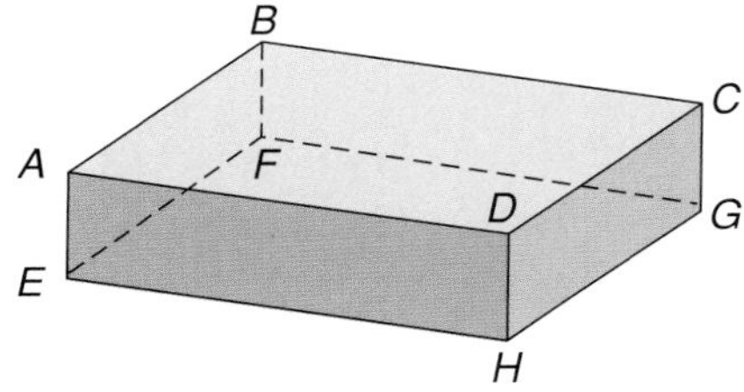

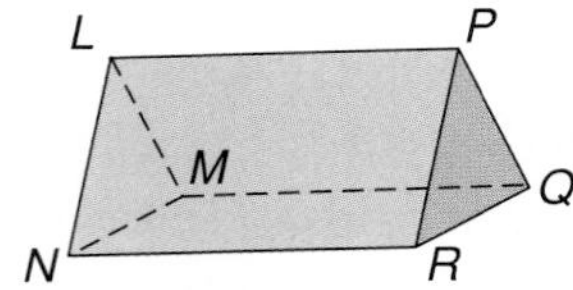

The parallel bases of the prism shown at the left are shaped like triangles, so it is called a **triangular prism**. Can you think of something in everyday life that has the shape of a triangular prism?

The amount of material that it would take to cover a geometric solid is the **surface area** of the solid. If you "open up" or "unfold" a prism, the result is a **net**. Nets help us see the regions or **faces** that make up the surface of the prism. The surface area of a prism is the area of the net.

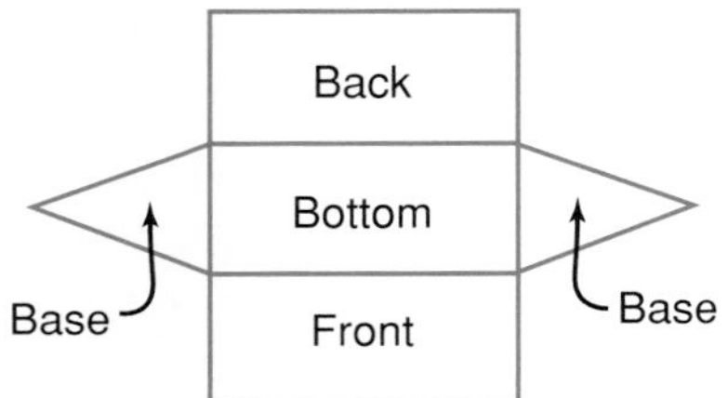

A triangular prism has five faces.

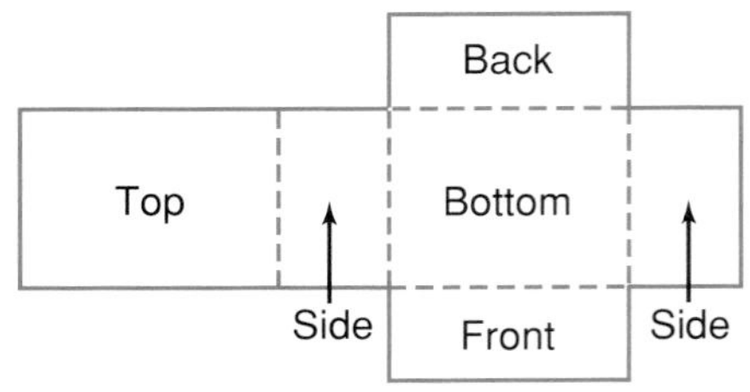

A rectangular prism has six faces.

Example 1 **Find the surface area of the 15 in. × 10 in. × 12 in. box in the application at the beginning of the lesson.**

Method 1

Sketch the box.
Use the formula $A = \ell w$ to find the areas of the faces.

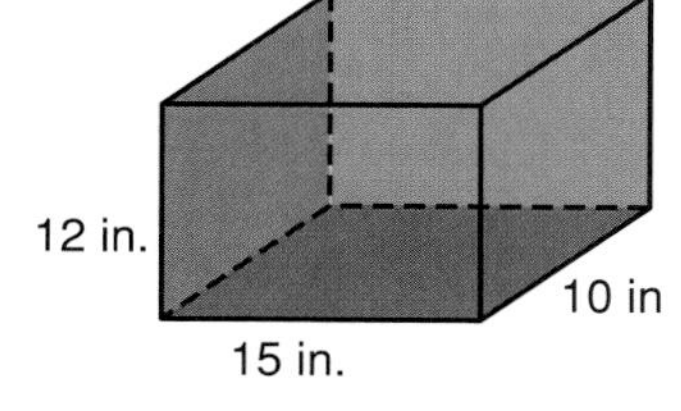

Front and Back	12 [×] 15 [×] 2 = 360
Top and Bottom	10 [×] 15 [×] 2 = 300
Two Sides	12 [×] 10 [×] 2 = 240

Add to find the total surface area. $360 + 300 + 240 = 900$
The surface area of a 15 in. × 10 in. × 12 in. box is 900 in^2.

Method 2

Draw a net for the box and label the dimensions of each face. Find the area of each face. Then add to find the total surface area.

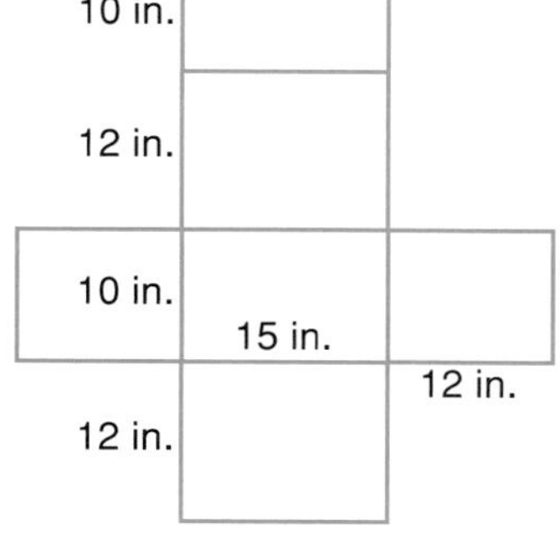

Bottom	10 [×] 15 [=] *150*
Side	12 [×] 10 [=] *120*
Top	10 [×] 15 [=] *150*
Side	12 [×] 10 [=] *120*
Back	12 [×] 15 [=] *180*
Front	12 [×] 15 [=] *180*
Total	150 [+] 120 [+] 150 [+] 120 [+] 180 [+] 180 [=] *900*

The surface area of the box is 900 in^2.

Example 2 **Find the surface area of the triangular prism shown below.**

Use the formula $A = \frac{1}{2}bh$ to find the area of the bases.

$\frac{1}{2} \cdot 6 \cdot 8 = 24$

Use the formula $A = \ell w$ to find the area of the sides.

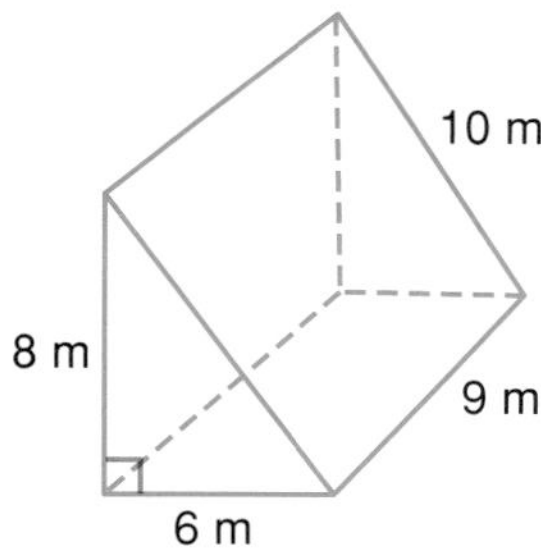

Front	$9 \cdot 10 = 90$
Bottom	$6 \cdot 9 = 54$
Back	$8 \cdot 9 = 72$

Add to find the total surface area. $2(24) + 90 + 54 + 72 = 264$
The surface area of the triangular prism is 264 m^2.
Check this answer by drawing a net of the prism.

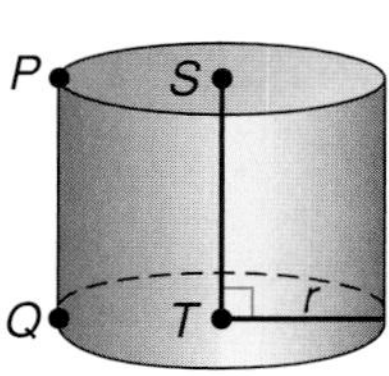

You probably have cans of food in your kitchen at home. These cans are examples of **cylinders**. The bases of a circular cylinder are two parallel, congruent circular regions.
An altitude is any perpendicular line segment joining the two bases. $\overline{ST}$ and $\overline{PQ}$ are altitudes.

If you open the top and bottom of a cylinder and then make a vertical cut in the curved surface, you could lay the cylinder flat in the same way you did with a rectangular prism. You would then have a pattern, or net, for making that cylinder.

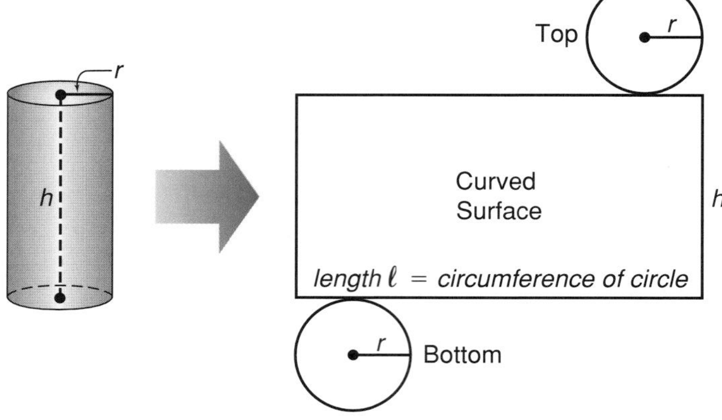

From the diagram above, you can see that:

1. The bases of the cylinder are congruent circular regions.
2. The curved surface of the cylinder opens to form a rectangular region.
3. The width of the rectangle is the height of the cylinder, or h.
4. The length of the rectangle is the circumference of a base, or $2\pi r$.

THINK ABOUT IT

Why is the length of the rectangle the same as the circumference of a base?

Connection to Algebra

The area of each base is πr^2. Since the area of a rectangle is ℓw, the area of the curved surface is $2\pi r \cdot h$. So, the surface area of a circular cylinder is $2(\pi r^2) + 2\pi r \cdot h$.

Example 3 **APPLICATION** **Manufacturing**

A Pringles™ can has a diameter of 3 inches and a height of 8.5 inches.

a. How much area must the label cover?

b. What is the total surface area of the can?

Explore You need to find the area of the label and the total surface area of the can. You know the diameter and the height of the can.

Plan The label covers the curved surface of the can. Use $A = \ell w$ or $A = 2\pi r \cdot h$ to find the area of the curved surface.

Estimate: $2 \cdot \pi \cdot 1\frac{1}{2} \cdot 8.5 \rightarrow 2 \cdot 3 \cdot \frac{3}{2} \cdot 9 = 81$

The top and bottom bases of the can are circles. Use $A = \pi r^2$ to find the area of each base.

Estimate: $\pi \cdot 1\frac{1}{2} \cdot 1\frac{1}{2} \rightarrow 3 \cdot 1 \cdot 2 = 6$

Then add to find the total surface area.

Estimate: $81 + 6 + 6 = 93$

Solve	**a.** $A = \ell w$	*Area of a rectangle*
	$A = 2\pi r \cdot h$	*Replace ℓ with the expression for the circumference of a circle.*
	$A = 2 \cdot \pi \cdot 1.5 \cdot 8.5$	*$d = 3$, so $r = 1.5$.*
	$A \approx 80.1$	

The label covers 80.1 in^2.

b. $A = \pi r^2$ *Formula for area of a circle*

$A = \pi \cdot 1.5^2$

$A \approx 7.1$ The area of each base is about 7.1 in^2.

The total surface area is about $80.1 + 2(7.1)$ or about 94.3 in^2.

Examine Compared to the estimates, the answers make sense.

Checking Your Understanding

Communicating Mathematics

Read and study the lesson to answer these questions.

1. How is surface area important when you giftwrap a package?
2. **Define** surface area in your own words.
3. **Explain** how to find the surface area of a circular cylinder.
4. **State** an example from your daily life of a rectangular prism, a triangular prism, and a circular cylinder.
5. A cube is a rectangular prism with square faces. Which size of box in the application at the beginning of the lesson is a cube? How do you know?

MODELING MATHEMATICS

MATERIALS

- small cereal boxes
- scissors
- ruler
- calculator

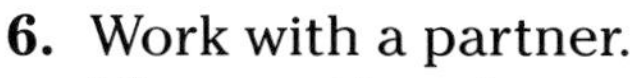

6. Work with a partner.
   Measure the edges of a small cereal box. Sketch the box and label its dimensions. Then find a way to cut the box along the seams so you can lay it out flat. Sketch the figure formed.
   a. The figure you formed is a net for the box. What shape is each section of the net?
   b. Find the area of the cardboard used to make the box.

Guided Practice

Name each solid. Then find the surface area. Round to the nearest tenth.

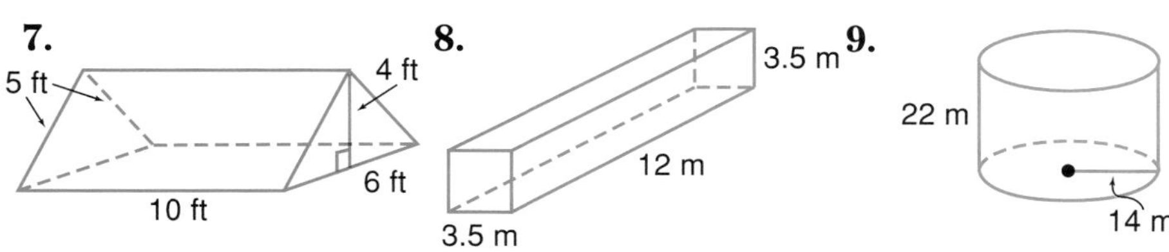

10. **Building Maintenance** The walls of a cylindrical tank that holds natural gas need to be painted. If the tank is 40 feet high and has a radius of 50 feet, what is the area that needs to be painted? If a gallon of paint covers 500 ft^2, how many gallons of paint are needed?

Exercises: Practicing and Applying the Concept

Independent Practice

Find the surface area of each solid. Round to the nearest tenth.

11.

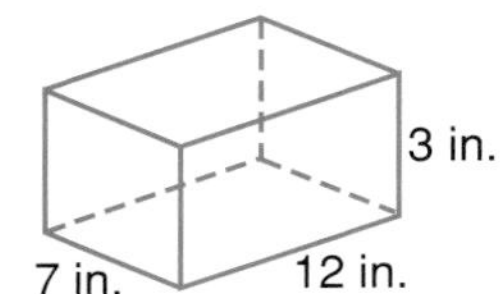

12.

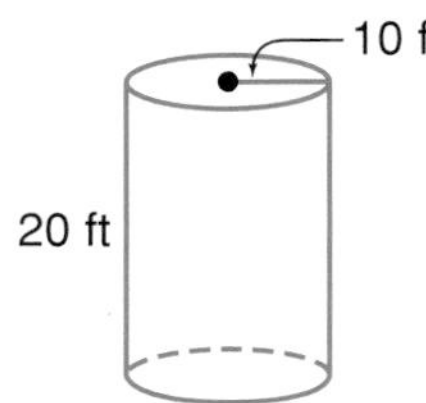

13.

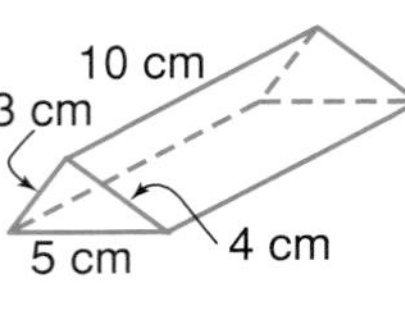

14.

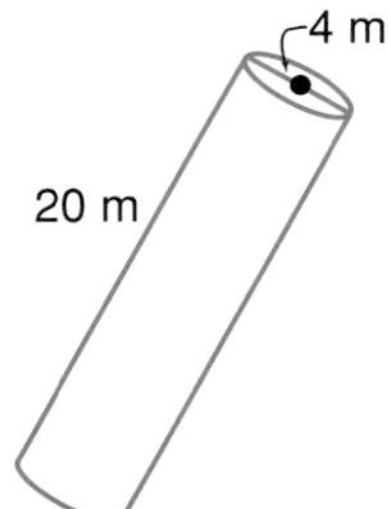

15.

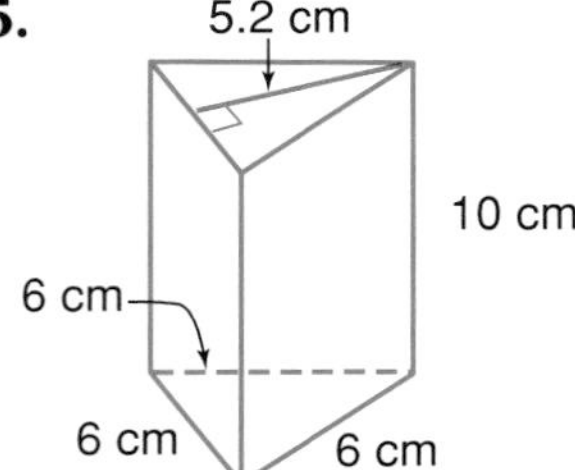

16.

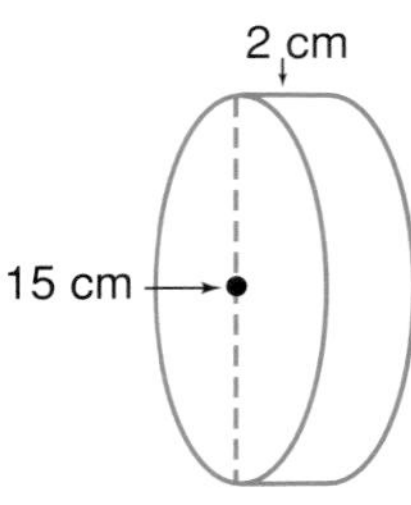

17.

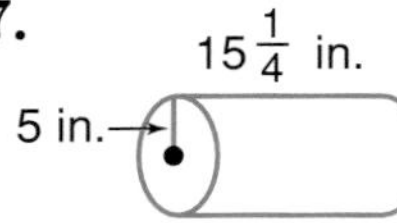

18.

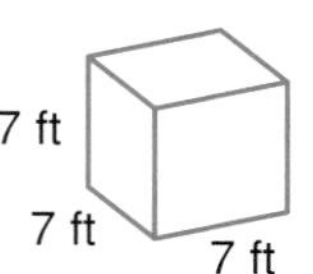

19.

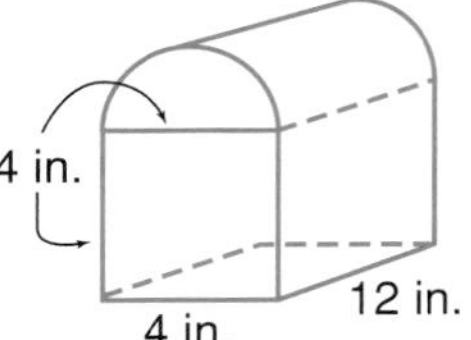

20. A cube has edges measuring a units. Find a formula for the surface area of the cube in terms of a.

21. If you add 1 cm to the length of each side of a cube, which of the following is true about the surface area?

a. It would increase by 1 cm^2. **b.** It would increase by 6 cm^2.

c. It would remain the same. **d.** need more information

22. Suppose you double the length of the sides of a cube. How is the surface area affected?

Critical Thinking

23. A cube with 12-inch sides is placed on a cube with 15-inch sides. Then a cube with 9-inch sides is placed on the 12-inch cube.

a. What is the surface area of the three-cube tower?

b. Does the way each cube is placed on the one below it affect the surface area? Why or why not?

Applications and Problem Solving

24. Business Claire works after school at her family's tent company. One of their best-selling tents is an A-frame tent that is 4 feet high and has a rectangular bottom 4 feet wide by 6 feet long. The sides of the tent are 4.5 feet long. How much canvas is needed to make the tent?

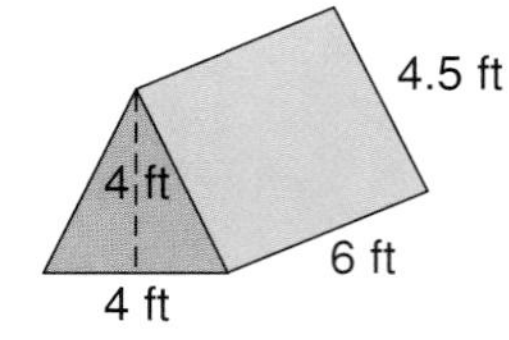

25. Manufacturing A carton of canned fruit cocktail holds 24 cans. Each can has a diameter of 7.6 cm and a height of 10.8 cm. How much paper is needed to make the labels for the 24 cans?

26. **Home Improvement** Juan's uncle wants to put drywall on his garage walls and ceiling. The garage is 24 feet by 20 feet, and the walls are $12\frac{1}{2}$ feet high. The drywall he will be using is sold in sheets that are 4 feet wide by 12 feet high by $\frac{1}{2}$ inch thick. How many sheets of drywall will he need to purchase?

27. **Postal Service** Refer to the application at the beginning of the lesson. How much cardboard does each box take if there are pieces that overlap completely at the top and bottom? Is the ratio of price to material about the same for each box?

28. **Architecture** A hallway in a Spanish-style house has a semi-circular arch above the walls. The width of the hall and the diameter of the arch are 6 feet. The walls are eight feet from the floor to the beginning of the arch. The hall is 20 feet long. If a gallon of paint covers about 400 square feet, how much paint will be needed to paint the hallway with two coats?

29. **Family Activity** Choose a room in your home. Sketch the room and measure the walls, windows, and doors.
    a. Find the area of the walls that would be painted or wallpapered.
    b. Visit a paint store and select a paint that would be appropriate for the room. Find the cost for two coats of paint for the room.

Mixed Review

30. **Farming** A farmer attaches a 25-foot rope to the corner of a 10-foot by 15-foot shed to tie up the family's pet goat. How much grazing area does the goat have? (Lesson 12-4)

31. **Geometry** Draw an angle that measures 75°. (Lesson 11-1)

32. Find the value C(12, 6). (Lesson 10-6)

33. Write 3.605×10^3 in standard form. (Lesson 6-9)

34. A house sold for more than 3 times its original purchase price of $32,000. Define a variable and translate the sentence into an inequality. Then solve. (Lesson 3-8)

Refer to the Investigation on pages 606–607.

Rooms in a home can be separated into *private*, *service*, and *social* zones. Bedrooms and bathrooms are part of the private zone. The service zone includes the kitchen, laundry, and garage. The social zone is an area where members of the household gather to spend time together.

The three zones don't have to be in different rooms. Many kitchen tables are used to do homework, thus making it a private zone. A kitchen becomes a social room when people gather to eat a meal.

Begin working on the Investigation by doing a group activity to design a model of a 2-bedroom house. The house you design must include all three zones.

Add the results of your work to your Investigation Folder.

12-6 Surface Area: Pyramids and Cones

Setting Goals: *In this lesson, you'll find surface areas of pyramids and cones.*

Modeling a Real-World Application: Architecture

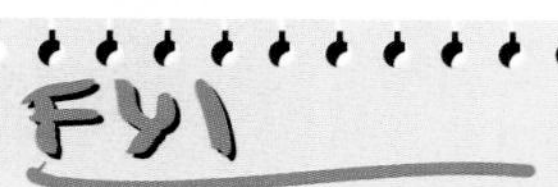

I.M. Pei also designed the Rock-and-Roll Hall of Fame in Cleveland, Ohio, which opened in September, 1995. About the Hall, Pei said, "Architecture is form and space. It's my interpretation of what the music is."

The pyramids of Egypt were built thousands of years ago as burial tombs for the pharaohs and their relatives. Pyramids have also inspired modern architects. The noted Chinese-American architect I. M. Pei completed renovation of the Louvre Museum in Paris in 1990. For the new entrance, he designed a huge square glass pyramid. It has a square base that is 116 feet on each side. How much glass did it take to cover the pyramid? *This problem will be solved in Example 2.*

Learning the Concept

In geometry, a **pyramid** is a solid figure that has a polygon for a base and triangles for the sides. We name pyramids by the shape of their bases. Notice that a pyramid has just one base. All the other faces of a pyramid, the *lateral* faces, intersect at a point called the **vertex**. The pyramid shown at the right is a **square pyramid**.

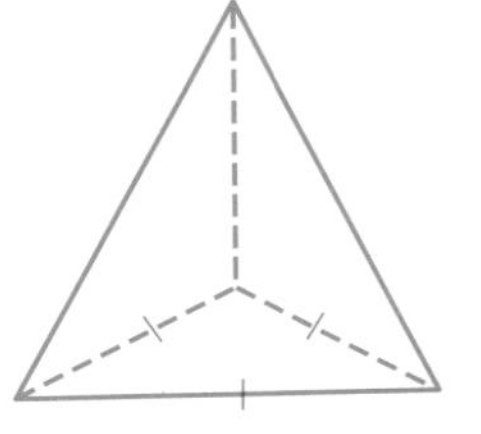

The pyramid shown at the left is a **triangular pyramid**. Its base is an equilateral triangle, so its lateral faces are congruent triangles.

Let's look at a net of a square pyramid. Notice that the lateral faces are triangles. Because the base is a square, the triangles are all congruent, and their altitudes all have the same length. The length of any one of these altitudes is called the **slant height** of the pyramid.

THINK ABOUT IT

If the base of a pyramid is a pentagon, how many faces would the pyramid have? Would the sides still be shaped like triangles?

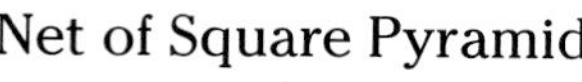

Square Pyramid

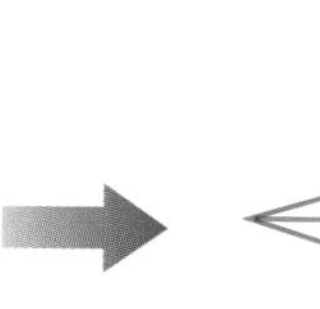

Net of Square Pyramid

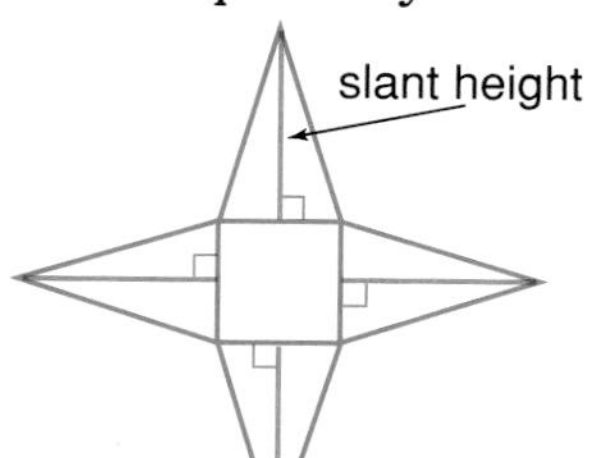

The surface area of a pyramid is the sum of the areas of its base and its lateral faces.

Example  **Find the surface area of the pyramid.**

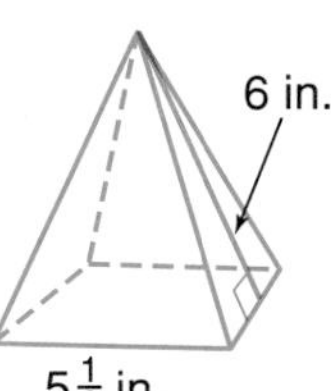

The base is a square. Use the formula $A = s^2$ to find the area. ***Estimate:*** *$5 \times 5 = 25$ and $6 \times 6 = 36$. The area is about 30 in^2.*

Use the formula $A = \frac{1}{2}bh$ to find the area of each triangular side.

Estimate: *$\frac{1}{2} \cdot 5 \cdot 6 = 15$. The area of the four sides is about 60 in^2.*

Base	**Each Triangular Side**
$A = s^2$	$A = \frac{1}{2}bh$
$A = \left(5\frac{1}{2}\right)^2$	$A = \frac{1}{2} \cdot 5\frac{1}{2} \cdot 6$
$A = 30\frac{1}{4}$ or 30.25	$A = 16\frac{1}{2}$ or 16.5

Add to find the total surface area.

area of base + *area of 4 triangular surfaces* = *surface area*

30.25 [+] 4 [×] 16.5 [=] 96.25

The surface area is 96.25 in^2. *Compare with the estimate.*

In the application at the beginning of the lesson, we need to find the **lateral surface** of the pyramid, the surface not including the base.

Example 2

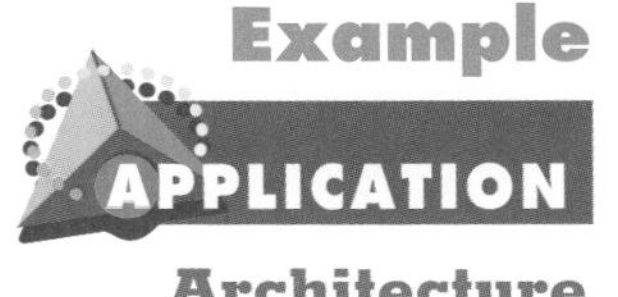

Architecture

The Louvre pyramid in the application at the beginning of the lesson has a slant height of about 92 feet. Its square base is 116 feet on each side. Find the amount of glass it took to cover the pyramid.

You are covering the lateral surface of the pyramid. Use the formula $A = \frac{1}{2}bh$ to find the area of each triangular face.

$A = \frac{1}{2}bh$

$A = \frac{1}{2} \cdot 116 \cdot 92$ *Replace b with 116 and h with 92.*

$A = 5336$

There are four triangular faces. Multiply to find the lateral surface area.

4 [×] 5336 [=] 21344

It took 21,344 ft^2 of glass to cover the pyramid.

In this textbook, when we say cone, assume we are talking about a circular cone.

A cone is another three-dimensional shape. You see cones in everyday life such as the cone for some ice cream cones, a holder for cotton candy, or some paper drinking cups. Most of the cones that you see are called **circular cones** because the base is a circle.

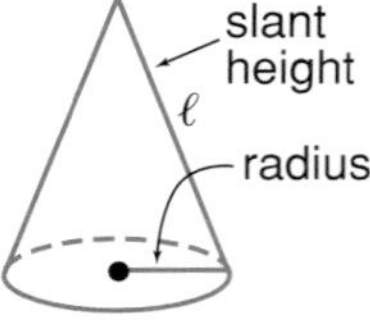

The net of a cone shows the regions that make up the cone. You can see that the base of the cone is a circle, with area πr^2. If you cut the lateral surface into sections, you can form a parallelogram. The base of the parallelogram is half the circumference of the cone, or $\frac{1}{2} \cdot 2\pi r$. Its height is the slant height of the cone, ℓ. So the area of the lateral surface is $A = \frac{1}{2} \cdot 2\pi r \cdot \ell$, or $\pi r\ell$. You can find the total surface area *SA* of a cone using the formula $SA = \pi r^2 + \pi r\ell$.

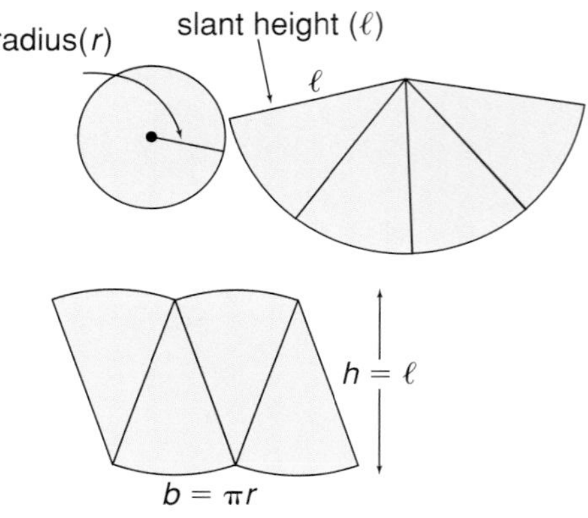

Example 3

Find the surface area of a circular cone with a slant height of 10 cm and a base with a radius of 5 cm. Round to the nearest tenth.

Base	**Lateral Surface**
$A = \pi r^2$	$A = \pi r\ell$
$A = \pi \cdot 5^2$	$A = \pi \cdot 5 \cdot 10$
$A \approx 78.5$	$A \approx 157.1$

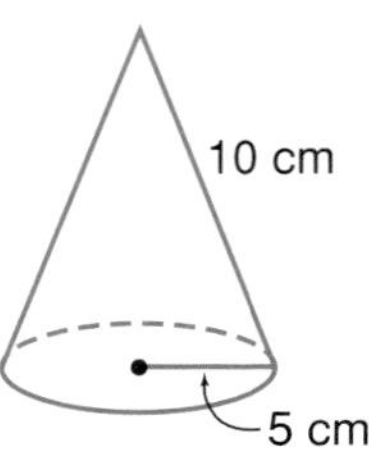

Add to find the total surface area.

78.5 [+] 157.1 [=] 235.6

The surface area is about 235.6 cm^2.

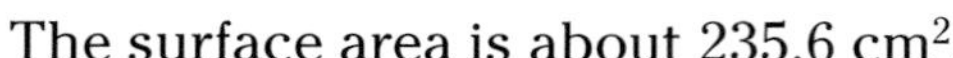

Checking Your Understanding

Communicating Mathematics

Read and study the lesson to answer these questions.

1. **Describe** the slant height of a pyramid and of a cone.
2. **Explain** why the expression $2\pi r$ can be used to find the perimeter of the base of a circular cone.
3. **State** how to find the area of the lateral surfaces of a pyramid.

4. **Assess Yourself** Write about the meaning of the term *surface area*. Then list some situations in everyday life when you might use surface area.

Guided Practice

Name each shape. Then find its surface area. Round to the nearest tenth.

5. 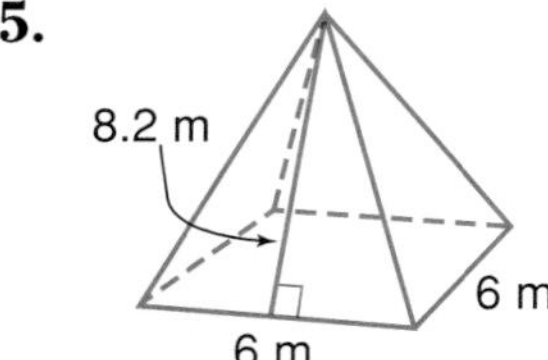

6. 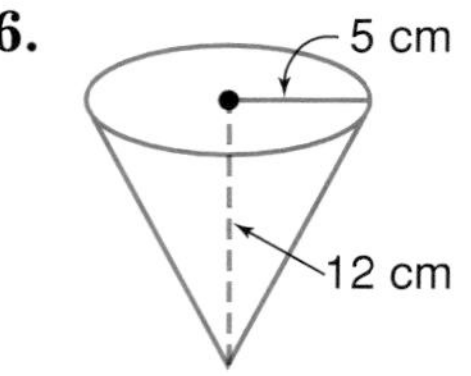

7. 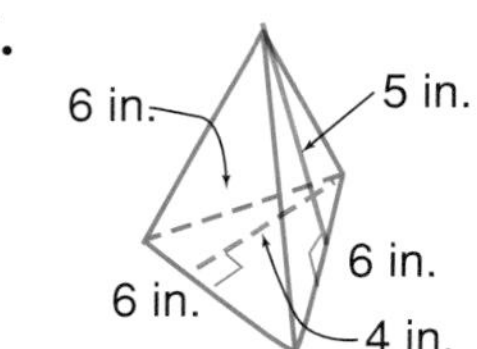

8. **Construction** A church spire is being built in the shape of a regular octagonal pyramid. Each edge of the base is 8 feet long. The slant height of the spire is about 75.3 feet. How much roofing will be needed to cover the spire?

Exercises: Practicing and Applying the Concept

Independent Practice

Find the surface area of each solid. Round to the nearest tenth.

9.

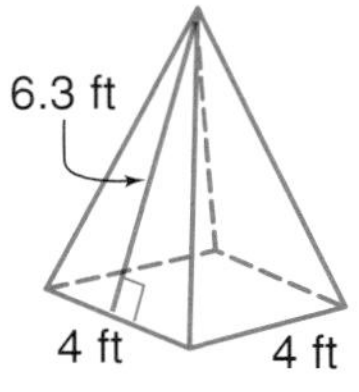

10.

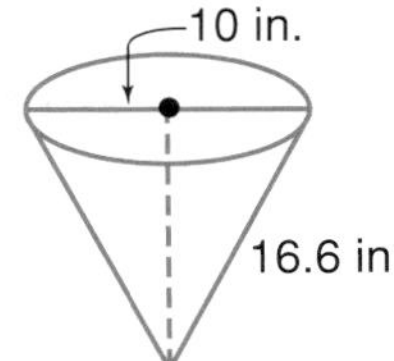

11.

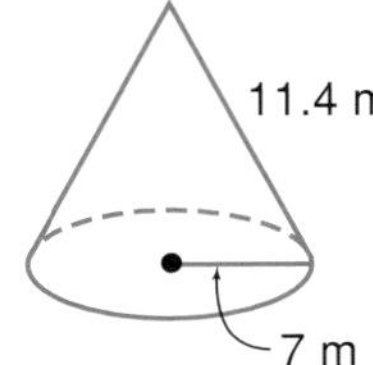

12.

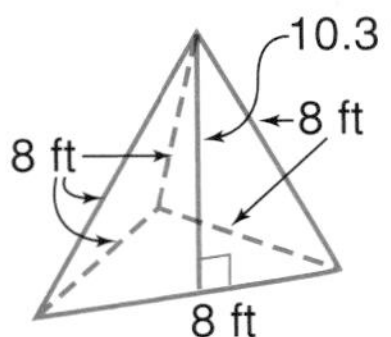

13.

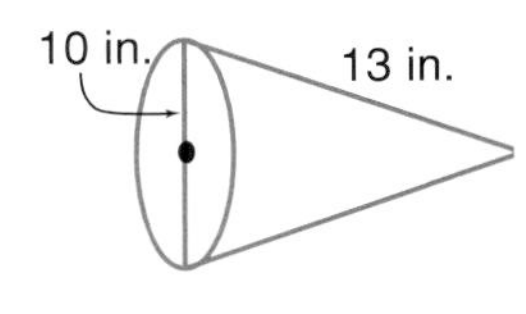

14.

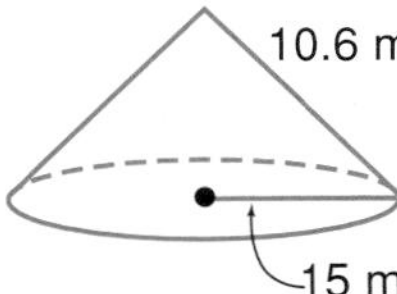

15.

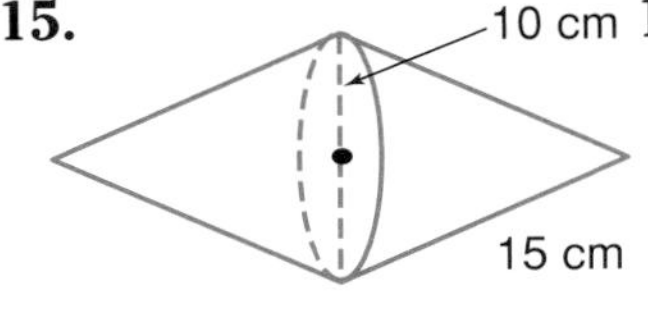

16.

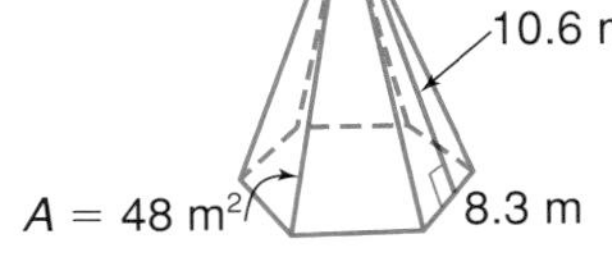

17.

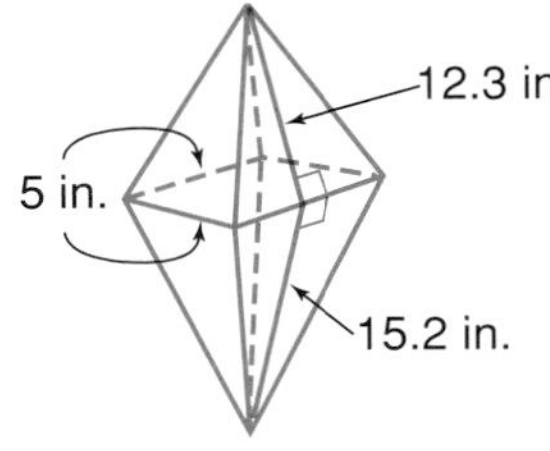

Critical Thinking

18. **Manufacturing** A bar of lead 13 inches by 2 inches by 1 inch is melted and recast into 100 conical fishing sinkers that have a diameter of 1 inch and a height of 1 inch. The slant height of each sinker is about 1.1 inches. How does the total surface area of all the sinkers compare to the surface area of the original lead bar?

Applications and Problem Solving

19. **History** The monument of Cestius in Rome is a square pyramid. It is 121.5 feet high and has a base 98.4 feet long on each side. The slant height is about 131 feet. What is its lateral surface area?

20. **Architecture** I. M. Pei also designed smaller square pyramids that stand near the entrance of the Louvre. These pyramids have bases 26 feet on a side and a slant height of about 20.6 feet. How much glass did it take to cover one of these pyramids?

21. **Tepees** The largest tepee in the United States has a diameter of 42 feet and a slant height of about 47.9 feet. The tepee belongs to Dr. Michael Doss of Washington, D.C. Dr. Doss is a member of Montana's Crow Tribe. How much canvas was used to make the cover of the tepee?

22. **Food** An ice cream shop uses sugar cones that are 5 inches long and have a diameter of 2 inches. The manager would like to order special napkins that would fit around the cone twice. Sketch the shape of the napkin and label its dimensions.

Mixed Review

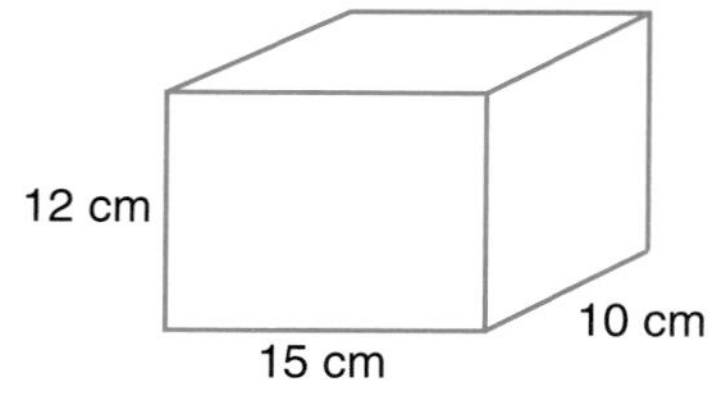

23. **Geometry** Find the surface area of the rectangular prism at the right. (Lesson 12-5)
24. **Probability** If two coins are tossed, what is the probability that one will land heads up and the other tails up? (Lesson 9-3)
25. Define a variable and write and solve an equation for the following situation. Four times a number plus eight is 20. What is the number? (Lesson 7-3)
26. Solve $w + \frac{1}{6} = 3\frac{2}{3}$. (Lesson 5-6)
27. Simplify $(5x)(-3y)$. (Lesson 2-7)

Self Test

Find the area of each figure. (Lessons 12-1 and 12-2)

1. 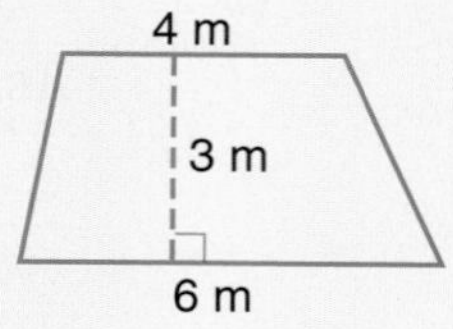

2. 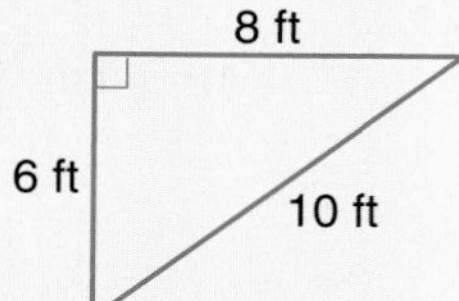

3. 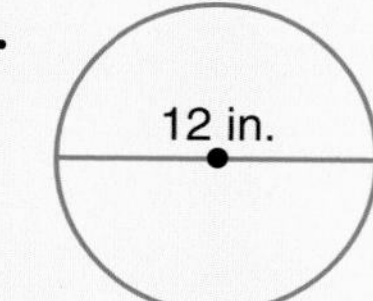

Name each shape. Then find its surface area. Round to the nearest tenth. (Lessons 12-5 and 12-6)

4. 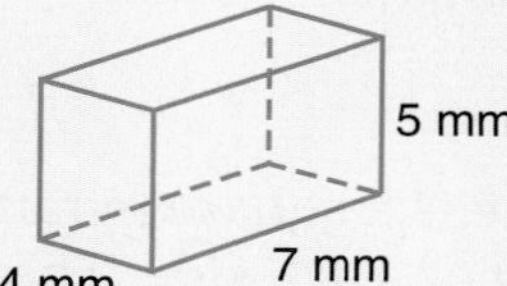

5. 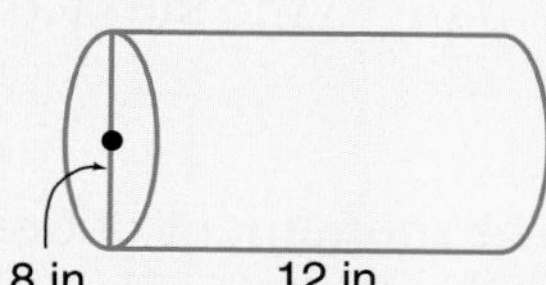

6. 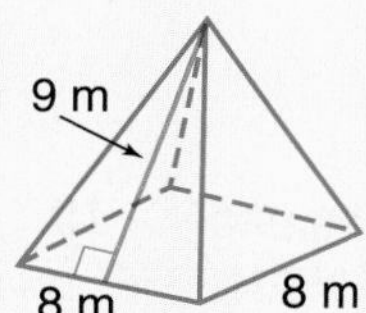

Solve.

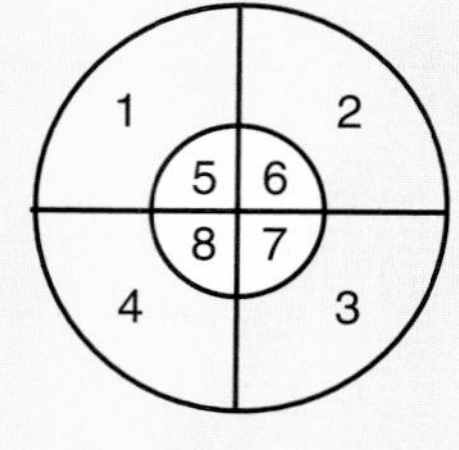

7. **Probability** In a dart game, you win points based on the region in which the dart lands. The diameter of the whole target is 24 inches. The diameter of the inner circle is 8 inches. What is the probability that the dart lands in the 8-point region? (Assume the dart lands on the target.) (Lesson 12-3)
8. **Make a Drawing** A circular fountain has a diameter of 8 feet. A flower garden is planted around the fountain. The garden extends 12 feet all around the fountain. If a pound of organic fertilizer covers about 300 square feet, how much fertilizer is needed for the garden? (Lesson 12-4)
9. How much paper would be used to make an open paper cone with a diameter of 5 inches and a slant height of 6 inches? (Lesson 12-6)

HANDS-ON ACTIVITY

12-7A Volume

A Preview of Lesson **12-7**

MATERIALS

- 5 × 8 inch index cards
- tape
- rice
- grid paper

In this activity, you will investigate volume by making containers of different shapes and comparing how much each container holds.

Your Turn

Work with a partner.

- Use three 5 × 8 cards to make three containers with a height of 5 inches as shown. Make one with a square base (with 2 in. sides.), one with a triangular base (with sides of 2 in., 3 in., and 3 in.) and one with a circular base.
- Then tape one end of each container to another card as a bottom, but leave the top open.
- Estimate which container would hold the most (have the greatest volume) and which would hold the least (have the least volume). Do you think each container would hold the same amount?
- Use rice to fill the container that you believe holds the least amount. Then put the rice into another container. Does the rice fill this container? Continue the process until you find out which, if any, container has the least volume and which has the greatest.

TALK ABOUT IT

1. Which container held the most? Which held the least?
2. How do the heights of the three containers compare? What is each height?
3. Compare the perimeters of the bases of each container. What is each base perimeter?
4. Trace the base of each container onto grid paper. Estimate the area of each base.
5. Which container has the greatest base area?
6. Does there appear to be a relationship between the area of the base and the volume? Explain.

12-7 Volume: Prisms and Cylinders

Setting Goals: *In this lesson, you'll find volumes of prisms and circular cylinders.*

Modeling with Technology

The program below computes the volume of a rectangular prism when you enter the dimensions. Use this program to discover how to calculate the volume of a container.

Your Turn Enter the program in a programmable calculator.

```
PROGRAM: VOLUME
: Input "LENGTH =", L
: Input "WIDTH =", W
: Input "HEIGHT =", H
: L*W→A
: L*W*H→V
: Disp "AREA OF BASE ="
: Disp A
: Disp "VOLUME ="
: Disp V
: Stop
```

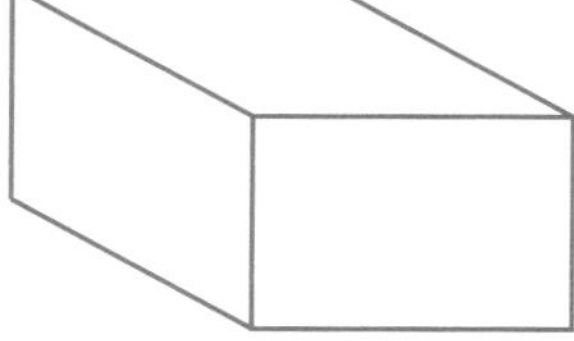

- ► Sketch a rectangular prism like the one shown at the right. Choose measurements for each side and label the sketch.
- ► To run the program, press PRGM. Use the down arrow to highlight the program. Press ENTER twice. Enter the data asked for in the program. Record the results on your sketch.
- ► Repeat for at least three different rectangular prisms. Sketch each prism and record the dimensions and the results of the program. (To run the program again, press ENTER.)

a. What dimensions did the program ask for in a rectangular prism?

b. What is the relationship between the dimensions and the volume?

c. How does the area of the base of the prism relate to its volume?

Learning the Concept

The amount a container will hold is called its capacity, or **volume**. Volume is usually measured in cubic units. Two common units of measure for volume are the cubic centimeter (cm^3) and the cubic inch (in^3). The volume of a prism can be found by multiplying the area of the base times the height.

Volume of a Prism

In words: If a prism has a base area of B square units and a height of h units, then the volume V is $B \cdot h$ cubic units.

In symbols: $V = Bh$

Example  1

Find the volume of each prism.

THINK ABOUT IT

How could you find the height of a prism if you know the area of the base and the volume?

a.

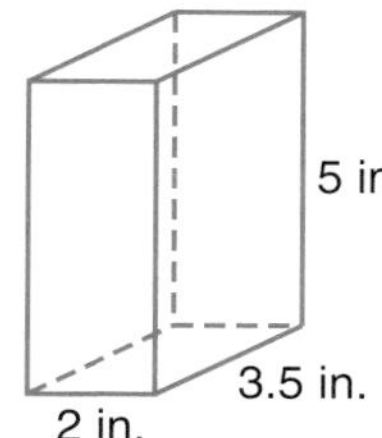

$V = Bh$ *Formula for volume of a prism*

$V = \ell wh$ *Since the base of the prism is a rectangle, $B = \ell w$.*

$V = 2 \cdot 3.5 \cdot 5$ *Replace ℓ with 2, w with 3.5, and h with 5.*

$V = 35$

The volume is 35 in^3.

b. A prism has bases that are right triangles as shown below. The height of the prism is 10 cm.

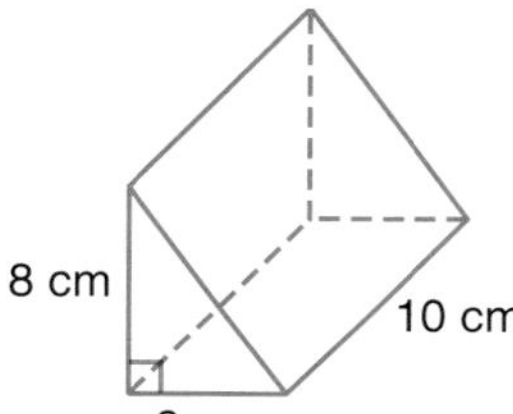

$V = Bh$

$V = \left(\frac{1}{2} \cdot 6 \cdot 8\right) \cdot 10$ *The base of the prism is a triangle. The area of the triangle is $\frac{1}{2} \cdot 6 \cdot 8$.*

$V = 24 \cdot 10$

$V = 240$

The volume is 240 cm^3.

Since the base of a circular cylinder is a circle and its area is equal to πr^2, we can substitute πr^2 for B to find the volume of a cylinder.

Volume of a Circular Cylinder

In words: If a circular cylinder has a base with a radius of r units and a height of h units, then the volume V is $\pi r^2 h$ cubic units.

In symbols: $V = \pi r^2 h$

Example  2

A can of tomato juice has a diameter of 14 cm and a height of 20 cm. What is its volume?

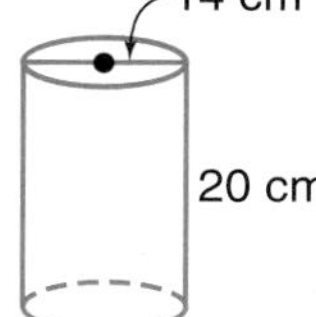

$V = \pi r^2 h$ *Formula for volume*

$V = \pi \cdot 7^2 \cdot 20$ *The diameter of the can is 14 cm; so the radius is 7 cm.*

$V \approx 3079$

The volume is about 3079 cm^3.

Connection to Algebra

If you know the volume of a cylinder, you can solve the formula for h or r to find the height or radius of the cylinder.

Engineering

A cylindrical natural gas storage tank is being manufactured to hold at least 1,000,000 cubic feet of natural gas and have a diameter of no more than 80 feet. What height should the tank be?

Explore The tank is a cylinder. You know the volume and the diameter. You must find the height.

Plan The formula for the volume of a cylinder is $V = \pi r^2 h$. Solve the formula for h to find the height.

Solve
$V = \pi r^2 h$
$1{,}000{,}000 = \pi \cdot (40)^2 \cdot h$ *$V = 1{,}000{,}000,\ r = 40$*
$1{,}000{,}000 = \pi \cdot 1600 \cdot h$
1,000,000 [÷] [π] [÷] 1600 [=] 198.9436789
$198.9 \approx h$

The tank should have a height of at least 199 feet.

Examine If the tank is 199 feet high with a diameter of 80 feet, will it have a volume of at least 1,000,000 cubic feet?

Is $\pi \cdot (40)^2 \cdot 199 \geq 1{,}000{,}000$?

π [×] 40 [x^2] [×] 199 [=] 1000283.101

The answer checks.

THINK ABOUT IT

Why should the result be rounded *up* in this case?

Checking Your Understanding

Communicating Mathematics

Read and study the lesson to answer each question.

1. **Explain** why you can use the same formula to find the volume of a prism or a cylinder.
2. **Tell** what each variable represents in the formula $V = Bh$.
3. **Compare and contrast** surface area and volume.
4. **You Decide** Jake says that doubling the length of each side of a cube doubles the volume. Carlos says the volume is eight times greater. Who is correct? Explain your answer.

5. Write a problem from an everyday situation in which you will need to find the volume of a cylinder or a rectangular prism. Explain how to solve the problem.

Guided Practice

Find the volume of each prism or cylinder. Round to the nearest tenth.

6.
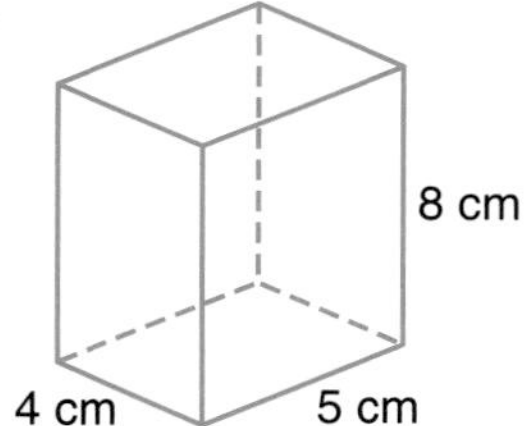

7.
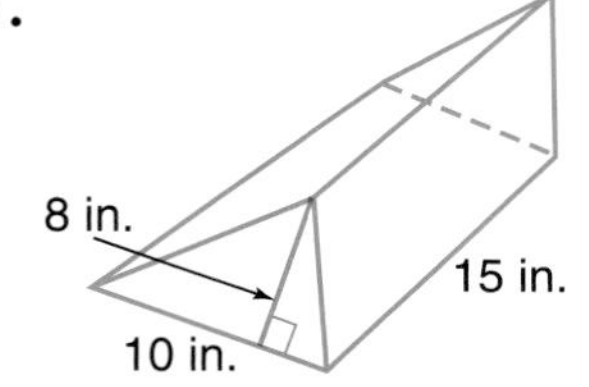

8.
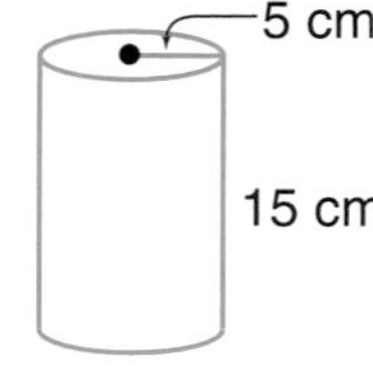

Find the volume of each prism or cylinder. Round to the nearest tenth.

9. rectangular prism: length 3 in.; width 5 in.; height 15 in.
10. octagonal prism; base area 25 m^2; height of 1.5 m
11. circular cylinder: radius 2 ft; height $2\frac{1}{4}$ ft
12. **Pet Care** Tina has an old fish tank that is a circular cylinder. The tank is 2 feet in diameter and 6 feet high. How many cubic feet of water does it hold?

Exercises: Practicing and Applying the Concept

Independent Practice

Estimation
Mental Math
Calculator
Paper and Pencil

Find the volume of each solid. Round to the nearest tenth.

13.
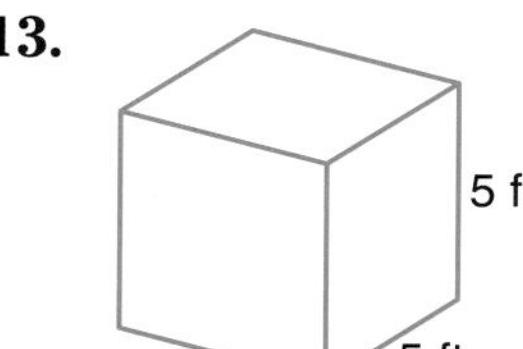

14.
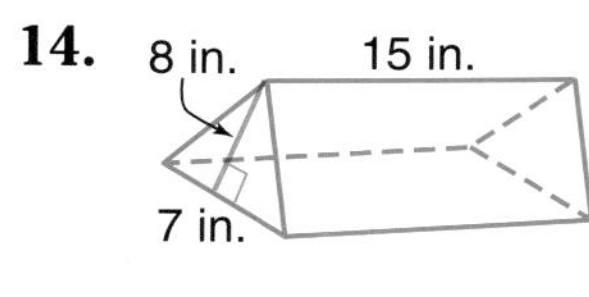

15.
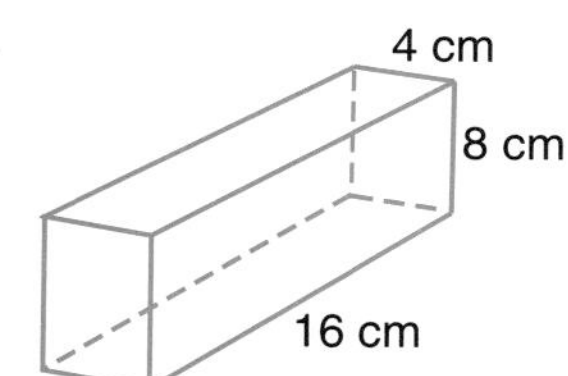

16.
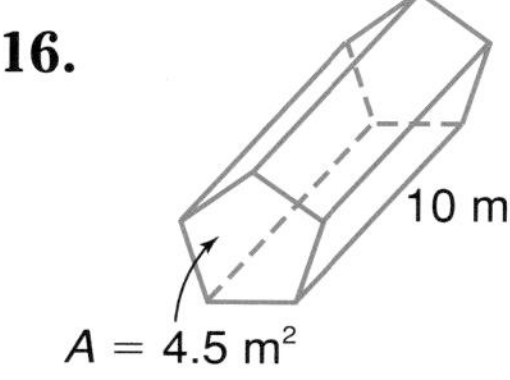

17.
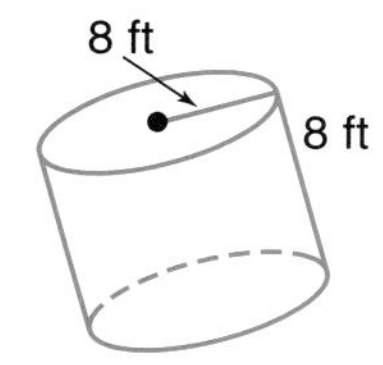

18.
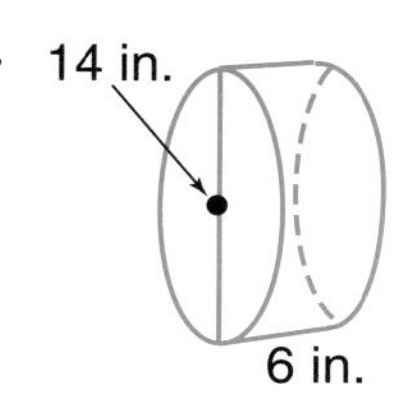

19.
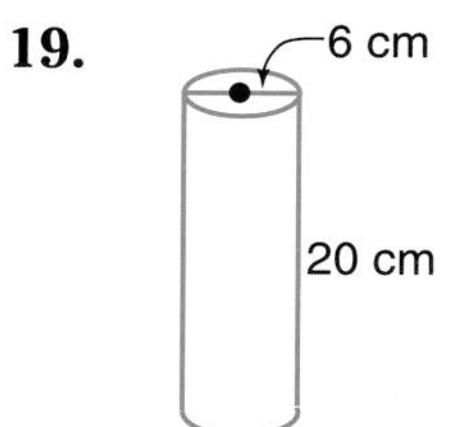

20.
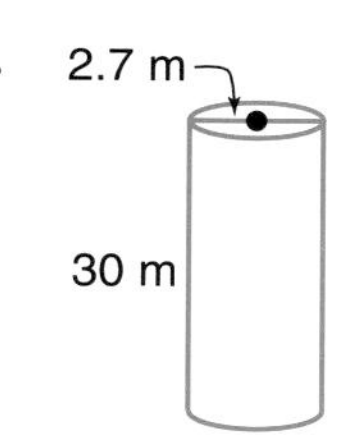

21.
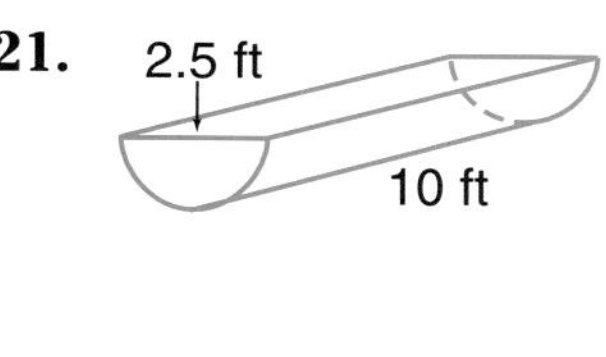

22. rectangular prism: length 6 in.; width 6 in.; height 6 in.
23. rectangular prism: length 4.2 cm; width 3.2 cm; height 6.2 cm
24. pentagonal prism: base area 45.2 cm^2; height 47.8 cm
25. circular cylinder: radius 10 yd; height 15 yd
26. triangular prism; base of triangle 8 in.; altitude of triangle 15 in.; height of prism 6.5 in.
27. circular cylinder: diameter 2.6 m; height 3.5 m

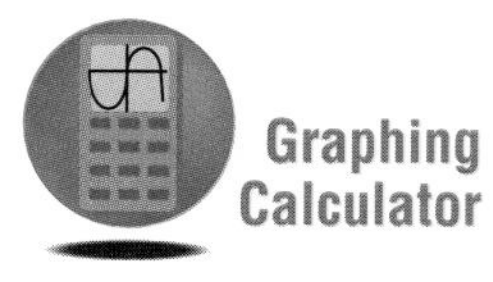

28. Refer to the program in the activity at the beginning of the lesson.
    a. Change the program to find the volume of a circular cylinder.
    b. Sketch three circular cylinders with the same height and different diameters. Run the program to find the volume of each cylinder.
    c. Graph the diameters and volumes on a coordinate plane. What pattern do you observe?

Critical Thinking

29. Suppose you roll an $8\frac{1}{2}$- by 11-inch piece of paper to form a cylinder. Will the volume be greater if you roll it so the height is $8\frac{1}{2}$ inches or 11 inches, or will the volumes be the same?

30. A cube has a side a units long. What is the formula for the volume of the cube in terms of a?

Applications and Problem Solving

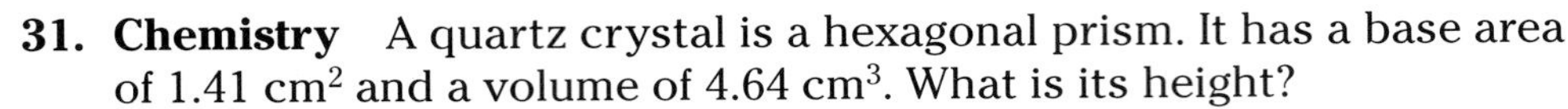

31. Chemistry A quartz crystal is a hexagonal prism. It has a base area of 1.41 cm^2 and a volume of 4.64 cm^3. What is its height?

32. Baking A rectangular cake pan is 30 cm by 21 cm by 5 cm. A round cake pan has a diameter of 21 cm and a height of 4 cm. Which will hold more batter, the rectangular pan or two round pans?

33. Recreation A cubic foot of water is approximately 7.481 gallons. Which swimming pool below requires about 45,000 gallons of water to fill?

a.

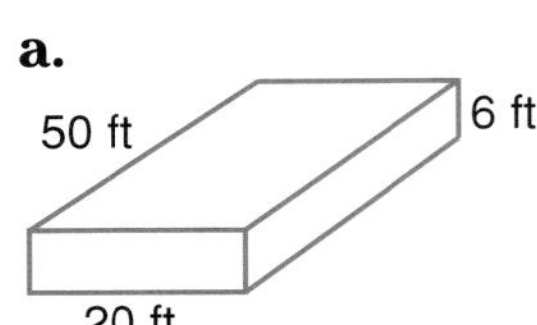

b.

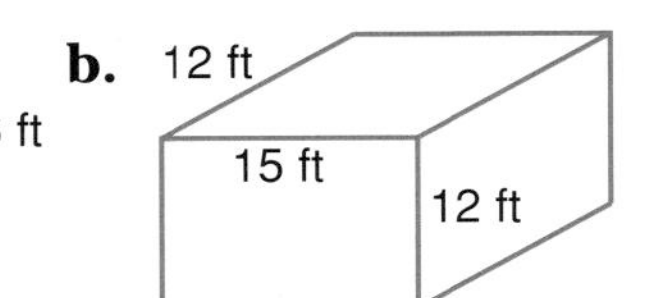

c.

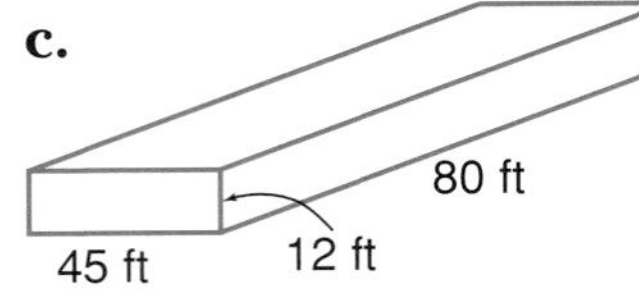

34. Make a Drawing A twelve-pack of ice tea cans is sold in a rectangular box. The cans are 6 inch high cylinders with a diameter of 3 inches.

a. What are the dimensions of the box if the cans are arranged in a 4 by 3 array?

b. How much wasted space is there in the box?

35. Consumer Awareness Firewood is usually sold by a measure known as a cord. A full cord may be a stack $8 \times 4 \times 4$ feet or $8 \times 8 \times 2$ feet.

a. What is the volume of a full cord?

b. A "short cord" is $8 \times 4 \times$ the length of the logs. What part of a full cord is a short cord of $2\frac{1}{2}$ foot logs?

Mixed Review

36. Find the surface area of a square pyramid with a base 14 yards long and a height of 3 yards. (Lesson 12-6)

37. Geometry Find the value of x in $\triangle ABC$ at the right. (Lesson 11-6)

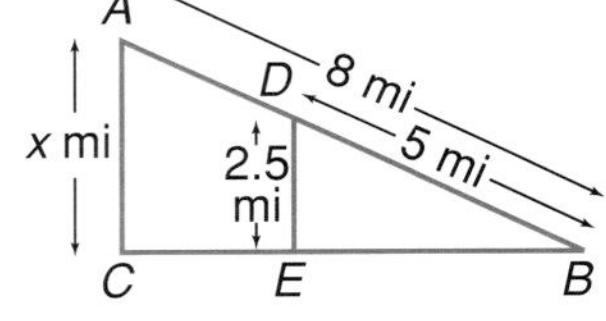

38. Geometry Angles X and Y are supplementary. If $m\angle X = 58°$, find $m\angle Y$. (Lesson 11-3)

39. If $f(x) = 5x - 3$, find $f(2)$. (Lesson 8-4)

40. Work Backward Drew has \$2.25. Half of the money he had when he left home this morning was spent on lunch. He loaned Cassie two dollars after lunch. How much money did Drew start with? (Lesson 7-1)

41. Games If you toss three 4-sided dice, how many different combinations of tosses are possible? (Lesson 4-3)

42. Write a problem that could be solved using the equation $3x = 25$. (Lesson 1-8)

12-8 Volume: Pyramids and Cones

Setting Goals: *In this lesson, you'll find volumes of pyramids and cones.*

Modeling with Manipulatives

MATERIALS

Look at the nets shown at the right. They will fold into models of a square prism (a cube) with one open side and a square pyramid. The base of both solids is the same size. If you fill each model with rice, you can compare their volumes.

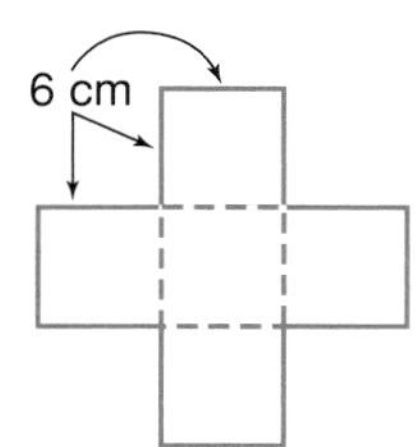

Your Turn Work with a partner.

- Copy the two nets on centimeter grid paper.
- Cut them out and fold on the dashed lines.
- Tape the edges together to form models of the solids.
- Estimate how much larger the volume of the cube is than the volume of the pyramid.
- Make an opening in the base of the pyramid so you can put rice into it.
- Fill the pyramid with rice. Then pour this rice into the cube. Repeat until the cube is filled.

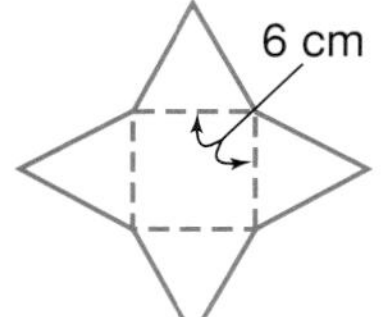

a. How many pyramids of rice did it take to fill the cube?
b. What do you know about the areas of the bases of each solid?
c. Compare the heights of the cube and the pyramid.
d. Compare the volume of the cube and the pyramid.

Learning the Concept

In the modeling activity, you found that the volume of the pyramid was $\frac{1}{3}$ the volume of the prism. Since the volume of a prism is Bh, the volume of the pyramid is $\frac{1}{3}Bh$. The height, h, of a pyramid is the length of a segment from the vertex, perpendicular to the base.

Volume of a Pyramid

In words: If a pyramid has a base of B square units and a height of h units, then the volume V is $\frac{1}{3} \cdot B \cdot h$ cubic units.

In symbols: $V = \frac{1}{3}Bh$

Example  Find the volume of the pyramid.

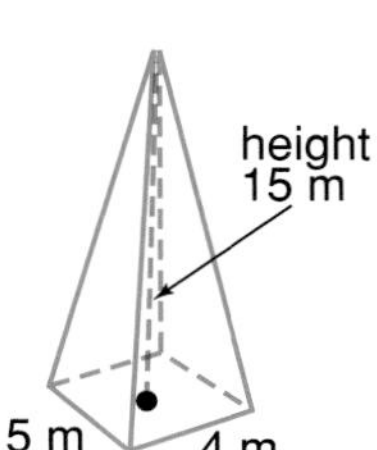

$V = \frac{1}{3}Bh$ *Formula for volume of a pyramid*

$V = \frac{1}{3}\ell wh$ *Replace B with ℓw.*

$V = \frac{1}{3} \cdot 4 \cdot 5 \cdot 15$

$V = 100$

The volume is 100 m^3.

The relationship of the volumes of a cone and a cylinder is similar to the relationship of the volumes of a pyramid and a prism. The volume of a cone is $\frac{1}{3}$ the volume of a cylinder. Since the volume of a cylinder is $\pi r^2 h$, the volume of a cone is $\frac{1}{3}\pi r^2 h$. *You'll verify this formula in Exercise 4.*

Volume of a Cone

In words: If a cone has a radius of r units and a height of h units, then the volume V is $\frac{1}{3} \cdot \pi \cdot r^2 \cdot h$.

In symbols: $V = \frac{1}{3}\pi r^2 h$

Example  Find the volume of the cone to the nearest whole number.

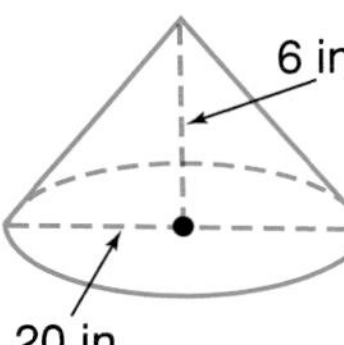

$V = \frac{1}{3}\pi r^2 h$ *Formula for volume of a cone*

$V = \frac{1}{3} \cdot \pi \cdot 10^2 \cdot 6$ *Since $d = 20$, $r = 10$; $h = 6$*

$V \approx 628$

The volume is about 628 in^3.

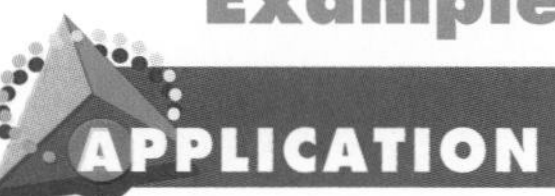

Example 3 APPLICATION

Highway Maintenance

Many communities use a salt and sand mixture on snowy or icy roads. When the mixture is dumped from a truck, it forms a mound with a conical shape.

a. If the mound has a diameter of 15 feet and a height of 10 feet, how much salt-sand mixture is there? *Estimate: $\frac{1}{3} \cdot 3 \cdot 8^2 \cdot 10 = 640$*

b. If the mixture is spread at a rate of 1 cubic foot per 500 square feet of roadway, how many square feet of roadway can be salted by the mixture in one mound?

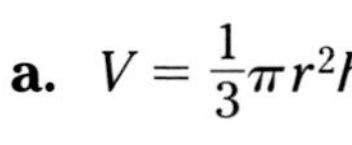

a. $V = \frac{1}{3}\pi r^2 h$ *Formula for volume of a cone*

$V = \frac{1}{3} \cdot \pi \cdot 7.5^2 \cdot 10$ *Since $d = 15$, $r = 7.5$; $h = 10$*

$V \approx 589$

The volume is 589 ft^3.

b. Each cubic foot of mixture will salt 500 square feet of roadway.
589 [×] 500 [=] 294000

294,000 square feet of roadway can be salted by the mixture in the mound.

Checking Your Understanding

Communicating Mathematics

Read and study the lesson to answer these questions.

1. **Explain** why you can use πr^2 to find the area of the base of a cone.
2. **State** the formula for finding the volume of a pyramid. Then tell what each variable represents.
3. **State** the formula for finding the volume of a cone. Then tell what each variable represents.
4. Work with a partner.

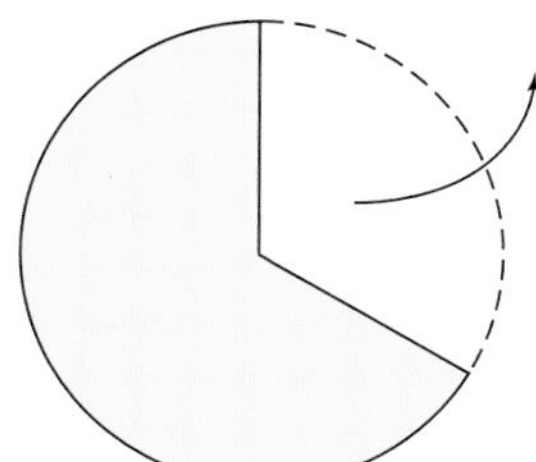

MODELING MATHEMATICS

MATERIALS

compass
scissors
tape
rice

- ▶ Draw a circle with a radius of at least 2 inches. Cut out the circle.
- ▶ Remove about $\frac{1}{3}$ of the circle as shown at the right. Tape the edges of the remaining part of the circle to make an open cone.
- ▶ Trace the base of the cone on another piece of paper and cut out the circle.

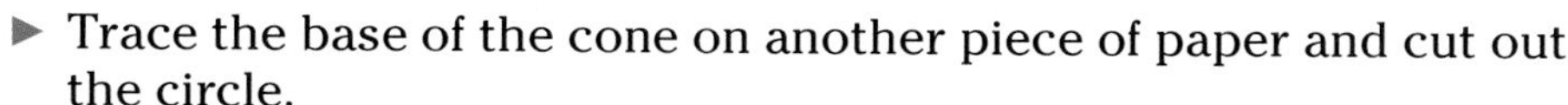

- ▶ Cut a strip of paper as wide as the cone is high. Use the strip and the circle to make a model of a cylinder that has the same size base and the same height as the cone.
- ▶ Estimate how many times larger the cylinder is than the cone. Use rice to check your estimate.

Guided Practice

Explain how to find the volume of each pyramid or cone. Then find the volume. Round to the nearest tenth.

5.

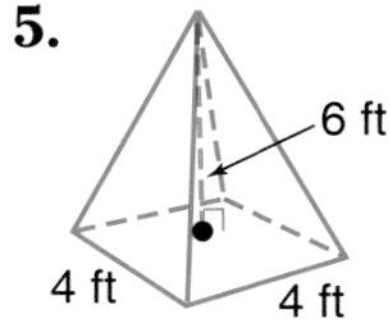

6.

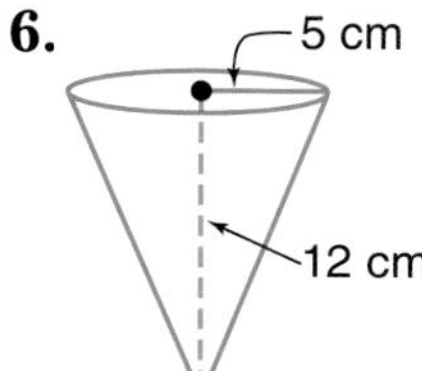

7.

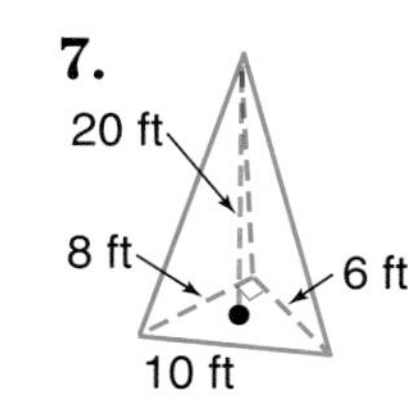

8. circular cone: diameter 12 m; height 15 m
9. hexagonal pyramid: base area 125 in^2; height $6\frac{1}{2}$ in.
10. **Historical Monuments** The top of the Washington Monument is a square pyramid 54 feet high and 34 feet long on each side. What is the volume of the room under the top?

Exercises: Practicing and Applying the Concept

Independent Practice

Find the volume of each solid. Round to the nearest tenth.

11. 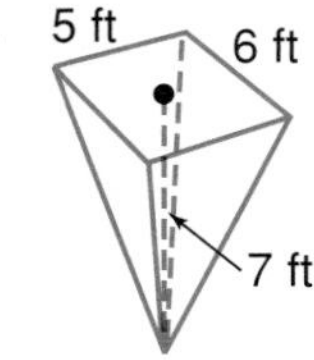

12. 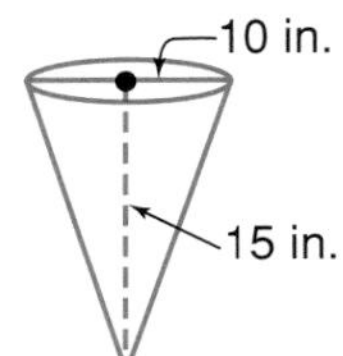

13. 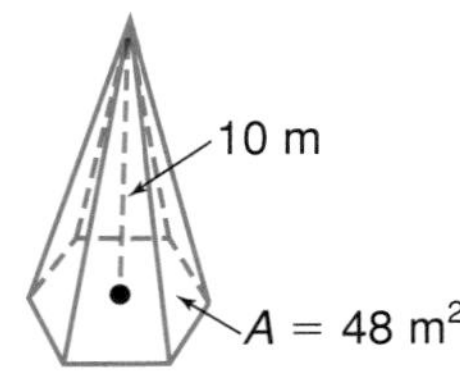

14. 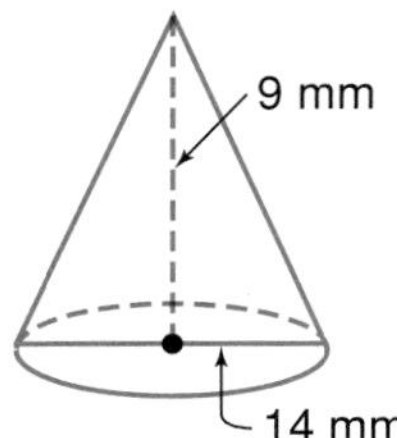

15. 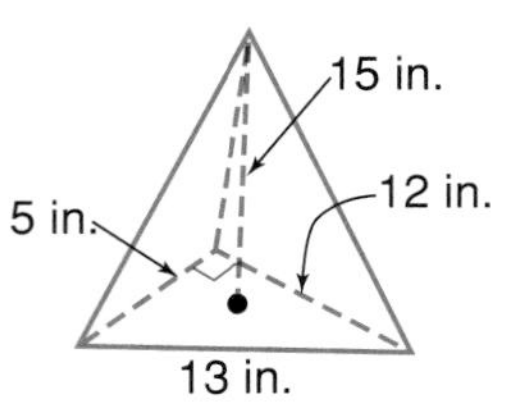

16. 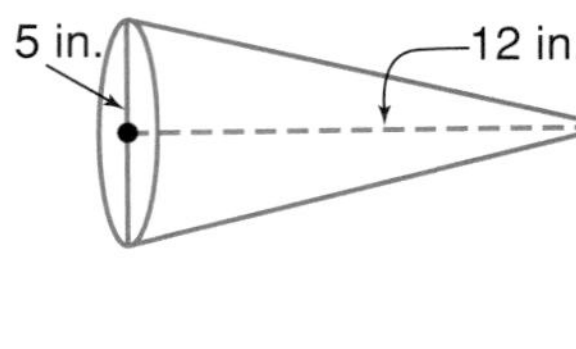

17. 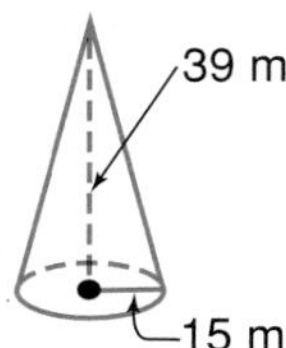

18. 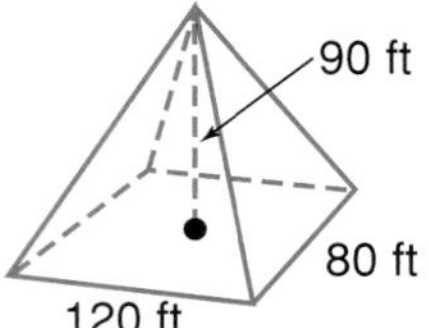

19. 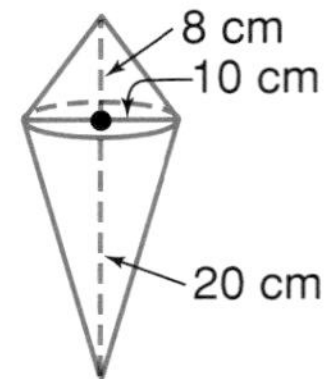

20. rectangular pyramid; length, 9 in.; width, 7 in.; height, 18 in.

21. square pyramid; length, 5 cm; height, 6 cm

22. circular cone; radius, 3 ft; height, 14 ft

23. circular cone; radius 10 m; height, 18 m

24. 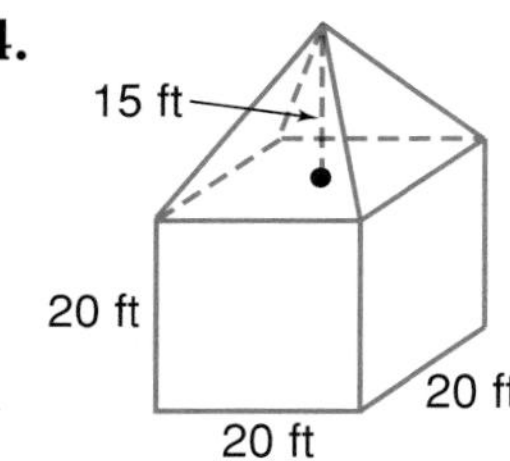

25. 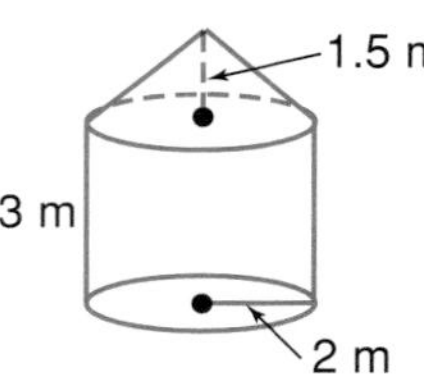

26. 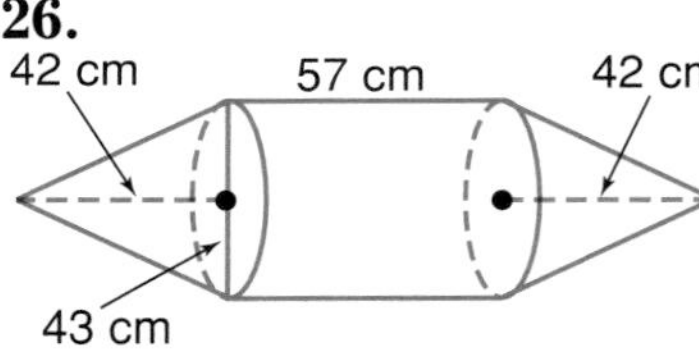

Critical Thinking

27. Consumer Awareness Popcorn at the Bijou Theatre used to be sold in rectangular prism containers. Recently the theater changed to a rectangular pyramid container. Both containers have the same height and the same size base. Popcorn in the new box sells for half the former price. Is this a good deal for the customer? Why or why not?

28. a. Suppose you double the height of a cone. How does the volume change?

b. Suppose you double the radius of the base of a cone. How does the volume change?

Applications and Problem Solving

29. History The Great Pyramid of Khufu in Egypt was originally 481 feet high and had a square base 756 feet on a side. What was its volume? Use an estimate to check your answer.

30. **Geology** A stalactite in the Endless Caverns in Virginia is shaped like a cone. It is 4 feet long and has a diameter at the roof of 1.5 feet.

   a. Find the volume of the stalactite to the nearest tenth.

   b. The stalactite is made of calcium carbonate, which weighs 131 pounds a cubic foot. What is the weight of the stalactite?

Mixed Review

31. **Construction** Marty's Masonry makes concrete steps for new homes. The steps for one home are to be 5 feet long, with three steps. Each step will be 0.5 foot high and 1 foot wide. How much concrete will be needed? (*Hint:* Make a drawing.) (Lesson 12-7)

32. **Probability** A die is tossed 30 times. How many times would you expect an outcome of a 1 or a 6? (Lesson 10-10)

33. Eighteen is what percent of 90? (Lesson 9-5)

34. Solve $x = 3\frac{3}{5} - 1\frac{3}{10}$. Write the solution in simplest form. (Lesson 5-5)

35. **Geometry** Find the perimeter and area of the rectangle at the right. (Lesson 3-5)

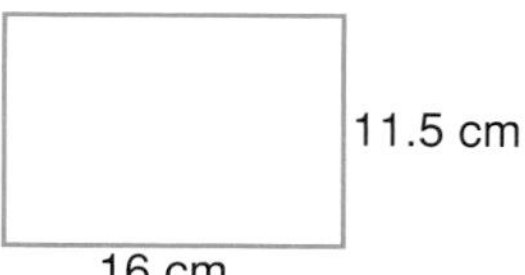

36. Solve the equation $\frac{-144}{6} = k$. (Lesson 2-8)

WORKING ON THE Investigation

Refer to the Investigation on pages 606–607.

Once your group has completed the basic layout of the rooms in your house, you are ready to prepare the *working drawings*. Working drawings are used to show specific information that is necessary to build the structure.

The working drawing that shows the location and size of the rooms is called the *floor plan*. A floor plan shows how the rooms would look if the ceiling were removed and you looked down from above.

Using your basic layout, design your house so that it is no larger than 1200 square feet. Each room should have at least 1 window. Sketch each floor plan on graph paper to a scale of $\frac{1}{4}$ inch = 1 foot.

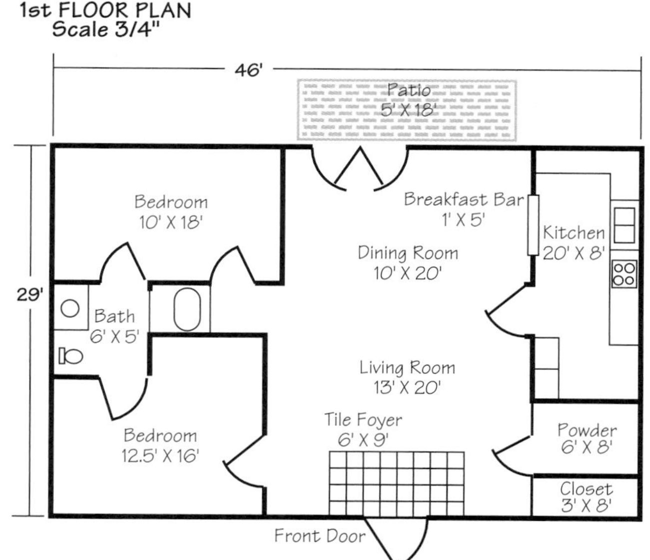

Add the results of your work to your Investigation Folder.

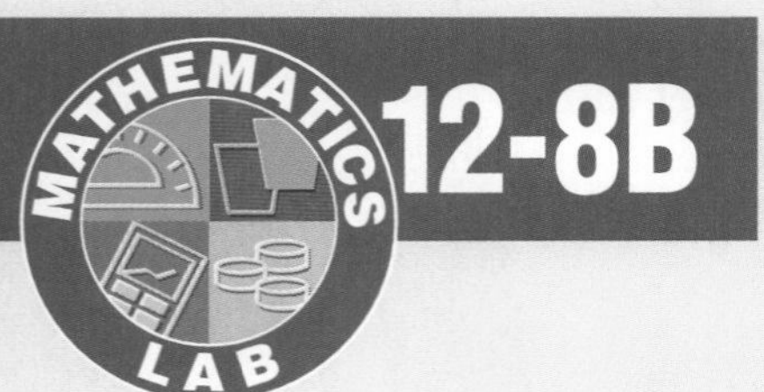

12-8B

HANDS-ON ACTIVITY

Similar Solid Figures

An Extension of Lesson **12-8**

MATERIALS

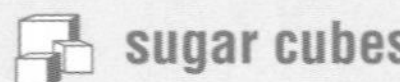

You can review similar figures in Lesson 11-6.

Have you or someone you know ever built a model car or boat? A scale model is an exact replica of the real object, but is larger or smaller than the original. The dimensions of the model and the original are proportional. The number of times that you increase or decrease the linear dimensions is called the **scale factor**. In this activity, you will find out how changing the dimensions of a solid affects the surface area and volume of the solid.

Activity  Work with a partner.

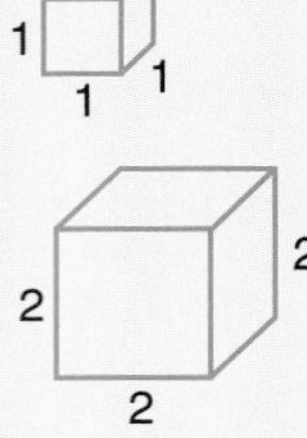

Your Turn

- If each edge of a sugar cube is 1 unit long, then each face is 1 square unit and the volume of the cube is 1 cubic unit.
- Use cubes to make a larger cube that has sides twice as long as the original cube.
- How many small cubes did you use?

TALK ABOUT IT

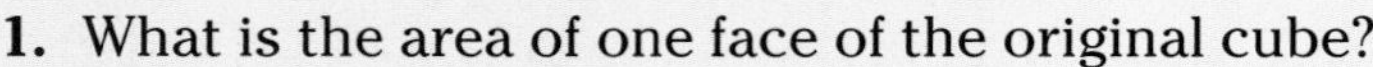

1. What is the area of one face of the original cube?
2. What is the area of one face of the cube that you built?
3. What is the volume of the original cube?
4. What is the volume of the cube that you built?

Activity  Work with a partner.

Your Turn

- Now use cubes to build a cube that has sides three times as long as the original cube.
- How many small cubes did you use?

5. What is the area of one face of the cube?
6. What is the volume of the cube?
7. Complete the table below.

Scale Factor	Length of a Side	Area of a Face	Volume
1			
2			
3			

8. Study the table. What happens to the area of a face when the length of a side is doubled? tripled?
9. If the scale factor is x, what is the area of the face?
10. What happens to the volume of a cube when the length of a side is doubled? tripled?
11. Let the scale factor be x. Write an expression for the cube's volume.
12. What would be the area of a face and the volume of a cube with sides 4 times as long as the original cube?

CHAPTER 12 Highlights

Vocabulary

After completing this chapter, you should be able to define each term, property, or phrase and give an example or two of each.

Geometry

altitude (pp. 612, 614, 633)
area of a circle (p. 620)
area of a parallelogram (p. 613)
area of a trapezoid (p. 614)
area of a triangle (p. 613)
base (pp. 614, 632)
circular cone (p. 639)
cylinder (p. 633)
face (p. 632)
lateral surface (p. 639)
net (p. 632)
prism (p. 632)
pyramid (p. 638)
rectangular prism (p. 632)
scale factor (p. 654)
slant height (p. 638)
square pyramid (p. 638)
surface area (p. 632)
triangular prism (p. 632)
triangular pyramid (p. 638)
vertex (p. 638)
volume (p. 644)
volume of a circular cylinder (p. 645)
volume of a cone (p. 650)
volume of a prism (p. 645)
volume of a pyramid (p. 649)

Probability

geometric probability (p. 623)

Problem Solving

make a model or a drawing (p. 629)

Understanding and Using Vocabulary

Choose the letter of the term or terms that best completes each statement.

A. altitude
B. cubic centimeter
C. perpendicular
D. prism
E. pyramid
F. slant height
G. square centimeter
H. surface area
I. volume

1. The _?_ is a common unit of measure for area.
2. The _?_ is a common unit of measure for volume.
3. The _?_ of a solid is the sum of the areas of its faces.
4. _?_ is the amount of space that a solid contains.
5. The _?_ of a parallelogram is a segment perpendicular to the bases with endpoints on the bases.
6. The _?_ of a pyramid is the length of an altitude of one of its lateral faces.
7. The height of a pyramid is the length of a segment from the vertex _?_ to the base of the pyramid.
8. A _?_ is named by the shape of its base.

CHAPTER 12 Study Guide and Assessment

Skills and Concepts

Objectives and Examples

Upon completing this chapter, you should be able to:

- **find areas of parallelograms, triangles, and trapezoids** (Lesson 12-1)

Find the area of the trapezoid.

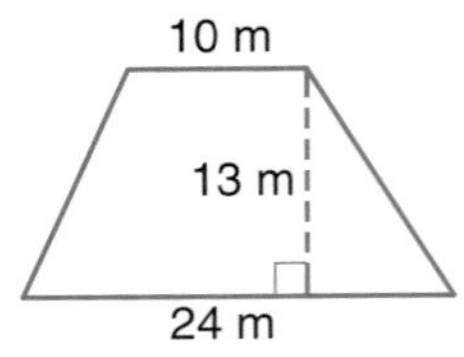

$A = \frac{1}{2}h(a + b)$

$A = \frac{1}{2} \cdot 13(10 + 24)$

$A = 221 \text{ m}^2$

Review Exercises

Use these exercises to review and prepare for the chapter test.

Find the area of each figure.

9. **10.**

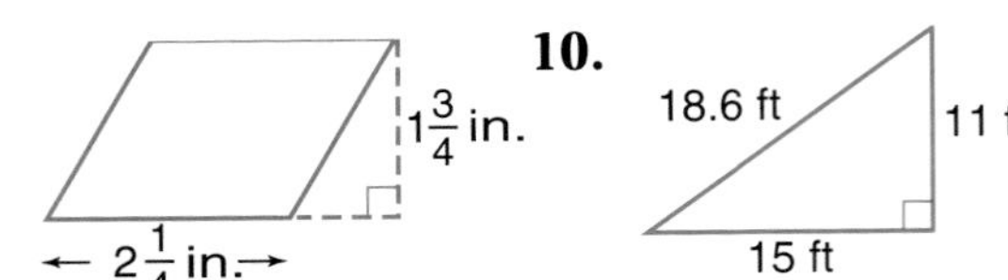

11.

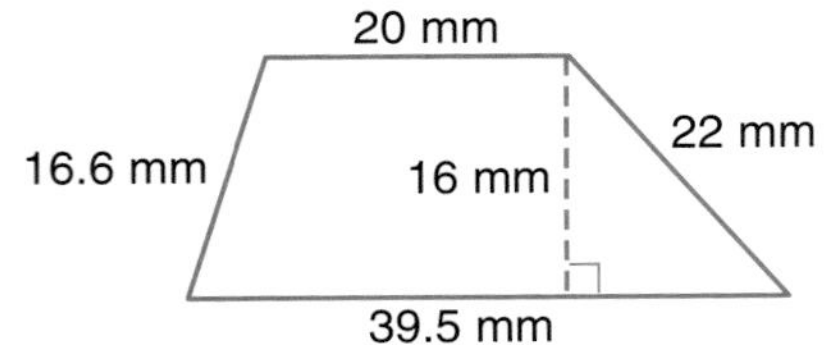

- **find the area of circles** (Lesson 12-2)

Find the area of the circle.

$A = \pi r^2$

$A = \pi \cdot (3)^2$

$A \approx 28.3 \text{ in}^2$

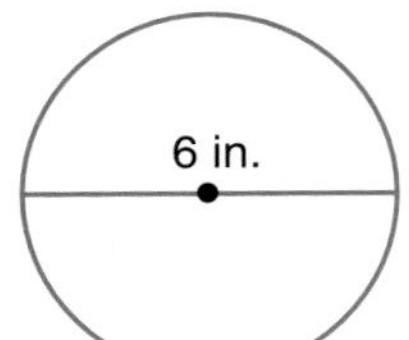

Find the area of each circle described below. Round answers to the nearest tenth.

12.

13. diameter, 4.2 m

- **find probabilities using area models** (Lesson 12-3)

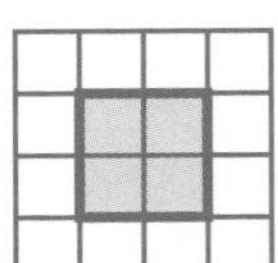

The probability of landing in the shaded region is $\frac{1}{4}$.

14. The figure below represents a dartboard. Find the probability of landing in the shaded region.

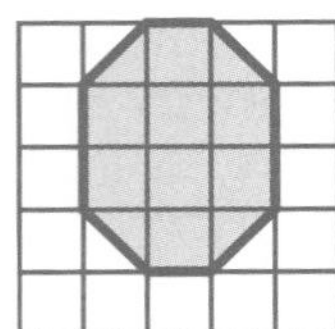

Objectives and Examples	Review Exercises

solve problems using a model or a drawing (Lesson 12-4)

Making a model or drawing can help you visualize a problem so you can solve it.

Solve by making a drawing.

15. The radius of a circular flower garden and its border is 2.8 meters. Without the border, the radius is 2.3 meters. Find the area of the border.

find the surface area of rectangular prisms and circular cylinders (Lesson 12-5)

Find the surface area of the prism.

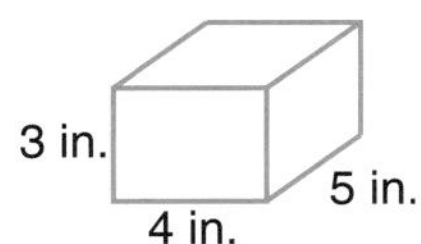

Top and bottom: $2 \times 4 \times 5 = 40$
Front and back: $2 \times 3 \times 4 = 24$
Sides: $2 \times 3 \times 5 = 30$
Surface area $= 40 + 24 + 30$ or 94 in^2

Find the surface area of each figure.

16.

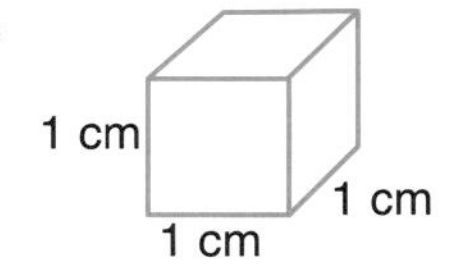

17.

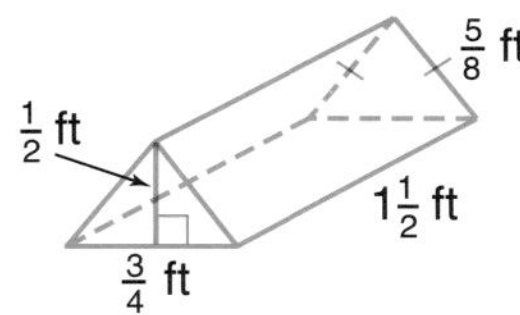

18. circular cylinder: height, 2.3 m; radius, 1 m
19. circular cylinder: height, 41 ft; diameter, 40 ft

find the surface areas of pyramids and cones (Lesson 12-6)

Find the surface area of a cone with radius 5 inches and height 8 inches.

Base:
$A = \pi r^2$
$A = \pi \cdot (5)^2$
$A \approx 78.5$

Curved Surface:
$A = \pi r \ell$
$A = \pi \cdot 5 \cdot 8$
$A \approx 125.7$

Total surface area: $78.5 + 125.7$ or 204.2 in^2

Find the surface area of each figure. Round decimal answers to the nearest tenth.

20.

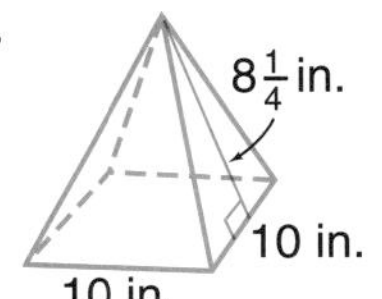

21.

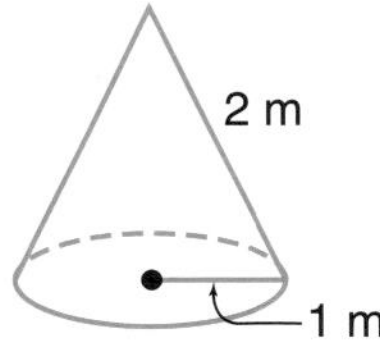

find the volumes of prisms and circular cylinders (Lesson 12-7)

Find the volume of the cylinder.

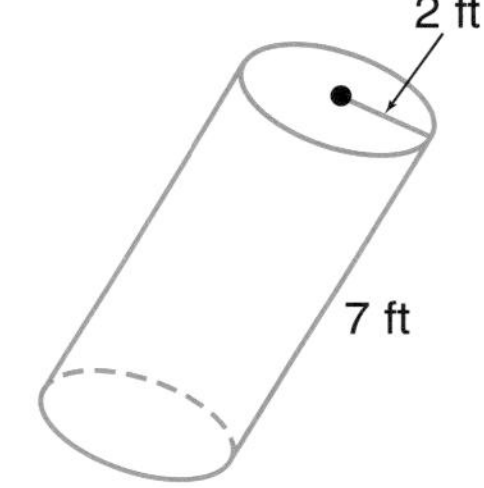

$V = \pi r^2 h$
$V = \pi \cdot (2)^2 \cdot 7$
$V = 88.0 \text{ ft}^3$

Find the volume of each figure described below. Round decimal answers to the nearest tenth.

22. triangular prism: base of triangle, 5 cm; height, 3.4 cm; prism height, 12 cm
23. hexagonal prism: area of base, 166.25 in^2; height, 20 in.
24. rectangular prism: length, 4 m; width, 4 m; height, 16 m
25. cylinder: radius, 7 ft; height, 18 ft
26. cylinder: diameter, 10 m; height, 22 m

Objectives and Examples	Review Exercises

- **find the volumes of pyramids and cones** (Lesson 12-8)

Find the volume of the pyramid.

$V = \frac{1}{3}Bh$

$V = \frac{1}{3} \cdot (3 \cdot 3) \cdot 4$

$V = 12\text{m}^3$

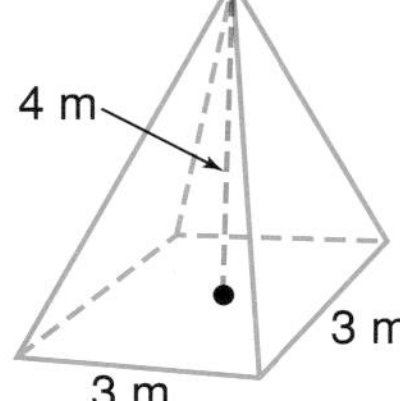

Find the volume of each figure. Round decimal answers to the nearest tenth.

27.

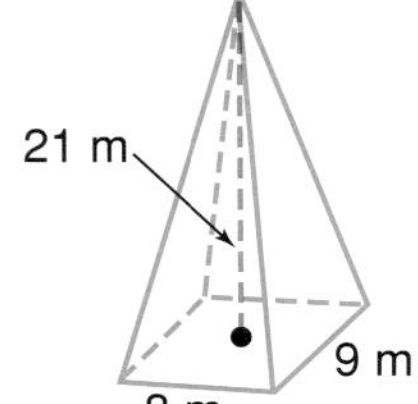

28.

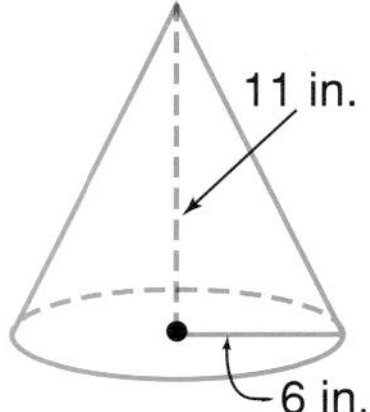

Applications and Problem Solving

29. Entertainment The Eastside High School Drill Team is having new flags made. Some of the flags will be shaped like triangular pennants. The base of the pennants will be 36 inches and the height will be 48 inches. What is the area of one of these flags? (Lesson 12-1)

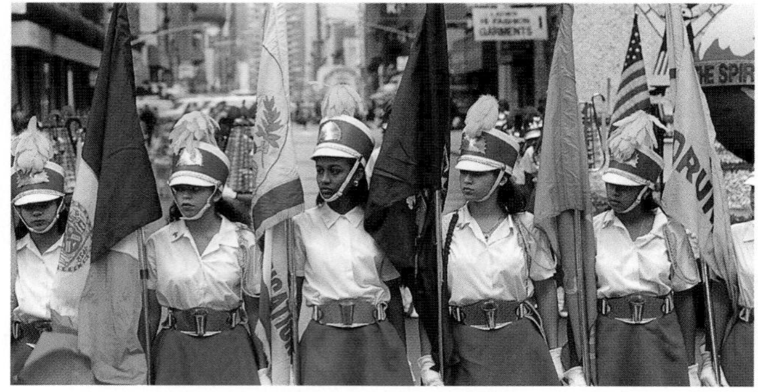

30. Agriculture A silo has a cylindrical shape. The radius of the base is about 20 feet, and the silo is 60 feet high. A company has been hired to paint the silo. (Lesson 12-5)

a. What is the area of the curved surface?

b. How many gallons of paint would be needed to apply one coat of paint to the silo if one gallon of paint covers about 450 square feet?

c. How much would it cost to paint the silo if one gallon costs $10?

d. How much would it cost to apply two coats of paint to the silo?

31. Manufacturing A leading manufacturer of sugar cubes packs a rectangular box so that there are six cubes along one edge, eleven cubes along a second edge, and three cubes along the third edge. How many sugar cubes are in the box? (Lesson 12-7)

32. Science Jenny knows that 1 mL of water has a volume of 1 cm^3. She needs 225 mL of water for an experiment. She has an unmarked cylindrical container with a diameter of 12 cm. To what height should she fill the container to measure the water for the experiment? (Lesson 12-7)

33. Agriculture A water trough is half of a circular cylinder with a radius of 6 feet and a height of 3 feet. If one gallon of water has a volume of about 230 in^3, about how many gallons of water will the trough hold? (Lesson 12-7)

34. Mining An open pit mine in the Elk mountain range is shaped like a cone. The mine is 420 feet across and 250 feet deep. What volume of material was removed? (Lesson 12-8)

A practice test for Chapter 12 is available on page 785.

Alternative Assessment

Cooperative Learning Project

Work with a group.

1. Research how Egyptian pyramids were constructed and what tools were used.
2. Choose one pyramid. Find the dimensions of that pyramid.
3. Plan a way to make a scale model of the pyramid. Include the following information in your plan.
   a. Decide how big to make the model.
   b. Determine the scale factor for the model.
   c. Select materials for making the model.
4. Construct the model.
5. Prepare a report on your findings, your procedures, and any problems you encountered. Include information on the surface area and volume of the original pyramid and of the model. You may wish to make a poster or brochure to accompany your model.

Thinking Critically

- The edge of a cube and the diameter and height of a cylinder all have the same measure. Which has the least surface area? Which has the least volume?
- Suppose you are designing a solid figure with two congruent parallel bases and the distance around the base and the height must be the same for whatever solid you design. What shape base would give the greatest surface area? the greatest volume?
- Design three different nets that could be used to make a rectangular prism 4 cm long, 3 cm wide, and 2 cm high. Which net will allow you to cut the greatest number of patterns from a 20-cm × 30-cm grid?

Portfolio

Select one of the assignments from this chapter that you found especially challenging and place it in your portfolio.

Self Evaluation

Good problem solvers often break a problem into several smaller steps that allow them to solve the entire problem.

Assess yourself. Are there any daily problems that you have solved by tackling them one step at a time? Write a paragraph about your experience. How can this strategy be applied to problems about surface area or volume?

CHAPTERS 1–12 Ongoing Assessment

Section One: Multiple Choice

There are eleven multiple-choice questions in this section. After working each problem, write the letter of the correct answer on your paper.

1. **Geometry** What is the area of the trapezoid?

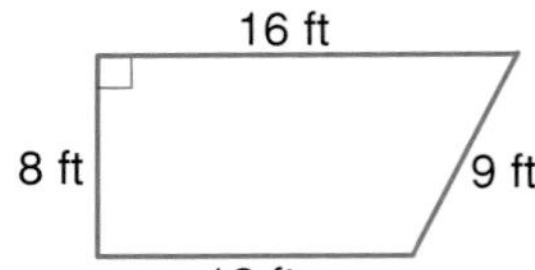

A. 45 ft^2
B. 64 ft^2
C. 96 ft^2
D. 112 ft^2

2. **Sports** Sandy's handicap score is found using the formula: $s = g + c$, where s is the handicap score, g is the game score, and c is the handicap. What was Sandy's game score if her handicap score is 186 and her handicap is 15?

A. 161 B. 171
C. 191 D. 201

3. **Geometry** A cylindrical can is 12 cm tall, and the ends each have a diameter of 10 cm. Which expression can be used to find the area the can label covers?

A. $\pi \times 10 \times 12$
B. $\pi \times 5 \times 5 \times 12$
C. $2(\pi \times 5^2) + \pi \times 10 \times 12$
D. $\pi \times 5^2 + 2(\pi \times 10 \times 12)$

4. Which is equivalent to $a^5 \cdot a^2$?

A. $2a^{10}$ B. $2a^7$
C. a^{10} D. a^7

5. **Geometry** What is the volume of the triangular prism?

A. 3600 cm^3
B. 4800 cm^3
C. 30,000 cm^3
D. 36,000 cm^3

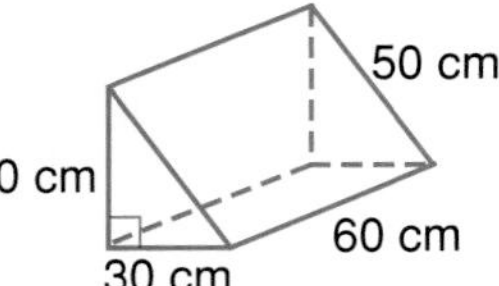

6. Which fractions and decimals are in order from least to greatest?

A. $0.5, \frac{1}{3}, \frac{1}{4}$ B. $\frac{3}{5}, 0.7, \frac{9}{12}$
C. $0.75, \frac{5}{8}, \frac{9}{10}$ D. $\frac{5}{6}, \frac{5}{8}, 0.5$

7. Lee is practicing for a 2500-meter run. He runs the distance twice a day for 5 days. How many kilometers has he run?

A. 25 km B. 25,000 km
C. 250 km D. none of these

8. **Geometry** Which shows a step in constructing an angle congruent to a given angle?

A.
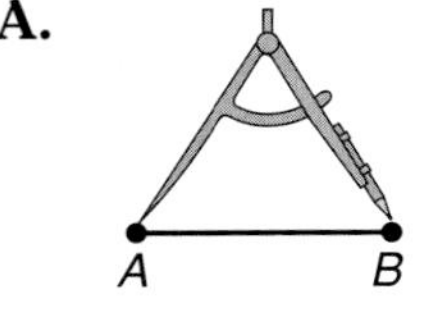

B.
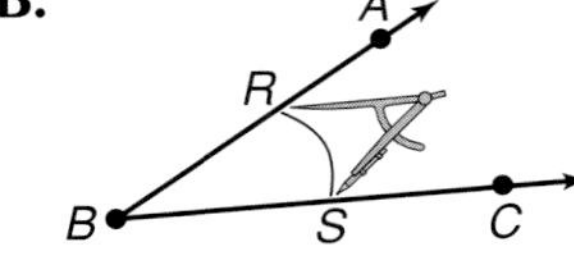

C.
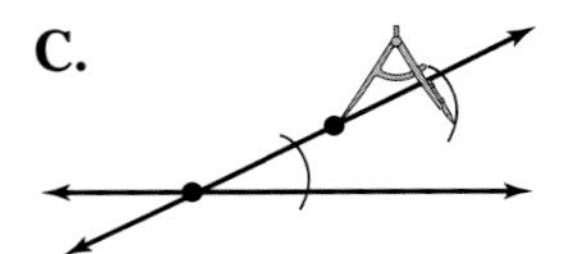

D.
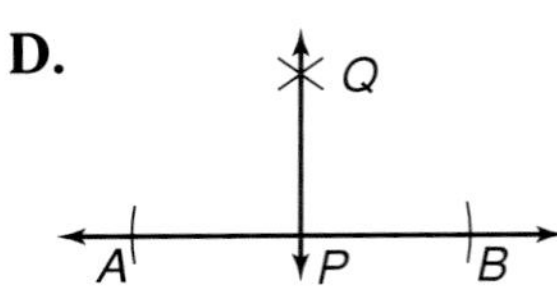

9. In how many different ways can all four students be arranged in a row?

Suki Janet Marcus Bill

A. 4 ways B. 12 ways
C. 24 ways D. 36 ways

10. **Geometry** $\triangle QRS$ is similar to $\triangle XYZ$. What is the length of $\overline{RQ}$?

A. 16 units
B. 18 units
C. 20 units
D. 22 units

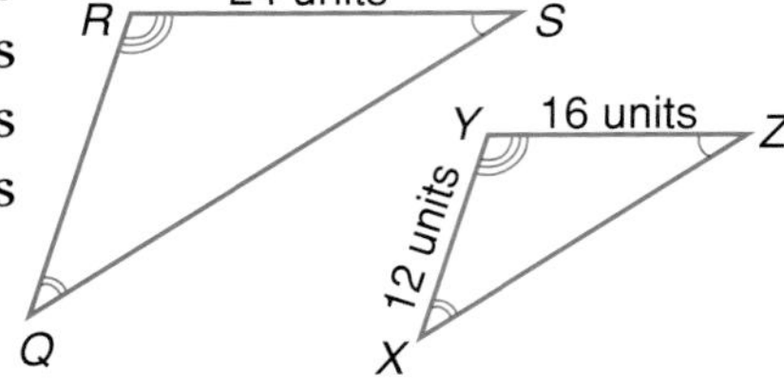

11. Find the circumference of a circle that has a diameter 1.2 meters long.

A. 4.5 meters
B. 7.5 meters
C. 3.8 meters
D. 1.1 meters

Section Two: Free Response

This section contains eight questions for which you will provide short answers. Write your answer on your paper.

12. Geometry Find the value of x. Then classify the triangle as acute, obtuse, or right.

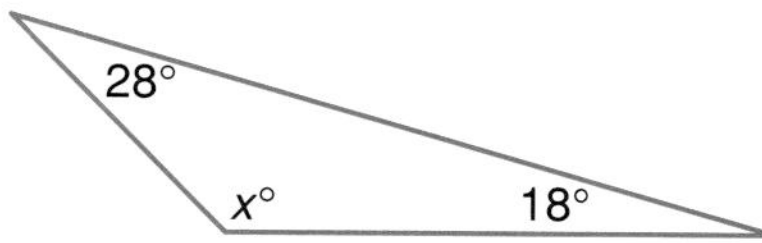

13. A sweatshirt is on sale for \$16. This is 80% of the regular price. What is the regular price?

14. A rectangular swimming pool is 6 meters wide and 12 meters long. A concrete walkway is poured around the pool. The walkway is 1 meter wide and 0.2 meters deep. What is the volume of the concrete?

15. On a trip, the Changs drive 104 miles in 2 hours. If they continue at the same rate, how long will it take them to complete a 550 mile trip?

16. From a 52-card deck, Susanna draws 2 cards. Find the probability they will both be kings.

17. Geometry In the figure below, ℓ is parallel to m. If the measure of $\angle 1$ is 108°, find the measure of the remaining angles.

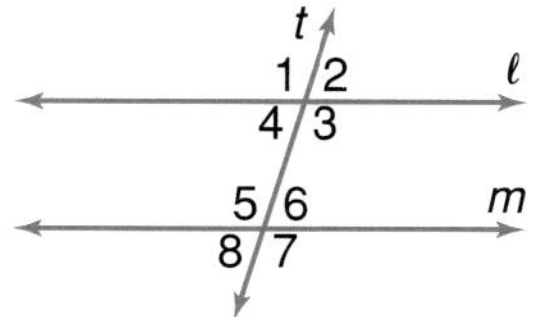

18. Geometry Find the value of x.

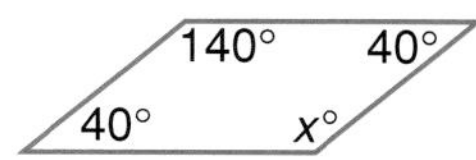

Test-Taking Tip

Although the basic formulas that you need to answer the questions on a standardized test are often given to you in the test booklet, it is a good idea to review the formulas beforehand.

Area of a circle = πr^2
Circumference of a circle = $2\pi r$
Area of a square = s^2
Perimeter of a rectangle = $2(\ell + w)$
Area of a rectangle = ℓw

19. Sports The Los Angeles Lakers scored 52 points in the first half of a recent game against the Phoenix Suns. Phoenix averages 106 points scored per game. Write and solve an inequality that shows how many points the Lakers need to score in the second half to beat the Suns' average points scored per game.

Section Three: Open-Ended

This section contains two open-ended problems. Demonstrate your knowledge by giving a clear, concise solution to each problem. Your score on these problems will depend upon how well you do the following.

- Explain your reasoning.
- Show your understanding of the mathematics in an organized manner.
- Use charts, graphs, and diagrams in your explanation.
- Show the solution in more than one way or relate it to other situations.
- Investigate beyond the requirements of the problem.

20. Probability Give two examples where geometric probability can be used.

21. Geometry Explain what the difference is between a line, a line segment, and a ray.

CHAPTER 13

Applying Algebra to Right Triangles

TOP STORIES in Chapter 13

In this chapter, you will:

- find squares and square roots,
- solve equations by finding square roots,
- use the Pythagorean theorem,
- use trigonometric ratios to solve problems, and
- use Venn diagrams to solve problems.

MATH AND TECHNOLOGY IN THE NEWS

My how time flies!

Source: New Scientist, April, 1995

The coming of the year 2000 is making people think about the changes that have taken place in the world in this century and the changes that will take place in the next one. But even though life is very different now than it was one hundred years ago, many computers and appliances with clocks will "think" it's 1900 again. At the stroke of midnight on January 1, 2000, these internal clocks will change from "99" to "00." And unless the programs are changed, that means that the systems will treat the year as 1900. That could mean chaos at banks that determine interest based on time, insurance companies that set rates using their customers' ages, and any other place where the data is affected by time. Experts estimate that the cost of making the necessary software changes before 2000 will cost up to $500 million!

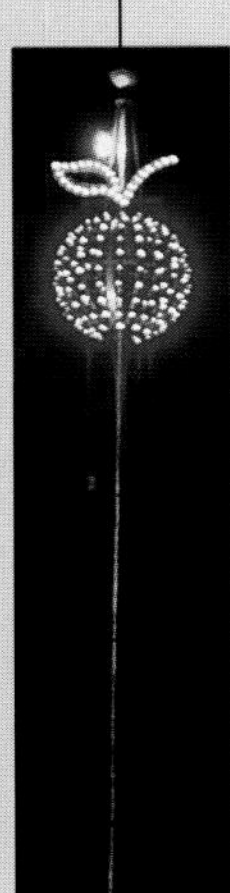

Putting It into Perspective

8000 B.C.
A calendar with 12 months of 30 days each is developed in Egypt.

1200
A calendar with 18 months of 20 days is developed in Mesomenia by the Olmecs of Mexico.

46
Julius Caesar reforms the Roman calendar to use 365 days each year and 1 additional day every four years to adjust the calendar.

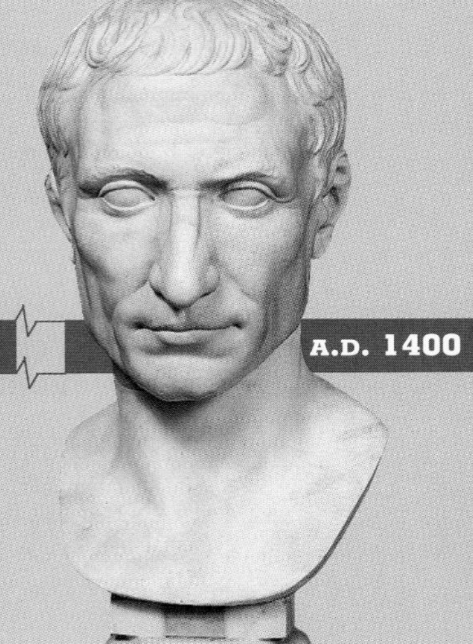

A.D. 1400

On the Lighter Side

"What time is it when the little hand is on two and the big hand is on the floor?"

interNET CONNECTION For up-to-date information on the calendar, visit:
www.glencoe.com/sec/math/prealg/mathnet

Statistical Snapshot

Comparative Scale of Time

Number of Seconds	Duration
10^{18}	Approximate age of the known universe (12 billion years)
10^{15}	Time since the age of the dinosaurs (135 million years)
10^{12}	Time for light from the center of our galaxy to reach Earth (about 30,000 years)
10^{9}	Human generation span (about 30 years)
10^{7}	Length of a school semester
10^{4}	Length of an average baseball game
10^{2}	Length of a television commercial break
10^{0}	One second
10^{-1}	Time of 1 vibration of lowest-pitched sound audible to humans
10^{-3}	Time for a midge to beat its wings once
10^{-7}	Time for an electron beam to go from source to screen of a television tube
10^{-11}	Time for a visible light wave to pass through a pane of glass
10^{-15}	Time for an electron to revolve once around proton in a Hydrogen molecule

1656
Dutch scientist Christian Huygens designs the first weight-driven clock controlled by a pendulum.

1948
Dr. Willard Frank Libby, professor of chemistry at the University of Chicago, invents the atomic clock.

1600 1800 2000

A.D. 1582
Pope Gregory XIII uses exact measurements for the solar year to revise the Julian calendar. The Gregorian calendar is the one in use in most of the world today.

1859
Big Ben, the tower clock famous for its accuracy and 13-ton bell, is installed.

1995
The computer industry addresses the year 2000.

13-1 Squares and Square Roots

Setting Goals: *In this lesson, you'll find and use squares and square roots.*

Modeling a Real-World Application: Ballooning

Have you ever wondered how far you can see out from an airplane or from the top of a hill? How far you can see depends on the curvature of Earth and your height above it. There is a formula that will help you to estimate how far you can see. The formula states that the view in miles V equals 1.22 times the square root of the altitude in feet A.

$$V = 1.22 \times \sqrt{A}$$

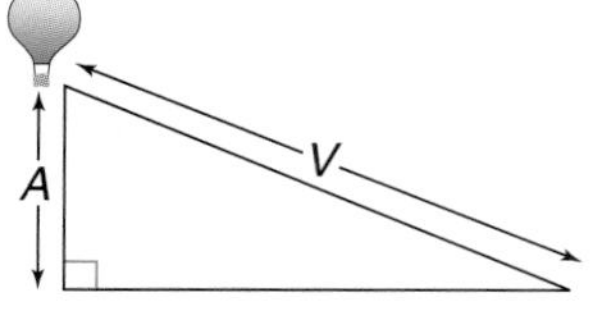

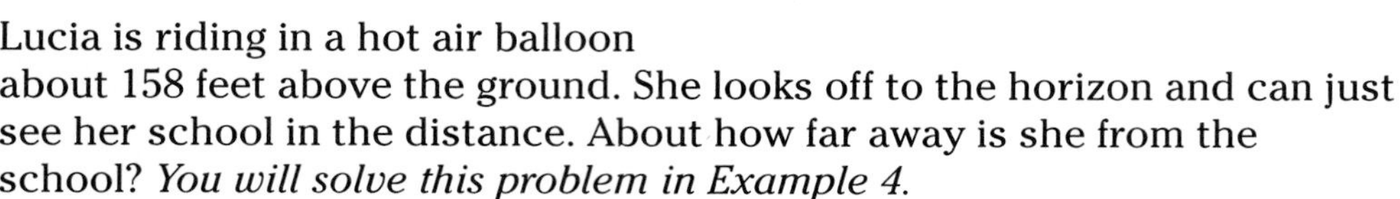

Lucia is riding in a hot air balloon about 158 feet above the ground. She looks off to the horizon and can just see her school in the distance. About how far away is she from the school? *You will solve this problem in Example 4.*

Learning the Concept

A **square root** is one of two equal factors of a number. For example, the square root of 49 is 7 since $7 \cdot 7$ or 7^2 is 49. It is also true that $-7 \cdot (-7) = 49$. So -7 is another square root of 49.

Definition of Square Root	
In words:	**The square root of a number is one of its two equal factors.**
In symbols:	**If $x^2 = y$, then x is a square root of y.**

The symbol $\sqrt{}$, called the **radical sign**, is used to indicate a nonnegative square root.

$\sqrt{49}$ indicates the nonnegative square root of 49.

$$\sqrt{49} = 7$$

$-\sqrt{49}$ indicates the negative square root of 49.

$$-\sqrt{49} = -7$$

Example  **1** **Find each square root.**

a. $\sqrt{25}$

The symbol $\sqrt{25}$ represents the nonnegative square root of 25. Since $5 \cdot 5 = 25$, $\sqrt{25} = 5$.

b. $-\sqrt{64}$

The symbol $-\sqrt{64}$ represents the negative square root of 64. Since $8 \cdot 8 = 64$, $-\sqrt{64} = -8$.

Squaring a number and finding the square root of a number are closely related to the square.

Connection to Geometry

The square shown at the right has sides that are 6 meters long. The area of the square is found by squaring *6*. If you know that the area of the square is 36 square meters, then the length of a side is found by finding the square root of 36.

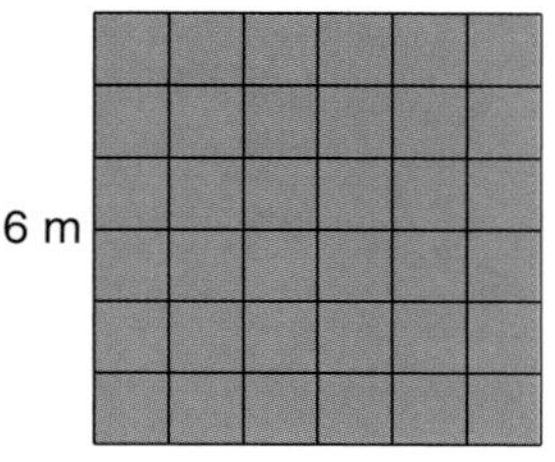

Example 2 **The area of a square is 100 square inches. Find its perimeter.**

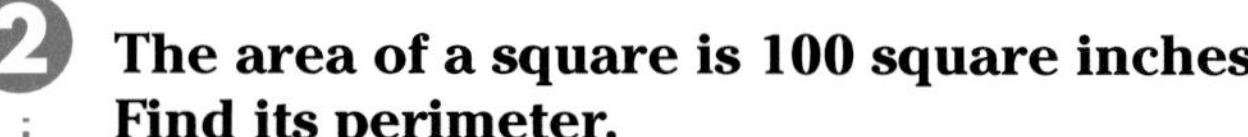

First, find the length of each side.

$A = s^2$

$100 = s^2$ *Replace A with 100.*

Area = 100 in^2

What number s when multiplied by itself is 100?

Both 10 and -10 when multiplied by themselves are 100. However, since length cannot be negative, s must be 10.

The length of each side is 10 inches.

Now find the perimeter.

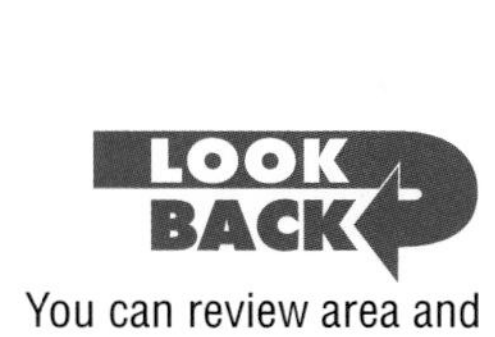

You can review area and perimeter in Lesson 3-5.

$P = 4 \cdot s$

$P = 4 \cdot 10$ *Replace s with 10.*

$P = 40$ The perimeter is 40 inches.

Numbers like 25, 49, and 64 are called **perfect squares** because when you take the square root you get an answer that is a whole number. What if the number is not a perfect square? How do you find the square root of a number like 125?

List some squares of numbers that are close to 125.

$10^2 = 100$ $11^2 = 121$ $12^2 = 144$

The number 125 is not a perfect square; that is, 125 has no square root that is a whole number. However, we know that 125 is greater than 121 or 11^2 and less than 144 or 12^2. So the square root of 125 should be greater than 11 and less than 12.

$$\sqrt{121} < \sqrt{125} < \sqrt{144}$$
$$\sqrt{11^2} < \sqrt{125} < \sqrt{12^2}$$
$$11 < \sqrt{125} < 12$$

Since 125 is closer to 121 than to 144, the best whole number estimate for $\sqrt{125}$ would be 11.

Example 3 Find the best integer estimate for each square root.

Some calculators have a *square root key* labeled  or $\sqrt{x}$. When you press this key, the number in the display is replaced by its nonnegative square root.

a. $\sqrt{59}$

49 and 64 are the closest perfect squares.

$\sqrt{49} < \sqrt{59} < \sqrt{64}$
$\sqrt{7^2} < \sqrt{59} < \sqrt{8^2}$
$7 < \sqrt{59} < 8$

Since 59 is closer to 64 than 49, the best integer estimate is 8.

b. $-\sqrt{38}$

36 and 49 are the perfect squares closest to 38.

$-\sqrt{49} < -\sqrt{38} < -\sqrt{36}$
$-\sqrt{7^2} < -\sqrt{38} < -\sqrt{6^2}$
$-7 < -\sqrt{38} < -6$

Since 36 is closer to 38 than 49, the best integer estimate is −6.

Many applications in science and other fields use square roots in important formulas.

Example 4 Refer to the application at the beginning of the lesson. How far away is Lucia from the school?

Explore You know that Lucia is about 158 feet above the ground. You also know that she can just see her school in the distance.

Plan Find the distance she is from the school. Use the formula $V = 1.22 \times \sqrt{A}$

Solve Substitute values into the formula.
$V = 1.22 \times \sqrt{A}$
$V = 1.22 \times \sqrt{158}$
Use your calculator.

Enter: 1.22 [×] 158 [$\sqrt{x}$] [=] 15.335162

Lucia is about 15.3 miles from the school.

Examine The closest perfect square to 158 is 169 or 13^2. $13 \times 1.22 = 15.86$. The answer is reasonable.

Checking Your Understanding

Communicating Mathematics

Read and study the lesson to answer each question.

1. **Explain** why a positive number has two different square roots.
2. **Draw** a square that has an area of 64 square centimeters.
3. **Explain** why 49 is a perfect square.
4. **Explain** why finding the square root and squaring are inverse operations.

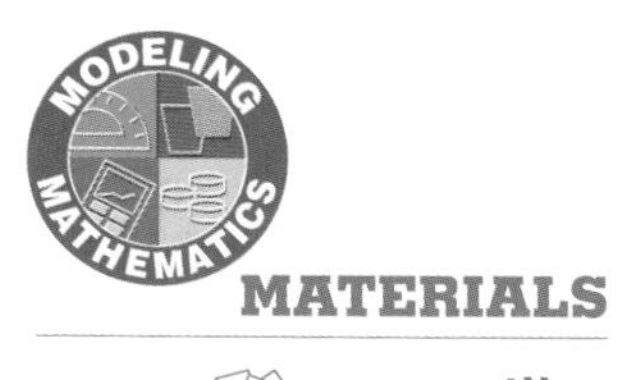

MATERIALS

square tiles

5. Arrange 40 tiles into the largest square possible.
   a. How many tiles did you use? How many are left over?
   b. Add tiles until you have the next larger square. How many tiles did you add?
   c. Between what two whole numbers is the square root of 40?

Guided Practice

Find each square root.

6. $-\sqrt{9}$ 7. $\sqrt{16}$ 8. $-\sqrt{4}$ 9. $\sqrt{81}$

Find the best integer estimate for each square root. Then check your estimate using a calculator.

10. $\sqrt{79}$ 11. $-\sqrt{53}$
12. $\sqrt{29}$ 13. $\sqrt{120}$
14. **Aviation** The British Airways Concorde flies from New York to London at an altitude of 60,000 feet. How far can the pilot see when he or she looks out the window?

Exercises: Practicing and Applying the Concept

Independent Practice

Find each square root.

15. $\sqrt{4}$ 16. $\sqrt{64}$ 17. $\sqrt{9}$ 18. $-\sqrt{25}$
19. $-\sqrt{16}$ 20. $-\sqrt{81}$ 21. $\sqrt{100}$ 22. $\sqrt{1}$
23. $-\sqrt{144}$ 24. $\sqrt{169}$ 25. $\sqrt{400}$ 26. $\sqrt{2.25}$

Choose

Estimation
Mental Math
Calculator
Paper and Pencil

Find the best integer estimate for each square root. Then check your estimate using a calculator.

27. $\sqrt{89}$ 28. $-\sqrt{44}$ 29. $\sqrt{200}$ 30. $-\sqrt{170}$
31. $-\sqrt{97}$ 32. $-\sqrt{6.76}$ 33. $\sqrt{2600}$ 34. $\sqrt{625}$
35. $\sqrt{118}$ 36. $\sqrt{13.69}$ 37. $-\sqrt{26.79}$ 38. $-\sqrt{156.25}$

Simplify.

39. $\sqrt{a^2}$ 40. $\sqrt{169} - (-\sqrt{121})$ 41. $\sqrt{\sqrt{81}}$

Critical Thinking

42. Enter a negative number into your calculator. Press the square root key. What is the result? Why?
43. In $x^2 = y$, could x be negative? Why or why not?

Applications and Problem Solving

44. **Sports** The ability of a hang glider to fly depends on the area of its wing. The relationship between wingspan and wing area is called the aspect ratio. The formula for the aspect ratio, R, is $R = \frac{s^2}{A}$, where s represents the wingspan and A represents the wing area.
   a. Solve $R = \frac{s^2}{A}$ for s to write a formula for wingspan when the aspect ratio and wing area are known.
   b. An advertisement for a hang glider says the aspect ratio is 2.7 and the wing area is 30 square feet. What is the wingspan of the glider?

45. **Construction** Bloomington city code requires that a party house must allow at least 4 square feet for each person on the dance floor. Reston's Hotel Restaurant is going to add a dance floor that is square and is large enough for 100 people. How long should it be on each side?

46. **Plumbing** Every water heater must have a pipe connecting it to the water meter supplied by the utility company. An engineer who is designing a new water heater has determined that the proper amount of water will be supplied if the opening of the pipe has an area of 0.442 square inches. Pipe is usually described in terms of its interior diameter. Use the formula for the area of a circle, $A = \pi r^2$, to find the proper size of pipe for the water heater.

Mixed Review

47. **Geometry** Find the volume of a rectangular pyramid with length 8 inches, width 8 inches, and height 9 inches. (Lesson 12-8)

48. How many ways can the letters of the word GYMNAST be arranged? (Lesson 10-6)

49. Express 28% as a fraction in simplest form. (Lesson 9-7)

50. **Carpentry** How many boards, each $2\frac{1}{2}$ feet long, can be cut from an 18-foot board? (Lesson 6-4)

51. Estimate 9.2×3.85. (Lesson 6-2)

52. Find the prime factorization of 2160. (Lesson 4-4)

53. Solve $a = -15 + (-5)$. (Lesson 2-4)

54. **Agriculture** Make a bar graph of the data about the fruit production in the table at the right. (Lesson 1-10)

interNET CONNECTION

For the latest U.S. fruit production, visit: **www.glencoe.com/sec/math/prealg/mathnet**

Annual Fruit Production

Fruit	Millions of Metric Tons
Grapes	60.7
Bananas	49.6
Apples	43.1
Coconuts	41.0
Plantains	26.8

Source: *Top Ten of Everything*

From the FUNNY PAPERS

1. What is wrong with Hobbes's reasoning?
2. Is the figure that Hobbes drew a square? Explain.
3. Write an explanation for Calvin and Hobbes of how to use square tiles to solve $6 + 3$.

13-2 Problem-Solving Strategy: Use Venn Diagrams

Setting Goals: *In this lesson, you'll use Venn Diagrams to solve problems.*

Modeling a Real-World Application: Surveys

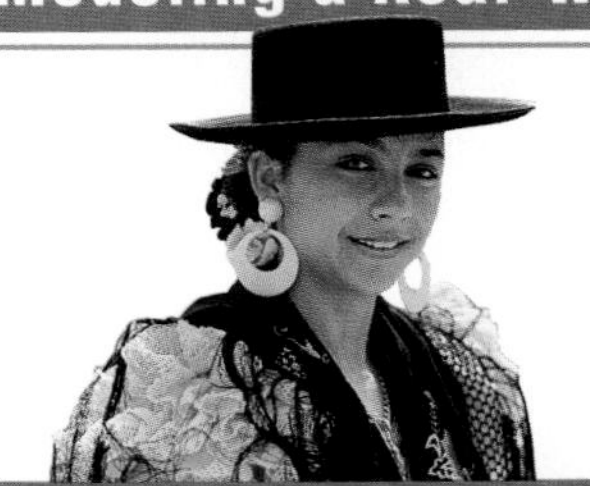

The guidance counselor surveyed 80 eighth-graders to find out their interest in learning Japanese or Spanish. Twenty-two students expressed an interest in both languages. Eighteen expressed interest in Japanese only, and five were not interested in learning any foreign language. How many of the students who were surveyed were interested in Spanish only?

Learning the Concept

You can use a **Venn diagram** to illustrate data. A rectangle is used to represent all the data. A circle inside the rectangle is used to represent one group of data. Data that is common to more than one group is represented by the region where intersecting circles overlap.

Foreign Language Preference

Spanish
Japanese
22
18
neither
5

Draw two intersecting circles in a rectangle to represent Japanese and Spanish. Write the number of students who are interested in both languages and the number of students who are interested in only Japanese in the appropriate regions. Write the number of students who are not interested in learning either language outside the circles but inside the rectangle. There are 22 + 18 + 5 or 45 students represented in the Venn diagram. There are 80 − 45 or 35 students that are interested in learning only Spanish.

Example

APPLICATION

Music

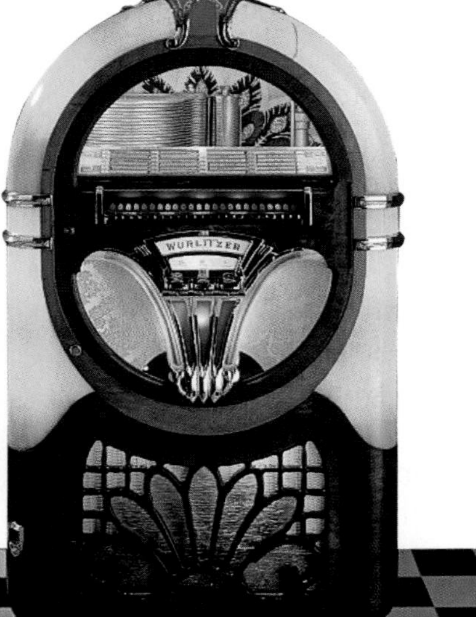

The Venn diagram at the right shows students' music preferences.

Music Preference

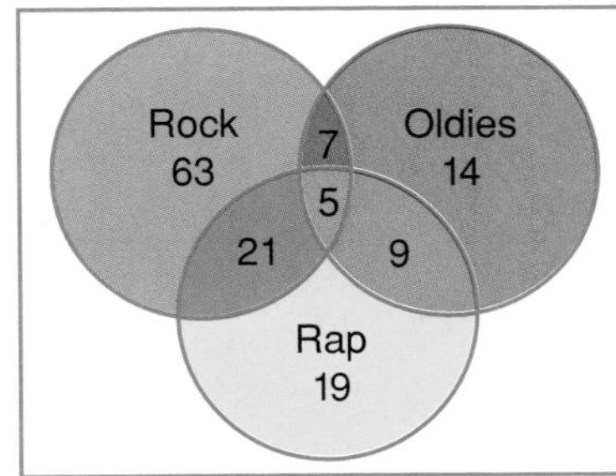

a. How many students like all three types of music?

The area where all three circles overlap contains a 5. There are 5 students who like all three types.

b. How many students like oldies?

There are 14 students who only like oldies, 9 who like oldies and rap, 7 who like oldies and rock, and 5 who like all three. There are 14 + 9 + 7 + 5 or 35 students who like oldies.

c. How many students like rap and rock?

The area where the circles for rap and rock overlap contains 5 and 21. There are 5 + 21 or 26 students who like rap and rock.

Checking Your Understanding

Communicating Mathematics

Read and study the lesson to answer each question. The Venn diagram at the right represents the states that produce more than 100 million bushels of corn, wheat, or soybeans per year.

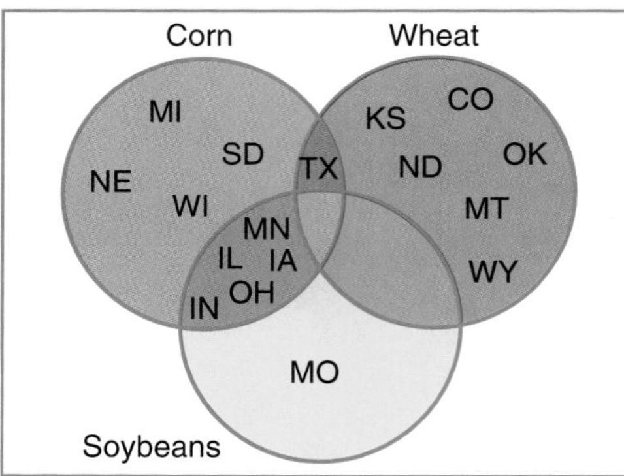

1. **Write** what the section where all three circles overlap represents.
2. **Determine** how many states produce more than 100 million bushels of corn or wheat.

Guided Practice

Solve using a Venn diagram.

3. At a birthday party, 9 people chose just cake for dessert, and 4 people chose just ice cream. Five people chose both cake and ice cream. If each person chose at least one dessert, how many people were at the birthday party?
4. **Recycling** The Venn diagram at the right shows the number of communities serviced by EarthCare Recycling that choose to have curbside recycling of glass, newspaper, and aluminum. How many communities does EarthCare service?

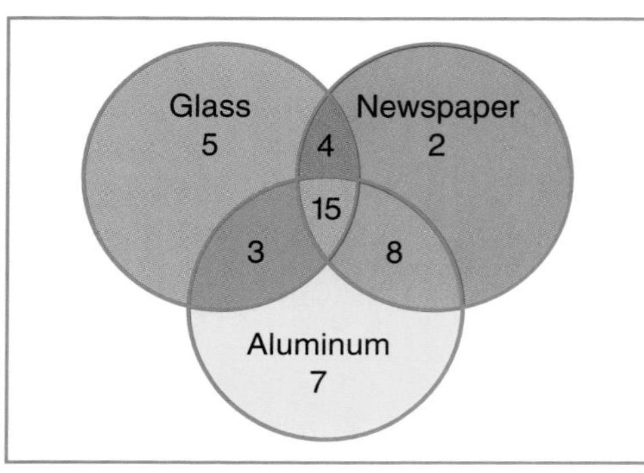

Exercises: Practicing and Applying the Concept

Independent Practice

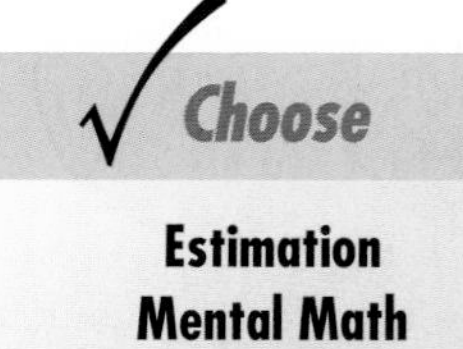

Estimation
Mental Math
Calculator
Paper and Pencil

Solve. Use any strategy.

5. **Geometry** There are 2 tiles. One is in the shape of a square, and the other is in the shape of an equilateral triangle. The sides of the square are congruent to the sides of the triangle. The tiles are put together to form a pentagon. If one side of the triangle is 20 cm long, what is the perimeter of the pentagon?
6. Of 150 TV viewers one evening, 64 watched MTV, 36 watched ESPN, and 10 watched both. How many did not watch either?
7. Chef Martino made a huge pan of lasagna for a banquet. He makes 6 cuts along the length of the rectangular pan and 10 cuts along the width. How many pieces does he have?
8. Jaria must read a book for a book report. She read half of her book on Tuesday. On Wednesday she read another 30 pages. On Thursday she read 6 more, and on Friday she read half of what was left. If she still has 20 pages yet to read, how many pages does the book have?
9. A school survey of 250 girls showed that 85 read *Seventeen* and 65 read *Elle*. 110 girls read neither. How many read both?

10. There are 26 students in a math class. The class takes a survey on pets and finds that 14 students have dogs, 10 students have cats, and 5 students have birds. Four students have dogs and cats, 3 students have dogs and birds, and 1 student has a cat and a bird. If no one has all three of these animals, how many students have none of these animals?

11. In a geography class of 30 students, 16 said they wanted to visit France, 16 wanted to visit Germany, and 11 wanted to visit Italy. Five said they wanted to visit both France and Italy, and of these, 3 wanted to visit Germany as well. Five wanted only Italy, and 8 wanted only Germany. How many students wanted France only?

Critical Thinking

12. **Geometry** A geometry teacher drew some quadrilaterals on the chalkboard. There were 5 trapezoids, 12 rectangles, 5 squares, and 8 rhombuses. What is the least number of figures the teacher could have drawn?

13. **Geometry** Draw a Venn diagram to show the relationship between squares and parallelograms. Let the rectangle represent polygons.

Mixed Review

14. Find $-\sqrt{121}$. (Lesson 13-1)

15. **Geometry** Draw an example of a rhombus and describe its characteristics. (Lesson 11-7)

16. **Finance** If \$800 is deposited in a savings account for six months at 5.75% annual interest, how much interest will it earn? (Lesson 9-9)

17. Solve $2x - 5 = 48$. (Lesson 7-2)

18. Solve $n - (-20) \geq 14$ and graph the solution on a number line. (Lesson 3-6)

WORKING ON THE Investigation

Refer to the Investigation on pages 606–607.

Look at the floor plan(s) of the house that you placed in your Investigation Folder at the end of Lesson 12-8. Use the floor plan(s) to make elevation drawings from four different directions. Be sure to use the same scale on the elevation drawing that you used on the floor plan. An example of an elevation drawing is shown at the beginning of Chapter 12.

Using your working drawings, construct a cardboard or foam board model of your house. It will probably be easier to work with a scale of $\frac{1}{2}$ inch = 1 foot instead of $\frac{1}{4}$ inch = 1 foot. (*Hint:* Cut doors and windows out before you erect the walls. You can also trace window openings on wax paper. Then draw in the frames and tape the wax paper on the inside wall.)

Make a roof for your model. Don't glue the roof on so that you can remove it to show the arrangement of the rooms.

Place your model on a base that is larger than the house. The excess can be used to show the yard, sidewalks, and driveway. You can also show trees and shrubs.

If possible, take a photograph of your model.

Add the results of your work to your Investigation Folder.

13-3 The Real Number System

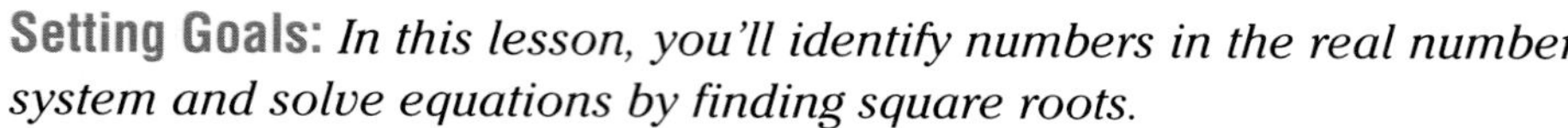

Setting Goals: *In this lesson, you'll identify numbers in the real number system and solve equations by finding square roots.*

Modeling a Real-World Application: Law Enforcement

HELP WANTED

Dedicated individuals are needed as law enforcement officers. If you are interested, contact:

National Fraternal Order of Police
2100 Gardiner Lane
Louisville, KY 40205

When investigating an accident, police officers often need to determine how fast a car was traveling before it skidded to a stop. They use the formula $s = \sqrt{30df}$, where s is the speed in miles per hour, d is the length of the skid marks in feet, and f is the coefficient of friction. What was the approximate speed of a car if the skid marks are 80 feet long and the coefficient of friction is estimated to be 1.2?

$s = \sqrt{30df}$

$s = \sqrt{30 \cdot 80 \cdot 1.2}$ *Replace d with 80 and f with 12.*

$s = \sqrt{2880}$

$s = 53.66563146\ldots$

The car was traveling at about 54 miles per hour.

Learning the Concept

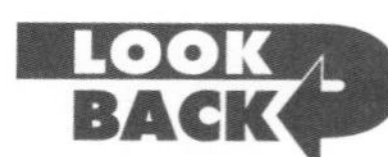

You can review rational numbers in Lesson 5-1.

Recall that numbers that can be expressed as $\frac{a}{b}$, where a and b are integers and b is not 0, are rational numbers. Rational numbers can always be expressed as decimals that terminate or repeat. For example, 7, $\frac{1}{3}$, and -4.9 are rationals. The number $\sqrt{2880}$, which is $53.66563146\ldots$, is not a repeating or terminating decimal. This kind of number is called an **irrational number**.

Definition of Irrational Number

An irrational number is a number that cannot be expressed as $\frac{a}{b}$, where a and b are integers and b does not equal 0.

Other examples of irrational numbers are shown below.

$\sqrt{3} = 1.7320508\ldots$ $\qquad \pi = 3.14159\ldots$

Example 

Determine whether each number is rational or irrational.

a. 0.22222 . . .

The three dots means that the 2s keep repeating. This is a repeating decimal, so it can be expressed as a fraction.

$0.22222\ldots = \frac{2}{9}$

Thus, it is a rational number.

b. 0.35

This is a terminating decimal. It can be expressed as $\frac{35}{100}$ or $\frac{7}{20}$. It is also a rational number.

c. 0.52522522252222 . . .

This decimal does not repeat or terminate and is therefore an irrational number. It does have a pattern to it, but since the 5s are separated by an increasing number of 2s, there is no exact repetition.

The set of rational numbers and the set of irrational numbers together make up the set of **real numbers**. The Venn diagram at the right shows the relationships among whole numbers, integers, rational numbers, irrational numbers, and real numbers.

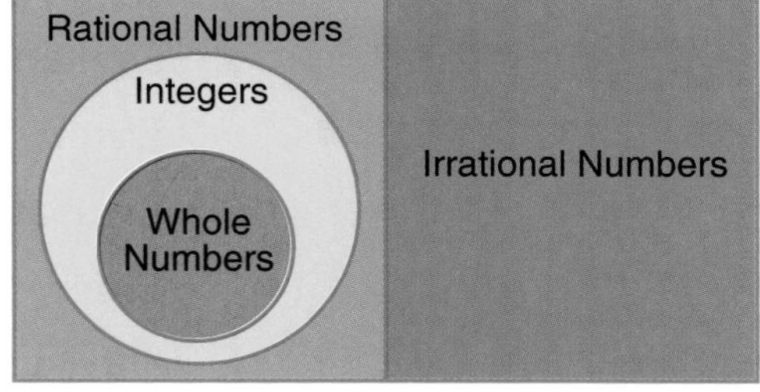

The following chart shows the sets of numbers to which several numbers belong.

Number	Whole Number	Integer	Rational	Irrational	Real
-9		✓	✓		✓
$\sqrt{35}$				✓	✓
$\sqrt{49}$	✓	✓	✓		✓
0.58585858 . . .			✓		✓
0.121231234 . . .				✓	✓
9	✓	✓	✓		✓

The equations you have encountered so far have had rational number solutions. Some equations have irrational number solutions.

You can solve some equations that involve squares by taking the square root of each side.

Example  **Solve each equation. Round decimal answers to the nearest tenth.**

a. $x^2 = 81$

$x^2 = 81$

$x = \sqrt{81}$ or $x = -\sqrt{81}$ *Take the square root of each side.*

$x = 9$ or $x = -9$

Check: $9^2 = 81$ and $(-9)^2 = 81$, so the solution checks.

b. $y^2 = 75$

$y^2 = 75$

$y = \sqrt{75}$ or $y = -\sqrt{75}$ *Take the square root of each side.*

75 [$\sqrt{x}$] 8.660254038

$y \approx 8.7$ or $y \approx -8.7$

The solutions to many real-world problems are irrational numbers.

Example 3

APPLICATION

Electricity

The voltage, V, in a circuit is given by the formula $V = \sqrt{PR}$. In this formula, V is in volts, P is the power in watts, and R is the resistance in ohms. An electrician has a circuit that produces 1500 watts of power. She wants the voltage in the circuit to be no more than 110 volts. Should she design the circuit with a resistance of a 7.5 ohms or 7.8 ohms?

For this circuit, $P = 1500$ and $V = 110$. To determine which resistance to use, you can use a calculator to evaluate $\sqrt{1500R}$ for $R = 7.5$ and for $R = 7.8$ to find out which one will produce no more than 110 volts.

Evaluate $\sqrt{1500R}$ for $R = 7.5$.

Enter: 1500 [×] 7.5 [=] [$\sqrt{x}$] 106.0660172

Evaluate $\sqrt{1500R}$ for $R = 7.8$.

Enter: 1500 [×] 7.8 [=] [$\sqrt{x}$] 108.1665383

She can design the circuit with *either* resistance. If she wants the voltage in the circuit to be as close to 110 volts as possible, then she should design the circuit with a resistance of 7.8 ohms.

Checking Your Understanding

Communicating Mathematics

Read and study the lesson to answer each question.

1. **Compare and contrast** rational and irrational numbers.
2. **You Decide** Jaya says that $\sqrt{16}$ can be both an integer and a rational number. Do you agree? Explain.
3. **Give an example** of an irrational number that is less than -10.

Guided Practice

Name the sets of numbers to which each number belongs: the whole numbers, the integers, the rational numbers, the irrational numbers, and/or the real numbers.

4. 0.3333 . . .
5. 0
6. $\sqrt{11}$
7. $-\frac{1}{2}$

Solve each equation. Round decimal answers to the nearest tenth.

8. $m^2 = 49$
9. $r^2 = 361$
10. $t^2 = 1$
11. $n^2 = 17$
12. **Hang Gliding** Using $R = \frac{s^2}{A}$, find the wingspan s of Kesse's hang glider if the aspect ratio R is 2.5 and the area of the wing A is 120 square feet.

Exercises: Practicing and Applying the Concept

Independent Practice

Name the sets of numbers to which each number belongs: the whole numbers, the integers, the rational numbers, the irrational numbers, and/or the real numbers.

13. 8
14. -5
15. $\frac{2}{3}$
16. $\sqrt{5}$
17. -3.5
18. $-\sqrt{25}$

19. 0.121121112... **20.** $\frac{5}{4}$ **21.** $\sqrt{18}$

22. $\frac{-9}{3}$ **23.** $2.\overline{5}$ **24.** 3.14359265...

Solve each equation. Round decimal answers to the nearest tenth.

25. $r^2 = 36$ **26.** $x^2 = 64$ **27.** $y^2 = 12$ **28.** $169 = m^2$

29. $n^2 = 120$ **30.** $f^2 = 200$ **31.** $180 = j^2$ **32.** $p^2 = 1.44$

33. $0.0004 = s^2$ **34.** $h^2 = 240$ **35.** $400 = q^2$ **36.** $a^2 = 90{,}000$

37. $(-b)^2 = 81$ **38.** $c^2 - 3^2 = \sqrt{16^2}$ **39.** $\sqrt{81} = d^2$

40. **Geometry** The length of a rectangle is three times its width. What are the dimensions of the rectangle if its area is 192 ft^2?

Critical Thinking

41. Can the product of two irrational numbers be rational? Support your answer with examples.

Applications and Problem Solving

FYI

In A.D. 60, Heron, an Egyptian mathematician and scientist, invented the *dioptra* for surveying and making astronomical observations. A dioptra could measure horizontal and vertical angles.

42. **Meteorology** You can use the formula $t^2 = \frac{d^3}{216}$ to estimate the amount of time that a thunderstorm will last. In this formula, t is the time in hours, and d is the diameter of the storm in miles. If a thunderstorm is 6 miles wide, how long will the storm last?

43. **Geometry** You can use Heron's formula to find the area of a triangle if you know the measures of its sides. If the measures of the sides are a, b, and c, the area A equals $\sqrt{s(s-a)(s-b)(s-c)}$, where s is one-half the perimeter. Suppose you have a triangle with sides 80 feet, 60 feet, and 100 feet long. Find the area of the triangle.

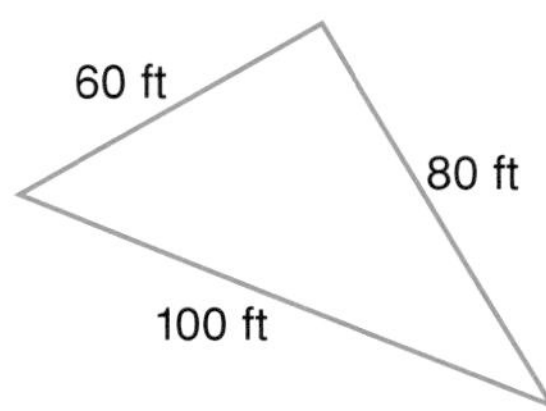

44. **Physics** If an object is dropped, the time t in seconds it takes to reach a given distance can be estimated by using the formula $d = 0.5gt^2$. In this formula, d is the free-fall distance, and g is the acceleration due to gravity, 32 ft/s^2. If Rachelle drops a ball from the top of a 55-foot building, how long does it take for the ball to hit the ground?

45. **Family Activity** Drop a ball from a second or third story window several times and time its fall to the ground. Find the average time. See how that time compares to the answer using the formula given in Exercise 44 by measuring the actual falling distance.

Mixed Review

46. **Statistics** A survey of 120 people showed that from 7:00 to 8:00 one evening, 46 people watched news programs, 34 watched game shows, and 15 watched both. How many did not watch either? (Lesson 13-2)

47. **Geometry** Find the volume of a circular cylinder with diameter 25 cm and height 10 cm. Round to the nearest tenth. (Lesson 12-7)

48. Graph $y = 2x + 3$. (Lesson 8-3)

49. Solve $v = \left(\frac{3}{7}\right)\left(-\frac{14}{15}\right)$. (Lesson 6-3)

50. Find the GCF of 36 and $-30ab$. (Lesson 4-5)

51. Evaluate $4cd$ if $c = -2$ and $d = -9$. (Lesson 2-7)

52. Simplify $7(a + b) - 2(3a + 4b)$. (Lesson 1-5)

13-4 The Pythagorean Theorem

Setting Goals: *In this lesson, you'll use the Pythagorean theorem to find the length of the side of a right triangle and to solve problems.*

Modeling with Manipulatives

MATERIALS

 dot paper

On dot paper, the horizontal or vertical distance between two adjacent dots is defined as one unit. Therefore, a square with sides 1 unit long has an area of 1 square unit.

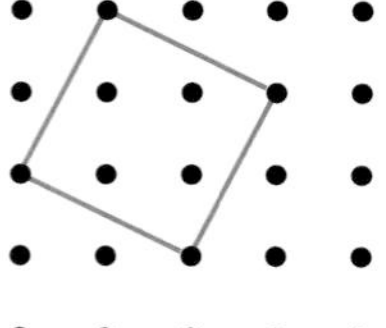

Your Turn **Work with a partner.**

▶ Draw a square like the one shown at the right.

TALK ABOUT IT

a. How can you prove that the area of the square is 5 square units?

Your Turn

▶ Use the side of the square to make a triangle. Then draw squares on the other two sides of the triangle.

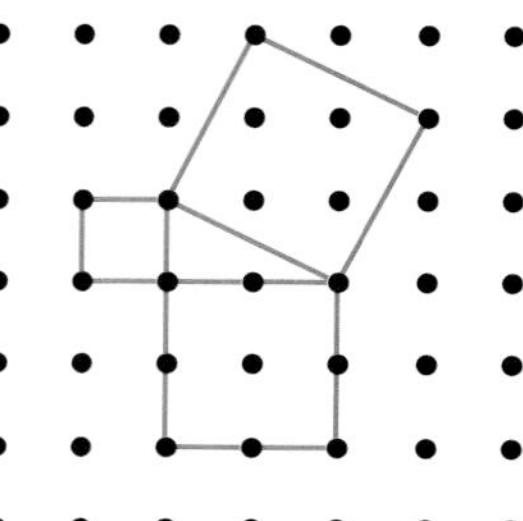

TALK ABOUT IT

b. Classify the triangle you drew.

c. What are the areas of the two smaller squares?

d. What is the relationship between the areas of the three squares?

e. If the area of the largest square is 5 square units, what is the length of the side of the triangle opposite the right angle?

Learning the Concept

In the movie *The Wizard of Oz*, with Judy Garland, when the scarecrow gets his brain he recites the Pythagorean theorem . . . but he says it incorrectly!

The sides of a right triangle that are adjacent to the right angle are called the **legs**. The side opposite the right angle is called the **hypotenuse**.

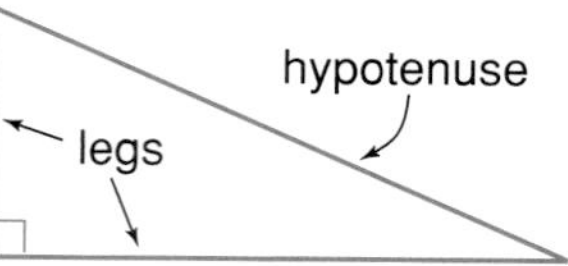

The relationship between the lengths of the legs and the hypotenuse that you noticed in the activity at the beginning of the lesson was proven in the fifth century B.C. by a Greek mathematician, Pythagoras, and his followers. This relationship is called the **Pythagorean theorem**. It is true for *any* right triangle.

Pythagorean Theorem

In words: In a right triangle, the square of the length of the hypotenuse is equal to the sum of the squares of the lengths of the legs.

In symbols: $c^2 = a^2 + b^2$

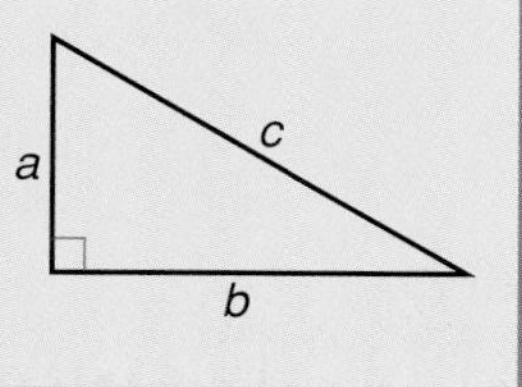

Example 1

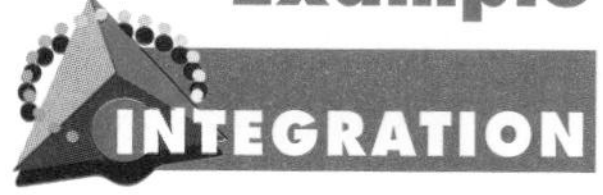

Geometry

A rectangle has sides of 6 and 8 inches. How long is a diagonal?

A diagonal and two adjacent sides of the rectangle form a right triangle. The legs of the right triangle are sides of the rectangle, and the hypotenuse of the right triangle is a diagonal of the rectangle.

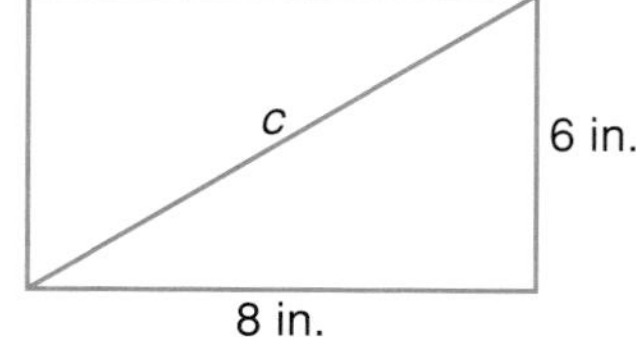

$c^2 = a^2 + b^2$ *Pythagorean theorem*
$c^2 = 6^2 + 8^2$ *Replace a with 6 and b with 8.*
$c^2 = 36 + 64$
$c^2 = 100$
$c = \sqrt{100}$ *Take the square root of each side.*
$c = 10$

The length of the diagonal is 10 inches.

THINK ABOUT IT

Why can you ignore the negative value of *c* in this problem?

The Pythagorean theorem can also be used to find the length of any side of a right triangle if the lengths of the other two sides are known.

Example 2 **Find the length of the third side of the right triangle.**

$c^2 = a^2 + b^2$
$27^2 = a^2 + 25^2$
$729 = a^2 + 625$
$729 - 625 = a^2 + 625 - 625$
$104 = a^2$
$\sqrt{104} = a$

104  10.19803903

The length of the leg is about 10.2 cm.

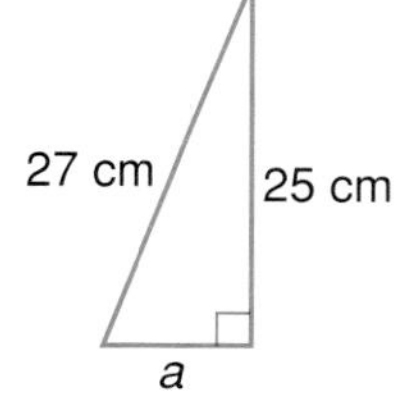

You can use the Pythagorean theorem to see whether a triangle is right.

Example 3 **The measurements of three sides of a triangle are given. Determine whether each triangle is a right triangle.** *Remember, the hypotenuse is the longest side.*

a. 5 ft, 7 ft, 8 ft

$c^2 = a^2 + b^2$
$8^2 \stackrel{?}{=} 5^2 + 7^2$
$64 \stackrel{?}{=} 25 + 49$
$64 \neq 74$

The triangle is *not* right.

b. 10 cm, 24 cm, 26 cm

$c^2 = a^2 + b^2$
$26^2 \stackrel{?}{=} 10^2 + 24^2$
$676 \stackrel{?}{=} 100 + 576$
$676 = 676$

The triangle is right.

The Pythagorean theorem can be applied to solve problems that occur in real life.

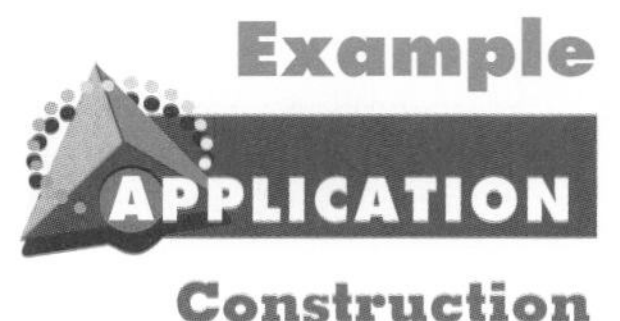

Example 4 **A Forest Service warehouse facility is used as a supply depot and a fire-fighting center. A security fence is needed to protect a helicopter, some tanker trucks, and other fire-fighting equipment that must be stored outdoors. A diagram of the region to be fenced is shown at the right. How much fencing must be purchased to surround the region?**

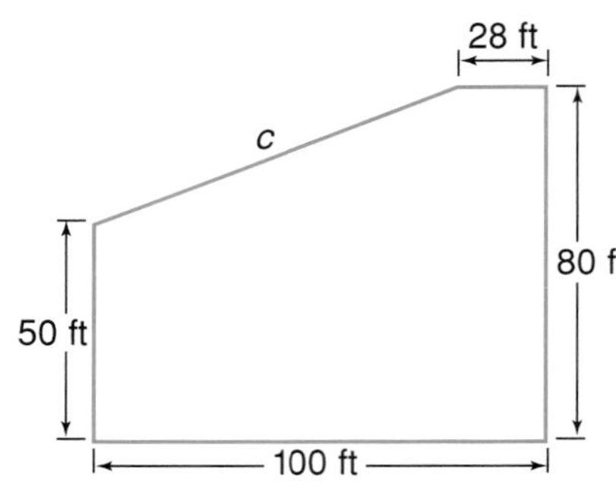

CULTURAL CONNECTIONS

Pythagoras is credited with the proof of the theorem that bears his name. However, the theorem may have been known to Egyptian, Babylonian, and Chinese scholars as much as one thousand years earlier.

Explore You know what the boundary of the region to be fenced looks like. You must find the perimeter of the region.

Plan To find the perimeter, you must find the unknown length, c. You can separate the region into more familiar shapes as shown below.

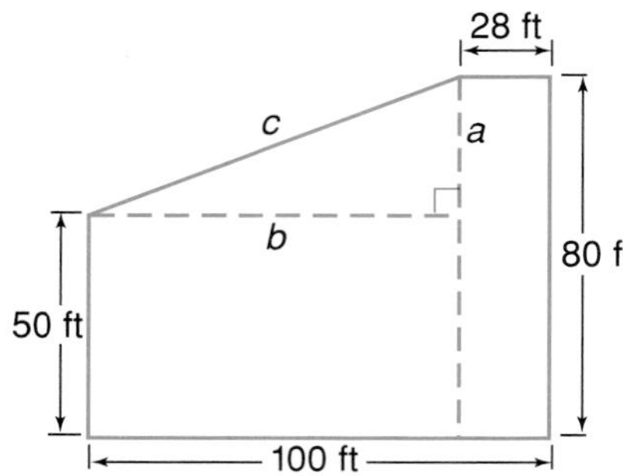

The hypotenuse of the right triangle is the side with length c. Find the measures of the legs of the right triangle and then use the Pythagorean theorem to find the length of the hypotenuse.

Solve First find the measures of the legs.

$a = 80 - 50$ $b = 100 - 28$
$a = 30$ $b = 72$

Next, apply the Pythagorean theorem.

$c^2 = a^2 + b^2$
$c^2 = 30^2 + 72^2$ *Replace a with 30 and b with 72.*
$c^2 = 900 + 5184$
$c^2 = 6084$
$c = \sqrt{6084}$ *Take the square root of each side.*
6084 [$\sqrt{x}$] 78
$c = 78$

Finally, find the perimeter.

$28 + 80 + 100 + 50 + 78 = 336$

So, 336 feet of fencing is needed.

Examine If the region included the corner area, it would be a rectangular shape. Its perimeter would be $2 \cdot 100 + 2 \cdot 80$ or 360 feet. The answer should be less than this. Thus, the answer is reasonable.

Checking Your Understanding

Communicating Mathematics

Read and study the lesson to answer each question.

1. **Explain** how to find the length of a leg of a right triangle if you know the length of the hypotenuse and the length of the other leg.
2. **Draw** and label a triangle with sides that measure 6 meters, 8 meters, and 10 meters. Is it a right triangle? Explain.
3. The roof of a house is 24 feet above the ground. You have a 26-foot ladder. Use the Pythagorean theorem to explain why the base of the ladder must be placed no more than 10 feet from the house in order to reach the roof.
4. Use dot paper to draw squares on each side of a triangle like the one at the right. Write an equation you could use to find the length of the hypotenuse.

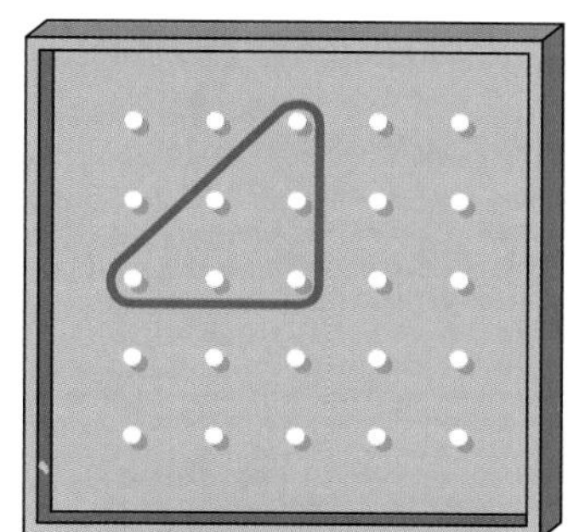

MODELING MATHEMATICS

MATERIALS

dot paper

Guided Practice

Write an equation you could use to solve for *x*. Then solve. Round decimal answers to the nearest tenth.

5.

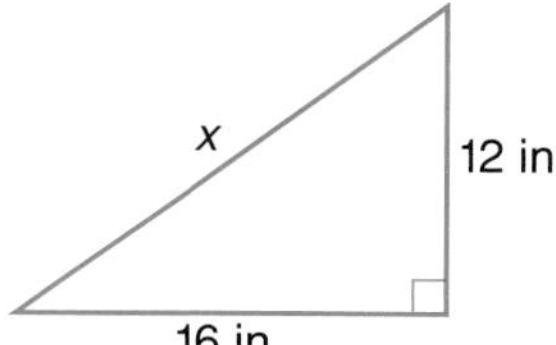

6.

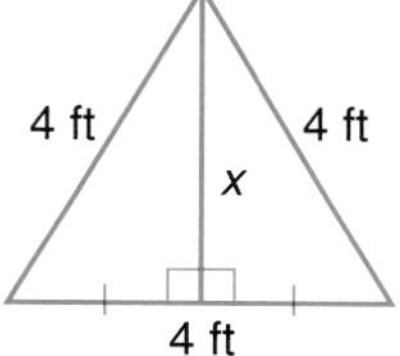

Write an equation that can be used to answer each question. Then solve. Round decimal answers to the nearest tenth.

7. How long is the lake?

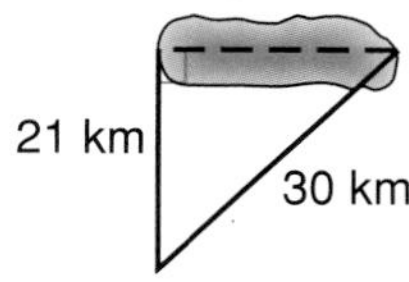

8. How high is the TV screen?

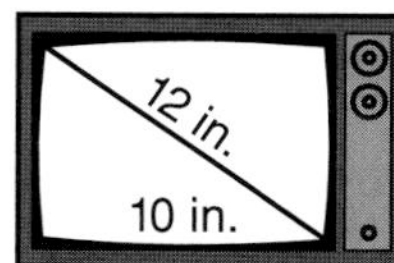

In a right triangle, if *a* and *b* are the measures of the legs and *c* is the measure of the hypotenuse, find each missing measure. Round decimal answers to the nearest tenth.

9. $a = 3, b = 4$
10. $a = 3, c = 7$
11. $b = 12, c = 35$
12. $a = 15, b = 16.7$

The measurements of three sides of a triangle are given. Determine whether each triangle is a right triangle.

13. 9 cm, 12 cm, 15 cm
14. 6 ft, 7 ft, 12 ft
15. **Baseball** A baseball diamond is actually a square. The distance between bases is 90 feet. When a runner on first base tries to steal second, a catcher has to throw from home to second base. How far is it between home and second base?

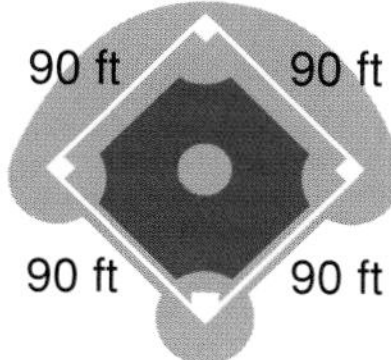

Exercises: Practicing and Applying the Concept

Independent Practice

Write an equation you could use to solve for x. Then solve. Round decimal answers to the nearest tenth.

16.

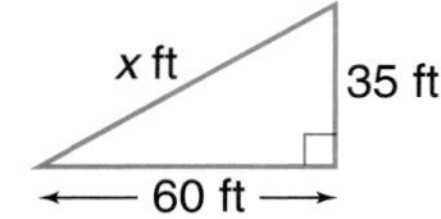

17.

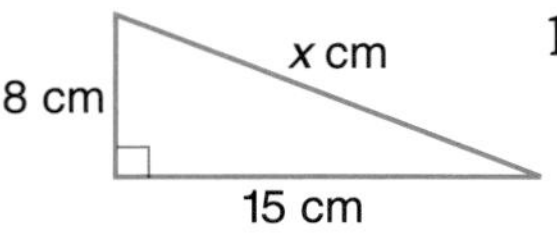

18.

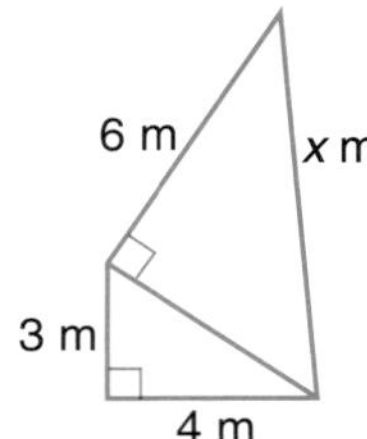

Write an equation that can be used to answer each question. Then solve. Round decimal answers to the nearest tenth.

19. How high is the kite?

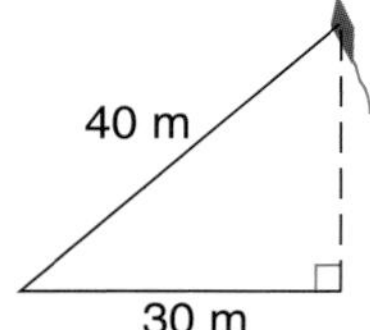

20. How long is a suspension line?

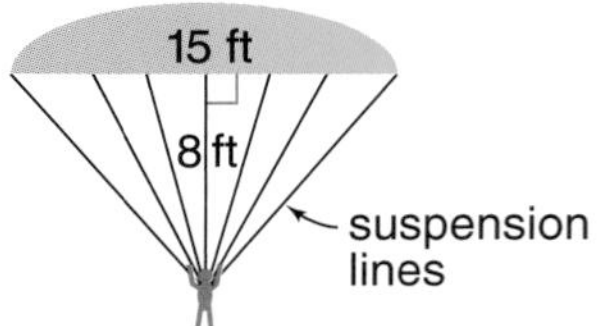

21. How high does the ladder reach?

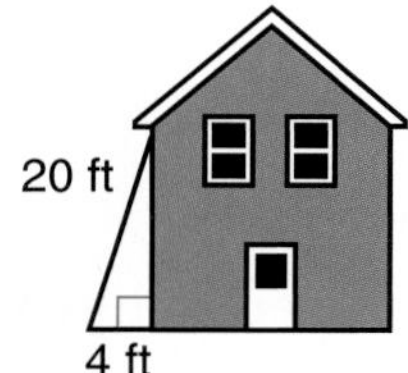

22. How far apart are the planes?

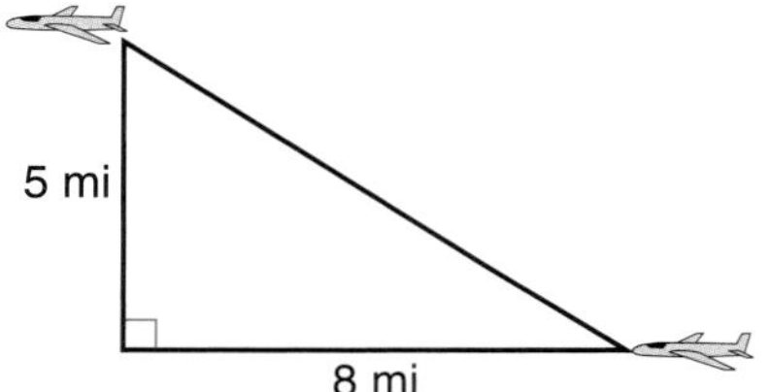

In a right triangle, if a and b are the measures of the legs and c is the measure of the hypotenuse, find each missing measure. Round decimal answers to the nearest tenth.

23. $a = 12, b = 16$

24. $b = 21, c = 29$

25. $a = 2, b = 5$

26. $a = 5, c = 10$

27. $a = 7, c = 9$

28. $b = 3, c = 7$

29. $a = 7, b = 7$

30. $b = 36, c = 85$

31. $a = 14, b = 15$

32. $a = 180, c = 181$

33. $a = \sqrt{11}, c = 6$

34. $b = 13, c = \sqrt{233}$

The measurements of three sides of a triangle are given. Determine whether each triangle is a right triangle.

35. 8 ft, 9 ft, 10 ft

36. 10 m, 24 m, 26 m

37. 15 in., 12 in., 9 in.

38. 6 mi, 7 mi, 8 mi

39. 18 cm, $\sqrt{24}$ cm, 30 cm

40. 15 ft, 16 ft, $\sqrt{31}$ ft

41. Geometry Find the length of a diagonal of a square if its area is 72 m^2.

42. Geometry Find the length of the diagonal of a cube if each side of the cube is 3 feet long.

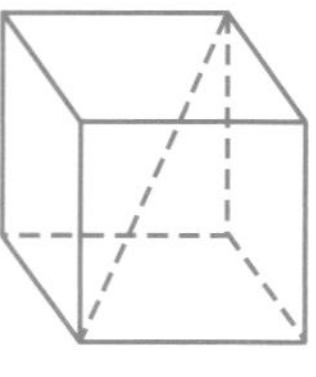

Critical Thinking

43. The points $C(5, 1)$ and $D(8, 5)$ are graphed at the right. Find the distance between C and D. (*Hint:* Notice the right triangle. First find the measures of $\overline{BC}$ and $\overline{BD}$.)

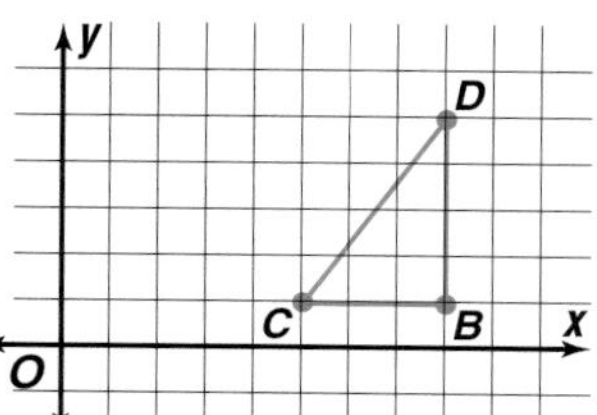

Applications and Problem Solving

Choose

Estimation
Mental Math
Calculator
Paper and Pencil

44. **Safety** The National Safety Council recommends placing the base of a ladder one foot from the wall for every three feet of the ladder's length.
    a. How far away from a wall should you place a 15-foot ladder?
    b. How high can the ladder safely reach?
45. **Hiking** Gracia hikes 10 kilometers south, 15 kilometers east, and then 10 km south again. How far is Gracia from the starting point of her hike?
46. **Sailing** A rope from the top of a mast on a sailboat is attached to a point 6 feet from the base of the mast. If the rope is 24 feet long, how high is the mast?

Mixed Review

47. Solve $x^2 = 2.5$ to the nearest tenth. (Lesson 13-3)
48. **Geometry** Draw and label a diagram to represent perpendicular lines k and m. (Lesson 11-1)
49. If $g(x) = -x + 4$, find $g(-1)$. (Lesson 8-4)
50. Solve $5 - 2n = 4n + 41$. (Lesson 7-5)
51. Solve $\left(\frac{2}{3}\right)^2 = x$. (Lesson 6-3)
52. **Sports** Willa swam 1 lap on the first day. She swam 2 laps on the second day, 4 laps on the third day, and 8 laps on the fourth day. To continue this pattern, how many laps should she swim on the seventh day? (Lesson 2-6)
53. Solve $6b = 120$ mentally. (Lesson 1-6)
54. Find the value of the expression $3[4(6 - 2) - 5]$. (Lesson 1-2)

Self Test

Find each square root. (Lesson 13-1)

1. $\sqrt{36}$
2. $\sqrt{121}$
3. $-\sqrt{100}$
4. In a recent survey of 120 students, 60 students said they play tennis and 50 students said they play softball. If 20 students play both sports, how many students do not play either tennis or softball? (Lesson 13-2)

Solve each equation. Round decimal answers to the nearest tenth. (Lesson 13-3)

5. $x^2 = 144$
6. $y^2 = 50$
7. $g^2 = 40$

In a right triangle, if a and b are the measures of the legs and c is the measure of the hypotenuse, find each missing measure. Round answers to the nearest tenth. (Lesson 13-4)

8. $a = 12, b = 16$
9. $b = 63, c = 65$
10. $a = 15, c = 39$

HANDS-ON ACTIVITY

13-4B Graphing Irrational Numbers

An Extension of Lesson **13-4**

MATERIALS

compass

straightedge

You already know how to graph integers and rational numbers on a number line. How would you graph an irrational number like $\sqrt{10}$?

Activity

- Draw a number line. At 3, construct a perpendicular line segment 1 unit in length. Draw the line segment shown in color. Label it c.
- Using the Pythagorean theorem, you can show that the hypotenuse is $\sqrt{10}$ units long.

$$c^2 = a^2 + b^2$$
$$c^2 = 1^2 + 3^2$$
$$c^2 = 10$$
$$c = \sqrt{10}$$

- Open the compass to the length of the segment you drew in color. With the tip of the compass at 0, draw an arc that intersects the number line at B. The distance from 0 to B is $\sqrt{10}$ units.

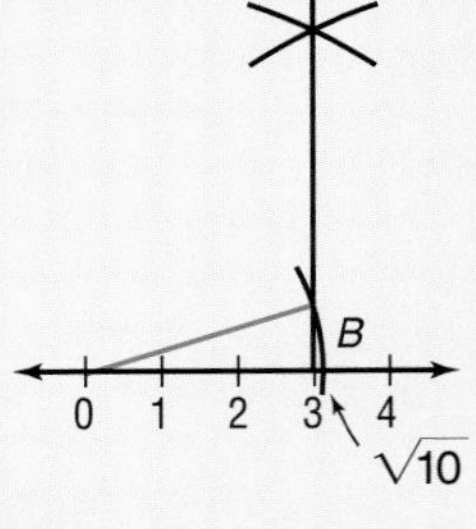

Your Turn Graph $\sqrt{8}$ on a number line. Think of $\sqrt{8}$ as $\sqrt{2^2 + 2^2}$.

TALK ABOUT IT

1. Explain how to graph $-\sqrt{10}$.
2. Explain how to graph $\sqrt{2}$.
3. Describe two different ways to graph $\sqrt{5}$.
4. Explain how the graph of $\sqrt{2}$ can be used to locate the point that represents $\sqrt{3}$.

Extension

5. Graph $\sqrt{12}$ on a number line. (*Hint:* Think of $\sqrt{12}$ as $\sqrt{4^2 - 2^2}$.)

13-5 Special Right Triangles

Setting Goals: *In this lesson, you'll find missing measures in 30°–60° and 45°–45° right triangles.*

Modeling with Manipulatives

MATERIALS

- compass
- protractor
- scissors
- ruler

Most people know that Alexander Graham Bell invented the telephone. But the drawings that Bell needed to get a patent were done by Lewis Howard Latimer (1848–1928), an African-American drafter and engineer.

Drafters and engineers often use a compass and straightedge to copy measurements accurately from one location to another.

Your Turn

Work with a partner.

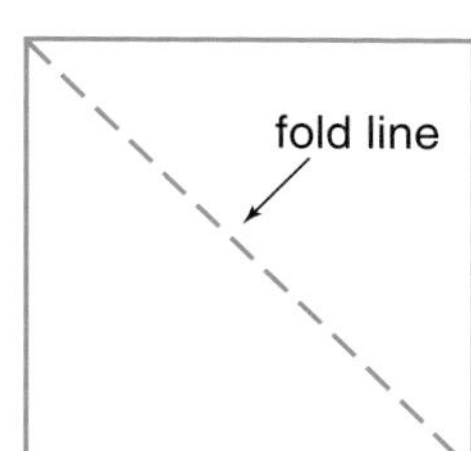

- ▶ Construct and cut out a square.
- ▶ Fold the square along the diagonal to form an isosceles right triangle.
- ▶ Measure each leg and each angle of the triangle.
- ▶ Use the Pythagorean theorem to determine the length of the hypotenuse.
- ▶ Use a calculator to divide the length of the hypotenuse by $\sqrt{2}$.
- ▶ Repeat the above steps for several other squares.

TALK ABOUT IT

a. Describe a method for finding the length of the hypotenuse of a right isosceles triangle if you know the length of its legs.

Your Turn

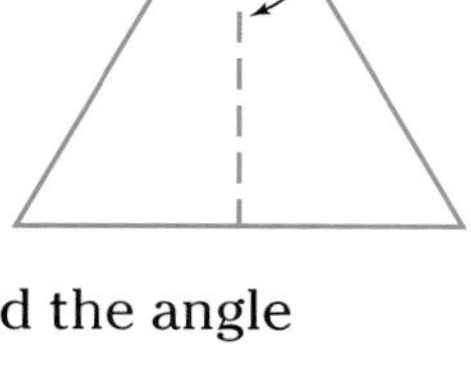

- ▶ Construct an equilateral triangle. Cut out the triangle. Fold the triangle to form a right triangle.
- ▶ Measure each side and each angle of the right triangle.
- ▶ Repeat the above steps for several other equilateral triangles.

TALK ABOUT IT

b. What is the relationship between the shortest side and the angle having the smallest measure?

c. What is the relationship between the length of the hypotenuse and the length of the shortest side?

Learning the Concept

The triangle formed in the first activity above is called a **45°–45° right triangle**. The triangle in the second activity is called a **30°–60° right triangle**. The relationships you discovered above are true for any 45°–45° or 30°–60° right triangle.

In a 45°–45° right triangle, you can find the length of the hypotenuse by multiplying the length of a leg by $\sqrt{2}$.

Example 

Find the length of $\overline{AC}$ in $\triangle ABC$.

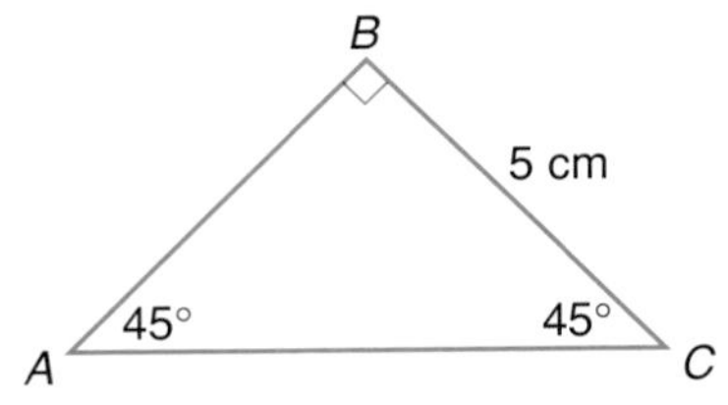

$c = a\sqrt{2}$

$c = 5\sqrt{2}$ *Replace a with 5.*

5 [×] 2 [√x] [=] 7.071067812

The length of $\overline{AC}$ is about 7.1 cm.

In a 30°–60° right triangle, the length of the side opposite the 30° angle is one-half the length of the hypotenuse.

Example 

Find the length of $\overline{PQ}$ in $\triangle PQR$.

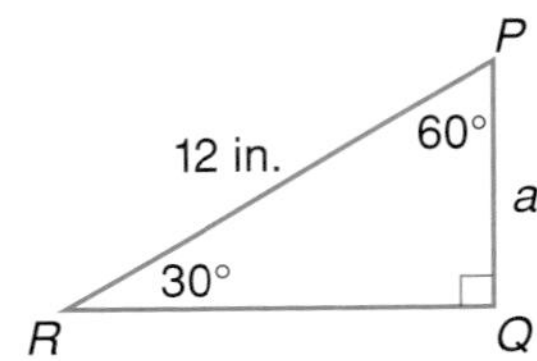

$a = \frac{1}{2}c$

$a = \frac{1}{2}(12)$ *Replace c with 12.*

$a = 6$

The length of $\overline{PQ}$ is 6 inches.

Also, in a 30°–60° right triangle, you can find the length of the side opposite the 60° angle by multiplying the length of the other leg by $\sqrt{3}$.

Example 3

APPLICATION

Construction

Large doors, like those on barns and airplane hangars, are often reinforced with a diagonal brace to prevent warping. A barn door is designed so that the brace forms a 30° angle as shown. If the width of the door is 7 feet, how high is the door?

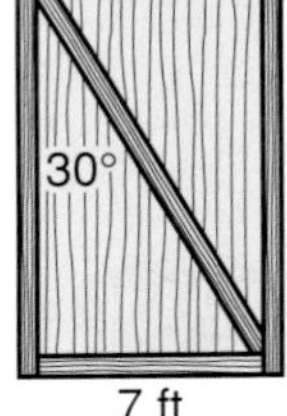

The brace forms two congruent right triangles. Draw a model of one triangle to find the measures.

Let a be the measure of the side opposite the 30° angle and b be the measure of the side opposite the 60° angle.

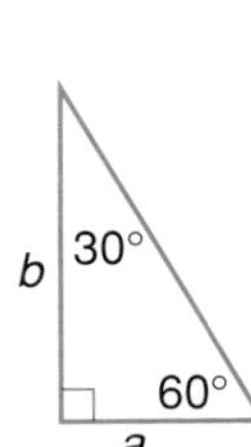

$b = a\sqrt{3}$

$b = 7\sqrt{3}$

$b \approx 12.1$

The door is about 12 feet high.

Checking Your Understanding

Communicating Mathematics

Read and study the lesson to answer each question.

1. **Explain** how to find the length of the hypotenuse of an isosceles right triangle without using the Pythagorean theorem.
2. **Draw** and label all the sides and angles of a 30°–60° right triangle with the shortest side having a measurement of 5 meters.

Guided Practice

The length of a leg of a 45°–45° right triangle is given. Find the length of the hypotenuse. Round decimal answers to the nearest tenth.

3. 1 yd

4. 9.5 m

The length of a hypotenuse of a 30°–60° right triangle is given. Find the length of the side opposite the 30° angle.

5. 18 m

6. $6\frac{1}{2}$ in.

Find the lengths of the missing sides in each triangle. Round decimal answers to the nearest tenth.

7.

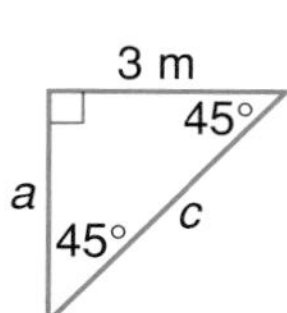

8.

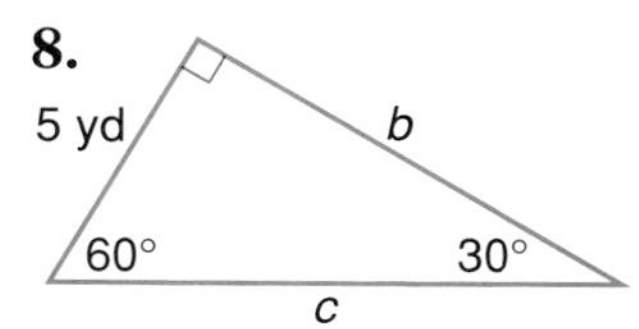

9.

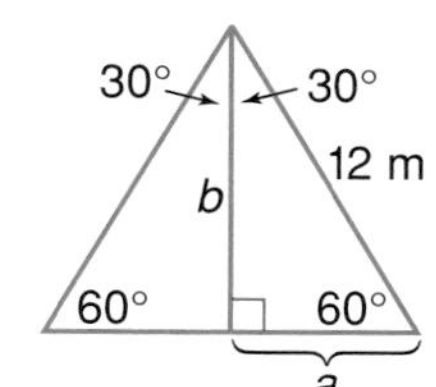

10. Camping A rope tied to the top of a tent pole makes a 60° angle with the ground and is anchored 4 feet from the base of the pole.

a. How long is the rope?

b. How tall is the tent pole?

Exercises: Practicing and Applying the Concept

Independent Practice

11. One leg of a 45°–45° right triangle is 15 centimeters long. What is the length of the other leg?

12. The shorter leg of a 30°–60° right triangle is 3 inches long.

a. What is the length of the hypotenuse?

b. What is the length of the other leg?

The length of a leg of a 45°–45° right triangle is given. Find the length of the hypotenuse. Round decimal answers to the nearest tenth.

13. 15 cm

14. 7 yd

15. 5.2 m

16. $2\frac{1}{2}$ ft

17. 6.9 in.

18. 4.1 mm

The length of a hypotenuse of a 30°–60° right triangle is given. Find the length of the side opposite the 30° angle.

19. 48 in.

20. 11 yd

21. $\frac{1}{4}$ mi

22. 3000 m

23. 4.63 cm

24. 13 mm

Find the lengths of the missing sides in each triangle. Round decimal answers to the nearest tenth.

25.

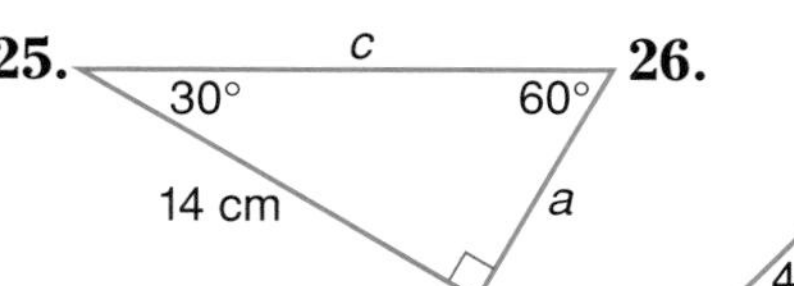

26.

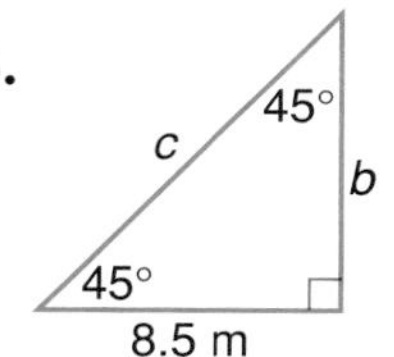

27. 60° a 12 in. 30° b

28.

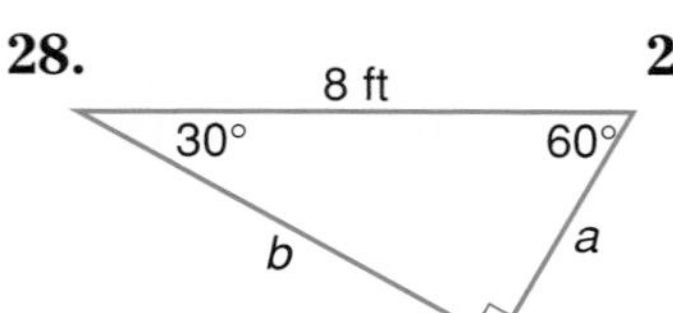

29.

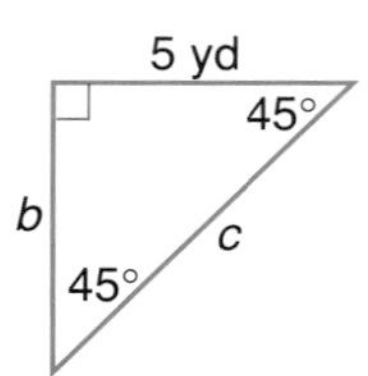

30.

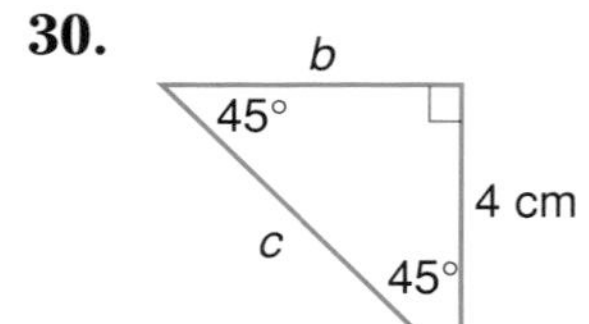

31.

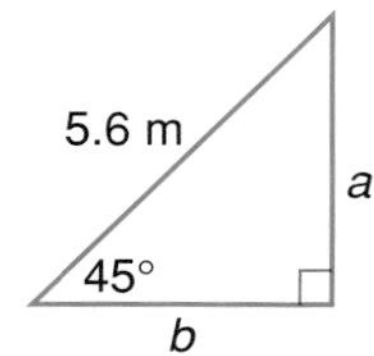

32.

a
60°
3 in.
b

33.

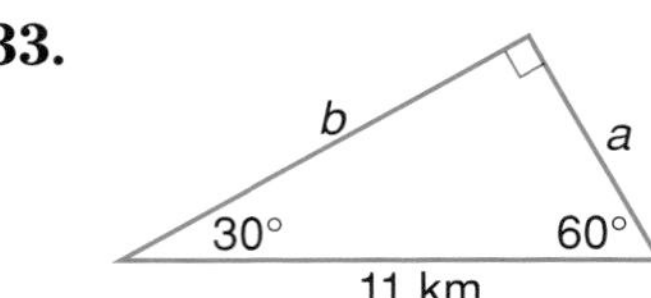

34. The hypotenuse of a 45°–45° right triangle is 12 cm long. Find the length of each of the other sides.

35. In triangle *JKL* shown at the right, the measure of $\angle LJK$ is 30°, and the altitude *h* is 5 yards long. Find the length of $\overline{KL}$.

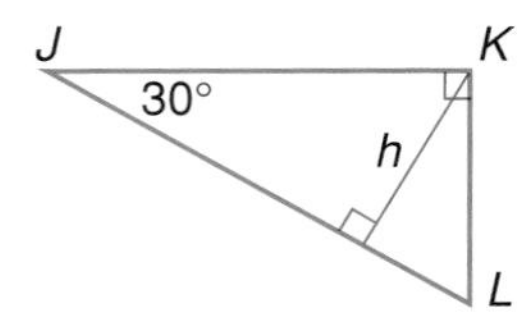

Critical Thinking

36. The area of a square is 900 square feet. Find the length of each of its diagonals.

Applications and Problem Solving

37. Recreation Two poles at the end of a swing set form a triangle with the ground. Each pole is 10 feet long and forms a 45° angle with the ground. What is the height of the swing set?

38. Construction An A-frame house is one that looks like the letter A when viewed from the front. An architect is designing an A-frame that has sides that are as long as the width of the front of the house at its base. If the building is to be 26 feet wide, what is the height of the house?

Mixed Review

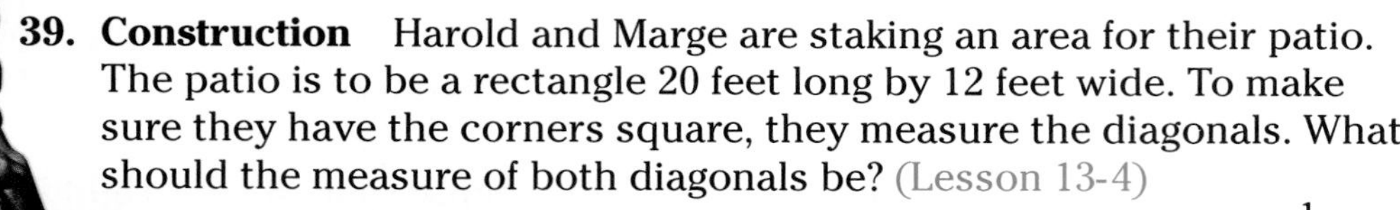

39. Construction Harold and Marge are staking an area for their patio. The patio is to be a rectangle 20 feet long by 12 feet wide. To make sure they have the corners square, they measure the diagonals. What should the measure of both diagonals be? (Lesson 13-4)

40. Geometry Find the area of a circle with a diameter of $8\frac{1}{2}$ inches. Round to the nearest tenth. (Lesson 12-2)

41. Architecture An architect is designing a gazebo in the shape of a regular pentagon. What will be the measure of each interior angle of the gazebo? (Lesson 11-8)

42. Express $\frac{5}{6}$ as a percent. (Lesson 9-7)

43. Evaluate $3c^2$ if $c = -5$. (Lesson 4-2)

44. Solve $f + 19 = 4$ and graph the solution on a number line. (Lesson 3-2)

45. Translate *a score not less than 50 is a win* into an inequality. (Lesson 1-9)

HANDS-ON ACTIVITY

13-6A Ratios in Right Triangles

A Preview of Lesson **13-6**

MATERIALS

- protractor
- metric ruler

Right triangles and the relationships among their sides have been studied for thousands of years.

When working with right triangles, the side **opposite** an angle is the side that is not part of the angle. In the triangle shown at the right, side r is opposite $\angle R$. The side that is not opposite an angle and not the hypotenuse is called the **adjacent** side. In the triangle at the right, side s is adjacent to $\angle R$.

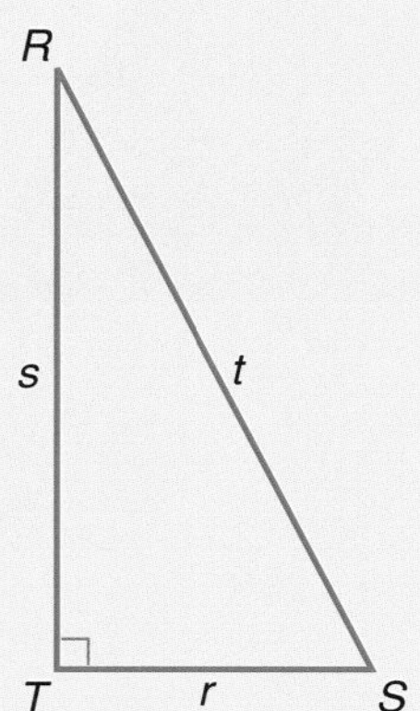

Your Turn

Work in groups of three.

- Each person should copy the table at the right.
- Each person in your group should draw a right triangle XYZ in which $m\angle X = 40°$, $m\angle Y = 50°$, and $m\angle Z = 90°$.
- Measure the leg opposite the 40° angle. Record the measurement to the nearest millimeter.

	40°angle
Length of leg opposite	
Length of leg adjacent	
Length of hypotenuse	
Ratio 1	
Ratio 2	
Ratio 3	

- Measure the leg adjacent to the 40° angle. Record the measurement to the nearest millimeter.
- Measure the hypotenuse and record the measurement to the nearest millimeter.
- Use your measurements and a calculator to find each ratio for the 40° angle. Record each ratio to the nearest hundredth.

$$\text{ratio 1} = \frac{\text{opposite leg}}{\text{adjacent leg}} \qquad \text{ratio 2} = \frac{\text{opposite leg}}{\text{hypotenuse}} \qquad \text{ratio 3} = \frac{\text{adjacent leg}}{\text{hypotenuse}}$$

- Add a column to your table and repeat the above procedure for the 50° angle.

1. Compare your ratios with the other members of your group. How do they compare?
2. Make a conjecture about the ratio of the sides of any 40°–50° right triangle.
3. Repeat the activity for 25°–75° right triangles. What do you find?

13-6 The Sine, Cosine, and Tangent Ratios

Setting Goals: *In this lesson, you'll find missing sides and angles of triangles using the sine, cosine, and tangent ratios.*

Modeling with Technology

If you know certain measures of a right triangle, ratios can be used to find the measures of the remaining parts. These ratios are used in the study of **trigonometry**.

Your Turn **Work with a partner.**

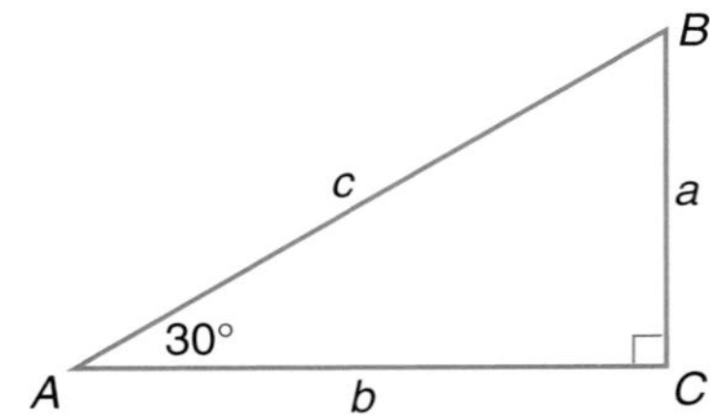

- ▶ In triangle ABC, the measure of side a is 1, the measure of side b is 1.732, and the measure of side c is 2.
- ▶ Find and record the ratio of a to c. Using a calculator, enter 30 SIN .
- ▶ Find and record the ratio of b to c. Using a calculator, enter 30 COS .
- ▶ Find and record the ratio of a to b. Using a calculator, enter 30 TAN .

a. The first ratio compares the leg opposite $\angle A$ to the hypotenuse. Describe the relationship between the measures of the second ratio and $\angle A$.

b. What is the relationship between the measures of the third ratio and $\angle A$?

c. How did each ratio compare to each calculator display that followed?

Learning the Concept

The word *trigonometry* means triangle measurement. The ratios of the measures of the sides of a right triangle are called **trigonometric ratios**.

Three common trigonometric ratios are defined below.

Definition of Trigonometric Ratios

In words: If $\triangle ABC$ is a right triangle and A is an acute angle,

$$\text{sine of } \angle A = \frac{\text{measure of the leg opposite } \angle A}{\text{measure of the hypotenuse}}$$

$$\text{cosine of } \angle A = \frac{\text{measure of the leg adjacent to } \angle A}{\text{measure of the hypotenuse}}$$

$$\text{tangent of } \angle A = \frac{\text{measure of the leg opposite } \angle A}{\text{measure of the leg adjacent to } \angle A}$$

In symbols: $\sin A = \frac{a}{c}$, $\cos A = \frac{b}{c}$, $\tan A = \frac{a}{b}$

Sine, cosine, and tangent are abbreviated as sin, cos, and tan, respectively.

The value of the trigonometric ratios depends only on the measure of the angle. The size of the triangle does not affect the value of the ratio.

Example  Find sin *S*, cos *S*, and tan *S* to the nearest thousandth.

$$\sin S = \frac{\text{measure of leg opposite } \angle S}{\text{measure of hypotenuse}}$$

$$= \frac{12}{13} \text{ or about } 0.923$$

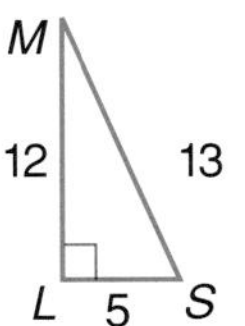

$$\cos S = \frac{\text{measure of leg adjacent to } \angle S}{\text{measure of hypotenuse}}$$

$$= \frac{5}{13} \text{ or about } 0.385$$

$$\tan S = \frac{\text{measure of leg opposite } \angle S}{\text{measure of leg adjacent to } \angle S}$$

$$= \frac{12}{5} \text{ or } 2.4$$

A calculator can be used to find the sine, cosine or tangent ratio for an angle with a given degree measure.

Example  Find each value to the nearest ten thousandth.

a. sin 37°

37 [SIN] 0.601815023

Rounded to the nearest ten thousandth, sin 37° ≈ 0.6018.

b. cos 56°

56 [COS] 0.559192903

Rounded to the nearest ten thousandth, cos 56° ≈ 0.5592.

c. tan 72°

72 [TAN] 3.077683537

Rounded to the nearest ten thousandth, tan 72° ≈ 3.0777.

You can also use a calculator to find the degree measure of an angle if the sine, cosine, or tangent ratio is known.

Example 

a. Find the measure of $\angle S$ given that sin *S* = 0.3569.

.3569 [SIN^{-1}] 20.909936

The measure of $\angle S$ is about 21°.

b. Find the measure of $\angle C$ given that cos *C* = 0.2831.

.2831 [COS^{-1}] 73.55469014

The measure of $\angle C$ is about 74°.

c. Find the measure of $\angle T$ given that tan *T* = 4.703.

4.703 [TAN^{-1}] 77.99596131

The measure of $\angle T$ is about 78°.

TECHNO TIP

To use [SIN^{-1}], [COS^{-1}], and [TAN^{-1}], you have to press the [2nd] or [INV] key and then [SIN], [COS], or [TAN].

You can use the trigonometric ratios to find missing measures.

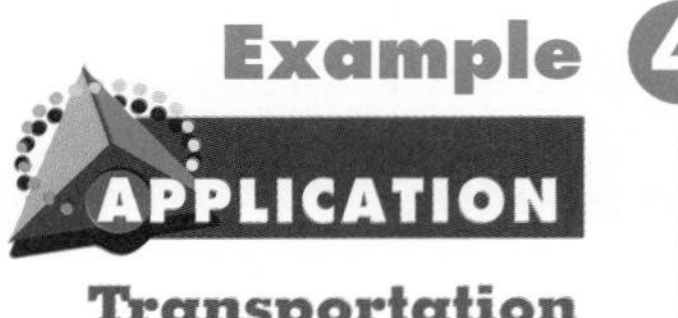

The end of an exit ramp from an interstate highway is 22 feet higher than the highway. If the ramp is 630 feet long, what angle does it make with the highway?

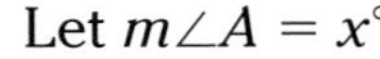
Let $m\angle A = x°$

leg opposite $\angle A$ = 22 feet

hypotenuse = 630 feet

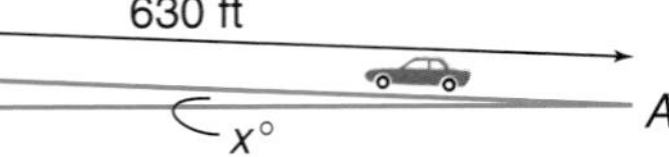

Now substitute these values into the definition of sine.

$\sin x° = \frac{22}{630}$ ← *side opposite* / ← *hypotenuse*

Use a calculator to find the value of x.

22 [÷] 630 [=] [SIN⁻¹] 2.0012119

The ramp makes an angle of about 2° with the road.

Checking Your Understanding

Communicating Mathematics

Read and study the lesson to answer each question.

1. **Write** a definition of the sine ratio.
2. **Compare and contrast** the sine ratio with the cosine ratio.
3. **Describe** the procedure for using a calculator to find the degree measure of the angle that corresponds to a given tangent ratio.
4. **Draw** a right triangle such that $\sin D = \frac{4}{5}$, $\cos D = \frac{3}{5}$, and $\tan D = \frac{4}{3}$.
5. **You Decide** Kelli says that the value of sin A is greater than the value of sin D. Derice says that sin A and sin D are equal. Who is correct and why?

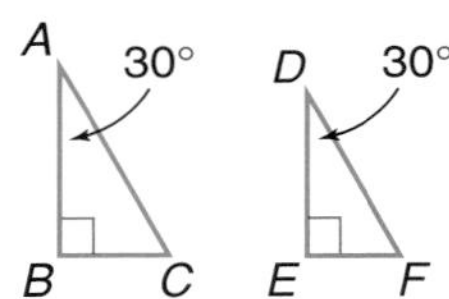

6. *Some Old Horse - Caught A Horse - Taking Oats Away* is a helpful mnemonic device for remembering the trigonometric ratios. S, C, and T represent sin, cos, and tan respectively, while O, H, and A represent opposite, hypotenuse, and adjacent respectively. Develop your own mnemonic device for remembering the ratios.

Guided Practice

Refer to $\triangle XYZ$. Express each ratio as a fraction in simplest form.

7. sin X
8. sin Y
9. cos X
10. cos Y
11. tan X
12. tan Y

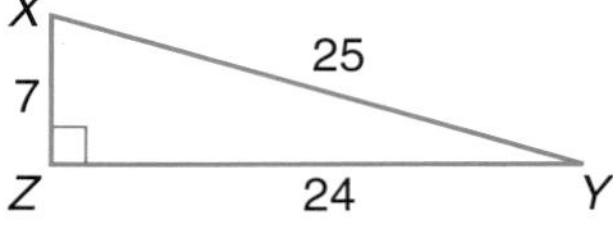

Use a calculator to find each ratio to the nearest ten thousandth.

13. cos 71°
14. tan 2°
15. sin 25°

Use a calculator to find the angle that corresponds to each ratio. Round to the nearest degree.

16. $\sin M = 0.4$
17. $\cos N = 0.18$
18. $\tan F = 0.64$

19. Public Access The angle that a wheelchair ramp makes with the ground cannot exceed 6°. If a ramp needs to rise 2 feet, how long does the ramp need to be?

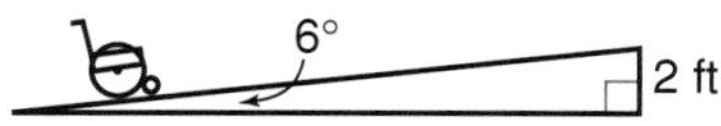

Exercises: Practicing and Applying the Concept

Independent Practice

For each triangle, find sin *B*, cos *B*, and tan *B* to the nearest thousandth.

20. 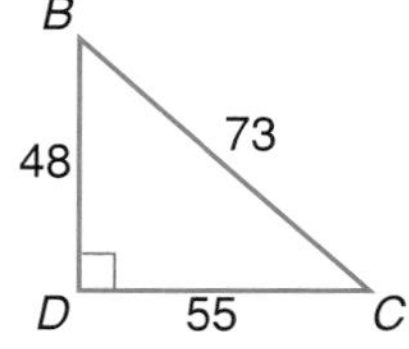

21. 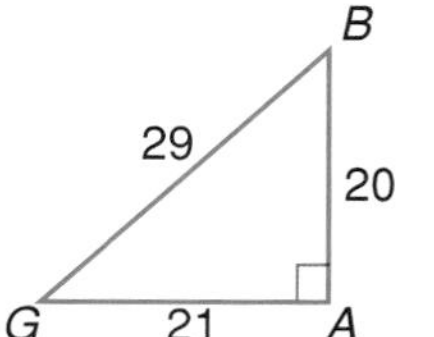

22. 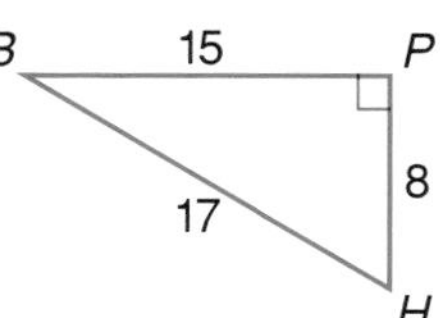

23. 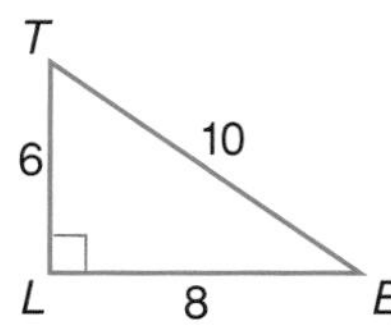

24. 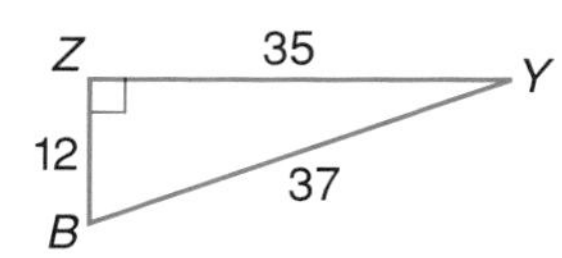

25. 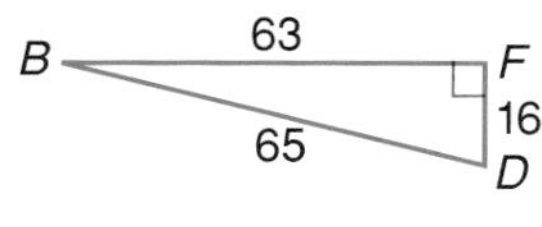

Use a calculator to find each ratio to the nearest ten thousandth.

26. tan 45° **27.** sin 30° **28.** cos 60°

29. cos 25° **30.** tan 31° **31.** sin 71°

Use a calculator to find the angle that corresponds to each ratio. Round answers to the nearest degree.

32. $\tan J = 0.6$ **33.** $\sin R = 0.8$ **34.** $\cos F = 0.866$

35. $\sin E = 0.6897$ **36.** $\cos B = 0.4706$ **37.** $\tan K = 1.8$

For each triangle, find the measure of the marked acute angle to the nearest degree.

38. 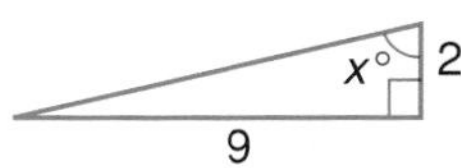

39. 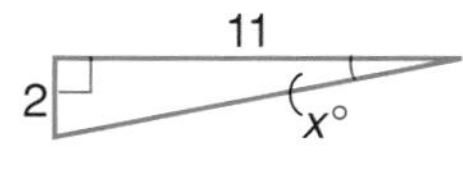

40. 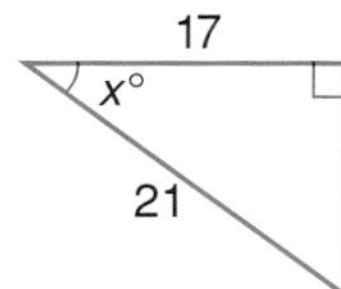

41. 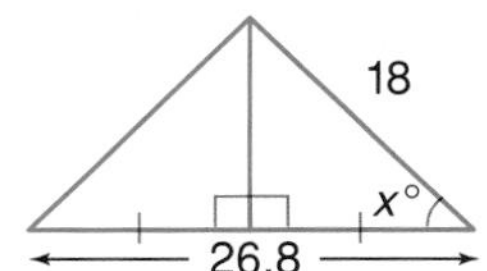

42. 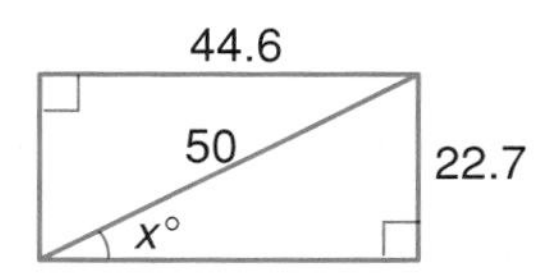

43. 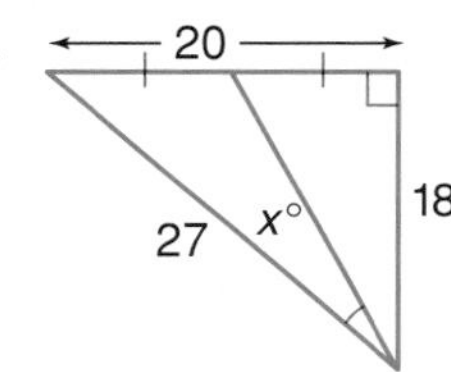

Critical Thinking

44. Write a conjecture about the relationship between the sine and cosine of complementary angles.

Applications and Problem Solving

45. **Scale Drawing** The Leaning Tower of Pisa is 55 meters tall and tilts 5 meters off the perpendicular. Maria wants to make a scale drawing for a project in history class. What should be the measure of the angle she draws to represent the tilt of the tower?

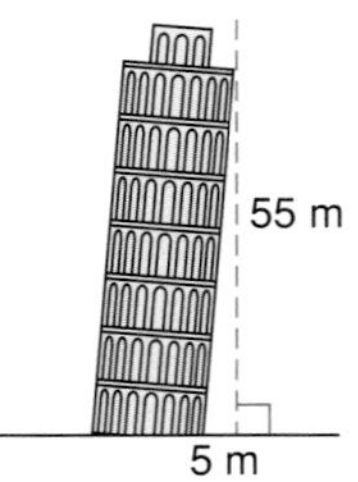

46. **Transportation** The steepest streets in the United States are Filbert Street and 22nd Street in San Francisco. They rise 1 foot for every 3.17 feet of horizontal distance. Find the angle these streets form with the horizontal.

Mixed Review

47. The hypotenuse of a $30°–60°$ right triangle is 20 yards long. What is the length of the side opposite the $30°$ angle? (Lesson 13-5)

48. **Construction** A rectangular patio is designed to have a concrete border around a tiled area. The tiled area is to be rectangular, and the border is to extend 2 feet around the tile on all sides. The entire patio will be 24 feet by 14 feet. What is the area to be covered by tile? (Lesson 12-4)

49. There are 3 routes between Jean's house and the school. How many ways can Jean go from her house to the school and back home again? (Lesson 10-5)

50. How many milligrams are in 2.5 grams? (Lesson 7-8)

51. **Geometry** If two angles of a triangle have the same measure, the triangle is an isosceles triangle. The triangle shown at the right is an isosceles triangle. Is this an example of inductive or deductive reasoning? (Lesson 5-8)

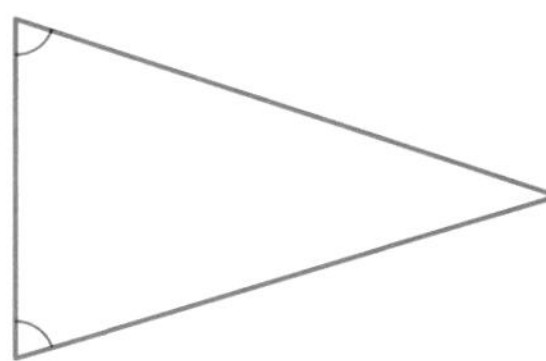

52. Find the product of $(2b^3)$ and $(-6b^4)$. (Lesson 4-8)

53. Solve $22 = \frac{n}{-11}$. Then graph the solution on a number line. (Lesson 3-3)

54. Is $4 > 16$ *true*, *false*, or *open*? (Lesson 1-9)

COOPERATIVE LEARNING PROJECT

THE SHAPE OF THINGS TO COME

Thought-Controlled Computer Technology

Have you ever heard the expression "It's the thought that counts?" A new computer accessory will make that expression more than an adage. Experimental work has been done to translate thoughts into digital impulses that can be read by a computer. A sensor is placed on your finger and then attached to a personal computer. Minute physiological responses are measured through the skin and the computer responds onscreen. Some applications of the technology include action games that can be played without a joystick, art and music software that adjust colors and tempos according to thoughts, and memory and learning games.

See for Yourself

Research thought-controlled computer technology.

- When will the technology be commercially available?
- Is research being done on applications of the technology for things other than games?

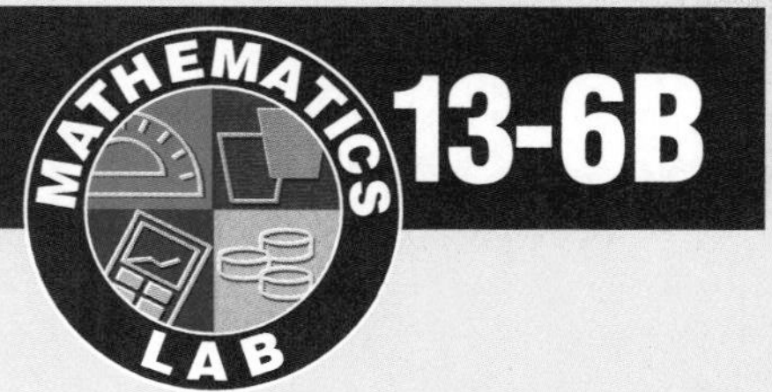

13-6B Slope and Tangent

An Extension of Lesson **13-6**

Be sure to turn off any functions in the Y= list and any statistical plots before starting to draw.

You can use a graphing calculator to draw line segments. The right triangle shown was drawn in the viewing window [0, 14] by [0, 10] with a scale factor of 1 on both axes. The vertices are (0, 3), (6, 3), and (6, 6).

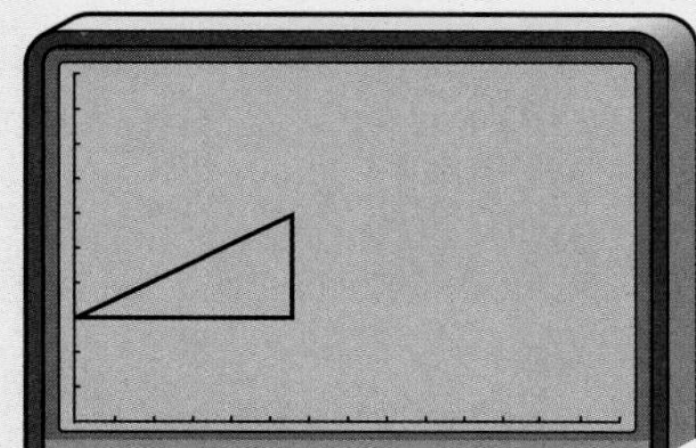

Enter: 2nd DRAW 2 0 , 3 , 6 , 3) ENTER CLEAR
2nd DRAW 2 6 , 3 , 6 , 6) ENTER CLEAR
2nd DRAW 2 6 , 6 , 0 , 3) ENTER

To find the measure of the angle formed by the horizontal line and the hypotenuse, you can use the tangent ratio. The leg opposite the angle is 3 units long, and the leg adjacent to the angle is 6 units long.

$$\tan a = \frac{\text{leg opposite}}{\text{leg adjacent}}$$
$$= \frac{3}{6} \text{ or } 0.5$$

Use a calculator to find a.

Enter: 2nd TAN⁻¹ .5 ENTER 26.56505118

$a \approx 26.6°$

Your Turn

- Find the slope of the hypotenuse of the triangle shown at the beginning of the lab.
- Draw the triangle shown above on a graphing calculator. Then graph the line that contains the hypotenuse, $y = 0.5x + 3$.

  Enter: Y .5 X,T,θ + 3 GRAPH

1. What is the slope of the line?
2. What is the slope of the hypotenuse?
3. How does the slope of the hypotenuse compare to the tangent of the angle formed by the hypotenuse and the horizontal leg?
4. Graph $y = 3x + 4$ on your graphing calculator. What is the slope of the line?
5. Sketch the right triangle formed by the graph of $y = 3x + 4$, the x-axis, and the y-axis. Label it $\triangle ABC$, so that $\angle A$ is between the x-axis and the graph of $y = 3x + 4$. What is $\tan A$?
6. What is the measure of the angle that the graph of $y = 3x + 4$ forms with the x-axis?

13-7 Using Trigonometric Ratios

Setting Goals: *In this lesson, you'll solve problems by using the trigonometric ratios.*

Modeling with Manipulatives

MATERIALS

- protractor
- index card
- straw
- paper clip (large)
- string
- tape

Many problems that can be solved using trigonometric ratios deal with angles of elevation. An **angle of elevation** is formed by a horizontal line and a line of sight above it.

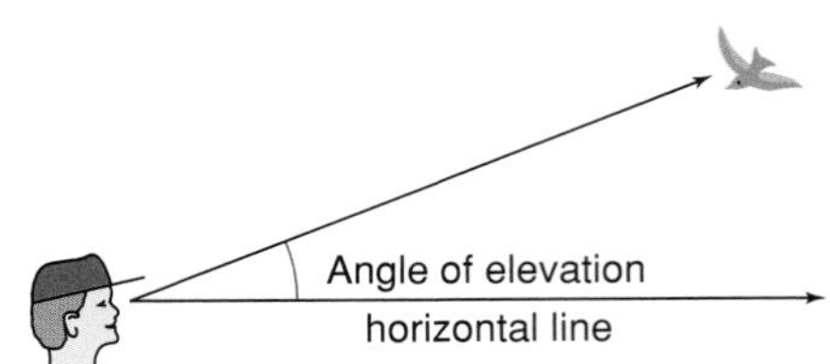

A hypsometer is an instrument that measures angles of elevation. You can make a model of a hypsometer.

Your Turn

Work with a partner.

- Tape a protractor on an index card so that both zero points align with the edge of the card. Mark the center of the protractor on the card and label it *C*. Then mark and label every 10° on the card like the one shown at the right.

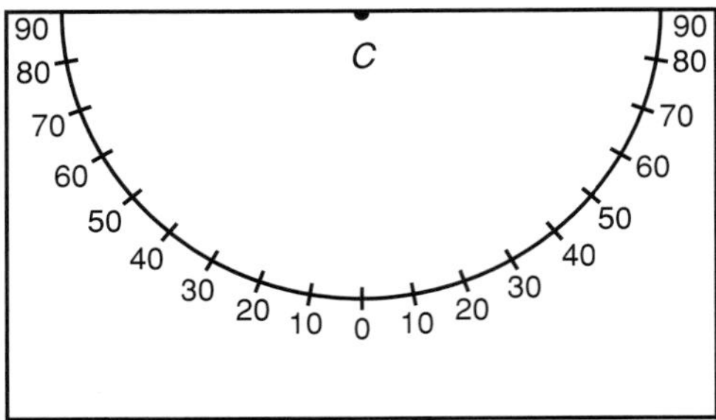

- Tie a piece of string to a large paper clip. Attach the other end of the string to the index card at *C*.
- Tape a straw to the edge of the index card that contains *C*.
- To use the hypsometer, look through the straw at the object. Have your partner read the angle from the scale.
- If possible, use your hypsometer to find the angle of elevation to a tree on your school property. Also measure the distance from the hypsometer to the ground and from the hypsometer to the base of the object.

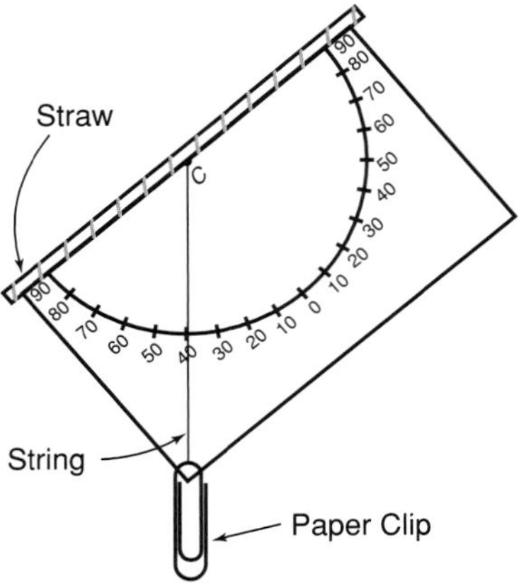

a. Draw a diagram of the situation with the hypsometer and the tree. If you do not have your own measurements, use 35° as the angle of elevation, 4 feet as the distance from the hypsometer to the ground, and 46 feet as the distance from the hypsometer to the tree base.

b. How could you find the height of a tree? Solve.

Learning the Concept

The trigonometric ratios and the Pythagorean theorem can be used to find the measure of any side or angle of a right triangle if the measure of one side and any other side or acute angle are known.

Example 1

Find the measure of $\angle J$ in $\triangle JKL$.

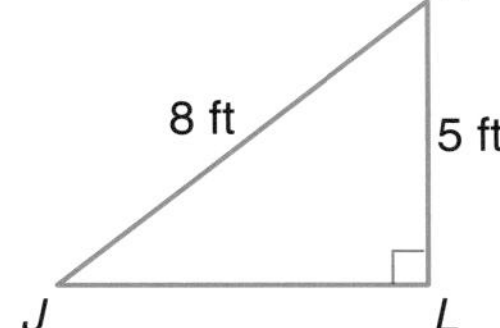

The measure of the leg opposite $\angle J$ and the measure of the hypotenuse are known. Use the sine ratio.

$$\sin J = \frac{\text{measure of leg opposite } J}{\text{measure of hypotenuse}}$$

$\sin J = \frac{5}{8}$ — *Replace the measure of the leg opposite J with 5 and the measure of the hypotenuse with 8.*

$\sin J = 0.625$

$J \approx 38.7$ — *Use a calculator.*

The measure of $\angle J$ is about 38.7°.

A person on a tower or a cliff must look down to see an object below. An **angle of depression** is formed by a horizontal line and another line of sight below it.

Example 2

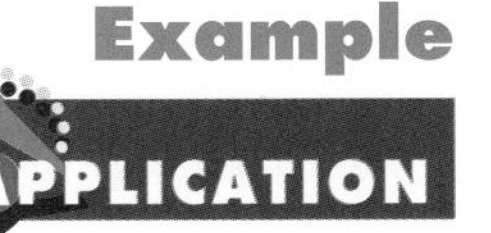

Fire Fighting

A fire is sighted from a fire tower in the Black Hills National Forest. Using a hypsometer, the ranger found the angle of depression to be 2°. If the ranger's eyes are 125 feet above the ground, how far is the fire from the base of the tower?

Let d represent the distance from the fire to the base of the fire tower.

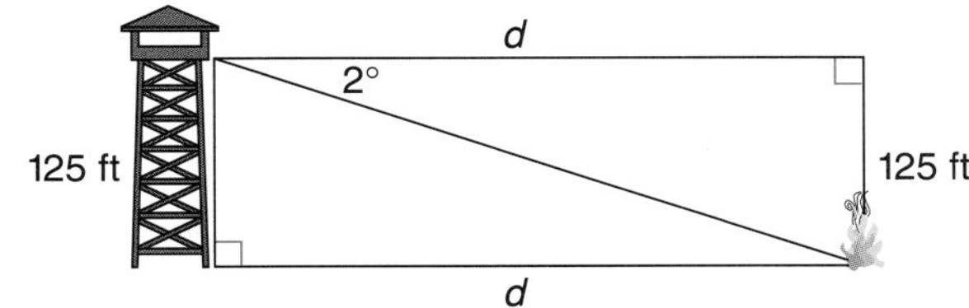

In the diagram, a line from the fire perpendicular to the horizontal line of sight forms a rectangle. Therefore, both lengths can be labeled d. Notice that 125 feet is the length of the leg opposite the 2° angle and d is the length of the leg adjacent to the 2° angle. The tangent ratio should be used to solve this problem.

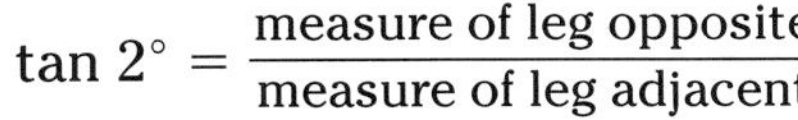

$$\tan 2^\circ = \frac{\text{measure of leg opposite}}{\text{measure of leg adjacent}}$$

$\tan 2^\circ = \frac{125}{d}$ — *Replace the measure of the leg opposite with 125 and the measure of the leg adjacent with d.*

$d \tan 2^\circ = 125$ — *Multiply each side by d.*

$d = \frac{125}{\tan 2^\circ}$ — *Divide each side by tan 2°.*

$d \approx 3580$ — *Use a calculator.*

The fire is about 3580 feet from the base of the tower.

Checking Your Understanding

Communicating Mathematics

Read and study the lesson to answer each question.

1. **Draw** a diagram and write a few sentences to explain what is meant by an angle of elevation.
2. **Explain** how to decide whether to use sine, cosine, or tangent when you are finding the measure of an acute angle in a right triangle.

3. **Assess Yourself** Describe a measure in your home or school that you could find using an angle of elevation. If possible, use the hypsometer you made at the beginning of the lesson to find the measure. Then write a few sentences explaining how you found the height.

Guided Practice

Write an equation that you could use to solve for x. Then solve. Round decimal answers to the nearest tenth.

4. 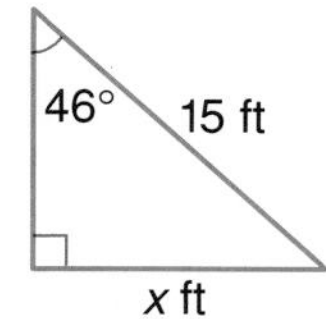

5. 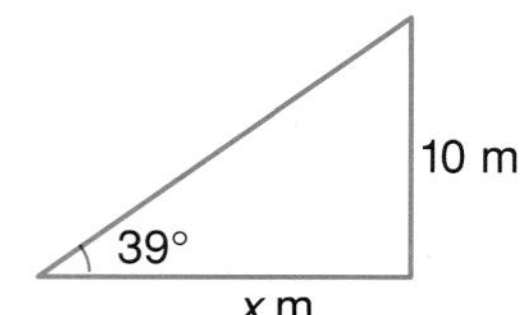

6. 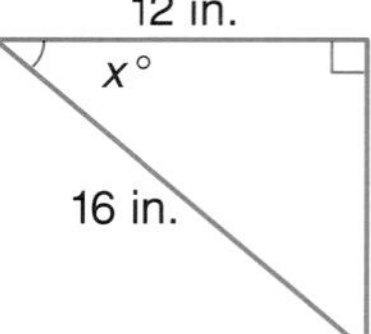

7. **Home Maintenance** A painter props a 20-foot ladder against a house. The angle it forms with the ground is 65°. How far up the side of the house does the ladder reach?

Exercises: Practicing and Applying the Concept

Independent Practice

Write an equation that you could use to solve for x. Then solve. Round decimal answers to the nearest tenth.

8. 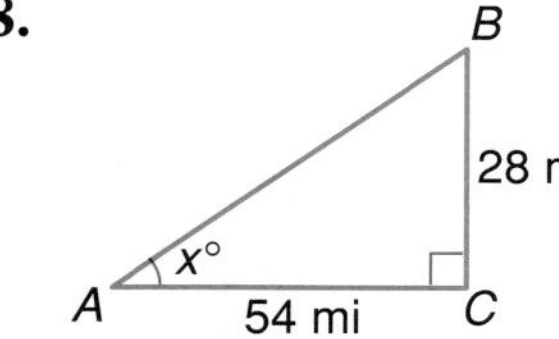

9. 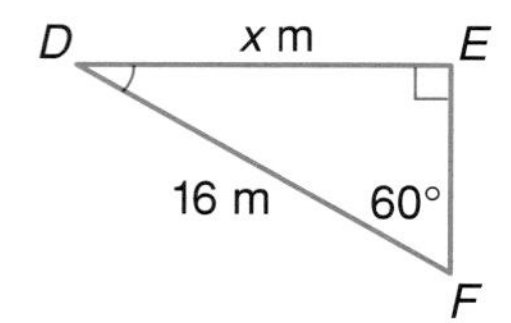

10. 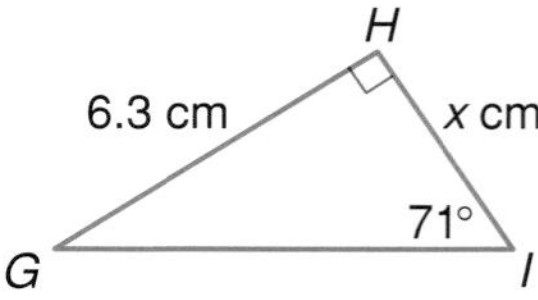

11. 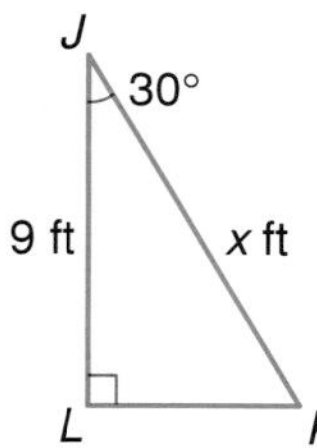

12. 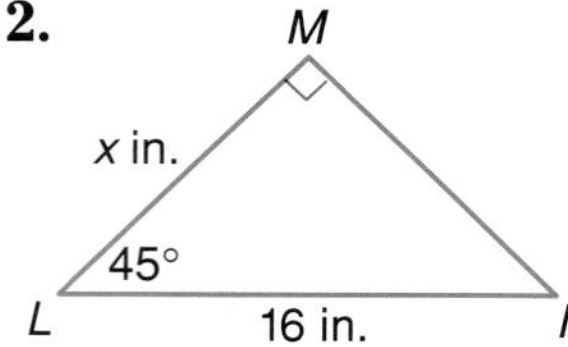

13. 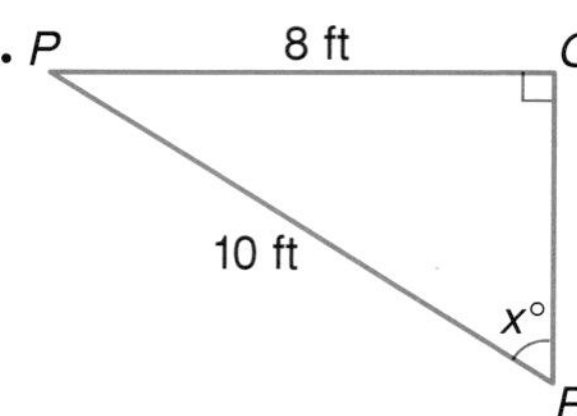

14. 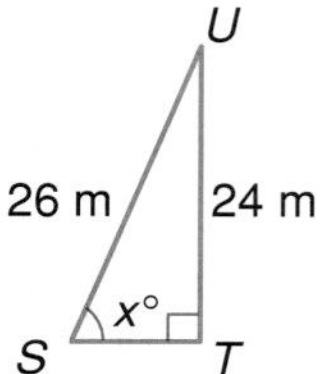

15. 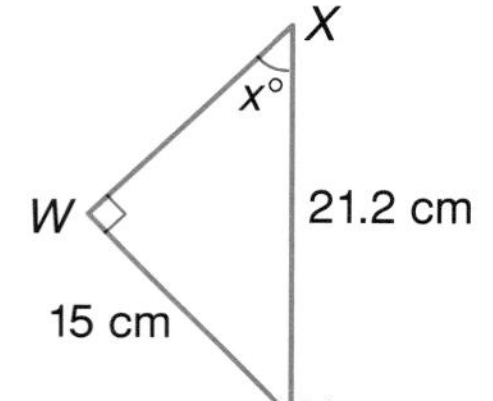

16. 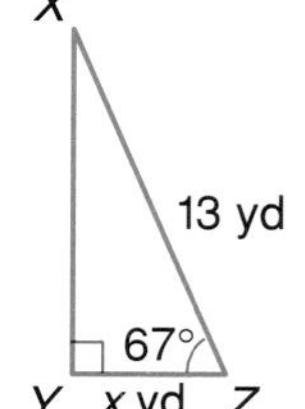

Use trigonometric ratios to solve each problem.

17. Martina is flying a kite on a 50-yard string. The string is making a 50° angle with the ground. How high above the ground is the kite?

18. A flagpole casts a shadow 25 meters long when the angle of elevation of the sun is 40°. How tall is the flagpole?

19. A guy wire is fastened to a TV tower 50 feet above the ground and forms an angle of 65° with the tower. How long is the wire?

20. Find the area of a right triangle in which one acute angle measures 35° and the leg opposite that angle is 80 cm long.

21. A surveyor is 85 meters from the base of a building. The angle of elevation to the top of the building is 26.5°. If her eye level is 1.6 meters above the ground, find the height of the building.

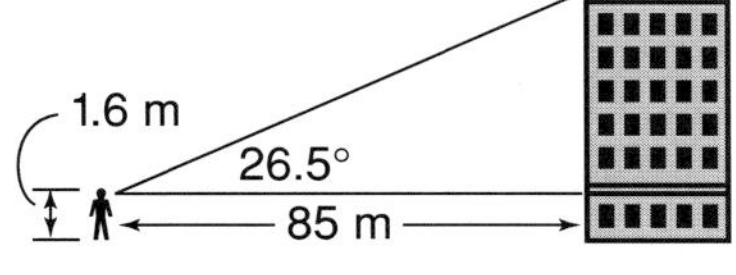

22. Danica is in the observation area of the Sears Tower in Chicago overlooking Lake Michigan. She sights two sailboats going due east from the tower. The angles of depression to the boats are 25° and 18°. If the observation deck is 1353 feet high, how far apart are the boats?

Critical Thinking

23. Find the missing measures in the triangle. (*Hint:* Draw an altitude from *B*.)

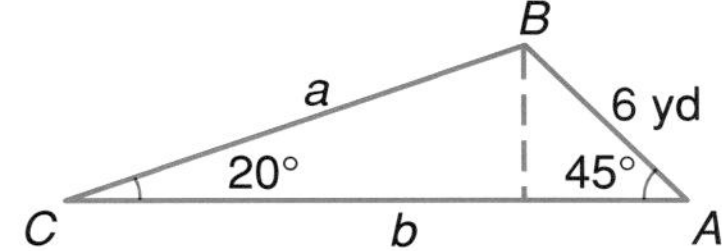

Applications and Problem Solving

24. Aviation A plane is 3 miles above the ground. The pilot sights the airport at an angle of depression of 15°. What is the distance between the plane and the airport?

25. Mountain Climbing A mountain climber is planning to scale a vertical rock wall. From a point 10 feet from the base, she estimates the angle of elevation to the top of the wall to be 75°. How high is the vertical wall to the nearest foot?

26. Meteorology The cloud ceiling is the lowest altitude at which solid cloud is present. To find the cloud ceiling at night, a searchlight is shone vertically onto the clouds. If a searchlight is 100 meters from a weather office and the angle of elevation to the spot on the clouds is 70°, how high is the cloud ceiling?

Mixed Review

27. Use a calculator to find the measure of $\angle B$ given that $\sin B = 0.8829$. (Lesson 13-6)

28. Collect Data Find out how many students are in each grade in your school. Make a circle graph to display your findings. (Lesson 11-2)

29. Find the range, median, upper and lower quartiles, and the interquartile range for the set of data {45, 62, 72, 51, 47, 68, 69, 50, 75}. (Lesson 10-2)

30. Taxes Use the information in the graph to find the probability that a randomly chosen taxpayer begins his or her tax return in March or April. (Lesson 9-3)

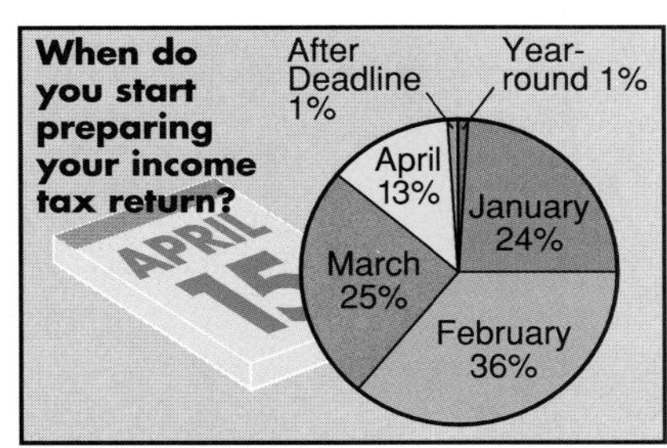

Source: RGA Survey

31. Determine whether {(2, −1), (3, −3), (4, −5)} is a function. (Lesson 8-1)

CLOSING THE

Investigation

OH GIVE ME A HOME

Architects often give presentations to potential buyers to show and explain their designs. Plan and give an oral presentation to describe what you know about architectural styles and designing homes. Your presentation should include the following.

- A brief architectural history of your community. You might include old and new photographs, drawings, or maps.
- Examples from your Investigation Folder to help you describe and define building terms you used most often in this Investigation.
- Floor plan(s) and elevation drawings of the house your group designed.
- The model of the house your group designed.
- The reasoning behind your design; that is, an explanation that tells why you designed the house the way you did.
- An explanation of what you would do differently if you had to do this Investigation over again.

Extension

Schedule an interview with an architect. Make a list of questions to ask him or her.

- Be sure to ask about the importance of mathematics to architecture.
- Include a question about what courses high school students interested in becoming architects should take.

PORTFOLIO ASSESSMENT

You may want to keep your work on this Investigation in your portfolio.

CHAPTER 13 Highlights

Vocabulary

After completing this chapter, you should be able to define each term, property, or phrase and give an example or two of each.

Algebra
irrational number (p. 672)
perfect squares (p. 665)
Pythagorean theorem (p. 676)
radical sign (p. 664)
real numbers (p. 673)
square root (p. 664)

Geometry
adjacent leg (p. 687)
angle of depression (p. 695)
angle of elevation (p. 694)
cosine (p. 688)
45°–45° right triangle (p. 683)
hypotenuse (p. 676)
leg (p. 676)
opposite side (p. 687)
sine (p. 688)
tangent (p. 688)
30°–60° right triangle (p. 683)
trigonometric ratios (p. 688)
trigonometry (p. 688)

Problem Solving
Venn diagram (p. 669)

Understanding and Using Vocabulary

Choose the correct term to complete each sentence.

1. The (square, square root) of 6 equals 36.
2. In the expression $\sqrt{25}$, the symbol $\sqrt{}$ is called a (radical sign, square).
3. The number $\sqrt{121}$ belongs to the set of (irrational, rational) numbers.
4. The longest side of a right triangle is the (hypotenuse, leg).
5. A right triangle with a leg that is half the length of the hypotenuse is a (30°–60°, 45°–45°) right triangle.
6. The angle formed by a horizontal line and another line of sight above it is an angle of (depression, elevation).

CHAPTER 13 Study Guide and Assessment

Skills and Concepts

Objectives and Examples

Upon completing this chapter, you should be able to:

- **find squares and square roots** (Lesson 13-1)

Find $-\sqrt{196}$.

The symbol $-\sqrt{196}$ represents the negative square root of 196. Since $14 \cdot 14 = 196$, $-\sqrt{196} = -14$.

Review Exercises

Use these exercises to review and prepare for the chapter test.

Find each square root.

7. $\sqrt{9}$
8. $\sqrt{121}$
9. $-\sqrt{25}$
10. $-\sqrt{81}$
11. $\sqrt{225}$
12. $-\sqrt{900}$

- **find and use squares and square roots** (Lesson 13-1)

Find the best integer estimate for $-\sqrt{29}$.
25 and 36 are the closest perfect squares.

$$-\sqrt{36} < -\sqrt{29} < -\sqrt{25}$$
$$-\sqrt{6^2} < -\sqrt{29} < -\sqrt{5^2}$$
$$-6 < -\sqrt{29} < -5$$

Since -29 is closer to -25 than -36, the best integer estimate is -5.

Find the best integer estimate for each square root. Then check your estimate using a calculator.

13. $\sqrt{83}$
14. $\sqrt{54}$
15. $\sqrt{220}$
16. $-\sqrt{900}$
17. $-\sqrt{39}$
18. $\sqrt{9.61}$

- **identify numbers in the real number system** (Lesson 13-3)

0.25 can be expressed as $\frac{1}{4}$ so it is a rational number.

The number $\sqrt{6} = 2.449489743\ldots$, which is not a repeating or terminating decimal, is an irrational number.

Name the sets of numbers to which each number belongs: the whole numbers, the integers, the rational numbers, the irrational numbers, and/or the real numbers.

19. 7
20. $\frac{3}{4}$
21. -2.6
22. $\sqrt{11}$

- **solve equations by finding square roots** (Lesson 13-3)

Solve $b^2 = 49$.

$$b^2 = 49$$
$$b = \sqrt{49} \quad \text{or} \quad b = -\sqrt{49}$$
$$b = 7 \quad \text{or} \quad b = -7$$

Solve each equation. Round decimal answers to the nearest tenth.

23. $x^2 = 196$
24. $n^2 = 160$
25. $t^2 = 15$
26. $a^2 = 0.04$

Objectives and Examples

▶ **use the Pythagorean theorem to find the length of the side of a right triangle and to solve problems** (Lesson 13-4)

Find the length of the third side of the right triangle.

$$c^2 = a^2 + b^2$$
$$8^2 = 5^2 + b^2$$
$$64 = 25 + b^2$$
$$39 = b^2$$
$$\sqrt{39} = b$$
$$6.2 \approx b$$

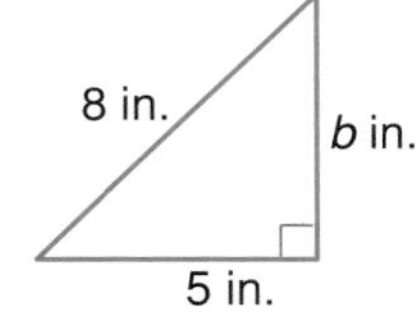

The length of the leg is about 6.2 inches.

Review Exercises

In a right triangle, if *a* and *b* are the measures of the legs and *c* is the measure of the hypotenuse, find each missing measure. Round answers to the nearest tenth.

27. $a = 12, b = 16$
28. $a = 14, b = 40$
29. $a = 30, b = 16$
30. $a = 8, c = 15$
31. $b = 63, c = 65$
32. $a = 15, c = 39$

Objectives and Examples

▶ **find missing measures in 30°–60° and 45°–45° right triangles** (Lesson 13-5)

In a 30°–60° right triangle, the hypotenuse is twice as long as the length of the side opposite the 30° angle.

In a 45°–45° right triangle, the hypotenuse is $\sqrt{2}$ times the length of a leg.

Review Exercises

Find the lengths of the missing sides in each triangle. Round decimal answers to the nearest tenth.

33.

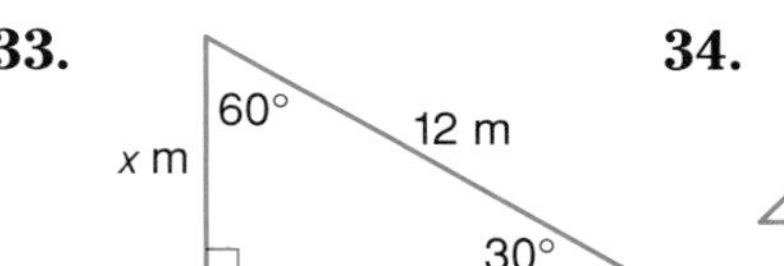

34.

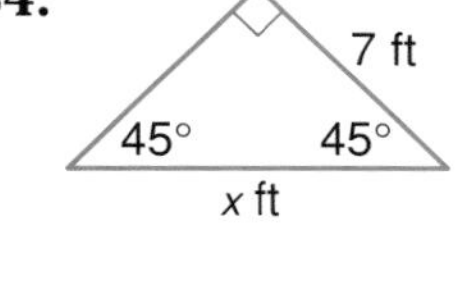

35.

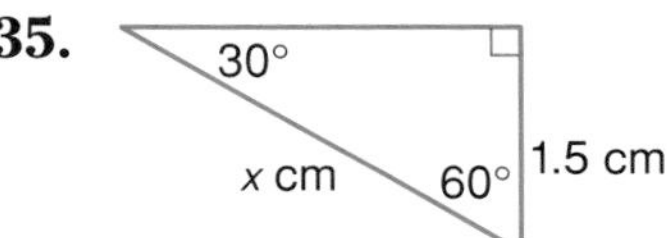

Objectives and Examples

▶ **find missing sides and angles of triangles using the sine, cosine, and tangent ratios** (Lesson 13-6)

$$\sin \angle A = \frac{\text{measure of the leg opposite } \angle A}{\text{measure of the hypotenuse}}$$

$$\cos \angle A = \frac{\text{measure of the leg adjacent to } \angle A}{\text{measure of the hypotenuse}}$$

$$\tan \angle A = \frac{\text{measure of the leg opposite } \angle A}{\text{measure of the leg adjacent to } \angle A}$$

Review Exercises

For each triangle, find sin *X*, cos *X*, and tan *X* to the nearest thousandth.

36.

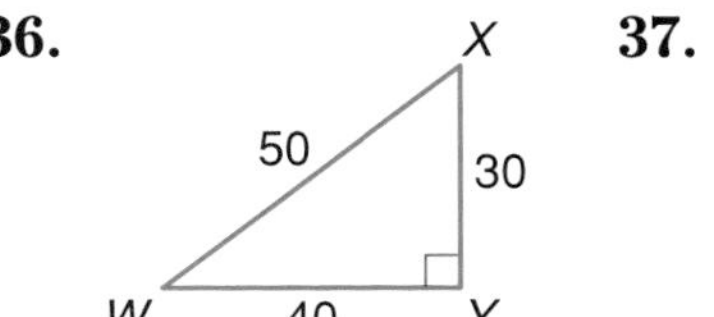

37.

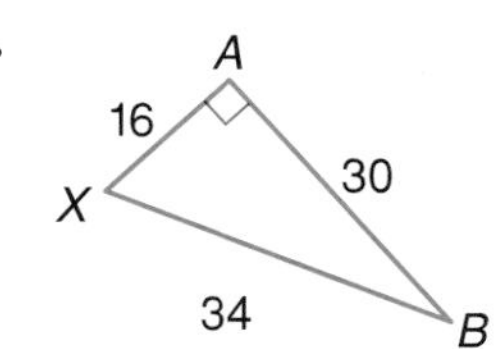

Use a calculator to find the angle that corresponds to each ratio. Round to the nearest degree.

38. $\tan C = 2.145$
39. $\sin H = 0.9945$
40. $\cos M = 0.2588$

Objectives and Examples

- **solve problems using the trigonometric ratios** (Lesson 13-7)

In $\triangle ABC$, find the length of $\overline{BC}$.

$$\sin A = \frac{\text{measure of leg opposite } \angle A}{\text{measure of the hypotenuse}}$$

$$\sin 28^\circ = \frac{x}{12}$$

$$(12)(\sin 28^\circ) = x$$

$$5.6 \approx x$$

The measure of $\overline{BC}$ is about 5.6 feet.

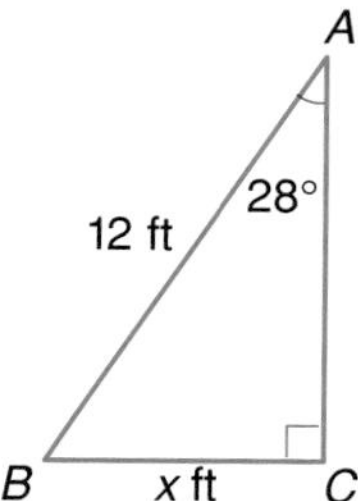

Review Exercises

Write an equation that you could solve for *x*. Round decimal answers to the nearest tenth.

41.

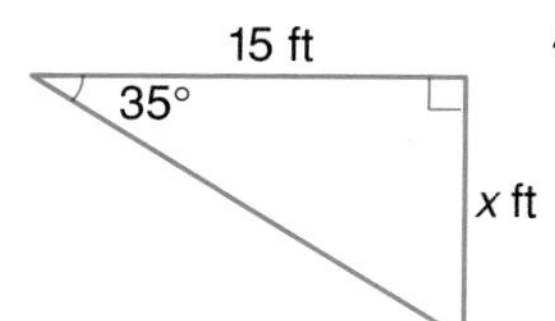

42.

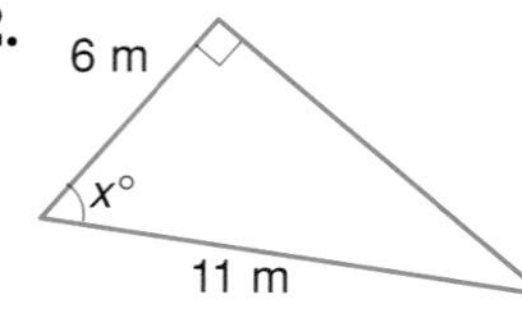

43.

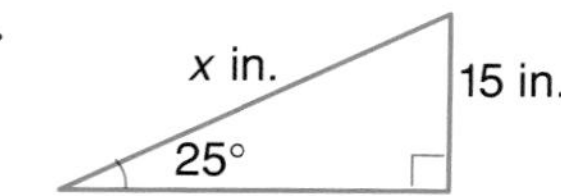

Applications and Problem Solving

44. Recreation A class survey of 32 students about favorite weekend activities showed that 20 students chose a movie and 15 students chose going to a shopping mall. There were 8 students who chose neither activity. How many students chose both activities? (Lesson 13-2)

45. Communication A telephone pole is 28 feet tall. A wire is stretched from the top of the pole to a point on the ground that is 5 feet from the bottom of the pole. How long is the wire? Round to the nearest tenth. (Lesson 13-3)

46. Geometry Write a formula for the length of the diagonal of a square if the sides are *s* units long. (Lesson 13-5)

47. Navigation From the top of a lighthouse, the angle of depression to a buoy is 25°. If the top of the lighthouse is 150 feet above sea level, find the distance from the buoy to the foot of the lighthouse. (Lesson 13-7)

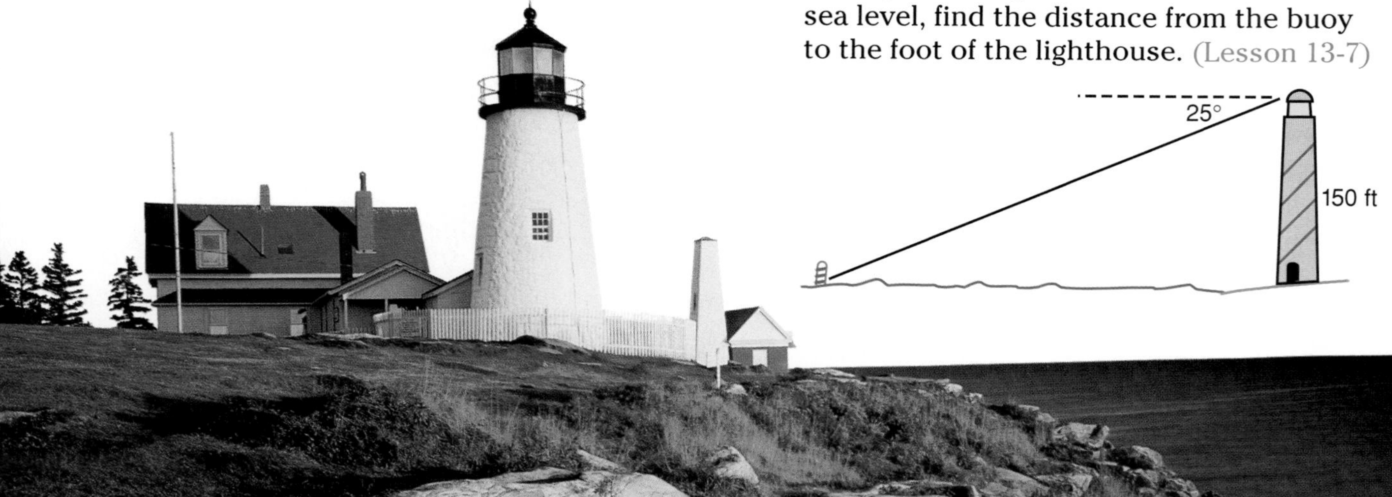

A practice test for Chapter 13 is available on page 786.

Alternative Assessment

Performance Task

Demonstrate your knowledge by giving a clear, concise solution to each problem. Be sure to include all relevant drawings and justify your answers. You may show your solutions in more than one way or investigate beyond the requirements of the problem.

1. You and your group are hiking in the woods. You leave camp and walk 15 kilometers due north. Then you walk 6 kilometers due east. Your job in the group is to chart the group's movement and to find how far you are from the starting point. Vectors are line segments with both magnitude (length) and direction. Draw vectors to represent the group's movement. Then use what you have learned in this chapter to find the direction and distance the hikers are from their starting point.

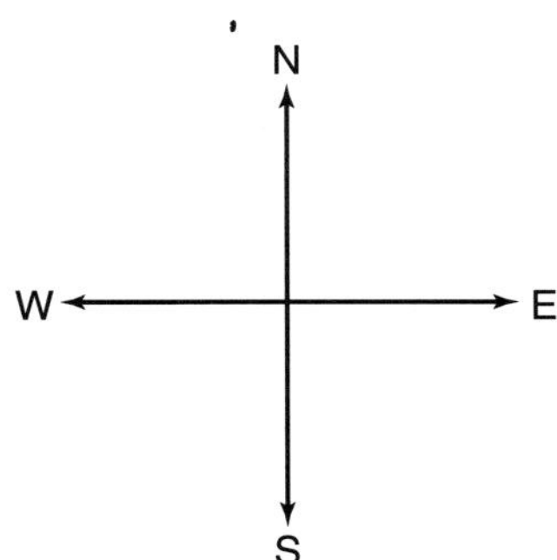

2. A soccer player kicks the ball 12 feet to the west. Another player kicks the ball 18 feet to the north. What is the direction and distance from the starting point of the ball? Round your answer to the nearest whole number.

Thinking Critically

Use a calculator to find the sine, cosine, and tangent of several acute angles.

- When is $\cos A > \sin A$?
- When is $\cos A = \sin A$?
- Describe the value of $\tan A$ when $m\angle A > 45°$.
- Describe the range of values for $\cos A$ and $\sin A$.

Portfolio

Place your favorite word problem from this chapter in your portfolio along with your solution and attach a note explaining why it is your favorite.

Self Evaluation

Euclid's *Elements* is the basis for much of the geometry that is studied today. It was written in about 300 B.C. and was a compilation of most of the geometry known at the time. Euclid taught mathematics at the Library of Alexandria in Alexandria, Egypt. When asked if there was a simpler way to learn geometry than by studying the *Elements*, Euclid replied "There is no 'royal road' to geometry."

Assess yourself. Learning anything well is a building process that takes time. Evaluate the time that you spend studying. Do you invest time each day on study or do you study for tests and forget what you learned? Keep a log of the time that you spend studying for one week, noting when tests and quizzes were given. If you find that you are studying ineffectively, make a new plan for your study time in the next week.

CHAPTER 14

Polynomials

TOP STORIES in Chapter 14

In this chapter, you will:

- identify and classify polynomials, and
- add, subtract, and multiply polynomials.

MATH AND GAMES IN THE NEWS

Bridge-building in China

Source: The Atlantic Monthly, February, 1995

"The Chinese are naturals for bridge, a game that brings people together," says Kathie Wei-Sender, a Tennessee-based businesswoman and former world champion bridge player. The Chinese did not begin to compete seriously in international bridge until the 1980s. Nevertheless, both the women's and men's national teams are making strong showings. Young Chinese players are being encouraged to become competitive, and bridge is a part of the curriculum in many schools. Working with a partner and careful bidding are key elements of a winning bridge player. Shanghai-born C. C. Wei invented the "precision system" of bidding that is based on statistics. This system is widely used in China and other countries. Experts on international competition think that it is only a matter of time before the Chinese team wins the championship.

Putting It into Perspective

1780 | 1920 | 1940

1925
American Harold Vanderbilt develops modern contract bridge while on a winter cruise in the Caribbean Sea.

1674
Whist, the forerunner of modern bridge, is described in Charles Cotton's book, "The Compleat Gamester."

1912
The People's Republic of China is established.

NORTH
♠ 952
♥ K8642
♦ K54
♣ 63

WEST	EAST
♠ AJ864	♠ 10 3
♥ Q5	♥ AJ10 7
♦ 10763	♦ AJ9
♣ 54	♣ Q10 97

SOUTH
♠ KQ7
♥ 93
♦ Q82
♣ AKJ82

interNET CONNECTION For up-to-date information on bridge, visit: **www.glencoe.com/sec/math/prealg/mathnet**

Statistical Snapshot

Chances of Suit Distributions

Thirteen cards are dealt in a hand of bridge. The distribution of cards that come from different suits is important to the bidding process. A suit distribution of 4-4-3-2 means that 4 cards come from one suit, 4 from a second suit, 3 from a third, and 2 from the fourth suit. The odds of different suit distributions are given in the table below.

Distribution	Approximate Odds Against
4–4–3–2	4 to 1
5–4–2–2	8 to 1
6–4–2–1	20 to 1
7–4–1–1	254 to 1
8–4–1–0	2211 to 1
13–0–0–0	188,753,389,899 to 1

Source: *The World Almanac*

On the Lighter Side

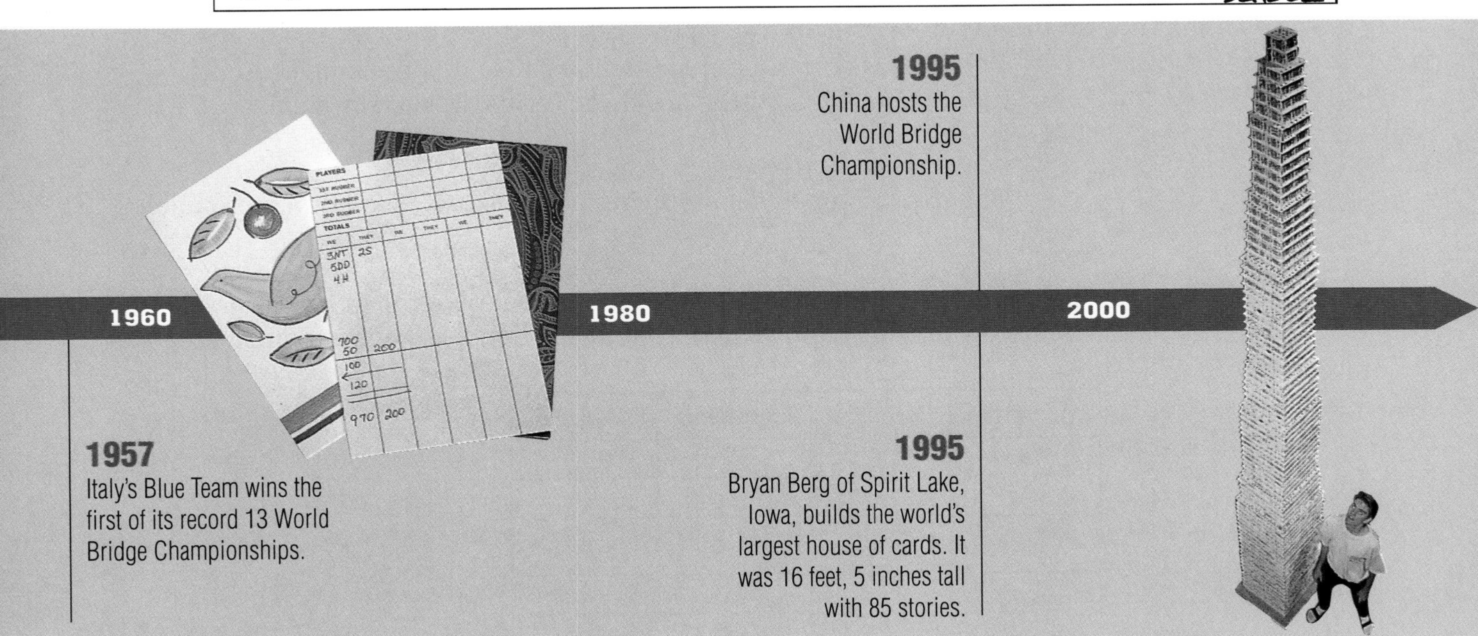

1957
Italy's Blue Team wins the first of its record 13 World Bridge Championships.

1995
China hosts the World Bridge Championship.

1995
Bryan Berg of Spirit Lake, Iowa, builds the world's largest house of cards. It was 16 feet, 5 inches tall with 85 stories.

14-1 Polynomials

Setting Goals: *In this lesson, you'll identify and classify polynomials and find their degree.*

Modeling a Real-World Application: Construction

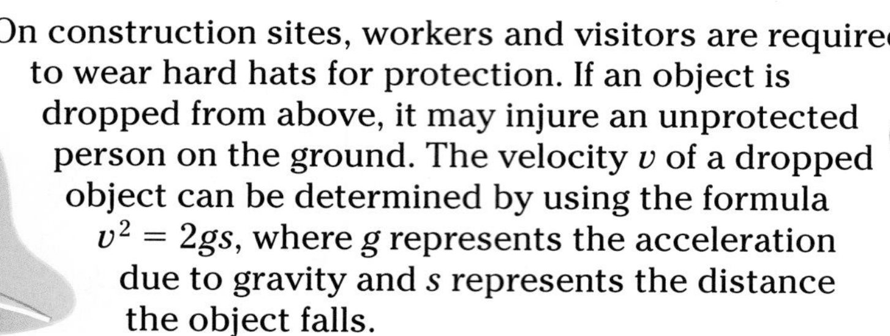

On construction sites, workers and visitors are required to wear hard hats for protection. If an object is dropped from above, it may injure an unprotected person on the ground. The velocity v of a dropped object can be determined by using the formula $v^2 = 2gs$, where g represents the acceleration due to gravity and s represents the distance the object falls.

Learning the Concept

You can review monomials in Lesson 4-1.

As you know, expressions like v^2 and $2gs$ are monomials. Monomials can be numbers, variables, or products of numbers and variables. Expressions that involve variables in the denominator of a fraction or variables with radical signs are not monomials. For example, $3x + 2$, $\frac{5}{m}$, and $\sqrt{ab}$ are not monomials.

An algebraic expression that contains one or more monomials is called a **polynomial**. A polynomial is a sum or difference of monomials. A polynomial with two terms is called a **binomial**, and a polynomial with three terms is called a **trinomial**.

Example 

Determine whether the given expression is a polynomial. If it is, classify it as a *monomial*, a *binomial*, or a *trinomial*.

a. $3x^2 + 7x - 5$

The expression is a polynomial because it is the sum of three monomials. Since there are three terms, it is a trinomial.

b. $3y - \frac{1}{y^2}$

The expression is *not* a polynomial because $\frac{1}{y^2}$ is not a monomial.

The degree of a monomial is the sum of the exponents of its variables. The following chart shows how to find the degree of a monomial.

Monomial	Variables	Exponents	Degree
z	z	1	1
$-3c^2$	c	2	2
$\frac{2}{5}m^3n^4$	m, n	3, 4	3 + 4 or 7
$0.62p^3q^3$	p, q	3, 3	3 + 3 or 6

Remember that $z = z^1$.

A monomial like 7 that does not have a variable associated with it is called a **constant**. The degree of a nonzero constant is 0. The constant 0 has *no* degree.

A polynomial also has a degree. The degree of a polynomial is the same as that of the term with the greatest degree.

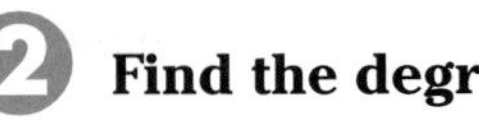

Example 2 **Find the degree of $x^2 + xy^2 + y^4$.**

First, find the degree of each term.

x^2 has degree 2.
xy^2 has degree $1 + 2$ or 3.
y^4 has degree 4. *Greatest degree*

So, the degree of $x^2 + xy^2 + y^4$ is 4.

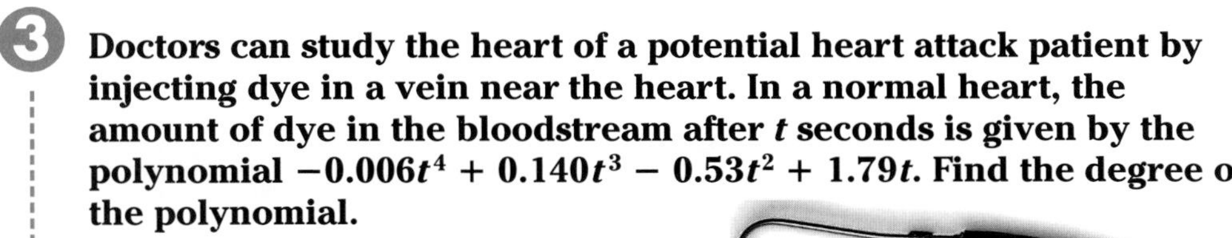

Example 3 APPLICATION Medicine

Doctors can study the heart of a potential heart attack patient by injecting dye in a vein near the heart. In a normal heart, the amount of dye in the bloodstream after t seconds is given by the polynomial $-0.006t^4 + 0.140t^3 - 0.53t^2 + 1.79t$. Find the degree of the polynomial.

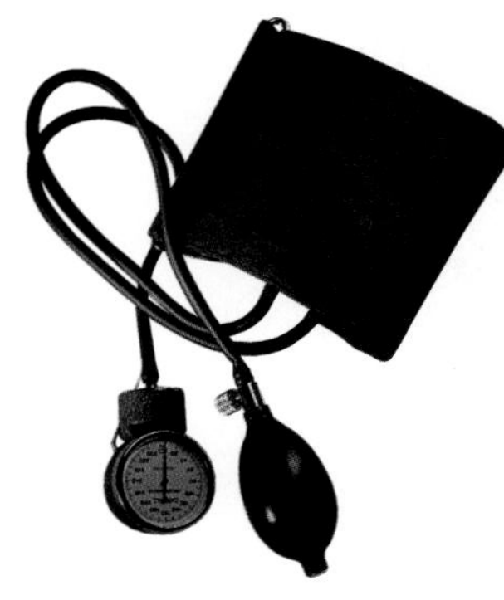

$-0.006t^4$ has degree 4.
$0.140t^3$ has degree 3.
$0.53t^2$ has degree 2
$1.79t$ has degree 1.

$-0.006t^4 + 0.140t^3 - 0.53t^2 + 1.79t$ has degree 4.

You can evaluate polynomials when you know the values of the variables involved.

Example 4 **Evaluate $6x^2 - 5xy$ if $x = -2$ and $y = -3$.**

$$
\begin{aligned}
6x^2 - 5xy &= 6 \cdot (-2)^2 - 5(-2)(-3) && \textit{Replace x with } -2 \textit{ and y with } -3.\\
&= 6 \cdot 4 - 5 \cdot 6 && \textit{Use the order of operations.}\\
&= 24 - 30\\
&= -6
\end{aligned}
$$

Checking Your Understanding

Communicating Mathematics

Read and study the lesson to answer these questions.

1. **Determine** if $\frac{3w^2}{5y}$ is a monomial. Explain your answer.
2. **Write** three binomial expressions. Explain why they are binomials.
3. **Explain** how to find the degree of a monomial and a polynomial.

4. Write about a way that you can remember the definitions of monomial, binomial, trinomial, and polynomial.

Guided Practice

State whether each expression is a polynomial. If it is, classify it as a *monomial*, *binomial*, or *trinomial*.

5. $\frac{x}{2}$
6. -5
7. $\frac{ab}{c} - c$
8. $3x^2 + 4x - 2$

Find the degree of each polynomial.

9. $-2xy^2$
10. $12x^3y^2$
11. $x^3 + 7x$
12. $x^2 + xy^2 - y^4$

Evaluate each polynomial if $x = 3$, $y = -5$, and $z = -1$.

13. $3x - 5y + z^2$
14. $x^3 + 2y^2 - z$
15. $x^3 - 2xy$
16. $2xy + 6yz^2$
17. **Business** The cost of an order of 12 burgers, 11 fries, and 5 shakes can be represented by the polynomial $12b + 11f + 5s$. Determine the cost of the order if burgers are \$1.79 each, fries are \$0.99 each, and shakes are \$1.29 each.

Exercises: Practicing and Applying the Concept

Independent Practice

State whether each expression is a polynomial. If it is, classify it as a *monomial*, *binomial*, or *trinomial*.

18. $-\frac{1}{9}r^2$
19. $c^2 + 3$
20. $x^2 - 4$
21. $1 + 3x + 5x^2$
22. $\frac{6}{a} + b$
23. $5xy$
24. $3x^2y$
25. $ab^2 + 3a - b^2$
26. $x + y$
27. $-\sqrt{49}$
28. $\frac{e}{6}$
29. $x^2 - \frac{1}{2}x + \frac{1}{3}$

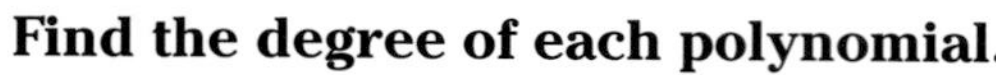

Find the degree of each polynomial.

30. $11c^2 + 4$
31. $3x + 5$
32. 121
33. $4x^3 + xy - y^2$
34. $x^6 + y^6$
35. $d^4 + c^4d^2$
36. $16y^2 + mnp$
37. $x^3 - x^2y^3 + 8$
38. $-7x^3y^4$
39. $-x^2yz^4$
40. $14c^5 - 16c^6d$
41. $2x^5 + 9x + 1$

Evaluate each polynomial if $a = 2$, $b = -3$, $c = 4$, and $d = -5$.

42. $a^3 - 2bc$
43. $ab + cd$
44. $2abc + 3a^2b$
45. $b^3 - 2ac + d^2$
46. $b^2 + a^2b$
47. $a^5 + bd^2$
48. $a^2 + b^2 - c^2 + d^2$
49. $abcd - 25 + a$
50. $5ad^2 - 2a + (bc)^2$
51. $-(-ac)^2 - cd$
52. $\sqrt{c} - b^2d$
53. $d - abc^2 - \sqrt{-27b}$

Critical Thinking

54. Find the degree of the polynomial $a^{x+3} + x^{x-2}b^3 + b^{x+2}$.

Applications and Problem Solving

55. **Landscaping** Lee's Garden Shop wants to place hosta plants along the perimeter of a garden.
    a. Write a polynomial that represents the perimeter of the garden in feet.
    b. What is the degree of the polynomial?

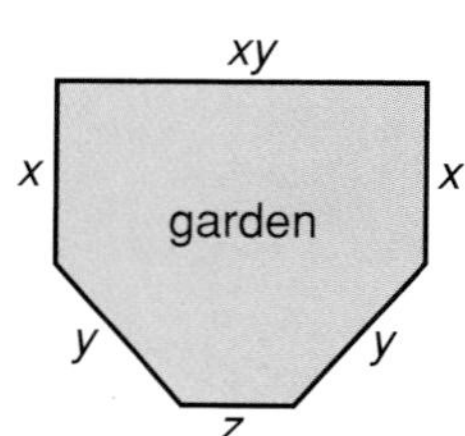

56. **Ecology** In the early 1900s, the deer population of the Kaibab Plateau in Arizona increased rapidly because hunters had reduced the number of predators. However, the food supply was not great enough to support the increased population. So, the population eventually began to decline. The deer population for the years 1905 to 1930 can be approximated by the polynomial $-0.125x^5 + 3.125x^4 + 4000$, where x is the number of years from 1900.
    a. Find the degree of the polynomial.
    b. Evaluate the polynomial to find the deer population in 1920.

57. **Family Activity** The amount of money in a savings account can be written as a polynomial. Suppose that your grandparents placed \$100 in your savings account each year on your birthday. On your fifth birthday, there would have been $100x^4 + 100x^3 + 100x^2 + 100x + 100$ dollars, where x is the annual interest rate plus 1. Research the interest rate at your family's bank. How much money would you have on your next birthday at their passbook savings rate?

Mixed Review

58. **Geometry** Find the length of $\overline{XY}$ in $\triangle WXY$. Round to the nearest tenth. (Lesson 13-7)

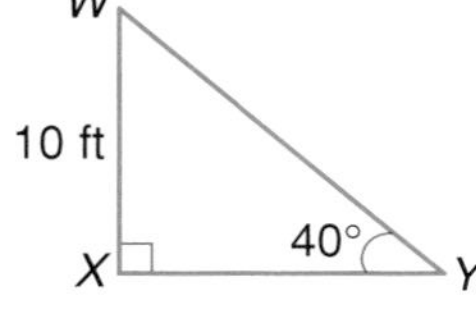

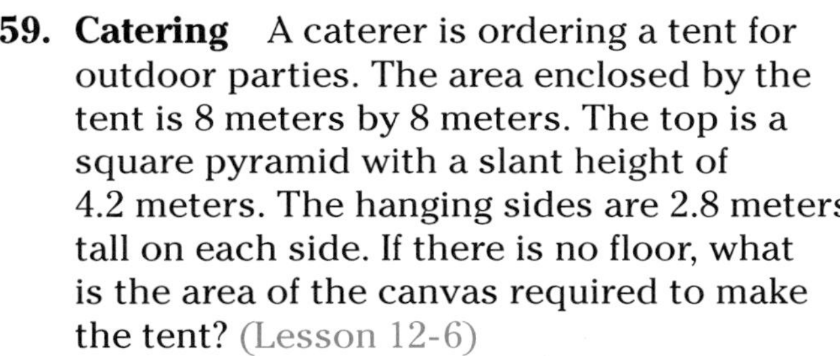

59. **Catering** A caterer is ordering a tent for outdoor parties. The area enclosed by the tent is 8 meters by 8 meters. The top is a square pyramid with a slant height of 4.2 meters. The hanging sides are 2.8 meters tall on each side. If there is no floor, what is the area of the canvas required to make the tent? (Lesson 12-6)

60. The stem-and-leaf plot shows the number of canned goods collected by each class for the food drive. (Lesson 10-2)
    a. What is the range?
    b. What is the median?
    c. What is the interquartile range?
    d. Are there any outliers?

Stem	Leaf
1	98
2	10 56 78 80 85
3	08 12 60 62 75 80
4	15
5	25

1 | 98 = 198 cans

61. **Retail Sales** The retail price of a camera is \$98. The sales tax rate is 6.5% of the price. What is the total cost of the camera? (Lesson 9-5)

62. Solve $\frac{2x}{5} - 3 > -5$. (Lesson 7-6)

63. State whether 1, 3, 9, 27, . . . is a geometric sequence. If so, state the common ratio and list the next three terms. (Lesson 6-8)

64. **Geometry** A rectangle has a length of 3.5 meters and a width of 1.75 meters. What is the perimeter of the rectangle? (Lesson 5-3)

65. Write an inequality for the solution set graphed below. (Lesson 3-6)

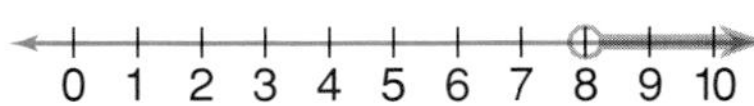

66. Find the quotient of 48 and -3. (Lesson 2-8)

67. Simplify $|-8| + |5|$. (Lesson 2-1)

14-1B

HANDS-ON ACTIVITY

Representing Polynomials with Algebra Tiles

An Extension of Lesson **14-1**

MATERIALS

algebra tiles

You can use the concept of area to model monomials. A model of a monomial can be made using tiles or drawings like the ones shown below.

Let [1 × 1 tile] be a 1×1 square. The square has an area of 1.

Let [1 × x tile] be a $1 \times x$ rectangle. The rectangle has an area of x.

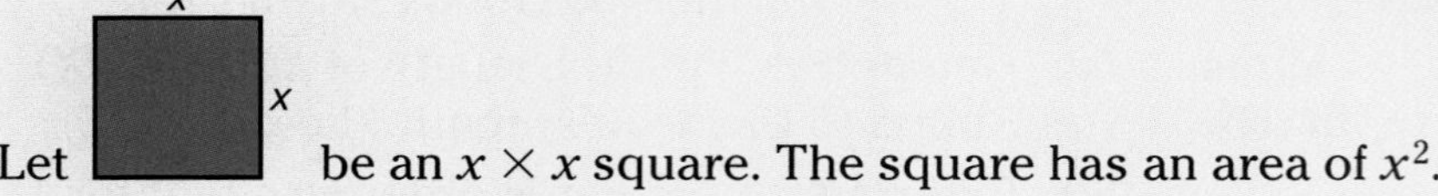

Let [x × x tile] be an $x \times x$ square. The square has an area of x^2.

Red tiles are used to represent -1, $-x$, and $-x^2$.

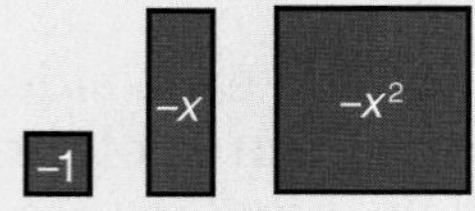

You can use these tiles to make a model of a polynomial.

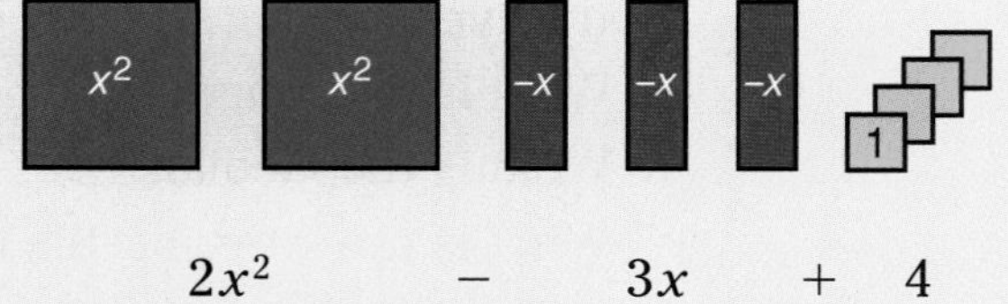

$2x^2 \quad - \quad 3x \quad + \quad 4$

Your Turn ▶ Model $-3x^2$, $5x + 3$, $4x^2 - x$, and $5x^2 + 2x - 3$ using algebra tiles.

1. Explain how you can tell if an expression is a monomial, binomial, or trinomial by looking at the area tiles.
2. Name the polynomial represented by the model below.

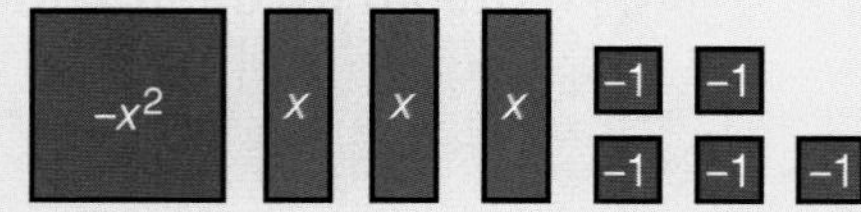

3. Explain how you would find the degree of a polynomial using algebra tiles.

14-2 Adding Polynomials

Setting Goals: *In this lesson, you'll add polynomials.*

Modeling with Manipulatives

MATERIALS

algebra tiles

Algebra tiles can be used to model addition of polynomials.

- The numerical part of a monomial is called the **coefficient**. For example, the coefficient of $5x^2$ is 5.
- When monomials are the same or differ only by their coefficients they are called **like terms**. For example, $2x$ and $7x$, and $-5x^2y$ and $9x^2y$ are like terms; $6a$ and $-5b$, and $7a^3b^2$ and $7a^2b$ are unlike terms. Like terms are represented by tiles that are the same shape and size.
- A zero pair is formed by pairing one tile with its opposite.
- You can remove or add zero pairs without changing the value of the polynomial.

The model below represents $(2x^2 + 2x - 4) + (x^2 + 3x + 6)$.

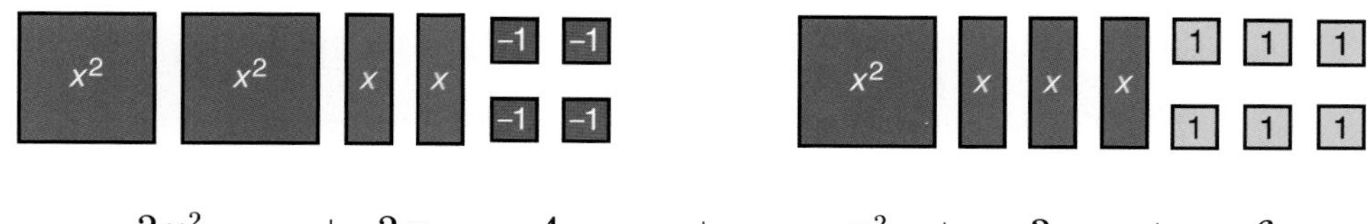

$2x^2 \quad + \quad 2x \quad - \quad 4 \quad + \quad x^2 \quad + \quad 3x \quad + \quad 6$

- To add the polynomials, combine like terms and remove all zero pairs.

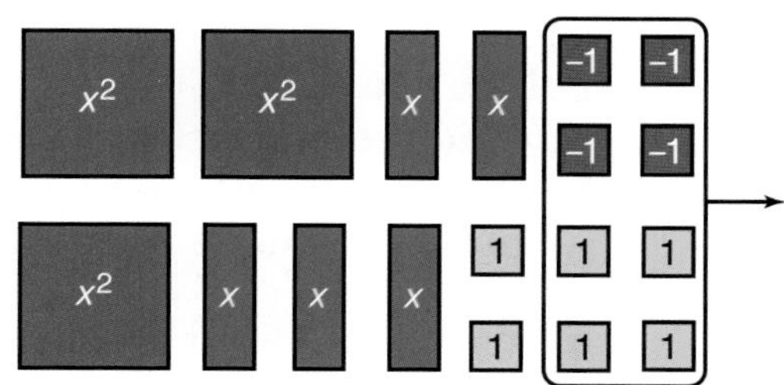

- Write the polynomial for the tiles that remain.
  Therefore, $(2x^2 + 2x - 4) + (x^2 + 3x + 6) = 3x^2 + 5x + 2$.

Your Turn Model $(x^2 + 4x + 2) + (7x^2 - 2x + 3)$ using algebra tiles.

a. What is the sum of $(x^2 + 4x + 2)$ and $(7x^2 - 2x + 3)$?

b. Compare and contrast finding sums of polynomials using algebra tiles to finding the sum of integers using counters.

Learning the Concept

You can use the principles and properties you learned about integers and rational numbers to add polynomials.

Example 1 **Find each sum.**

a. $(5x + 7) + (3x + 3)$

Method 1: Add vertically

$$\begin{array}{r} 5x + 7 \\ (+)\ 3x + 3 \\ \hline 8x + 10 \end{array}$$

Notice that the terms are aligned and added.

Method 2: Add horizontally

$(5x + 7) + (3x + 3)$

$= (5x + 3x) + (7 + 3)$ *Associative and commutative properties of addition*

$= (5 + 3)x + (7 + 3)$ *Distributive property*

$= 8x + 10$

The sum is $8x + 10$.

b. $(3x^2 + x + 2) + (x^2 + 2x + 4)$

$$(3x^2 + x + 2) + (x^2 + 2x + 4) = (3x^2 + x^2) + (x + 2x) + (2 + 4)$$
$$= 4x^2 + 3x + 6$$

The sum is $4x^2 + 3x + 6$.

c. $(2x^2 + 5xy + 7y^2) + (4x^2 - 3y^2)$

$$\begin{array}{r} 2x^2 + 5xy + 7y^2 \\ (+)\ 4x^2 \qquad\quad - 3y^2 \\ \hline 6x^2 + 5xy + 4y^2 \end{array}$$

The sum is $6x^2 + 5xy + 4y^2$.

Polynomials are often used to represent measures of geometric figures.

Example 2

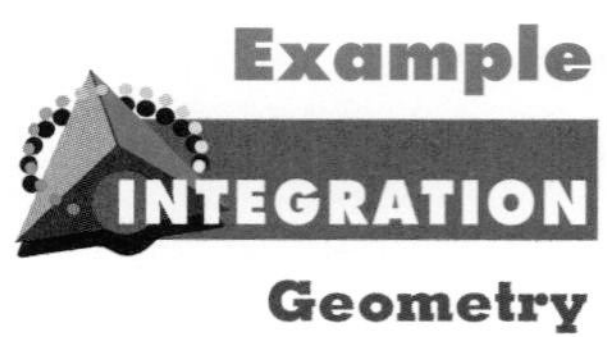

Geometry

The lengths of the sides of golden rectangles are in a special ratio. The rectangle at the right shows the lengths of the sides of a general golden rectangle.

a. Find the perimeter of the rectangle.

b. Find the length, the width, and the perimeter of the rectangle if x is 5.2.

a. $x + 1.618x + x + 1.618x = (1 + 1.618 + 1 + 1.618)x$
$= 5.236x$

b. If x is 5.2, then the width is 5.2 units, the length is 1.618(5.2) units, and the perimeter is 5.236(5.2) units.

Estimate the length and perimeter first and then use a calculator.

Length: (1.618)(5.2) is about $2 \cdot 5$ or 10 units.

1.618 [×] 5.2 [=] 8.4136

Perimeter: (5.236)(5.2) is about $5 \cdot 5$ or 25 units.

5.236 [×] 5.2 [=] 27.2272

Checking Your Understanding

Communicating Mathematics

Read and study the lesson to answer these questions.

1. **Name** the like terms in $(x^2 + 7x + 1) + (2x^2 - 3x + 5)$.
2. **Demonstrate** how to find the sum of $5x^2 + 6x + 4$ and $2x^2 + 3x + 1$.
3. **You Decide** Caroline says that $4xyz$ and $3zyx$ are like terms. Nikita says that they are not. Who is correct? Explain.

MATERIALS

algebra tiles

Use algebra tiles to find each sum.

4. $(3x^2 - 2x + 1) + (x^2 + 5x - 3)$
5. $(-2x^2 + x - 5) + (x^2 - 3x + 2)$
6. $(x^2 - 3x + 6) + (2x^2 + 5x - 4) + (x^2 + x + 1)$

Guided Practice

Find each sum.

7. $(x + 3) + (2x + 5)$
8. $(4x + 3) + (x - 1)$
9. $(3x - 5) + (x + 9)$
10. $(2x - 3) + (x - 1)$
11. $\begin{array}{r} 2x^2 + 4x + 5 \\ (+)\ x^2 - x - 3 \\ \hline \end{array}$
12. $\begin{array}{r} 2x^2 - 5x + 4 \\ (+)\ 3x^2 + 8x - 1 \\ \hline \end{array}$

13. **Geometry** Find the perimeter of the rectangle shown at the right.

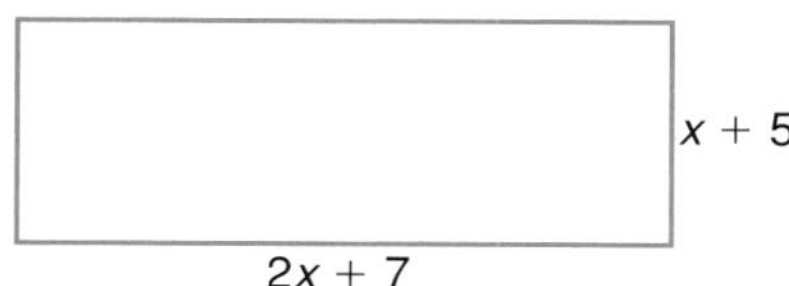

Exercises: Practicing and Applying the Concept

Independent Practice

Find each sum.

14. $(6y - 5r) + (2y + 7r)$
15. $(3x + 9) + (x + 5)$
16. $(11x + 2y) + (x - 5y)$
17. $(8m - 2n) + (3m + n)$
18. $(13x - 7y) + 3y$
19. $(2x^2 + 5x) + (9 - 7x)$
20. $(3r + 6s) + (5r - 9s)$
21. $(5m + 3n) + 12m$
22. $(3x^2 - 9x + 5) + (5x^2 + 5x - 11)$
23. $(5x^2 - 7x + 9) + (3x^2 + 4x - 6)$
24. $(6x^2 + 15x - 9) + (5 - 8x - 8x^2)$
25. $(a^3 - b^3) + (3a^3 + 2a^2b - b^2 + 2b^3)$
26. $\begin{array}{r} -6y^2 + 7b - 5 \\ (+)\ 2y^2 - 9b + 8 \\ \hline \end{array}$
27. $\begin{array}{r} 3a + 5b - 4c \\ 2a - 3b + 7c \\ (+)\ -a + 4b - 2c \\ \hline \end{array}$
28. $\begin{array}{r} 3x^2 - 7x + 9 \\ -2x^2 + x - 4 \\ (+)\ x^2 + 3x - 1 \\ \hline \end{array}$
29. $(3a + 5ab - 3b^2) + (7b^2 - 8ab) + (2 - 5a)$
30. $(x^2 + x + 5) + (3x^2 - 4x - 2) + (2x^2 + 2x - 1)$
31. $(3x^2 + 7) + (4x - 2) + (x^2 - 3x - 6)$

Find each sum. Then, evaluate if $a = -3$ and $b = 4$.

32. $(3a + 5b) + (2a - 9b)$
33. $(a^2 - 3ab + b^2) + (3a^2 - 2b - 5b^2)$
34. $(a^2 + 7b^2) + (5 - 3b^2) + (2a^2 - 7)$

Critical Thinking

35. If $(4r + 3s) + (6r - 5s) = (10r - 2s)$, find $(10r - 2s) - (6r - 5s)$.

Applications and Problem Solving

36. **Geometry** The measures of the angles of a triangle are shown in the figure at the right.

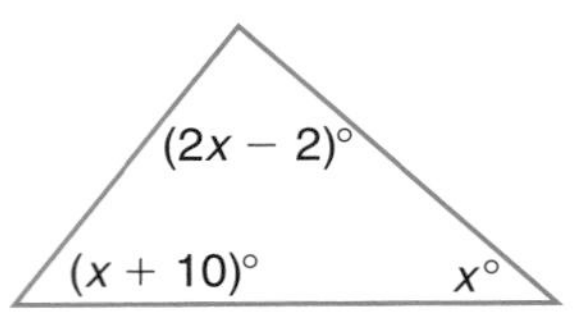

a. Find the sum of the measures of the angles.

b. The sum of the measures of the angles in any triangle is 180°. Find the value of x.

c. Find the measure of each angle.

37. **Construction** A standard measurement for a window is the *united inch*. You can find the united inches of a window by adding the length of the window to the width. If the length of a window is $3x - 5$ inches and the width is $x + 7$, what is the size of the window in united inches?

Mixed Review

38. State whether $a^2 - \frac{1}{4}$ is a polynomial. If it is, classify it as a *monomial*, *binomial*, or *trinomial*. (Lesson 14-1)

39. **Geometry** Find the area of the triangle at the right. (Lesson 12-1)

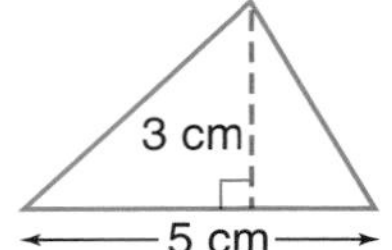

40. **Consumer Awareness** At a 20%-off sale, Antjuan bought a book that regularly sells for \$15.98. How much is the discount? (Lesson 9-9)

41. **Weather** When the sum of the temperature (in °F) and humidity exceeds 130, people generally find the weather to be uncomfortable. (Lesson 8-9)

a. Write an inequality to represent this situation. Then graph the inequality.

b. Which area of the graph indicates uncomfortable weather?

c. Humidity cannot exceed 100%. How does this affect the graph?

42. Consider the sequence 2, 6, 11, 17, 24, . . . (Lesson 5-9)

a. Is the sequence arithmetic?

b. Write the next three terms of the sequence.

43. **Geometry** A rectangle has a perimeter of 50 feet and an area of 150 square feet. Its width is 10 feet. What is its length? (Lesson 3-5)

44. **Statistics** The table at the right shows the percent of American homes that have the most popular pets. Make a bar graph of the data. (Lesson 1-10)

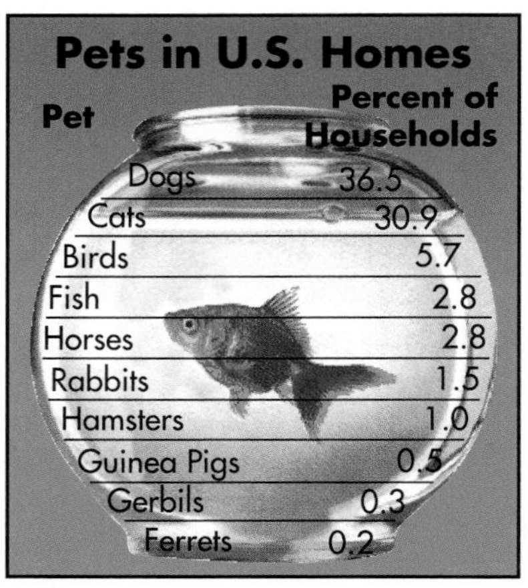

Pets in U.S. Homes

Pet	Percent of Households
Dogs	36.5
Cats	30.9
Birds	5.7
Fish	2.8
Horses	2.8
Rabbits	1.5
Hamsters	1.0
Guinea Pigs	0.5
Gerbils	0.3
Ferrets	0.2

Source: American Veterinary Medical Association

45. **Sports** Two hundred thirty-six women were members of the United States Olympic team for 1992. That number grew to 298 in 1996. How many more women were on the 1996 team than were on the 1992 team? (Lesson 1-6)

14-3 Subtracting Polynomials

Setting Goals: *In this lesson, you'll subtract polynomials.*

Modeling a Real-World Application: Running

FYI

German Silva of Mexico was the 1995 men's winner with a time of 2 hours 11 minutes. He also won the race in 1994.

Tegla Loroupe of Kenya was the first woman to cross the finish line of the 1994 New York Marathon. She completed the 26.2-mile course in 2 hours 27 minutes 37 seconds. In 1995, she won again with a time of 2 hours 28 minutes 6 seconds. How much faster was her 1994 time?

To find the difference of the times you must subtract.

2 hours 28 minutes 6 seconds	→	2 hours 27 minutes 66 seconds
− 2 hours 27 minutes 37 seconds		− 2 hours 27 minutes 37 seconds
		29 seconds

Miss Loroupe's 1994 time was 29 seconds faster.

Learning the Concept

In the application above, like units were subtracted from each other. Similarly, in algebra, when one polynomial is subtracted from another, only like terms can be subtracted.

Algebra tiles were used to help you understand how to add polynomials. They can also be used to help you understand how to subtract polynomials.

Example  **Find $(4x^2 + 7x + 4) - (x^2 + 2x + 1)$ by using algebra tiles.**

Make a model of $4x^2 + 7x + 4$ as shown in the diagram below. Then remove 1 x^2-tile, 2 x-tiles, and 1 1-tile.

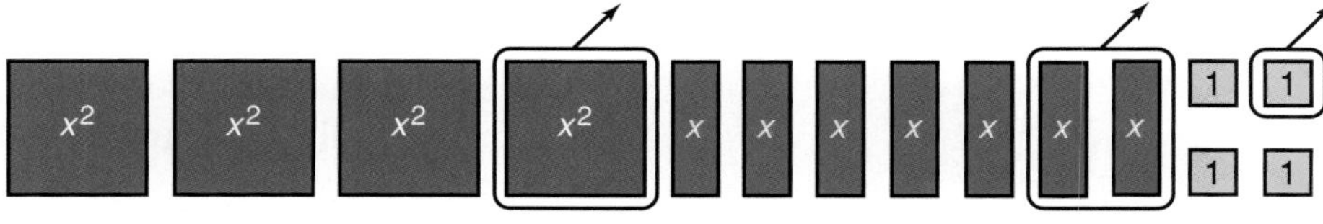

To find the difference, count the number of x^2-tiles, x-tiles, and 1-tiles that remain. There are 3 x^2-tiles, 5 x-tiles, and 3 1-tiles remaining.

$(4x^2 + 7x + 4) - (x^2 + 2x + 1) = 3x^2 + 5x + 3.$

To subtract polynomials, you can use the same methods you used to add them.

Example 2 **Find $(5x + 9) - (3x + 6)$.**

Method 1: Subtract Vertically

$$\begin{array}{r} 5x + 9 \\ (-)\ 3x + 6 \\ \hline 2x + 3 \end{array}$$

The difference is $2x + 3$.

Method 2: Subtract Horizontally

$$\begin{aligned}(5x + 9) - (3x + 6) &= 5x + 9 - 3x - 6 \\ &= (5x - 3x) + (9 - 6) \\ &= (5 - 3)x + (9 - 6) \\ &= 2x + 3\end{aligned}$$

Recall that you can subtract a rational number by adding its additive inverse. You can also subtract a polynomial by adding its additive inverse.

To find the additive inverse of a polynomial, it is necessary to multiply the entire polynomial by -1. Study the examples in the table below.

Polynomial	Multiply by -1	Additive Inverse
$-y$	$-1(-y)$	y
$4x - 3$	$-1(4x - 3)$	$-4x + 3$
$2xy + 5y$	$-1(2xy + 5y)$	$-2xy - 5y$
$2x^2 - 3x - 5$	$-1(2x^2 - 3x - 5)$	$-2x^2 + 3x + 5$

Example  **Find each difference.**

a. $(7x + 5) - (3x + 2)$

$$\begin{aligned} &= (7x + 5) + (-1)(3x + 2) && \textit{Add the additive inverse of } 3x + 2. \\ &= 7x + 5 + (-3x - 2) && (-1)(3x + 2) = -3x - 2 \\ &= 7x + 5 - 3x - 2 \\ &= 7x - 3x + 5 - 2 && \textit{Commutative property of addition} \\ &= 4x + 3 \end{aligned}$$

b. $(3x^2 - 5xy + 7y^2) - (x^2 - 3xy + 4y^2)$

Align like terms and add the additive inverse of $x^2 - 3xy + 4y^2$.

$$\begin{array}{r} 3x^2 - 5xy + 7y^2 \\ (-)\ x^2 - 3xy + 4y^2 \\ \hline \end{array} \rightarrow \begin{array}{r} 3x^2 - 5xy + 7y^2 \\ (+)\ -x^2 + 3xy - 4y^2 \\ \hline 2x^2 - 2xy + 3y^2 \end{array}$$

Example 

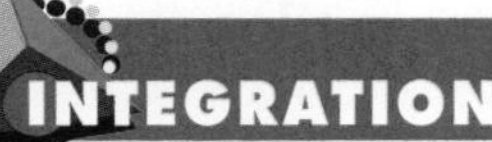

Geometry

Lashonda plans to mat and frame a picture. The area inside the frame is $3x^2 - 5$ square inches. The area of the picture is $x^2 + 5$ square inches. How much matting will Lashonda need?

Explore You know the area inside the frame and the area of the picture.

Plan The area of the picture is smaller than the area inside the frame. Subtract the area of the picture from the area inside the frame to find the area needed for the matting.

Solve

$$\begin{aligned}(3x^2 - 5) - (x^2 + 5) &= 3x^2 - 5 + (-1)(x^2 + 5) \\ &= 3x^2 - 5 + (-x^2 - 5) \\ &= 3x^2 - 5 - x^2 - 5 \\ &= 3x^2 - x^2 - 5 - 5 \\ &= 2x^2 - 10\end{aligned}$$

Lashonda will need $2x^2 - 10$ square inches of matting.

Examine Check by aligning like terms and subtracting.

Checking Your Understanding

Communicating Mathematics

Read and study the lesson to answer each question.

1. **Describe** how subtraction and addition of polynomials are related.
2. **Explain** how to find the additive inverse of a given polynomial. What is the additive inverse of $3x^2 - 5x + 7$?
3. **Write** the sum of a polynomial and its additive inverse.
4. **Write** a polynomial subtraction problem with a difference of $3m^2 - 5m + 4$. Check your work by adding.

MATERIALS

algebra tiles

Use algebra tiles to find each difference.

5. $(5x^2 + 6x + 4) - (3x^2 + 2x + 1)$
6. $(2x^2 + 5x + 3) - (x^2 - 2x + 3)$
7. $(3x^2 + 2x - 5) - (2x^2 + x - 4)$
8. $(8x^2 + 2x - 1) - (2x^2 + x + 5)$

Guided Practice

State the additive inverse of each polynomial.

9. $2abc$
10. $3x + 2y$
11. $x^2 + 5x + 1$
12. $3x^2 - 2x + 5$
13. $-8m + 7n$
14. $-4h^2 - 5hk - k^2$

Find each difference.

15. $(3x + 4) - (x + 2)$
16. $(4x + 5) - (2x + 3)$
17. $(5x - 7) - (3x - 4)$
18. $(2x - 5) - (3x + 1)$
19. $\begin{array}{r} 5x^2 + 4x - 1 \\ (-)\ 4x^2 + x + 2 \\ \hline \end{array}$
20. $\begin{array}{r} 3x^2 + 5x + 4 \\ (-)\ x^2 \qquad - 1 \\ \hline \end{array}$

21. **Geometry** Find how much longer the hypotenuse is than the shorter leg of the triangle shown at the right.

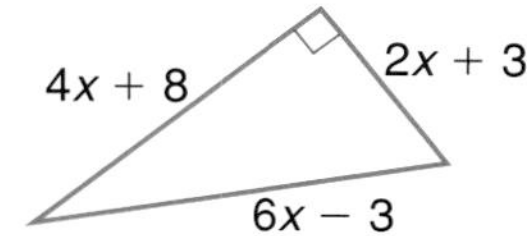

Exercises: Practicing and Applying the Concept

Independent Practice

Find each difference.

22. $(9x + 5) - (4x + 3)$
23. $(2x + 5) - (x + 8)$
24. $(3x - 2) - (5x - 4)$
25. $(6x - 5) - (4x + 3)$
26. $(2x + 3y) - (x - y)$
27. $(9x - 4y) - (12x - 9y)$
28. $(5x^2 - 3) - (2x^2 - 7)$
29. $(x^2 + 6x) - (3x^2 + 7)$
30. $\begin{array}{r} 5a^2 + 7a + 9 \\ (-)\ 3a^2 + 4a + 1 \\ \hline \end{array}$
31. $\begin{array}{r} 6m^2 - 5m + 3 \\ (-)\ 5m^2 + 2m - 7 \\ \hline \end{array}$
32. $\begin{array}{r} 5x^2 - 4xy \qquad \\ (-)\ \qquad - 3xy + 2y^2 \\ \hline \end{array}$
33. $\begin{array}{r} 9m^2 \qquad + 7 \\ (-)\ -6m^2 + 2m - 3 \\ \hline \end{array}$
34. $\begin{array}{r} 15x^2y^2 + 11xy - 9 \\ (-)\ 9x^2y^2 - 13xy + 6 \\ \hline \end{array}$
35. $\begin{array}{r} 14a + 10b - 18c \\ (-)\ 5a + 7b - 11c \\ \hline \end{array}$
36. $(10x^2 + 8x - 6) - (3x^2 + 2x - 9)$
37. $(5y^2 + 9y - 12) - (-3y^2 + 5y - 7)$
38. $(6a^2 + 7ab - 3b^2) - (2a^2 + 3ab - b^2)$
39. $(x^3 - 3x^2y + 4xy^2 + y^3) - (7x^3 - 9xy^2 + x^2y + y^3)$

Critical Thinking

40. Suppose that A and B represent polynomials. If $A + B = 3x^2 + 2x - 2$ and $A - B = -x^2 + 4x - 8$, find A and B.

Applications and Problem Solving

interNET CONNECTION

For the latest basketball scores, visit: **www.glencoe.com/sec/math/prealg/mathnet**

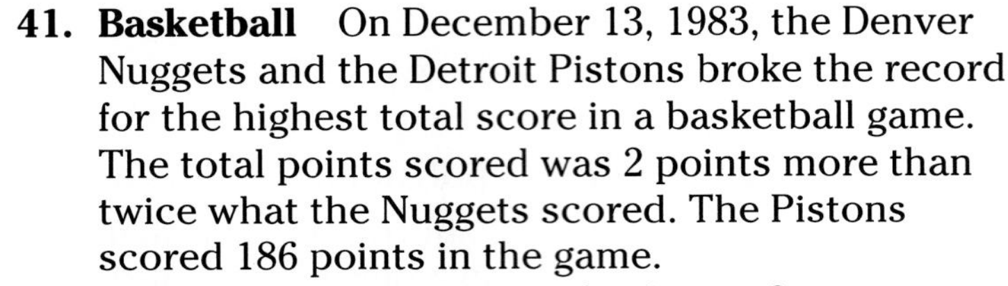

41. Basketball On December 13, 1983, the Denver Nuggets and the Detroit Pistons broke the record for the highest total score in a basketball game. The total points scored was 2 points more than twice what the Nuggets scored. The Pistons scored 186 points in the game.

a. What was the Nuggets final score?

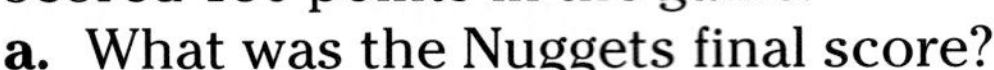

b. How many points did the two teams score?

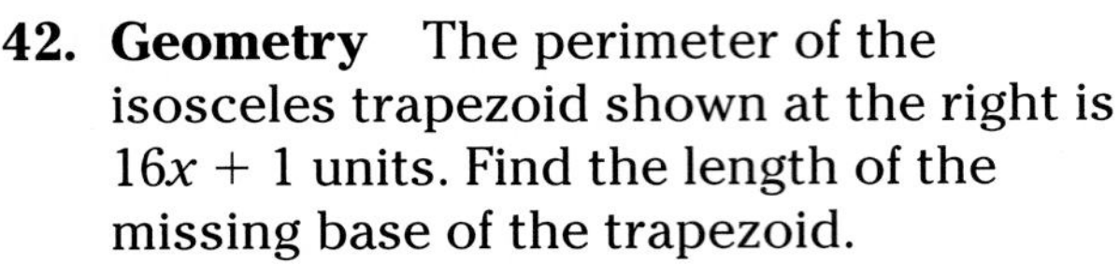

42. Geometry The perimeter of the isosceles trapezoid shown at the right is $16x + 1$ units. Find the length of the missing base of the trapezoid.

Mixed Review

43. Find the sum $(2a^2 + 3a - 4) + (6a^2 - a + 5)$. (Lesson 14-2)

44. Solve $n^2 = 18$ to the nearest tenth. (Lesson 13-3)

45. Geometry In $\triangle CDE$, $m\angle C = 35°$ and $m\angle D = 55°$. Find the measure of $\angle E$ and classify the triangle as *acute*, *right*, or *obtuse*. (Lesson 11-4)

46. Graph $y = -x + 5$ by using the x-intercept and the y-intercept. (Lesson 8-7)

47. Patterns Two groups share the same meeting room. If the debate team meets every 3rd school day and the dance committee meets every 4th school day, how often will both groups need the room on the same day? (Lesson 4-7)

48. Solve $x = 32 + 56 + (-18)$. (Lesson 2-4)

49. Geometry In which quadrant does the point at $(-5, 3)$ lie? (Lesson 2-2)

50. Solve $8c = 72$ mentally. (Lesson 1-6)

Self Test

State whether each expression is a polynomial. If it is, classify it as a *monomial*, a *binomial*, or a *trinomial*. Then find the degree of each polynomial. (Lesson 14-1)

1. $ax^2 + 6x$ **2.** $4b^2 + c^3d^4 + x$ **3.** $a^3b^4c^5$

Find each sum. (Lesson 14-2)

4. $(4x + 5y) + (7x - 3y)$ **5.** $(2x^2 + 5) + (-3x^2 + 7)$ **6.** $(9m - 3n) + (10m + 4n)$

7. Geometry The perimeter of the triangle shown at the right is $7x + 2y$ units. Find the length of the third side of the triangle. (Lessons 14-1 and 14-2)

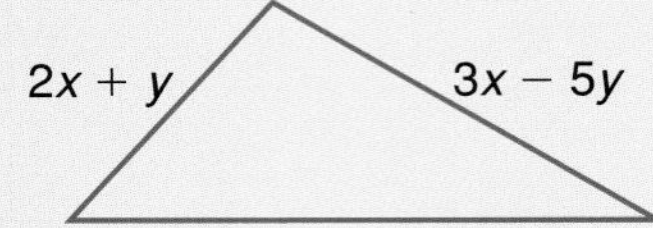

Find each difference. (Lesson 14-3)

8. $(11p + 5r) - (2p + r)$ **9.** $(7a + 6d) - (6a - 7d)$ **10.** $(4t + 11r) - (t + 2r)$

14-4 Powers of Monomials

Setting Goals: *In this lesson, you'll find powers of monomials.*

Modeling with Technology

Recall that when you multiply powers that have the same base, you add the exponents. For example, $3^2 \cdot 3^3 = 3^{2+3}$ or 3^5. But how do you find the power of a power like $(3^2)^3$? You can use a calculator to discover a pattern.

Your Turn

The [^] or [y^x] key allows you to evaluate expressions with exponents.

Copy the table below. Use a calculator to find each value and complete the table.

Power	Value	Power	Value
3^1	3	$(3^2)^1$	9
3^2		$(3^2)^2$	
3^3		$(3^2)^3$	
3^4		$(3^2)^4$	
3^5		$(3^2)^5$	
3^6		$(3^2)^6$	

a. Compare the values of 3^6 and $(3^2)^3$. What do you observe?
b. Compare the exponents of 3^6 and $(3^2)^3$. What do you observe?
c. Guess the value of $(3^4)^2$. Use a calculator to check your answer.
d. Write a rule for finding the value of a power of a power.

Learning the Concept

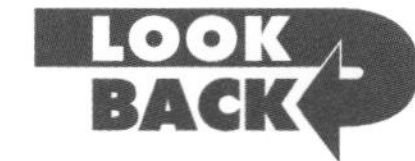

You can review multiplying powers in Lesson 4-8.

Consider the following powers.

$$
\begin{aligned}
(5^4)^2 &= 5^4 \cdot 5^4 && \textit{Definition of exponent} \\
&= 5^{4+4} && \textit{Product of powers} \\
&= 5^8 && \textit{Substitution}
\end{aligned}
\qquad
\begin{aligned}
(x^2)^3 &= (x^2) \cdot (x^2) \cdot (x^2) \\
&= x^{2+2+2} \\
&= x^6
\end{aligned}
$$

These and other similar examples suggest that you can find a *power of a power* by multiplying the exponents.

Power of a Power — **For any number a and positive integers m and n, $(a^m)^n = a^{mn}$.**

Example 1 **Simplify $(p^6)^2$.**

$(p^6)^2 = p^{6 \cdot 2}$ *Power of a power*

$= p^{12}$

Check: $(p^6)^2 = p^6 \cdot p^6$

$= p^{6+6}$ or p^{12} ✓

Connection to Geometry

The volume of a cube can be found by multiplying its length, width, and height. The volume of the small cube shown at the right is $x \cdot x \cdot x$ or x^3. If the length of each edge of the cube is doubled, the volume of the new cube would be $(2x)^3$.

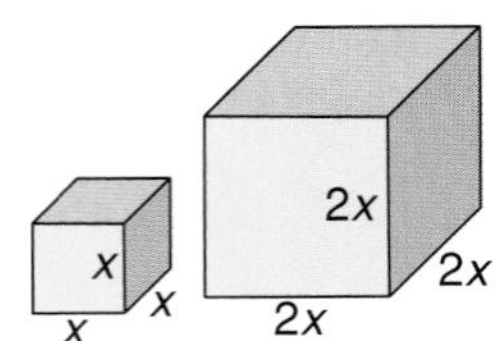

$(2x)^3 = (2x)(2x)(2x)$ *Definition of exponent*

$= (2 \cdot 2 \cdot 2)(x \cdot x \cdot x)$ *Commutative and associative properties*

$= 2^3x^3$ *Exponential notation*

$= 8x^3$ *$2^3 = 8$*

Notice that when $2x$ was raised to the third power, *both* the 2 and the x were raised to the third power. This example suggests that the power of a product can be found by multiplying the individual powers.

Power of a Product — **For all numbers a and b and positive integer m, $(ab)^m = a^mb^m$.**

Example 2

Geometry

The formula for the volume of a pyramid is $V = \frac{1}{3}\ell wh$.

a. Find the measure of the volume of the pyramid.

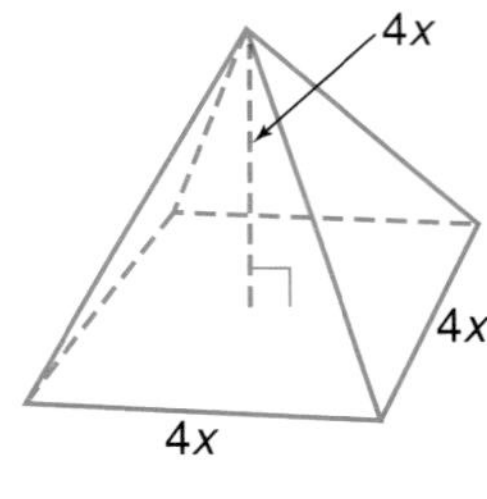

$V = \frac{1}{3}\ell wh$

$= \frac{1}{3}(4x)(4x)(4x)$

$= \frac{1}{3}(4 \cdot 4 \cdot 4)(x \cdot x \cdot x)$

$= \frac{1}{3}(4^3)(x^3)$

$= \frac{1}{3}(4x)^3$ *Power of a product*

The measure of the volume of the pyramid is $\frac{1}{3}(4x)^3$.

b. If x is 5 cm, what is the volume of the pyramid?

$\frac{1}{3}(4x)^3 = \frac{1}{3}(4 \cdot 5)^3$ *Replace x with 5.*

$= \frac{1}{3}(20)^3$ *Estimate: $\frac{1}{3} \cdot 20 \cdot 20 \cdot 20 = 7 \cdot 20 \cdot 20$ or 2800*

20 [y^x] 3 [÷] 3 [=] 2666.6666

The volume is about 2667 cm^3. *Compare to the estimate.*

Sometimes the rules for the *power of a power* and the *power of a product* are combined into one rule.

Power of a Monomial For all numbers a and b and positive integer m, n, and p, $(a^m b^n)^p = a^{mp} b^{mp}$.

Example 3 Simplify $(x^2y^3)^4$.

$$(x^2y^3)^4 = (x^2)^4(y^3)^4 \quad \textit{Power of a monomial}$$
$$= x^{2\cdot 4}y^{3\cdot 4}$$
$$= x^8y^{12}$$

Check:
$$(x^2y^3)^4 = (x^2y^3)(x^2y^3)(x^2y^3)(x^2y^3)$$
$$= (x^2 \cdot x^2 \cdot x^2 \cdot x^2)(y^3 \cdot y^3 \cdot y^3 \cdot y^3)$$
$$= (x^{2+2+2+2})(y^{3+3+3+3})$$
$$= (x^8)(y^{12}) \quad \text{or} \quad x^8y^{12} \ \checkmark$$

These rules can also be used with negative exponents.

Example 4 Simplify $(x^3y^2z)^{-2}$.

$$(x^3y^2z)^{-2} = x^{3(-2)}y^{2(-2)}z^{1(-2)} \quad \textit{Power of a monomial}$$
$$= x^{-6}y^{-4}z^{-2}$$

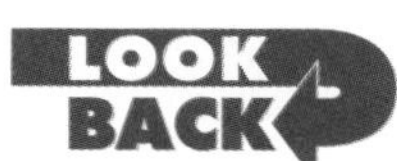

You can review negative exponents in Lesson 4-9.

Check:
$$(x^3y^2z)^{-2} = \frac{1}{(x^3y^2z)^2}$$
$$= \frac{1}{(x^3y^2z)(x^3y^2z)}$$
$$= \frac{1}{(x^3 \cdot x^3)(y^2 \cdot y^2)(z \cdot z)}$$
$$= \frac{1}{x^{3+3}y^{2+2}z^{1+1}}$$
$$= \frac{1}{x^6y^4z^2} \quad \text{or} \quad x^{-6}y^{-4}z^{-2} \ \checkmark$$

Checking Your Understanding

Communicating Mathematics

Read and study the lesson to answer each question.

1. **You Decide** Nelia says that $5x^3$ and $(5x)^3$ are equivalent expressions. Julie says they are not. Who is correct? Explain.
2. **Analyze** the square shown at the right. What is its area? What would its area be if the length of each side were doubled or tripled?

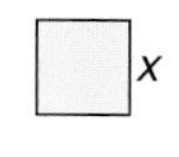

3. **Assess Yourself** How well do you understand the rules given in this lesson? For each of the following, write the rules in your own words and then give an example of it.
   a. Power of a power
   b. Power of a product
   c. Power of a monomial

Guided Practice

Simplify.

4. $(y^5)^3$
5. $(2^4)^3$
6. $(2m)^4$
7. $(3xy)^3$
8. $(a^2b^3)^2$
9. $(-xy)^2$

Evaluate each expression if $a = -1$ and $b = 3$.

10. a^2b
11. $-ab^2$
12. $(-4a^3b)^2$

13. **Geometry** Find the volume of the rectangular prism at the right.

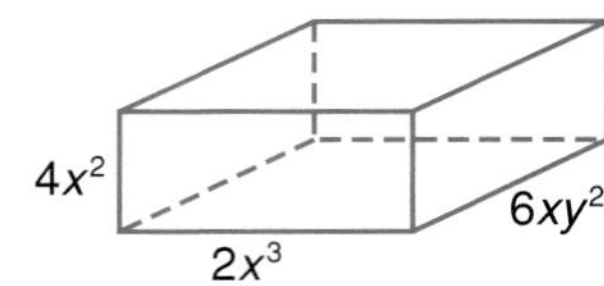

Exercises: Practicing and Applying the Concept

Independent Practice

Simplify.

14. $(7^3)^2$
15. $(yz)^4$
16. $(-3w)^3$
17. $(-2rs)^4$
18. $[(-4)^2]^2$
19. $(m^2)^5$
20. $(-y^3)^6$
21. $(-2x^2)^5$
22. $(x^2y)^3$
23. $(-xy^3)^4$
24. $(2a^3b)^5$
25. $3x(2x)^2$
26. $-5x(x^3)^2$
27. $(4x^2y^3)^2$
28. $[(-3)^2]^3$
29. $(t^3)^{-4}$
30. $(-w^2)^{-5}$
31. $-2(x^3y)^{-2}$

Evaluate each expression if $a = -2$ and $b = 3$.

32. $2ab^2$
33. $-3a^2b$
34. $-(ab^2)^2$
35. $(-2ab^3)$
36. $(a^3b)^2$
37. $-(3a^2b)^2$
38. $-3b(2a)^{-2}$
39. $2b^2(-3ab)^{-3}$
40. $2(3a^{-2})^2$

Calculators

41. You can use a calculator to confirm the power of a product rule. Copy and complete the following table using a calculator.

Monomial	Value	Monomial	Value
$(2 \cdot 3)^2$		$2^2 \cdot 3^2$	
$(4 \cdot 5)^3$		$4^3 \cdot 5^3$	
$(6 \cdot 7)^4$		$6^4 \cdot 7^4$	
$(8 \cdot 3)^5$		$8^5 \cdot 3^5$	

a. Compare the values in each row. What do you observe?
b. Compare the monomials in each row. What do you observe?
c. Write a rule for finding the power of a product in your own words.

Critical Thinking

42. Are $(2^6)^4$ and $(4^6)^2$ equal? Explain why or why not. (*Hint:* Write both expressions as a power of 2.)

Applications and Problem Solving

43. **Finance** Nuna opens a savings account with \$1000. It earns 1% interest every 3 months. So, in one year, the balance is $1000(1.01)(1.01)(1.01)(1.01)$ or $1000(1.01)^4$. If she keeps the money in the account for 3 years, the deposit is worth $1000(1.01^4)^3$.
   a. Write an expression that represents what the deposit is worth in 6 years.
   b. What is the deposit worth in 6 years?

44. **Biology** It takes 2 hours for a culture with 1 bacterium to grow to two bacteria. To grow to 32, or 2^5, bacteria, the culture must double five times. If the culture doubles another five times, there will be $(2^5)^2$ bacteria.

   a. Write $(2^5)^2$ as a power of 2.

   b. $2^5 = 32$. Use this information to estimate $(2^5)^2$.

45. **Chemistry** Chemicals like salt and magnesium can be extracted from sea water. To determine if such a venture would be profitable, a company would like to know how much salt or magnesium could be extracted from a cubic mile of sea water. How many cubic feet of sea water are in a cubic mile of sea water? (*Hint:* There are 5280 feet in a mile.)

Mixed Review

46. **Geometry** The perimeter of the triangle at the right is $5x + 2$ units. Find the length of the base of the triangle. (Lesson 14-3)

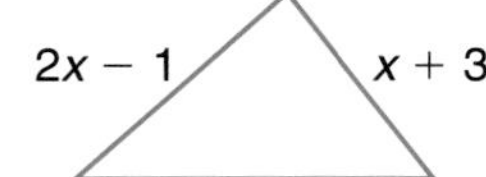

47. **Geometry** In a right triangle, if the measures of the legs are 5 inches and 12 inches, find the measure of the hypotenuse. (Lesson 13-4)

48. **Geometry** Write a congruence statement for the pair of congruent triangles at the right. Then name the corresponding parts. (Lesson 11-5)

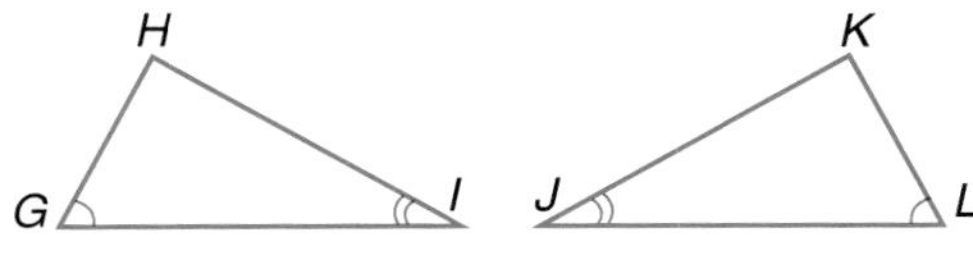

49. **Probability** Betsy plays on a basketball team. Her free throw average is 60%. How could you simulate the probable results of her next five free throw attempts? (Lesson 10-8)

50. Solve $\frac{2x}{-5} = -24$. (Lesson 7-2)

51. On graph paper, draw coordinate axes. Then graph and label each point. Name the quadrant in which each point is located. (Lesson 2-2)

   a. $X(-4, 6)$ b. $Y(2, 5)$ c. $Z(-3, 0)$

52. Evaluate each expression if $x = -2$, $y = 4$, and $z = -1$. (Lesson 2-1)

   a. $|2x| + yz$ b. $\frac{4(x + y)}{2z}$ c. $\frac{xyz}{8}$

53. Write an expression that means the same as $\frac{xy}{5}$. (Lesson 1-3)

EARTH WATCH

BUILDING WITH RECYCLED MATERIALS

It seems like a house built out of newspaper, sawdust, plastic bags, and old cars would not be the nicest one on the block. But the National Association of Home Builders made a beautiful new home out of recycled materials in Bowie, Maryland. The framing material was steel from old cars and the insulation was made from recycled newspapers.

See for Yourself
Research recycled building materials.

- New siding, decking, and asphalt products have been invented that make use of recycled material. What recycled materials do these products use?
- What advantages and disadvantages do products made of recycled materials have as compared to traditional building products?

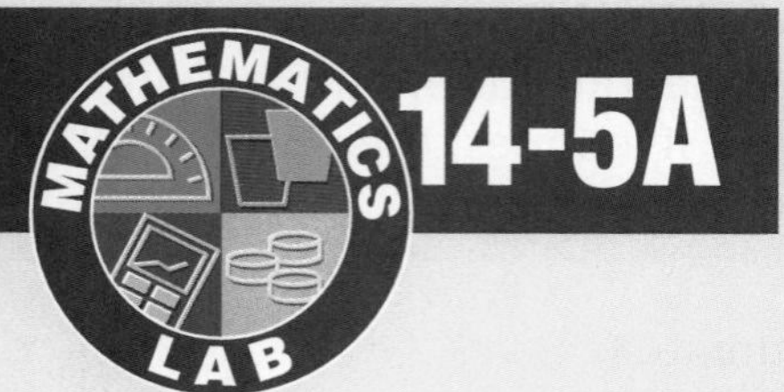

HANDS-ON ACTIVITY

14-5A Multiplying Polynomials

A Preview of Lesson **14-5**

MATERIALS

algebra tiles

product mat

Algebra tiles are named based on the area of the rectangle. The area of the rectangle is the product of the width and length.

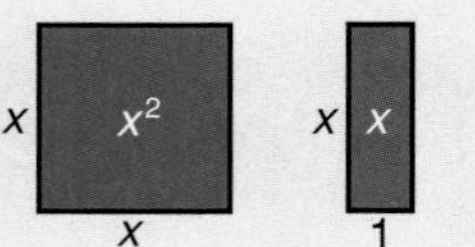

You can use algebra tiles to model more complex rectangles. These rectangles will help you understand how to find the product of simple polynomials. The width and length each represent a polynomial being multiplied. The area of the rectangle represents their product.

Your Turn

Work with a partner to find $x(x + 2)$.

- Make a rectangle with a width of x and a length of $x + 2$. Use algebra tiles to mark off the dimensions on a product mat.

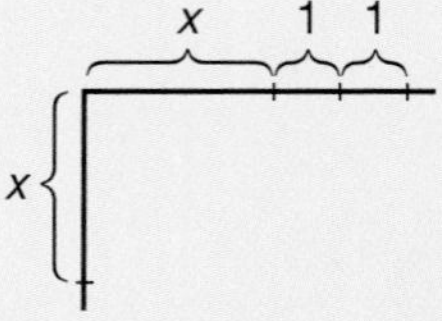

- Using the marks as a guide, fill in the rectangle with algebra tiles.

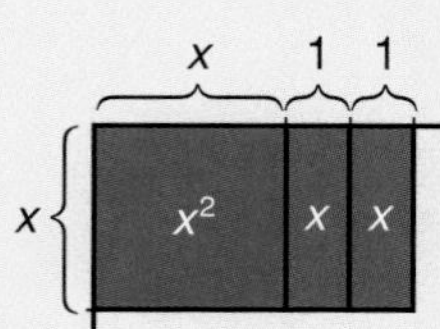

- The area of the rectangle is $x^2 + x + x$. In simplest form, the area is $x^2 + 2x$. Therefore, $x(x + 2) = x^2 + 2x$.

TALK ABOUT IT

Tell whether each statement is *true* or *false*. Justify your answer with algebra tiles.

1. $x(2x + 3) = 2x^2 + 3x$
2. $2x(3x + 4) = 6x^2 + 4x$

Find each product using algebra tiles.

3. $x(x + 5)$
4. $2x(x + 2)$
5. $3x(2x + 1)$
6. Suppose you have a square garden plot that measures x feet on a side.
   a. If you double the length of the plot and increase the width by 3 feet, how large will the new plot be? Write two expressions for the area of the new plot.
   b. If the original plot was 10 feet on a side, what is the area of the new plot?

14-5 Multiplying a Polynomial by a Monomial

Setting Goals: *In this lesson, you'll multiply a polynomial by a monomial.*

Modeling a Real-World Application: Construction

A chain-link fence is going to be installed around the school tennis courts. To determine how much fencing is needed, a representative from the fencing company measures the length and the width of the region that has to be enclosed. The total length of fencing can be determined by finding the perimeter P of the rectangular region.

$P = \ell + w + \ell + w$
$= 2\ell + 2w$ *Combine like terms.*
$= 2(\ell + w)$ *Distributive property*

In the expression $2(\ell + w)$, a polynomial is being multiplied by a monomial. In general, the distributive property can be used to multiply a monomial and a polynomial.

Learning the Concept

As you saw in Lesson 14-5A, you can use algebra tiles to multiply a polynomial and a monomial.

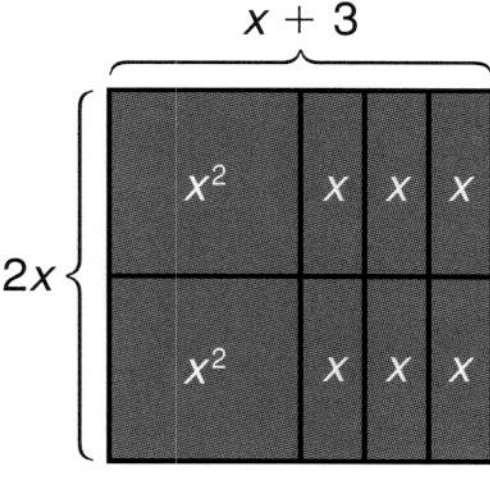

The figure at the right is a rectangle whose length is $x + 3$ and whose width is $2x$.

The area of any rectangle is the product of its length and its width. The area of this rectangle can also be found by adding the areas of the tiles.

Area: Formula	**Area: Algebra tiles**
$A = \ell w$	$A = x^2 + x^2 + x + x + x + x + x + x$
$A = (x + 3)2x$	$A = 2x^2 + 6x$

Therefore, $(x + 3)2x = 2x^2 + 6x$.

Example 1 **Find each product.**

a. $7(2x + 5)$

$7(2x + 5) = 7(2x) + 7(5)$ *Distributive property*
$= 14x + 35$

b. $(3x - 7)4x$

$(3x - 7)4x = (3x)4x - (7)4x$ *Distributive property*
$= 12x^2 - 28x$

c. $2ab(-9a^2 + 5ab - 4b^2)$

$2ab(-9a^2 + 5ab - 4b^2)$
$= 2ab(-9a^2) + 2ab(5ab) - 2ab(4b^2)$ *Distributive property*
$= -18a^3b + 10a^2b^2 - 8ab^3$ *Multiply monomials.*

Sometimes problems can be solved by simplifying polynomial expressions.

Example 2 INTEGRATION Algebra

The world's largest swimming pool is the Orthlieb Pool in Casablanca, Morocco. It is 30 meters longer than 6 times its width. If the perimeter of the pool is 1110 meters, what are the dimensions of the pool?

Explore You know the perimeter of the pool. You want to find the dimensions of the pool.

Plan Let w represent the width of the pool. Then $6w + 30$ represents the length. Then write an equation.

Perimeter equals twice the sum of the length and width.

$$\underbrace{P}_{\text{Perimeter}} \quad \underbrace{=}_{\text{equals}} \quad \underbrace{2}_{\text{twice}} \quad \underbrace{(\ell + w)}_{\text{the sum of the length and width}}$$

Solve

$P = 2(\ell + w)$

$1110 = 2((6w + 30) + w)$ *Replace P with 1110 and ℓ with 6w + 30.*

$1110 = 2(7w + 30)$ *Combine like terms.*

$1110 = 14w + 60$ *Distributive property*

$1050 = 14w$ *Subtract 60 from each side.*

$\frac{1050}{14} = w$

$75 = w$

The width is 75 meters and the length is $6w + 30$ or 480 meters.

Examine If the width is 75 meters, the length should be 30 meters longer than 6 times 75, or 480 meters. The perimeter is 2(75 + 480) or 1110 meters. The solution checks.

Checking Your Understanding

Communicating Mathematics

Read and study the lesson to answer these questions.

1. **Explain** how you would find the product of x and $2x - 1$.
2. **State** the product of x and $2x + 3$ using the rectangle at the right.
3. **Explain** why $x(2x + 3)$ and $(2x + 3)x$ are equivalent expressions.

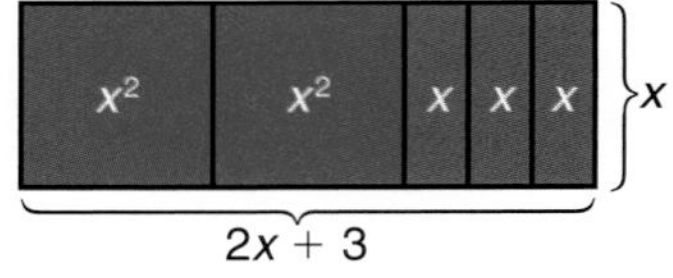

MATERIALS

algebra tiles

product mat

Use algebra tiles to find each product.

4. $5(x + 2)$
5. $x(x + 4)$
6. $2x(x - 1)$

Guided Practice

Find each product.

7. $3(x + 4)$
8. $x(x + 5)$
9. $3(x - 2)$
10. $2x(x - 8)$
11. $4x(3x + 7)$
12. $x(5x - 12)$

13. **Sports** The perimeter of a football field is 1040 feet. The length of the field is 120 feet less than 3 times the width. What are the dimensions of the football field?

Exercises: Practicing and Applying the Concept

Independent Practice

Find each product.

14. $7(3x + 5)$
15. $-2(x + 8)$
16. $y(y - 9)$
17. $3x(2x - 1)$
18. $-3x(x - 5)$
19. $c(a^2 + b)$
20. $4m(m^2 - m)$
21. $pq(pq + 8)$
22. $-3x(x^2 - 7x)$
23. $-2a(9 - a^2)$
24. $7(-2a^2 + 5a - 11)$
25. $-5(3x^2 - 7x + 9)$
26. $-3y(6 - 9y + 4y^2)$
27. $4c(c^3 + 7c^2 - 10)$
28. $-5x^2(3x^3 - 8x - 12)$
29. $6x^2(-2x^3 + 8x^2 - 7)$
30. $2x(5x^3 - 4x^2 + 6x - 9)$
31. $-x^2(x^3 - x^2 + 3x - 5)$

Solve each equation.

32. $-3(2a - 12) + 48 = 3a + 3$
33. $2(5w - 12) = 6(-2w + 3) + 2$

Critical Thinking

34. Use algebra tiles to make a model of $2x + 6$. Then, form a tile rectangle and find the factors of $2x + 6$. Finally, outline a procedure for finding the factors of any binomial.

Applications and Problem Solving

35. **Manufacturing** The figure at the right shows a pattern for a cardboard box before it has been cut and folded.
    a. Find the area of each rectangular region and add to find a formula for the number of square inches of cardboard needed.
    b. Find the surface area if x is 2.5 inches.

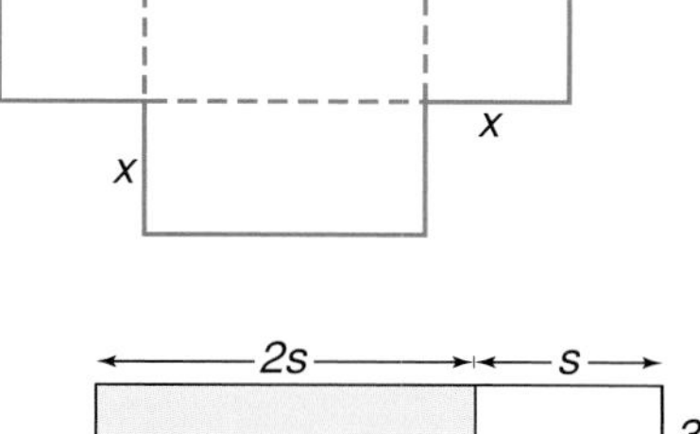

36. **Geometry** Find the measure of the area of the shaded region in simplest terms.

Mixed Review

37. Simplify $-4b(2b)^3$. (Lesson 14-4)
38. Find the best integer estimate for $-\sqrt{150}$. (Lesson 13-1)
39. **Statistics** Describe two ways a graph of sales of several brands of cereal could be misleading. (Lesson 10-4)
40. Solve the system of equations $y = x - 3$ and $y = x - 4$ by graphing. (Lesson 8-8)
41. **Draw a Diagram** The Avarillo's backyard is 130 by 90 feet. If their sprinkler can water an area 30 by 30 feet, how many times will it need to be moved to water the entire yard? (Lesson 4-3)
42. Is 528 divisible by 2, 3, 5, 6, or 10? (Lesson 4-1)
43. Find the distance traveled if you drive at 55 mph for $3\frac{1}{2}$ hours. (Lesson 3-4)

14-6 Multiplying Binomials

Setting Goals: *In this lesson, you'll multiply binomials.*

Modeling a Real-World Application: Genetics

Punnett squares are used to show possible ways that genes can combine at fertilization. In a Punnett square, *dominant* genes are shown with capital letters. *Recessive* genes are shown with the lowercase of the same letter.

The Punnett square below represents a cross between two hybrid tall pea plants. A hybrid trait is the result of a combination of a dominant and a recessive gene. The parent plants are tall because the dominant trait masks the recessive trait.

	T hybrid tall	t
T (hybrid tall)	TT	Tt
t	Tt	tt

Letters representing the parents' genes are placed on the outer sides of the Punnett square.

Letters inside the boxes of the square show the possible gene combinations for their offspring.

Let T represent the dominant gene for tallness and t represent the recessive gene for shortness.

When the parents' genes are combined, you see all of the possible combinations of the genes of the offspring.

$$\begin{aligned}(T + t)(T + t) &= TT + Tt + Tt + tt \\ &= TT + 2Tt + tt\end{aligned}$$

Learning the Concept

$(T + t)(T + t)$ is an example of the product of two binomials. Algebra tiles can also be used to help you understand how to multiply two binomials.

Consider the binomials $x + 3$ and $x + 2$. To multiply these binomials, mark off a rectangle on a product mat that has dimensions $x + 3$ and $x + 2$. Then fill in the rectangle with algebra tiles.

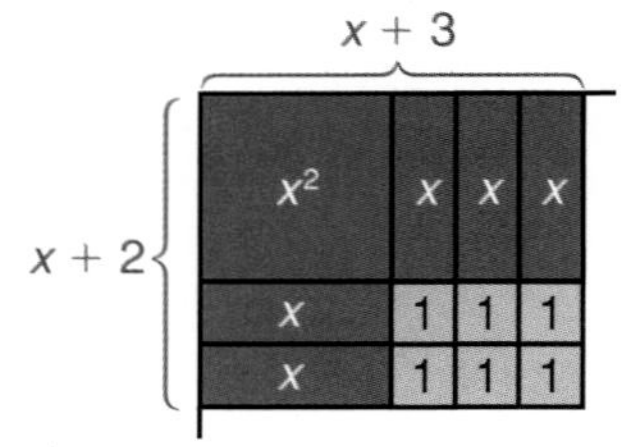

Area: Formula	**Area: Algebra Tiles**
$A = \ell w$	$A = x^2 + x + x + x + x + x + 1 + 1 + 1 + 1 + 1 + 1$
$= (x + 3)(x + 2)$	$= x^2 + 5x + 6$

The product $(x + 3)(x + 2)$ represents the area of the rectangle. The area is also the sum of the areas of the algebra tiles, $x^2 + 5x + 6$. Therefore, $(x + 3)(x + 2) = x^2 + 5x + 6$.

You can also use the distributive property to multiply binomials.

Example 

Find each product.

a. $\mathbf{(2x + 3)(3x + 5)}$

$$\begin{aligned}(2x + 3)(3x + 5) &= (2x + 3)3x + (2x + 3)5 && \textit{Distributive property}\\ &= (2x)3x + (3)3x + (2x)5 + (3)5\\ &= 6x^2 + 9x + 10x + 15 && \textit{Multiply polynomials.}\\ &= 6x^2 + 19x + 15 && \textit{Combine like terms.}\end{aligned}$$

b. $\mathbf{(2x + 1)(5x - 3)}$

$$\begin{aligned}(2x + 1)(5x - 3) &= (2x + 1)5x + (2x + 1)(-3) && \textit{Distributive property}\\ &= (2x)5x + (1)5x + (2x)(-3) + (1)(-3)\\ &= 10x^2 + 5x + (-6x) + (-3) && \textit{Multiply polynomials.}\\ &= 10x^2 - x - 3 && \textit{Combine like terms.}\end{aligned}$$

Example 2

APPLICATION

Gardening

A rectangular garden is 5 feet longer than twice its width. It has a sidewalk 3 feet wide on two of its sides. The area of the sidewalk is 213 square feet. Find the dimensions of the garden.

Explore Use algebra tiles to model the situation. Let x represent the width of the garden. Then, $2x + 5$ represents the length of the garden, $x + 3$ the width of the garden and sidewalk, and $2x + 8$ the length of the garden and sidewalk.

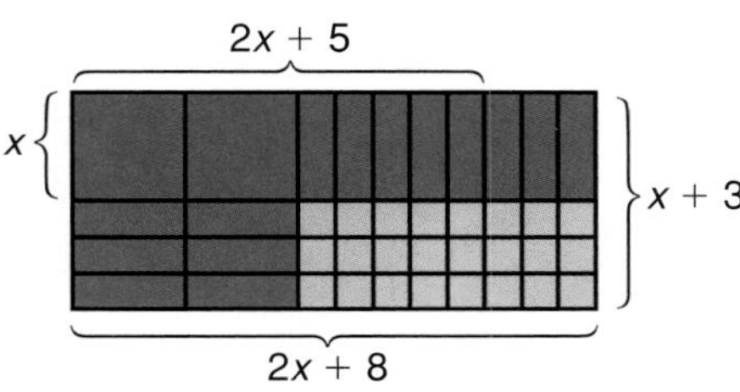

Then, $x(2x + 5)$ = the area of the garden and $(x + 3)(2x + 8)$ = the area of the garden and the sidewalk.

Plan *area of garden and sidewalk* − *area of garden* = *area of sidewalk*

$$(x + 3)(2x + 8) \;-\; x(2x + 5) = 213$$

Solve

$$\begin{aligned}2x^2 + 6x + 8x + 24 - 2x^2 - 5x &= 213 && \textit{Count the tiles.}\\ (2x^2 - 2x^2) + (6x + 8x - 5x) + 24 &= 213\\ 9x + 24 &= 213 && \textit{Simplify.}\\ 9x &= 189\\ x &= 21\end{aligned}$$

The width is 21 feet, and the length is $2x + 5 = 2(21) + 5$ or 47 feet.

Examine

Total Area	−	area of garden	$\stackrel{?}{=}$	area of sidewalk
(24)(50)	−	(21)(47)	$\stackrel{?}{=}$	213
1200	−	987	$\stackrel{?}{=}$	213
		213	=	213 ✓

Checking Your Understanding

Communicating Mathematics

MATERIALS

- algebra tiles
- product mat

Read and study the lesson to answer these questions.

1. **Draw** a rectangular region that represents the product of $x + 3$ and $2x + 1$. Then find the product.
2. **Explain** how the procedure used for multiplying two binomials is similar to the procedure for multiplying a binomial and a monomial. Then explain how it is different.

Use algebra tiles to find each product.

3. $(x + 1)(x + 2)$
4. $(x + 3)(x + 4)$
5. $(2x + 3)(x + 2)$

Guided Practice

For each model, name the two binomials being multiplied and then give their product.

6.

7. 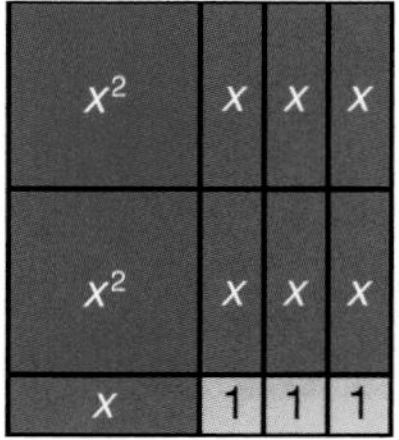

Find each product.

8. $(x + 3)(2x + 5)$
9. $(2x + 3)(x + 1)$
10. $(2x + 3)(3x + 2)$
11. $(5x - 3)(2x + 1)$

12. **Sports** A shuffleboard court is 10 feet longer than it is wide. A brick path 3 feet in width surrounds the court. Express the total area of the court and path algebraically if the court is w feet wide.

Exercises: Practicing and Applying the Concept

Independent Practice

For each model, name the two binomials being multiplied and then give their product.

13.

14. 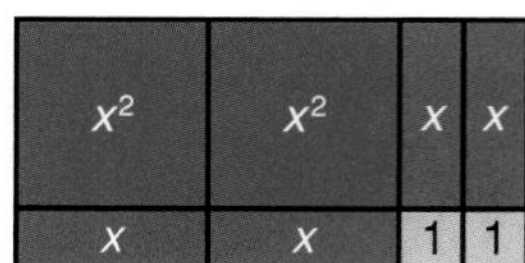

15. 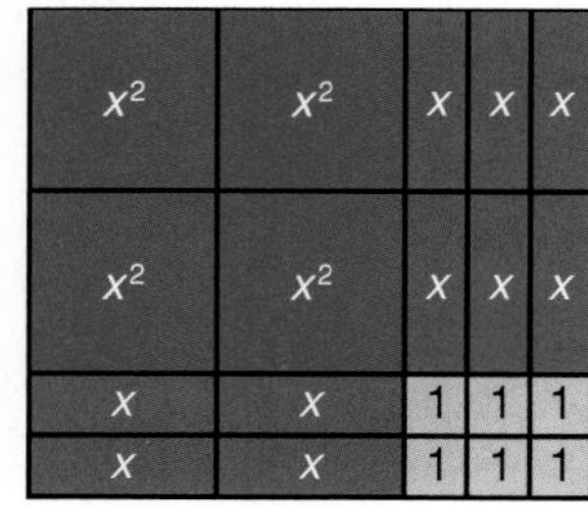

16.

Find each product.

17. $(x + 4)(x + 3)$
18. $(x + 5)(x + 2)$
19. $(x - 6)(x + 2)$
20. $(x + 7)(x - 5)$
21. $(x - 9)(x + 4)$
22. $(x + 2)(x + 2)$
23. $(2x + 3)(x - 4)$
24. $(3x - 1)(x + 8)$
25. $(x + 3)(x + 3)$
26. $(2x + 5)(3x + 1)$
27. $(5x + 2)(2x - 3)$
28. $(x - 5)(x + 5)$
29. $\left(3x - \frac{1}{4}\right)\left(6x - \frac{1}{2}\right)$
30. $(x - 2)(x^2 + 2x + 4)$

Critical Thinking

31. Write the multiplication problem modeled by the figure at the right. Name the product.

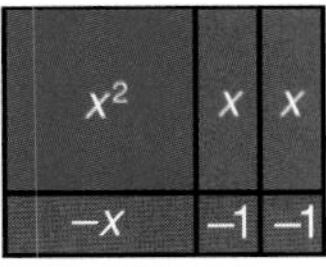

Applications and Problem Solving

32. **Volunteer Work** As a project, a senior citizens group makes baby quilts and full-size quilts for victims of natural disasters. The baby quilts are $x + 2$ feet long by x feet wide. The full-size quilts are 4 feet longer and 3 feet wider than the baby quilts.

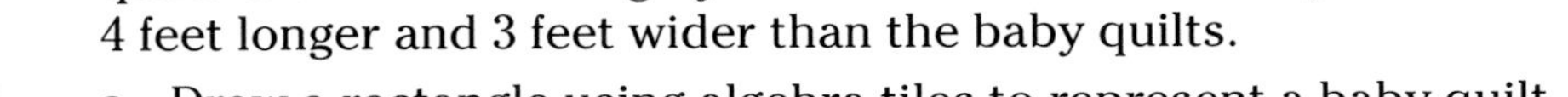

a. Draw a rectangle using algebra tiles to represent a baby quilt.
b. Draw additional tiles to increase the rectangular region to represent a full-size quilt.
c. Write a product for the full-size quilt and another for the baby quilt.
d. What is the difference between the area of the full-size quilt and the area of the baby quilt?

33. **Geometry** The length of the smaller rectangle at the right is 1 inch less than twice its width. Both the dimensions of the larger rectangle are 2 inches longer than the smaller rectangle. The area of the shaded region is 86 inches.

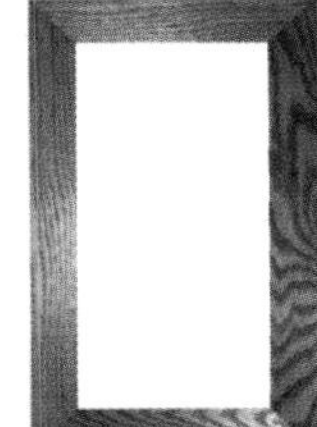

a. What are the dimensions of the smaller rectangle?
b. What is the area of the smaller rectangle?
c. What is the area of the larger rectangle?

Mixed Review

34. Find the product $-2(4c^2 - 3c + 5)$. (Lesson 14-5)

35. **Measurement** On a sunny day, a tree casts a shadow 60 feet long. At the same time, a yardstick casts a shadow 2 feet long. How tall is the tree? (Lesson 11-6)

36. **Business** One way businesses judge their success is by the ratio of their sales to the total sales of similar products. This is called their market share. The circle graph at the right shows the market shares of ready-to-drink teas. Find the measure of the angle to represent Arizona® brand tea. (Lesson 11-2)

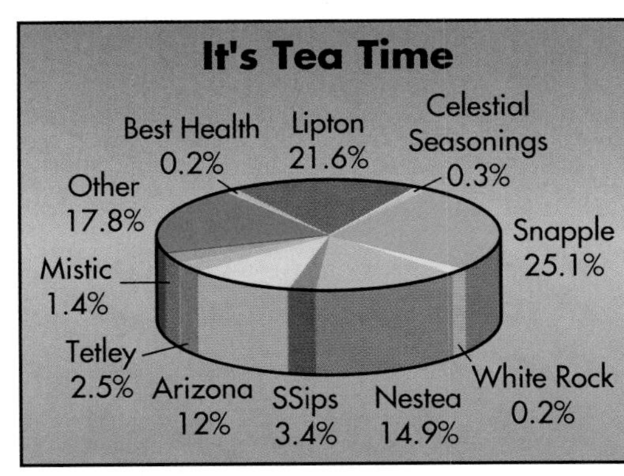

Source: Beverage Marketing Corporation

37. **Geometry** The radius of a circle is 0.5 m. Find its circumference. (Lesson 7-4)

38. What is the GCF of 60 and 25? (Lesson 4-5)

39. Solve $6x < -84$. (Lesson 3-7)

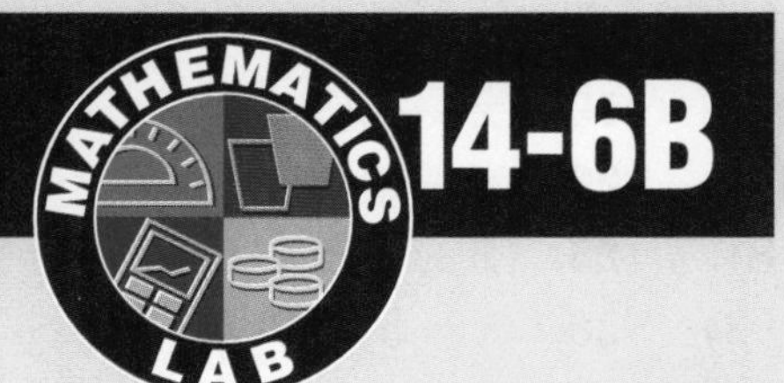

HANDS-ON ACTIVITY

14-6B Factoring

An Extension of Lesson **14-6**

MATERIALS

- algebra tiles
- product mat

Consider a rectangle with a length of $x + 1$ and a width of $x + 3$. From the previous lesson, you know that $(x + 1)(x + 3) = x^2 + 4x + 3$. The binomials $(x + 1)$ and $(x + 3)$ are **factors** of the trinomial $x^2 + 4x + 3$.

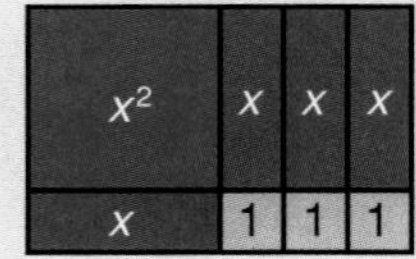

You can use algebra tiles as a model for factoring simple trinomials. If a rectangle with a length and width greater than 1 cannot be formed to represent the trinomial, then the trinomial is not factorable.

Your Turn

Work with a partner to factor $x^2 + 3x + 2$.

► Model the polynomial.

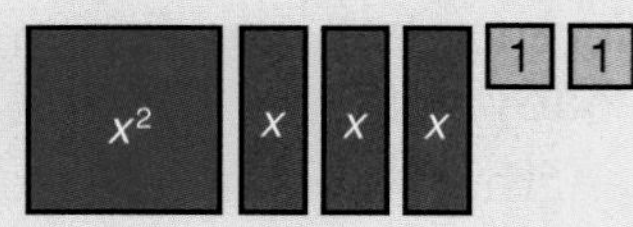

► Place the x^2-tile and the 1-tiles as shown.

► Complete the rectangle with the x-tiles.

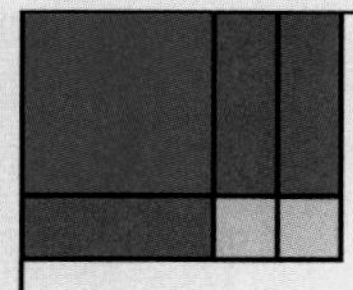

► Since a rectangle can be formed, $x^2 + 3x + 2$ is factorable. The rectangle has a width of $x + 1$ and a length of $x + 2$. So the factors of $x^2 + 3x + 2$ are $x + 1$ and $x + 2$.

TALK ABOUT IT

Tell whether each polynomial is factorable. Justify your answers with algebra tiles.

1. $x^2 + 6x + 8$
2. $x^2 + 5x + 6$
3. $x^2 + 7x + 3$
4. $3x^2 + 8x + 5$
5. $5x^2 - x + 16$
6. $8x^2 - 31x - 4$
7. Write a paragraph that explains how you can determine whether a trinomial can be factored. Include an example of one trinomial that can be factored and one that cannot.

CHAPTER 14 Highlights

Vocabulary

After completing this chapter, you should be able to define each term, property, or phrase and give an example or two of each.

Algebra

binomial (p. 706)

coefficient (p. 711)

constant (p. 707)

like terms (p. 711)

polynomial (p. 706)

power of a monomial (p. 721)

power of a power (p. 719)

power of a product (p. 720)

trinomial (p. 706)

Understanding and Using Vocabulary

Choose the correct term to complete each sentence.

1. The degree of a monomial is the (sum, product) of the exponents of its variables.
2. A (binomial, trinomial) is the sum or difference of three monomials.
3. The degree of a polynomial is the same as that of the term that has the (least, greatest) degree.
4. The numerical part of a monomial is called the (coefficient, variable).
5. To subtract two polynomials, you add the (additive inverse, multiplicative inverse) of the second polynomial.
6. The (commutative property, distributive property) is used in multiplying a polynomial by a monomial.

Determine whether each statement is *true* or *false*.

7. Every polynomial is a monomial.
8. The coefficient of $6ab^2$ is 2.
9. The additive inverse of $(3x + 6y)$ is $(-3x - 6y)$.
10. The power of a product property says that $(8m)^3 = 8^3m^3$.

CHAPTER 14 Study Guide and Assessment

Skills and Concepts

Objectives and Examples

Upon completing this chapter, you should be able to:

- **identify and classify polynomials** (Lesson 14-1)

  The expression $5a^2 - 3a + 4$ is a polynomial. It is a trinomial because it is the sum of three monomials.

  The expression $\frac{2}{a}$ is not a polynomial because it has a variable in the denominator.

Review Exercises

Use these exercises to review and prepare for the chapter test.

State whether each expression is a polynomial. If it is, identify it as a *monomial*, *binomial*, or *trinomial*.

11. $3x^4 - x$
12. $\frac{4}{ax}$
13. ax^2
14. $9b^2 + b - 1$
15. $8x - \frac{5}{3}$
16. $k^3 - \frac{2}{k}$

- **find the degree of polynomials** (Lesson 14-1)

  Find the degree of $x^3y + 3xy - 6y - 1$.

  Find the degree of each term.

  x^3y has degree $3 + 1$ or 4.
  $3xy$ has degree $1 + 1$ or 2.
  $6y$ has degree 1.
  1 has degree 0.

  So, the degree of $x^3y + 3xy - 6y - 1$ is 4.

Find the degree of each polynomial.

17. $4x$
18. $5a^2b$
19. $3x + y^2$
20. $19m^2n^3 - 14mn^4$
21. $x^2 - 6xy + xy^2$
22. $12rs^2 + 3r^2s + 5r^4s$

- **add polynomials** (Lesson 14-2)

  Find the sum $(b^2 + 3b + 8) + (2b^2 + 5b - 5)$.

$$\begin{array}{r} b^2 + 3b + 8 \\ (+)\ 2b^2 + 5b - 5 \\ \hline 3b^2 + 8b + 3 \end{array}$$

Find each sum.

23. $(2x^2 - 5x) + (3x^2 + x)$
24. $(a^2 - 6ab) + (3a^2 + ab)$
25. $(x^2 - 5x + 3) + (4x - 3)$
26. $(-3y^2 + 2) + (4y^2 - 5y - 2)$
27. $\begin{array}{r} 4x^2 + 3x + 2 \\ (+)\ x^2 \quad\quad - 1 \\ \hline \end{array}$
28. $\begin{array}{r} 16x^2y - 2xy + xy^2 \\ (+)\ 4x^2y + 6xy - 8xy^2 \\ \hline \end{array}$

Objectives and Examples

▸ **subtract polynomials** (Lesson 14-3)

Find the difference
$(4y^2 - 2y + 3) - (y^2 + 2y - 4)$.

Align like terms and add the additive inverse of $y^2 + 2y - 4$.

$$\begin{array}{r} 4y^2 - 2y + 3 \\ (-)\ y^2 + 2y - 4 \\ \hline \end{array} \rightarrow \begin{array}{r} 4y^2 - 2y + 3 \\ (+)\ -y^2 - 2y + 4 \\ \hline 3y^2 - 4y + 7 \end{array}$$

Review Exercises

Find each difference.

29. $(7a - 11b) - (3a + 4b)$

30. $(6y - 8z) - (6y + 4z)$

31. $(3a^2 - b^2 + c^2) - (a^2 + 2b^2)$

32. $(14a^2 - 3a) - (6a^2 + 5a + 17)$

33. $$\begin{array}{r} 18x^2 + 3x - 1 \\ (-)\ \ 2x^2 + 4x + 6 \\ \hline \end{array}$$

34. $$\begin{array}{r} 12m^2 - \ \ mn + 9n^2 \\ (-)\ \ 7m^2 + 2mn - 4n^2 \\ \hline \end{array}$$

▸ **find powers of monomials** (Lesson 14-4)

$$\begin{aligned}(x^5)^3 &= x^{5 \cdot 3} \\ &= x^{15}\end{aligned}$$

$$\begin{aligned}(2d)^4 &= 2^4 \cdot d^4 \\ &= 16d^4\end{aligned}$$

$$\begin{aligned}(a^3b^2)^3 &= a^{3 \cdot 3} \cdot b^{2 \cdot 3} \\ &= a^9b^6\end{aligned}$$

Simplify.

35. $(a^2)^3$

36. $(-2x)^3$

37. $(p^2q)^3$

38. $-5c(2cd)^3$

39. $4y(y^2z)^3$

40. $(12a^5)^2b^3$

41. $6a(-ab)^7$

42. $(-2d^2)^6(-3)^3$

▸ **multiply a polynomial by a monomial** (Lesson 14-5)

$$\begin{aligned}-5y(2y^2 - 4) &= -5y(2y^2) - (-5y)(4) \\ &= -10y^3 - (-20y) \\ &= -10y^3 + 20y\end{aligned}$$

Find each product.

43. $4d(2d - 5)$

44. $x(-5x + 3)$

45. $a^2(2a^3 + a - 5)$

46. $3y(-y^2 - 8y + 4)$

47. $-2g(g^3 + 6g + 3)$

48. $-3az(2z^2 + 4az + a^2)$

Objectives and Examples	Review Exercises

▶ **multiply binomials** (Lesson 14-6)

$$\begin{aligned}(x - 4)(2x + 3) &= (x - 4)2x + (x - 4)(3)\\ &= x(2x) + (-4)(2x) + x(3) + (-4)(3)\\ &= 2x^2 - 8x + 3x - 12\\ &= 2x^2 - 5x - 12\end{aligned}$$

49. For the model below, name the two binomials being multiplied and then give their product.

Find each product.

50. $(x + 3)(x + 1)$

51. $(2x + 1)(x + 1)$

52. $(3x + 2)(2x + 2)$

Applications and Problem Solving

53. Architecture Norma Merrick Sklarek was the first African-American woman registered as an architect in the United States. Suppose Ms. Sklarek drew the floor plan for the first floor of a house shown below. Using the measurements given, write a polynomial that represents the total area of the first floor. (Lesson 14-1)

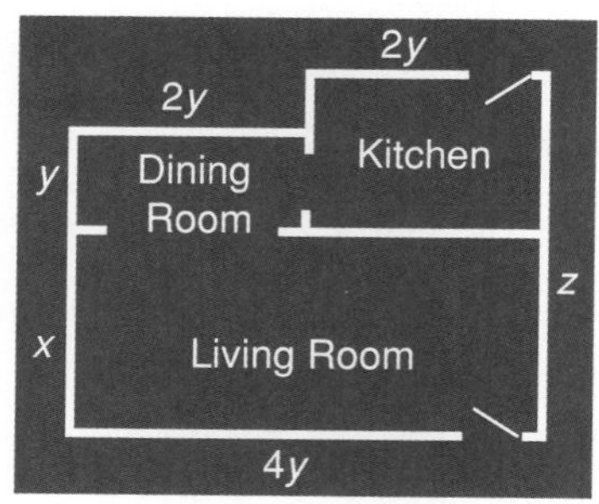

54. Geometry The perimeter of the triangle shown below is $4x - 2y$ units. Find the length of the third side of the triangle. (Lessons 14-1 and 14-2)

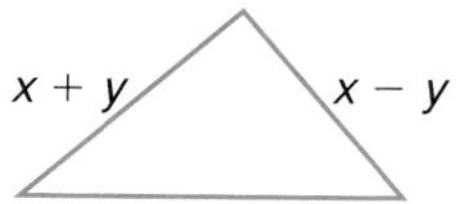

55. Geometry Find the volume of the triangular prism below. (Lesson 14-4)

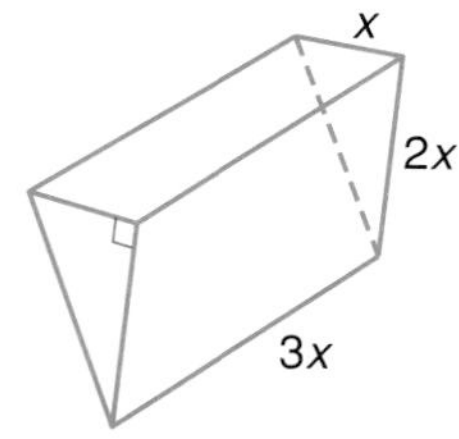

56. Sports The perimeter of a soccer field is 1040 feet. The length of the field is 40 feet more than 2 times the width. What are the dimensions of the field? (Lesson 14-5)

A practice test for Chapter 14 is available on page 787.

Alternative Assessment

Cooperative Learning Project

The triangular arrangement of numbers below is called Pascal's Triangle. Many patterns exist in Pascal's Triangle. One of the patterns is that the rows in the triangle can be used to write powers of binomials.

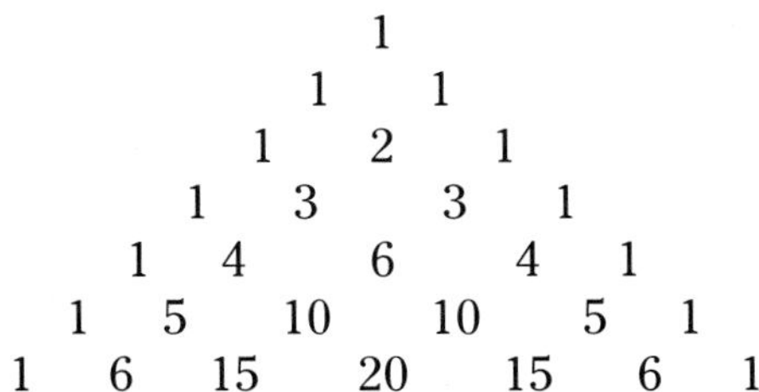

Work with a partner to find the powers of the binomial $(x + y)$ and complete the table below.

Power	Product	Coefficients of the terms
$(x + y)^2$		1, 2, 1
$(x + y)^3$		
$(x + y)^4$		
$(x + y)^5$	$x^5 + 5x^4y + 10x^3y^2 + 10x^2y^3 + 5xy^4 + y^5$	

Compare the last column of the table to the rows of Pascal's Triangle. Describe the relationship you find. Predict the product of $(x + y)^6$. Then multiply to verify your product.

Thinking Critically

- For all numbers a and b such that $b \neq 0$, and any integer m, is $\left(\frac{a}{b}\right)^m = \frac{a^m}{b^m}$ a true sentence?
- A trapezoid has an area of 425 square inches and a height of 10 inches. The lower base is 5 inches less than twice the upper base. Find the length of each base.

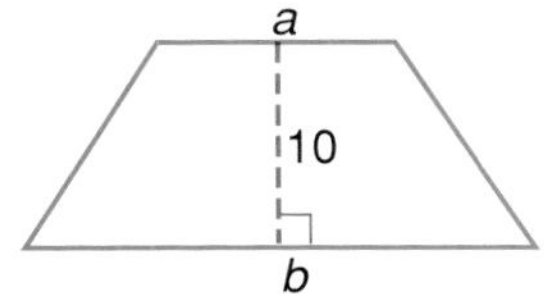

Portfolio

Review the items in your portfolio. Make a list of the items, noting why each item was chosen. Replace any items that are no longer appropriate.

Self Evaluation

Assess yourself. Look back over your performance in Pre-Algebra this year. Did you meet the goals you set for yourself at the beginning of the course? What could you have done differently to make the year more successful? Write at least one goal that you would like to accomplish in the next mathematics course you take.

CHAPTERS 1–14 Ongoing Assessment

Section One: Multiple Choice

There are ten multiple-choice questions in this section. After working each problem, write the letter of the correct answer on your paper.

1. Which is equivalent to $\frac{3^6}{3^2}$?

   A. 3^3 **B.** 3^4
   C. 3^8 **D.** 3^{12}

2. In a right triangle, the tangent ratio is the measure of the leg opposite an angle to the measure of the leg adjacent to the angle. What is the tangent ratio of angle D?

   A. $\frac{3}{5}$
   B. $\frac{4}{5}$
   C. $\frac{4}{3}$
   D. $\frac{5}{3}$

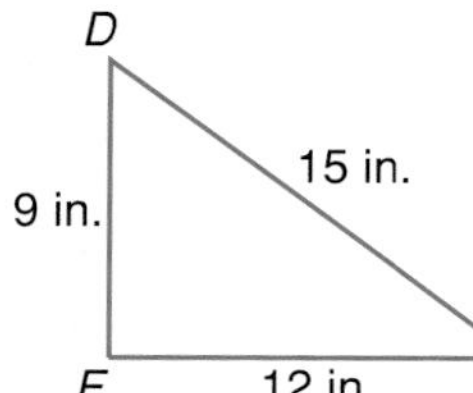

3. Paul Olsen is designing a cereal box. The height of the box is 12 inches, the width is 2 inches, and the length is 6 inches. What is the surface area of the cardboard he needs for one box?

   A. 40 in^2
   B. 144 in^2
   C. 216 in^2
   D. 288 in^2

4. Which is equivalent to $(7x^2 + 3y) - (3x^2 + 5y)$?

   A. $4x^2 - 2y$ **B.** $4x^4 - 2y^2$
   C. $4x^2 + 8y$ **D.** $4x^4 + 8y^2$

5. A sweater that normally sells for \$35 is on sale at 25% off. Which is the best estimate of the sale price?

   A. \$9. **B.** \$32.
   C. \$26. **D.** \$43.

6. A rectangular window is 40 inches high and 30 inches wide. What is the length of a diagonal?

   A. 60 in. **B.** 35 in.
   C. 50 in. **D.** 25 in.

7. Find the surface area of a cone with a diameter of 12 meters and a slant height of 18 meters. Round your answer to the nearest tenth.

   A. 75.4 m^2 **B.** 452.2 m^2
   C. 94.2 m^2 **D.** 678.2 m^2

8. Which is equivalent to $(-2x)^3$?

   A. $-8x^3$ **B.** $-2x^3$
   C. $8x^3$ **D.** $2x^3$

9. $\triangle EFG$ is similar to $\triangle KML$. What is the measure of $\angle M$?

   A. 40°
   B. 60°
   C. 70°
   D. 80°

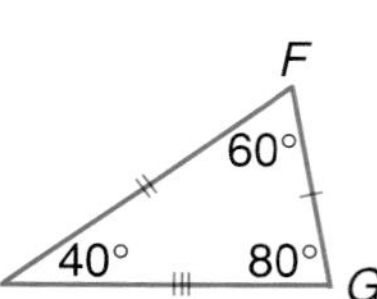

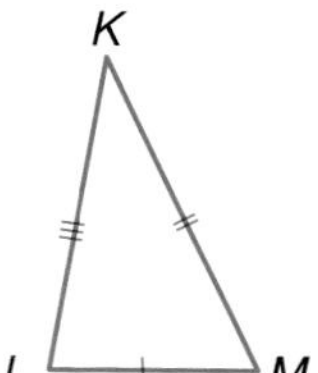

10. $\triangle ABC$ is a right triangle. What is the length of the hypotenuse?

    A. 21 in.
    B. 26 in.
    C. 30 in.
    D. 52 in.

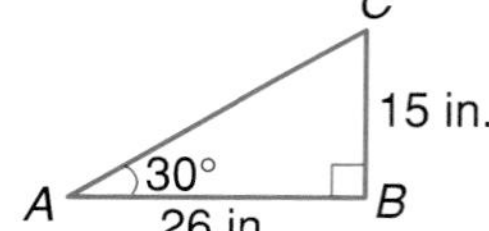

Section Two: Free Response

This section contains twelve questions for which you will provide short answers. Write your answer on your paper.

11. Find the perimeter of a regular hexagon with sides 12.6 meters long.

12. A nickel, a dime, a quarter, and a half dollar are in a purse. Without looking, Alex picks up two coins. What are the different amounts he could choose?

13. Joan and Miguel are flying kites in the park. When Joan is 70 feet from Miguel, her kite forms an angle of 56° with the ground. If the kite is directly over Miguel, how high is the kite to the nearest tenth of a foot?

14. Evaluate $a^4 - 2a^3 + 4b^2 - c$ if $a = 2$, $b = -6$, and $c = -2$.

15. Find the area of the circle shown below.

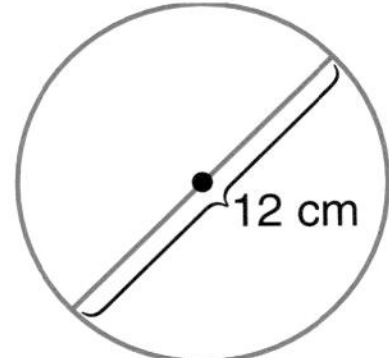

16. There are 25 students in a math class. The class takes a survey and finds that 15 people like pepperoni, 12 people like sausage, and 7 people like both equally. How many people like neither pepperoni nor sausage?

17. In a survey of favorite music types, 15 out of 50 people preferred country. In a group of 1000 people, how many would you expect to like country?

18. Find the sum of $(4x^3 + 2x^2) + (-2x^3 - 7x^2)$.

19. Approximate the value of $\sqrt{45}$.

20. The largest country in the world is Russia, with an area of 6,590,876 square miles. The second largest country is Canada. The area of Russia is 110,416 square miles less than twice the area of Canada. What is the area of Canada?

21. Write 2.25 as a mixed number.

22. Find the sum of the measures of the interior angles of a trapezoid.

Test-Taking Tip

On standardized tests, guessing can improve your score if you make educated guesses. First, find out if there is a penalty for incorrect answers. If there is no penalty, a guess can only increase your score, or at worst, leave your score the same. If there is a penalty, try to eliminate enough choices to make the probability of a correct guess greater than the penalty for an incorrect answer. That is, if there is a quarter point penalty for an incorrect response and you have a choice of three responses, the probability of making a correct guess is 1 out of 3, which is better than the penalty.

Section Three: Open-Ended

This section contains three open-ended problems. Demonstrate your knowledge by giving a clear, concise solution to each problem. Your score on these problems will depend on how well you do the following.

- Explain your reasoning.
- Show your understanding of the mathematics in an organized manner.
- Use charts, graphs, and diagrams in your explanation.
- Show the solution in more than one way or relate it to other situations.
- Investigate beyond the requirements of the problem.

23. Show that $4n(-2n^2 + n - 12) = -8n^3 + 4n^2 - 48n$. Use values for n to justify your answer.

24. Explain the difference between integers and whole numbers.

25. Write a problem that can be simulated by rolling a die.

Extra Practice

EXTRA PRACTICE

Lesson 1-1

1. **Postal Service** The U.S. Postal Service offers airmail service to other countries. The rates for International Air Mail letters and packages are shown in the table at the right. Determine the air mail rate for a package that weighs 5.5 ounces.
   a. Write the *Explore* step. What do you know and what do you need to find?
   b. Write the *Plan* step. What strategy will you use? What do you estimate the answer to be?
   c. *Solve* the problem using your plan. What is your answer?
   d. *Examine* your solution. Is it reasonable? Does it answer the question?

Weight not over (ounces)	Rate
0.5	\$0.50
1.0	\$0.95
1.5	\$1.34
2.0	\$1.73
2.5	\$2.12
3.0	\$2.51
3.5	\$2.90
4.0	\$3.29

2. **Postal Service** In 1995, the state of Florida celebrated the 150th anniversary of its statehood. The U.S. Postal Service issued a stamp, the first to bear the 32-cent price, to honor the occasion. Ninety million of the commemorative stamps were issued. About how much postage did the stamps represent?
   a. Which method of computation do you think is most appropriate for this problem? Justify your choice.
   b. Solve the problem using the four-step plan. Be sure to examine your solution.

Lesson 1-2

Find the value of each expression.

1. $8 + 7 + 12 \div 4$
2. $20 \div 4 - 5 + 12$
3. $(25 \cdot 3) + (10 \cdot 3)$
4. $36 \div 6 + 7 - 6$
5. $30 \cdot (6 - 4)$
6. $(40 \cdot 2) - (6 \cdot 11)$
7. $\frac{86 - 11}{11 + 4}$
8. $\frac{12 + 84}{11 + 13}$
9. $\frac{5 \cdot 5 + 5}{5 \cdot 5 - 15}$
10. $(19 - 8)4$
11. $75 - 5(2 \cdot 6)$
12. $81 \div 27 \times 6 - 2$

Lesson 1-3

Evaluate each expression if $a = 2$, $b = 4$, and $c = 3$.

1. $ba - ac$
2. $4b + a \cdot a$
3. $11 \cdot c - ab$
4. $4b - (a + c)$
5. $7(a + b) - c$
6. $8a + 8b$
7. $\frac{8(a + b)}{4c}$
8. $36 - 12c$
9. $\frac{9(b + a)}{c - 1}$
10. $abc - bc$
11. $28 - bc + a$
12. $a(b - c)$

Translate each phrase into an algebraic expression.

13. nine more than a
14. eleven less than k
15. three times p
16. the product of some number and five
17. twice Shelly's score decreased by 18
18. the quotient of 16 and n

Lesson 1-4

Name the property shown by each statement.

1. $1 \cdot 4 = 4$
2. $6 + (b + 2) = (6 + b) + 2$
3. $9(6n) = (9 \cdot 6)n$
4. $8t \cdot 0 = 0 \cdot 8t$
5. $0(13n) = 0$
6. $7 + t = t + 7$

Find each sum or product mentally.

7. $6 + 8 + 14$
8. $5 \cdot 18 \cdot 2$
9. $0(13 \cdot 6)$
10. $8 + 4 + 12 + 16$
11. $8 \cdot 20 \cdot 10$
12. $4 \cdot 14 \cdot 5$

13. History Many flags have represented the United States since the first colonists arrived. The current flag, called the Stars and Stripes, was originated by the Marine Committee of the Second Continental Congress in a resolution made on June 14, 1777. The flag now has fifty stars in the field to represent the fifty states of the Union. Each row of the flag has either 5 or 6 stars in each row. Rows of 5 stars alternate with rows of 6 stars. How many rows are there with 6 stars?

Lesson 1-5

Simplify each expression.

1. $8k + 2k + 7$
2. $3 + 2b + b$
3. $t + 2t$
4. $9(3 + 2x)$
5. $4(xy + 2) - 2$
6. $(6 + 3e)4$
7. $4 + 9c + 3(c + 2)$
8. $5(7 + 2s) + 3(s + 4)$
9. $9(f + 2) + 14f$

10. Business The Columbus Dispatch newspaper can be ordered for delivery on weekdays or Sundays. A weekday paper is 35 cents and the Sunday Edition is $1.50. The Stadlers ordered delivery of the weekday papers. The month of March had 23 weekdays and April had 20. How much should the carrier charge the Stadlers for those two months?

Lesson 1-6

Identify the solution to each equation from the given list.

1. $16 - f = 7$; 5, 7, 9
2. $9 = \frac{72}{m}$; 8, 9, 11
3. $4b + 1 = 17$; 3, 4, 5
4. $17 + r = 25$; 6, 7, 8
5. $9 = 7n - 12$; 3, 5, 7
6. $67 = 98 - q$; 21, 26, 31

Solve each equation mentally.

7. $131 - u = 120$
8. $88 = 11d$
9. $\frac{84}{h} = 12$
10. $5t = 0$
11. $\frac{x}{2} = 8$
12. $88 + y = 96$
13. $13g = 39$
14. $23 = w + 6$
15. $9z = 45$

EXTRA PRACTICE

Lesson 1-7

Use the grid at the right to name the point for each ordered pair.

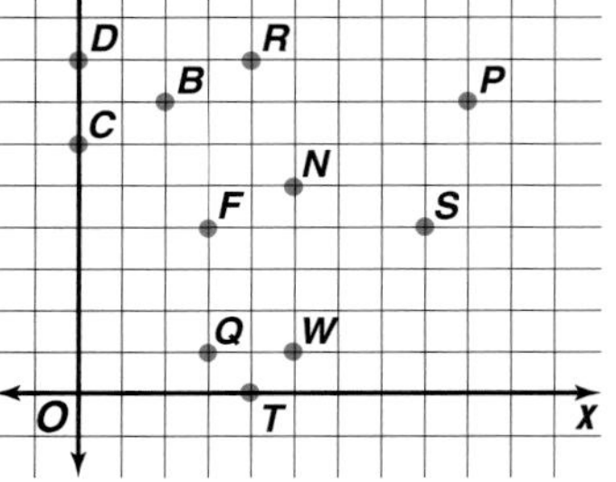

1. (9, 7)
2. (5, 5)
3. (3, 1)
4. (2, 7)
5. (8, 4)
6. (4, 0)

Use the grid at the right to name the ordered pair for each point.

7. *R*
8. *P*
9. *W*
10. *C*
11. *D*
12. *F*

Lesson 1-8

Solve each equation using the inverse operation. Use a calculator when needed.

1. $37 = b + 22$
2. $6 + u = 14$
3. $p - 19 = 3$
4. $6x = 30$
5. $3v = 99$
6. $48c = 192$
7. $r \div 7 = 4$
8. $12 = \frac{z}{3}$
9. $6.2 = 5.8 + g$

Translate each sentence into an equation.

10. The sum of a number and 8 is 14.
11. Twelve less than a number is 50.
12. The product of a number and ten is seventy.
13. A number divided by three is nine.

Lesson 1-9

State whether each inequality is *true*, *false*, or *open*.

1. $8 > 0$
2. $18 \leq 18$
3. $10 < x$
4. $24 > 2w$
5. $9 < 7$
6. $13 \leq 55$

State whether each inequality is *true* or *false* for the given value.

7. $5 \geq 2t - 12; t = 11$
8. $7 + n < 25; n = 4$
9. $6r - 18 > 0; r = 3$
10. $3n + 2 < 26, n = 3$
11. $h - 19 < 13; h = 28$
12. $20m \geq 10; m = 0$

13. **Literature** William Shakespeare, who lived from 1564 to 1616, was England's best known poet, actor, and playwright. Each year several productions of Shakespeare's plays are performed around the world. Some of his plays have been made into films, including a 1993 version of *Much Ado About Nothing* and a 1990 version of *Hamlet*. There are currently more than 50 versions of *Hamlet* on film. Write an inequality for the number of versions of Hamlet.

Lesson 1-10

The table at the right shows the albums that have stayed on the U.S. music charts the longest as of December 31, 1993.

Artist/title	Weeks on chart
Carole King, *Tapestry*	302
Johnny Mathis, *Heavenly*	295
Johnny Mathis, *Johnny's Greatest Hits*	490
Pink Floyd, *Dark Side of the Moon*	741
Tennessee Ernie Ford, *Hymns*	277
Various, Original cast recording of *Camelot*	265
Various, Original cast recording of *My Fair Lady*	480
Various, Original cast recording of *The Sound of Music*	276
Various, Soundtrack of *The King and I*	277
Various, Soundtrack of *Oklahoma!*	305

1. Which album has been on the charts the greatest number of weeks?
2. How many weeks was *Tapestry* by Carole King on the charts?
3. How much longer was Johnny Mathis' *Greatest Hits* on the charts than his *Heavenly* album?
4. Make a bar graph of the data on albums.

Lesson 2-1

Simplify.

1. $|-3| + |9|$	2. $|-18| - |5|$	3. $|12 + 7|$
4. $-|6|$	5. $|-8| + |4|$	6. $-|-20|$
7. $|15 - 12|$	8. $|8 + 9|$	9. $-|4| \cdot |-5|$
10. $|-6| \cdot |8|$	11. $-|12| \cdot |9|$	12. $-||-16| + |-22||$

Lesson 2-2

Use the coordinate grid at the right to name the point for each ordered pair.

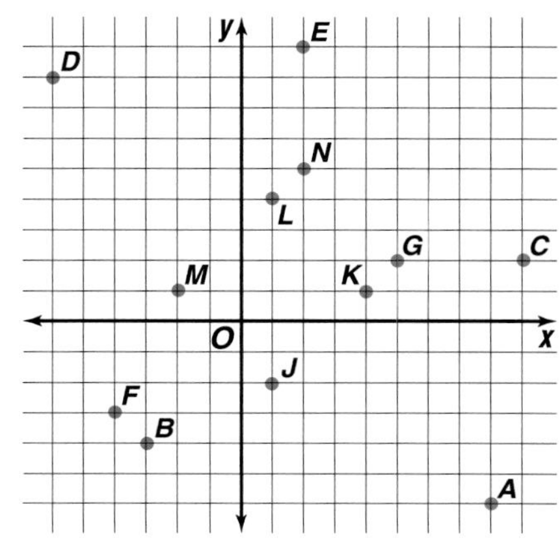

1. $(-6, 8)$
2. $(1, -2)$
3. $(9, 2)$
4. $(1, 4)$
5. $(-3, -4)$
6. $(2, 5)$

On graph paper, draw coordinate axes. Then graph and label each point. Name the quadrant in which each point is located.

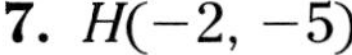

7. $H(-2, -5)$
8. $P(1, 5)$
9. $R(-3, 1)$
10. $M(4, -2)$
11. $K(-4, 5)$
12. $G(3, -5)$

EXTRA PRACTICE

Lesson 2-3

Order the integers in each set from least to greatest.

1. $\{-1, 2, -5\}$
2. $\{0, -2, 8, 5, -9\}$
3. $\{100, -34, -86, 21, 0\}$
4. $\{-1, 16, -43, 8, 27, -40\}$
5. $\{0, -23, 75, -15, 24\}$
6. $\{-6, 6, -5, 18\}$

Write an inequality using the numbers in each sentence. Use the symbols $<$ or $>$.

7. **World** Finland covers 130,119 square miles and is 3 times greater than Ohio which covers 40,953 square miles.
8. **Sports** Atlanta, Georgia hosted the 1996 Summer Olympics. The point with the highest altitude in the city is 1050 feet above sea level. The lowest point is 940 feet above sea level.
9. **Statistics** The oldest dog lived to be 24 years old. The oldest cat lived to be 26 years old.
10. **Statistics** The slowest animal in the world is the garden snail that moves 0.03 mph. The fastest animal is the cheetah that can run 70 mph.
11. **Astronomy** Mars is 227.9 million kilometers from the Sun. Earth is 149.6 million kilometers from the Sun.

Lesson 2-4

Solve each equation.

1. $n = 5 + (-6)$
2. $-17 + 24 = y$
3. $k = 15 + (-29)$
4. $m = -6 + 13$
5. $50 + (-14) = x$
6. $w = -21 + (-4)$
7. $30 + (-7) = t$
8. $z = (-3) + (-10)$
9. $-15 + 26 = q$
10. $b = -17 + 4 + (-2)$
11. $d = 50 + (-16) + (-11)$
12. $-17 + 8 + (-14) = f$

Lesson 2-5

Solve each equation.

1. $8 - 17 = f$
2. $-15 - 3 = r$
3. $10 - 21 = a$
4. $20 - (-5) = m$
5. $5 - (-9) = g$
6. $-12 - (-7) = v$
7. $-19 - (-6) = k$
8. $-16 - (-23) = b$
9. $-56 - 32 = z$
10. $-49 - (-52) = h$
11. $-6 - 9 - (-7) = d$
12. $6 - (-10) - 7 = n$

Lesson 2-6

Solve. Look for a pattern.

1. Find the next two integers in the pattern 2, 3, 5, 9, 17, ? , ? .
2. If FOG = 6 − 15 − 7 and LOG = 12 − 15 − 7; what does JOG equal by the same logic?
3. The digits 0 to 9 have been used to make the multiplication problem below. Find the missing numbers that make the multiplication correct.

```
  ? 0 2
×   ? 9
-------
1 ? ? ? 8
```

Lesson 2-7

Solve each equation.

1. $n = -4(2)$
2. $-8(-5) = h$
3. $r = 13(-4)$
4. $-5 \cdot 6 \cdot 10 = c$
5. $-6(-2)(-14) = w$
6. $u = 18(-3)(6)$

Evaluate each expression.

7. $-6t$, if $t = 15$
8. $7p$, if $p = -9$
9. $-4k$, if $k = -16$
10. aw, if $a = 0$ and $w = -72$
11. dk, if $d = -12$ and $k = 11$
12. st, if $s = -8$ and $t = -10$
13. $3hp$, if $h = 9$ and $p = -3$
14. $-5bc$, if $b = -6$ and $c = 2$
15. $-4wx$, if $w = -1$ and $x = -8$

Lesson 2-8

Divide.

1. $-36 \div 9$
2. $112 \div (-8)$
3. $-72 \div 2$
4. $-26 \div (-13)$
5. $-144 \div 6$
6. $-180 \div (-10)$
7. $304 \div (-8)$
8. $-216 \div (-9)$
9. $80 \div (-5)$
10. $-105 \div 15$
11. $120 \div (-30)$
12. $-200 \div (-8)$

Lesson 3-1

Solve by eliminating possibilities.

1. Lindsay, Lee, Ann, and Marcos formed a study group. Each one has a favorite subject that is different from the others. The subjects are art, math, music, and physics. Use the following information to match each person with his or her favorite subject.
   A. Lindsay likes subjects where she can use her calculator.
   B. Lee does not like music or physics.
   C. Ann and Marcos prefer classes in the cultural arts.
   D. Marcos plans to be a professional cartoonist.
2. Five friends decided to run in a 5-kilometer run to benefit an education association. Nancy beat Pete. Jeff was not last. Gina was beaten by Tom and Jeff. Jeff crossed the finish line just after Tom did. Tom lost to Pete. Choose the correct order in which the friends finished the race.
   a. Nancy, Pete, Gina, Tom, Jeff
   b. Tom, Jeff, Nancy, Pete, Gina
   c. Nancy, Pete, Tom, Jeff, Gina
   d. Nancy, Gina, Jeff, Tom, Pete

Lesson 3-2

Solve each equation and check your solution. Then graph the solution on a number line.

1. $y + 49 = 26$
2. $d - (-31) = -24$
3. $q - 8 = 16$
4. $x - 16 = 32$
5. $40 = a + 12$
6. $b + 12 = -1$
7. $21 = u - (-6)$
8. $-52 = p + 5$
9. $-14 = 5 - g$
10. $121 = k + (-12)$
11. $-234 = m - 94$
12. $110 = x + 25$
13. $f - 7 = 84$
14. $y - 864 = -652$
15. $475 + z = -18$

EXTRA PRACTICE

Lesson 3-3

Solve each equation and check your solution. Then graph the solution on a number line.

1. $-y = -32$

2. $7r = -56$

3. $\frac{t}{-3} = 12$

4. $4 = \frac{s}{-14}$

5. $\frac{b}{47} = -2$

6. $64 = -4n$

7. $-144 = 12q$

8. $\frac{r}{11} = -132$

9. $-5g = -385$

10. $-16x = -176$

11. $-21 = \frac{y}{-4}$

12. $-372 = 31k$

13. $84 = \frac{k}{5}$

14. $-b = 19$

15. $\frac{y}{112} = -9$

Lesson 3-4

Solve by replacing the variables with the given values.

1. $A = \pi \cdot r^2$, if $\pi = 3.14$ and $r = 5$

2. $\frac{5}{9}(F - 32) = C$, if $F = 86$

3. $d = r \cdot t$, if $d = 366$ and $t = 3$

4. $S = (n - 2) \cdot 180$, if $n = 8$

5. $A = \frac{1}{2}bh$, if $A = 36$ and $h = 12$

6. $P = 4s$, if $P = 108$

7. $V = \frac{1}{3}(B \cdot H)$, if $B = 27$ and $H = 5$

8. $h = 69 + 2.2F$, if $F = 42$

Lesson 3-5

Find the perimeter and area of each rectangle.

1. a rectangle 23 centimeters long and 9 centimeters wide

2. a 16-foot by 14-foot rectangle

3. a rectangle with a length of 31 meters and a width of 3 meters

4. a square with sides 7.5 meters long

Find the missing dimension of each rectangle.

	Length	Width	Area	Perimeter
5.	9 ft		126 ft^2	46 ft
6.		18 in.	108 in^2	48 in.
7.	13 yd		273 yd^2	68 yd
8.		12 cm	168 cm^2	52 cm
9.		3 m	162 m^2	114 m
10.	18 mi		720 mi^2	116 mi

Lesson 3-6

Solve each inequality and check your solution. Then graph the solution on a number line.

1. $m + 9 < 14$
2. $k + (-5) < -12$
3. $-15 < v - 1$
4. $-7 + f \geq 47$
5. $r > -15 - 8$
6. $18 \geq s - (-4)$
7. $38 < r - (-6)$
8. $z - 9 \leq -11$
9. $-16 + c \geq 1$
10. $24 < a + -3$
11. $-52 \geq d + (-6)$
12. $24 < -2 + b$
13. $n - (-17) \geq 12$
14. $31 < x - 24$
15. $-20 \leq n + (-3)$
16. $p + (-11) < -37$
17. $-40 \leq -72 + w$
18. $72 > a + 88$

Lesson 3-7

Solve each inequality and check your solution. Then graph the solution on a number line.

1. $6p < 78$
2. $\frac{m}{-3} > 24$
3. $-18 < 3b$
4. $-5k \geq 125$
5. $-75 > \frac{a}{5}$
6. $\frac{w}{6} < -5$
7. $\frac{y}{-13} > -20$
8. $14t < 266$
9. $\frac{g}{-25} \geq 8$
10. $-216 \leq 9h$
11. $\frac{g}{-9} < -8$
12. $-18d > 108$
13. $2268 < -63a$
14. $52 \leq \frac{p}{4}$
15. $42n \leq -210$
16. $-60 > -4d$
17. $\frac{k}{-3} \geq -21$
18. $\frac{v}{-8} < 0$

Lesson 3-8

Define a variable and translate each sentence into an equation or inequality. Then solve.

1. Three times a number is equal to thirty-six.
2. The sum of a number and 5 is less than 12.
3. **Sports** In the 1993–1994 NBA season, David Robinson of the San Antonio Spurs was the leading scorer. Shaquille O'Neal of the Orlando Magic had 6 points less than David Robinson. If O'Neal had 2377 points for the season, how many points did David Robinson get?
4. **College** The enrollment at the Columbus campus of The Ohio State University is more than six times as large as the enrollment at Stanford University. If there are 38,958 students enrolled at Ohio State, how many students are currently enrolled at Stanford?

EXTRA PRACTICE

Lesson 4-1

Using divisibility rules, state whether each number is divisible by 2, 3, 5, 6, or 10.

1. 98 **2.** 243 **3.** 800
4. 252 **5.** 105 **6.** 210
7. 1001 **8.** 2016 **9.** 2475

Determine whether each expression is a monomial. Explain why or why not.

10. $-6h$ **11.** $9 - v$ **12.** $4g^3$

Lesson 4-2

Write each multiplication expression using exponents.

1. $(-6)(-6)(-6)(-6)(-6)$
2. $(y \cdot y \cdot y) \cdot (y \cdot y \cdot y \cdot y)$
3. 9
4. $3q \cdot 3q \cdot 3q \cdot 3q \cdot 3q \cdot 3q$
5. $\underbrace{n \cdot n \cdot n \cdot \ldots \cdot n}_{\text{17 factors}}$
6. $8 \cdot 8 \cdot 8 \cdot 8$

Write each power as a multiplication expression of the same factor.

7. 13^3
8. $(-4)^5$
9. k^9
10. $(2 - w)^2$
11. $(3m)^{15}$
12. $(-x)^4$

Lesson 4-3

Solve. Use a diagram.

1. Four people are eligible to be officers in the National Honor Society. The positions available are President, Vice President, Treasurer, and Secretary. How many different ways can the offices be filled?
2. Copy the magic square at the right. Use the numbers 0, 4, 8, and 12 to make all the rows, columns, and diagonals add up to 24.
3. Beverly and Pam are painting a fence. For every 3 sections Pam paints, Beverly paints 5. If Beverly paints 20 sections in 2 hours how long will it take Pam to paint 18 sections?
4. Determine how many games have to be played to determine a champion from 32 teams in a single elimination softball tournament.

0		12	0
4			4
			12

Lesson 4-4

Determine whether each number is prime or composite.

1. 59　　**2.** 369　　**3.** 116

Factor each number or monomial completely.

4. 40　　**5.** $630a$　　**6.** 187
7. 310　　**8.** 510　　**9.** 1589
10. $-18ab^2$　　**11.** $-117x^3$　　**12.** $435j^2k^5$

EXTRA PRACTICE

Lesson 4-5

Find the GCF of each set of numbers or monomials.

1. 112, 216　　**2.** 120, 245　　**3.** $84k$, $108k^2$
4. $135ab$, $-171b$　　**5.** $185fg$, $74f^2g$　　**6.** $44m$, $60n$
7. $90gh$, $225k$　　**8.** 8, $-28h$　　**9.** $-16w$, $-28w^3$
10. $24a$, $30ab$, $66a^2$　　**11.** $-13z$, $39yz$, $52y$　　**12.** $60a^3b$, $150a^2b^2$, $36a^2b$

Lesson 4-6

Write each fraction in simplest form. If the fraction is already in simplest form, write *simplified*.

1. $\frac{3}{54}$　　**2.** $\frac{3}{16}$　　**3.** $\frac{6}{58}$
4. $\frac{15}{55}$　　**5.** $\frac{10}{90}$　　**6.** $\frac{20}{49}$
7. $\frac{8}{20}$　　**8.** $\frac{99}{9}$　　**9.** $\frac{18}{54}$
10. $\frac{21}{64}$　　**11.** $\frac{40}{76}$　　**12.** $\frac{49}{56}$

Lesson 4-7

Find the least common multiple (LCM) of each set of numbers or algebraic expressions.

1. 30, 18　　**2.** $3m$, 12　　**3.** $6a$, $17a^5$　　**4.** 2, 5, 7

Find the least common denominator (LCD) for each pair of fractions.

5. $\frac{2}{5}, \frac{6}{25}$　　**6.** $\frac{3}{12}, \frac{4}{5}$　　**7.** $\frac{4}{6}, \frac{7}{9}$　　**8.** $\frac{1}{4}, \frac{5}{6}, \frac{2}{9}$

Replace each ● with $<$ or $>$ to make a true statement.

9. $\frac{5}{11}$ ● $\frac{6}{13}$　　**10.** $\frac{15}{34}$ ● $\frac{4}{8}$　　**11.** $\frac{8}{18}$ ● $\frac{5}{16}$　　**12.** $\frac{3}{14}$ ● $\frac{5}{20}$

EXTRA PRACTICE

Lesson 4-8

Find each product or quotient. Express your answer in exponential form.

1. $r^4 \cdot r^2$ **2.** $\frac{2^9}{2^3}$ **3.** $\frac{b^{18}}{b^5}$ **4.** $12^3 \cdot 12^8$

5. $x \cdot x^9$ **6.** $(2s^6)(4s^2)$ **7.** $w^3 \cdot w^4 \cdot w^2$ **8.** $(-2)^2(-2)^5(-2)$

9. $\frac{4^7}{4^6}$ **10.** $3(f^{17})(f^2)$ **11.** $(5k)^2 \cdot k^7$ **12.** $\frac{6m^8}{3m^2}$

Find each missing exponent.

13. $(2^{\bullet})(2^2) = 2^6$ **14.** $y(y^3)(y^4) = y^{\bullet}$ **15.** $\frac{5^7}{5^3} = 5^{\bullet}$

16. $12^9 \cdot 12^{\bullet} = 12^{14}$ **17.** $\frac{b^{10}}{b^{\bullet}} = b$ **18.** $\frac{2^7}{8^2} = 2^{\bullet}$

Lesson 4-9

Write each expression using positive exponents.

1. y^{-9} **2.** $3m^{-4}$ **3.** $5^{-3} \cdot (-a)$ **4.** $\left(\frac{2}{x}\right)^{-7}$

Write each fraction as an expression using negative exponents.

5. $\frac{1}{p^4}$ **6.** $\frac{6a}{b^9}$ **7.** $\frac{2}{27}$ **8.** $\frac{17}{a}$

Evaluate each expression.

9. 5^n if $n = -2$ **10.** $(2a^{-2}b)^2$ if $a = 3$ and $b = 6$

11. $15n^{-3}$ if $n = 5$ **12.** $9x^{-6}$ if $x = -3$

Find each product or quotient. Express using positive exponents.

13. $(x^8)(x^{-2})$ **14.** $(g^{-12})(g^9)$ **15.** $\frac{r^{14}}{r^{-2}}$ **16.** $\frac{y^7}{y^{10}}$

Lesson 5-1

Name the set(s) of numbers to which each number belongs.

1. $\frac{7}{10}$ **2.** 41 **3.** -19

Replace each ● with <, >, or = to make each sentence true. Use a number line if necessary.

4. $\frac{3}{4} \bullet \frac{2}{5}$ **5.** $\frac{-13}{25} \bullet \frac{-3}{5}$ **6.** $\frac{9}{10} \bullet \frac{7}{8}$

7. $5\frac{2}{9} \bullet \frac{47}{9}$ **8.** $\frac{5}{26} \bullet \frac{4}{13}$ **9.** $\frac{-11}{4} \bullet -2\frac{1}{2}$

Express each decimal as a fraction or mixed number in simplest form.

10. 0.38 **11.** 2.346 **12.** $-0.\overline{4}$

Lesson 5-2

Round to the nearest whole number.

1. 6.5
2. 5.193
3. 42.09

Round each fraction to 0, $\frac{1}{2}$, or 1.

4. $\frac{9}{10}$
5. $\frac{1}{14}$
6. $\frac{89}{101}$

Estimate each sum or difference.

7. $9.9 - 3.2$
8. $16.2 + 9.31$
9. $1.3 + 2.8 - 3.4$
10. $8\frac{3}{5} - 2\frac{1}{6}$
11. $19\frac{4}{10} + 13\frac{8}{11}$
12. $863\frac{1}{12} - 241\frac{189}{221}$

Lesson 5-3

Solve each equation.

1. $b = 5.8 + 9.3$
2. $s = 12.4 - 4.52$
3. $-4.9 + 8.4 = k$
4. $-5.2 - 7.8 = q$
5. $p = 14.8 - 29.46$
6. $-4.25 + 11.2 = t$
7. $21.4 - 9.2 = z$
8. $45.26 - (-6.1) = y$
9. $x = -9.27 - 8.1$
10. $-28.94 + 3.48 = u$

Lesson 5-4

Solve each equation. Write the solution in simplest form.

1. $\frac{2}{7} + \frac{3}{7} = g$
2. $\frac{8}{15} - \frac{4}{15} = y$
3. $\frac{3}{7} + \frac{4}{7} = w$
4. $\frac{5}{6} - \frac{1}{6} = t$
5. $\frac{7}{12} - \frac{5}{12} = r$
6. $\frac{5}{12} + \frac{11}{12} = w$

Simplify each expression.

7. $12\frac{7}{8}s - 7\frac{3}{8}s + 2\frac{5}{8}s$
8. $-6\frac{4}{9}t - \left(-4\frac{5}{9}t\right) + 3\frac{2}{9}t$
9. $6\frac{1}{4}g + \left(-6\frac{3}{4}g\right)$
10. $7\frac{2}{5}n - \left(-4\frac{2}{5}n\right)$

Lesson 5-5

Solve each equation. Write each solution in simplest form.

1. $\frac{1}{5} + \frac{2}{7} = d$
2. $a = \frac{4}{5} + \frac{7}{9}$
3. $\frac{1}{9} - \frac{7}{12} = n$
4. $z = \frac{8}{11} - \frac{4}{5}$
5. $\frac{7}{12} - \left(\frac{-4}{11}\right) = y$
6. $\ell = -\frac{9}{14} + \frac{15}{16}$
7. $3\frac{2}{5} + 2\frac{4}{7} = k$
8. $r = -4\frac{1}{8} + 2\frac{5}{9}$
9. $-3\frac{3}{7} - 5\frac{1}{14} = g$

Lesson 5-6

Solve each equation. Check your solution.

1. $a - 4.86 = 7.2$
2. $n + 6.98 = 10.3$
3. $87.64 = f - (-8.5)$
4. $x - \frac{2}{5} = -\frac{8}{15}$
5. $3\frac{3}{4} + m = 6\frac{5}{8}$
6. $4\frac{1}{6} = r + 6\frac{1}{4}$
7. $7\frac{1}{3} = c - \frac{4}{5}$
8. $-4.62 = h + (-9.4)$
9. $w - 1\frac{1}{5} = \frac{2}{9}$

Lesson 5-7

Solve each inequality and check your solution. Graph the solution on a number line.

1. $h + 5.7 > 21.3$
2. $78.26 \leq v - (-65.854)$
3. $\frac{2}{3} \leq a - \frac{5}{6}$
4. $-13.2 > w - 4.87$
5. $a + \frac{5}{12} \geq \frac{7}{18}$
6. $7\frac{1}{2} < n - \left(-\frac{7}{8}\right)$
7. $t - 8.5 > -4.2$
8. $-7.42 \leq d - 5.9$
9. $m - (-18.4) < -17.6$
10. $s - \frac{2}{3} \geq 9\frac{4}{5}$

Lesson 5-8

State whether each is an example of *inductive* or *deductive* reasoning. Explain your answer.

1. Every day that Felicia takes off from work it rains. She is going to take her birthday off, so she thinks it will rain on her birthday.
2. The local radio station calls monthly with a survey. There is less than a week left in the month. They will be calling this week.
3. If you have a C or better in all of your classes, you are eligible to play sports. Hank has an A or a B in all of his classes. Hank is eligible to play sports.
4. The credit card company charges finance charges if the customer doesn't pay the balance in full. Carla could only afford to pay half of her bill this month. Carla will be charged a finance charge.

Lesson 5-9

State whether each sequence is arithmetic. Then write the next three terms of each sequence.

1. 3.5, 4.3, 5.1, . . .
2. 5, 10, 20, . . .
3. $\frac{1}{2}, \frac{5}{6}, 1\frac{1}{6}, \ldots$
4. $\frac{1}{4}, \frac{1}{2}, 1, 2, \ldots$
5. 23, 18, 13, . . .
6. 45, 43, 39, 33, . . .

Lesson 6-1

Write each fraction as a decimal. Use a bar to show a repeating decimal.

1. $\frac{6}{10}$
2. $-4\frac{7}{12}$
3. $\frac{8}{11}$
4. $3\frac{4}{18}$
5. $-\frac{3}{16}$
6. $8\frac{36}{44}$
7. $\frac{9}{37}$
8. $\frac{6}{15}$

Replace each ● with <, >, or = to make a true sentence.

9. $\frac{7}{8}$ ● $\frac{5}{6}$
10. 0.04 ● $\frac{5}{9}$
11. $\frac{1}{3}$ ● $\frac{2}{7}$
12. $\frac{3}{5}$ ● $\frac{12}{20}$
13. $\frac{1}{2}$ ● 0.75
14. 0.3 ● $\frac{1}{3}$
15. $\frac{2}{3}$ ● 0.64
16. $\frac{2}{20}$ ● 0.10

Lesson 6-2

Estimate each product or quotient.

1. 16.38×1.5
2. $35.54 \div 4.1$
3. $6\frac{4}{9} \cdot 7.09$
4. $18.24 \cdot 3.25$
5. $\frac{6}{13} \times 150$
6. $\left(\frac{1}{4}\right)(15)$
7. $78 \div 1\frac{11}{12}$
8. $1\frac{7}{8} \cdot 40$
9. $\frac{1}{3} \cdot 37$
10. $75 \div 1\frac{7}{16}$
11. $88 \div \frac{3}{8}$
12. $71.99 \div 5.7$
13. Estimate $\frac{4}{9}$ times 20.
14. Estimate the quotient of 65.46 and 5.6.
15. Estimate 32 times $5.49.
16. Estimate the quotient of 45 and $\frac{6}{13}$.

Lesson 6-3

Solve each equation. Write each solution in simplest form.

1. $d = \frac{2}{5} \cdot \frac{3}{16}$
2. $u = 3\frac{1}{4} \cdot \frac{2}{11}$
3. $\left(\frac{3}{5}\right)\left(-\frac{5}{12}\right) = g$
4. $t = \left(\frac{4}{5}\right)^3$
5. $s = 2\frac{2}{6} \cdot 6\frac{2}{7}$
6. $2\left(-\frac{7}{12}\right) = r$
7. $1\frac{3}{7} \cdot \left(-9\frac{4}{5}\right) = c$
8. $h = \left(-\frac{6}{7}\right)^2$

Evaluate each expression if $r = -\frac{1}{5}$, $s = \frac{2}{3}$, $x = 1\frac{1}{4}$, and $y = -2\frac{1}{8}$.

9. rx
10. $5r^2$
11. $s(x + y)$
12. $8y + 12x$
13. $x^2(s + 2)$
14. $-x(x - 2s)$

EXTRA PRACTICE

EXTRA PRACTICE

Lesson 6-4

Name the multiplicative inverse for each rational number.

1. $-\frac{5}{9}$ **2.** $5\frac{3}{8}$ **3.** 0.7

4. 2.35 **5.** -18 **6.** $\frac{a}{b}$

Estimate the solution to each equation. Then solve. Write the solution in simplest form.

7. $w = \frac{3}{4} \div \frac{15}{16}$ **8.** $16 \div 1\frac{7}{8} = m$ **9.** $q = 2\frac{1}{6} \div 1\frac{1}{5}$

10. $y = -11 \div 3\frac{1}{7}$ **11.** $a = \frac{8}{45} \div \frac{10}{27}$ **12.** $220 \div \left(-5\frac{1}{2}\right) = p$

Lesson 6-5

Solve each equation.

1. $7.3017 \div 0.57 = a$ **2.** $13.42 \div 67.1 = d$ **3.** $x = 80 \div (-3.2)$

4. $m = -2.016 \div (-0.13)$ **5.** $3.8 \cdot 2.9 = k$ **6.** $85 \cdot 0.07 = w$

7. $r = 15.32(0.0015)$ **8.** $(16.2)(0.013) = b$ **9.** $c = 5.4 \cdot 9.7$

Evaluate each expression.

10. $4y^3$, if $y = 0.6$ **11.** xy, if $x = 0.348$ and $y = -6.4$

12. $\frac{3a}{w}$, if $a = 0.4$ and $w = 2$

Lesson 6-6

Find the mean, median, and mode for each set of data. When necessary, round to the nearest tenth.

1. 82, 79, 93, 91, 95

2. 88, 85, 76, 94, 85, 97

3. 0.57, 12.81, 12.6, 0.96, 6.1, 14.3, 4.1, 12.81, 0.96

Use the table to answer Exercises 4–6.

4. What is the mean percent of students who plan to attend college?

5. What is the median percent of students who plan to attend tech school?

6. What is the mode percent of students who plan to start a career after high school?

Percent of Independence High School Seniors with After-High School Plans

Year	College	Tech School	Career
1990	67.7	23.9	11.5
1991	68.2	20.3	11.5
1992	71.1	22.7	6.2
1993	70.4	24.1	5.5
1994	63.9	20.5	10.2
1995	67.4	22.2	10.4

Lesson 6-7

Solve each equation or inequality. Check your solution.

1. $3.5a = 7$
2. $0.8 = -0.8b$
3. $8 < \frac{2}{3}c$
4. $\frac{m}{13} \geq 0.5$
5. $-9 = \frac{3}{4}g$
6. $0.4y > -2$
7. $-\frac{1}{2}d \leq -5\frac{1}{2}$
8. $-3.5 = 0.07z$
9. $-\frac{1}{6}s = 15$

Solve each equation or inequality and graph the solution on a number line.

10. $\frac{2}{7}t < 4$
11. $8.37 = 2.7d$
12. $\frac{1}{5}m \geq 4\frac{3}{5}$

EXTRA PRACTICE

Lesson 6-8

State whether each sequence is a geometric sequence. If so, state the common ratio and list the next three terms.

1. 2, 4, 8, 16, . . .
2. 125, 75, 45, . . .
3. 100, 75, 50, . . .
4. $\frac{1}{5}$, 1, 5, 25, . . .
5. 2401, 49, 7, . . .
6. $-\frac{4}{5}$, 2, -5, $12\frac{1}{2}$, . . .

7. Write the first five terms in a geometric sequence with a common ratio of 3. The first term is -4.
8. Write the first five terms of a geometric sequence if $a = -12$ and $r = \frac{1}{2}$.
9. In a certain geometric sequence, $a = 1.8$ and $r = -3$. Write the first five terms of the sequence.
10. Use the expression $ar^{(n-1)}$ to find the eighth term in the geometric sequence -8, 12, -18, ...

Lesson 6-9

Write each number in scientific notation.

1. 6,184,000
2. 27,210,000
3. 0.00004637
4. 0.00546
5. 500,300,100
6. 0.00000321

Write each number in standard form.

7. 9.562×10^{-3}
8. 8.2453×10^{-7}
9. 8.2×10^{4}
10. 9.102040×10^{2}
11. 2.41023×10^{6}
12. 4.21×10^{-5}

EXTRA PRACTICE

Lesson 7-1

Solve by working backward.

1. Doug brought some milk bottle caps to Antonio's house to trade. Doug traded Antonio a third of his caps in exchange for 2 designer caps. Then Doug gave Antonio's little sister Maria 6 caps as a gift. If Doug left Antonio's house with 42 milk caps, how many caps did he bring to Antonio's house?
2. Kwan had some cookies. She gave a fourth of her cookies to Justin. Justin then gave half of his cookies to Simon. Simon gave a third of his cookies to Sandy. If Sandy has 3 cookies, how many cookies did Kwan have in the beginning?
3. A certain bacteria doubles its population every 12 hours. After 3 full days, there are 1600 bacteria in a culture. How many bacteria were there at the beginning of the first day?
4. Sam and Akita are playing a game. On five turns, Sam's piece advanced 5 spaces, moved back 4 spaces, advanced 2 spaces, moved back 8 spaces, and advanced two spaces. How far did Sam's piece move from its original place on the board in these turns?

Lesson 7-2

Solve each equation. Check your solution.

1. $3t - 13 = 2$
2. $-8j - 7 = 57$
3. $9d - 5 = 4$
4. $6 - 3w = -27$
5. $\frac{k}{6} + 8 = 12$
6. $-4 = \frac{q}{8} - 19$
7. $15 - \frac{n}{7} = 13$
8. $7.25 = 3r - 6.25$
9. $21.63 - h = -32.7$
10. $-19 = 11b - (-3)$
11. $6 = 20 + \frac{x}{3}$
12. $8.12 + 3a = -3.25$

Lesson 7-3

Define a variable and write an equation for each situation. Then solve.

1. The sum of 29 and 3 times a number is 44. What is the number?
2. The opposite of twice a number less 3 is 17. What is the number?
3. Find five consecutive numbers whose sum is 95.
4. The lengths of the sides of a quadrilateral are consecutive even integers. The perimeter of the quadrilateral is 108 yards. What are the lengths of the sides of the quadrilateral?

Lesson 7-4

Copy and complete the table below by finding the radius, diameter, or circumference of the circle using the information provided.

	Radius	Diameter	Circumference
1.	5 cm		
3.		14 in.	
5.			≈ 56.52 ft
7.		17 m	
9.	6 yds		
11.	20 mm		

	Radius	Diameter	Circumference
2.		25 mi	
4.			≈ 65.94 mi
6.	8 in.		
8.			≈ 100.48 ft
10.		22 cm	
12.			≈ 81.64 in.

EXTRA PRACTICE

Lesson 7-5

Solve each equation. Check your solution.

1. $-7h - 5 = 4 - 4h$

2. $5t - 8 = 3t + 12$

3. $w + 6 = 2(w - 6)$

4. $m + 2m + 1 = 7$

5. $3.21 - 7y = 10y - 1.89$

6. $3(b + 1) = 4b - 1$

7. $\frac{5}{9}g + 8 = \frac{1}{6}g + 1$

8. $\frac{s - 3}{7} = \frac{s + 5}{9}$

Lesson 7-6

Solve each inequality and check your solution. Graph the solution on a number line.

1. $2m + 1 < 9$

2. $-3k - 4 \leq -22$

3. $-2 > 10 - 2x$

4. $-6a + 2 \geq 14$

5. $3y + 2 < -7$

6. $\frac{d}{4} + 3 \geq -11$

7. $\frac{x}{3} - 5 < 6$

8. $-5g + 6 < 3g + 20$

9. $-3(m - 2) > 12$

10. $\frac{r}{5} - 6 \leq 3$

11. $\frac{3(n + 1)}{7} \geq \frac{n + 4}{5}$

12. $\frac{n + 10}{-3} \leq 6$

Lesson 7-7

Define a variable and write an inequality for each situation. Then solve.

1. If 8 times a number is decreased by 2, the result is less than 15. What is the number?

2. George plans to spend at most $40 for shirts and ties. He bought 2 shirts for $13.95 each. How much can he spend for ties?

3. Mia is buying a boat. She can make a down payment of $2,200. She wants to pay off the boat in 3 years. She cannot afford monthly payments over $350. What is the most she can pay for the boat?

Lesson 7-8

Complete each sentence.

1. 8.2 mm = __?__ cm
2. 6.7 km = __?__ m
3. 8.4 kg = __?__ mg
4. 18 cm = __?__ m
5. 250 ml = __?__ L
6. 4 g = __?__ mg

Write which metric unit you would probably use to measure each item.

7. length of a car
8. weight of a boy
9. capacity of a can of motor oil
10. weight of a container of deodorant

Lesson 8-1

Write the domain and range of each relation. Then determine whether each relation is a function.

1. {(3, 6),(35, 64),(1, 1),(21, 7)}
2. {(32, 24),(27, 24),(36, 24),(45, 24)}
3. {(2, 9),(3, 18),(4, 27),(2, 36)}
4. $\left\{\left(\frac{1}{2}, 3\right),\left(\frac{1}{4}, 5\right),\left(\frac{1}{6}, 7\right),\left(\frac{1}{8}, 9\right),\left(\frac{1}{10}, 11\right)\right\}$
5. {(1, 0),(1, 9),(1, 18)}
6. {(5, 5),(6, 6),(7, 7),(8, 7)}
7.

x	8	15	22	29
y	8	8	51	22

8. 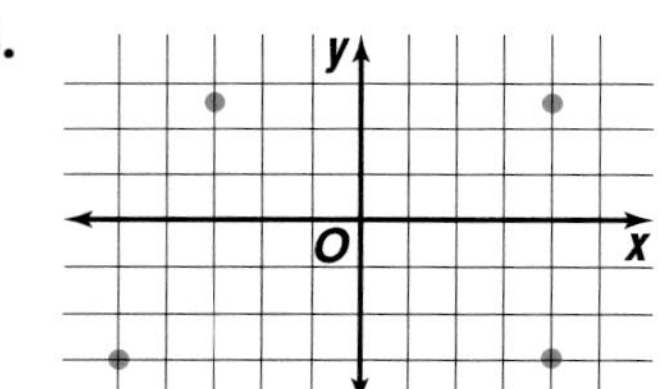

Lesson 8-2

Determine whether a scatter plot of the data for the following might show a *positive*, a *negative*, or *no* relationship. Explain your answer.

1. Size of the bag, number of potato chips
2. Number of team members, amount of playing time
3. Number of flowers in a garden, number of bees
4. Hair color, number of dates
5. Candy consumed, number of cavities
6. Number of meals, weight
7. Number of classes, hours studied

Lesson 8-3

Find four solutions for each equation. Write your solutions as ordered pairs.

1. $x = 4$
2. $y = 0$
3. $x + y = 2$
4. $y = 2x - 6$
5. $x - y = 5$
6. $3x - y = 8$
7. $y = \frac{1}{2}x - 3$
8. $y = \frac{1}{3}x + 1$
9. $2x + y = -2$
10. $2x + 3y = 12$
11. $x + 2y = -4$
12. $2x - 4y = 8$

Lesson 8-4

Determine whether each equation is a function.

1. $y = 2x - 3$
2. $x = 17$
3. $y = |x| - 3$
4. $x + |y| = 2$

Given $f(x) = 3x + 2$ and $g(x) = 2x^2 - x$, determine each value.

5. $f(3)$
6. $g(-5)$
7. $g\left(\frac{1}{2}\right)$
8. $f\left(\frac{2}{3}\right)$
9. $g(-1)$
10. $g(2.5)$
11. $f(-2)$
12. $2[f(3)]$

Lesson 8-5

Use a graph to solve each problem. Assume that the rate is constant in each problem.

1. Climate The average high temperature in Vancouver, Canada in January is 41°. In July, the average high temperature is 74°. What would the average high temperature be in April?

2. Economy A 23 minute phone call from Columbus, Ohio to Dayton, Ohio costs \$1.65. A similar phone call lasting 18 minutes costs \$1.30. How much would a 20 minute phone call cost?

3. Measurement A distance of 5 miles corresponds to the distance of 8 kilometers. A distance of 30 miles corresponds to the distance of 48.2 kilometers. If you traveled 50 miles, how many kilometers would you have traveled.

Lesson 8-6

Find the slope of the line that contains each pair of points.

1. $P(3, 8)$, $Q(4, -3)$
2. $D(4, 5)$, $E(-3, -9)$
3. $L(-1, 2)$, $M(0, 5)$
4. $J(6, 2)$, $K(6, -4)$
5. $B(8, -3)$, $C(-4, 1)$
6. $D(1, 5)$, $E(3, 10)$
7. $H(7, 2)$, $I(-2, -2)$
8. $K(2, -4)$, $L(5, -19)$
9. $G(5, 6)$, $H(7, 6)$
10. $A(-6, -3)$, $B(-9, 4)$
11. $P(-1, -6)$, $Q(-5, -10)$
12. $B(5, 9)$, $C(-4, -5)$

Lesson 8-7

Use the *x*-intercept and the *y*-intercept to graph each equation.

1. $2x + y = 6$
2. $-4x + y = 8$
3. $3x - 3y = -12$
4. $x + 2y = -4$
5. $y = -\frac{1}{2}x - 6$
6. $y = \frac{5}{2}x - 1$

Graph each equation using the slope and *y*-intercept.

7. $y = 3x - 2$
8. $x - 3y = 9$
9. $y = \frac{1}{2}x + 4$
10. $y = -\frac{2}{3}x - 1$
11. $x - y = -4$
12. $2x + 4y = -4$

Lesson 8-8

Use a graph to solve each system of equations.

1. $x + 2y = 6$
$y = -0.5x + 3$

2. $y = -2$
$4x + 3y = 2$

3. $2x = 3y$
$y = 4 + \frac{2}{3}x$

4. $y = x - 2$
$y = -\frac{1}{3}x + 2$

5. $x + y = 4$
$y = 2$

6. $2x + y = 8$
$2x - y = 0$

7. $x = 3$
$y = 4$

8. $y = x + 2$
$2y + x = 1$

Lesson 8-9

Graph each inequality.

1. $y > 2x - 2$
2. $y \geq x$
3. $y < 1$
4. $x + y \leq -1$
5. $y + 3x \leq 0$
6. $x < -3$
7. $2x + 3y \geq 12$
8. $-2x + y > -1$

Lesson 9-1

Express each ratio or rate as a fraction in simplest form.

1. 15 out of 240
2. 140:12
3. $\frac{98}{14}$
4. 6 pens to 14 pens
5. 30 cats to 6 cats
6. 18 out of 45
7. 321 to 96
8. 3 cups to 3 quarts

Express each ratio as a unit rate.

9. 343.8 miles on 9 gallons
10. $7.95 for 5 pounds.
11. $52 for 8 tickets
12. $43.92 for 4 CDs

Lesson 9-2

Solve by making a table.

1. Susan Lee has $1.15 made up of six United States coins. However, she cannot make change for a dollar, a half-dollar, a dime, or a nickel. What six coins does Susan have?
2. How many ways can you add eight prime numbers to get a sum of 20? You may use a number more than once.
3. Anne Zody buys a car for $100, sells it for $110, buys it back for $120, and sells it again for $130. How much does Ms. Zody make or lose?
4. Dwight bought the books he needed for the fall semester for $230. He dropped a class and sold the book back for $18. He added a new class and bought a book for $35. At the end of the semester, he sold his books back for $62. How much did Dwight make or lose?
5. A new computer company begins with 48 employees. After the first year, 8 employees leave for more stable jobs, 2 are transferred to other offices, and 17 new employees are hired. Over the next five years the company's profits increase, resulting in the hiring of 21 more people with the loss of only 3 employees. How many employees are employed by the company after 6 years?

EXTRA PRACTICE

Lesson 9-3

There are 4 blue marbles, 5 red marbles, and 3 green marbles in a bag. Suppose you select one marble at random. Find each probability.

1. P(green)
2. P(blue)
3. P(red)
4. P(not green)
5. P(white)
6. P(blue or red)
7. P(neither red nor green)
8. P(not orange)

Lesson 9-4

Solve each proportion.

1. $\frac{7}{k} = \frac{49}{63}$
2. $\frac{s}{4.8} = \frac{30.6}{28.8}$
3. $\frac{6}{11} = \frac{19.2}{g}$
4. $\frac{8}{13} = \frac{b}{65}$

Write a proportion that could be used to solve for each variable. Then solve the proportion.

5. 6 plums at \$1
   10 plums at d
6. 8 gallons at \$9.36
   f gallons at \$17.55
7. 3 packages at \$53.67
   7 packages at m
8. 10 cards at \$7.50
   p cards at \$18

Lesson 9-5

Use the percent proportion to solve each problem.

1. What is 81% of 134?
2. 52.08 is 21% of what number?
3. 11.18 is what percent of 86?
4. What is 120% of 312?
5. 140 is what percent of 400?
6. 430.2 is 60% of what number?

Lesson 9-6

William Wonker wants to find out favorite candy bars. He surveyed 100 people at a candy store.

1. Give some reasons why this is a valid sample.
2. Give some reasons why this is not a valid sample.
3. What percent of the people preferred Giggles Bar?
4. If 600 customers were surveyed, how many would prefer a Venus Bar?
5. How many would prefer a Galaxy Bar?
6. What is the mode?

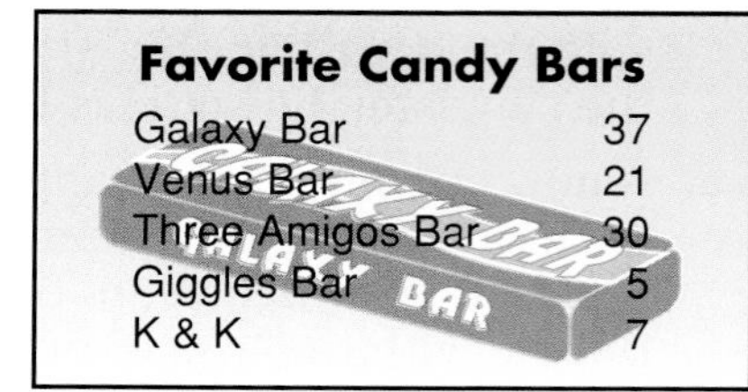

Favorite Candy Bars	
Galaxy Bar	37
Venus Bar	21
Three Amigos Bar	30
Giggles Bar	5
K & K	7

Lesson 9-7

Express each decimal as a percent.

1. 0.06 **2.** 0.374 **3.** 0.0095 **4.** 56.71

Express each fraction as a percent.

5. $\frac{3}{4}$ **6.** $3\frac{1}{4}$ **7.** $\frac{45}{50}$ **8.** $\frac{3}{1000}$

Express each percent as a fraction.

9. 17% **10.** 0.8% **11.** 5268% **12.** $\frac{15}{4}\%$

Lesson 9-8

Choose the best estimate.

1. 28% of 500 **a.** 1.5 **b.** 15 **c.** 150

2. 96% of 900 **a.** 90 **b.** 900 **c.** 9000

3. 148% of 350 **a.** 5.25 **b.** 52.5 **c.** 525

4. $\frac{1}{3}\%$ of 360 **a.** 1.08 **b.** 10.8 **c.** 108

Estimate.

5. 72% of 250 **6.** 47% of 198

7. 0.8% of 380 **8.** 98% of 32

9. $12\frac{1}{2}\%$ of 130 **10.** 122% of 84

Estimate each percent.

11. 12 out of 20 **12.** 14 out of 40 **13.** 3 out of 75

14. 75 out of 179 **15.** 19 out of 96 **16.** 1.6 out of 88

17. 5 out of 9 **18.** 6 out of 210 **19.** 12 out of 2250

Lesson 9-9

Solve each problem by using the percent equation, $P = R \cdot B$.

1. 9.28 is what percent of 58? **2.** What number is 43% of 110?

3. 88% of what number is 396? **4.** What number is 61% of 524?

5. 126 is what percent of 90? **6.** 52% of what number is 109.2?

Find the discount or interest to the nearest cent.

7. $64.98 dress at 32% off. **8.** $1000 at 4% annually for 2 years.

9. $589 sofa at 20% off. **10.** $500 at 2.5% a month for 18 months.

11. $800 at $7\frac{1}{2}\%$ annually for 15 months. **12.** $2148 bedroom set at 40% off.

Lesson 9-10

State whether each percent of change is a percent of increase or a percent of decrease. Then find the percent of increase or decrease. Round to the nearest whole percent.

1. old: $56
   new: $42
2. old: $26
   new: $29.64
3. old: $22
   new: $37.18
4. old: $137.50
   new: $85.25
5. old: $455
   new: $955.50
6. old: $3
   new: $15
7. old: 750.75
   new: $765.51
8. old: $953
   new: $476.5
9. old: $101.25
   new: $379.69
10. old: $836
    new: $842.27

Lesson 10-1

Make a stem-and-leaf plot of each set of data.

1. 37, 44, 32, 53, 61, 59, 49, 69
2. 3, 26, 35, 8, 21, 24, 30, 39, 35, 5, 38
3. 15.7, 7.4, 0.6, 0.5, 15.3, 7.9, 7.3
4. 172, 198, 181, 182, 193, 171, 179, 186, 181
5. 101.6, 101.8, 100.5, 102.1, 101.0, 100.1, 100.5, 100.5

Lesson 10-2

Find the range, median, upper and lower quartiles, and the interquartile range for each set of data.

1. 44, 37, 23, 35, 61, 95, 49, 96
2. 30, 62, 35, 80, 12, 24, 30, 39, 53, 38
3. 7.15, 4.7, 6, 5.3, 30.1, 9.19, 3.2
4. 271, 891, 181, 193, 711, 791, 861, 818

Lesson 10-3

Use the stem-and-leaf plot at the right to answer each question.

1. Make a box-and-whisker plot of the data.
2. What is the range?
3. What is the median?
4. What is the upper quartile?
5. What is the lower quartile?
6. What is the interquartile range?
7. What are the extremes?
8. Are there any outliers? If so, what are they?
9. What are the limits for outliers?

Stem	Leaf
1	2
2	
3	
4	3
5	0
6	2 3 7 9
7	0 3
8	0 2 2 4 5 7
9	0 7

4 | 3 = 43

Lesson 10-4

Use the graphs to answer the following questions.

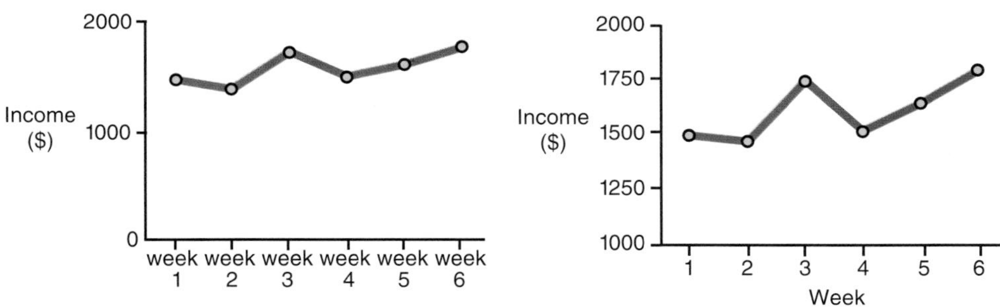

1. Explain why these graphs made from the same data look different.
2. Which graph shows that the family income is consistent? Why?
3. Which graph shows why a son can't have a raise in his allowance? Explain.

Lesson 10-5

Find the number of possible outcomes for each event.

1. Engagement rings come in silver, gold, and white gold. The diamond can weigh $\frac{1}{2}$ karat, $\frac{1}{3}$ karat, or $\frac{1}{4}$ karat. The diamond can have 4 possible shapes.
2. A dress can be long, tea-length, knee-length, or mini. It comes in 2 colors and the dress can be worn on or off the shoulders.
3. The first digit of a 7 digit phone number is a 2. The last digit is a 3.
4. A chair can be a rocker, recliner, swivel, straight back, or a combination of any of the first three features.

Lesson 10-6

1. Seven people are running for student council. There will be four people on the board and one alternate. How many ways can the students be elected?
2. How many ways can the letters of the word ISLAND be arranged?
3. How many ways can five candles be arranged in three candlesticks?

Find each value.

4. 7! **5.** $P(3, 2)$ **6.** $C(9, 4)$

7. $P(10, 5)$ **8.** 10! **9.** $\frac{6!2!}{5!}$

EXTRA PRACTICE

Lesson 10-7

Find the odds of each outcome if the spinner below is spun.

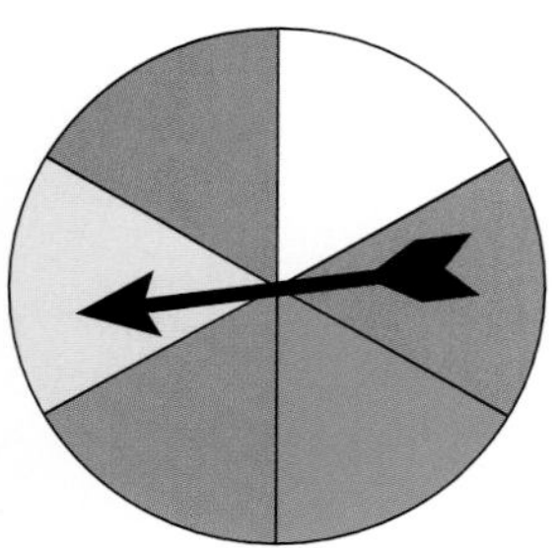

1. blue
2. a color with less than 5 letters
3. a color that begins with a consonant
4. red, yellow, or blue

Find the odds of each outcome if a 10-sided die is rolled.

5. number less than 7
6. odd number
7. composite number
8. number divisible by 3

Lesson 10-8

Solve. Use a simulation.

1. Laneeda is a forward on the freshman basketball team. She usually makes $\frac{2}{3}$ of her shots from the field and $\frac{5}{6}$ of her free throws. If she averages 10 shots a game and 5 free throws, describe a simulation that could give her probable number of points in the next game.
2. A certain restaurant gives out game cards. One out of six cards wins a small soft drink. Describe a simulation to predict how many soft drinks Ted will win with 14 game cards.

Lesson 10-9

Determine whether the events are independent or dependent.

1. go to the movies on Monday and then going back to the same theater on Tuesday
2. rolling a die and spinning a spinner
3. watching a 1 p.m. football game and watching a 4 p.m. football game
4. choosing a person to be quarterback from a roster of 45 people and then choosing a person to be wide receiver
5. selecting a name for your daughter and then selecting a name for your son

A deck of Euchre cards consists of 4 nines, 4 tens, 4 jacks, 4 queens, 4 kings, and 4 aces. Once a card is selected, it is not replaced. Find the probability of each outcome.

6. 3 nines in a row
7. a black jack and then a red queen
8. a nine of clubs, a black king, and a red ace
9. 4 face cards in a row
10. 2 cards lower than a jack

Lesson 10-10

Determine whether each event is *mutually exclusive* or *inclusive*. Then find the probability. A number from 6 to 19 is drawn.

1. P(12 or even)

2. P(13 or less than 7)

3. P(even or odd)

4. P(14 or greater than 20)

5. P(even or less than 10)

6. P(odd or greater than 10)

7. P(divisible by 3 or even)

8. P(prime or even)

Lesson 11-1

In the figure at the right, use a protractor to find the measure of each angle. Then classify the angle as *acute*, *right*, or *obtuse*.

1. $m\angle PQW$

2. $m\angle VQW$

3. $m\angle TQW$

4. $m\angle SQW$

5. $m\angle SQR$

6. $m\angle VQR$

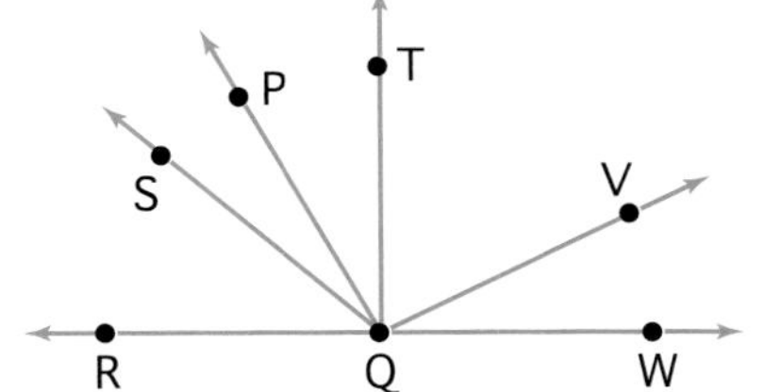

Lesson 11-2

Make a circle graph to display each set of data.

1.

Items Donated for Food Drive	
Canned Goods	169
Pasta	86
Cereal	70
Peanut Butter	42
Condiments	18

2.

Davis Family Monthly Expenditures	
Housing	\$800
Taxes	\$600
Food	\$350
Clothing	\$200
Insurance	\$150
Savings	\$100

Lesson 11-3

In the figure at the right, ℓ is parallel to m. If the measure of $\angle 2$ is 38°, find the measure of each angle.

1. $\angle 1$

2. $\angle 3$

3. $\angle 4$

4. $\angle 5$

5. $\angle 6$

6. $\angle 8$

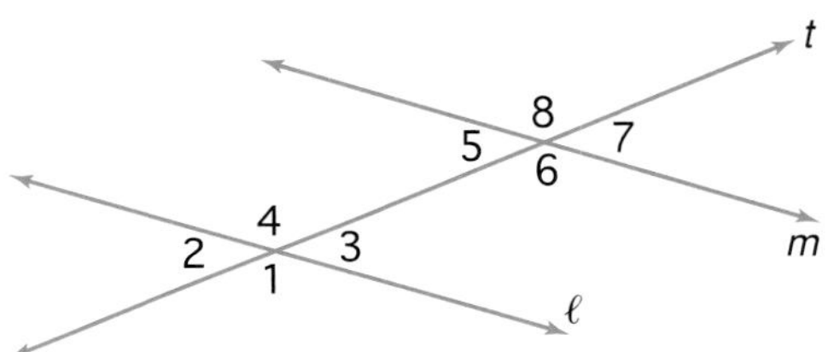

EXTRA PRACTICE

Lesson 11-4

1. Find the measure of each angle of a triangle if they are in the ratio 1:2:3.
2. Find the measure of each angle of a triangle if they are in the ratio 1:1:2.
3. Find the measure of each angle of a triangle if they are in the ratio 1:9:26.

Lesson 11-5

If $\triangle JKL \cong \triangle DGW$, name the part congruent to each angle or segment given.

1. $\angle K$ 2. $\overline{WG}$ 3. $\angle D$ 4. $\overline{KL}$ 5. $\overline{DG}$

Lesson 11-6

Write a proportion to find each missing measure x. Then find the value of x.

1. 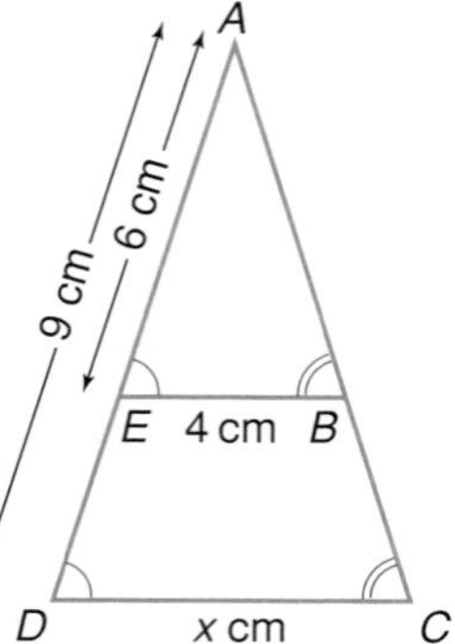

2. 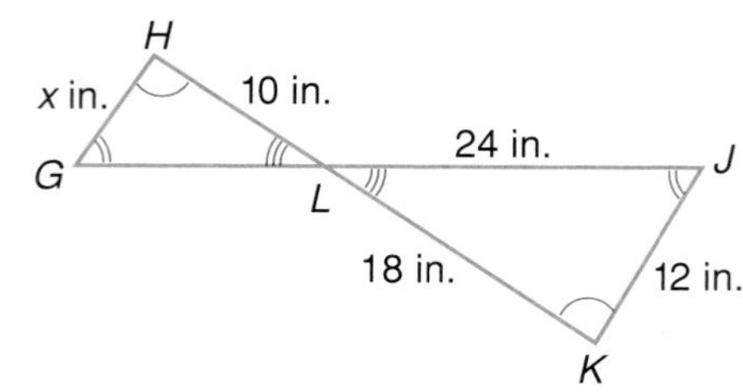

Lesson 11-7

Find the value of x. Then find the missing measures.

1. 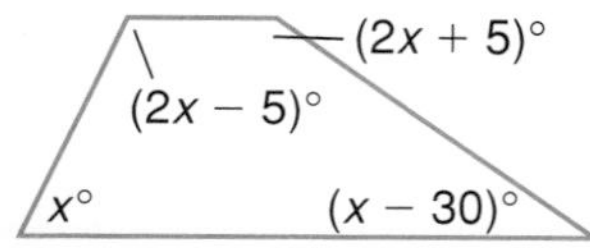

2. 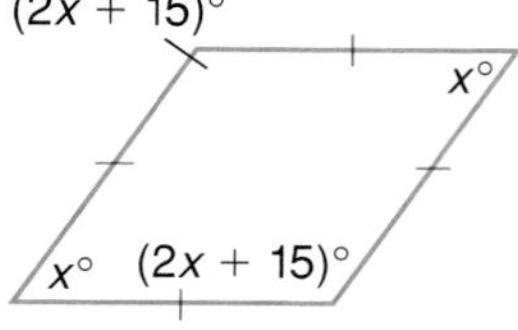

3. $(\frac{1}{2}x + 10)°$ $x°$ $(x + 10)°$

Lesson 11-8

Find the measure of each exterior angle and each interior angle of each regular polygon.

1. equilateral triangle
2. regular pentagon
3. regular octagon
4. regular nonagon

Lesson 11-9

Trace each figure. Draw all lines of symmetry.

1. 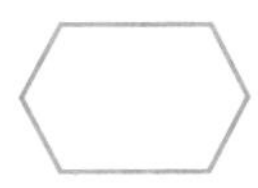

2. 

3. 

4. 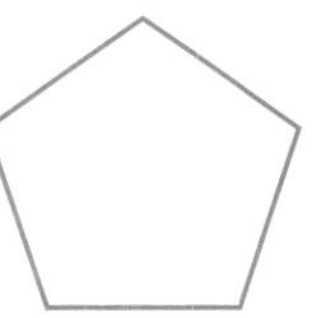

EXTRA PRACTICE

Lesson 12-1

Find the area of each figure.

1. 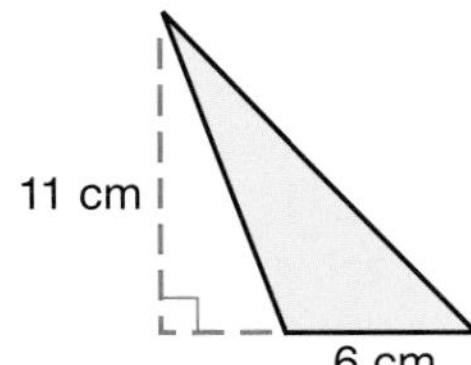

2. 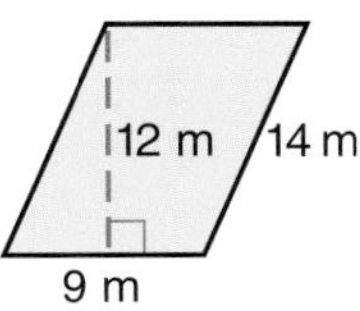

3. 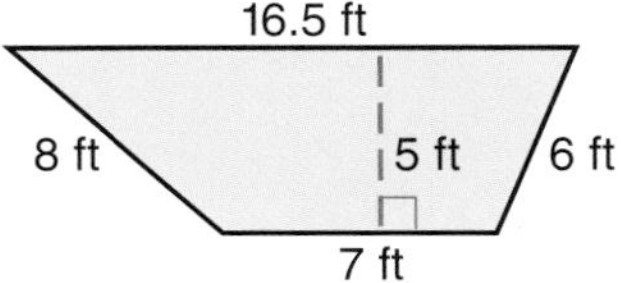

Lesson 12-2

Find the area of each circle. Round to the nearest tenth.

1. diameter, 10 in. **2.** radius, 8 mm **3.** diameter, 15.4 m

Lesson 12-3

Each figure represents a dart board. Find the probability of landing in the shaded region.

1. 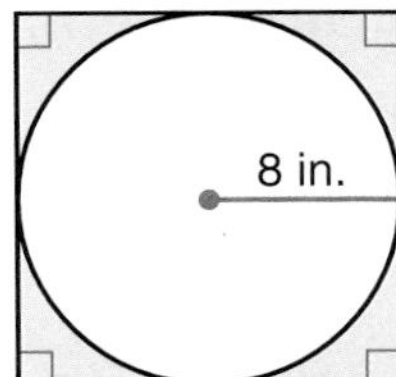

2. 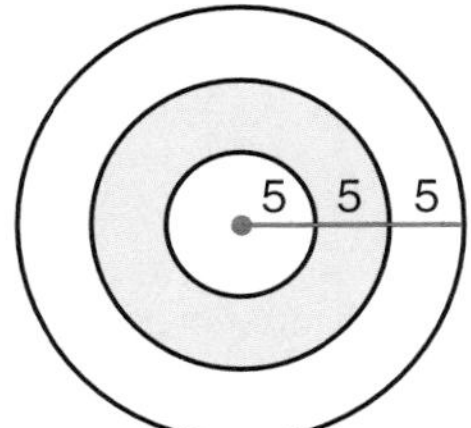

3. 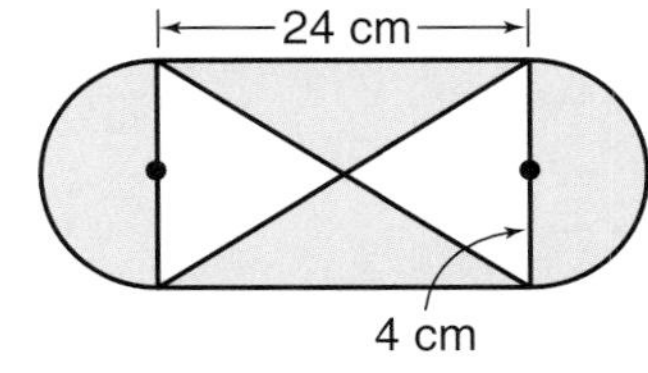

Lesson 12-4

Solve by making a model or drawing.

1. A theater section is arranged so that each row has the same number of seats. Andrew is seated in the fifth row from the front and the third row from the back. His seat is sixth from the left and second from the right. How many seats are in this section?
2. Sixteen players are competing in a bowling tournament. Each player in the competition bowls against another opponent and is eliminated after one loss. How many games does the winner of the tournament play?
3. An artist is creating a pyramid piece of art using differently-colored marble cubes. There is one cube on the top, two cubes in the next layer, three cubes in the next layer, and so on until 10 layers are completed. How many cubes did the artist use?

Lesson 12-5

Find the surface area of each solid. Round decimal answers to the nearest tenth.

1.
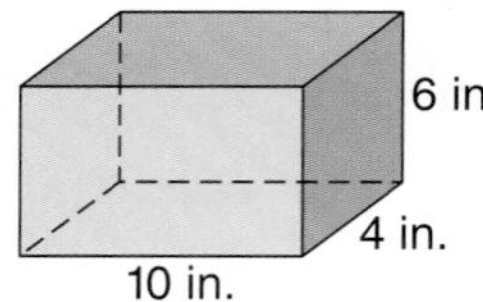

2.
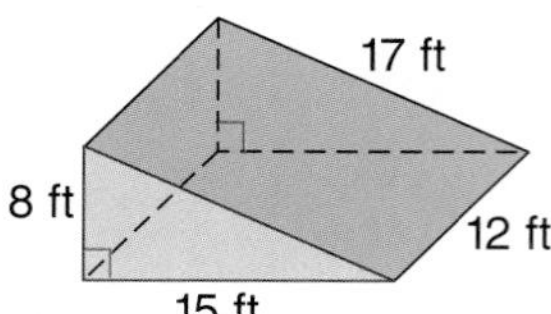

3.
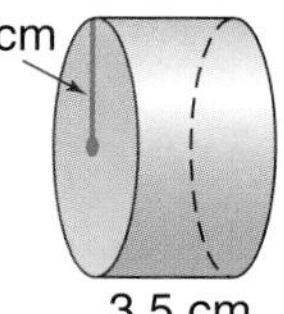

Lesson 12-6

Find the surface area of each pyramid or cone. Round decimal answers to the nearest tenth.

1.
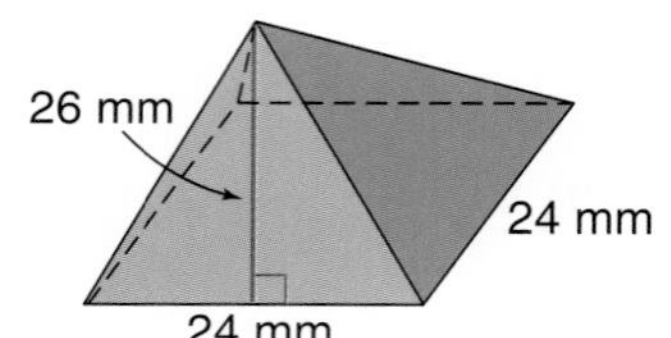

2.
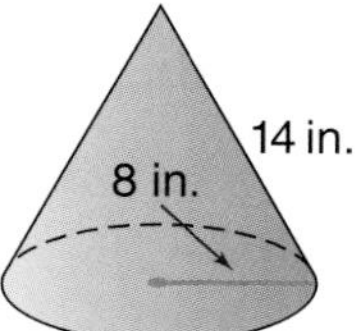

3.
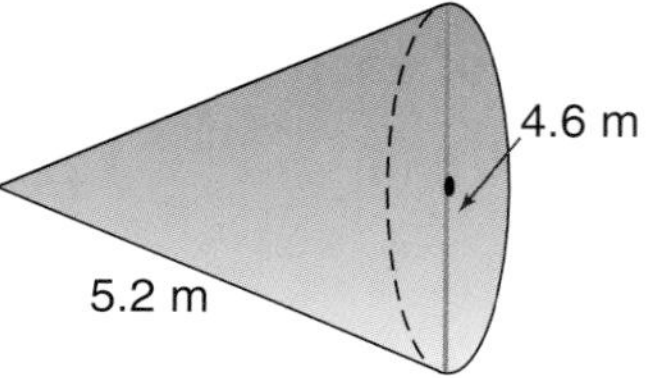

Lesson 12-7

Find the volume of each prism or cylinder. Round decimal answers to the nearest tenth.

1.
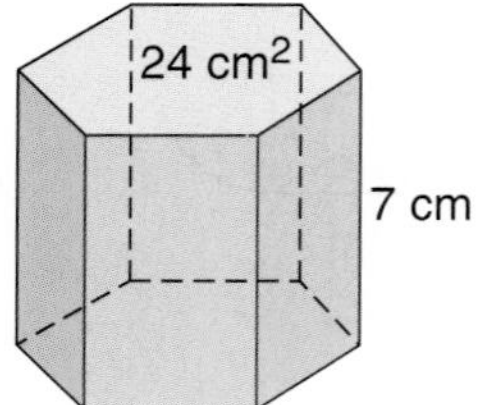

2.
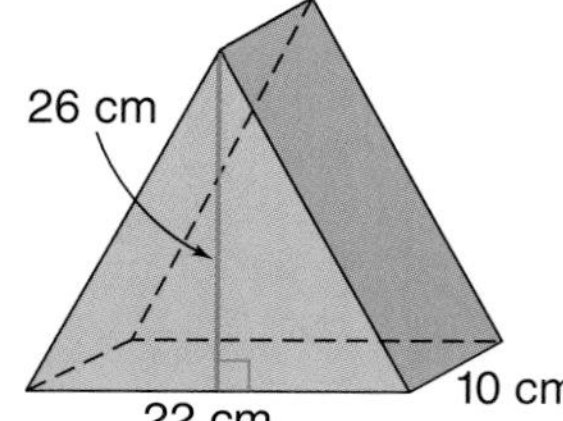

3.
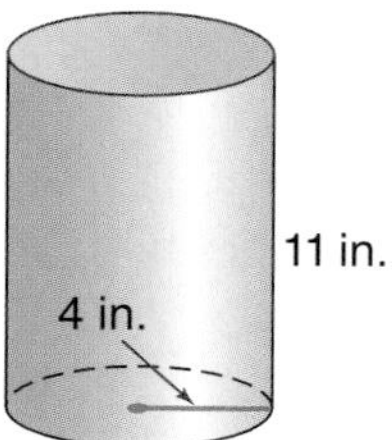

Lesson 12-8

Find the volume of each pyramid or cone. Round decimal answers to the nearest tenth.

1.
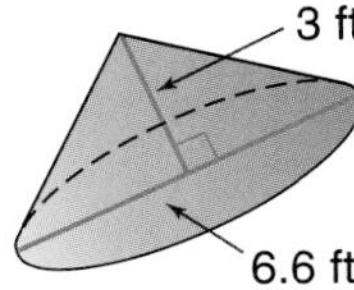

2.
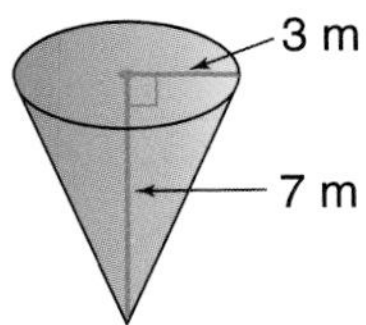

3.
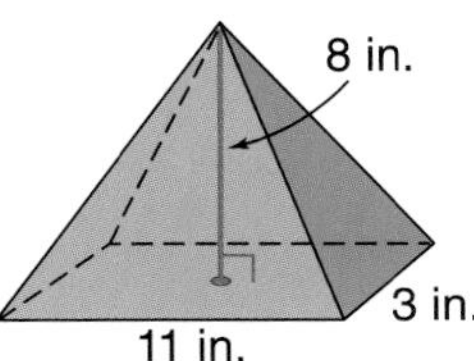

EXTRA PRACTICE

Lesson 13-1

Find the best integer estimate for each square root. Then check your estimate with a calculator.

1. $\sqrt{21}$ **2.** $-\sqrt{85}$ **3.** $\sqrt{7.3}$ **4.** $\sqrt{1.99}$
5. $-\sqrt{62}$ **6.** $\sqrt{74.1}$ **7.** $\sqrt{810}$ **8.** $-\sqrt{88.8}$

Lesson 13-2

Solve using a Venn diagram.

1. A supermarket survey showed that 83 customers chose wheat cereal, 83 chose rice, and 20 chose corn. Six customers bought a box of corn cereal and a box of wheat cereal and 10 others bought a box of rice cereal and a box of corn cereal. Four customers bought all three. How many customers bought a box of rice cereal and a box of wheat cereal if 61 bought just a box of wheat cereal.
2. Of the thirty members in a cooking club, 20 like to mix salads, 17 prefer baking desserts, and 8 like to do both.
   a. How many like to do salads, but not bake desserts?
   b. How many do not like either baking desserts or mixings salads?

Lesson 13-3

Solve each equation. Round decimal answers to the nearest tenth.

1. $x^2 = 14$ **2.** $y^2 = 25$ **3.** $34 = p^2$ **4.** $55 = h^2$
5. $225 = k^2$ **6.** $324 = m^2$ **7.** $d^2 = 441$ **8.** $r^2 = 25{,}000$

Lesson 13-4

In a right triangle, If *a* and *b* are the measures of the legs and *c* is the measurement of the hypotenuse, find each missing measure.

1. *a*, 7 m; *b*, 24 m **2.** *a*, 18 in.; *c*, 30 in. **3.** *b*, 10 ft; *c*, 20 ft
4. *a*, 3 cm; *c*, 9 cm **5.** *b*, 8 m; *c*, 32 m **6.** *a*, 32 yd; *c*, 65 yd

Lesson 13-5

Find the lengths of the missing sides in each right triangle. Round decimal answers to the nearest tenth.

1. 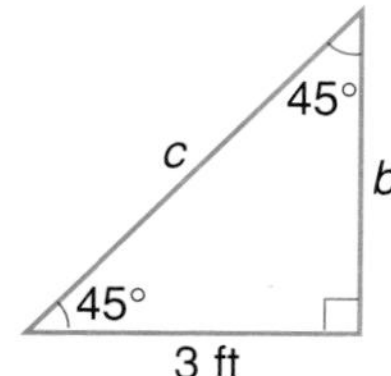

2. 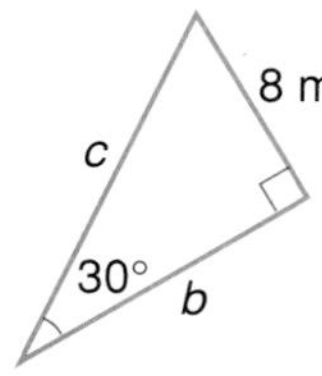

3. 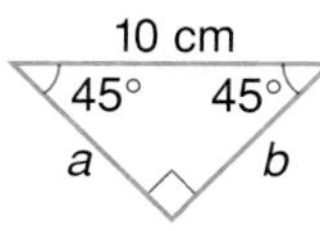

4. 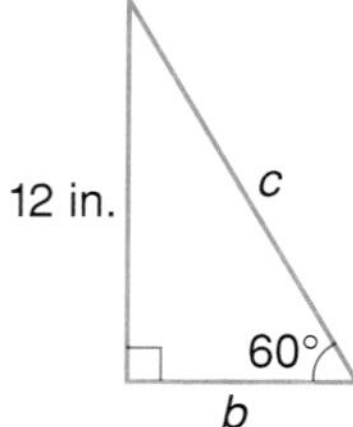

Lesson 13-6

For each triangle, find the measure of the marked acute angle to the nearest degree.

1. 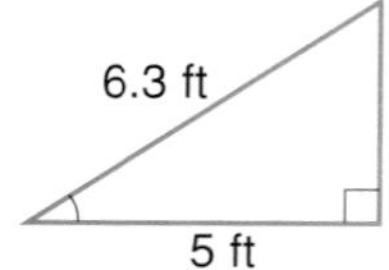

2. 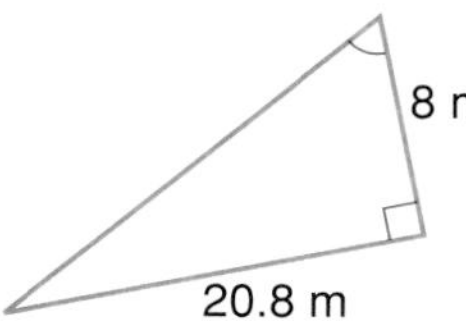

3. 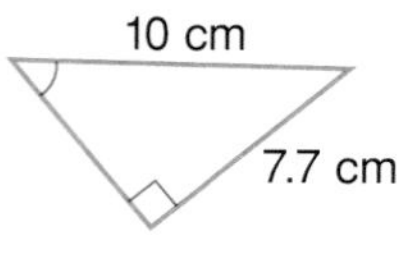

4. 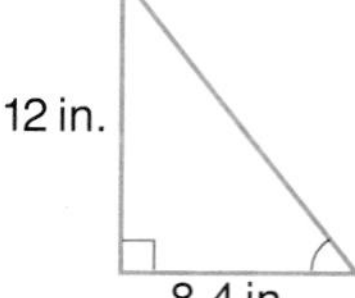

Lesson 13-7

Use trigonometric ratios to solve each problem.

1. In a sightseeing boat near the base of Horseshoe Falls at Niagara Falls, a passenger estimates that angle of elevation to the top of the falls to be 30°. If Horseshoe Falls is 173 feet high, what is the distance from the boat to the base of the falls?
2. While picnicking in San Jacinto Battlefield Park, a student 800 feet from the base of the monument estimates the angle of elevation to the top of the monument to be 25°. From this information, estimate the height of the monument.
3. From a boat in the ocean, a cliff is sighted through the fog. The angle of elevation is 42° and the height of the cliff is 135 meters. How far is the boat from the cliff?

Lesson 14-1

Find the degree of each polynomial.

1. $4x$	**2.** $a^2 - 6$	**3.** $11r + 5s$
4. $3y^2 4y - 2$	**5.** $9cd^3 - 5$	**6.** $-5p^3 + 8q^2$
7. $w^2 + 2x - 3y^3 - 7z$	**8.** $\frac{x^3}{6} - x$	**9.** $-17n^2p - 11np^3$

Lesson 14-2

Find each sum.

1. $(3a + 4) + (a + 2)$
2. $(8m - 3) + (4m + 1)$
3. $(5x - 3y) + (2x - y)$
4. $(8p^2 - 2p + 3) + (-3p^2 - 2)$
5. $(-11r^2 + 3s) + (5r^2 - s)$
6. $(3a^2 + 5a + 1) + (2a^2 - 3a - 6)$

EXTRA PRACTICE

Lesson 14-3

Find each difference.

1. $(3n + 2) - (n + 1)$
2. $(-3c + 2d) - (7c - 6d)$
3. $(4x^2 + 1) - (3x^2 - 4)$
4. $(5a - 4b) - (-a + b)$
5. $\begin{array}{r} 6x^2 - 4x + 11 \\ \underline{(-)\ 5x^2 + 5x - 4} \end{array}$
6. $\begin{array}{r} 8n^2 + 3mn \\ \underline{(-)\ 4n^2 + 2mn - 9} \end{array}$

Lesson 14-4

Simplify.

1. $(a^2)^3$
2. $(-2x)^3$
3. $(p^2q)^3$
4. $-5c(2cd)^3$
5. $4y(y^2z)^3$
6. $6a(-ab)^7$

Lesson 14-5

Find each product.

1. $4n(5n - 3)$
2. $-3x(4 - x)$
3. $6m(-m^2 + 3)$
4. $-5x(2x^2 - 3x + 1)$
5. $7r(r^2 - 3r + 7)$
6. $-3az(2z^2 + 4az + a^2)$

Lesson 14-6

Find each product.

1. $(2x + 2)(3x + 1)$
2. $(x + 4)(3x + 1)$
3. $(7x + 4)(3x - 11)$
4. $(x + 3)(x + 1)$
5. $(2x + 1)(x + 1)$
6. $(3x + 2)(2x + 2)$

Chapter 1 Practice Test

Solve using the four-step plan.

1. The distance between Jackie's house and Roberta's house is 90 feet. If it takes Jackie 3 seconds to walk 10 feet, how long will it take her to walk to Roberta's house?

Find the value of each expression.

2. $7 \cdot 4 + 6 \cdot 5$
3. $8 + 3(16 - 12)$
4. $6[4 \cdot (72 - 63) \div 3]$
5. $6[5 \times (41 - 36) - (8 + 14)]$

Evaluate each expression if $a = 8$, $b = 4$, and $c = 3$.

6. $2c + 6bc - 7$
7. $15 \div c + 7b$
8. $6b \div a + 5c$

Name the property shown by each statement.

9. $4 \cdot (6 \cdot 0) = 4 \cdot (0 \cdot 6)$
10. $5 + (9 + 3) = (5 + 9) + 3$
11. $0 + h = h$
12. $8(a + 4) = 8a + 8(4)$
13. $8(6 \cdot 9) = (8 \cdot 6)9$
14. $r \cdot 1 = r$

Simplify each expression.

15. $(b + 12) + 15$
16. $6(v \cdot 2)$
17. $18yz + 13yz$
18. $5(2w + 3) + 7(w + 13)$

Solve each equation mentally.

19. $5x = 25$
20. $8 + g = 8$
21. $\frac{s}{14} = 3$
22. $22 - k = 17$

Name the ordered pair for each point graphed on the coordinate plane at the right.

23. B
24. E
25. F
26. H
27. N
28. K

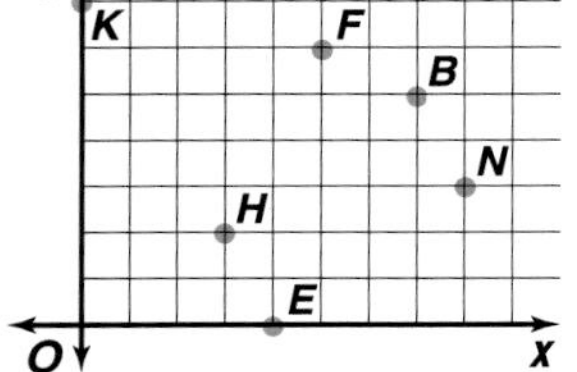

Solve each equation using the inverse operation. Use a calculator when needed.

29. $x + 48 = 55$
30. $v - 57 = 72$
31. $\frac{b}{2} = 18$
32. $672 = 21t$

State whether each inequality is *true* or *false* for the given value.

33. $16 - t > 7, t = 14$
34. $4p - 6 \geq 9, p = 4$
35. $2v + v > 21, v = 7$
36. $8m - 2m \leq 7m, m = 0$

The frequency table at the right contains data about the number of dogs of different breeds registered in the American Kennel Club in 1992.

Breed	Dogs (thousands)	Breed	Dogs (thousands)
Beagle	61	Maltese	18
Chow Chow	43	Newfoundland	3
Collie	17	Shar-Pei	90
Dalmatian	39	Springer Spaniel	22
Labrador Retriever	121	Weimaraner	5

Source: American Kennel Club

37. Which breed has the most dogs registered?
38. Write an equation to represent the difference between the number of Maltese and the number of springer spaniels registered. Then find the difference.
39. How many dogs of these ten breeds are registered?
40. Make a bar graph of the data.

Chapter 2 Practice Test

PRACTICE TEST

Write an integer for each situation.

1. a gain of 15 pounds
2. a rent increase of \$10
3. a bank overdraft of \$25
4. a loss of 6 yards

Simplify.

5. $|-8|$
6. $|-3| + |2|$
7. $|15| - |-3|$
8. $|-20| + |-19|$

Name the ordered pair for each point graphed on the coordinate plane at the right.

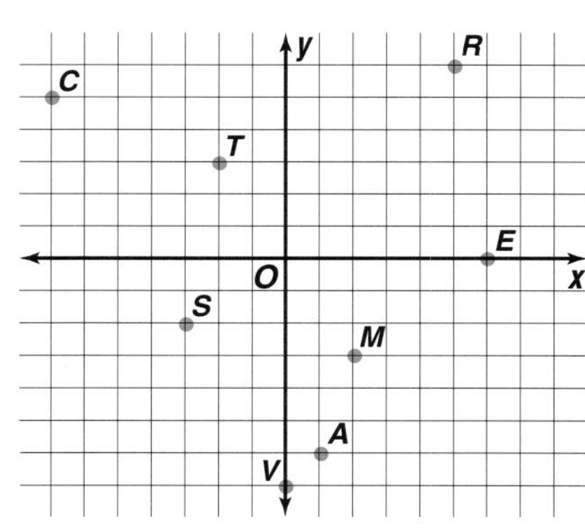

9. *A*
10. *C*
11. *E*
12. *R*
13. *S*
14. *M*
15. *T*
16. *V*

On graph paper, draw coordinate axes. Then graph and label each point. Name the quadrant in which each point is located.

17. $A(8, -5)$
18. $J(-4, -18)$
19. $W(-1, 6)$
20. $F(6, 17)$
21. $M(-8, 5)$
22. $S(-8, -5)$

Replace each ● with <, >, or =.

23. -6 ● -12
24. 18 ● 18
25. -5 ● -4
26. $|-7|$ ● 7

Simplify each expression.

27. $3x + 5x$
28. $6p - (-4p)$
29. $-9m - (-7m)$
30. $-8k + (-12k)$
31. $-15a - 6a$
32. $10d + (-3d)$

Solve. Look for a pattern.

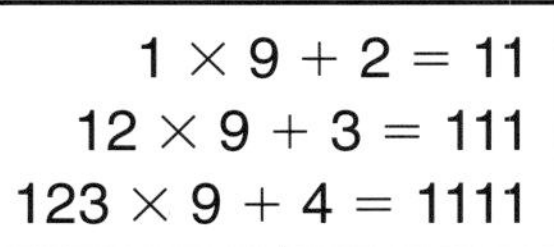
$1 \times 9 + 2 = 11$
$12 \times 9 + 3 = 111$
$123 \times 9 + 4 = 1111$

33. Use the pattern at the right to find $123{,}456 \times 9 + 7$
34. The following is referred to as an alphametric problem. Each letter stands for one digit 0–9.

```
  TRIED
+ DRIVE
-------
  RIVET
```

Solve each equation.

35. $q = -13(10)$
36. $-40 \cdot (-6) = h$
37. $-5(-8)(3) = r$
38. $b = \frac{-15}{3}$
39. $\frac{-12}{-6} = a$
40. $115 \div (-5) = k$

Chapter 3 Practice Test

Solve by eliminating possibilities.

1. Dave is younger than Angie, but older than Bob and Dwight. Angie is older than Dwight. Dwight is younger than Andy. Donna is older than Angie, but younger than Andy. Andy is older than Dave. Who is the oldest?
2. Three people are standing in a straight line as follows: The teenager is to the right of the baby and the adult. The blonde is to the left of the baby. The brunette is to the right of the redhead. State the order of the people and their hair color.

Solve each equation and check your solution. Then graph the solution on a number line.

3. $19 = f + 5$
4. $-15 + z = 3$
5. $x - (-27) = -40$
6. $j - 7 = -13$
7. $-8y = 72$
8. $-9k = -162$
9. $-42 = \frac{p}{3}$
10. $\frac{n}{-30} = -6$

Solve by replacing the variables with the given values.

11. $P = 2(s + b)$, if $s = 4$ and $b = 16$
12. $S = 2\ell w + 2wh + 2\ell h$, if $\ell = 3$, $w = 5$, and $h = 8$
13. $E = mc^2$, if $E = 325$ and $c = 5$
14. $C = \frac{5}{9}(F - 32)$, if $F = 212$

Find the perimeter and area of each rectangle.

15. 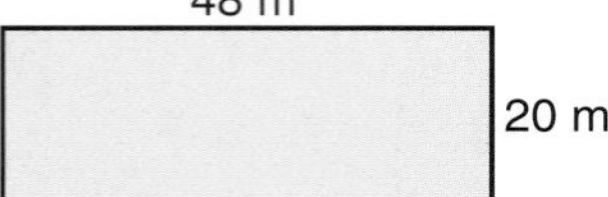

16. 

17. 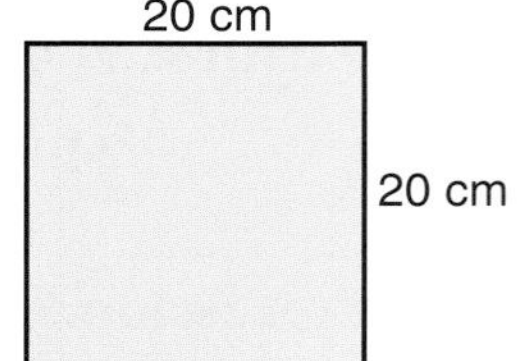

Solve each inequality and check your solution. Then graph the solution on a number line.

18. $g - 8 > -5$
19. $-61 \leq w + 50$
20. $17 \geq a - (-19)$
21. $3t \geq -42$
22. $\frac{p}{-12} > 6$
23. $-5c > 30$

Define a variable and write an equation or inequality. Then solve.

24. **Animals** According to the International Union for Conservation of Nature and Natural Resources, there are just under 6000 types of endangered species. If 2574 are invertebrates, what is the maximum number of vertebrates?
25. **Statistics** In 1990, there were 308,000 Cherokee Indians and 44,000 Creek Indians. How many times more Cherokee Indians are there than Creek Indians?

Chapter 4 Practice Test

PRACTICE TEST

Using divisibility rules, state whether each number is divisible by 2, 3, 5, 6, or 10.

1. 315 **2.** 400 **3.** 216

Write each multiplication expression using exponents.

4. $(-7)(-7)(-7)(-7)(-7)$ **5.** 40 **6.** $x \cdot x \cdot x \cdot x \cdot x \cdot x \cdot x \cdot x$

Solve. Use a diagram.

7. Six teams are to play each other in a round-robin volleyball tournament, that is, each team will play all five of the other teams. How many games will be played?

Factor each number or monomial completely.

8. 280 **9.** $882x^3$ **10.** $-696f^2g$

Find the GCF of each set of numbers or monomials.

11. 63, 231 **12.** $9xy$, $27x^3y^4$ **13.** $12z^4$, $72z^3$, $120z^2$

Write each fraction in simplest form. If the fraction is already in simplest form, write *simplified*.

14. $\frac{95}{155}$ **15.** $\frac{120}{210}$ **16.** $\frac{65}{104}$

Replace each ● with <, >, or = to make a true statement.

17. $\frac{3}{7}$ ● $\frac{6}{10}$ **18.** $\frac{5}{11}$ ● $\frac{3}{5}$ **19.** $\frac{7}{15}$ ● $\frac{1}{3}$

Find each product or quotient. Express your answer in exponential form.

20. $(2x^4)(-15x^5)$ **21.** $\frac{q^6}{q}$ **22.** $18 \cdot 18^7$

Write each expression using positive exponents.

23. $3^{-2}xy^{-3}$ **24.** $(st)^{-1}$ **25.** $\frac{3^{-4}}{2^{-2}}$

Chapter 5 Practice Test

Express each decimal as a fraction or mixed number in simplest form.

1. -9.8
2. 0.125
3. $1.\overline{6}$

Name the set(s) of numbers to which each number belongs.

4. $\frac{2}{3}$
5. -7
6. 51

Estimate each sum or difference.

7. $\$10.03 + \$5.84$
8. $44.03 - 32.9$
9. $20.3 + 59.7 + 62.8$

Simplify each expression.

10. $4.75m - 6.2m + 9$
11. $6.16q - 19 - (-7.32q)$

Solve each equation. Write the solution in simplest form.

12. $a = \frac{7}{9} + \frac{5}{9}$
13. $\frac{27}{18} - \frac{9}{18} = h$
14. $2\frac{3}{8} - 1\frac{5}{8} = p$
15. $\frac{1}{6} + \frac{5}{24} = p$
16. $5\frac{1}{2} - 2\frac{2}{3} = r$
17. $\frac{-6}{7} - \left(-2\frac{10}{21}\right) = y$

Solve each equation or inequality. Check your solution.

18. $u - 7.28 = 14.9$
19. $a + \frac{2}{3} = 2$
20. $-9.3 > y - (-4.8)$
21. $j + \frac{7}{12} \leq 7\frac{1}{2}$

State whether each is an example of *inductive* or *deductive* reasoning. Explain your answer.

22. For the past 3 years of school when Joe wears his black shoes to play football, his team wins. On the nights Joe wears his white shoes, his team loses. Sherry noticed that Joe was wearing his white shoes, so she figured they were going to lose.
23. At 32°F water freezes. It is 15° outside. Kelly knows that it is going to snow instead of rain.

State whether each sequence is arithmetic. Then write the next three terms of each sequence.

24. $\frac{1}{2}, \frac{3}{4}, 1 \ldots$
25. 5040, 720, 120, 24, . . .

Chapter 6 Practice Test

Write each fraction as a decimal. Use a bar to show a repeating decimal.

1. $\frac{3}{8}$ **2.** $\frac{-6}{11}$ **3.** $4\frac{8}{9}$

Estimate each product or quotient.

4. $14.34 \cdot 5.8$ **5.** $81.2 \div 15.57$ **6.** $\frac{3}{7} \cdot 98$

Solve each equation. Write each solution in simplest form.

7. $k = -5\left(\frac{-4}{3}\right)$ **8.** $1\frac{4}{5} \cdot -1\frac{5}{6} = n$ **9.** $t = \left(\frac{2}{5}\right)^3$

Evaluate each expression.

10. $a \div b$, if $a = \frac{-8}{9}$ and $b = \frac{-4}{3}$ **11.** $r \div s$, if $r = -1\frac{1}{2}$ and $s = \frac{21}{30}$

Solve each equation.

12. $(2.01)(0.04) = d$ **13.** $2.13 \div (-0.3) = a$ **14.** $a = (81)(0.02)(1.5)$
15. $x = 27.9 \div 0.31$ **16.** $y = 2.3(-0.004)$ **17.** $-51.408 \div (-5.4) = g$

Find the mean, median, and mode for each set of data. When necessary, round to the nearest tenth.

18. 36, 37, 41, 43, 43 **19.** 0.2, 0.4, 0.1, 0.6, 1.2, 1.1
20. 2, 8, 16, 21, 3, 8, 9, 6, 7 **21.** 44, 48, 55, 56, 55, 68, 70

Solve each equation or inequality. Check your solution.

22. $\frac{7}{8}p \geq -4\frac{1}{8}$ **23.** $0.25w = 6\frac{4}{5}$ **24.** $-0.5h < 12.5$

State whether each sequence is a geometric sequence. If so, state the common ratio and list the next three terms.

25. 384, 96, 24, . . . **26.** −7, 0, 7, 14, . . . **27.** 20, 16, 12.8, . . .

Write each number in scientific notation.

28. 0.0021 **29.** 87,500,000 **30.** 0.00000743

Write each number in standard form.

31. 3.9×10^3 **32.** 5.32×10^{-4} **33.** 7.02×10^0

Chapter 7 Practice Test

Solve.

1. Diana makes a salary of \$250 a week. In addition to her weekly salary, she receives a bonus of \$50 for every 25 boxes of books she sells. If she earned a total of \$450 in salary and bonuses this week, how many boxes did she sell?
2. Frank walks 10 blocks to work every day. It takes him between 15 and 20 minutes to get to work. At this rate, what is the least amount of time it would take for Frank to walk 50 blocks?

Solve each equation. Check your solution.

3. $25 = 2d - 9$
4. $4w + (-18) = -34$
5. $\frac{p}{6} - (-21) = 8$
6. $-7 = \frac{d}{-5} + 1$

Define a variable and write an equation for each situation. Then solve.

7. The quotient of a number and 8, decreased by 17, is -15.
8. Find the two consecutive odd integers whose sum is 76.

Find the circumference of each circle.

9. 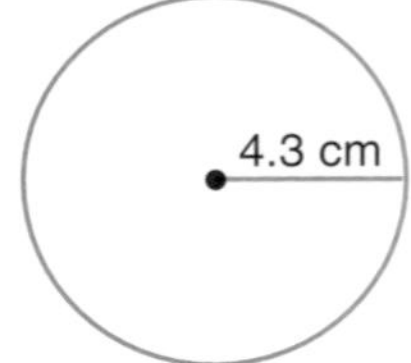

10. 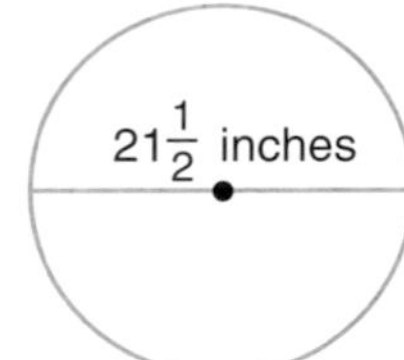

Solve each equation or inequality. Check your solution.

11. $7y + 6 = 3y - 14$
12. $3(x + 2) - 6 = 3x$
13. $\frac{7}{9}d - 7 = \frac{5}{9}d - 3$
14. $-3j - 4 < -22$
15. $-3.2 + 14s > 15s$
16. $3(y - 2) \geq 5(y - 7)$

Define a variable and write an inequality for each situation. Then solve.

17. Linda plans to spend at most \$85 on jeans and shirts. She bought 2 shirts for \$15.30 each. How much can she spend on jeans?

Complete each sentence.

18. 6 m = __?__ cm
19. 8.3 L = __?__ ml
20. 0.7 kg = __?__ g

Chapter 8 Practice Test

Write the domain and range of each relation. Then determine whether each relation is a function.

1. $\{(8, 9), (7, 6.5), (10, 6)\}$

2. $\{(3, 4), (5, 6), (5, 7), (8, 3)\}$

3. $\{(4, 6), (18, 40), (1.6, 4), (6, 6)\}$

4. $\{(10, 9), (4, 8.5), (17, 3), (4, 7)\}$

5.

x	5	8	10	2
y	1	3	3	1

6. $\{(-4, 30), (-4, -5), (6, 10)\}$

PRACTICE TEST

Determine whether a scatter plot of the data for the following might show a *positive*, *negative*, or *no* relationship. Explain your answer.

7. income, amount of clothes owned

8. hair color, weight

9. month of birth, years of college

10. outside temperature, frostbite

Find four solutions for each equation. Write your solution as ordered pairs.

11. $x + y = 8$

12. $x = -2$

13. $y = \frac{1}{3}x - 2$

For each equation,
a. solve for the domain {−4, −2, 0, 2, 4}, and
b. determine if the equation is a function.

14. $y = 4x + 5$

15. $x = 9$

16. $x^2 - y = 6$

Solve by using a graph. Assume the rate is constant.

17. Of the top 25 collegiate football teams on Nov. 26, 1995, Duke University was rated the number 1 academic football college having a 95% graduation rate. Penn State University was rated 8th having a 77% rate. The Ohio State University was ranked 15th. What was OSU's graduation rate?

Find the slope of the line that contains each pair of points.

18. $K(3, 5), L(7, 4)$

19. $G(9, -3), H(-1, 6)$

20. $L(-3, -4), M(-5, 2)$

21. $J(-6, 7), K(25, 4)$

Use the *x*-intercept and *y*-intercept to graph each equation.

22. $y = x - 6$

23. $y = -3 + x$

24. $y = -2x - 9$

Graph each equation using the slope and *y*-intercept.

25. $y = \frac{1}{4}x - 3$

26. $2x - y = -1$

27. $x + 3y = 6$

Use a graph to solve each system of equations.

28. $y = 0.4x$
$y = 0.4x - 3$

29. $x + y = 2$
$2x - y = -2$

30. $y = 3x - 6$
$-6x + 2y = -12$

Graph each inequality.

31. $y \geq 6$

32. $x + y < 4$

33. $y > \frac{1}{3}x + 3$

Chapter 9 Practice Test

Express each ratio or rate as a fraction in simplest form.

1. 16 pencils to 72 pencils
2. 2 pints out of 1 gallon
3. 88 out of 564 students

Solve by making a table.

4. How many ways can you make change for a \$50-bill using only \$5-, \$10-, and \$20-bills?

A standard deck of playing cards is placed on the table. Suppose you select one card at random. Find each probability.

5. P(a red card)
6. P(a six)
7. P(a face card)

Write a proportion that could be used to solve for each variable. Then solve the proportion.

8. 13 cans for \$2.99
   x cans for \$4.83
9. 3 teaspoons for 24 cookies
   7 teaspoons for y cookies

Use the percent proportion to solve each problem.

10. What is 40% of 60?
11. Find 37.5% of 80.
12. Twenty-one is 35% of what number?
13. Seventy-five is what percent of 250?
14. Fifty-two is what percent of 80?
15. Thirty-six is 45% of what number?

Solve.

16. Eight of the 72 students surveyed at Perry Middle School said they would like to take a music appreciation class. The school's policy says there must be at least 30 interested students in the class. If there are 567 students in the school, predict how many students would be interested. Is this enough for the class?

Express each decimal or fraction as a percent.

17. $\frac{85}{120}$
18. 0.42
19. $\frac{9}{5}$
20. 0.086

Express each percent as a decimal and as a fraction.

21. 67.5%
22. $33\frac{1}{3}\%$
23. 10.17%

Choose the best estimate.

24. 26% of 49 **a.** 13 **b.** 1.3 **c.** 130
25. 47% of 550 **a.** 26 **b.** 2.6 **c.** 260

Find the discount or interest to the nearest cent.

26. \$1550 at 4% for 2 years
27. \$138 dress, 15% off
28. \$143 suit, 10% off
29. \$1250 at 10.5% for 3 months

State whether each percent of change is a percent of increase or a percent of decrease. Then find the percent of increase or decrease. Round to the nearest whole percent.

30. old: \$32
    new: \$19
31. old: \$245
    new: \$315
32. old: \$178
    new: \$201
33. old: \$1010
    new: \$1075

Chapter 10 Practice Test

The ages of Miss Rebar's seventh grade class were 13.3, 12.5, 12.8, 12.5, 12.8, 13.1, 13.2, 12.2, 12.9, 13.3, and 12.6.

1. Make a stem-and-leaf plot of the data.

2. How young was the youngest student?

Use the data in the stem-and-leaf plot shown at the right.

3. What is the range?

4. Find the median.

5. What is the upper quartile?

6. What is the lower quartile?

7. Find the interquartile range.

8. What are the outliers, if any?

9. Make a box-and-whisker plot of the data.

Stem	Leaf
2	5 8 8
3	2 3 3 6 9
4	2 3 5 7
5	1 1 3

5 | 1 = 51

The graphs below show how many sports cards Doug buys each month.

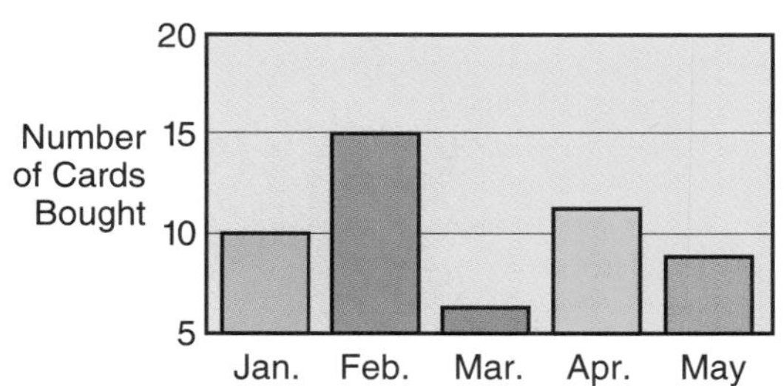

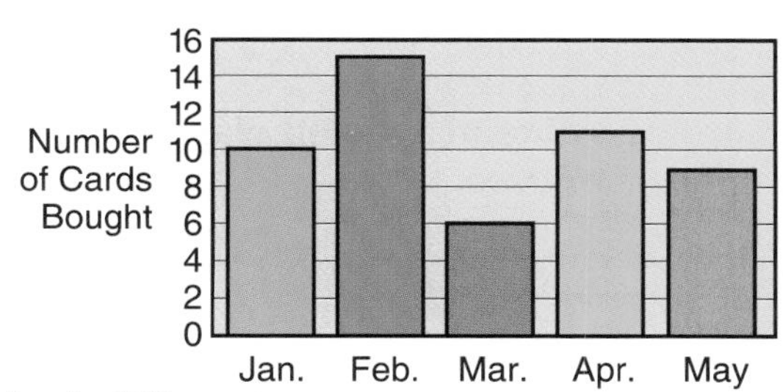

10. Explain why these graphs made from the same data look different.

11. If Doug wanted to convince his wife that he isn't buying very many cards which graph would he use? Explain.

Find the number of possible outcomes for each event.

12. A die is rolled and a coin is tossed.

13. Four dice are rolled.

Tell whether each situation represents a *permutation* or a *combination*.

14. eight notes in a song

15. eight new CDs from a group of thirty

16. five outfits from 15 to be used for a window display

Find the odds of each outcome if a fair eight-sided die is rolled.

17. A number less than five.

18. A composite number.

Heather makes three out of five 10-foot putts. To simulate her chances of making a putt, she puts 40 marbles in a box. A red marble represents a putt made, and a blue marble represents a putt missed. After a marble is drawn, it is replaced in the box.

19. How many blue marbles does Heather need?

20. In 120 drawings, how many red marbles can she expect?

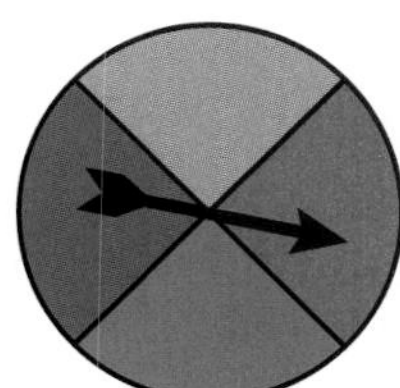

An eight-sided die is rolled and the spinner is spun. Find each probability.

21. *P*(green and 6)

22. *P*(orange and an odd number)

Determine whether each event is *mutually exclusive* or *inclusive*. Then find the probability. A letter of the alphabet is chosen.

23. *P*(a vowel or a letter following R)

24. *P*(a consonant or the letter E)

25. *P*(a vowel or a consonant)

Chapter 11 Practice Test

Draw and label a diagram to represent each of the following.

1. point C
2. $\overline{XY}$
3. plane GHI
4. $\overleftrightarrow{RS}$
5. Make a circle graph to display the data in the chart at the right.

Water Usage in the United States	
Agriculture	36%
Public water	8%
Utilities	33%
Industry	23%

In the figure at the right, $\overline{KG} \parallel \overline{AE}$. If the measure of $\angle HFG$ is 45°, find the measure of each angle.

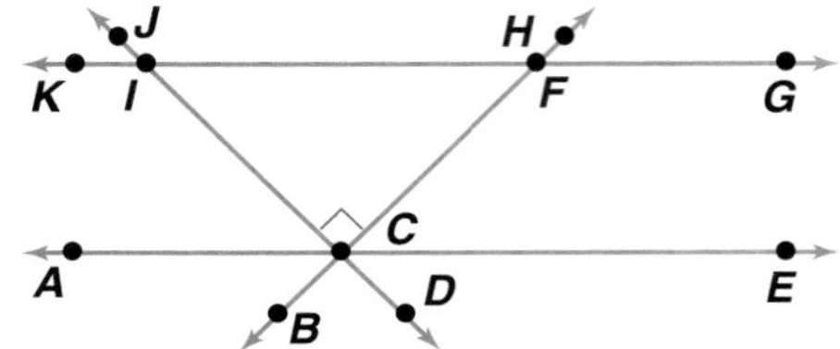

6. $\angle FCE$
7. $\angle BCD$
8. $\angle JIF$
9. $\angle KIC$

Find the value of x. Then classify the triangle as *acute*, *right*, or *obtuse*.

10. 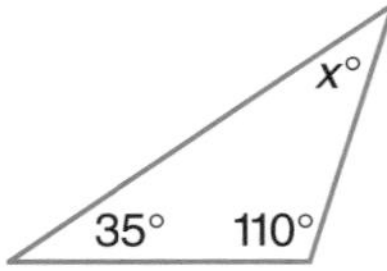

11. 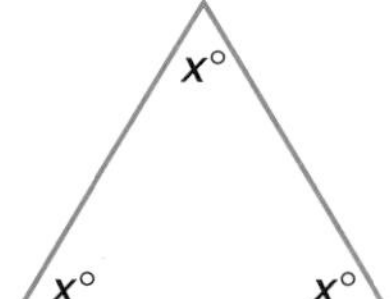

12. 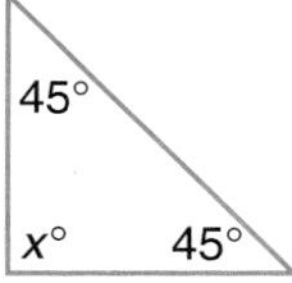

If $\triangle XYZ \cong \triangle ABC$, name the part congruent to each angle or segment.

13. $\angle X$
14. $\angle B$
15. $\overline{YZ}$
16. $\overline{AC}$

Write a proportion to find the missing measure x. Then find the value of x.

17. 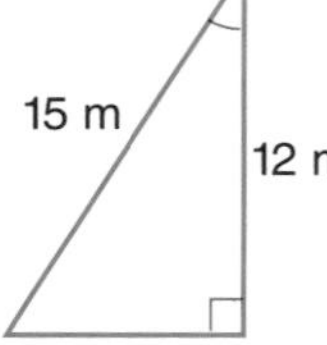

18. 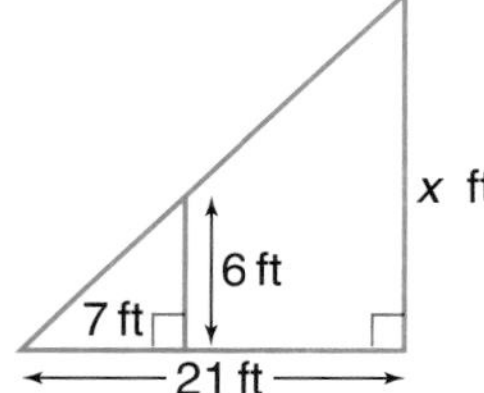

Find the value of x.

19. 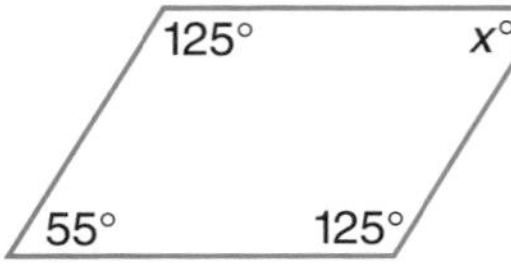

20. 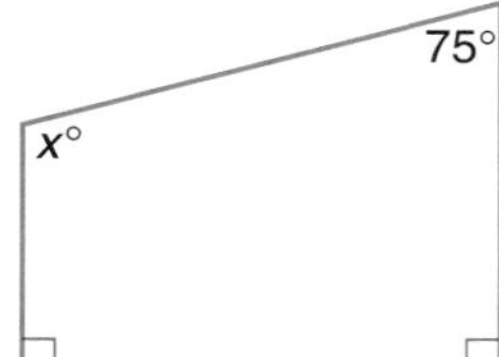

21. 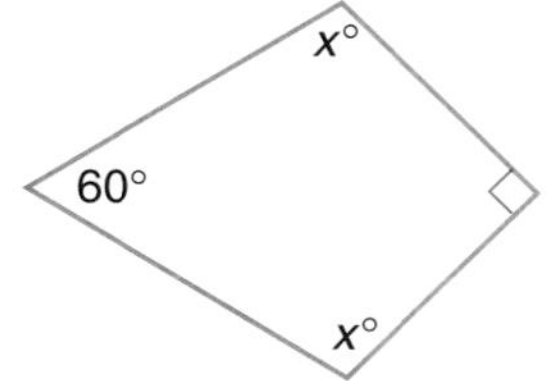

Find the sum of the measures of the interior angles of each polygon.

22. hexagon
23. 20-gon
24. A translation moves $\triangle CAT$ 6 units to the right and 3 units up to form $\triangle C'A'T'$. Write the coordinates of each vertex of $\triangle C'A'T'$.
25. Draw $\triangle C'A'T'$ on a coordinate plane and graph its reflection if the x-axis is the line of reflection.

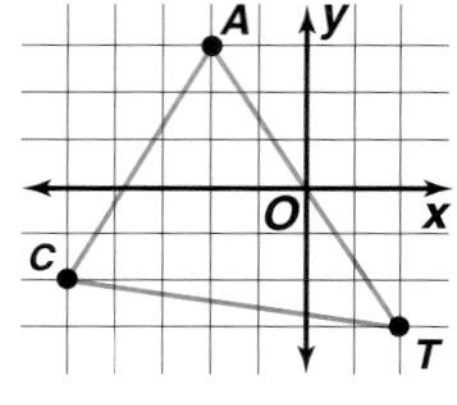

Chapter 12 Practice Test

Find the area of each figure.

1. 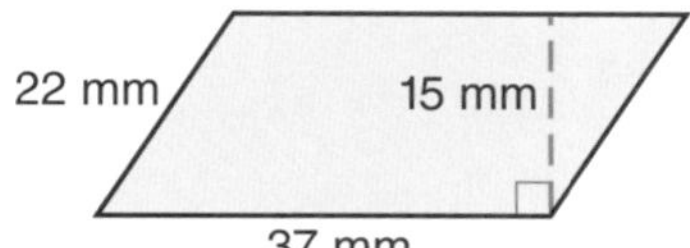

2. 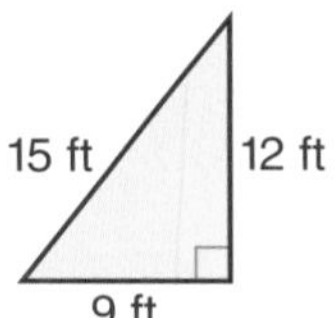

3. 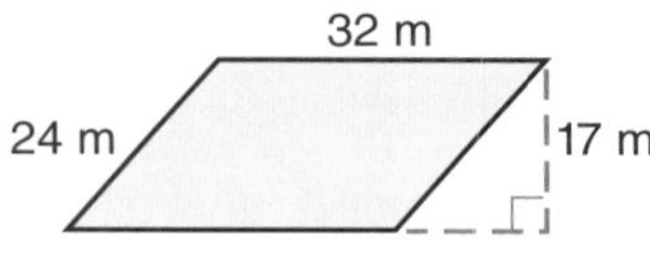

Find the area of each circle. Round to the nearest tenth.

4. radius, 16 cm

5. diameter, 42 in.

6. radius, $3\frac{1}{2}$ in.

7. The figure at the right represents a dartboard. Find the probability of a dart landing in the shaded region.

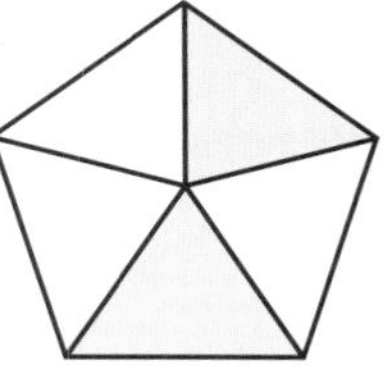

Solve by making a model or drawing.

8. Half of a garden is planted in corn. Half of the remaining garden is planted in strawberries. A third of the part not planted in corn or strawberries is planted in tomatoes. If 6 m^2 is planted in tomatoes, what is the area planted in corn?

Find the surface area of each solid.

9. 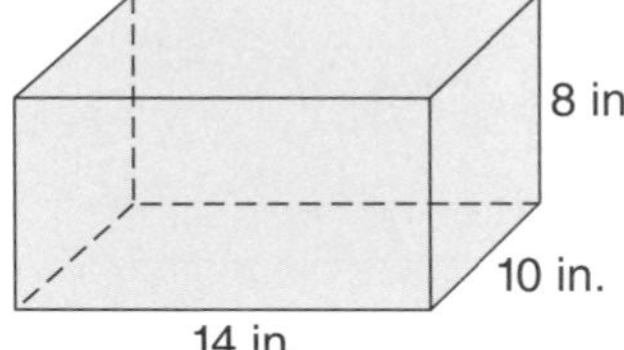

10. 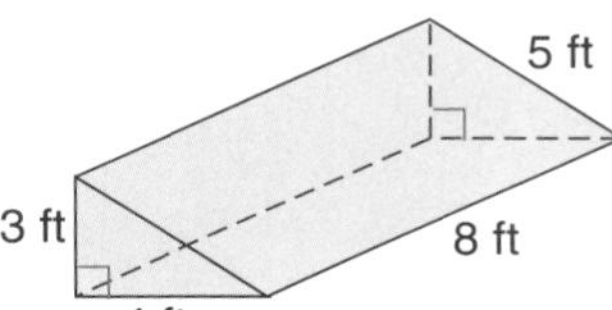

11. 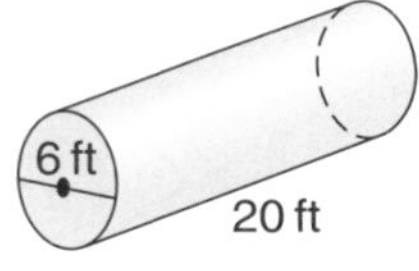

12. 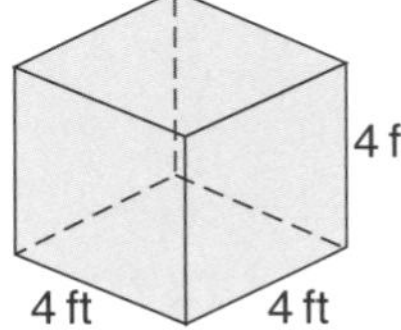

13. 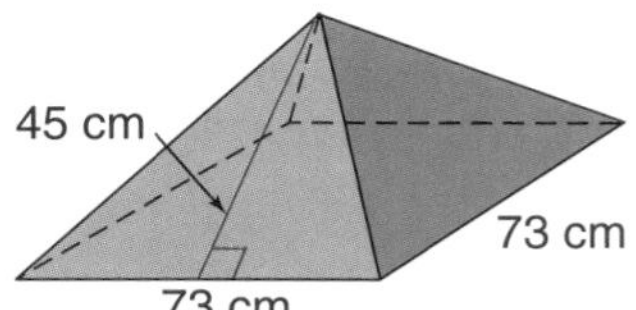

14. 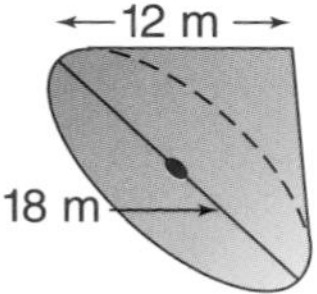

Find the volume of each solid. Round to the nearest tenth.

15. rectangular prism: length, 4 m; width, 7 m; height, 9 m

16. triangular prism: base of triangle, $2\frac{1}{2}$ ft; altitude, 1 ft; prism height, 2 ft

17. cylinder: diameter, 12 m; height, 9 m

18. rectangular pyramid: length, 18 cm; width, 15 cm; height, 20 cm

19. square pyramid: length, 8 in.; height, 9 in.

20. circular cone: radius, 9 cm: height, 16 cm

Chapter 13 Practice Test

PRACTICE TEST

Find the best integer estimate for each square root. Then check your estimate using a calculator.

1. $\sqrt{90}$ **2.** $\sqrt{20.25}$ **3.** $-\sqrt{62}$ **4.** $-\sqrt{71}$

Solve using a Venn diagram.

5. Exquisite Interiors has 40 sample floor tiles in the shapes of various polygons. Twenty tiles are regular polygons, 14 tiles are quadrilaterals, and 5 tiles are squares. How many tiles are not regular polygons?

Solve each equation. Round decimal answers to the nearest tenth.

6. $s^2 = 121$ **7.** $g^2 = 40$ **8.** $f^2 = 11$ **9.** $y^2 = 60$

In a right triangle, if *a* and *b* are the measures of the legs and *c* is the measure of the hypotenuse, find each missing measure. Round decimal answers to the nearest tenth.

10. b, 77 m; c, 85 m **11.** a, 15 ft; c, 17 ft **12.** b, 91 yd; c, 109 yd

Find the lengths of the missing sides in each triangle. Round decimal answers to the nearest tenth.

13.

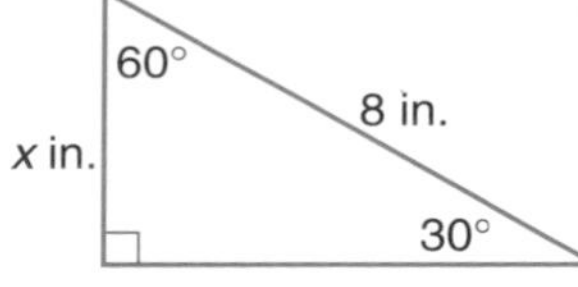

14.

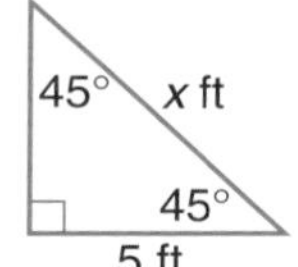

15.

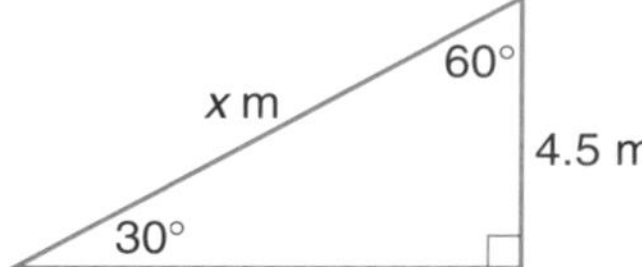

For each triangle, find the measure of the marked acute angle to the nearest degree.

16.

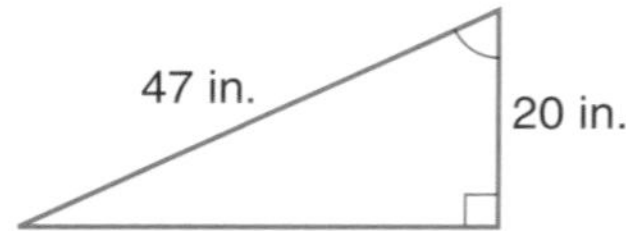

17.

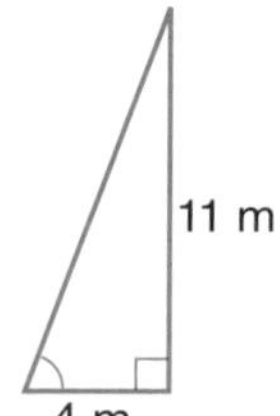

18.

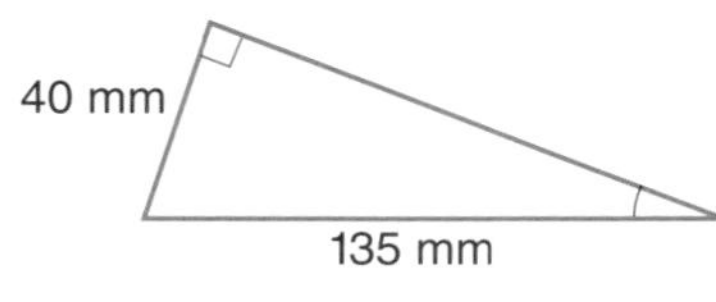

Use trigonometric ratios to solve each problem.

19. What is the angle of elevation of the sun when a 100-foot water tower casts a shadow 165 feet long?

20. **Building Design** A ramp is designed to help people in wheelchairs move more easily from one level to another. If a ramp 16 feet long forms an angle of 6° with the level ground, what is the vertical rise?

Chapter 14 Practice Test

State whether each expression is a polynomial. If it is, identify it as a *monomial*, *binomial*, or *trinomial*.

1. $4a^2$
2. $-2y + 3$
3. $\frac{a}{5}$
4. $r^3 - s^2$
5. $-\frac{3}{(x + y)}$
6. $7mt^2 + 3m - 2t$

Find the degree of each polynomial.

7. $5a^2b$
8. $12rs^2 + 3r^2s + 5r^4s$
9. $x^2 + 6xy + xy^2$

Find each sum or difference.

10. $(2x^2 + 5x) + (3x^2 + x)$
11. $(7a - 11b) - (3a + 4b)$
12. $(x^2 - 5x + 3) + (4x - 3)$
13. $(3a^2 - b^2 + c^2) - (a^2 + 2b^2)$
14. $\begin{array}{r} -3y^2 \qquad + 2 \\ \underline{(+)\ 4y^2 - 5y - 2} \end{array}$
15. $\begin{array}{r} 14a^2 - 3a \qquad \\ \underline{(-)\ 6a^2 + 5a + 17} \end{array}$

Simplify.

16. $(5c)^3$
17. $(4x^2y^3)^2$
18. $-3x(4xy)^2$

Find each product.

19. $x(-5x + 3)$
20. $a^2(2a^3 + a - 5)$
21. $-2g(g^3 + 6g + 3)$
22. $(2x + 1)(3x + 2)$
23. $(3x + 3)(x - 1)$
24. $(-x + 3)(3x - 2)$

Solve.

25. **Geometry** For the rectangular prism at the right, compute the area of the top, bottom, front, back, left, and right sides. Then add to find the total surface area.

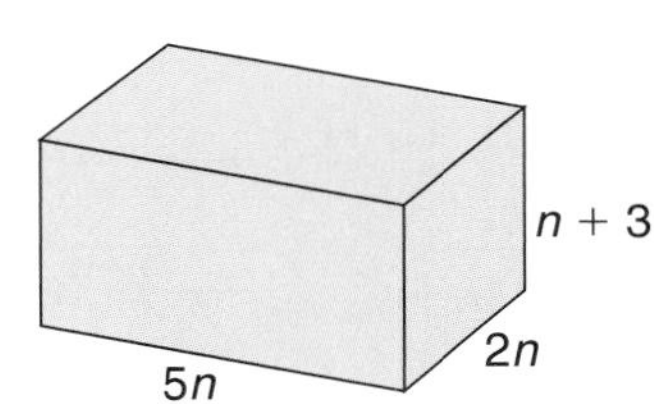

GLOSSARY

absolute value (67) The number of units a number is from zero on the number line.

acute angle (550) An angle with a measure greater than 0° and less than 90°.

acute triangle (569) A triangle that has three acute angles.

addition property of equality (125) If you add the same number to each side of an equation, the two sides remain equal. For any numbers a, b, and c, if $a = b$, then $a + c = b + c$.

addition property of inequality (147) Adding the same number to each side of an inequality does not change the truth of the inequality. For all numbers a, b, and c
1. If $a > b$, then $a + c > b + c$.
2. If $a < b$, then $a + c < b + c$.

additive inverses (89) An integer and its opposite. The sum of an integer and its additive inverse is zero.

adjacent angles (561) Two angles that have a common side and the same vertex, but they do *not* overlap.

adjacent leg (687) The leg of a right triangle that is not opposite an angle and not the hypotenuse.

algebraic expression (16) A combination of variables, numbers, and at least one operation.

algebraic fraction (197) A fraction with variables in the numerator or denominator.

alternate exterior angles (563) Nonadjacent exterior angles found on opposite sides of the transversal.
$\angle 1$ and $\angle 7$, and $\angle 2$ and $\angle 8$ are alternate exterior angles.

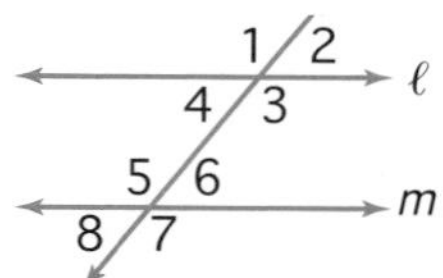

alternate interior angles (563) Nonadjacent interior angles found on opposite sides of the transversal.
$\angle 4$ and $\angle 6$, and $\angle 3$ and $\angle 5$ are alternate interior angles.

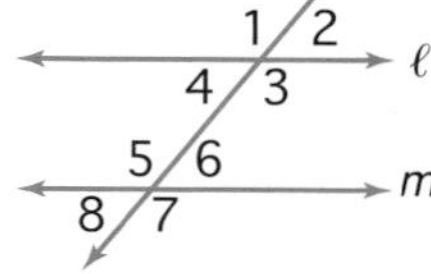

altitude (612, 614, 633) A line segment that is perpendicular to the base of a figure with endpoints on the base and the side opposite the base.

angle (549) Two rays with a common endpoint form an angle. The rays and vertex are used to name an angle.

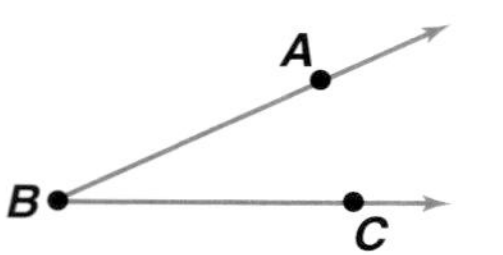

angle ABC or $\angle ABC$

angle of depression (695) An angle of depression is formed by a line of sight along the horizontal and another line of sight below it.

angle of elevation (694) An angle of elevation is formed by a line of sight along the horizontal and another line of sight above it.

area (138) The number of square units needed to cover a surface enclosed by a geometric figure.

arithmetic sequence (258) A sequence in which the difference between any two consecutive terms is the same.

associative property of addition (22, 234) For any rational numbers a, b, and c,
$$(a + b) + c = a + (b + c).$$

associative property of multiplication (22, 296) For any numbers a, b, and c,
$$(ab)c = a(bc).$$

axes (73) Two intersecting number lines that form a coordinate system.

back-to-back stem-and-leaf plot (487) Used to compare two sets of data. The leaves for one set of data are on one side of the stem and the leaves for the other set of data are on the other side.

bar graph (52) A graphic form using bars to make comparisons of statistics.

bar notation (225) In repeating decimals the line or bar placed over the digits that repeat. For example, $2.\overline{63}$ indicates the digits 63 repeat.

base (175) In 5^3, the base is 5. The base is used as a factor as many times as given by the exponent (3). That is,
$$5^3 = 5 \times 5 \times 5.$$

base (449) In a percent proportion, the number to which the percentage is compared.
$$\frac{\text{percentage}}{\text{base}} = \text{rate}$$

base (614) The base of a rectangle, a parallelogram, or a triangle is any side of the figure. The bases of a trapezoid are the parallel sides.

base (632) The bases of a prism are the two parallel congruent sides.

binomial (706) A polynomial with exactly two terms.

boundary (418) A line that separates a graph into half-planes.

box-and-whisker plot (495) A diagram that summarizes data using the median, the upper and lower quartiles, and the extreme values. A box is drawn around the quartile value and whiskers extend from each quartile to the extreme data points.

center (341) The given point from which all points on the circle are the same distance.

circle (341) The set of all points in a plane that are the same distance from a given point called the center.

circle graph (497, 556) A type of statistical graph used to compare parts of a whole.

circular cone (639) A shape in space that has a circular base and one vertex.

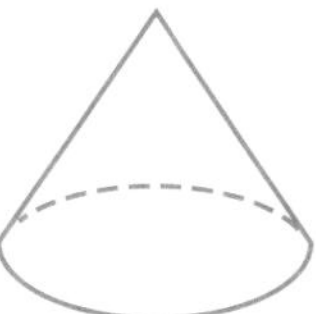

circumference (341) The distance around a circle.

coefficient (711) The numerical part of a monomial.

combination (516) An arrangement or listing in which order is not important.

commission (470) A fee paid to a sales associate based on a percent of the total sales.

common difference (258) The difference between any two consecutive terms in an arithmetic sequence.

common multiples (200) Multiples that are shared by two or more numbers. For example, some common multiples of 2 and 3 are 6, 12, and 18.

common ratio (312) The ratio between any two successive terms in a geometric sequence.

commutative property of addition (22, 234) For any rational numbers a and b,

$$a + b = b + a.$$

commutative property of multiplication (22, 296) For any rational numbers a and b, $ab = ba$.

comparative graph (498) A graph used to compare results of similar groups to show trends.

compass (554) An instrument used for drawing circles or parts of circles.

compatible numbers (280) Numbers that have been rounded so, when the numbers are divided by each other, the remainder is zero.

complementary (562) Two angles are complementary if the sum of their measures is $90°$.

composite number (184) Any whole number greater than 1 that has more than two factors.

congruent (554, 561, 573) Line segments that have the same length, or angles that have the same measure, or figures that have the same size and shape.

constant (707) A monomial that does not contain a variable.

coordinate (67, 126) A number associated with a point on the number line.

coordinate system (36, 73) A coordinate plane separated into four equal regions by two intersecting number lines called axes.

corresponding angles (563) Angles that have the same position on two different parallel lines cut by a transversal. $\angle 1$ and $\angle 5$, $\angle 2$ and $\angle 6$, $\angle 3$ and $\angle 7$, and $\angle 4$ and $\angle 8$ are corresponding angles.

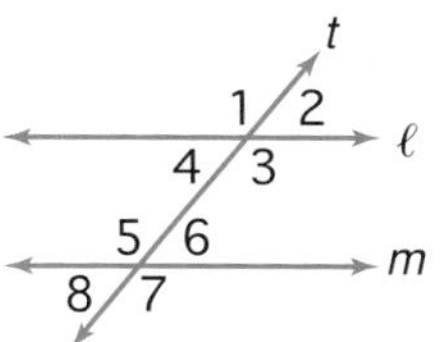

corresponding parts (573) Parts on congruent or similar figures that match.

cosine (688) If $\triangle ABC$ is a right triangle and A is an acute angle,

$$\text{cosine } \angle A = \frac{\text{measure of the side adjacent to } \angle A}{\text{measure of the hypotenuse}}$$

$$\cos A = \frac{b}{c}$$

cylinder (633) A 3-dimensional shape that has two parallel, congruent bases (usually circular) and a curved side connecting the bases.

data (50) Numerical information gathered for statistical purposes.

defining a variable (42) Choosing a variable and a quantity for the variable to represent in an equation.

degree (549) The most common unit of measure for angles.

dependent events (532) Two or more events in which the outcome of one event does affect the outcome of the other event(s).

diagonal (590) A line segment that joins two nonconsecutive vertices of a polygon.

diameter (341) The distance across a circle through its center.

discount (468) The amount deducted from the original price.

discrete mathematics (258) A branch of mathematics that deals with discontinuous or finite sets of numbers. Some fields that deal with discrete mathematics are logic and statistics.

distributive property (26) For any numbers a, b, and c, $a(b + c) = ab + ac$ and $(b + c)a = ba + ca$.

divisible (170) A number is divisible by another if, upon division, the remainder is zero.

division property of equality (130) For any numbers a, b, and c, with $c \neq 0$, if $a = b$, then $\frac{a}{c} = \frac{b}{c}$.

division properties of inequality (152) For all integers a, b, and c, where $c > 0$, if $a > b$, then $\frac{a}{c} > \frac{b}{c}$.
For all integers a, b, and c, where $c < 0$, if $a > b$, then $\frac{a}{c} < \frac{b}{c}$.

domain (372) The set of all first coordinates from each ordered pair.

empty set (347) A set with no elements shown by the symbol { } or $\varnothing$.

equation (32) A mathematical sentence that contains the equals sign, =.

equiangular (591) All angles are congruent.

equilateral (591) All sides are congruent.

equivalent (275) The same value represented in different forms. For example, $\frac{3}{5}$ and 0.6 are equivalent.

equivalent equations (124) Two or more equations with the same solution. For example, $x + 3 = 5$ and $x = 2$ are equivalent equations.

evaluating an expression (11) Replacing the variables with numbers and finding the numerical value of the expression.

expanded form (176) A number expressed using place value to write the value of each digit in the number.

experimental probability (529) An estimated probability based on the relative frequency of positive outcomes occurring during an experiment.

exponent (175) In 5^3, the exponent is 3. The exponent tells how many times the base, 5, is used as a factor.

$$5^3 = 5 \times 5 \times 5$$

exterior angle (591) The angle formed when a side of a polygon is extended.

face (632) A face of a prism is any surface that forms a side or a base of the prism.

factorial (515) The expression $n!$ is the product of all counting numbers beginning with n and counting backward to 1.

factors (170) Numbers that divide into a whole number with a remainder of zero.

factor trees (185) A diagram showing the prime factorization of a composite number. The factors branch out from the previous factors until all the factors are prime numbers.

Fibonacci sequence (263) A list of numbers in which the first two numbers are both 1 and each number that follows is the sum of the previous two numbers, 1, 1, 2, 3, 5, 8, . . .

formula (134) An equation that states a rule for the relationship between certain quantities.

45°–45° right triangle (683) A right triangle that has two angles that each measure 45°.

frequency table (51) A chart that indicates the number of values in each interval.

function (373) A function is a relation in which each element of the domain is paired with exactly one element in the range.

functional notation (393) Equations that represent functions can be written in *functional notation*, $f(x)$. The symbol $f(x)$ is read "f" of "x" and represents the value in the range that corresponds to the value of x in the domain.

Fundamental counting principle (509) If event M can occur in m ways and is followed by event N that can occur in n ways, then the event M followed by event N can occur in $m \cdot n$ ways.

geometric probability (623) Probability that uses area to find the probability of an event.

geometric sequence (312) A sequence in which the ratio between any two successive terms is the same.

gram (359) The basic unit of mass in the metric system.

graph (73, 126) A dot marking a point that represents a number on a number line or an ordered pair on a coordinate plane.

greatest common factor (GCF) (190) The GCF of two or more integers is the greatest factor that is common to each of the integers.

histogram (502) A bar graph whose columns are next to each other to show how the data are distributed.

hypotenuse (676) The side opposite the right angle in a right triangle.

identity property of addition (23, 234) For any number a, $a + 0 = a$ and $0 + a = a$.

identity property of multiplication (23, 296) For any number a, $a \times 1 = a$.

inclusive (536) When two events can happen at the same time.

independent events (531) Two or more events in which the outcome of one event does *not* affect the outcome of the other event(s).

indirect measurement (580) Finding a measurement by using similar triangles and writing a proportion.

inequality (46) A mathematical sentence that contains $<$, $>$, $\neq$, $\leq$, or $\geq$.

integers (66, 224) The whole numbers and their opposites.

$$\ldots, -3, -2, -1, 0, 1, 2, 3, \ldots$$

intercepts (406) The points where a graph crosses the axes.

interior angle (591) An angle inside a polygon.

interquartile range (491) The range of the middle half of a set of numbers.

$$\text{Interquartile range} = \text{UQ} - \text{LQ}.$$

intersect (552) Lines that cross or touch each other at a point.

inverse operations (41) Operations that undo each other, such as addition and subtraction.

inverse property of addition (234) For every rational number a, $a + (-a) = 0$.

inverse property of multiplication (289) For every nonzero number $\frac{a}{b}$, where $a, b \neq 0$, there is exactly one number $\frac{b}{a}$ such that $\frac{a}{b} \times \frac{b}{a} = 1$.

irrational number (672) A number that cannot be expressed as $\frac{a}{b}$, where a and b are integers and b does not equal 0. For example, $\sqrt{3} = 1.7320508\ldots$ is an irrational number.

isosceles triangle (569) A triangle that has at least two congruent sides.

lateral surface (639) The lateral surface of a prism, cylinder, pyramid, or cone is all the surface of the figure except the base or bases.

least common denominator (LCD) (201) The least common multiple of the denominators of two or more fractions.

least common multiple (LCM) (200) The least of the nonzero common multiples of two or more numbers. The LCM of 2 and 3 is 6.

leaves (486) The next greatest place value forms the leaves in a stem-and-leaf plot.

leg (676) Either of the two sides that form the right angle of a right triangle.

like terms (27, 711) Expressions that contain the same variables, such as $3ab$ and $7ab$.

line (548) A never-ending straight path. A representation of a line CD ($\overleftrightarrow{CD}$) is shown below.

linear equation (386) An equation for which the graph is a straight line.

line graph (497) A type of statistical graph used to show how values change over a period of time.

line segment (548) Two endpoints and the straight path between them. A line segment is named by its endpoints. A representation of line segment ST ($\overline{ST}$) is shown below.

liter (359) The basic unit of capacity in the metric system.

lower quartile (491) The median of the lower half of a set of numbers indicated by LQ.

mean (300) The sum of the numbers in a set of data divided by the number of pieces of data.

measures of central tendency (300) Numbers or pieces of data that can represent the whole set of data.

median (300) In a set of data, the median is the number in the middle when the data are organized from least to greatest.

meter (359) The basic unit of length in the metric system.

metric system (360) A system of measurement using the basic units: meter for length, gram for mass, and liter for capacity.

mixed number (239) The sum of a whole number and a fraction. For example, $3\frac{1}{2}$.

mode (300) The number or item that appears most often in a set of data.

monomial (172) An expression that is either a real number, a variable, or a product of real numbers and variables.

multiple (200) The product of the number and any whole number.

multiplication properties of inequality (152)

1. For all integers a, b, and c, where $c > 0$, if $a > b$, then $ac > bc$.
2. For all integers a, b, and c, where $c < 0$, if $a > b$, then $ac < bc$.

multiplication property of equality (130) For any numbers a, b, and c, if $a = b$, then $ac = bc$.

multiplicative inverses (289) Two numbers whose product is 1. The multiplicative inverse of $\frac{3}{4}$ is $\frac{4}{3}$.

GLOSSARY

multiplicative property of zero (23) For any number a, $a \times 0 = 0$.

mutually exclusive (535) Two or more events such that no two events can happen at the same time.

negative (66) A value less than zero.

negative relationship (380) If points in a scatter plot appear to suggest a line that slants downward to the right, there is a *negative relationship*.

net (632) The shape that is formed by "unfolding" a prism. The net shows all the faces that make up the surface area of a prism.

no relationship (380) Points that seem to be random in a scatter plot are said to have no relationship.

null set (347) A set with no elements shown by the symbol { } or ∅.

obtuse angle (550) Any angle that measures greater than 90° but less than 180°.

obtuse triangle (569) A triangle with one obtuse angle.

odds (520) A way to describe the chance of an event occurring.

odds in favor = number of successes: number of failures

odds against = number of failures: number of successes

open sentence (32) An equation that contains variables.

opposite side (687) The side that is not part of the angle in a right triangle.

ordered pair (37) A pair of numbers used to locate a point in the coordinate system.

order of operations (12, 176)

1. Simplify the expressions inside grouping symbols.
2. Evaluate all powers.
3. Then do all multiplications and divisions from left to right.
4. Then do all additions and subtractions from left to right.

origin (36) The point of intersection of the x-axis and y-axis in a coordinate system.

outcomes (509) Possible results of a probability event. For example, 4 is an outcome when a die is rolled.

outliers (496) Data that are more than 1.5 times the interquartile range from the quartiles.

parallel (551) Lines in the same plane that do not intersect. The symbol ∥ means parallel.

parallelogram (585) A quadrilateral with two pairs of parallel sides.

percent (449) A ratio with a denominator of 100. For example, 77% and $\frac{77}{100}$ name the same number.

percentage (449) In a percent proportion, a number that is compared to another number called the base.

$$\frac{\text{percentage}}{\text{base}} = \text{rate}$$

percent of change (472) The ratio of the amount of change to the original amount.

percent of decrease (473) The ratio of an amount of decrease to the previous amount, expressed as a percent.

percent of increase (472) The ratio of an amount of increase to the previous amount, expressed as a percent.

percent proportion (450)

$$\frac{\text{percentage}}{\text{base}} = \text{rate or } \frac{P}{B} = \frac{r}{100}$$

perfect squares (665) Rational numbers whose square roots are whole numbers. 25 is a perfect square because $\sqrt{25} = 5$.

perimeter (138) The distance around a geometric figure.

permutation (515) An arrangement or listing in which order is important.

perpendicular (551) Two lines that intersect to form a right angle.

pi, π (341) The ratio of the circumference of a circle to the diameter of a circle. Approximations for π are 3.14 and $\frac{22}{7}$.

pictograph (497) A type of statistical graph that uses pictures or illustrations to show how specific quantities compare. Each symbol represents a specific number.

plane (550) A flat surface that has no edges or boundaries. A plane can be named by a single uppercase script letter or by using any three noncollinear points of the plane.

point (548) A specific location in space.

polygon (589) A simple closed figure in a plane formed by three or more line segments.

polynomial (706) A monomial or the sum or difference of two or more monomials.

positive relationship (380) In a scatter plot, if the points appear to suggest a line that slants upward to the right, there is a *positive relationship*.

power (175) A number that can be written using an exponent.

power of a monomial (721) For all numbers a and b, and positive integer m, n, and p, $(a^m b^n)^p = a^{mp} b^{np}$.

power of a power (719) For any number a and positive integers m and n, $(a^m)^n = a^{mn}$.

power of a product (720) For all numbers a and b, and positive integer m, $(ab)^m = a^m b^m$.

prime factorization (185) Expressing a composite number as the product of its prime factors.

prime number (184) A whole number greater than 1 that has exactly two factors , 1 and itself.

principal (467) The amount of money in an account.

prism (632) A figure in space that has two parallel and congruent bases in the shape of polygons.

probability (440) The chance that some event will happen. It is the ratio of the number of ways a certain event can occur to the number of possible outcomes.

$$\text{probability} = \frac{\text{number of favorable outcomes}}{\text{number of possible outcomes}}$$

probability of inclusive events (536) The probability of one or the other of two inclusive events can be found by adding the probability of the first event to the probability of the second event and subtracting the probability of both events happening.
$P(A \text{ or } B) = P(A) + P(B) - P(A \text{ and } B)$

probability of mutually exclusive events (535) The probability of one or the other of two mutually exclusive events can be found by adding the probability of the first event to the probability of the second event.
$P(A \text{ or } B) = P(A) + P(B)$

probability of two dependent events (532) If two events A and B are dependent, then the probability of both events occurring is the product of the probability of A and the probability of B after A occurs.
$P(A \text{ and } B) = P(A) \times P(B \text{ following } A)$

probability of two independent events (531) The probability of two independent events can be found by multiplying the probability of the first event by the probability of the second event.
$P(A \text{ and } B) = P(A) \times P(B)$

product of powers (205) For any number a and positive integers m and n, $a^m \times a^n = a^{m+n}$.

property of proportions (444)
If $\frac{a}{b} = \frac{c}{d}$, then $ad = bc$.
If $ad = bc$, then $\frac{a}{b} = \frac{c}{d}$.

proportion (444) A statement of equality of two or more ratios.

protractor (549) An instrument used to measure angles.

pyramid (638) A solid figure that has a polygon for a base and triangles for sides.

Pythagorean theorem (676) In a right triangle, the square of the length of the hypotenuse is equal to the sum of the squares of the lengths of the legs. $c^2 = a^2 + b^2$

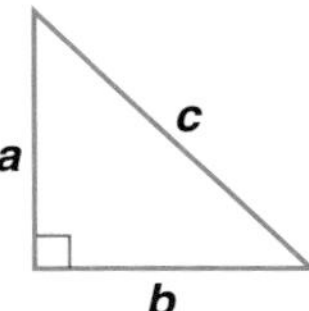

quadrants (73) The four regions into which two perpendicular number lines separate the plane.

quadrilateral (584, 585) A polygon having four sides.

quartile (491) One of four equal parts of data from a large set of numbers.

quotient of powers (206) For any nonzero number a and whole numbers m and n, $\frac{a^m}{a^n} = a^{m-n}$.

radical sign (664) The symbol used to indicate a nonnegative square root. $\sqrt{\ }$

radius (341) The distance from the center of a circle to any point on the circle.

range (372) The range of a relation is the set of all second coordinates from each ordered pair.

range (490) The difference between the least and greatest numbers in the set.

rate (433) A ratio of two measurements having different units.

rate (449) In a percent proportion the ratio of a number to 100.

ratio (196, 432) A comparison of two numbers by division. The ratio of 2 to 3 can be stated as 2 out of 3, 2 to 3, 2:3, or $\frac{2}{3}$.

rational numbers (224) Numbers of the form $\frac{a}{b}$ where a and b are integers and $b \neq 0$.

ray (548) A part of a line that extends indefinitely in one direction.

real numbers (673) The set of rational numbers together with the set of irrational numbers.

reciprocal (289) Another name for a multiplicative inverse.

rectangle (585) A quadrilateral with four congruent angles.

rectangular prism (632) A prism with rectangular bases.

reflection (595) A type of transformation where a figure is flipped over a line.

regular polygon (591) A polygon having all sides congruent and all angles congruent.

relation (372) A set of ordered pairs.

repeating decimal (225) A decimal whose digits repeat in groups of one or more. Examples are 0.181818 . . . and 0.8333. . . .

rhombus (585) A parallelogram with four congruent sides.

right angle (550) An angle that measures 90°.

right triangle (569) A triangle with one right angle.

rotation (595) When a figure is turned around a point.

GLOSSARY

sample (50, 51) A group that is used to represent a whole population.

sample space (441) The set of all possible outcomes.

scale drawing (585) A drawing that is similar but either larger or smaller than the actual object.

scalene triangle (569) A triangle with no congruent sides.

scatter plot (378) In a scatter plot, two sets of data are plotted as ordered pairs in the coordinate plane.

scientific notation (317) A way of expressing a number as the product of a number that is at least 1 but less than 10 and a power of 10. For example, $687{,}000 = 6.87 \times 10^5$.

sequence (258) A list of numbers in a certain order, such as, 0, 1, 2, 3, or 2, 4, 6, 8.

sides (549) The two rays that form an angle are called the sides.

sides (589) The line segments that form a polygon.

similar (578) Figures that have the same shape but not necessarily the same size.

simplest form (28) An expression in simplest form has no like terms and no parentheses.

simplest form (196) A fraction is in simplest form when the GCF of the numerator and the denominator is 1.

simulation (524) The process of acting out a problem.

sine (688) If $\triangle ABC$ is a right triangle and A is an acute angle,

$$\text{sine } \angle A = \frac{\text{measure of the side opposite } \angle A}{\text{measure of the hypotenuse}}$$

$$\sin A = \frac{a}{c}$$

skew lines (551) Two lines that do not intersect and are not in the same plane.

slant height (638) The length of the altitude of a lateral face of a regular pyramid.

slope (400) The slope of a line is the ratio of the change in y to the corresponding change in x.

$$\text{slope} = \frac{\text{change in } y}{\text{change in } x}$$

slope-intercept form (407) The slope-intercept form of a linear equation of a line is $y = mx + b$. The slope of the line is m, and the y-intercept is b.

solution (32) A value for the variable that makes an equation true. The solution for $12 = x + 7$ is 5.

solution of system (412) The ordered pair that is the solution of both equations.

solving an equation (32) The process of finding a solution to an equation.

square (585) A parallelogram with all sides congruent and all angles congruent.

square pyramid (638) A pyramid with a square base.

square root (664) One of the two equal factors of a number. A square root of 144 is 12 since $12^2 = 144$.

standard form (176) The standard form for seven hundred thirty-nine is 739.

statistics (51) Information that has been collected, analyzed, and presented in an organized fashion.

stem-and-leaf plot (486) A system used to condense a set of data where the greatest place value of the data forms the stem and the next greatest place value forms the leaves.

stems (486) The greatest place value common to all the data values is used for the stem of a stem-and-leaf plot.

straightedge (554) Any object that can be used to draw a straight line.

substitution property of equality (17) For all numbers a and b, if $a = b$, then a may be replaced with b.

subtraction property of equality (124) If you subtract the same number from each side of an equation, the two sides remain equal. For any numbers a, b, and c, if $a = b$, then $a - c = b - c$.

subtraction property of inequality (147) Subtracting the same number from each side of an inequality does not change the truth of the inequality.
For all numbers a, b, and c:
1. If $a > b$, then $a - c > b - c$.
2. If $a < b$, then $a - c < b - c$.

supplementary (562) Two angles are supplementary if the sum of their measures is 180°.

surface area (632) The sum of the areas of all the surfaces (faces) of a 3-dimensional figure.

symmetric (596) A figure created by a shape and its reflection.

system of equations (412) A set of equations with the same variables. For example, $c = 10 + 0.1n$ and $c = 5 + 0.2n$ are called a *system of equations*.

tangent (688) If $\triangle ABC$ is a right triangle and A is an acute angle,

$$\text{tangent } \angle A = \frac{\text{measure of the side opposite } \angle A}{\text{measure of the side adjacent to } \angle A}$$

$$\tan A = \frac{a}{b}$$

term (27) A number, a variable, or a product of numbers and variables.

term (258) Each number within a sequence is called a term.

terminating decimal (225, 275) A decimal whose digits end. Every terminating decimal can be written as a fraction with a denominator of 10, 100, 1000, and so on.

tessellation (594) A repetitive pattern of polygons that fit together with no holes or gaps.

30°–60° right triangle (683) A right triangle that has one angle measuring 30° and another angle measuring 60°.

transformation (595) A movement of a geometric figure.

translation (595) One type of transformation where a figure is slid horizontally, vertically, or both.

transversal (563) A line that intersects two parallel lines to form eight angles.

trapezoid (585) A quadrilateral with exactly one pair of parallel sides.

tree diagram (509) A diagram used to show the total number of possible outcomes in a probability experiment.

triangle (568) A polygon having three sides.

triangular prism (632) A prism with triangular bases.

triangular pyramid (638) A pyramid with an equilateral triangle as a base.

trigonometric ratios (688) Ratios that involve the measures of the sides of right triangles. The tangent, sine, and cosine ratios are three trigonometric ratios.

trigonometry (688) The study of triangle measurement.

trinomial (706) A polynomial with three terms.

truncate (275) An answer being cut off by a calculator at a certain place-value position, ignoring the digits that follow.

unit rate (433) A rate with a denominator of 1.

upper quartile (491) The median of the upper half of a set of numbers.

variable (16) A placeholder in a mathematical expression or sentence.

Venn diagram (669) A diagram consisting of circles inside a rectangle which is used to show the relationships of sets.

vertex (549) A vertex of an angle is the common endpoint of the rays forming the angle.

vertex (75, 589) A vertex of a polygon is a point where two sides of the polygon intersect.

vertex (638) The vertex of a pyramid is the point where all the faces except the base intersect.

vertical angles (561) Congruent angles formed by the intersection of two lines. In the figure, the vertical angles are $\angle 1$ and $\angle 3$, and $\angle 2$ and $\angle 4$.

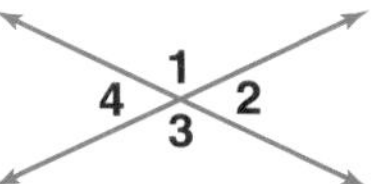

vertical line test (373) If any vertical line drawn on the graph of a relation passes through no more than one point of its graph, then the relation is a function.

volume (644) The number of cubic units needed to fill a container.

whole numbers (224) The set of numbers $\{0, 1, 2, 3, \ldots\}$. It also includes any number that can be written as a whole number, such as $\frac{5}{5}$ or $\frac{9}{1}$.

***x*-axis** (36) The horizontal number line which helps to form the coordinate system.

***x*-coordinate** (37) The first number of an ordered pair.

***x*-intercept** (406) The x-coordinate of the point where the graph crosses the x-axis.

***y*-axis** (36) The vertical number line which helps to form the coordinate system.

***y*-coordinate** (37) The second number of an ordered pair.

***y*-intercept** (406) The y-coordinate of the point where the graph crosses the y-axis.

zero pair (82) The result of pairing one positive counter with one negative counter.

SPANISH GLOSSARY

absolute value/valor absoluto (67) El número de unidades que un número dista de cero en la recta numérica.

acute angle/ángulo agudo (550) Un ángulo cuya medida es mayor que $0°$ y menor que $90°$.

acute triangle/triángulo agudo (569) Un triángulo que tiene tres ángulos agudos.

addition property of equality/propiedad de adición de la igualdad (125) Si sumas el mismo número a ambos lados de una ecuación, los dos lados permanecen igual. Para cualquiera de los números a, b y c, si $a = b$, entonces $a + c = b + c$.

addition property of inequality/propiedad de adición de la desigualdad (147) El sumar el mismo número a cada lado de una desigualdad, no altera la validez de la desigualdad. Para cualquiera de los números a, b y c:
1. Si $a > b$, entonces $a + c > b + c$.
2. Si $a < b$, entonces $a + c < b + c$.

additive inverses/inversos aditivos (89) Un número entero y su opuesto. La suma de un número entero y su inverso aditivo es cero.

adjacent angles/ángulos adyacentes (561) Dos ángulos que tienen un lado común y el mismo vértice, pero que *no* se traslapan.

adjacent leg/cateto adyacente (687) El cateto de un triángulo rectángulo que no es el opuesto a un ángulo y que no es la hipotenusa.

algebraic expression/expresión algebraica (16) Una combinación de variables, números y por lo menos una operación.

algebraic fraction/fracción algebraica (197) Una fracción con variables en el numerador o en el denominador.

alternate exterior angles/ángulos externos alternos (563) Ángulos exteriores no adyacentes que se encuentran en lados opuestos de la transversal. $\angle 1$ y $\angle 7$, y $\angle 2$ y $\angle 8$ son ángulos externos alternos.

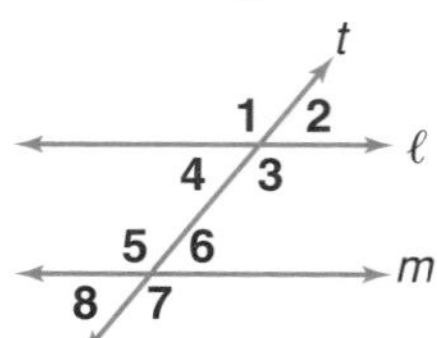

alternate interior angles/ángulos internos alternos (563) Ángulos internos no adyacentes que se encuentran en lados opuestos de la transversal. $\angle 4$ y $\angle 6$, y $\angle 3$ y $\angle 5$ son ángulos internos alternos.

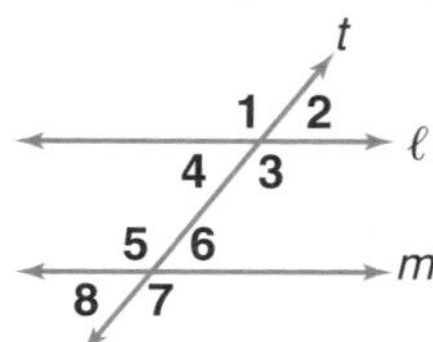

altitude/altitud (612, 614, 633) Un segmento de recta perpendicular a la base de una figura con extremos en la base y el lado opuesto a la base.

angle/ángulo (549) Dos rayos con un punto en común forman un ángulo. Los rayos y los vértices se usan para nombrar el ángulo.

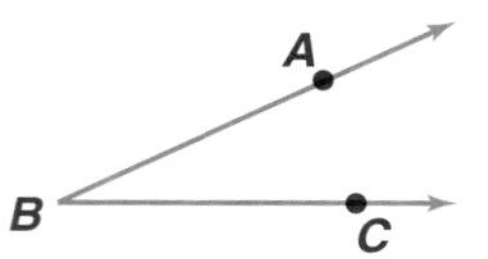

ángulo ABC o $\angle ABC$.

angle of depression/ángulo de depresión (695) El ángulo de depresión se forma por una línea visual a lo largo de la horizontal y otra línea visual por debajo de la misma.

angle of elevation/ángulo de elevación (694) Un ángulo de elevación se forma por una línea visual a lo largo de la horizontal y otra línea visual por encima de la misma.

area/área (138) El número de unidades cuadradas necesarias para cubrir una superficie no encerrada por una figura geométrica.

arithmetic sequence/secuencia aritmética (258) Una secuencia en que la diferencia entre cualquier par de términos consecutivos es la misma.

associative property of addition/propiedad asociativa de la adición (22, 234) Para cualquiera de los números racionales a, b y c,

$$(a + b) + c = a + (b + c).$$

associative property of multiplication/propiedad asociativa de la multiplicación (22, 296) Para cualquiera de los números a, b y c,

$$(ab)c = a(bc).$$

axes/ejes (73) Dos rectas numéricas que se intersecan forman un sistema de coordenadas.

back-to-back stem-and-leaf plot/diagrama de tallo y hojas consecutivo (487) Se usa para comparar dos conjuntos de datos. Las hojas de uno de los conjuntos de datos están a un lado del tallo y las hojas para el otro conjunto de datos están en el otro lado.

bar graph/gráfica de barra (52) Una forma gráfica que usa barras para comparar estadísticas.

bar notation/notación de barra (225) En decimales que se repiten, la línea o barra que se coloca sobre los dígitos que se repiten. Por ejemplo, $2.\overline{63}$ indica que los dígitos 63, se repiten.

base/base (175) En 5^3, la base es 5. La base se usa como un factor tantas veces como lo indique el exponente (3). Es decir,

$$5^3 = 5 \times 5 \times 5.$$

base/base (449) En una proporción de porcentaje, el número con el cual se compara el por ciento.

$$\frac{\text{porcentaje}}{\text{base}} = \text{proporción}$$

base/base (614) La base de un rectángulo, un paralelogramo o un triángulo es cualquier lado de la figura. Las bases de un trapecio son los lados paralelos.

base/base (632) Las bases de un prisma son los dos lados congruentes paralelos.

binomial/binomio (706) Un polinomio con exactamente dos términos.

boundary/frontera (418) Una línea que separa una gráfica en mitades de planos.

box-and-whisker plot/diagrama de caja y patillas (495) Una representación gráfica que resume datos usando la mediana, los cuartillos superior e inferior y los valores extremos. Se dibuja una caja alrededor del valor cuartílico y las patillas se extienden desde cada cuartillo hasta los puntos extremos de los datos.

center/centro (341) El punto dado desde el cual todos los puntos en un círculo están equidistantes.

circle/círculo (341) El conjunto de todos los puntos en un plano que están equidistantes de un punto dado llamado centro.

circle graph/gráfica de círculo (497, 556) Un tipo de gráfica estadística que se usa para comparar partes de un todo.

circular cone/cono circular (639) Una figura en el espacio que tiene una base circular y un vértice.

circumference/circunferencia (341) La distancia alrededor de un círculo.

coefficient/coeficiente (711) La parte numérica de un monomio.

combination/combinación (516) Un arreglo o lista en el cual el orden no es importante.

commission/comisión (470) Un estipendio que se le paga a un vendedor o a una vendedora basado en un porcentaje del total de las ventas.

common difference/diferencia común (258) La diferencia entre cualquiera par de términos consecutivos en una secuencia aritmética.

common multiples/múltiplos comunes (200) Múltiplos compartidos por dos o más números. Por ejemplo, algunos múltiplos comunes de 2 y 3 son 6, 12 y 18.

common ratio/proporción común (312) La proporción entre cualquier par de términos consecutivos en una secuencia geométrica.

commutative property of addition/propiedad conmutativa de la adición (22, 234) Para cualquiera de los números racionales a y b, $a + b = b + a$.

commutative property of multiplication/propiedad conmutativa de la multiplicación (22, 296) Para cualquiera de los números racionales a y b, $ab = ba$.

comparative graph/gráfica comparativa (498) Una gráfica que se usa para comparar resultados de grupos similares y mostrar tendencias.

compass/compás (554) Un instrumento que se usa para dibujar círculos o partes de círculos.

compatible numbers/números compatibles (280) Números que han sido redondeados de forma que cuando se dividen uno entre otro, el residuo es cero.

complementary/complementario (562) Dos ángulos son complementarios si la suma de sus medidas es 90°.

composite number/número compuesto (184) Cualquier número entero mayor de uno que posee más de dos factores.

congruent/congruente (554, 561, 573) Segmentos de recta que tienen la misma longitud, o ángulos que tienen la misma medida, o figuras que tienen la misma forma y tamaño.

constant/constante (707) Un monomio que no contiene una variable.

coordinate/coordenada (67, 126) Un número asociado con un punto en la recta numérica.

coordinate system/sistema de coordenadas (36, 73) Un plano de coordenadas separado en cuatro regiones iguales por dos rectas numéricas que se intersecan, llamadas ejes.

corresponding angles/ángulos correspondientes (563) Ángulos que tienen la misma posición en dos rectas paralelas diferentes cortadas por una transversal. $\angle 1$ y $\angle 5$, $\angle 2$ y $\angle 6$, y $\angle 3$ y $\angle 7$, y $\angle 4$ y $\angle 8$ son ángulos correspondientes.

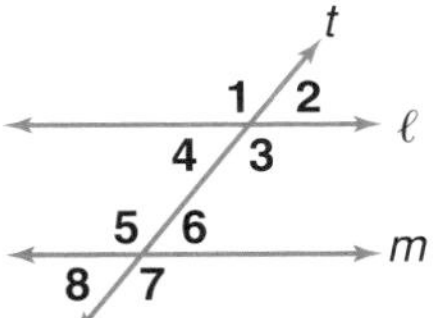

corresponding parts/partes correspondientes (573) Partes en figuras congruentes o similares que encajan.

cosine/coseno (688) Si $\triangle ABC$ es un triángulo rectángulo y A es un ángulo agudo,

$$\text{coseno } \angle A = \frac{\text{medida del lado adyacente a } \angle A}{\text{medida de la hipotenusa}}$$

$$\cos A = \frac{b}{c}$$

cylinder/cilindro (633) Una figura tridimensional que posee dos bases congruentes paralelas (generalmente circulares) y un lado curvo que conecta las bases.

data/datos (50) Información numérica que se reúne con propósitos estadísticos.

defining a variable/definir una variable (42) El escoger una variable y una cantidad para la variable para representarla en una ecuación.

degree/grado (549) La unidad de medida más común para ángulos.

dependent events/eventos dependiente (532) Dos o más eventos en los cuales el resultado de un evento afecta el resultado de otro(s) evento(s).

diagonal/diagonal (590) Un segmento de recta que une dos vértices no consecutivos de un polígono.

diameter/diámetro (341) La distancia a través del centro de un círculo.

discount/descuento (468) La cantidad que se rebaja del precio original.

discrete mathematics/matemáticas discretas (258) Una rama de las matemáticas que tiene que ver con los conjuntos de números finitos o interrumpidos. Algunas disciplinas que tratan con las matemáticas discretas son la lógica y la estadística.

distributive property/propiedad distributiva (26) Para todos los números reales a, b y c, $a(b + c) = ab + ac$ y $(b + c)a = ba + ca$.

divisible/divisible (170) Un número es divisible entre otro si el residuo de la división es cero.

division property of equality/propiedad de división de la igualdad (130) Para cualquiera de los números a, b y c, con $c \neq 0$, si $a = b$, entonces $\frac{a}{c} = \frac{b}{c}$.

division properties of inequality/propiedades de división de la desigualdad (152) Para todos los números enteros a, b y c, en que $c > 0$, si $a > b$, entonces $\frac{a}{b} > \frac{b}{c}$.
Para todos los números enteros a, b y c, en que $c < 0$, si $a > b$, entonces $\frac{a}{c} < \frac{b}{c}$.

domain/dominio (372) El conjunto de todas las coordenadas de cada par ordenado.

empty set/conjunto vacío (347) Un conjunto sin ningún elemento, el cual se representa con el símbolo { } o $\varnothing$.

equation/ecuación (32) Un enunciado matemático abierto que contiene el signo de igualdad, =.

equiangular/equiangular (591) Todos los ángulos son congruentes.

equilateral/equilátero (591) Todos los lados son congruentes.

equivalent/equivalente (275) El mismo valor representado de diferentes formas. Por ejemplo, $\frac{3}{5}$ y 0.6 son equivalentes.

equivalent equations/ecuaciones equivalentes (124) Dos o más ecuaciones con la misma solución. Por ejemplo, $x + 3 = 5$ y $x = 2$ son ecuaciones equivalentes.

evaluating the expression/evaluando una expresión (11) Reemplazar las variables con números y hallar el valor numérico de la expresión.

expanded form/forma desarrollada (176) Un número que se expresa usando el valor de posición para escribir el valor de cada dígito en el número.

experimental probability/probabilidad experimental (529) Una posibilidad estimada con base en la frecuencia relativa de los resultados positivos que ocurren durante un experimento.

exponent/exponente (175) En 5^3, el exponente es 3. El exponente indica cuántas veces la base, 5, se usa como factor.

$$5^3 = 5 \times 5 \times 5$$

exterior angle/ángulo externo (591) El ángulo que se forma cuando se extiende un lado de un polígono.

face/cara (632) La cara de un prisma es cualquier superficie que forma un lado o la base del prisma.

factorial/factorial (515) La expresión $n!$ es el producto de todos los números contables comenzando con n y contando al revés hasta 1.

factors/factores (170) Números que se dividen entre un número entero y cuyo residuo es cero.

factor trees/árboles de factores (185) Un diagrama que muestra la factorización prima de un número compuesto. Los factores se ramifican a partir de los factores previos hasta que todos los factores son números primos.

Fibonacci sequence/secuencia de Fibonacci (263) Un listado de números en el que los dos primeros números son ambos 1 y cada número que sigue es la suma de los dos números previos, 1, 1, 2, 3, 5, 8, . . .

formula/fórmula (134) Una ecuación que expresa una regla para la relación entre ciertas cantidades.

45°–45° right triangle/triángulo rectángulo de 45°–45° (683) Un triángulo rectángulo que posee dos ángulos que miden 45° cada uno.

frequency table/tabla de frecuencia (51) Una tabla que indica el número de valores en cada intervalo.

function/función (373) Una función es una relación en que cada elemento del dominio se aparea exactamente con un elemento de la amplitud.

functional notation/notación funcional (393) Las ecuaciones que representan funciones se pueden escribir en *forma de notación*, $f(x)$. El símbolo $f(x)$ se lee "f" de "x" y representa el valor en la amplitud que corresponde al valor de x en el dominio.

fundamental counting principle/principio fundamental de contar (509) Si el evento M puede ocurrir de m maneras y es seguido por el evento N que puede ocurrir de n maneras, entonces el evento M seguido del evento N puede ocurrir en $m \cdot n$ maneras.

geometric probability/probabilidad geométrica (623) Probabilidad que usa el área para hallar la probabilidad de un evento.

geometric sequence/secuencia geométrica (312) Una secuencia en que la proporción entre cualquier par de números consecutivos es la misma.

gram/gramo (359) La unidad básica de masa del sistema métrico.

graph/gráfica (73, 126) Un puntito que representa un número en una recta numérica o un par ordenado en un plano de coordenadas.

greatest common factor (GCF)/máximo factor común (MFC) (190) El MFC de dos o más números enteros es el máximo factor que es común a cada uno de los números.

histogram/histograma (502) Una gráfica de barra cuyas columnas están una al lado de la otra para mostrar cómo están distribuidos los datos.

hypotenuse/hipotenusa (676) El lado opuesto al ángulo recto en un triángulo rectángulo.

identity property of addition/propiedad de identidad de la adición (23, 234) Para cualquier número a, $a + 0 = a$ y $0 + a = a$.

identity property of multiplication/propiedad de identidad de la multiplicación (23, 296) Para cualquier número a, $a \times 1 = a$.

inclusive/inclusivo (536) Cuando dos eventos pueden suceder al mismo tiempo.

independent events/eventos independientes (531) Dos o más eventos en los cuales el resultado de uno de ellos *no* afecta el resultado de ninguno de los otros.

indirect measurement/medida indirecta (580) Hallar una medida mediante el uso de triángulos similares y escribiendo una proporción.

inequality/desigualdad (46) Un enunciado matemático que contiene $<$, $>$, $\neq$, $\leq$ o $\geq$.

integers/números enteros (66, 224) Los números enteros y sus opuestos.

$$\ldots, -3, -2, -1, 0, 1, 2, 3, \ldots$$

intercepts/intersecciones (406) Los puntos en donde una gráfica cruza los ejes.

interior angles/ángulos internos (591) Un ángulo dentro de un polígono.

interquartile range/amplitud intercuartílica (491) La amplitud de la mitad central de un conjunto de datos.

$$\text{Amplitud intercuartílica} = UQ - LQ,$$

intersect/intersecar (552) Líneas que se cruzan o se tocan unas a otras en un punto.

inverse operations/operaciones inversas (41) Operaciones que se anulan entre sí, tales como la adición y la sustracción.

inverse property of addition/propiedad del inverso de la adición (234) Para cualquier número racional a, $a + (-a) = 0$.

inverse property of multiplication/propiedad del inverso de la multiplicación (289) Para todo número no cero $\frac{a}{b}$, donde $a, b \neq 0$, hay exactamente un número $\frac{b}{a}$, tal que $\frac{a}{b} \times \frac{b}{a} = 1$.

irrational number/número irracional (672) Un número que no se puede expresar como $\frac{a}{b}$, donde a y b son números enteros y b no es igual a 0. Por ejemplo, $\sqrt{3} = 1.7320508\ldots$ es un número irracional.

isosceles triangle/triángulo isósceles (569) Un triángulo que tiene por lo menos dos lados congruentes.

lateral surface/superficie lateral (639) La superficie lateral de un prisma, cilindro, pirámide o cono es toda la superficie de la figura excepto la base o las bases.

least common denominator (LCD)/mínimo común denominador (MCD) (201) El mínimo común múltiplo de los denominadores de dos o más fracciones.

least common multiple (LCM)/mínimo común múltiplo (MCM) (200) El menor número diferente de cero de los múltiplos comunes de dos o más números. El MCM de 2 y 3 es 6.

leaves/hojas (486) El valor de posición que le sigue al valor mayor forma las hojas en una gráfica de tallo y hojas.

leg/cateto (676) Cualquiera de los dos lados que forman el ángulo recto de un triángulo rectángulo.

like terms/términos semejantes (27, 711) Expresiones que contienen las mismas variables, tales como $3ab$ y $7ab$.

line/recta (548) Una trayectoria recta sin fin. Una representación de la recta CD ($\overleftrightarrow{CD}$) se muestra a continuación.

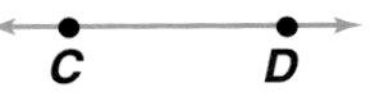

linear equation/ecuación lineal (386) Una ecuación cuya gráfica es una recta.

line graph/gráfica lineal (497) Un tipo de gráfica estadística que se usa para mostrar cómo los valores cambian a través de un período de tiempo.

line segment/segmento de recta (548) Dos extremos y la trayectoria recta entre ellos. Un segmento de recta se denomina por sus extremos.

liter/litro (359) La unidad básica de capacidad en el sistema métrico.

lower quartile/cuartillo inferior (491) La mediana de la mitad inferior de un conjunto de números indicados por LQ.

mean/media (300) La suma de los números en un conjunto de datos dividida entre el número de datos.

measures of central tendency/medidas de tendencia central (300) Datos o grupos de datos que pueden representar el conjunto total de datos.

median/mediana (300) En un conjunto de datos, la mediana es el número en el centro cuando los datos se organizan en orden de menor a mayor.

meter/metro (359) La unidad básica de longitud en el sistema métrico.

metric system/sistema métrico (360) Un sistema de medida que usa las unidades básicas: metro para longitud, gramo para masa y litro para capacidad.

mixed numbers/números mixtos (239) La suma de un número entero y una fracción. Por ejemplo, $3\frac{1}{2}$.

mode/modal (300) El número o artículo que aparece con más frecuencia en un conjunto de datos.

monomial/monomio (172) Una expresión algebraica que es un número real, una variable o el producto de números reales y variables.

multiple/múltiplo (200) El producto del número y cualquier número entero.

multiplication properties of inequality/ propiedades multiplicativas de la desigualdad (152) **1.** Para todos los números enteros a, b y c, en que $c > 0$, si $a > b$, entonces $ac > bc$. **2.** Para todos los números enteros a, b y c, en que $c < 0$, si $a < b$, entonces $ac > bc$.

multiplication property of equality/propiedad multiplicativa de la igualdad (130) Para cualquiera de los números reales a, b y c, si $a = b$, entonces $ac = bc$.

multiplicative inverses/inversos multiplicativos (289) Dos números cuyo producto es 1. El inverso multiplicativo de $\frac{3}{4}$ es $\frac{4}{3}$.

multiplicative property of zero/propiedad multiplicativa de cero (23) Para cualquier número a, $a \times 0 = 0$.

mutually exclusive/exclusivo mutuamente (535) Dos o más eventos tales que ningún par de eventos pueden suceder al mismo tiempo.

negative/negativo (66) Un valor menor que cero.

negative relationship/relación negativa (380) Si los puntos en un diagrama de dispersión parecen sugerir una recta que se inclina hacia abajo y hacia la derecha, entonces existe una *relación negativa*.

net/red (632) La figura que se forma al "desdoblar" un prisma. La red muestra todas las caras que componen el área de un prisma.

no relationship/no relacionado (380) Se dice que los puntos que parecen ser aleatorios en un diagrama de dispersión son no relacionados.

null set/conjunto vacío (347) Un conjunto sin elementos y el cual se representa con el símbolo { } o $\varnothing$.

obtuse angle/ángulo obtuso (550) Cualquier ángulo que mide más de 90° pero menos de 180°.

obtuse triangle/triángulo obtuso (569) Un triángulo con un ángulo obtuso.

odds/posibilidades (520) Una manera de describir la posibilidad de que un evento ocurra.
posibilidades en favor = número de éxitos: número de fracasos
posibilidades en contra = número de fracasos: número de éxitos

open sentence/enunciado abierto (32) Una ecuación que contiene variables.

opposite side/lado opuesto (687) El lado que no forma parte del ángulo en un triángulo rectángulo.

ordered pair/par ordenado (37) Un par de números utilizados para ubicar un punto en el sistema de coordenadas.

order of operations/orden de operaciones (12, 176)
1. Simplifica las expresiones dentro de los símbolos de agrupación.
2. Evalúa todas las potencias.
3. Luego, realiza todas las multiplicaciones y las divisiones de izquierda a derecha.
4. Finalmente, realiza todas las sumas y las restas de izquierda a derecha.

origin/origen (36) El punto de intersección del eje x y del eje y en un sistema de coordenadas.

outcomes/resultados (509) Posibles resultados de un evento de probabilidad. Por ejemplo, 4 es un resultado cuando se lanza un dado.

outliers/valores atípicos (496) Datos que están a más de 1.5 veces que la amplitud intercuartílica de los cuartillos.

parallel/paralelo (551) Rectas en el mismo plano que no se intersecan. El símbolo $\parallel$ quiere decir paralelo.

parallelogram/paralelogramo (585) Un cuadrilátero con dos pares de lados paralelos.

percent/por ciento (449) Una proporción con un denominador de 100. Por ejemplo, 77% y $\frac{77}{100}$ son el mismo número.

percentage/porcentaje (449) En una proporción de por ciento, un número que se compara con otro llamado base.

$$\frac{\text{porcentaje}}{\text{base}} = \text{tasa}$$

percent of change/porcentaje de cambio (472) La proporción de la cantidad de cambio comparada con la cantidad inicial.

percent of decrease/porcentaje de disminución (473) La proporción de una cantidad de disminución comparada con la cantidad previa, expresada en forma de porcentaje.

percent of increase/porcentaje de aumento (472) La proporción de una cantidad de aumento comparada con la cantidad previa, expresada en forma de porcentaje.

percent proportion/proporción de porcentaje (450)

$$\frac{\text{porcentaje}}{\text{base}} = \text{tasa o } \frac{P}{B} = \frac{r}{100}$$

perfect squares/cuadrados perfectos (665) Números racionales cuyas raíces cuadradas son números enteros. 25 es un cuadrado perfecto porque $\sqrt{25} = 5$.

perimeter/perímetro (138) La distancia alrededor de una figura geométrica.

permutation/permutación (515) Un arreglo o listado en que el orden es importante.

perpendicular/perpendicular (551) Dos rectas que se intersecan para formar un ángulo recto.

pi, π/pi, π (341) La proporción de la circunferencia de un círculo con el diámetro del círculo. Aproximaciones para π son 3.14 y $\frac{22}{7}$.

pictograph/pictografia (497) Un tipo de gráfica estadística que usa láminas o ilustraciones para mostrar comparación entre cantidades específicas. Cada símbolo representa un número específico de artículos.

plane/plano (550) Una superficie plana sin aristas o fronteras. Un plano se puede nombrar con una sola letra mayúscula o mediante el uso de tres puntos no colineales del plano.

point/punto (548) Una ubicación específica en el espacio.

polygon/polígono (589) Una figura cerrada simple en un plano formado por tres o más segmentos de recta.

polynomial/polinomio (706) Un monomio o la suma o diferencia de dos o más monomios.

positive relationship/relación positiva (380) En una gráfica de dispersión, si los puntos parecen sugerir una recta que se inclina hacia arriba y a la derecha, entonces existe una *relación positiva*.

power/potencia (175) Un número que se puede escribir usando un exponente.

power of a monomial/potencia de un monomio (721) Para todos los números a y b, y los números enteros positivos m, n y p, $(a^m b^n)^p = a^{mp} b^{np}$.

power of a power/potencia de una potencia (719) Para cualquier número a y cualquiera de los números enteros m y n, $(a^m)^n = a^{mn}$.

power of a product/potencia de un producto (720) Para todos los números a y cualquier número entero positivo m, $(ab)^m = a^m b^m$.

prime factorization/factorización prima (185) El expresar un número compuesto como un producto de sus factores primos.

prime number/número primo (184) Un número entero mayor que 1 cuyos únicos factores son 1 y el número mismo.

principle/capital (467) La cantidad de dinero en una cuenta.

prism/prisma (632) Figura en el espacio que tiene dos bases paralelas y congruentes en forma de polígonos.

probability/probabilidad (440) La posibilidad de que algún evento pueda suceder. Es la proporción del número de veces que cierto evento puede ocurrir comparada con el número de resultados posibles.

$$\text{Probabilidad} = \frac{\text{número de resultados favorables}}{\text{número de resultados posibles}}.$$

probability of inclusive events/probabilidad de eventos inclusivos (536) La probabilidad de que uno u otro de dos eventos inclusivos ocurra se puede hallar sumando la probabilidad del primero a la probabilidad del segundo evento y restando la probabilidad de que ambos eventos ocurran.

$$P(A \text{ o } B) = P(A) + P(B) - P(A \text{ y } B)$$

probability of mutually exclusive events/probabilidad de eventos exclusivos mutuamente (535) La probabilidad de que el primero o el segundo de dos eventos mutuamente exclusivos ocurra se puede hallar sumando la probabilidad del primero a la probabilidad del segundo evento.

$$P(A \text{ o } B) = P(A) + P(B)$$

probability of two dependent events/probabilidad de dos eventos dependientes (532) Si dos eventos A y B son dependientes, entonces la probabilidad de que ambos eventos ocurran es el producto de la probabilidad de A y de la probabilidad de B después de que A ocurra.

$$P(A \text{ y } B) = P(A) \times P(B \text{ seguida de } A)$$

probability of two independent events/probabilidad de dos eventos independientes (531) La probabilidad de dos eventos independientes se puede hallar multiplicando la probabilidad del primero por la probabilidad del segundo evento.

$$P(A \text{ y } B) = P(A) \times P(B)$$

product of powers/producto de potencias (205) Para cualquiera de los números a y los números enteros positivos m y n, $a^m \times a^n = a^{m+n}$.

property of proportions/propiedad de las proporciones (444)

Si $\frac{a}{b} = \frac{c}{d}$, entonces $ad = bc$.

Si $ad = bc$, entonces $\frac{a}{b} = \frac{c}{d}$.

proportion/proporción (444) Un enunciado de la igualdad de dos o más razones.

protractor/transportador (549) Un instrumento que se usa para medir ángulos.

pyramid/pirámide (638) Una figura sólida que tiene un polígono por base y triángulos por lados.

Pythagorean theorem/teorema de Pitágoras (676) En un triángulo rectángulo, el cuadrado del largo de la hipotenusa es igual a la suma de los cuadrados de los largos de los catetos.

$$c^2 = a^2 + b^2$$

quadrants/cuadrantes (73) Las cuatro regiones en que dos rectas numéricas perpendiculares separan el plano.

quadrilateral/cuadrilátero (584, 585) Un polígono con cuatro lados.

quartile/cuartillo (491) Una de las cuatro partes iguales de datos de un conjunto grande de números.

quotient of powers/cociente de potencias (206) Para todo número a diferente de cero y los números enteros m y n, $\frac{a^m}{a^n} = a^{m-n}$.

radical sign/signo radical (664) El símbolo que se usa para indicar una raíz cuadrada no negativa. $\sqrt{\ }$

radius/radio (341) La distancia desde el centro de un círculo hasta cualquier punto en el círculo.

range/amplitud (372) La amplitud de una relación es el conjunto de todas las segundas coordenadas de cada par ordenado.

range/amplitud (490) La diferencia entre los valores mayor y menor en un conjunto de datos.

rate/razón (433) La proporción de dos medidas que tienen diferentes unidades.

rate/tasa (449) En una proporción de por ciento, la razón de un número a 100.

ratio/razón (196, 432) Una comparación de dos números mediante división. La razón de 2 a 3 se puede expresar como 2 de 3, 2 a 3 ó $\frac{2}{3}$.

rational numbers/números racionales (224) Números que se pueden expresar como $\frac{a}{b}$, en que a y b son números enteros y $b \neq 0$.

ray/rayo (548) Una parte de una recta que se extiende infinitamente en una dirección.

real numbers/números reales (673) El conjunto de números racionales unido con el conjunto de números irracionales.

reciprocal/recíproco (289) Otro nombre para el inverso multiplicativo.

rectangle/rectángulo (585) Un cuadrilátero con cuatro ángulos congruentes.

rectangular prism/prisma rectangular (632) Un prisma con bases rectangulares.

SPANISH GLOSSARY

reflection/reflexión (595) Un tipo de transformación en que una figura se voltea alrededor de una línea.

regular polygon/polígono regular (591) Un polígono que tiene todos los lados y todos los ángulos congruentes.

relation/relación (372) Un conjunto de pares ordenados.

repeating decimal/decimal periódico (225) Un decimal cuyos dígitos se repiten en grupos de uno o más. Por ejemplo: 0.181818 . . . y 0.8333. . . .

rhombus/rombo (585) Un paralelogramo con cuatro lados congruentes.

right angle/ángulo recto (550) Un ángulo que mide 90°.

right triangle/triángulo rectángulo (569) Un triángulo con un ángulo recto.

rotation/rotación (595) Cuando se hace girar una figura alrededor de un punto.

sample/muestra (50, 51) Un grupo que se usa para representar una población entera.

sample space/espacio muestral (441) El conjunto de todos los resultados posibles.

scale drawing/dibujo a escala (585) Un dibujo que es similar, pero que es o más grande o más pequeño que el objeto en sí.

scalene triangle/triángulo escaleno (569) Un triángulo sin lados congruentes.

scatter plot/diagrama de dispersión (378) En un diagrama de dispersión, se grafican dos conjuntos de datos como pares ordenados en el plano de coordenadas.

scientific notation/notación científica (317) Una manera de expresar un número como el producto de un número que es por lo menos 1, pero menos de 10, y una potencia de 10. Por ejemplo, $687{,}000 = 6.87 \times 10^5$.

sequence/secuencia (258) Una lista de números en cierto orden, como por ejemplo: 0, 1 2 3 ó 2, 4, 6, 8.

sides/lados (549) Los dos rayos que forman un ángulo se llaman lados.

sides/lados (589) Los segmentos de recta que forman un polígono.

similar/similar (578) Figuras que tienen la misma forma, pero no necesariamente el mismo tamaño.

simplest form/forma reducida (28) Una expresión en forma reducida no tiene términos semejantes ni paréntesis.

simplest form/forma reducida (196) Una fracción está en forma reducida cuando el máximo factor común (MFC) del numerador y del denominador es 1.

simulation/simulación (524) El proceso de representar un problema.

sine/seno (688) Si $\triangle ABC$ es un triángulo rectángulo y A es un ángulo agudo,

$$\text{seno } \angle A = \frac{\text{medida del lado opuesto a } \angle A}{\text{medida de la hipotenusa}}$$

$$\text{sen } A = \frac{a}{c}.$$

skew lines/rectas alabeadas (551) Dos rectas que no se intersecan y no están en el mismo plano.

slant height/altura inclinada (638) El largo de la altura de una cara lateral de una pirámide regular.

slope/pendiente (400) La pendiente de la recta es la razón del cambio en y al cambio correspondiente en x.

$$\text{Pendiente} = \frac{\text{cambio en } y}{\text{cambio en } x}$$

slope-intercept form/forma de pendiente-intersección (407) La forma de pendiente-intersección de una ecuación lineal de una recta es $y = mx + b$. La pendiente de la recta es m y la intersección con el eje y es b.

solution/solución (32) El valor de una variable que hace que una ecuación sea válida. La solución de $12 = x + 7$ es 5.

solution of system/solución de un sistema (412) El par ordenado que resuelve ambas ecuaciones en el sistema.

solving an equation/resolviendo una ecuación (32) El proceso de hallar una solución a una ecuación.

square/cuadrado (585) Un paralelogramo con todos los lados congruentes y todos los ángulos congruentes.

square pyramid/pirámide cuadrada (638) Una pirámide con una base cuadrada.

square root/raíz cuadrada (664) Uno de los dos factores iguales de un número. La raíz cuadrada de 144 es 12 puesto que $12^2 = 144$.

standard form/forma estándar (176) La forma estándar de setecientos treinta y nueve es 739.

statistics/estadística (51) Información que ha sido reunida, analizada y presentada de una manera organizada.

stem-and-leaf plot/gráfica de tallo y hojas (486) Un sistema que se usa para condensar un conjunto de datos en el cual el mayor valor de posición de los datos forma el tallo y el siguiente valor de posición forma las hojas.

stems/tallos (486) El mayor valor de posición común a todos los valores de los datos y que forma el tallo en una gráfica de tallo y hojas.

straightedge/regla (554) Cualquier objeto que se puede usar para dibujar una recta.

substitution property of equality/propiedad de sustitución de la igualdad (17) Para todos los números a y b, si $a = b$, entonces, a se puede reemplazar por b.

subtraction property of equality/propiedad de sustracción de la igualdad (124) Si restas el mismo número de cada lado de una ecuación, los dos lados permanecen iguales. Para cualquiera de los números a, b y c, si $a = b$, entonces $a - c = b - c$.

subtraction property of inequality/propiedad de sustracción de la desigualdad (147) Si restas el mismo número de cada lado de una desigualdad, la validez de la desigualdad no cambia.
Para todos los números a, b y c:
1. si $a > b$, entonces $a - c > b - c$;
2. si $a < b$, entonces $a - c < b - c$.

supplementary/suplementario (562) Dos ángulos son suplementarios si la suma de sus medidas es 180°.

surface area/área de superficie (632) La suma de las áreas de todas las superficies (caras) de una figura tridimensional.

symmetric/simétrico (596) Una figura creada por la figura y su reflexión.

system of equations/sistema de ecuaciones (412) Un conjunto de ecuaciones con las mismas variables. Por ejemplo: $c = 10 + 0.1n$ y $c = 5 + 0.2n$ se llaman un *sistema de ecuaciones.*

tangent/tangente (688) Si $\triangle ABC$ es un triángulo rectángulo y A es un ángulo agudo,

$$\text{tangente } \angle A = \frac{\text{medida del lado opuesto a } \angle A}{\text{medida del lado adyacente al } \angle A}$$

$$\tan A = \frac{a}{b}.$$

term/término (27) Un número, una variable o un producto de números y variables.

term/término (258) Cada número en una secuencia se llama un término.

terminating decimal/decimal terminal (225, 275) Un decimal cuyos dígitos terminan. Cada decimal terminal se puede escribir en forma de fracción con un denominador de 10, 100, 1000 y así sucesivamente.

tessellation/teselación (594) Un patrón repetitivo de polígonos que encajan juntos sin dejar espacios vacíos.

30°–60° right triangle/triángulo rectángulo de 30°–60° (683) Un triángulo rectángulo que tiene un ángulo que mide 30° y otro que mide 60°.

transformation/transformaciones (595) Movimientos de figuras geométricas.

translation/traslación (595) Un tipo de transformación en la cual una figura se desliza horizontal, verticalmente o en ambas direcciones.

transversal/transversal (563) Una recta que interseca dos líneas paralelas para formar ocho ángulos.

trapezoid/trapezoide (585) Un cuadrilátero con exactamente un par de lados paralelos.

tree diagram/diagrama de árbol (509) Un diagrama que se usa para mostrar el número total de posibles resultados en un experimento de probabilidad.

triangle/triángulo (568) Un polígono con tres lados.

triangular prism/prisma triangular (632) Un prisma con bases triangulares.

triangular pyramid/pirámide triangular (638) Una pirámide cuya base es un triángulo equilátero.

trigonometric ratios/razones trigonométricas (688) Razones que involucran las medidas de los lados de triángulos rectángulos. La tangente, el seno y el coseno son tres razones trigonométricas.

trigonometry/trigonometría (688) El estudio de la medición de triángulos.

trinomial/trinomio (706) Un polinomio con tres términos.

truncate/truncado (275) Una respuesta que la calculadora interrumpe en una posición determinada, ignorando los dígitos que siguen.

unit rate/razón unitaria (433) Una razón cuyo denominador es 1.

upper quartile/cuartillo superior (491) La mediana de la mitad superior de un conjunto de datos.

variable/variable (16) Símbolo que ocupa un lugar en expresiones matemáticas.

Venn diagram/diagrama de Venn (669) Un diagrama que consiste en círculos dentro de un rectángulo y el cual se usa para mostrar las relaciones de conjuntos.

vertex/vértice (549) El vértice de un ángulo es el extremo común de los rayos que forman el ángulo.

vertex/vértice (75, 589) El vértice de un polígono es un punto en donde se intersecan dos lados del polígono.

vertex/vértice (638) El vértice de una pirámide es el punto en donde se intersecan todas las caras, excepto la base.

vertical angles/ángulos verticales (561) Ángulos congruentes formados por la intersección de dos rectas. En la figura, los ángulos verticales son $\angle 1$ y $\angle 3$, y $\angle 2$ y $\angle 4$.

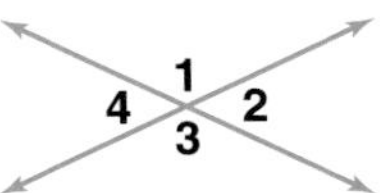

vertical line test/prueba de recta vertical (373) Si cualquier recta vertical dibujada en la gráfica de una relación pasa por un solo punto de esa gráfica, entonces la relación es una función.

volume/volumen (644) El número de unidades cúbicas que se necesitan para llenar un recipiente.

whole numbers/números enteros (224) El conjunto de números $\{0,1,2,3,\ldots\}$. También incluye cualquier número que se pueda escribir como un número entero, por ejemplo, $\frac{5}{5}$ ó $\frac{9}{1}$.

x-axis/eje x (36) La recta numérica horizontal que ayuda a formar el plano de coordenadas.

x-coordinate/coordenada x (37) El primer número en un par ordenado.

x-intercept/intersección con el eje x (406) La coordenada x del punto sobre el cual una gráfica cruza el eje x.

y-axis/eje y (36) La recta numérica vertical que ayuda a formar el plano de coordenadas.

y-coordinate/coordenada y (37) El segundo número en un par ordenado.

y-intercept/intersección con el eje y (406) La coordenada y del punto sobre el cual una gráfica cruza el eje y.

zero pair/par cero (82) El resultado de aparear un contador positivo con un contador negativo.

SELECTED ANSWERS

Chapter 1 Tools for Algebra and Geometry

Pages 9–10 Lesson 1-1

5a. We know the bonus for 2, 4, 6, and 8 packages over plan. We need to know the bonus for 16 packages over plan. **5b.** Extend the pattern. The bonus should be about $250.

5c.

10	$200
12	$225
14	$250
16	$275

5d. The answer is reasonable and answers the question. **7.** yes **9.** $2.07

11a.

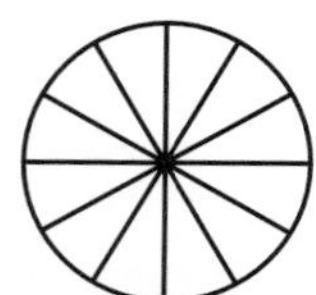

11b. 2, 4, 6, 8, 10, 12 **11c.** 1, 2, 3, 4, 5, 6 **11d.** The number of cuts is half the number of pieces.

Pages 14–15 Lesson 1-2

7. ÷, 61 **9.** ×, 5 **11.** −, 3 **13.** 8 **15.** 9 **17.** 12 **19.** 4 **21.** 4 **23.** 1 **25.** 145 **27.** 60 **29.** 128 **31.** $71 - (17 + 4) = 50$ **33.** $18 \div (3 + 6) + 12 = 14$ **35a.** 3($330) + 2($247) + 4($229) **35b.** $2400 **37.** $252 **39.** 2024, 3024, 7224

Pages 19–20 Lesson 1-3

5. 19 **7.** 12 **9.** 10 **11.** $3 + v$ **13.** $6n$ **15a.** $6x$ **15b.** 204 pounds **17.** 2 **19.** 14 **21.** 4 **23.** 12 **25.** 3 **27.** 10 **29.** $92 \div c$ **31.** $p - 5$ **33.** $u - 10$ **35.** $88 \div b$ **37.** $3k$ **39.** $2s + 2$ **41.** 16 less some number **43.** the quotient of some number and 5 **45.** twice some number plus 1 **47a.** $c \div 4 + 37$ **47b.** 68° **49.** 27 **51.** 29

Page 21 Lesson 1-3B

1. 14 **3.** 7 **5.** 55 **7.** Use REPLAY then change the expression.

Pages 24–25 Lesson 1-4

5. associative addition **7.** commutative addition **9.** 29 **11.** 48 **13.** $(7 \cdot 6)z$; $42z$ **15.** $8.35 **17.** commutative multiplication **19.** multiplicative zero **21.** commutative addition **23.** commutative addition **25.** commutative addition **27.** associative addition **29.** 360 **31.** 0 **33.** 130 **35.** 660 **37.** $9 + 5$ **39.** $18w + 9$ **41.** $b(6 \cdot 5)$; $30b$ **43.** $(13 + 11) + m$; $24 + m$ **45.** $p(7 \cdot 4)$; $28p$ **47.** $15 + (4w + w)$; $15 + 5w$ **49b.** Division is not commutative. **53.** 6 **55a.** $w \div 16$ **55b.** 9 pints

Pages 28–30 Lesson 1-5

5. 5(7) + 5(8) **7.** $4(x) + 4(3)$ **9.** $13a + 7$ **11.** $7c + 7$ **13.** $20ab$ **15.** 3(11) + 3(12) **17.** $t(6 + 11)$ **19.** $2(r + 6s)$ **21.** $9(v) + 8(v)$ **23.** $2(4x) + 2(9y)$ **25.** $38a + 45$ **27.** $25c + 16$ **29.** $33y + 35$ **31.** $32b + 60$ **33.** $72 + 25f$ **35.** $11x + 15y$ **37.** 12,000 cubic meters **39.** 8 + (19 + 17) **41.** $n \div 9$ **43.** $362.65 **45.** yes

Page 30 Self Test

1.
1 5 10 10 5 1
1 6 15 20 15 6 1

3. 59 **5.** 2 **7.** commutative addition **9.** $62.50

Page 31 Lesson 1-5B

1. F **3.** T **5.** F

Pages 34–35 Lesson 1-6

7. 13 **9.** 5 **11.** 50 **13a.** $220 + d = 360$ **13b.** 140 people **15.** 39 **17.** 32 **19.** 1 **21.** 6 **23.** 120 **25.** 8 **27.** 9 **29.** 11 **31.** 7 **33.** 7 **35a.** $3973.05 + c = 4003.33$ **35b.** 30.28 **37.** $7x + 1$ **39.** $1943 + n$ **41.** 5

Pages 38–40 Lesson 1-7

5. (3, 2) and (3, 3) **7.** State Capitol **9.** no **11.** M **13.** L **15.** T **17.** J **19.** (6, 4) **21.** (5, 0) **23.** (9, 2) **25.** (10, 8) **27a.** $(a, 0)$ **27b.** $(0, b)$ **27c.** origin

29a.

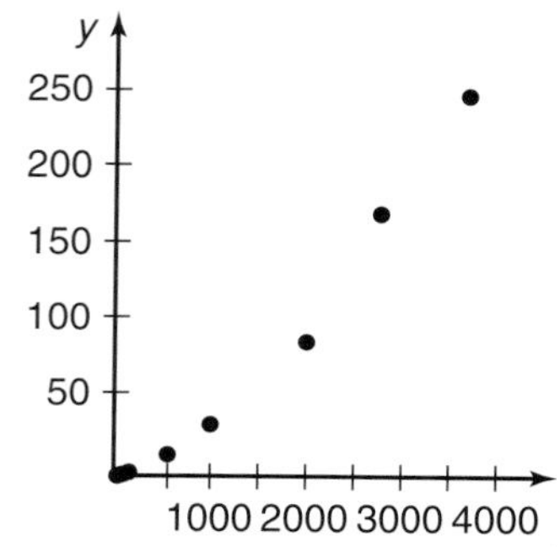

29b. about 78 years **31a.** 50 cars are waiting at the light at 7:00 A.M. and 10 cars are waiting at 9:00 A.M. **31b.** The workday started at most businesses in the area. **31c.** Sample answer: Add point (5:15, 50). About the same number of cars will take people out from work that took them there in the morning. **33.** $10m + 1$ **35.** 12 **37.** 1

Pages 43–45 Lesson 1-8

5. Sample answer: $12 - 9 = 3$ **7.** Sample answer: $4 + 12 = 16$ **9.** 14 **11.** 11.88 **13.** $n + 27 = 31$; $n = 4$ **15.** Sample answer: Evan missed 7 questions on the quiz. If his score was 28, how many points were possible? **17.** $14 = 11 + 3$ **19.** $14 = 28 \div 2$ **21.** $h = 48 \div 6$ **23.** 22 **25.** 88 **27.** 120 **29.** 0.83 **31.** 3.77 **33.** 0.6 **35.** $3x = 18$; 6 **37.** $n - 7 = 22$; 29 **39.** $n \div 4 = 14$; 56 **41.** Sample answer: Jack scored 16 points in this week's game. That is two more than he scored last week. How

many points did Jack score last week? **43.** Sample answer: The product of any number and 0 is 0 according to the multiplicative property of zero. Thus, we could rewrite $5 = a \cdot 0$ as $5 = 0.5 \neq 0$, so division by zero does not make sense.
45. 46 times **47.** 19 **49.** associative property of addition **51.** $5 \cdot t$ **53a.** Sample answer: calculator **53b.** about 317 miles

Pages 48–49 Lesson 1-9

5. false **7.** true **9.** true **11.** $d \geq 27$ **13.** true **15.** open **17.** true **19.** true **21.** open **23.** true **25.** true **27.** true **29.** $t \geq \$100$ **31.** $\ell \geq 3$ **33.** $p < 15$ **35.** $f > 80{,}000$ **37a.** $t \leq 10$ years **37b.** $t \leq 1$ year **39.** 11 **41.** $(8 \cdot 7)c$; $56c$ **43.** 9

Page 50 Lesson 1-10A

1. 9–11 hours

Pages 54–55 Lesson 1-10

7. Miami Metrozoo and Toledo **9.** 13.6 million **11.** 15 **13.** yes, the total of the states is 51 **15.** Most states will issue a driver's license to a 16-year-old, all to an 18-year-old.

17.

Height	Buildings
0–100	0
101–200	0
201–300	0
301–400	4
401–500	8
501–600	4
601–700	1
701–800	3
801–900	0
901–1000	1

19.

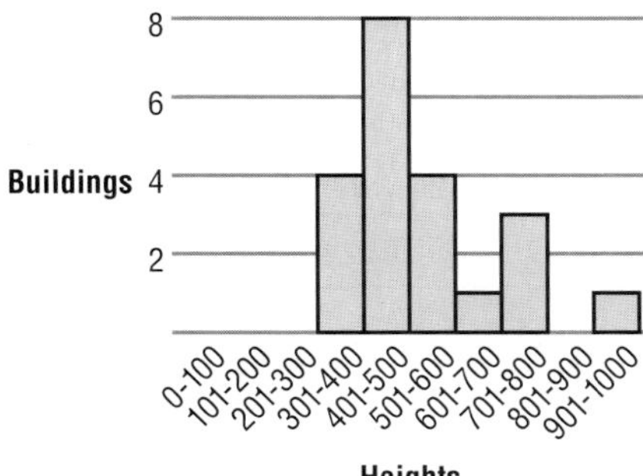

21a. 41 **21b.** 72–74 in. **21c.** You cannot tell from the frequency table how many presidents were exactly six feet tall. The table shows how many presidents fell in different height ranges, not exact heights. **21d.** Leadership is associated with tallness. **23.** $s \geq \$210{,}070$

25.

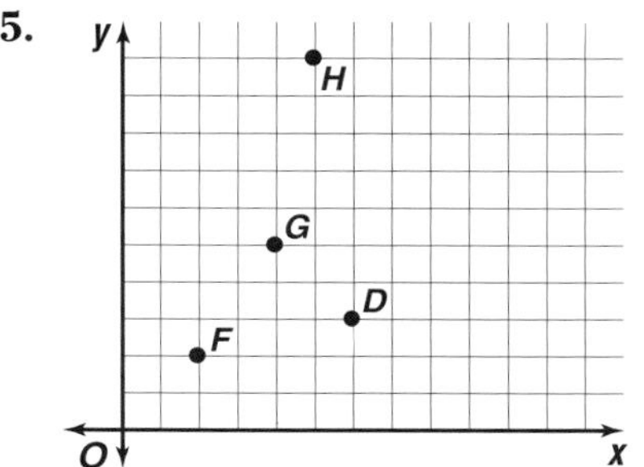

27a. 6 **27b.** 12

Page 57 Chapter 1 Highlights

1. algebraic expression **3.** in simplest form **5.** addition **7.** like terms

Pages 58-61 Study Guide and Assessment

9. 9 **11.** 16 **13.** 6 **15.** 60 **17.** 4 **19.** 10 **21.** 1 **23.** 24 **25.** $10n$ **27.** $b + 10$ **29.** $r - 12$ **31.** commutative, + **33.** associative, × **35.** multiplicative identity **37.** $8n$ **39.** $18x + 8$ **41.** $10x + 26y$ **43.** 6 **45.** 9 **47.** 3 **49.** 5 **51.** N **53.** (5, 6) **55.** (0, 4) **57.** 12 **59.** 28.35 **61.** 108 **63.** false **65.** false **67.** $p < 60$

69.

Representatives	States
1–10	38
11–20	7
21–30	3
31–40	1
41–50	0
51–60	1

71a. Sample answer: Estimation because an exact answer is not needed. **71b.** about 28,000 miles

Chapter 2 Exploring Integers

Pages 69–70 Lesson 2-1

7. −5 **9.** +13 **11.** 5 **13.** 10 **15.** 12 **17a.** −15, −24

17b. −25 −20 −15

19. −5 −4 −3 −2 −1 0 1 2 3 4 5

21. −9 −8 −7 −6 −5 −4 −3 −2 −1 0 1

23. +400 **25.** −3 **27.** +5 **29.** 11 **31.** 0 **33.** 8 **35.** 9 **37.** 12 **39.** −36 **41.** 2 **43.** 6 **45.** 2 **47.** 8, 0, −8 **49.** Sample answer: −1.5

51. −5 −4 −3 −2 −1 0 1 2 3 4 5

53. 17–18

55. 

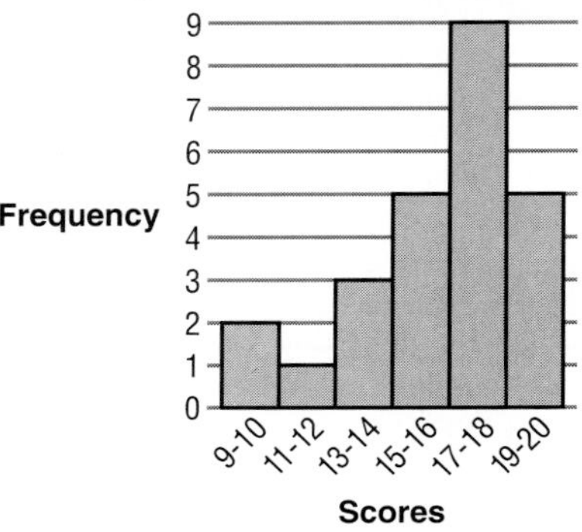

57. 180 square feet **59.** $3x - 8$

Page 71 Lesson 2-1B

1. -1 **3.** Sample answer: There seems to be a cluster between -2 and 1. **5.** Sample answer: Northern states are losing House members while southern states are gaining House members.

Pages 74–76 Lesson 2-2

5. B **7.** D **9.** IV **11.** none **13a.** I **13b.** III **13c.** II **15.** $(4, -2)$ **17.** $(1, -4)$ **19.** $(4, 3)$ **21.** $(2, 2)$ **23.** $(-1, 5)$ **25.** II **27.** III **29.** IV **31.** I **33.** III **35.** none

37.

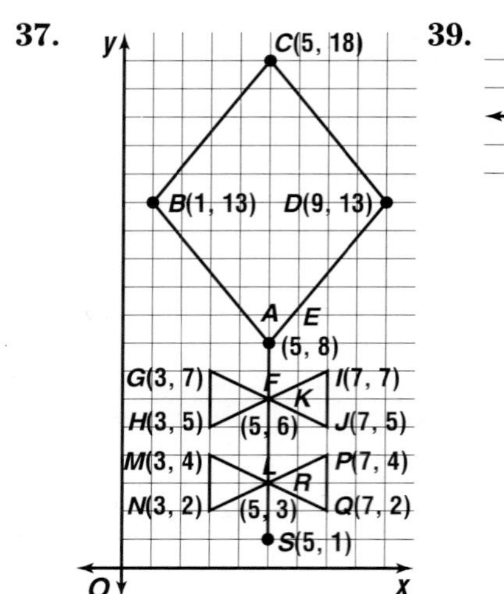

39.

41a. $P(2, 2)$, $Q(2, 5)$, $R(7, 5)$, $S(7, 2)$ **41b.** Rectangle *PQRS* shifted 4 units down **41c.** A rectangle shaped like *PQRS* in quadrant III **43.** 12 **45.** $s < 250{,}000$ **47a.** 22 **47b.** 4 **49.** eleven 2-point; eight 1-point

Pages 77–79 Lesson 2-2B

1.

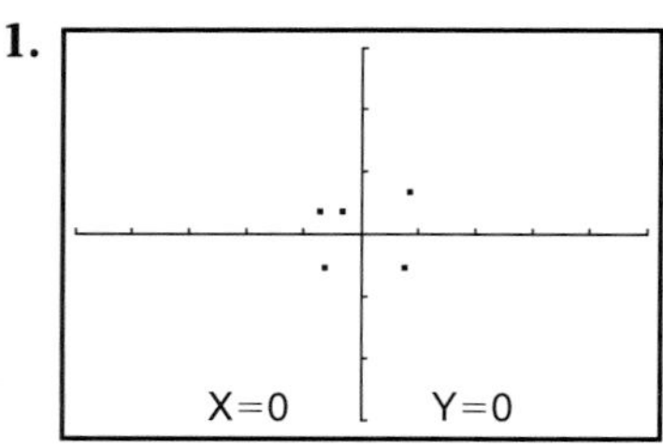

3.

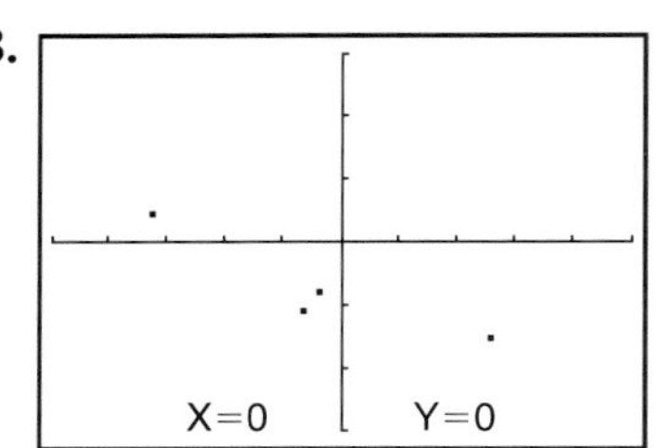

5a. $[-10, 10]$ by $[-10, 10]$ **5b.** $[-47, 47]$ by $[-31, 31]$

Pages 79–81 Lesson 2-3

5. $2 > -5$ **7.** $<$ **9.** $=$ **11.** $\{-88, -9, -4, 0, 3, 43, 234\}$ **13.** $-2 < 3$, $3 > -2$ **15.** $>$ **17.** $=$ **19.** $>$ **21.** $<$ **23.** $<$ **25.** $<$ **27.** $3 > 2$ **29.** $60 < 75$ **31.** $565 > 344$ **33.** Sample answer: $66 < 265$ **35.** $-8 < -3$; $-3 > -8$ **37.** $0 > -8$; $-8 < 0$ **39.** $\{-11, -3, 8\}$ **41.** $\{-65, -6, 1, 29, 56\}$ **43.** $\{-65, -53, 48, 87, 199\}$ **45b.** Lead **45c.** Helium

47–48.

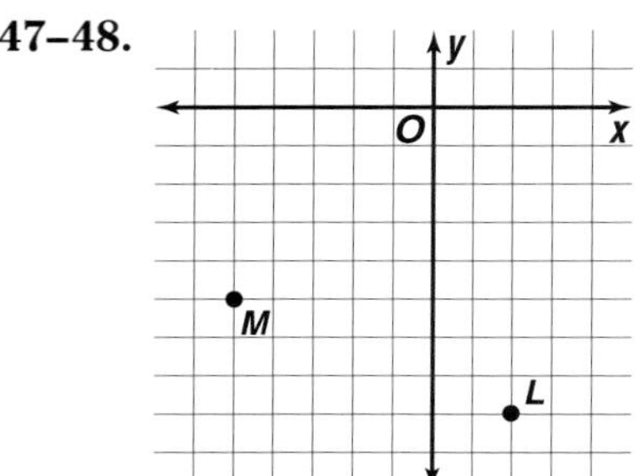

49. -6; 120 **51.** Sample answer: $(6 \times 20) + (6 \times 6)$ **53.** 4 adults

Page 82 Lesson 2-4A

1. 5 **3.** -1 **5.** -3 **7.** -8 **9.** Sample answer: First, place the appropriate number and kind of counters on the mat. Then match all positive counters with all the possible negative counters. These pairs can then be removed because zero does not affect the value of the set on the mat. The remaining counters give you the sum.

Pages 85–87 Lesson 2-4

7. $8 + (-5) = 3$ **9.** $-4 + 9 = 5$ **11.** positive, 9 **13.** negative, -4 **15.** -111 **17.** $-6a$ **19.** 13 **21.** 7 **23.** -4 **25.** 12 **27.** -66 **29.** 26 **31.** -9 **33.** -7 **35.** $-7 + 12$; 5 **39.** 21 **41.** -27 **43.** 63 **45.** $3y$ **47.** $-15m$ **49.** $-20d$ **51.** 7666 **53.** false

55a.

Breed	Change
Akita	−191
Beagle	−390
Chow Chow	+8846
Dachshund	+1473
Labrador Retriever	−4020
Pug	+286

55b. +6004 **57.** false **59.** 3 m/s **61.** $r - 9 = 15$; $24 **63.** $n + (8 + 9)$; $n + 17$ **65.** false

Page 88 Lesson 2-5A
1. 5 **3.** 1 **5.** −15 **7.** 4 **9.** Sample answer: The answers to each set of exercises are the same. The exercises differ in that the second term of the subtraction exercises is the additive inverse of the second term of the addition exercises.

Pages 91–93 Lesson 2-5
7. 8 **9.** $-9x$ **11.** $-7 + (-11) = x$; −18 **13.** $8 + 3 = b$; 11 **15.** $-14 + 19 = y$; 5 **17.** $-27x$ **19.** −12 **21.** −8 **23.** −14 **25.** −3 **27.** 32 **29.** 9 **31.** 37 **33.** −29 **35.** −63 **37.** −46 **39.** −20 **41.** −27 **43.** −5 **45.** 26 **47.** $-10x$ **49.** $15p$ **51.** $24cd$ **53.** $-8a$ **55.** Sample answer: Every integer has an additive inverse because every integer has an opposite. The sum of an integer and its opposite is zero. Zero is its own additive inverse because $0 + 0 = 0$. **57.** −$2133 **59.** $36m$ **61.** $4 < 6$ **63.** triangle **65.** $3y + 23$ **67.** Sandy Koufax **69.** $k + 3$

Page 93 Self Test
1. 8 **3.** −3, 0, 2, 5

4–6.
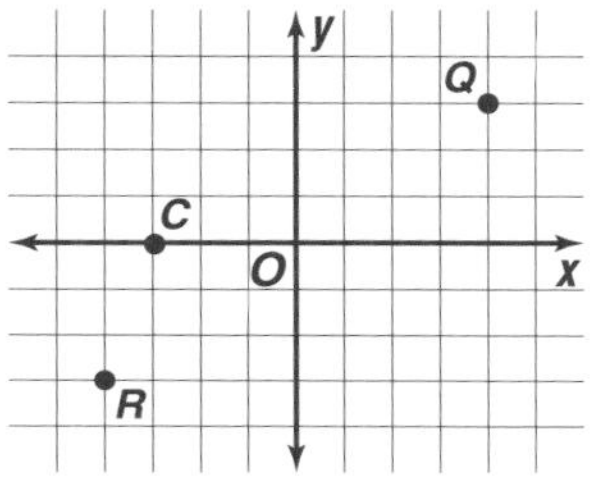

7. > **9.** = **11.** 10 **13.** −15

Pages 95–97 Lesson 2-6
5. 120, 720 **7.** 204 **9.** 123,454,321 **11.** 4 touchdowns **13.** 1:33 P.M.

15a.
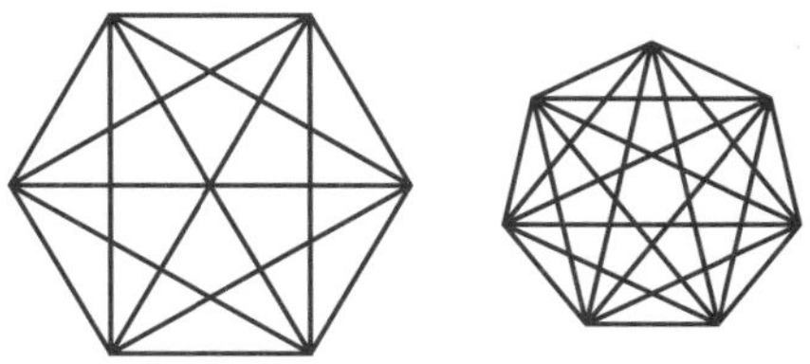

15b. 2, 5, 9, 14; 27 **17.** 36 segments **19.** 55 **21.** 1001 **23.** 6 **25.** $11y + 21$

Page 98 Lesson 2-7A
1. −6 **3.** 6 **5.** −6 **7.** −16 **9.** Sample answer: -2×4 means to remove 2 sets of 4 positive 1-tiles after putting in 8 zero pairs.

Pages 102–103 Lesson 2-7
5. −, −56 **7.** +, 518 **9.** +, 54 **11.** −27 **13.** −320 **15.** −132 **17.** $-40x$ **19.** −56 feet **21.** 28 **23.** −26 **25.** 75 **27.** −585 **29.** 168 **31.** −308 **33.** 56 **35.** −48 **37.** −36 **39.** $-12b$ **41.** $48b$ **43.** $-50rs$ **45.** $-3ab$ **47.** If there is an even number of negative numbers being multiplied, the product will be positive. If there is an odd number of negative numbers being multiplied, the product will be negative. **49.** −36,000 feet or about −6.8 miles **51.** 49°F **53.** Sample answer: 65 > 34 **55.** 39,177 **57.** commutative, +

Pages 107–108 Lesson 2-8
5. −; −7 **7.** −; −6 **9.** −; −15 **11.** 40 **13.** 7 **15.** 13 **17.** −3 **19.** $-60 \div 4 = y$; −15 **21.** −4 **23.** 6 **25.** 13 **27.** −111 **29.** −9 **31.** −12 **33.** −9 **35.** 4 **37.** −19 **39.** −5 **41.** 7 **43.** −9 **45.** −17 **47.** 12 **49.** −7 **51.** −3 **53.** Sample answer: $x = -144$; $y = 12$; $z = -12$ **55.** about $1096 **57.** 10:31, 10:35, 11:21 **59.** 41 **61.** 1014 **63.** 6 **65.** 7 **67.** 27

Page 109 Chapter 2 Highlights
1. F **3.** C **5.** E **7.** A

Pages 110–112 Study Guide and Assessment

9.
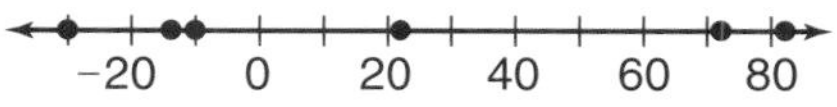

11.
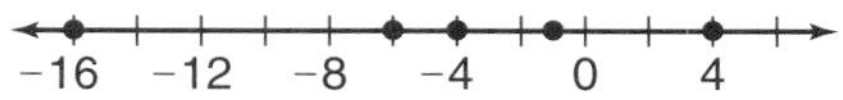

13. 24 **15.** −40 **17.** 29 **19.** 12 **21.** III **23.** II **25.** IV **27.** = **29.** > **31.** < **33.** 17 < 35 or 35 > 17 **35.** −16, −4, 2, 3 **37.** −300, −33, 9, 124, 210 **39.** −3 **41.** −6 **43.** −15 **45.** −10 **47.** −30 **49.** 21 **51.** $-7b$ **53.** $6r$ **55.** −8 **57.** 65 **59.** $36cd$ **61.** $-300yz$ **63.** 7 **65.** 15 **67.** −16 **69.** 19 **71.** −18°F **73.** 7623

Chapter 3 Solving One-Step Equations and Inequalities

Pages 120–122 Lesson 3-1
5. 6087 **7.** 24 **9.** 5 people **11.** 501 **13.** You would buy 4 pairs of $70 shoes at a cost of $280 before the $200 pair wore out. It is less expensive to buy the $200 pair of shoes that lasts longer. But, if the shoes are uncomfortable or go out of style, it may make more sense to buy the less expensive shoes. **15.** blue **17.** 21 **19.** $24k + 3$

Page 123 Lesson 3-2A
1. 4 **3.** 3 **5.** 7 **7.** 12 **9.** 11

Pages 127–128 Lesson 3-2
7. −25
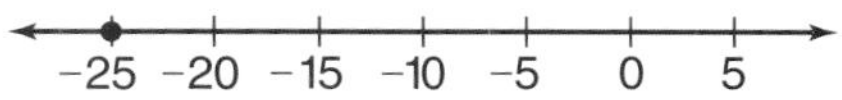

9. 37
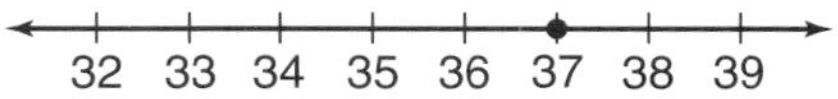

11. −1
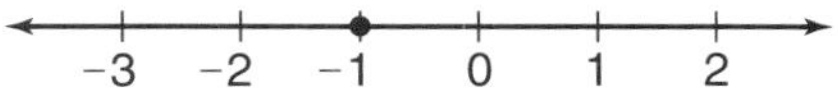

13. $322 = 264 + e$, $58 million

15. 38
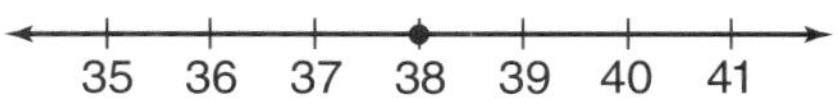

19. −32
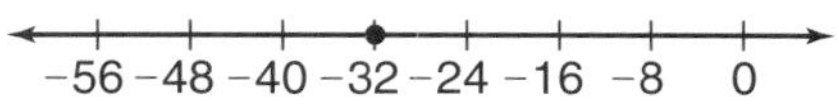

SELECTED ANSWERS

23. -4

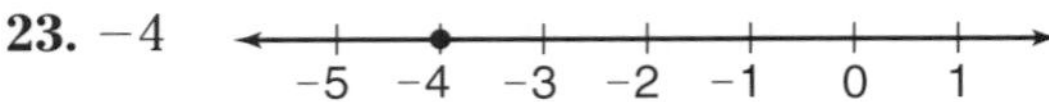

27. 50

31. -1384

33. a **35.** 5 **37.** -8 **39.** 12 **41.** 25 years old **43.** -162 **45.** $-3, -1, 3, 15$ **47a.** v = Calories burned in an hour of volleyball **47b.** $v + 120 = 384$ **47c.** $v = 264$

Pages 131–133 Lesson 3-3

5. 9

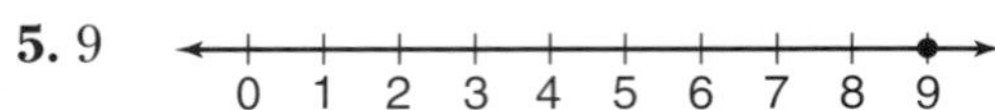

7. -12

9. 17

11. 24 dollars

13. -19

17. 168

21. 8

25. 936

29. 128

33. $-13{,}432$

39a. $6x = 300{,}000{,}000$ **39b.** $15x = 300{,}000{,}000$ **39c.** Most species lay between 20 and 50 million eggs. **41.** 18 cm **43.** d **45.** 17 **47.** 10 **49a.** Estimation; an exact answer is not needed. **49b.** Yes, $7 is enough.

Pages 135–137 Lesson 3-4

5. 45 **7.** 165 miles **9.** 360 **11.** 13 **13.** 3.4 **15.** 14 mph **17.** 10.9 hrs **19.** $s = \ell - d$ **21a.** 30 sq in. **21b.** 6 cm **23a.** 0.8 hours or 48 minutes **23b.** 8.9 mph **25a.** $a = \frac{f - s}{t}$ **25b.** 2 m/s^2 **25c.** -2 m/s^2 **25d.** negative **27.** -27 **29.** $x + 11 = 25$; 14

Page 137 Self Test

1. Stiers–Ratcliffe, Gibson–Smith, Means–Powhatan, Hunt–Willow, Bedard–Pocahontas

3. 6

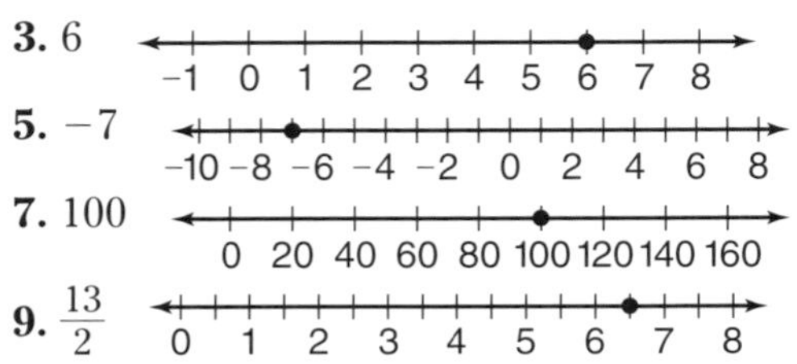

5. -7

7. 100

9. $\frac{13}{2}$

Page 138 Lesson 3-5A

1. 18 units

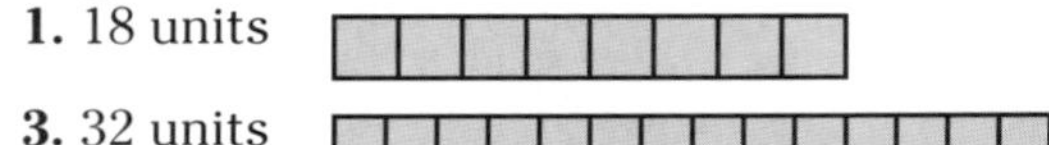

3. 32 units

5. The two figures have the same perimeter because the squares still share the same number of sides.

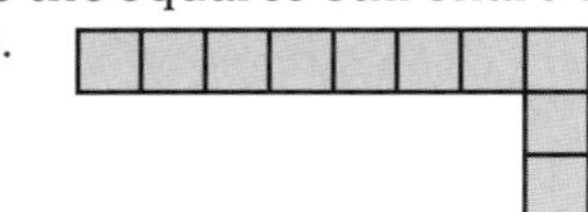

Pages 142–144 Lesson 3-5

5. 14 cm, 12 sq cm **7.** 42 ft; 90 sq ft **9.** 12 m **11.** 70 ft; 264 sq ft **13.** 34 km; 30 sq km **15.** 15 m; 9.86 sq m **17.** 3.6 m; 0.81 sq m **19.** 12.4 cm; 9.61 sq cm **21.** 19.2 mm, 23.04 sq mm **23.** 5 m **25.** 11 yd **27.** 39 ft **29.** 225 sq in.

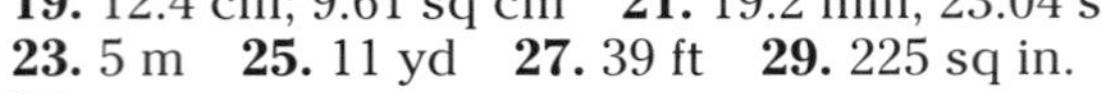

31a.

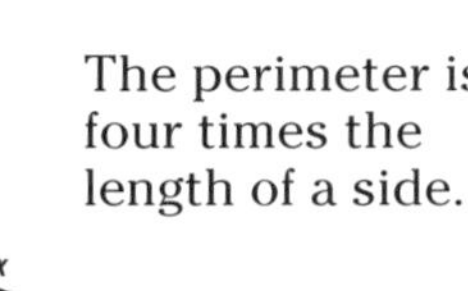

The perimeter is four times the length of a side.

31b.

The area is the square of the length of a side.

31c. perimeter: $8x$ units; area: $4x^2$ square units **33a.** 150 ft **33b.** 900 sq ft

37. 20

39. $375 > 210$ **41.** $20d + 28$

Page 145 Lesson 3-5B

1. 9 sq units **3.** 32 sq units **5.** Both give the same result.

Pages 148–150 Lesson 3-6

7. $y \leq -1$ **9.** $d < 75$

11. $m < -13$

13. $y < -4$

15. $a \leq -31$

17. $x < 6$ **19.** $y > 1$

21. $t \leq 14$

25. $m < 19$

29. $x > 27$

33. $m > -9$

37. $k < -26$

41. $c \geq -25$

45. $x + 285 \geq 375$; $x \geq \$90$ **47.** $x \leq 137$ minutes or 2 hours 17 minutes **49.** 100 **51.** $-2a$
53a. 49 million

53b.

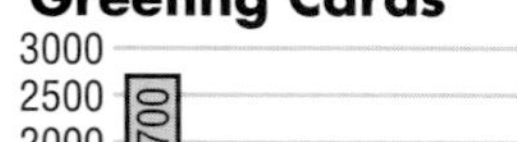
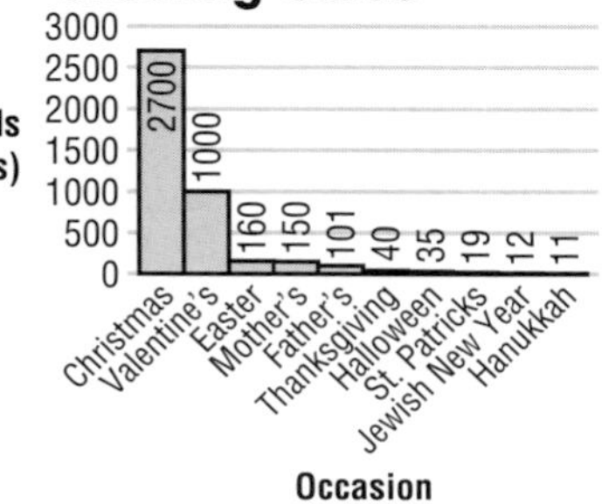

55. Identity, ×

Pages 154–155 Lesson 3-7

5. 5, no **7.** −3, yes

9. $x \leq 2$

11. $p < 105$

15. $y > 9$

19. $p \geq -14$

23. $m \geq 14$

27. $r > -21$

31. $x < -12$

33. 5 **35.** 800 **37.** $t > 6$ years **39.** $m < 28$ **41.** d
43. $-56y$

45a. $[(3 \times 4) + (2 \times 3) + (1 \times 2) + (1 \times 1)] \div 7$
45b. 3.0

Pages 158–159 Lesson 3-8

5. n = number; $4n < 96$; $n < 24$ **7.** Sample answer: The total cost of a car repair was \$187. If the labor cost \$169, how much did the parts cost?
9. n = number; $\frac{n}{-3} > 5$; $n < -15$ **11.** a = account balance; $a + 50 > 400$; $a > \$350$ **13.** p = original purchase price; $4p \leq 85{,}000$; $p \leq 21{,}250$
15. Sample answer: The low temperature was 16° F on Tuesday. This is at least 9° more than the Wednesday low. What was the low temperature on Wednesday?
17. $|x| < 4$; −3, −2, −1, 0, 1, 2, 3
19. a = average; $35a < 600$; $a < 17.1$ points per game **21.** c = changes in price; $8c = 32$; $c = 4$ times the 1971 price/oz. **23.** h = height; $5(8)h = 440$; $h = 11$ centimeters **25.** x = sample scores; $80 - 6 \leq x \leq 80 + 6$; $74 \leq x \leq 86$ strokes
27. $h > 13$;

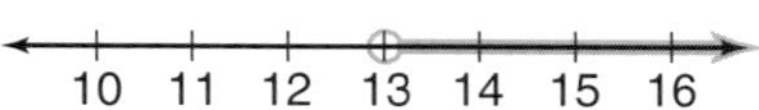

29. $-7y$ **31.** 24

Page 161 Chapter 3 Highlights

1. true **3.** false **5.** true **7.** false

Pages 162–164 Study Guide and Assessment

9. 51

13. 19

17. 126

21. 330

25. 24 **27.** 56.52 **29.** 36 in., 72 sq in. **31.** 40 ft; 75 sq ft

33. $x < -12$

37. $f > -4$

41. $t < 132$

45. $f \geq 15$

47. $3x < 36$; less than \$12 **49.** more than 224 boxes **51.** 103

Chapter 4 Exploring Factors and Fractions

Pages 173–174 Lesson 4-1

7. 2 **9.** 2, 3, 5, 6, 10 **11.** yes **13.** yes **15.** 3 **17.** 2, 5, 10 **19.** 5 **21.** 3 **23.** none **25.** 2 **27.** yes **29.** yes **31.** no **33.** no **35.** 0, 4, or 8 **37.** 1035 **39.** 5555 **41b.** 1996, 2008, 2020, 2032, 2044 **43a.** 6 **43b.** 28 **45.** −267 **47.** −8 **49.** 33

Pages 177–179 Lesson 4-2

7. m^3 **9.** $2 \cdot 2 \cdot 2$ **11.** $10 \cdot 10 \cdot 10 \cdot 10 \cdot 10 \cdot 10$ **13.** 54 **15a.** $6(5^2)$ or 150 in^2 **15b.** 5^3 or 125 in^3 **15c.** The surface area is multiplied by 4 and the volume is multiplied by 8. **17.** n^4 **19.** 14^1 **21.** 3^{15} **23.** b^4 **25.** $12 \cdot 12$ **27.** $(-7)(-7)(-7)$ **29.** $\underbrace{6 \cdot 6 \cdot \ldots \cdot 6 \cdot 6}_{\text{20 factors}}$ **31.** $\underbrace{1 \cdot 1 \cdot \ldots \cdot 1 \cdot 1}_{\text{55 factors}}$ **33.** $(-1)(-1)(-1)(-1)$ **35.** 31 **37.** 162 **39.** −78 **41.** $(1 \times 10^2) + (4 \times 10^1) + (9 \times 10^0)$ **43.** 23,405 **45a.** −1 **45b.** $\frac{1}{2}$ **45d.** $\frac{1}{3}$ **47.** The volume of a cube with sides s units long is s^3. **49a.** $2^0, 2^1, 2^2$ **49b.** 2^3 **49c.** 2^{29} cents **49d.** 536,870,912 **49e.** $2^0 + 2^1 + 2^2 + \ldots + 2^{29}$ **49f.** No, the reward will cost $10,737,418.23 over the thirty-day period.

51. $x > -12$

−14 −12 −10 −8 −6 −4 −2 0 2 4

53. 18°F **55.** $-8p$ **57.** $b + 4$

Page 180 Lesson 4-2B

1. 62 **3.** 2 **5.** 1024 **7.** The values are not the same because the order of operations is different. The value of the first expression is −24 and the value of the second expression is 0.

Pages 182–183 Lesson 4-3

5.

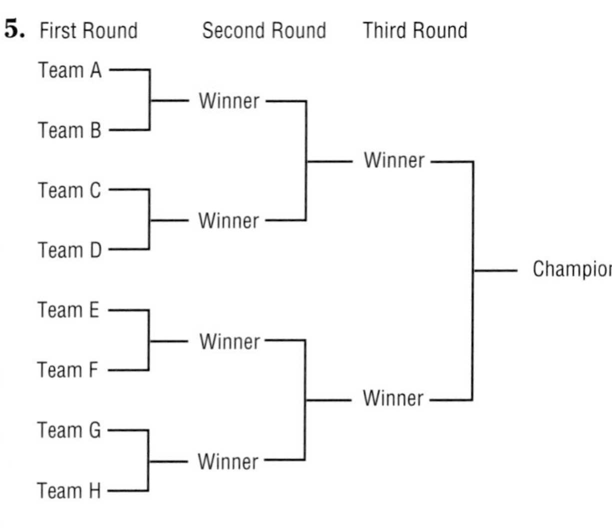

7. $(4 + 3) \times 6 + 3 = 45$ **9.** 1:23 P.M.

11.

4	15	1	14
9	6	12	7
16	3	13	2
5	10	8	11

13. about $243 **15.** 30 feet **17.** $x \leq -9$ **19.** 4

Pages 187–188 Lesson 4-4

5. composite **7.** $2 \cdot 19$ **9.** $2 \cdot 2 \cdot 2 \cdot 7$ **11.** $-1 \cdot 2 \cdot 3 \cdot 7 \cdot a \cdot b \cdot c$ **13.** Damon is correct that 3067 is prime, but it has two factors, 1 and 3067. **15.** prime **17.** prime **19.** $3 \cdot 17$ **21.** $1 \cdot 41$ **23.** $2 \cdot 5 \cdot 11$ **25.** $3 \cdot 3 \cdot 3 \cdot 3$ **27.** $1 \cdot 13$ **29.** $2 \cdot 3 \cdot 7 \cdot x \cdot y \cdot y$ **31.** $3 \cdot 7 \cdot x \cdot y \cdot y \cdot y$ **33.** $2 \cdot 2 \cdot 7 \cdot f \cdot f \cdot g$ **35.** $2 \cdot 3 \cdot 5 \cdot 7 \cdot m \cdot n \cdot n \cdot n$ **37.** $3 \cdot 5 \cdot 5 \cdot m \cdot m \cdot k$ **39.** 211 **41.** 31 and 53 **43a.** 23 **43b.** yes **45.** 16 **47.** 52 inches; 153 square inches **49.** $>$ **51.** $x + 15$

Page 189 Lesson 4-4B

1. Sample answer: 67, 121, 55, 81, 63, 66

Pages 192–194 Lesson 4-5

5. 2 **7.** 1 **9.** b **11.** 12 inches **13.** 8 **15.** 12 **17.** 36 **19.** $8x$ **21.** $14b$ **23.** 9 **25.** 6 **27.** 6 **29.** $6a$ **31.** yes **33.** no **35.** no **37.** 15 **39.** 18 shelves **41.** $2 \cdot 2 \cdot 2 \cdot 5 \cdot 7 \cdot 11$ **43.** 10 **45.** 135 **47.** 1728

Page 194 Self Test

1. 3 **3.** 3 **5.** 2^3 **7.** 121 **9.** $-1 \cdot 2 \cdot 13$ **11.** 14 **13.** 9

Page 195 Lesson 4-6A

1. Shade 2 sections. **3.** They are equal in size. **5.** Yes; sample answer $\frac{1}{4}$ **7.** yes

Pages 198–199 Lesson 4-6

7. $\frac{1}{7}$ **9.** $\frac{11}{15}$ **11.** $\frac{5}{7}$ **13.** $\frac{9}{22}$ **15.** $\frac{1}{11t}$ **17.** simplified **19.** $\frac{1}{7}$ **21.** $\frac{17}{19}$ **23.** $\frac{1}{3}$ **25.** $\frac{5}{8}$ **27.** simplified **29.** simplified **31.** $\frac{62}{111}$ **33.** $\frac{4x}{5y}$ **35.** $\frac{41}{7}$ **37.** $\frac{z^3}{x}$ **39.** $\frac{10p}{13q}$ **41.** simplified **43.** simplified **45.** $\frac{22}{60 \cdot 24}$ or $\frac{11}{720}$ **47a.** 2:6 **47b.** $\frac{1}{3}$ **47c.** Sample answer: 1 oz. alcohol, 3 oz. water **49.** 108 **51.** $68d$ **53.** 12

Pages 202–204 Lesson 4-7

5. 70 **7.** $60a$ **9.** 630 **11.** 10 **13.** 100 **15.** $<$ **17a.** East Central **17b.** West Central **17c.** New England **19.** 18 **21.** 288 **23.** 84 **25.** $84y$ **27.** 40 **29.** 180 **31.** $105t^2$ **33.** 16 **35.** 35 **37.** 220 **39.** $25b$ **41.** $>$ **43.** $<$ **45.** $<$ **47.** $<$ **49.** $\frac{3}{8} < \frac{5}{12}$ **51.** dog, $\frac{7815}{12} > \frac{4723}{15}$ **53a.** video game **53b.** VCR **53c.** on-line services **55.** $x > 576$ **57.** −98 **59.** Utah

61. 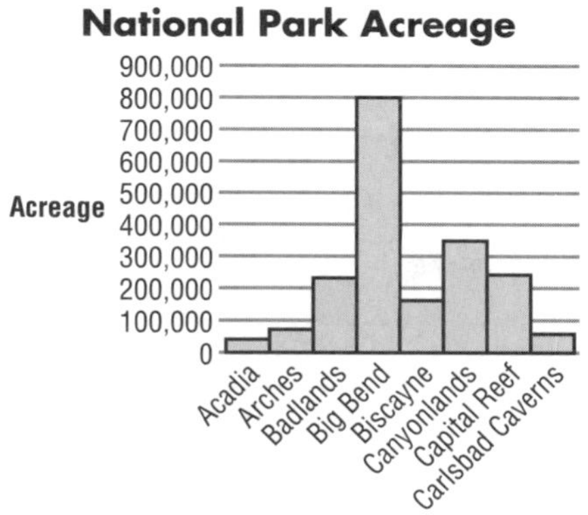

Pages 207–209 Lesson 4-8

5. 10^7 **7.** 3^{10} **9.** 8^6 **11.** y^2 **13.** about 10 times more intense **15.** w^6 **17.** b^7 **19.** 10^7 **21.** a^4 **23.** $20x^7$ **25.** f^{12} **27.** $12a^5$ **29.** a^3b^4 **31.** t **33.** k^2m **35.** w^4y^2 **37.** $3n^8$ **39.** 9 **41.** 3 **43.** 1 **45a.** 100 times **45b.** base **47.** $<$ **49.** 2

51. -7 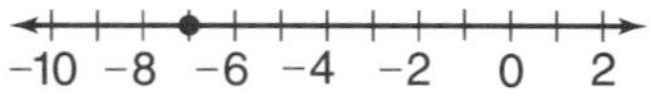

53. II **55.** $2x - 4$

Pages 212–214 Lesson 4-9

5. $\frac{1}{15^5}$ **7.** 10^{-7} **9.** 3^{-2} **11.** $\frac{3}{16}$ **13.** $\frac{1}{4}$ **15.** $\frac{1}{(-2)^5}$ **17.** $\frac{1}{s^2t}$ **19.** $\frac{2}{(xy)^2}$ **21.** 5^{-3} **23.** 2^{-4} or 16^{-1} **25.** $2 \cdot 3^{-2}$ **27.** fg^{-3} **29.** $\frac{1}{27}$ **31.** $\frac{2}{3}$ **33.** a^3 **35.** c^7 **37.** a^{2b-a} **39a.** $\frac{6.67}{10^{11}}\text{Nm}^2/\text{kg}^2$ **39b.** $0.0000000000667\ \text{Nm}^2/\text{kg}^2$ **41.** $-50x^4$ **43.** 7^5 **45.** $2230 + x = 4750$; \$2520 **47.** $n - 8 = 3$; 11

Page 215 Chapter 4 Highlights

1. greatest common factor **3.** factors **5.** multiple **7.** least common denominator **9.** ratio

Pages 216–218 Study Guide and Assessment

11. 5 **13.** 3 **15.** 2, 3, 6 **17.** 2, 3, 6 **19.** $(c + 2)^3$ **21.** 1^3 **23.** $k \cdot k \cdot k \cdot k \cdot k \cdot k \cdot k \cdot k \cdot k$ **25.** $32 \cdot 32 \cdot 32 \cdot 32 \cdot 32 \cdot 32 \cdot 32 \cdot 32 \cdot 32 \cdot 32 \cdot 32$ **27.** $-1 \cdot 2 \cdot 3 \cdot 19$ **29.** $-1 \cdot 3 \cdot 5 \cdot 5 \cdot 11 \cdot x \cdot x \cdot y$ **31.** $-1 \cdot 3 \cdot 3 \cdot 7 \cdot 29 \cdot j \cdot k \cdot k \cdot k$ **33.** $2 \cdot 5 \cdot 5 \cdot 11 \cdot m \cdot n \cdot q \cdot q$ **35.** 13 **37.** $35a^2b^2$ **39.** 5 **41.** 1 **43.** $\frac{19}{20}$ **45.** $\frac{8}{11}$ **47.** $\frac{1}{2}$ **49.** simplified **51.** 60 **53.** $30jk$ **55.** 180 **57.** $36abcd$ **59.** $>$ **61.** $>$ **63.** d^5 **65.** $(-4)^4$ **67.** $2x^{22}$ **69.** $110a^{10}b^3$ **71.** $\frac{4d}{c^2}$ **73.** $\frac{4x}{(yz)^2}$ **75.** $3f^9$ **77.** 21 games **79.** home improvement

Chapter 5 Rationals: Patterns in Addition and Subtraction

Pages 227–228 Lesson 5-1

7. $\frac{1}{20}$ **9.** $\frac{2}{9}$ **11.** rationals **13.** none **15.** $=$ **17a.** $\frac{1}{16}$ **17b.** Rational; it is a fraction. **19.** $-\frac{7}{10}$ **21.** $\frac{57}{100}$ **23.** $\frac{3}{100}$ **25.** $-\frac{1}{3}$ **27.** $\frac{25}{99}$ **29.** $2\frac{34}{99}$ **31.** rationals **33.** whole numbers, integers, rationals **35.** none **37.** rationals **39.** rationals **41.** none **43.** $>$ **45.** $=$ **47.** $>$ **49.** $<$ **51.** $\frac{2}{7}$ **53.** Yes; $\frac{6}{2.4} = \frac{5}{2}$ or $\frac{60}{24}$. **55.** $\frac{1}{40}$ **57a.** $\frac{1}{30}$ **57b.** Yes; it can be written as a fraction. **59.** $60x^2$ **61.** $b < -18$ **63.** -19 **65.** 8 servings

Pages 231–233 Lesson 5-2

5. 6 **7.** 803 **9.** 11 **11.** 1 **13.** $\frac{1}{2}$ **15.** a **17.** 40 **19.** 9 **21.** 29 **23.** 60 **25.** 21 **27.** 1 **29.** $7\frac{1}{2}$ **31.** 1 **33.** 5 **35.** 110 **37.** $28\frac{1}{2}$ **39.** 50 **41.** \$5 **43.** \$3 **45.** No, the change should be about \$7. **47.** Sample answer: $11\frac{1}{10} + 1\frac{7}{15}$ **49.** \$59 **51.** more than half **53.** $\frac{13}{50}$ **55.** -16 **57.** false **59a.** $2(6.25 + 2.25)$ or $2(6.25) + 2(2.25)$ **59b.** \$17

Pages 236–238 Lesson 5-3

5. 3.68 **7.** -4.91 **9.** 77.33 **11.** -6.4 **13.** 43.9 **15.** $-4.1m$ **17.** 14.1 **19.** 1.5 **21.** 63.431 **23.** 2.3 **25.** -26.4 **27.** -51.73 **29.** $12.3m$ **31.** $6.9y$ **33.** $11.5x - 5$ **35.** 29.4 **37.** 12.56 **39.** 12.67 **41.** 24.45 left; no; third side must be less than 20.55 feet **43.** 378.5 **45a.** 24.3 bags per 10,000 passengers **45b.** Sample answer: Bad weather causes changes in flight schedules. **47.** yes; $2 + 4 + 1 + 2 = \$9$ **49.** $2 < a$ **51.** -11 **53.** 12

Pages 241–243 Lesson 5-4

5. $1\frac{2}{7}$ **7.** $\frac{1}{3}$ **9.** 1 **11a.** $\frac{9}{25}$ **11b.** $\frac{1}{5}$ of each dollar **13.** $\frac{1}{5}$ **15.** $\frac{4}{9}$ **17.** $1\frac{1}{3}$ **19.** $1\frac{1}{3}$ **21.** $2\frac{1}{3}$ **23.** 1 **25.** $\frac{1}{2}$ **27.** $\frac{2}{3}$ **29.** $-\frac{1}{6}$ **31.** $-2r$ **33.** $\frac{2}{5}a$ **35.** $\frac{1}{3}b$ **37.** Sample answer: $\frac{2x}{5} - \frac{1}{5}$ **39.** $5\frac{3}{8}$ inches **41.** $3\frac{1}{2}$ yards **43a.** $\frac{25}{170}$ or $\frac{5}{34}$ **43b.** $\frac{1}{28}$ **45.** $c \geq -13$ **47.** $-22a$

Pages 246–247 Lesson 5-5

7. $\frac{1}{8}$ **9.** $-\frac{1}{8}$ **11.** $\frac{11}{14}$ **13.** $\frac{5}{9}$ **15.** $-\frac{1}{12}$ **17.** $\frac{15}{26}$ **19.** $1\frac{5}{6}$ **21.** $3\frac{1}{6}$ **23.** $4\frac{1}{4}$ **25.** $16\frac{10}{21}$ **27.** $10\frac{1}{15}$ **29.** $-\frac{17}{90}$ **31.** $3\frac{5}{24}$ **33.** $-1\frac{3}{8}$ **35.** $1\frac{5}{24}$ **37.** Sample answer: $\frac{1}{18} + \frac{1}{6}$ **39.** $\frac{11}{102}$ **41.** $\frac{t^3}{s^2}$ **43.** -5 **45.** 13 **47.** $18x$

Page 247 Self Test

1. $-\frac{1}{6}$ **3.** $1\frac{4}{9}$ **5.** 1 **7.** 14.52 **9.** 30.6 **11.** $5\frac{1}{7}$ **13.** $8\frac{3}{4}$ inches

Pages 249–250 Lesson 5-6

5. $-\frac{1}{15}$ **7.** -5.28 **9.** 3.25 **11.** -8.7 **13.** 21.9 miles

15. 6.4 **17.** $15\frac{3}{4}$ **19.** 13.4 **21.** $-\frac{31}{36}$ **23.** $1\frac{39}{40}$ **25.** −2.8 **27.** 8.01 **29.** $12\frac{17}{18}$ **31.** $\frac{19}{312}$ **33.** 688.2 million barrels **35.** 0.39 million or 390,000 metric tons **37.** $4\frac{11}{12}$ **39.** $d \geq -84$ **41.** 18 **43.** 15

Pages 253–254 Lesson 5-7

5. $x \leq 2\frac{23}{30}$

9. $y < 1.92$

13. $y < -\frac{1}{6}$

17. $d < 2\frac{5}{6}$

19. $r \leq 1\frac{7}{8}$

23. $h < 15.96$

27. $t > \frac{5}{108}$

29. x is between 0 and 1 or x is less than −1. **31.** $x > 9.7$ **33a.** Sample answer: The average salary of a high school graduate is slightly less than half of the average salary of a college graduate. **33b.** Sample answer: No, if it was that low the inequality would be written differently. The inequality means that it is less than but close to one-half. **35.** $3\frac{3}{4}$ **37.** 2 **39.** $-30abc$ **41a.** Los Angeles **41b.** 20 cents per mile

Pages 256–257 Lesson 5-8

5. inductive **7.** inductive **9.** 18, 21 **11.** $1111^2 = 1{,}234{,}321$ **13a.** ground you; deductive **13b.** No; there could be another reason. **15a.** 1010 **15b.** 110 **15c.** Inductive reasoning does not always work, so it would not be wise to use it to prove something. **17.** pink **19.** $z \leq 0.91$ **21.** 1 **23.** 10

Pages 260–262 Lesson 5-9

5. yes; 17, 20, 23 **7.** no; 2, −4, −11 **9.** no; 23, 30, 38 **11a.** 9.95, 12.90, 15.85, 18.80, 21.75, 24.70 **11b.** $36.50 **13.** yes; 2.5, 2.0, 1.5 **15.** no; 25, 36, 49 **17.** no; 32, 64, 128 **19.** yes; 55, 46, 37 **21.** yes; 3, −1, −5 **23.** yes; 9.55, 10.57, 11.59 **25.** no; 46, 59, 74 **27.** no; $\frac{7}{8}, \frac{8}{9}, \frac{9}{10}$ **29.** 7; 11 **31.** 100, 125, 150, 175, 200, 225 **33a.** 50 s **33b.** $d = \frac{1}{5}t$ **35.** 320 feet/second **37.** deductive **39.** 6

41.

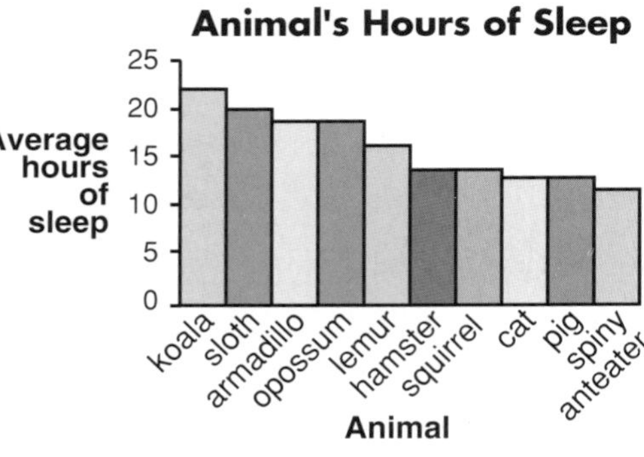

Page 263 Lesson 5-9B

1. 13 **3.** 2, 5, 13, 34, ... The sequence is the 3rd, 5th, 7th, . . . terms of the Fibonacci sequence.

Page 265 Chapter 5 Highlights

1. B **3.** I **5.** C **7.** A

Pages 266–268 Study Guide and Assessment

9. integers, rationals **11.** none **13.** $-9\frac{9}{20}$ **15.** $-\frac{13}{100}$ **17.** $4\frac{8}{33}$ **19.** $8\frac{89}{100}$ **21.** 8 **23.** 10 **25.** 1 **27.** $2\frac{1}{2}$ **29.** 11.23 **31.** −7.68 **33.** −17.95 **35.** $13\frac{7}{9}$ **37.** $\frac{1}{4}$ **39.** $-\frac{3}{5}$ **41.** $-\frac{1}{15}$ **43.** $6\frac{11}{12}$ **45.** $5\frac{1}{36}$ **47.** −8.8 **49.** $1\frac{13}{16}$ **51.** $1\frac{5}{6}$

53. $f \leq 4\frac{1}{8}$

55. $q > 5.3$

57. $m > -6.151$

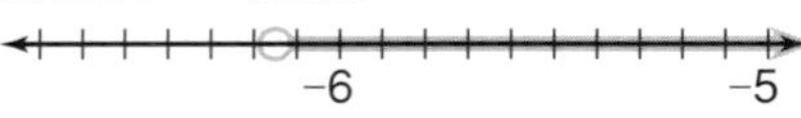

59. $a \leq \frac{9}{10}$

61. yes; 13, 5, −3 **63.** no; $\frac{4}{5}, \frac{2}{3}, \frac{0}{1}$ **65a.** $\frac{35}{1200}$ or $\frac{7}{240}$ **65b.** Rational, it is a fraction. **67.** $6\frac{5}{8}$ inches **69a.** 25 units **69b.** The lengths are the squares of 1, 2, 3, ... It is not an arithmetic sequence because there is no common difference.

Chapter 6 Rationals: Patterns in Multiplication and Division

Pages 277–279 Lesson 6-1

7. $2.\overline{6}$ **9.** 0.125 **11.** $-\frac{3}{4}$ **13.** $-\frac{3}{4}$ **15.** 0.7 **17.** 0.5 **19.** 0.7 **21.** 0.3125 **23.** $-3.\overline{36}$ **25.** −4.3125 **27.** > **29.** < **31.** < **33.** < **35.** > **37.** < **39.** .7777777 **41.** 0.7777778 **43.** $0.\overline{3}$, 0.25, 0.7, 0.8, $0.41\overline{6}$, $0.\overline{571428}$, 0.12 **43a.** terminating: $\frac{1}{4}, \frac{7}{10}, \frac{4}{5}, \frac{3}{25}$;

repeating $\frac{1}{3}, \frac{5}{12}, \frac{4}{7}$ **43b.** $3 = 1 \times 3$; $4 = 2^2$; $10 = 2 \times 5$; $5 = 1 \times 5$; $12 = 2^2 \times 3$; $7 = 1 \times 7$; $25 = 5^2$ **43c.** 2 and 5 **45.** 0.33 **47.** \$82.38; \$46.88; \$36.75 **49.** 64, 58, 52 **51.** b^{11} **53.** 20 feet **55.** $r > 7$ or $r \geq 8$ **57.** 27

Pages 282–283 Lesson 6-2

5. c **7.** d **9.** $18 \div 3 = 6$ **11.** $8 \div 1 = 8$ **13.** $20 \times \frac{1}{4} = 5$ **15.** 160 **17.** 4 **19.** 5 **21.** 2 **23.** 6 **25.** 20 **27.** 150 **29.** 5 **31.** 10 **33.** If the numbers were rounded up the actual answer is less than the estimate. If the numbers were rounded down the product is greater than the estimate. **35.** about 15,000,000 **37.** 5 mph **39.** $\frac{17}{14}$ or $1\frac{3}{14}$ **41.** 27 **43.** 2

Pages 286–288 Lesson 6-3

5. d **7.** $\frac{1}{2}$ **9.** $\frac{4}{9}$ **11.** -24 **13.** $\frac{3}{8}$ **15.** $\frac{9}{25}$ **17a.** 3.8¢ **17b.** 28 years, 1.5 months **19.** $\frac{1}{7}$ **21.** $-1\frac{1}{2}$ **23.** $-4\frac{1}{2}$ **25.** -3 **27.** -24 **29.** $\frac{9}{25}$ **31.** $6\frac{12}{25}$ **33.** $\frac{4}{15}$ **35.** $1\frac{1}{18}$ **37.** $\frac{4}{5}$ **39.** $\frac{11}{36}$ **41.** $\frac{7}{2} \times \frac{6}{3}$ or $\frac{6}{2} \times \frac{7}{3}$; use the greatest factors as numerators and the least as denominators. **43.** $3\frac{3}{4}$ hours **45a.** 6 picas **45b.** $4\frac{1}{2}$ picas **45c.** about 91 characters **47.** 15 **49.** $2^3 \cdot 3^2$ **51.** $-20xy$ **53.** 15

Pages 291–293 Lesson 6-4

7. no **9.** yes **11.** -1 **13.** $-3\frac{3}{4}$ **15.** $\frac{20}{57}$ **17.** $\frac{11}{18}$ **19.** $-\frac{3}{7}$ **21.** $\frac{5}{8}$ **23.** $\frac{2}{3}$ **25.** $\frac{y}{x}$ **27.** $-1\frac{2}{25}$ **29.** -10 **31.** $-1\frac{1}{2}$ **33.** 27 **35.** $34\frac{2}{7}$ **37.** $-1\frac{1}{2}$ **39.** $6\frac{2}{3}$ **41.** $-2\frac{5}{8}$ **43.** $-2\frac{2}{7}$ **45.** $12\frac{2}{3}$ **47.** 21 uniforms plus extra fabric remaining **49.** 8 days **51.** $\frac{2}{5}$ **53.** 6 **55.** II **57.** $13h + 23$

Page 294 Lesson 6-5A

1. 24 squares; 24 hundredths or 0.24 **3.** $0.4 \times 0.2 = 0.08$, each square represents one one-hundredth **5.** 5

Pages 297–299 Lesson 6-5

5. 1.44 **7.** 0.7 **9.** 0.62 **11.** -90 **13.** 25.12 **15.** Inverse, multiplication **17.** 70 pounds **19.** 1.4147 **21.** -5.5875 **23.** -6.592 **25.** 0.32 **27.** -17 **29.** 1460 **31.** 5.6 **33.** -15 **35.** 1.21 **37.** -1.16 **39.** Identity, multiplication **41.** Associative, multiplication **43.** Commutative, multiplication **45.** -2.5; -2.6 **47.** -10; -10.2 **49.** 0.1; 0.2 **51.** Muesli, muesli 18.1¢ cents per oz; frosted corn flakes 18.4¢ per oz. **53.** \$3.2 billion **55.** $2\frac{7}{12}$ **57.** integers, rationals **59.** \$314 **61.** Commutative, addition

Page 299 Self Test

1. 0.875 3. $-0.\overline{72}$ 5. 16 7. 3 9. -24

Pages 304–306 Lesson 6-6

5. 29.4, 33, 12 **7.** 100.4, 105, 78 **9.** 49.9, 45, 45 **11.** 6.0, 6.5, 3.6 and 7.2 **13.** 104.4, 105, 113 **15.** 0.92, 0.85, none **17.** 85.4 **19.** 86 **21a.** 86 **21b.** no effect **23a.** 90 **23b.** no effect **25.** 91 **27.** Sample answer: 2, 4, 6, 8, 8, 10, 12, 14 **31.** -12.9824 **33.** $y \leq -24$ **35.** Multiplication property of zero

Page 307 Lesson 6-6B

1. 5.55, 5.75 **3.** -16.36, -15 **5.** 73.6, 49.7 **7.** The mean is affected more by very large or very small numbers. So the median is a better representation if the data has extreme values. If the data are more closely clustered, the mean is a good representation.

Pages 310–311 Lesson 6-7

5. -15

−20 −10 0 10

7. $-9 < f$

−10 −8 −6 −4 −2 0 2

9. $\frac{18}{25}$

0 $\frac{1}{5}$ $\frac{2}{5}$ $\frac{3}{5}$ $\frac{4}{5}$ 1

11. 6 cm **13.** -0.5 **15.** $16\frac{1}{2}$ **17.** $x < -14.4$ **19.** $-6\frac{1}{4}$ **21.** -3.55 **23.** $z \leq -1.3$ **25.** $\frac{-7}{10}$ **27.** 19 **29.** $x < \frac{1}{5}$

31. $y < \frac{15}{32}$

0 $\frac{1}{4}$ $\frac{1}{2}$ $\frac{3}{4}$ 1

35. \$16.66 **37.** 88.8 **39.** $1\frac{3}{14}$ **41.** -36

Pages 315–316 Lesson 6-8

7. no **9.** yes, $\frac{1}{3}$; $1, \frac{1}{3}, \frac{1}{9}$ **11.** yes, 2; 8, 16, 32 **13.** 2048 **15.** yes, $\frac{1}{2}$; $\frac{3}{2}, \frac{3}{4}, \frac{3}{8}$ **17.** yes, $\frac{1}{12}$; $\frac{1}{144}, \frac{1}{1728}, \frac{1}{20736}$ **19.** yes, $\frac{1}{2}$; $\frac{1}{32}, \frac{1}{64}, \frac{1}{128}$ **21.** yes, -2; 112, -224, 448 **23.** no **25.** no **27.** yes, $\frac{1}{4}$; $\frac{3}{32}, \frac{3}{128}, \frac{3}{512}$ **29.** 2, -4, 8, -16, 32 **31.** 80, 100, 125 **33.** 1 **35a.** 1.3 **35b.** 37,129,300 **37.** 40 **39.** 27 **41.** $\frac{1}{3^4}$ or $\frac{1}{81}$ **43.** 5 **45.** 34

Pages 319–320 Lesson 6-9

5. yes **7.** 8.490×10^3 **9.** 8.479×10^2 **11.** 61,000 **13.** 31,557,600 seconds in a year; (31,557,600)(3.00×10^5) or about 9.47×10^{12} km **15.** 5.9×10^{-3} **17.** 1.5×10^{-4} **19.** 4.98×10^{-8} **21.** 7.01×10^8 **23.** 0.0000056 **25.** 0.009001 **27.** 0.00005985 **29.** 1454 **31.** 0.00005 **33.** 700,080,003 **35.** 3.14 **37a.** Bezymianny **37b.** Ngauruhoe **37c.** Bezymianny; Santa Maria; Agung; Mount St. Helens tied with Hekla, 1947;

Hekla, 1970; Ngauruhoe **39.** $x \geq 2\frac{1}{3}$ **41.** -13 **43.** $b + 3$

Page 321 Highlights
1. true **3.** false; Each term in a geometric sequence increases or decreases by a common ratio. **5.** false **7.** true **9.** false; It is possible to have a set of numerical data with no mode.

Pages 322–324 Study Guide and Assessment
11. 0.625 **13.** 0.425 **15.** $0.2\overline{6}$ **17.** 21 **19.** 9 **21.** $-\frac{8}{45}$ **23.** $-1\frac{1}{5}$ **25.** $\frac{64}{125}$ **27.** $\frac{3}{4}$ **29.** $-\frac{1}{9}$ **31.** 28.28 **33.** 0.0629 **35.** $0.291\overline{6}$ **37.** -47.7 **39.** 20.6; 20; 18, 21, and 25 **41.** 14.17; 14; 14 **43.** 9 **45.** -12 **47.** $-4 \geq n$ **49.** 5.3 **51.** no **53.** no **55.** no **57.** 6.74×10^{-6} **59.** 5.81×10^3 **61.** 57,200 **63.** 202,000,000 **65.** 132¢ or \$1.32 **67.** 15.2 times

Chapter 7 Solving Equations and Inequalities

Pages 331–332 Lesson 7-1
5. 3 ounces **7.** \$16,200 **9.** \$5 **11.** 729 people **13.** 29.8.92 and 2.9.92 **15.** 0.0000006789 **17.** whole numbers, integers, rationals **19.** 31

Page 333 Lesson 7-2A
1. -3 **3.** 4 **5.** 2 **7.** Sample answer: You undo operations to find the value of x.

Pages 336–337 Lesson 7-2
5. 13 **7.** -4 **9.** -3 **11.** 6 ounces **13.** 4 **15.** 17 **17.** -1.2 **19.** 84 **21.** 33 **23.** 51 **25.** -153 **27.** 22.5 **29.** -93 **31.** 16 **33.** Sample answer: $6x + 3 = 2$ **35.** 4, 5 **37.** 90 **39.** $>$ **41.** y^3 **43.** false

Pages 339–340 Lesson 7-3
3. $17 - 2x = 5$; 6 **5a.** $2682 = 4d - 354$ **5b.** 759 thousand **7.** C **9.** E **11.** $\frac{n+6}{7} = 5$; 29 **13.** $\frac{c}{-4} - 8 = -42$; 136 **15.** $5x - 10 = 145$; 31 **17.** 7 yd, 9 yd, 11 yd **19a.** $2w + 50 = 130$ **19b.** 40 feet **21.** 50 **23.** 72 **25.** associative, addition

Pages 343–344 Lesson 7-4
7. 11.31 cm **9.** 10.99 m **11.** 12.57 cm **13.** 34.56 ft **15.** 43.96 mm **17.** 59.032 m **19.** 8.164 yd **21.** $7\frac{1}{3}$ ft **23.** C **25.** B **27.** 0.65 m **29.** 2 units **31.** about 238.76 m **33a.** The perimeter of the square is longer. **33b.** Sample answer: The perimeter of a square is 4 times the length of one side. The circumference of a circle is approximately 3.14 times the diameter. $4 > 3.14$ **35.** 3.7 million **37.** $2 \cdot 2 \cdot 5 \cdot 11 \cdot p \cdot q \cdot q$ **39.** $>$

Page 345 Lesson 7-5A
1. -8 **3.** 4 **5.** 1 **7.** Sample answer: No; the result is the same.

Pages 348–350 Lesson 7-5
5. Sample answer: Add 1 to each side. Subtract $3x$ from each side. The solution is 3. **7.** Sample answer: Simplify the left side to $3a + 66$. Subtract $3a$ from each side. The solution is 4. **9.** Sample answer: Add n to each side. Subtract 4 from each side. The solution is 3. **11.** 21 **13.** 2 **15.** -0.5 **17.** 5 **19.** all numbers **21.** $\varnothing$ **23.** 2 **25.** all numbers **27.** 2.48 **29.** 100 m by 130 m **31.** 25 ft by 30 ft **33.** $3y - 14 = y$; 7 **35a.** $199{,}000 - 700x$ **35b.** $165{,}000 + 3700x$ **35c.** $199{,}000 - 700x = 165{,}000 + 3700x$ **35d.** about 8 years **37a.** \$20 **39.** 32 **41.** $\frac{2}{7}$ **43.** 52

Page 350 Self Test
1. \$569.19 **3.** -12 **5.** $4x - 12 = 18$; 7.5 **7.** 1.02 yd **9.** 65 yards by 120 yards

Pages 353–354 Lesson 7-6
5. $x \leq 7$
0 5 10

7. $k > 2$
−4 −3 −2 −1 0 1 2 3 4

9. $m < -2$
−4 −3 −2 −1 0 1 2 3 4

11. Sample answer: If Heather can find CDs that cost no more than \$12.50 each, she can buy the shoes and two CDs.

13. $u \geq -4$
−6 −5 −4 −3 −2 −1 0 1 2 3

17. $j \geq -3$
−5 −4 −3 −2 −1 0 1 2 3 4

21. $1.17 > t$
1.10 1.12 1.14 1.16 1.18 1.20

25. $x \leq 4.5$
4.0 4.2 4.4 4.6 4.8 5.0

29. $c < -4$
−8 −6 −4 −2 0

31. $-5 < x < 1$ **33.** $s \geq 97$ **35.** 258 **37.** 8 **39.** $n + 3$

Pages 356–357 Lesson 7-7
5. $2n - 9 > 11$; $n > 10$ **7.** $55c + 35 \leq 200$; $c \leq 3$ **9.** $4x + 4 \geq 16$; $x \geq 3$ **11.** Sample answer: A stove and a freezer weigh at least 260 kg. The stove weighs 115 kg. What is the weight of the freezer? **13.** $3(h - 1) \geq 18$; $h \geq 7$ **15.** 17, 18 **17.** $t \geq \frac{11}{3}$ **19.** 3, 4, or 5 **21.** $50 > c > 6$ **23.** $x > -5$ **25a.** $\frac{1.7}{10^{24}}$ **25b.** 0.0000000000000000000000017 **27.** $-126b$

Pages 360–361 Lesson 7-8
5. mL **7.** kg **9.** 0.040 **11.** 12.8 km **13.** cm or m **15.** L **17.** 0.4 **19.** 9.4 **21.** 6 **23.** 8000 **25.** 1000

27. 316 **29.** 946 mL **31.** 603,200 cm **33.** 13 cm **35.** 65 cm, 55 cm **37.** 0.035 m **41.** $b > 7$ **43.** 17.25 **45.** $-20\frac{4}{5}$

Page 363 Highlights
1. false **3.** true **5.** true **7.** true

Pages 364–366 Study Guide and Assessment
9. 12 **11.** 60 **13.** -20 **15.** 0.2 **17.** 13 **19.** \$250 **21.** 22.0 in. **23.** 30.8 m **25.** -2 **27.** -1 **29.** 14

31. $n > 6$

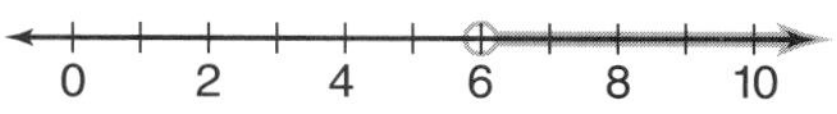

33. $r \leq 98$

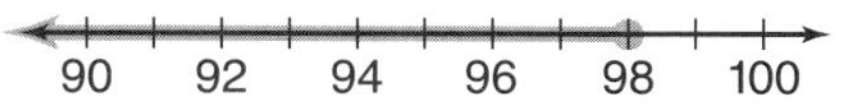

35. $c < 7.5$

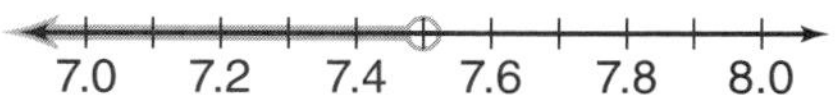

37. $8n + 2 \geq 18$; $n \geq 2$ **39.** $x + (x + 1) > 47$; $x > 23$ **41.** 0.006 **43.** 43,000 **45.** 300,000 **47.** 8800 **49.** 140.8 beats per minute **51.** less than 16.67 minutes

Chapter 8 Functions and Graphing

Pages 374–377 Lesson 8-1
5. $\{-1.3, 4, -2.4\}$; $\{1, -3.9, 3.6\}$ **7.** $\{(-1, 3), (0, 6), (4, -1), (7, 2)\}$; D = $\{-1, 0, 4, 7\}$, R = $\{3, 6, -1, 2\}$ **9.** $\{(-2, 1), (0, -1), (2, 2)\}$; D = $\{-2, 0, 2\}$, R = $\{-1, 1, 2\}$ **11.** yes **13.** no **15.** $\{(8, 4), (9, 3), (10, 2), (11, 1), (12, 0)\}$ **17.** $\{-1, 4, 2, 1\}$; $\{6, 2, 36\}$ **19.** $\{1.4, -2, 4, 6\}$; $\{3, 9.6, 4, -2.7\}$
21. $\left\{-\frac{1}{2}, 4\frac{2}{3}, -12\frac{3}{8}\right\}$; $\left\{\frac{1}{3}, -17, 66\right\}$ **23.** $\{(-4, -2), (-2, 1), (0, 2), (1, -3), (3, 1)\}$; D = $\{-4, -2, 0, 1, 3\}$, R = $\{-2, 1, 2, -3\}$ **25.** $\{(5, 4), (2, 8), (-7, 9), (2, 12), (5, 14)\}$; D = $\{5, 2, -7\}$, R = $\{4, 8, 9, 12, 14\}$ **27.** $\{(-3, -2), (-2, -1), (0, 0), (1, 1)\}$; D = $\{-3, -2, 0, 1\}$, R = $\{-2, -1, 0, 1\}$ **29.** yes **31.** no **33.** yes **35.** no **37.** no **39.** no **41.** yes **43a.** \$6.95 **43b.** more than \$30 and less than or equal to \$70 **43c.** Yes, each x-value (total price) has exactly one y-value (shipping cost).
45. $-2 > b$ **47.** $z \geq -1\frac{3}{5}$ **49.** 16 miles
51. Sample answer: two times the sum of a number and three

Page 378 Lesson 8-2A
1. Sample answer: Generally, as shoe length increases, armspan increases also. **3.** Sample answer: Because points appear to be scattered around a general area.

Pages 381–384 Lesson 8-2
5. negative **7.** positive **9.** positive **11.** none **13.** negative **15.** positive **17.** no **19.** negative **21.** no **23a.** The data show a positive relationship. **23b.** Sample answer: about 62 wpm **23c.** Sample answer: between 3 and 4 weeks **23d.** The more experience a student has, the more words per minute he or she can key.

25a.

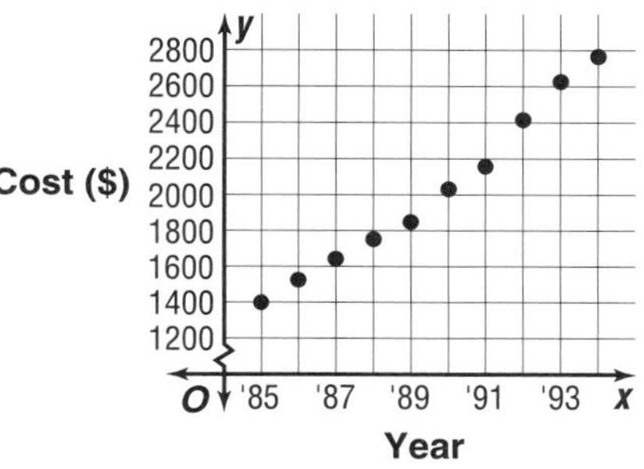

25b. Yes; positive.
27a.

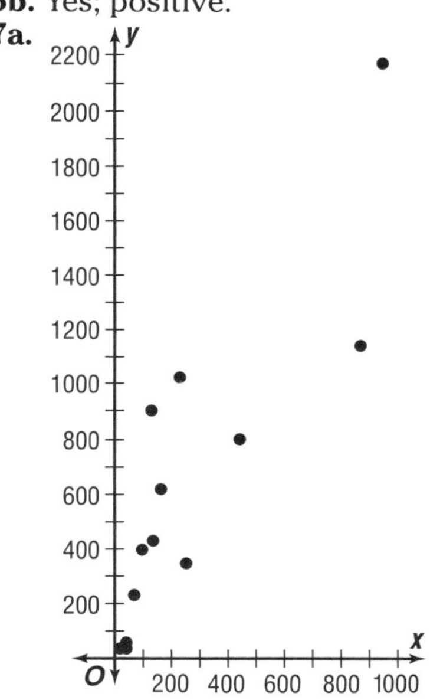

27b. yes; positive **27c.** Sample answer: yes; you could estimate by using the points in the scatter plot. **27d.** Sample answer: Yes; some players have more opportunities to get rebounds because of the position they play for the team. **27e.** Sample answer: Yes, since more playing time provides a player with more opportunity to score points.
29. $\{8, 4, 6, 5\}$; $\{1, 2, -4, -3, 0\}$ **31.** 127.7; 129; 130 **33.** 2075

Pages 387–390 Lesson 8-3
7. c, d **9.** $(-2, -5)$, $(0, -2)$, $(2, 1)$, $(4, 4)$ **11.** Sample answer: $(0, 2.8)$, $(-2, -7.2)$, $(2, 12.8)$, $(-1, -2.2)$. **13.** no

15.

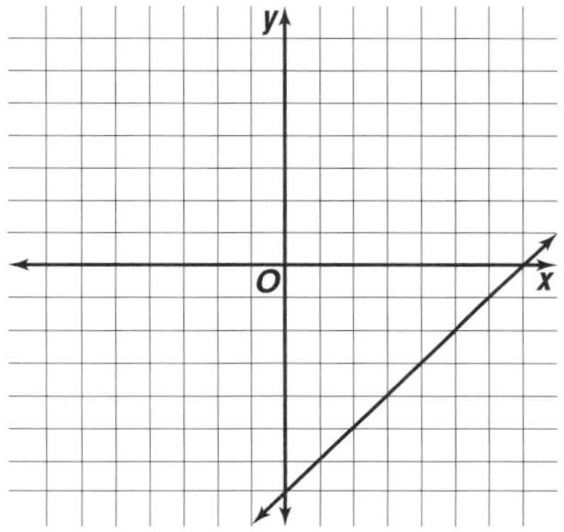

17.

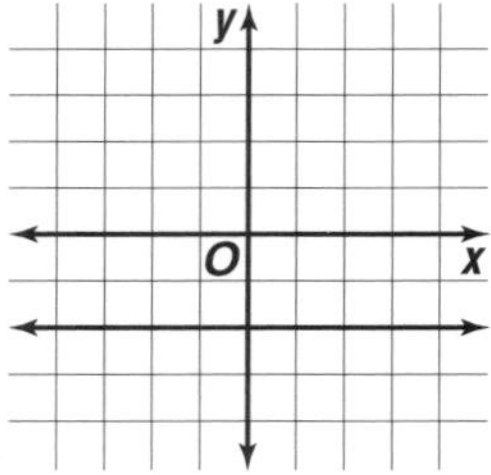

19a. $y = 0.5x$

19b.

x	$y = 0.5x$	y	(x, y)
0	$y = 0.5(0)$	0	(0, 0)
1	$y = 0.5(1)$	0.5	(1, 0.5)
2	$y = 0.5(2)$	1	(2, 1)
3	$y = 0.5(3)$	1.5	(3, 1.5)

19c. Sample answer: There is no such thing as a negative number of seedlings.

19d.

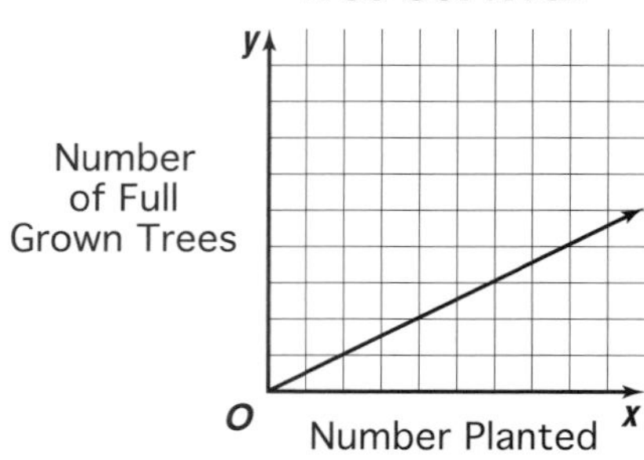

21. a, d **23.–37.** Sample answers given.
23. {(0 ,0), (1, 2.5), (2, 5), (3, 7.5)}
25. {(0, 7), (1, 10), (2, 13), (3, 16)}
27. {(0, −4), (1, −5), (2, −6), (3, −7)}
29. {(0, 1), (1, 0), (2, −1), (3, −2)}
31. {(1, 2), (3, 4), (5, 6), (6, 7)}
33. {(0, 3), (3, 0.5), (4, 1), (6, 0)}
35. {(1, −1), (2, −2), (3, −3), (4, −4)}
37. {(−2, 0), (−2, 1), (−2, 2), (−2, 3)} **39.** no

41.

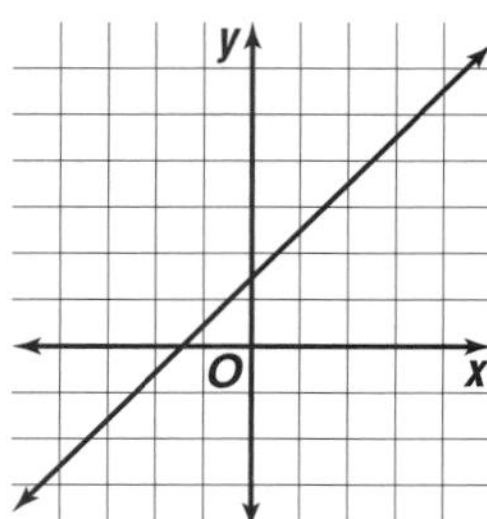

45.

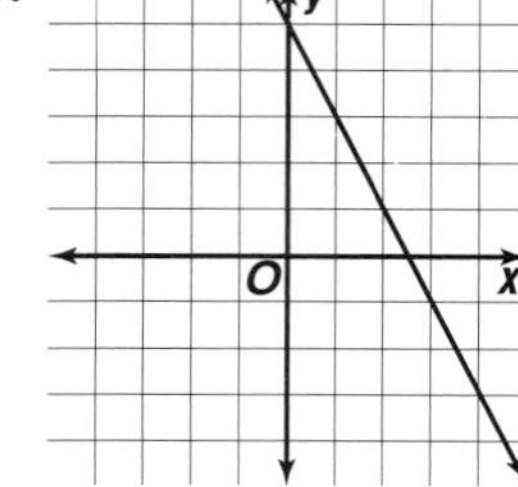

49.

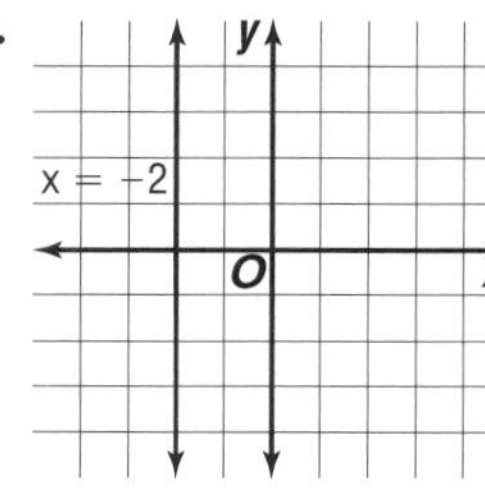

51. $x = \frac{1}{4}y$; {(0, 0), (1, 4), (2, 8), (3, 12)};

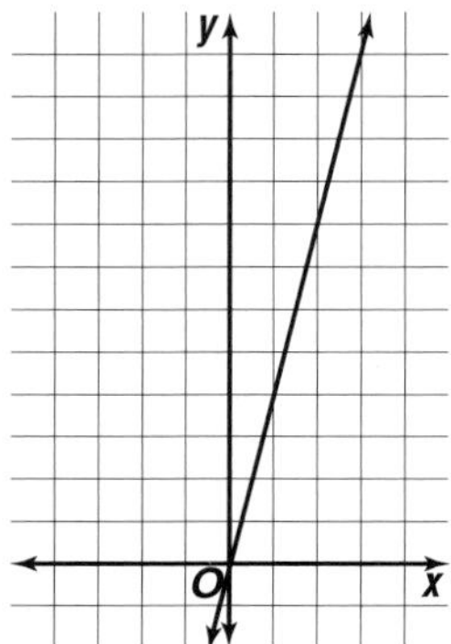

53. 2 **55.** $y = -3x$ or $3x + y = 0$ **57a.** Sample answer: (1, 4), (2, 8), (4, 16), (6, 24), (8, 32)

57b.

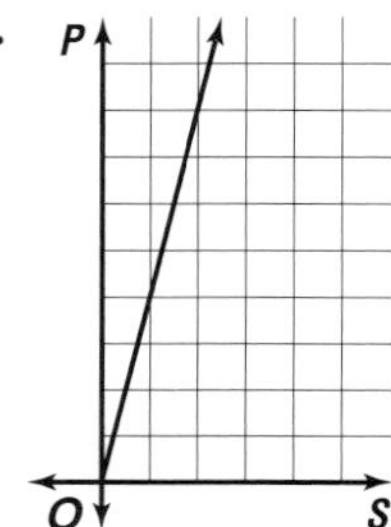

57c. Sample answer: Length cannot have a negative value. **59a.** (0, 32), (100, 212)

59b.

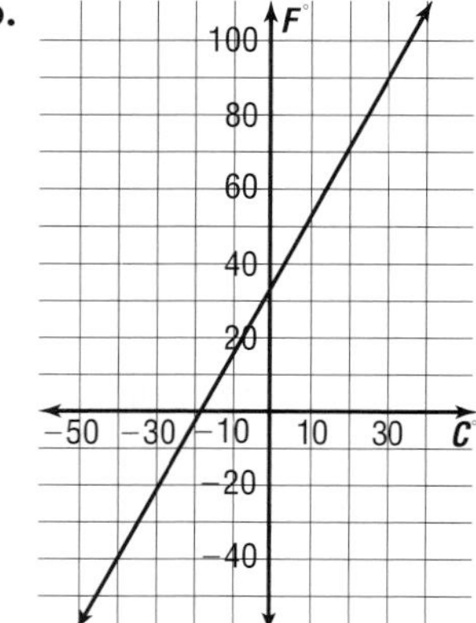

59c. Find the coordinates of other points on the graph. **59d.** Sample answer: (20, 68) **61.** $a > 11$
63. 4 **65.** 7^2

67.

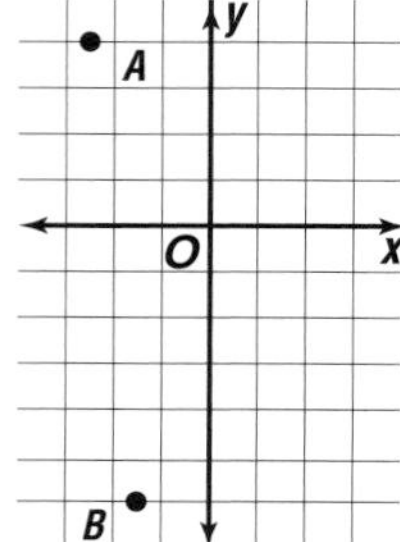

69. $14n$

Page 391 Lesson 8-3B

1. Each distance is the same.
3a. {7, 2, −1, −2, −1, 2, 7}
3b. {(−3, 7), (−2, 2), (−1, −1), (0, −2), (1, −1), (2, 2), (3, 7)};

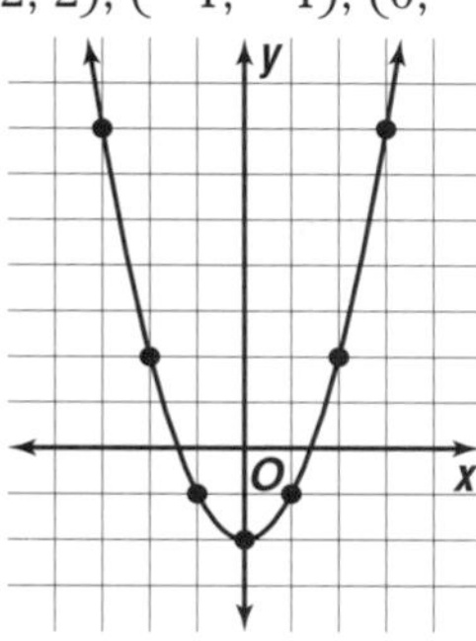

3c. Sample answer: They both open upward and have their vertex on the *y*-axis. The graphs pass through different points on the *y*-axis. The graph of $y = x^2 - 2$ is narrower.

Pages 394–395 Lesson 8-4

5. 10, 1, 2, 17; yes **7.** 3 **9.** 6 **11.** 9 weeks **13.** 8, 16, 28, 36; yes **15.** 2, 2, 2, 2; yes **17.** 5, 4, 2.5, 1.5; yes **19.** 7, 11, 2, −14; yes **21.** −47 **23.** $-1\frac{3}{4}$ **25.** 35 **27.** 4.25 **29.** $4b^2 - 2$ **31.** 118
33a. $\left\{-17\frac{7}{9}, 0, 22\frac{2}{9}, 37\right\}$
33b.

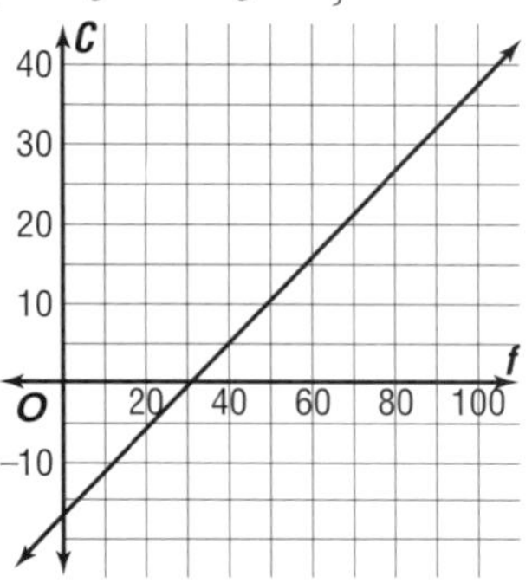

33c. Yes; No member of the domain is paired with more than one member of the range.
35a. Sample answer:

x	y = 18x + 66.5	y	(x, y)
1	y = 18(1) + 66.5	84.5	(1, 84.5)
5	y = 18(5) + 66.5	156.5	(5, 156.5)
10	y = 18(10) + 66.5	246.5	(10, 246.5)
15	y = 18(15) + 66.5	336.5	(15, 336.5)
20	y = 18(20) + 66.5	426.5	(20, 426.5)

35b.

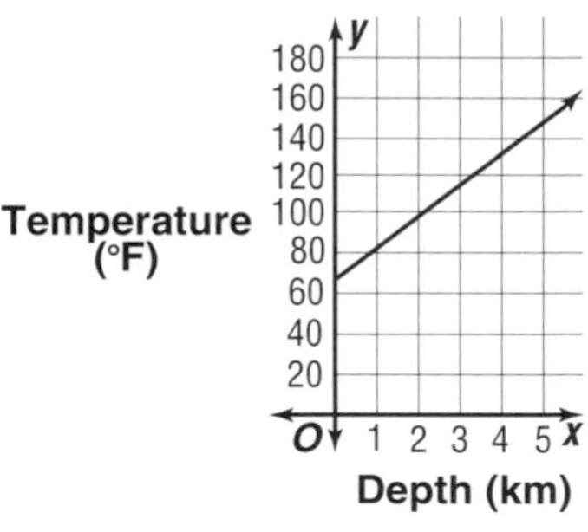

35c. 131.3°F **37.** $\frac{9}{16}$
39. $a \leq -6$;

Pages 398–399 Lesson 8-5

3a. about 78.2 m^3 **3b.** about $12.76 **5.** about 4500 lb/in^2 **7.** $1.11 a gallon **9.** about 150 cubic meters **11.** 156 pounds **13.** 4 **15.** $-\frac{5}{12}$
17. negative

Page 399 Self Test

1. D = {4, 1, 3, 6}, R = {2, 3, 4}; yes **3.** positive
5. Sample answer: (2, 6), (1, 2), (0, −2), (−1, −6)

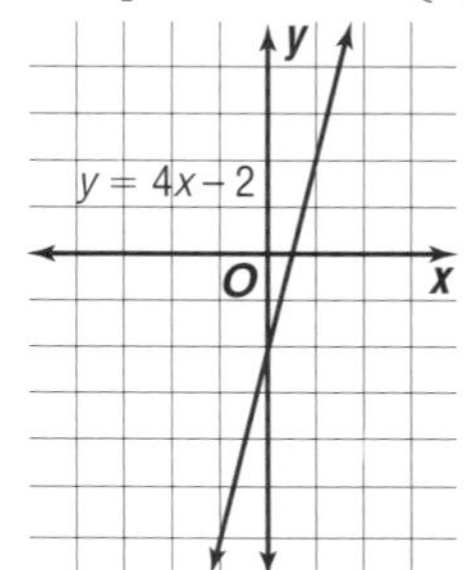

7. 50 **9.** −31

Pages 402–404 Lesson 8-6

7. 1 **9.** $\frac{1}{2}$ **11.** −1 **13.** 4 **15.** −1 **17.** 0 **19.** $-\frac{1}{5}$ **21.** −2 **23.** Sample answer: Both graphs slant up to the right. **25.** $-\frac{3}{2}$ **27.** $\frac{7}{4}$ **29.** 1 **31.** $\frac{3}{8}$ **33.** $-\frac{3}{4}$ **35.** $\frac{2}{3}$ **37.** 11,080 feet **39.** 89 minutes or 1 hour 29 minutes **41.** composite **43.** 12

Page 405 Lesson 8-6B

3. The point should be on the line.

Pages 409–410 Lesson 8-7

9. 1.5, 3 **11.** −6, 6

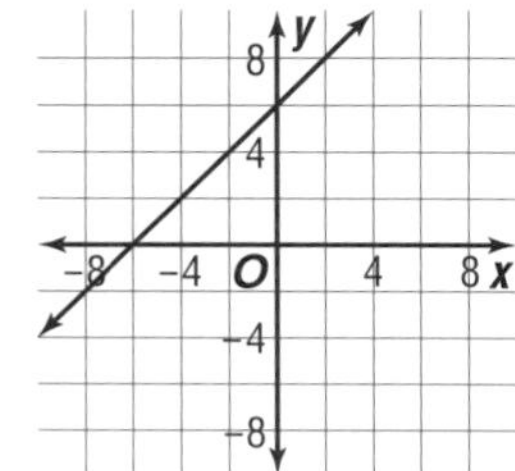

13. $\frac{2}{3}$, 6

SELECTED ANSWERS

15.

17. 5, −5 19. −2, −1 21. 2, −4

23.

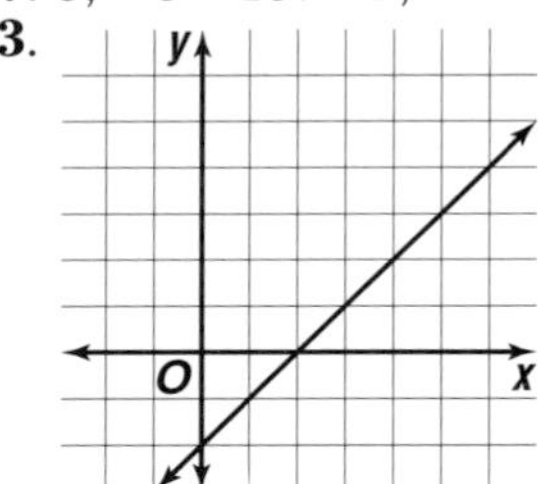

27.

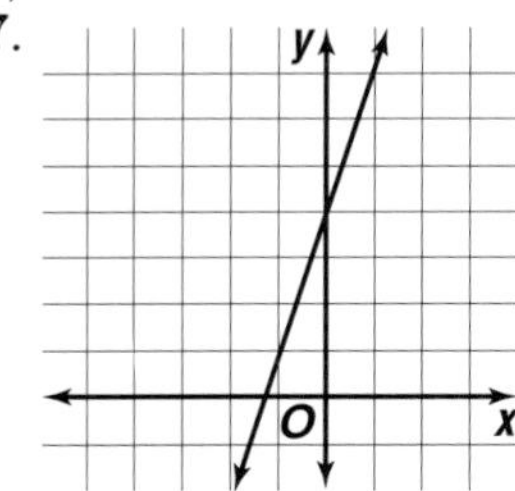

31.

35.

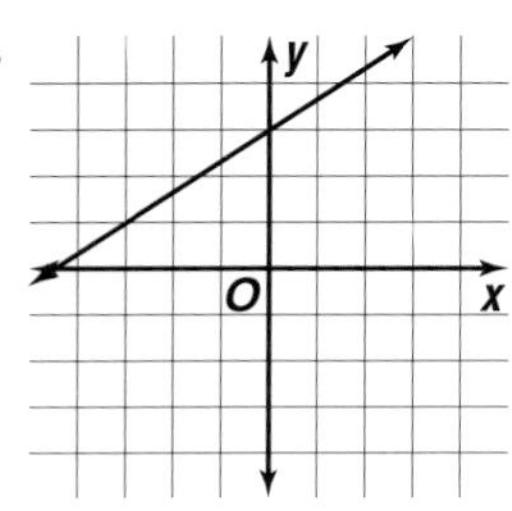

39.

43. The x-intercept and y-intercept are both zero. The line, therefore, passes through the origin. Since two points are needed to graph a line, $y = 2x$ cannot be graphed using only the intercepts.

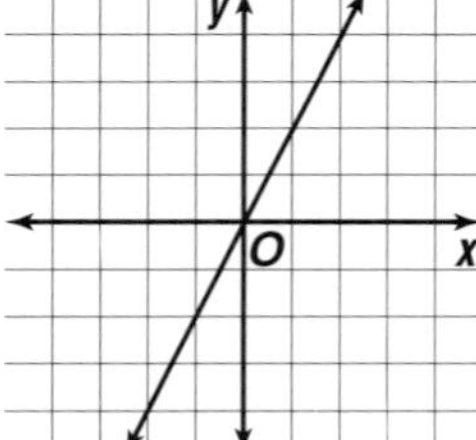

45a.

Altitude (1000 ft)

45 40 35 30 25 20 15 10 5

O 1 2 3 4 5 6 7 8 9

Time (minutes)

45b. 16 **45c.** The x-intercept is the time when the plane lands. **47.** Gettysburg **49.** 13 **51.** $-5a$ **53.** associative, addition

Page 411 Lesson 8-7B

1. The family of graphs is lines having a slope of 1.5. All are parallel and have different x- and y-intercepts. **3.** Sample answer: $y = -5x$

Pages 414–416 Lesson 8-8

7. (1, 1) **9.** (2, −6) **11.** (2, 1) **13a.** $2\ell + 2w = 12$; $\ell = 2w$ **13b.** (4, 2) **13c.** The length of the frame will be 4 feet and the width will be 2 feet. **15.** (0, 0) **17.** (2, 1) **19.** no solution **21.** no solution **23.** (5, 7) **25.** (−1, 3) **27.** (3, 9) **29.** (−2, 8) **31.** (5, 7) **33.** infinitely many **35.** $A = 3, B = 4$ **37a.** (300, 1500) **37b.** The break-even point would be at point (300, 1500) because that is the solution when the cost equals the income. When 300 items are sold, the income is \$1500. **39a.** $y = 15x$, $y = 5x + 60$ **39b.** (6, 90) **39c.** After 6 seconds, the dog will catch up to the prowler at 90 meters away from where the dog started. **41.** 3, 4, or 5 **43.** \$78.51 **45.** 14

Page 417 Lesson 8-9A

1. Both graphs are shaded regions and the boundary is the same line. **3.** The graph of $y = 2x + 5$ would not be shaded at all.

Pages 420–422 Lesson 8-9

7. yes **9.** b, c **11.** The region above the boundary. **13.** The region to the right of the boundary.

15.

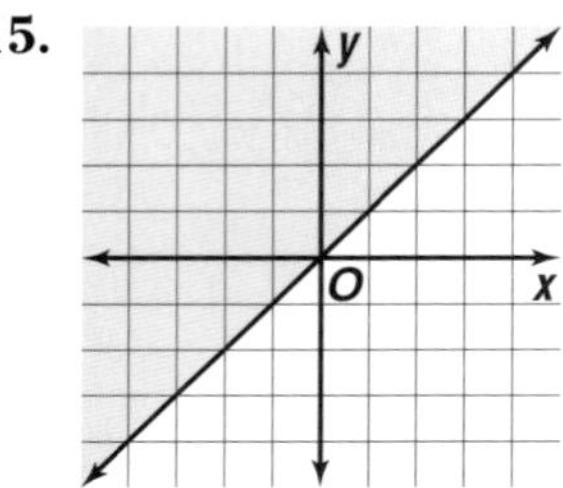

17.

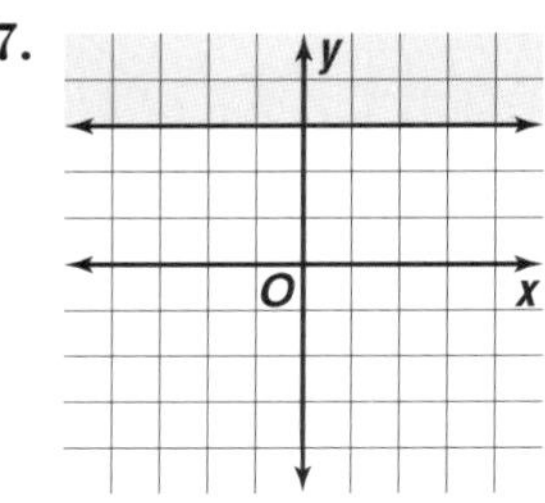

19. a **21.** c **23.** a, b, c **25.** The region to the left of the boundary **27.** The region to the left of the boundary.

29.

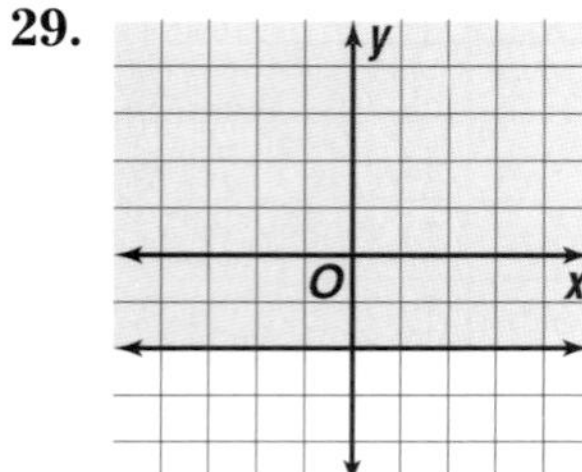

33.

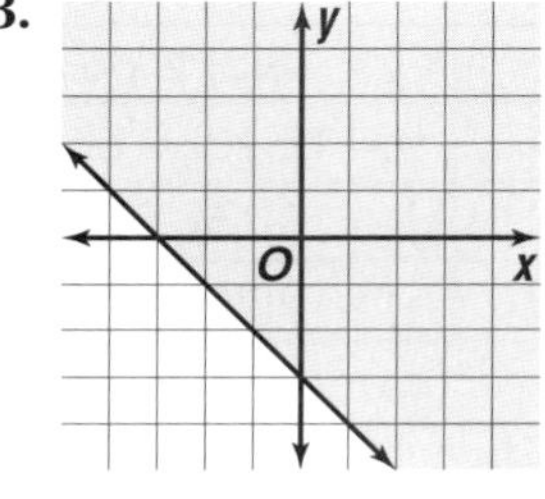

37.

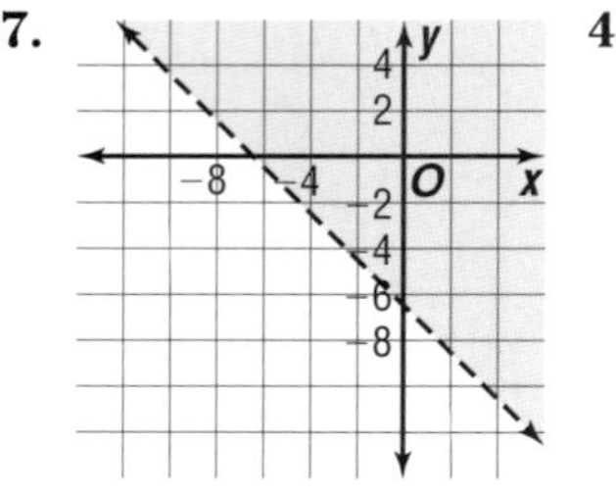

41.

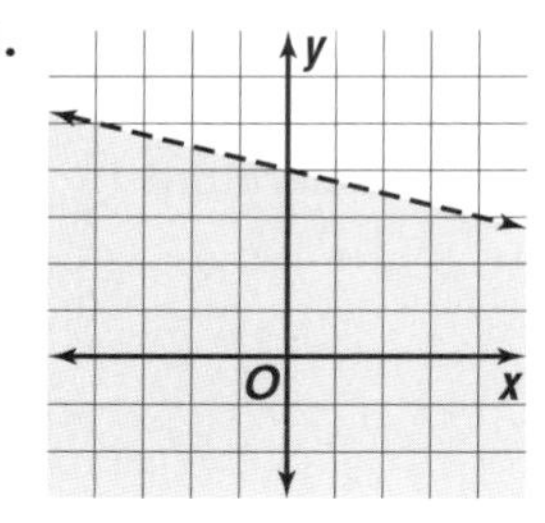

45a. $O < S - 3000$

45b.

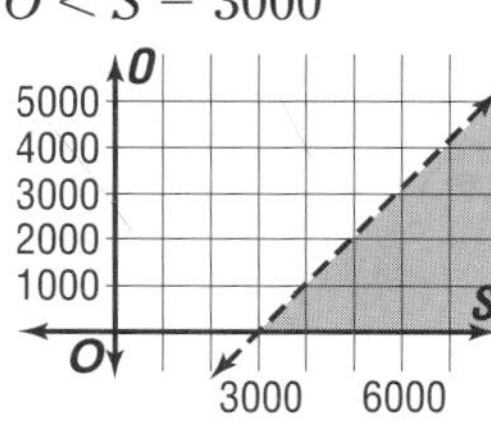

45c. below the line 45d. No; this would mean that your cost was greater than your sales, which doesn't happen in a successful business.
45e. Sample answer: (6600, 3600), (7000, 4000)

47a.

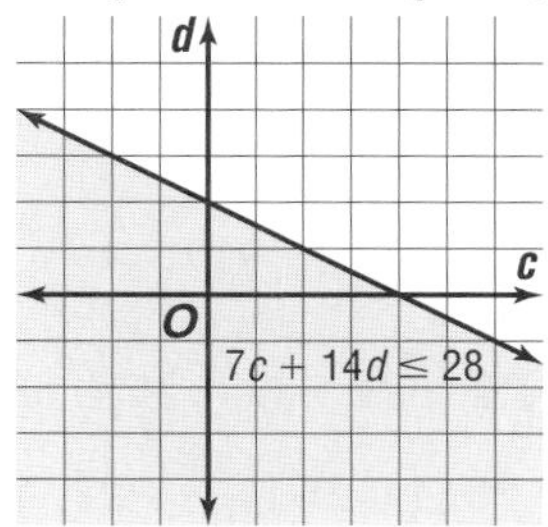

47b. Sample answer: (1, 1) means purchasing 1 cassette and 1 CD for a total cost of $21. (2, 1) means purchasing 2 cassettes and 1 CD for a total of $28. 49. $(-1, -1)$ 51. $\frac{3}{5}$, 6.02×10^{-1}, 0.63
53. 2 55. -12

Page 423 Highlights
1. function 3. y-intercept 5. relation
7. x-intercept

Pages 424–426 Study Guide and Assessment
9. {0, 6, 7, 9}; {17, 18, 19, 40}; yes 11. {6, 8, 15}; {2}; yes 13. $\left\{-2, -\frac{1}{2}, 4, 5\right\}$; {6, 3, 1}; yes
15. no 17. negative 19. no

21.

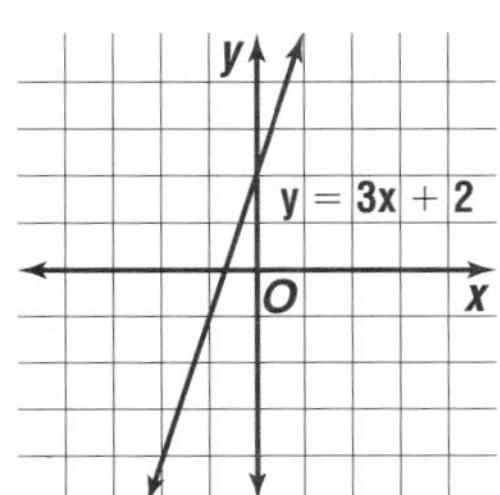

(0, 2), (1, 5), (−1, −1), (2, 8)

25.

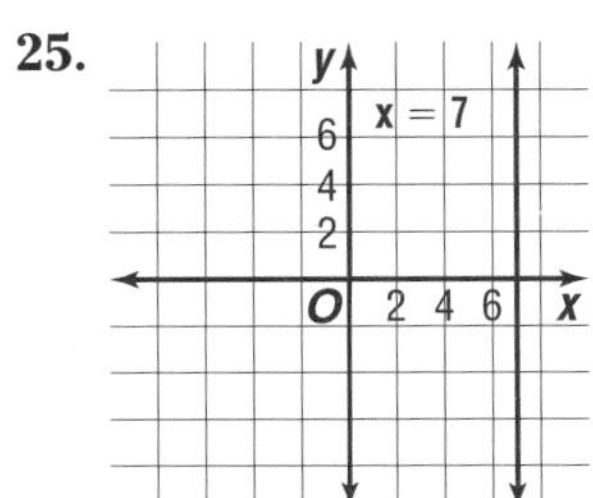

(7, 0), (7, 2), (7, 4), (7, 6)
29. 5 31. 4 33. $\frac{2}{5}$ 35. -1 37. 0

39.

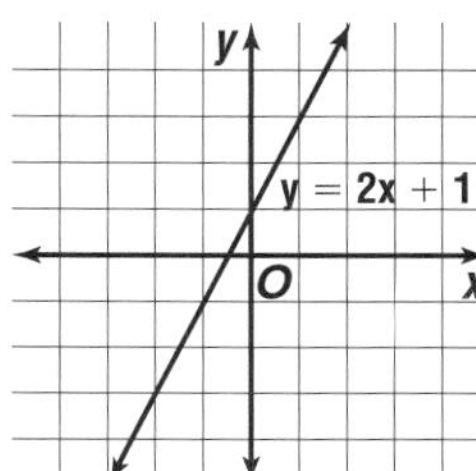

43.

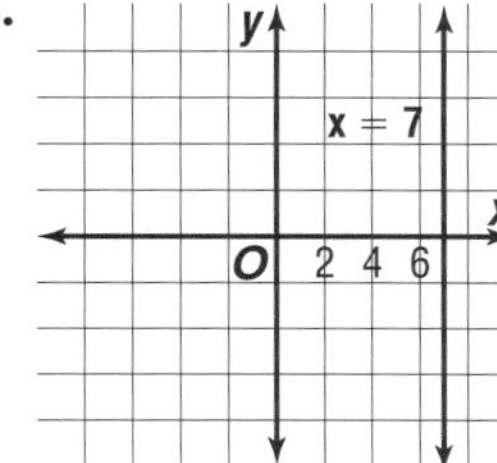

45.

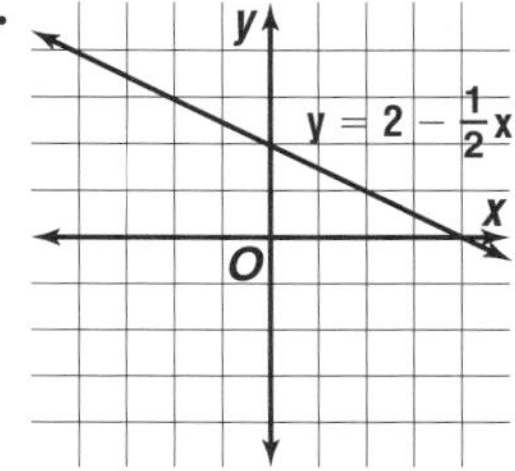

47. (0, 8) 49. empty set 51. empty set

53.

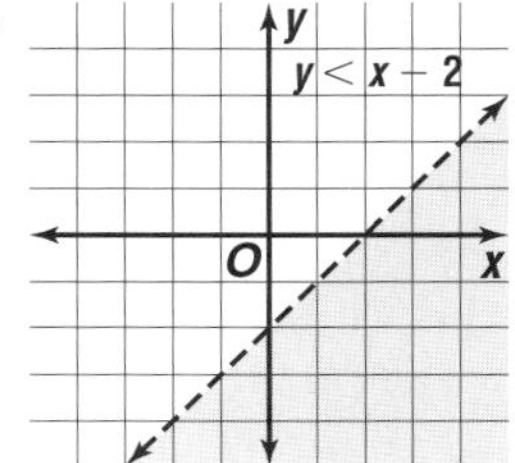

55.

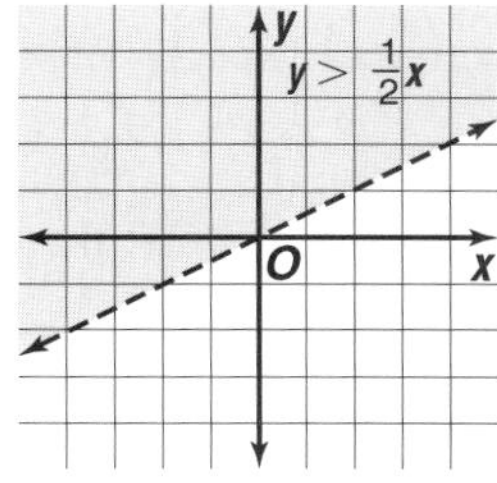

57.

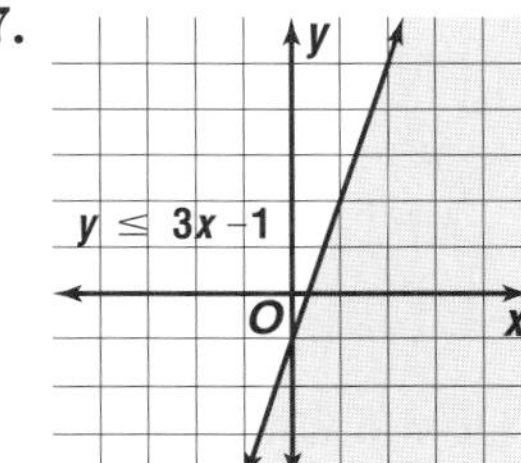

59a.

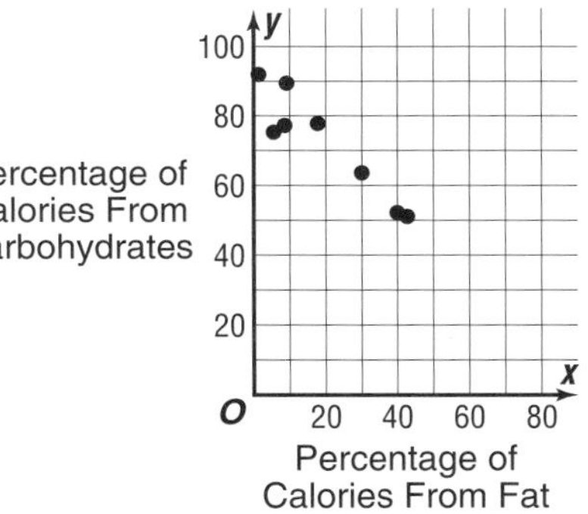

59b. negative 61. approximately 36 goals scored

Chapter 9 Ratio, Proportion, and Percent

Pages 434–436 Lesson 9-1
5. $\frac{1}{5}$ 7. $\frac{25}{1}$ 9. $\frac{43}{46}$ 11. \$0.38/can 13. 0.25 in./hour

15. $\frac{1}{13}$ **17.** $\frac{4}{1}$ **19.** $\frac{13}{7}$ **21.** $\frac{2}{15}$ **23.** $\frac{5}{7}$ **25.** $\frac{3}{7}$ **27.** 24.2 miles/gallon **29.** 9.52 m/s **31.** $7/ ticket **33.** Sample answer: 2697:13,485 **35a.** the 32 oz bag at $3.69 **35b.** Sample answer: When the quality of the less expensive product is not as good as that of the more expensive product, it may be better to buy the more expensive product. **37a.** All ratios are 2. **37b.** The ratios approach the golden ratio. **39a.** 34.8 mph **39b.** 5 ft/s **41.** −11 **43.** 2 · 2 · 2 · 3 · 19 **45.** −13

Page 437–438 Lesson 9-2

3. 10 **5.** 21 **7.** 6 **9.** She gained $800. **11.** −21 **13.** 5 **15.** 249.5 mph **17.** Sample answer: 180 **19.** 24 > 7

Page 439 Lesson 9-3A

1. Game A is fair. It is fair because there are the same amount of odd and even sums. Game B is unfair because there are many more even products than odd products. **3.** Yes; there are the same number of blue and red marbles.

Pages 442–443 Lesson 9-3

7. $\frac{1}{12}$ **9.** $\frac{1}{8}$ **11.** $\frac{1}{2}$ **13.** $\frac{7}{8}$ **15.** $\frac{5895}{7424}$ **17.** 0 **19.** $\frac{17}{24}$ **21.** 1 **23.** $\frac{1}{2}$ **25.** 0 **27.** $\frac{1}{4}$ **29.** $\frac{1}{36}$ **31.** 0 **33.** $\frac{7}{10}$ **35.** 1–humans have been to the Moon; 0–the Sun is too hot. **37.** 4 **39.** −2 **41.** 15.4 m; 13.26 sq meters

Pages 446–447 Lesson 9-4

7. = **9.** 7.5 **11.** 2.1 **13.** $\frac{2.7}{m} = \frac{3}{7}$; 6.3 **15.** ≠ **17.** ≠ **19.** = **21.** 7 **23.** 42 **25.** 2.5 **27.** 16 **29.** 2.1 **31.** 10.2 **33.** $\frac{5}{6.15} = \frac{x}{8.00}$; 6.5 **35.** $\frac{8}{2} = \frac{z}{5}$; 20 **37.** $\frac{25}{5} = \frac{m}{25}$; 125 **39.** 1 **41.** 15 **43.** 6 and 8 **45.** 3.24 inches **47.** $\frac{8}{3}$ **49.** 9.412×10^6 **51.** $\frac{1}{4^3}$ **53.** $b + 200$

Page 448 Lesson 9-4B

1. Sample answer: A larger sample will lead to a more reliable prediction and result.

Pages 451–453 Lesson 9-5

7. 44% **9.** 3.2% **11.** 9 **13.** 12.5% **15.** 25% **17.** 60% **19.** 225% **21.** 75.5% **23.** 106.72 **25.** 45% **27.** 35 **29.** 16 **31.** 80 **33.** 1200 **35.** 100 **37a.** 50 g **37b.** about 2375 mg or 2.375 g **39.** 31,813; 8238; 18,779; 7279; 35,071 **41.** $\frac{3}{4}$ **43.** $4.22 **45.** 5

Pages 455–457 Lesson 9-6

5. 2% **7.** Sample answer: High school cafeteria; because teen's tastes are not representative of all ages. **9.** 35% **11.** 125 **13.** No; because the sample is too small and it is not random. **15.** Sample answer: No; many of the common two-letter words like *at*, *in*, *of*, and *on* do not contain the letter E. **17a.** 52% **17b.** 60% **19.** 58% **21.** D **23.** true

Page 457 Self Test

1. 60 ft/min **3.** 12 **5.** $\frac{9}{16}$ **7.** 1.8 **9.** 50 **11.** 105

Pages 460–461 Lesson 9-7

5. 37% **7.** 103% **9.** 7.2% **11.** $\frac{1}{4}$ **13.** $\frac{3}{1000}$ **15.** 0.245 **17a.** $\frac{2}{5}$ **17b.** 40% **19.** 3% **21.** 237% **23.** 132% **25.** 0.04% **27.** $58\frac{1}{3}$% **29.** $233\frac{1}{3}$% **31.** $55\frac{5}{9}$% **33.** 7.5% **35.** $\frac{5}{4}$ **37.** $\frac{69}{200}$ **39.** $\frac{3}{8}$ **41.** $\frac{49}{300}$ **43.** 0.75 **45.** 0.398 **47.** 0.004 **49.** 2.354 **51.** 22% **53.** 0.067, 16%, $\frac{1}{4}$ **55.** $\frac{1}{3}$ **57.** $\frac{4}{5}$ **59.** 1.5, −12 **61.** $\frac{5}{27}$ **63.** n^{14} **65.** C **67.** 8

Pages 464–466 Lesson 9-8

7. c **9.** $\frac{4}{5}$ **11.** 215 **13.** 1.5 **15.** 50% **17a.** 60% **17b.** 40% **19.** c **21.** b **23.** 1 **25.** $\frac{3}{100}$ **27.** $\frac{1}{10}$ **29.** $1\frac{1}{5}$ **31.** 120 **33.** 98 **35.** 5.15 **37.** 20 **39.** 80% **41.** 50% **43.** 10% **45.** 10% **47.** 0.3% **49.** 10%, 20%, 60%, 45%, 40% **51.** Sample answers: Connecticut 10%; Delaware 1%; Georgia 2%; Kentucky 1%; Maine 2%; Maryland 10%; New Hampshire 3%; New Jersey 5%; New York 10%; North Carolina 10%; Pennsylvania 15%; Plymouth 10%; Rhode Island 2%; South Carolina 10%; Tennessee 1%; Vermont 2%; Virginia 20% **53.** $\frac{16}{25}$ **55.** geometric; 1.5; 60.75, 91.125, 136.6875 **57.** 21;

15 20 25 30

59.

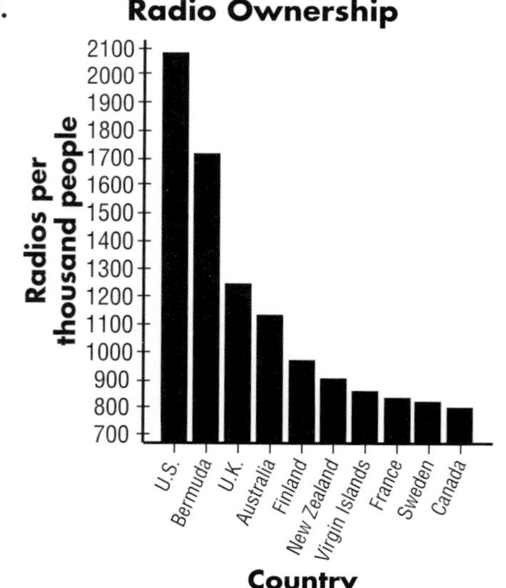

Pages 469–471 Lesson 9-9

7. 200 **9.** 17.28 **11.** $47.25 **13.** 3.36 million **15.** 27 **17.** 70 **19.** 31.5 **21.** 300% **23.** $1.\overline{6}$% **25.** 120 **27.** 20 **29.** $70 **31.** $187.50 **33.** $42.50 **35.** $41.34 **37.** about 93 million **39.** They would be paying the customers. **41a.** $264.72 **41b.** No; the price is the same either way. **41c.** You could

not add the percents first. For example, if the price was \$100, 10% + 33% = 43% off. Using this method the price would be 100 − 43 or \$57. Taking 33% off and then 10% would have the price as \$60.30.
43. The statements are not equivalent. For example, suppose an item costs \$1.00 on day 1. According to the first statement, it would cost \$2.08, not \$1.00 × 2 or \$2.00, on the fifth day.

Day 2 $1.2 \times 1.0 = \$1.20$
Day 3 $1.2 \times 1.20 = \$1.44$
Day 4 $1.2 \times 1.44 = 1.728$ or \$1.73
Day 5 $1.2 \times 1.73 = 2.076$ or \$2.08

45. positive **47a.** $3v - 298 = 1916$ **47b.** 738 **49.** 15 **51.** 7

Pages 474–475 Lesson 9-10

7. decrease; 12% **9.** decrease; 7% **11.** decrease; 20% **13.** increase; 13% **15.** increase; 13% **17.** increase; 10% **19.** increase; 7% **21.** increase; 136% **23a.** no; Suppose the price of the outfit is \$100. The price would be \$50 after the first reduction. After the second reduction, the price would be $50 - 0.3(50) = 50 - 15$ or \$35. A final price of \$35 is $\frac{100 - 35}{100}$ or 65% off, not 80% off. **23b.** 65% **25.** 20% **27.** \$3.20 **29.** 125 **31.** Yes; it passes the vertical line test. **33.** 12 outfits

Page 477 Highlights

1. true **3.** false **5.** true **7.** false

Pages 478–480 Study Guide and Assessment

9. $\frac{41}{45}$ **11.** $\frac{3}{5}$ **13.** \$0.89 per pound **15.** 57 miles/hour **17.** $\frac{13}{21}$ **19.** $\frac{8}{21}$ **21.** $\frac{11}{21}$ **23.** 2.5 **25.** $\frac{88.4}{3.4} = \frac{161.2}{g}$; 6.2 gal. **27.** 76% **29.** 4 **31.** 32 **33.** 0.56% **35.** 56.5% **37.** $\frac{13}{20}$ **39.** 2.35 **41.** a **43.** a **45.** b **47.** 20 **49.** 4000 **51.** 31.25% **53.** decrease; 1% **55.** decrease; 22% **57.** decrease; 12% **59.** 28

Chapter 10 More Statistics and Probability

Pages 488–489 Lesson 10-1

7. 5, 6, 7, 8

Stem	Leaf
5	4 7
6	3 7 8 9
7	1 5 7
8	5

5|4 = 5.4

9.

Stem	Leaf
6	0 1 4 5 7 8
7	0 0 2 3
8	0 0

6|0 = 60

11.

Stem	Leaf
0	5 6 7 9 9
1	1 2 4
2	1 3
3	
4	0 2 5

4|0 = 40

13.

Stem	Leaf
9	2
10	4 8
11	1 2 2 7
12	3 6 8 9
13	3 8
14	7

9|2 = 9.2

15.

Women's		Men's
5	4	5
5 5	5	
7 5 0	6	
5 2 0 0 0	7	0 0 0 0 0 5
5 0	8	2 2 3
0	9	
	10	0
	11	0
4	12	0 5
	13	3

5|4 = 45 *4|5 = 45*

15b. Answers will vary. Sample answer: Men's running shoes are more expensive than women's running shoes. The typical running shoe costs \$70. **17.** Sample answer: Yes; Since the data can be read from a stem-and-leaf plot, you can estimate their sum and then divide by the number of leaves to estimate the mean.

19a.

Distress		No Distress
8 7 3	5	
3	6	6 7 7 7 8 9
5 0 0	7	0 0 2 3 5 6 6 8 9
	8	1

3|5 = 53 *7|0 = 70*

21. $\frac{9}{20}$ **23.** d

Pages 492–494 Lesson 10-2

5. 13 **7.** 52

9a.

Stem	Leaf
3	8 9
4	0 2 6
5	8
6	1 2 3 3
7	0
8	0 4
9	4

3|8 = 38

9b. 61.5 **9c.** 70, 42 **9d.** 28 **11.** 70; 65; 85, 45; 40 **13.** 21; 7; 13.5, 3; 10.5 **15.** 49; 208; 213, 200; 13 **17.** 4.1; 9.2; 9.75, 8.85; 0.9 **19.** 37,600,000; 43,100,000; 52,600,000; 33,100,000; 19,500,000 **21.** American League: 39; 42.5; 46, 38; 8; National League: 21; 40.5; 46.5, 38; 8.5 **23c.** The first set of data has a smaller interquartile range, thus the data in the first set are more tightly clustered around the median and the data in the second set are spread out over the range. **25a.** participants 60; observers 71 **25b.** 51; 35 **25c.** Sample answer: The participants' ages are between 12 and 72. Their ages are spread out over the range of the data. The median age of the participants is 25 years. **25d.** Sample answer: The observers are both younger and older than the participants, but the range of ages is very close to that of the participants. The median age of the observers is 46.5 while the median ages of the participants is 25. **27.** > **29.** cd^4

Pages 499–501 Lesson 10-3

7a. 40 **7b.** 50 **7c.** 30 **7d.** 20 **7e.** 25, 55 **7f.** none **9a.** 15 **9b.** 68 **9c.** 71 **9d.** 65 **9e.** 6 **9f.** no **11a.** 50% **11b.** 50% **11c.** 25% **11d.** 25% **11e.** Scores were closer together between 80–83. **11f.** The scores were spread further apart. **13.** box-and-whisker plot **15.** box-and-whisker plot **17.** line graph or comparative graph **19.** The box did not change. The weight 200 is an outlier.

21a.

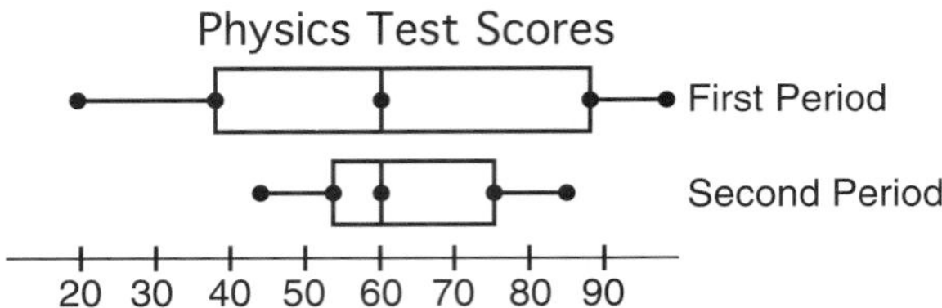

21b. The medians are the same. **21c.** The range and interquartile range are greater for the first period class.

23a.

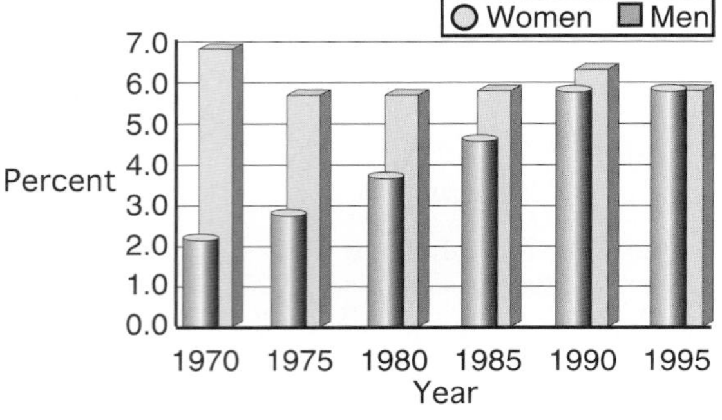

23b. Men's employment decreased from 1970 to 1975, but since 1975 men's employment has been relatively consistent. **23c.** Women's employment increased steadily from 1970 to 1990 and has remained steady from 1990–95. **23d.** Women are as likely as men to have two jobs. **25.** {5, −1, 4.5}; {−3, 4, 0} **27.** Sample answer: $7

Pages 502–503 Lesson 10-3B

1.

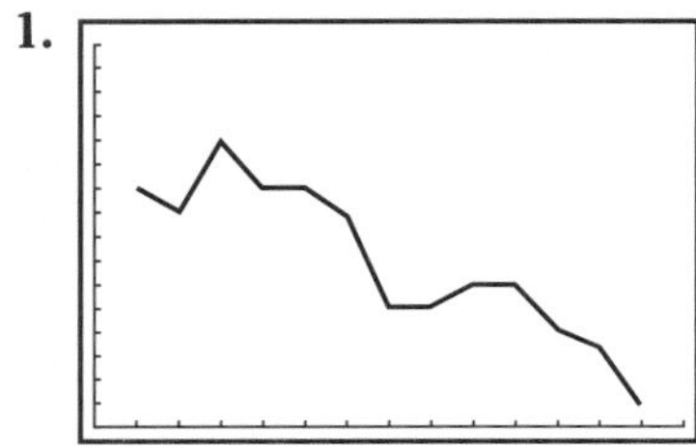

3.

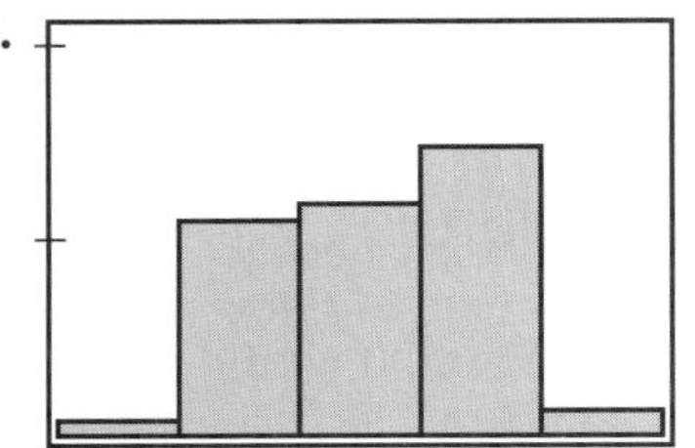

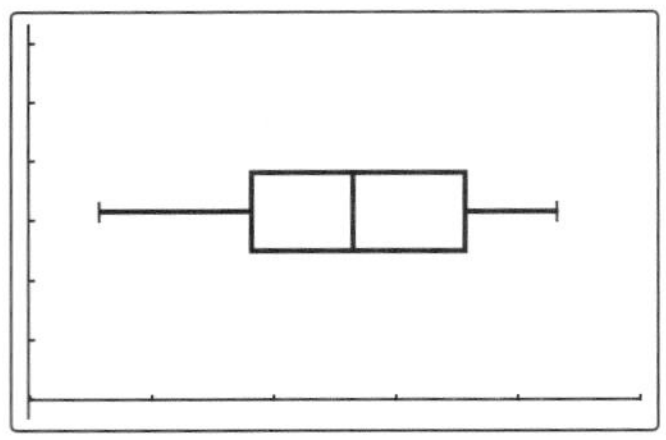

Pages 506–508 Lesson 10-4

5a. about 1.1 **5b.** The second car is shorter in length and width. **5c.** Yes; the actual number of cars sold in 1992 is 1.1 times greater than the number sold in 1991. However, the graph appears to be longer and wider, but both dimensions of the car should not have changed. **7a.** Median; there are large and small values. **7b.** Mean; the high values will raise the center of the data. **9.** 172.6; mean **11.** sports; mode

13.

	Company A	Company B
mean	$22,413.79	$21,935.48
median	$15,000.00	$20,000.00
mode	$15,000.00	$20,000.00
range	$85,000.00	$30,000.00

You would probably prefer to work for Company B. Both the median and mode salaries are greater, so the salary is greater for more employees.
15a. Sample answer: No; it only shows the population increases for 6 years.

15b.

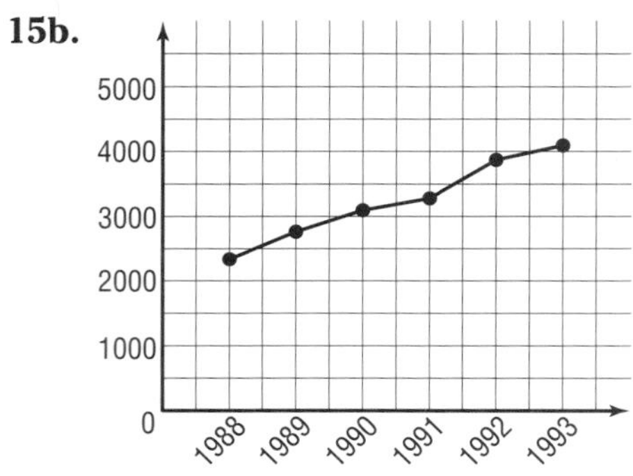

17a. $58 **17b.** $25 **17c.** $35 **17d.** $15 **17e.** $20 **17f.** $12, $70 **17g.** $0 and $65 **17h.** $70 **19a.** $\frac{8}{11}$ **19b.** 1,965,955 **21.** −6, 6

Pages 511–513 Lesson 10-5

5a. 12 outcomes **5b.** $\frac{1}{12}$ **7.** 18 outcomes **9.** 24 outcomes **11.** 25 ways **13.** 36 outcomes

15a. 8 outcomes **15b.** $\frac{1}{8}$ **17.** 32 outcomes

19a. 2^{30} **19b.** $\frac{1}{2^{30}}$ **21a.** 10,000 extensions
21b. 4,860,000 **21c.** 2,880,000 new numbers
23. The median; it is close to three of the four actual prices. **25.** -6 **27.** h = hours, $3.50h < 20$, $h < 5.7$

Page 513 Self Test

1.

2	8
3	6 7 9
4	3 7 9
5	0 1 1 3
6	2

6|2 = 6.2

3. 2.7, 4.7 **5.** A comparative graph is probably the best way. **7.** Brian would use the bottom graph because it makes the increases he has received appear smaller. **9.** 24 outcomes

Page 514 Lesson 10-6A

1. 120 words

Pages 517–519 Lesson 10-6

5. combination **7.** 2 **9.** 24 **11.** 120 **13.** 252 **15.** 56 programs **17.** combination **19.** permutation **21.** combination **23.** 720 **25.** 720 **27.** 362,880 **29.** 39,916,800 **31.** 35 **33.** 1 **35.** 45 **37.** 252 **39.** $1\frac{2}{3}$ **41.** 5005 teams **43.** 6 ways **45.** 10 students **47.** 5005 combinations **49.** 140,400,000 license plates **51.** about $-15°$ C **53.** -2.5 **55.** $\frac{5}{4}$ **57.** $-9x$

Pages 521–523 Lesson 10-7

5. 1:1 **7.** 5:1 **9.** 1:4 **11.** 4:1 **13.** 1:1 **15.** 1:2 **17.** 7:5 **19.** 5:1 **21.** 1:17 **23.** 7:11 **25.** 9:4 **27.** 1:9999 **29a.** less than 1 or none **29b.** about 89 people **31.** 40,320 **33.** \$0.16, \$0.31, \$0.36, \$0.40, \$0.51, \$0.55, \$0.60 **35.** >

Pages 526–528 Lesson 10-8

5a. Sample answer: Select a marble out of a bag with 4 marbles the same color and one marble a different color. Then repeat three more times. **7a.** Roll a die. **9.** \$25 **11a.** 15 games **11b.** 240 **13b.** Change the second line to "For (N, 1, 50)".
15. 1:5 **17a.** $110 = \frac{1}{2}x + 20$ **17b.** \$180 **19.** 35°F

Page 529 Lesson 10-8B

1. The most frequently drawn color is probably the one with the greatest number of marbles in the bag. **3.** yes; It is more likely to happen with fewer draws.

Pages 533–534 Lesson 10-9

5. dependent **7.** $\frac{21}{380}$ **9.** $\frac{9}{190}$ **11.** $\frac{11}{625}$ or 1.76% **13.** independent **15.** independent **17.** $\frac{1}{30}$ **19.** $\frac{2}{15}$ **21.** 0 **23.** $\frac{10}{91}$ **25.** $\frac{3}{91}$ **27.** $\frac{1}{91}$ **29a.** $\frac{3}{8}$ **29b.** $\frac{3}{10}$ **31.** $\frac{4993}{5000}$ or 99.86% **33.** \$10; about \$40

35. 2;
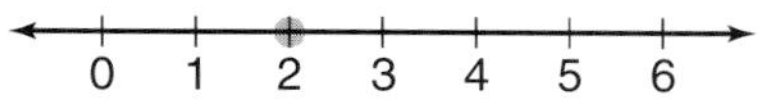

37. $2(n + 3)$

Pages 537–538 Lesson 10-10

5a. $\frac{1}{3}$ **5b.** $\frac{1}{3}$ **5c.** $\frac{2}{3}$ **7a.** inclusive; $\frac{3}{4}$ **7b.** exclusive; $\frac{5}{8}$ **7c.** exclusive; $\frac{5}{8}$ **7d.** inclusive; $\frac{1}{2}$ **9.** 0 **11.** $\frac{2}{15}$ **13.** $\frac{2}{221}$ **15.** $\frac{4}{9}$ **17.** about 10 times **19.** about 35 times **21.** about 20 times **23.** 53% **25.** independent

27.
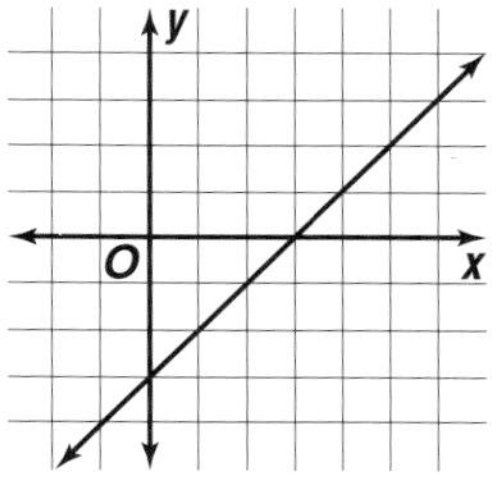

29. 300

Page 539 Highlights

1. C **3.** I **5.** A **7.** G

Pages 540–542 Study Guide and Assessment

9.

9	5 8
10	7 7 8
11	0 5 9

9|5 = 9.5

11.

11	1 2 7
12	0 1
13	3
14	6 8

11|1 = 111

13. 30; 18; 10, 28; 18 **15.** 98; 217; 205, 240; 35

17a.
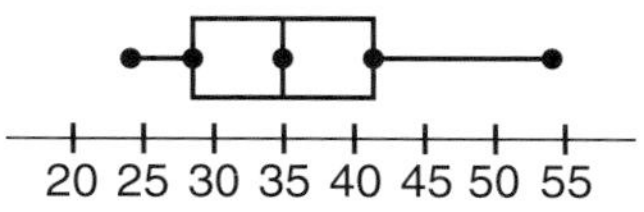

17b. 35 **17c.** none **19.** 6.75; mean **21.** favorite sport: soccer; 6.75 classes; 6.5 family members **23.** 1024 **25.** 104 **27.** C **29.** 120 **31.** 90 **33.** 10:3 **35.** 1:1 **37.** dependent **39.** $\frac{3}{22}$ **41.** $\frac{3}{22}$ **43.** inclusive; $\frac{27}{36}$ **45.** $\frac{1}{2}$ **47.** 64

Chapter 11 Applying Algebra to Geometry

Pages 552–553 Lesson 11-1

7. plane **9.** A **11.** $\overrightarrow{CD}$ **13.** 63°, acute
15.
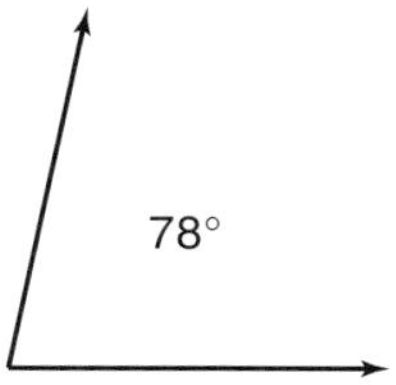

SELECTED ANSWERS

17. The three "pods" determine a plane. **19.** plane **21.** ray

23.

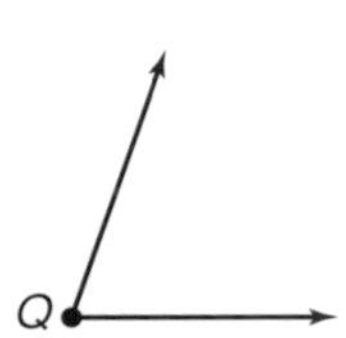

25.

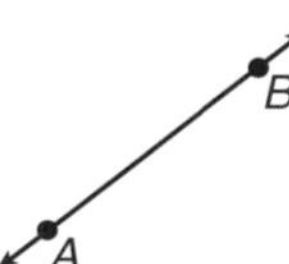

27.

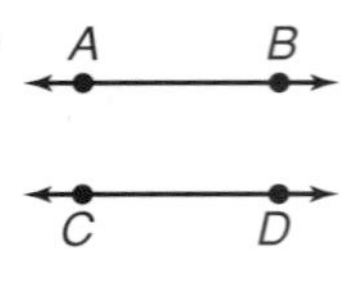

29.

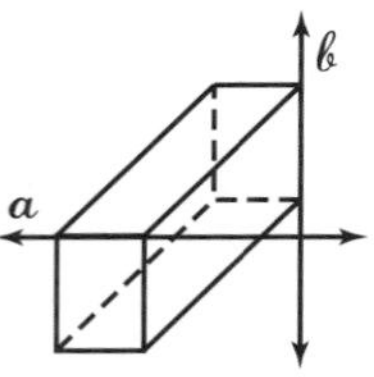

31. A B C D **33.** a ℓ

35. 155°; obtuse **37.** 65°; acute **39.** 15°; acute

41. acute;

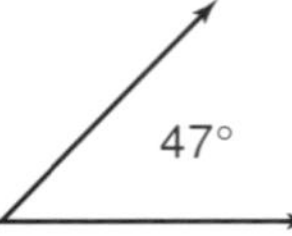

43. acute;

45. right;

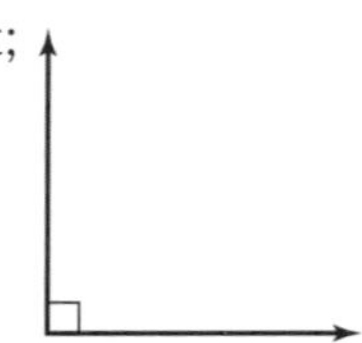

47. 3° **49.** 22 times **51.** = **53.** 50, 52 **55.** 15 **57.** −12

Page 555 Lesson 11-1B
3. Cut in two equal pieces. **5.** They are the same.

Pages 558–560 Lesson 11-2
5. silicon **7a.** 100% **7b.** 360° **7c.** Housing: 31%, 111.6°; Food:19%, 68.4°; Transportation: 22%, 79.2°; Insurance: 7%, 25.2°; Savings: 4%, 14.4°; Misc.: 17%, 61.2°; Totals, \$2080, 100%, 360.0°

7d.

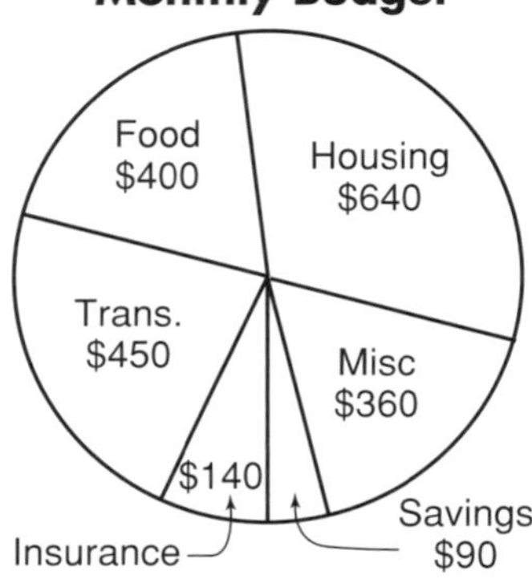

9.

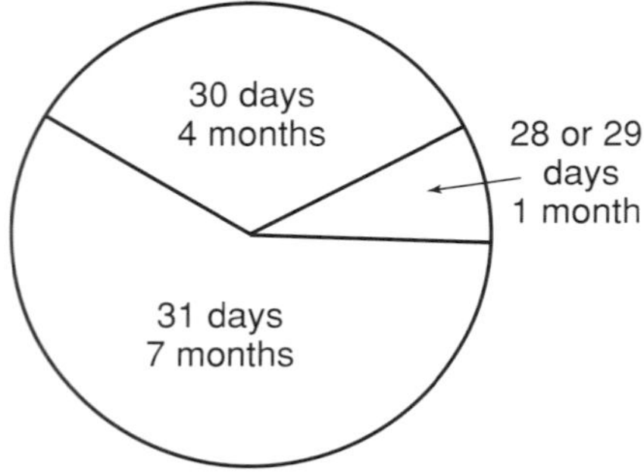

11a. $\frac{3}{8}$

11b.

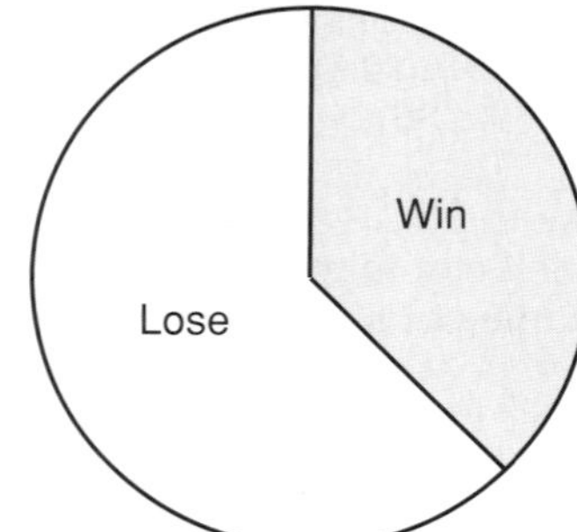

13.

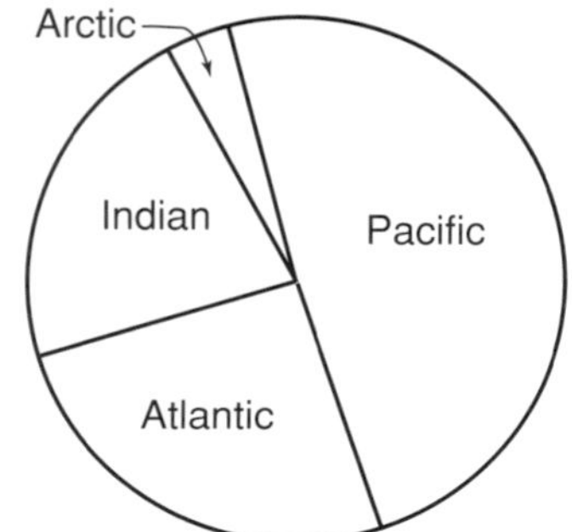

15a. **Black Hills' Reclamation Mix**

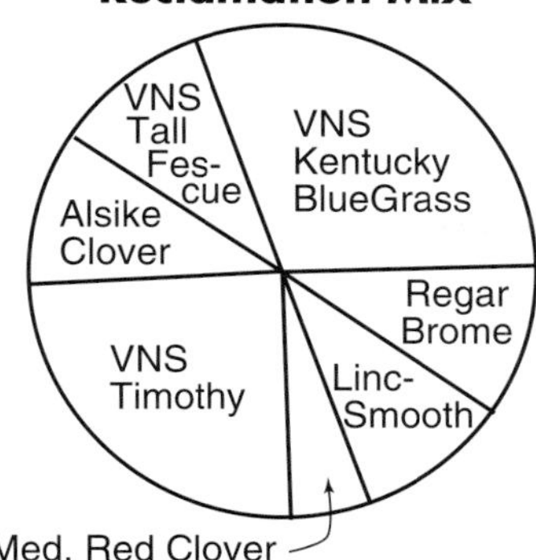

15b. 75 lb. VNS Kentucky Bluegrass; 25 lb each of Regar Bromegrass, Lincoln Smooth Brome, Alsike Clover, and VNS Tall Fescue; 12.5 lb Medium Red Clover; 62.5 lb VNS Timothy

17.

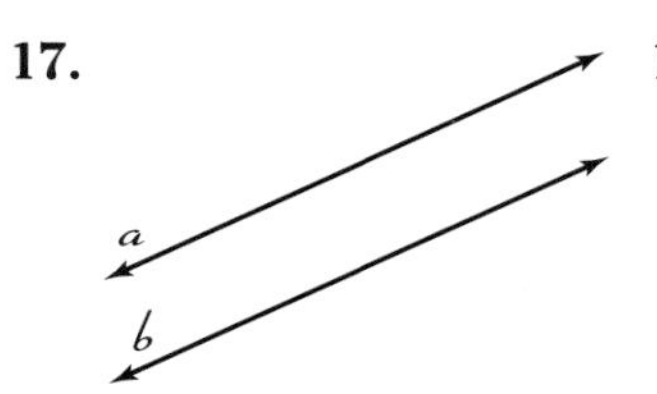

19.

21. Sample answer: Joe earned 3 times as much as Sue. If Joe earned \$25, how much did Sue earn?

Pages 564–566 Lesson 11-3

7. 40 **9.** 15° **11.** $m\angle X = 60°$, $m\angle Y = 120°$ **13.** $\angle AFG$, $\angle GFB$, $\angle CGF$, $\angle FGD$ **15.** $m\angle AFE = 130°$; $m\angle EFB = 50°$; $m\angle BFG = 130°$; $m\angle AFG = 50°$; $m\angle CGF = 130°$; $m\angle FGD = 50°$; $m\angle DGH = 130°$; $m\angle CGH = 50°$ **17.** 105 **19.** 66 **21.** 55 **23.** 120°, 60° **25.** 18°, 72° **27.** 47° **29.** 133° **31.** 133° **33.** $m\angle R = 32°$; $m\angle S = 58°$ **35.** $m\angle F = 118°$; $m\angle G = 62°$ **37.** 30 **39.** 55° **41.** Yes; all the angles formed are right angles.

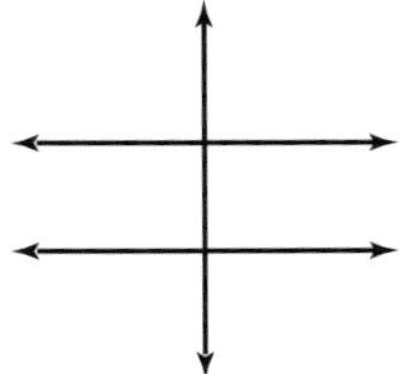

43. 135° **45.** $\frac{5}{18}$ **47.** 5.678×10^4 **49.** $-7b$ **51.** 5

Page 567 Lesson 11-3B

1. It divides the difference of the y-coordinates by the difference of the x-coordinates. **3.** yes **5.** Draw two lines with the same slope.

Pages 570–572 Lesson 11-4

7. 117; obtuse **9.** equilateral **11.** 119° **13.** 50; right **15.** 36°; 54°; 90° **17.** isosceles **19.** equilateral **21.** obtuse **23.** 85° **25.** 25° **27.** 85° **29.** 45°, 80°, 55° **33.** 60° **35.** 25° **37.** $\frac{5}{13}$ **39.** $59\frac{1}{4}$ inches **41.** $4x + 12$

Page 572 Self Test

1. acute **3.** acute **5.** 34° **7.** 34° **9.** $m\angle F = 100°$; $m\angle G = 80°$ **11.** 100; obtuse

Pages 575–577 Lesson 11-5

5a. $\triangle ZYX$ **5b.** $\overline{ZY}$ **5c.** $\angle B$ **5d.** $\angle X$ **5e.** $\overline{YX}$ **5f.** $\overline{CA}$ **5g.** $\angle A$ **7.** c **9.** $\triangle ABC \cong \triangle FED$; $\angle A \cong \angle F$, $\angle B \cong \angle E$, $\angle C \cong \angle D$, $\overline{AB} \cong \overline{FE}$, $\overline{BC} \cong \overline{ED}$, $\overline{AC} \cong \overline{FD}$ **11.** $\triangle BAD \cong \triangle ABC$; $\angle BAD \cong \angle ABC$, $\angle D \cong \angle C$, $\angle ABD \cong \angle BAC$, $\overline{BA} \cong \overline{AB}$, $\overline{AD} \cong \overline{BC}$, $\overline{BD} \cong \overline{AC}$ **13.** $\triangle DEF \cong \triangle HGF$ **15.** $\overline{MA}$ **17.** $\overline{RT}$ **19.** $\angle M$ **21.** 3 **23.** 8 **25.** $\triangle ABH \cong \triangle IJG$; $\triangle ACG \cong \triangle IEF$ **27.** 5 ft **29.** 95°; obtuse **31.** 125 **33.** Liz–pack; Renee–snack bar; Pablo–plate lunch

Pages 581–583 Lesson 11-6

7a. $\frac{m\overline{AB}}{m\overline{AC}} = \frac{m\overline{BE}}{m\overline{CD}} = \frac{m\overline{AE}}{m\overline{AD}}$ **7b.** $\angle CAD$ **7c.** $\overline{CD}$ **7d.** 10 **9.** $\frac{1}{x} = \frac{2}{6}$; 3 **11.** 121 m **13.** $\frac{3}{2} = \frac{x}{3}$; 4.5 **15.** $\frac{4}{5} = \frac{8}{x}$; 10 **17.** $\frac{3}{5} = \frac{4.5}{x}$; 7.5 **19a.** Yes; the ratio of the corresponding sides is 1:1. **19b.** No; only those where the ratio of the corresponding sides is 1:1. **21.** 2.7 ft **23.** 32 m **25.** 2^5 **27.** 47.9; 42.5; 54

Pages 586–588 Lesson 11-7

5. 115 **7.** quadrilateral, trapezoid; trapezoid **9.** 6 m **11.** 30° **13.** $x = 60$; 60°, 120° **15.** $x = 60$; 60°, 120° **17.** square **19.** rectangle **21.** quadrilateral **23.** sometimes **25.** never **27.** always **29.** Yes; a rhombus is equilateral but may not be equiangular. **29b.** Yes; a rectangle is equiangular but may not be equilateral. **35.** 3 cm **37.** 22.5 miles per gallon **39.** inductive

Pages 592–593 Lesson 11-8

7. hexagon, not regular **9.** 40°; 140° **11.** 900° **13.** 1800° **15.** 2340° **17.** 60°, 120° **19.** 30°, 150° **21.** 60° **23.** 72 cm **25.** 27 in.

27.

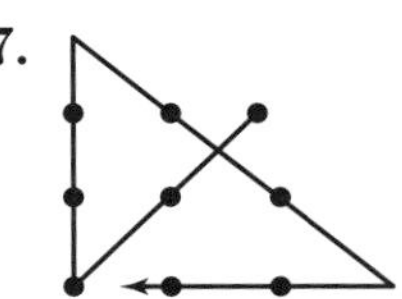

29a. trapezoid **29b.** 45°, 45°, 135°, 135° **29c.** square **31.** $\frac{1}{5}$ **33.** $(1 \times 10^3) + (6 \times 10^2) + (4 \times 10^0)$

Page 594 Lesson 11-8B

1. It will tessellate if 360 is a multiple of the angle's measure.

Pages 597–599 Lesson 11-9

5. translation

7.

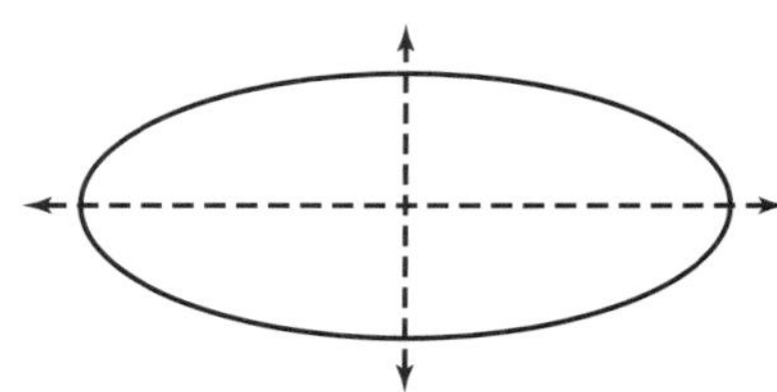

9. reflection **11.** rotation **13.** b

15.

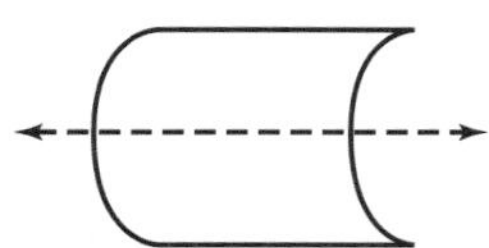

17.

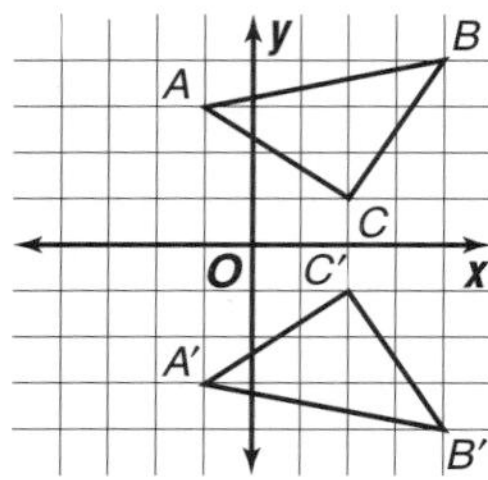

19.

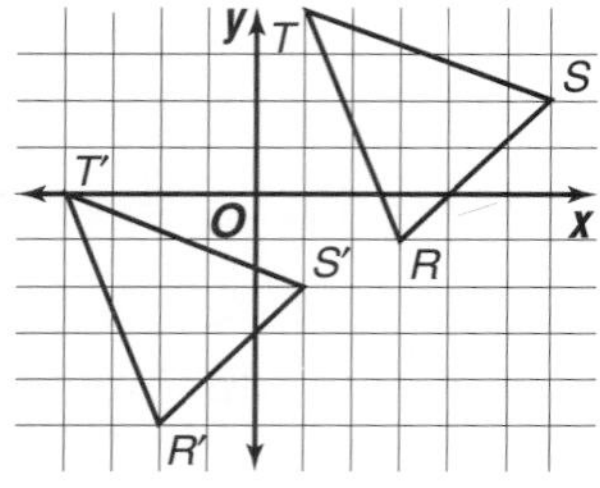

21. top to bottom: *all except* E L N O U Y Z
side to side: C D E F I J M P Q R S V W X
diagonal: F I L M N P Q R S U V W X Y

25. 200% increase

27. $t < -21$

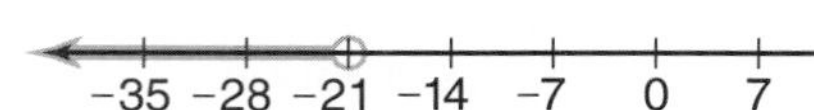

Page 601 Highlights

1. b **3.** g **5.** a **7.** i

Pages 602–604 Study Guide and Assessment

9.

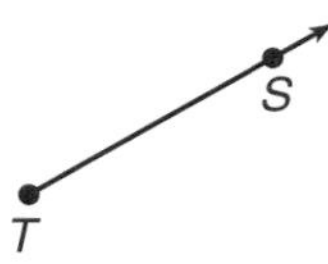

11.

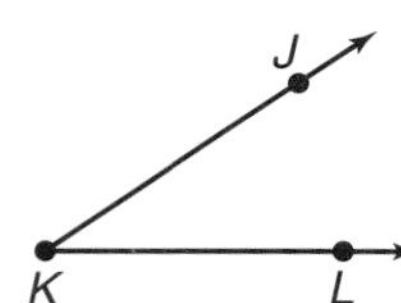

13.

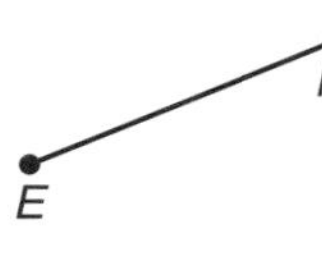

15.

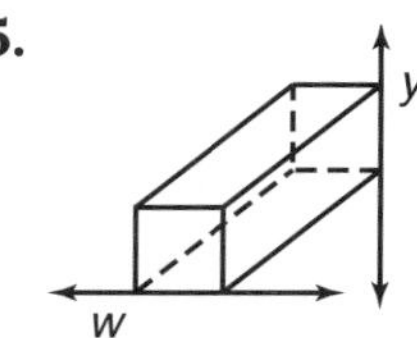

19. 48° **21.** 48° **23.** 90° **25.** 92°; obtuse
27. $\angle Y$ **29.** $\overline{XZ}$ **31.** 2 in. **33.** 70°; 110° **35.** 144°

37.

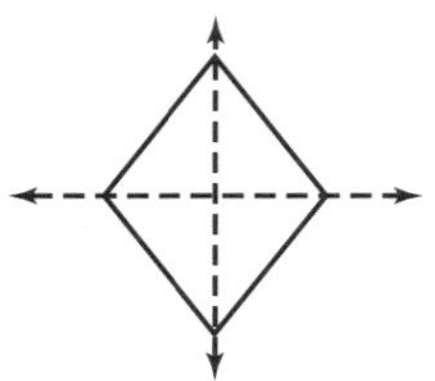

39. reflection

41.

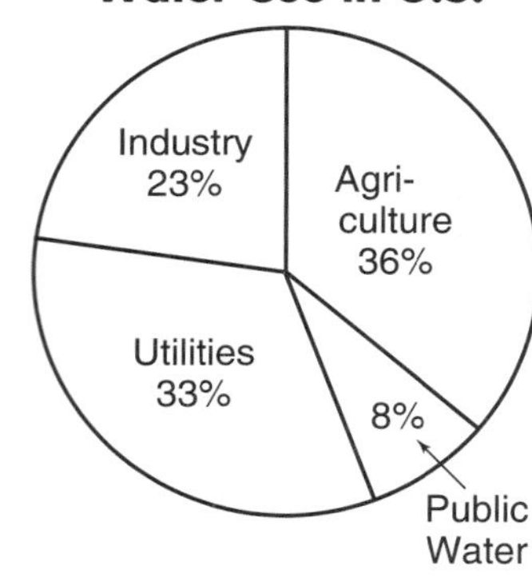

Chapter 12 Measuring Area and Volume

Pages 610–611 Lesson 12-1A

1. Sample answer: Count squares and half-squares. **3.** Sample answer: Find the area of the rectangle and divide by 2.

Pages 615–617 Lesson 12-1

5. 3; 4; 6 sq cm **7.** 15 m and 5 m; 6 m; 60 m^2 **9.** 216 cm^2 **11a.** about 33,600 mi^2 **11b.** The actual area is 35,936 mi^2. **13.** 6 cm^2 **15.** 90 cm^2 **17.** 21.6 in^2 **19.** 57 ft^2 **21.** 9 in^2 **23.** $6\frac{1}{2}$ ft^2 **25.** 36.75 m^2 **27.** $A_1 = \frac{1}{2}h_1d$; $A_2 = \frac{1}{2}h_2d$; $A_1 + A_2 = \frac{1}{2}d(h_1 + h_2) = \frac{1}{2}d(d) = \frac{1}{2}d^2$ **29a.** 308 ft^2 **29b.** 15-foot width **31.** A shape as close as possible to a square gives the greatest area. **33.** 5 **35.** 2.1 **37.** $x + 7$

Page 618 Lesson 12-1B

1. Each section is a replica of the whole curve. **3.** infinite

Pages 621–622 Lesson 12-2

5. 452.4 in^2 **7.** 49,087.4 ft^2 **9.** 50.3 cm^2 **11.** 120.8 m^2 **13.** 475.3 cm^2 **15.** 167.4 in^2 **17.** 39.3 ft^2 **19.** 7854.0 ft^2 **21.** about 707 m^2 **23.** about 26 times **25.** about 13 mi^2 **27.** 18 **29.** $\frac{4}{9}$

Pages 625–627 Lesson 12-3

5. $\frac{9}{50}$ **7.** about 9 **9.** $\frac{3}{8}$ **11.** $\frac{1}{6}$ **13.** $\frac{3}{4}$

15. about 38 **17.** 0.16 **19.** $\frac{1}{35}$ **21.** about 450 cm^2 **23.** 0.1 **25.** $32 < w < 212$

Page 628 Lesson 12-3B

1. The ratios should be similar. **3.** More data makes the ratios closer.

Pages 630–631 Lesson 12-4

5. 875 cm^2
7b. 10 units; 10 units; 8 units; 10 units; 10 units
7c. Shapes where two squares share one side with another square and the rest of the squares share two sides with other squares are the greatest in perimeter. **7d.** Square shapes have the least perimeter. **7e.** 12; yes **9.** 4 m by 12 m
11. $3 \times 3 \times 4$ **13.** $700 gain **15.** A **17.** $\frac{2}{5}$
19. 11 hours **21a.** Nevada **21b.** 17 million acres

Pages 635–637 Lesson 12-5

7. triangular prism; 184 ft^2 **9.** cylinder; 3166.7 m^2 **11.** 282 in^2 **13.** 132 cm^2 **15.** 211.2 cm^2 **17.** 636.2 in^2 **19.** 264 in^2 **21.** d **23a.** 2250 in^2 **23b.** No, as long as the entire surface of the face of the smaller cube is touching a face of the larger cube, the surface area of the tower will be the same. **25.** about 6189 cm^2 **27.** 512 in^2, 1200 in^2, 1800 in^2; no

31.

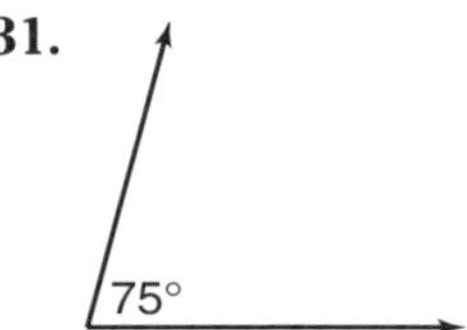

33. 3,605

Pages 641–642 Lesson 12-6

5. pyramid; 134.4 m^2 **7.** triangular pyramid; 57 in^2 **9.** 66.4 ft^2 **11.** 404.6 m^2 **13.** 282.7 in^2 **15.** 471.2 cm^2 **17.** 275 in^2 **19.** $\approx$ 25,780.8 ft^2 **21.** 3160.1 sq ft **23.** 900 cm^2 **25.** $4x + 8 = 20$; 3 **27.** $-15xy$

Page 642 Self Test

1. 15m^2 **3.** 113.1 in^2 **5.** cylinder, 402.1 in^2
7. $\approx$ 0.028 or about $\frac{3}{100}$ **9.** 47.1 in^2

Page 643 Lesson 12-7A

1. triangular prism; cylinder **3.** same; 8 in. **5.** cylinder

Pages 646–648 Lesson 12-7

7. 600 in^3 **9.** 225 in^2 **11.** 28.3 ft^3 **13.** 125 ft^3 **15.** 512 cm^3 **17.** 1608.5 ft^3 **19.** 565.5 cm^3 **21.** 24.5 ft^3 **23.** 83.3 cm^3 **25.** 4712.4 yd^3
27. 18.6 m^3 **29.** The volume is greater with $8\frac{1}{2}$ as the height. **31.** 3.29 cm **33.** a **35a.** 128 ft^3
35b. $\frac{5}{8}$ **37.** 4 mi **39.** 7 **41.** 64

Pages 651–653 Lesson 12-8

5. 32 ft^3 **7.** 160 ft^3 **9.** 270.8 in^3 **11.** 70 ft^3 **13.** 160 m^3 **15.** 150 in^3 **17.** 9189.2 m^3 **19.** 733 cm^3 **21.** 50 cm^3 **23.** 1885.0 m^3 **25.** 44 m^3
27. No; the customer gets only $\frac{1}{3}$ as much popcorn for $\frac{1}{2}$ the price. **29.** 91,636,272 ft^3 **31.** 15 ft^3 **33.** 20% **35.** 55 cm; 184 cm^2

Page 654 Lesson 12-8B

1. 1 unit^2 **3.** 1 unit^3 **5.** 9 unit^2

7.

Scale Factor	Length of a Side	Area of a Face	Volume
1	1	1	1
2	2	4	8
3	3	9	27

9. x^2 **11.** x^3

Page 655 Highlights

1. g **3.** h **5.** a **7.** c

Pages 656–658 Study Guide and Assessment

9. $3\frac{15}{16}$ in^2 **11.** 476 mm^2 **13.** 13.9 m^2 **15.** 8.0 m^2
17. $3\frac{3}{8}$ ft^2 **19.** 7665 ft^2 **21.** 9.4 m^2 **23.** 3325 in^3
25. 2770.9 ft^3 **27.** 504 m^3 **29.** 864 in^2 **31.** 198 **33.** about 1275 gallons

Chapter 13 Applying Algebra to Right Triangles

Pages 667–668 Lesson 13-1

7. 4 **9.** 9 **11.** -7 **13.** 11 **15.** 2 **17.** 3 **19.** -4 **21.** 10 **23.** -12 **25.** 20 **27.** 9 **29.** 14 **31.** -10 **33.** 51 **35.** 11 **37.** -5 **39.** a **41.** 3 **43.** Yes; because when a negative number is squared the result is positive. **45.** 20 feet **47.** 192 in^3 **49.** $\frac{7}{25}$ **51.** 36 **53.** -20

Pages 670–671 Lesson 13-2

3. 18 people **5.** 100 cm **7.** 77 pieces **9.** 10 girls **11.** 7 students

13.

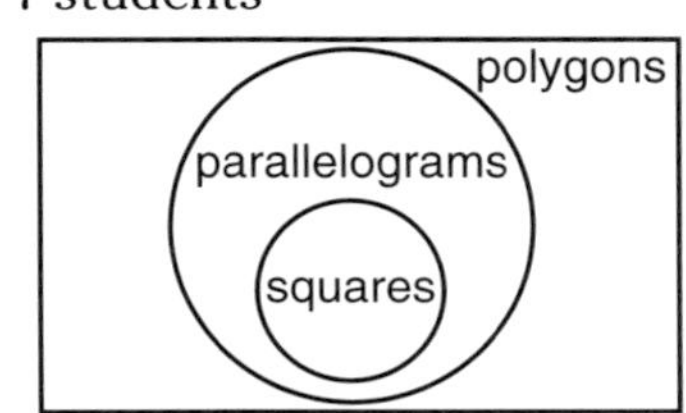

15. A rhombus is a quadrilateral with four congruent sides. **17.** 26.5

Pages 674–675 Lesson 13-3

5. whole, integer, rational, real **7.** rational, real **9.** 19, -19 **11.** 4.1, -4.1 **13.** whole number, integer, rational, real **15.** rational, real

17. rational, real **19.** irrational, real **21.** irrational, real **23.** rational, real **25.** 6, −6 **27.** 3.5, −3.5 **29.** 11.0, −11.0 **31.** 13.4, −13.4 **33.** 0.02, −0.02 **35.** 20, −20 **37.** 9, −9 **39.** 3, −3 **41.** Yes; sample answer: $\sqrt{2} \times \sqrt{2} = 2$ **43.** 2400 ft^2 **47.** 4908.7 cm^3 **49.** $-\frac{2}{5}$ **51.** 72

Pages 679–681 Lesson 13-4
5. $x^2 = 12^2 + 16^2$; 20 in. **7.** $30^2 = 21^2 + b^2$; 21.4 km **9.** 5 **11.** 32.9 **13.** yes **15.** about 127.3 ft **17.** $x^2 = 8^2 + 15^2$; 17 cm **19.** 26.5 m **21.** 19.6 ft **23.** 20 **25.** 5.4 **27.** 5.7 **29.** 9.9 **31.** 20.5 **33.** 5 **35.** no **37.** yes **39.** no **41.** 12 m **43.** 5 **45.** 25 km **47.** 1.6 **49.** 5 **51.** $\frac{4}{9}$ **53.** 20

Page 681 Self Test
1. 6 **3.** −10 **5.** 12, −12 **7.** 6.3, −6.3 **9.** 16

Page 682 Lesson 13-4B
1. Use the procedure in the activity, drawing a right triangle with legs 1 and 3 units long. Then use a compass to find the distance $\sqrt{10}$ on the x-axis. **3.** Use the procedure in the activity, drawing a right triangle with legs 1 unit and 2 units long; or use the procedure in the activity to find the lengths $\sqrt{2}$ and $\sqrt{3}$ and use those lengths as the legs of a right triangle. **5.** Draw a triangle with one leg 2 units long and hypotenuse 4 units long. Transfer $\sqrt{12}$ to the number line.

Pages 685–686 Lesson 13-5
3. 1.4 yd **5.** 9 m **7.** $a = 3$ m, $c \approx 4.2$ m **9.** $a = 6$ m, $b = 10.4$ m **11.** 15 cm **13.** 21.2 cm **15.** 7.4 m **17.** 9.8 in. **19.** 24 in. **21.** $\frac{1}{8}$ mi **23.** 2.315 cm **25.** $a \approx 8.1$ cm, $c \approx 16.2$ cm **27.** $a = 6$ in., $b \approx 10.4$ in. **29.** $b = 5$ yd, $c \approx 7.1$ yd **31.** $a \approx 4.0$ m, $b \approx 4.0$ m **33.** $a = 5.5$ km, $b \approx 9.5$ km **35.** about 5.8 yards **37.** about 7.07 feet **39.** 23.3 feet **41.** 108° **43.** 75 **45.** $s \geq 50$

Page 687 Lesson 13-6A
1. The ratios are the same. **3.** The ratios of the given angles are the same no matter how long the sides are.

Pages 690–692 Lesson 13-6
7. $\frac{24}{25}$ **9.** $\frac{7}{25}$ **11.** $\frac{24}{7}$ **13.** 0.3256 **15.** 0.4226 **17.** 80° **19.** about 19 feet **21.** 0.724, 0.690, 1.05 **23.** 0.6, 0.8, 0.75 **25.** 0.246, 0.969, 0.254 **27.** 0.5000 **29.** 0.9063 **31.** 0.9455 **33.** 53° **35.** 44° **37.** 61° **39.** 10° **41.** 42° **43.** 19° **45.** about 84.8° **47.** 10 yd **49.** 9 ways **51.** deductive
53. -242;
−250 −245 −240

Page 693 Lesson 13-6B
1. $\frac{1}{2}$ **3.** They are the same. **5.** 3

Pages 696–697 Lesson 13-7
5. $\tan 39° = \frac{10}{x}$; 12.3 **7.** about 18.1 feet **9.** $\sin 60° = \frac{x}{16}$; 13.9 **11.** $\cos 30° = \frac{9}{x}$; 10.4 **13.** $\sin x° = \frac{8}{10}$; 53.1 **15.** $\sin x° = \frac{15}{21.2}$; 45.0 **17.** 38.3 yards **19.** about 118 feet **21.** about 44 m **23.** $a \approx 12.4$ yd, $b \approx 15.9$ yd **25.** 37 feet **27.** about 62° **29.** 30; 62; 70.5; 48.5; 22 **31.** yes

Page 699 Highlights
1. square **3.** rational **5.** 30°–60°

Pages 700–702 Study Guide and Assessment
7. 3 **9.** −5 **11.** 15 **13.** 9 **15.** 15 **17.** −6 **19.** whole, integer, rational, real **21.** rational, real **23.** 14, −14 **25.** 3.9, −3.9 **27.** 20 **29.** 34 **31.** 16 **33.** 6 m **35.** 3 cm **37.** $\sin X = 0.882$, $\cos X = 0.471$, $\tan X = 1.875$ **39.** 84° **41.** $\tan 35° = \frac{x}{15}$; 10.5 ft **43.** $\sin 25° = \frac{15}{x}$; 35.5 in. **45.** 28.4 ft **47.** about 321.68 ft

Chapter 14 Polynomials

Pages 708–709 Lesson 14-1
5. yes; monomial **7.** no **9.** 3 **11.** 3 **13.** 35 **15.** 57 **17.** \$38.82 **19.** yes, binomial **21.** yes, trinomial **23.** yes, monomial **25.** yes, trinomial **27.** yes, monomial **29.** yes, trinomial **31.** 1 **33.** 3 **35.** 6 **37.** 5 **39.** 7 **41.** 5 **43.** −26 **45.** −18 **47.** −43 **49.** 97 **51.** −44 **53.** 82 **55a.** $2x + 2y + z + xy$ **55b.** 2 **59.** 156.8 m^2 **61.** \$104.37 **63.** yes; 3; 81, 243, 729 **65.** $x > 8$ **67.** 13

Page 710 Lesson 14-1B
1. A monomial uses just one size of tiles, a binomial uses two different sizes, and a trinomial uses three different sizes. **3.** The degree of the polynomial is determined by the size of the largest algebra tile.

Pages 713–714 Lesson 14-2
7. $3x + 8$ **9.** $4x + 4$ **11.** $3x^2 + 3x + 2$ **13.** $6x + 24$ units **15.** $4x + 14$ **17.** $11m - n$ **19.** $2x^2 - 2x + 9$ **21.** $17m + 3n$ **23.** $8x^2 - 3x + 3$ **25.** $4a^3 + 2a^2b - b^2 + b^3$ **27.** $4a + 6b + c$ **29.** $-2a - 3ab + 4b^2 + 2$ **31.** $4x^2 + x - 1$ **33.** $4a^2 - 3ab - 2b - 4b^2$; 0 **35.** $(4r + 3s)$ **37.** $4x + 2$ united inches **39.** 7.5 cm^2 **41a.** $t + h > 130$

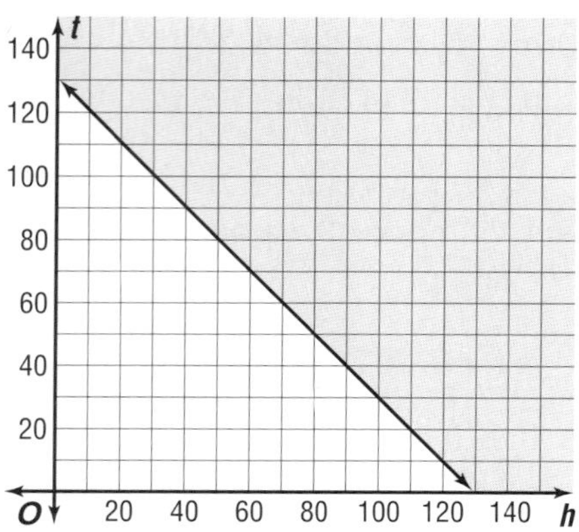

41b. the area above the boundary line

41c. No points with $h > 100$ are included in the graph.
43. 15 ft

Pages 717–718 Lesson 14-3
9. $-2abc$ **11.** $-x^2 - 5x - 1$ **13.** $8m - 7n$ **15.** $2x + 2$ **17.** $2x - 3$ **19.** $x^2 + 3x - 3$ **21.** $4x - 6$ **23.** $x - 3$ **25.** $2x - 8$ **27.** $-3x + 5y$ **29.** $-2x^2 + 6x - 7$ **31.** $m^2 - 7m + 10$ **33.** $15m^2 - 2m + 10$ **35.** $9a + 3b - 7c$ **37.** $8y^2 + 4y - 5$ **39.** $-6x^3 - 4x^2y + 13xy^2$ **41a.** 184 **41b.** 370 **43.** $8a^2 + 2a + 1$ **45.** 90°; right **47.** every 12 days **49.** II

Page 718 Self Test
1. yes, binomial; 3 **3.** yes, monomial; 12 **5.** $-x^2 + 12$ **7.** $2x + 6y$ units **9.** $a + 13d$

Pages 722–723 Lesson 14-4
5. 2^{12} **7.** $3^3x^3y^3$ or $27x^3y^3$ **9.** x^2y^2 **11.** 9 **13.** $48x^6y^2$ **15.** y^4z^4 **17.** $16r^4s^4$ **19.** m^{10} **21.** $-32x^{10}$ **23.** x^4y^{12} **25.** $12x^3$ **27.** $16x^4y^6$ **29.** t^{-12} **31.** $-2x^{-6}y^{-2}$ **33.** -36 **35.** 108 **37.** -1296 **39.** $\frac{1}{324}$ **41a.** The numbers in the values columns are the same. **41b.** The exponents are distributed across the factors. **41c.** When you raise a product to a power, distribute the exponent over each factor. **43a.** $1000(1.01^4)^6$ **43b.** $1269.73 **45.** 5280^3 **47.** 13 in. **49.** A spinner divided into 10 sections, six of them being successes.

51.

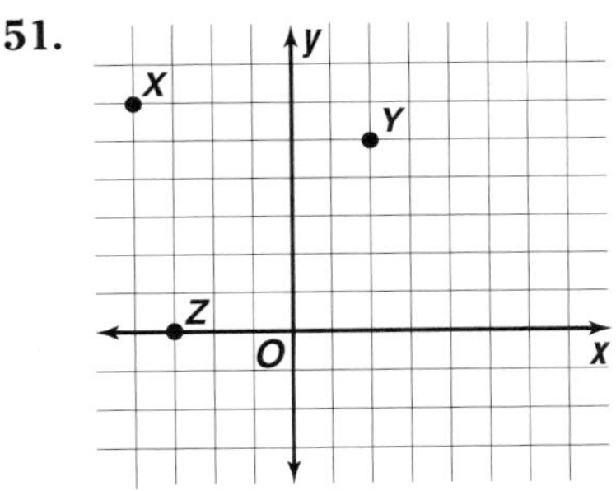

51a. II **51b.** I **51c.** none **53.** $xy \div 5$

Page 724 Lesson 14-5A
1. true **3.** $x^2 + 5x$ **5.** $6x^2 + 3x$

Pages 726–727 Lesson 14-5
7. $3x + 12$ **9.** $3x - 6$ **11.** $12x^2 + 28x$ **13.** 160 ft by 360 ft **15.** $-2x - 16$ **17.** $6x^2 - 3x$ **19.** $a^2c + bc$ **21.** $p^2q^2 + 8pq$ **23.** $-18a + 2a^3$ **25.** $-15x^2 + 35x - 45$ **27.** $4c^4 + 28c^3 - 40c$ **29.** $-12x^5 + 48x^4 - 42x^2$ **31.** $-x^5 + x^4 - 3x^3 + 5x^2$ **33.** 2 **35a.** $12x^2 + 8x + 1$ **35b.** 96 in² **37.** $-32b^4$ **39.** Sample answer: Choose a scale that makes one product look much better than the other; omit some information or labels. **41.** 13 times **43.** 192.5 miles

Pages 730–731 Lesson 14-6
7. $x + 3, 2x + 1; 2x^2 + 7x + 3$ **9.** $2x^2 + 5x + 3$ **11.** $10x^2 - x - 3$ **13.** $2x + 1, x + 2; 2x^2 + 5x + 2$ **15.** $2x + 3, 2x + 2; 4x^2 + 10x + 6$ **17.** $x^2 + 7x + 12$ **19.** $x^2 - 4x - 12$ **21.** $x^2 - 5x - 36$ **23.** $2x^2 - 5x - 12$ **25.** $x^2 + 6x + 9$ **27.** $10x^2 - 11x - 6$ **29.** $18x^2 - 3x + \frac{1}{8}$ **31.** $(x + 2)(x - 1); x^2 + x - 2$ **33a.** 14 in. by 27 in. **33b.** 378 in² **33c.** 464 in² **35.** 90 ft **37.** 3.14 m **39.** $x < -14$

Page 732 Lesson 14-6B
1. yes **3.** no **5.** no

Page 733 Highlights
1. sum **3.** greatest **5.** additive inverse

Pages 734–736 Study Guide and Assessment
11. yes; binomial **13.** yes; monomial **15.** yes; binomial **17.** 1 **19.** 2 **21.** 3 **23.** $5x^2 - 4x$ **25.** $x^2 - x$ **27.** $5x^2 + 3x + 1$ **29.** $4a - 15b$ **31.** $2a^2 - 3b^2 + c^2$ **33.** $16x^2 - x - 7$ **35.** a^6 **37.** p^6q^3 **39.** $4y^7z^3$ **41.** $-6a^8b^7$ **43.** $8d^2 - 20d$ **45.** $2a^5 + a^3 - 5a^2$ **47.** $-2g^4 - 12g^2 - 6g$ **49.** $2x + 1, x + 3; 2x^2 + 7x + 3$ **51.** $2x^2 + 3x + 1$ **53.** $2xy + 2y^2 + 2yz$ **55.** $3x^3$

PHOTO CREDITS

2, 10, 35, 56 (t) Vince Streano/Tony Stone Images; **3** (t) Charles O'Rear/Westlight, (b) Chris Sorensen; **4** (bl) Lois Gervais/Westlight, (bc) Nikolai Zurek/FPG International, (br) Bettmann; **5** (t) Bob Daemmrich/Stock Boston, (bl) Beverly Johnson by Francesco Scavullo August 1974 Vogue. Courtesy Vogue. ©1974 by the Conde Nast Publications, Inc.; (br) Carl Corey/Westlight; **6** (t) Duomo, (b) Mark Burnett; **7** Stacy Pick/Stock Boston; **8** Custom Medical Stock Photo; **9** Mark Steinmetz; **11** Mark Burnett; **13** Joe Rimkus Jr./The Miami Herald; **14** Lester Lefkowitz/Tony Stone Images; **15** W. Cody/Westlight; **17** Don Smetzer/Tony Stone Images; **19** (t) Jack Zehrt/FPG International, (b) Dr. E.R. Degginger/Color-Pic; **20** (t) Wide World Photos, (b) FOXTROT ©Bill Amend. Reprinted with permission of UNIVERSAL PRESS SYNDICATE. All rights reserved.; **23** (l) David Allen/LGI Photo Agency, (r) David Allen/LGI Photo Agency; **27** Michael S. Yamashita/Westlight; **28** Katrina De Leon/Still Life Stock; **30** (t) Mark Burnett, (b) E. Sampers/Gamma Liaison; **32** (t) David Young Wolff/Tony Stone Images, (b) Mark Burnett; **33** Doug Martin; **34** Frank Spinelli/Liaison International; **36** Matt Meadows; **38** Mark Burnett; **39** National Optical Astronomy Observatories; **41** Bettmann; **43** Aaron Haupt; **44** Ben Van Hook/Duomo; **45** Photofest; **47** Mark Burnett; **48** Owen Franken/Stock Boston; **50** Matt Meadows; **51** Murray & Associates/Tony Stone Images; **52** Mark Reinstein/FPG International; **54** Lynn Stone; **55** Alexander Gardner, 1863/Bettmann Archive; **56** Chris Sorensen; **61-63** Mark Burnett; **64** (t) Mark Burnett, (bl) Eberhard E. Otto/FPG International, (br) Chromosohm/Joe Sohm/Photo Researchers; **65** FRANK & ERNEST, reprinted by permission of Newspaper Enterprise Association, Inc.; **65** (tl) Steve Gottlieb/FPG International, (b) Bettmann; **66** Buddy Mays/FPG International; **69** Mark Conlin; **70** D.R. Goff, courtesy CAPA; **73** (t) Michael Keller/FPG International, (b) J. Scower/FPG International; **76** Robin Smith/Tony Stone Images; **78** (t) David Scharf/Peter Arnold, Inc., (b) Mark Burnett; **79** Mark Steinmetz; **80** Jose Carrillo; **81** Steve Seiller/Ken Graham Agency; **84** AP/World Wide Photos; **86** Stephen Rose/Gamma International; **87** (t) G. Randall/FPG International, (b) Stan Osolinski/FPG International; **90** Stan Osolinski/FPG International; **92** AP/Wide World Photos; **93** R. Y. Kaufman/Yogi, Inc.; **94** Biblioteca Universitaria, Bologna, Italy/Art Resource; **96** Comstock Inc./Gary Benson; **97** Mark Conlin; **99** (t) Mark Conlin, (l) Painting by Ken Marshall from the book *The Discovery of the Titanic* by Dr. Robert Ballard, (b) Stephen Rose/The Gamma Liaison Network; **101** Josef Beck/FPG International; **102** Ken Graham/Ken Graham Agency; **103** Joseph Pobereski/Tony Stone Images; **106** Remi Benali/Gamma International; **107** John Terence Turner/FPG International; **108** (t) courtesy NASA, (b) Comstock Inc./Jack Clark; **112** Carl R. Sams II/Peter Arnold, Inc.; **113** Wally Dallenbach/Peter Arnold, Inc.; **116** (t) Hoffines/Young, (br) courtesy Perkins School for the Blind, (bl) Bettmann; **117** (t) Ron Rovtar/FPG International, (b) Bettmann; **118** Matt Meadows; **120** Nawrocki Stock Photo; **121** (l) John Elk III/Stock Boston, (r) J. McGuire/Washington Stock Photo; **122** (t) Henry Horenstein/Viesti, (b) Mark Steinmetz; **125** Photofest; **127** Bonnie Sue/Photo Researchers; **128** Photofest; **129, 131** Mark Burnett; **132** Nawrocki Stock Photo; **133** (t) Sharon M. Kurgis, (b) Deborah Denker/Liaison International; **134** Frank Whitney/The Image Bank; **135** Fred Bavendam/Peter Arnold, Inc.; **136** Nawrocki Stock Photo; **137** (t) Joe Viesti/Viesti Associates, (b) Glencoe files; **138** Mark Burnett; **141** Michael Newman/PhotoEdit; **142** Marcel Ehrhard/Liaison International; **144** (t) Gerard Lacz/Peter Arnold, Inc., (b) ZIGGY ©1978 ZIGGY & FRIENDS, INC. Dist. by UNIVERSAL PRESS SYNDICATE. Reprinted with permission. All rights reserved.; **145** Nawrocki Stock Photo; **148** Mark Burnett; **149** Uniphoto; **153** Frank Spinelli/Liaison International; **154** Charlie Trainor/The Miami Herald; **155** (t) Michael Simpson/FPG International, (b) Cliff Beaver; **159** John Lei/Stock Boston; **160** Mark Burnett; **164** Bettmann; **165** Robert Burke/Liaison International; **167** Mark Steinmetz; **168** (br) Frank Driggs Collection, (bl) Painting by Abraham Anderson, c. 1889, National Portrait Gallery, Smithsonian/Bettmann Archives; **169** (t) Comstock Inc./Comstock Inc., (c) Ken Frick, (br) Photofest, (bl) Charles O'Rear/Westlight; **172** Bridgeman/Art Resource, NY; **173** (t) Myrleen Ferguson/PhotoEdit, (b) Nawrocki Stock Photo; **174** G. Randall/FPG International; **175** Aaron Haupt; **177** Comstock Inc./Contrib. Library; **178** Comstock Inc./Comstock Inc.; **179** (t) Alinari/SEAT/Art Resource, NY, (b) THE FAR SIDE ©1985 FARWORKS, INC. Dist. by UNIVERSAL PRESS SYNDICATE. Reprinted with permission. All rights reserved.; **181** (t) Pam Francis/Liaison International, (b) Lawrence Migdale; **183** Lynn Stone; **185** Rosemary Weller/Tony Stone Images; **186** courtesy Dr. Arjen Lenstra, Bell Communications Research; **187** Photofest; **188** (t) Everett Collection, (b) Dick Luria/FPG International; **190** S. Feld/H. Armstrong Roberts, Inc.; **191** Elizabeth Simpson/FPG International; **192** Aaron Haupt; **194** Comstock Inc./Stuart Cohen; **196** courtesy GCI/Atlanta; **199** (t) Matthew Neal McVay/Tony Stone Images, (b) James Blank/FPG International; **200** Stuart Westmorland/Tony Stone Images; **201** NASA/Peter Arnold, Inc.; **203** Matt Meadows; **204** (t) Mike Theiler/Reuters/Bettmann Archive, (b) Larry Ulrich/Tony Stone Images; **207** Paul Richards/UPI/Bettmann Newsphotos; **208, 211** Manfred Kage/Peter Arnold, Inc.; **212** Howard Hughes Medical Institute/Peter Arnold, Inc.; **214** (t) Arthur Tilley/Tony Stone Images, (b) Stock Montage; **219** Comstock Inc./Robert Houser; **222** *American Heritage* magazine; **223** (t)*The Saturday Evening Post* © 1994, (c) Ken Huang/The Image Bank, (br) Mark Steinmetz, (bl) Studiohio; **224** (l) Photofest, (r) Everett Collection; **225** Bob Eighmie/The Miami Herald; **227** Steven Alexander/Uniphoto Picture Agency; **228** (l) courtesy Ingrid Proctor-Fridia, (r) Leonard Lessin/Peter Arnold, Inc.; **229, 230** Mark Burnett; **232** courtesy N.O.A.A.; **233** Robert Landau/Westlight; **236** Eugene Gebhardt/FPG International; **237** Bettmann; **241** G. Schwartz/FPG International; **242** (t) Mark Burnett, (b) David W. Hamilton/The Image Bank; **243** Mark Burnett; **245** Frank Saragnese/FPG International; **248** NASA/Peter Arnold, Inc.; **249, 250** Bettmann; **252** Mark Burnett; **254** (t) William R. Sallaz/Duomo, (b) Mark Steinmetz; **255** PEANUTS reprinted by permission of United Feature Syndicate, Inc.; **256** (t) Stephen Dalton/Photo Researchers, (b) Mark Steinmetz;

257 ©1966 *The Saturday Review,* reproduced by permission of Ed Fisher; **258** A. T. Willett/The Image Bank; **260** Mark Steinmetz; **261** Studiohio; **262** (t) Science and Society Picture Library, (b) Shinichi Kanno/FPG International; **264** Mark Steinmetz; **268** First Image; **269** David M. Grossman; **271** Mark Burnett; **272** (t) Uniphoto, (bl, bc) Bettmann, (br) courtesy International Business Machines Corporation; **273** (t) Drawing by W. Miller, 1994 *The New Yorker* Magazine, Inc., (c) Arthur Tilley/FPG International, (bl) Photofest, (br) Kazu Studio Ltd./FPG International; **274** Mark Burnett; **276** Tim Courlas; **277** courtesy Wendy's International, Inc.; **278** (l) David Zosel, (r) NASA; **279** Raphael. *The Triumph of Galatea.* Palazzo della Farnesina, Rome, Italy./Art Resource; **281** Michal Heron/The Stock Market; **283** (l) Mark Sennet/Onyx, (r) Laurel Canty, Dolphin Research Center, Grassy Key, Florida; **285** Richard Bowditch; **287** Mark Steinmetz; **288** Rudy VonBriel; **290** Bob Daemmrich; **292** Aaron Haupt; **293** Dave Bartruff; **295** Aaron Haupt; **297** Michael Nelson/FPG International; **298** Elaine Shay; **299** Aaron Haupt; **300** Will Vinton/Photofest; **301** U.S. Representative/Westlight; **303** (l) James Schnepf/The Gamma Liaison Network, (r) Thomas Ives/The Stock Market; **304** Mark Steinmetz; **305** Chromosohm/Sohm/The Stock Market; **306** (l) Jim Pickerell/Westlight, (r) Dan Lecca/FPG International; **308** (t) Terence Gili/FPG International, (b) Mark Steinmetz; **310** Photofest; **312** Matt Meadows; **314** Long Photography; **315** Mark Burnett; **317** Manfred Gottschalk/Westlight; **319** (l) Paul W. Nesbit, (r) Stan Osolinski/FPG International; **325** Richard Pasley/Viesti Associates; **328** (t, br) FPG International, (bl) Science & Society Picture Library; **329** (t) THE FAR SIDE ©1990 FARWORKS, INC. Dist. by UNIVERSAL PRESS SYNDICATE. Reprinted with permission. All rights reserved., (b) Photofest; **331** Matthew Klein/Photo Researchers; **332** (t) James Blank/FPG International, (b) Bob Daemmrich; **334** E. Nagele/FPG International; **336** Mark Steinmetz; **337** Aaron Haupt; **338** Focus on Sports; **339** Photofest; **340** from *Not Strictly By the Numbers* ©Carolina Biological Supply Company; **342-3** Jack Zehrt/FPG International; **343** Mark Steinmetz; **344** Ben Simmons/The Stock Market; **349** (t) Elaine Shay, (b) Alan Pitcairn from Grant Heilman; **350** (t) Aaron Haupt, (b) Comstock Inc./Stuart Cohen; **352** Larry Lefever from Grant Heilman; **353** Mark Steinmetz; **354** Warren Morgan/Westlight; **355** Mark Steinmetz; **357** courtesy John Deere Precision Farming; **358** Joachim Messerschmidt/FPG International; **359** Mark Burnett; **360** (t) Bob Daemmrich, (bl) Dr.E.R. Degginger/Color-Pic; **361** (l) Mark Steinmetz, (r) Aaron Haupt; **362** James Marshall/The Stock Market; **366** (l) The Smithsonian Institution, (r) From the Al Rose Collection, Hogan Jazz Archive, Tulane University; **369** (t) Bob Daemmrich, (b) Flip Chalfant/The Image Bank; **370** (t) Ben Van Hook/Black Star Publishing, (b) Archive Photos; **371** (t) *The Saturday Evening Post* ©1995, (c) Cindy Lewis, (bl) Bettmann, (br) Michael R. Brown/Gamma-Liaison; **372** Comstock Inc./Comstock Inc.; **374** MAK-I; **375** Fred Hirschmann/Ken Graham Agency; **377** (t) Henry Groskinsky/Peter Arnold, Inc., (b) Telegraph Colour Library/FPG International; **378** Aaron Haupt; **381** Long Photography; **382** Uniphoto; **383** Brian Drake/Rim Light Photography; **384** J.R. Schnelzer; **388** Comstock Inc./M & C Werner; **390** ©1989 by Sidney Harris - *Einstein Simplified,* Rutgers University Press; **394** Jose Carrillo; **395** (l) J&L Weber/Peter Arnold, Inc., (r) Mark Stephenson/Westlight, (b) Steve Lissau; **397** Ken Skalski/Liaison International; **398** Elaine Shay; **400** B. Busco/The Image Bank; **402** Ken Graham/Ken Graham Agency; **404** Bob Daemmrich; **405** Mark Steinmetz; **406** Aaron Haupt; **408** Penny Gentieu/Tony Stone Images; **410** Warren Motts; **412** Peter Timmerans/Tony Stone Images; **416** James Shaffer/PhotoEdit; **419** Photofest; **421** William J. Weber; **422** (t) Mark Conlin, (b) Bob Daemmrich; **430** (l) Francois Gohier/Photo Researchers, (r) Larry Hamill; **431** (t) Anthro-Photo, (l) John Reader/Science Photo Library/Photo Researchers, (r) Francois Gohier/Photo Researchers; **432** Mark Burnett; 433 Aaron Haupt; **434** Uniphoto; **435** Eberhard Otto/FPG International; **436** John Mitchell/Photo Researchers; **437** Bettmann; **438** (t) Mark Steinmetz, (c) *People Weekly* 1995 Steve Liss, (b) Focus on Sports; **439** Fred Lyon/Photo Researchers; **440** courtesy WSNY; **441** (l) Matt Meadows, (r) Mark Steinmetz; **443** courtesy NASA; **444** (l) P. Le Segretain/Sygma, (r) De Keerle/Sygma; 445 Mark Steinmetz; **446** Doug Martin; **447** Mark Steinmetz; **448** D. Robert Franz/Masterfile; **449** Mark Burnett; **452** (t) Ralph Clevenger/Westlight, (b) Mark Steinmetz; **454** Mark Burnett; **455** W. Hodges/Westlight; **456** Doug Martin; **459** Rosemary Weller/Tony Stone Images; **461** (l)Uniphoto, (r) Paul Morse/LGI Photo Agency; **462, 463** Aaron Haupt; **466** (t) Cindy Lewis Photography, (b) THE BORN LOSER reprinted by permission of Newspaper Enterprises Association, Inc.; **467** Bob Daemmrich; **468** courtesy The Nature Conservancy; **469** Lisa Quinones/Black Star; **470** (t) Larry Mulvehill/Photo Researchers, (b) Aldo Pavan/Gamma Liaison Network; **471** Auguste Renoir. *The Reader,* 1874-1876. Musee d'Orsay, Paris, France. Erich Lessing/Art Resource, NY; **475** Geof du Feu/Masterfile; **476** Brian Bahr/Allsport; **480** courtesy Vehicle Research & Testing Center; **481** Mark Steinmetz; **482, 508, 528** Mark Burnett; **483** KS Studio; **483** Rod Joslin; **483** (t) Mark Steinmetz, (b) Mark Burnett; **484** Bettmann; **485** (t) Gaslight Advertising Archives, (l) Glencoe files, (c) Bettmann, (r) Mark Steinmetz; **486** (l) Mark Burnett, (r) Glencoe files; **488** (l) Mark Burnett, (r) Ed Bock/The Stock Market; **490** Mark Steinmetz; **492** Everett Collection; **492** Mark Burnett; **493** David R. Frazier Photolibrary; **494** James Carlson/David R. Frazier Photolibrary; **495** Trent Steffler/David R. Frazier Photolibrary; **498, 500** Mark Steinmetz; **502** Matt Meadows; **503** Nawrocki Stock Photo; **504** Barbara Campbell/Gamma Liaison International; **505** Matt Meadows; **507** Daniel J. Cox/Gamma Liaison International; **508, 509** Mark Burnett; **510** (l) Mark Steinmetz, (r) David M. Grossman; **513** Comstock Inc./Comstock Inc.; **514** Conway Trefethen/FPG International; **515** Tim Davis/Photo Researchers; **516** Larry Kunkel/FPG International; **517** Nawrocki Stock Photo; **518** Mark Burnett; **519** courtesy Dr. Carol Malloy; **520, 521** Mark Steinmetz; **521** (t) Mark Burnett from the Paul Baumann collection; **523** (t) Cecil Fox/Science Source/Photo Researchers, (b) PEANUTS reprinted by permission of United Feature Syndicate, Inc.; **524** Corbis Media; **525** Mary Ann Evans; **526** James Levin/FPG International; **527** Ken Biggs/The Stock Market; **529** Mark Burnett from the Paul Baumann collection; **530** Thayer Syme/FPG International; **532** Mark Steinmetz; **533** Index Stock Photography; **536** Mark Burnett; **538** Mark Steinmetz; **543** Philippe Poulet/Gamma Liaison International; **546** (l) Bettmann, (c) Stock Newport/Daniel Forster, (r) Thomas Zimmermann/FPG International; **547** (t) G. Whiteley/Photo Researchers, (bl) Stock Newport/Bob Grieser, (br) Stock Newport/Daniel Forster; **548** David Zalubowski/AP Colorado; **550** Bettmann; **551** Mathias

Oppersdorff/Photo Researchers; **552** Mark Burnett; **556** Tracy Borland; **557** Bettmann; **559** (t) Comstock Inc./Comstock Inc., (b) Harald Sund/The Image Bank; **560** (t) Mark Burnett, (b) Russell D. Curtis/Photo Researchers; **564** Mark Steinmetz; **565** David R. Frazier Photolibrary; **566** CNRI/Science Photo Library/Photo Researchers; **567, 568** David R. Frazier Photolibrary; **570** David Gorman; **576** Carl Purcell/Photo Researchers; **577, 578** Mark Steinmetz; **580** V.A.P. Neal/Photo Researchers; **583** Mark C. Burnett/Photo Researchers; **584** Janet Adams; **591** Defense Dept.; **594** Pictures Unlimited; **595** David R. Frazier Photolibrary; **596** (l) Phil Degginger, (lc) Matt Meadows, (c) Aaron Haupt, (rc) Mark Steinmetz, (r) Mark Burnett; **597** (l) Dr.E.R. Degginger/Color-Pic, (r) Harald Sund/Image Bank; **599** (t) Audrey Gibson/Mark E. Gibson Photography, (b) FOXTROT ©Bill Amend. Reprinted with permission of UNIVERSAL PRESS SYNDICATE. All rights reserved.; **600** Mark Burnett; **605** Scott Camazine/Photo Researchers; **606** Jose Luis Banus-March/FPG International; **607** David A. Barnes/FPG International; **608** (l) Bridgeman/Art Resource, NY, (r) Scala/Art Resource, NY; **609** (t) Uniphoto, (l) Bettmann, (c) Superstock, (r) Photofest; **612** (l) Glencoe files, (r) Mark Burnett; **615** (l) Mark Steinmetz, (r) Mark Burnett; **617** Joseph Nettis/Photo Researchers; **618, 619** Mark Steinmetz; **621** Mark Segal/Tony Stone Images; **622** (t) Howard Bluestein/Photo Researchers, (b) FRANK & ERNEST, reprinted by permission of Newspaper Enterprise Association, Inc.; **623** Mark Steinmetz; **624** Mark Burnett; **625** Heinz Fischer/The Image Bank; **626** David R. Stoecklein/TexStock; **629** Werner Muller/Peter Arnold, Inc.; **630** Ben Mitchell/The Image Bank; **631** Navaswan/FPG International; **632** Mark Steinmetz; **634** Mark Burnett; **636** Gary Cralle/The Image Bank; **638** Marvin E. Newman/The Image Bank; **640** Jules Zalon/The Image Bank; **641** courtesy Dr. Michael R. Doss; **645** Gerard Photography; **646** Ellis Herwig/Stock Boston; **647** E.R.Degginger/Photo Researchers; **648** Herve Berthoule Jacana/Photo Researchers; **650** Bernard Roussel/The Image Bank; **651** Mark Burnett; **653** William E. Ferguson; **654** Mark Steinmetz; **658** David M. Grossman; **659** Ancient Art & Architecture Collection; **662** (t) George Obremski/The Image Bank, (l) Glencoe files, (r) Scala/Art Resource, NY; **663** (t) Reprinted from *The Saturday Evening Post* ©1995, (bl) Peter Beney/The Image Bank, (br) Bettmann; **664** Superstock; **665** Mark Burnett; **667** Cliff Feulner/The Image Bank; **668** CALVIN AND HOBBES ©1990 Watterson. Dist. by UNIVERSAL PRESS SYNDICATE. Reprinted with permission. All rights reserved.; **669** (t) Travelpix/FPG International, (b) Telegraph Colour Library/FPG International; **670, 672** David R. Frazier Photolibrary; **674** Index; **675** Doug Martin; **678** Leverett Bradley/FPG International; **681, 682** Mark Burnett; **683** Stock Montage; **684** Pictures Unlimited; **685** Mark E. Gibson/Mark E. Gibson Photography; **686** Janine Herron/FPG International; **687** Morton & White Photographic; **690** James Blank/FPG International; **692** John Elk III/Stock Boston; **695** Bill Lea; **697** Aaron Haupt/David R. Frazier Photolibrary; **698** (t) Courtesy Sullivan Gray Bruck Architects, Inc., (b) courtesy Darlene Leshnock; **702** Ron Thomas/FPG International; **704** (l) Mark Burnett, (c) Bettmann, (r) Superstock; **705** (t) PEANUTS reprinted by permission of United Feature Syndicate, Inc., (l) Mark Burnett, (r) courtesy Bryan Berg; **706** Tom Carroll/FPG International; **707** (l) Will & Deni McIntyre/Photo Researchers, (r) Photo Researchers; **708** (t) Mark Burnett, (b) John Kaprielian/Photo Researchers; **709** Audrey Gibson/Mark E. Gibson Photography; **709** Craig K. Lorenz/Photo Researchers; **712** Ronald Sheridan/Ancient Art & Architecture Collection; **713** Glencoe files; **714** (t) Index, (b) Spencer Grant/Photo Researchers; **715** Steve Allen/Gamma Liaison; **716** Beth Ava/FPG International; **718** John McDonough/Focus on Sports; **720** Superstock; **723** (l) Porterfield-Chickering/Photo Researchers, (r) Platinum Productions; **724** Index; **725** Glencoe files; **726** Franc Shor/©National Geographic Society; **727** (t) Mark Burnett, (b) Mark Steinmetz; **728** Index; **729** Ed Braverman/FPG International; **730** Superstock; **731** Superstock; **736** Tony Quinn Photography; **Cover** Gus Chan; **Cover** Tony Stone Images; **Cover** Tony Stone Images

APPLICATIONS INDEX

Academics, 336
Accounting, 92
Advertising, 127, 470, 617
Aerospace, 108
Aerospace Design, 534
Agriculture, 357, 361, 658, 668
Air Travel, 159, 396
Airports, 250
Ambulance Service, 534
Anthology, 395
Architecture, 434, 435, 488, 591, 622, 637, 639, 641, 686
Art, 20, 209, 471
Astronomy, 39, 87, 200, 213, 324
Automotive, 344
Aviation, 92, 366, 410, 438, 553, 697

Baking, 648
Banking, 299
Basketball, 718
Biology, 133, 208, 283, 384, 559, 566, 723
Boating, 247
Business, 15, 29, 93, 121, 150, 194, 218, 233, 245, 279, 293, 306, 311, 398, 410, 416, 422, 461, 470, 513, 526, 536, 560, 636, 731

Careers, 133
Carpentry, 193, 243, 268, 390, 403, 564, 566, 668
Cartography, 73
Catering, 709
Census, 349
Charity, 49, 468
Chemistry, 80, 208, 336, 648, 723
Civics, 48, 174
Civil Defense, 571
Communication, 150, 598, 702
Computers, 494, 626
Conservation, 507
Construction, 144, 247, 575, 577, 583, 588, 653, 668, 684, 686, 692, 714
Consumer Awareness, 70, 81, 353, 359, 377, 422, 435, 475, 534, 538, 577, 648, 652, 714
Consumerism, 121, 230, 288, 336, 433, 501, 617
Cooking, 144, 288, 350, 446
Cosmetics, 459, 461
Cryptography, 186, 188
Cycling, 237

Decorating, 283
Demographics, 45, 188, 199, 279, 443
Desktop Publishing, 290
Driver's License Exam, 593
Driving, 256

Ecology, 361, 709
Economics, 108, 243, 299, 324, 445, 470, 489, 533, 538, 709
Education, 118-119, 179, 311, 383, 525
Electricity, 178, 674
Employment, 254, 281, 306, 501, 507
Energy, 397, 398, 473, 475
Engineering, 29, 228, 489, 646
Entertainment, 35, 45, 125, 157, 179, 233, 241, 332, 344, 353, 455, 466, 504, 522
Exercise, 128, 188

Family Activity, 144, 188, 254, 306, 361, 422, 435, 443, 523, 559, 637, 675, 709
Family Management, 55
Farming, 637
Fashion, 257, 456
Figure Skating, 254
Finance, 35, 49, 133, 513, 521, 671, 723
Fire Fighting, 695
Food, 228, 233, 276, 283, 293, 311, 337, 466, 457, 642
Food Preparation, 404
Forestry, 196, 324, 453
Fund Raising, 233, 238

Games, 441, 447, 480, 531, 542, 599, 626, 648
Game Show, 70
Geography, 76, 87, 92, 174, 316, 383, 395, 436, 508, 559, 588, 617
Geology, 39, 320, 653
Geometry, 10, 30, 35, 39, 45, 55, 93, 159, 177, 178, 183, 188, 243, 288, 293, 340, 342, 344, 357, 354, 357, 361, 377, 395, 399, 403, 414, 415, 438, 443, 457, 518, 519, 523, 528, 577, 583, 593, 599, 617, 627, 631, 637, 642, 653, 668, 670, 671, 675, 681, 686, 692, 702, 709, 714, 718, 723, 727, 731
Grades, 149, 155

Health, 49, 60, 155, 288, 293, 366, 390, 399, 452, 461, 463, 553, 599
Highway Maintenance, 650
Highway Safety, 401
Hiking, 681
History, 183, 188, 193, 199, 465, 583, 622, 641, 653
Hobbies, 148, 191, 254, 268
Hockey, 527
Home Improvements, 636
Home Projects, 243
Horse Riding, 122
Horticulture, 658
Housekeeping, 457
Human Growth, 416

Jewelry, 453
Jobs, 136

Kennels, 143
Keyboarding, 382

Landscaping, 144, 150, 559, 615, 622, 708
Law Enforcement, 416
License Plates, 519
Life Science, 350
Lifestyles, 155
Literature, 179, 218
Living Expenses, 426

Manufacturing, 133, 228, 475, 593, 634, 637, 658, 727
Measurement, 228, 288, 501, 508, 731
Medicine, 25, 40, 199, 250, 252, 336, 470, 522, 707
Merchandising, 194, 475
Meteorology, 19, 79, 127, 179, 261, 262, 299, 489, 494, 528, 675, 697
Mining, 395, 626
Models, 447
Money, 340
Money Management, 10
Mountain Climbing, 697
Music, 366, 456

Navigation, 702
Number Theory, 571
Nutrition, 15, 247, 300, 426, 465, 487

Oceanography, 93, 306, 393
Oil Production, 250

Paper Recycling, 154
Patterns, 384, 435, 447, 457, 466, 508, 553, 718
Personal Finance, 108, 153, 214, 236, 238, 293, 330, 337, 416, 421, 453
Pet Care, 617
Pets, 87, 89, 155, 422, 436, 475, 714
Photography, 361, 447, 583
Physical Science, 137, 213, 228, 268, 361, 398
Physical Therapy, 136
Physics, 112, 164, 262, 278, 319, 377, 389, 566, 675
Plumbing, 668
Politics, 204, 268
Population, 283, 324, 471, 480
Postage, 159, 260, 475
Postal Service, 10, 637
Probability, 447, 528, 538, 553, 558, 559, 566, 572, 593, 617, 631, 642, 653, 723
Public Safety, 622
Publishing, 416, 588

Real Estate, 141, 513, 622
Records, 480
Recreation, 76, 87, 97, 422, 577, 648, 686, 702
Recycling, 232, 237, 297
Research, 344, 465
Retail, 398, 470
Retail Sales, 20, 233, 461, 489, 566, 637

Safety, 480, 496, 681
Sailing, 681
Sales, 350, 419
Scale Drawing, 692
Science, 25, 159, 357, 461, 627, 658
Seismology, 207
Sequences, 293
Sewing, 243, 279, 572
Shipping Rates, 377
Shopping, 133, 150, 243
Social Studies, 55, 173
Space Exploration, 10, 60, 81, 84, 443
Space Flight, 177
Space Travel, 553
Sports, 15, 25, 45, 49, 70, 76, 96, 103, 122, 137, 159, 194, 214, 218, 237, 254, 279, 299, 314, 324, 383, 410, 426, 487, 523, 525, 534, 542, 588, 622, 630, 667
Statistics, 121, 154, 262, 288, 354, 385, 403, 410, 452, 471, 494, 501, 506, 521, 583, 604, 631, 675, 727
Survey, 583

Taxes, 150, 697
Teaching, 501
Technology, 204
Teepees, 642
Telecommunication, 512
Television, 522
Theater, 310
Tourism, 344, 352
Tournaments, 159
Traffic, 255
Traffic Control, 40, 376
Transportation, 159, 199, 236, 542, 690, 692
Travel, 128, 144, 237, 404, 435, 470
TV Game Show, 534

United Nations, 518

Veterinary Medicine, 204
Volunteer Work, 731

Wages, 631
Weather, 103, 112, 390, 422, 714
World Cultures, 494

Zoology, 406

INDEX

A

Absolute value, 67, 76, 78–80, 110
Acute angle, 550, 552–553
Acute triangle, 569, 653
Addition
  associative property of, 234
  commutative property of, 234
  of decimals, 234–237
  of fractions, 239–247
  identity property of, 234
  of integers, 83–84
  inverse property of, 234
  phrases, 18
  property of equality, 125
  property of inequality, 147
Additive inverses, 89, 125–126, 716
Adjacent angles, 561–562, 602
Adjacent leg, 687–691
Algebra connections, *see* Connections
Algebraic expression, 16–19, 21, 240–242
  absolute value, 76
  evaluating, 11–15, 68–69, 76, 102, 106, 112, 176–178, 180, 212–213, 286, 287, 291, 297–298
  simplifying, 27–29, 68–69, 85, 86–87, 90–93
Algebraic fraction, 197
Alternate angles, 563–565
Altitude, 612, 614, 634
Angle of depression, 695–697
Angle of elevation, 694–697
Angles, 549–550, 552–554, 602–603
  acute, 550, 552–553
  adjacent, 561, 602
  alternate exterior, 563
  alternate interior, 563
  complementary, 562, 602
  congruent, 561
  corresponding, 563
  of depression, 695–697
  of elevation, 694–697
  exterior, 591
  interior, 591
  measuring, 549, 584, 590–592, 603
  obtuse, 553, 569
  of polygons, 590–592
  of quadrilaterals, 584, 603
  right, 550, 552–553
  supplementary, 562, 602
  vertical, 561, 602
Applications (refer to Applications Index; *see also* Modeling)
Area, 138
  of a circle, 545, 620–622, 629, 656
  of a parallelogram, 612–613, 620, 656
  of a rectangle, 140–143, 150, 634–635
  of a square, 624, 665
  of a trapezoid, 614–616, 656
  of a triangle, 610–611, 613–614, 616, 656
Area model, 286
Arithmetic sequence, 258–261
Assessment
  Alternative, 61, 113, 165, 219, 269, 325, 367, 427, 481, 543, 605, 659, 703, 737
  Chapter Tests, 774–787
  Cooperative Learning Project, 113, 481, 543, 605, 659
  math journal, 53, 106, 148, 177, 231, 282, 319, 356, 408, 446, 505, 592, 640, 696, 722
  Ongoing Assessment, 114–115, 326–327, 428–429, 544–545, 660–661, 738–739
  performance task, 61, 165, 219, 269, 325, 367, 427, 703, 737
  portfolio, 61, 113, 165, 219, 269, 325, 367, 427, 481, 543, 605, 659, 703, 737
  self evaluation, 61, 165, 219, 269, 325, 367, 427, 481, 543, 605, 659, 703, 737
  Self Test, 30, 93, 137, 194, 247, 301, 350, 399, 457, 513, 572, 642, 681, 718
  Study Guide and Assessment, 58–60, 110–112, 162–164, 216–218, 266–268, 322–326, 364–366, 424–426, 478–480, 540–542, 602–604, 656–658, 700–702, 734–736
  thinking critically, 61, 113, 165, 219, 269, 325, 367, 427, 481, 543, 605, 659, 703, 737
Associative property
  of addition, 22, 24–25, 234, 236, 712
  of multiplication, 22, 24–25, 30, 101
Axes, 73–75

B

Back-to-back stem-and-leaf plot, 487–489
Bar graph, 52–54, 60, 150
Bar notation, 225
Base, 175, 205, 449–451, 614, 632
Binomials, 706–708
  multiplying, 728–731, 732, 735
Boundary, 418–420
Box-and-whisker plot, 495–496, 498–501, 503, 540
Brackets, 12
Break-even point, 416

C

Cake method, 185
Capital gain, 166, 264
Capture-recapture, 448
Careers
  employment opportunities
    design, 547
    engineer, 547
    fashion, 5
    field archaeologist, 609
    marketing research and management, 485
    multimedia development, 169
    naval architect, 547
    paleontologist, 431
    for visually impaired, 117
  help wanted
    artist, 297
    automotive service industry, 134
    civil engineer, 568
    landscape architect, 615
    law enforcement officer, 672
    meteorologist, 249
    oceanographer, 393
    physical therapist, 449
    science-related careers, 358
    travel industry, 26
Cell, 209
Center, 341
Central angle, 344
Central tendency, 505
Change in *x*, 400–408
Change in *y*, 400–408
Circle, 341–344
  area of 545, 620–622
Circle graph, 497, 556–559, 602
Circular cone, 640–641, 650–652, 657
Circumference, 341–344, 364
Closing price, 264
Coefficient, 711
Combination, 516, 517, 541
Commission, 470
Common difference, 258, 258–259
Common multiples, 200–204
Common ratio, 312
Commutative property
  of addition, 22–25, 234, 712
  of multiplication, 22, 24–25
Comparative graph, 498, 500
Compass, 554–555
Compatible numbers, 280–282
Complementary angle, 562–565, 602
Composite number, 184–187
Compound inequality, 357
Cones
  surface area, 640–641
  volume of, 650–651
Congruent, 554, 561, 563, 573, 579
  angles, 561, 573–574
  line segments, 554
  regular polygon, 591
  sides, 570–571, 573–575
  triangles, 573–577
Connections
  algebra, 23, 27, 68, 85, 90, 101, 106, 172, 176, 186, 191, 197, 201, 202,

206, 211, 225, 235, 240, 259, 286, 291, 297, 313, 510, 562, 569, 590, 620, 634, 646, 726
geometry, 134, 401, 405, 434, 445, 677, 712, 716, 720
interdisciplinary
biology, 211, 225
construction, 678
cultural, 32, 66, 72, 146, 172, 308, 342, 385, 441, 459, 524, 593, 629, 678, 712
gardening, 729
geography, 90
health, 463
health and exercise, 7
history, 580
meteorology, 106
music, 669
science, 318
social studies, 52, 172
Consecutive numbers, 340
Constant, 707
Construct, 554
Consumer price index (CPI), 489
Continuous review, *see* Mixed review
Cooperative learning projects, *see also* The Shape of Things to Come, 113, 481, 543, 605, 659, 737
Coordinate, 67, 126
Coordinate system, 36–40, 73
Corresponding angles, 563–564, 574, 579
Corresponding parts, 573–576, 579, 603
Cosine, 688–691, 695–697, 701
Cross products, 444–447
Cumulative review, *see* Ongoing Assessments
Cylinder, 634–635, 645–648, 657

D

Data, 50–55
box-and-whisker plot, 495–501
cluster, 71
displaying, 495–501
gathering, 50–55
graphing, 379–383, 432
interpretation, 233
measures of variation, 490–494
scatter plots, 379–383
stem-and-leaf plot, 486–489
Data collection, 50, 51–53, 56, 128, 160, 174, 209, 238, 264, 337, 362, 384, 404, 439, 453, 471, 486, 508, 524, 588
Decimals, 296–299, 301–307, 309–311, 317–326
adding, 234–237
bar notation, 275, 277
dividing, 294–298
estimating differences, 229–232
estimating sums, 229–232
to fractions, 225–228
multiplying, 294–298
to percents, 458–461
repeating, 225, 275, 277
subtracting, 234–237
terminating, 225, 275, 277
Deductive reasoning, 255–256
Defining a variable, 42
Degree, 549
of temperature, 128
Demographics, 50, 154, 164
Dependent events, 530–533, 542
Diagonal, 586, 590, 677
of square, 702
of polygon, 96
Diameter, 341
Diastolic pressure, 16
Dilation, 578
Dioptra, 675
Discount, 468–470, 475
Discrete mathematics
arithmetic sequence, 258–261
combination, 516–517, 541
common difference, 258–259
factorial, 515–517
Fibonacci sequence, 263
Fundamental counting principle, 509–511, 515, 541
geometric sequences, 313, 315
outcomes, 509–512
permutation, 514–517
term, 258
tree diagram, 509, 511, 541
sequence, 258–262, 313, 315
Distributive property, 26–29, 31, 712, 725–726, 729–731
Dividend, 166
Divisibility rules, 170–173, 186, 216
Division
of decimals, 296–298
of fractions, 290–292
of integers, 104–106
properties of inequality, 152
property of equality, 130
Domain, 372–376

E

Earth Watch, 80, 214, 232, 376, 436, 518, 723
Eliminating possibilities, 164, 577
Employment Opportunities, *see* Careers
Empty set, 348
Equation, 32–35
graphing, 126–127, 130–132, 137, 162
inverse operations, 41–45
one-step, 124–127, 129–132, 134–136
percent, 467–470
with rational numbers, 248–250
solving, 32–35, 41–45, 84, 85–87, 90–93, 100–103, 105–108, 111, 123–127, 129–132, 134–137, 162, 285, 287, 291, 295–299, 308–311
two-step, 333–340
with variables on each side, 345–349, 365
writing, 338–340, 364
Equiangular, 587, 591
Equilateral, 587, 591
Equilateral triangle, 569–571
Equivalent
equations, 124
fractions, 195
rational numbers, 275
***Escherichia coli* (*e. coli*),** 208, 320
Estimation
capture-recapture, 448
of differences, 229–232
experimental probability, 529
hints, 8, 156–157, 177, 233, 281–283, 285, 295–299, 398, 473, 620, 629, 634, 639, 666–667, 700, 712, 720
of percents, 462–465, 479
of products, 280–283
of quotients, 280–283
of sums, 229–232
Euclid, 703
Evaluating an expression, 16–21, 24
Expanded form, 176
Experimental probability, 511, 529
Experiments, *see* Data collection
Exponent, 175–178, 180, 706, 719–723
Exterior angle, 591–592, 604
Extra practice, 740–773

F

Face, 632–633
Factorial, 515
Factor patterns, 189
Factors, 170–173, 175–179, 184–193
divisibility rule, 170–173
evaluating expressions, 180
greatest common factor, 190–193, 196–199
and monomials, 170–173
patterns, 189
powers and exponents, 175–178
prime factors, 184–187, 196–199
Factor trees, 185–187
Family of graphs, 411
Fibonacci sequence, 95, 263
Floor plan, 653
Formulas, 134–137, 177, 278, 314
area of circle, 620–622, 629, 635, 661
area of parallelogram, 612–613, 620
area of rectangle, 70, 140–143, 163, 661, 725, 728
area of square, 624, 639, 661
area of trapezoid, 335, 399, 614–616
area of triangle, 311, 613–314, 639
aspect ratio, 667
car speed according to skid, 672
circumference, 267, 661
diagonal of a square, 702
distance, 35
energy, 179
finding Celsius, 346, 394
finding Fahrenheit, 346
45°–45° right triangle, 683–686, 701
Heron's, 675
interest, 467
men's shoe size, 257
perimeter of rectangle, 139–140, 142–143, 163, 661, 725
perimeter of square, 389
Pythagorean theorem, 676–680
safe heart rate, 366

sequence, 314
slope, 400–401
slope-intercept form, 407
surface area of a cone, 640–641
temperature, 385
using, 135–137
velocity of dropped objects, 706
voltage in a circuit, 674
volume of circular cylinder, 645–646
volume of a cone, 650–651
volume of a pyramid, 649–650, 720

Fractals, 618

Fractions
- **adding**, 239–247
- **to decimals**, 274–278
- **dividing**, 290–292
- **equivalent**, 195, 276–277
- **estimating differences**, 230–231
- **estimating products**, 280
- **estimating quotients**, 280
- **estimating sums**, 230–231
- **least common denominator**, 201-203
- **like denominators**, 239–243
- **mixed**, 239–243, 245–247
- **multiplying**, 284–287
- **negative exponents**, 210–213
- **to percents**, 459–461
- **simplifying**, 217
- **unlike denominators**, 244–247

Frequency table, 51–55, 60, 70, 79–80, 108, 506

From the Funny Papers, 20, 144, 179, 257, 340, 390, 466, 523, 599, 622

Function
- **family of graphs**, 411
- **graphing**, 373–377, 399
- **intercepts**, 406–410
- **linear equations**, 392–394, 399
- **slope**, 400–403
- **systems of equations**, 412–415

Functional notation, 393–394

Fundamental counting principle, 509–511, 515, 541

FYI, 6, 38, 41, 46, 51, 52, 71, 73, 81, 85, 131, 186, 225, 248, 274, 285, 303, 312, 336, 339, 380, 388, 400, 412, 434, 463, 489, 492, 507, 584, 599, 638

G

Geometric probability, 623–626, 628, 656

Geometric sequence, 312–315

Geometry
- **acute angle**, 552–553
- **acute triangle**, 569
- **adjacent angles**, 561–562, 602
- **adjacent leg**, 687–691
- **alternate exterior angles**, 563–565
- **alternate interior angles**, 563–565
- **altitude**, 612, 614, 634
- **angle**, 549–550, 554, 602
- **angle of depression**, 695–697
- **angle of elevation**, 694–697
- **area**, 138
- **area model**, 286
- **area of circle**, 620–622, 629
- **area of parallelogram**, 612–613, 620, 656
- **area of a rectangle**, 140, 282
- **area of a trapezoid**, 614–616, 656
- **area of triangle**, 613–614, 656
- **axes**, 73–75
- **base**, 614, 632
- **center**, 341
- **change in x**, 400–408
- **change in y**, 400–408
- **circle**, 341
- **circular cone**, 640–641, 650–652, 657
- **circumference**, 341–344
- **compass**, 554–555
- **complementary**, 562–565, 602
- **congruent**, 554, 561, 563, 573
- **construct**, 554
- **coordinate**, 67, 126
- **coordinate system**, 36–40, 73
- **corresponding angles**, 563–564, 574
- **corresponding parts**, 573–576
- **cosine**, 688–691, 695–697, 701
- **cylinder**, 634–635, 645–648, 657
- **degree**, 549
- **diagonal**, 586, 590
- **diameter**, 341
- **equiangular**, 587, 591
- **equilateral**, 587, 591
- **equilateral triangle**, 569–571
- **exterior angle**, 591–592
- **face**, 632–633
- **family of graphs**, 411
- **45°–45° right triangle**, 683–686, 701
- **golden ratio**, 434
- **graph**, 73, 126
- **hypotenuse**, 676–680, 683–686, 693, 701
- **indirect measurement**, 580
- **intercepts**, 406–409
- **interior angle**, 591–592
- **intersect**, 550–551
- **isosceles triangle**, 569–571
- **lateral surface**, 639–640
- **leg**, 676–681, 683–686
- **line**, 548
- **line segment**, 548–549, 551–552, 554–555
- **net**, 632
- **obtuse angle**, 550, 552
- **obtuse triangle**, 553, 569–571
- **opposite side**, 687
- **ordered pair**, 36–40
- **origin**, 36–40
- **parallel**, 550–551, 561, 563–565, 602
- **parallel lines**, 401
- **parallelogram**, 585–587
- **perimeter**, 138
- **perimeter of a rectangle**, 140
- **perpendicular**, 551
- **pi (π)**, 343
- **plane**, 550, 552, 602
- **plotting a point**, 73
- **point**, 548–549, 552, 602
- **polygon**, 589–593
- **prism**, 632–633, 645–648, 657
- **protractor**, 549–552
- **pyramid**, 638–642, 649–652, 657
- **quadrants**, 73
- **quadrilateral**, 584, 603
- **radius**, 341
- **ray**, 548–549, 552, 555, 602
- **rectangle**, 585
- **rectangular prism**, 632–633, 657
- **reflection**, 595–598, 604
- **regular polygon**, 591–592
- **rhombus**, 585–587
- **right angle**, 550, 552–553, 569–571, 603
- **rotation**, 595–598, 604
- **scale**, 224
- **scale drawing**, 585
- **scale factor**, 654
- **scalene triangle**, 569–571
- **sides**, 549, 589
- **similar**, 578
- **sine**, 688–691, 695–697, 701
- **skew lines**, 551, 602
- **slant height**, 638, 640
- **slope**, 400–403
- **slope-intercept form**, 407–408
- **square**, 585–587
- **square pyramid**, 638–639
- **straightedge**, 554–555
- **supplementary**, 562–565, 602
- **surface area**, 632–636, 657
- **symmetric**, 596
- **tangent**, 688–691, 695–697, 701
- **tessellation**, 594
- **30°–60° right triangle**, 683–686, 701
- **transformation**, 595–598
- **transversal**, 563–566, 602
- **trapezoid**, 585, 587
- **triangle**, 568–572
- **triangular prism**, 632–633
- **triangular pyramid**, 638
- **trigonometric ratios**, 688–691, 694–697, 701
- **trigonometry**, 688
- **vertex**, 549, 589, 638
  - **of a polygon**, 75
- **vertical angles**, 561–563, 602
- **volume**, 643–648
  - **of a circular cylinder**, 645–648, 657
  - **of a cone**, 650–652, 658
  - **of a prism**, 645–648, 657
  - **of a pyramid**, 649–652, 658
- ***x*-axis**, 36–40
- ***x*-coordinate**, 37–40
- ***x*-intercept**, 406–409, 425
- ***y*-axis**, 36–40
- ***y*-coordinate**, 37–40
- ***y*-intercept**, 406–409, 425

Geometry connections, *see* Connections

giga, 461

Golden ratio, 434, 435

Golden rectangles, 712

Gram, 359–361

Graphing calculator
- **activities**
  - **area and coordinates**, 145
  - **evaluating expressions**, 21
  - **exploring factors and fractions**, 180

factor patterns, 189
families of graphs, 411
finding mean and median, 307
geometric probability, 628
graphing inequalities, 417
slope and tangent, 693
slopes of parallel lines, 567
applications, energy cost, 397
exercises
graphing linear relations, 387
measures of variation, 493
probability, 443
scatter plot, 382
solving one-step equations, 132
use a simulation, 527
volume, 647
introduction, xx, 1
modeling
congruent triangles, 573
functions and graphing, 412
plotting points, 77
probability, 440
scatter plots, 379–380
solving equations, 39, 124
statistics and probability, 490, 502–503, 530
programming, 21, 34, 77, 124, 132, 145, 180, 387, 397, 411, 417, 440, 490, 502–503, 567, 573, 628, 693
Graphs, 73, 126
bar, 52–54, 150, 395
box-and-whisker plot, 495–496, 498–501, 503, 540
circle, 556–557
coordinate systems, 36–40, 74–76, 93, 110–128, 583
data, 80, 432
of equations, 412–415
functions, 372–376
of inequalities, 75, 147–149
latitude, 76
longitude, 76
on number line, 78–80, 110, 112–113, 126–127, 130–132, 162
parabola, 391–394
points, 67–69, 71, 73–76, 78–80
problem solving, 396–399
rectangle, 145
relations, 372–376
Greatest common factor (GCF), 190–193, 196–197, 217, 285

Hands-on activities, *see* Mathematics Lab
Heat index, 97, 299
Helge von Koch, 618
Help wanted, *see* Careers
Histogram, 502–503
Hypotenuse, 676–680, 686–686, 693, 701
Hypsometer, 694

Identity property
of addition, 23, 234
of multiplication, 23
Inclusive, 536–538, 542
Independent events, 530–533, 542
Indirect measurement, 580–583
Inductive reasoning, 255–256
Inequalities, 46–49, 78–80, 110, 146–159, 203, 217, 227–228
checking solutions, 147–149, 152–154
compound, 357
graphing, 147–149, 152–154, 251–253, 351–353, 417–421, 426
solving multi-step, 351–354, 355–357, 365
solving one-step, 146–149, 151–154, 156–158, 163, 309–311
writing, 355–357, 365
Integers, 66–69, 224
absolute value, 67–69
adding, 82–87, 111
comparing, 78–81
dividing, 104–107
graphing, 73–75, 78–81
multiplying, 98–103, 111
ordering, 78–81
subtracting, 88, 89–92, 104–106, 112
Integration
assessment, *see* Assessment
discrete mathematics
arithmetic sequences, 258–261
geometric sequences, 312–315
geometry, 128, 133
area and perimeter, 139–143
circles and circumference, 341–343
coordinate system, 72–75
ordered pairs, 36–40
Look Back, *see* Look Back
measurement, using the metric system, 358–361
number theory, 15, 30, 95, 174, 188, 218, 287, 437, 443, 553, 571
patterns, *see* Patterns
probability, 374, 440–443, 510
statistics, 53, 93, 121, 164, 262, 288, 384, 403, 452, 471, 494
gathering and recording data, 51–54
measures of central tendency, 301–305
technology, *see* Graphing calculator, Spreadsheets, Technology Tips
Intercepts, 406–409
Interdisciplinary, *see* Connections
Interior angle, 591–592
Interquartile range, 491–494, 540
Intersecting lines, 550–551
Inverse operations, 41–45
of addition, 234
of multiplication, 289
Investigations (long-term projects)
Bon Voyage, 62–63, 70, 97, 128, 160
It's Only a Game, 482–483, 508, 528, 588, 600
Oh, Give Me a Home, 606–607, 637, 653, 671, 698
Roller Coaster Math, 270–271, 279, 293, 337, 362
Stadium Stampede, 368–369, 384, 404, 453, 471, 476
Taking Stock, 166–167, 174, 209, 238, 264
Up, Up, and Away, 2–3, 10, 35, 56
Irrational numbers, 672–675, 700
Isosceles triangle, 569–571

Journal, *see* Math Journal

Koch curve, 618

L

Lateral surface, 639–640
Least common denominator (LCD), 201–203
Least common multiple (LCM), 200–204, 217
Leaves, 486
Leg, 676–681, 683–686
Like terms, 27, 711, 716
Linear equation, 386–389
Line graph, 497–500
Line plot, 71
Lines, 548, 550–554
intersecting, 550–551
parallel, 561, 563, 567, 602
perpendicular, 551, 602
of symmetry, 550–551, 596–598
transversal, 563
Line segment, 548–549, 551–552, 554–555
bisecting, 555
congruent, 554
perpendicular, 613–614
Liter, 359–361
Logic
matrix, 119–121
reasoning, 70, 74
Long-term project, *see* Investigations
Look Back, 73, 89, 123, 125, 225, 251, 274, 287, 309, 317, 351, 355, 358, 379, 432, 459, 498, 502, 505, 510, 556, 579, 620, 630, 654, 665, 672, 706, 719, 721
Lower quartile, 491–494, 540

Manipulatives, *see* Modeling, Mathematics Lab
Market price, 166
Mass, 359–361
Math Journal, 9, 28, 53, 79, 106, 148, 153, 177, 212, 231, 241, 256, 260, 282, 291, 319, 343, 356, 360, 402, 408, 420, 446, 464, 474, 505, 516, 521, 558, 570, 592, 630, 640, 646, 690, 696, 708, 722
Mathematical modeling
charts, 6–7, 11, 51, 71, 78, 118, 139, 175, 189, 200, 205, 210, 234, 251, 258, 280, 301, 308, 330, 341, 372, 385, 396, 432, 437, 444, 462, 502, 535, 719, 728

INDEX

diagrams, 22, 31, 36, 72, 82, 88, 89, 98, 104, 123, 129, 138, 139, 181, 195, 244, 284, 351, 509, 520, 584, 594, 595, 610, 612, 618, 664, 669, 694
equations, 21, 34, 83, 104, 123, 124, 295, 346, 400, 411, 412, 672, 725
graphs, 21, 71, 77, 145, 151, 180, 229, 248, 274, 372, 385, 391, 405, 406, 411, 417, 418, 454, 495, 502, 504, 556, 567, 573, 693, 694
scatter plots, 379–380
Mathematics Labs, 21, 31, 50, 71, 77, 82, 88, 98, 123, 138, 145, 180, 189, 195, 263, 294, 300, 307, 333, 345, 378, 391, 405, 411, 417, 439, 448, 502, 514, 529, 554–555, 567, 594, 610–611, 618, 628, 643, 654, 682, 687, 693, 710, 724, 732
Mean, 20, 300–301, 303–307, 505
Measurement
gram, 359–361
liter, 359–361
meter, 358–361
metric system, 358–361
Measures of central tendency, 300–307
mean, 300–307, 505
median, 300–307, 491–494, 501, 505, 540
mode, 300–306, 505
Measures of variation, 490–494
Median, 300–307, 491–494, 501, 505, 540
Meter, 358–361
Metric system, 358–361
Mixed number, 239–243
adding, 240–243, 245–247
subtracting, 240–243, 245–247
Mixed review, 20, 25, 29, 35, 40, 45, 49, 55, 70, 75, 80, 87, 93, 97, 103, 108, 122, 128, 133, 137, 144, 150, 155, 159, 174, 179, 194, 199, 204, 209, 214, 228, 233, 238, 243, 247, 250, 254, 257, 262, 279, 283, 288, 299, 306, 316, 320, 332, 337, 340, 344, 350, 354, 357, 361, 377, 384, 395, 399, 404, 410, 416, 422, 436, 438, 447, 453, 457, 461, 466, 471, 475, 494, 501, 507, 519, 523, 528, 534, 553, 560, 566, 572, 577, 588, 593, 617, 622, 627, 631, 637, 648, 653, 668, 671, 681, 686, 692, 697, 709, 714, 718, 723, 727, 731
Mode, 300–306, 505
Modeling
with manipulatives, 22, 31, 36, 89, 98, 104, 123, 126, 138, 139, 146, 170, 184, 284, 289, 333, 341, 345, 351, 392, 405, 418, 432, 458, 520, 524, 535, 561, 578, 595, 612, 619, 649, 676, 694, 711
real-world applications
advertising, 509
air travel, 396
architecture, 638
backpacking, 355
ballooning, 664
baseball, 338
business, 229, 437
cartography, 72–73
child care, 407
comics, 255
construction, 400, 548, 706, 725
consumerism, 334, 454, 486
ecology, 372
economics, 444
education, 118, 504
electricity rates, 472
engineering, 568
entertainment, 32, 51, 181, 339
food, 274
forestry, 196
games, 175, 623
geography, 66
geology, 317
health, 46, 78, 462
health and exercise, 6
history, 190
law enforcement, 672
medicine, 16
meteorology, 248, 301
movies, 224
oceanography, 99
pet care, 629
physical therapy, 449
postal service, 632
quilting, 584
recycling, 151
safety, 495
science, 385
sports, 251, 406
swimming, 515
toys, 129
transportation, 41
travel, 26, 134, 358
weather, 556
world records, 156
with technology, 11, 83, 94, 124, 205, 210, 234, 258, 280, 295, 330, 346, 412, 440, 467, 490, 530, 573, 589, 644, 688, 719
Modeling Mathematics, 14, 21, 22, 31, 36, 47, 50, 54, 62, 68, 71, 77, 82, 88, 89, 91, 98, 102, 104, 120, 123, 135, 138, 139, 145, 146, 170, 173, 180, 182, 184, 189, 192, 195, 198, 202, 227, 239, 244, 263, 277, 284, 288, 289, 294, 297, 300, 304, 307, 315, 331, 333, 335, 341, 345, 348, 351, 353, 378, 387, 391, 392, 404, 405, 411, 414, 417, 418, 432, 434, 439, 442, 448, 451, 458, 460, 488, 498, 502, 511, 514, 520, 524, 529, 535 554–555, 561, 567, 575, 578, 581, 586, 594, 595, 597, 601–611, 618, 619, 625, 628, 635, 643, 649, 651, 654, 667, 676, 679, 682, 687, 693, 694, 710, 711, 713, 717, 724, 726, 730, 732
Monomials, 172–173, 186–187, 205–208, 706, 708, 719–723
dividing, 206–208
factoring, 186–187
multiplying, 205–208
power of, 721–723, 735
power of a power, 719–723
power of a product, 720–723
power of ten, 205, 207
Multiple, 200–204
Multiplication
associative property of, 22
of binomials, 728–731, 732, 735
commutative property of, 22
of decimals, 294–298
of fractions, 284–287
of integers, 98–103
of monomials, 205–208
phrases, 18
polynomials by monomials, 725–727
of powers, 719–723
property of equality, 130
properties of inequality, 152
Multiplicative inverse, 289–292
Multiplicative property of zero, 23
Multiplying integers with different signs, 99
Multiplying integers with the same sign, 100
Mutually exclusive, 535–538, 542

Negative, 66
Negative exponents, 210–213
Negative relationship, 380–384, 424
Net, 632
Net change, 71
No relationship, 380–384, 424
Null set, 348
Number lines
graphing points, 67, 69, 71, 78–80
graphing solutions to equations, 126–127, 130, 132, 162
graphing solutions to inequalities, 147–149, 152–154, 163
integers, 682
irrational numbers, 682
ordering integers, 78–80
parabolas, 391–394
rational numbers, 682
Numbers, 673–674
composite, 184–187
consecutive, 340
integers, 224–228, 673–674
irrational 673–674
mixed, 239–243, 245–247
perfect squares, 665–667
pi (π), 343
prime, 184–187
rational, 224–228, 672–674, 700
real, 673–674, 700
sequence, 20
whole, 224, 228, 673–674

Obtuse angle, 550–553, 569, 603
Obtuse triangle, 553, 569–571
Octagon, 255
Odds, 520–522, 541
Ongoing Assessments, 114–115, 326–327, 428–429, 544–545, 660–661, 738–739
Open-ended problems, 115
Open sentence, 32

Opposite side, 687
Ordered pair, 37, 36–40, 73–77, 103
Order of operations, 11–15, 21, 76
Origin, 36–40
Outcomes, 509–512
Outliers, 71, 496

Parabola, 391–394
Parallel lines, 401, 550–551, 561, 563–565, 602
Parallelogram, 585–587
  area of, 612–613, 616
Parentheses, 12–15
Pascal's Triangle, 30
Patterns: problem solving, 20, 25, 35, 49, 95, 96, 103, 122, 178, 183, 189, 193, 205, 210, 256, 312–316, 354, 366, 384, 435, 438, 457, 466, 508, 553
Percent, 449–452
  of change, 472–475
  to decimals, 459–461
  of decrease, 472–475, 480
  equations, 467–470
  estimating, 462–465, 479
  to fractions, 459–461
  of increase, 472–475, 480
Percent proportion, 450–451, 499
Perfect number, 174
Perfect squares, 665–667
Performance task, 61, 165, 219, 269, 325, 367, 427, 703
Perimeter, 138, 678
  of a rectangle, 140
  of a square, 665
Permutation, 514–517
Perpendicular lines, 551, 602
Pi (π), 255, 341–344
Pictograph, 497, 500
Plane, 550, 552, 602
Plotting a point, 73–75, 77
Point, 548–549, 552, 602
  graphing, 77
Polygons, 589–593
  angle measures, 590–592
  area of, 610–611
  exterior angles, 591–592
  heptagon, 590
  hexagon, 590, 591
  interior angles, 591–592
  octagon, 590
  pentagon, 590–591
  quadrilaterals, 590
  regular, 591–593
  tessellations, 594
Polynomial, 706–708, 710, 734
  adding, 711–714, 734
  additive inverse, 716–717
  binomials, 728–731, 732, 735
  multiplying, 724, 725–731, 732, 735
  subtracting, 715–718, 734
Portfolio, 61, 113, 160, 165, 219, 269, 325, 427, 476, 481, 543, 600, 605, 659, 698, 703
Positive relationship, 380–384
Powers, 175–178
  of a monomial, 706, 720–723, 735
  of a power, 719–723
  of a product, 720–723
Prime day, 187
Prime factorization, 185–187, 190–191, 196–197, 201–202, 211, 216
Prime number, 545
Principal, 467
Prism, 632–633, 645–648
Probability, 440–443, 478
  capture-recapture, 448
  dependent events, 530–533, 542
  experimental, 511, 529
  geometric, 623–626, 628, 656
  inclusive, 536–538, 542
  independent events, 530–533, 542
  mutually exclusive, 535–538, 542
  odds, 520–522, 541
  relative frequency, 529
  sample space, 441, 479
  simulation, 524–526
  theoretical, 511
Problem-solving strategies
  acting it out, *see* Simulations
  deductive reasoning, 255–256
  draw a diagram, 181–183
  draw a graph, 396–398
  eliminate possibilities, 118–121
  guess and check, 186
  inductive reasoning, 255–256
  look for a pattern, 94–96, 171
  make a drawing, 629–631
  make a model, 629–631, 657
  make a table, 437, 441
  matrix logic, 119–121
  method of computation, 8
  patterns, 20, 25, 35, 49, 94–97
  problem-solving plan, 6–10, 43, 148, 156–157, 171, 181, 290, 314, 330–331, 339, 342, 355–356, 359, 396–397, 407–408, 437, 445, 454, 510, 516, 525, 557, 580–581, 629, 634–635, 646, 678, 726, 729
  use a simulation, 524–526
  Venn diagram, 669–671, 673
  work backward, 314, 330
Product of powers, 205–208
Programming, graphing calculator
  area and coordinates, 145
  congruent triangles, 573
  energy cost, 397
  evaluating expressions, 21
  exploring factors and fractions, 180
  families of graphs, 411
  geometric probability, 628
  graphing inequalities, 417
  graphing linear relations, 387
  plotting points, 77
  probability, 440
  slope and tangent, 693
  slopes of parallel lines, 567
  solving equations, 34, 124, 132
  statistics and probability, 490, 502–503
Projects
  cooperative, 113, 481, 543, 605, 659
  long-term, *see* Investigations
Properties
  of addition, 22, 24–25, 30, 234
  additive inverse, 89–91
  associative, 22, 24–25, 101, 296, 712
  commutative, 22–25, 296, 712
  distributive, 26–29, 31, 712
  of equality, 17, 28, 124, 125
  identity, 23, 25, 30, 296
  inverse, 296
  of multiplication, 22, 24–25, 296
  of proportions, 444
  substitution, 17, 28
  of zero, 23, 25
Proportions, 444–447, 478, 579–583, 585
Protractor, 549–552, 694
Punnett squares, 728
Putting It into Perspective, 4–5, 64–65, 116–117, 168–169, 222–223, 272–273, 328–329, 370–371, 430–431, 484–485, 546–547, 608–609, 662–663, 704–705
Pyramid, 638–642, 649–652, 657
Pythagoras, 676
Pythagorean theorem, 676–680, 682, 694–697, 701

Quadrants, 73–75
Quadrilateral, 584–588, 603
  angles, 584, 603
  parallelogram, 585–587
  rectangle, 585–587
  rhombus, 585–587
  similar, 585
  square, 585–587
  trapezoid, 585–587, 614–616
Quartile, 491–494
Quotient of powers, 206–208

Radical sign, 664–667, 672–675
Radius, 341
Range, 372–376, 490–494, 540
Rate, 433, 449–452
Ratio, 196–199, 432–435, 478, 687, 688–691
  cosine, 688–691
  sine, 688–691
  tangent, 688–691
  trigonometric, 688–691, 694–697, 701
Rational numbers, 224–228, 672, 700
  adding, 234–237
  equivalent, 275
  solving equations, 248–250
  subtracting, 234–237
Ray, 548–549, 552, 555, 602
Real numbers, 673–675, 700
Reciprocal, 289
Rectangle, 585
  area of, 140–143, 150, 165, 615, 634–635
  graphing, 145
  golden, 712
  line of symmetry, 596
  perimeter, 139–143, 165
  volume, 159

Rectangular prism, 632–633
Reflection, 595–598, 604
Regular polygon, 591–593
Relation, 372–376
  graphing linear relations, 385–387
  inverse, 376
  with two variables, 425
Relative frequency, 529
Relatively prime, 193, 218
Repeating decimals, 225–226, 277
Review
  Chapter Tests, 774–787
  Extra Practice, 740–773
  Highlights, 57, 109, 161, 215, 265, 321, 363, 423, 477, 539, 601, 655, 699, 733
  Mixed review, *see* Mixed review
  Self Test, 30, 93, 137, 194, 247, 299, 350, 399, 457, 513, 572, 642, 681, 718
  Study Guide, 58–60, 110–112, 162–164, 216–218, 266–268, 322–326, 364–366, 424–426, 478–480, 540–542, 602–604, 656–658, 700–702, 734–738
Rhombus, 585–587
Right angle, 550, 552–553, 569–571, 603
Right triangle, 569–571, 676–680, 683–686, 701
  angles, 569
  cosine ratio, 688–692, 695–697
  hypotenuse, 676–681
  legs, 676–681
  Pythagorean theorem, 676–681
  ratios, 687–691, 694–697
  sine ratio, 688–692, 695–697
  tangent ratio, 688–697
Rotation, 595–598, 604
Rounding
  truncate, 275, 281, 296, 298

Sales, 383
Sample, 50, 51
Sample space, 441, 479
Scale, 224
Scale drawing, 585, 605
Scale factor, 654
Scalene triangle, 569–571
Scatter plot, 379–383, 387–384, 424
Scientific notation, 317–319, 324–325
Security, 166
Self Evaluation, 165, 219, 269, 325, 367, 427, 481, 543, 605, 659, 703
Self Test, 30, 93, 137, 194, 247, 299, 350, 399, 457, 513, 572, 642, 681, 718
Sequence, 258–262
  geometric, 312–316
Shadow reckoning, 580
Share, 166, 264
Sides, 549, 589
Sieve of Eratosthenes, 193
Similar, 578, 654
Simplest form, 28, 196–199
Simulation, 524–525
Sine, 688–691, 695–697, 701
Skew lines, 551, 602
Slant height, 638, 640
Slope, 400–403, 693
Slope-intercept form, 407–408
Solution, 32–35
  graphing, 126–127, 130–132, 147, 152–154
  of system, 412–415
Solving an equation, 32
Spatial reasoning
  angle relationships and parallel lines, 561–566
  area and perimeter of rectangles, 139–143, 163
  congruent triangles, 573–577
  construction, 554–555
  polygons, 589–593
  quadrilaterals, 584–587
  similar triangles, 578–583
  surface area: prisms and cylinders, 632–636
  surface area: pyramids and cones, 638–642
  tessellations, 594
  transformations, 595–598
  triangles, 568–572
  volume, 643–653, 720
Spiral review, *see* Mixed review
Spreadsheet, 209, 238, 467
Square, 585–587
  area, 624, 665
  diagonals, 702
  Punnett, 728
Square pyramid, 638–639
Square root, 664–667, 672–674
Standard form, 176
Statistical Snapshots, 5, 65, 117, 169, 223, 273, 329, 371, 431, 485, 547, 609, 663, 705
Statistics, 51–55
  back-to-back stem-and-leaf plot, 487–489
  bar graph, 52–54, 60, 503
  box-and-whisker plot, 495–496, 498–501
  circle graph, 497, 556–559
  comparative graph, 498, 500
  data, 50–55
  demographics, 50
  frequency table, 51–55
  histogram, 502–503
  interquartile range, 491–494, 540
  leaves, 486
  line graph, 497, 500
  line plot, 71
  lower quartile, 491–494, 540
  making graph, 502–503
  mean, 300–301, 303–307
  measures of central tendency, 300–307
  median, 300–307
  misleading, 504–506, 541
  mode, 300–307
  negative relationship, 380–384
  no relationship, 380–384
  odds, 520–522
  outliers, 496
  pictograph, 497, 500
  positive relationship, 382–384, 424
  quartile, 491–494
  range, 490–494, 540
  sample, 50, 51
  scatter plot, 378–384, 424
  stem-and-leaf plot, 486–489, 540
  stems, 486
  surveys, 50
  upper quartile, 491–494, 540
  using to make predictions, 455–456
Stem-and-leaf plot, 486–489, 540
Stems, 486
Stock exchange, 166
Straightedge, 554–555
Substitution property of equality, 17, 28
Subtraction
  of integers, 89
  like fractions, 240–243
  phrases, 18
  unlike fractions, 244–247
Subtraction property of equality, 124
Subtraction property of inequality, 147
Supplementary, 288, 562–565, 602
Surface area, 632–636, 657
  of cones, 640–641, 657
  of cubes, 635–636
  of cylinders, 634
  of prisms, 632–636, 657
  of pyramids, 639–641, 657
Surveys, 50
Symbols
  multiplication, 13
  division, 13
Symmetric, 596
System of equations, 412–415, 426
Systolic pressure, 16

T

Tangent, 688–691, 695–697, 701
Technology, *see* Graphing calculator; Modeling; Spreadsheet; Technology Tips
Technology Tips, 12, 83, 177, 235, 276, 318, 450, 491, 666, 689, 719
Terminating decimal, 225, 275
Terms, 27, 258
Tessellation, 594
Tests, *see* Assessment
Test-taking tips, 115, 327, 429, 545, 661
Theoretical probability, 511
The Shape of Things to Come
  digital cameras, 627
  the global positioning system, 40
  living batteries, 320
  technology in sports, 354
  thought-controlled computer technology, 692
  virtual reality, 560
Think About It, 7, 12, 17, 18, 23, 24, 27, 33, 37, 42, 43, 52, 67, 104, 171, 172, 197, 201, 202, 226, 252, 281, 341, 352, 372, 386, 401, 406, 458, 536, 548, 556, 562, 563, 574, 580, 589, 596, 613, 614, 629, 632, 634, 638, 645

30°–60° right triangle, 683–686, 701
Timelines, *see* Putting It into Perspective
Transformation, 595–598
Translation, 595, 604
Transversal, 563–566, 602
Trapezoid, 585, 587
  area of, 614–616
Tree diagram, 509, 511, 541
Triangle, 568–572
  acute, 569–571, 603
  angles, 569–577
  area of, 311, 610–611, 613–614, 616
  congruent, 573–577, 603
  equilateral, 569–571, 596
  isosceles, 569–571
  obtuse, 569–571, 603
  right, 569–571, 603
  scalene, 569–571
  similar, 578–583, 603
  special, 683–686
Triangular prism, 632–633
Triangular pyramid, 638
Trigonometric ratios, 688–691, 694–697, 701
Trigonometry, 688
Trinomials, 706–708, 734
Truncate, 275
Twin primes, 188

Unit rate, 433–435, 478
Upper quartile, 491–494, 540

Variable, 16–18, 32–35
  defining, 42
Venn diagram, 669, 671, 673
Vertex, 75, 549, 589, 638
Vertical angles, 561–563, 602
Vertical line test, 373–374
Volume, 643–648
  of a circular cylinder, 645–648, 657
  of a cone, 650–652, 658
  of a cube, 177–178
  of a prism, 645–648, 657
  of a pyramid, 649–652, 658

Whole numbers, 224
Windchill factor, 97
Working drawings, 653

***x*-axis,** 36–40
***x*-coordinate,** 37–40
***x*-intercept,** 406–409

***y*-axis,** 36–40
***y*-coordinate,** 37–40
***y*-intercept,** 406–409

Zero pair, 82

INDEX